Robot Control

IEEE PRESS
445 Hoes Lane, PO Box 1331
Piscataway, NJ 08855-1331

IEEE Control Systems Society, *Sponsor*

CS-S Liaison to the IEEE PRESS
Steven Yurkovich

Technical Reviewer for CS-S
T. J. Tarn
Washington University

Robot Control

Dynamics, Motion Planning, and Analysis

Edited by

Mark W. Spong
University of Illinois at Urbana-Champaign

F. L. Lewis
The University of Texas at Arlington

C. T. Abdallah
University of New Mexico

IEEE
PRESS

A Selected Reprint Volume
IEEE Control Systems Society, *Sponsor*

The Institute of Electrical and Electronics Engineers, Inc., New York

This book may be purchased at a discount from the publisher when ordered in bulk quantities. For more information contact:

IEEE PRESS Marketing
Attn: Special Sales
PO Box 1331
445 Hoes Lane
Piscataway, NJ 08855-1331
Fax: (908) 981-8062

Printed in the United States of America

10 9 8 7 6 5 4 3 2 1

ISBN 0-7803-0404-7

IEEE Order Number: PC0299-8

Library of Congress Cataloging-in-Publication Data

Robot control / edited by Mark W. Spong, Frank L. Lewis, Chaouki T. Abdallah.
p. cm.
"A selected reprint volume."
"IEEE Control Systems Society, sponsor."
Includes bibliographical references and index.
ISBN 0-7803-0404-7
1. Robots—Control systems. I. Spong, Mark W. II. Lewis, Frank L. III. Abdallah, Chaouki T. IV. IEEE Control Systems Society.
TJ211.35.R63 1993
670.42'72—dc20 92-5764

Contents

Preface

ROBOTICS is an extremely large and interdisciplinary field and it is therefore difficult for one person to become an expert in the entire field. When the idea for an IEEE Press book of reprints was first discussed in the summer of 1989, it was decided to solicit the advice and opinions of colleagues from around the world as to which papers they felt would serve as useful guides for students and researchers entering the field of robotics. The response to our initial inquiry was overwhelming. Nearly four hundred papers in all aspects of robotics were suggested for inclusion. We quickly realized the futility of trying to put together a book of reprints covering more than a single aspect of robotics, and decided instead to limit the scope of the book to robotic motion planning and control and to some aspects of force control. Toward this end we solicited the help of several colleagues, each of whom was asked to suggest a list of papers covering a specific topic and to write a short introduction. The result is the present volume. We believe that this collection of papers will be of use to anyone wishing to have a broad overview of the main ideas and results in robot motion and force control, as well as to students and researchers in the field. The large number of additional references in all of the chapters should be of particular benefit in this regard.

We are first of all indebted to the chapter organizers, Vladimir Lumelsky, Jean-Jacques Slotine, Suguru Arimoto, Kang Shin, Harris McClamroch, Steve Yurkovich, Shankar Sastry, Ping Hsu, Alessandro DeLuca, and Dan Koditschek for the excellent job they have done in organizing the chapters and writing the introductions. This was truly a group effort. Second, we regret that there was not room in this single volume to include more topics and more papers. The large number of papers excluded from this volume are certainly no less interesting or important than the small number of papers that we were able to include. In fact, we hope that one of the consequences of this volume will be to inspire other researchers to put together similar volumes in other aspects of robotics and robotic manipulation.

MARK W. SPONG
FRANK LEWIS
CHAOUKI ABDALLAH

Introduction

1. Robotics

INDUSTRIAL robots are basically positioning and handling devices. A useful robot is one that is able to control its movement and the forces it applies to its environment. This reprint volume is concerned with the control aspect of robotics. It is meant to be a supplement to works that describe other important aspects of industrial robots [1]–[4]. To control requires the knowledge of a mathematical model and of some sort of intelligence to act on the model. The mathematical model is obtained from the basic physical laws governing the robot's dynamics. Intelligence requires sensory capabilities and means for acting and reacting to the sensed variables. This book is not generally concerned with the modeling of robots except in the case of flexible joints and links. The reason for this omission is that rigid robots have well-known dynamics that may be found elsewhere [12], [17], [21], [23], [26]. We will, however, summarize the rigid robot model in section 3 of this introduction and discuss some of its properties. On the other hand, the problem of controlling robots (both in motion and force) still offers many theoretical and practical challenges and the papers collected in this book address different aspects of this problem. It may be argued that industrial robotics have been effective at performing many tasks without the body of knowledge collected in this book. In fact, robot manufacturers do not even attempt to obtain dynamic models of their robots, let alone measure their physical parameters. The theoretical issues addressed in this volume are interesting, challenging; but more important, their study is necessary to achieve higher performance than is possible with the existing controllers provided on commercial robot arms.

2. Robot Control: A Historical Perspective

It is quite remarkable that industrial robots in use today are still controlled using classical concepts [15]. In this section we present a historical perspective of control methodologies as used in robotics. This perspective includes both theoretical and practical highlights. We also reference other works that review the history of other aspects of robots such as mechanical designs, drive methods, sensing mechanisms, programming languages, and others [1]–[4].

The first theoretical development that had a direct effect on the control of robots was the convention for assigning frames to kinematic linkages developed in 1955 by Denavit and Hartenberg [5]. In 1958, the first industrial robot was built by Unimation and was a five-axis, hydraulic manipulator [6]. As is typical of the earlier robots, no servo-control was provided. In fact, the robot was taught a sequence of movements that were then stored in memory and repeated as desired. Such point-to-point open-loop robots were slightly improved in the following years but retained the same basic architecture. The first robotics patent was granted in 1961 to George Devol in the U.S. [6]. In 1963, AMF, Inc., introduced a cylindrical-configuration, hydraulically actuated robot that provided a continuous-path following ability in addition to the point-to-point movement. In 1965, the D-H convention was applied to robotics [7]. This had a direct effect on the mechanical design of robotic structures and their control. In particular Pieper [8] analyzed the inverse kinematics problem that led to the design of robotic wrist mechanisms with intersecting joint axes. The control of these robots was still effected in an open-loop fashion. In fact, the trajectory planning problem was considered in great detail by many authors, with proposed solutions including techniques for minimizing the time [9] or the energy [10] needed to complete a certain task. The Boston arm and the Stanford arms were developed in 1969 to demonstrate some advanced robotics concepts [11]. In 1973, Cincinnati Milacron introduced their T^3 robot. This was a large hydraulic robot capable of tracking moving parts. This robot marked the introduction of a servo-controller to industrial robotics. Bejczy in 1974 [12] implemented a computed-torque controller on the Stanford arm and Hohn in 1976 presented a version of a computed-torque controller for a Cincinnati Milacron robot [13]. In 1975, Rosen and Nitzan [14] combined many sensory capabilities (vision, force, proximity) to improve the performance of their robot controller.

Most robot controllers in use today are direct descendants of the original linear controllers developed during this "classical" period. The simplest of these is the *independent joint control* method [15], which attempts to control each robot joint with a simple proportional-plus-derivative controller while treating the joint interactions as a disturbance. The operator can interface to this controller using a high-level language through a master computer. Different control objectives may be specified at this level. The master computer then translates the control objectives to subtasks to be performed by the microprocessors at the joint level.

It was shown that in point-to-point applications, such controllers are globally asymptotically stable [16]. In fact, even in continuous-path applications, such simple controllers will perform reasonably well in an indirect-drive robot moving slowly along its path and actuated by large motors [17], [18]. In the old stability versus performance trade-off, current industrial robots are tilted toward stability. It is only the attempt to extract better performance that has led researchers to investigate more advanced control schemes for robots. Such schemes include the computed-torque method [17], which can theoretically control the motion of rigid robots to any desired degree of performance. Its practical implementation is unfortunately hindered by its complexity, which is of the order of the robot's dynamic complexity, as well as by the requirement for perfect knowledge of the robot dynamics. The trend has therefore shifted to robust controllers, which have a simpler structure than the computed-torque controller and which can operate under uncertain conditions. In another trend, adaptive controllers have been implemented without exact knowledge of the robot dynamics. Note that adaptive controllers are more performance oriented since they improve with time as opposed to the fixed-structure robust controllers. For a survey of the robust control of robots see [19] and for one of adaptive control see [20]. Both robust and adaptive controllers have yet to find their way into industrial robotics, although simple proportional-plus-integral controllers have been shown to be somewhat robust [21].

The control designs presented in this reprint volume have one thing in common. They all attempt to push the limits of performance while decreasing the required prior knowledge of the robot's dynamics. Better performance may be obtained by including higher-order effects (such as flexibilities), controlling more variables (such as force), or designing better end-effectors (dexterous hands). As robots become more accepted, more demanding requirements are placed on them, and that is why more advanced controllers are needed.

3. Modeling

For servo-control design purposes, and to design better controllers, it is necessary to reveal the dynamic behavior of the robot via a mathematical model obtained from basic physical laws. In this section we use Lagrangian dynamics [22] to obtain the describing mathematical equations. Let us caution, however, that the model obtained in this section does not account for the flexibility of the joints and links. Such complex effects will be included in some reprints in parts 2 and 7. Often, the rigid modeling approach given here is general enough.

We begin our development with the general Lagrange equations of motion [22]. Consider then Lagrange's equations for a conservative system as given by

$$\frac{d}{dt}\left(\frac{\delta L}{\delta \boldsymbol{q}}\right) - \frac{\delta L}{\delta \boldsymbol{q}} = \boldsymbol{\tau} \tag{1}$$

where $\boldsymbol{q}$ is an n-vector of generalized coordinates q_i, $\boldsymbol{\tau}$ is an n-vector of generalized force τ_i, and the Lagrangian L is the difference between the kinetic and potential energies

$$L = K - P \tag{2}$$

It can then be shown [21] that the robot dynamics are given by

$$\boldsymbol{M}(q)\ddot{q} + \boldsymbol{V}(q,\dot{q}) + \boldsymbol{G}(q) = \tau \tag{3}$$

where $\boldsymbol{M}(q)$ is an n × n, symmetric, positive-definite inertia matrix, $\boldsymbol{V}(q,\dot{q})$ is a vector containing the effects of the Coriolis and centripetal torques, and $\boldsymbol{G}(q)$ is an n-vector of gravity torques.

3.1 Properties

In reality, a rigid robot arm is always affected by friction and disturbances. Therefore, let us generalize the model (3) by writing the dynamics as

$$\boldsymbol{M}(q)\ddot{q} + \boldsymbol{V}(q,\dot{q}) + \boldsymbol{F}(\dot{q}) + \boldsymbol{G}(q) + \boldsymbol{\tau}_d = \boldsymbol{\tau} \tag{4}$$

where $\boldsymbol{F}(\dot{q})$ represents friction and τ_d represents external disturbances. The *friction term* can often be assumed to have the form [23]

$$\boldsymbol{F}(\dot{q}) = \boldsymbol{F}_v\dot{q} + \boldsymbol{F}_s \tag{5}$$

with $\boldsymbol{F}_v$ the coefficient matrix of *viscous friction* and $\boldsymbol{F}_s$ a *static friction* term. The *disturbance torque*, τ_d, represents unmodeled dynamics and other unknown terms. In this section we examine the structure and the properties of the terms in (4) that will be useful in designing control algorithms [17]. We should stress again that this formulation pertains to rigid robots and that some of the following properties may be lost if joint or link flexibilities are present.

Properties of the Inertia Matrix. $\boldsymbol{M}(q)$ is a symmetric, positive definite matrix. Another vital property of $\boldsymbol{M}(q)$ is that it is bounded above and below. That is,

$$\mu_1(q)I \le \boldsymbol{M}(q) \le \mu_2(q)I \tag{6}$$

where I is the identity matrix, $\mu_1(q) \neq 0$ and $\mu_2(q)$ are scalar constants for a revolute arm and generally scalar functions of q for an arm containing prismatic joints. The prismatic-joints case is not stressed in most robot textbooks but may be easily seen from a simple two-link cylindrical robot [24]. It is easy to see then that $\boldsymbol{M}^{-1}(q)$ is also bounded since

$$0 \le 0\frac{1}{\mu_2(q)}I \le \boldsymbol{M}^{-1}(q) \le \frac{1}{\mu_1(q)}I \tag{7}$$

Properties of the Coriolis and Centripetal Terms. The first observation one can make about the term $V(q,\dot{q})$ is that it is quadratic in the generalized velocity $\dot{q}$. In fact, $V(q,\dot{q})$ is bounded as follows:

$$\|\boldsymbol{V}(q,\dot{q})\| \le v_b(q)\|\dot{q}\|^2 \tag{8}$$

where $v_b(q)$ is a scalar constant for an all-revolute arm and a scalar function of q for arms containing prismatic joints. In addition there exist numerous factorizations of $V(q,\dot{q})$, one of which is given by

$$V(q,\dot{q}) = V_m(q,\dot{q})\dot{q} \tag{9}$$

This factorization is particularly important in controllers that exploit the passivity of the robot, where V_m is selected to satisfy the following property:

$$\dot{M}(q) - 2V_m(q, \dot{q}) \text{ is skew-symmetric} \tag{10}$$

Properties of the Gravity Terms. The gravity term $G(q)$ is bounded as follows

$$\|G(q)\| \le g_b(q) \tag{11}$$

where $g_b(q)$ is a scalar constant for revolute arms and a scalar function of q for arms containing prismatic joints.

Properties of the Friction Terms. Since friction is a local effect, we may assume that the friction terms $F(\dot{q})$ are uncoupled among the joints so that

$$F(\dot{q}) = \begin{bmatrix} f_1(\dot{q}_1) \\ f_2(\dot{q}_2) \\ \vdots \\ f_n(\dot{q}_n) \end{bmatrix} \tag{12}$$

where each $f_i(\dot{q}_i)$ is a known scalar function. In fact, we shall assume the following forms for the viscous and static frictions:

$$F_v = \operatorname{diag}\{v_i \dot{q}_i\}$$

$$F_s(\dot{q}) = \operatorname{diag}\{k_i\} \operatorname{sgn}(\dot{q}) \tag{13}$$

One can then bound the friction terms as follows

$$\|F(\dot{q})\| \le v\|\dot{q}\| + k \tag{14}$$

Properties of the Disturbance Torques. The disturbance torque τ_d is by definition unknown. It is, however, reasonable to assume that it is bounded by a known function

$$\|\tau_d\| \le d(q, \dot{q}) \tag{15}$$

In addition, equation (4) has the following properties:

Property 1: For the rigid robot model (4), there is an independent control input for each generalized coordinate.

This result is true for redundant and nonredundant robots. It fails to hold for robots possessing joint flexibilities.

Property 2: The dynamics equation (4) define a passive mapping from the input torque τ to the generalized velocity $\dot{q}$.

This property may be shown using Hamilton's equations of motion. In fact let the Hamiltonian H be defined by

$$H = \text{Kinetic Energy} + \text{Potential Energy} \tag{16}$$

It is then easy to show that [17]

$$\frac{dH}{dt} = \dot{q}^T \tau \tag{17}$$

from which it follows that

$$\int_0^t \dot{q}^T(u)\tau(u)\, du = H(t) - H(0) \ge -H(0) \tag{18}$$

which proves the passivity of the mapping between τ and $\dot{q}$ [25]. This property is intimately related to (10).

Property 3: The equation of motion (4) is linear in a suitably defined set of parameters.

In other words, one can write equation (4) as

$$Y(q, \dot{q}, \ddot{q})\theta = \tau \tag{19}$$

where $Y(q, \dot{q}, \ddot{q})$ is a matrix of known time functions and θ is a vector containing known functions of the robot's physical parameters such as link masses, moments of inertia, and other factors.

4. Control

Once a satisfactory model of the robot dynamics is obtained, control theory may be used to modify the actions and reactions of the robot to different stimuli. The particular controller used will depend on the complexity of the mathematical model, the application at hand, the available resources, and a host of other criteria. To control includes the following steps:

1. Planning
2. Sensing
3. Comparing
4. Reacting
5. Go to 1.

The first step is modified when other objects or robots are present. The motion is also constrained by kinematic and dynamic constraints (Part 1). Once motion is planned, the control loop 2–3–4–5 is activated. For robots, feedback linearization and passivity properties are used at this stage (Part 2).

4.1 Control Objectives

A useful robot is one whose trajectory in its workspace may be specified, and the forces it exerts on its environment may be controlled. Looking back at equation (4), our control objectives will usually fall into one of the following categories:

Motion Control. Given a desired trajectory, specified by the vector time functions $q_d(t)$ and $\dot{q}_d(t)$, design and implement a controller whose output $\tau(t)$ will drive the actual trajectory $\{q(t), \dot{q}(t)\}$ to $\{q_d(t), \dot{q}_d(t)\}$ asymptotically. Note that the desired trajectory is usually specified in the task space and some preprocessing is required to obtain a desired joint space trajectory. The alternative would be to obtain the dynamics of the robot in the task space where the desired trajectory is specified. The fine motion control is then accomplished using precision movements of the end-effector.

Force Control. When the robot comes in contact with its environment, the contact forces and reactions of the robot need to be regulated. It is important, for example, that a robot holding a load should exert enough force to carry the load but not so much as to crush it. In another example, the robot may come in contact with a very stiff environment that may require some compliance on the part of the robot. The force control requirements are specified by a desired force vector $f_d(t)$ in the task space. This force vector is the trajectory in the force space that the end-effector should follow, when a force controller is properly designed. We should caution here that forces felt by the robot at the moment of impact with its environment are highly impulsive and as such are not addressed in this book.

Motion and Force Control. In some cases, the robot is required to follow a desired motion trajectory while exerting a certain force on its environment. In this case, both previous control objectives are combined to design a suitable controller.

4.2 Control Techniques

The dynamic model of rigid robots is well understood as mentioned at the beginning of this chapter, and although the describing equations are highly nonlinear, many theoretically sound controllers can be designed and implemented when all physical parameters are accurate. In other words, because the structure of rigid robots is well understood, if one can eliminate the parametric uncertainties, many control techniques are available. Among such techniques we list the Inverse-Dynamics (or computed-torque) approach [27], which is a particular case of the feedback-linearization approach for nonlinear systems [27]. This will lead to a linear multivariable control problem where many standard performance measures are meaningful [28], and where pole-placement and optimal control methods are readily available.

Since models are inherently uncertain, we require either robustification or adaptive control (Parts 3 and 4). Such control methods have been used in deriving (and sometimes implementing) performance-oriented robot controllers. In repetitive tasks, learning control is a suitable alternative and one step beyond adaptive control (Part 5). It may even be argued that most robots are in effect employed to do repetitive or even periodic tasks. Finally, the time-optimal control of robots is studied (Part 6), along with the force-control problem (Part 7).

To account for mechanical flexibilities, a reexamination of the basic dynamic model (3) is required (Part 8). Further control performance is obtained with the introduction of dexterous end-effectors (Part 9) and redundant manipulators (Part 10). Finally, advanced robotic applications are presented (Part 11).

5. Organization of the Book

Because of the vast research in robot control and the space limitations of a reprint volume, we have selected ten topics that represent what we see as major areas of research. Within each topic we have attempted with the help of the section editors to choose papers that, in our opinion, are seminal or final. In the remainder of this section, we briefly describe the organization of the enclosed topics, leaving to the section editors the difficult task of introducing their particular sections.

The first part is concerned with the motion-planning problem, which is the first step in designing a robot controller. As discussed in the introduction to that part, motion planning may also be a part of the control strategy such as in the Act-While-Thinking strategy.

The second part is concerned with theoretically established robot control methodologies that have yet to be applied in industrial robots. These controllers exploit the structure of the robot's dynamics, and are for the most part motion controllers.

The third part reviews different robustification schemes of robot controllers. It concentrates on motion control and presents three separate philosophies of robust control, (1) the sliding-mode approach, (2) the use of functional analytic methods, and (3) a Lyapunov-based variable-structure approach. These methodologies represent the major different robust controllers in robotics.

The fourth part deals with the adaptive motion control of robots. This particular part is doubly informative because it shows that systems described by nonlinear and multivariable equations may be successfully controlled with adaptive controllers.

In the fifth part, learning control strategies are discussed. Since most robots are used to repeat a certain task, the idea of learning through practice is very applicable in robotics.

The sixth part concludes what we may call the classical control objectives by addressing the time-optimal control problem. With the mounting concerns of efficiency and productivity, it is appropriate that robots should be made to move faster as well as more accurately.

In Part 7 a new era of robot control is described, because a useful robot must interact with its environment in a controlled manner. In fact, there are those who contend that the motion-control problem of rigid robots is completely solved and all interesting problems remain in the arena of force control.

Part 8 is concerned with the modeling and control of flexible-link robots. The dynamics of these robots are highly complex and their control difficult. One may argue that all robots are rigid only to a first approximation and that their joint and link flexibilities impose important constraints on their performance. In addition, certain environments will require the use of specialty robots designed for maximum flexibility.

In Part 9 the control of robotic hands is described. The kinematics and dynamics of the hand as it grasps an object are presented. Moreover, a robotic hand is often used to effect fine motion and its control poses unique problems as described in the selected papers.

Redundant robots are discussed in Part 10. These robots are often used to provide a higher level of performance than that possible with nonredundant robots. The redundancy is often used to advantage, but the control algorithms are more complicated.

Finally, Part 11 describes some of the more advanced applications of robotics. In fact, the robots described in this part provide us with a vision of the future of robotics.

REFERENCES

[1] C. S. G. Lee, R. C. Gonzales, and K. S. Fu, *Tutorial on Robotics*, 2nd ed., Silver Springs, MD: IEEE Computer Society Press, 1986.

[2] M. Brady, J. H. Hollerbach, T. L. Johnson, T. Lozano-Perez, and M. T. Mason, Eds., *Robot Motion: Planning and Control*, Cambridge, MA: MIT Press, 1982.

[3] A. J. Critchlow, *Introduction to Robotics*, New York: Macmillan, 1985.

[4] G. Beni and S. Hackwood, Eds., *Recent Advances in Robotics*, New York: Wiley, 1985.

[5] J. Denavit and R. S. Hartenberg, "A kinematic notation for lower pair mechanisms based on matrices," *J. of Appl. Mech.*, vol. 22, pp. 215–221, June 1955.

[6] J. F. Engelberger, *Robotics in Practice*, New York: AMACOM, American Management Association, 1980.

[7] L. G. Roberts, "Homogeneous matrix representation and manipulation of n-dimensional constructs," *Document No. MS1045*, Lincoln Laboratories, Cambridge, MA: MIT Press, 1965.

[8] D. Pieper, "The kinematics of manipulators under computer control," Ph.D. Thesis, Stanford, CA: Stanford University, 1968.

[9] M. E. Kahn and B. Roth, "The near minimum-time control of open-loop articulated kinematic chains," *J. Dyn. Sys., Meas., and Cont.*, vol. 93, pp. 164–172, 1971.

[10] D. E. Whitney, "Force feedback control of manipulator fine motions," *IEEE Proc. American Control Conf.*, San Francisco, CA, pp. 694–699, 1976.

[11] V. C. Scheinman, "Design of a computer-controlled manipulator," *Memo AIM-92* Artificial Intelligence Laboratory, Stanford, CA: Stanford University, 1969.

[12] A. K. Bejczy, "Robot arm dynamics and control," *Technical Memo 33-669*, Pasadena, CA: NASA Jet Propulsion Laboratory, 1974.

[13] R. E. Hohn, "Applications flexibility of a computer-controlled industrial robot," in *Industrial Robots*, 2nd ed., vol. 1, W. R. Tanner, Ed, Dearborn, MI: Robotics International of SME, 1981, pp. 224–242.

[14] C. A. Rosen and D. Nitzan, "Developments in programmable automation," in *Industrial Robots*, 2nd ed., vol. 1, W. R. Tanner, Ed., Dearborn, MI: Robotics International of SME, 1981, pp. 254–258.

[15] J. Y. S. Luh, "Conventional controller design for industrial robots: A Tutorial," *IEEE Trans. Sys., Man, and Cybernetics*, vol. SMC-13, no. 3, pp. 298–316, May/June 1983.

[16] S. Arimoto and F. Miyazaki, "Stability and robustness of PID feedback control for robot manipulators of sensory capability," *Proc. 1st Int. Symp. Robotics Res.*, pp. 783–799, 1983.

[17] M. W. Spong and M. Vidyasagar, *Robot Dynamics and Control*, New York: Wiley, 1989.

[18] H. Asada and K. Youcef-Toumi, *Direct-Drive Robots: Theory and Practice*, Cambridge, MA: MIT Press, 1987.

[19] C. Abdallah, D. Dawson, P. Dorato, and M. Jamshidi, "Survey of robust control for rigid robots," *IEEE Cont. Sys. Magazine*, vol. 11, no. 2, pp. 24–30, Feb. 1991.

[20] R. Ortega and M. W. Spong, "Adaptive motion control of rigid robots: a tutorial," *Proc. IEEE Conf. Dec. and Cont.*, Austin, TX, pp. 1575–1584, 1988.

[21] J. J. Craig, *Introduction to Robotics: Mechanics and Control*, 2nd ed., Reading, MA: Addison-Wesley, 1989.

[22] H. Goldstein, *Classical Dynamics*, Reading, MA: Addison-Wesley, 1950.

[23] R. J. Schilling, *Fundamentals of Robotics: Analysis and Control*, Englewood Cliffs, NJ: Prentice-Hall, 1990.

[24] F. Lewis, C. Abdallah, and D. Dawson, *Robot Control*, New York: Macmillan, to appear 1992.

[25] C. Desoer and M. Vidyasagar, *Feedback Systems: Input-Output Properties*, New York: Academic Press, 1975.

[26] R. Paul, *Robot Manipulators: Mathematics, Programming, and Control*, Cambridge, MA: MIT Press, 1981.

[27] K. Kreutz, "On manipulator control by exact linearization," *IEEE Trans. Auto. Cont.*, vol. 34, no. 7, pp. 763–767, July 1989.

[28] M. W. Spong and M. Vidyasagar, "Robust linear compensator design for nonlinear robotic control," *IEEE J. Robotics and Autom.*, vol. RA-3, no. 4, pp. 345–351, Aug. 1987.

C. ABDALLAH AND F. L. LEWIS

Part 1
Motion Planning

VLADIMIR J. LUMELSKY
University of Wisconsin-Madison

GENERATING collision-free motion of acceptable quality is one of the main concerns in robotics. A typical robot presents an arm manipulator with a fixed base operating in three-dimensional space, or a mobile vehicle operating in two-dimensional space, or a combination of the two. Whatever form it takes, the robot is expected to move purposely and safely in an often complex environment filled with known or unknown obstacles.

There are two distinct models and categories of approaches—global and local—into which all existing techniques for robot motion planning largely fall. Each category includes provable (that is, nonheuristic) as well as heuristic approaches, and each has its own advantages and difficulties. The important assumption in the first category—some names for it in literature are *Piano Movers model*, *model-based approach*, and *motion planning with complete information*—is that full information about the geometry of the robot and the obstacles in the environment is given beforehand, so path planning becomes a one-time off-line operation. Information about the environment is usually presented algebraically (e.g., obstacles are assumed to be polygons).

This approach follows the paradigm of human cognitive process pioneered by A. Newell and H. Simon in the 1950s and 1960s [1], which is in essence as follows: First, information necessary for the task is collected, then a world model as accurate as possible is built, then the complete solution to the task is found, and only after that does the task execution begin. In other words, task execution is seen as a preprogrammed rigid process with little information processing; the approach could be called *Act-After-Thinking*. Its main pluses and minuses are all related to the notion of complete information: In principle, most general cases can be handled and optimal performance can be delivered; in terms of applications, the approach is at its best in tightly structured tasks such as an assembly work cell. On the other hand, computational bottlenecks are the approach's major curse, and its open-loop nature makes it brittle in unstructured and time-changing environments where sensory feedback is essential.

The second category of approaches is often called *sensor-based motion planning* or *motion planning with incomplete information*; as a companion to the catchy term Piano Movers model, call it the *South Pole Search model* (see below). The main distinction of this model is that information about obstacles is assumed to be unknown or at least partial, and this is compensated by local on-line information coming, for example, from sensory feedback. No detailed model of the environment is assumed; planning is done continuously, based on whatever partial information is available at the moment. This paradigm of the cognitive process could be called *Act-While-Thinking*. Its obvious strength is an ability to deal with uncertainty typical of unstructured environments. Since relatively little information is used at a single step, the computational and memory requirements here are low. On the negative side, generality is an elusive goal, optimality is ruled out, and handling multidimensional cases and assuring convergence are difficult algorithmic issues.

A good part of motion planning by humans and animals seems to follow the Act-While-Thinking paradigm. Consider this example: You are a Robert Scott, and you set out to reach the South Pole. At every moment of your journey you know more or less where you are and where the Pole is. A straight line motion that you would prefer is not feasible: Your actual path will be full of intricate wiggles and loops and dead-end retreats caused by crevices, walls, hills, and valleys. The decisions on where to go next will be dictated by local information from sensors—perhaps your vision. Optimality is certainly not a strong side of your strategy: In retrospect, you might think of a better, more streamlined path, but it would be too late; even on the way back it would be of limited use because the environment would change.

The seeming incompatibility of these two models has not stopped researchers from attempting to combine their advantages while minimizing their drawbacks—which produces a relatively small body of work between the two main categories. Similar to the tunnel *projet du siecle* between France and England, the work progresses simultaneously from both directions. On the side of the Piano Movers model, the main thrust is on computationally efficient algorithms for the general and more restricted cases. On the side of the South Pole Search model, the emphasis is on extending the approach to more general cases. Besides, more direct global-local approaches have been attempted in which the planning job is carefully separated into subtasks requiring respectively global and local information.

The collection of four papers presented in this part addresses some of the algorithmic issues at the center of the aforementioned approaches. Obviously, the collection is too small to do justice to the whole area of robot motion planning. The first two papers relate to the Piano Movers model, the third to the South Pole Search model, and the fourth to the global-local approach.

It has been shown by J. Reif in his 1979 paper [2] that the Piano Movers problem is PSPACE-hard, that is, at least exponential, in the number of degrees of freedom in the system. The first provable algorithms for the general case, notably the 1983 work by J. T. Schwartz and M. Sharir [3], are double-exponential. Whether a single-exponential algorithm indeed exists has been an open question. The paper by Canny presented here resolves the issue—the answer is "yes." To achieve this result and to build the required algorithm, the developed method makes use of the notion of stratified sets in singularity theory.

The paper by Hopcroft and Wilfong addresses the following issue: Can planning with complete information be successfully done in some subspace of the configuration space (i.e., the space of the available degrees of freedom) instead of the whole space? A positive answer would promise an additional resource in the search for computationally manageable algorithms. In simple terms, the presented result says that if the configuration space at hand satisfies certain topological conditions, then instead of analyzing the free space, the search for a path among obstacles can be limited to obstacle surfaces only, that is, to a space of smaller dimensionality than the free space. One is left with a feeling that some algorithmic consequences of this result are yet to be realized.

The problem of motion planning for simple arm manipulators in the context of the South Pole Search model is considered in the paper by Lumelsky. It is shown that purely local information about obstacles in the arm vicinity, such as data from a skinlike tactile or proximity sensor covering the arm body, is sufficient for planning collision-free motion in a complex environment with unknown obstacles of arbitrary shapes. The resulting provable algorithms are surprisingly inexpensive computationally, and the motion they prescribe makes the arm cling to the obstacles, instead of exploring the free space—one can see a connection to the Hopcroft and Wilfong paper. The fact that nonheuristic motion planning can be done with so little input information poses a fundamental question: What is the amount of input information that is necessary and sufficient for motion planning?

Finally, the paper by Barraquand and Latombe explores the global-local approach: The authors show that by using the known technique of potential functions, motion planning can be reduced in some rather nontrivial cases to a simpler problem of following one-dimensional roadmaps. First, an appropriate potential function is computed off-line based on complete information about the environment. It is shown then that in many cases the resultant potential field is characterized by one-dimensional valleys, which can easily be explored via local inexpensive techniques.

References

[1] A. Newell and H. Simon, "GPS, a program that simulates human thought," in *Computers and Thought*, E. Feigenbaum and J. Feldman, Eds., New York: McGraw-Hill, pp. 279–293, 1963.

[2] J. Reif, "Complexity of the mover's problem and generalizations," *Proc. 20th Symp. Foundations of Computer Science*, 1979.

[3] J. T. Schwartz and M. Sharir, "On the "piano movers" problem. II. General techniques for computing topological properties of real algebraic manifolds," *Adv. Appl. Math.*, vol. 4, pp. 298–351, 1983.

A New Algebraic Method for Robot Motion Planning and Real Geometry

John Canny

Artificial Intelligence Laboratory,
Massachusetts Institute of Technology

and

Computer Science Division,
University of California, Berkeley

Abstract We present an algorithm which solves the findpath or generalized movers' problem in single exponential sequential time. This is the first algorithm for the problem whose sequential time bound is less than double exponential. In fact, the combinatorial exponent of the algorithm is equal to the number of degrees of freedom, making it worst-case optimal, and equaling or improving the time bounds of many special purpose algorithms. The algorithm accepts a formula for a semi-algebraic set S describing the set of free configurations and produces a one-dimensional skeleton or "roadmap" of the set, which is connected within each connected component of S. Additional points may be linked to the roadmap in linear time. Our method draws from results of singularity theory, and in particular makes use of the notion of stratified sets as an efficient alternative to cell decomposition.

We introduce an algebraic tool called the *multivariate resultant* which gives a necessary and sufficient condition for a system of homogeneous polynomials to have a solution, and show that it can be computed in polynomial parallel time. Among the consequences of this result are new methods for quantifier elimination and an improved gap theorem for the absolute value of roots of a system of polynomials.

1 Introduction

This work is motivated by the abstract path planning problem in robotics. Given a way of specifying the configuration of a moving object, constraints on its motion define a subset of its configuration space within which the object is clear of obstacles, which is called "free-space" [Loz] The abstract motion planning problem is to find a continuous path between two specified configurations such that the path is collision-free i.e. it lies entirely in the set of safe configurations.

In virtually all cases the constraints on motion have an algebraic description, and the free-space can be defined as a semi-algebraic set [SS], [Ca86]. In principle, algorithms based on cellular algebraic decomposition [Col], [BKR] can then be used to decompose free space, and the existence of a path demonstrated by computing cell adjacency [SS], [KY], [Pri]. However, these algorithms are also capable of deciding the theory of reals with addition, which is hard for exponential time with a polynomial number of alternations [Ber]. The average case performance of cell decomposition algorithms is not much better than worst case and they basically run in double-exponential time. Here the crucial input parameter is the number of variables, or the dimension of the configuration space. Thus cell decomposition seems completely impractical for even simple problems. Reif [Rei] first showed that the general robot motion planning problem is PSPACE-hard, which suggests that single exponential time may be required in the worst case.

Acknowledgements. This report describes research done at the Artificial Intelligence Laboratory of the Massachusetts Institute of Technology. Support for the laboratory's artificial intelligence research is provided in part by the Office of Naval Research under Office of Naval Research contract N00014-81-K-0494 and in part by the Advanced Research Projects Agency of the Department of Defense under Office of Naval Research contract N00014-85-K-0124. John Canny is supported by an IBM Fellowship.

1.1 Summary of Results

On the other hand, there do exist many efficient algorithms for restricted classes of motion planning problems [SY], [SHS]. This suggests that it may be possible to solve the general problem in exponential time. This is quite useful in practice, where the number of degrees of freedom is best thought of as a small constant rather than a input parameter. We present the such an algorithm here. In fact the running time of our algorithm equals or improves the worst-case performance of many of the special purpose algorithms that have appeared over the last few years, especially with regard to algebraic complexity. From appendix I, the running time of the algorithm is

$$T_{R_1} = \left(n^r(\log n(2d)^{O(r)}\log^3 w + (2dr)^{O(r^2)}\log^2 w)\right) \quad (1)$$

where r is the number of degrees of freedom, n is the number of constraints (product of the number of edges, vertices etc. in the obstacles and the moving object), and d and w are respectively the degree and coefficient magnitude of the constraint polynomials.

The bound may be weakened to a product of "geometric" part $O(n^r \log n)$, measuring growth with environment complexity, and an "algebraic" part $O((2dr)^{O(r^2)}\log^3 w)$ measuring growth with degree and coefficient size. Previous algorithms for the general problem exhibit double exponential growth in both parts. There are two new tools which make single exponential bounds possible. The first of these is the notion of stratification, which is a partition of S into smooth pieces (manifolds). We use the coarsest possible stratification to limit the geometric complexity of the algorithm.

Reprinted from *Proc. IEEE COMPUS '87*, Los Angeles, CA, pp. 39–48, Oct. 1987.

The second tool is the multivariate resultant, which allows simultaneous elimination of variables in single exponential sequential time, or polynomial parallel time. We also derive from bounds on the size of the resultant an improved gap theorem. This theorem is used throughout the roadmap algorithm to reduce computation with algebraic numbers to computation with finite precision binary numbers.

1.2 Outline of Paper

Some definitions and basic results from singularity theory are given in section 2. Stratifications are described in section 3, where we also give a constructive definition of the notion of a tube around a variety. Section 4 gives a simple mathematical definition of the roadmap, while the algorithm itself is described in appendix I. Section 5 describes the multivariate resultant, and gives a polynomial parallel time algorithm for its computation. We also give a gap theorem which gives a tight bound on the rate of exponential growth of the number of bits required to describe and algebraic number.

2 Preliminaries

Let Q_r be the ring of polynomials in r variables with rational coefficients, $Q_r = Q[x_1, \ldots, x_r]$.

Definition. A *semi-algebraic set* $S \subseteq \Re^r$ defined by the polynomials $F_1, \ldots, F_n \in Q_r$ is a set derived from the sets

$$S_i = \{x \in \Re^r \mid F_i(x) > 0\} \tag{2}$$

by union, intersection and complement.

Definition For a map $f : \Re^m \to \Re^n$ the *differential* of f relates differential changes in $\mathrm{dom}(f)$ to the corresponding changes in the value of f. The differential of f at x is denoted df_x, and is given by the Jacobian:

$$df_x = \begin{pmatrix} \frac{\partial f_1}{\partial x_1}(x) & \cdots & \frac{\partial f_1}{\partial x_m}(x) \\ \vdots & & \vdots \\ \frac{\partial f_n}{\partial x_1}(x) & \cdots & \frac{\partial f_n}{\partial x_m}(x) \end{pmatrix} \tag{3}$$

The notion of differential generalizes to maps between manifolds. Let f be a smooth map of manifolds $f : M \to N$, where M and N have dimension m and n respectively, and let T_xM denote the tangent space to M at x. We define the differential of f at $x \in M$, (denoted df_x) as the induced map of tangent spaces $df_x : T_xM \to T_{f(x)}N$. Using local coordinates about x and $f(x)$, the differential is again given by the Jacobian (3).

Definition. A point x in M is a *critical point* of $f : M \to N$ if the differential df_x is not surjective at x, i.e. if the Jacobian (3) has rank less than $\dim(N)$ at x. We will call the set of all critical points of f its *critical set*, and denote it by $\Sigma(f)$. Points where df_x is surjective are called *regular points.*

The image of a critical point $x \in M$ is a critical *value* of f. Values $v \in \Re^n$ which are not critical values are called *regular* values. So a value v is regular if and only if all the points in $f^{-1}(v)$ are regular.

A generalization of the regularity of a map is the notion of transversality.

Definition. Let $f : M \to N$ be a smooth map, and let V be a submanifold of N. We say that f is *transversal* to V if

$$df_x(T_xM) + T_{f(x)}V = T_{f(x)}N \tag{4}$$

at every point $x \in f^{-1}(V)$, and we denote this relationship by $f \pitchfork V$.

If $f \pitchfork V$, then the set $f^{-1}(V)$ is either empty or a manifold whose codimension is $\mathrm{codim}(V)$. (The codimension of a submanifold $V \subset N$ is $\mathrm{codim}(V) = \dim(N) - \dim(V)$). We will make use of the following result which is derived from Sard's theorem. A proof of both theorems may be found in [GG].

Generic Map Lemma Let M, N, B be smooth manifolds, and let V be a submanifold of N. Suppose $\Phi : M \times B \to N$ is a smooth map such that for every $b \in B$, $\Phi_b : M \to N$ is smooth, where $\Phi_b(x) = \Phi(x, b)$. If $\Phi \pitchfork V$, then $\Phi_b \pitchfork V$ for almost all $b \in B$ (all but a measure zero subset).

In other words, if a parametrized class of mappings is transversal to V, then almost every map in the class is transversal to V.

3 Stratifications

Stratifications (or stratified sets) have their origin in a paper of Whitney [Whi] on the geometry of real algebraic sets. He was able to show that singular algebraic (and semi-algebraic) sets of dimension d may be partitioned into manifolds of dimensions ranging from 0 to d. This result has been extended to semi- and sub-analytic sets, and such a partition has come to be known as a stratification. The later addition of certain regularity conditions (also due to Whitney) lead to isotopy theorems analogous to those for manifolds. Stratifications enjoy many of the useful properties of manifolds, and in particular they can be triangulated [Joh]. More recently, stratifications have played an important role in proofs of some fundamental stability theorems for maps between manifolds [GWD].

Stratifications are highly applicable to computational geometry and robotics, since the class of sets which are easily described as stratified sets includes polyhedra, constructive solid geometric (CSG) models, and the transformed obstacles in the configuration space [Loz] of virtually any robot. The arrangement algorithm for hyperplanes of [EOS] may be viewed as a stratification procedure. So can the cell decomposition algorithm for semi-algebraic sets of [Col]. The latter is however extremely fine, and ill-suited to geometric applications, since it produces a double exponential (in the dimension of the space) number of strata. We will be here considering a coarsest possible stratification of semi-algebraic sets, which has a single exponential number of strata.

Definition A *stratification* $\underline{\underline{S}}$ of a set $S \subset \Re^r$ is partition of S into a finite number of disjoint subsets S_i called *strata* such that each S_i is a manifold.

The Whitney Conditions Let U and V be any two strata of a stratification $\underline{\underline{S}}$, such that $U \subset \bar{V}$. Let $x \in U$ and let (x_i) and (y_i) be any two sequences in U and V respectively which converge to x. The following are called the Whitney conditions:

(A) If $(T_{y_i}V) \to \tau$, then $T_xU \subset \tau$.
(B) If $(T_{y_i}V) \to \tau$ and $(\overline{x_iy_i}) \to l$, then $l \subset \tau$.

where $\overline{x_iy_i}$ is the line passing through x_i and y_i. Note that the second condition implies the first. A stratification which satisfies condition (B) is called a *Whitney stratification.*

We next show that the stratification into sign-invariant sets of a collection of polynomials in general position satisfies the Whitney conditions. First we define $R^- = (-\infty, 0)$, $R^+ = (0, \infty)$, and let $\underline{\underline{\Re}}$ denote the set $\{R^-, \{0\}, R^+\}$. Then $\underline{\underline{\Re}}$ is a Whitney stratification of the real line. It is simple to show that a product of Whitney stratifications is also a Whitney stratification, so that $\underline{\underline{\Re}}^n = (\underline{\underline{\Re}})^n$ is a Whitney stratification of $\Re^n$.

Definition. Let $F : \Re^r \to \Re^n$ be a polynomial map. A *sign sequence* σ is an element of $\underline{\underline{\Re}}^n$, so each σ is a manifold. Each set $F^{-1}(\sigma)$ is called a *sign-invariant set* of F. The map F is said to *define* the semi-algebraic set $S \subset \Re^r$ if S can be written as a union of sign-invariant sets of F.

We now generalize the transversality notation and write $F \pitchfork \underline{\underline{A}}$ where $\underline{\underline{A}}$ is a collection of manifolds, and take this to mean that $F \pitchfork U$ for every $U \in \underline{\underline{A}}$. Now we can show that almost any perturbation in the constant terms of all the polynomials that define S causes the sign-invariant sets to be a Whitney stratification of $\Re^r$.

General position lemma #1.
Let $F : \Re^r \to \Re^n$ be a polynomial map, $\epsilon \in \Re^n$ a parameter. Define $F_\epsilon : \Re^r \to \Re^n$ as $F_\epsilon(x) = F(x) + \epsilon$, then for almost all $\epsilon \in \Re^n$, $F_\epsilon \pitchfork \underline{\underline{\Re}}^n$, and $F_\epsilon^{-1}(\underline{\underline{\Re}}^n)$ is a Whitney stratification of $\Re^r$.

The proof uses basic properties of stratifications which can be found in [GWD], section 1. Notice that if $F : \Re^r \to \Re^n$ is a polynomial map that defines the semi-algebraic set S, and $F \pitchfork \underline{\underline{\Re}}^n$, then the sign invariant sets contained in S constitute a Whitney stratification of S. In future we will assume that S can be stratified this way, and that $\underline{\underline{S}}$ denotes the stratification into sign-invariant sets.

Once we have $S \subset \Re^r$ represented as a stratification, for any map $f : \Re^r \to \Re^k$, we can define the critical set of $f|_{\underline{\underline{S}}}$ as the union of the critical sets of all maps $f|_U$ (this denotes the restriction of the domain of f to U) where $U \in \underline{\underline{S}}$. Critical values of $f|_{\underline{\underline{S}}}$ are the images of critical points, and all other points in $\Re^k$ are regular values. Since there are only finitely many strata, the set of critical values of $f|_{\underline{\underline{S}}}$ has measure zero in $\Re^k$. In particular, if f is real-valued and S is semi-algebraic, $f|_{\underline{\underline{S}}}$ has a finite number of critical values. A crucial property of stratifications is that the topological type of sections $f|_{\underline{\underline{S}}}^{-1}(v)$ changes only at critical values of $f|_{\underline{\underline{S}}}$.

Isotopy Theorem Let $f : \Re^r \to \Re$ be a smooth mapping, and let S be a compact subset of $\Re^r$ with Whitney stratification $\underline{\underline{S}}$. If (c, d) is an open interval in $\Re$ containing no critical values of $f|_{\underline{\underline{S}}}$, then all sections $f|_{\underline{\underline{S}}}^{-1}(v)$ with $v \in (c, d)$ are isotopic, and $f|_{\underline{\underline{S}}}^{-1}(c, d)$ is homeomorphic to $f|_{\underline{\underline{S}}}^{-1}(v) \times (c, d)$.

This result was first proved by Thom [Th69], or can be inferred from [GWD] section II, theorem 5.2. From now on we will use the slice notation $S|^B = f|_{\underline{\underline{S}}}^{-1}(B)$ where B is either a real value or an interval in $\Re$. The result can be extended to:

Retraction Lemma. Let $(c, d) \subset \Re$ be an interval containing no critical values of $f|_{\underline{\underline{S}}}$, with c possibly a non-degenerate critical value. Then $S|^{(-\infty, c]}$ is a deformation retract of $S|^{(-\infty, d)}$.

Later we will need to be able to choose a projection function a such that $\Sigma(a|_U)$ is a 1-manifold for any $U \in \underline{\underline{S}}$. This will be the case if the map a is sufficiently "generic". In order to formalize this and prove the result, we use jet transversality. A description of jets and one-genericity is given in [GG]. Using these notions we obtain:

General position lemma #2.
Let $a : \Re^r \to \Re^2$ be a linear map, and $M \subset \Re^r$ a manifold of $\dim(M) \geq 1$. Then for almost all $a \in \Re^{2r}$, the critical set $\Sigma(a|_M)$ is a one-dimensional manifold.

3.1 Tubes Around Varieties

We can reduce the problem of computing critical points on varieties of arbitrary dimension to critical point analysis for certain hypersurfaces. Suppose we wish to find the critical points of the map a restricted to the non-singular algebraic set M defined by $f_1 = 0, \ldots, f_n = 0$. We define a new polynomial

$$g_\epsilon = \sum_i f_i^2 + \epsilon \tag{5}$$

This construction was used by both Thom [Th65] and Milnor [Mil] in their bounds for Betti numbers of varieties. Now let $(\epsilon_i) \to 0$ be a sequence, there is a critical point of $a|_{\ker(g_{\epsilon_i})}$ converging to every critical point of $a|_M$.

4 The Roadmap

We now have the hardware necessary to construct a one dimensional semi-algebraic subset of S which we call the roadmap of S and which we denote $R(S)$. This subset serves as a skeleton for motion planning and provides candidate paths for motion between any two configurations. In order to guarantee that solutions may be found along the roadmap, it must satisfy the following condition:

Definition A subset R of a set S satisfies the *roadmap condition* if every connected component of S contains a single connected component of R.

Let $\pi_{ij} : \Re^r \to \Re^2$ be projection on the i and j coordinates. We define the silhouette Σ as $\Sigma(\pi_{12}|_{\underline{\underline{S}}})$. Before describing the roadmap, we define a refinement of the stratification $\underline{\underline{S}}$ of S, to give a new Whitney stratification $\underline{\underline{S}}'$, called the *Silhouette compatible stratification* of S. This stratification includes as a subset, a stratification of the silhouette Σ.

Let $\underline{\underline{\Sigma}} = \{\Sigma(\pi_{12}|_U) | U \in \underline{\underline{S}}\}$ be a stratification of Σ. This is not a Whitney stratification, since a boundary point of one curve may lie inside another. If however, we define $B = \{\partial C | C \in \underline{\underline{\Sigma}}\}$ as the set of such points, we can construct a finer stratification of Σ which is a Whitney stratification. Let $\underline{\underline{B}}$ be the stratification of B into one point sets, then

$$\underline{\underline{\Sigma}}' = \underline{\underline{B}} \cup \{U - B | U \in \underline{\underline{\Sigma}}\} \quad (6)$$

is a Whitney stratification of Σ. If now we define

$$\underline{\underline{S}}' = \underline{\underline{\Sigma}}' \cup \{U - \Sigma | U \in \underline{\underline{S}}\} \quad (7)$$

then $\underline{\underline{S}}'$ is a Whitney stratification of S, which includes $\underline{\underline{\Sigma}}'$ as a subset. In what follows we will use the notation $\underline{\underline{A}} \cap V = \{U \cap V | U \in \underline{\underline{A}}\}$ and the slice notation $\underline{\underline{A}}|^B = \underline{\underline{A}} \cap \pi_1^{-1}(B)$ where B is a point or an interval in $\Re$.

4.1 The Basic Roadmap

Definition. Let $\underline{\underline{S}}$ be a Whitney stratification of a compact set. Then the roadmap $R_0(\underline{\underline{S}})$ is defined inductively as follows:

1. If $\dim(\underline{\underline{S}}) = 1$, then $R_0(\underline{\underline{S}}) = S$.
2. Otherwise let $\Sigma = \Sigma(\pi_{12}|_S)$ be the silhouette of S and let $\underline{\underline{S}}'$ be the silhouette compatible stratification of S. Let $P_c \subset \Re$ be the (finite) set of critical values of $\pi_1|_{\underline{\underline{S}}'}$. The roadmap $R_0(\underline{\underline{S}})$ is given by

$$R_0(\underline{\underline{S}}) = \Sigma \cup \bigcup_{v \in P_c} R_0(\underline{\underline{S}}'|^v) \quad (8)$$

That is, the roadmap of a one dimensional set is the set itself. Otherwise the roadmap is the union of the silhouette Σ and the roadmaps of slices through $\underline{\underline{S}}'$ at critical values of $\pi_1|_{\underline{\underline{S}}'}$. The construction is well defined because if $\dim(\underline{\underline{S}}) = k$ then the slice $\underline{\underline{S}}'|^v$ has dimension at most $k-1$. We claim that $R_0(\underline{\underline{S}})$ satisfies the roadmap condition, but to prove this we need the following inductive lemma:

Lemma. If for each slice $S|^v$ through a critical value $v \in P_c$, $R_0(\underline{\underline{S}}'|^v)$ satisfies the roadmap condition, then $R_0(\underline{\underline{S}})$ satisfies the roadmap condition.

The proof makes use of the retraction lemma of the last section. With this lemma, we can show by induction on dimension:

Theorem. $R_0(\underline{\underline{S}})$ satisfies the roadmap condition.

In fact there is no need to compute the complete roadmap of a slice through a critical value, because only those strata in the neighborhood of the critical point are affected. This observation leads to a more efficient algorithm, and locally constructed curves are described in the next section.

4.2 Linking Curves

We have set ourselves the subgoal of finding a one-dimensional set which satisfies the roadmap condition within the closure of every stratum in $\underline{\underline{S}}$. The first step is to define a link curve for a stratum U which joins any point p in $\bar{U}$ to a distinct point q which is contained in the silhouette of $\bar{U}$. This curve should lie entirely within the slice $\bar{U}|^v$.

Now U is contained in the kernel of a set of n polynomials, with $n = \text{codim}(U)$. Similarly, $U|^v$ is contained in $\hat{U}$, which is the intersection of this kernel with the set $\pi_1^{-1}(v)$. In fact $\hat{U} - \{p\}$ is a manifold of codimension $n+1$, (even if p is a critical point). So the set $\underline{\underline{\hat{U}}} = \{\hat{U} - \{p\}, \{p\}\}$ is a Whitney stratification of $\hat{U}$.

Definition We define $L_0(p, U)$ as the connected component of $R_0(\underline{\underline{\hat{U}}}) \cap \bar{U}$ which contains p, where the silhouette of $\underline{\underline{\hat{U}}}$ is computed using the linear map π_{23}.

Now the link $L_0(p, U)$ contains a point which is extremal wrt π_2. This point, call it p_1, must either lie on the silhouette of U, or it must be at the endpoint of a closed curve segment, which implies that it lies in some stratum U_1, which is adherent to U. Then the link $L_0(p_1, U_1)$ must contain a point p_2 in the silhouette of U_1, or p_2 must lie in some stratum adherent to U_1 etc. Now we set

$$L_1(p, U) = L_0(p, U) \cup \bigcup_{i<k-1} L_0(p_i, U_i) \quad (9)$$

where k is the least i s.t. $p_i \in \Sigma(\bar{U})$. This is well defined because $\dim(U_i) < \dim(U_{i-1})$. Then $L_1(p, U)$ is a connected one-dimensional set which contains both p and a point on the silhouette of $\bar{U}$. This set suffices as the link curve as long as p is a regular point of $\pi_1|_U$. However, if p is a critical point, several components of U may "meet" at p for the first time. That is, let C be the component of $U|^v$ which contains p. $C - \{p\}$ may not be connected, however it is important that we link p to the silhouette of every connected component of $C - \{p\}$. So it is neccessary to link every connected component of $L_0(p, U) - \{p\}$ to the silhouette. Let q_i denote an extremal point wrt π_2 of the ith connected component of $L_0(p, U) - \{p\}$, and let V_i be the stratum which contains q_i. Then we define

$$L_2(p, U) = L_0(p, U) \cup \bigcup L_1(q_i, V_i) \quad (10)$$

where $L_1(q_i, V_i)$ is empty if q_i is already a silhouette point. One other case where the link curve L_0 is insufficient occurs when the connected component of $U|^v$ which contains p consists of the single point p. This occurs when p is a critical point of $\pi_1|_U$ of index 0 (see [Mil]). In this case there is no point trying to link p to the silhouette of U, since it consists only of the point p itself. Instead we link p to the silhouettes of all strata to which p is adherent in $\pi_1^{-1}(v)$. In fact it suffices to do this only for the strata of dimension $\dim(U)+1$ which we denote by W_i. So we set

$$L_3(p, \underline{\underline{S}}) = \bigcup_i L_1(p, W_i) \quad (11)$$

and given an arbitrary point p we define:

$$L(p, \underline{\underline{S}}) = \begin{cases} L_1(p, U) & \text{if } p \text{ is a regular point of } \pi_1|_U; \\ L_2(p, U) & \text{if } p \text{ is a critical point of } \pi_1|_U \text{ of index} \neq 0; \\ L_3(p, \underline{\underline{S}}) & \text{otherwise} \end{cases} \quad (12)$$

4.3 The Full Roadmap

Let $\Sigma = \Sigma(\pi_{12}|_S)$ be the silhouette of S, and let $\underline{\underline{S}}'$ be the silhouette-compatible stratification of S. Let P_c denote the set of critical points of $\pi_1|_{\underline{\underline{S}}'}$ *which are not contained in the set of boundary points B* defined earlier. That is

$$P_c = \Sigma(\pi_1|_{\underline{S}'}) - \{\partial C | C \in \underline{\Sigma}\} \tag{13}$$

then the full roadmap $R_1(\underline{S})$ is defined by:

$$R_1(\underline{S}) = \Sigma \cup \bigcup_{p \in P_c} L(p, \underline{S}) \tag{14}$$

and we have our main result:

Theorem $R_1(\underline{S})$ satisfies the roadmap condition.

The proof follows by induction on critical values, and by verifying that each type of link curve defined in the last subsection actually links p to the silhouette.

5 The Multivariate Resultant

Necessary and sufficient conditions for the solution of a system of polynomials were known over a century ago [Sal], [Mac], [Wae] and still appear in modern treatments of algebraic geometry, although usually in non-constructive form. For a system of n homogeneous polynomials in n variables, there is a single resultant polynomial (in the coefficients of the input polynomials) whose vanishing is necessary and sufficient for the system to have a solution over an algebraically closed field. This resultant is generically irreducible, so that it is the smallest such condition. The term resultant has come to be used exclusively for the Sylvester resultant (for two polynomials), so we will distinguish the more general case by calling it the multivariate resultant. The multivariate resultant includes both the Sylvester resultant and the determinant of a matrix as special cases.

Our resultant algorithm shares some similarities with the first step of Lazard's method [Laz] for solving systems of equations, although his method does not compute a resultant. Lazard's method is also used in the single-exponential time algorithm of [CG] for the existential theory of complex numbers. The new method, however, is considerably simpler, does not involve division or branching, and can be implemented in polynomial time in parallel.

While the multivariate resultant R of n polynomials not readily expressible as the determinant of a single matrix, it can be defined as the greatest common divisor of the determinants of n matrices $A^1, \ldots, A^n$. As we shall see below, it is also possible to define the resultant as a rational function of certain subdeterminants of the A^k thereby avoiding the GCD computation. The key is to introduce enough indeterminates that the matrix determinants can be easily factored. The matrices A^k are constructed as follows. First we define the degree d of monomials that index the columns of A^k:

$$d = 1 + \sum_{i=1,\ldots,n} (d_i - 1) \tag{15}$$

Let $\alpha \in Z^n$ represent the exponents of a monomial in $x_1, \ldots, x_n$, i.e. if $\alpha = (\alpha_1, \alpha_2, \ldots, \alpha_n)$ we use the notation x^α for the monomial

$$x^\alpha = x_1^{\alpha_1} x_2^{\alpha_2} \ldots x_n^{\alpha_n}$$

Then the set of monomials of degree d, denoted X^d is

$$X^d = \{x^\alpha \mid \alpha_1 + \alpha_2 + \cdots + \alpha_n = d\} \tag{16}$$

and we observe that the cardinality of X^d is

$$N = |X^d| = \binom{d+n-1}{d} \tag{17}$$

We now partition X^d into n subsets as follows:

$$\begin{aligned} X_1^d &= \{x^\alpha \in X^d \mid \alpha_1 \geq d_1\} \\ &\vdots \qquad \vdots \\ X_n^d &= \{x^\alpha \in X^d \mid \alpha_n \geq d_n \text{ and } \alpha_i < d_i, \text{ for } i < n\} \end{aligned} \tag{18}$$

and it is readily shown that every element of X^d is contained in exactly one of these subsets. Now for each X_i^d we define a set of polynomials F_i as

$$F_i = \frac{X_i^d}{x_i^{d_i}} f_i \tag{19}$$

that is, $F_i = \{x^\beta f_i \mid x^\beta x_i^{d_i} \in X_i^d\}$. Then the polynomials in F_i, for $i = 1, \ldots, n$, have degree d. The union F of the F_i is a collection of N polynomials of degree d. Since there are N monomials of degree d, we can construct a square matrix whose columns correspond to the N monomials in X^d, and each row of which contains the coefficients of those monomials in some polynomial in F. Specifically, we choose any ordering of the N elements of X^d, and then construct an $N \times N$ matrix A^n where A_{ij}^n is the coefficient of the ith monomial of X^d in the jth polynomial of F. The determinant of A^n has degree $d_1 d_2, \ldots, d_{n-1}$ in the coefficients of f_n, because the number of power products in X_n^d is $d_1 d_2, \ldots, d_{n-1}$.

The A^k for $k = 1, \ldots, n-1$ are defined similarly, but the power products X_i^d are different. In the definition (18) each X_i^d is the set of elements of X^d which are multiples of $x_i^{d_i}$, which are not contained in previous X_i^d's. For A^k, we reorder the X_i^d so that X_k^d appears last in the definition, so that it is the set of all multiples of $x_k^{d_k}$ which are not multiples of $x_i^{d_i}$ for any $i \neq k$. Thus X_k^d has $d_1, \ldots, d_{k-1} d_{k+1}, \ldots, d_n$ elements, and this number is also the degree of det A^k in the coefficients of f_k.

From now on, we will let a^k denote the determinant of A^k. Now the resultant of $f_1, \ldots, f_n$ is the greatest common divisor of the a^k *as polynomials in the indeterminate coefficients of* $f_1, \ldots, f_n$ (see [Wae], section 82). This is a crucial point. We are interested in the value of the resultant for a certain specialization of the coefficients. By a specialization of the coefficients, we mean the introduction of polynomial relations on the indeterminate coefficients, e.g. $u_1 = 5$ or $u_1 = 2u_2$. Specialization commutes with addition and multiplication and therefore with determinant computation. However, specialization does not commute with the greatest common divisor step. (e.g. consider the GCD of ax and bx, where a and b specialize to zero). The most obvious alternative, which is to compute the determinants symbolically, and then compute the GCD of the a^k as polynomials in *all* the symbolic coefficients of the f_i., is a massive task, and generates terms of size double exponential in d.

5.1 Partial Coefficient Specialization

There is a second alternative, which allows the resultant to be computed for any specialization of the coefficients, but which can be computed in time only single exponential in the input size. The key is to specialize *almost all* of the coefficients of the f_i, but to leave n coefficients indeterminate.

The determinants a^k are computed as polynomials in the n indeterminates, and their GCD is guaranteed to be the partial specialization of the resultant R. In fact, a true polynomial GCD computation is never necessary, because the GCD can be computed directly from the a^k by a sequence of $2n$ polynomial divisions.

We define a new system of polynomials $\hat{f}_i$ from the f_i as follows:

$$\hat{f}_i(x_1, \ldots, x_n) = f_i(x_1, \ldots, x_n) + u_i x_i^{d_i} \tag{20}$$

where the u_i are indeterminates. Notice that the u_i appear on the leading diagonal of every matrix A^k. In fact u_i occupies the leading diagonal position in the columns indexed by monomials in X_i^d. The product of the diagonal u_i's appears as the leading term of a^k, treating a^k as a polynomial in $u_1, \ldots, u_n$, and its coefficient is 1. It follows that a^k has degree $d_1, \ldots, d_{k-1}d_{k+1}, \ldots, d_n$ in u_k. But the resultant R also has degree $d_1, \ldots, d_{i-1}d_{i+1} \ldots, d_n$ in u_i for all i. In fact the resultant (before specialization) has a single term of maximal degree in all the u_i, and this term has coefficient 1 ([Wae], section 82). Since the resultant divides all the a^k, we can write

$$a^k(u_1, \ldots, u_n) = b^k(u_1, \ldots, u_{k-1}, u_{k+1}, \ldots, u_n) R(u_1, \ldots, u_n) \tag{21}$$

where b^k is a polynomial independent of u_k because R and a^k have the same degree in u_k. Knowing the a^k, and knowing that the leading coefficients of both the a^k and R are 1, we can actually solve for all the coefficients of R (and b^k if desired). First we need:

Definition A multivariate polynomial $p(x_1, \ldots, x_2)$ of degree d_i in x_i is said to be *rectangular* if its leading term is of maximal degree in all the x_i, i.e. the leading term is of the form $cx_1^{d_1}x_2^{d_2} \ldots, x_n^{d_n}$. The product or quotient of two rectangular polynomials is also rectangular.

In particular $R(u_1, \ldots, u_n)$ is rectangular and monic, i.e. its leading coefficient is one. All the $a^k(u_1, \ldots, u_n)$ are rectangular and monic, so the $b^k(u_1, \ldots, u_{k-1}, u_{k+1}, \ldots, u_n)$ are rectangular and monic. To compute the resultant we can use:

The Basic Recurrence

Let $R_j(u_1, \ldots, u_j)$ be the leading coefficient of R considered as a polynomial in $u_{j+1}, \ldots, u_n$. Since R is rectangular, it follows that the leading coefficient of R_j equals the leading coefficient of R, so that R_j is monic. Let $b_j^k(u_1, \ldots, u_j)$ be the leading coefficient of b^k, considered as a polynomial in $u_{j+1}, \ldots, u_n$. Then b_j^k is also monic. Finally, let $a_j^k(u_1, \ldots, u_j)$ be the leading coefficient of a^k considered as a polynomial in $u_{j+1}, \ldots, u_n$. The following relationships hold:

$$R_{k-1} b_{k-1}^k = a_{k-1}^k \qquad \text{for } k = 1, \ldots, n$$

$$R_k b_{k-1}^k = a_k^k \qquad \text{for } k = 1, \ldots, n \tag{22}$$

Where the second expression exploits the fact that $b_k^k = b_{k-1}^k$ since b^k is independent of u_k. So if we know R_{k-1}, we can compute R_k by first dividing a_{k-1}^k by R_{k-1} to get b_{k-1}^k, and then dividing b_{k-1}^k into a_k^k. Since we know $R_0 = 1$, after $2n$ divisions we obtain $R_n = R$. The constant term of this polynomial is the resultant we seek.

Method 1 The recurrence relation (22) implies corresponding identities on the constant coefficients of all the polynomials in the recurrence. The constant coefficient of a_j^k is just the determinant of a submatrix A_j^k of A^k. So this method involves $2n$ divisions of determinants of matrices whose elements are the coefficients of the original polynomials. Unfortunately, one of these determinants may vanish, even if the resultant does not, and this methods fails.

Method 2 The recurrence relation in (22) is valid for any specialization of the $u_1, \ldots, u_n$. In particular we can take $u_1 = u_2 = \cdots = u_n$. For this specialization, the polynomials $a_j^k(u_1)$ are monic, in fact they are characteristic polynomials of submatrices A_j^k of the A^k. Thus, by computing $2n$ characteristic polynomials, and performing $2n$ divisions of univariate monic polynomials, we obtain the resultant. Now division of monic polynomials does not require division over the base field, and can be computed using subdeterminants of a certain matrix. Using Csanky's [Csa] algorithms for parallel matrix computations, all polynomials (which have exponential size) can be computed in polynomial parallel time.

5.2 Degree bounds and Gap Theorem

For the algebraic algorithms described in Appendix I, we will be making use of finite precision binary numbers to approximate algebraic numbers. Although this only allows us to evaluate expressions approximately, if we are careful, we can still determine the values of certain expressions exactly using "Gap" theorems. Specifically, these theorems state that for any collection of integral polynomials of a certain size (bounded number of bits in each coefficient and bounded degree) there is a finite lower bound on the spacing between any two distinct roots of the polynomials in the collection.

Thus if the binary approximations to two algebraic numbers of a certain size are sufficiently close (their difference is smaller than the gap), then the two algebraic numbers must be identical. In particular, if the binary approximation to an algebraic number of a certain size is sufficiently small, then the algebraic number must be zero. It turns out that manipulation of these binary approximations is much more efficient than the manipulation of "exact" algebraic numbers as described in [SS]. Roughly speaking, the approximate computations can be done in exponential time, while the exact arithmetic representation grows doubly exponentially with the number of arithmetic operations. Such approximate techniques have been used by Prill [Pri], and Hong [Hon] who gave a single exponential bound for certain systems of polynomials. The gap bound given below improves on the bounds of these authors, and we show that the rate of exponential growth of required number of bits is tight.

Definition The *weight* of an integral multivariate polynomial f is the sum of the magnitudes of the coefficients of f. We denote the weight of f as $w(f)$, and it is readily shown that $w(f+g) \leq w(f) + w(g)$ and $w(fg) \leq w(f)w(g)$.

Weight is a useful measure for most types of polynomial, but we will also be manipulating characteristic polynomials quite a lot and these have a special structure which makes weight a very poor measure of size. Instead we use:

Definition. The *height* of an integral monic polynomial $f(s)$ of degree m is the smallest h such that $w(f_{m-i}) \leq v^i$, where f_j is the coefficient of s^j in f. We denote the height of f as $h(f)$.

Then it is readily seen that the characteristic polynomial of an $N \times N$ matrix has height Nw, where w is the maximum weight of any element of the matrix. The utility of height as a measure of polynomial size is emphasized by the following result:

Lemma. Let $a(s)$, $b(s)$ be monic polynomials. Then $h(ab) \leq h(a) + h(b)$ and $h(a/b) \leq h(a) + 2h(b)$.

Lemma. If $p(x)$ is an integral polynomial of weight $w(p)$, ξ a non-zero root of $p(x) = 0$, then ξ satisfies:

$$\frac{1}{w(p)} < |\xi| < w(p) \tag{23}$$

These and other bounds are given in [Mig]. Now for the gap theorem, we consider the class $\wp(d, c)$ of integral polynomials of degree d and maximum coefficient magnitude c.

Gap Theorem Let $\wp(d,c)$ be the class of polynomials of degree d and coefficient magnitude c. Let $f_1(x_1, \ldots, x_n), \ldots, f_n(x_1, \ldots, x_n) \in \wp(d,c)$ be a collection of n polynomials in n variables which has only finitely-many solutions when homogenized. Then if $(\alpha_1, \ldots, \alpha_n)$ is a solution of the system, then for any j either

$$\alpha_j = 0 \quad \text{or} \quad |\alpha_j| > (3dc)^{-n(n+d)d^{(n-1)}} \tag{24}$$

Proof We homogenize the system by adding the homogenizing variable x_0. We then add a linear form $\sum v_i x_i$ and compute the u-resultant [Wae], which is a polynomial in $v_0, \ldots, v_n$. We assume for simplicity that $v_i = 0$ for $i \neq 0, j$, so that the resultant is an integral polynomial in v_0, v_j. Setting $v_0 = 1$, we obtain a polynomial whose roots are the values of α_j in the solution set. If we can bound the weight of the resultant, using the corollary to the last lemma, we obtain an immediate lower bound on $|\alpha_j|$. From method 2 we see that the resultant is the product of $\bar{a}_k^k$ for $k = 1, \ldots, n+1$ divided by the product of the $\bar{a}_{k-1}^k$ for $k = 1, \ldots, n+1$. Now the matrices A^k all have size:

$$N = \binom{nd+1}{n} < \frac{(dn+1)^n}{n!} < (de)^n \tag{25}$$

where e is the base of natural logarithms. This follows from Stirling's approximation for factorial. Now $\bar{a}_k^k$ is the characteristic polynomial of matrix of size at most $Nk/(n+1)$, while $\bar{a}_{k-1}^k$ comes from a matrix of size $N(k-1)/(n+1)$. From the expression for height of a characteristic polynomial we have $h(\bar{a}_k^k) = Nck/(n+1)$ and $h(\bar{a}_{k-1}^k) = Nc(k-1)/(n+1)$. From the linear growth of height with multiplication and division we obtain:

$$h(\bar{R}) < \sum_{k=1,\ldots,n+1} \frac{3Nck}{(n+1)} < \frac{3}{2}Ncn \tag{26}$$

and since $\bar{R}(u_1)$ has degree $nd^{(n-1)} + d^n = (n+d)d^{(n-1)}$ in u_1, the weight of the resultant (the constant term of $\bar{R}$) is

$$(3/2Ncn)^{(n+d)d^{(n-1)}} < (3dc)^{n(n+d)d^{(n-1)}} \tag{27}$$

and so the non-zero roots α_j of the resultant satisfy $|\alpha_j| > (3dc)^{-n(n+d)d^{(n-1)}}$. □

This result implies that we can represent algebraic numbers defined by such a system using $O(n(n+d)d^{(n-1)}\log(dc))$ bits, since the weight of the resultant also bounds the maximum size of a solution. In fact, the bound on the base of exponential growth, d, is tight because there exist systems having gaps of $d^{n-1}\log c$ bits. e.g. Consider $Lx_1 - 1 = 0$, L a large integer, and the equations $x_i = x_{i-1}^d$ for $i = 2, \ldots, n$. Then $x_n = (1/L)^{d^{(n-1)}}$ is a solution.

References

[BKR] Ben-Or M., Kozen D., and Reif J., "The Complexity of Elementary Algebra and Geometry", J. Comp. and Sys. Sciences, Vol. 32, (1986), pp. 251-264.

[Ber] Berman L., "Precise Bounds for Presburger Arithmetic and the Reals with Addition," Proc. 18th IEEE Symp. FOCS, (1977).

[Ca85] Canny J. F., "A Voronoi Method for the Piano-Movers Problem," Proc. IEEE Int. conf. Robotics and Automation, (March 1985).

[Ca86] Canny J. F., "Collision Detection for Moving Polyhedra," IEEE trans. PAMI, vol 8, no 2, (March 1986).

[Csa] Csanky L., "Fast Parallel Matrix Inversion Algorithms" SIAM J. Comp., Vol. 5, No. 4, (Dec. 1976).

[Col] Collins G.E. "Quantifier Elimination for Real Closed Fields by Cylindrical Algebraic Decomposition" Lecture Notes in Computer Science, No. 33, Springer-Verlag, New York, (1975).

[CG] Chistov A. L. and Grigoryev D. Y., "Complexity of quantifier elimination in the theory of algebraically closed fields", Lect. Notes Comp. Sci. 176, Springer Verlag, (1984).

[EOS] Edelsbrunner H., O'Rourke J., Seidel R., "Constructing Arrangements of Lines and Hyper-planes with Applications" Proc. 24th Symp. on Foundations of Computer Science, (1983).

[GWD] Gibson C. G., Wirthmüller K., Du Plessis A. A., Looijenga E. J. N., "Topological Stability of Smooth Mappings", Lecture Notes in Mathematics, No. 552, Springer-Verlag, New York, (1976).

[GG] Guilleman V., and Golubitsky M., "Stable Mappings and Their Singularities", GTM-14, Springer Verlag, New York, (1973).

[Hon] Hong J. W., "Proving by Example and Gap Theorem", Proc. 27th IEEE Symp. FOCS, (1986), pp. 107-116.

[Joh] Johnson F. E. A., "On the Triangulation of Stratified Sets and Singular Varieties", Trans. Amer. Math. Soc. 275, (1983).

[KY] Kozen D., and Yap C. "Algebraic Cell Decomposition in NC", Proc IEEE symp. FOCS, (1985), pp. 515-521.

[Laz] Lazard D., "Résolution des Systèmes d'Équations Algébriques", Theor. Comp. Sci. vol 15, (1981).

[Loz] Lozano-Pérez T., "Spatial Planning: A Configuration Space Approach," IEEE Trans. Computers, C-32, No 2 (Feb 1983), pp 108-120.

[Mac] Macaulay F. S., "Some Formulae in Elimination" Proc. London Math. Soc. (1) 35 (1902) pp. 3-27.

[Mig] Mignotte M., "Some Useful Bounds", in "Computer Algebra, Symbolic and Algebraic Computation", Buchberger et al. ed., Springer-Verlag, New York, (1982).

[Mil] Milnor J., "On the Betti Numbers of Real Varieties", Proc. Amer. Math. Soc. 15, (1964).

[Pri] Prill D., "On Approximations and Incidence in Cylindrical Algebraic Decompositions", SIAM Jour. Comp., Vol. 15, No. 4, (Nov. 1986), pp 972-993.

[Rei] Rief J., "Complexity of the Mover's Problem and Generalizations," Proc. 20th IEEE Symp. FOCS, (1979).

[Sal] Salmon G., "Modern Higher Algebra", G. E. Stechert and Co., New York, reprinted 1924, (1885).

[SHS] Schwartz J., Hopcroft J., and Sharir M. (eds.) "Planning, Geometry and Complexity of Robot Motion Planning", Albex Publishing Co., New Jersey, (1987).

[SS] Schwartz J. and Sharir M., "On the 'Piano Movers' Problem, II. General Techniques for Computing Topological Properties of Real Algebraic Manifolds," Computer Science Department, New York University report 41, (1982).

[SY] Schwartz J. and Yap C.K., "Advances in Robotics," Lawrence Erlbaum associates, Hillside New Jersey, (1986).

[Th65] Thom R., "Sur L'Homologie des Variétés Algébriques Réelles", in "Differential and Combinatorial Topology", S. Cairns ed., Princeton University Press, New Jersey, (1965).

[Th69] Thom R., "Ensembles et Morphismes Stratifiés", Bull. AMS. vol 75, (1969), pp 240-284.

[Wae] van der Waerden B. L., "Modern Algebra", (third edition) F. Ungar Publishing Co., New York (1950).

[Whi] Whitney H., "Elementary Structure of Real Algebraic Varieties", Annals of Math., Vol. 66, No. 3, (Nov. 1957).

Appendix I. The Roadmap Algorithm

We define the algorithm bottom-up, describing first the algebraic algorithms for computing equations for curves and points, and later the geometric algorithms for construction of the adjacency graph.

Algorithm 1
Input: Homogeneous polynomials $f_1, \ldots, f_{r-1}$ of degree d in r variables $x_1, \ldots, x_r$ which have a finite number of distinct solutions. The solutions are assumed to have distinct x_1 coordinates (this occurs with probability one after a change of coordinates).
Output: A list of the real root vectors $(s_1, \ldots, s_r)$, $s_r \neq 0$, which are solutions of the system.

Description We use the u-resultant defined in section 5. However, rather than computing the complete u-resultant as a polynomial in $v_1, \ldots, v_r$, we compute several resultants in three indeterminates, and then match solutions by v_1-coordinate. Each resultant includes the original polynomials and one of the linear forms:

$$l_i = v_1 x_1 + v_r x_r + v_i x_i \qquad \text{for } i = 2, \ldots, r-1 \tag{28}$$

and let the corresponding resultants be denoted R_i. The R_i are homogeneous in v_1, v_r, v_i. Now by setting $v_r = 1, v_i = 0$ in R_i, R_i becomes a univariate polynomial in v_1. We can isolate the real roots of this polynomial to any precision using Sturm sequences. Let $v_1 = \alpha_j$ be one such approximate root. Now let $v_r = 1$, $v_1 = 0$, we can isolate a root of R_i in the same way and we get an approximation to a solution $v_i = \beta_{k_i}$. Now for each α_j, there is a corresponding β_{k_i} such that $v_1 = \alpha_j$, $v_i = \beta_{k_i}$, $v_r = 1$ is a solution of $R_i = 0$. We check this by substitution, making use of the gap theorem to bound the numerical precision we require. Doing this for all i, we obtain all the solution vectors $(\alpha_j, \beta_{k_2}, \ldots, \beta_{k_{r-1}}, 1)$ by identify α_j values. (Note that this works because what we are really doing is examining the terms of highest power in v_1, v_r, v_i in the *full* u-resultant)

Using the defining polynomials for α_j, β_k and R_i, the gap for values of R_i is $O(r^2 d^{3r} \log dw)$ bits, where w is the maximum absolute value of any coefficient of any f_i. By simple error analysis, using the upper bounds for α_j and β_k, and the fact that R_i has degree d^r, we need $O(r^2 d^{3r} \log dw)$ bits of accuracy in the value of each root to ensure that the error in the value of R_i is within the gap. Once we have a sufficiently good approximation to a root, Newton's method can be used to isolate the root to this accuracy in time $O(r^4 d^{3r} \log^3 dw)$. Evaluation of R_i for each solution β_k takes time $O(r^3 d^{3r} \log^2 dw)$. (Both of the previous two bounds make use of a simple divide-and-conquer scheme for polynomial evaluation, and rounding to $O(r^2 d^{3r} \log dw)$ bits). To find all three-element solution rays (solution rays for some R_i) takes time $O(r^3 d^{5r} \log^2 dw)$. Once we have these for each R_i, we can immediately obtain all n-element real root vectors by comparing their v_1-coordinates. Since the cost for evaluation dominates the time for computation of the resultants themselves, the overall cost for this algorithm is $O(r^4 d^{5r} \log^2 dw)$ binary operations.

Algorithm 2
Input: Polynomials $f_1, \ldots, f_n$ in r variables $x_1, \ldots, x_r$ of degree d defining a manifold M, and a linear map $a : \Re^r \rightarrow \Re^2$.
Output: a single polynomial $h(x_1, x_2)$ s.t. $\ker(h) \supset \pi_{12}(\Sigma(a|_M))$

Description
We first define a polynomial whose kernel is the boundary of a tube around the variety.

$$g = \sum_{i=1,\ldots,n} f_i^2 + \sum_{i=1,\ldots,r} \epsilon x_i^{2d} \quad (29)$$

Let $N = \ker(g)$. Now for ϵ small there are critical points of $a|_N$ near all the critical points of $a|_M$. We make a change of coordinates if necessary so that a corresponds to the projection π_{12}. Then the critical points of $a|_N$ are the points where the system

$$g = 0, \quad \frac{\partial g}{\partial x_3} = 0, \cdots \frac{\partial g}{\partial x_r} = 0 \quad (30)$$

has a common zero. We can compute the resultant (after homogenizing) of this system as a polynomial in x_1, x_2 and ϵ. For a generic choice of the coefficients of a, the critical set of $a|_N$ is one-dimensional in projective space, and so the resultant is non-vanishing. Now as $\epsilon \to 0$, the zero set of the resultant (in the x_1-x_2 plane) converges to the projection of the critical set of $a|_M$. Arranging the resultant according to powers of ϵ, the limit of the zero set is the zero set of the coefficient of the lowest power of ϵ. This is the polynomial h which we are after.

Computation of h requires the computation of the resultant of a system of $r-2$ polynomials of degree $2d-1$ and one of degree $2d$. Because of the introduced ϵ, the constant terms in the recurrence (22) are non-vanishing (in fact most are monic polynomials in ϵ), so we can use method 1 for resultant computation. Each matrix in the computation has size at most $O((6d)^r)$ and we need compute only the highest $(2d)^r$ terms of the matrix coefficients as polynomials in ϵ. This can be done in $O(r^2(12d^2)^{3r} \log dw)$ binary operations.

Algorithm 3
Input: A linear map $a : \Re^r \to \Re^2$ and polynomials $f_1, \ldots, f_n$ and f_k in $x_1, \ldots, x_r$ of degree d. The zero set of $f_1, \ldots, f_n$ is assumed to be a manifold M.
Output: The intersection points of the curve $C = \Sigma(a|_M)$ and the hypersurface $f_k = 0$.

Description This algorithm is a hybrid of the previous two algorithms. First we form a polynomial g from $f_1, \ldots, f_n$ using (29). The solution set of the system (30) is a smooth one-dimensional curve, and the intersection of this curve with the hypersurface $f_k = 0$ generically consists of a finite number of points, and as $\epsilon \to 0$ these points converge to the solution points we seek. Once again, we find these points using the u-resultant. We introduce a linear form in three variables as in (28), and find the resultant of the polynomial system (30) and $f_k = 0$ and one of the forms (28). This is a system of $r+1$ equations in $r+1$ unknowns (once the homogenizing variable is added). We arrange the resultant in powers of ϵ, and take the coefficient of the lowest power.

Once again, the intermediate polynomials in the resultant computation are non-vanishing, so we can use method 2 to compute them. This step takes $O(r^3(12d^2)^{3r} \log dw)$ binary operations. We use the substitution method described in algorithm 1 to find solution vectors, and this phase dominates the running time of the algorithm. We can apply the analysis algorithm 1 to algorithm 3, but where the degrees of the polynomials are doubled. This gives an $O(r^4(2d)^{5r} \log^2 dw)$ bound for the number of binary operations.

Algorithm 4
Input: A linear map $a : \Re^r \to \Re^2$, and polynomials $f_1, \ldots, f_n$, in r variables $x_1, \ldots, x_r$ of degree d defining a semi-algebraic set S, such that the first $m < r$ polynomials define a manifold M.
Output: The adjacency graph of the smooth curve $C = \Sigma(a|_M)$ and all the intersection points of this curve with the hypersurfaces $f_i = 0$ for $i = m+1, \ldots, n$. Each point and curve segment is labelled with the signs of all the f_i.

Description We first call algorithm 2 to compute the projection of the curve C. Then we use a variation of a line sweep method to order intersection points along each connected component of C. The algorithm is essentially the same as that described in [Ca85]. Algorithm 2 returns a polynomial h in two variables say x_1 and x_2. We first ensure that h is square-free. Then we compute the Sturm sequence of h as a polynomial in x_1. The last term in this sequence is a polynomial $D(x_2)$ in x_2 only, called the discriminant of h. If h is square-free, then the discriminant is a polynomial in x_2 having only a finite number of zeros. For x_2 in between these zeros, the curve $h = 0$ has a fixed number of solutions in x_1, and these form smooth curves as x_2 is varied. They may therefore be uniquely ordered by x_1 coordinate in this range of x_2.

Given the x_1, x_2-coordinates of an intersection point p found using algorithm 3, we can determine from the signs of the Sturm sequence at p whether the projection of the point lies on a curve, and if it does, the number of the curve in the x_1-ordering. We can check by substitution into the original system of polynomials whether the point actually lies on the curve. Thus we can completely construct the adjacency graph of points and curve segments in between zeros of $D(x_2)$. At the zeros of $D(x_2)$ we can make use of the fact that the projection of a smooth curve into the plane is generically an immersion with normal crossings. This means that there are only two types of event that can occur at a zero α of x_2: (i) two curve segments meet smoothly, in which case there will be either be two more or two less zeros of $h(x_1)(x_2)$ for $x_2 > \alpha$ than for $x_2 < \alpha$. (ii) two curve segments cross, in which case there will be the same number of curves on either side of α. Both these cases are easily recognised, and the Sturm sequence indicates which curve segments join or cross. Thus we can unambiguously identify curve segments across zeros of $D(x_2)$, and therefore compute the complete adjacency graph of the curve and all intersection points.

Since the graph has pure degree two (C is a smooth curve), there is a linear ordering of points along each component of the graph. Exactly one polynomial changes sign at each point, so by using a persistent data structure, (e.g. an AVL-tree without mutation), we can store the signs of all n polynomials at each point at $O(\log n)$ cost per point, and retrieve them in $O(\log n)$ time.

We need $n - m$ calls to algorithm 3, a cost of $O(nr^4(2d)^{5r} \log^2 dw)$. We use approximate arithmetic once again to evaluate the Sturm sequences, and in fact the gap required, $O(r^2(2d)^{3r} \log dw)$ is the same as that used in algorithm 3, and so the values supplied by that algorithm are accurate enough. Evaluation of the entire sequence at a point takes time $O(r^3(2d)^{4r} \log^2 dw)$, and there are less than $n(2d)^r$ points. The Sturm evaluation and point ordering phase therefore takes $O((n \log n) r^4 (2d)^{5r} \log^3 dw)$ binary operations, including time to generate the query structure. This dominates the running time of the algorithm.

Algorithm 5 (Basic Roadmap)

Input: Polynomials $f_1, \ldots, f_n$ in r variables of degree d defining a semi-algebraic set S. The f_i are assumed to be in general position, so that the stratification $\underline{\underline{S}}$ of S into sign-invariant sets is a Whitney stratification.
Output: The adjacency graph and point coordinates of the basic roadmap $R_0(\underline{\underline{S}})$.

Description. We first choose a one-generic projection map a. Then the basic roadmap consists of the silhouette $\Sigma(a|_{\underline{\underline{S}}})$ and the basic roadmap of slices through critical points on the silhouette. To compute the silhouette, we call algorithm 4 on all subsets of $m < r$ polynomials. The adjacency graph of the silhouette is readily constructed from the graphs of individual curves in two steps: (i) identifying corresponding points on curves on strata of the same dimension, and (ii) find the position of points on curves of lower dimensional strata. Step (i) can be accomplished in time linear in r if points are stored lexicographically ordered by coordinates, and in arrays indexed by the polynomials that are zero at the point. For step (ii) we simply locate the point on the curve as though it were an intersection point for that curve. Overall, the adjacency graph of the silhouette can be constructed in $O(\binom{n}{r-1}(n \log n)r^4(2d)^{5r} \log^3 dw)$ binary operations.

Now for the recursive step, we need to count the number of critical points on silhouette curves. First of all there are $\binom{n}{r-1}$ curves and each will have at most $(2d)^{2r}$ critical points. This follows from the maximum degree of the curves. So if $T_{R_0}(k)$ is the number of binary operations needed in the basic roadmap in k dimensions, then we have the following recurrence:

$$T_{R_0}(k) = \binom{n}{k-1}(2d)^{2k}T_{R_0}(k-1) + O(n^k \log n \frac{k^4}{k!}(2d)^{5k} \log^3 dw) \tag{31}$$

and solving, we find that the recursive step dominates, and that the basic roadmap in r dimensions requires

$$T_{R_0} = O(n^{\frac{1}{2}r(r-1)}(2d)^{r(r+1)}) \tag{32}$$

binary operations.

Algorithm 6 (Point Linking)
Input: The coordinates of a point $p \in \Re^r$, polynomials $f_1, \ldots, f_n$ of degree d in r variables, a linear map a and the silhouette (or partial roadmap) of $\underline{\underline{S}}$ (including adjacency graph).
Output: The link curve $L(p, \underline{\underline{S}})$ defined in section 5.2, and a modified adjacency graph which includes this curve.

Description. We must first determine whether p is a critical point. We test the signs of all the f_i at p, and those which are zero (say m of them) define the local geometry of the stratum U containing p. If the projection of the tangent space of this stratum is not surjective, then p is a critical point. To find the index of p, we examine the local direction of the flow induced by the map a_1. The flow at each point is a vector whose inner product with a vector in the a_1 direction is always non-negative, and which is zero only at critical points. By solving a simple $m \times m$ system of linear equations, we can explicitly define (giving the components as polynomials) a vector field parallel to the flow which also vanish only at critical points, and has non-negative inner product with a_1. This vector field has degree $3dm$. Taking partial derivatives of the vector field with respect to the x_i, we obtain an $r \times r$ matrix H, which describes the local change in the flow field near p. We can transform this matrix to an $(r-m) \times (r-m)$ matrix H' using a local basis for the tangent space to U at p. Then p is a critical point of index 0 iff all the eigenvalues of H' are positive.

By the above tests we can determine which type of linking curve is needed at p. We analyse here only the most complicated case, which is where p is a critical point of index 0. Thus we must link p to the silhouettes of all strata adherent to p of dimension one greater than U. First we call algorithm 5 on all subsets of size $m-1$ of the m polynomials which define U. Then to each adjacency graph returned by one of these calls, we add the intersections with all constraints $f_k = 0$. However, rather than maintaining a sorted structure as in algorithm 4, we keep track only of those intersection points which are closest to p in the linear ordering along some curve through p (in fact we need only keep track of the closest point on one such curve). If there is a direct path from p to a silhouette point (i.e. no intermediate intersection point with some $f_k = 0$) then we are done for that graph, and we add it to the main graph. Otherwise, we must compute the roadmap of a lower-dimensional stratum. We repeat this at most $r - m$ times until we obtain a silhouette point.

Thus there are $m(r-m)$ calls to the basic roadmap algorithm in dimension $r-1$, each with at most m polynomials. From algorithm 5, these calls have a total cost of $O(r^2 r^{\frac{1}{2}(r-1)(r-2)}(2d)^{r(r-1)})$. The other step is addition of new nearest intersection points with some $f_k = 0$, which from the description of algorithm 4 takes time $O(r^3(2d)^{4r} \log^2 dw)$ operations per point. Counting the number of possible intersections using the recurrence (31) we find that there will be at most $O(nr^2 r^{\frac{1}{2}(r-1)(r-}$ [illegible] $^{-1)})$ such intersection points. This phase dominates the running time, so that linking critical points of index 0 takes time:

$$T_L = O(nr^5 r^{\frac{1}{2}r(r-1)}(2d)^{r(r+3)} \log^2 dw) \tag{33}$$

binary operations. All other types of link require less time than this.

Algorithm 7 (Full Roadmap)
Input: A Formula defining a semi-algebraic set S, with defining polynomials in general position.
Output: The full roadmap $R_1(\underline{\underline{S}})$ of the stratification $\underline{\underline{S}}$ of S into sign-invariant sets.

Description
The algorithm has two phases. First we choose a generic linear map a and compute the silhouette Σ of $\underline{\underline{S}}$. From the description of algorithm 5, this requires $O(n^r \log n \frac{r^4}{r!}(2d)^{5r} \log^3 dw)$ binary operations.

Then we call algorithm 6 (linking) on all critical points of $a_1|_\Sigma$. Now there are $\binom{n}{r-1}$ silhouette curves, and each may have at most $(2d)^{2r}$ critical points. So the total time for linking is $O((n^r)\frac{r^4}{r!}r^{\frac{1}{2}r(r-1)}(2d)^{r(r+5)} \log^2 dw)$ binary operations. So overall the full roadmap algorithm requires

$$T_{R_1} = O\Big((n^r)\frac{r^4}{r!}(\log n(2d)^{5r} \log^3 dw + r^{\frac{1}{2}r(r-1)}(2d)^{r(r+5)} \log^2 dw\Big) \tag{34}$$

binary operations.

John Hopcroft
Gordon Wilfong*
Department of Computer Science
Cornell University
Ithaca, New York 14853

Motion of Objects in Contact

Abstract

The use of computers in the design, manufacture, and manipulation of physical objects is increasing. An important aspect of reasoning about such actions concerns the motion of objects in contact. The study of problems of this nature requires not only the ability to represent physical objects but also the development of a framework or theory in which to reason about them. In this paper such a development is investigated and a fundamental theorem concerning the motion of objects in contact is proved. The simplest form of this theorem states that if two objects in contact can be moved to another configuration in which they are in contact, then there is a way to move them from the first configuration to the second configuration such that the objects remain in contact throughout the motion. This result is proved when translation and rotation of objects are allowed. The problem of more generalized types of motion is also discussed. This study has obvious applications in compliant motion and in motion planning.

Introduction

The increasing use of computers in the design, manufacture, and manipulation of physical objects underscores the need for a theory to provide a framework for reasoning about transformations of objects. In this paper we take a first step toward developing such a theory.

One can define an object such as a shaft or a connecting rod without instantiating its position or orientation. More generally, one can define a rectangular solid without specifying its dimensions. In fact, one can define an object without instantiating its shape. For example, the shape of an ellipsoid parametrized by the ratio of its axes is determined only when the parameter is fixed. To obtain a specific instance of the ellipsoid, one provides the ratio of the major to the minor axis, the length of the major axis, and its orientation with respect to some coordinate system, along with the position in 3-space of the center of the ellipsoid. In this framework an object is the image of an instance of a parametrized homeomorphism from a canonical region in $\mathbf{R}^3$ to $\mathbf{R}^3$. Thus, all instances of an object are homeomorphic and therefore topologically equivalent. A sphere and an ellipsoid can be instances of the same object, but a sphere and a torus cannot. A sphere might be given by the parametrized mapping

$$f(x, y, z) = (xr + a, yr + b, zr + c)$$

from a unit sphere centered at the origin in $\mathbf{R}^3$, where a, b, and c give the coordinates of the center and r is the radius. Corresponding to a particular instantiation of the sphere is a point in the four-dimensional parameter space.

In this generalized setting, a motion is a continuous mapping from [0,1] into the appropriate parameter space. Thus, a motion is a path in the parameter space. A motion can be a combination of translation, rotation, and growth, or it can be a more complicated continuous deformation of shape. It is our hope that this view will be useful in defining and manipulating generic objects as well as deformable or nonrigid objects. For our purposes, we limit ourselves to motions where the transformation can be parametrized by a finite number of parameters. The parameter space is called configuration space (Lozano-Pérez and Wesley 1979).

Although it is traditional to think of objects in terms of their shape and dimension and then to deduce functionality from the shape, it is enticing to think of

---

* Author's present address: AT&T Bell Laboratories, 600 Mountain Avenue, Murray Hill, NJ 07974

This work was supported in part by NSF grant ECS-8312096 and an NSERC graduate scholarship.

The International Journal of Robotics Research, Vol. 4, No. 4, Winter 1986

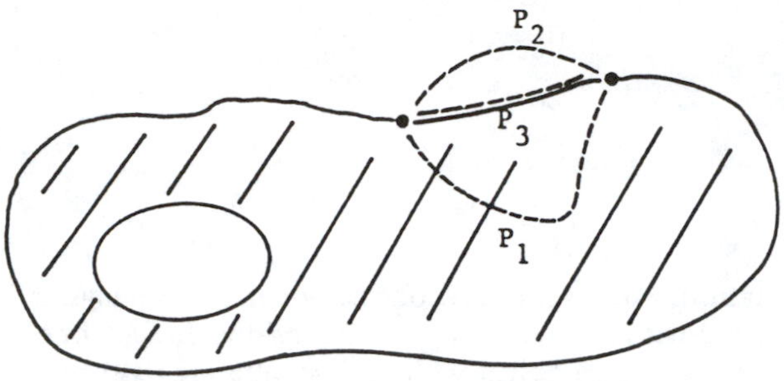

Fig. 1. Configuration space.

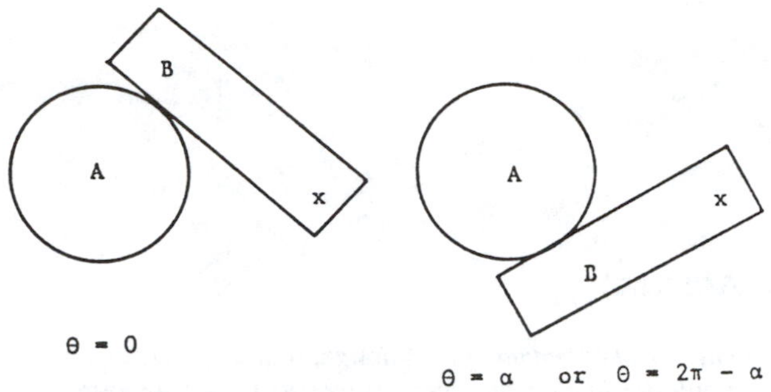

Fig. 2. Restriction to rotational motion.

representing objects by functionality and then deducing shape. In designing for automatic assembly one is normally free to modify objects for ease of assembly. Thus, designing for functionality and allowing the functionality and the assembly process to determine shape and size is a desirable goal. Furthermore, parametrized design provides additional advantages. For example, instead of designing a driveshaft for a particular torque, it would be preferable to design the driveshaft with torque as a parameter. This allows changes in design specifications without necessitating the redesign of components. The study of problems of this nature will require substantial advances in the representation of physical objects and in our ability to reason about them. In this paper we begin with a modest step: establishing a fundamental theorem concerning the motion of objects in contact.

In the special case where motion is restricted to rotations and translations, the theorem states that if there is a way to move a set of objects from an initial configuration where the objects form a connected component to a final configuration where the objects form a connected component then there is a way to move the objects from the initial to the final configuration such that at all times the objects form a connected component. To understand the theorem in a more general setting, consider the motion of two objects A and B relative to one another. Normally one would consider A fixed and B moving relative to A. For ordinary motions such as translations or rotations, there is of course no loss in generality in fixing A. However, the fact that A may be changing shape makes it more desirable to view both objects as moving. A point in configuration space represents the values for the parameters of A and B. Certain points correspond to positions and orientations where B overlaps A. In the situation where configuration space is contractible to a point, the theorem states that the existence of a path (P_1 in figure 1) from initial configuration to final configuration where A and B always intersect and of a path (P_2 in figure 1) where A and B do not overlap implies the existence of a path (P_3 in figure 1) where A and B touch at all times but are not overlapping. The point of contact need not be a continuous function even though motion is continuous.

Care must be exercised in applying the theorem. For example, in figure 2 there are two objects A and B; A is fixed and B is permitted only to rotate about x. Rotating B $2\pi - \alpha$ in the clockwise direction results in the same apparent configuration as rotating B α radians in the counterclockwise direction. However, these two configurations are the same in configuration space only if we identify points that differ by a rotation of 2π. This results in a cylindrical space that is not contractible to a point, and hence the hypothesis of our theorem is not valid. In the above example, the only motion from $\theta = 0$ to $\theta = 2\pi - \alpha$ is a motion where the objects do not overlap, and thus there is no motion that keeps the objects intersecting.

The most obvious applications of the results of this paper are in the area of compliant motion. In compliant motions, an object (e.g. a robot) is constrained to move along the surface of another object. Using compliant motion allows the search for a path to be restricted to a lower-dimensional search space. It also allows the robot to use force sensing as in Mason 1981. Compliant motion is useful, for example, when one wishes to mate two parts whose exact dimensions and positions are not known.

The main result of this paper shows that if there is

any motion of objects between given initial and final configurations of the objects, then there is a motion where the objects remain in contact. That is, using compliant motion for automated assembly does not limit the types of assemblies that can be performed, because we know that if there is any way to achieve the desired product then there is a way to do it while the moving robot keeps parts in contact.

Another useful concept is the notion that an object is just an instance of a parametrized canonical object. Using this definition of an object, one should be able to reason about an object in terms of parameters that can then be instantiated for any particular instance of the canonical object.

In addition to the obvious applications in compliant motion, the theorem has potential applications in motion planning and in complexity theory. In the planning of coordinated motion, not only trajectories must be determined but also the relative timing of objects as they move on their individual trajectories. A path in configuration space contains this information about relative timing. Thus, searching the paths in configuration space simplifies the problem conceptually. In general, configuration space is of very high dimension. The above theorem reduces the search from this high-dimensional space to a lower-dimensional surface in the space. The surface can be thought of as composed of faces that intersect in lower-dimensional faces that in turn intersect in still lower-dimensional faces. Under suitable restrictions, the surface of contact in configuration space will have edge-connected vertices. In order to move a set of objects from one configuration to another, we first push the objects together in both the initial and final configurations. Then we move the objects along faces until they reach lower-dimensional faces. We continue this process until the initial and final configurations have been converted to vertices of the surface in configuration space at which the objects are in contact. This reduces the problem to a graph-searching problem. In general the number of vertices of the graph will be astronomical. We need not construct the entire graph; we need only generate vertices and edges as the search progresses. With a suitable heuristic, it may be possible in practical situations to find the desired path having generated only a tiny fraction of the graph. The knowledge of such a path could be used in constructing a path where the objects do not touch one another.

In complexity theory it is often important to show that if a certain motion exists, then a canonical motion exists. In the case of linkages (Hopcroft, Joseph, and Whitesides 1982), for example, it is important from the viewpoint of complexity theory that various joints need not be moved to locations that are algebraically independent in order for a motion to take place. Our theorem establishes that if a motion exists, then a motion that follows features of the surface exists and hence a canonical motion exists.

In section 1 some general properties of the space of all configurations are developed. These properties are used in section 2 to show that certain regions of the space of all configurations are path connected or contractible to a point. From this it is shown that if there is a motion between configurations in which two objects touch, then there is a motion between them such that at all times two objects touch. In section 3 an inductive argument is developed to show the main result. That is, it is shown that if there is a motion of rotations and translations between two configurations in which the objects form a connected component, then there is a motion that keeps the objects in a connected component. In section 4 we discuss the case where more general motions are allowed.

1. Basic Properties of Configuration Space

Let A be a set in $\mathbf{R}^n$. The *interior* of A, denoted int(A), is the union of all open sets of $\mathbf{R}^n$ contained in A. A point x is a *limit point* of A if there exists a sequence $\{x_i\}$ of points in A such that

$$\lim_{i \to \infty} x_i = x.$$

The *closure* of A, denoted by cl(A), is the set of all limit points of A.

An *object* is a convex, compact region of $\mathbf{R}^n$ that is the closure of its interior (Requicha and Tilove 1978) and is bounded by a finite number of algebraic surfaces. (The limitation of convexity will be removed later by the introduction of composite objects.) Each object contains a designated point, called the *origin,* at

Fig. 3. Two objects and the corresponding configuration space.

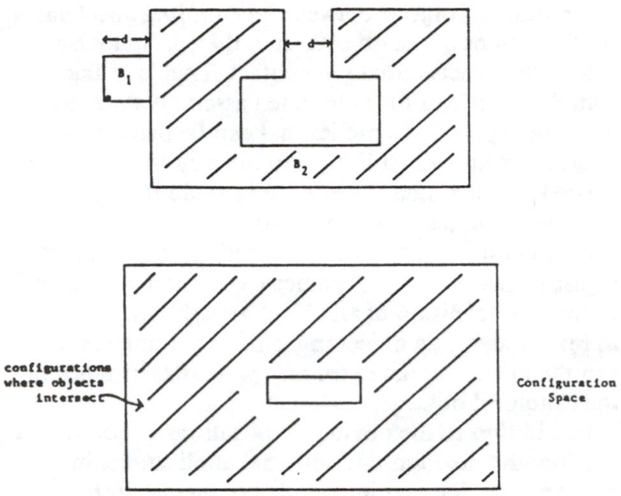

which the origin of a coordinate system, affixed to the object, is located. The position and the orientation of an object are specified by the location of the origin of the object in $\mathbf{R}^n$ and the orientation of the affixed coordinate system relative to the coordinate system of $\mathbf{R}^n$. Given a set of objects, a *configuration* is a vector whose components specify the position and the orientation of each object. The space of all such vectors is called *configuration space.* Given a set of objects and a point x in the corresponding configuration space, we let $B(x)$ denote the region in $\mathbf{R}^n$ occupied by object B in the given configuration. If b is a point on object B, then let $b(x)$ be the point in $\mathbf{R}^n$ occupied by b when B is in the position and orientation specified by x.

Objects B_i and B_j *intersect* in configuration x if $B_i(x) \cap B_j(x) \neq \emptyset$. The objects *overlap* if their interiors intersect. If the objects intersect but do not overlap, then we say that they *touch* in configuration x.

It is convenient to partition the set of objects into subsets called *composite objects.* A composite object is intended to be a single object made up of smaller objects. Associate with each composite object a graph whose vertices are the objects and whose edges are pairs of objects that intersect. A composite object is *connected* if the associated graph is connected. A configuration is *proper* if each composite object is connected. A configuration is *valid* if, in addition to being proper, the interiors of each pair of objects in a composite object do not intersect. Let VALID denote the set of valid configurations.

Composite objects are used in an inductive argument in section 3. By considering two or more objects to be a single composite object, we are able to establish a motion for n objects from a motion for $n - 1$ objects, one of which is a composite object. We introduce the notion of valid so that individual objects in a composite object will touch but not overlap throughout the motion.

Let OVERLAP denote the set of valid configurations in which two or more composite objects overlap. Let TOUCH denote the set of valid configurations in which two or more composite objects touch and no two composite objects overlap. Let NONOVERLAP be the complement of OVERLAP with respect to the set of valid configurations.

Figure 3 shows two objects, B_1 and B_2. Object B_2 is stationary and object B_1 is allowed only translational motion. Given two configurations in which the two objects are touching and a motion between the configurations in which the objects do not overlap, we wish to show that there is a motion where the objects are always in contact. The graphic representation of configuration space in figure 1 suggests that the boundary of the region where the objects intersect corresponds to the configurations where the objects touch. Figure 3 shows that this is not exactly the case. The configurations where B_1 is in the opening of B_2 are not in the boundary of the space of configurations where the objects intersect.

The first goal of this section is to show that the configurations where at least two composite objects touch is exactly

$$\mathrm{cl(OVERLAP)} - \mathrm{OVERLAP}.$$

That is,

$$\mathrm{TOUCH} = \mathrm{cl(OVERLAP)} \cap \mathrm{NONOVERLAP}.$$

The second aim of this section is to prove a lemma concerning TOUCH that will aid us in the next section in proving that certain motions in TOUCH exist. Toward this end, we designate one object in each composite object as the *base* object. We will call the origin of the base object the origin of the composite

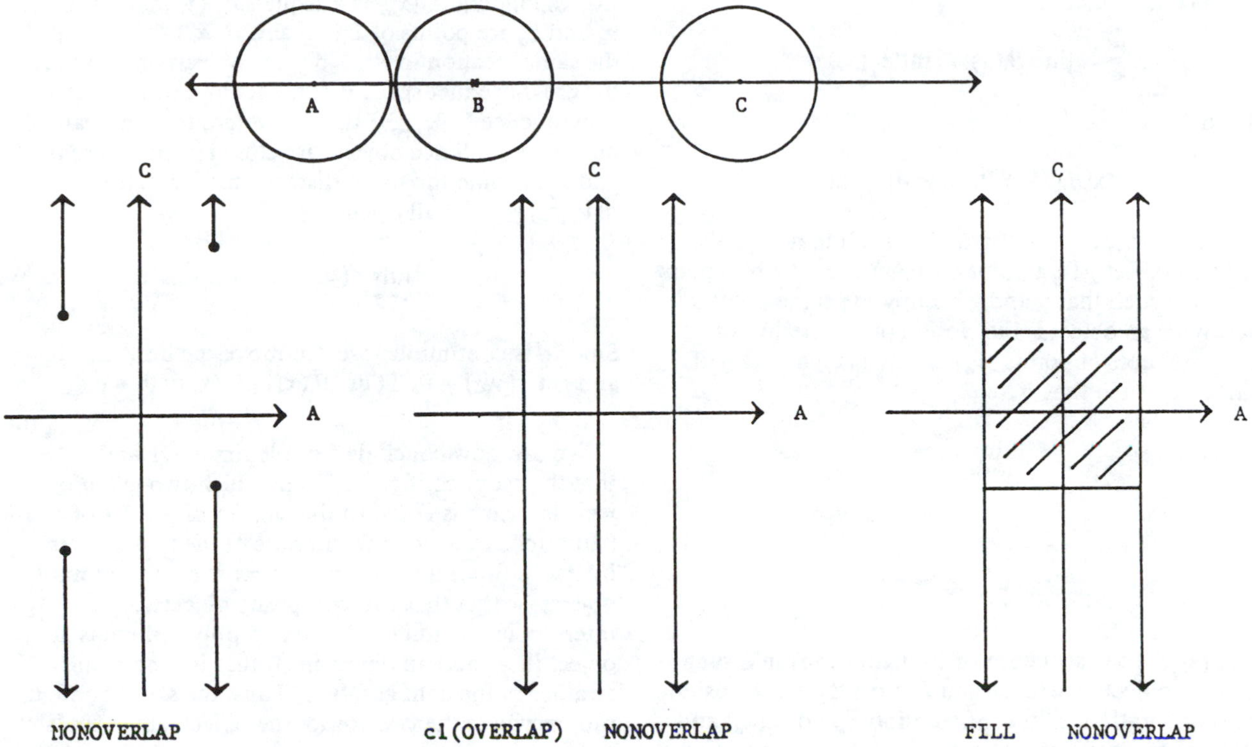

Fig. 4. Three circles constrained to horizontal motion. B is fixed. A and B form a composite object with B as base.

object. Let BASE be the set of proper configurations in which the base object of some composite object intersects the base object of the nth composite object. Let

$$\text{FILL} = \text{cl(OVERLAP)} \cup \text{BASE}.$$

In section 2 we will need the fact that FILL ∪ NONOVERLAP is contractible to a point. Suppose we had not included BASE in FILL. Consider the example shown in figure 4. Here we have three circles that are allowed to move along a line. Object B is fixed, and objects A and B form one composite object in which B is the base object. In this case, NONOVERLAP is four rays and cl(OVERLAP) ∪ NONOVERLAP is two parallel lines. Thus, not only is cl(OVERLAP) ∪ NONOVERLAP not contractible to a point; it also is not path-connected. However, when we include BASE, the set FILL ∪ NONOVERLAP becomes contractible to a point. Thus, the points of FILL fill in the holes of NONOVERLAP so that the union of FILL and NONOVERLAP is contractible to a point. We will show that

$$\text{TOUCH} = \text{FILL} \cap \text{NONOVERLAP}.$$

This will be used in the next section to show that if there is a path in NONOVERLAP between two configurations in TOUCH then there is a path in TOUCH between them. Throughout this section, B_i will denote an object and A_i will denote a composite object.

We now proceed with a series of lemmas. Lemmas 1.1 through 1.4 are used to establish in Theorem 1.5 that TOUCH = cl(OVERLAP) ∩ NONOVERLAP and in Theorem 1.6 that TOUCH = FILL ∩ NONOVERLAP. This latter theorem is used in the next section to prove that certain motions in TOUCH exist.

First we show that for any configuration in which two objects intersect there is an arbitrarily close configuration in which the two objects overlap.

Lemma 1.1: Let

$$S = \{x|\mathrm{int}(B_i(x)) \cap \mathrm{int}(B_j(x)) \neq \emptyset\}.$$

Then

$$\{x|B_i(x) \cap B_j(x) \neq \emptyset\} \subseteq \mathrm{cl}(S).$$

Proof: Let x be a configuration such that $B_i(x) \cap B_j(x) \neq \emptyset$. Let y be a point on $B_i(x)$ and let z be a point on $B_j(x)$ such that y and z occupy the same point b in $\mathbf{R}^n$. Since an object is the closure of its interior, there are sequences of points $\{y_\alpha\}$ of $\mathrm{int}(B_i(x))$ and $\{z_\alpha\}$ of $\mathrm{int}(B_j(x))$ such that

$$\lim_{\alpha \to \infty} y_\alpha = y$$

and

$$\lim_{\alpha \to \infty} z_\alpha = z.$$

Let $\{x_\alpha\}$ be the sequence of configurations in S such that all objects except B_i and B_j have the same position and orientation as in configuration x and $B_i(x_\alpha)$ and $B_j(x_\alpha)$ have the same orientation as in configuration x but y_α and z_α are at position b in $\mathbf{R}^n$. Thus,

$$\lim_{\alpha \to \infty} x_\alpha = x$$

and so x is in $\mathrm{cl}(S)$. ■

Next we show that the closure of the set of configurations in which two objects overlap is contained within the set of configurations in which the two objects intersect. This result, when combined with lemma 1.1, establishes that these two sets are equal.

Lemma 1.2: $\mathrm{cl}(\{x|\mathrm{int}(B_i(x)) \cap \mathrm{int}(B_j(x)) \neq \emptyset\}) \subseteq \{x|B_i(x) \cap B_j(x) \neq \emptyset\}.$

Proof: Let $S = \{x|\mathrm{int}(B_i(x)) \cap \mathrm{int}(B_j(x)) \neq \emptyset\}$. Let x be in $\mathrm{cl}(S)$. Then there exists a sequence of configurations $\{x_\alpha\}$ in S such that

$$\lim_{\alpha \to \infty} x_\alpha = x.$$

Corresponding to $\{x_\alpha\}$ is a sequence $\{\langle y_\alpha, z_\alpha\rangle\}$, where y_α and z_α are points of $B_i(x_\alpha)$ and $B_j(x_\alpha)$ that occupy the same location in $\mathbf{R}^n$. Since the objects are compact, the cross-product space is compact and so there is a subsequence $\{\langle \hat{y}_\alpha, \hat{z}_\alpha\rangle\}$ that converges to some pair of points $\langle y,z\rangle$. Since objects are closed, y and z are in B_i and B_j. Define the usual distance metric d. Since $d(y_\alpha, z_\alpha) = 0$ for all α, clearly

$$\lim_{\alpha \to \infty} d(\hat{y}_\alpha, \hat{z}_\alpha) = 0.$$

Since d is continuous, we can move the limit inside and get $d(y,z) = 0$. Thus, $B_i(x) \cap B_j(x) \neq \emptyset$. ■

We can now conclude from lemmas 1.1 and 1.2 that the set of configurations in which two given objects intersect is equal to the closure of the set of configurations in which the interiors of the two objects intersect. Since a composite object is some union of objects, we get that two composite objects, A_1 and A_2, intersect in a configuration if and only if there is an object in A_1 and an object in A_2 that intersect one another in the configuration. Thus, the set of configurations where the two composite objects intersect is some union of sets of configurations where two objects intersect. Similarly, the set of configurations where the interiors of two given composite objects intersect is some union of sets of configurations in which the interiors of two objects intersect. Since the closure of the union is the union of the closure, the closure of the set of configurations in which the interiors of two given composite objects intersect is some union of the closure of sets of configurations in which the interiors of two objects intersect. Thus, we can conclude that the set of configurations in which two given composite objects intersect is equal to the closure of the set of configurations in which the interiors of the two composite objects intersect. By the same argument, we can show that the set of configurations in which there are at least two composite objects which intersect is equal to the closure of the set of configurations where the interiors of at least two composite objects intersect.

The next step is to show that a composite object also has the property that it is the closure of its interior. Note that interior points of a composite object may not be interior points of any object.

Lemma 1.3: For composite object A, $A = \mathrm{cl}(\mathrm{int}(A))$.

Proof: Let $A = \cup_i B_i$. Suppose $y \in A$. Then $y \in B_i$ for some i. Since $B_i = \mathrm{cl}(\mathrm{int}(B_i))$, there is a sequence $\{y_\alpha\}$ in $\mathrm{int}(B_i)$ such that $\lim_\alpha y_\alpha = y$. But each $y_\alpha \in \mathrm{int}(B_i)$ implies each $y_\alpha \in \mathrm{int}(A)$. Thus, $y \in \mathrm{cl}(\mathrm{int}(A))$ and so $A \subseteq \mathrm{cl}(\mathrm{int}(A))$.

Suppose $y \in \mathrm{cl}(\mathrm{int}(A))$. Then y is the limit point of a sequence in $\mathrm{int}(A)$ and hence the limit point of a sequence in A. Since A is closed, y must be in A. Thus, $\mathrm{cl}(\mathrm{int}(A)) \subseteq A$. ■

Since VALID is a closed set, we can compute the closure of OVERLAP by taking the closure of all configurations where two composite objects overlap and then intersecting with VALID.

Lemma 1.4: $\mathrm{cl}(\mathrm{OVERLAP}) = \mathrm{cl}(\{x | \exists\, i,j \quad i \neq j \quad \mathrm{int}(A_i(x)) \cap \mathrm{int}(A_j(x)) \neq \emptyset\} \cap \mathrm{VALID}) = \{x | \exists\, i,j \quad i \neq j \quad A_i(x) \cap A_j(x) \neq \emptyset\} \cap \mathrm{VALID}$.

Proof: Let

$$F = \{x | \exists\, i,j \quad i \neq j \quad \mathrm{int}(A_i(x)) \cap \mathrm{int}(A_j(x)) \neq \emptyset\}.$$

Then

$$\begin{aligned}\mathrm{cl}(\mathrm{OVERLAP}) &= \mathrm{cl}(F \cap \mathrm{VALID}) \subseteq \mathrm{cl}(F) \cap \mathrm{cl}(\mathrm{VALID}) \\ &= \mathrm{cl}(F) \cap \mathrm{VALID}\end{aligned}$$

because VALID is closed.

Let $x \in \mathrm{cl}(F) \cap \mathrm{VALID}$. We want to show $x \in \mathrm{cl}(\mathrm{OVERLAP})$. Since $x \in \mathrm{cl}(F)$, we know by the remark after lemma 1.2 that for some i and j, $i \neq j$, $A_i(x) \cap A_j(x) \neq \emptyset$. By a construction similar to that in lemma 1.3 we create a sequence $\{x_\alpha\}$ with limit point x such that $\mathrm{int}(A_i(x_\alpha)) \cap \mathrm{int}(A_j(x_\alpha)) \neq \emptyset$ and $x_\alpha \in \mathrm{VALID}$. Thus,

$$x \in \mathrm{cl}(F \cap \mathrm{VALID}) = \mathrm{cl}(\mathrm{OVERLAP}). \ ■$$

We can now establish the result that TOUCH = cl(OVERLAP) ∩ NONOVERLAP.

Theorem 1.5: TOUCH = cl(OVERLAP) ∩ NONOVERLAP.

Proof: By definition,

$$\begin{aligned}\mathrm{TOUCH} = \{x | \exists\, i,j, \quad i \neq j \quad A_i(x) \cap A_j(x) \neq \emptyset\} \\ \cap \{x | \forall\, i,j, \quad i \neq j \quad \mathrm{int}(A_i(x)) \\ \cap\, \mathrm{int}(A_j(x)) = \emptyset\} \cap \mathrm{VALID}.\end{aligned}$$

Therefore,

$$\begin{aligned}\mathrm{TOUCH} &= \mathrm{NONOVERLAP} \cap \{x | \exists\, i,j, \\ &\qquad i \neq j \quad A_i(x) \cap A_j(x) \neq \emptyset\} \\ &= \mathrm{NONOVERLAP} \cap \mathrm{cl}(\mathrm{OVERLAP})\end{aligned}$$

by lemma 1.4. ■

Theorem 1.6: TOUCH = FILL ∩ NONOVERLAP.

Proof: By definition, FILL = cl(OVERLAP) ∪ BASE. Since TOUCH = cl(OVERLAP) ∩ NONOVERLAP by theorem 1.5, we get

$$\begin{aligned}\mathrm{TOUCH} &= \mathrm{cl}(\mathrm{OVERLAP}) \cap \mathrm{NONOVERLAP} \\ &\subseteq \mathrm{FILL} \cap \mathrm{NONOVERLAP}.\end{aligned}$$

Let $x \in$ FILL ∩ NONOVERLAP. If $x \in$ cl(OVERLAP), then $x \in$ TOUCH. Suppose $x \in$ BASE. Since $x \in$ NONOVERLAP, we have $x \in$ VALID. Also, $x \in$ BASE implies that two composite objects intersect, and so $x \in$ cl(OVERLAP). Thus, $x \in$ TOUCH. Therefore, FILL ∩ NONOVERLAP ⊆ TOUCH. ■

2. Requiring Two Objects to Touch throughout a Motion

In this section we show the following intermediate result: Given two configurations x and y with n objects, at least two of which are touching in each configuration, if it is possible to move the objects from configuration x to configuration y then it is possible to do so by a motion such that two objects are always touching. To do this, we make use of the Mayer-Vietoris theorem from algebraic topology to show that the path-connected components of TOUCH are in one-to-one correspondence with the path-connected components of NONOVERLAP.

Fig. 5. TOUCH not a retract of NONOVERLAP.

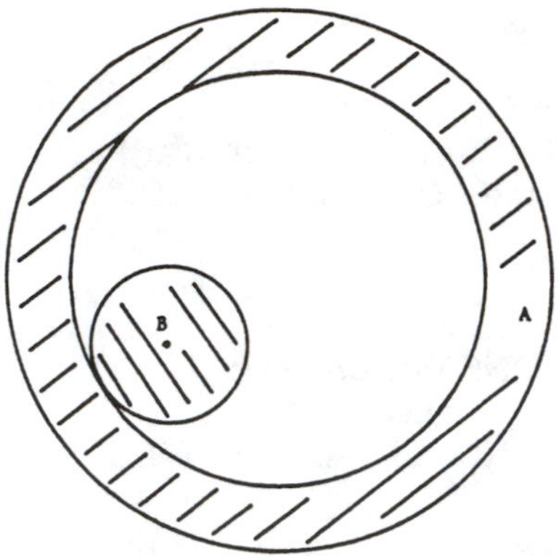

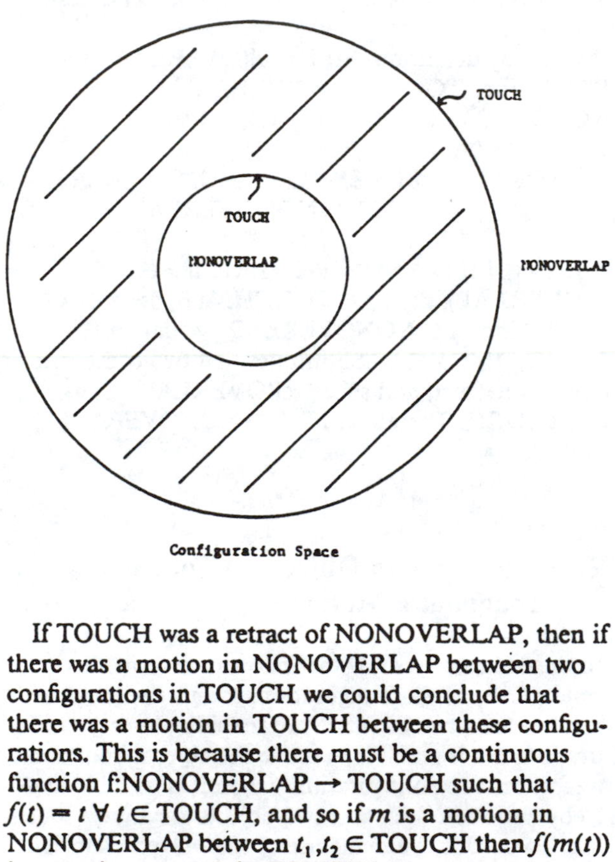

If TOUCH was a retract of NONOVERLAP, then if there was a motion in NONOVERLAP between two configurations in TOUCH we could conclude that there was a motion in TOUCH between these configurations. This is because there must be a continuous function f:NONOVERLAP → TOUCH such that $f(t) = t \; \forall \; t \in$ TOUCH, and so if m is a motion in NONOVERLAP between $t_1, t_2 \in$ TOUCH then $f(m(t))$ is a continuous path in TOUCH between t_1 and t_2.

However, we cannot guarantee that TOUCH will be a retract of NONOVERLAP. Consider figure 5, where there is one stationary object A and another object B which is free to move about. Configuration space is just $\mathbf{R}^2$, where a configuration consists of the position of the center of the disk. NONOVERLAP is the unshaded region in the figure. Thus, in this case NONOVERLAP is not retractible to TOUCH, the boundaries of NONOVERLAP in configuration space.

We begin by defining a motion. A movement of the objects corresponds in an obvious manner to a path in configuration space. Thus, a *motion* is a continuous function from [0,1] to configuration space. If m is a motion, then the *reversal* of the motion m^r is defined as $m^r(t) = m(1 - t)$. If m_1 and m_2 are motions where $m_1(1) = m_2(0)$, then the composition $m = m_1 \| m_2$ is a motion defined by

$$m(t) = \begin{cases} m_1(2t), & 0 \leq t \leq \frac{1}{2} \\ m_2(2t-1), & \frac{1}{2} \leq t \leq 1. \end{cases}$$

At certain times we shall be concerned with motions where the orientation of each object is maintained while each object is moving along a straight line at a constant rate. Thus, in configuration $x(t)$ the location of point b of an object is given by

$$b(x_t) = b(x_0) + [b(x_1) - b(x_0)]\,t.$$

When we talk about a motion in which objects move in a straight line, we are referring to a motion of the above type. When we talk about moving a composite object in a straight line, the objects making up the composite object maintain their relative spacing.

In many of the following results a straight-line motion is used. In lemma 2.1 we show that a straight-line motion of two objects keeps the objects intersecting if they intersect at the beginning and at the end of the straight-line motion.

Lemma 2.1: Let x and y be configurations and B_1 and B_2 be objects. Suppose that b_1 and c_1 are points of B_1 and that b_2 and c_2 are points of B_2 such that $b_1(x) = b_2(x)$ and $c_1(y) = c_2(y)$. Then moving B_1 and B_2 along straight lines so that c_1 and c_2 are positioned at $c_1(y)$ keeps the objects intersecting. See figure 6.

Proof: At time t during the motion, the point of B_1 on the line between b_1 and c_1 given by $b_t = b_1 + (c_1 -$

Fig. 6. Straight line motion.

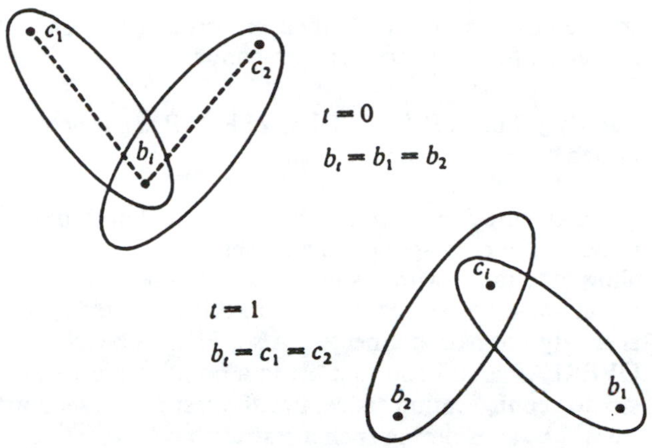

$b_1)t$ occupies the same point as the point of B_2 on the line between b_2 and c_2 given by $b_t = b_2 + (c_2 - b_2)t$. ■

Now it is shown that there is a motion between any two configurations in PROPER in which the base object of some composite object intersects the base object of the nth composite object such that during the motion all configurations have that property. This is done by showing that there is some fixed configuration with the property such that there is a motion, which keeps the property true, between any configuration with the property and the fixed configuration.

Lemma 2.2: BASE is path-connected.

Proof: Fix some configuration y in BASE such that in y the origin of every object has the same location. Let $x \in$ BASE. Move the objects in a straight line from configuration x to the configuration z that has the same orientations as x but in which the location of the origins of the objects are as in y. By lemma 2.1 any objects that intersect in x will intersect throughout the motion. Now rotate the objects about their origins to the orientations given by y. Thus, there is a path in BASE from any $x \in$ BASE to y. Since motions are reversible, there is a path in BASE from y to any $x \in$ BASE, and so BASE is path-connected. ■

In the following theorem we show that FILL is path-connected by constructing a motion from any configuration in FILL to some configuration in BASE. BASE is path-connected, by the previous result, and since BASE is contained in FILL we conclude that FILL is path-connected.

Theorem 2.3: FILL is path-connected.

Proof: Let $x \in$ FILL. We will show that there is a path in FILL from x to some configuration in BASE, and since BASE is path-connected and motions are reversible we will conclude that FILL is path-connected.

By definition, FILL = cl(OVERLAP) ∪ BASE. If $x \in$ BASE, then we are done. Suppose $x \in$ cl(OVERLAP). Then $A_i(x) \cap A_j(x) \neq \emptyset$ for some i and j $(i \neq j)$ and $x \in$ VALID. Let b be the origin of the base object of A_i and let b' be the origin of the base object of the nth composite object. Move A_i and A_j in a straight line (considering A_i and A_j as one composite object) to the configuration where the location of b is $b'(x)$. All other objects remain stationary. Thus, the motion is in cl(OVERLAP) and hence in FILL, and the resulting configuration is in BASE. Therefore FILL is path-connected. ■

The next lemma will be used when we show that FILL ∪ NONOVERLAP is contractible to a point. We show that there is a path in VALID from any configuration in VALID to a configuration in BASE in which the origins of all the composite objects have the same location. The same construction can be used to construct a path in BASE from any configuration in BASE to some configuration in BASE in which the origins of all the composite objects have the same location.

Lemma 2.4: From every configuration in VALID (BASE) there is a path in VALID (BASE) to some configuration in BASE in which the origins of the composite objects coincide.

Proof: Let $x \in$ VALID (BASE). Move the composite objects in a straight line from x to the configuration where all the origins have location equal to the location in x of the origin of the nth composite object. The resulting configuration is in BASE, and the motion described is as desired by lemma 2.1. ■

We now use the motions constructed in lemmas 2.2 and 2.4 to show that FILL ∪ NONOVERLAP is contractible to a point. That is, we show that there is a configuration $y \in S = \text{FILL} \cup \text{NONOVERLAP}$ and a continuous function $f: S \times [0,1] \rightarrow S$ such that

$$\left.\begin{array}{l} f(x,0) = x \\ f(x,1) = y \end{array}\right\} \forall x \in S$$
$$f(y,t) = y \quad \forall t \in [0,1].$$

In order that FILL ∪ NONOVERLAP be contractible to a point, we cannot identify a rotation of 2π with no rotation at all, as is done in Schwartz and Sharir 1982. Thus, in configuration space a dimension corresponding to a rotation is infinite even though every 2π radians the object returns to its apparent initial position.

Theorem 2.5: FILL ∪ NONOVERLAP is contractible to a point.

Proof: Let $S = \text{FILL} \cup \text{NONOVERLAP}$. Then $S = \text{BASE} \cup \text{VALID}$. Let y be the fixed configuration in BASE as in lemma 2.2. Define $f: S \times [0,1] \rightarrow S$.

$$f(x,t) = \begin{cases} m_1(2t), & 0 \leq t \leq ½ \text{ where } m_1 \text{ is the motion described in lemma 2.4 and } m_1(0) = x \\ m_2(2t-1), & ½ \leq t \leq 1 \text{ where } m_2 \text{ is the motion described in lemma 2.2 and } m_2(0) = m_1(1). \end{cases}$$

Then

$$\left.\begin{array}{l} f(x,0) = x \\ f(x,1) = y \end{array}\right\} x \in S$$
$$f(y,t) = y \quad t \in [0,1].$$

By the construction of m_1 and m_2, f is continuous. Thus, f is a homotopy between the retraction $r: S \rightarrow \{y\}$ and the identity $i: S \rightarrow S$. That is, $\{y\}$ is a deformation retract of S and so S is contractible to a point. ■

The above construction gives a motion in $S = \text{FILL} \cup \text{NONOVERLAP}$ from any configuration in S to the fixed configuration $y \in S$. Thus, if x_1 and x_2 are two configurations in S and if m_i is the motion constructed from x_i to y in S, then $m_1 \| m_2^r$ is a motion in S from x_1 to x_2 and so S is path-connected.

Corollary 2.6: FILL ∪ NONOVERLAP is path-connected.

If configuration space is restricted so that each parameter that corresponds to an orientation is only allowed to range within some closed and bounded interval, then the portion of FILL in this restricted space is clearly still path-connected. Also, FILL ∪ NONOVERLAP is still contractible to a point in this restricted configuration space. For the rest of this section we will be considering such a restricted configuration space. Thus, when we speak of some set such as NONOVERLAP we will mean the part of the set that is in the restricted configuration space.

Theorem 2.7: NONOVERLAP consists of a finite number of path-connected components.

Proof: As in Schwartz and Sharir 1982, we divide configuration space into finitely many cells such that the set of polynomials that describe the relative positions of the objects are sign-invariant within each cell. Then NONOVERLAP is the finite union of some of these cells, and for any two points x,y in a cell there is a path within the cell between them. Thus, there must be a finite number of path-connected components in NONOVERLAP. ■

The next result makes use of the concept of an exact sequence. An *exact sequence*

$$A \xrightarrow{h_1} B \xrightarrow{h_2} C \xrightarrow{h_3} \ldots \xrightarrow{h_n} \{0\}$$

is such that the image of h_i, $\text{Im}(h_i)$, is the kernel of h_{i+1}, $\ker(h_{i+1})$. Let **Z** be the group of integers under addition. The notation H_0 and H_1 in the theorem means the following:

$H_0(S)$ is the zeroth homology group of S where $H_0(S) \simeq Z \oplus Z \oplus \ldots \oplus Z$ (m copies) if S has m path-connected components.

$H_1(S)$ is the first homology group of S where $H_1(S) \simeq \{0\}$ if $H_1(S)$ is contractible to a point.

Fig. 7. Mayer-Vietoris sequence.

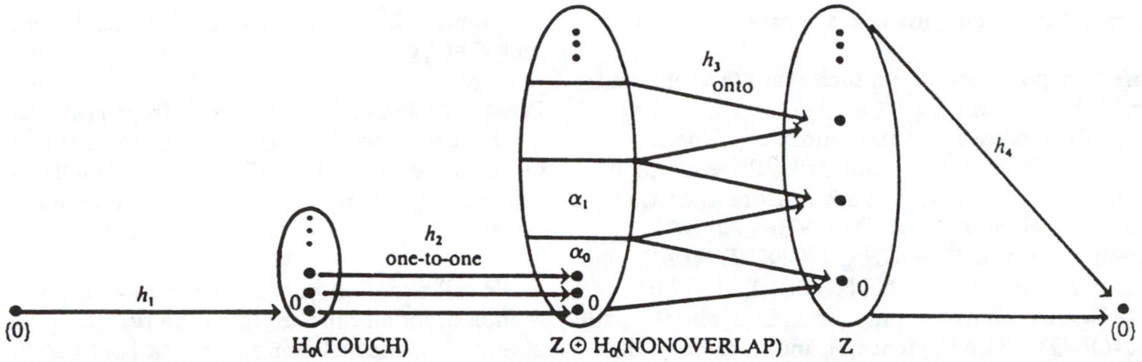

Theorem 2.8 (Mayer-Vietoris): The sequence

$$H_1(A \cup B) \rightarrow H_0(A \cap B) \rightarrow H_0(A) \oplus H_0(B) \rightarrow H_0(A \cup B) \rightarrow \{0\}$$

is an exact sequence.

Proof: See Massey 1978. ∎

In the following we will use the Mayer-Vietoris sequence with A = FILL and B = NONOVERLAP to show that in our restricted configuration space TOUCH and NONOVERLAP have the same number of path-connected components.

Theorem 2.9: H_0(NONOVERLAP) ≃ H_0(FILL ∩ NONOVERLAP) = H_0(TOUCH).

Proof: By theorem 2.5, S = FILL ∪ NONOVERLAP is contractible to a point and so $\Pi_1(S) = \{0\}$ (see Massey 1967). Since $\Pi_1(S)$ is abelian, $H_1(S) = \Pi_1(S) = \{0\}$. Also, we have $H_0(S) = \mathbf{Z}$, because by corollary 2.6 S is path-connected. By theorem 2.3 FILL is path-connected, so H_0(FILL) = Z. Therefore the sequence in theorem 2.8 is as follows:

$$\{0\} \xrightarrow{h_1} H_0(\text{TOUCH}) \xrightarrow{h_2} \mathbf{Z} \oplus H_0(\text{NONOVERLAP}) \xrightarrow{h_3} \mathbf{Z} \xrightarrow{h_4} \{0\}.$$

Since the sequence is exact, $\text{Im}(h_3) = \ker(h_4) = \mathbf{Z}$ and so h_3 is onto. Also, $\ker(h_2) = \text{Im}(h_1) = \{0\}$ and so h_2 is one-to-one. Thus we have the situation shown in figure 7. Therefore, H_0(TOUCH) ≃ Im(h_2). The α_i's are the cosets of Im(h_2), and so there is a one-to-one correspondence between any α_i and Im(h_2). Hence,

$$\mathbf{Z} \oplus H_0(\text{NONOVERLAP}) \simeq \mathbf{Z} \oplus \text{Im}(h_2) \simeq \mathbf{Z} \oplus H_0(\text{TOUCH}).$$

Then, since H_0(NONOVERLAP) ≃ Z ⊕ · · · ⊕ Z (n copies of Z) and H_0(TOUCH) ≃ Z ⊕ · · · ⊕ Z (m copies of Z) by theorem 2.8, it must be that $m = n$ and so H_0(NONOVERLAP) ≃ H_0(TOUCH). ∎

Thus, it has been shown that the number of path-connected components of NONOVERLAP, denoted by #NONOVERLAP, equals the number of path-connected components of TOUCH, denoted by #TOUCH. Let m = #TOUCH = #NONOVERLAP, and let $t_1, \ldots, t_m$ be the path-connected components of TOUCH and $n_1, \ldots, n_m$ be the path-connected components of NONOVERLAP.

Now we wish to show that each path-connected component of NONOVERLAP contains exactly one path-connected component of TOUCH. The reason that this property is desired is that, in this case, if there is a path in NONOVERLAP between two configurations in TOUCH (i.e., the two configurations are in the same path-connected component of NONOVERLAP) then there must be a path in TOUCH between them, because they must be in the one path-connected component of TOUCH that is contained in the path-connected component of NONOVERLAP.

Lemma 2.10: Each t_i intersects at most one n_k.

Proof: Suppose there is a t_i such that $t_i \cap n_k \neq \emptyset$ and $t_i \cap n_j \neq \emptyset$. Let $x_1 \in n_j$, $x_2 \in n_k$, $x_3 \in t_i \cap n_j$, and $x_4 \in t_i \cap n_k$. Then, since n_j is path-connected, there is a path P_1 in n_j from x_1 to x_3. Similarly, there is a path P_2 in n_k from x_4 to x_2 and a path P_3 from x_3 to x_4 in t_i. Since TOUCH = FILL $\cap$ NONOVERLAP by theorem 1.6, we get TOUCH $\subseteq$ NONOVERLAP, and so a path in t_i is a path in NONOVERLAP. Thus, $P_1 \| P_3 \| P_2$ is a path from x_1 in n_j to x_2 in n_k and the path is in NONOVERLAP. Hence, n_j and n_k must be the same path-connected component of NONOVERLAP. ■

Thus, for each i, $t_i \cap n_k \neq \emptyset$ for at most one k, and since TOUCH $\subseteq$ NONOVERLAP we know that for every t_i there is one n_k such that $t_i \subseteq n_k$. Now it will be shown that each n_k contains at most one t_i and so each n_k contains exactly one t_i.

Lemma 2.11: For each n_k there is a t_j such that $t_j \subseteq n_k$.

Proof: We must show that for any $x \in$ NONOVERLAP there is motion in NONOVERLAP from x to some configuration $y \in$ TOUCH. Let $x \in$ NONOVERLAP and suppose $x \notin$ TOUCH. Let a_1 and a_2 be the origins of A_1 and A_2 and let $s(t) = ta_2(x) + (1-t)a_1(x)$. Let $m(t)$ be the motion such that $a_1(m(t)) = s(t)$ and everything else stays constant. Let

$$t_0 = \min_t \{m(t) \mid m(t) \notin \text{NONOVERLAP} - \text{TOUCH}\}.$$

Thus, for some i and j,

$$A_i(m(t_0)) \cap A_j(m(t_0)) \neq \emptyset.$$

If $a \in A_i(m(t_0)) \cap A_j(m(t_0))$, then

$$a \in A_i(m(t_0)) - \text{int}(A_i(m(t_0)))$$

and

$$a \in A_j(m(t_0)) - \text{int}(A_j(m(t_0)));$$

otherwise we would contradict the definition of t_0. Therefore, $m(t_0) \in$ TOUCH, as required. ■

Theorem 2.12: For each n_k there is exactly one t_j such that $t_j \subseteq n_k$.

Proof: By lemma 2.11 we know there is at least one $t_j \subseteq n_k$. By lemma 2.10, each t_i is contained in at most one n_k. Since #NONOVERLAP = #TOUCH, we conclude that there is exactly one t_j contained in each n_k. ■

We call a motion m of n objects a *k-component motion* if, for all t in the closed interval [0,1], $m(t)$ has at most k connected components. A configuration x is said to be a *k-component configuration* if x has at most k connected components. We call a 1-component configuration a *connected configuration*. Thus, theorem 2.12 can be restated as follows: If there is an n-component motion from x to y (i.e., $x,y \in n_k$) and x and y are $(n-1)$-component configurations, then there is an $(n-1)$-component motion from x to y (i.e., $x,y \in t_j$).

3. The Existence of a Motion in Contact

In this section we establish our main result. Suppose that we have n objects and that x and y are two configurations in which the n objects form a connected composite object. Suppose further that there is a motion from configuration x to configuration y such that no two objects overlap. Then there is a motion such that all configurations throughout the motion are also connected. (For the remainder of the section, all configurations will be in NONOVERLAP. Thus, if we say that a configuration x has k components, we mean that x is in NONOVERLAP and has k components.)

Let $P_1, P_2, \ldots, P_P$ be the partitionings of the n objects into $k+1$ or fewer connected components. We say that a configuration x satisfies P_i if the connected components of P_i are contained in the connected components of x. Let x_1 and x_2 be configurations of n composite objects both of which have k $(1 \leq k \leq n)$ or fewer connected components, and let m be a $(k+1)$-component motion from x_1 to x_2. Partition NONOVERLAP into regions so that all x satisfying a given P_i are in one region. Further partition NONOVERLAP into path-connected components. Without loss of generality, we assume that m never returns to a path-connected component satisfying P_i once it has left it. Let $T_i = \{t|m(t)$ satisfies $P_i\}$, where $1 \leq i \leq p$. Since $m(t)$ is a continuous function and T_i is the set of all t such that $m(t)$ is a closed region of configuration space, each T_i is a closed set. Furthermore, each T_i is a finite union of closed intervals. Partition the interval [0,1] into a finite number of closed subintervals $J_i = [a_i, a_{i+1}]$ $(1 \leq i \leq t)$ such that for each J_i there are $k+1$ sets of composite objects where each set remains as a connected composite object during the motion m on interval J_i. We can assume that the J_i's are maximal with respect to the above conditions. Thus, $m(a_i)$, where $1 \leq i \leq t+1$, is a k-component configuration and the motion m_i, which is m on J_i, is a $(k+1)$-component motion. Notice that during m_i the composite objects in each of the $k+1$ sets that define J_i remain connected composite objects and so we can think of m_i as a motion of $k+1$ composite objects rather than a motion of n composite objects.

We will use such a partitioning of the interval [0,1] in the following result, where we show that it is possible to reduce the number of components during a motion to the number of components in the initial and final configurations if there is a motion during which there is one more component.

Lemma 3.1: Let n be the number of composite objects and let k be such that $1 \leq k < n$. If there is a $(k+1)$-component motion between two k-component configurations, then there is a k-component motion between them.

Proof: The proof is by induction n, the number of composite objects.

Base step: If $n = 2$, then $k = n - 1$ and the result follows from theorem 2.12.

Induction step: Assume that the result holds when there are fewer than n composite objects. Suppose we have n composite objects. For $k = n - 1$, the lemma is true by theorem 2.12.

Let k be such that $1 \leq k < n - 1$. Partition [0,1] into $J_1, \ldots, J_t$ as above. Then m_i is a $(k+1)$-component motion of $k+1$ composite objects between k-component configurations a_i and a_{i+1}. Since $k+1 < n$, the induction hypothesis holds for each m_i, and so there is a k-component motion m'_i between a_i and a_{i+1}. Thus, $m' = m'_1 \| m'_2 \| \cdots \| m'_t$ is a k-component motion between the given k-component configurations. ∎

Now an immediate corollary to lemma 3.1 for the case of connected configurations is stated.

Corollary 3.2: If there is a $(k+1)$-component motion between two connected configurations of composite objects $(1 \leq k < n)$, then there is a k-component motion between them.

It is now possible to show that, if there is any motion between two connected configurations such that during the motion no two composite objects overlap, then there is a motion between the configurations such that all configurations during the motion are connected and no two composite objects overlap.

Theorem 3.3: If there is any motion between two connected configurations x and y, then there is a motion between them such that throughout the motion the configurations are connected.

Proof: Suppose there is a $(k+1)$-component motion between x and y for $k \geq 1$. Then by corollary 3.2 there is a k-component motion between x and y. Thus, by induction, there is a 1-component motion between x and y. ∎

4. Generalizations

In the previous sections we restricted motion to translation and rotation. It was shown that, if there was a motion between two connected configurations such that throughout the motion no two objects overlapped, then there was a motion between the configurations such that throughout the motion the objects formed a connected configuration. This result depended on the fact that certain subsets of configuration space were path-connected and the fact that one subset, namely FILL ∪ NONOVERLAP, was contractible to a point. These facts used translational motions in their proofs. Notice that the results hold if the only motions allowed are translations. However, as noted above, if

Fig. 8. A nonconvex object.

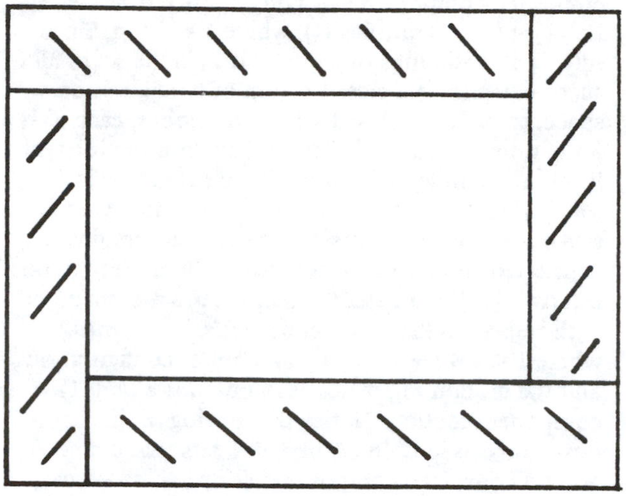

only rotations are permitted then the result does not hold as stated.

For more general motion that allows continuous deformation of the objects, such as stretching or radial growth about some point of an object, we must make sure that FILL and FILL ∪ NONOVERLAP are again path-connected and that FILL ∪ NONOVERLAP is contractible to a point. If so, we can again conclude that, if there is a generalized motion that keeps the objects from overlapping one another between two connected configurations, then there is a generalized motion between these configurations that keeps the configurations connected throughout.

Suppose that motions consist of translations and any kind of continuous deformation of objects such that the objects remain convex and the deformation has a fixed point. The motions described in theorems 2.3 and 2.5 can be extended in the obvious way to include the type of motion described above. Thus, FILL ∪ NONOVERLAP is contractible to a point and FILL is path-connected for these motions, and hence we can conclude that, if there is a motion in NONOVERLAP between two connected configurations, then there is a motion that keeps the objects connected throughout.

For some types of generalized motions, FILL will not be path-connected. Suppose that FILL is not path-connected but that it is the case that the set consisting of all the configurations of one path-connected component of FILL and all the configurations in a path-connected component of NONOVERLAP that intersects the path-connected component of FILL is contractible to a point. Then, by taking A to be the path-connected component of FILL and B to be the path-connected component of NONOVERLAP in the Mayer-Vietoris sequence of section 2, we can conclude that, for two configurations that are in both the path-connected component of FILL and the path-connected component of NONOVERLAP (and hence in TOUCH), there is a motion in TOUCH between these configurations.

If we strengthen the definition of VALID such that VALID remains a closed set in configuration space, the results of the preceding sections still hold. For example, instead of just requiring that a composite object be connected, we could insist that the objects of the composite object touch each other in a specific manner. In this way we could have nonconvex objects by dividing them into convex pieces and then defining VALID so that a configuration is in VALID only if the convex pieces form the nonconvex object that is required. An example of this is shown in figure 8, where the four rectangular objects form a nonconvex composite object and where a configuration must have these objects touching in this way for it to be in VALID.

Acknowledgments

The authors would like to thank Peter Kahn for pointing out the use of the Mayer-Vietoris sequence.

References

Hopcroft, J. E., D. Joseph, and S. Whitesides. 1982. Movements problems for 2-dimensional linkages. *SIAM J. Comput.* 13(3):610–629.

Lozano-Pérez, T., and M. A. Wesley. 1979. An algorithm for planning collision-free paths among polyhedral objects. *Commun. ACM* 22(10):560–570.

Mason, M. T. 1981. Compliance and force control for computer controlled manipulators. *IEEE Trans. Systems, Man, and Cybernetics SMC-11(6).*

Massey, W. S. 1967. *Algebraic Topology: An Introduction.* New York: Springer.
Massey, W. S. 1978. *Homology and Cohomology Theory.* New York: Dekker.
Requicha, A. G., and R. B. Tilove. 1978. Mathematical foundations of constructive solid geometry: General topology of closed regular sets. Technical memo 27, Production Automation Project, University of Rochester.
Schwartz, J. T., and M. Sharir. 1982. On the piano mover's problem II: General techniques for computing topological properties of real algebraic manifolds. Technical report 41, Computer Science Department, New York University.

Effect of Kinematics on Motion Planning for Planar Robot Arms Moving Amidst Unknown Obstacles

VLADIMIR J. LUMELSKY, SENIOR MEMBER, IEEE

Abstract—An approach of dynamic path planning (DPP) was introduced elsewhere, and nonheuristic algorithms were described for planning collision-free paths for a point automaton moving in an environment filled with unknown obstacles of arbitrary shape. The DPP approach was further extended to a planar robot arm with revolute joints; in this case, every point of the robot body is subject to collision. Under the accepted model, the robot, using information about its immediate surroundings provided by the sensory feedback, continuously (dynamically) generates its path. Various kinematic configurations of planar arms with revolute and sliding joints are analyzed in this paper from the standpoint of applying the same strategy. It is shown that, depending on the arm kinematics, specific modifications must be introduced in the path planning algorithm to preserve convergence. The approach presents an attractive method for robot motion planning in unstructured environments with uncertainty.

I. Introduction

PLANNING a path for a robot system moving in an environment with obstacles amounts to finding continuous and mutually compatible trajectories for all its parts, such that the resulting motion is collision-free. Typically, the robot system is either a mobile robot or a multilink robot arm manipulator with a fixed base. In this paper, we are concerned with the latter type.

There are two distinct categories into which all existing approaches to robot motion planning fall. The important assumption in the first category (often called the "piano movers" model) is that complete information on the geometry of the robot and the obstacles in the scene is given beforehand so that path planning becomes a one-time operation. The main difficulty in this formulation is in designing algorithms efficient enough to handle the overwhelming complexity of the input data.

In the second category of works, the information on the obstacles is incomplete, and this is compensated by additional on-line sources of information such as maps or sensory feedback. Computationally, handling the input information in this case is usually not difficult, since the path is being built continuously, step-by-step, and relatively little information has to be processed at every single step. The main issue with the planning algorithms in this category is to show how to use the provided partial information either to assure reaching the target or to conclude that no path to the target exists.

Additionally, in each category one needs to distinguish between various heuristics and approaches where convergence is guaranteed (nonheuristic or exact methods). With heuristic procedures, although they can provide satisfactory results in simpler applications, the chances for the robot "to get lost" increase as the applications become more complex (such as in cases with nonconvex obstacles or with three-dimensional motion). Since human intuition is often of little help in more or less complex cases of space orientation, providing theoretical assurances of the algorithm convergence becomes an important task.

Historically, the piano mover's model has received most attention; both exact and heuristic approaches were developed [1]–[9]. Works on motion planning with uncertainty have so far concentrated on various heuristics, either for mobile robots [13]–[17] or for arm manipulators [18].

This work is a further extension of a nonheuristic approach, developed in [10], [11], [23], to planar robot arms of different kinematic configurations with revolute or sliding joints. In terms of the classification given, the approach, called dynamic path planning (DPP), belongs to the second category (i.e., motion planning with uncertainty). In [10], path planning algorithms were developed for a point automaton traveling in an environment which presents a surface homeomorphic to a (two-dimensional) plane. The environment may contain unknown obstacles of arbitrary shapes. The automaton knows its current position, the target position, and is provided with sensory feedback which allows it to detect (physical or distant) contact with an obstacle. Using a DPP algorithm, the automaton plans its path continuously; that is, at any point of its path, the automaton is generating its next position, based on its current position and the target's position and on the feedback information about its immediate surroundings.

In [11], the DPP approach has been extended to a planar arm of a specific kinematic configuration—the so-called articulated arm (see Fig. 1(a)). A representation space, called *image space* or *I* space, has been introduced, in which the arm becomes a point moving along the (two-dimensional) surface of some manifold; for said arm, this manifold is a common torus. Modifications were introduced in the basic DPP algorithm to account for specifics of the torus topology. Examples were given of the algorithm performance in rather complicated environments where, even if full information

Manuscript received May 1986; revised November 1986. The material in this paper has been partially presented at the 24th IEEE Conference on Decision and Control, Fort Lauderdale, FL, December 1985. This research has been supported in part by National Science Foundation under Grant DMC-8519542.

The author is with the Department of Electrical Engineering, Yale University, New Haven, CT 06520, USA.

IEEE Log Number 8714997.

Reprinted from *IEEE J. of Robot. Automat.*, vol. RA-3, no. 3, pp. 207–223, June 1987.

about the obstacles is provided, path planning presents a formidable task.

Although some limited direct applications of planar arm motion planning can be envisioned (e.g., with SCARA arms which are typically used in plane-oriented applications, such as assembly on the table or on the conveyer belt), the main motivation behind this work is to develop the groundwork for handling three-dimensional (3D) robot arms of various kinematics. For the general case of a body moving in 3D space, the main property that is being exploited in the DPP approach—the unique choice for following a simple curve in an orientable two-dimensional manifold—does not hold. However, for a 3D robot arm, there are natural physical constraints (such as the fact that the arm links follow in succession and that one of the links is attached to the floor), which can be exploited in the design of 3D algorithms. Although these constraints do not produce a unique choice for maneuvering around obstacles, they do give some "preferable" directions—for example, moving in the direction of the minimum value of a joint variable, or moving toward the boundary surfaces of the *work space* (*W* space). The resulting strategies and the proofs of their convergence are based on the theory of planar motion, and use, as their components, algorithms for planar arms developed in this paper. Dynamic path planning algorithms for simple 3D robot arms are described in [19], [21].

Because of the uncertainty involved, no optimality of a path generated by a DPP algorithm can be guaranteed; instead, the objective is to generate a reasonable path or to conclude that the target cannot be reached. It has been shown in [11], [23] that the local feedback information is, indeed, sufficient to guarantee reaching a global objective.[1] The length of the produced path and the amount of required computations are proportional to the total of the distance start-to-target and of perimeters of the obstacles met by the automaton on its way to the target. In [10], [23], the lower bound on the length of paths generated by any imaginable algorithm operating with uncertainty has been produced. The bound shows the best performance that can be expected from an algorithm in the worst case and is a powerful means for comparing different algorithms.

Briefly, one of the *basic procedures*, described in [10] and also used in this paper, works as follows. A simple and easily recognizable line, called *main line* or *M* line, is defined which connects the start and the target points; for example, this may be a straight line. The *M* line is the desirable trajectory of the automaton's motion to the target. Normally, the automaton moves along the *M* line toward the target. If the *M* line crosses no obstacles, the automaton simply reaches the target without ever interacting with obstacles; in this case, the *M* line becomes the actual path.

If, on the other hand, the automaton hits an obstacle, it has only two options for passing around it: turning either right or left along the obstacle boundary. Since the available local information is not sufficient to decide which of the two options is better, the direction of turning around an obstacle is fixed. Using this direction, the automaton starts walking along the obstacle boundary until it meets the *M* line again, at a distance from the target shorter than that from the hit point. Next, the automaton continues moving along the *M* line toward the target until it meets another (or, maybe, another part of the same) obstacle. The procedure repeats until termination. The algorithm effectively uses the fact that the motion takes place along a two-dimensional orientable manifold in which any obstacle presents a simple closed curve.

The algorithm can be shown to converge. Its behavior is rather similar to what a human might do in an uncertain environment. In terms of the length of generated paths, the performance of the algorithm is quite complex and depends on the mutual position of the start and target points relative to the obstacles in the scene. In most cases, the generated paths are quite efficient and are shorter than the lower bound mentioned earlier. In some cases, however, the paths may include (a finite number of) cycles, which make the automaton pass the same segment of the path more than once.

The upper bound on the length of the paths [10], [23] generated by the basic procedure, also applies to the algorithms presented below. For each of the planar arms, the main idea is to introduce an appropriate image space in which the motion of the arm would be reduced to the motion of a point along a two-dimensional surface filled with simple closed curves; the curves represent the images of the actual obstacles.

The list of planar arms studied in this work is based on the classification of kinematic configurations of three-dimensional major linkages with revolute and sliding joints given in [12]. Out of 36 possible configurations, 12 are listed as kinematically useful and distinct. The rest are not admissible, either because they degenerate into one- or two-dimensional cases, or because they are equivalent to some others. Of these 12 types, only five are meaningful for the case of a planar major linkage. These are as follows: two revolute joints (an articulated arm, Fig. 1(a)); two sliding joints (typically referred to as a Cartesian arm, Fig. 1(b)); a revolute joint followed by a parallel sliding joint (Fig. 1(c)); a revolute joint followed by a perpendicular sliding joint (Fig. 1(d)); and a sliding joint followed by a revolute joint (Fig. 1(e)). We study these five planar arm configurations and show that the approach described in [11] is indeed applicable if important modifications of the path planning procedure are introduced for each of the configurations; with these modifications, the procedures converge.

Following [12], we use the notion of a separable arm—that is, an arm which can be naturally divided into the major linkage responsible for the position planning (or the gross motion) of the arm, and the minor linkage responsible for the orientation planning of the arm end effector. As a rule, existing arm manipulators are separable. In the 3D case, the major linkage includes three links (and three joints), and for 2D it includes two links. In the DPP approach, motion planning is limited to the gross motion. This can be interpreted as planning the major linkage motion only, with the implicit assumption that the minor linkage motion can be planned

[1] Besides being important for robot motion planning, this interesting result continues the discussion of the value of local means in reaching a global objective, which has been an area of active study in the theory of games and the theory of automata during the last three decades (see, e.g., [22]). In the geometric context, it has also been discussed in [26].

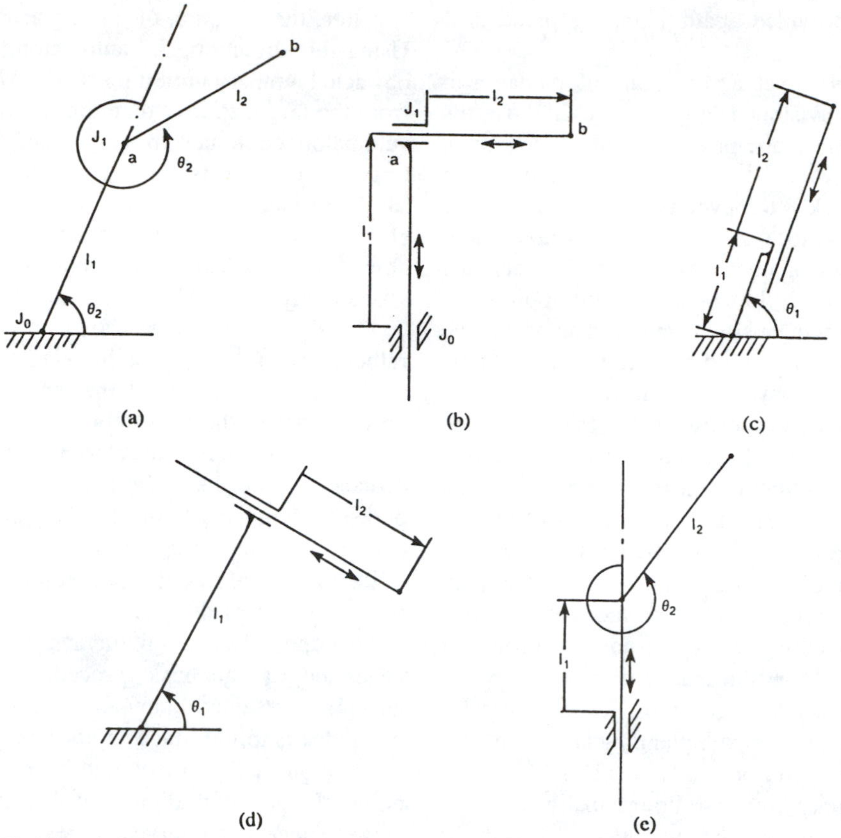

Fig. 1. Five kinematically distinct planar robot arms. (a) Arm 1. (b) Arm 2. (c) Arm 3. (d) Arm 4. (e) Arm 5.

separately, after the arm arrives in the vicinity of the target position. For all but very unusual applications, this assumption should be plausible. Since no constraints are imposed by the approach on the shape of the arm links (see below), the joints of the minor linkage can be, for example, "frozen" during the gross motion, thus effectively making the minor linkage a part of one of the links of the major linkage.

In the next section, a general model of the arm and of the environment in which the arm operates is formulated, and relevant definitions are given. This is followed by five sections, each dealing with one of the planar arms of Fig. 1. In each of these sections, an appropriate I space is introduced, the interaction between the arm and the obstacles is analyzed, and a modified path planning procedure formulated.

II. Model and Definitions

The arm's objective is to move from the position "start" (or S) to the position "target" (or T). Only continuous motion of the robot links is allowed. The arm and its *work space* (W space) are defined in the plane. The *boundaries* of the W space are defined by the arm configuration and by the arm links dimensions. Both positions S and T are assumed to lie within the W space. The arm body consists of two links l_1 and l_2 and two joints J_0 and J_1 (Fig. 1(a)–(e)). Joint J_0 is fixed and is the origin of the reference system. Assume, for presentation purposes, that each link is a straight line segment; the lengths of the links are l_1 and l_2, respectively.

Depending on the arm configuration, the length of a link may be constant or variable. An *arm solution* (equivalent terms to be used are "arm position," "arm coordinates," and "link positions") corresponding to a given point P in the W space is defined by a pair of variables, *joint values,* which are either angles (as in Fig. 1(a)), or linear translations (as in Fig. 1(b)), or both (as in Fig. 1(c), (d), (e)). An equivalent presentation for the same solution P is given by the coordinates of the link endpoints, a_p and b_p; b_p also designates the position of the arm endpoint at point P. Positive directions and zero positions for all the joint values of the five arms are shown in Fig. 1(a)–(e).

The W space may contain obstacles. Each obstacle is a simple closed curve of finite size; there may be only a finite number of obstacles present in the W space. Formally, this means that the boundary of any obstacle is homeomorphic to a circle and that any circle of a limited radius or a straight line passing through W space will intersect with a finite set of obstacles. Being rigid bodies, obstacles cannot intersect each other; two or more obstacles may touch each other, in which case for the arm they effectively present one obstacle. At any position of the arm with respect to a set of obstacles, at least some arm motion is assumed to be feasible.

We emphasize that for the considered class of the path planning algorithms, the shape of arm links and of the obstacles (e.g., the fact of their convexity or concavity) is of no importance. Solely for better visualization and material presentation, line segment links and circular obstacles are used in most of this text.

The only information available to the arm are its current position and the target position T. The arm position S is known to be feasible. Because of the obstacles, position T may or may not be feasible. Or, even if position T is feasible, it may or may not be reachable from the position S. The arm is capable of performing the following actions:

1) moving the arm endpoint through a prescribed simple curve (called main line or M line) that connects S and T;
2) when the arm body hits an obstacle, identifying the point(s) of the arm body that is in contact with the obstacle;
3) following the obstacle boundary.

The first of these operations implies that the arm is capable of computing coordinates of consecutive points along M line and, if necessary, transforming them into the corresponding joint values (using, for example, appropriate procedures of inverse kinematics [20]).

The sole purpose of the second operation is to provide local information needed to pass around the obstacle. At any moment, when at least one point of the arm body is in contact with an obstacle, the arm identifies coordinates of the points of contact on the arm body, relative to the arm internal reference system. Note that such identification is a local operation that does not require information about the environment. (Recall that a blindfolded person can easily indicate a point of his body that touches an object.) We do not discuss here specific ways in which this capability can be realized. For our purpose, assume that the arm body is covered with sensors such that when one or more sensors contact an obstacle, the points of contact on the arm body are known.

For the third operation, imagine that, while being in contact with an obstacle, the arm follows the obstacle boundary while a (weak) constant force pushes it against the obstacle. (This situation is similar to a blindfolded person walking around a building while keeping his finger on the wall.) At any given moment during the motion, there is a variable point of contact between the obstacle boundary and the arm body. In the algorithm, the arm will plan its next step along the obstacle boundary in such a way that, after the step has been made, the arm is still in contact with the obstacle. Again, we will not discuss here how this important capability can be realized. Note that if the arm endpoint follows an obstacle boundary up to the W space boundary, it is not clear whether on the boundary the arm is still in contact with the obstacle. To avoid this limit case, assume that no point of the W space boundary may be a point of contact between an obstacle and the arm endpoint.

Passing around an obstacle is a continuous motion of the arm during which the arm is in constant contact with some obstacle. Because of the arm/obstacle interaction, some areas of the W space not occupied by the actual obstacle may be inaccessible by the arm endpoint. Such areas create a *shadow* of the obstacle; for the arm endpoint, a shadow presents as real an obstacle as points of the actual obstacle. The actual obstacle and its shadow(s) constitute the *virtual obstacle* of a given obstacle. When the arm is passing around the actual obstacle, the arm endpoint follows the *virtual line*, which presents the boundary of the virtual obstacle.

The *image space* (I space) is a representation space in which the arm is shrunk to a point.[2] Every path and every virtual obstacle has its corresponding image in the I space. A combination of the virtual line with the corresponding arm solutions defines the *virtual boundary* of the obstacle. The virtual boundary is a curve which forms the boundary of the obstacle image in I space. The transformation from I space to W space is unique, and the transformation from I space to W space may or may not be unique, depending on the arm configuration (see below). For all the arms of Fig. 1, the I space presents a two-dimensional manifold.

To be able to use the basic path planning procedure, the problem of motion planning for a planar arm has to be reduced to moving a point around simple closed curves. In the I space, the arm becomes a point. The key problem is, therefore, to show that, in the case of the arm of a given configuration, no matter how complex the arm motion around an actual obstacle is, the corresponding virtual boundary in I space presents a curve with no self-intersections or double points—that is, a simple curve. Indeed, this turns out to be true for each of the arms of Fig. 1.

According to the path planning procedure, the arm image point moves in the I space along (a simple) M line from point S to point T. During this motion, when some point of the arm body meets an obstacle, in the I space this corresponds to the M line image intersecting the obstacle virtual boundary. Such a point of intersection is said to define a *hit point* H_j. Because both the M line and the virtual boundary are simple curves, at the hit point the arm has a choice of walking along the virtual boundary in one of two directions: when facing the obstacle at the hit point, these are "right" and "left." Since the arm has no information which would help it choose between the two directions, the direction for passing around an obstacle can be fixed as well; this is called the *local direction*.

While following the virtual boundary, the arm may meet the M line again. If this occurs at a distance from T shorter than that from the hit point to T (as measured along the M line), then the arm is said to define a *leave point* L_j. Hit and leave points play an important role in the path planning procedure. One will see below that they come in pairs (H_j, L_j), $j = 1, 2, \cdots$. For the sake of convenience, denote L_0 = start, with no corresponding H_0.

III. Arm 1: Two Revolute Joints

This kinematic configuration (Fig. 1(a)) is very common in the major linkages of industrial arm manipulators; PUMA series and various robots of SCARA type would be typical examples. This arm has been studied extensively in [11]; here, only some distinctive characteristics of the arm motion and the text of the path planning algorithm are given. The outer boundary of the W space of arm 1 presents a circle whose radius is equal to the sum of link lengths $(l_1 + l_2)$. The W space inner boundary creates a circular "dead zone" around

[2] To avoid possible confusion, the term "image space" is used instead of "configuration space" employed in a number of works (see, e.g., [8]); configuration space is meant to represent some Euclidean space, whereas, as one will see in the sequel, I space represents a k-dimensional manifold in a Euclidean space whose dimension is higher than k. For example, the metric in I space may be quite different from that of the configuration space.

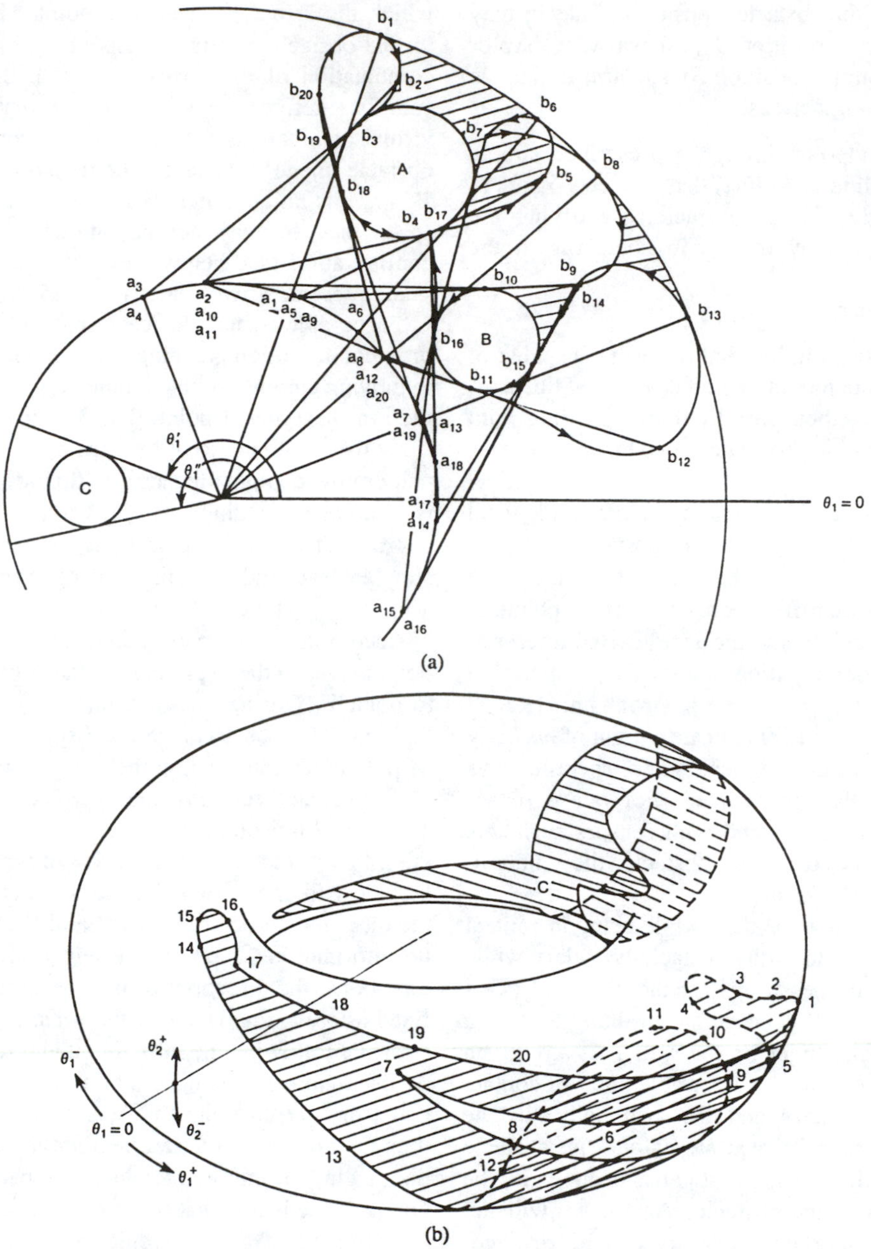

Fig. 2. Arm 1. (a) W space. A, B, C: actual obstacles. (b) I space images of same obstacles.

the origin, of radius $|l_1 - l_2|$; if $l_1 = l_2$, there is no dead zone, and the W space becomes a disk. Both joint values, angles θ_1 and θ_2, can increase or decrease indefinitely, with an obvious constraint that $\theta_i = \theta_i \pm n \cdot 2\pi$; $n = 0, 1, \cdots$; $i = 1, 2$. In general, for any position of the arm endpoint in the W space, except for points along the W space boundaries, there are two arm solutions.

Although the variables θ_1 and θ_2 are independent, from the standpoint of path planning it is important that there is certain asymmetry between them regarding the link motion, due to the fact that the links are positioned sequentially relative to the fixed origin. In general, for a given value θ_2, some value θ_1 can be found such that this θ_2 would be realizable. The opposite is not true: for example, because of the obstacle C in Fig. 2(a), values of θ_1 in the range $\theta_1' \leq \theta_1 \leq \theta_1''$ cannot be realized for any value of θ_2.

The combination of the arm kinematics and the obstacles creates a complicated pattern of areas inaccessible to the arm endpoint. Even with complete information about the obstacles, these areas would present difficulties for direct analysis. When the arm is passing around an obstacle, it may create a single shadow or more than one subshadow. The latter are shown in Fig. 2(a) as two disconnected shaded areas behind the circular obstacle B. Note that with the same position but a smaller diameter of B, or with B of the same diameter positioned slightly further from the origin, both subshadows would merge into a single shadow.

Actual obstacles may interact; in Fig. 2(a), the shadow behind the circular obstacle A is asymmetric because of the influence of obstacle B. Note that if the arm endpoint moves in the figure through the positions (a_1, b_1), (a_2, b_2), $\cdots$, (a_{20}, b_{20}), (a_1, b_1), at all moments of this motion the arm is in contact with some obstacle. Here, point b_i is the end of a (straight line) segment of the length l_2 which starts at a_i and is tangential to the obstacle. Therefore, as far as the arm is concerned, two actual obstacles A and B present one virtual obstacle. The virtual line of this obstacle is defined by the points $b_1, b_2, \cdots, b_{20}$, and its virtual boundary is defined by the corresponding link positions (a_i, b_i). It is easy to see that a set of even simple actual obstacles may produce extremely complicated virtual obstacles.

A set of two independent angular variables can be represented by a common torus, with a pair of values of the variables uniquely representing a point on the surface of the torus. We represent I space of arm 1 by the surface of a torus whose two independent variables are θ_1 and θ_2. Since one point in W space corresponds, in general, to two arm solutions, this creates two image points in I space.

Theorem 1: Any virtual boundary in the I space of arm 1 can consist only of simple closed curves.

Lemma 1: Any virtual boundary in the I space of arm 1 can be formed by no more than two simple closed curves.

The proofs of these statements [11] are based on the Jordan curve theorem and on the topology of a common torus (specifically, on the *first connectivity number* of the torus [25]). In Fig. 2(b), images of the obstacles of Fig. 2(a) are shown. Note that in I space the obstacles A and B form a single closed curve (these are called Type I obstacles) whereas the obstacle C forms a band-like structure limited by two simple closed curves (these are called Type II obstacles). The path planning procedure has to be able to recognize and appropriately handle each type of obstacle. This is necessary because if the arm, while passing around an obstacle, completes a full circle without ever meeting the M line, it may mean that the target cannot be reached, or that "somewhere" there is another closed curve belonging to the same virtual boundary, which has not yet been investigated but which has to be investigated before any conclusions are drawn.

To do the obstacle type recognition, two counters, C_1 and C_2, corresponding to angles θ_1 and θ_2, respectively, are used. When the arm is traveling in free space, the contents of the counters are zeros. Once the arm hits an obstacle, both counters are turned on, and, while the arm follows a closed curve of the virtual boundary, each counter integrates its angle, taking the sign into account. After completing a closed curve, the content of the counter C_i must be $n_i \cdot 2\pi$, $|n_i| = 0, 1, 2, \cdots$ (see Fig. 2(b)). For a given closed curve, the resulting values of the pair (C_1, C_2) define its *arm joints range* (or simply *range*) (n_1, n_2). The range defines the type of the obstacle as follows: the range of a Type I obstacle is (0, 0), and the range of a Type II obstacle is (n_1, n_2), with either $n_i = 0$ and $|n_{3-i}| = 1$, or $|n_i| = 1$ and $|n_{3-i}| = 1, 2, \cdots; i = 1, 2$.

To locate the second closed curve of a Type II obstacle, a notion of *complementary* M lines is introduced. At this stage, it is useful to be more specific about the type of the M line being used. Although no restrictions are imposed on the choice of the M line, we use, for specificity, a straight line in the plane of variables θ_1 and θ_2. The advantage of such an M line is that it produces a monotonic and economical pattern of change of each of the joint values on the way from S to T. On the I space torus, a model of this M line is an imaginary tight thread connecting points S and T.

There are, however, three other ways to have a tight thread between S and T on the torus surface (Fig. 2(c)). These are obtained by switching the direction of change of one or both angles θ_1 and θ_2 compared to the first M line. The first M line corresponds to the global minimum, and the other three correspond to local minima of the Euclidean distance (in the plane (θ_1, θ_2)) between S and T as a function of a route from S to T (e.g., all four become equal if S and T are located at opposite points of the torus outer equator). Denote $\delta_i = \text{sgn}(\theta_i^T - \theta_i^S)$, $i = 1, 2$, where sgn $(\cdot)$ takes values $+1$ or -1, depending on the sign of the argument. Then, each of these M_k lines, $k = 1, 2, 3, 4$, is defined by a pair of its endpoints, S and T_k, which are presented as follows (for details, see [11]):

$$
\begin{aligned}
&M_1\colon S=(\theta_1^S, \theta_2^S), && T_1=(\theta_1^T, \theta_2^T)\\
&M_2\colon S=(\theta_1^S, \theta_2^S), && T_2=(\theta_1^T, \theta_2^T-2\pi\cdot\delta_2)\\
&M_3\colon S=(\theta_1^S, \theta_2^S), && T_3=(\theta_1^T-2\pi\cdot\delta_1, \theta_2^T)\\
&M_4\colon S=(\theta_1^S, \theta_2^S), && T_4=(\theta_1^T-2\pi\cdot\delta_1, \theta_2^T-2\pi\cdot\delta_2).
\end{aligned} \tag{1}
$$

We order these four segments according to (1), with M_1 being the shortest of the four segments. Then, for each of the angles θ_i, $i = 1, 2$, and each of the segments M_k, define an interval

$$\delta\theta_{i,k}=|\theta_i^T-\theta_i^S| \quad \text{or} \quad \delta\theta_{i,k}=|\theta_i^T-\theta_i^S-2\pi|. \tag{2}$$

Apparently, $\delta\theta_{i,k} \leq 2\pi$. Any two segments, M_k and M_m, $k, m = 1, 2, 3, 4$, $k \neq m$, are complementary in the sense that, for one or both angles θ_i, intervals $\delta\theta_i$ add to 2π. Two segments M_k and M_m are said to be *complementary over the angle* θ_i if $\delta\theta_{i,k} + \delta\theta_{i,m} = 2\pi$. To find the second closed curve of a Type II obstacle, an M line complementary to the current M line over an appropriately chosen angle θ_i is selected (see the algorithm below). This way, not more than two M lines are ever used to produce the path.

For a Type I obstacle, completing a closed curve of the obstacle virtual boundary without ever meeting the M line means that the target is "inside" the obstacle (e.g., the obstacle forms a ring on the torus surface in I space, with the target point inside the ring) and thus cannot be reached. A similar condition is true in the case of a Type II obstacle, if two closed curves have been passed without ever meeting the M line. These two facts are both necessary and sufficient, and they form the basis for the test for target reachability incorporated in the procedure below.

It is known that sometimes the relative arrangement of the positions S and T and of the set of obstacles is such that the path planning procedure produces *local cycles* [10], [11], [23]. A local cycle is created when the arm image point on the torus comes back to a previously defined hit point and at this moment the content of one or both of its counters C_i is different from $n_i \cdot 2\pi$, $|n_i| = 0, 1, 2, \cdots, i = 1, 2$. The number of local cycles is always finite, and so convergence of the algorithm is preserved; they do result, however, in longer paths. For an example and a study on the effect of local cycles, refer to [11].

In the path planning procedure formulated below, the following entities are used. The hit points H_j and the leave points L_j are numbered in the order of their occurrence. If a complementary M line is introduced, the arm starts again at the point S, and the numbering starts over. The distance $d(P, Q)$ between points P and Q is Euclidean distance in the plane of variables θ_1 and θ_2; the length of an M line segment is defined similarly. A flag is used to indicate that, in case of a Type II obstacle, one of the two closed curves of the current virtual boundary has been processed. The procedure consists of the following steps.

1) All four complementary M lines are defined according to (1), and then ordered as follows: M_1 is the shortest of the segments M_k, $k = 1, 2, 3, 4$; M_2 complements M_1 over the angle θ_2 (as in (1)); M_3 complements M_1 over θ_1; M_4 complements M_1 over both θ_1 and θ_2. Go to Step 2.

2) The M_1 line is designated as M line. The flag is set down. Set $j = 1$. Go to Step 3.

3) Counters C_1 and C_2 are set to zero. From point L_{j-1}, the arm moves along M line until one of the following occurs.

a) The target is reached. The procedure stops.
b) An obstacle is encountered and a hit point H_j is defined. Go to Step 4.

4) Counters C_1 and C_2 are turned on. The arm follows the virtual boundary until one of the following occurs.

a) The target is reached. The procedure stops.
b) M line is met at a distance d from T such that $d < d(H_j, T)$; point L_j is defined. Increment j. Go to Step 3.
c) The arm returns to H_j (i.e., a closed curve along the virtual boundary has been completed) without ever meeting M line. Go to Step 5.

5) Examine the obstacle range accumulated in the counters C_1 and C_2. One of the following takes place.

a) The range is (0, 0) (i.e., this is a Type I obstacle). The target cannot be reached. The procedure stops.
b) The range is not (0, 0) and the flag is up (i.e., this is the second closed curve of the virtual boundary of a Type II obstacle). The target cannot be reached. The procedure stops.

The remaining three events relate to the case when the range is not (0, 0) and the flag is down (i.e., the first closed curve of the virtual boundary of a Type II obstacle is being processed).

c) The range is $(0, n_2)$; $|n_2| \geq 1$, an integer; designate the shorter of M_3 and M_4 as M line. Go to Step 6.
d) The range is $(n_1, 0)$; $|n_1| \geq 1$; designate the shorter of M_2 and M_4 as M line. Go to Step 6.
e) The range is (n_1, n_2); $|n_1|, |n_2| \geq 1$; designate the shortest of M_2, M_3, and M_4 as M line. Go to Step 6.

6) The arm moves back to Start. Set the flag up. Set $j = 1$. Go to Step 3.

IV. Arm 2: Two Sliding Joints

The joint values of this arm (Fig. 1(b)) are l_1 and l_2. The boundaries of its W space are limited by a rectangle whose sides are equal to the maximum lengths of the corresponding links (Fig. 3). Assume that no obstacles outside the W space can interfere with the arm motion; hence the path planning problem is limited to the W space rectangle. The M line is defined as a straight line segment connecting the points S and T.

Observe that, for any obstacle, no points of the W space to the right of the obstacle can ever be reached by the arm endpoint (Fig. 3). Therefore, from the path planning standpoint, the arm endpoint actually represents the whole arm; this makes the I space identical to the W space. For the same reason, the boundary of a virtual obstacle, which includes the actual obstacle and the corresponding shadow, is always a simple curve. For example, in Fig. 3, the virtual boundary of the obstacle A is a simple curve passing through the points 8, 2, 3, 4, 5, 6, 7. This produces one important condition necessary for applying the basic path planning procedure and for assuring its convergence.

As one can see, the second condition—that the virtual boundary be a closed curve—does not hold; e.g., in Fig. 3, points of the segment (7–8) cannot be reached by the arm endpoint. In general, a virtual boundary consists of four distinct segments: the left curve corresponding to the arm endpoint following the actual obstacle boundary (in Fig. 3, this segment passes through the points 3-4-5); two mutually parallel straight line segments corresponding to the points of the arm body (other than the arm endpoint) touching the actual obstacle (lines 3-2-8 and 5-6-7); and the straight line segment which presents a part of the W space boundary (line 7-8). The first three segments form a simple open curve each point of which can be reached by the arm endpoint. The fact that the fourth segment cannot be reached by the arm endpoint poses no algorithmical problems since the endpoints of the specific simple open curves (in this example, points 7 and 8) can easily be recognized from the fact that they correspond to the maximum value of the joint value l_2. With this modification, the convergence of the path planning procedure is thus guaranteed.

The path planning procedure becomes simpler and more efficient if the link l_2 is assumed to present a rectangle whose sides are parallel to the joint axes l_1 and l_2, respectively. When the actual link l_2 has a more complicated shape, the algorithm replaces the link with a minimum rectangle which contains the link and whose sides are parallel to the axes l_1 and l_2. In these terms, the link l_2 in the example of Fig. 3 presents a rectangle of zero width. Then an important statement, which helps simplify the path planning procedure, follows directly from Fig. 3.

Lemma 2: In the case of arm 2, if T is reachable from S then a path exists from S to T such that it corresponds to a monotonic change of the joint value l_1.

Such a path can be found as follows. Depending on whether the difference $(l_1^S - l_1^T)$ is positive, zero, or negative, establish the M line direction of change of l_1 (positive, zero, or

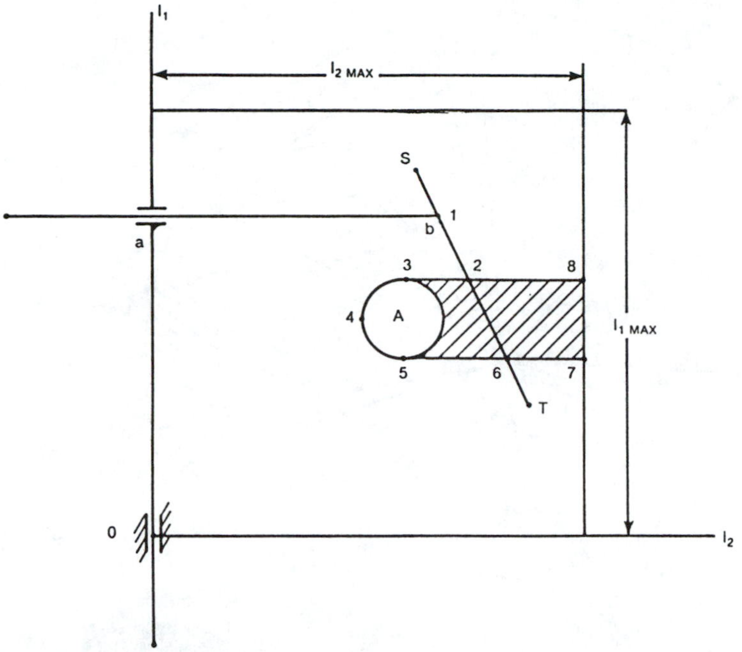

Fig. 3. Arm 2. Virtual obstacle A includes actual obstacle and its shadow (shaded).

negative, respectively). Note that this can be done before the motion takes place, using the coordinates of the points S and T alone. If, during its motion from S to T along the M line, the arm meets the virtual boundary of an obstacle, such a local direction is chosen for passing around the obstacle for which the corresponding change in l_1 coincides with the M line direction of change of l_1. If l_1 is constant in the vicinity of the hit point then the local direction should correspond to decreasing values of l_2; otherwise, the arm will not be able to pass around the obstacle.

If, while passing around an obstacle, the current value of l_1 moves outside the interval (l_1^S, l_1^T), then, clearly, T lies in the shadow of the obstacle and cannot be reached. If the M line direction of change of l_1 is 0 (i.e., the M line is parallel to the link l_2) and an obstacle is met along the way, then, again, T cannot be reached because it is in the shadow of the obstacle. Now, the whole path planning procedure is formulated; L_0 = start.

1) Establish the M line direction of change of l_1 (see above). Set $j = 1$. Go to Step 2.

2) From point L_{j-1}, the arm moves along the M line until one of the following occurs.

a) The target is reached. The procedure stops.
b) An obstacle is encountered and a hit point H_j is defined. Choose the local direction using the M line direction of change of l_1. Go to Step 3.

3) The arm follows the obstacle virtual boundary until one of the following occurs:

a) The target is reached. The procedure stops.
b) The M line is met at a distance d from T such that $d < d(H_j, T)$; point L_j is defined. Increment j. Go to Step 2.
c) Current l_1 is outside the interval (l_1^S, l_1^T). The target cannot be reached. The procedure stops.

V. Arm 3: Revolute Link Followed by a Parallel Sliding Link

This kinematic configuration (Fig. 1(c)) is also quite common in the major linkages of existing arm manipulators; Stanford manipulator and some Unimation robots would be typical examples. Joint values of this arm are the angle θ_1 and the variable length l_2 of the second link; $0 \le l_2 \le l_{2max}$ (Fig. 4(a)). The outer boundary of the W space is a circle whose radius is $(l_1 + l_{2max})$; its inner boundary, which defines the dead zone inaccessible by the arm endpoint, is a circle of radius l_1. For simplicity, no dead zone is shown in Fig. 4(a); that is, $l_1 = 0$. When the arm endpoint b is positioned at some point (say, S) of the W space, the position of the rear end of that link a_S can be found by passing a line segment of the length l_{2max} from b_S through the origin O.

For specificity, consider an M line which presents in the W space a straight line segment connecting points S and T; denote it M_1 line. Observe (Fig. 4(a)) that if, because of the obstacles, the arm cannot reach T by following the M_1 line, it might be able to reach T "from the other side," by changing θ_1 in the direction opposite to that of the M_1 line. As for arm 1, a complementary M line, the M_2 line, can be introduced. The M_2 line is defined as consisting of three parts: two straight line segments (S, S') and (T, T') continuing the M_1 line outwards till their intersection with the W space outer boundary, and a segment of the W space outer boundary corresponding to the interval of θ_1 complementing that of the M_1 line to 2π. (The choice of the complementary M line is largely arbitrary; any M_2 line will do, as long as it is uniquely defined, is computationally simple, and complements the θ_1 interval of the M_1 line to 2π.)

Obstacles can interfere with the arm 3 motion in two different ways. Observe that if the basic path planning procedure is applied to the example of Fig. 4(a), then the arm

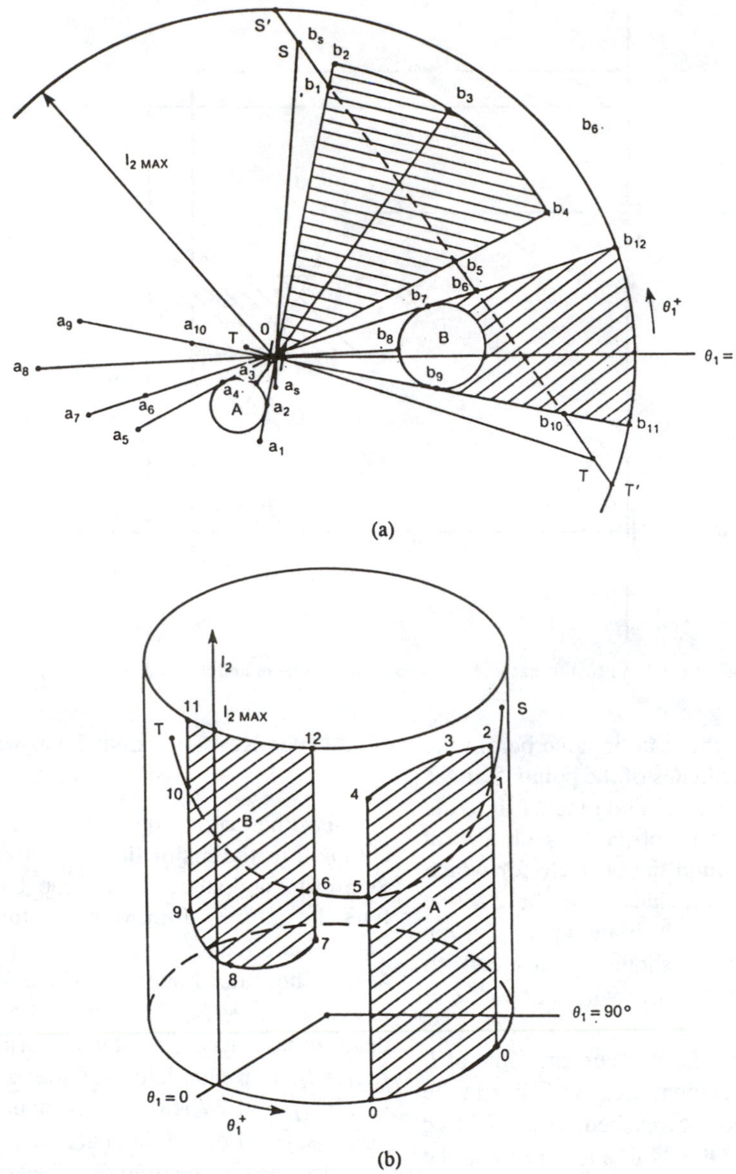

Fig. 4. Arm 3. (a) *W* space, with obstacles *A* and *B*. Shadows (shaded areas) resulting from interaction of rear end of link l_2 with *A* and of front end of l_2 with *B* are shown. (b) *I* space, with images of corresponding virtual obstacles. Line (*S*, *T*) is image of straight line (*S*, *T*) in *W* space.

endpoint, while traveling along the *M* line, will be forced to leave it two times: first, when the rear part of link l_2 interferes with the circular obstacle *A*, and, second, when the front part of link l_2 interferes with the circular obstacle *B*. The example shows that, as a rule, each obstacle forms two shadows in the *W* space of arm 3. (In Fig. 4(a), only one shadow per obstacle is shown for better visualization.) As before, a shadow contains those points (areas) of *W* space which cannot be accessed by the arm endpoint, due to a given obstacle.

Define a *front contact* of the link l_2 as a situation in which a part of the link between its front end (which is the arm endpoint) and the origin is in contact with the obstacle. A *rear contact* of the link refers to a situation when a part of the link between its rear end and the origin is in contact with the obstacle. Correspondingly, a front contact forms a *front shadow* of the obstacle, and a rear contact forms a *rear shadow*. In Fig. 4(a), the front shadow of the obstacle *B* (shaded) is limited by the line $b_6 - b_7 - b_9 - b_{10} - b_{11} - b_{12} - b_6$; the rear shadow of the obstacle *A* is limited by the line $b_1 - O - b_5 - b_4 - b_3 - b_2 - b_1$.

Therefore, any virtual obstacle consists of two parts, one due to the front shadow and the other due to the rear shadow, with one of the parts including the actual obstacle itself. Unlike the more complicated situation in the case of arm 1, the virtual lines of the obstacles here are always simple curves. One part of a virtual obstacle, due to the front shadow, extends from the obstacle to the outer boundary of the *W* space (e.g., the one limited by the line $b_6 - b_7 - b_8 - b_9 - b_{10} - b_{11} - b_{12} - b_6$, Fig. 4(a)), whereas the other, due to the rear shadow, extends from the origin into the periphery, but not to the boundary, of the *W* space (e.g., the one limited by the line $b_1 - O - b_5 - b_4 - b_3 - b_2 - b_1$). In a special case, when the M line crosses the dead zone, the latter can be treated simply as an obstacle interfering with the rear end of the link l_2.

Two independent variables, one angular and the other linear displacement, can be represented by the surface of a cylinder. The I space of arm 3 is represented by a cylinder whose *base circles* (its flat sides) correspond to the first joint value θ_1 and whose height corresponds to the second joint value l_2. In Fig. 4(b) images of the M line and of those parts of the virtual obstacles A and B that are depicted in Fig. 4(a) are shown.

Since each point in the W space has only one arm solution, there is a one-to-one mapping between the W space and the I space. Because of this, and also because the virtual lines in the W space are always simple curves, the virtual boundaries in the I space are also simple curves. These factors constitute the necessary conditions for the basic path planning procedure. In general, in the I space each virtual boundary consists of a combination of three distinct segments: 1) a curve formed when the front or the rear end of link l_2 follows the boundary of the actual obstacle; for the virtual boundary of obstacle B in Fig. 4(b), this segment passes through points 7–8–9; 2) a vertical straight line segment formed when points of the arm body, other than the arm endpoint, touch the obstacle while passing around it (e.g., lines 7–6–12 and 9–10–11, Fig. 4(b)); and 3) a segment which is a part of one of the base circles (e.g., line 11–12). The latter cannot be reached by the arm.

To apply the basic path planning procedure to this arm, the algorithm has to be modified to reflect specifics of moving along the I space cylinder. Observe that a ringlike actual obstacle, positioned in the W space so that it separates the arm from the outer boundary of the W space, would produce a bandlike virtual obstacle in the I space. One simple closed curve of the band can be reached by the arm, whereas the other, formed by one of the base circles, is inaccessible to the arm. Because of this and in spite of the fact that the virtual boundary has two closed curves, from the standpoint of path planning, this is an example of a Type I obstacle (using the terminology introduced in the section on the arm 1).

If an actual obstacle extends from the W space inside the dead zone, it produces a swathlike virtual obstacle, whose virtual boundary in the I space includes two separate simple curves each connecting the opposite base circles of the I space cylinder. This is a Type II obstacle. Similar to arm 1, if during the arm motion one of such curves of a Type II obstacle has been investigated without ever meeting the M line, the second curve has to be looked at. To do that, the *complementary* M_2 line is used.

As with arm 2, the choice of local direction for following the virtual obstacle by arm 3 is actually unique. Once the arm meets an obstacle, one of two cases takes place. If the contact is a front contact then only such local direction as corresponds to decreasing values of l_2 is meaningful; one can see (Fig. 4(a)) that the opposite local direction will never bring the arm any closer to the target. If, on the other hand, the contact is a rear contact then only such local direction as corresponds to increasing values of l_2 should be chosen. Because of the unique choice of the local direction, there is no need to investigate the whole curve of the virtual boundary. If, while passing around the obstacle in the chosen local direction, the arm reaches one of the limits of l_2, it can safely conclude that it is dealing with a Type II obstacle, and start looking for the second curve of the virtual boundary using the complementary M line. The procedure can be simplified further using the following statement similar to the one in Section IV.

Lemma 3: For arm 3, if T is reachable from S then a path exists from S to T such that it corresponds to a monotonic change of the joint value θ_1.

In the procedure, a flag is used to indicate processing of each of the two curves of a Type II virtual boundary. When the complementary M line is introduced, the numbering of hit and leave points starts over; L_0 = start. Distance is a Euclidean distance in W space. The procedure consists of the following steps.

1) Establish M_1 line as the M line. Set the flag down. Set $j = 1$. Go to Step 2.

2) From point L_{j-1}, the arm moves along the M line until one of the following occurs.

a) The target is reached. The procedure stops.
b) An obstacle is encountered and a hit point H_j is defined. In case of a front contact, choose the local direction such that it corresponds to decreasing values of l_2. In case of a rear contact, choose the local direction such that it corresponds to increasing values of l_2. Go to Step 3.

3) The arm follows the virtual boundary until one of the following occurs.

a) The target is reached. The procedure stops.
b) Current θ_1 is outside the interval (θ_1^S, θ_1^T). The target cannot be reached. The procedure stops.
c) The M line is met at a distance d from T such that $d < d(H_j, T)$. Point L_j is defined. Increment j. Go to Step 2.
d) The value l_2 approaches one of its limits, and the flag is down (i.e., the first curve of the virtual boundary of a Type II obstacle has been processed). Set the flag up. Set $j = 1$. Establish M_2 line as the M line. Move the arm back to S. Go to Step 2.
e) The value l_2 approaches one of its limits, and the flag is up (i.e., the second curve of the virtual boundary of a Type II obstacle has been processed). The target cannot be reached. The procedure stops.

VI. Arm 4: Revolute Link Followed by a Perpendicular Sliding Link

From the standpoint of path planning, arm 4 (Fig. 1(d)) is very similar to arm 3. For the algorithm itself and for a discussion of its various details, refer to the previous section. An example of interaction between the arm and two circular obstacles A and B is shown in Fig. 5(a). Also shown are the M line, the rear shadow of the obstacle A, and the front shadow of the obstacle B. The images of the corresponding virtual obstacles and of the M line are shown in Fig. 5(b); the generated path passes through the points S, 1, 2, $\cdots$, 9, T.

VII. Arm 5: Sliding Link Followed by a Revolute Link

A more detailed sketch of the arm of Fig. 1(e) is shown in Fig. 6. The first joint value is the variable length of the first

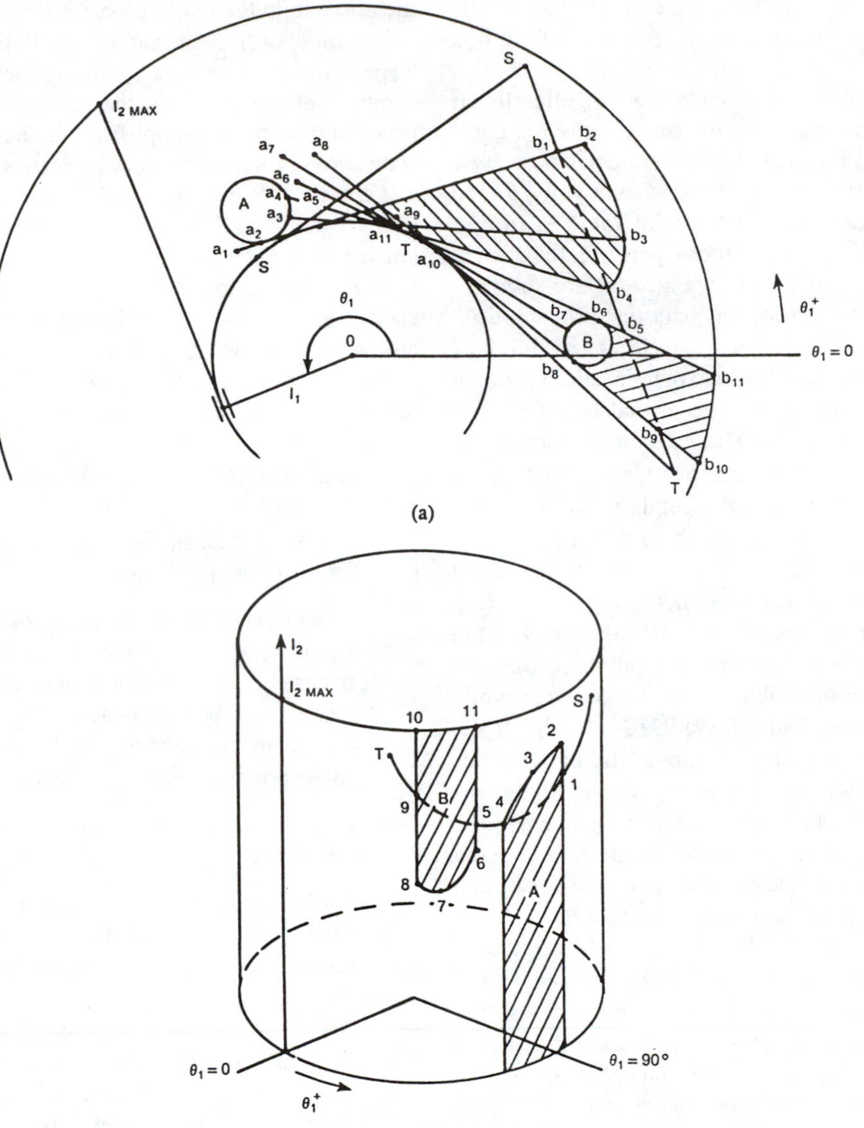

Fig. 5. Arm 4. (a) *W* space, with obstacles *A* and *B*; shadows (similar to Fig. 4) are shown. (b) *I* space images of corresponding virtual obstacles.

link l_1; $0 \le l_1 \le l_{1\max}$. The second joint value is the angle θ_2. The length of the second link l_2 is constant. The boundary of the *W* space of this arm is a combination of a rectangle whose sides are equal to $l_{1\max}$ and $2l_2$, respectively, and of two semicircles of radius l_2 attached to the rectangle as shown in Fig. 6. As before, assume that no obstacles may interfere with the parts of the arm outside of the *W* space.

Two circles of radius l_2, centered at the limit positions O and O_1 of the first link, are called the *limit areas* of the *W* space. Note that any point belonging to a limit area has only one corresponding arm solution, and any point outside the limit areas has two possible arm solutions. For example, in Fig. 6 a single arm solution exists for point P, and two arm solutions exist for point P_1. For path planning purposes, this peculiarity makes the arm distinct from the previous arms and leads to a combination of some features of the algorithms for those arms.

For a given actual obstacle, the corresponding virtual obstacle, virtual line, and virtual boundary are defined as before. An obstacle is considered to be *inside* the limit area if only one arm solution exists for any point of the obstacle virtual line. It is considered to be *outside* the limit area if two arm solutions exist for any point of the obstacle virtual line. An intermediate situation, when some points of the virtual line have one solution, and some other points have two solutions, is referred to as being *partially inside* the limit area.

Consider three circular obstacles, *A*, *B*, and *C* (Fig. 7(a)), positioned outside, partially inside, and fully inside the limit areas of the arm *W* space, respectively. Positions of the link endpoints at some points of the corresponding virtual lines are indicated in the figure by pairs of numbers, (1, 1), (2, 2), etc. Note that the virtual line of obstacle *A* is a closed curve: starting at some position of contact with *A* (say, point 1), the arm can pass around the obstacle, while keeping in contact with it (points 2, 3, $\cdots$, 12), and eventually return to the starting position. On the other hand, the virtual line of obstacle

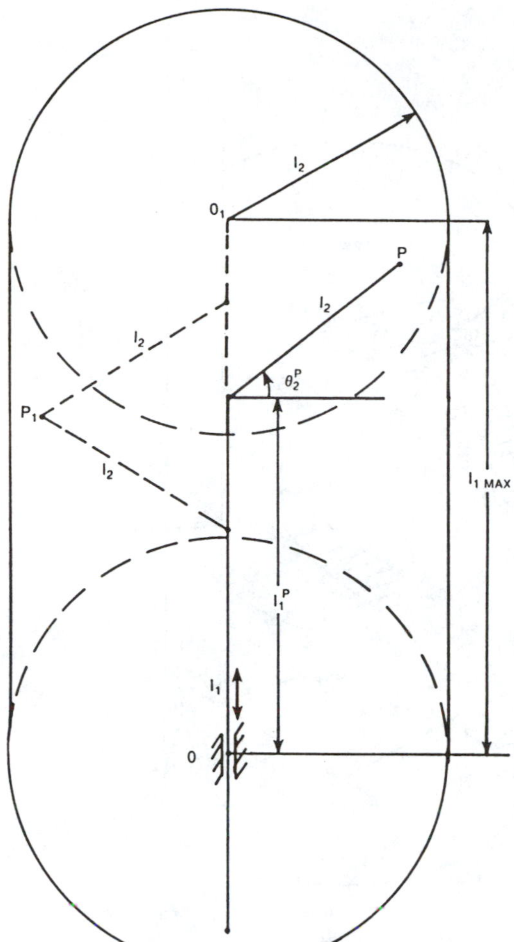

Fig. 6. W space of arm 5. Any point (such as P) within circles whose centers are 0 and 0_1, and whose radius is l_2, has one corresponding arm solution. Any point (such as P_1) outside these circles has two corresponding arm solutions, except on border of W space.

B is an open curve whose endpoints, 19 and 23, lie inside the W space but on the boundary of a limit area. Finally, the virtual line of obstacle C is an open curve whose endpoints, 24 and 30, lie on the boundary of the W space.

The I space of arm 5 presents the surface of a cylinder whose height is equal to the upper limit of the first joint value $l_{1\max}$ and whose base circles correspond to the second joint value θ_2. Any virtual boundary in the I space presents a curve which may or may not include as its part a segment of the base circle, depending on whether it corresponds to an inside, partially inside, or an outside obstacle (Fig. 7(b)). No point of a base circle segment of a virtual boundary can be accessed by the arm.

Similar to the arm 1 [11], it can be shown that in the case of an outside obstacle, even if the virtual line in the W space has self-intersections or double points, the corresponding virtual boundary in the I space presents a single simple closed curve, and this curve can be traced by the arm fully. The obstacle A in Fig. 7 gives an example of this type. Continuing, one can show that, similar to arms 3 and 4, for inside and partially inside obstacles the corresponding virtual boundaries are simple curves that include two types of segments: 1) one or two open segments, which can be followed by the arm, and 2) one or two segments of the base circles of the I space cylinder, which cannot be accessed by the arm. Obstacles B and C in Fig. 7 provide relevant examples of this type. That the curves are not closed in case 1 produces no algorithmic difficulties, because the endpoints of the open curves correspond to one or both limit values of l_1 and hence are easily recognizable. Together with simplicity of the virtual boundaries, this fact assures conditions necessary for convergence of the path planning procedure for arm 5.

From the practical standpoint, using a straight line for the M line in W space is not convenient because continuous motion of both links often cannot be maintained. Try, for example, to follow a straight line between the points S and T in Fig. 7(a); somewhere a discontinuity takes place in the motion of one or both links. As for arm 1, define the M line as a straight line in the plane of variables l_1 and θ_2—that is, as a function $\theta_2 = p \cdot l_1 + q$, with the coefficients p and q determined from the coordinates of S and T. The image of this M line, denoted as M_1 line in Fig. 7(a) and (b), is a geodesic line between the points S and T on the surface of the I space cylinder. If, because of the obstacles, T cannot be reached from S using this M line, a complementary M line (denoted as M_2 line, Fig. 7) is used; the M_2 line is defined similarly to the M_1 line, with the coordinate of the point T being $(\theta_2 - 2\pi)$ instead of θ_2.

To complete the study of interaction between the arm and the obstacles, some special cases have to be considered. For the sake of specificity, assume that every time the arm meets an obstacle and defines on it a hit point, it starts with the local direction left. In the case shown in Fig. 8, the virtual obstacle A forms a swath along the I space cylinder. This means that the virtual boundary includes two separate open curves (2–3–4 and 5–6–1, Fig. 8(b)). After leaving S, following M line and meeting the obstacle A at point 7, the arm turns left along the virtual boundary and goes through points 3 and 2. For the arm, point 2 is the first endpoint of the first open curve of the virtual boundary. To complete the investigation of this curve, the arm will now return to the hit point and try to pass around the obstacle using the local direction right. Along this segment of the path, point 4, which is another endpoint of the same open curve, will be reached.

The difference between this case and, say, that of the obstacle B in Fig. 7 is that both endpoints of the virtual boundary B correspond to the same (here, upper) limit of the joint value l_1, whereas the endpoints of the virtual boundary A in Fig. 8 correspond to both limits of l_1. The consequence of this is that in the former case it is possible to pass around the obstacle while being in constant contact with it, whereas in the latter case this is not possible. The fact that both endpoints of the open curve are located at the opposite base circles of the I space cylinder (Fig. 8(b)) indicates that the arm is dealing with a Type II obstacle, and, therefore, the second curve of the virtual boundary has to be found and investigated before a conclusion about the target reachability can be drawn. Because of the topology of a common cylinder, the second curve must be similar to the first one in that both its endpoints correspond to both base circles of the I space cylinder. Indeed, both curves would not separate the virtual obstacle from the rest of

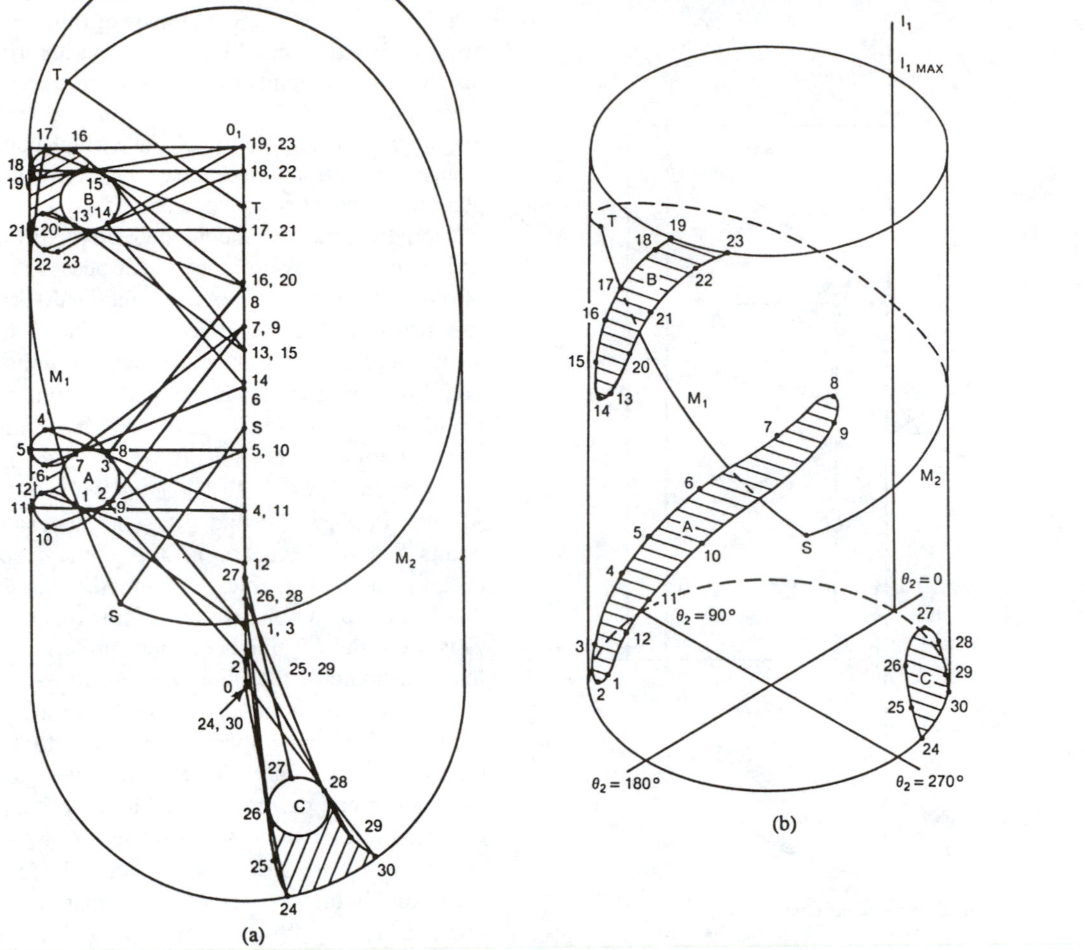

Fig. 7. Arm 5. (a) W space; shown are M lines M_1 and M_2, and link positions during passing around obstacles A, B, and C. (b) I space images of same M lines and of virtual obstacles.

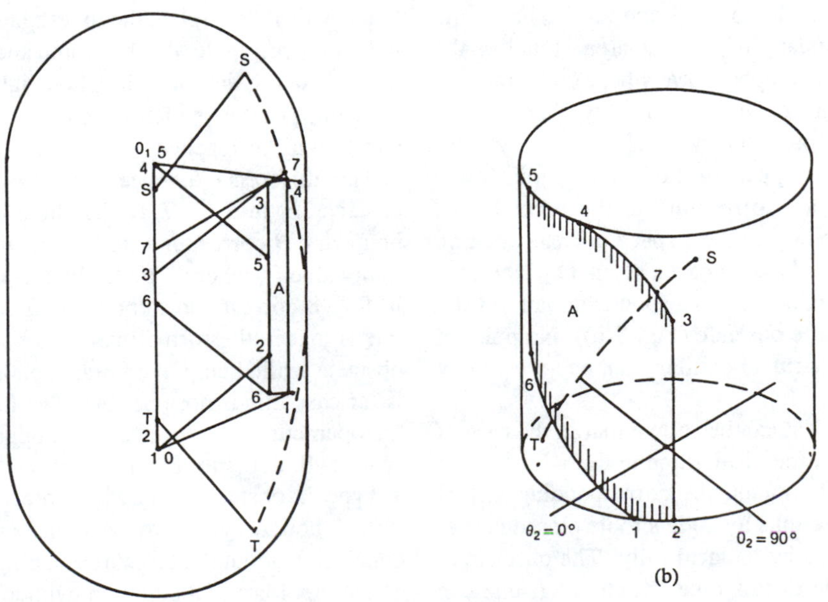

Fig. 8. Arm 5. Barlike obstacle A in W space forms swathlike virtual obstacle in I space.

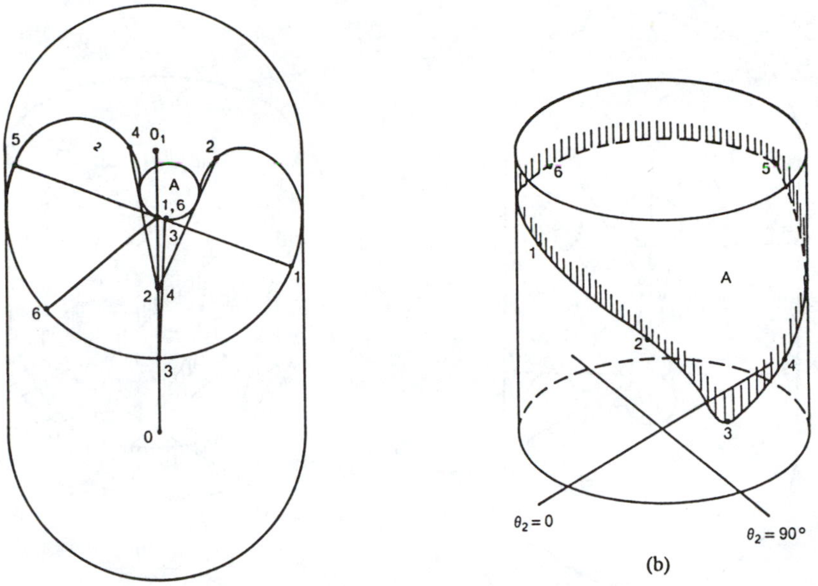

(a) (b)

Fig. 9. Arm 5. If obstacle A interferes in W space with motion of link l_1, in I space this creates bandlike virtual obstacle.

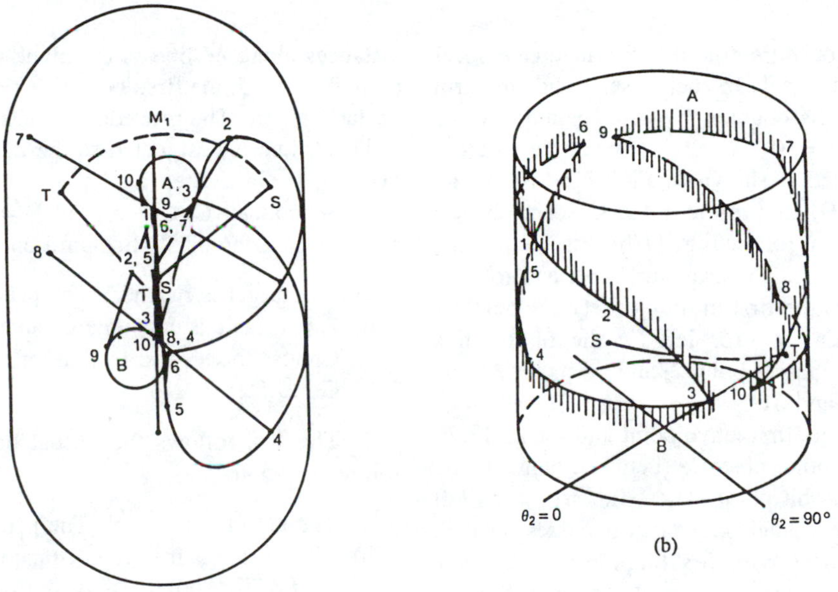

(a) (b)

Fig. 10. Arm 5. Because of two obstacles (A and B), two bandlike virtual obstacles appear in I space, such that two free areas are formed unaccessible from each other. Here, T cannot be reached from S. (Virtual obstacles are partially shaded.)

the I space cylinder otherwise. To reach the second curve, a complementary M line will be used. If, after following the whole second curve, neither the M line nor the target is met, then the target cannot be reached.

If an obstacle interferes with the first link (Fig. 9(a)), the resulting virtual boundary forms a band around the I space cylinder (Fig. 9(b)). While following such a virtual boundary, the arm would make a full circle and return to the hit point. The fact of completing a full circle is indicated by a counter C_2, which, starting at the hit point, integrates the angle θ_2, considering the sign. If a full circle is made without ever meeting the M line (that is, the value of C_2 is 2π), then the target cannot be reached.

If two or more obstacles interfere with the first link, two bands are formed on the surface of the I space cylinder (Fig. 10). If the distance between the obstacles along the line OO_1 is longer than l_2, then in the I space the bands are disconnected; this case is similar to that of Fig. 9(a). If the distance is shorter than l_2, then the I space is divided into disconnected free areas. For example, in Fig. 10(a) arm position T cannot be reached from arm position S. In such a case, if the arm starts at S, follows the M line, hits the obstacle (between points 2 and 3; Fig. 10(b)) and tries to pass around the obstacle, it will return to the hit point without ever meeting the M line. However, unlike the example of Fig. 9, the counter C_2 will contain in this case zero. Obviously, the target cannot be reached. This is quite different from the outcome in a similar situation with arm 1 (refer to Section III or, for more detail, to [11]).

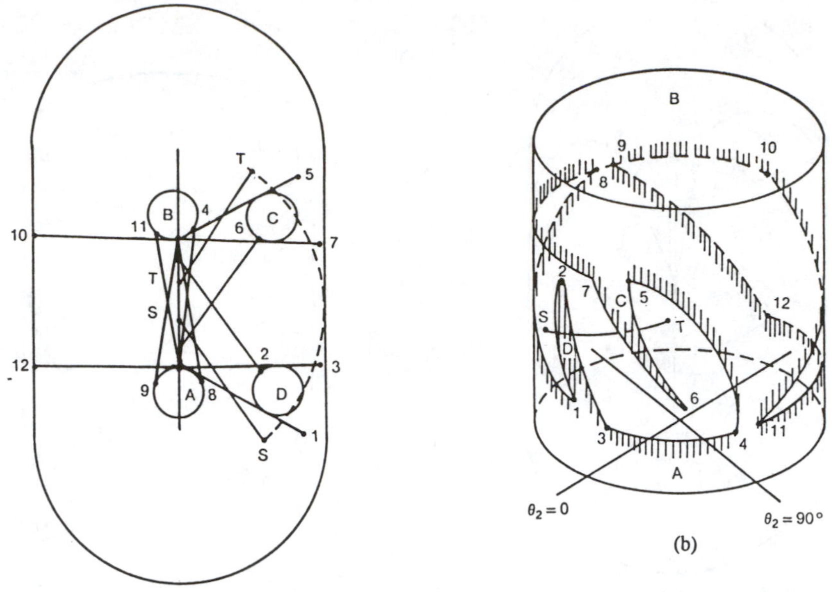

Fig. 11. Arm 5. Here local cycle is created when moving from S to T and following basic path planning algorithm.

The path planning procedure for arm 5 can create local cycles similar to those of arm 1. In such cases, when the arm returns to the hit point, its counter C_2 will contain a value different from $|n|\cdot 2\pi$, $n = 0, 1, \cdots$. An example where a local cycle will be created is shown in Fig. 11; four actual obstacles, A, B, C, and D, are present in the W space. To see how the local cycle appears, follow Fig. 11(b); assume that the chosen local direction is left. The arm starts at S and follows the M line until it defines the first hit point (between points 1 and 2, Fig. 11(b)). Then, it turns left (in the figure, this corresponds to going up), passes point 2, and meets the M line again (between points 2 and 3).

Here, the arm defines its first leave point and then follows the M line until it hits another obstacle (between points 6 and 7) and defines the second hit point. Then, the arm turns left and, while following the virtual boundary, it passes through points 7, 8, and 1 and returns to the first hit point; a local cycle has been completed. Since counter C_2 was last turned on at the second hit point, at the completion of the local cycle the content of the counter will be different from $|n|\cdot 2\pi$. Therefore, the arm proceeds further along the virtual boundary, passes through points 2, 3, 4, and 5, meets the M line again (between points 5 and 6) and defines there the third hit point. Finally, the arm proceeds along the M line directly to T.

In the path planning procedure to follow, a parameter F is used to handle an open curve of a Type II obstacle. It is set to $+1$ when the arm, while following a virtual boundary, reaches the upper limit of the joint value l_1, to -1 when, while following a virtual boundary, it reaches the lower limit of the joint value l_1, and to 0 at the first hit point of the virtual boundary. A flag is used to distinguish between the first and the second open curve of the virtual boundary of a Type II obstacle. A counter C_2 is used to handle closed curves of virtual boundaries. The complementary M lines, M_1 and M_2, are defined as before; L_0 = start. Distances are Euclidean distances along M line in the plane (l_1, θ_2). Every time a hit point is defined, the first local direction for passing around the obstacle is left. The procedure consists of the following steps.

1) M_1 line is designated as the M line. Set the flag down. Set $j = 1$. Go to Step 2.

2) Set counter C_2 to zero. Set $F = 0$. From point L_{j-1}, the arm moves along the M line until one of the following occurs.

- a) The target is reached. The procedure stops.
- b) An obstacle is encountered and a hit point H_j is defined. Choose the local direction left. Turn on the counter C_2. Go to Step 3.

3) The arm follows the virtual boundary until one of the following occurs

- a) The target is reached. The procedure stops.
- b) The M line is met at a distance d from T such that $d < d(H_j, T)$; point L_j is defined. Set the flag down. Increment j. Go to Step 2.
- c) The arm reaches one of the limits of l_1 (this corresponds to an endpoint of one open curve of a virtual boundary) without ever meeting the M line. Go to Step 4.
- d) The arm returns to H_j (i.e., a closed curve along the virtual boundary has been completed) without ever meeting the M line. The target cannot be reached. The procedure stops.

4) Depending on the value F, on the flag condition, and on the current arm position, one of the following occurs.

- a) $F = 0$. Set F to $+1$ or -1 according to the rule above. Return to the last hit point. Choose the local direction right. Go to Step 3.
- b) The value F does correspond to the current curve endpoint (i.e., $+1$ for the upper limit and -1 for the lower limit of l_1); this means the whole obstacle has been

investigated. The target cannot be reached. The procedure stops.

c) The value F does not correspond to the current curve endpoint, and the flag is down; this means the first open curve of a Type II obstacle has been investigated. Set $j = 1$. Set the flag up. Designate M_2 line as the M line. Return to S. Go to Step 2.

d) The value F does not correspond to the current curve endpoint, and the flag is up; this means the second open curve of a Type II obstacle has been investigated. The target cannot be reached. The procedure stops.

VIII. Concluding Remarks

The results indicate that, at least in the case of simpler robot arms of various kinematics, the approach of dynamic path planning exhibits impressive power: indeed, it allows one to do real-time motion planning of the arm manipulator in a complicated environment with unknown arbitrarily shaped obstacles. In the examples given, even if complete information about the obstacles were available, processing it along the lines of the piano movers approach (let alone obtaining, coding, storing, and presenting this information in an algebraic form) would present a tremendous computational burden. One source of this power lies in the notion of feedback utilized in the approach. The other source relates to natural parallel processing capability offered by an array of sensors; in fact, an unlimited increase in the number of sensors—up to a continuous sensing capability of the robot "skin"—should not affect the computational burden of the algorithms.

In its current form, the approach is subject to some constraints, one of which is that it is limited to the gross motion (position) planning. This does not seem to be a high price, since in most applications gross motion planning is the main issue in the robot collision avoidance problem, with the orientation planning for the robot end effector being a secondary and simpler problem. Another constraint—on the problem dimensionality—is more essential. All the cases studied are of planar robot arms. The work on dynamic path planning for three-dimensional arm manipulators may change this situation [19], [21].

There are, however, other difficulties, outside of the realm of algorithms. Practical realization of the DPP approach requires availability of robot arms whose skin is sensitive to contact with an obstacle. In case of physical contact, the arm's sensitivity will be similar to that of a human arm. In case of a nonphysical contact—which is certainly preferable for arm manipulators—the type of sensitivity goes beyond that of a human arm. Infrared or sonar sensors covering the whole body of the robot arm present viable alternatives. Solving technical problems of development, installation, and information processing of such large multisensor systems presents an interesting area of research in robotics. Some issues of simulating such systems have been discussed in [24].

One byproduct of this study is the realization that the effect of the same obstacle on the arm motion flexibility and on the area of the work space accessible to the arm is greatly influenced by the arm kinematics. For example, typically more area of the work space around an obstacle remains accessible to the arm endpoint for an articulated arm, arm 1, than for any of the other four arms studied in this work. In this sense, the variance in behavior between arms with different kinematics is much smaller in the free work space than it is in the work space filled with obstacles. Therefore, more care should be exercised in choosing the arm kinematics if, in the intended application, the arm is expected to interact with obstacles.

References

[1] D. L. Pieper, "The kinematics of manipulators under computer control," Ph.D. dissertation, Stanford University, Stanford, CA, 1972.

[2] S. M. Udupa, "Collision detection and avoidance in computer controlled manipulators," in *Proc. 5th Joint Int. Conf. Artificial Intelligence*, Cambridge, MA, Aug. 1977.

[3] J. T. Schwartz and M. Sharir, "On the 'piano movers' problem. I. The case of a two-dimensional rigid polygonal body moving amidst polygonal barriers," *Comm. Pure Appl. Math.*, vol. 34, 1983.

[4] C. O'Dunlaing, M. Sharir, and C. K. Yap, "Retraction: A new approach to motion planning," in *Proc. 15th Symp. Theory of Computing*, May 1984.

[5] J. Reif, "Complexity of the mover's problem and generalizations," in *Proc. 20th Symp. Foundations of Computer Science*, 1979.

[6] J. Hopcroft, D. Joseph, and S. Whitesides, "On the movement of robot arms in 2-dimensional bounded regions," in *Proc. IEEE Conf. Foundations of Computer Science*, Chicago, IL, Nov. 1982.

[7] J. T. Schwartz and M. Sharir, "On the 'piano movers' problem. II. General techniques for computing topological properties of real algebraic manifolds," in *Advances in Applied Mathematics*. New York: Academic, 1983, pp. 298–351.

[8] T. Lozano-Perez, "Automatic planning of manipulator transfer movements," *IEEE Trans. Syst., Man, Cybern.*, vol. SMC-11, Oct. 1981.

[9] R. A. Brooks, "Planning collision-free motions for pick-and-place operations," *Int. J. Robotics Res.*, vol. 2, no. 4, 1983.

[10] V. J. Lumelsky and A. A. Stepanov, "Dynamic path planning for a mobile automaton with limited information on the environment," *IEEE Trans. Automat. Contr.*, vol. AC-31, Nov. 1986.

[11] V. J. Lumelsky, "Dynamic path planning for a planar articulated robot arm moving amidst unknown obstacles," *Automatica, J. Int. Fed. Automatic Control* (IFAC). Pergamon Press, Sept. 1987.

[12] V. Milenkovic and B. Huang, "Kinematics of major robot linkage," in *Proc. 13th Int. Symp. Industrial Robots and Robots 7 Conf.*, Chicago, IL, Apr. 1983.

[13] H. Moravec, "Obstacle avoidance and navigation in the real world by a seeing robot rover," Stanford AIM-340, Sept. 1980.

[14] B. Bullock, D. Keirsey, J. Mitchell, T. Nussmeier, and D. Tseng, "Autonomous vehicle control: An overview of the Hughes project," in *Proc. IEEE Conf. Trends and Applications, 1983: Automating Intelligent Behavior*, Gaithersburg, MD, May 1983.

[15] A. M. Thompson, "The navigation system of the JPL robot," in *Proc. 5th Joint Int. Conf. Artificial Intelligence*, Cambridge, MA, Aug. 1977.

[16] L. Gouzenes, "Collision avoidance for robots in an experimental flexible assembly cell," in *Proc. IEEE Int. Conf. Robotics*, Atlanta, GA, Mar. 1984.

[17] B. Faverjon, "Obstacle avoidance using an octree in the configuration space of a manipulator," in *Proc. the IEEE Int. Conf. Robotics*, Atlanta, GA, Mar. 1984.

[18] A. A. Petrov and I. M. Sirota, "Control of a robot manipulator with obstacle avoidance under little information about the environment," in *Proc. VIII Congress of IFAC*, Kyoto, Japan, 1981, v.XIV.

[19] V. J. Lumelsky, "Continuous motion planning in unknown environment for a 3D Cartesian robot arm," in *Proc. IEEE Int. Conf. Robotics and Automation*, San Francisco, CA, Apr. 1986.

[20] R. Paul, *Robot Manipulators: Mathematics, Programming, and Control*. Cambridge, MA: MIT Press, 1981.

[21] V. Lumelsky and K. Sun, "Gross motion planning for a simple 3D articulated robot arm moving amidst unknown arbitrarily shaped obstacles," in *Proc. IEEE Int. Conf. Robotics and Automation*, Raleigh, NC, Apr. 1987.

[22] E. F. Moore, "The firing squad synchronization problem," in *Sequential Machines: Selected Papers*. E. F. Moore, Ed. Reading, MA: 1964.

[23] V. J. Lumelsky and A. A. Stepanov, "Path planning strategies for a point automaton moving amidst unknown obstacles of arbitrary shape," *Algorithmica*, to appear, 1987.
[24] K. Sun and V. Lumelsky, "Computer simulation of sensor-based robot collision avoidance in an unknown environment," *Robotica*, to appear, 1987.
[25] W. S. Massey, *Algebraic Topology*. New York: Harcourt, Brace, & World, 1967.
[26] H. Abelson and A. diSessa, *Turtle Geometry*. Cambridge, MA: MIT Press, 1981.

Jérôme Barraquand
Jean-Claude Latombe
Robotics Laboratory
Department of Computer Science
Stanford University
Stanford, California 94305

Robot Motion Planning: A Distributed Representation Approach

Abstract

We propose a new approach to robot path planning that consists of building and searching a graph connecting the local minima of a potential function defined over the robot's configuration space. A planner based on this approach has been implemented. This planner is considerably faster than previous path planners and solves problems for robots with many more degrees of freedom (DOFs). The power of the planner derives both from the "good" properties of the potential function and from the efficiency of the techniques used to escape the local minima of this function. The most powerful of these techniques is a Monte Carlo technique that escapes local minima by executing Brownian motions. The overall approach is made possible by the systematic use of distributed representations (bitmaps) for the robot's work space and configuration space. We have experimented with the planner using several computer-simulated robots, including rigid objects with 3 DOFs (in 2D work space) and 6 DOFs (in 3D work space) and manipulator arms with 8, 10, and 31 DOFs (in 2D and 3D work spaces). Some of the most significant experiments are reported in this article.

1. Introduction

In this article we propose a new approach to robot path planning that is based on the systematic use of a hierarchical bitmap to represent the robot work space. This "distributed" representation, which strongly differs from the "centralized" semialgebraic representations used in most path-planning algorithms so far, makes it possible to define simple and powerful numeric potential field techniques. Using such techniques, we implemented a planner that is considerably faster than previous path planners and solves problems for robots with many more degrees of freedom (DOFs).

The principle of our approach is to construct *collision-avoiding attractive potential fields* over the work space. Each of these potentials applies to a selected point in the robot, called a *control point*, and pulls this point toward its goal position among the obstacles. The work space potentials are then combined into another potential function defined over the configuration space of the robot. This new potential attracts the whole robot toward its goal configuration.

Each work space potential is computed over the bitmap representation of the work space (at some resolution) in such a way that it has no other local minimum than the desired final position of the robot's control point it applies to. Therefore it can be regarded as a numeric navigation function as introduced in Koditschek (1987). Such a "perfect" potential tends to keep the robot outside the work space concavities created by the obstacles. When the work space potentials are combined together (i.e., when they are applied concurrently to the various control points), the resulting configuration space potential may have (and indeed has) local minima other than the goal. However, it is usually possible to define the combination in such a way that either the number of minima or their domains of attraction remain relatively small. The idea is then to build a graph connecting the local minima and to perform a search of this graph until the goal is attained. If the goal cannot be achieved, the planner repeats the whole process at a finer level of resolution in the work space pyramid. It stops when either a path has been found (success) or the maximal resolution has been attained (failure).

In order to build the local minima graph, our approach requires efficient techniques for escaping

---

Barraquand is now with the Paris Research Laboratory of Digital Equipment Corporation (DEC), 85 Avenue Victor Hugo, 92563 Rueil-Malmaison Cedex, France.

The International Journal of Robotics Research,
Vol. 10, No. 6, December 1991,

the local minima. In this article we describe two such techniques. One is a brute force technique that exhaustively explores the local minimum wells in the discretized configuration space. The other is a Monte Carlo technique that escapes local minima by executing Brownian motions. The second technique turns out to be very powerful and has solved tricky path-planning problems for robots with many DOFs. Furthermore, it is highly parallelizable. However, for robots with few degrees of freedom, the first technique is faster on a sequential computer. In addition, it is *deterministically resolution complete*, whereas the Monte Carlo technique is only *probabilistically resolution complete*.

We have implemented the planner in a program written in C language and run on a DEC 3100 MIPS-based workstation. (All the running times given in this article refer to this implementation.) We have experimented with the implemented planner using several computer-simulated robots, including rigid objects ("mobile robots") with 3 DOFs (in two-dimensional work spaces) and 6 DOFs (in three-dimensional work spaces) and articulated objects ("manipulator arms") with 8, 10, and 31 DOFs (in two- and three-dimensional work spaces). Some of the most significant experiments are reported in this article. Our planner demonstrated the following capabilities:

It is much faster than any previous planner. For instance, it generates paths for a holonomic 3-DOF mobile robot in nontrivial work spaces in about 1 s of computation, as opposed to minutes or even tens of minutes for many other planners.

It generates paths for robots with many DOFs. In particular, within a few minutes of computation, it constructs complex paths for a 10-DOF non-serial manipulator arm with both revolute and prismatic joints.

It solves path-planning problems for multiple robots. For example, without domain-specific heuristics, it can generate the coordinated paths of two 3-DOF mobile robots in a work space made of narrow corridors.

In addition, the algorithms are highly parallelizable. This allows us to envision an implementation of the planner using specific hardware for generating paths in real time, even for robots with many DOFs. Another advantage of the planner is that it accepts goals defined by specifying the desired positions of one or several points in the robot. This feature is essential when robots have many DOFs, as specifying the goal configuration of the robot (i.e., a collision-free placement of the various bodies of the robot) is a difficult task in itself. It also allows the easy handling of any kind of redundancy of the robot. Finally, a version of the planner (not described in this article) generates paths for nonholonomic robots—i.e., robots with nonintegrable kinematic constraints such as a car and a car towing a trailer; this version of the planner is described in Barraquand and Latombe (1989a; 1990).

This article is organized as follows. In section 2 we relate our work to previous research. In section 3 we describe the distributed representations of the work space and the configuration space and propose various techniques for computing the work space and configuration space potentials. In section 4 we present a simplified version of the planner that is applicable to robots with few DOFs (four or less). In section 5 we show how the use of Brownian motions for escaping local minima allows us to extend the planner and solve path-planning problems with robots having many DOFs.

2. Relation to Other Work

Much research has been devoted to robot motion planning during the past 10 years (Latombe 1990). Most of this research has focused on path planning (i.e., the geometric problem of finding a collision-free path between two given configurations of a robot). Today the mathematic and computational structures of the general problem (when stated in algebraic terms) are reasonably well understood (Schwartz and Sharir 1983b; Canny 1988). In addition, practical algorithms have been implemented in more or less specific cases (Brooks and Lozano-Pérez 1983; Gouzènes 1984; Laugier and Germain 1985; Faverjon 1986; Lozano-Pérez 1987; Faverjon and Tournassoud 1987; Barraquand et al. 1989; Zhu and Latombe 1989).

One of the most widely studied path-planning approaches is the "cell decomposition" approach (Schwartz and Sharir 1983a; Brooks and Lozano-Pérez 1983). It consists of first decomposing (exactly or approximately) the set of free configurations of the robot into a finite collection of cells and then searching a connectivity graph representing adjacency relation among these cells. However, in this approach, the number of cells to be generated is a function of (1) the number of polynomial constraints used to model the robot and the obstacles and (2) the degree of these constraints. This function also grows exponentially with the number of DOFs (n) of the robot, as the volume of the configuration space (locally diffeomorphic to $\mathbf{R}^n$) increases exponentially with n. Thus the approach is intractable even for reasonably small values of n. To our knowledge, no effective planner has been implemented using this approach with $n > 4$. In fact, this is also true of the other so-called "global" methods—e.g., retraction (Ó'Dúnlaing et al. 1983)—which also represent the connectivity of free space in the form of a graph before actually starting the search for a path.

Some approximate cell decomposition methods proceed hierarchically by decomposing the configuration space into rectangloid cells organized at several levels of resolution. For example, Faverjon (1984) uses an "octree" to represent a three-dimensional configuration space. Each cell is labeled as EMPTY if it intersects no configuration space obstacle, FULL if it lies entirely in configuration space obstacles, and MIXED otherwise. Although there may be a loose resemblance between such a tree and the hierarchical bitmap representations used in our planner, the two approaches are very different. In particular, our planner does not attempt to explicitly approximate the configuration space obstacles as collections of cells.

Because the intractability of the cell decomposition approach—and more generally of the other global path-planning methods—is caused in part by the precomputation of a connectivity graph representing the "global" topology of the robot's free space, "local" methods to path planning have been developed for handling more DOFs, and some successful systems have been implemented (Donald 1984; Faverjon and Tournassoud 1987). A local path-planning method consists of placing a grid (at some resolution) over the robot configuration space and searching this grid. Heuristics computed from partial information about the geometry of the configuration space are used to guide the search. Thus a local method requires no expensive precomputation step before starting the search of a path. In favorable cases, it runs substantially faster than any global method. However, because the search graph (i.e., the grid) is considerably larger than the connectivity graph searched by global methods, it may require much more time than global methods in less favorable cases.

In order to deal efficiently with the large size of the grid, local methods need powerful heuristics to guide the search. However, such known heuristics have the drawback of eventually leading the search to dead-ends from which it is difficult to escape. For example, a widely used heuristics consists of guiding the robot along the negated gradient of an artificial potential field (Khatib 1986). However, this technique may get stuck at local minima of the potential and provides no systematic way to escape these minima. The problem of defining an analytic "navigation function" (i.e., a potential field with a unique minimum at the goal configuration in the connected component of the free space containing the goal configuration of the robot) has been investigated with only limited success so far. Solutions have been proposed only in Euclidean configuration spaces when all the configuration space obstacles are spherical or star-shaped objects (Rimon and Koditschek 1989). Furthermore, if such a navigation function could be defined in the general case, its computation would probably be expensive and would constitute a precomputation step before search, similar in drawback to the construction of the connectivity graph in the cell decomposition approach. Along another line of research, paths for an 8-DOF manipulator have been generated with a variant of the potential field method, called the *constraint method* (Faverjon and Tournassoud 1987). Although impressive, this result has been obtained in specific work spaces where all the obstacles are vertical cylindrical pipes, with interactive human assistance for moving the robot out of the encountered local minima.

Recently, we have developed a potential-based approach using a numeric valley-tracking algorithm (Barraquand et al. 1989) to escape the local minima of the potential function. The planner was able to generate paths for a 10-DOF manipulator arm with a nonserial kinematic chain. (An example with the same arm will be given in this article.) However, the planner implemented using this approach was quite slow and not very reliable. In particular, it failed to solve several problems that the planner described in the present article has been able to solve. Nevertheless, the approach described later derives from this earlier work.

The problem of generating paths for multiple robots has attracted some attention (Schwartz and Sharir 1983c; Kant and Zucker 1986; Erdmann and Lozano-Pérez 1986; O'Donnell and Lozano-Pérez 1989). Implemented systems rely on a simple paradigm with multiple variants—e.g., "velocity tuning" and "prioritizing"—which allows one to consider the individual robots separately or sequentially. This paradigm makes it possible to build planners whose time complexity is "only" exponential in the maximum of the numbers of DOFs of the individual robots, rather than in their sum. However, the paradigm is incomplete and is unable to solve problems in which robots "strongly" interact. An example of such a problem is when two mobile robots have to interchange their positions in a work space made of narrow corridors. We successfully run our planner on several examples of this kind.

Since Reif's early paper (Reif 1979), the computational complexity of path planning has been analyzed by many authors when the problem is stated in semi-algebraic form (Schwartz and Sharir 1988). The bitmap representations used in our planner may open new perspectives on some computational complexity aspects. The worst-case complexity of our planner is still exponential in the number of DOFs (i.e., the dimension of the configuration space). However, although planners using semi-algebraic representations are time polynomial in the number of polynomial constraints and their maximal degrees, our planner is polynomial in the inverse of the maximal resolution of the bitmap description.[1] One may argue that, unlike semi-algebraic models, the bitmap representation is not "exact." However, when compared with the real world, neither of these representations is exact, and both can be made as precise as one wishes by increasing the resolution of the bitmap, for one representation, and the number and degrees of the semi-algebraic constraints, for the other representation.

3. Distributed Representation and Potential Fields

3.1. Work Space Representation

Let $\mathcal{A}$ denote the robot, $\mathcal{W}$ its work space, and $\mathcal{C}$ its configuration space. A configuration of the robot (i.e., a point in $\mathcal{C}$) completely specifies the position of every point in $\mathcal{A}$ with respect to a coordinate system attached to $\mathcal{W}$ (Lozano-Pérez 1983). Let n be the dimension of $\mathcal{C}$. We represent a configuration $q \in \mathcal{C}$ by a list of n parameters $(q_1, \ldots, q_n)$, with appropriate modulo arithmetic for the angular parameters (Latombe 1990). The subset of $\mathcal{C}$ consisting of all the configurations where the robot has no contact or intersection with the obstacles in $\mathcal{W}$ is called the *free space* and is denoted by $\mathcal{C}_{free}$.

The work space $\mathcal{W}$ is modeled as a multiscale pyramid of bitmap arrays, each of which is N-dimensional, with $N = 2$ or 3 being the dimension of $\mathcal{W}$. At any given resolution level, the array is defined by the following function BM:

$$BM : \mathcal{W} \rightarrow \{1, 0\}$$
$$x \mapsto BM(x)$$

1. Different complexity measures with nonalgebraic models were previously given in Lumelsky (1987).

in such a way that the subset of points x such that $BM(x) = 1$ represents the work space obstacles, and the subset of points x such that $BM(x) = 0$ represents the empty part of the work space. We write: $\mathcal{W}_{empty} = \{x \mid BM(x) = 0\}$. Figure 1 displays the bitmap representation of a particular two-dimensional work space at the 256^2 resolution (1 = black; 0 = white).

For each point $p \in \mathcal{A}$, one can consider the geometric application that maps any configuration $q = (q_1, \ldots, q_n) \in \mathcal{C}$ to the position $x \in \mathcal{W}$ of p in the work space. This map:

$$X : \mathcal{A} \times \mathcal{C} \rightarrow \mathcal{W}$$
$$(p, q) \mapsto X(p, q) = x$$

is called the *forward kinematic map*.

The work space representation is given to the planner at the finest level of resolution, typically 512^2 or 256^2 for a two-dimensional work space and 128^3 for a three-dimensional work space. The other levels are automatically derived from it in a conservative fashion. The scaling factor between two successive levels of resolution is 2, but a different factor could have been chosen.

The planner iteratively considers each level of resolution in the pyramid, from the coarsest to the finest, until it generates a path or exits with failure. At each level of resolution, it computes the various potential functions in the same fashion, using the bitmap array at the current level of resolution. We now describe these computations.

Fig. 1. 256^2 bitmap representation of a work space.

3.2. Extraction of the Work Space Skeleton

Let $p_1, \ldots, p_s$ be s given points in $\mathcal{A}$, called *control points*. For each point p_i we construct a function:

$$V_{p_i} : x \in \mathcal{W}_{empty} \mapsto V_{p_i}(x) \in \mathbf{R}$$

called the *work space potential*. Next, we combine these potentials into another potential function U defined over the configuration space:

$$U : q \in \mathcal{C}_{free} \mapsto U(q)$$
$$= G(V_{p_1}(X(p_1, q)), \ldots, V_{p_s}(X(p_s, q))) \in \mathbf{R}.$$

U is called the *configuration space potential*. The construction of the V_{p_i}s is described in this subsection and the next. The construction of U is described in section 3.5.

In constructing the work space potentials V_{p_i}, we have two goals:

1. We want each function V_{p_i} to have a single minimum at the goal position of the point p_i. This is a major heuristic step toward the construction of a configuration space potential with few or small spurious local minima.
2. We want the path obtained by following the negated gradient of V_{p_i}, from any initial position of p_i, to lie as far away as possible from the obstacles in order to maximize the maneuvering space of the robot.

To achieve these goals, we compute each work space potential in two steps. First, we compute the discrete L^1 (Manhattan) distance $d_1(x)$ from every point $x \in \mathcal{W}_{empty}$ to the obstacles and we simultaneously extract a subset $\mathcal{S}$ of $\mathcal{W}$ of co-dimension 1. We call this subset the *work space skeleton*. In two dimensions, it is a network of one-dimensional curves similar to the skeleton widely used in mathematic morphology (Serra 1982) and to the generalized Voronoi diagram (Lee and Drysdale 1981). Second, we compute the potential functions V_{p_i} using both the d_1 map and the skeleton $\mathcal{S}$. The first step is detailed later in this section. The second is described in the next subsection.

The distance $d_1(x)$ between every point $x \in \mathcal{W}_{empty}$ and the obstacles is computed as follows: First, the points in the boundary of the obstacles are identified, and the value of d_1 at these points is set to zero (we also include the points in the frame bounding the bitmap as boundary points). Then, starting from these boundary points, we apply a wavefront expansion procedure that recursively labels all the points in $\mathcal{W}_{empty}$. More precisely, the values of d_1 at all the neighbors of the boundary points are set to 1; the value of d_1 at the neighbors of these points, if not yet computed, is set to 2; etc. The procedure is repeated until all the points in $\mathcal{W}_{empty}$ have been attained. Figure 2 displays contours of the d_1 map thus computed for the work space shown in Figure 1.

In parallel, we compute the work space skeleton $\mathcal{S}$ as the set of points where the "waves" issued from the boundary points of $\mathcal{W}_{empty}$ meet. This is done by propagating not only the values of d_1, but also the points in the boundary of $\mathcal{W}_{empty}$ that are at the origin of the propagation. Figure 3 displays the skeletons computed in several work spaces, including the work space of Figure 1. Every connected component of the work space yields a connected component of the skeleton $\mathcal{S}$.

The computation of both d_1 and $\mathcal{S}$ is not local and therefore must be done prior to the execution of the rest of our path-planning algorithm. However, the time complexity of the algorithm is linear in the number of points in $\mathcal{W}$ and constant in the number and the shape of obstacles. Its implementation is quite fast (a fraction of a second for a 256^2 bitmap array).

From a conceptual point of view, neither the choice of the L^1 metric nor the precise definition of the skeleton is very important for the rest of our path-planning approach, as the potential functions are only used as heuristics. Instead, we could have used the more classic L^2 distance and computed the generalized Voronoi diagram for that metric. However, as mentioned earlier, the construction of the work space potentials over $\mathcal{W}_{empty}$ is a necessary preliminary step before the rest of our path-planning

Fig. 2. Contours of the d_1 *map in the work space of Fig. 1.*

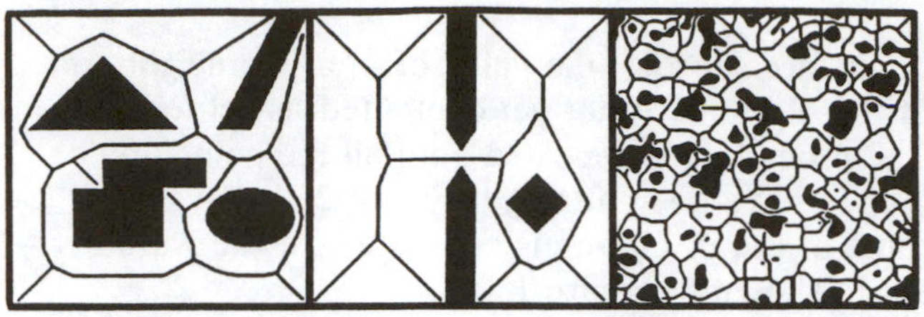

Fig. 3. Examples of work space skeletons.

Fig. 4. Equipotential contours of the work space potential computed by NF1.

algorithm can be executed. The computation of the L^1 distance is faster than the computation of the L^2 distance. Notice also that unlike the earlier computation of $\mathcal{S}$, the time complexity of constructing the algebraic description of the L^2 Voronoi diagram of a polygonal work space increases with the number of vertices of the obstacles.

3.3. Work Space Potential Fields Without Local Minima

The bitmap description of the work space allows us to compute a numeric navigation function—i.e., a collision-avoiding attractive potential field defined over the work space bitmap that has no other local minima than the goal. Such a navigation function happens to be very helpful for avoiding concavities of the work space obstacles.

We propose two algorithms, NF1 and NF2, for computing numeric navigation functions. NF1 is simpler and does not make use of the map d_1 and the skeleton $\mathcal{S}$. NF2 is slightly more involved but has the advantage of producing work space potentials that keep the control points as far away as possible from the obstacles.

NF1 computes the work space potential V_p for every control point p by using a wavefront expansion technique starting at the goal position x_{goal} of p. The value of V_p is first set to 0 at x_{goal}. Next, the value of V_p at the neighbors of x_{goal} in $\mathcal{W}_{empty}$ is set to 1, the value of V_p at the neighbors of these neighbors in $\mathcal{W}_{empty}$ is set to 2 (if not previously computed), etc. This procedure is recursively repeated until the connected subset of $\mathcal{W}_{empty}$ containing x_{goal} has been completely explored. The complexity of NF1 is linear in the number of points of the bitmap description and constant in the number and shape of the obstacles. Equipotential contours of the resulting work space potential field for the two-dimensional work space of Figure 1 are displayed in Figure 4. This computation was performed in a fraction of a second.

A property of the function V_p thus computed is that by following the flow of the negated gradient from any initial point x_{init}, we obtain a path between x_{init} to x_{goal} (if one exists) that is the shortest one for the L^1 distance (at the resolution of the bitmap array) over all the paths connecting x_{init} to x_{goal}. In a three-dimensional work space, this computation may be preferable to exact methods, as the problem of computing the exact shortest distance in a polyhedral space is known to be NP-hard in the number of vertices under any L^p metric (Canny 1988).

Notice that NF1 computes V_p only in the connected subset of $\mathcal{W}_{empty}$ that contains the goal position x_{goal}. Hence if the initial position x_{init} is not in the same connected component as the goal, the value of V_p is not computed at this point, and we can immediately infer that there is no collision-free path for the robot.

However, a significant drawback of this work space potential is that it induces paths that typically graze the obstacles in the work space (i.e., it only achieves the first of the two goals stated in section 3.2). Because several potentials will have to be combined into a configuration space potential U attracting the robot toward a goal configuration, the individual work space potentials may compete in such a way that they produce local minima of U. To reduce the risk of such a competition and to enlarge the maneuvering space of the robot, our planner makes use of a more involved work space potential function V_p computed by the algorithm NF2.

NF2 computes V_p in three steps:

1. The first step consists of tracing a line σ following the gradient of the distance map d_1 from the goal point x_{goal} to the nearest point in the

skeleton $\mathcal{S}$. Once this line is computed, we include it in $\mathcal{S}$, yielding the "extended skeleton" $\mathcal{S}_p = \mathcal{S} \cup \sigma$.

2. The second step of the algorithm consists of labeling all the points of $\mathcal{S}_p$ starting from x_{goal}. The label 0 is assigned to x_{goal}, the label 1 is assigned to its neighbors in $\mathcal{S}_p$, and these neighbors are inserted into a list Q. The point x_1 in Q that is at maximum distance d_1 from the obstacles is considered next and removed from Q; the label $l(x_1) = U(x_1) + 1$ is assigned to its neighbors in $\mathcal{S}_p$ that have not been labeled yet, and these neighbors are inserted into Q. This operation is repeated until the list Q is empty (i.e., the subset of $\mathcal{S}_p$ accessible from x_{goal} has been completely explored). At each step of the recursion, the list of points in Q is stored in a balanced tree sorted according to d_1 so that each insertion of a new point and extraction of a point at maximum distance from the obstacles takes logarithmic time in the size of Q (Aho et al. 1983). At the end of the second step, all the points $x \in \mathcal{S}_p$ connected to x_{goal} in $\mathcal{S}_p$ have a label $l(x)$.
3. The third step is a wavefront expansion starting from the subset S of $\mathcal{S}_p$ labeled at the previous step. NF2 first gives the label $l(x) + 1$ to every neighbor of every point $x \in S$. It then computes the unlabeled neighbors of these neighbors and increments the labels iteratively by 1 until the connected component of $\mathcal{W}_{empty}$ containing x_{goal} has been completely explored.

Figure 5 shows equipotential contours of the work

Fig. 5. Equipotential contours of the work space potential computed by NF2.

space potential computed by NF2 for the work space of Figure 1. This potential has no other local minima than the goal. Following its negated gradient from an initial position leads to generating a path that lies as far away from the obstacles as possible by following the safest portion on the work space skeleton. Like NF1, NF2 only computes V_p over the connected subset of $\mathcal{W}_{empty}$ that contains x_{goal}.

The complexity of NF2 is slightly higher than that of NF1. Let a be the number of points in the bitmap array and b the number of points in the augmented skeleton $\mathcal{S}_p$. The complexity of NF2 (including the cost of computing d_1 and the skeleton) is $O(a + b \log b)$, instead of $O(a)$. For reasonable work spaces, however, because the skeleton has codimension 1 in the work space, we have

$$b \propto a^{N-1/N},$$

with $N = 2$ or 3 being the dimension of the work space. For these work spaces, the complexity of NF2 reduces to

$$O(a + a^{N-1/N} \log a)$$

and hence is linear in a. For the 256^2 work space of Figure 1, it took about 2s for our implementation of NF2 to compute the work space potential shown in Figure 5. This time includes the computation of d_1 and $\mathcal{S}$, which does not have to be repeated if several work space potentials are computed.

Variants of NF2 can easily be imagined to compute work space potentials with slightly different properties. For example, in the third step, we could increment the potential at each iteration by $1/d_1$, rather than by 1, and obtain a potential V_p that becomes infinite in the boundary of $\mathcal{W}_{empty}$. We will not detail the cosmetics of the computation of numeric potential fields further. Our point is simply to show that a large family of potential functions with various properties can be built within our distributed representation approach.

3.4. Discretization of the Configuration Space

Because we represent the work space as a pyramid of bitmap arrays, it is consistent to discretize the configuration space $\mathcal{C}$ into a multiresolution grid pyramid. The configuration space pyramid has as many levels of resolution as the work space pyramid, and the resolutions at each level of the two pyramids are tightly related, as described later.

Let δ denote the distance between two adjacent points along the same coordinate axis in a work space bitmap. In the work space pyramid, δ varies between δ_{min} and δ_{max}. For example, if the work space is represented by a pyramid of arrays whose sizes are ranging between 16^2 and 512^2 and if distances are measured in percentage of the work space diameter ("normalized" distances), we have $\delta_{min} = 1/512$ and $\delta_{max} = 1/16$.

We can define the resolution of a grid as the logarithm of the inverse of the distance between two discretization points in the base defined by the scaling factor between two successive resolution levels (2 in our implementation). Hence, in our example, the resolution r varies between $r_{min} = -\log_2(\delta_{max}) = 4$ and $r_{max} = -\log_2(\delta_{min}) = 9$.

We represent the configuration space $\mathscr{C}$ as a rectangloid subset of the n-dimensional Cartesian space $\mathbf{R}^n$ by representing a configuration q as a list $(q_1, \ldots, q_n)$ of n independent parameters. For any given work space resolution, say $r = -\log_2(\delta)$, the corresponding resolution $R_i = -\log_2(\Delta_i)$ of the discretization along the q_i axis of $\mathscr{C}$ is chosen in such a way that a modification of q_i by $\Delta_i = 2^{-R_i}$ generates a "small motion" of the robot in the work space. By "small motion" we mean that any point $p \in \mathscr{A}$ moves by less than $nbtol \times \delta$, where $nbtol$ is a small number (typically, 1 or 2).

The relation between the position of a robot point in the work space and a robot configuration is given by the forward kinematic map $X(p, q)$. For every point $p \in \mathscr{A}$, a modification of q_i ($i \in [1, n]$) by Δ_i results in a modification of each coordinate x_j ($j \in [1, N]$) of p by:

$$\frac{\partial x_j}{\partial q_i}(p, q)\Delta_i.$$

If we impose the work space motions to be less than $nbtol \times \delta$, we must have:

$$\Delta_i = nbtol \times \delta / \sup_{p\in\mathscr{A}, q\in\mathscr{C}, j\in[1,N]} \left(\frac{\partial x_j}{\partial q_i}(p, q)\right)$$

$$= nbtol \times \delta / J^i_{\text{sup}}.$$

For a given robot, the numbers J^i_{sup} are generally straightforward to compute. This leads us to compute the resolution R_i as:

$$R_i = r + \log_2(J^i_{\text{sup}}) + \log_2(nbtol).$$

For example, consider a bar of length L moving freely in a two-dimensional work space. A configuration of the bar can be represented as (x_G, y_G, θ), with x_G and y_G being the coordinates of the center of mass of the bar in the fixed Cartesian coordinate system embedded in the work space and θ being the orientation of the bar relative to an axis of this system. Let us normalize x_G, y_G, and θ so that their values range between 0 and 1. We have:

$$J^{x_G}_{\text{sup}} = J^{y_G}_{\text{sup}} = 1 \quad \text{and} \quad J^{\theta}_{\text{sup}} = \pi L/2.$$

If we set $nbtol$ to 2, we get:

$$R_{x_G} = R_{y_G} = r - 1 \quad \text{and}$$

$$R_\theta = r + \log_2(\pi L/2) - 1.$$

This means that we need $2^1 = 2$ times less samples for x_G and y_G than for the work space representation at each level of resolution and $1/(\pi L)$ less samples for θ. In our implementation, the J^i_{sup}s are input by hand.

3.5. *Configuration Space Potential*

In our planner, the goal configurations of a robot are specified by the goal positions of one or several points in the robot. By definition, the robot is at a goal configuration whenever all the control points are at their goal positions. For instance, if $\mathscr{A}$ is a two-dimensional object that can both translate and rotate in the plane (three-dimensional configuration space), the specification of the goal positions of two points uniquely determines the goal configuration of the robot. If $\mathscr{A}$ is, say, a 10-DOF manipulator arm, then specifying the desired positions of some points in the end effector determines a goal region in configuration space.

It is important that a path planner allows specification of a goal region in configuration space. Indeed, for many tasks, the goal configuration is incompletely specified. Arbitrarily selecting one would possibly result in a more difficult or impossible path-planning problem. Furthermore, if the robot has many degrees of freedom, specifying a unique goal configuration is a difficult task in itself, as it requires collision-free placement of the various bodies of the robot to be found.

The points used to specify the goal configurations of a robot are exactly those that are later used by the planner as the control points. Let $p_1, \ldots, p_s$ be these points. The configuration space potential U is defined as a combination:

$$U(q) = G\{V_{p_1}[X(p_1, q)], \ldots, V_{p_s}[X(p_s, q)]\}$$

of the work space potentials V_{p_i}, $i = 1, \ldots, s$, defined for the s control points p_i. This combination concurrently attracts the different points p_i toward their respective goal positions. G is called the *arbitration function*. This terminology reflects the facts that the various control points may compete to attain their goal positions and that the function G arbitrates this competition.

In most of the previous collision-avoidance systems using artificial potential fields, G was chosen as a linear combination of the work space potentials (Khatib 1986); i.e.,

$$G(y_1, \ldots, y_s) = \sum_{i=1}^{i=s} \lambda_i y_i.$$

This simple choice seems natural, because it does not favor one control point over the others. However, precisely for that same reason, it tends to increase the number of conflicts among the control points, thus producing numerous undesired local minima.

The choice of the function G is important, as it highly influences the number of local minima of the potential U. With our "perfect" work space potentials, the work space concavities do not directly create local minima. It is the concurrent attraction of the different control points toward their respective goal positions that creates these local minima.[2] This results from the fact that these points do not move independently. As suggested earlier, the function G precisely defines the way in which the competition between the different points is to be regulated.

The choice of G that seems to minimize the number of local minima is:

$$G(y_1, \ldots, y_s) = \min_{i=1}^{i=s} y_i.$$

Indeed, this competition function favors the attraction of the point that is already in the best position to reach its goal. However, when one point has reached its goal position, the potential field is identically zero, and it does not attract the other points toward their goal positions. A solution to avoid this problem is simply to add another term to the arbitration function:

$$G(y_1 \ldots, y_s) = \min_{i=1}^{i=s} y_i + \varepsilon \max_{i=1}^{i=s} y_i \quad (1)$$

where ε is a small real number. In our experiments with robots with few degrees of freedom (see section 4), we obtained the best results with $\varepsilon = 0.1$. However, the best value of ε may depend on the robot.

Another choice for G is:

$$G(y_1, \ldots, y_s) = \max_{i=1}^{i=s} y_i. \quad (2)$$

This choice tends to increase the number of competitions between the control points and, therefore, the number of local minima. However, it can be a good choice for robots with many DOFs. As a matter of fact, the number of local minima is not the only measure for the quality of the potential, as it might be much more difficult to escape a local minimum with a small attractive well than a minimum with a large well. The above competition function increases the number of local minima, but experiments show that in general it also reduces their volumes. This is the function that gave the best results for planning the paths of manipulator arms with many DOFs and multiple control points (see section 5).

Unlike the work space potentials V_{p_i}, the configuration space potential U does not have to be precomputed before actually searching for a path. In fact, in high-dimensional configuration spaces, this precomputation would be intractable. The function U is only computed during the search of a path at those configurations that are attained by the search algorithm.

4. Fast Path Planning for Robots With Few DOFs

We have implemented two versions of our planner, called the *Best-First Planner* (BFP) and the *Randomized Path Planner* (RPP). These two versions[3] differ mainly in the way the local minima of the configuration space potential function are escaped. In this section we present the simplest version of the planner, BFP. This version is fast for robots with few DOFs (less than five), but it is only applicable to these robots.

BFP iteratively considers each level of resolution in the work space pyramid, from the coarsest to the finest. It terminates with success as soon as it has generated a path. It terminates with failure if, after having considered the finest bitmap, it still has not generated a path.

At each level, BFP performs a best-first search (Nilsson 1980) of the collision-free subset $\mathscr{C}_{free}$ of the configuration space grid using the potential U defined in formula (1) as the heuristic function. If r is the resolution of the work space bitmap, $R_i = r - \log_2 (J^i_{sup}) - \log_2 (nbtol)$ is the resolution of the discretization along each q_i axis in $\mathscr{C}$, for every $i = 1$ to n. The successors of a configuration in the search graph are all its neighbors lying in $\mathscr{C}_{free}$. In our implementation, we chose the n-neighborhood, which means that two configurations are neighbors iff 1 to n of their coordinates differ by a single incre-

2. As we will make explicit later, local minima may also appear in the boundary of $\mathscr{C}_{free}$ at configurations where the gradient of the potential function U is not zero.

3. In fact, the two versions are blended in a single program that offers two options. In this article we distinguish between BFP and RPP to make the presentation clearer.

ment, the others being the same. Hence each configuration may have up to $3^n - 1$ neighbors. The size of this neighborhood is reasonable because, for other reasons, n has to be small.

As long as the best-first search process does not reach a local minimum of the function U, the search reduces to following the negated gradient of the potential (fastest descent procedure). When a local minimum is reached, the search algorithm simply fills up the well of the local minimum until it reaches a saddle point. Then the search proceeds again along the negated gradient of U. It stops when the goal configuration is attained.

At this stage, there is an important aspect of the algorithm that we must make precise. The use of the potential U to guide the search does not guarantee that there will be no collision of the robot with the obstacles. Therefore whenever the planner considers a new configuration q in $\mathcal{C}$, it should check that it lies in the free space. Because the planner does not represent the configuration space obstacles explicitly, the verification is done in the work space using a simple divide-and-conquer technique. To illustrate the idea, let the robot be a line segment of length L in a two-dimensional work space. We assume that the motion increments in the configuration space grid are small enough so that if the robot is in free space at one discretized configuration, it cannot lie entirely inside an obstacle at a neighboring configuration. Using the precomputed d_1 map, we obtain the distances $d_1(begin)$ and $d_1(end)$ of the two end points of the segment representing the robot at the configuration q. If the minimum of these two distances is smaller than the length L of the segment, then we are certain that the robot does not hit any of the obstacles at q. Otherwise, we divide the segment into two segments of equal lengths and repeat the computation recursively for the two segments. In two-dimensional work spaces, this computation can easily be generalized to robots whose boundaries consists of straight edges and/or curve segments of higher degrees (e.g., circular and elliptical arcs) by treating each segment separately. In three-dimensional work spaces, the computation can be extended as follows to robots made up of three-dimensional bodies: Assume first that the robot is a triangle whose vertices are v_1, v_2, and v_3. If the maximum of the distances $d_1(v_1)$, $d_1(v_2)$, and $d_1(v_3)$ is greater than the maximum of the lengths of the edges of the triangle, then we are certain that the configuration is collision-free. Otherwise, we divide every edge of the triangle in two equal segments and repeat the test recursively for the four triangles whose vertices are v_1, v_2, v_3, and the edge mid-points. This computation is extended to robots described as collections of polyhedra by triangulating the faces of the polyhedra and considering the generated triangles separately.

Remark. One may argue that the need for collision checking would be eliminated if we used a potential tending toward infinity in the boundary of $\mathcal{C}_{free}$. However, it must be noticed that the computation of such a potential at any configuration q includes the computation of the shortest distance between the robot at q and the obstacles. Hence it includes the computational cost of making the above collision test.

□

Note here that two types of local minima can be attained: the natural minima of U (where the gradient is zero) and the minima located in the boundary of $\mathcal{C}_{free}$ (where the gradient is nonzero in general). Both types of minimum are escaped in the same fashion.

We have implemented BFP in a program written in C language and run on a DEC 3100 MIPS-based workstation. We have experimented with this program using a "mobile robot" with two DOFs of translation and one DOF of rotation—namely, a long rectangle in a two-dimensional work space. Figure 6 shows a path generated by the planner that demonstrates the ability of the planner to produce complex maneuvers. In this example, the configuration space potential was computed using formula (1) with two control points located at the two extremities of the bar. The path was generated in a 256^2 work space bitmap. The running time of the planner

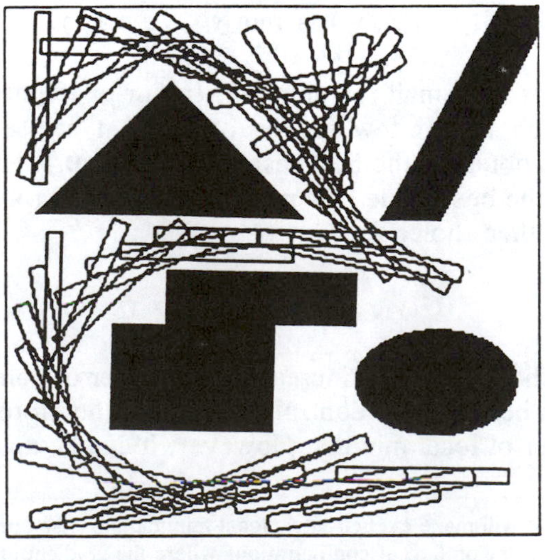

Fig. 6. Path generated for a 3-DOF mobile robot.

was 1 s. This is three orders of magnitude faster than the running times reported in Brooks and Lozano-Pérez (1983) for similar (though apparently simpler) path-planning problems. Roughly one order of magnitude is a result of the faster computer that we used. The other two seem to be the product of our algorithm.

Figure 7 shows another path generated by BFP for the same mobile robot. The path was generated in less than 5 s within a 512^2 bitmap representing a work space with more than 70 randomly constructed obstacles of arbitrary shapes. This example would be very difficult (at best) to run with a planner using a semi-algebraic representation of the work space and illustrates the power of the distributed representation approach used in our planner.

BFP is practical only for robots with a small number n of DOFs—typically, $n < 5$—because the number of discrete configurations in a local minimum well increases exponentially with the dimension of the configuration space. For such robots, it has two major advantages:

1. As illustrated earlier, it is extremely fast. Simpler but nontrivial problems than those of Figures 6 and 7, requiring less complex maneuvers, were solved in about ⅕ s (at a coarse level of resolution), which can almost be considered real time.
2. It is deterministically resolution complete (i.e., the algorithm generates a path to the goal whenever a solution exists at the maximal resolution and returns failure if there is no solution). In both cases, BFP returns control within some bounded amount of time.

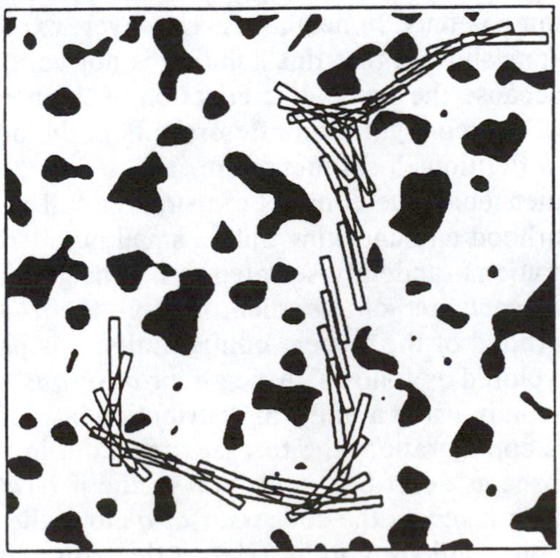

Fig. 7. Path generated among randomly distributed obstacles.

At first sight, the experimental efficiency of such a brute force technique is surprising. In fact, the potential functions computed in the work space are designed to be "perfect" potential fields for a point robot, in the sense that they have no other minimum than the goal. Therefore complex-shaped obstacles do not directly create local minima of the potential. Local minima occur only when the work space is so cluttered that the solution path has to come very close to the obstacles. However, in such cases, because the work space potentials computed by NF2 guide the robot along a path where the maneuvering space is maximized, the local minima usually have a reduced domain of attraction—i.e., the number of discrete configurations contained in the local minimum wells is usually small. The best-first search algorithm therefore fills the wells very quickly.

Remark. Whenever the planner fails to find a path at some resolution level, it forgets about it and considers the next level of resolution. Because the time required to explore a coarser grid is small relative to the time needed for a finer grid, this naive coarse-to-fine approach is adequate. Nevertheless, one could imagine another approach where the work done at a resolution level would be used to guide the work at the next resolution level. However, the solution of a path-planning problem is often so versatile with respect to the resolution of the representation that the information passed from one resolution level to the next might be more misleading than helpful. Moreover, having no interaction among the searches at the various resolution levels makes it possible to perform them concurrently on a parallel machine. □

5. Path Planning for Robots With Many DOFs

The BFP planner presented in the previous section cannot efficiently plan motions for robots with many DOFs. On the other hand, human beings are able to solve motion-planning problems with a high number of DOFs. Sometimes redundancy among the DOFs even seems to simplify planning. Hence there is a "divorce" between the exponential-time complexity of the general path-planning problem and our everyday life experience. In order to design an efficient path planner applicable to robots with many DOFs, it seems that we have to drop the completeness requirement. In this section we describe the RPP version of our planner that has demonstrated its ability to solve many complex planning problems, some being nontrivial for humans. The new algorithm differs from the one of the previous section in the way it escapes local minima. Rather than filling up a local minimum, it applies a Monte Carlo procedure that consists of generating Brownian motions until the minimum is escaped. The resulting planning algorithm is probabilistically (rather than deterministically) resolution complete. For robots with many DOFs, we used the configuration space potential defined by formula (2) in our experiments, rather than the one defined by formula (1). However, the Monte Carlo procedure itself does not depend on the particular potential function that is used.

5.1. Overview

Starting from the initial configuration q_{init} of the robot, RPP first applies a best-first algorithm (i.e., it descends the potential U until it reaches a local minimum q_{loc}; see section 5.2). We call such a motion a *gradient motion*. Let $U_{loc} = U(q_{loc})$. If $U_{loc} = 0$, the problem is solved, and the planner returns the constructed path. Otherwise, it attempts to escape the local minimum by executing a series of *random motions* issued from q_{loc}. These random motions are approximations of Brownian motions described in section 5.3.

At the terminal configuration of every random motion, the algorithm executes a gradient motion until it reaches a (hopefully) new local minimum. From each local minimum, if none of them is the goal, it performs another series of random motions. The graph of the local minima is thus incrementally built, the path joining two "adjacent" local minima being the concatenation of a random motion and a gradient motion. A randomized depth-first search of this graph is performed (see section 5.4) until the goal configuration is reached or the planner gives up. If the search terminates successfully, the generated path is transformed into a smoother path (section 5.5).

An interesting property of this planning algorithm is that all the random motions starting from a given local minimum can be performed concurrently on a parallel machine, as there is no need for communication among the different processing units.

Because the algorithm uses a random procedure to build the graph of the local minima, it is not guaranteed to find a path whenever one exists. In other words, the algorithm is not complete. However, the properties of Brownian motions make it possible to prove that when the number of Brownian motions executed from every local minimum is unbounded (the computation time may then tend toward infinity), the probability to reach the goal converges toward 1. Hence we say that the algorithm is *probabilistically resolution complete*. However, this convergence-in-distribution property, which is well known for the so-called "simulated annealing" algorithms (Geman and Hwang 1986), is a very weak one. Indeed, the totally uninformed algorithm that executes a Brownian motion from the initial configuration q_{init} and terminates when it enters a small neighborhood of the goal configuration is also probabilistically resolution complete! Despite the weakness of this theoretical result, our experiments show that RPP is quite efficient.

We describe the various components of RPP in more detail in the following sections. We also give experimental results obtained with the implemented planner.

5.2. Gradient Motions

The planner operates in the configuration space grid over which U is defined. In principle, a gradient motion consists of searching this grid in a best-first fashion as described in section 4. However, when the dimension n of the configuration space becomes large, say $n \geq 6$, the n-neighborhood used in section 4 to determine the direction of motion is much too large to be fully explored at each step. For example, if $n = 10$, a configuration has approximately 60,000 n-neighbors; if $n = 31$, it has over 600 thousands of billions n-neighbors.

One way to deal with this difficulty is simply to use a smaller neighborhood—for example, the 1-neighborhood (two configurations are 1-neighbors iff only one of their coordinates differs and if it differs by a single increment of the grid). Then each configuration has only $2n$ neighbors. However, experimentations showed that this solution is not very good, because the crude discretization of the neighborhood of a configuration often results in the detection of a fictitious local minimum.

Another technique consists of using the full n-neighborhood and checking only a small number of configurations randomly selected in this neighborhood. At each step of a gradient motion, the n-neighborhood of the current configuration q is partially explored as follows. A neighbor q' of q is chosen randomly using a uniform distribution law. If it is a free configuration (the test is carried out in the work space as described in section 4) and if $U(q') < U(q)$, q' is taken as the successor of q along the path of the gradient motion. (Hence the path may only follow a rough approximation of the negated gradient flow.) If q' is not free or if $U(q') \geq U(q)$, another neighbor q' is randomly chosen. The number of iterations is limited to a few tens to a few hundreds (depending on the value of n). If none of them generates a successor of q, q is considered to be a local minimum.

Both techniques mentioned above have been implemented, and the second one gave much better results, generating almost no fictitious minima. As a matter of fact, if the curvature of the surface $U = U(q)$ at q is small and if q is not a minimum of U, then at every iteration, the probability of guessing a configuration q' such that $U(q') < U(q)$ is approximately equal to 0.5. The probability of finding such a configuration q' in 10 independent guesses is of the order of 0.999. Although the number of guesses may be increased with n, it does not have to be increased proportionally.

5.3. Random Motions

When the algorithm reaches a local minimum of the potential field U, we consider that there is no more information that we can extract locally from U in order to guess the direction of motion that will lead us to the goal. Then *if we do not make any assumption on the statistics of the obstacle distribution*, we have no additional information for helping us to reach the goal. RPP continues the search by executing random motions issued from the current local minimum q_{loc}.

The most uninformed type of motion is known to be the Brownian motion (Papoulis 1965). Because a Brownian motion is a continuous stochastic process, the random motions performed by RPP are approximations of Brownian motions and are defined as discrete random walks. A random walk in the configuration space consists of executing a certain number t of steps (the "duration" of the random walk). Each step corresponds to a "unit" of time and projects into every q_i axis, $i \in [1, n]$, as an increment $+v_i$ or $-v_i$ of fixed amplitude, each with the constant probability 0.5 (hence independent of the previous steps). The amplitude of the increment, v_i, is the "velocity" of the walk along the q_i axis. This random walk is known to converge almost surely toward a Brownian motion when the amplitude v_i of every increment tends toward 0 (Papoulis 1965).

Without lack of generality, let us take the current local minimum, q_{loc}, as the origin of the coordinates of $\mathscr{C}$. The configuration attained by a Brownian motion of duration t and velocity v_i along each q_i axis is a random variable $Q(t) = (Q_1(t), \ldots, Q_n(t))$ with the following properties (Papoulis 1965):

- The density p_i of $Q_i(t)$ is the Gaussian distribution given by:

$$p_i(q_i, t) = \frac{1}{v_i\sqrt{2\pi t}} \exp\left(-\frac{q_i^2}{2v_i^2 t}\right).$$

- The standard deviation $D_i(t)$ of the difference $Q_i(t + t_0) - Q_i(t_0)$ increases proportionally to the square root of t; i.e.,

$$D_i^2(t) = E\{[Q_i(t_0 + t) - Q_i(t_0)]^2\} = v_i^2 t.$$

- The two processes $Q(t') - Q(t)$ and $Q(t'') - Q(t')$ are independent, for any (t, t', t'') such that $t < t' < t''$.

The Brownian motion (also called the *Wiener-Levy process*) is well defined as long as it does not encounter any obstacle in configuration space. When the process $Q(t)$ hits the boundary of an obstacle, the Brownian motion has to be adapted so that it remains in the free space. The classic generalization of a Brownian motion when the space is bounded consists of reflecting the motion that would have taken place in the absence of boundary, symmetrically to the tangent hyperplane of the boundary at the collision configuration (Brownian motion with "reflective boundary"). The mathematic consistency of this adaptation is discussed in detail in Anderson and Orey (1976). Our planner, which does not construct an explicit representation of the configuration space obstacles, does not know the orientation of the tangent hyperplane at the collision configuration. Hence whenever a random motion step leads to colliding with an obstacle, instead of reflecting the motion on the boundary, the planner guesses another random step and substitutes it for the previous one. Collisions along a Brownian motion are checked in the work space using the divide-and-conquer technique presented in section 4.

We still have to select the velocities v_i and the duration t of every random motion. Because we approximate a Brownian motion as a random walk in a grid where the increment along each q_i axis is Δ_i, we would like the standard deviation of each step to be equal to Δ_i. This leads to choosing $v_i = \Delta_i$. Regarding the duration t, we should choose it such that the generated random motion take the robot out of the current local minima of U. Let us define the *attractive radius* $a_{R_i}(q_{loc})$ of any local minimum q_{loc} of U along the q_i axis as the distance along q_i between q_{loc} and the nearest saddle point of U in that direction. In order to escape the local minimum q_{loc}, the minimum distance that the robot must travel in each direction q_i from q_{loc} is $a_{R_i}(q_{loc})$.

If we were able to estimate the statistics of a_{R_i}, the property $D_i(t) = \Delta_i\sqrt{t}$ would give us a clue for choosing t. The duration of the motion would then be:

$$t \approx \max_{i\in[1,n]} \left(\frac{a_{R_i}(q_{loc})}{\Delta_i}\right)^2. \qquad (3)$$

However, as we make no assumption on the obstacle distribution, we cannot infer any strong statistical property about U and a_{R_i}. However, in general, we may assume that the distance a_{R_i} for each parameter q_i does not exceed the distance that would provoke a motion of the robot longer than the work space diameter (defined by the input bitmap). This diameter being equal to 1 (with the normalized L^1 distance previously used), we obtain the following estimate of a_{R_i} for any local minimum q_{loc}:

$$a_{R_i}(q_{loc}) \approx 1/J^i_{\text{sup}}.$$

On the other hand, we have $\Delta_i \approx \delta/J^i_{\text{sup}}$, where δ is the distance between two consecutive points along the same coordinate axis of the work space. Combining these two formulas with (3), we obtain:

$$t \approx \frac{1}{\delta^2}.$$

We could take t equal to this value. However, this choice would mean that we implicitly assume that all the attraction radii are the same, which is not the case. Instead, we take a_{R_i} as the value of a strictly positive random variable A_{R_i} whose expected value is $1/J^i_{sup}$. The most uninformed distribution (i.e., the one that maximizes entropy) of a positive random variable of given expected value is the truncated Laplace distribution. Therefore we define the density of A_{R_i} as:

$$p(a_{R_i}) = J^i_{\text{sup}} \exp(-J^i_{\text{sup}} a_{R_i}).$$

This leads to choosing the duration t of a random motion as the value of a random variable T. The above density of A_{R_i}, combined with the relation (3), entails the following density for T:

$$p(t) = \frac{\delta}{2\sqrt{t}} \exp(-\delta\sqrt{t}). \qquad (4)$$

One can verify that the expected value of this distribution is indeed $1/\delta^2$.

In fact, a value of T gives a maximal duration of the random motion. After each step of the motion, the planner checks the value of the potential at the current configuration against $U(q_{loc})$. If it is smaller, the planner terminates the random motion.

Remark. As mentioned at the beginning of this subsection, executing Brownian motions when the algorithm reaches a local minimum of the potential U other than the goal configuration corresponds to assuming that there is no more local information that we can extract from U in order to guess the direction of motion that will lead us toward the goal. However, higher order derivatives could provide useful additional information. As a matter of fact, in Barraquand et al. (1989), we used the concept of valley (which is based on the first and second derivatives) to escape local minima. The resulting planner was not very reliable, but the idea of tracking valleys for escaping local minima could be reused here to generate more informed random motions.

□

5.4. *Searching the Local Minima Graph*

By combining best-first motions and random motions, RPP can incrementally construct a graph of the local minima of the potential function U. The search of this graph can be done using a best-first strategy. This simply consists of iteratively generating the successors of the pending local minimum having the smallest potential value and limiting the number of random motions issued from the same minimum to a prespecified number M. A drawback of this strategy is that the same minimum may be attained several times, which is difficult and costly to detect. The strategy may also waste time exploring a local minimum well containing smaller local minimum wells imbricated in one another.

Another search strategy is the following depth-first strategy. When a local minimum q_{loc} is attained, a maximum of M random motions are generated. Each random motion is immediately followed by a best-first motion that attains a local minimum q'_{loc}. If $U(q'_{loc}) \geq U(q_{loc})$, then the planner forgets q'_{loc}; otherwise, if q'_{loc} is not a goal configuration (in which case the problem is solved), its successors are generated in the same fashion. If none of the M motions issued from q_{loc} allow the planner to attain a lower local minimum than q_{loc}, the latter is considered to be a dead-end, and the search is resumed at the most recent local minimum whose M successors have not all been generated yet (chronologic backtracking).

This second search strategy gives better experimental results than the first, but it may still waste time exploring imbricated minima. Moreover, if a low local minimum has been attained (this is the case in the example shown in Figure 20 later), it may have difficulty attaining a lower one from it. (It is not difficult to construct examples where all the free paths between a very low local minimum and a goal configuration are quite long and unlikely to be generated by a single random motion.)

The above strategy can be slightly modified as follows: Rather than memorizing the whole graph G, the planner only memorizes the constructed path τ connecting the initial configuration to the current one (τ is represented at a list of configurations in the configuration space grid). At every local minimum, the planner iteratively generates a maximum of M (typically $M \approx 20$) random/best-first motions as described earlier. As soon as one of them reaches a lower minimum q'_{loc}, it inserts the path from q_{loc} to q'_{loc} at the end of the current path τ and continues the search from q'_{loc}. If the M motions are performed and none of them attains a lower minimum, the planner randomly selects a configuration q_{back} in the subset of τ formed by random motions, using a uniform distribution law over that subset, and backtracks to q_{back}. The search is resumed at q_{back} by executing a best-first motion. Because q_{back} belongs to the path of a random motion, this best-first motion may terminate at a new local minimum that has not been attained so far. (If the first local minimum q_{loc} encountered by the planner turns out to be a dead-end, the above backtracking mechanism cannot be applied. Then RPP randomly selects one of the local minima q'_{loc} attained from q_{loc} and resumes the search with a best-first motion starting at a configuration randomly selected along the path of the random motion leading to q'_{loc}.)

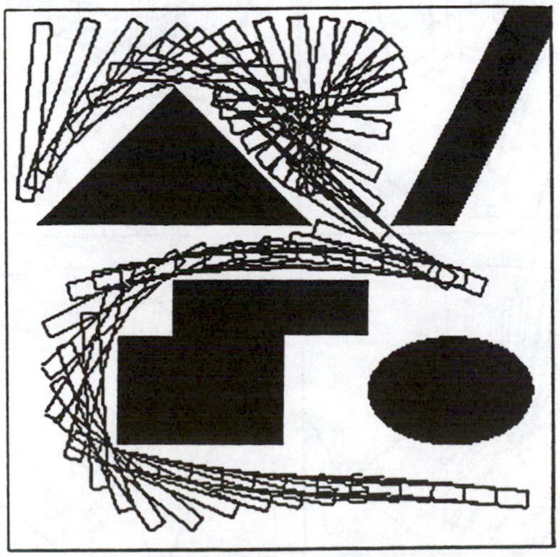

Fig. 8. Path generated by RPP for a 3-DOF rectangular robot.

A more formal expression of the search carried out by RPP is given in the following procedure:

```
procedure RPP-SEARCH;
begin
  τ ← GRADIENT-PATH(q_init); q_loc ← LAST(τ);
  while q_loc ∉ GOAL do
    begin
      ESCAPE ← false;
      for i = 1 to M until ESCAPE do
        begin
          t ← RANDOM-TIME;
          τ_i ← RANDOM-PATH(q_loc, t);
          q_rand ← LAST(τ_i);
          τ_i ← PRODUCT(τ_i, GRADIENT-
            PATH(q_rand));
          q'_loc ← LAST(τ_i);
          if U(q'_loc) < U(q_loc) then
            begin
              ESCAPE ← true;
              τ ← PRODUCT(τ, τ_i);
            end;
        end;
      if ¬ ESCAPE then
        begin
          τ ← BACKTRACK(τ, τ_1, . . . , τ_M);
          q_back ← LAST(τ);
          τ ← PRODUCT(τ, GRADIENT-
            PATH(q_back));
        end;
      q_loc ← LAST(τ);
    end;
end;
```

where:

- τ is a list of configurations representing the path constructed so far;
- LAST(τ) returns the last configuration in a path τ;
- PRODUCT(τ_1, τ_2) returns the list of configurations representing the concatenation $\tau_1 \bullet \tau_2$ of two paths τ_1 and τ_2;
- GRADIENT-PATH(q) returns the path generated by a gradient motion starting at q;
- RANDOM-PATH(q, t) returns the path generated by a random motion of duration t starting at q;
- RANDOM-TIME returns a random duration computed with the distribution defined in formula (4);
- BACKTRACK(τ, τ_1, . . . , τ_M) selects a backtracking configuration and returns the path from q_{init} to this configuration; if τ includes a subpath generated by a random motion, then the

returned path is a subpath of τ; otherwise it is a subpath of $\tau \bullet \tau_i$, with i randomly chosen in $[1, M]$.

The above search techniques have all been implemented, and the technique described in the procedure RPP-SEARCH gave the best experimental results.

5.5. Path Optimization

The path produced by the search of the local minima graph consists of a succession of gradient and random motions, both of them containing a large spectrum of spatial frequencies. To enable the robot to execute a graceful motion, the resulting path has to be smoothed. The smoothing procedure can be theoretically described as an optimization problem: Given an initial path τ_{init}, find a new path τ in the homotopy class of τ_{init} that minimizes the functional:

$$J(\tau) = \int_0^T K(\dot{\tau}(t))dt,$$

where K is the quadratic form of the kinetic energy under the obstacle avoidance constraint, and the two conditions $\tau(0) = q_{init}$ and $\tau(T) = q_{goal}$. To reduce the amount of computation, we use a simplified diagonal form for K, which corresponds to artificially decoupling the different degrees of freedom. The geodesics of the corresponding Riemannian metric are simply straight line segments in the generalized coordinate system $(q_1, \ldots, q_n)$.

Because any spatial frequency may be present in the initial path, it is highly preferable to use a multiscale technique for minimizing J. Our optimization procedure consists of iteratively modifying the path τ_{init} by replacing subpaths of decreasing lengths with straight line segments in the configuration space. It is necessary to check each of the straight line segments to ensure that it is collision free. The algorithm first checks long segments of the order of the total length of the path, and then smaller and smaller ones until the resolution of the configuration space grid is attained. The final path generated by this algorithm generally lies in the same homotopy class as the initial one.

5.6. Experimental Results

We have implemented RPP in a program written in C language that runs on a DEC 3100 MIPS-based workstation. Interestingly, the program only consists of about 1500 lines of code (but it does not include fancy inputs/outputs). We have experimented successfully with RPP using a variety of robot structures. We have also run the planner to generate paths for a PUMA robot in our laboratory and for a dual-arm system in the Stanford Aerospace Robotics Laboratory. We present here some of the most significant experiments. Because the algorithm contains several random components, neither the running time of the planner nor the generated solution are constant across several runs for the same problem. The times given below are typical times. It is not unusual for two running times for the same example to differ by a ratio of 5. The execution times would be both smaller and much more stable on a parallel architecture allowing the concurrent execution of several random motions.

3-DOF Rectangular Robot in 2D Work Space. Figure 8 shows a path generated by the planner for a holonomic rectangular robot in the plane. The resolution of the work space bitmap was 256^2. The computation time for this example was approximately 10 s, whereas the best-first planner (BFP) of section 4 takes only a second to solve this same problem.

6-DOF U-Shaped Robot in 3D Work Space. Figure 9 shows snapshots along a path generated by RPP for a U-shaped robot that can translate and rotate freely in a three-dimensional work space. The obstacles consist of a parallelepipedic block and a lattice. Because the work space is bounded, the robot must "maneuver" among the bars of the lattice.

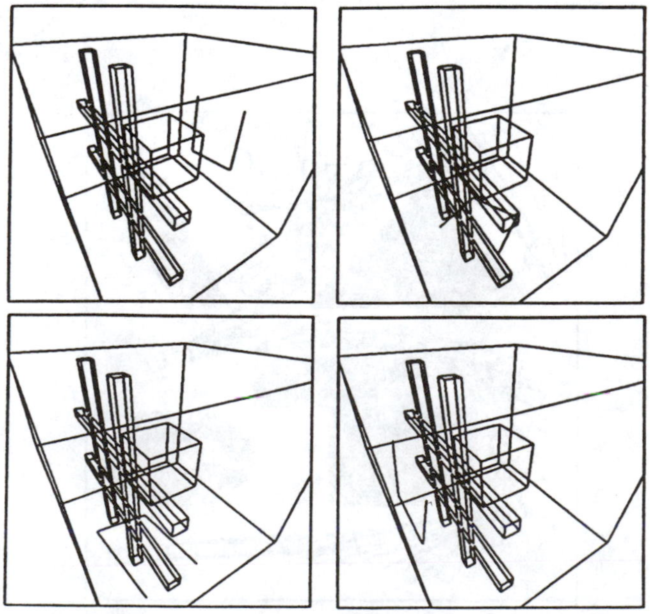

Fig. 9. Path generated by RPP for a 6-DOF U-shaped robot.

8-Revolute-DOF Serial Manipulator Robot in 2D Work Space. We ran RPP with the planar serial manipulator shown in Figure 10, which has eight revolute joints. Figures 11 and 12 show two paths generated by RPP. In both examples, the goal was defined by the position of the end point of the last link of the robot. The potential U was computed using this point as the only control point. The paths were generated in approximately 2 minutes for the first example and 30 seconds for the second, with a 216^2 work space bitmap. Collisions among the links were forbidden and checked by the planner.

10-DOF Nonserial Manipulator Robot in 2D Work Space. We also applied RPP to the planar nonserial manipulator robot shown in Figure 13, which includes three prismatic joints (telescopic links) and seven revolute joints. Figures 14 and 15 illustrate two different paths of the robot. In both examples, we used a potential U computed with two control points located at the end points of the two kinematic chains. Overlapping of the links was forbidden in the first example, but not in the second. The first example was solved in 3 minutes, whereas the second was solved in 2 minutes. Both paths were constructed with a 256^2 work space bitmap. The size of the corresponding configuration space grid is of the order of 10^{20} configurations.

31-DOF Nonserial Manipulator Robot in 3D Work Space. Continuing with our manipulator series, we experimented with RPP using the 31-DOF manipulator illustrated in Figure 16. This manipulator consists of 10 telescopic links connected by 10 spherical joints. The bar at the end of the manipulator is connected to the last link by a revolute joint. A path generated by RPP is illustrated in Figure 17. The potential was computed with two control points located at the end points of the bar. The computation time was of the order of 15 minutes. The size of the work space bitmap was 128^3. The size of the corresponding configuration space grid is of the order of 10^{62} configurations.

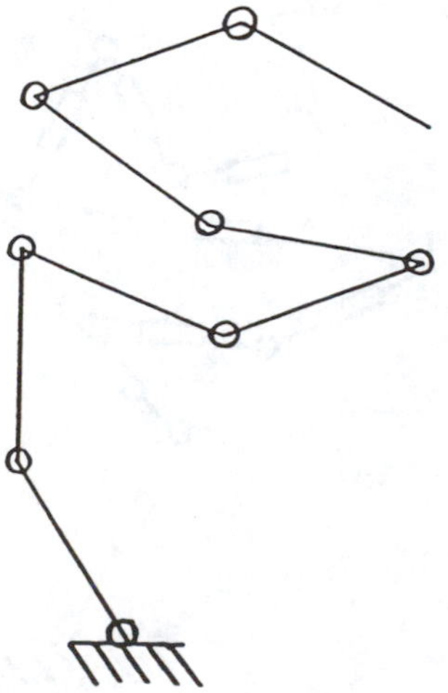

Fig. 10. Structure of the 8-DOF serial manipulator.

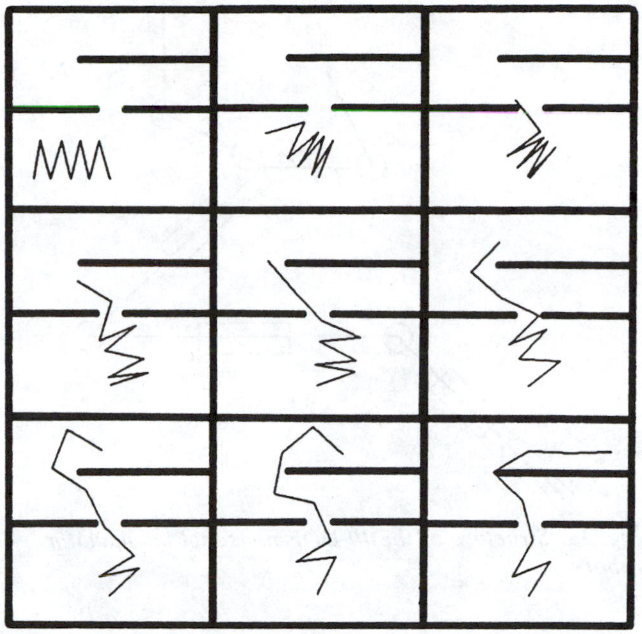

Fig. 11. Path generated by RPP for the 8-DOF serial manipulator (example 1).

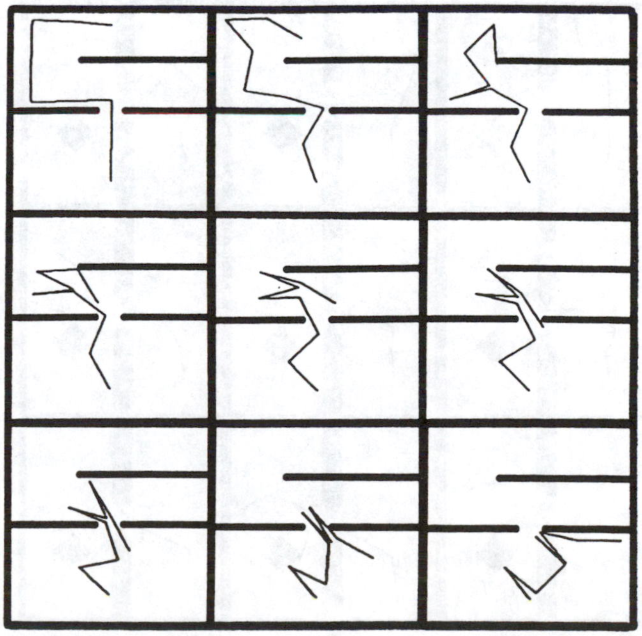

Fig. 12. Path generated by RPP for the 8-DOF serial manipulator (example 2).

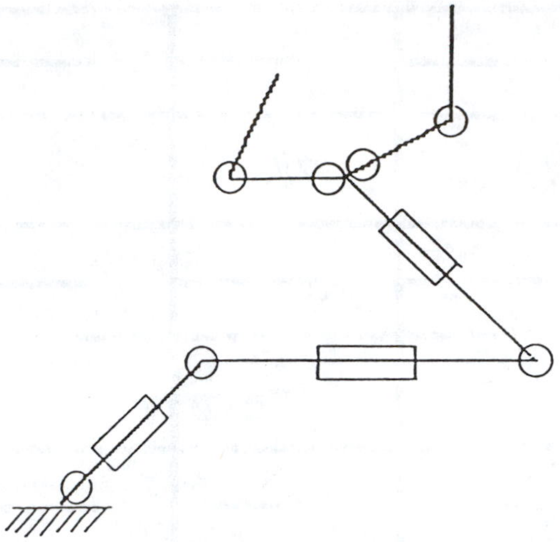

Fig. 13. Structure of the 10-DOF nonserial manipulator robot.

Coordination of Two 3-DOF Mobile Robots. The same planner was applied to problems requiring the coordination of two 3-DOF mobile robots in a two-dimensional work space made of several corridors. These are narrow enough so that the two robots cannot pass each other in the same corridor (Figure 18). The two robots are treated by RPP as a single two-body robot with 6 DOFs. Figures 19 and 20 display two paths generated by RPP for two different problems in the same environment. The second

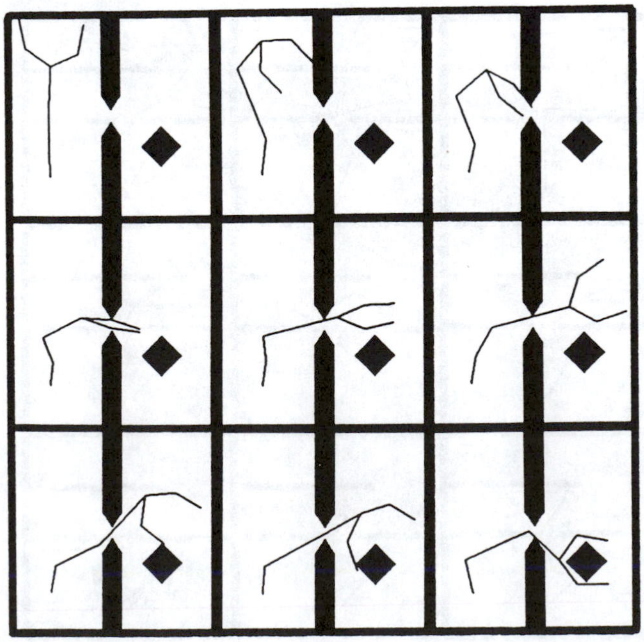

Fig. 14. Path generated by RPP for the 10-DOF nonserial manipulator (example 1).

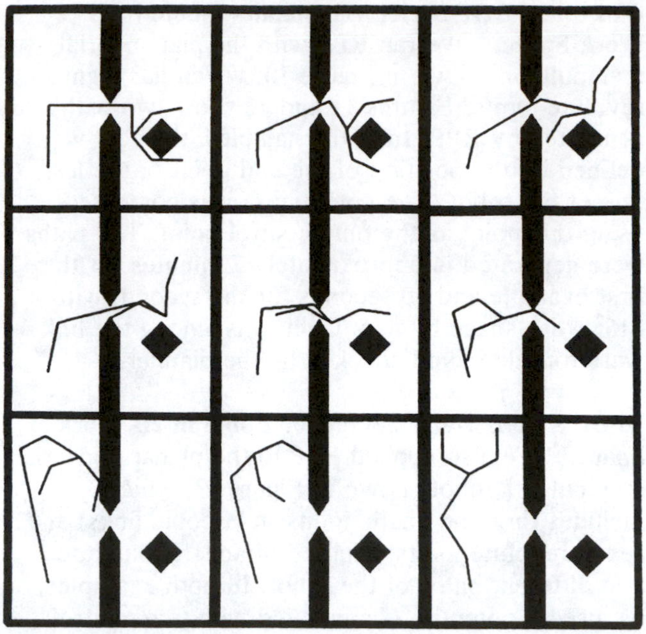

Fig. 15. Path generated by RPP for the 10-DOF nonserial manipulator (example 2).

problem is particularly difficult, because the two robots have to interchange their positions in the central corridor; hence both of them must first move to an intermediate position in order to allow the permutation. Notice that in the initial configuration, both robots are rather close to their respective goal configurations, despite the fact that the paths to move there are not short. This example illustrates the

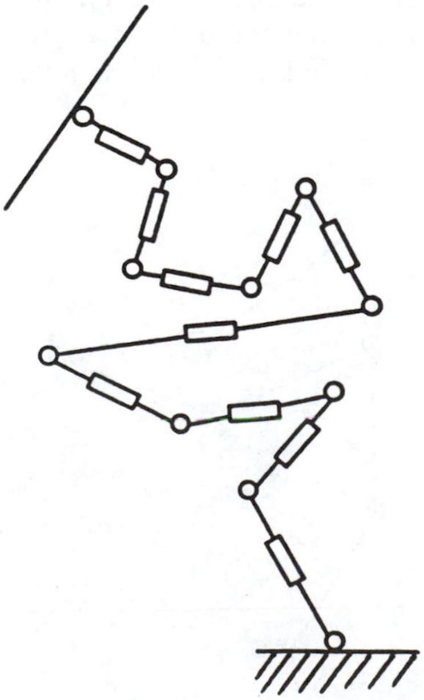

Fig. 16. Structure of the 31-DOF serial manipulator robot.

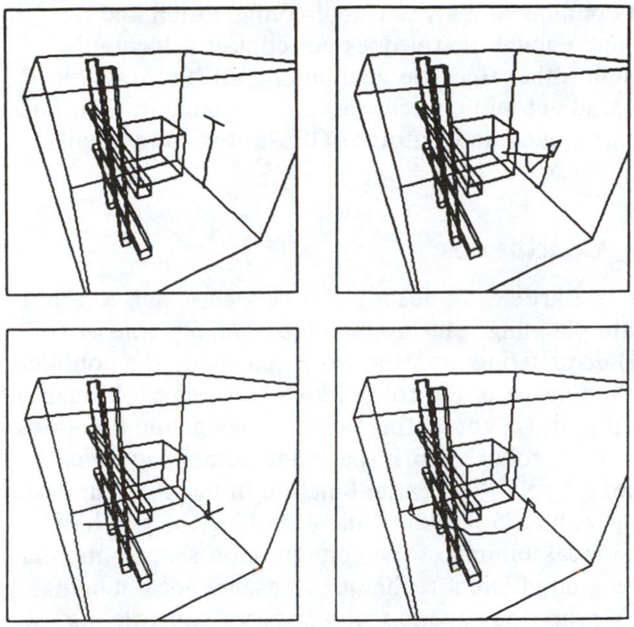

Fig. 17. Path generated by RPP for the 31-DOF serial manipulator.

power of the random techniques used in RPP. The two paths were generated in about 30 s.

5.7. Discussion

The experiments with RPP have shown that randomized planning is both efficient and reliable. The efficiency of RPP results from the fact that a typical path-planning problem has many solutions, so that a globally random search procedure can find one if it is well informed most of the time (by the potential

Fig. 18. The corridor problem for two 3-DOF mobile robots.

Fig. 19. Coordinated path for the corridor problem (example 1).

function). In fact, similar randomized techniques (e.g., "simulated annealing") have proven to be useful for solving other NP-hard problems—e.g., the traveling salesman problem (Cerny 1985) and VLSI placement and routing (Kirkpatrick et al. 1983; Sechen 1988). In these problems the very large search space is associated with a large number of "good" suboptimal solutions.

Fig. 20. Coordinated path for the corridor problem (example 2).

Monte Carlo procedures for optimization have also been used more or less successfully in Computer Vision. In Geman and Geman (1984), a simulated annealing approach is applied for restoring images blurred with nonlinear filters. Several authors have implemented edge detection algorithms based on the same paradigm. Recently, simulated annealing has also been applied to higher level problems in Computer Vision. In Barnard (1988) the stereo-matching problem is addressed using a hierarchical pixel-level simulated annealing algorithm. A stochastic optimization approach for the three-dimensional reconstruction of stratigraphic layers and the detection of geologic faults in seismic data is proposed in Barraquand (1988).

Nevertheless, RPP behaves very differently from the classic simulated annealing procedures. On a sequential computer, simulated annealing procedures perform a kind of breadth-first search of the graph of local minima of the function to be optimized, whereas RPP performs a depth-first search of this graph using the potential U as the heuristic function. See Barraquand and Latombe (1989b) for a more detailed comparison of BFP and simulated annealing.

Randomized planning has some drawbacks, however. The planner typically generates different paths if it is run several times with the same problem, and the running time varies from one run to another. Furthermore, if the input path-planning problem admits no solution, the planner usually has no way to recognize it, even after a large amount of computation. Hence a limit on the running time of the algorithm has to be imposed. However, if this limit is attained and no path has been generated, there is no guarantee that no paths exist. However, in practice, experiments have shown that it is not difficult, for a given class of problem (e.g., an object moving in a three-dimensional work space) and a given size of the configuration space grid, to determine a time limit (through a series of preliminary trials) such that if no path has been found by the end of the time limit, there is little chance that one actually exists.

In Barraquand and Latombe (1989b), we proved that the randomized planning algorithm implemented in RPP is probabilistically resolution complete. This result is based on a general property of the Wiener-Levy process: Whenever the free space is connected and relatively compact, the probability for a Brownian motion ω with reflective boundary starting at any initial configuration $q_{init} \in \mathscr{C}_{free}$ to reach any given open subset B of $\mathscr{C}_{free}$ at least once during the interval of time $[0, t]$ converges toward 1 when the duration t tends toward infinity. We can choose B so that it contains at least one goal configuration and is small enough that it does not contain a local minimum (other than the goal minimum). From within B, a gradient motion achieves a goal configuration. See Barraquand and Latombe (1989b) for more details.

6. Conclusions

In this article we described a new approach to robot path planning. This approach essentially consists of (1) discretizing both the work space and the configuration space of the robot into a hierarchical bitmap and grid; (2) computing numeric navigation functions over the robot's work space and combining them into a "good" potential function in the configuration space; and (3) building and searching the graph of the local minima of the configuration space potential using an efficient technique to escape local minima. We proposed several techniques for constructing the potential function and two techniques for escaping local minima. The most powerful of these two techniques, which is applicable to robots with many DOFs, is a stochastic process technique based on the execution of Brownian motions with reflective boundary.

We have implemented our approach and the various techniques presented in this article in two programs, BFP and RPP, which we ran successfully on many different examples. BFP has solved 3-DOF robot problems also solved by previous planners, but several orders of magnitude faster. It has also solved problems with may obstacles of arbitrary shapes that were never attempted before. On the other hand, RPP has solved a large variety of problems that fall far outside the range of the capabilities of any other previous planner (e.g., problems with 8-DOF, 10-DOF, and 31-DOF robots and multirobot problems).

The algorithms implemented in RPP are highly parallelizable. A preliminary investigation of the parallelization of the planner has been conducted in Barraquand and Latombe (1989b) and in Métivier and Urbschat (1990). We envision an implementation of RPP using a specific hardware system, which should permit real-time path planning. Such a system would open new perspectives on some key issues in robotics related to the interaction of planning and execution in partially known and dynamically changing environments.

In addition to implementing a real-time planner, we currently conduct research aimed at using the randomized planning approach for planning manipulation tasks (Alami et al. 1989) involving grasping and regrasping operations on movable objects with multiple robots.

Acknowledgments

This research was funded by DARPA contract DAAA21-89-C0002 (U.S. Army), DARPA contract N00014-88-K-0620 (Office of Naval Research), SIMA (Stanford Institute of Manufacturing and Automation), CIFE (Center for Integrated Facility Engineering), and Digital Equipment Corporation. The authors also thank Professor J. M. Harrison (Stanford Graduate School of Business) and J. Chang (Stanford Statistical Consulting Service) for their helpful advice on Brownian motions with reflective boundaries. Bruno Langlois also provided useful suggestions.

References

Aho, A. V., Hopcroft, J. E., and Ullman, J. D. 1983. *Data Structures and Algorithms*. Reading, MA: Addison-Wesley.

Alami, R., Siméon, T., and Laumond, J. P. 1989. A geometrical approach to planning manipulation tasks—the case of discrete placements and grasps. In Miura, H., and Arimoto, S. (eds.): *Robotics Research 5*. Cambridge, MA: MIT Press, pp. 453–463.

Anderson, R. F., and Orey, S. 1976. Small random perturbations of dynamical systems with reflecting boundary. *Nagoya Math. J.* 60:189–216.

Barnard, S. T. 1988. Stochastic stereo matching over scale. *Int. J. Computer Vision*. 2(4).

Barraquand, J. 1988. Markovian random fields in computer vision: Applications to seismic data understanding. Ph.D. dissertation. Dept. of Computer Vision, INRIA, Sophia-Antipolis, France. In French.

Barraquand, J., Langlois, B., and Latombe, J. C. 1989. Robot motion planning with many degrees of freedom and dynamic constraints. In Miura, H., and Arimoto, S. (eds.): *Robotics Research 5*. Cambridge, MA: MIT Press, pp. 435–444.

Barraquand, J., and Latombe, J. C. 1989a. On nonholonomic mobile robots and optimal maneuvering. *Revue d'Intelligence Artificielle* 3(2):77–103. (Also in *Proc. of the 4th IEEE Int. Symp. on Intelligent Control*, Albany, NY: pp. 340–347.)

Barraquand, J., and Latombe, J. C. 1989b. *Robot Motion Planning: A Distributed Representation Approach*. Report no. STAN-CS-89-1257. Dept. of Computer Science, Stanford University.

Barraquand, J., and Latombe, J. C. 1990. Controllability of mobile robots with kinematic constraints. Rep. no. STAN-CS-90-1317. Dept. of Computer Science, Stanford University.

Brooks, R. A., and Lozano-Pérez, T. 1983 (Karlsruhe). A subdivision algorithm in configuration space for findpath with rotation. *Proc. of the 8th Int. Joint Conf. on Artificial Intelligence*, pp. 799–806.

Canny, J. F. 1988. The Complexity of Robot Motion Planning. Cambridge, MA: MIT Press.

Cerny, V. 1985. Thermodynamical approach to the traveling salesman problem: An efficient simulation algorithm. *J. Optimization Theory Applications*. 45(1):41–51.

Donald, B. R. 1984. Motion planning with six degrees of freedom. Tech. rep. 791. Artificial Intelligence Laboratory, MIT.

Erdmann, M., and Lozano-Pérez, T. 1986. On multiple moving objects. AI memo no. 883. Artificial Intelligence Laboratory, MIT.

Faverjon, B. 1984 (Atlanta). Obstacle avoidance using an octree in the configuration space of a manipulator. *Proc. of the IEEE Int. Conf. on Robotics and Automation*, pp. 504–512.

Faverjon, B. 1986 (San Francisco). Object level programming of industrial robots. *Proc. of the IEEE Int. Conf. on Robotics and Automation*, pp. 1406–1412.

Faverjon, B., and Tournassoud, P. 1987 (Raleigh, NC). A local based approach for path planning of manipulators with a high number of degrees of freedom. *Proc. of the IEEE Int. Conf. on Automation and Robotics*, pp. 1152–1159.

Geman, D., and Geman, S. 1984. Stochastic relaxation, Gibbs distributions, and the Bayesian restoration of images. *IEEE Trans. Pattern Analysis Machine Intelligence* PAMI-6:721–741.

Geman, S., and Hwang, C. R. 1986. Diffusions for global optimization. *SIAM J. Control Optimization* 24(5).

Gouzènes, L. 1984. Strategies for solving collision-free trajectories problems for mobile and manipulator robots. *Int. J. Robot. Res.* 3(4):51–65.

Kant, K., and Zucker, S. W. 1986. Toward efficient trajectory planning: Path velocity decomposition. *Int. J. Robot. Res.* 5:72–89.

Khatib, O. 1986. Real-time obstacle avoidance for manipulators and mobile robots. *Int. J. Robot. Res.* 5(1):90–98.

Kirkpatrick, S., Gelatt, C. D., and Vecchi, M. P. 1983. Optimization by simulated annealing. *Science* 220:671–680.

Koditschek, D. E. 1987 (Raleigh, NC). Exact robot navigation by means of potential functions: Some topological considerations. *Proc. of the IEEE Int. Conf. on Robotics and Automation*, pp. 1–6.

Latombe, J. C. 1990. *Robot Motion Planning*. Boston: Kluwer Academic Publishers.

Laugier, C., and Germain, F. 1985 (Tokyo). An adaptative collision-free trajectory planner. *Int. Conf. on Advanced Robotics*.

Lee, D. T., and Drysdale, R. L. 1981. Generalization of Voronoi diagrams in the plane. *SIAM J. Computing* 10:73–87.

Lozano-Pérez, T. 1983. Spatial planning: A configuration space approach. *IEEE Trans. Computers* C-32(2):108–120.

Lozano-Pérez, T. 1987. A simple motion-planning algorithm for general robot manipulators. *IEEE J. Robot. Automat.* RA-3(3):224–238.

Lumelsky, V. 1987 (Los Angeles). Algorithmic issues of sensor-based robot motion planning. *Proc. of the 26th IEEE Conf. on Decision and Control*, pp. 1796–1801.

Métivier, C., and Urbschat, R. 1990. Run-time statistical analysis of a robot motion planning algorithm. Internal technical note. Robotics Laboratory, Dept. of Computer Science, Stanford University.

Nilsson, N. J. 1980. *Principles of Artificial Intelligence*. Los Altos, California: Morgan Kaufmann.

O'Donnell, P. A., and Lozano-Pérez, T. 1989 (Scottsdale, AZ). Deadlock-free and collision-free coordination of two robot manipulators. *Proc. of the IEEE Int. Conf. on Robotics and Automation*, pp. 484–489.

Ó'Dúnlaing, C., Sharir, M., and Yap, C. K. 1983 (Boston). Retraction: A new approach to motion planning. *Proc. of the 15th ACM Symp. on the Theory of Computing*, pp. 207–220.

Papoulis, A. 1965. *Probability, Random Variables, and Stochastic Processes*. New York: McGraw-Hill.

Reif, J. H. 1979. Complexity of the mover's problem and generalizations. *Proc. of the 20th Symp. on the Foundations of Computer Science*, pp. 421–427.

Rimon, E., and Koditschek, D. E. 1989 (Scottsdale, AZ). The construction of analytic diffeomorphisms for exact robot navigation on star worlds. *Proc. of the IEEE Int. Conf. on Robotics and Automation*, pp. 21–26.

Schwartz, J. T., and Sharir, M. 1983a. On the piano movers' problem: I. The case of a two-dimensional rigid polygonal body moving amidst polygonal barriers. *Comm. Pure Applied Math.* 36:345–398.

Schwartz, J. T., and Sharir, M. 1983b. On the piano movers' problem: II. General techniques for computing topological properties of real algebraic manifolds. *Adv. Applied Math.* 4:298–351.

Schwartz, J. T., and Sharir, M. 1983c. On the piano movers' problem: III. Coordinating the motion of several independent bodies: The special case of circular bodies moving amidst polygonal barriers. *Int. J. Robot. Res.* 2(3):46–75.

Schwartz, J. T., and Sharir, M. 1988. A survey of motion planning and related geometric algorithms. *Artificial Intelligence* 37(1–3):157–169.

Schwartz, J. T., Sharir, M., and Hopcroft, J. 1987. *Planning, Geometry, and Complexity of Robot Motion*. Norwood, NJ: Ablex.

Sechen, C. 1988. *VLSI Placement and Global Routing Using Simulated Annealing*. Boston, MA: Kluwer Academic Publishers.

Serra, J. 1982. *Image Analysis and Mathematical Morphology*. New York: Academic Press.

Zhu, D., and Latombe, J. C. 1989. New heuristic algorithms for efficient hierarchical path planning. Rep. no. STAN-CS-89-1279. Dept. of Computer Science, Stanford University.

Part 2
Geometric and Asymptotic Methods in Control

MARK W. SPONG
University of Illinois at Urbana-Champaign

THE HIGHLY nonlinear nature of robot dynamics means that standard techniques from linear control theory usually give local results at best. In order to obtain global results, more advanced and quite different techniques from nonlinear control theory are required. This part deals with such techniques, chiefly differential geometric methods and singular perturbation techniques.

Historically, one of the first results in nonlinear control of robot manipulators was the so-called *method of computed torque* [1], [2], [3]. More recently this concept has been used interchangeably with the concepts of *feedback linearization* and *inverse dynamics* [4]. In these techniques a nonlinear control law is computed either in a feedforward (computed torque) or a feedback (feedback linearization, inverse dynamics) form with the intent of linearizing and decoupling the nonlinear robot dynamics. In equation form, if we express the Euler-Lagrange equations for a rigid robot manipulator as

$$\boldsymbol{M}(q)\ddot{q} + \boldsymbol{C}(q,\dot{q})\dot{q} + \boldsymbol{g}(q) = \boldsymbol{\tau} \tag{1}$$

then the inverse dynamics control law for $\boldsymbol{\tau}$ is given as

$$\boldsymbol{\tau} = \boldsymbol{M}(q)a + \boldsymbol{C}(q,\dot{q}) + \boldsymbol{g}(q) \tag{2}$$

where a is an additional (sometimes called outer loop [4]) signal to be designed. Substituting (2) into (1) yields

$$\boldsymbol{M}(q)\ddot{q} = \boldsymbol{M}(q)a \tag{3}$$

which, since the inertia matrix $\boldsymbol{M}(q)$ is invertible, results in the linear second-order system

$$\ddot{q} = a \tag{4}$$

Equation (4) represents n decoupled linear second-order systems each with transfer function

$$G(s) = \frac{1}{s^2} \tag{5}$$

and is therefore called the *double integrator system*. It is now obvious that the signal a whose design completes the design of the overall control law (2) can be designed using standard linear methods based on the linear model (4). Related to this notion have been several other results such as *resolved acceleration* [5], *operational space control* [6], and differential geometric control, which attempt through nonlinear feedback to linearize the manipulator dynamics with respect to a task frame. The double integrator model can also be used as a basis for designing robust control algorithms that rely on the satisfaction of so-called *matching conditions* [7], [8]. More discussion of this can be found in the papers in this part and in the references and bibliography that follow.

The first paper in this part, "On Manipulator Control by Exact Linearization," by K. Kreutz, generalizes the idea of computed torque and serves to synthesize the various results from the research literature into a unified framework. The paper is important because it shows in a clear and readable style how all of the various algorithms that have been developed to linearize the nonlinear rigid robot dynamics are fundamentally the same or at least derivable as special cases of a single unifying algorithm.

The second paper, "Nonlinear Feedback in Robot Arm Control," by T. J. Tarn et al., is included for the background it gives in differential geometric methods applied to robot control. As the paper by Kreutz points out, an important design concept is emphasized in this paper; the separation of the control algorithm into a feedback linearization/decoupling loop, a nominal pole assignment loop, and a final feedback control term to enhance robustness to disturbances and unmodeled dynamics.

The third paper, "Dynamic Control of Robots with Joint Elasticity," by A. DeLuca illustrates several important features of both the robot control problem and feedback linearization. First, the paper shows that even a small amount of flexibility at the joints can destroy the feedback linearizability property of rigid robots (when only a static state feedback law is employed). This is an important observation in practice because, in fact, joint flexibility, whether caused by harmonic drive transmissions, elasticity in cables, hydraulic lines, or other factors, is known to be a significant factor in the dynamics of many industrial robot arms. The paper also shows that it is still possible to linearize the dynamics of the robot globally, provided *dynamic feedback control* is employed. The resulting controller is considerably more complex than the linearizing controllers designed for rigid robots.

The fourth paper, "Modeling and Control of Elastic Joint Robots," by M. W. Spong, illustrates the importance of modeling as a first step in control system design. It shows that by making a few modeling assumptions, justified on physical grounds, it is possible to derive a model of elastic joint robot dynamics whose complexity is nearly the same as that of rigid robots and that is once again feedback linearizable using static-state feedback together with a nonlinear coordinate transformation. The paper also shows the usefulness of so-called *asymptotic* methods of singular perturbation theory for the flexible-joint control problem. In practice, the joint-stiffness parameters of a robot arm are considerably larger than the other parameters representing such things as inertias. This means that the rigid robot can be viewed as a singular perturbation of the flexible-joint robot in a sense made precise by singular perturbation theory. Algorithms designed for rigid robots can therefore be incorporated as part of so-called *composite* and *corrective* control algorithms for flexible-joint robots, which greatly simplifies both the design and complexity of controllers for such systems.

REFERENCES

[1] B. R. Markiewicz, "Analysis of the computed-torque drive method and comparison with conventional position servo for a computer-controlled manipulator," *Jet Propulsion Lab. Rep.* JPL TM 33-669, Feb. 1974.

[2] R. C. Paul, "Modeling, trajectory calculation, and servoing of a computer-controlled arm," *Stanford A. I. Lab. Memo 177*, Nov. 1972.

[3] A. K. Bejczy, "Robot arm dynamics and control," *Jet Propulsion Lab. Rep.* JPL TM 33-669, Feb. 1974.

[4] M. W. Spong and M. Vidyasagar, *Robot Dynamics and Control,* New York: Wiley, 1989.

[5] J. Y. S. Luh, M. W. Walker, and R. P. Paul, "Resolved-acceleration control of mechanical manipulators," *IEEE Trans. Automat. Contr.,* vol. AC-25, pp. 468–474, 1980.

[6] O. Khatib, "Dynamic control of manipulators in operational space," in *Proc. 6th IFTOMM Congress on Theory of Machines and Mechanisms,* New Delhi, pp. 1128–1131, 1983.

[7] J.-J. E. Slotine, "Robust control of robot manipulators," *Int. J. Robotics Res.,* vol. 4, no. 2, pp. 49–64, 1985.

[8] M. J. Corless and G. Leitman, "Continuous state feedback guaranteeing uniform ultimate boundedness for uncertain dynamic systems," *IEEE Trans. Automat. Contr.,* vol. AC-26, pp. 1139–1144, 1981.

BIBLIOGRAPHY

1. C. Abdallah, D. Dawson, P. Dorato, and M. Jamshidi, "Survey of robust control for rigid robots," *IEEE Cont. Syst. Magazine,* vol. 11, no. 2, pp. 24–30, Feb. 1991.
2. C. Abdallah and R. Jordan, "A positive-real design for robotic manipulators," *Proc. IEEE American Cont. Conf.,* San Diego, CA, May 1990.
3. R. J. Anderson, "Passive computed-torque algorithms for robots," *Proc. IEEE Conf. Dec. and Cont.,* Tampa, FL, pp. 1638–1644, Dec. 1989.
4. S. Arimoto and F. Miyazaki, "Stability and robustness of PID feedback control for robot manipulators of sensory capability," *Proc. 1st Int. Symp. Robotics Res.,* pp. 783–799, 1983.
5. E. Bailey and A. Arapostathis, "Simple sliding mode control scheme applied to robot manipulators," *Int. J. Contr.,* vol. 45, pp. 1197–1209, 1987.
6. N. Becker and W. M. Grimm, "On L_2 and $L_i nf$ stability approaches for the robust control of robot manipulators," *IEEE Trans. Automat. Contr.,* vol. 33, pp. 118–122, Jan. 1988.
7. G. Cesareo and R. Marino, "On the controllability properties of elastic robots," presented at the 6th Int. Conf. on Anal. and Optimiz. Syst., INRIA, Nice, June 1984.
8. Y.-F. Chen, T. Mita, and S. Wakui, "A new and simple algorithm for sliding mode trajectory control of the robot arm," *IEEE Trans. Automat. Contr.,* vol. AC-35, pp. 828–829, July 1990.
9. M. J. Corless, "Tracking controllers for uncertain systems: application to a Manutec R3 robot," *J. Dyn. Sys., Meas., and Cont.,* vol. 111, pp. 609–618, Dec. 1989.
10. V. Cvetkovic and M. Vukobratovic, "One robust, dynamic control algorithm for manipulation systems," *Int. J. Robotics Res.,* vol. 1, pp. 15–28, 1982.
11. E. J. Davison and A. Goldenberg, "The robust control of a general servomechanism problem: the servo compensator," *Automatica,* vol. 11, pp. 461–471, 1975.
12. R. A. DeCarlo, S. Zak, and G. P. Matthews, "Variable structure control of nonlinear multivariable systems," *IEEE Proc.,* vol. 76, pp. 212–232, March 1988.
13. P. Dorato, Ed., *Robust Control,* New York: IEEE Press, 1987.
14. M. S. Fadali, M. Zohdy, and B. Adamczyk, "Robust pole assignment for computed-torque robotic manipulators control," *Proc. IEEE American Cont. Conf.,* pp. 37–41, June 1989.
15. E. Freund, "Fast nonlinear control with arbitrary pole-placement for industrial robots and manipulators," *Int. J. Robotics Res.,* vol. 3, pp. 76–86, 1982.
16. E. G. Gilbert and I. J. Ha, "An approach to nonlinear feedback control with applications to robotics," *IEEE Trans. Sys., Man, and Cyber.,* vol. SMC-14, pp. 879–884, Nov./Dec. 1984.
17. S. Gutman, "Uncertain dynamic systems—A Lyapunov min-max approach," *IEEE Trans. Automat. Contr.,* vol. AC-24, pp. 437–443, 1979.
18. H. Hemami and P. C. Camana, "Nonlinear feedback in simple locomotion systems," *IEEE Trans. Automat. Contr.,* vol. AC-21, pp. 855–860, 1976.
19. J. M. Hollerbach and G. Sahar, "Wrist-partitioned inverse kinematic accelerations and manipulator dynamics," *Int. J. Robotics Res.,* vol. 2, pp. 61–76, 1983.
20. L. R. Hunt, R. Su, and G. Meyer, "Global transformations of nonlinear systems," *IEEE Trans. Automat. Contr.,* vol. AC-28, pp. 24–31, 1983.
21. D. Koditschek, "Natural control of robot arms," *Yale University Center for Syst. Sci. Rep.* 1985.
22. C. S. G. Lee, R. C. Gonzales, and K. S. Fu, *Tutorial on Robotics,* 2nd ed., Silver Spring, MD: IEEE Computer Society Press, 1986.
23. J. Y. S. Luh, M. W. Walker, and R. P. Paul, "On-line computational scheme for mechanical manipulators," *J. Dyn. Syst. Meas. and Cont.,* vol. 102, pp. 69–76, 1980.
24. R. Ortega and M. W. Spong, "Adaptive motion control of rigid robots: a tutorial," *Proc. IEEE Conf. Dec. and Cont.,* pp. 1575–1584, Dec. 1988.
25. C. Samson, "Robust nonlinear control of robotic manipulators," *Proc. 22nd IEEE Conf. Dec. Cont.,* pp. 1211–1216, Dec. 1983.
26. S. N. Singh, "Adaptive model following control of nonlinear robotic systems," *IEEE Trans. Automat. Contr.,* vol. AC-30, pp. 1099–1100, Nov. 1985.
27. J-J. E. Slotine and S. S. Sastry, "Tracking control of nonlinear systems using sliding surfaces with applications to robot manipulators," *Int. J. Cont.,* vol. 38, pp. 465–492, 1983.
28. M. W. Spong, "Control of flexible-joint robots: a survey," *Rep. UILU-ENG-90-2203, DC-116,* Coordinated Science Laboratory, University of Illinois at Urbana-Champaign, Feb. 1990.
29. M. W. Spong, J. S. Thorp, and J. M. Kleinwaks, "Robust micropro-

cessor control of robot manipulators," *Automatica*, vol. 23, pp. 373–379, 1987.

30. M. W. Spong and M. Vidyasagar, "Robust linear compensator design for nonlinear robotic control," *Proc. IEEE Int. Conf. on Robotics and Automation*, pp. 954–959, March, 1985.
31. M. W. Spong and M. Vidyasagar, "Robust linear compensator design for nonlinear robotic control," *IEEE J. Robotics and Automation*, vol. RA-3, pp. 345–351, Aug. 1987.
32. M. Takegaki and S. Arimoto, "A new feedback method for dynamic control of manipulators," *J. Dyn. Sys., Meas., and Cont.*, vol. 102, pp. 119–125, June 1981.
33. V. I. Utkin, "Variable structure systems with sliding modes," *IEEE Trans. Automat. Contr.*, vol. AC-22, pp. 212–222, April 1977.
34. M. Vidyasagar, *Control Systems Synthesis: A Factorization Approach*, Cambridge, MA: MIT Press, 1985.
35. K. S. Yeung and Y. P. Chen, "A new controller design for manipulators using the theory of variable structure systems," *IEEE Trans. Automat. Contr.*, vol. AC-33, pp. 200–206, Feb. 1988.
36. K-K. D. Young, "Controller design for a manipulator using theory of variable structure systems," *IEEE Trans. Syst., Man, and Cyber.*, vol. SMC-8, pp. 210–218, Feb. 1978.

On Manipulator Control by Exact Linearization

KENNETH KREUTZ

Abstract—**Comments on the application to rigid link manipulators of geometric control theory, resolved acceleration control, operational space control, and nonlinear decoupling theory are given, and the essential unity of these techniques for externally linearizing and decoupling end effector dynamics is discussed. Exploiting the fact that the mass matrix of a rigid link manipulator is positive definite, and the fact that there is an independent input for each degree of freedom, it is shown that a necessary and sufficient condition for a locally externally linearizing and output decoupling feedback law to exist is that the end effector Jacobian matrix be nonsingular.**

I. Introduction

An "exactly linearizing" control makes a nonlinear system behave as if it has linear and decoupled dynamics. It has been known at least since the early 1970's [1]–[5] that exact linearization of manipulators in joint space is accomplished by the so-called inverse or computed torque technique. Efforts to accomplish decoupled linearization of end effector (EE) motions directly in task space began soon thereafter [6]–[14].

The work [6] is concerned with controlling the tip location of a three-link manipulator in the plane, and proceeds by the three explicit steps of: 1) decoupled linearization of tip behavior; 2) stabilization of the resulting tip dynamics; followed by 3) trajectory control of the now linearly behaving tip. Such clarity of approach will only be retrieved in the latter work of [19]–[22]. Reference [6] also presages future work in its dealing with the problems of manipulator redundancy and actuator saturation.

Resolved acceleration control (RAC) is developed in [7], [8]. RAC essentially extends the work of [6] to a six dof manipulator yielding linearized EE positional dynamics and "almost" linearized EE attitude dynamics (see Section III). This work did not make clear the three steps of [6] and consequently appears not to have been appreciated as a technique for exact linearization of EE motions. The fact that the attitude dynamics are not completely linearized also helped to obscure the appreciation of RAC as an exactly linearizing control technique.

References [9]–[11] apply nonlinear decoupling theory (NDT) to obtain decoupled linearization of a manipulator EE with simultaneous pole placement of the linearized EE dynamics. The simultaneous pole placement and linearization of EE dynamics is a blurring of the distinct steps 1) and 2) described above for the approach [6]. In [23] correspondences of NDT to RAC and the computed torque technique are discussed.

In [12]–[14], manipulator dynamics are expressed in the task space, or operational space of the EE. The resulting nonlinear end effector dynamics are then linearized by the computed torque method. Thus, the operational space control (OCS) of [12]–[14] can be viewed as a generalized computed torque technique. In [12] correspondences to RAC and the computed torque technique are noted.

Recently, geometric control theory (GCT) based techniques for exactly externally linearizing and decoupling general affine-in-the-input nonlinear systems have been developed [15]–[19]. References [15]–[19] extend the idea of feedback linearization for linearizing system state equations (the original problem considered by GCT) to include exact linearization of the input–output equations (hence, "external" linearization). These references give constructive sufficient conditions for local decoupled external linearization which produce the linearizing feedback law. Applications to manipulator control are given in [19]–[22], along with a clear control perspective which keeps the following steps distinct: 1) exactly linearize and decouple end effector dynamics to a canonical decoupled double integrator form, i.e., to Brunovsky canonical form (BCF); 2) effect a stabilizing loop (pole placement step); 3) perform feedforward precompensation to obtain nominal model following performance; 4) institute an error correcting feedback loop.

RAC, NDT, OCS, and GCT can be shown to give the same linearizing control law for exact external linearization and decoupling of EE motions [24]. (This equivalence is specific to the nonlinear systems considered here, viz. systems dynamically similar to rigid link manipulators. NDT and GCT apply to a much larger class than this, and so the equivalence to RAC and OSC holds for systems restricted to this class but not in general.) For example, the approaches discussed above are recovered within the GCT framework of [15]–[22] by making an appropriate choice of states and outputs and applying an output linearizing control. With q giving manipulator joint variables and y the EE position and orientation, the following associations can be made.

a) Computed torque method: state $x = (q, \dot{q})$, output q.

b) NDT, GCT, and (in a sense discussed in Section III) RAC: state $x = (q, \dot{q})$, output $y = h(q)$, where $h(\cdot)$ gives the manipulator forward kinematics.

c) OSC: state $x = (h(q), J\dot{q}) = (y, \dot{y})$, output $y = h(q)$, where J is the manipulator Jacobian matrix.

In this note these equivalences are discussed and a simple form for the linearizing control is given.

II. Dynamics of Finite-Dimensional Natural Systems

Many physical systems have finite-dimensional nonlinear dynamics of the form [25], [26]

$$M(q)\ddot{q}+C(q, \dot{q})=\tau; \qquad \dot{q}, \ddot{q} \in R^n;$$
$$M(q) \in R^{n\times n}; \qquad M(q)=M^T(q)>0, \forall q \tag{1}$$

where q evolves on a manifold of dimension n, $q \in \mathfrak{M}^n$. For example, $q \in R^n$ for a Cartesian manipulator, while $q \in T^n$ for a revolute manipulator. Typically (1) arises as a solution to the Lagrange equations $(d/dt)(\partial L/\partial \dot{q})L - (\partial L/\partial q) = Q^T$ where $L = T - U$, $T = 1/2\dot{q}^T M(q)\dot{q}$ is positive definite and autonomous, U is a conservative potential function, $Q = \tau + F$ are generalized forces, and F are dissipative or constraint forces. Manipulator dynamics can be obtained in this way and hence have the form (1). Such systems are known as natural systems [25], [26]. Not only is $M(q)$ positive definite for these systems, but $C(q, \dot{q})$ of (1) has terms which depend on $M(q)$ in a very special way [27]–[29]. In fact, natural systems are nongeneric in the class of all affine-in-the-input nonlinear systems [38], [39]. In addition to (1) describing the dynamics of a natural system, in (1) τ provides a direct independent input for each configuration degree of freedom—this is an additional assumption which we have made since our concern is with rigid link manipulators for which every joint can be actuated. Although we only exploit the fact that $M(q)$ is invertible for any q, and the fact that τ directly influences the system configuration degrees of freedom, it should be noted that the nongeneric structure of (1) has recently enabled important statements to be made on the existence of time optimal control laws [38]–[40], on the existence of globally stable control laws [27]–[33], on the existence of robust exponentially stable control laws [34], and on the existence of stable adaptive control laws [35]–[37] for the natural system (1). Recognizing the special properties of (1), it is not surprising that results yielding external linearizing behavior can be obtained much more easily than by application of NDT or GCT—theories which apply to the whole general class of smooth affine-in-the-input nonlinear systems.

Manuscript received May 28, 1987; revised May 26, 1988. Paper recommended by Associate Editor, T. J. Tarn. This work was supported by the National Aeronautics and Space Administration.

The author is with the Jet Propulsion Laboratory, California Institute of Technology, Pasadena, CA 91109.

IEEE Log Number 8927765.

Reprinted from *IEEE Trans. Automat. Contr.*, vol. 34, no. 7, pp. 763–767, July 1989.

III. End Effector Kinematics and Control After Linearization

The system (1) is assumed to have a read-out map of either the form

$$y=h(q) \in R^m, \; V=\dot{y}=J(q)\dot{q} \in R^m, \; J(q)=\frac{\partial h}{\partial q} \in R^{m\times n} \tag{2}$$

or the more general form

$$y=h(q) \in \mathfrak{M}^m, \; V=J_0(q)\dot{q} \in R^m, \; J_0(q) \in R^{m\times n} \tag{3}$$

where in general $J_0 dq = V\,dt$ is not an exact differential [25], [26], $h(\cdot)$ is c^2 [44], [47] and defined on the manipulator configuration manifold $\mathfrak{M}^n$, $\mathfrak{M}^m$ is some m-dimensional output manifold, J or J_0 is c^1, and m and n can have different values. Often $h(\cdot)$ is smooth (i.e., c^∞) or even a diffeomorphism when the domain is suitably restricted. In subsequent discussion $V = \bar{J}\dot{q}$ will mean that $\bar{J}$ can be either J or J_0. Let the state of system (1) be $(q, \dot{q})$. Then for $y = h(q)$, $V = \bar{J}(q)\dot{q}$ will be called the "velocity associated with the output y." Note that (2) is a special case of (3) where V, the velocity associated with y, is just $V = \dot{y}$ and $\mathfrak{M}^m = R^m$ giving $\bar{J} = J = \partial h/\partial q$. Also note that for the case (3), since h is c^2, it is still meaningful to talk about $\dot{y} = J\dot{q}$ and $J = \partial h/\partial q$, $J(q): T_q\mathfrak{M}^n = R^n \to T_{h(q)}\mathfrak{M}^m$, but now $V \neq \dot{y}$ and $J \neq J_o$ is admitted as a possibility.

For rigid link manipulators moving in Euclidean three-space, typically $V = \text{col}(\dot{x}, \omega) \in R^6$, where $x \in R^3$ gives the EE location, $\dot{x}$ the EE linear velocity, and $\omega \in R^3$ the EE angular rate of change. It is well known that ω is not the time derivative of any minimal (i.e., three-dimensional) representation of attitude, so that $V = \text{col}(\dot{x}, \omega) = J_0(q)\dot{q}$ as in (3). In this case, we call $J_0(q)$ the "standard Jacobian" ([13] refers to J_0 as the "basic Jacobian"). It is also common to represent EE attitude by a proper orthogonal matrix $A \in SO(3) = \{A \mid A^TA = AA^T = I, \det A = +1\} \subset R^{3\times 3}$, where the columns of A determine EE fixed body axes in the usual way. It is well known that $\dot{A} = \tilde{\omega}A$ where $\tilde{\omega}v := \omega \times v$ for all $v \in R^3$. Thus, EE location and kinematics are often given by

$$y=(x, A)=h(q) \in R^3\times SO(3), \; V=\begin{pmatrix}\dot{x}\\ \omega\end{pmatrix}=J_0(q)q \in R^6$$

$$\dot{A}=\tilde{\omega}A, \; A \in SO(3), \; \tilde{\omega}^T=-\tilde{\omega}, \; J_0(q) \in R^{6\times n} \tag{4}$$

which should be compared to (3). Alternatively, we can take [cf. (2)]

$$y=\begin{pmatrix}x\\ \beta\end{pmatrix}=h(q) \in R^6, \; V=\dot{y}=\begin{pmatrix}\dot{x}\\ \dot{\beta}\end{pmatrix}=J(q)\dot{q} \in R^6, \; \beta \in \Omega \subset R^3. \tag{5}$$

$\beta \in \Omega \subset R^3$ is a minimal representation of EE attitude [i.e., of the rotation group $SO(3)$]. In general, $\beta = f(A)$ for some function $f(\cdot)$ which is many-to-one or undefined if the domain of $f(\cdot)$ on $SO(3)$ is not properly restricted. That is, because $SO(3)$ cannot be covered by a single coordinate chart, β is not valid for all possible EE orientations and there will be singularity of attitude representation unless we restrict EE attitude to some subregion of $SO(3)$ [25], [41], [42]. This restriction then forces β to be defined in the image of admissible attitudes, namely in some $\Omega \subset R^3$. (It may be true, however, that $\Omega = R^3$, as in the case of Euler-Rodriquez parameters where singularity of attitude representation corresponds to $\|\beta\| = \infty$ [42].) Typical β's are roll-pitch-yaw angles, axis/angle variables, Euler angles, Euler parameters, and Euler-Rodriquez parameters [25], [41]-[43]. The kinematical relationship between β and ω is given by

$$\dot{\beta}=\Pi(\beta)\omega \tag{6}$$

where $\Pi \in R^{3\times 3}$ will lose rank, i.e., become singular, precisely when β becomes a singular representation of EE attitude.

Note from (3)-(6) that $J = \text{diag}(I, \Pi)J_0$. Generally, the standard Jacobian matrix J_0 will become singular only at a manipulator kinematic singularity, in which case J will also be singular. However, J can also be singular when $\beta = \beta(q)$ gives a singularity of EE attitude representation—this compounds the trajectory planning problem for EE motions, since now we must plan trajectories which avoid manipulator kinematic singularities and also ensure that $\beta(q) \in \Omega$.

Henceforth the system (1), (2) or (1), (3) will be said to be exactly externally linearized and decoupled if

$$\dot{V}=u \in R^6. \tag{7}$$

This is somewhat of an abuse of notation, as a consideration of the system (1), (4) shows; for $u = \text{col}(u_1, u_2)$, $\dot{V} = u$ yields

$$\ddot{x}=u_1 \in R^3, \; \dot{\omega}=u_2 \in R^3, \; \dot{A}=\tilde{\omega}A. \tag{8}$$

Although EE positional dynamics are decoupled and linearized to $\ddot{x} = u_1$, attitude dynamics are nonlinear and given by $\dot{\omega} = u_2$, $\dot{A} = \tilde{\omega}A$. Equation (8) is precisely the sense in which RAC can be said to almost "exactly externally linearize and decouple" attitude dynamics, as was discussed in the Introduction. In the case of the system (1), (5), $\dot{V} = \text{col}(u_1, u_2)$ gives

$$\ddot{x}=u_1 \in R^3, \; \ddot{\beta}=u_2 \in R^3, \; \beta \in \Omega \subset R^3 \tag{9}$$

which can indeed be said to be exactly externally linearized and decoupled. Drawbacks to using (9) are that β must always be controlled to remain in Ω, trajectories involving β may be difficult to visualize, and the generalized force u_2, which drives β, may be nonintuitive. On the other hand, it is obvious how to obtain stable attitude tracking from (9). The advantage to using (8) is that ω and A are easily visualized entities, while u_2 is the ordinary torque that we are all familiar with. Fortunately, despite the nonlinear attitude dynamics, it is possible to use (8) to perform EE attitude tracking with asymptotically vanishing attitude error [7], [8].

Note that once (8) is obtained, it is easy to get (9) by use of the relationship (6). If we have $\dot{\omega} = u$, $\dot{\beta} = \Pi(\beta)\omega$, and $\beta \in \Omega$ so that $\Pi^{-1}(\beta)$ exists, then use of

$$\dot{\omega}=u, \; u=\Pi^{-1}(\beta)[\bar{u}-\dot{\Pi}(\beta)\omega] \tag{10}$$

gives

$$\ddot{\beta}=\Pi(\beta)\dot{\omega}+\dot{\Pi}(\beta)\omega=\bar{u}. \tag{11}$$

Therefore, having (8), we can perform attitude control directly on $\dot{\omega} = u_2$, $\dot{A} = \tilde{\omega}A$ or we can transform to $\ddot{\beta} = \bar{u}_2$ and then control.

IV. Comparison of GCT, NDT, and OSC

For brevity, we consider the nonredundant manipulator case, taking $n = 6$ in (1), and we omit derivations. A more detailed discussion is given in [24]. Note that the system (1), (5) can be written as

$$\frac{d}{dt}\begin{pmatrix}q\\ \dot{q}\end{pmatrix}=\begin{pmatrix}\dot{q}\\ -M^{-1}C\end{pmatrix}+\begin{pmatrix}0\\ M^{-1}\end{pmatrix}\tau, \qquad y=h(q) \tag{12}$$

or, taking $Z = \text{col}(q, \dot{q})$,

$$\frac{d}{dt}Z=A(Z)+B(Z)\tau, \qquad y=H(Z) \tag{13}$$

where the definitions of A, B, and H are obvious. GCT asks: Does there exist i) a nonlinear feedback $\tau = Q(Z) + R(Z)u$; and ii) a nonlinear change of basis $x = X(Z)$ such that (12) is placed into BCF?

$$\frac{d}{dt}\begin{pmatrix}y\\ \dot{y}\end{pmatrix}=\begin{pmatrix}0 & I\\ 0 & 0\end{pmatrix}\begin{pmatrix}y\\ \dot{y}\end{pmatrix}+\begin{pmatrix}0\\ I\end{pmatrix}u \Leftrightarrow \ddot{y}=u. \tag{14}$$

The constructive sufficient conditions of [19]-[22] can be applied and give the linearizing and decoupling feedback law $\tau = -MJ^{-1}\partial J\dot{q} + C + MJ^{-1}u$, where $\partial J = [\Sigma_{k=1}^n (\partial J_{ik}/\partial q_j)\dot{q}_k]$. Although $\partial J \neq \dot{J}$, it is true that $\partial J\dot{q} = \dot{J}\dot{q}$ giving

$$\tau=-MJ^{-1}\dot{J}\dot{q}+MJ^{-1}u+C. \tag{15}$$

Note that J must be nonsingular for (15) to exist. This is consistent with the theory of [19]-[21] which provides sufficient conditions for *local* linearization. Note also that to implement (15), explicit expressions for M, J^{-1}, J, and C are required.

The NDT approach of [11] constructs the linearizing feedback in the following way. For the system (13) define

$$G(Z)=\frac{\partial}{\partial Z}\left(\left[\frac{\partial}{\partial Z}H(Z)\right]A(Z)\right),\ D^*(Z)=G(Z)B(Z),$$
$$H^*(Z)=G(Z)A(Z). \quad (16)$$

The use of

$$\tau=-D^{*-1}(Z)H^*(Z)+D^{*-1}(Z)u \quad (17)$$

will transform (12), (13) to $\ddot{y}=u$, i.e., to (14). It is straightforward to show that, for A, B, and H as in (12) and (13), (17) is precisely (15). Note that in (13) we take $Z=\text{col}\,(q,\dot{q})$ and *not* $Z=(q_1,\dot{q}_1,\cdots,q_n,\dot{q}_n)^T$. The latter choice for Z is taken in [11] and serves to obscure the final result—namely that (17) and (15) are equivalent.

Now consider the OSC approach of [12]–[14]. In this approach the EE coordinates y are viewed as generalized coordinates for the manipulator, and a change of basis $(q,\dot{q})\rightarrow(y,\dot{y})$ is made. In (1) let $C=b-g$ where b are the Coriolis forces and g the gravity forces. Restrict the domain of the system (1), (5) to ensure that $h(\cdot)$ is a bijection (and consequently det $J(q)\neq 0$ on this restriction). This restriction means that, as for DGC and NDT, OSC gives a local result for external linearization. In [12]–[14], the effective EE dynamics are determined to be

$$\Lambda(y)\ddot{y}+c(y,\dot{y})=F,\ \tau=J^TF,\ \Lambda=J^{-T}MJ,\ c=U-P,$$
$$P=J^{-1}g,\ U=J^{-T}b-\Lambda\dot{J}\dot{q},\ q=h^{-1}(y),\ \dot{q}=J^{-1}\dot{y}. \quad (18)$$

Recall that for the system (1), $M\ddot{q}+C=\tau$, the computed torque technique is to take $\tau=Mu+C$, yielding $M(\ddot{q}-u)=0\Rightarrow\ddot{q}=u$, since $M(q)>0$, $\forall q$. Similarly, in (18) $\Lambda(y)>0$ for every $y=h(q)$ where q is in the restricted domain. The choice of

$$F=\Lambda(y)u+c(y,\dot{y}),\qquad \tau=J^TF \quad (19)$$

in (18) then yields $\Lambda(y)(\ddot{y}-u)=0\Rightarrow\ddot{y}=u$. In this sense the work in [12]–[14] can be viewed as a generalized computed torque technique. From (18) and (19) it is straightforward to determine that τ of (19) is exactly τ of (15).

$JM^{-1}=D^*$ of (15), (16) and Λ^{-1} of (19) are the "decoupling matrices" [19], [20] for (13) and (18), respectively. The difficulty in feedback linearizing (13) is seen to arise from the decoupling matrix becoming singular, while the difficulty in feedback linearizing (19) is due to the change of basis $(q,\dot{q})\rightarrow(y,\dot{y})$ (since if y is a good generalized coordinate for the manipulator Λ^{-1} must be nonsingular). In either interpretation we see that the difficulty is due to the Jacobian matrix J becoming singular.

V. Derivation of a Feedback Law for Local Exact Decoupled External Linearization and Its Relationship to RAC and GCT

Recall that the system (1), (2) or (1), (3) is of the form

$$M(q)\ddot{q}+C(q,\dot{q})=\tau;\ q\in\mathfrak{M}^n;\ \dot{q},\ddot{q}\in R^n$$
$$y=h(q)\in\mathfrak{M}^m;\ h(\cdot)\text{ is }c^2;\ V=\bar{J}(q)\dot{q}\in R^m;\ \bar{J}(q)\text{ is }c^1;$$
$$M(q)\in R^{n\times n};\ M(q)=M(q)^T>0,\ \forall q\in\mathfrak{M}^n \quad (20)$$

where in general, it may be that $m\neq n$, $\mathfrak{M}^m\neq R^m$, $V\neq\dot{y}$, and $\bar{J}\neq J=\partial h/\partial q$. It is assumed that a necessary and sufficient condition for $h(q)$ to be onto some neighborhood of $y=h(q)$ in $\mathfrak{M}^m$ is that the mapping $\bar{J}(q)$ be onto R^m, i.e., we assume that $\bar{J}(q)$ is onto R^m if and only if $J(q)=\partial h(q)/\partial q$ is onto $T_{h(q)}\mathfrak{M}^m\simeq R^m$. This is a reasonable assumption; for example, when $\mathfrak{M}^m=R^m$, $V=\dot{y}=J(q)\dot{q}\in R^m$, and $\bar{J}=J=\partial h/\partial q$ this is trivially true. For the case $y=h(q)=(x,A)\in\mathfrak{M}^6=R^3\times SO(3)$ and $\bar{J}=J_0$ where $V=\text{col}\,(\dot{x},\omega)=J_0(q)\dot{q}$, the fact that $\dot{x}\in T_xR^3$ and $\dot{A}=\tilde{\omega}A\in T_ASO(3)$ means that for $J_0(q)$ onto, we can fill out a neighborhood of (x,A) and otherwise we cannot. (A general element of $T_ASO(3)$ is precisely of the form $\tilde{\omega}A$, $\tilde{\omega}\in R^{3\times3}$ skew-symmetric, so that if $\omega=\omega(q)$ can be mapped onto R^3, $\tilde{\omega}A$ can be mapped onto $T_ASO(3)$ [44], [47].)

Definition LEL: The system (20) can be locally exactly externally linearized and decoupled (LEL) over an open neighborhood $B^m(y')\subset\mathfrak{M}^m$ of $y'\in h(\mathfrak{M}^n)\subset\mathfrak{M}^m$ with the arm in the configuration $q'\in h^{-1}(y')$ if there is an open neighborhood of q', $B^n(q')\subset\mathfrak{M}^n$, such that $B^m(y')=h[B^n(q')]$ and if for any $u\in R^m$, $q\in B^n(q')$, and $\dot{q}\in T_q\mathfrak{M}^n\simeq R^n$ there exists a nonlinear feedback $\tau=F(q,\dot{q},u)$ such that V, the velocity associated with $y=h(q)\in B^m(y')$, obeys $\dot{V}=u$. ●

Note that for an EE to be LEL at y' it must be true that y' is in the range of $h(\cdot)$, i.e., y' must be a physically attainable EE position. Also, for a given EE location, $y\in h(\mathfrak{M}^n)$, a manipulator can physically be in only one of the possible configurations $h^{-1}(y')$. Thus, we can interpret $q'\in h^{-1}(y')$ to be the actual physical configuration of a manipulator. If the system (20) is not LEL at y' in the configuration $q'\in h^{-1}(y')$, it may be LEL at a different configuration $q\in h^{-1}(y')$.

Theorem LEL: A necessary and sufficient condition for (20) to be LEL at $y'\in h(\mathfrak{M}^n)$ in the configuration $q'\in h^{-1}(y')$ is that $\bar{J}(q')\in R^{m\times n}$ be onto, which is true iff $m\leq n$ and rank $\bar{J}(q')=m$. Furthermore, the locally exactly linearizing and decoupling feedback is given by

$$\tau=M(q)\xi+C(q,\dot{q}) \quad (21)$$

where ξ is any solution to

$$\bar{J}(q)\xi=-\dot{\bar{J}}(q)\dot{q}+u. \quad (22)$$

When $m=n$ this gives

$$\tau=-M(q)\bar{J}(q)^{-1}\dot{\bar{J}}(q)\dot{q}+M(q)\bar{J}(q)^{-1}u+C(q,\dot{q}). \quad (23)$$

●

Proof. Necessity: Suppose that $\dot{V}=\bar{J}(q')\ddot{q}'+\dot{\bar{J}}(q')\dot{q}'=u$ can be made to hold regardless of the values of $u\in R^m$ and $\dot{q}'$. This means that there must exist $\ddot{q}'\in R^n$ such that

$$\bar{J}(q')\ddot{q}'=-\dot{\bar{J}}(q')\dot{q}'+u. \quad (24)$$

If $\bar{J}(q')$ is not onto, then $Im\bar{J}(q')\subset R^m$ and $Im\bar{J}(q')\neq R^m$. Let u and $\dot{q}'$ be such that $-\dot{\bar{J}}(q')\dot{q}'+u\notin Im\bar{J}(q')$. Then there is no $\ddot{q}'$ for which (24) holds, yielding a contradiction. *Sufficiency:* By assumption $\bar{J}(q')$ is full rank and onto $\Leftrightarrow J(q')=\partial h(q')/\partial q$ is full rank and onto. Since $\bar{J}$ and J are c^1, there exist neighborhoods $B^m(y')$ and $B^n(q')$, $y'=h(q')$, such that $B^m(y')=h[B^n(q')]$ and such that $\bar{J}$ is full rank and onto when restricted to $B^n(q')$ [44]. Now consider any $q\in B^n(q')$ and its associated $y=h(q)\in B^m(y')$. Then, $V=\bar{J}(q)\dot{q}\Rightarrow$

$$\dot{V}=\bar{J}(q)\ddot{q}+\dot{\bar{J}}(q)\dot{q}. \quad (25)$$

Let ξ be any solution to (22). ξ is guaranteed to exist since $Im\bar{J}(q)=R^m$. Take τ to be (21), then

$$M\ddot{q}+C=\tau=M\xi+C\Rightarrow M(\ddot{q}-\xi)=0\Rightarrow\ddot{q}=\xi$$

which with (22) and (25) gives $\dot{V}=u$. ●

Comments:

1) Note that this result applies to all systems of the form (20), of which rigid link manipulators are a special case.

2) Note that with $y\in\mathfrak{M}^m$ and $\tau\in R^n$, the fact that we need $m\leq n$ can be interpreted to mean that there must be at least as many inputs as outputs.

3) When $\bar{J}=J=\partial h/\partial q$, $V=\dot{y}$, and $m=n$ we have that $\tau=-MJ^{-1}\dot{J}\dot{q}+MJ^{-1}u+C\Rightarrow\ddot{y}=u$ when det $J\neq 0$. This is the same result provided by GCT, NDT, and OCS as seen in the last section [cf. (15)].

4) Note that in the proof we force $\ddot{q}=\xi$ precisely like $\ddot{q}=u$ is forced to happen in the computed torque method. In fact, for $y=q$ we have $J=I$ and $\dot{J}=0$ giving $\xi=u$. Thus, the exact linearizing control of (21), (22)

is seen to be a generalization of the computed torque method in a somewhat different, and perhaps more illuminating, way than OCS.

5) In addition to the invertibility of the manipulator mass matrix, a key reason that a relatively simple solution to the output linearizing problem can be found is that an independent input exists for every system configuration degree of freedom. In the case of manipulators with elastic joints, this condition may be violated, and it is not possible in general to feedback linearize using only static feedback of measured joint variables and rates. See [50], [51].

Consider the case of EE control given by the system (1), (4). Here $J = J_0$ where $V = \text{col}(\dot{x}, \omega) = J_0\dot{q}$. In this case, when $m = n$, (23) is

$$\tau = -MJ_0^{-1}\dot{J}_0\dot{q} + MJ_0^{-1}u + C. \tag{26}$$

When $\det J_0 \neq 0$, the use of (26) yields $\text{col}(\ddot{x}, \dot{\omega}) = \text{col}(u_1, u_2)$. This is precisely RAC [7], [8]. Theorem LEL can be interpreted as an extension of RAC to the redundant arm case which allows for the use of a minimal representation of EE attitude [24]. The more general case $m \leq n$ is given by

$$\tau = Mu + C, \quad J_0\xi = -\dot{J}_0\dot{q} + u. \tag{27}$$

By using the indirect form (27), τ can be obtained, after ξ has been found, by use of the Newton-Euler recursion [45]. Furthermore, ξ can be obtained recursively—either directly [46], or by first recursively obtaining J_0 and $\dot{J}_0$ and then solving for u by Gaussian elimination. The point to be drawn here is that (27) shows us how to perform exact external linearization without the need for an explicit manipulator model. After exactly linearizing to $\text{col}(\ddot{x}, \dot{\omega}) = \text{col}(u_1, u_2)$ one can perform EE tracking at this stage [7], [8], or one can continue to the form (11) by the use of (10).

When using (26) or (27), the only way that rank $J_0 < m$ can occur for $m \leq n$ is when the manipulator is at a mechanically singular configuration. Recall (Section IV) that in the case when a minimal representation of EE attitude is used, the resulting Jacobian matrix J will be rank deficient not just for a manipulator singularity, but at a configuration which leads to a singularity of attitude presentation. Thus, rank deficiency of J_0 is kinematically cleaner to understand. The necessity that rank $J_0 = m$ in order to use (26) or (27) allows two obvious, but important statements to be made: i) for a manipulator with a workspace boundary (ignoring joint stops), as in the case of a PUMA-type manipulator, exact linearization at the boundary is impossible; ii) for a nonredundant (6 dof) manipulator with workspace interior singularities, there cannot be exact linearization throughout the workspace interior. For a redundant manipulator with workspace interior singularities, it may be possible to avoid workspace interior configurations which cannot be exactly linearized by the use of self motions as described in [48], [49]. This is related to the multiplicity of solutions available for ξ in (27).

How does the control (23) fulfill the aim of GCT as stated in (12)-(14)? The nonlinear feedback (taking $V = \dot{y}$ and $J = \bar{J})\tau = Q(Z) + R(Z)u = (C - MJ^{-1}\dot{J}\dot{q}) + (MJ^{-1})u$ applied to (12), (13) gives

$$\frac{d}{dt}\begin{pmatrix} q \\ \dot{q} \end{pmatrix} = \begin{pmatrix} 0 & I \\ 0 & -J^{-1}\dot{J} \end{pmatrix}\begin{pmatrix} q \\ \dot{q} \end{pmatrix} + \begin{pmatrix} 0 \\ J^{-1} \end{pmatrix} u. \tag{28}$$

Consider the local nonlinear change of basis given by

$$\begin{pmatrix} y \\ \dot{y} \end{pmatrix} = \begin{pmatrix} h(q) \\ J\dot{q} \end{pmatrix}; \begin{pmatrix} q \\ \dot{q} \end{pmatrix} = \begin{pmatrix} h^{-1}(y) \\ J^{-1}\dot{y} \end{pmatrix}.$$

The fact that $\dot{y} = J\dot{q}$ and $\ddot{y} = J\ddot{q} + \dot{J}\dot{q}$ gives

$$\frac{d}{dt}\begin{pmatrix} y \\ \dot{y} \end{pmatrix} = \begin{pmatrix} J & 0 \\ \dot{J} & J \end{pmatrix}\frac{d}{dt}\begin{pmatrix} q \\ \dot{q} \end{pmatrix}.$$

Writing (28) as

$$\begin{pmatrix} J & 0 \\ \dot{J} & J \end{pmatrix}\frac{d}{dt}\begin{pmatrix} q \\ \dot{q} \end{pmatrix} = \begin{pmatrix} J & 0 \\ \dot{J} & J \end{pmatrix}\begin{pmatrix} 0 & I \\ 0 & -J^{-1}\dot{J} \end{pmatrix}$$

$$\cdot \begin{pmatrix} h^{-1} & (y) \\ J^{-1} & \dot{y} \end{pmatrix} + \begin{pmatrix} J & 0 \\ \dot{J} & J \end{pmatrix}\begin{pmatrix} 0 \\ J^{-1} \end{pmatrix} u$$

we obtain the BCF

$$\frac{d}{dt}\begin{pmatrix} y \\ \dot{y} \end{pmatrix} = \begin{pmatrix} 0 & I \\ 0 & 0 \end{pmatrix}\begin{pmatrix} y \\ \dot{y} \end{pmatrix} + \begin{pmatrix} 0 \\ I \end{pmatrix} u \Leftrightarrow \ddot{y} = u.$$

VI. Concluding Remarks

Recognizing the fundamental unity of RAC, GCT, OSC, and NDT [7]-[22] for exact linearization of manipulators, we can focus on their true differences—namely, differences in implementation detail and design philosophy. With the awareness that they all produce essentially the same linearizing feedback, we can ask why this particular feedback form is appropriate for manipulator-like systems.

OCS and RAC exploit the specific structure of such systems. Not surprisingly, the solutions arrived at, reflecting the philosophies and implementation perspectives of the researchers involved, are quite distinct in their flavor and presentation. Yet, since the properties specific to manipulator dynamics ultimately forced the solution, they are fundamentally the same. The important point here is that researchers consciously exploited the specific properties of a system of interest, but without pinpointing precisely what these properties were which made the system amenable to linearizing control.

GCT and NDT provide techniques for exactly linearizing general smooth affine-in-the-input dynamical systems. These techniques ignore any specific nongeneric structural properties that a system might have and as a consequence the solutions obtained are much less transparent than those of OCS or RAC. The strength of these approaches, particularly GCT, is that they can provide necessary and sufficient conditions for a system to be exactly linearizable and constructive sufficient conditions which produce the linearizing feedback when satisfied. Interestingly, when applied to the problem of manipulator exact linearization, the solutions obtained can be shown to be equivalent to those of RAC and OCS. Again, the structural properties of the system forced the solution. Certainly, once a solution is known to exist, it is reasonable to attempt to produce it from more physical arguments knowing now that the search is not fruitless. This leads to a reexamination of OCS and RAC.

The work of [17]-[22] stresses a perspective which serves to enable a clearer comparison between competing techniques for external linearization: Place the system in a standard linear canonical form before additional control efforts are made—this ensures that the process of linearizing the system is not mixed up with, and confused with, the process of stabilizing and controlling it. This perspective greatly aided the comparison of GCT, OCS, RAC, and NDT which resulted in [24]. In turn, this comparison focuses attention on the structural properties of manipulators.

Much current research makes it apparent that systems dynamically similar to rigid link manipulators have important structural properties which can be exploited to achieve results which are quite strong when compared to those available for general smooth affine in the inputs nonlinear systems [25]-[40]. Here we have seen that exploiting the nongeneric second-order form of system (1) which has an everywhere positive definite mass matrix, direct independent control of every configuration degree of freedom, and a c^2 locally onto readout map enables a simple form for the linearizing feedback.

One can also approach the problem of hybrid force/position control from the framework of the approaches discussed in this note. Unfortunately, lack of space precludes discussion here. For hybrid control from the OCS perspective see [52], for the RAC perspective see [53], and for approaches based on GCT see [54], [55].

Acknowledgment

This work was performed at the Jet Propulsion Laboratory, California Institute of Technology, Pasadena, CA.

References

[1] R. C. Paul, "Modeling, trajectory calculation and servoing of a computer controlled arm," Stanford A.I. Lab., Stanford Univ., Stanford, CA, A.I. Memo 177, Nov. 1972.

[2] B. R. Markiewicz, "Analysis of the computed torque drive method and

comparison with conventional position servo for a computer-controlled manipulator," Jet Propulsion Lab. Rep. JPL TM 33-601, Mar. 1973.

[3] A. K. Bejczy, "Robot arm dynamics and control," Jet Propulsion Lab. Rep. JPL TM 33-669, Feb. 1974.

[4] H. Hemami and P. C. Camana, "Nonlinear feedback in simple locomotion systems," *IEEE Trans. Automat. Contr.*, vol. AC-21, pp. 855-860, 1976.

[5] M. H. Raibert and B. K. Horn, "Manipulator control using the configuration space method," *Indust. Robot*, vol. 5, pp. 69-73, 1978.

[6] J. R. Hewit and J. Padovan, "Decoupled feedback control of robot and manipulator arms," in *Proc. 3rd CISM-IFToMM Symp. Theory and Practice of Robot Manipulators*, Udine, Italy, Sept. 1976, pp. 251-266.

[7] R. Paul, J. Luh *et al.*, "Advanced industrial robot control systems," Purdue Univ., West Lafayette, IN, Rep. TR-EE 78-25, May 1978.

[8] J. Luh, M. Walker, and R. Paul, "Resolved—Acceleration control of mechanical manipulators," *IEEE Trans. Automat. Contr.*, vol. AC-25, pp. 468-474, 1980.

[9] E. Freund, "A nonlinear control concept for computer controlled manipulators," in *Proc. IFAC Symp. Multivariable Technol. Syst.*, 1977, pp. 395-403.

[10] E. Freund and M. Syrbe, "Control of industrial robots by means of microprocessors," in *IRIA Conf. Lecture Notes on Information Sciences*. New York: Springer-Verlag, 1976, pp. 167-85.

[11] E. Freund, "Fast nonlinear control with arbitrary pole-placement for industrial robots and manipulators," *Int. J. Robotics Res.*, vol. 3, pp. 76-86, 1982.

[12] O. Khatib, "Commande dynamique dans l'espace operational des robots manipulators en presence d'obstacles," Engineering Doctoral dissertation 37, ENSAE, Toulouse, France, 1980.

[13] ——, "Dynamics control of manipulators in operational space," in *Proc. 6th CISM-IFToMM*, 1983.

[14] ——, "The operational space formulation in the analysis, design, and control of robot manipulators," in *Proc. 3rd Int. Symp. Robotics Res.*, 1985, pp. 103-110.

[15] A. Isidori and A. Ruberti, "On the synthesis of linear input-output responses for nonlinear systems," *Syst. Contr. Lett.*, vol. 4, pp. 17-22, 1984.

[16] A. Isidori, "The matching of a prescribed linear input-output behavior in a nonlinear system," *IEEE Trans. Automat. Contr.*, vol. AC-30, pp. 258-265, 1985.

[17] D. Cheng, T. J. Tarn, and A. Isidori, "Global external linearization of nonlinear systems via feedback," *IEEE Trans. Automat. Contr.*

[18] A. Isidori, *Nonlinear Control Systems: An Introduction*. New York: Springer-Verlag, 1985.

[19] Y. Chen, "Nonlinear feedback and computer control of robot arms," D.Sc. dissertation, Dep. Syst. Sci. and Math, Washington Univ., St. Louis, MO, 1984.

[20] T. J. Tarn, A. K. Bejczy *et al.*, "Nonlinear feedback in robot arm control," in *Proc. 23rd Conf. Decision Contr.*, 1984, pp. 736-751.

[21] A. K. Bejczy, T. J. Tarn, and Y. L. Chen, "Robot arm dynamic control by computer," in *Proc. IEEE ICRA*, 1985, pp. 960-970.

[22] T. J. Tarn, A. K. Bejczy, and X. Yun, "Coordinated control of two robot arms," in *Proc. IEEE ICRA*, 1986, pp. 1193-202.

[23] M. Brady *et al., Robot Motion Planning and Control*. Cambridge, MA: M.I.T. Press, 1984.

[24] K. Kreutz, "On nonlinear control for decoupled exact external linearization of robot manipulators," in *Recent Trends in Robotics: Modeling, Control, and Education*, M. Jamshidi *et al.*, Eds. Amsterdam, The Netherlands, North-Holland, 1986, pp. 199-212.

[25] L. Meirovitch, *Methods of Analytical Dynamics*. New York: McGraw-Hill, 1970.

[26] F. Gantmacher, *Lectures in Analytical Mechanics*. Moscow: MIR, 1975.

[27] D. Koditschek, "Natural control of robot arms," Dep. Elec. Eng., Yale Univ., New Haven, CT, Center for Syst. Sci. Rep. 1985.

[28] D. Koditschek, "High gain feedback and telerobotic tracking," in *Proc. Workshop on Space Telerobotics*, Pasadena, CA, Jan. 20-22, 1987, vol. 3, pp. 355-364.

[29] H. Asada and J. Slotine, *Robot Analysis and Control*. New York: Wiley, 1986.

[30] R. Pringle, Jr., "On the stability of a body with connected moving parts," *AIAA J.*, vol. 4, pp. 1395-1404, 1966.

[31] M. Takegaki and S. Arimoto, "A new feedback method for dynamic control of manipulators," *J. Dynam. Syst. Meas. Contr.*, vol. 102, pp. 119-125, 1981.

[32] S. Arimoto and F. Miyazaki, "Stability and robustness of PID feedback control for robot manipulators of sensory capacity," in *Proc. 1st Int. Symp. Robotics Res.*, 1983, pp. 783-99.

[33] ——, "Stability and robustness of PD feedback control with gravity compensation for robot manipulators," *Robotics: Theory and Practice*, DSC-vol. 3, pp. 67-72, 1986.

[34] J. T. Wen and D. S. Bayard, "Simple robust control laws for robotic manipulators—Part I: Nonadaptive case," in *Proc. Workshop on Space Telerobotics*, Pasadena, CA, Jan. 1987, JPL publication 87-13, vol. 3, pp. 215-230.

[35] D. S. Bayard and J. T. Wen, "Simple robust control laws for robotic manipulators—Part II: Adaptive case," in *Proc. Workshop on Space Telerobotics*, Pasadena, CA, Jan. 1987, JPL publication 87-13, vol. 3, pp. 231-244.

[36] J. Slotine and W. Li, "On the adaptive control of robot manipulators," *Robotics: Theory and Practice*, DSC-vol. 3, pp. 51-56, 1986.

[37] B. Paden and D. Slotine, "PD + robot controllers: Tracking and adaptive control," presented at the *1987 IEEE Int. Conf. Robotics Automat.*, 1987.

[38] E. Sontag and H. Sussman, "Time-optimal control of manipulators," in *Proc. IEEE 1986 Int. Conf. Robotics Automat.*, San Francisco, CA, 1986, pp. 1692-1697.

[39] ——, "Remarks on the time-optimal control of two-link manipulators," in *Proc. 24th IEEE Conf. Decision Contr.*, 1985, pp. 1643-1652.

[40] J. Wen, "On minimum time control for robotic manipulators," in *Recent Trends in Robotics: Modeling, Control, and Education*, M. Jamshidi *et al.* Eds. Amsterdam, The Netherlands: North-Holland, 1986, pp. 283-292.

[41] J. Stuelpnagel, "On the parameterization of the three dimensional rotation group," *SIAM Rev.*, vol. 6, pp. 422-430, 1964.

[42] P. C. Hughes, *Spacecraft Attitude Dynamics*. New York: Wiley, 1986.

[43] J. Craig, *Introduction to Robotics*. Reading, MA: Addison-Wesley, 1986.

[44] W. Boothby, *An Introduction to Differentiable Manifolds and Riemannian Geometry*, 2nd ed. New York: Academic, 1986.

[45] J. Y. S. Luh, M. W. Walker, and R. P. Paul, "On-line computational scheme for mechanical manipulators," *J. Dynam. Syst. Measure, Contr.*, vol. 102, pp. 69-76, 1980.

[46] J. M. Hollerbach and G. Sahar, "Wrist-partitioned inverse kinematic accelerations and manipulator dynamics," *Int. J. Robotics Res.*, vol. 2, pp. 61-76, 1983.

[47] C. Von Westenholz, *Differential Forms in Mathematical Physics*. Amsterdam, The Netherlands: North-Holland, 1986.

[48] J. Hollerbach, "Optimum kinematic design for a seven degree of freedom manipulator," presented at the *2nd Int. Symp. Robotics Res.*, 1984.

[49] J. Baillieul *et al.*, "Programming and control of kinematically redundant manipulators," in *Proc. 23rd Conf. Decision Contr.*, 1986, pp. 768-774.

[50] G. Cesareo and R. Marino, "On the controllability properties of elastic robots," presented at the *6th Int. Conf. on Anal. and Optimiz. Syst.*, INRIA, Nice, June 1984.

[51] A. De Luca, "Dynamic control of robots with joint elasticity," in *Proc. 1988 IEEE Int. Conf. Robotics Automat.*, Philadelphia, PA, Apr. 24-29, 1988, pp. 152-158.

[52] O. Khatib and J. Burdick, "Motion and force control of robot manipulators," in *Proc. IEEE 1986 Int. Conf. Robotics Automat.*, San Francisco, CA, 1986, pp. 1381-1386.

[53] Z. Li and S. Sastry, "Hybrid velocity/force control of a robot manipulator," Univ. California, Berkeley, Eng. Res. Lab. Rep. M87/9, Mar. 3, 1987.

[54] T. J. Tarn, A. K. Bejczy, and X. Yun, "Nonlinear feedback control of multiple robot arms," in *Proc. Workshop on Space Telerobotics*, Pasadena, CA, JPL publication 87-13, vol. 3, pp. 179-192.

[55] ——, "Robot arm force control through system linearization by nonlinear feedback," in *Proc. IEEE Int. Conf. Robotics Automat.*, Philadelphia, PA, Apr. 24-29, 1988, pp. 1618-1625.

Nonlinear Feedback in Robot Arm Control*

T. J. TARN,† A. K. BEJCZY,††, A. ISIDORI,††† Y. CHEN†

† Department of Systems Science and Mathematics, Box 1040, Washington University, St. Louis, Missouri 63130, U.S.A.
†† JPL, California Institute of Technology, 4800 Oak Grove Drive, Pasadena, California 91109, U.S.A.
††† Dipartimento di Informatica e Sistemistica, Universita di Roma 'La Sapienza,' 18 Via Eudossiana, 00184 Roma, Italy.

Abstract

Nonlinear feedback control is proposed for implementation of an advanced dynamic control strategy for robot arms. Using differential geometric system theory we obtained necessary and sufficient conditions for the existence of a nonlinear feedback control for a general nonlinear system to be externally linearized and simultaneously output decoupled. An algorithm is given for the construction of the required nonlinear feedback. To design a dynamic control for robot arms we apply the above result to the JPL-Stanford arm and propose a new control strategy, which also contains an optimal error-correcting feedback. Simulation results show great promise for the obtained dynamic control strategy.

1. Introduction

In recent years the use of industrial robots throughout industry has increased significantly. In many industrial applications such as painting and arc welding, programming of a sequence of movements is done by a 'teaching by doing' process in which a human operator physically takes the robot hand through the desired sequence. When operating the robot the controller reads the memory by a play-back method. In such a situation, however, the robot can only repeat what it has been taught.

In some cases, where the kinematic design is simple enough and the dynamic control demands are low, on-board computing power is used to perform the necessary coordinate transformation between the joint coordinates, which are controlled directly, and the task coordinates, which are convenient to the task description. In such cases robot motions can be programmed from a computer keyboard, and the motions of robot arm joints can be controlled through the kinematic transformation of coordinates.

When the performance requirements for industrial robots are increased and require accurate, fast and versatile manipulations, dynamic effects become significant. This requires the control of a multi-input multi-output system described by a set of highly nonlinear, strongly coupled differential equations.

During the last decade, many papers have been written that discuss the control of robots through dynamic effects. Pieper [1], Whitney [2], Kahn and Roth [3], Popov [4] and Dubowsky and Forges [5], suggest using linearized system models as the basis for control. However, it should be noted that the implementation of a linear regulator in a robot system leads to many problems because the complex robot control has been synthesized based on the approximate linear model. This was demonstrated by a digital computer simulation of various control methods in [6].

The 'computed torque' technique, investigated by Markiewicz [7] and Bejczy [8] is called the 'inverse problem' technique by Paul [9] and Raibert and Horn [10]. Basically this technique has a feedforward and a feedback path. The feedforward path compensates the coupled forces among all the various joints and the feedback path computes the correction torques to compensate for any deviations from the desired tasks. The performance of this control technique can vary a great deal because linear corrections are imposed on a nonlinear system. This technique becomes less effective when high speed is required with high-position accuracy.

The use of nonlinear feedback to decouple the nonlinear coupling terms in a nonlinear system is not new to control engineers. The earlier work was done by Hemami and Camana [11]. They applied the nonlinear feedback control technique to a simple locomotion system and obtained decoupled subsystems. In 1980 Hewit and Burdess [12] applied a method of active force control to the problem of obtaining dynamic decoupling of the motion trajectories of a robot arm. However, they assumed that one part of the dynamic equation is completely unknown and introduced accelerometers that provide information about the accelerations of the manipulator joints. They calculate on-line the inertia matrix of the system in order to estimate gravitational, Coriolis and friction forces, which makes the control law unnecessarily complicated. Almost at the same time, an attractive idea of how to compensate for coupling and nonlinearities through a suitable feedback was introduced by Lentini, Nicolo, Nicosia and Vecettio [13-15]. For robot systems whose number of inputs equals the number of outputs, they obtained parallel chains of double integrators by means of nonlinear feedback. However, the paper discusses formal manipulations and gives no general results. In a recent paper, Freund [16] presented suitable nonlinear feedback methods for the control of industrial robots. The methods presented are based on partitioning the dynamic equations of the robot. The overall system behavior that results from the application of this nonlinear control strategy is characterized by decoupled, linear, second-order equations. But the methods suffer from the same restrictions as in

* This research was supported in part by the National Science Foundation under grants ECS-8017184-01, ECS-8309527 and INT-8201554.

Reprinted from *Proc. IEEE DECC*, Las Vegas, NV, Dec. 1984, pp. 736–751.

[13-15]: the number of inputs equals the number of outputs. Moreover, the linearization in general is not in terms of the state of the system but rather in terms of the output. Thus it is difficult to investigate the stability of the system for a broad class of states; and, as stated by the author, they were developed from practical experience with the design of control strategies for a few types of industrial robots.

There are other dynamic control strategies that exist in the literature [17-22]. Each has its own advantages and disadvantages. For an efficient implementation of advanced feedback control of robotic systems, a solid foundation is needed. However, most of the research that has been done on the dynamic control of robot arms has either fallen short of a general mathematical frame and justification or else has bypassed important structural details and complexities.

For the purpose of dynamic control of robot arms, we will use tools from differential geometric system theory. The objectives of this paper are then:

1. To develop for the first time the necessary and sufficient conditions for the existence of a nonlinear feedback and a diffeomorphic transformation for a given nonlinear system to be externally linearized and simultaneously output decoupled.
2. To derive an algorithm for constructing the required nonlinear feedback.
3. To apply the above results to the control of robot arms.
4. To propose a new strategy for dynamic control of robot arms.
5. To design and implement a digital simulation to evaluate this new control technique.

To the authors' knowledge, researchers working in the field of differential geometric control theory have considered the external linearization and output decoupling (noninteracting) problems separately. There has been no study of external linearization and simultaneous output decoupling in the literature. Thus the thrust of this paper is the development of a mathematical foundation for robotic systems control, taking into account also the dynamics of robots. The proposed control strategy appears to be intuitive and the simulation results obtained by us are very encouraging.

It is expected that with the rapidly expanding interest in the dynamic behavior of robots combined with the realization that when the performance requirements (high speed, handling varying loads and accuracy) for the robots are increased, the dynamic control of robots will become a necessity. We believe that the results obtained in this paper will contribute significantly to the dynamic control of robot arms.

In Section 2 we present the dynamic control model of rigid robot arms in state equation form. The main theoretical results of our control, using a differential geometric approach, are presented in Section 3. The application of our theoretical results to the design of a dynamic control strategy for the robot arms, exemplified by the JPL-Stanford arm, is presented in Section 4, followed by the presentation of simulation results in Section 5.

2. Dynamic Control Model of Robot Arm

In this paper we consider an active, rigid robot arm mechanism with m degrees of freedom. A mathematical model of the complete mechanical system comprises the mechanical part of the system and m actuators that actuate one degree of freedom each. Using Lagrangian techniques to express the dynamical behavior of a rigid (nonflexing) robot arm with m joints results in an equation of the form [8, 23]:

$$\tau_i = \sum_{j=1}^{m} D_{ij}(q)\ddot{q}_j + \sum_{j=1}^{m}\sum_{k=1}^{m} D_{ijk}(q)\dot{q}_j\dot{q}_k + D_i(q),$$

$$i=1,\ldots,m, \tag{1}$$

Where τ_i is the generalized force generated by or required for the motion of the ith mechanical joint, q_i is the coordinate of the ith mechanical joint, $q = (q_1,\ldots,q_m)$, $D_{ij}(q)$ is the inertia load projections to joint i related to the acceleration of the joint variables, $D_{ijk}(q)$ is the inertia load projections to joint i related to the velocity of the joint variables, and $D_i(q)$ is the gravity load at joint i. The function definitions of the dynamical coefficients or dynamical projection functions D_i, D_{ij}, and D_{ijk} can be found in [8,23]. Actuator models are given in the form [24]

$$\ddot{q}_i = a_i\dot{q}_i + f_i\tau_i + b_iu_i, i=1,\ldots,m, \tag{2}$$

where a_i, f_i and b_i are constants, and u_i is the input for the ith actuator, $|u_i| \le u_i^m$ (a positive constant).

Let $x_i = q_i, x_{m+i} = \dot{q}_i$, and denote $(x_1,\ldots,x_{2m})$ by x and $(x_1,\ldots,x_m)$ by $\tilde{x}$. From (1) and (2) we then obtain the state equation of the system:

$$\begin{bmatrix} \dot{x}_1 \\ \vdots \\ \dot{x}_m \\ \dot{x}_{m+1} \\ \vdots \\ \dot{x}_{2m} \end{bmatrix} = \left[\begin{array}{c|c} 0 & I \\ \hline 0 & 0 \end{array}\right] \begin{bmatrix} x_1 \\ \vdots \\ x_m \\ x_{m+1} \\ \vdots \\ X_{2m} \end{bmatrix} + \left[\begin{array}{c} 0 \\ \hline E^{-1}(\tilde{x})\bar{p}(x) \end{array}\right] + \left[\begin{array}{c} 0 \\ \hline E^{-1}(\tilde{x})B \end{array}\right] u \tag{3}$$

where

$$E = \begin{bmatrix} 1-f_1D_{11} & -f_1D_{12} & \cdots & -f_1D_{1m} \\ -f_2D_{21} & 1-f_2D_{22} & \cdots & -f_2D_{2m} \\ \vdots & & & \\ -f_mD_{m1} & -f_mD_{m2} & \cdots & 1-f_mD_{mm} \end{bmatrix},$$

$$
\bar{p} = - \begin{bmatrix} a_1 x_{m+1} + f_1 (\sum_{j=1}^{m} \sum_{k=1}^{m} D_{1jk} x_{m+j} x_{m+k} + D_1) \\ \cdot \\ \cdot \\ \cdot \\ a_m x_{2m} + f_m (\sum_{j=1}^{m} \sum_{k=1}^{m} D_{mjk} x_{m+j} x_{m+k} + D_m) \end{bmatrix}
$$

$$
B = \begin{bmatrix} b_1 & 0 & \dots & 0 \\ 0 & b_2 & \dots & 0 \\ \cdot & & & \\ \cdot & & & \\ \cdot & & & \\ 0 & 0 & \dots & b_m \end{bmatrix}
$$

For simplicity in writing, the dependence of D_i, D_{ij} and D_{ijk} on $\tilde{x}$ has been omitted.

With the task description y solved in the joint space, we then have the following compact representation for the robot arm control dynamics:

$$
\dot{x} = f(x) + \sum_{i=1}^{m} g_i(x) u_i ,
$$
$$
y_i = h_i(x), i=1,\dots r, \tag{4}
$$

where the first equation is the state equation and the second is the output or task equation.

The parameters in (4) can contain many thousands of algebraic terms should the robot arm consist of six joints. This makes the overall system very complex. To have a general idea about its complexity, nonlinearity, and coupling effect we refer the readers to [8].

The robot arm control model (3) is highly nonlinear and strongly coupled. Therefore, in the case of simultaneous motion of several joints, the motion and the torque (or force) applied at one joint have a dynamic effect on the motion at other joints. Since the dynamic coefficients or projection functions are dependent on the values of the joint variables, the effect of dynamic coupling between motions at different joints will depend on the actual link configuration during motion. Furthermore, we must also notice that, as the velocity increases, we cannot neglect the centripetal and Coriolis forces.

With the above discussion in mind, to have an effective dynamic control of robot arms, we ask the following questions:

1. For a nonlinear and coupled system with certain task, does there exist a feedback such that in a suitable coordinate representation the system can be converted to a linear and output decoupled system? What are the necessary and sufficient conditions?
2. If the answer to the above question is positive, then is it possible to construct such a feedback?
3. Is it possible to construct the feedback described above for the robot system (4)?

To answer these questions, we present in the next section our general theoretical results using a differential geometric approach.

3. External Linearization and Simultaneous Output Decoupling for Nonlinear Systems

The decomposition of nonlinear system via a geometric approach was studied by Krener [25]. He considered the problem when there exists a globally decomposed system that is homomorphic rather than equivalent to the given one. Later, Mikhail and Wonham [26] dealt with the local decomposability and the disturbance decoupling problem for nonlinear autonomous systems. In 1981 Isidori, Krener, Gori-Giorgi and Monaco [27] studied the problem of nonlinear decoupling via feedback. The key tools are the nonlinear generalizations of the notion of (A,B) and (C,A) invariant subspaces introduced by Basile and Marro [28] and by Wonham and Morse [29] for linear systems. For a nonlinear system with the same number of input and output channels, they obtained necessary and sufficient conditions for solvability of the static, state-feedback, noninteracting control problem. A similar result is given by Respondek [30].

It has been known for many years that there are nasty-looking systems that are disguised versions of simple systems: they can be made linear [31] or bilinear [32] by a suitable change of coordinates. Brockett [33] introduced the first definition of feedback-transformation, which includes state space change of coordinates, additive state feedback and control space change of coordinates, linear over the reals. Jakubczyk-Respondek [34] and Su, Hunt and Meyer [35,36] generalized Brockett's definition allowing control space change of coordinates, linear over the ring of smooth functions. Nijmeijer-VanderSchaft [37,38] gave a coordinate-free description of the same concept.

Feedback-equivalence is an equivalence relation among systems, and it generalizes the concept of a linear feedback group, which plays a role in linear system theory [39] leading, among other things, to the Brunovsky [40] canonical form and the definition of controllability indices.

For nonlinear systems the appeal of the feedback-equivalence concept relies on the same motivation: one can study equivalence classes of systems containing representations with special structures or with a canonical form of interest for the analysis and synthesis of control systems. In other words, we look for structures that are left invariant or, even more important, that can be destroyed or induced by feedback transformation.

A first and important step in this direction has been accomplished by Brockett [33], Jakubczyk-Respondek [34] and Su, Hunt and Meyer [35,36] who demonstrated necessary and sufficient conditions for a given nonlinear system to be locally, around a given point, feedback equivalent to a linear controllable system.

So far, investigators working in the differential geometric control theory are considered output decoupling (or noninteracting) and external linearization problems separately. However, in practice such as in the dynamic control of robots we often wish to accomplish both the output decoupling and the external linearization together. In this section we will present conditions that assure that for a given nonlinear system we can accomplish output decoupling and external linearization via a nonlinear feedback and a diffeomorphic transformation. Because of lack of space, we will only present the results, omitting the proofs, which may be found in [41].

3.1. Preliminaries

In this subsection we introduce some notations and definitions. We take M (or N) to be a paracompact connected C^∞ manifold of dimension n, V(M) the set of C^∞ vector fields on M and $C^\infty(M)$ the set of C^∞ functions on M. T_xM is the tangent to M at $x \in M$ and TM is a tangent bundle over M; pointwise it is composed of T_xM. A distribution Δ on M is a mapping, which assigns to each $x \in M$ a subspace $\Delta(x)$ of T_xM in a smooth (analytic) fashion. $Sp\{X_1,\ldots,X_k\}$ is the free submodule of V(M) over the ring $C^\infty(M)$ generated by $X_1,\ldots,X_k \in V(M)$. $\phi_t^X(p)$ is the integral curve of $X \in V(M)$ with initial condition

$$\phi_o^X(p) = p, \text{i.e.,} \frac{d}{dt}(\phi_t^X(p)) = X(\phi_t^X(p)), \ \phi_o^X(p) = p.$$

A C^∞ mapping $F: M \to N$ between C^∞ manifolds is a diffeomorphism if it is a homeomorphism and F^{-1} is C^∞. We will denote the differential of a C^∞ mapping F by F_*. If $F:M \to N$ is a diffeomorphism, then $F_*:T_pM \to T_{F(p)}N$ is an isomorphism.

Furthermore, $L_f^{k+1}\lambda \overset{\Delta}{=} \langle dL_f^k\lambda, f\rangle$ with $L_f^o\lambda = \lambda$ for $\lambda \in C^\infty(M), f \in V(M)$. $L_f^{k+1}g \overset{\Delta}{=} L_f(L_f^k g)$ with

$$L_f^o g = g, L_f g = [f,g] = \frac{\partial g}{\partial x} f - \frac{\partial f}{\partial x} g, \text{ for } f,g \in V(M).$$

Definitions

1. A set of C^∞ vector fields $\{X_1,\ldots,X_\ell\}$ on M is involutive if there exist C^∞ functions $\gamma_{ijk}(X)$ such that

$$[X_i,X_j](x) = \sum_{k=1}^{\ell} \gamma_{ijk}(x)X_k(x), \quad 1\le i,j\le \ell, i\ne j.$$

2. A distribution Δ is nonsingular on U, an open subset of M, if there exists an integer d such that $\dim\Delta(p) = d$ for all p U.

3. Distributions $\Delta_1,\Delta_2,\ldots,\Delta_\ell$ on an n-dimensional manifold M with dimensions equal respectively to $d_1,d_2,\ldots,d_\ell$ are simultaneously integrable if in a neighborhood of any $p_o \in M$ there exists a coordinate system $(x_1,\ldots,x_n)$ in which the integral manifolds of $\Delta_i, j = 1,\ldots,\ell$ are of the form $x_i = c_i$ = constant for any $i = 1,\ldots,n$ except $i = r_{j-1}+1,\ldots,r_j$, where $r_p = r_o + \sum_{i=1}^{p} d_i$ for $p = 1,\ldots,\ell$, $r_o = n - \sum_{j=1}^{\ell} d_j$.

3.2 Main Result

Consider a nonlinear system described by

$$\begin{aligned} \dot{x} &= f(x) + \sum_{j=1}^{m} g_j(x)u_j \\ y_i &= h_i(x), i=1,\ldots,r, \end{aligned} \tag{5}$$

where $x \in M, f, g_1,\ldots,g_m \in V(M)$, $h_1,\ldots,h_r \in C^\infty(M)$, and we assume $f(o) = 0$.

Definition 3.1. System (5) is said to be locally externally linearizable or simply externally linearizable on a neighborhood U around an equilibrium point, say $0 \in M$ by (α,β,T), if there exist α, invertible β and a diffeomorphism $T:M \to U$ such that on U the system is described by

$$\begin{aligned} \dot{z} &= \hat{f}(z) + \sum_{j=1}^{m} \hat{g}_j(z)u_i \\ y_i &= \hat{h}_i(z), \quad i = 1,2,\ldots,r, \end{aligned} \tag{6}$$

where

$$\hat{f} = T_*(f) + (T_*(g_1),\ldots,T_*(g_m))\alpha,$$

$$\hat{g} = (T_*(g_1),\ldots,T_*(g_m))\beta,$$

has the Brunovsky canonical form [40] and $\hat{h}_i = h_i \circ T^{-1}$.

Now let $(\hat{g}_1,\ldots\hat{g}_m)$ and $(\hat{h}_1,\ldots,\hat{h}_r)$ be partitioned to k blocks respectively, i.e., $(\hat{g}_1,\ldots\hat{g}_m) = (\hat{g}^1,\ldots,\hat{g}^k)$, each

$$\hat{g}_i = [\hat{g}_1^i,\ldots,\hat{g}_{m_i}^i], \sum_{i=1}^{k} m_i = m;$$

similarly $(\hat{h}_1,\ldots,\hat{h}_r)^T = (\hat{h}^1,\ldots,\hat{h}^k)$, each $\hat{h}^i = [\hat{h}_1^i,\ldots,\hat{h}_{r_i}^i]$, $\sum_{i=1}^{k} r_i = r$.

Definition 3.2. The system (5) is said to be an output block decoupled by (α,β,T) if there exist α, β and T, as in definition 3.1, such that for system (6) the ith block input only affects the ith block output, i.e.,

$$L_{\hat{g}_k^i} L_{\hat{\tau}_1} \cdots L_{\hat{\tau}_s} \hat{h}_t^j = 0.$$

$\forall i\ne j$, $k=1,\ldots,m_i$, $t=1,\ldots,r_j$, where $\hat{\tau}_1,\ldots,\hat{\tau}_s \in \{\hat{f},\hat{g}_1,\ldots,\hat{g}_m\}$.

We next introduce an important subdistribution called controllability distribution [42].

Definition 3.3. A controllability distribution R is a distribution with the following properties

(i) Let $D^\circ \overset{\Delta}{=} R\cap G$ and define D^ℓ inductively as $D^\ell = R\cap([f,D^{\ell-1}] + \sum_{i=1}^{m}[g_i,D^{\ell-1}] + G)$. Then there is a ℓ_* such that $R = D^{\ell_*} = D^{\ell_*+1}$, where $G = Sp\{g_1,\ldots,g_m\}$.

(ii) R is (f,g) - invariant [43].

Note that for each involutive distribution Δ, there exists a unique maximal controllability distribution denoted by mcd (Δ) contained in Δ [42,44]. Now since

$$\Big(\bigcap_{\substack{j=1\\ j\ne i}}^{k} \ker(dh^j)\Big)$$

is an involutive distribution, so let

$$R_i^* = \text{mcd}\Big(\bigcap_{\substack{j=1\\ j\ne i}}^{k} \ker(dh^j)\Big), i=1,2,\ldots,k.$$

We define a sequence of distributions as

$$D_i^{\circ} = R_i^* \cap G$$

$$D_i^{\ell+1} = R_i^* \cap ([f, D_i^{\ell}] + \sum_{j=1}^{m} [g_j, D_i^{\ell}] + G), \; i=1,\dots,k$$

where $G = Sp\{g_1, \dots, g_m\}$.

Now we present the main result.

Theorem 3.4. Suppose $R_1^*, R_2^*, \dots, R_k^*$ are linearly independent; then the system (5) can be externally linearized and simultaneously output decoupled by (α, β, T) on U an open neighborhood around $0 \in M$ if and only if:

(i) $R_1^*, R_2^*, \dots, R_k^*$ are simultaneously integrable and $R_1^* + R_2^* + \dots + R_k^* = T(U)$.

(ii) $G \cap R_1^* + \dots + G \cap R_k^* = G$ and $G \cap R_i^* \neq \{o\}$,

$\forall i=1,2,\dots,k.$

(iii) $D_i^{\ell}, i=1,\dots,K, \ell=1,\dots,n-1$, are nonsingular, involutive and g-invariant.

All conditions stated in the above theorem are verifiable except simultaneous integrability. The following proposition provides an easy criterion to check whether or not a set of distributions are simultaneously integrable.

Proposition 3.5. Let $R_1, \dots, R_k$ be a set of linearly independent distributions and let $\{X_1^i, \dots, X_{m_i}^i\}$ be a basis of $R_i, i=1,\dots,k$, then $R_1, \dots, R_k$ are simultaneously integrable if and only if

$$[X_s^i, X_t^j] \in R_i + R_j$$

for $i,j=1,\dots,k, s=1,\dots,n_i$ and $t=1,\dots,n_j$.

3.3. An Algorithm

In this section we present a sufficient condition for the external linearization and simultaneous output decoupling that yields an algorithm. In spite of its restrictiveness, it is useful in practice. First we introduce some notations.

For the system (5), we define a set of constant numbers

$$\rho_i = \max_{\rho}\{\rho \,|\, L_{g_j} L_f^k h_i(x) = o \quad \forall k<\rho, \forall j, \forall x \in U\}, \; i=1,2,\dots,r.$$

Define an rxm matrix

$$A(x) = (a_{ij}(x))$$

where $a_{ij}(x) = L_{g_j} L_f^{\rho_i} h_i(x)$, $i=1,\dots,r$, $j=1,\dots,m$ and ρ_i are well defined.

To present the algorithm, we need two lemmas.

Lemma 3.6. Suppose $\Delta_1, \Delta_2, \dots, \Delta_k$ are involutive distributions satisfying (i) $\Delta_1 + \Delta_2 + \dots + \Delta_k = T(U)$, and (ii) there exist $\{X_s^i | s=1,\dots,n_i\}$ $i=1,\dots k$, commutative bases of Δ_i, i.e., $[X_s^i, X_\ell^j] = o \quad \forall i \neq j. i,j=1,\dots,k$ and let

$$\phi: (z_1^1, \dots, z_{n_1}^1, \dots, z_1^k, \dots, z_{n_k}^k) \rightarrow \phi_{z_1^1}^{X_1^1} \circ \phi_{z_2^1}^{X_2^1} \circ \dots \circ \phi_{z_{n_k}^k}^{X_{n_k}^k}(o)$$

be a mapping from V, a neighborhood of $o \in R^n$ to U; then (ϕ^{-1}, U) is a coordinate chart of U and on this new coordinate

$$\Delta_i = Sp\{ \frac{\partial}{\partial z_s^i} | \; s=1,\dots,n_i\}, \; i=1,\dots,k.$$

Lemma 3.7. Let $r_i \leq m_i$, $i=1,\dots,k$, and the matrix $A(x)$, defined above, has rank r; then there exist $\alpha(x)$ and invertable $\beta(x)$, satisfying

$$A(x)\alpha(x) = \begin{bmatrix} -L_f^{\rho_1+1} h_1(x) \\ \cdot \\ \cdot \\ \cdot \\ -L_f^{\rho_r+1} h_r(x) \end{bmatrix},$$

$$A(x)\beta(x) = \begin{bmatrix} I_1 & | & 0 \end{bmatrix} \dot{+} \begin{bmatrix} I_2 & | & 0 \end{bmatrix} \dot{+} \dots \dot{+} \begin{bmatrix} I_k & | & 0 \end{bmatrix}.$$

Where $[I_i | 0]$ is an $r_i \times m_i$ matrix with I_i, $r_i \times r_i$ identity, $\dot{+}$ is the direct sum.

Now assume that the conditions of Lemma 3.7 hold. Let $\tilde{f} \overset{\Delta}{=} f + g\alpha$, $\tilde{g} \overset{\Delta}{=} g\beta$, α, β are as in Lemma 3.7; then we have

Theorem 3.8. The system (5) can be externally linearized to the Brunovsky canonical form and simultaneously output decoupled on V if

(i) $[L_{\tilde{f}}^s \tilde{g}^i, L_{\tilde{f}}^t \tilde{g}^j] = 0 \quad \forall i \neq j, s,t=0,1,\dots,n-1,$

(ii) the distributions $D_i^{\ell} \overset{\Delta}{=} Sp\{\tilde{g}^i, \dots, L_{\tilde{f}}^{\ell-1} \tilde{g}^i\}$, $\ell = 0,1,\dots,n-1$, are nonsingular and involutive,

(iii) $\bigoplus_{i=1}^{k} D_i^{n-1} = T(U)$, where U is an open neighborhood of $o \in M$.

Lemma 3.6, Lemma 3.7, and Theorem 3.8 provide a complete algorithm for computing the nonlinear feedback law $u = \alpha(x) + \beta(x)v$ and the diffeomorphic transformation T for a given nonlinear system to be externally linearized and simultaneously output decoupled by (α, β, T).

4. Application of Nonlinear Feedback to Robot Arm Control

The hypotheses and theorems for external linearization and simultaneous output decoupling presented in the previous section can be tested for any nonlinear system. In this paper we are interested in the application of our theoretical results to the control of robot arms. In order to have a realistic quantitative illustration for our theoretical results as applied to robot arm control, we take a specific robot arm as an example. Our example is the JPL-Stanford robot arm since (i) its dynamic model is quantitatively well defined and (ii) its kinematic configuration is a good representative example for a large and interesting class of industrial robots. Here we refer to those robot arms that have two rotary joints at the base followed by a linear joint that replaces a rotary elbow action (see, e.g., the Unimate robot arm).

First we introduce and summarize the dynamic model of the JPL-Stanford robot arm. Then we present the nonlinear feedback and diffeomorphic transformation that convert the overall dynamically controlled robot arm system to a new linear system in the Brunovsky canonical form with outputs simultaneously decoupled. Thereafter we present a new control strategy that also uses optimal control techniques to render control performance robust versus a reasonably large percentage of machine and task model parameter uncertainties.

4.1 JPL-Stanford Arm Dynamic Model

This robot arm has six degrees of freedom in the following sequence starting from the base: 2R-P-3R, where R denotes rotary joint and P denotes prismatic (linear or sliding) joint. The first two rotary joints provide shoulder azimuth and elevation actions, and the last three rotary joints provide the wrist actions: roll-pitch/yaw-roll. The wrist joints are at the end of the linear (sliding boom) joint, as shown in Figure 1.

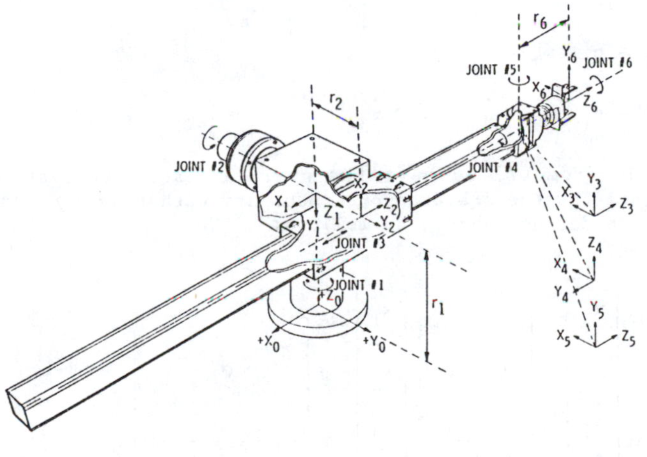

Fig. 1. JPL-Stanford Arm Kinematic Configuration

The dynamic model of the JPL-Stanford robot arm is described in detail in [8,45] including the numerical values of the nonzero geometric and inertia parameters of the arm. In order to keep the mathematical writing within reasonable limits in this paper so that our application results stand out not only clearly but also precisely, we restrict the motion of the six arm joints to the motion of three joints. In our subsequent treatment of the control problem we freeze the motion of the last three (wrist) joints and consider only the dynamically interacting or coupled motion of the first three joints, that is, the rotary motion of the two shoulder joints and the linear motion of the sliding boom. In that case the numeric values of the inertia parameters of the last three (wrist) joints can be added to the inertia parameters of the third (linear) joint by considering the frozen geometric configuration of the last three (wrist) joints relative to the third (linear) joint. Following the general symbolism introduced in Section 2 of this paper, the dynamic model of the JPL-Stanford robot arm restricted to motions of the first three joints becomes:

$$w_i = \sum_{j=1}^{3} D_{ij}(q)\ddot{q}_j + \sum_{j=1}^{3}\sum_{k=1}^{3} D_{ijk}(q)\dot{q}_j\dot{q}_k + D_i(q) ,$$

$$i = 1,2,3.$$

We may denote the above equation as

$$D(q)\ddot{q} = p(q,\dot{q}) + w \qquad (7)$$

Where

$$q = \begin{bmatrix} q_1 \\ q_2 \\ r_3 \end{bmatrix}, \quad w = \begin{bmatrix} w_1 \\ w_2 \\ w_3 \end{bmatrix}, \text{ the generalized force}$$

$$D(q) = \begin{bmatrix} D_{11}(q) & D_{12}(q) & D_{13}(q) \\ D_{21}(q) & D_{22}(q) & D_{23}(q) \\ D_{31}(q) & D_{32}(q) & D_{33}(q) \end{bmatrix}$$

$$p(q,\dot{q}) = - \begin{bmatrix} \sum_{j=1}^{3}\sum_{k=1}^{3} D_{1jk}(q)\dot{q}_j\dot{q}_k + D_1(q) \\ \sum_{j=1}^{3}\sum_{k=1}^{3} D_{2jk}(q)\dot{q}_j\dot{q}_k + D_2(q) \\ \sum_{j=1}^{3}\sum_{k=1}^{3} D_{3jk}(q)\dot{q}_j\dot{q}_k + D_3(q) \end{bmatrix}$$

From [8,45] we have the following functions for the dynamic coefficients or projection functions in this restricted model:

$$D_{11} = m_1 k_{122}^2 + m_2[k_{211}^2 \sin^2 q_2 + k_{233}^2 \cos^2 q_2 + r_2(2y_2 + r_2)] + m_3^*$$
$$[k_{322}^{2*} \sin^2 q_2 + k_{333}^{2*} \cos^2 q_2 + r_3 \sin^2 q_2 (2z_3^* + r_3) + r_2]$$

$$D_{22} = m_2 k_{222}^2 + m_3^*[k_{311}^{2*} + r_3(2z_3^* + r_3)]$$

$$D_{33}^* = M_3^*$$

$$D_{12} = D_{21} = -[m_2 z_2 r_2 + m_3^* r_2(z_3^* + r_3)]\cos q_2$$

$$D_{13} = D_{31} = -m_3^* r_2 \sin q_2$$

$D_{23} = D_{32} = 0$

$D_1 = 0$

$D_2 = g[m_2 z_2 + m_3^*(z_3^*+r_3)]\sin q_2$

$D_3 = -m_3^* g\cos q_2$

$D_{111} = 0$

$D_{122} = [m_2 z_2 + m_3^*(z_3+r_3)]r_2\sin q_2$

$D_{133} = 0$

$D_{211} = [m_2(k_{233}^2-k_{211}^2) + m_3^{*2} -$

$(k_{333}^{2*}-2z_3^* r_3)]\sin q_2\cos q_2$

$D_{222} = 0$

$D_{233} = 0$

$D_{311} = -m_3^*(z_3^*+r_3)\sin^2 q_2$

$D_{322} = -m_3^*(z_3^*+r_3)$

$D_{333} = 0$

$D_{112} = D_{121} = 2\ \{m_2(k_{211}^2-k_{233}^2) +$

$m_3^*[k_{322}^{2*}-k_{333}^{2*} + r_3(2z_3^*+r_3)]\}$

$\sin q_2\cos q_2$

$D_{113} = D_{131} = 2m_3^*(z_3+r_3)\sin^2 q_2$

$D_{123} = D_{132} = -2m_3^*(r_2\cos q_2$

$D_{212} = D_{221} = 0$

$D_{213} = D_{231} = 0$

$D_{223} = D_{232} = 2m_3^*(z_3^*+r_3)$

$D_{312} = D_{321} = 0$

$D_{313} = D_{331} = 0$

$D_{323} = D_{332} = 0$

Where g is the gravity acceleration and

$m_3^* = m_3 + m_4 + m_5 + m_6$

$z_3^* = [m_3 z_3 - m_4 y_4 + m_5 z_5 + m_6(z_6 + r_6)]\ /m_3^*$

$k_{322}^{2*} = [m_3 k_{322}^2 + m_4 k_{433}^2 + m_5 k_{522}^2 + m_6 k_{622}^2]\ /m_3^*$

$k_{333}^{2*} = [m_3 k_{333}^2 + m_4 k_{422}^2 + m_5 k_{533}^2 + m_6 k_{633}^2]\ /m_3^*$

$k_{311}^{2*} = [m_3 k_{311}^2 + m_4 k_{411}^2 + m_5 k_{511}^2 + m_6 k_{611}^2]\ /m_3^*$

Note that, following the notations of [8,45], we put superscript * for those dynamic parameters that belong to the dynamic model restricted to motions at the first three joints. These parameters are different from those that belong to the full model when all joints can be in motion.

We now want the end effector to move on a prescribed trajectory in the work space described by Cartesian coordinates. We have the following general kinematic solution for the end effector position in the work space as a function of joint variables [9,23]:

$$y = \begin{cases} y_x = \cos q_1 \sin q_2 r_3 - \sin q_1 r_2 \\ y_y = \sin q_1\ \sin q_2 r_3 + \cos q_1 r_2 \\ y_z = \cos q_2 r_3 \end{cases} \tag{8}$$

(8) is the 'general positioning task equation' or system output.

Now let

$x_1 = q_1,\ x_2 = q_2,\ x_3 = r_3,$

$x_4 = \dot{q}_1,\ x_5 = \dot{q}_2,\ x_6 = \dot{r}_3.$

Then from (7) we have the overall restricted system model for the JPL-Stanford robot arm with tasks y and D, as defined above, as follows:

$$\dot{x} = \begin{bmatrix} \dot{x}_1 \\ \dot{x}_2 \\ \dot{x}_3 \\ \dot{x}_4 \\ \dot{x}_5 \\ \dot{x}_6 \end{bmatrix} = \left[\begin{array}{c|c} 0 & I \\ \hline 0 & 0 \end{array}\right] \begin{bmatrix} x_1 \\ x_2 \\ x_3 \\ x_4 \\ x_5 \\ x_6 \end{bmatrix} + \begin{bmatrix} 0 \\ \hline [D(\tilde{x})]^{-1}p(x) \end{bmatrix} + \tag{9}$$

$$\begin{bmatrix} 0 \\ \hline [D(\tilde{x})]^{-1} \end{bmatrix} \begin{bmatrix} w_1 \\ w_2 \\ w^2 \end{bmatrix} = f(x) + \sum_{i=1}^{3} g_i(\tilde{x}) w_i$$

$$y(\tilde{x}) = \begin{bmatrix} h_1(x_1,x_2,x_3) \\ h_2(x_1,x_2,x_3) \\ h_3(x_1,x_2,x_3) \end{bmatrix} = \begin{bmatrix} \sin x_1 \sin x_2 x_3 + \cos x_1 r_2 \\ \sin x_1 \sin x_2 x_3 + \cos x_1 r_2 \\ \cos x_2 x_3 \end{bmatrix} \quad (10)$$

where $x = (x_1,\ldots,x_6)$ and $\tilde{x} = (x_1,x_2,x_3)$.

Note that, for simplicity in writing, in (9) we did not include the actuator model and, therefore we use the generalized force w as our control vector.

4.2 Nonlinear Feedback Applied to JPL-Stanford Arm

Given the dynamic model of JPL-Stanford robot described by (9), with the task described by (10), our question is, Does there exist a nonlinear feedback and a diffeomorphic transformation such that the given nonlinear and coupled robot arm system is converted to a linear system where the outputs y are decoupled, i.e., w_i affects only y_i? To answer this question, we apply the external (or exact) linearization and simultaneous output decoupling theorem stated in Section 3 of this paper. The application of this theorem requires to compute R_i^*, the largest controllability distribution contained in $\bigcap_{j\neq i}^{k} (dh^j)^\perp$. To do so, we need first to find Δ_i^*, the largest (f,g) invariant distribution contained in $\bigcap_{j\neq i}^{k} (dh^j)^\perp$ [46], where i = 1, 2, 3, and k = r = 3.

The mathematical details of the above computational procedures can be found in [41]. The concluding result of the computations is that the dynamic model of the JPL-Stanford robot arm given by (9) with the task given by (10) satisfies the necessary and sufficient conditions for external (or exact) linearization and simultaneous output decoupling for all $x \in U$, an open neighborhood of equilibrium point x^o. Therefore there must exist:

1. A nonlinear feedback $w = \alpha(x) + \beta(x)v$ where $\beta(x)$ is invertable.
2. A diffeomorphic transformation T(x) such that, after the nonlinear feedback and diffeomorphic transformation, the new system is linear in the Brunovsky canonical form and the outputs are decoupled.

Now, the next step is: given the dynamic system described by (9) with the 'positioning' task described by (10), apply the algorithm described in Subsection 3.3 of this paper (Lemma 3.6, Lemma 3.7, and Theorem 3.8) to find the required nonlinear feedback $w = \alpha(x) + \beta(x)v$ and diffeomorphic transformation T(x) by which the dynamic system (9) is externally (or exactly) linearized and simultaneously output decoupled in suitable coordinates around the equilibrium point. The mathematical details of applying the algorithm can be found in [41]. Here we only present the results.

1. The required nonlinear feedback is

$$w = \alpha(x) + \beta(x)v$$

$$= -D(\tilde{x})J_h^{-1}\partial J_h \begin{bmatrix} x_4 \\ x_5 \\ x_6 \end{bmatrix} - p(x) + D(\tilde{x})J_h^{-1}v \quad (11)$$

Where J_h is the Jacobian matrix of $h(x_1,x_2,x_3)$, $D(\tilde{x})$ and p(x) are defined previously, and where we introduced the following notations:

$$\partial J_h \triangleq \left[\frac{\partial J_h}{\partial x_1}\begin{bmatrix} x_4 \\ x_5 \\ x_6 \end{bmatrix} \quad \frac{\partial J_h}{\partial x_2}\begin{bmatrix} x_4 \\ x_5 \\ x_6 \end{bmatrix} \quad \frac{\partial J_h}{\partial x_3}\begin{bmatrix} x_4 \\ x_5 \\ x_6 \end{bmatrix} \right]$$

with

$$\frac{\partial J_h}{\partial x_i} = \begin{bmatrix} \frac{\partial J_{11}}{\partial x_i} & \frac{\partial J_{12}}{\partial x_i} & \frac{\partial J_{13}}{\partial x_i} \\ \frac{\partial J_{21}}{\partial x_i} & \frac{\partial J_{22}}{\partial x_i} & \frac{\partial J_{23}}{\partial x_i} \\ \frac{\partial J_{31}}{\partial x_i} & \frac{\partial J_{32}}{\partial x_i} & \frac{\partial J_{33}}{\partial x_i} \end{bmatrix}$$

2. The required diffeomorphic transformation is:

$$z = T(x) = \begin{bmatrix} h_1(x) \\ L_f h_1(x) \\ h_2(x) \\ L_f h_2(x) \\ h_3(x) \\ L_f h_3(x) \end{bmatrix} \quad (12)$$

Where L_f is defined in Subsection 3.1.

By using the nonlinear feedback and diffeomorphic transformation given above, the nonlinear dynamic system (9) with task (10) is converted into the following Brunovsky canonical form and simultaneously output decoupled (see also Fig. 2):

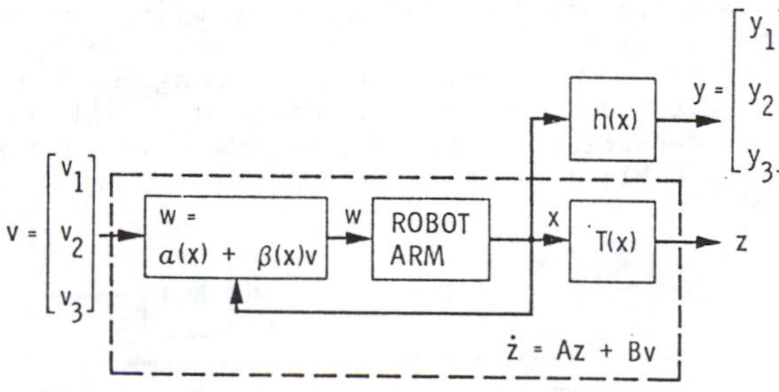

Fig. 2, Externally Linearized and Output Decoupled System when Nonlinear Feedback and Diffeomorphic Transformation Are Introduced.

$$\dot{z} = Az + Bv \quad (13)$$

$$y = Cz \quad (14)$$

where

$$z = \begin{bmatrix} z_1 \\ z_2 \\ z_3 \\ z_4 \\ z_5 \\ z_6 \end{bmatrix} \qquad A = \begin{bmatrix} 0&1&0&0&0&0 \\ 0&0&0&0&0&0 \\ 0&0&0&1&0&0 \\ 0&0&0&0&0&0 \\ 0&0&0&0&0&1 \\ 0&0&0&0&0&0 \end{bmatrix}$$

$$B = \begin{bmatrix} 0&0&0 \\ 1&0&0 \\ 0&0&0 \\ 0&1&0 \\ 0&0&0 \\ 0&0&1 \end{bmatrix}, y = \begin{bmatrix} y_1 \\ y_2 \\ y_3 \end{bmatrix}, C = \begin{bmatrix} 1&0&0&0&0&0 \\ 0&0&1&0&0&0 \\ 0&0&0&0&1&0 \end{bmatrix}$$

As a matter of fact, this system consists of three ($i=1,2,3$) independent subsystems of the following form

$$\dot{z}_i = \begin{bmatrix} 0&1 \\ 0&0 \end{bmatrix} z_i + \begin{bmatrix} 0 \\ 1 \end{bmatrix} v_i$$

$$y_i = [\ 1\ 0\ ]\ z_i$$

where

$$z_i = \begin{bmatrix} z_{2i-1} \\ z_{2i} \end{bmatrix}$$

Since each subsystem has double poles at the origin, the overall system in unstable. To render it stable we can add a linear feedback loop F to the system and thus assign the poles arbitrarily by using the fact that the linear system is completely controllable. It is not difficult to see that, as long as F is a constant block diagonal matrix, the system will remain an output decoupled linear system. We call this new system L & D block, as shown in Figure 3.

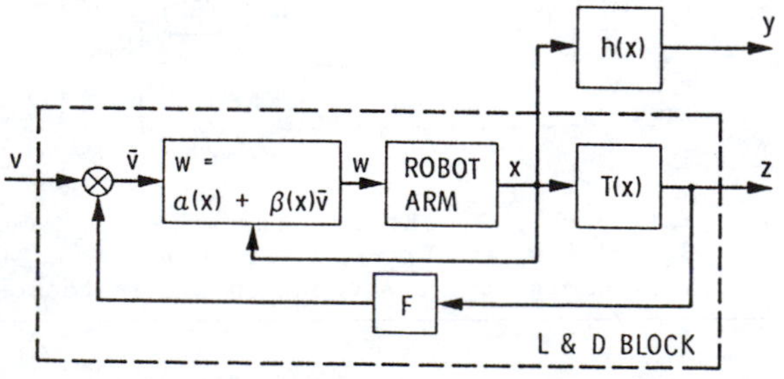

Fig. 3. A Linear Feedback Loop with Constant Gains to Stabilize the Externally Linearized and Output Decoupled System

Remarks

1. When we add the actuator model (see (2)) to the system equation (7), it can be shown that the actuator-expanded system also satisfies the necessary and sufficient conditions for external (or exact) linearization and simultaneous output decoupling. The required nonlinear feedback law including the actuator model becomes then:

$$u = -B^{-1}E(\tilde{x})J_h^{-1}\partial J_h \begin{bmatrix} x_4 \\ x_5 \\ x_6 \end{bmatrix} - B^{-1}\bar{p}(x) + B^{-1}E(\tilde{x})J_h^{-1}v \qquad (15)$$

where J_h, ∂J_h are as defined following (11), the corresponding diffeomorphic transformation $T(x)$ remains the same as (12), and B, E and $\bar{p}(x)$ are as defined in Section 2 of this paper.

2. The control results stated in (11,12,13,14,15) are not restricted to the control of the JPL-Stanford arm only. They are much more general. They can be applied to the control of any robot arm as long as the robot arm dynamic model and task description satisfies the characteristics used in deriving (11,12,15). This is apparent when one considers the details of proofs elaborated in [41].

4.3 A New Strategy for Dynamic Control of Robot Arms

The design of a new dynamic control strategy for robot arms proposed in this paper has three major steps.

(i) Convert the original nonlinear robot arm control dynamics into an externally linearized and simultaneously output decoupled system by using the required nonlinear feedback and diffeomorphic transformations. (See (11,12), the Brunovsky canonical form, (13,14), and Figure 2.)

(ii) Stabilize the externally linearized and simultaneously output decoupled system by designing a linear error correction PD (or PID) controller for each individual (decoupled) subsystem.

(iii) Render the control robust versus uncertainties in machine and task model parameter values by adding an optimal error-correcting loop to each individual (decoupled and linear) subsystem.

Step (i) has been discussed above in Subsection 4.2. Here we discuss steps (ii) and (iii).

First, to stabilize the externally linearized and simultaneously output decoupled system by pole assignment, we add a linear PD feedback controller to the system, as indicated in Figure 3. That is,

$$\bar{v} = v - Fz$$

where

$$F = \begin{bmatrix} f_{11} & f_{12} & 0 & 0 & 0 & 0 \\ 0 & 0 & f_{21} & f_{22} & 0 & 0 \\ 0 & 0 & 0 & 0 & f_{31} & f_{32} \end{bmatrix}$$

(13,14) become then:

$$\dot{z} = Az + B(v-Fz) = (A - BF)z + Bv \quad (16)$$

$$= \begin{bmatrix} 0 & 0 & 0 & 0 & 0 & 0 \\ -f_{11} & -f_{12} & 0 & 0 & 0 & 0 \\ 0 & 0 & 0 & 1 & 0 & 0 \\ 0 & 0 & -f_{21} & -f_{22} & 0 & 0 \\ 0 & 0 & 0 & 0 & 0 & 1 \\ 0 & 0 & 0 & 0 & -f_{31} & -f_{32} \end{bmatrix} z + \begin{bmatrix} 0 & 0 & 0 \\ 1 & 0 & 0 \\ 0 & 0 & 0 \\ 0 & 1 & 0 \\ 0 & 0 & 0 \\ 0 & 0 & 1 \end{bmatrix} \begin{bmatrix} v_1 \\ v_2 \\ v_3 \end{bmatrix}$$

and $y = Cz$ (17)

Thus the new system (L & D block in Figure 3) can be considered to be three independent subsystems. Each has the following second order form:

$$\begin{bmatrix} \dot{z}_{2i-1} \\ \dot{z}_{2i} \end{bmatrix} = \begin{bmatrix} 0 & 1 \\ -f_{i1} & -f_{i2} \end{bmatrix} \begin{bmatrix} z_{2i-1} \\ z_{2i} \end{bmatrix} + \begin{bmatrix} 0 \\ 1 \end{bmatrix} v_i \quad (18)$$

$$y_i = [\ 1\ \ 0\ ] \begin{bmatrix} z_{2i-1} \\ 2i \end{bmatrix} \quad \text{for } i = 1, 2, 3. \quad (19)$$

It can be computed easily that the poles of this subsystem are

$$s_{1,2} = -\xi\omega_n \pm \omega_n j\sqrt{1-\xi^2}.$$

where ξ = damping ratio,

ω_n = natural frequency

and $\omega_n^2 = f_{i1}$, $2\xi\omega_n = f_{i2}$.

At the same time the mathematical model of the real system should correspond to the following form:

$$\begin{bmatrix} \dot{z}^d_{2i-1} \\ \dot{z}^d_{2i} \end{bmatrix} = \begin{bmatrix} 0 & 1 \\ -f_{i1} & -f_{i2} \end{bmatrix} \begin{bmatrix} z^d_{2i-1} \\ z^d_{2i} \end{bmatrix} + \begin{bmatrix} 0 \\ 1 \end{bmatrix} v^d_i \quad (20)$$

$$y^d_i = [\ 1\ \ 0\ ] \begin{bmatrix} z^d_{2i-1} \\ z^d_{2i} \end{bmatrix}, \quad i = 1, 2, 3. \quad (21)$$

where y^d_i is the desired path, and 'd' superscript denotes that these equations are the 'model' equations for a robust (or optimal) control system design.

Now let's denote the output error:

$$e_i = \begin{bmatrix} e_{i1} \\ e_{i2} \end{bmatrix} = \begin{bmatrix} y_i - y^d_i \\ \dot{y}_i - \dot{y}^d_i \end{bmatrix}$$

Thus from (18,19) and (20,21) we have

$$(v_i - v^d_i) = (\ddot{y}_i - \ddot{y}^d_i) + f_{i2}(\dot{y}_i - \dot{y}^d_i) + f_{i1}(y_i - y^d_i)$$

That is

$$\Delta v_i = \dot{e}_{i2} + f_{i2}e_{i2} + f_{i1}e_{i1}$$

or

$$\dot{e}_i = \begin{bmatrix} 0 & 1 \\ -f_{i1} & -f_{i2} \end{bmatrix} e_i + \begin{bmatrix} 0 \\ 1 \end{bmatrix} (v_i - v^d_i) \overset{\Delta}{=} A_i e_i + b_i \Delta v_i \quad (22)$$

Next, as the third step, we introduce an optimal error-correcting loop by optimizing the following cost function for Δv_i:

$$J(\Delta v_i) = \int_0^T \Delta v_i' R \Delta v_i \, dt + \int_0^T e(t)' Q e(t) dt + e(T)' S e(T) \quad (23)$$

Where R = positive definite matrix

Q = semipositive definite matrix

S = semipositive definite matrix

T = terminal time

From the well-known optimal linear control theory (see, e.g., [47]), the optimal correction is

$$\Delta v^0_i = -R^{-1} b_i' P(t) e(t) \quad (24)$$

where

$$P(t) = \begin{bmatrix} P_{11} & P_{12} \\ P_{12} & P_{22} \end{bmatrix}$$

is a positive solution of the Riccati equation:

$$\dot{P}(t) = -P(t)A_i - A_i'P(t) + P(t)b_i R^{-1} b_i' p(t) - Q$$

with

$$p(T) = S$$

If we consider the steady state solution ($t \to \infty$) of the above equation then $\dot{P}(t) = 0$ and the Riccati equation becomes an algebraic equation

$$-PA_i - A_i'P + Pb_i R^{-1} b_i' P - Q = 0$$

That is

$$2f_{i2}P_{12} + R^{-1}P_{12}^2 - Q_{11} = 0$$

$$-P_{11} + f_{i2}P_{12} + f_{i1}P_{22} + R^{-1}P_{12}P_{22} - Q_{12} = 0$$

$$-2(P_{12} - f_{i2}P_{22}) + R^{-1}P_{22}^2 - Q_{22} = 0$$

Thus (24) becomes

$$\Delta v_i^0 = -R^{-1}b_i'Pe(t) \tag{25}$$

that is,

$$\Delta v_i^0 = - R^{-1} [P_{12} (y_i - y_i^d)] + P_{22}(\dot{y}_i - \dot{y}_i^d)]$$

Where

$$P_{12} = - R f_{i2} \pm R\sqrt{f_{i2}^2 + Q_{11}R^{-1}}$$

$$P_{22} = - 2Rf_{i2} \pm R\sqrt{4f_{i2}^2 - (2f_{i2} \mp 2\sqrt{f_{i2}^2 + Q_{11}} - Q_{22})R^{-1}}$$

The proposed new strategy for dynamic control of robot arms is summarized in Figure 4 and has been evaluated by a digital simulation technique described in the next section.

5. Simulation of New Dynamic Control Strategy

An extensive digital simulation program was developed to answer the following main question: How stable and robust is the new overall dynamic control scheme versus model imperfections and other causal uncertainties? Note that the nonlinear feedback is essentially based on robot arm model (geometric and inertia) parameters. In particular, we were interested in evaluating the following cases or in finding quantitative answers to the following questions:

1. If the initial condition is away from the desired path, how does the real trajectory track (or converge to) the desired path?
2. In general, how far away are the real trajectory and velocity from the desired path and velocity; that is, what is the overall position accuracy of the control scheme?
3. What is the torque (or force) that is needed for each joint?
4. Considering the perturbation because of the actuators, gears and amplifiers, if we change by a certain percentage the torques (or forces) that are applied to the robot arm relative to the required torque (or force), how does this difference affect the behavior of the overall system? This question is equivalent to the situation when the actuator model is missing from the nonlinear feedback.
5. When F (linear PD feedback gain), R, Q, S (in the optimal controller) and other factors change, how do they affect the answers to the above questions?

5.1 Simulated Task

To make the simulation more realistic dynamically, instead of some straight lines or movement on a horizontal plane, we chose the desired path of the end effector as a circle C with the center at the origin, radius R_d (= 2.0 feet) and constant angular velocity ω on the plane X'OY', which is tilted by 45° relative to the XOY plane (see Fig. 5). From Figure 5 the explicit expression of the desired end effector path and velocity can be easily derived, referenced to the task space (that is, to the XYZ frame).

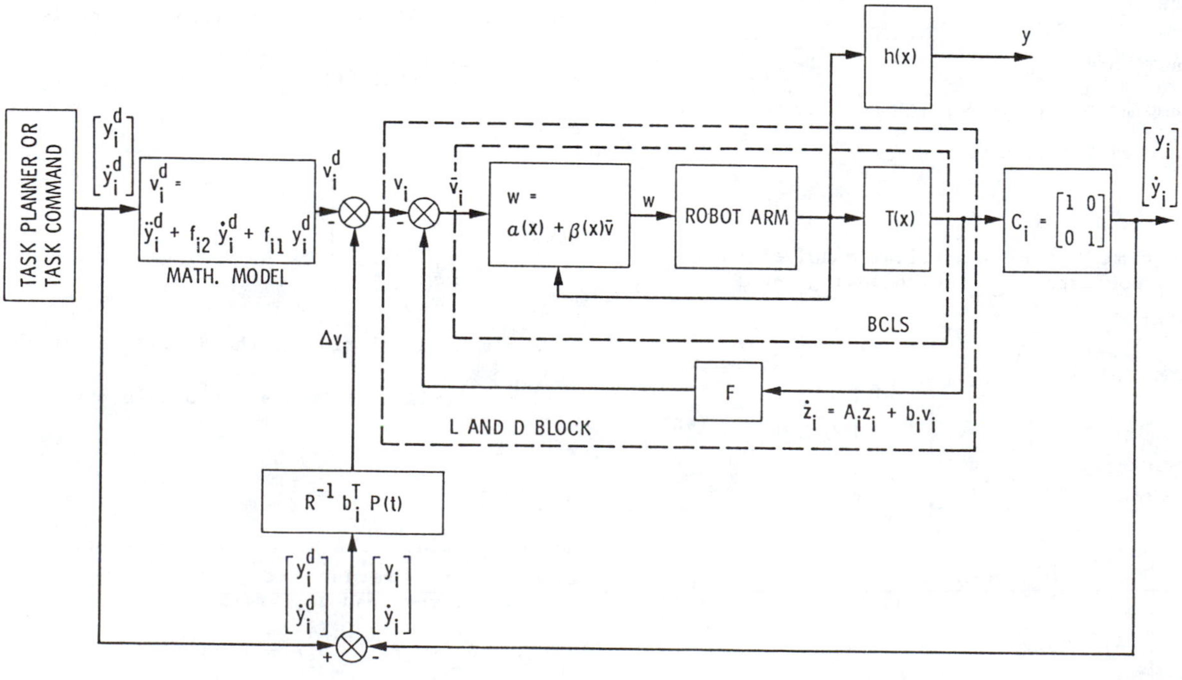

Fig. 4. New Strategy for Dynamic Control of Robot Arms, Introducing an Optimal Error-Correcting Loop

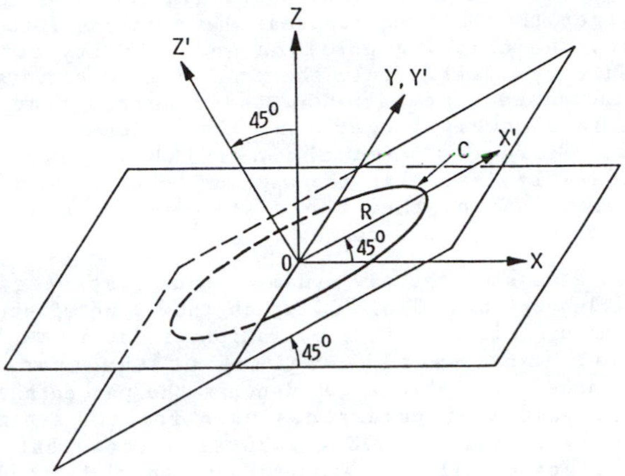

Fig. 5. Desired Circular Tilted Path in Cartesian Coordinate Frame for Performance Simulation Study

The desired path is:

$$\begin{bmatrix} y_1^d \\ y_2^d \\ y_3^d \end{bmatrix} = \begin{bmatrix} x \\ y \\ z \end{bmatrix} = \begin{bmatrix} \frac{\sqrt{2}}{2}(x'+z') \\ y' \\ \frac{\sqrt{2}}{2}(-x'+z') \end{bmatrix} = \begin{bmatrix} \frac{\sqrt{2}}{2}(R_d\cos\omega t+0) \\ R_d\sin\omega t \\ \frac{\sqrt{2}}{2}(-R_d\cos\omega t+0) \end{bmatrix}$$

$$= \begin{bmatrix} \frac{\sqrt{2}}{2}R_d\cos\omega t \\ R_d\sin\omega t \\ -\frac{\sqrt{2}}{2}R_d\cos\omega t \end{bmatrix} \tag{26}$$

The desired velocity is:

$$\begin{bmatrix} y_4^d \\ y_5^d \\ y_6^d \end{bmatrix} = \begin{bmatrix} \dot{y}_1^d \\ \dot{y}_2^d \\ \dot{y}_3^d \end{bmatrix} = \begin{bmatrix} \frac{-\sqrt{2}}{2}R_d\omega\sin\omega t \\ R_d\omega\cos\omega t \\ \frac{\sqrt{2}}{2}R_d\omega\sin\omega t \end{bmatrix} \tag{27}$$

where $R_d = 2$ feet. We assume that the period T is 2, 4, 6, and 8s, respectively, yielding the following angular velocities in the task space:

$\omega = \pi, \frac{\pi}{2}, \frac{\pi}{3}, \frac{\pi}{4}$ (rad/sec). (See Fig. 6.)

The output of the digital simulation program includes:

1. The values of the real position and velocity of the end effector in Cartesian coordinates.
2. The position and velocity errors between the real and desired path in Cartesian coordinates.
3. The torque (or force) of each joint for each time instant. The computer printer also gives graphic plots for the position and velocity errors.

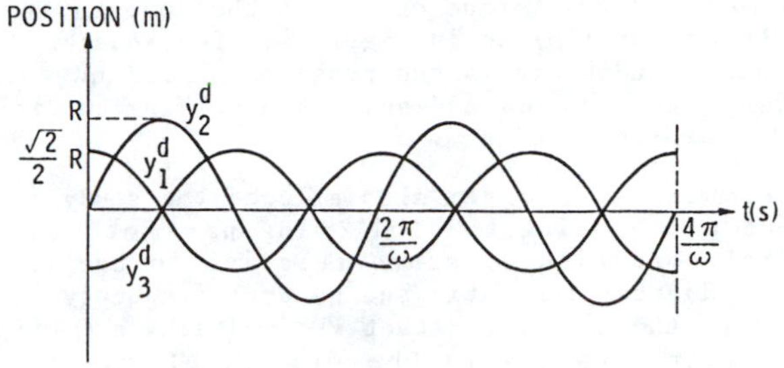

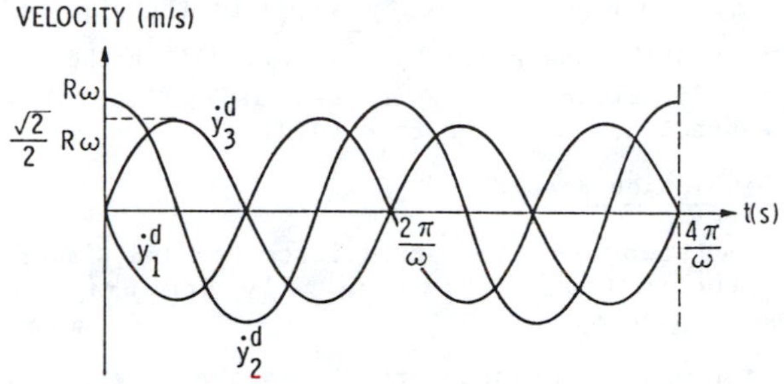

Fig. 6. Cartesian Components of Desired Path and Velocity of End Effector

5.2 Simulation Program and Procedures

The details of the simulation program and procedures can be found in [41]. Here we only summarize the main features.

Essentially the simulation program implemented the new dynamic control strategy shown in the block diagram in Figure 4. Two sets of values were used for the dynamic parameters of the JPL-Stanford arm: 'nominal' and 'assigned' values. The nominal values were used in the design and computation of the nonlinear feedback and diffeomorphic transformation. The nominal values are the 'best known values' and have been taken from [8]. These are the 'ideal' values that would have been used in a real-time implementation of the nonlinear feedback. The assigned values were used in the computational implementation of the 'Robot Arm' block in Figure 4. The assigned values differed from the nominal values with varying percentages in order to simulate the difference between a 'real arm' and its 'ideal' parameters. The difference between the assigned and nominal values automatically takes care of the fact that we omitted the actuator model from the model of the arm dynamics; that is, the actuator model was omitted from the design and calculation of the nonlinear feedback.

For each simulation the following 'base assignments' were employed:

1. Values of all parameters of robot arm were 10% more than the nominal values.
2. The load held by the hand was 100% more than the nominal load, which is 2 pounds.
3. Because of the anticipated perturbation originating from the actuators, gears and amplifiers the produced torques (or forces) were 20% different from the torques (or forces) that were required to track the desired path.

Each time we only changed one assigned value, while all other assigned values remained unchanged. This means, for example, that when we evaluated the effect of changing arm parameter values then, in addition to all these parameter variations, we also had at least 100% load and 20% torque errors in the computation of the 'Robot Arm' block in Figure 4. Consequently, our assignment procedure in the program simulated rather large perturbations and quite bad working conditions for the JPL-Stanford arm.

The simulation program also allowed the study of the effect of changing the following parameters: (i) initial condition of state relative to the desired state, (ii) damping ratio and natural frequency in the design of the linear constant PD feedback, and (iii) R and Q matrix values in the design of the linear optimal error-correcting feedback loops.

The 'task command' input (y_i^d, $\dot{y}_i^d$) to the simulation program was the one given by (26,27). Altogether more than 60 simulation cases were run, using FORTRAN and some subroutines modified from [48].

5.3 Simulation Results

The simulation results are illustrated and discussed in detail in [41]. Here we only summarize and illustrate the main points.

The first major simulation result is that the optimal error-correcting feedback loop Δv, based on the steady state Riccati equation, plays an important role in eliminating position and velocity errors in the task space during tracking of the desired path. Moreover, it also reduces the servo power required in the tracking. These facts are illustrated in Table 1 and Figure 7. Note that the 'no Δv' results shown in Table 1 and Figure 7 still include the use of the linear constant F (PD) feedback loop shown in Figure 4. Motivated by this fundamental result, all other simulations were performed by using the optimal steady state error-correcting feedback, Δv.

Table 1. The Effect of Optimal Error Correction Δv

	pos. error (m)	vel. error (m/s)	Tmax (steady state) Joint1 (n.m)	Joint2 (n.m)	Joint3 (n)	conv. time T_c (s)
With ΔV	0.0022	0.0033	1.80	114.5	68.8	0.5
No ΔV	0.4200	0.7700	2.23	191.8	94.6	/

The simulations have also shown that a suitable selection of S, Q, and R matrices in computing the optimal steady state error-correcting feedback gains also plays an important role in reducing position and velocity errors in tracking a desired path. On the other hand, the selection of damping ratio has practically no effect on tracking performance; only the selection of natural frequency influences to some extent the tracking performance.

The variations in the initial conditions of the state affects only the torques (or force) in the beginning of the tracking task, and has no effect on position and velocity errors. Of course, some 'bad' initial conditions may cause the use of excessive (i.e., not available) torques and forces, and therefore will increase the convergent time.

In general, the simulations have shown that as long as the available 'real' control power (actuator output) is larger than the required maximum tracking force or torque, the tracking position and velocity errors remain very small. Only the transient or convergent time increases somewhat when the required tracking power is somewhat larger than the maximum available power. However, when the available power is considerably less than the maximum power required in the steady state, then the system will output big errors or might even diverge.

The robustness of the new dynamic control strategy is best illustrated in Table 2, which shows the effect of assigned dynamic parameter variations relative to the nominal (or 'ideal') values on the tracking performance. In Table 2 COE denotes the percentage of nominal values of parameters used for the assigned values. For example, COE = 1.30 simulates that the real values of all the parameters of the JPL-Stanford robot arm are 30% more than the nominal values. Table 2 shows that COE has only a minor effect on the position and velocity errors and mainly affects the maximum torques TMAX of each joint during tracking the desired path. The overall system has a good robustness. It can be seen from Table 2 that even when COE is 30% different from the nominal value (COE = 1.0), the system is still stable and the errors are still in an acceptable minimum range. Note that, in addition to the 30% difference in the dynamic parameter values, the load in the hand was 100% heavier (4 lb instead of 2 lb) and the real servo power output was 20% less than the assumed nominal (or 'ideal') output.

In the simulation program the integration step size was equal to the sampling period. It turned out that the critical integration step size was 0.017 sec, corresponding to 60 Hz sampling frequency. For larger integration step size (or, for lower sampling frequencies), the system became unstable or divergent. In the simulation program a fourth order Runge-Kutta integration procedure was employed. It took about 15 ms CPU time on an IBM 370 computer to compute the required control torques (or forces) for each step.

6. Summary and Conclusions

The results presented in this paper touch both control theory and control applications.

Theory. Using differential geometric system theory, necessary and sufficient conditions were obtained for the existence of a nonlinear feedback and a diffeomorphic transformation, which convert a given nonlinear system to a linear one and provide simultaneous output decoupling. An algorithm was also derived (Lemma 3.6, Lemma 3.7 and Theorem 3.8) for finding the required nonlinear feedback and diffeomorphic transformation for a given nonlinear system.

The general theoretical results presented here extend the previous theoretical results on nonlinear feedback decoupling for the dynamic control of robot arms obtained by Freund [16,6]. The extensions are:

(i) Our results provide both necessary and sufficient conditions, and the output can be block decoupled, and each block can be a vector. Freund gives only sufficient condition for a single (not block) output decoupling.

(ii) Our algorithm for finding the required nonlinear feedback can be applied to systems in which the dimension of output is less than or equal to the dimension of input. In Freund's result the dimension of output must be equal to the dimension of input.

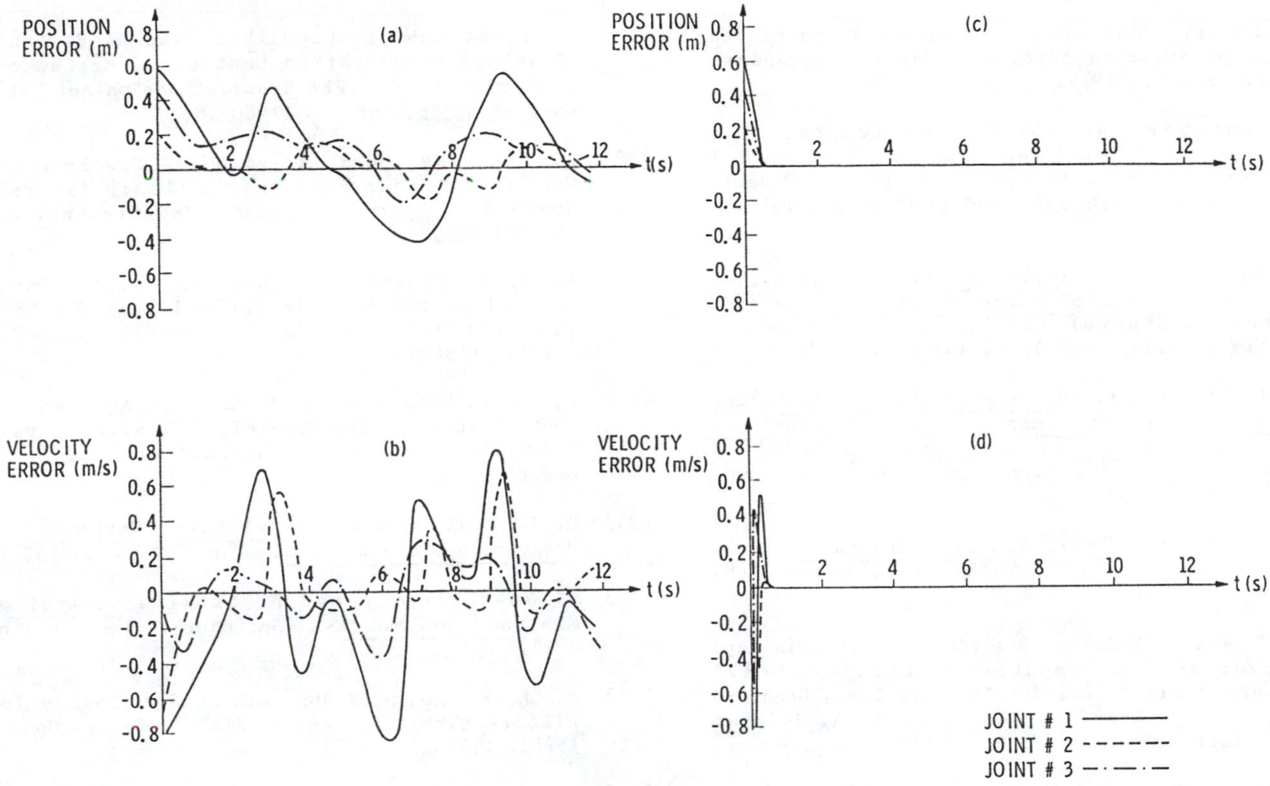

Fig. 7. Errors With and Without Optimal Error Correction Feedback Δv. [(a)(b): Without Δv; (c)(d): With Δv.]

Table 2. The Effect of Dynamic Model Parameter Deviations from Nominal Values

COE	pos. error (m)	vel. error (m/s)	Tmax Joint1 (n.m)	Tmax Joint2 (n.m)	Tmax Joint3 (n)	conv. time T_c (s)
0.70	0.0045	0.0051	1.41	51.0	43.6	0.5
0.80	0.0030	0.0023	1.36	64.6	49.8	0.5
0.90	0.0014	0.0013	1.33	79.7	56.1	0.5
1.00	0.00072	0.00094	1.38	64.5	56.5	0.5
1.10	0.0022	0.0033	1.80	114.5	68.8	0.5
1.20	0.0042	0.0059	2.23	134.2	75.3	0.5
1.30	0.0063	0.0087	2.66	155.4	81.8	0.5

(iii) Our results provide nonlinear feedback decoupling in terms of the system's state. Freund provides nonlinear feedback decoupling in terms of system output; that is, the new system is not necessarily a linear system.

Application. The developed general theoretical results were applied to the derivation of the required nonlinear feedback and diffeomorphic transformation for external (or exact) linearization and simultaneous output decoupling in the dynamic control of robot arms that perform trajectory tracking (position and velocity control) tasks in the work space. This has been illustrated by an example related to a restricted (three-dimensional) motion of the JPL-Stanford robot arm performing trajectory tracking.

Based on nonlinear feedback and simultaneous output decoupling, a new control strategy was proposed for dynamic control of robot arms. First, the externally linearized and output decoupled system was stabilized by a constant gain linear feedback. Then an optimal error-correcting feedback loop was added to the system based on the steady state Riccati equation. The simulation studies performed for the JPL-Stanford robot arm gave very encouraging results. They have shown that the proposed new control strategy is indeed robust. It can cope with considerable causal uncertainties in model and task parameter values. This suggests that the complex dynamic model equations for robot arms can be simplified if real time computational burden warrants the simplification. Otherwise, for the simulated example, the computational burden was not overwhelming in view of the capabilities of today's computers.

References

[1] D. L. Pieper, The Kinematic of Manipulators Under Computer Control. Ph.D. Dissertation, Stanford University Memo 72. Stanford, California: Stanford University Artificial Intelligence Laboratory (1968).

[2] D. E. Whitney, 'Resolved Motion Rate Control of Manipulators and Human Prostheses.' IEEE Transactions MMS-10, 47-53 (1969).

[3] M. E. Kahn and B. Roth, 'The Near-Minimum-Time Control of Open-Loop Articulated Kinematic Chains.' Trans. ASME, Journal of Systems, Measurement, and Control 93, 164-172 (1971).

[4] E. P. Popov et al., 'Synthesis of Control System of Robots Using Dynamic Models of Manipulation

Mechanism.' (in Russian) Proceedings of 6th IFAC Symposium on Automatic Control in Space, Nauka, Moscow, 1974.

[5] S. Dubowsky and D. T. Des Forges, 'The Application of Model-Referenced Adaptive Control To Robotic Manipulators.' Journal of Dynamic Systems, Measurement, and Control 101 193-200 (1979).

[6] E. Freund and M. Syrbe, Control of Industrial Robots by Means of Microprocessors. Lecture Notes in Control and Information Sciences, Springer-Verlag, New York, 167-185, 1977.

[7] B. R. Markiewicz, Analysis of the Compute Torgue Drive Method and Comparison with Conventional Position Servo for a Computer-Controlled Manipulator. Jet Propulsion Lab., Tech. Memo. 33-601, Mar. 1973.

[8] A. K. Bejczy, Robot Arm Dynamics and Control. Jet Propulsion Lab., Tech. Memo. 33-669, Feb. 1974.

[9] R. C. Paul, 'Modeling Trajectory Calculation and Servoing of a Computer Controlled Arm.' Stanford Artificial Intelligence Lab., Stanford University, Stanford, California, A. I. Memo 177, Sept. 1972.

[10] M. H. Raibert and B. K. P. Horn, 'Manipulator Control Using the Configuration Space Method.' Industrial Robot, 5, 69-73 (1978).

[11] H. Hemami and P. C. Camana, 'Nonlinear Feedback in Simple Locomotion System.' IEEE Transactions on Automatic Control, AC-19, 855-860 (1976).

[12] J. R. Hewit and J. R. Burdess, 'Fast Dynamic Decoupled Control for Robotics, Using Active Force Control.' Mechanism and Machine Theory Journal (1980).

[13] D. Lentini, F. Nicolo, S. Nicosia, and L. Veccttio, 'Decoupling and Nonlinearities Compensation in Robot Control Systems.' April 1980.

[14] F. Nicolo and S. Nicosia, 'Dynamic Control of Industrial Robots.' 8th Italian Conference on Systems Theory in Economics Man. Tech. (1980).

[15] S. Nicosia, F. Nicolo, and D. Lentini, 'Dynamical Control of Industrial Robots with Elastic and Dissipative Joints.' International Federation of Automatic Control 8th Triennial World Congress, Kyoto, Japan, August 24-28, 1981.

[16] E. Freund, 'Fast Nonlinear Control with Arbitrary Pole-placement for Industrial Robots and Manipulators.' The International Journal of Robotics Research 1(1) (1982).

[17] G-L. Luo and G. N. Saridis, 'Robust Compensation of Optimal Control for Manipulators.' Proceedings of the 21st IEEE Conference on Decision and Control, Orlando, Florida 1, 351-56 (1982).

[18] G. N. Saridis and C. S. G. Lee, 'An Approximation Theory of Optimal Control for Trainable Manipulators.' IEEE Transactions on Systems, Man, and Cybernetics SMC-9(3), 152-159 (1979).

[19] J. Y. S. Luh, M. W. Walker, and R. P. Paul, 'Resolved Acceleration Control of Mechanical Manipulators.' IEEE Transactions on Automatic Control AC-25, 468-474 (1980).

[20] M. Takegake and S. Arimoto, 'A New Feedback Method for Dynamic Control of Manipulators.' Journal of Dynamic Systems, Measurement, and Control 102, 119-125, (1981).

[21] A. J. Koivo and T. H. Guo, 'Adaptive Linear Controller for Robotic Manipulators.' IEEE Transactions on Automatic Control AC-28(2), 162-171 (1983).

[22] C. S. G. Lee and M. J. Chung, 'An Adaptive Control Strategy for Mechanical Manipulators.' IEEE Transactions on Automatic Control, in press.

[23] R. P. Paul, Robot Manipulators: Mathematics, Programming, and Control. The MIT Press, 1981.

[24] M. Vukobratovic and D. Stokic, Control of Manipulation Robots. Springer-Verlag, Berlin, 1982.

[25] A. J. Krener, 'A Decomposition Theory for Differentiable Systems.' SIAM J. Contr. Optim. 15(5), 813-829 (1977).

[26] S. H. Mikhail and W. H. Wonham, 'Local Decomposability and the Disturbance Decoupling Problem in Nonlinear Autonomous Systems.' Proceedings of the 16th Allerton Conf. Commun. Contr. Comput., 664-669, Oct. 1978.

[27] A. Isidori, A. J. Krener, C. Gori-Giorgi, and S. Monaco, 'Nonlinear Decoupling via Feedback: A Differential Geometric Approach.' IEEE Transactions on Automatic Control AC-26(2), (1981).

[28] G. Basile and G. Marro, 'Controlled and Conditioned Invariant Subspaces in Linear System Theory.' J. Optimiz. Theory Appl. 3(5), 306-315 (1969).

[29] W. M. Wonham and A. S. Morse, 'Feedback Invariants of Linear Multivariable Systems.' Automatica 8, 93-100 (1972).

[30] W. Respondek, 'On Decomposition of Nonlinear Control Systems.' Systems and Control Letters 1, 301-308 (1982).

[31] A. J. Krener, 'On the Equivalence of Control Systems and Linearization of Nonlinear Systems.' SIAM J. Contr. and Opt. 11 (1973).

[32] J. L. Sedwick and D. L. Elliott, 'Linearization of Analytic Vector Fields in the Transitive Case.' J. Diff. Eqs. 25(3), 377-390 (1977).

[33] R. W. Brockett, 'Feedback Invariants for Nonlinear Systems.' IFAC Congress, Helsinki, 1978.

[34] B. Jakubczyk and W. Respondek, 'On Linearization of Control Systems.' Bull. Acad. Pol. Sci., Ser. Sci. Math. Astronom. Phys. 28 (1980).

[35] R. Su, 'On the Linear Equivalence of Nonlinear Systems.' Systems and Control Letters 2(1) (1982).

[36] L. R. Hunt, R. Su, and G. Meyer, 'Design for Multi-Input Nonlinear Systems.' Differential Geometric Control Theory, Proceedings of the Conference held at MTU, 268-298, June 28-July 2, 1982. (Published by Birkhauser.)

[37] H. Nijmeijer and A. Van der Schaft, 'Controlled Invariance for Nonlinear Systems.' IEEE AC-27(4), 904-914 (1982).

[38] H. Nijmeijer and A. Van der Schaft, 'Controlled Invariance by Static Output Feedback for Nonlinear Systems.' System and Control Letters 2(1), 39-47 (1982).

[39] W. M. Wonham, 'Linear Multivariable Control, A Geometric Approach.' 2nd ed., Applic. of Math 10. Springer-Verlag, Berlin, 1979.

[40] P. Brunovsky, 'A Classification of Linear Controllable Systems.' Kibernetica (Praha) 6, 173-188 (1970).

[41] Y. Chen, Nonlinear Feedback and Computer Control of Robot Arms. D.Sc. Dissertation, Department of Systems Science and Mathematics, Washington University, St. Louis, Missouri, December 1984.

[42] A. J. Krener and A. Isidori, '(Adf, G) Invariant and Controllability Distributions.' Lecture Notes in Control and Information Sciences 39, 157-164 (1982).

[43] A. Isidori, A. J. Krener, C. Gori-Giorgi, and S. Monaco, 'Locally (f,g)--Invariant Distributions.' System and Control Letters 1, 12-15 (1981).

[44] H. Nijmeijer, 'Controllability Distributions for Nonlinear Control Systems.' Systems and Control Letters 2(2), 122-129 (1982).

[45] A. K. Bejczy, Dynamic Models and Control Equations for Manipulators, JPL Report, 715-719, November 30, 1979.

[46] A. Isidori, 'Lecture Notes on Differential Geometric Control Theory,' to appear in 1985.

[47] T. Kailath, Linear Systems, Prentice-Hall, Inc., Englewood Cliffs, N.J., 1980.

[48] J. L. Melsa and S. K. Jones, Computer Programs for Computational Assistance in the Study of Linear Control Theory. McGraw-Hill Book Company, New York, 1973.

DYNAMIC CONTROL OF ROBOTS WITH JOINT ELASTICITY

Alessandro De Luca

Dipartimento di Informatica e Sistemistica
Università di Roma "La Sapienza"
Via Eudossiana 18 , 00184 Roma, Italy

Abstract. The problem of controlling the dynamic behavior of robots with rigid links but in presence of joint elasticity is much more complex than the one in which the manipulator is assumed to be rigid in both the links and the transmission elements. When the full nonlinear and interacting dynamics is taken into account in the model, it is known that the resulting system may not be feedback linearizable using nonlinear *static* state-feedback. It is shown here that use of the more general class of *dynamic* nonlinear state-feedback allows to solve both the feedback linearization and the input-output decoupling problems. A constructive procedure for the decoupling and linearizing feedback is given which is based on generalized system inversion and on the properties of the so-called zero-dynamics of the system. A case study of a planar two-link robot with elastic joints is included. The role of dynamic feedback for this class of robots is discussed.

1. Introduction

In the process of modeling the dynamic behavior of robot manipulators, some physical phenomena like backlash, stiction, link flexibility and joint elasticity are usually neglected. In general, when a reliable model of such disturbances is available, one can reduce the control effort and achieve better performances. Various experimental and simulation studies [1,2,3] have shown that elasticity of the transmission elements between actuators and links may have a relevant influence on robot dynamics. In particular, robots using transmission belts, long shafts or harmonic drives show a typical resonant behavior in the same range of frequencies used for control; this effect can be reconducted to the presence of elasticity at the joints between the rigid links of the arm.

From the modeling point of view, joint elasticity implies that the position of the actuator (i.e. the angle of the motor shaft) is not uniquely related to the position of the driven link. This *internal* deflection is taken into account by inserting a linear torsional spring at each joint. As a consequence, the rigid arm dynamic model has to be modified in order to describe completely the relation between applied torques and links motion. For quasi-static applications, simplified models which consider only the dynamics of the drive system have been used widely, see e.g. [4]. Models including the full nonlinear dynamic interactions among joint elasticities and inertial properties of links and actuators have been first introduced in [5]. There exist programs that generate automatically the dynamic equations for these arms, using symbolic manipulation languages [6].

Control of robot arms with joint elasticity has recently become an active area of research. Several advanced control approaches have been proposed using singular perturbation techniques [7], integral manifold design [8,9], sliding mode [10], pseudo-linearization [11], and model reference adaptive control [12]. It is worth to point out that none of these methods achieves an *exact* design, in the sense that is able to mimic the results obtained in the nominal case for fully rigid manipulators.

The reason for such an outgrow of methods stands in the peculiar control theoretic feature that distinguishes robots with joint elasticity from rigid manipulators: in fact, this class of nonlinear systems does not satisfy the necessary conditions for obtaining linearization and noninteraction using nonlinear static state-feedback [13,14]. *In general*, the so-called inverse dynamics or computed torque method, which is the standard nominal trajectory control for rigid arms, cannot be extended directly. There are however simple types of robots with elastic joints, like the single-link [15] and the two-link cylindric arm [16], where use of nonlinear static feedback still leads to a linear system; in these cases, the closed-loop equivalent behavior is that of, respectively, one or two independent chains of *four* integrators each.

A detailed analysis of several kinematic types of arms with elastic joints is contained in [17,18]. The resulting picture is quite intriguing. Feedback linearization may or may not be achieved, depending on the specific kinematic arrangement (i.e. the set of Denavit-Hartenberg parameters) of the robot. In particular, it is never possible for structures in which two or more elastic joints have parallel axes of rotation, like for the common case of a two-link planar arm. As an effort to achieve feedback linearization properties, a simplified model which neglects the inertial couplings between actuators and links has been investigated in [19].

Only recently a unifying perspective has been found for extending exact methods of nonlinear control to the full model of robots with joint elasticity. The keypoint is to enlarge the class of allowed feedback rules to the use of *dynamic* nonlinear state-feedback. Exploiting this concept, input-output decoupling and exact state linearization have been obtained in the closed-loop control of two arms with joint elasticity, a two-link arm with gravity [20] and a three-link anthropomorphic one [21], both of which cannot be linearized using only static feedback.

These results can be applied to the whole class of robots with elasticity at the joints [17] and were obtained by extensive use of nonlinear differential-geometric concepts such as the properties of the maximal controlled invariant distribution [22] associated to the system. The complexity of the techniques and of the computations involved has limited to a certain extent the understanding of the generality of such a design.

Reprinted from *Proc. IEEE ROBOT '88*, Philadelphia, PA, April 1988, pp. 152–158.

In this paper, the introduction of dynamic compensation in the control scheme follows from a more natural approach. The control design is based on the capability of properly recovering the inputs to the system from the knowledge of the outputs and of their time derivatives, an issue which is related to system inversion. In particular, a full linearizing controller can be obtained if the system has *no zero-dynamics* [23], that is if there is no dynamics left in the system once the output is forced to be zero. This concept is a quite useful extension to nonlinear systems of the notion of transfer function zeros for linear systems. The *generalized inverse dynamics* approach that is presented here incorporates the use of either static or dynamic state-feedback. In the multi-input multi-output case, the need of the latter arises only if the so-called decoupling matrix [22] of the system is singular.

Dynamic modeling of robot arms is briefly reviewed next. Section 3 outlines how to construct a full linearizing controller for nonlinear systems without zero-dynamics [24,25], a result which is used in Section 4 for a planar two-link robot with joint elasticity. Discussion of the obtained results and comparison with other approaches completes the paper.

2. Dynamic model of robots with joint elasticity

As a whole, a robot arm with N elastic joints can be seen as a set of 2N elastically coupled rigid bodies, the N actuators and the N links. In face of 2N mechanical degrees of freedom, only N independent control inputs are available, the motor torques acting on the actuator side of the elastic joints.

The dynamic equations of motion are obtained following a Lagrangian approach. Two variables are associated to the generic i-th elastic joint (see Figure 1): q_{2i-1}, the rigid position of the i-th actuator with respect to the (i-1)-th link, and q_{2i}, the elastic position of the i-th link with respect to the previous one. The 2N-dimensional vector $\mathbf{q}$ of generalized coordinates is partitioned into two N-vectors $\mathbf{q}_r$ and $\mathbf{q}_e$ containing respectively the odd and the even components.

The potential energy U(q) and the kinetic energy T(q,q') are computed in the usual way, considering the arm as an open chain of 2N rigid bodies, links and actuators (see the example in Appendix). Beside the gravitational contribution $U_g(q)$, U(q) contains also the elastic energy stored in the joints:

$$U_e(q) = \sum_{i=1}^{N} U_{e,i}(q_{2i-1}, q_{2i}) = \sum_{i=1}^{N} \frac{1}{2} K_i \left[q_{2i} - \frac{q_{2i-1}}{N_i} \right]^2$$

K_i is the elastic constant and $N_i \geq 1$ is the transmission ratio of the i-th drive. No damping is given to the springs modeling joint elasticity. The Euler-Lagrange equations particularize as:

$$\frac{d}{dt}\left[\frac{\partial T}{\partial \dot{q}_j}\right] - \frac{\partial T}{\partial q_j} + \frac{\partial U_g}{\partial q_j} - \frac{K_i}{N_i}\left[q_{2i} - \frac{q_{2i-1}}{N_i}\right] = \tau_i \qquad \text{for } j = 2i-1$$

$$\frac{d}{dt}\left[\frac{\partial T}{\partial \dot{q}_j}\right] - \frac{\partial T}{\partial q_j} + \frac{\partial U_g}{\partial q_j} + K_i\left[q_{2i} - \frac{q_{2i-1}}{N_i}\right] = 0 \qquad \text{for } j = 2i$$

with i =1,...,N. τ_i is the torque supplied by the actuator at the i-th joint. Rhs zeros in the even equations mean lack of local control action at the link side of elasticity.

The dynamic model for robot arms with joint elasticity can be rewritten as:

$$B_E(q)\,\ddot{q} + c_E(q,\dot{q}) + e_E(q) + r_E(q) = \tau_E$$

In vector form the model looks similar to the one of a rigid arm but there are twice the number of second order nonlinear differential equations. The 2N-vector forcing term τ_E has even components equal to zero and odd components equal to the motor torques τ. The 2Nx2N generalized inertia matrix $B_E(q)$ is still symmetric and positive definite for all $\mathbf{q}$. $c_E(q,q')$ collects the centrifugal and Coriolis terms and is related to the elements of the matrix $B_E(q)$ by the same relations holding for the rigid model [6]. $e_E(q)$ contains gravitational forces while the elastic terms are grouped in $r_E(q)$.

Assume now a symmetric mass distribution of the motor around its rotation axis (this implies also that its center of mass is located on the motor shaft). Under this mild assumption it can be shown that B_E, c_E and e_E depend only on link variables:

$$B_E = B_E(q_e), \qquad c_E = c_E(q_e, \dot{q}_e), \qquad e_E = e_E(q_e)$$

The motor joint variables $\mathbf{q}_r$ enter the nonlinear equations only through r_E, that is in a linear way like the input motor torques τ_i.

To obtain state and output equations, define the state as $\mathbf{x} = (\mathbf{x}_1,\mathbf{x}_2) = (\mathbf{q},\mathbf{q}')$ and the input as $\mathbf{u} = \tau$. For compactness set $n_E(q,q') = c_E(q,q') + e_E(q) + r_E(q)$ and define a 2NxN odd-columns selection matrix B_S = block diag { $[1\ 0]^T$} and a Nx2N even-rows selection matrix C_S = block diag { [0 1] }. Then

$$\dot{x} = \begin{bmatrix} x_2 \\ -B_E(x_1)^{-1} n_E(x) \end{bmatrix} + \begin{bmatrix} 0 \\ B_E(x_1)^{-1} B_S \end{bmatrix} u$$

which is of the form

$$\dot{x} = f(x) + g(x)\,u \qquad x \in \mathbb{R}^n, \quad u \in \mathbb{R}^m$$

with n = 4N and m = N, where N is the number of elastic joints. Limiting ourself to joint space strategies, the outputs will be

$$y = h(x) = C_S\, x_1 \qquad y \in \mathbb{R}^p, \quad p = N$$

or, in a scalar notation, $y_i = x_{1,2i} = q_{2i}$, i = 1,...,N. The proper "joint" variables to be set under control are indeed the ones that specify the link positions. Notice that a nonlinear system of the square type (p = m = N) is obtained.

3. Linearization of systems without zero-dynamics

Consider a nonlinear system described by the equations:

$$\dot{x} = f(x) + \sum_{i=1}^{m} g_i(x)\, u_i \qquad y_i = h_i(x), \quad i = 1,\ldots,m$$

with state $\mathbf{x}$ in R^n, input $\mathbf{u}$ and output $\mathbf{y}$ in R^m. All the functions are assumed to be smooth. The problem of transforming a nonlinear system into a linear one by means of state-feedback and change of coordinates has been studied by several authors; necessary and sufficient conditions exist when static state-feedback is used [26,27]. If dynamic state-feedback laws are considered, the necessary conditions for linearization may be relaxed. In particular, the problem of full linearization for systems *with outputs* is of special interest, like in the present robotic application. It will be briefly described how constructive conditions for full linearization and input-output decoupling can be derived in the generalized case of dynamic state-feedback. Further details can be found in [24,25].

Some notation is needed first. Given a scalar function $\lambda(x)$ and a vector function f(x), the Lie derivative of $\lambda(x)$ along f(x), denoted by $L_f\lambda(x)$, is a new scalar function defined as:

$$L_f\lambda(x) = \frac{\partial \lambda}{\partial x} \cdot f(x) = \left(\frac{\partial \lambda}{\partial x_1}, \ldots, \frac{\partial \lambda}{\partial x_n}\right) \cdot f(x)$$

$L_f^k \lambda(x)$ is defined iteratively as $L_f(L_f^{k-1}\lambda(x))$, with $L_f^0 \lambda(x) = \lambda(x)$. This operation can be applied repeatedly and is used to define the relative degrees associated to the outputs of the system. An output y_i is said to have *relative degree* r_i (at $x°$) if:

$$L_{g_j} L_f^k h_i(x) = 0 \quad \text{for } j = 1,\dots,m \text{ and } k = 0,\dots,r_i - 2$$

for all x in a neighborhood of x°, and if for at least an index j:

$$L_{g_j} L_f^{r_i - 1} h_i(x°) \neq 0.$$

The relative degree of an output is connected to the integration structure of the given system; r_i is exactly the number of times one has to differentiate $y_i(t)$ at $t = t°$ in order to let some of the input values $u_j(t°)$ appear explicitly. Moreover, the set of relative degrees is used also to define the elements $a_{ij}(x)$ of the so-called decoupling matrix A(x) associated to the system:

$$a_{ij}(x) = L_{g_j} L_f^{r_i - 1} h_i(x)$$

Another relevant control concept which is intrinsic to a given nonlinear system with outputs is the so-called *zero-dynamics* [23]. This is the internal dynamics of the system obtained using as input a state-feedback law $u = u^*(x)$ which forces the output y to be constantly zero.

For SISO systems it is easy to see that, in order to have $y(t) \equiv 0$ for all times, it is required that:

$$h(x(t)) = L_f h(x(t)) = L_f^2 h(x(t)) = \dots = L_f^{r-1} h(x(t)) = 0$$

and

$$u(t) = -\frac{L_f^r h(x(t))}{L_g L_f^{r-1} h(x(t))} = u^*(x(t))$$

The first set of conditions defines an hypersurface $M^*(x)$ of dimension n-r in the state space which is an *invariant set* of the closed-loop system obtained using $u(t) = u^*(x(t))$. The zero-dynamics is exactly the dynamical behavior of this particular closed-loop system on $M^*(x)$. Note that the control $u^*(x(t))$ is defined also *outside* M^*.

For a description of the problem of "zeroing the output" in a multivariable nonlinear system the reader is referred to [23,25]. However, the essential steps of the zero-dynamics algorithm can be easily recovered directly from the robot application described in the next section. Roughly speaking, it will be shown how it is possible to recover the input $u(t)=u^*(x(t))$ from the output functions of the system and from other properly chosen functions which are *x-dependent* combinations of the output derivatives. If the algorithm terminates successfully, the system is said to be *invertible*. The set M^* is then found directly by zeroing the above functions.

It is well-known [22] that any system for which:
- the decoupling matrix A(x) is nonsingular at x°,
- the sum of the relative degrees equals the dimension of x,

can be transformed, locally near x°, into a linear system by means of a suitable change of coordinates and of the feedback:

$$u(x) = -A(x)^{-1} \begin{bmatrix} L_f^{r_1-1} h_1(x) \\ \dots \\ L_f^{r_m-1} h_m(x) \end{bmatrix}$$

The first is a necessary and sufficient condition for the existence of *noninteracting control* via static state-feedback. From the previous discussion, the second condition (i.e. $r_1 + \dots + r_m = n$) can be interpreted as the condition that the system has *no zero-dynamics* ; the hypersurface $M^*(x)$ degenerates into a single point, x°. The above result can be rephrased by saying that a system "without zeros" for which noninteracting control exists can always be transformed into a linear system via static state-feedback. This is the classic case of rigid manipulators.

In this perspective, milder sufficient conditions for full linearization can be obtained when dynamic state-feedback is allowed. More specifically, the assumption on the existence of noninteracting control can be relaxed; assuming only that the system is without zeros, it is possible to obtain linearity via feedback. The constructive technique that makes this possible follows the philosophy of adding a suitable number of integrators on the input channels until a system is obtained which has a nonsingular decoupling matrix (see [24]). Note that the input $u^*(x)$ that displays the zero-dynamics can always be written as a product where the inverse of a matrix Q(x) appears:

$$u^*(x) = -Q^{-1}(x)\, b(x)$$

(compare with the equation above for the linearizing controller). Whenever the system has a nonsingular decoupling matrix A(x), then Q(x) = A(x).

Consider the following *pre-processing* of the system:
- apply the static state feedback

$$u = u^*(x) + Q^{-1}(x)\, w$$

i.e. the control displaying the zero-dynamics of the system, with an additional external input w;
- perform a dynamic extension, i.e. add to each of the inputs w a proper number of integrators, or

$$\overline{w}_i = \frac{d^{\mu_i}}{dt^{\mu_i}} w_i \qquad i = 1,\dots, m$$

where the integer μ_i is the difference between the highest and the lowest order of derivation of the output y_i, which appears in the computation of $u^*(x)$.

Under the *invertibility* hypothesis on the original system, it is possible to show that the extended system obtained with the above pre-processing has a nonsingular decoupling matrix. Moreover, its zero-dynamics is left unchanged. Thus, if the original system had no zero-dynamics, then the extended system can be transformed into a fully linear one by means of static-state feedback and change of coordinates. In other words, denoting by z the states of the added integrators, the extended system will have relative degrees (with overbars) such that

$$\bar{r}_1 + \bar{r}_2 + \dots + \bar{r}_m = \bar{n} \qquad \text{where } \bar{n} = \dim \bar{x} = \dim \begin{bmatrix} x \\ z \end{bmatrix}$$

Therefore, the standard static noninteracting feedback will also be a linearizing one for the extended system. By composing this feedback from the extended state with the pre-processing, the full linearizing *dynamic* controller is obtained.

4. Controller design for a two-link arm

To illustrate the above methodology, the control problem for a two-link planar robot arm with the two rotary joints being elastic (see Figure 2) is considered here as a case study. The derivation of the dynamic model and the state-equations are reported in Appendix.

For this arm, which has a kinematic structure of two rotational joints with parallel axes of rotation, neither the conditions for exact linearization nor the ones for input-output decoupling are satisfied [7,17], if *static* state-feedback is used. In the following it will be shown how to construct a linearizing and decoupling *dynamic* state-feedback law for this example.

From the output and state equations, since $L_g h(\mathbf{x}) = 0$, one gets

$$y = h(\mathbf{x}) = \begin{bmatrix} x_2 \\ x_4 \end{bmatrix} \qquad \dot{y} = L_f h(\mathbf{x}) = \begin{bmatrix} x_6 \\ x_8 \end{bmatrix}$$

$$\ddot{y} = L_f^2 h(\mathbf{x}) + L_g L_f h(\mathbf{x})\, \mathbf{u} \quad \Rightarrow \quad A(\mathbf{x}) = \begin{bmatrix} 0 & g_{62}(x_4) \\ 0 & g_{82}(x_4) \end{bmatrix}$$

and hence $r_1 = r_2 = 2$. The decoupling matrix $A(\mathbf{x})$ is singular and so input-output decoupling is not possible using static state-feedback. The input $\mathbf{u}$ cannot be recovered only from the knowledge of $\mathbf{y}$, $\mathbf{y}'$ and $\mathbf{y}''$. To proceed further, one has to avoid the introduction of the time derivative of the second input u_2 at the next step of the algorithm. This is done as follows. Find the coefficient of linear dependence (on the field of smooth analytic functions) of the rows of $A(\mathbf{x})$

$$\begin{bmatrix} \gamma(\mathbf{x}) & 1 \end{bmatrix} \begin{bmatrix} 0 & g_{62}(x_4) \\ 0 & g_{82}(x_4) \end{bmatrix} = \begin{bmatrix} 0 & 0 \end{bmatrix} \quad \Rightarrow \quad \gamma(x_4) = -\frac{g_{82}(x_4)}{g_{62}(x_4)}$$

and define a function $\lambda(\mathbf{x})$ as

$$\lambda(\mathbf{x}) = \begin{bmatrix} \gamma(\mathbf{x}) & 1 \end{bmatrix} \begin{bmatrix} L_f^2 h_1(\mathbf{x}) \\ L_f^2 h_2(\mathbf{x}) \end{bmatrix} = \begin{bmatrix} -\dfrac{g_{82}(x_4)}{g_{62}(x_4)} & 1 \end{bmatrix} \begin{bmatrix} f_6(\mathbf{x}) \\ f_8(\mathbf{x}) \end{bmatrix}$$

or, using the system expressions in the Appendix,

$$\lambda(\mathbf{x}) = f_8(\mathbf{x}) + \left(1 + \frac{A_3}{A_2}\cos x_4\right) f_6(\mathbf{x}) =$$

$$= -\frac{K_2}{A_2 N_2}(N_2 x_4 - x_3) - \frac{A_3}{A_2} x_6^2 \sin x_4 = \lambda(x_3, x_4, x_6)$$

This function $\lambda(\mathbf{x})$ of the state will be treated as a new "output" of the system. Note that, dropping $\mathbf{x}$-dependence, this function can be alternatively rewritten as $\lambda = \gamma y''_1 + y''_2$. At the next step, its time derivative is

$$\dot{\lambda} = L_f \lambda + L_{g_1}\lambda\, u_1 + L_{g_2}\lambda\, u_2 =$$

$$= \left[\frac{\partial\lambda}{\partial x_3} x_7 + \frac{\partial\lambda}{\partial x_4} x_8 + \frac{\partial\lambda}{\partial x_6} f_6\right] + [0]\, u_1 + \left[\frac{\partial\lambda}{\partial x_6} g_{62}\right] u_2$$

Since the input u_1 still does not appear, the above reasoning has to be repeated. Solving

$$\begin{bmatrix} \delta(\mathbf{x}) & 1 \end{bmatrix} \begin{bmatrix} L_g L_f h_1(\mathbf{x}) \\ L_g \lambda(\mathbf{x}) \end{bmatrix} = \begin{bmatrix} \delta(\mathbf{x}) & 1 \end{bmatrix} \begin{bmatrix} 0 & g_{62} \\ 0 & \dfrac{\partial\lambda}{\partial x_6} g_{62} \end{bmatrix} = \begin{bmatrix} 0 & 0 \end{bmatrix}$$

for $\delta(\mathbf{x})$ gives

$$\delta(\mathbf{x}) = -\frac{\partial\lambda}{\partial x_6} = \frac{2A_3}{A_2} x_6 \sin x_4 = \delta(x_4, x_6)$$

Another function $\mu(\mathbf{x})$ is built as

$$\mu(\mathbf{x}) = \begin{bmatrix} \delta(\mathbf{x}) & 1 \end{bmatrix} \begin{bmatrix} f_6 \\ L_f \lambda \end{bmatrix} =$$

$$= -\frac{K_2}{A_2 N_2}(N_2 x_8 - x_7) - \frac{A_3}{A_2} x_6^2 x_8 \cos x_4 = \mu(x_4, x_6, x_7, x_8)$$

Again, μ can be given an alternative expression in terms of $\mathbf{x}$-dependent combinations of time derivatives of the original outputs, and in which the control $\mathbf{u}$ does not appear. Taking the time derivative of this function

$$\dot{\mu} = L_f \mu + L_{g_1}\mu\, u_1 + L_{g_2}\mu\, u_2 =$$

$$= \left[\frac{\partial\mu}{\partial x_4} x_8 + \frac{\partial\mu}{\partial x_6} f_6 + \frac{\partial\mu}{\partial x_7} f_7 + \frac{\partial\mu}{\partial x_8} f_8\right] + \left[\frac{\partial\mu}{\partial x_6} g_{62} + \frac{\partial\mu}{\partial x_7} g_{72} + \frac{\partial\mu}{\partial x_8} g_{82}\right] u_2$$

gives again no information on u_1. The coefficient $\varepsilon(\mathbf{x})$ of linear dependence between the rows of the matrix multiplying $\mathbf{u}$ at this step is found by solving

$$\begin{bmatrix} \varepsilon(\mathbf{x}) & 1 \end{bmatrix} \begin{bmatrix} 0 & g_{62} \\ 0 & L_{g_2}\mu \end{bmatrix} = \begin{bmatrix} 0 & 0 \end{bmatrix} \quad \Rightarrow \quad \varepsilon(\mathbf{x}) = -\frac{L_{g_2}\mu}{g_{62}}$$

that gives

$$\varepsilon(\mathbf{x}) = -\frac{K_2}{A_2 N_2}\left(N_2 \frac{g_{82}}{g_{62}} - \frac{g_{72}}{g_{62}}\right) + \frac{A_3}{A_2} x_6 \left(2x_8 + \frac{g_{82}}{g_{62}} x_6\right) \cos x_4$$

so that $\varepsilon(\mathbf{x}) = \varepsilon(x_4, x_6, x_8)$ since

$$\frac{g_{82}}{g_{62}} = -\left(1 + \frac{A_3}{A_2}\cos x_4\right) \qquad \frac{g_{82}}{g_{62}} = \frac{G_2}{A_2}\left(A_3^2 \cos^2 x_4 + A_4\right) - 1$$

The coefficient $\varepsilon(\mathbf{x})$ is used for defining a third function $\nu(\mathbf{x})$ as

$$\nu(\mathbf{x}) = \begin{bmatrix} \varepsilon(\mathbf{x}) & 1 \end{bmatrix} \begin{bmatrix} f_6(\mathbf{x}) \\ L_f \mu(\mathbf{x}) \end{bmatrix} =$$

$$= \frac{\partial\mu}{\partial x_4} x_8 + \frac{\partial\mu}{\partial x_7}\left[f_7 - f_6 \frac{g_{72}}{g_{62}}\right] + \frac{\partial\mu}{\partial x_8}\left[f_8 - f_6 \frac{g_{82}}{g_{62}}\right] =$$

$$= \frac{A_3}{A_2}(x_6 x_8)^2 \sin x_4 + \frac{G_2 K_2}{A_2 N_2}\Big[\frac{K_1}{N_1}(N_1 x_2 - x_1) - \frac{K_2}{N_2}(N_2 x_4 - x_3)\left(1 + \frac{A_3}{A_2}\cos x_4\right)$$

$$- A_3 \sin x_4 \left(\frac{A_3}{A_2} x_6^2 \cos x_4 + (x_6 + x_8)^2\right)\Big]$$

$$+ \frac{K_2 + A_3 x_6^2 \cos x_4}{A_2}\left[\frac{K_2}{A_2 N_2}(N_2 x_4 - x_3) + \frac{A_3}{A_2} x_6^2 \sin x_4\right] =$$

$$= \nu(x_1, x_2, x_3, x_4, x_6, x_8)$$

Note that this "output" function does not depend on x_5 ; this implies that its time derivative

$$\dot{\nu} = L_f \nu + L_{g_1}\nu\, u_1 + L_{g_2}\nu\, u_2 =$$

$$= \left[\frac{\partial\nu}{\partial x_1} x_5 + \frac{\partial\nu}{\partial x_2} x_6 + \frac{\partial\nu}{\partial x_3} x_7 + \frac{\partial\nu}{\partial x_4} x_8 + \frac{\partial\nu}{\partial x_6} f_6 + \frac{\partial\nu}{\partial x_8} f_8\right] + \left[\frac{\partial\nu}{\partial x_6} g_{62} + \frac{\partial\nu}{\partial x_8} g_{82}\right] u_2$$

is still independent from input u_1. An $\mathbf{x}$-dependent coefficient $\zeta(\mathbf{x})$ is found as

$$\zeta(\mathbf{x}) = -\frac{L_{g_2}\nu}{g_{62}} = \zeta(x_3, x_4, x_6, x_8)$$

where the functional dependencies follow from the expression of $L_{g_2}\nu$. A fourth function $\xi(\mathbf{x})$ is computed as

$$\xi(\mathbf{x}) = \begin{bmatrix} \zeta(\mathbf{x}) & 1 \end{bmatrix} \begin{bmatrix} f_6(\mathbf{x}) \\ L_f \nu(\mathbf{x}) \end{bmatrix} = \zeta(\mathbf{x})\, f_6(\mathbf{x}) + L_f \nu(\mathbf{x}) = \frac{\partial\nu}{\partial x_1} x_5 + \phi(\mathbf{x})$$

where $\phi(\mathbf{x})$ contains all terms which do not depend on x_5. From the expression of $\nu(\mathbf{x})$

$$\frac{\partial\nu}{\partial x_1} = -\frac{K_1 K_2}{A_2 N_1 N_2} \neq 0$$

so that x_5 appears linearly in $\xi(\mathbf{x})$. The time derivative of the function $\xi(\mathbf{x})$ yields

$$\dot{\xi} = L_f\xi + L_{g_1}\xi\ u_1 + L_{g_2}\xi\ u_2 = L_f\xi + [\frac{\partial \xi}{\partial x_5} g_{51}]\ u_1 + L_{g_2}\xi\ u_2$$

where finally the control input u_1 is explicitly present. The full input $\mathbf{u}$ can be recovered from the equation

$$\begin{bmatrix} \dddot{y}_1 \\ \dot{\xi} \end{bmatrix} = \begin{bmatrix} L_f^2 h_1(\mathbf{x}) \\ L_f \xi(\mathbf{x}) \end{bmatrix} + \begin{bmatrix} L_{g_1} L_f h_1(\mathbf{x}) & L_{g_2} L_f h_1(\mathbf{x}) \\ L_{g_1}\xi(\mathbf{x}) & L_{g_2}\xi(\mathbf{x}) \end{bmatrix} \mathbf{u} = \mathbf{b}(\mathbf{x}) + \mathbf{Q}(\mathbf{x})\,\mathbf{u}$$

where $Q(\mathbf{x})$ is a nonsingular (here, triangular) matrix. Since this equation is solvable in $\mathbf{u}$, then the robotic system is invertible. As already mentioned, all functions introduced in the algorithm may be expressed in terms of y_1, y_2 and of their time derivatives. In particular, ξ' contains combinations of derivatives of the first output, from $y_1^{(2)}$ to $y_1^{(6)}$, but only the sixth derivative of the second one (i.e. $y_2^{(6)}$).

Setting the lhs of the above equation to zero yields the input $\mathbf{u} = -\,Q^{-1}(\mathbf{x})\ \mathbf{b}(\mathbf{x}) = \mathbf{u}^*(\mathbf{x})$ which displays the zero-dynamics of the system. It is easy to see that the given robot with joint elasticity has no zero-dynamics: the hypersurface $M^*(\mathbf{x})$ is reduced to a single point, $\mathbf{x}^\circ = \mathbf{0}$. In fact, M^* is defined by imposing a zero-constrained output behavior. This is equivalent to find all $\mathbf{x}$'s which solve the following set of eight equations:

$$\begin{bmatrix} \mathbf{y}(\mathbf{x}) & \dot{\mathbf{y}}(\mathbf{x}) & \lambda(\mathbf{x}) & \mu(\mathbf{x}) & \nu(\mathbf{x}) & \xi(\mathbf{x}) \end{bmatrix} = \mathbf{0} \quad \Rightarrow \quad M^* \equiv \mathbf{0}$$

The above implication is checked by looking at the triangular structure of these equations. Using the unique solution of the first four ($x_2 = x_4 = x_6 = x_8 = 0$) into the fifth equation (λ=0) yields $x_3 = 0$; using these values into the sixth one (μ=0) gives $x_7 = 0$, and so on.

In order to construct the linearizing and decoupling control for the given two-link robot arm, first apply the *static* state-feedback law $\mathbf{u} = Q^{-1}(\mathbf{x})\ [\mathbf{w} - \mathbf{b}(\mathbf{x})] = \mathbf{u}^*(\mathbf{x}) + Q^{-1}(\mathbf{x})\ \mathbf{w}$, or

$$\mathbf{u} = \begin{bmatrix} u_1 \\ u_2 \end{bmatrix} = \begin{bmatrix} 0 & \dfrac{A_2}{A_3^2\cos^2 x_4 + A_4} \\ -\dfrac{G_1K_1K_2}{A_2N_1N_2} & L_{g_2}\xi(\mathbf{x}) \end{bmatrix}^{-1} \left\{ \begin{bmatrix} w_1 \\ w_2 \end{bmatrix} - \begin{bmatrix} f_6(\mathbf{x}) \\ L_f\xi(\mathbf{x}) \end{bmatrix} \right\} =$$

$$= \alpha(\mathbf{x}) + \beta(\mathbf{x})\ \mathbf{w}$$

which is the feedback derived directly from the computation of the zero-dynamics. The obtained system can be represented in a nice form as in Figure 3. On this graph it is easy to verify that the system has no zero-dynamics; it is also simple to see which additional steps have to be taken in order to get decoupling and linearization.

The system is then extended by adding a proper number of integrators to the inputs. In this case, $\mu_1 = 4$ while $\mu_2 = 0$. Note that for two-input two-output systems, one input is *always* left unchanged. So, *four* integrators have to be added on input w_1

$$w_1 = z_1 \quad \dot{z}_1 = z_2 \quad \dot{z}_2 = z_3 \quad \dot{z}_3 = z_4 \quad \dot{z}_4 = \overline{w}_1$$

$$w_2 = \overline{w}_2$$

As a result of this preprocessing, a system is obtained with state

$$\overline{\mathbf{x}} = \begin{bmatrix} \mathbf{x}^T & \mathbf{z}^T \end{bmatrix}^T \in \mathbb{R}^{12}$$

and described by the following equations (see also Figure 3):

$$\dot{\overline{\mathbf{x}}} = \overline{\mathbf{f}}(\overline{\mathbf{x}}) + \overline{\mathbf{g}}(\overline{\mathbf{x}})\,\overline{\mathbf{w}} \qquad \mathbf{y} = \overline{\mathbf{h}}(\overline{\mathbf{x}}) = \mathbf{h}(\mathbf{x})$$

where

$$\overline{f}(\overline{\mathbf{x}}) = \begin{bmatrix} x_5 & x_6 & x_7 & x_8 & \overline{f}_5 & z_1 & \overline{f}_7 & \overline{f}_8 & z_2 & z_3 & z_4 & 0 \end{bmatrix}^T$$

$$\overline{g}(\overline{\mathbf{x}}) = \begin{bmatrix} 0 & 0 & 0 & 0 & 0 & 0 & 0 & 0 & 0 & 0 & 0 & 1 \\ 0 & 0 & 0 & 0 & \overline{g}_{52} & 0 & 0 & 0 & 0 & 0 & 0 & 0 \end{bmatrix}^T$$

and

$$\overline{f}_5(\overline{\mathbf{x}}) = f_5(\mathbf{x}) + \frac{L_{g_2}\xi(\mathbf{x})}{L_{g_1}\xi(\mathbf{x})}\frac{g_{51}}{g_{62}(\mathbf{x})}(f_6(\mathbf{x}) - z_1) - \frac{L_f\xi(\mathbf{x})}{L_{g_1}\xi(\mathbf{x})} g_{51}$$

$$\overline{f}_7(\overline{\mathbf{x}}) = f_7(\mathbf{x}) - \frac{g_{72}(\mathbf{x})}{g_{62}(\mathbf{x})}(f_6(\mathbf{x}) - z_1)$$

$$\overline{f}_8(\overline{\mathbf{x}}) = f_8(\mathbf{x}) - \frac{g_{82}(\mathbf{x})}{g_{62}(\mathbf{x})}(f_6(\mathbf{x}) - z_1) \qquad \overline{g}_{52}(\overline{\mathbf{x}}) = \frac{g_{51}}{L_{g_1}\xi(\mathbf{x})}$$

It is easy to check that the decoupling matrix of this extended system is

$$\overline{A}(\overline{\mathbf{x}}) = L_{\overline{g}} L_{\overline{f}}^5 \overline{h}(\overline{\mathbf{x}}) = \begin{bmatrix} 1 & 0 \\ \gamma(x_4) & 1 \end{bmatrix}$$

which is *globally* nonsingular. The relative degrees are both equal to six and their sum gives the dimension of the extended state. Hence, the standard noninteracting control [22]

$$\overline{\mathbf{w}} = \overline{A}(\overline{\mathbf{x}})^{-1}[\,\mathbf{v} - L_{\overline{f}}^6\overline{h}(\overline{\mathbf{x}})\,] = \overline{\alpha}(\overline{\mathbf{x}}) + \overline{\beta}(\overline{\mathbf{x}})\ \mathbf{v}$$

will also linearize the extended system. The composition of the preprocessing and of this additional static feedback yields the linearizing *dynamic* state-feedback controller.

The input-state-output behavior between the reference inputs v_1, v_2 and the original outputs y_1 and y_2 is decoupled and linear. This is made evident once a change of coordinates is performed [22]. In the new coordinates θ specified by

$$\theta_i = L_{\overline{f}}^{i-1}\overline{h}_1(\overline{\mathbf{x}}), \quad i = 1,\ldots,6 \qquad \theta_i = L_{\overline{f}}^{i-1}\overline{h}_2(\overline{\mathbf{x}}), \quad i = 7,\ldots,12$$

the system takes on the canonical form of two strings of *six* integrators each. This set of linear coordinates is given just by the system outputs, namely the angular positions of the two links of the robot, together with their time derivatives up to the sixth order. As usual, the design of a linear stabilizing feedback for trajectory control has to be done in terms of this set of coordinates.

5. Discussion

It has been shown that a planar two-link robot with joint elasticity can be fully linearized and input-output decoupled using dynamic rather than only static feedback. control laws are allowed. As a matter of fact, this is a more general result: the *whole* class of robots with joint elasticity can be linearized once the larger class of dynamic state-feedback control laws is considered. The control problem for a series of robots with different kinematic types, including structures with mixed sets of rigid and elastic joints, has been solved in [17,18]. Similar linearization results hold also if the outputs are chosen in the task space, except for the presence of kinematic singularities.

Like for any rigid manipulator, the resulting closed-loop system is linear and decoupled. However, the main difference is in the length of the input-output chains of integrators which, for each elastic joint, is variable but is never less than four, as opposed to the constant double-integration structure of the rigid case. For the two-link planar robot, this length is six and is obtained with the help of the states of the compensator. Note that the lower bound of four input-output integrations is intrinsic to the purely elastic coupling between bodies. This can be checked by analyzing the simple one-dimensional case of two rigid masses elastically coupled, with a force acting at one side while observations are taken on the other.

The variability of this integration structure is a rather surprising aspect [28]. The reason of such behavior is related to the physics of the problem, namely to the kind of interactions that arise between the elastic and the rigid degrees of freedom of a multi-jointed arm. Including joint elasticity in the model creates two sources of dynamic interaction among links and actuators: elastic reaction forces $r_E(\mathbf{q})$, and inertial couplings (i.e. the off-diagonal terms of the inertia matrix $B_E(\mathbf{q})$). The actuator torques affect the motion of the robot links using both these dynamic pathways. Usually, the one going through inertial couplings has lower control authority. In any case, the paths which are faster, measured in terms of integration structure, dominate over the others. It may happen that one of the input torques is "felt" at different links before all of the other ones. No *instantaneous* feedback action performed on the system is capable of uncoupling these interaction effects. Moreover, the possibility of full compensation of model nonlinearities is lost. This happens whenever there exist inertial coupling paths which are faster than the ones due to elastic forces, as in the examined planar case. There are particular kinematic structures in which these inertial paths are not present, already at a *mechanical* level. This is true, for example, for pairs of elastic joints having orthogonal axes. This explains the results obtained for a two-link cylindric [16] and a polar [11,17] arm. It is also evident that the "racing" situation among torques does not occurr for one-dimensional systems. This idea is consistent with the result of feedback linearizability obtained in the single-joint elastic case [15].

The above discussion helps to understand the role of dynamic feedback in the control of robots with joint elasticity. In particular, dynamic compensation *delays* the contributions of the inertial interacting effects so that each input torque affects the relative output (the link position) after the same number of integrations. By slowing down these fast but "weak" actions, the high authority control paths are brought into play and cancellation of the nonlinearities becomes possible. Of course, if the inertial interaction paths are neglected as in the approximate modeling proposed in [19], decoupling and linearization are again possible using only static feedback, although a more complex one than in the rigid case. Roughly speaking, dynamic compensation gives robustness w.r.t. these parasitic effects. Moreover, the *balancing* role of dynamic feedback becomes even clearer when arms with mixed types of joints are considered (see [18]). In any case, there is essentially a trade-off between modeling accuracy and use of more sophisticated control laws.

Some remarks are in order on the choice of reference trajectories for this class of robot arms. The length of the linear chains of closed-loop integrators puts obvious smoothness constraints on the class of desirable time evolutions. If the outputs have to reproduce exactly a given reference trajectory, this should possess continuous time derivatives at least up to the third order. For the two-link planar arm this requirement increases up to the fifth order; the reference inputs $\mathbf{v}$ may be, for instance, the piecewise constant sixth derivative of the desired link motion. If the initial state is properly set, this allows exact output tracking of the trajectory. Otherwise, only asymptotic tracking is guaranteed, once the system has been externally stabilized. As intuition suggests, robots with joint elasticity have to be driven with very smooth reference commands.

Finally, it should be mentioned that the whole approach requires availability of the full state of the system. The measure of position and velocity at both sides of the elastic joint is tecnically feasible, although additional instrumentation is needed. Except for [12], full-state availability is a common request of all the proposed control methods [7-11]. Some work is in progress also on the design of exact linear observer for this class of robots [29].

References

[1] Good,M.C.,Sweet,L.M.,Strobel,K.L., Dynamic Models for Control System Design of Integrated Robot and Drive Systems,*ASME J. Dyn.Systems, Meas, and Control*, **107**, 1985.

[2] Rivin, E.I., Zeid, A., Rastgu-Ghamsari, A., Dynamic Effects Associated with Joint Compliance in Cartesian-Frame Manipulators, *1st IFAC Symp. Robot Control*, Barcelona, 1985.

[3] Sweet, L.M., Good, M.C., Redefinition of the Robot Motion Control Problem,*IEEE Control Systems Mag.*, **5**, 3,1985.

[4] Kuntze, H.B., Jacubasch, A.H.K., Control Algorithms for Stiffening an Elastic Industrial Robot, *IEEE J. Robotics and Automation*, **RA-1**, 2, 1985.

[5] Nicosia, S., Nicolò, F., Lentini, D., Dynamical Control of Industrial Robots with Elastic and Dissipative Joints, *8th IFAC World Congress*, Kyoto, 1981.

[6] Cesareo, G., Nicolò, F., Nicosia, S., DYMIR: A Code for Generating Dynamic Model of Robots, *1st IEEE Int. Conf. Robotics and Automation* , Atlanta, 1984.

[7] Marino, R, Nicosia, S., On the Feedback Control of Industrial Robots with Elastic Joints: A Singular Perturbation Approach, *Università di Roma "Tor Vergata"*, Rap.84.01, 1984.

[8] Khorasani, K., Spong, M.W., Invariant Manifolds and Their Application to Robot Manipulators with Flexible Joints, *2nd IEEE Int. Conf. Robotics and Automation*, St.Louis, 1985.

[9] Khorasani, K., Kokotovic, P.V., Feedback Linearization of a Flexible Manipulator Near its Rigid Body Manifold, *Systems & Control Lett.*, 6, 1985.

[10] Slotine, J.J., Hong, S., Two-Time Scale Sliding Control of Manipulators with Flexible Joints, *American Control Conference*, Seattle, 1986.

[11] Nicosia, S.,Tomei, P., Tornambé, A., Feedback Control of Elastic Robots by Pseudo-Linearization Techniques, *25th IEEE Conf. Decision and Control*, Athens, 1986.

[12] Tomei, P., Nicosia, S., Ficola, A., An Approach to the Adaptive Control of Elastic at Joints Robots; *3rd IEEE Int. Conf. Robotics and Automation*, S.Francisco, 1986.

[13] Cesareo, G., Marino, R., On the Controllability Properties of Elastic Robots, *6th Int. Conf. Analysis and Optimization of Systems*, Nice, 1984.

[14] De Simone, C., Nicolò, F., On the Control of Elastic Robots by Feedback Decoupling, *IASTED Int. J. Robotics and Automation*, **1**, 2, 1986.

[15] Marino,R., Spong, M.W.,Nonlinear Control Techniques for Flexible Joint Manipulators: A Single Link Case Study, *3rd IEEE Int. Conf. Robotics and Automation*, S.Francisco, 1986.

[16] Forrest-Barlach, M.G., Babcock, S.M., Inverse Dynamics Position Control of a Compliant Manipulator, *3rd*

IEEE Int. Conf. Robotics and Automation, S.Francisco, 1986.

[17] De Luca A., Control of Robot Arms with Joint Elasticity: A Differential-Geometric Approach, Ph.D.Thesis (in Italian), *Università di Roma "La Sapienza"*, December 1986.

[18] De Luca A., Control Properties of Robot Arms With Joint Elasticity, *Symp.Math.Theo.of Networks Systems*, Phoenix, 1987

[19] Spong M.W., Modeling and Control of Elastic Joint Robots, *ASME Winter Annual Meeting*, Anaheim, 1986.

[20] De Luca, A., Isidori, A., Nicolò, F., An Application of Nonlinear Model Matching to the Control of Robot Arm with Elastic Joints, *1st IFAC Symp. Robot Control*, Barcelona, 1985.

[21] De Luca, A., Isidori, A., Nicolò, F., Control of Robot Arms with Elastic Joints Via Nonlinear Dynamic Feedback, *24th IEEE Conf. Decision and Control*, Ft. Lauderdale, 1985.

[22] Isidori, A., Nonlinear Control Systems: An Introduction, *Lect. Notes in Control and Information Sci.*, **72**, Springer, 1985.

[23] Isidori, A., Moog, C.H., On the nonlinear equivalent of the notion of transmission zeros, *Modeling and Adaptive Control* (C.I.Byrnes, A.H.Kurszanski Eds.), Springer, 1987.

[24] Isidori, A., Moog, C.H., De Luca, A., A Sufficient Condition for Full Linearization Via Dynamic State Feedback, *25th IEEE Conf. Decision and Control*, Athens, 1986.

[25] De Luca, A., Isidori, A., Feedback Linearization of Invertible Systems, *Colloquium on Robotics and Automation*, Duisburg, 1987.

[26] Jakubczyk, B., Respondek, W., On Linearization of Control Systems, *Bull.Acad.Pol.Sci.Ser.Math.Phys.*, **28**, 1980.

[27] Hunt, L.R., Su, R., Meyer, G., Design for Multi-Input Nonlinear Systems, in: *Differential Geometric Control Theory* (R.W.Brockett, R.S.Millman, H.Sussmann Eds.), Birkhauser, 1983.

[28] Vidyasagar, M., Systems Theory and Robotics, *IEEE Control Systems Mag.*, April 1987.

[29] Nicosia, S., Tomei, P., Tornambé, A., A Nonlinear Observer for Elastic Robots, to appear on *IEEE J. Robotics and Automation.*

Appendix. A two-link robot with joint elasticity

The kinetic energy T of the arm in Figure 2 is the sum of the terms:

$$T_{mot1} = \frac{1}{2} JR_1 \dot{q}_1^{\,2} \qquad T_{link1} = \frac{1}{2} m_1 d_1^2 \dot{q}_2^{\,2} + \frac{1}{2} J_1 \dot{q}_2^{\,2}$$

$$T_{mot2} = \frac{1}{2} mr_2 l_1^2 \dot{q}_2^{\,2} + \frac{1}{2} JR_2 (\dot{q}_2 + \dot{q}_3)^2$$

$$T_{link2} = \frac{1}{2} m_2 [l_1^2 \dot{q}_2^{\,2} + d_2^2 (\dot{q}_2 + \dot{q}_4)^2 + 2 d_2 l_1 \dot{q}_2 (\dot{q}_2 + \dot{q}_4) \cos q_4] + \frac{1}{2} J_2 (\dot{q}_2 + \dot{q}_4)^2$$

The expressions of f(**x**), g(**x**) and h(**x**) are given below:

$$f(\mathbf{x}) = \begin{bmatrix} x_5 & x_6 & x_7 & x_8 & f_5(\mathbf{x}) & f_6(\mathbf{x}) & f_7(\mathbf{x}) & f_8(\mathbf{x}) \end{bmatrix}^T$$

$$g(\mathbf{x}) = \begin{bmatrix} 0 & 0 & 0 & 0 & g_{51}(\mathbf{x}) & 0 & 0 & 0 \\ 0 & 0 & 0 & 0 & 0 & g_{62}(\mathbf{x}) & g_{72}(\mathbf{x}) & g_{82}(\mathbf{x}) \end{bmatrix}^T$$

$$h(\mathbf{x}) = \begin{bmatrix} x_2 & x_4 \end{bmatrix}^T$$

with

$$g_{51} = G_1 \qquad g_{62}(x_4) = \frac{A_2}{A_3^2 \cos^2 x_4 + A_4}$$

$$g_{72}(x_4) = G_2 - g_{62}(x_4) \qquad g_{82}(x_4) = -\frac{A_3 \cos x_4 + A_2}{A_3^2 \cos^2 x_4 + A_4}$$

$$f_5(x_1, x_2) = \frac{G_1 K_1}{N_1^2} [N_1 x_2 - x_1]$$

$$f_6(x_1, x_2, x_3, x_4, x_6, x_8) = \frac{1}{A_3^2 \cos^2 x_4 + A_4} \Big\{ \frac{K_1 A_2}{N_1} [N_1 x_2 - x_1]$$

$$+ \frac{K_2 [A_2 - N_2 (A_3 \cos x_4 + A_2)]}{N_2^2} [N_2 x_4 - x_3] - A_3 \sin x_4 [A_3 x_6^2 \cos x_4 + A_2 (x_6 + x_8)^2] \Big\}$$

$$f_7(x_1, x_2, x_3, x_4, x_6, x_8) = -f_6(\mathbf{x}) + \frac{G_2 K_2}{N_2^2} [N_2 x_4 - x_3]$$

$$f_8(x_1, x_2, x_3, x_4, x_6, x_8) = \frac{1}{A_3^2 \cos^2 x_4 + A_4} \Big\{ -\frac{K_1 (A_3 \cos x_4 + A_2)}{N_1} [N_1 x_2 - x_1]$$

$$+ \frac{K_2 [(N_2 - 1)(A_3 \cos x_4 + A_2) + N_2 (A_3 \cos x_4 - \frac{A_4}{A_2})]}{N_2^2} [N_2 x_4 - x_3]$$

$$+ A_3 \sin x_4 [(A_3 \cos x_4 - \frac{A_4}{A_2}) x_6^2 + (A_3 \cos x_4 + A_2)(x_6 + x_8)^2] \Big\}$$

Besides $G_i = 1/JR_i$, the following constants were introduced:

$$A_1 = (J_1 + m_1 d_1^2) + (J_2 + m_2 d_2^2) + JR_2 + (m_2 + mr_2) l_1^2$$

$$A_2 = J_2 + m_2 d_2^2 \qquad A_3 = m_2 l_1 d_2 \qquad A_4 = A_2 (\frac{1}{G_2} + A_2 - A_1)$$

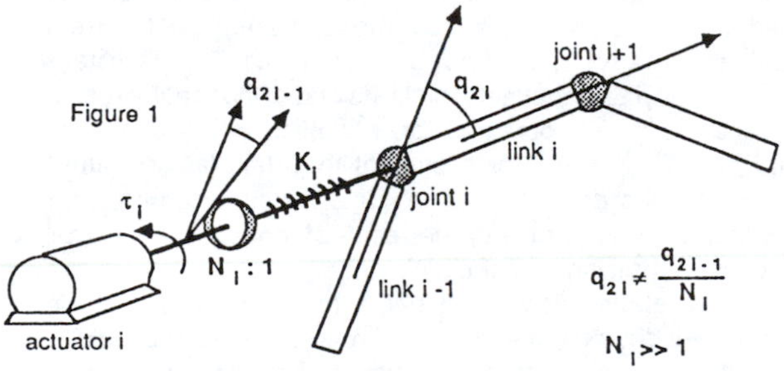

Figure 1

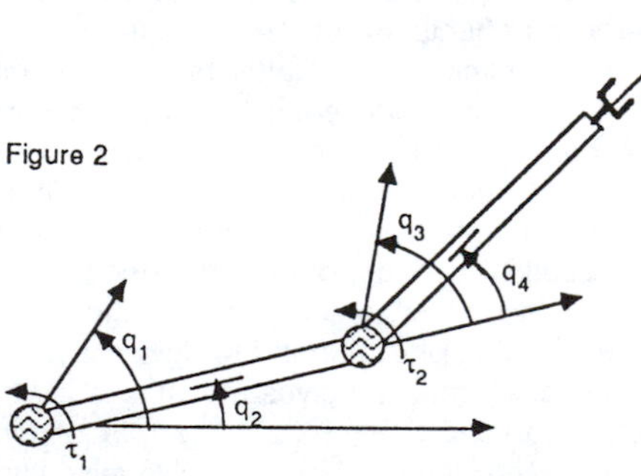

Figure 2

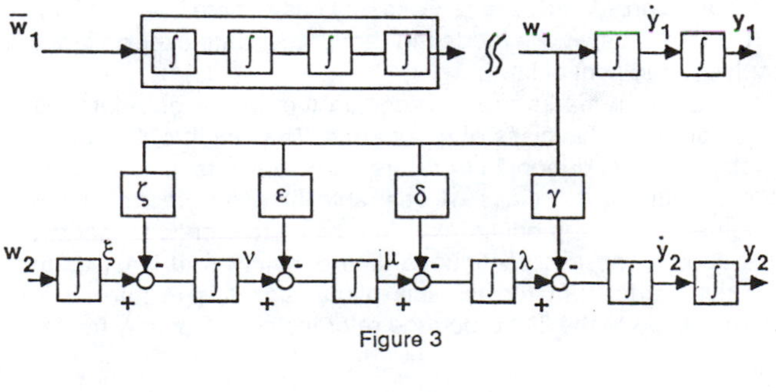

Figure 3

Modeling and Control of Elastic Joint Robots

M. W. Spong
Coordinated Science Laboratory,
University of Illinois at Urbana-Champaign,
Urbana, Ill. 61801

In this paper we study the modeling and control of robot manipulators with elastic joints. We first derive a simple model to represent the dynamics of elastic joint manipulators. The model is derived under two assumptions regarding dynamic coupling between the actuators and the links, and is useful for cases where the elasticity in the joints is of greater significance than gyroscopic interactions between the motors and links. In the limit as the joint stiffness tends to infinity, our model reduces to the usual rigid model found in the literature, showing the reasonableness of our modeling assumptions. We show that our model is significantly more tractable with regard to controller design than previous nonlinear models that have been used to model elastic joint manipulators. Specifically, the nonlinear equations of motion that we derive are shown to be globally linearizable by diffeomorphic coordinate transformation and nonlinear static state feedback, a result that does not hold for previously derived models of elastic joint manipulators. We also detail an alternate approach to nonlinear control based on a singular perturbation formulation of the equations of motion and the concept of integral manifold. We show that by a suitable nonlinear feedback, the manifold in state space which describes the dynamics of the rigid manipulator, that is, the manipulator without joint elasticity, can be made invariant under solutions of the elastic joint system. The implications of this result for the control of elastic joint robots are discussed.

1 Introduction

The proper choice of mathematical models for control system design is a crucial stage in the development of control strategies for any system. This is particularly true for robot manipulators due to their complicated dynamics. For simulation purposes one would like as detailed a model as possible, while for control design and implementation one would like to retain only the most significant dynamic effects in the model in order to simplify the analysis and minimize on-line computational requirements.

Because of the extreme complexity of the dynamic equations of motion for n-link manipulators with joint elasticity, most existing results on the control of such manipulators have relied either on computer programs to generate the equations [17] or have treated special configurations [26] and/or single link examples [27]. However, one generally obtains relatively little insight from symbolically generated equations, and an understanding of the physics underlying the model is of prime importance in understanding the control problem. For this reason we first investigate the problem of modeling the dynamics of elastic joint manipulators. For notational simplicity we treat the case of revolute joints driven by DC-motors whose rotors are elastically coupled to the links. It turns out that by making two rather simple approximating assumptions it is possible to derive a model of the system that is much more amenable to analysis and control than previous models.

Specifically, we assume

(A1) That the kinetic energy of the rotor is due mainly to its own rotation. Equivalently, the motion of the rotor is a pure rotation with respect to an inertial frame. We further assume.

(A2) The the rotor/gear inertia is symmetric about the rotor axis of rotation so that the gravitational potential of the system and also the velocity of the rotor center of mass are both independent of the rotor position.

Assumption (A2) hardly needs any justification and Assumption (A1) is easy to justify for a large class of robots, since roughly speaking it amounts to neglecting terms of order at most $1/m$ where m:1 is the gear ratio. In fact, most existing models of rigid manipulators are derived under precisely these same assumptions; see for example Paul [4], equation (6.49). The important point is to model the dynamic effects which are dominant, in this case the joint elasticity.

2 Modeling

We now consider an n-link manipulator with revolute joints actuated by DC-motors, and model the elasticity of the ith joint as a linear torsional spring with stiffness k_i. For notational simplicity we take $k_i \equiv k$ for all i. Because of the additional degrees of freedom introduced by the elastic coupling of the motor shaft to the links we model the rotor of each actuator as a "fictitious link," that is, as an additional rigid body in the chain with its own inertia. Thus the manipulator consists of n "actual" links and n "fictitious" or rotor links.

[1]This research was partially supported by the National Science Foundation under grant DMC-8616091

Contributed by the Dynamic Systems and Control Division for publication in the JOURNAL OF DYNAMIC SYSTEMS, MEASUREMENT, AND CONTROL. Manuscript received by the Dynamic Systems and Control Division July 1986.

Reprinted with permission from *ASME J. Dyn. Sys., Measurement, Contr.*, vol. 109, p. 310–319, Dec. 1987.

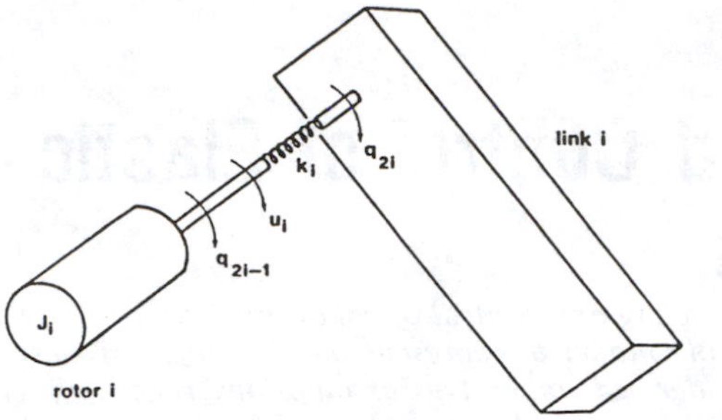

Fig. 1 Elastic Joint

The specific design of the manipulator will dictate the manner in which the actuators are coupled to the links. For simplicity we discuss the case in which the rotor is directly coupled to the link that it actuates as shown in Fig. 1. Other configurations, for example when the motors are located on link 1 and drive the distal links through cables, etc., can be handled by finding the corresponding transformation between "actuator space" and "joint space" as in [34]. The details are omitted.

Referring to Fig. 1, let $\mathbf{q}=(q_1, \ldots q_{2n})^T$ be a set of generalized of generalized coordinates for the system where

$$q_{2i} = \text{the angle of link } i,\ i=1, \ldots, n \tag{2.1}$$

$$q_{2i-1} = -\frac{1}{m_i}\theta_i,\ i=1, \ldots, n \tag{2.2}$$

where θ_i is the angular displacement of rotor i and m_i is the gear ratio. In this case then $q_{2i}-q_{2i-1}$ is the elastic displacement of link i.

Lagrangian Dynamics The rotor, as an intermediate link, now has its own coordinate frame and inertia tensor associated with it. We shall model the "rotor" link as a right circular cylinder of radius a and length b. From symmetry consideration we may establish the coordinate frame at the center of mass and assume that coordinate axes are principal axes of the cylinder, with the rotor angle θ_i measured about the z_{2i} axis. The inertia tensor of the rotor is then given by

$$I_i = \begin{bmatrix} I_{xx_i} & 0 & 0 \\ 0 & I_{yy_i} & 0 \\ 0 & 0 & I_{zz_i} \end{bmatrix} \tag{2.3}$$

where I_{xx}, I_{yy}, I_{zz} are the moments of inertia of the rotor about the principal axes. The kinetic energy of the rotor is

$$K_{r_i} = \frac{1}{2}M_i v_i^T v_i + \frac{1}{2}\ \omega_i^T I_i \omega_i \tag{2.4}$$

where v_i represents the velocity of the center of mass of the rotor, M_i is the rotor mass, and ω_i is the vector of angular velocities about the principal axes.

Now by the symmetry assumption (A2) the velocity v_i of the center of mass of the rotor can written as a function only of the link variables $q_2, \ldots, q_{2i-2}$. If we therefore include the rotor mass as part of link $2i-2$ for the purposes of calculating the inertia tensor of link $2i-2$ then the first term in (2.4) will be included with the kinetic energy of link $2i-2$.

We now invoke Assumption (A1) and model only the kinetic energy of the rotor about its principal axis of rotation, i.e., we assume that the second term in (2.4) above is given as

$$\frac{1}{2}\omega_i^T I_i \omega_i = \frac{1}{2}I_{zz_i}\dot{\theta}_i^2 \tag{2.5}$$

$$= \frac{1}{2}I_{zz_i} m_i^2 \dot{q}_{2i-1}^2$$

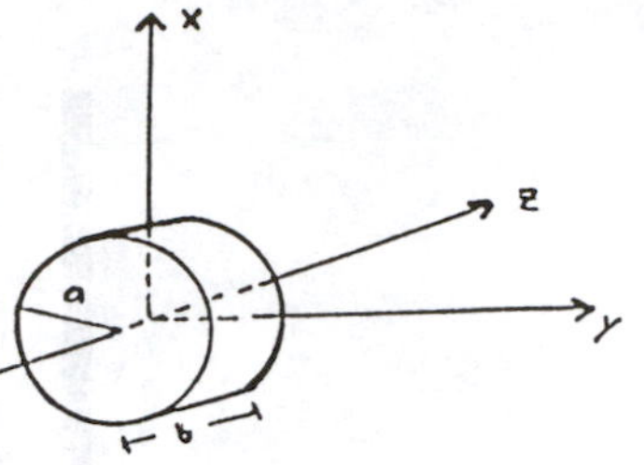

Fig. 2

The following example gives a simple illustration of the effect of the above assumption.

Example Consider the cylinder shown in Fig. 2. The rotational kinetic energy is then

$$K = \frac{1}{2}(I_{xx}\omega_x^2 + I_{yy}\omega_y^2 + I_{zz}\omega_z^2) \tag{2.6}$$

The principal moments of inertia of the cylinder with respect to the coordinate system shown are given by

$$I_{xx} = \frac{1}{4}Mb^2 = I_{yy} \tag{2.7}$$

$$I_{zz} = \frac{1}{2}Ma^2 \tag{2.8}$$

Due to the gear ratio m:1 the angular velocity ω_z will generally be a factor of m larger than the angular velocities about the other two axes. If we take therefore $\omega_z = m\omega_x = m\omega_y$ for the purposes of illustration, the kinetic energy becomes

$$K = \frac{1}{4}M\omega_z^2(a^2 + b^2/m^2) \tag{2.9}$$

We now approximate according to (A1) the kinetic energy K as

$$\bar{K} = \frac{1}{2}M\omega_z^2 a^2 \tag{2.10}$$

The percent error in the kinetic energy incurred by using the expression (2.10) instead of the true kinetic energy (2.9) is then

$$Error = \frac{K-\bar{K}}{K} \times 100 \tag{2.11}$$

$$= \frac{b^2}{b^2+m^2a^2} \times 100 \tag{2.12}$$

For example if $a=1$, $b=1/2$, and $m=100$, the percent error in kinetic energy is 0.01 percent.

Let us now partition the generalized coordinate vector $\mathbf{q}$ as $(\mathbf{q}_1, \mathbf{q}_2)^T$ where

$$\mathbf{q}_1 = (q_2, q_4, \ldots, q_{2n})^T \tag{2.13}$$

$$\mathbf{q}_2 = (q_1, q_3, \ldots, q_{2n-1})^T \tag{2.14}$$

In other words $\mathbf{q}_1$ is the vector of link variables and $\mathbf{q}_2$ is the vector of actuator variables (divided by the gear ratio).

We have shown by the previous discussion then that the kinetic energy of the system under our modeling assumption (A1) is

$$K = \frac{1}{2}\dot{\mathbf{q}}_1^T D(\mathbf{q}_1)\dot{\mathbf{q}}_1 + \frac{1}{2}\dot{\mathbf{q}}_2^T J\dot{\mathbf{q}}_2 \tag{2.15}$$

where $D(\mathbf{q}_1)$ is the inertia of the "rigid" robot

$$D(\mathbf{q}_1)=(d_{ij}(\mathbf{q}_1)) \tag{2.16}$$

which can be calculated using standard techniques, (e.g., formula 6.66 in [4]) once the rotor masses are included as part of the proximal links for the calculation of the latter's inertia tensor. The $n \times n$ matrix J is given by

$$J=diag\left[m_1^2 I_{zz_1}, \ldots, m_n^2 I_{zz_n}\right] \tag{2.17}$$

where the diagonal elements are the motor inertias about their principal axes of rotation multiplied by the square of the respective gear ratios.

We now invoke our second assumption (A2) again that the rotor inertia is symmetric about its axis of rotation. This implies that the gravitational potential is a function only of $\mathbf{q}_1$. Therefore the total potential energy of the system is

$$P=P_1(\mathbf{q}_1)+P_2(\mathbf{q}_1-\mathbf{q}_2) \tag{2.18}$$

where, as in the case of the kinetic energy, the potential energy term P_1 is found from standard formulae for rigid robots (e.g., formula 6.54 in [4]). The second term above is due to the elastic potential of the spring and is given as

$$P_2=\frac{1}{2}k(\mathbf{q}_1-\mathbf{q}_2)^T(\mathbf{q}_1-\mathbf{q}_2). \tag{2.19}$$

The Lagrangian $L=K-P$ of the system is now given by

$$L=\frac{1}{2}\dot{\mathbf{q}}_1^T D(\mathbf{q}_1)\dot{\mathbf{q}}_1+\frac{1}{2}\dot{\mathbf{q}}_2 J\dot{\mathbf{q}}_2-P_1(\mathbf{q}_1)$$

$$-\frac{1}{2}k(\mathbf{q}_1-\mathbf{q}_2)^T(\mathbf{q}_1-\mathbf{q}_2) \tag{2.20}$$

and the equations of motion are found from the Euler-Lagrange equations [4] using (2.20) to be

$$D(\mathbf{q}_1)\ddot{\mathbf{q}}_1+c(\mathbf{q}_1,\dot{\mathbf{q}}_1)+k(\mathbf{q}_1-\mathbf{q}_2)=0 \tag{2.21}$$

$$J\ddot{\mathbf{q}}_2-k(\mathbf{q}_1-\mathbf{q}_2)=u. \tag{2.22}$$

The $n \times n$ matrix $D(\mathbf{q}_1)$ is symmetric, positive definite for each $\mathbf{q}_1$. The vector $C(\mathbf{q}_1, \dot{\mathbf{q}}_1)$ contains coriolis, centripetal, and gravitational forces and torques, and can be expressed as

$$c(\mathbf{q}_1,\dot{\mathbf{q}}_1)=\dot{D}\dot{\mathbf{q}}_1-\frac{1}{2}\dot{\mathbf{q}}_1^T\frac{\partial D}{\partial \mathbf{q}_1}\dot{\mathbf{q}}_1-\frac{\partial P_1}{\partial \mathbf{q}_1} \tag{2.23}$$

We note however that the gyroscopic forces between each rotor and the other links are not included in this expression as a result of Assumption (A1). It is interesting to compare the simplicity of our model (2.21)-(2.22) with other models of elastic joint manipulators that have been derived in the literature [13, 15, 18, 20, 23, 29]. It turns out that (2.21)-(2.22) is structurally similar to the models used in [26] and [27]. Thus our model can be viewed as the general n-degree-of-freedom extension to the models in the latter references.

Interestingly enough, our model is also the direct extension to the case of elastic joints of the familar rigid models that have become standard in the literature. In fact, we can easily see that the usual rigid model can be recovered from (2.21)-(2.22) as the joint stiffness parameters k_i tend to infinity. To see this we assume that in the limit as $\mathbf{k}\to\infty$ there is no elastic deformation, so that

$$\mathbf{q}_1=\mathbf{q}_2, \dot{\mathbf{q}}_1=\dot{\mathbf{q}}_2. \tag{2.24}$$

On the other hand the force $k(\mathbf{q}_1-\mathbf{q}_2)$ transmitted through the coupling between the rotor and link remains finite in the rigid case, i.e., as $k\to\infty$, and it follows that the potential energy P_2 in (2.18) satifies

$$\frac{1}{2}k(\mathbf{q}_1-\mathbf{q}_2)^T(\mathbf{q}_1-\mathbf{q}_2)\to 0 \tag{2.25}$$

as $k\to\infty$ and $\mathbf{q}_2-\mathbf{q}_1\to 0$. Therefore the Lagrangian of the rigid system L_r is obtained from (2.20) as

$$L_r=\frac{1}{2}\dot{\mathbf{q}}_1^T(D(\mathbf{q}_1)+J)\dot{\mathbf{q}}_1-P_1(\mathbf{q}_1) \tag{2.26}$$

which leads to the equations of motion for the rigid system by applying the Euler-Lagrange equations to (2.26)

$$D(\mathbf{q}_1+J)\ddot{\mathbf{q}}_1+c(\mathbf{q}_1,\dot{\mathbf{q}}_1)=u \tag{2.27}$$

The interesting implication of this is that the usual textbook model of rigid robots is subject to the same assumptions (A1) and (A2) that we use to derive the elastic joint model. Gyroscopic forces due to the rotation of the actuators are thus not considered in most existing rigid models. See [32] for an exception to this statement which does consider the modeling of these gyroscopic terms in the case of the rigid joints. It is of interest to note that [32] concluded that these gyroscopic terms can indeed by neglected in most cases.

3 Feedback Linearization

It is well known that the rigid robot equations (2.27) may be globally linearized and decoupled by nonlinear feedback. This is just the familiar inverse dynamics control scheme which transforms (2.27) into a set of double integrator equations which can then be controlled by adding an "outer loop" control [31].

The above technique of inverse dynamics control is now understood as a special case of a more general procedure for transforming a nonlinear system to a linear system, known as *external* or *feedback linearization*.

Definition 3.1: A nonlinear system

$$\dot{x} = f(x)+\sum_{i=1}^{n} g_i(x)u_i \tag{3.1}$$

$$= f(x)+G(x)u$$

is said to be *feedback linearizable* in a neighborhood U_o of the origin if there is a diffeomorphism $T{:}U_o\to R^n$ and nonlinear feedback

$$u=\alpha(x)+\beta(x)v \tag{3.2}$$

such that the transformed state

$$y=T(x) \tag{3.3}$$

satisfies the linear system

$$\dot{y}=Ay+Bv \tag{3.4}$$

where (A, B) is a controllable linear system.

Necessary and sufficient conditions for a system of the form (3.1) to be feedback linearizable are given in [3]. In the case of elastic joint robots, the feedback linearization property was investigated in [13] using computer generated models of the manipulator dynamics. These models are sufficiently complex, even for two link examples that another computer program was used in [13] to check the conditions for feedback linearization. The answer was negative, i.e., the elastic joint model derived in [13] is not general linearizable in this fashion. In this section we show that the new model (2.21)-(2.22) is always globally feedback linearizable according to Definition 3.1. Moreover we do not need symbolic programs to check linearizability or to compute the required state space change of coordinates or the nonlinear feedback law. These can be found by inspection.

We first write the system (2.21)-(2.22) in state space by setting

$$x_1 = q_1 \qquad x_2 = \dot{q}_1$$
$$x_3 = q_2 \qquad x_4 = \dot{q}_2 \tag{3.5}$$

Then we have from (2.21)-(2.22)

$$\dot{x}_1 = x_2 \tag{3.6}$$

$$\dot{x}_2 = -D(x_1)^{-1}\{c(x_1, x_2)+k(x_1-x_3)\} \tag{3.7}$$

$$\dot{x}_3 = x_4 \tag{3.8}$$

$$\dot{x}_4 = J^{-1}k(x_1-x_3)+J^{-1}u \tag{3.9}$$

Since the nonlinearities enter into the second equation above, while the control appears only in the last equation, it is not obvious that the system is linearizable nor can u immediately be chosen to cancel the nonlinearities as in the case of the rigid equations (2.24).

In order to check feedback linearizability of the above system one needs, in principle, to check rank conditions and involutivity of certain sets of vector fields formed by taking Lie brackets of the vector fields defining the state equations (3.6)-(3.9). Our model is simple enough, however, that we can show global feedback linearizabilty by directly computing the required change of coordinates and nonlinear feedback law. Moreover the new coordinates themselves turn out to have physical significance for the control problem at hand.

Consider now the nonlinear state space change of coordinates.

$$y_1 = T_1(x)=x_1 \tag{3.10}$$

$$y_2 = T_2(x)=\dot{T}_1=x_2 \tag{3.11}$$

$$y_3 = T_3(x)=\dot{T}_2 \tag{3.12}$$
$$= -D(x_1)^{-1}\{c(x_1, x_2)+k(x_1-x_3)\}$$

$$y_4 = T_4(x)=\dot{T}_3 \tag{3.13}$$
$$= -\frac{d}{dt}D(x_1)^{-1}\{c(x_1, x_2)+k(x_1-x_3)\}$$
$$-D(x_1)^{-1}\Big\{\frac{\partial c}{\partial x_1}x_2$$
$$+\frac{\partial c}{\partial x_2}(-D(x_1)^{-1}(c(x_1, x_2)+k(x_1-x_3)))$$
$$+k(x_2-x_4)\Big\}$$
$$:= f_4(x_1, x_2, x_3)+D(x_1)^{-1}kx_4$$

where for simplicity we define the function f_4 to be everything in the definition of y_4 above except the last term, which is $D^{-1}kx_4$. Note that x_4 appears only in this last term so that f_4 depends only on x_1, x_2, x_3.

The above mapping is actually a global diffeomorphism. Its inverse is likewise found by inspection to be

$$x_1 = y_1 \tag{3.14}$$

$$x_2 = y_2 \tag{3.15}$$

$$x_3 = y_1+k^{-1}(D(y_1)y_3+c(y_1, y_2)) \tag{3.16}$$

$$x_4 = k^{-1}D(y_1)(y_4-f_4(y_1, y_2, y_3)). \tag{3.17}$$

The linearizing control law can now be found from the condition

$$\dot{y}_4 = \nu \tag{3.18}$$

where ν is a new control input. Computing $\dot{y}_4$ from (3.13) and suppressing function arguments for brevity yields

$$\nu=\frac{\partial f_4}{\partial x_1}x_2-\frac{\partial f_4}{\partial x_2}D^{-1}(c+k(x_1-x_3)) \tag{3.19}$$
$$+\frac{\partial f_4}{\partial x_3}x_4+\Big(\frac{d}{dt}D^{-1}\Big)kx_4+D^{-1}k(J^{-1}k(x_1-x_3)+J^{-1}u)$$
$$:= F(x_1, x_2, x_3, x_4)+D(x_1)^{-1}kJ^{-1}u$$

where $F(x_1, x_2, x_3, x_4)=F(x)$ denotes all the terms in (3.19) but the last term, which involves the input u.

Solving the above expression for u yields

$$u=Jk^{-1}D(x_1)(\nu-F(x)) \tag{3.20}$$

With the nonlinear change of coordinates (3.10)-(3.13) and nonlinear feedback (3.20) the system (3.6)-(3.9) now has the linear block form

$$\dot{y} = \begin{bmatrix} 0 & I & 0 & 0 \\ 0 & 0 & I & 0 \\ 0 & 0 & 0 & I \\ 0 & 0 & 0 & 0 \end{bmatrix} y+\begin{bmatrix} 0 \\ 0 \\ 0 \\ I \end{bmatrix}\nu \tag{3.21}$$

where $I=n\times n$ identity matrix, $0=n\times n$ zero matrix, $y^T=(y_1^T, y_2^T, y_3^T, y_4^T)\epsilon R^{4n}$, and $\nu\epsilon R^n$.

The nonlinear control law (3.20) is not completely determined until the function ν is specified. We will detail next one design scheme for ν which guarantees robust tracking for the above system.

Closed Loop Performance and Robustness It is easy to determine from the linear system (3.21) with linear feedback control (3.22) what the response of the system in the y_i coordinate system will be. The corresponding response of the original coordinates x_i is not necessarily easy to determine since the nonlinear coordinate transformation (3.10)-(3.13) must be inverted to find the x_i. However, in this case the transformed coordinates y_i are themselves physically meaningful. Inspecting (3.10)-(3.13) we see that the variables y_1, y_2, y_3, y_4 are n-vectors representing, respectively, the link positions, velocities, accelerations, and jerks (derivative of the acceleration). Since the motion trajectory of the manipulator is typically specified in terms of these quantities [4], they are natural variables to use for control.

The issue of robustness to parameter uncertainty is an important one at this point. In order to control the linear system (3.21) either the y_i coordinates must be physically measurable, or the y_i must be computed from the measured x_i variable according to (3.10)-(3.13), or a robust observer for these variables must be constructed. In the first case, the required measurements may be difficult to obtain, although solid state accelerometers are now available which could greatly simplify the problem. In the second case, that of computing the y_i via (3.10)-(3.130), one needs accurate estimates of the parameters in the manipulator model. In the third case there are results on the design of nonlinear observers which could be applied to this problem [35].

The computation of the overall nonlinear control law (3.20) also requires knowledge of the model parameters. In what follows we assume that both the original variables x_i and the transformed variables y_i may be used for feedback, and we consider the robust tracking problem.

Following [33] we consider the transformed system

$$\dot{y} = y_2 \tag{3.23}$$

$$\dot{y}_2 = y_3$$

$$\dot{y}_3 = y_4$$

$$\dot{y}_4 = F(x) + D^{-1}kJ^{-1}u$$

$$:= -\beta(x)^{-1}\alpha(x) + \beta(x)^{-1}u$$

that is, $\beta(x) = Jk^{-1}D(x)$ and $\alpha = \beta(x)F(x)$.

Now the control law

$$u = \alpha(x) + \beta(x)\nu \tag{3.24}$$

that is, (3.20), which ideally linearizes the system is unachievable in practice due to parameter uncertainty, computational roundoff, unknown disturbances, etc. It is more reasonable to assume a control law of the form

$$u = \hat{\alpha}(x) + \hat{\beta}(x)\nu \tag{3.25}$$

where $\hat{\alpha}(x)$ and $\hat{\beta}(x)$ are estimated or computed values of $\alpha(x)$ and $\beta(x)$, respectively. In addition, the functions α and β are extremely complicated so that $\hat{\alpha}$ and $\hat{\beta}$ may represent intentional model simplification to facilitate real-time computation. In what follows $\|x\|$ denotes the usual L_2-norm or Euclidean norm of a vector $x \epsilon R^n$ and, for any matrix M, $\|M\|$ is the corresponding induced matrix norm, i.e.,

$$\|M\| = \sqrt{\lambda_{\max}(M^T M)}$$

where $\lambda_{\max}(\cdot)$ denotes the largest eigenvalue of a matrix. We make the following assumptions on the functions α, $\hat{\alpha}$, β, $\hat{\beta}$.

(A3) There exist positive constants $\bar{\beta}$ and β such that

$$\beta \le \|\beta^{-1}(x)\| \le \bar{\beta} \tag{3.26}$$

(A4) There is a positive constant $a < 1$ such that

$$\|\beta^{-1}\hat{\beta} - I\| \le a \tag{3.27}$$

(A5) There is a known function $\phi(x,, t)$ such that

$$\|\hat{a} - \alpha\| \le \phi < \infty \tag{3.28}$$

We note that (A4) can always be satisfied by suitable choice of $\hat{\beta}$. For example, the choice $\hat{\beta} = 1/cI$, where I is the identity matrix and the constant c is $1/2\,(\bar{\beta} + \beta)$ results in [36]

$$\|\beta^{-1}\hat{\beta} - I\| \le \frac{\bar{\beta} - \beta}{\bar{\beta} + \beta} < 1.$$

Now we substitute the control law (3.25) into (3.23) which results in

$$\dot{y}_4 = \beta^{-1}\hat{\beta}\nu + \beta^{-1}\Delta\alpha \tag{3.29}$$

$$= \nu + E\nu + \beta^{-1}\Delta\alpha$$

where $\Delta\alpha = \hat{\alpha} - \alpha$, and $E := \beta^{-1}\hat{\beta} - I$. Note that $\|E\| \le a < 1$ from assumption (A4).

To track a desired trajectory $y_i^d(t)$ we first find K such that $\bar{A} := A + BK$ is stable, where A and B are defined by (3.21), and we set $\nu = \dot{y}_4^d(t) + Ke + \Delta\nu$, where e is the vector tracking error.

$$e(t) = \begin{bmatrix} y_1 - y_1^d \\ y_2 - y_2^d \\ y_3 - y_3^d \\ y_4 - y_4^d \end{bmatrix}$$

The above system may now be written in "error space" as

$$\dot{e} = Ae + B\{\Delta\nu + \Psi\} \tag{3.30}$$

where Ψ is the nonlinear function (hereafter referred to as the ("uncertainty") defined by

$$\Psi = E(\dot{y}_4^d + Ke + \Delta\nu) + \beta^{-1}\Delta\alpha \tag{3.31}$$

The problem of robust trajectory tracking now reduces to the problem of stabilizing the system (3.30) by suitable choice of the additional input $\Delta\nu$. The above formulation is valid for any system that is feedback linearizable as is shown in [33] and any number of techniques can now be used to design the input $\Delta\nu$. However, the problem of stabilizing (3.30) in nontrivial since Ψ is a function of both e and $\Delta\nu$ and hence Ψ cannot be treated merely as a disturbance to be rejected by $\Delta\nu$. A more sophisticated analysis and design is required to guarantee stability of (3.30).

Approaches that can be used to design $\Delta\nu$ in (3.30) to guarantee robust tracking include Lyapunov and sliding mode designs [6], [5], high gain [8]and other approaches. We shall outline one approach to robust stabilization of feedback linearizable systems based on Lyapunov's second method. See [33] for the details and proofs.

First we note that from our assumptions on the uncertainty we have

$$\|\Psi\| \le a(\|\dot{y}_4^d\| + \|Ke\| + \|\Delta\nu\|) + \bar{\beta}\phi \tag{3.32}$$

$$\le \bar{\phi} + a\|\Delta\nu\|$$

where $\bar{\phi} := a(\|\dot{y}_4^d\| + \|Ke\|) + \bar{\beta}\phi$. Suppose that we can simultaneously satisfy the inequalities

$$\|\Psi\| \le \rho(e, t) \tag{3.33}$$

$$\|\Delta\nu\| \le \rho(e, t) \tag{3.34}$$

for a known function $\rho(e, t)$. The function ρ can be determined as follows. First suppose that $\Delta\nu$ satisfies (3.34). Then from (3.32) we have

$$\|\Psi\| \le \bar{\phi} + a\rho := \rho \tag{3.35}$$

This definition of ρ is well-defined since $a < 1$ and we have

$$\rho = \frac{1}{1-a}\bar{\phi}. \tag{3.36}$$

It now follows from [33] that the null solution of (3.30) is uniformly asymptotically stable (in a generalized sense) if $\Delta\nu$ is chosen as

$$\Delta\nu = \begin{cases} -\rho\dfrac{B^T Pe}{\|B^T Pe\|}; & \text{if } \|B^T Pe\| \ne 0 \\ 0; & \text{if } \|B^T Pe\| = 0 \end{cases} \tag{3.37}$$

where P is the unique positive definite solution to the Lyapunov equation

$$\bar{A}^T P + P\bar{A} = -Q \tag{3.32}$$

for a given symmetric, positive definite Q. The argument is completed by noting that indeed $\|\Delta\nu\| \le \rho$.

4 Integral Manifold Approach

The above feedback linearizing control scheme requires measurement of the link positions and velocities, the motor positions and velocities as well as the link accelerations and jerks for successful implementation. In this section we present a differnt approach based on a reformulation of the dynamic equations (2.21)-(2.22) as a singularly perturbed system and the concept of integral manifold. In the case of weakly elastic joints, such as arise in harmonic drive gear elasticity, this approach has the advantage that it may be applied even when only the link position and velocity are available for feedback, provided that the system has a degree of natural damping at the joints. We will make this precise later.

Returning to the original system, we set

$$z = k(\mathbf{q}_2 - \mathbf{q}_1);\ \mu = \frac{1}{k} \tag{4.1}$$

Then z is the elastic force at the joints. If we now choose coordinates z and $\mathbf{q}_1$ we have from (2.21) and (2.22)

$$\ddot{\mathbf{q}}_1 = -D(\mathbf{q}_1)^{-1}c(\mathbf{q}_1, \dot{\mathbf{q}}_1) - D(\mathbf{q}_1)^{-1}z \tag{4.2}$$

$$: = a_1(\mathbf{q}, \dot{\mathbf{q}}) + A_1(\mathbf{q})z$$

where we henceforth drop the subscript on $\mathbf{q}$ for convenience. Likewise,

$$\mu\ddot{z} = \ddot{\mathbf{q}}_1 - \ddot{\mathbf{q}}_2 \tag{4.3}$$

$$= -D(\mathbf{q}_1)^{-1}c(\mathbf{q}_1, \dot{\mathbf{q}}) - (D(\mathbf{q}_1)^{-1} + J^{-1})z - J^{-1}u$$

$$= a_2(\mathbf{q}, \dot{\mathbf{q}}) + A_2(\mathbf{q})z + B_2u$$

In this case not that $a_2 = a_1$ and B_2 is constant and invertible. The model (4.2)-(4.3) is singularly perturbed. In the limit as $\mu \to 0$(4.2)-(4.3) reduces to the rigid equations of motion. In other words, by formally setting $\mu = 0$ in (4.3) and eliminating z from the equations, one obtains the rigid equations (2.24), as we now show.

Setting $\mu = 0$ in (4.3) and solving for z yields

$$z = -(D^{-1} + J^{-1})^{-1}(D^{-1}c + J^{-1}u) \tag{4.4}$$

which, when substituted into (4.2) yields

$$\ddot{q} = -D^{-1}c + D^{-1}(D^{-1} + J^{-1})^{-1}(D^{-1}c + J^{-1}u) \tag{4.5}$$

$$= -D^{-1}c + D^{-1}(D^{-1} + J^{-1})D^{-1}c + D^{-1}(D^{-1}J^{-1})^{-1}J^{-1}u$$

Now a straightforward calculation shows that the first two term in (4.5) above may be combined to yield

$$-D^{-1}c + D^{-1}(D^{-1} + J^{-1})^{-1}D^{-1}c \tag{4.6}$$

$$= -D^{-1}c + D^{-1}J(D+J)^{-1}c$$

$$= (-D^{-1}(D+J) + D^{-1}J)(D+J)^{-1}c$$

$$= -(D+J)^{-1}c$$

Likewise the second term in (4.5) can be simplified as

$$D^{-1}(D^{-1} + J^{-1})^{-1}J^{-1}u \tag{4.7}$$

$$= D^{-1}J(D+J)^{-1}DJ^{-1}u$$

$$= [J\,D^{-1}(D+J)J^{-1}D]^{-1}u = (D+J)^{-1}u$$

and so the reduced order system (4.5) simplifies to

$$\ddot{\mathbf{q}} = -(D+J)^{-1}c + (D+J)^{-1}u \tag{4.8}$$

which is just the rigid sytem (2.24).

Integral Manifold In the $4n$-dimensional state space of (4.2)-(4.3), a $2n$ dimensional manifold M_μ may be defined by the expressions,

$$z = h(\mathbf{q}, \dot{\mathbf{q}}, u, \mu) \tag{4.9}$$

$$\dot{z} = \dot{h}(\mathbf{q}, \dot{\mathbf{q}}, u, \mu) \tag{4.10}$$

The manifold M_μ is said to be an *integral manifold* (4.2)-(4.3) if it is invariant under solutions of the system. In other words, given an admissible input function $t \to u(t)$, if $\mathbf{q}(t)$, $z(t)$ are solutions of (4.2)-(4.3) for $t > t_0$ with initial conditions $\mathbf{q}(t_0) = \mathbf{q}^0$, $\dot{\mathbf{q}}(t_0) = \bar{\mathbf{q}}^0$, z^0, $\dot{z}(t_0) = \bar{z}^0$ then

$$z^0 = h(\mathbf{q}^0, \bar{\mathbf{q}}^0, u(t_0), \mu) \tag{4.11}$$

$$\bar{z}^0 = h(\mathbf{q}^0, \bar{\mathbf{q}}^0, u(t_0), \mu)$$

implies that for $t > t_0$

$$z(t) = h(\mathbf{q}(t), \dot{\mathbf{q}}(t), u(t), \mu) \tag{4.12}$$

$$\dot{z}(t) = \dot{h}(\mathbf{q}(t), \dot{\mathbf{q}}(t), u(t), \mu) \tag{4.13}$$

In otherwords, if the system lies initially on the manifold M_μ, then the solution trajectory remains on the manifold M_μ for $t > t_0$.

The integral manifold M_μ is characterized by the following partial differential equation, formed by substituting the expression (4.9) into the equation (4.3)

$$\mu\ddot{h} = a_2(\mathbf{q}, \dot{\mathbf{q}}) + A_2(\mathbf{q})h + B_2u \tag{4.14}$$

In otherwords, if the system lies initially on the manifold M_μ, then the solution trajectory remains on the manifold M_μ for $t > t_0$.

$$\dot{h} = \frac{\partial h}{\partial \mathbf{q}}\dot{\mathbf{q}} + \frac{\partial h}{\partial \dot{\mathbf{q}}}(a_1 + A_1z) + \frac{\partial h}{\partial u}\dot{u}$$

and $\ddot{h}$ is to be similarily expanded. Although the p.d.e. (4.14) is seemingly difficult we shall actually find an explicit solution.

Once h is determined from (4.14), the dynamics of the system (4.2)-(4.3) on the integral manifold are given by a reduced order system referred as the *reduced flexible system* formed by replacing z by h in (4.2)

$$\ddot{\mathbf{q}} = a_1(\mathbf{q}, \dot{\mathbf{q}}) + A_1(\mathbf{q})h(\mathbf{q}, \dot{\mathbf{q}}, u, \mu) \tag{4.15}$$

Equation (4.15) is of the same order as the rigid system, but as shown in [18] is a more accurate approximation of the flexible system than is the rigid model (2.24). We leave it to the reader to verify that the reduced flexible system reduces to the rigid system (2.24) as the perturbation parameter μ tends to zero.

We now utilize the concept of composite control [10] and choose the control input u of the form

$$u = u_s(\mathbf{q}, \dot{\mathbf{q}}, \nu, \mu) + u_f(\eta, \dot{\eta}) \tag{4.16}$$

where ν represents a new input to be specified. We also specify $u_\nu(0,0) = 0$ so that $u = u_s$ on the integral manifold. The variable η represents the deviation of the fast variables from the integral manifold, i.e.,

$$\eta = z - h(\mathbf{q}, \dot{\mathbf{q}}, u_s, \mu) \tag{4.17}$$

$$\dot{\eta} = \dot{z} - \dot{h}(\mathbf{q}, \dot{\mathbf{q}}, u_s, \mu).$$

Since $u = u_s$ on the integral manifold we may combine (4.3) and (4.14) to obtain

$$\mu\eta = \mu\ddot{z} - \mu\ddot{h}$$

$$\mu\ddot{\eta} = a_2 + A_2z + B_2u - (a_2 + A_2h + B_2u_s)$$

$$= A_2\eta + B_2u_f$$

Therefore, in terms of the variables $\mathbf{q}$ and η, the system (4.2)-(4.3) is rewritten as

$$\ddot{\mathbf{q}} = a_1 + A_1h(\mathbf{q}, \dot{\mathbf{q}}, u_s, \mu) + A_1\eta \tag{4.18}$$

$$\mu\ddot{d} = A_2(\mathbf{q})\eta + B_2u_f \tag{4.19}$$

In order to solve the P.D.E. (4.14) defining the integral manifold, we expand the function h in terms of μ as

$$h(\mathbf{q}, \dot{\mathbf{q}}, u_s, \mu) = h_0(\mathbf{q}, \dot{\mathbf{q}}, u_0) + \mu h(\mathbf{q}, \dot{\mathbf{q}}, u_0 + \mu u_1) + \ldots \tag{4.20}$$

and we choose u_s as

$$u_s = u_0 + \mu u_1 \tag{4.21}$$

Substituting these expressions into the manifold condition (4.14) yields

$$\mu\{\ddot{h}_0 + \mu\ddot{h}_1 + \ldots\} = a_2 + A_2(h_0 + \mu h_1 + \ldots) \tag{4.22}$$

$$+ B_2(u_0 + \mu u_1).$$

Equating coefficients of μ^k we obtain the sequence of equalities

$$0 = a_2 + A_2 h_0 + B_2 u_0 \quad (4.23)$$

$$\ddot{h} = A_2 h_1 + B_2 u_1 \quad (4.24)$$

$$\ddot{h}_{k-1} = A_2 h_k,\ k>1. \quad (4.25)$$

Equation (4.23) may be solved for h_0 to yield

$$h_0 = -A_2^{-1}(a_2 + B_2 u_0) \quad (4.26)$$

The derivation now proceeds iteratively. The control u_0 is first computed at $\mu = 0$, that is, based on the rigid model, and can be any one of the many schemes that have been derived for control of rigid manipulators. Given u_0 then h_0 is computable from (4.26). From this, with the given u_0 we can compute $\ddot{h}_0$ and so we can write (4.24) as

$$A_2 h_1 = \ddot{h}_0 - B_2 u_1. \quad (4.27)$$

where the right-hand side contains only known quantities and the control. Since both A_2, and B_2 are invertible, we see that by setting

$$u_1 = B_2^{-1}\ddot{h}_0 \quad (4.28)$$

it follows that

$$h_1 \equiv 0 \text{ and therefore } \ddot{h}_1 \equiv 0 \quad (4.29)$$

From this it follows iteratively from (4.25) and the invertibility of A_2 that $h_k = 0$ for $k > 1$.

We have show therefore that the choice of control input

$$u_s = u_0 + \mu u_1 \quad (4.30)$$

with u_1 given by (4.28) results in $h = h_0$. Thus, on the integral manifold, i.e., when $\eta = 0$, the dynamics of the system are described by the reduced order system.

$$\ddot{\mathbf{q}} = a_1 - A_1 A_2^{-1} a_2 - A_1 A_2^{-1} B_2 u_0 \quad (4.31)$$

$$= -(D(\mathbf{q}) + J)^{-1}\{c(\mathbf{q}, \dot{\mathbf{q}}) + u_0\}$$

which is of course just the rigid system.

We see that we have produced a solution h_0 of the manifold condition (4.14). The fact that h_0, given by (4.26), satisfies (4.14) is significant. What this implies is that by adding the corrective control $\mu\mu_1$ the integral manifold h becomes the rigid manifold h_0. To put it another way, the rigid manifold h_0 is made an invariant manifold for the flexible system by the corrective control.

If the control u_0 is chosen to be the feedback linearizing control for the rigid system

$$u_0 = (D(\mathbf{q}) + J)\nu + c(\mathbf{q}, \dot{\mathbf{q}}) \quad (4.32)$$

we obtain the overall system

$$\ddot{\mathbf{q}} = \nu + A_1(\mathbf{q})\eta \quad (4.33)$$

$$\mu\ddot{\eta} = A_2(\mathbf{q})\eta + B_2 u_f \quad (4.34)$$

Since B_2 is nonsingular the fast subsystem (4.34), which is a linear system in η parmeterized by $\mathbf{q}$, is controllable for each $\mathbf{q}$. Thus there exists a fast control $u_f(\eta, \dot{\eta})$ to place the poles of (4.34) arbitrarily. Note that we have not explicitly included damping in the model. Thus for each $\mathbf{q}$, since $-A_2$ is a positive definite matrix, the open loop poles of (4.34) are on the $j\omega$-axis. This shows clearly the resonance phenomonon whereby the elastic oscillations from (4.34) drive the slow variables through (4.33). Since A_2 is a function of $\mathbf{q}$ the resonant modes will be configuration dependent, a fact that was experimentally verified in [25]. In case the system (4.34) has some inherent natural damping one can show that the fast subsystem is of the form

Table 1 Parameters used for simulation

Mass	$m =$	1
Stiffness	$k =$	100
Length (2L)	$L =$	1
Gravity	$g =$	9.8
Inertias	$I =$	1
	$J =$	1

$$\mu\ddot{\eta} = A_2(\mathbf{q})\eta + \sqrt{\mu}A_3(\mathbf{q})\dot{\eta} + B_2 u_f \quad (4.35)$$

in which the fast variables, represented by η, decay to zero with $u_f = 0$. In other words the integral manifold, which in this case is the rigid manifold becomes an attracting set. Solutions off the manifold rapidily converge to the manifold after which the system equations are just the rigid equations. In this case the control u_s consisting of the rigid control plus the corrective control achieves the desired result. The point to note that this slow control is a function only of $\mathbf{q}$, $\dot{\mathbf{q}}$. Thus the corrective control compensates for the elasticity using a limited set of state measurements.

If there is no damping in the fast variables, or if the damping is insufficient then the fast control u_f must be added. Note that the choice

$$u_f = B_2^{-1}(\zeta - A_2(\mathbf{q})\eta) \quad (4.36)$$

where ζ is a new input, when applied to (4.34) results in

$$\mu\ddot{\eta} = \zeta \quad (4.37)$$

Inspecting (4.33), (4.37) we see that we have for all practical purposes produced an alternate feedback linearization of the original system, which exploits the two-time scale property of the elastic system. A linear control scheme can now be employed, for example

$$\nu = \alpha_1 \cdot \mathbf{q} + \alpha_2 \cdot \dot{\mathbf{q}} + r \quad (4.38)$$

$$\zeta = \beta_1 \cdot \eta + \sqrt{\mu}\beta_2 \cdot \dot{\eta} \quad (4.39)$$

to place the poles of the system arbitrarily. Note that implementation of the above control scheme requires either direct measurement of the fast variables, which in this case are the elastic forces at the joints and their time derivatives or else accurate knowledge of the system paramters in order that η and $\dot{\eta}$ can be computed from (4.17), which is an issue similar to that which arises in the feedback linearization approach of section 3.

5 An Example

For illustrative purposes consider the single link with the flexible joint of Fig. 1 with the parameters shown in Table 1. The equations of motion for this system in state space are easily computed to be

$$\begin{aligned}\dot{x}_1 &= x_2\\ \dot{x}_2 &= -MgL/I \sin x_1 - k/I(x_1 - x_3)\\ \dot{x}_3 &= x_4\\ \dot{x}_4 &= k/J(x_1 - x_3) + 1/Ju\end{aligned} \quad (5.1)$$

where $x_1 = q_1$, $x_3 = q_2$, etc.

In the limit as $k \to \infty$ the resulting rigid system is

$$\begin{aligned}\dot{x}_1 &= x_2\\ \dot{x}_2 &= -Mgl/(I+J)\sin x_1 + 1/(I+J)u\end{aligned}$$

where we take here $x_1 = {}_1 = q_1 = q_2$.

The feedback linearizing control law for (5.2) may be chosen as

$$u = (I+J)(\nu + Mgl \sin x_1) \quad (5.3)$$

with ν given as a a simple linear control term

$$\nu = \dot{x}_2^d - a_1(x_1 - x_1^d) - a_2(x_2 - x_2^d) \quad (5.4)$$

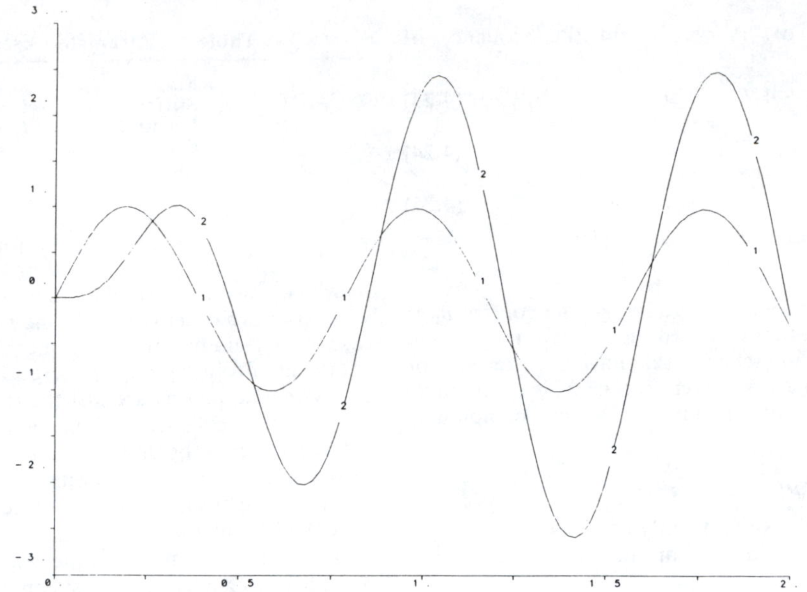

Fig. 3 Rigid control applied to flexible joint. 1 = reference trajectory; 2 = link angle.

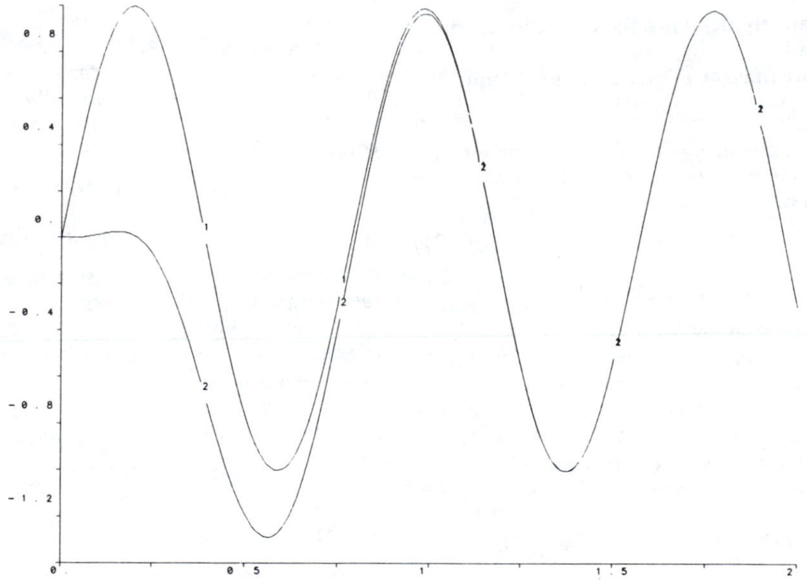

Fig. 4 Feedback linearization control. 1 = reference trajectory; 2 = link angle

designd to track a desired trajectory $t \to x_1^d(t)$.

It is interesting to see the response of the flexible joint system (5.1) to this "rigid control." At this point one must make a choice whether to use the motor variable q_2 or the link variable q_1 in this control law. Figure 3 shows the response of the link variable q_1 in the flexible joint system (5.1) using the motor variable $x_1 = q_2$ in (5.3)-(5.4), with a desired tracjectory $x_1^d = \sin 8t$. It is interesting to note that if one tries to feedback instead the link variable q_1 in (5.3)-(5.4) the system becomes unstable.

Feedback Linearization Control. The feedback linearizing transformation for this system is given by (3.10)-(3.13) as

$$\begin{aligned} y_1 &= x_1 \\ y_2 &= x_2 \\ y_3 &= -MgL/I \sin x_1 - k/I(x_1 - x_3) \\ y_4 &= -Mgl/I\cos x_1 \cdot x_2^2 - k/I(x_2 - x_4) \end{aligned} \tag{5.5}$$

The feedback linearizing control law computed from (3.19) and (3.20) turns out to be

$$u = \frac{IJ}{k}(\nu - F(x_1, x_2, x_3, x_4)) \tag{5.6}$$

where

$$F(x_1, x_2, x_3, x_4) = MgL/I \sin x_1 \cdot x_2^2 \tag{5.7}$$

$$+ (MgL/I\cos x_1 + k/I)(MgL/I \sin x_1 + k/I(x_1 - x_3))$$

$$+ k^2/IJ(x_1 - x_3)$$

A simple linear control law for ν designed to track a desired trajectory $t \to y_i^d(t)$ can be expressed as

$$\nu = \dot{y}_4^4 - \sum_{i=1}^{3} a_i(y_i - y_i^d) \tag{5.8}$$

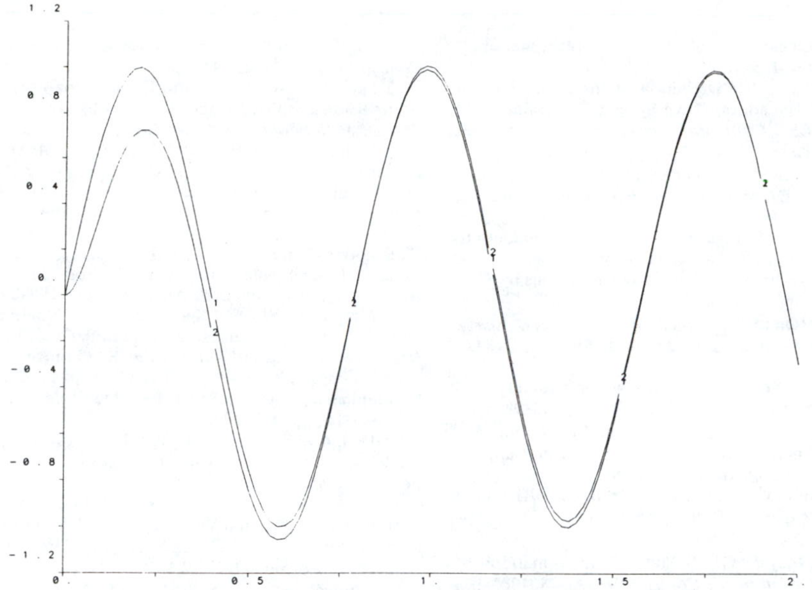

Fig. 5 Corrective control based on the integral manifold. 1 = reference trajectory; 2 = link angle

The zero-state response of the above system with a given desired trajectory $y_1 = \sin 8t$ is shown in Fig. 4. The gains in the control law (5.8) where chosen for simplicity to achieve a closed loop characteristic polynomial for the linearized system of $(s+10)^4$. This response illustrates the improved tracking resulting from basing the control design on the fourth-order flexible joint model rather than on the rigid model. The robust version of this control law, given by (3.37) is omitted.

Integral Manifold Control. In terms of the variables $q = q_1$ and $z = k(q_1 - q_2)$ the equations of motion for the system of Fig. 1 in singularly perturbed form with $\mu = 1/k$ are

$$\ddot{q} = -MgL \sin q - 1/I\, z \tag{5.5}$$

$$\mu \ddot{z} = -MgL \sin q - (1/I + 1/J)z - 1/Ju \tag{5.6}$$

In terms of the variables h and η the system (4.18)-(4.19) is

$$\ddot{q} = -MgL \sin q - 1/I\, h - 1/I \eta \tag{5.8}$$

$$\mu \ddot{\eta} = (1/I + 1/J)\eta - 1/Ju \tag{5.9}$$

where $\eta = z - h$ and h is determined from the manifold condition

$$\mu \ddot{\eta} = -MgL \sin q - (1/I + 1/J)h - 1/Ju \tag{5.10}$$

The detailed calculation of the asymptotic expansion of the function h and the corrective control are carried out in [23]. The interested reader is referred to that paper for the details and also more simulation results. In Fig. 5 the composite and corrective control law thus derived is applied to track the same desired trajectory $q^d = \sin 8t$.

6 Conclusions

In this paper we have rigorously derived a simple and rather intuitive model to represent the dynamics of elastic joint manipulators and presented two attractive control techniques for the resulting system. The first new result that we present is the global feedback linearization of the flexible joint system by nonlinear coordinate transformation and static state feedback. The importance of the property of feedback linearization is not necessarily that the nonlinearities in the system can by computed and exactly cancelled by feedback as this is never achievable in practice. Rather its significance is that once the proper coordinates are found in which to represent the system, the so-called matching conditions are satisfied, which is to say that the nonlinearities are all in the range space of the input. This property allows the design of control laws which are highly robust to parametric uncertainty. The second new result is based on the integral manifold formulation of the equations of motion. We have shown using the corrective control concept that the manifold in state space describing the dynamics of the rigid manipulator can be made invariant under solutions of the flexible joint system, independent of the joint stiffness. This result holds in general only for the model derived here. In previous models of elastic joint manipulators, as shown in [23], the results here hold up to $0(\mu^l)$ by applying a corrective control

$$u_s = u_0 + \mu u_1 + \ . \ . \ . \ + \mu^l u_l. \tag{6.1}$$

With the present model the result is exactly achieved to any order in μ and is done so only with a first order correction term μu_1.

There are several interesting research issues that arise at this point. Among are the design of robust state estimators to realize the feedback linearization control using only the joint positions and velocities and also the computational issues associated with computing in real-time what amounts to a very complicated nonlinear control algorithm. Also, the robustness of the integral manifold based corrective control strategy needs to be investigated.

References

1 Bejczy, A. C., "Robot Arm Dynamics and Control," JPL Technical Memo., 33-369, Feb. 1974.

2 Freund, "Fast Nonlinear Control with Arbitrary Pole-Placement for Industrial Robots and Manipulators," *Int. J. Robotics Res.*, Vol. 1, No. 1, 1982, pp. 65–78.

3 Hunt, L. R., Su, R., and Meyer, G., "Design for Multi-input Nonlinear Systems," *Differential Geometric Control Theory Conf.*, Birkhauser, Boston, 1983, pp. 268–298.

4 Paul, R. P., *Robot Manipilators: Mathematics, Programming and Control*, MIT Press, Cambridge, Mass., 1982.

5 Slotine, J. J., "Robust Control of Robot Manipulators," *Int. J. Robotics Res.*, Vol. 4, No. 2, 1985.

6 Spong, M. W., Thorp, J. S., and Kleinwaks, J. W., "The Control of Robot Manipulators with Bounded Control Part II: Robustness and Disturbance Rejection," *Proc. 23nd IEEE CDC*, Las Vegas, NV, Dec. 1984.

7 Tarn, T. J., Bejczy, A. K., Isidori, A., and Chen, Y., "Nonlinear Feedback in Robot Arm Control." *Proc. 23rd IEEE CDC*, Las Vegas, NV, Dec. 1984.

8 Samson, C., "Robust Non Linear Control of Robotic Manipulators," *Proc. 22nd IEEE CDC*, San Antonio, Dec. 1983.

9 Sweet, L. M., and Good, M. C., "Re'Definition of the Robot Motion Control Problem: Effects of Plant Dynamics, Drive System Constraints, and User Requirement," *Proc. 23rd IEEE Conf. on Decision and Control*, Las Vegas, NV, Dec. 1984, pp. 724-731.

10 Chow, J. H., and Kokotovic, P. V., "Two Time Scale Feedback Design of a Class of Nonlinear Systems," *IEEE Trans. on Automatic Control*, AC-23, 1978, pp. 438-443.

11 Kokotovic, P. V., "Applications of Singular Perturbation Techniques to Control Problems," *SIAM Review*, Vol. 26, No. 4, 1984, pp. 396-410.

12 Erzerber, H., "Analysis and Design of Model Following Systems by State Space Techniques," *Proc. JACC*, 1968, pp. 572-581.

13 Cesareo, G. and Marino, R., "On the Controllability Properties of Elastic Robots," *Sixth Int. Conf. on Analysis and Optimization of Systems*, INRIA, Nice, 1984.

14 Fenichel, "Geometric Singular Perturbation Theory for Ordinary Differential Equations," *J. Differential Equations*, Vol. 3, 1979, pp. 53-98.

15 Ficola, A., Marino, R., and Nicosia, S., "A Singular Peturbation Approach to the Control of Elastic Robots," *Proc. 21st. Annual Allerton Conf. on Communication, Control, and Computing*, Univ. of Illinois, 1983.

16 Hoppensteadt, F., "Properties of Solutions of Ordinary Differential Equations with Small Parameters," *Comm. Pure Appl. Math.*, Vol 34, 1971, pp. 807-840.

17 Hunt, L. R., Su. R., and Meyer, G., "Global Transformations of Nonlinear Systems," *IEEE Trans. on Automatic Control*, Vol. Ac-28, 1983, pp. 24-30.

18 Khorasani, K., and Spong, M. W., "Invariant Manifolds and their Applications to Robot Manipulators with Flexible Joints," *Proc. 1985 IEEE Int. Conf. on Robotics and Automation*, St. Louis, Mar. 1985.

19 Kokotovic, P. V., "Control Theory in the 80's: Trends in Feedback Design," *Proc. 9th World Congress of IFAC*, July 1984.

20 Marino, R., and Nicosia, S., "On the Feedback Control of Industrial Robots with Elastic Joints: A Singular Perturbation Approach," University of Rome, R-84.01, 1984.

21 Sobolev, V. A., "Integral Manifolds and Decompositions of Singularly Perturbed Systems," *Systems and Control Letters*, Vol. 5, 1984, pp. 1169-179.

22 Su, R., "On the Linear Equivalents of Nonlinear Systems," *Systems and Control Letters*, Vol 2, 1981, pp. 48-52.

23 Spong, M. W., Khorasani, K., and Kokotovic, P. V., "A Slow Manifold Approach to Feedback Control of Flexible Joint Robots," to appear in *IEEE Journal of Robotics and Automation*, 1986.

24 Rivin, E. I., "Effective Rigidity of Robot Structure: Analysis and Enhancement," *Proc. 1985, ACC*, Boston, June 1985.

25 Good, M., Strobel, K., and Sweet, L. M., "Dynamics and Control of Robot Drive Systems," General Electric Company, Corporate Research and Development, 1983.

26 Forrest-Barlach, M. G., and Babcock, S. M., "Inverse Dynamics Position Control of a Compliant Manipulator," *Proc. IEEE Conference of Robotics and Automation*, San Francisco, Apr. 1986, pp. 196-205.

27 Marino, R. W., and Spong, M. W., "Nonlinear Control Techniques for Flexible Joint Manipulators: A Single Link Case Study," *Proc. 1986 IEEE Conference on Robotics and Automation*, San Francisco, Apr. 1986, pp. 1030-1036.

28 Khorasani, K., and Kokotovic, P. V., "Feedback Linearization of a Flexible Manipulator Near its Rigid Body Manifold," *Systems and Control Letters*, Vol. 6, 1985, pp. 187-192.

29 De Luca, A., Isidori, A., and Nicolo, F., "An Application of Nonlinear Model Matching to the Dynamic Control of Robot Arm with Elastic Joints," preprint, 1985.

30 Bejczy, A. K., personal communication, San Francisco, Apr. 1986.

31 Spong, M. W., and Vidyasagar, M., "Robust Nonlinear Control of Robot Manipulators," *Proc. 24th IEEE Conf. on Decision and Control*, Fort Lauderdale, Dec. 1985, pp. 1767-1772.

32 Springer, H., Lugner, P., and Desoyer, "Equations of Motion for Manipulators Including Dynamic Effects of Active Elements," *Proc. IFAC Symposium and Robot Control*, SYROCO' 85, Barcelona, Nov. 1985.

33 Spong, M. W., "Robust Stabilization For a Class of Nonlinear Systems," *Proc. 7th Int. Symp. on the Math. Theory of Network and Systems*, (MTNS), Stockholm, June 1985, Springer-Verlag.

34 Craig, J., *Introduction to Robotics*, Addison-Wesley, 1985.

35 Krener, A. J., and Isidori, A., "Linearization by Output Injection and Nonlinear Observers," *Systems and Control Letters*, Vol. 3, 1983, pp. 47-52.

36 Shoureshi, R., Roesler, M. D., and Corlis, M. J., "Control of Industrial Manipulators with Bounded Uncertainty," *Japan-USA Symposium of Flexible Automation*, Osaka, Japan. 1986.

Part 3
Robust Control

MARK W. SPONG
University of Illinois at Urbana–Champaign

THIS PART treats the problem of robust control of robot manipulators. Robust control refers to the design of controllers with low sensitivity to parameter variations, disturbances, unmodeled dynamics, and other sources of uncertainty. Because of the nonlinearity and complexity of robot dynamics, the robust-control problem is extremely important and has received considerable attention in the research literature.

We distinguish robust control from adaptive control, considered in a later part, although both attempt to accomplish basically the same objective, namely, control under uncertainty. We follow the basic philosophy that an adaptive-control algorithm incorporates some sort of online parameter estimation, while a robust control is usually a fixed controller designed to satisfy performance specifications over a given range of uncertainty. Although the dividing line between robust and adaptive control is not always distinct, as the fourth paper in this part shows, the differences between the two design philosophies are often important. An adaptive controller, for example, can learn from experience in the sense that parameters are changed on-line, whereas a robust controller does not usually learn from past performance. In a repetitive motion, for example, the tracking errors produced by a fixed robust controller tend to be repetitive as well, while a well-designed adaptive controller should produce ever-decreasing tracking errors from trial to trial. On the other hand, adaptive controllers that perform well in the face of parametric uncertainty alone are known to be highly sensitive to other types of uncertainties such as unmodeled dynamics and external disturbances. An understanding of these trade-offs is important in deciding whether to use a robust nonadaptive or an adaptive controller in a given setting.

Robust controllers fall into several broad categories. Traditional methods of overcoming parametric uncertainty are high-gain [1], or switching or variable-structure controllers [2]. The method of sliding modes [3] and the Lyapunov-based guaranteed-stability approach [4] are two such variable-structure design approaches included here. A third approach to robust design is dynamic compensation [5], where a dynamic compensator is designed to minimize some measure of sensitivity to disturbances and uncertainty. Functional analytic methods, such as H_∞ optimal control, fall into this class of robust control.

The first paper in this part, "A Survey of Robust Control for Rigid Robots," by C. Abdallah et al., gives a more detailed discussion and classification of robust controllers than that provided in this introduction and surveys the major contributions in robust control of robots up to 1991. The list of references at the end of this introduction is therefore intended as a supplement to the extensive list of references from the first paper.

The second paper, "Robust Control of Robot Manipulators," by J-J. E. Slotine, is representative of the sliding-mode approach to robust design. The first application in robotics of the sliding-mode approach was in the work of Young [6], which has been reprinted in at least two other collections [7], [8]. The paper by Slotine reprinted here contains some important improvements over earlier sliding-mode designs. First, the switching portion of the controller is incorporated into an outer loop in conjunction with a nominal inverse dynamics inner loop. This control structure utilizes the available information about the parameters to reduce the apparent uncertainty that the switching controller needs to overcome. The second improvement presented in this work is the smoothing of the discontinuous controller within a so-called boundary layer about the switching surface. This smoothing has the effect of eliminating the chattering that is inherent in discontinuous control architectures. The paper presents clearly the design trade-offs between tracking performance and chattering when the continuous approximation to the discontinuous controller is utilized.

The third paper, "On L_2 and L_∞ Stability Approaches for the Robust Control of Robot Manipulators," by N. Becker and W. M. Grimm, is a study of the functional analytic approaches to robust design based on dynamic compensation. This paper was motivated by earlier work of Spong and Vidyasagar [5] on robust inverse dynamics control. In this approach, as in Slotine's paper, a nominal inverse-dynamics inner loop is followed by an outer loop control to achieve robustness. The difference lies in the design of the outer loop. Since the nominal inverse-dynamics control produces a nominally linear system, in effect a linear system perturbed by the nonlinear uncertainty, linear compensation techniques can be employed. The class of all linear dynamic compensators that stabilize the nominally linear system can be computed using the so-called Youla parametrization. Then, assuming known bounds on the uncertainty perturbing the linear system, the small-gain theorem is used to derive a sufficient condition for stability and a subset of stabilizing compensators can be shown to exist that satisfy this sufficient condition for stability. The paper by Becker and Grimm contains an analysis of this approach.

The fourth paper, "Adaptive Model Following Control of Nonlinear Robotic Systems," by S. N. Singh, while containing the word "adaptive" in the title, is included as a readable application of the so-called uncertain-systems or guaranteed-stability approach. This approach is sometimes referred to as the Leitmann or Corless-Leitmann approach in recognition of the pioneering work of Corless and Leitmann [4] in this area. In this approach, which is a Lyapunov-based variable-structure approach, an outer-loop-saturating nonlinear controller is designed based on a worst case bound on the uncertainty. The paper by Singh includes an update law to estimate this uncertainty bound. In this sense, the control algorithm is adaptive and it achieves asymptotic stability. In the case that a fixed uncertainty bound is used, only ultimate boundedness of the tracking error would be achievable, in general. This paper then combines features of the robust design based on variable structure systems with parameter adaptive control.

References

[1] S. Jayasuriya and C. N. Hwang, "Tracking controllers for robot manipulators: a high gain perspective," *J. Dyn. Sys. Meas., and Cont.*, vol. 110, pp. 39–45, March 1988.

[2] R. DeCarlo, S. H. Zak, and G. P. Matthews, "Variable structure control of nonlinear multivariable systems," *IEEE Proc.*, vol. 76, pp. 212–232, March 1988.

[3] V. I. Utkin, "Variable structure systems with sliding modes," *IEEE Trans. Autom. Cont.*, vol. AC-22, pp. 212–222, April 1977.

[4] M. Corless and G. Leitmann, "Continuous-state feedback guaranteeing uniform ultimate boundedness for uncertain dynamic systems," *IEEE Trans. Autom. Cont.*, vol. AC-26, pp. 1139–1144, 1981.

[5] M. W. Spong and M. Vidyasagar, "Robust linear compensator design for nonlinear robotic control," *IEEE J. Robotics and Automation*, vol. RA-3, pp. 345–351, Aug. 1987.

[6] K-K. D. Young, "Controller design for a manipulator using the theory of variable structure systems," *IEEE Trans. Syst., Man, and Cyber.*, vol. SMC-8, pp. 210–218, Feb. 1978.

[7] M. Brady, J. M. Hollerbach, T. L. Johnson, T. Lozano-Perez, and M. T. Mason, *Robot Motion: Planning and Control*, Cambridge, MA: MIT Press, 1982.

[8] C. S. G. Lee, R. C. Gonzales, and K. S. Fu, *Tutorial on Robotics, 2nd ed.*, Silver Springs, MD: IEEE Computer Society Press, 1986.

Bibliography

1. S. Cetinkunt and R. L. Tsai, "Accurate, robust, adaptive control in contour tracking applications of robotic manipulators," *Robotics and Computer-Integrated Manufacturing*, vol. 8, no. 1, pp. 45–51, 1991.
2. F. J. Chang, S. H. Twu, and S. Chang, "Adaptive chattering alleviation of variable structure systems control," *IEEE Proc. Part D, Cont. Theory and Appl.*, vol. 137, pp. 31–29, Jan. 1990.
3. Y. H. Chen, "Robust computed-torque schemes for mechanical manipulators: nonadaptive versus adaptive," *J. Dyn. Sys., Meas., and Cont.*, vol. 113, pp. 324–327, June, 1991.
4. J. J. Craig, *Adaptive Control of Mechanical Manipulators*, New York: Addison-Wesley, 1988.
5. C. Y. Kuo and S. P. T. Wang, "Robust position control of robotic manipulator in Cartesian coordinates," *IEEE Trans. Robotics and Automation*, vol. 7, pp. 653–659, Oct. 1991.
6. G. Leitmann, "On the efficacy of nonlinear control in uncertain linear systems," *J. Dyn. Sys. Meas., and Cont.*, vol. 102, pp. 95–102, June 1981.
7. J. K. Mills, and A. A. Goldenberg, "Robust control of robotic manipulators in the presence of dynamic parameter uncertainty" *J. Dyn. Sys., Meas., and Cont.*, vol. 111, pp. 444–51, Sept. 1989.
8. T. Narikiyo and T. Izumi, "On model feedback control for robot manipulators," *J. Dyn. Syst., Meas., and Cont.*, vol. 113, pp. 371–378, Sept. 1991.
9. Z. Novakovic and L. Zlajpah, "An Algorithm for robust tracking control of robots," *Robotica*, vol. 9, pp. 53–62, Jan. 1991.
10. Z. Novakovic, "Lyapunov-like methodology for robot tracking control synthesis," *Int. J. Cont.*, vol. 51, no. 3. pp. 567–583, Mar. 1990.
11. Z. Qu and J. Dorsey, "Robust tracking control of robots by a linear feedback law," *IEEE Trans. Autom. Cont.*, vol. 36, pp. 1081–1084, Sept. 1991.
12. Z. Qu and J. Dorsey, "Robust control of generalized dynamic systems without the matching conditions," *J. Dyn. Sys. Meas., and Cont.*, vol. 113, pp. 582–589, Dec. 1991.
13. M. W. Spong, "On the robust control of robot manipulators," *IEEE Trans. Autom. Cont.*, to appear, Aug. 1992.
14. C. Y. Su, T. P. Leung, and Q. J. Zhou, "A novel variable structure control scheme for robot trajectory control," *Proc. 11th IFAC World Congress*, vol. 9, pp. 121–124, Aug. 1990.
15. T. Sugie, T. Yoshikawa, and T. Ono, "Robust controller design for robot manipulators," *J. Dyn. Syst. Meas., and Cont.*, vol. 110, pp. 94–96, Mar. 1988.
16. A. Swarup and M. Gopal, "Robust trajectory control of a robot manipulator," *Int. J. Syst. Sci.*, vol. 22, no. 11, pp. 2185–2194, 1991.
17. P. L. Tin, J. Z. Qi, and Y. S. Chun, "An adaptive variable structure model following control design for robot manipulators," *IEEE Trans. Autom. Cont.*, vol. 36, pp. 347–352, Mar. 1991.

Survey of Robust Control for Rigid Robots

C. Abdallah, D. Dawson, P. Dorato, and M. Jamshidi

This survey discusses current approaches to the robust control of the motion of rigid robots and summarizes the available literature on the subject. The five major designs discussed are the "Linear-Multivariable" approach, the "Passivity" approach, the "Variable-Structure" approach, the "Saturation" approach and the "Robust-Adaptive" approach.

Introduction

There are basically two underlying philosophies to the control of uncertain systems: the adaptive control philosophy, and the robust control philosophy. In the adaptive approach, one designs a controller which attempts to "learn" the uncertain parameters of the particular system and, if properly designed will eventually be a "best" controller for the system in question. In the robust approach, the controller has a fixed-structure which yields "acceptable" performance for a given plant-uncertainty set. In general, the adaptive approach is applicable to a wider range of uncertainties , but robust controllers are simpler to implement and no time is required to "tune" the controller to the plant variations. More recently, researchers have attempted to "robustify" certain adaptive controllers in order to combine the advantages of both approaches.

We review here different robust control designs used in controlling the motion of robots. A discussion of adaptive controllers in robotics may be found in [1]. A comprehensive survey of robust control theory is available in [3],[18]. The techniques discussed in this survey belong to one of five categories. The first is the linear-multivariable or feedback-linearization approach [2] where the inverse dynamics of the robot are used in order to globally linearize and decouple the robot's equations. Since one does not have access to the exact inverse dynamics, the linearization and the decoupling will not be exact. This will be manifested by uncertain feedback terms that may be handled using multivariable linear robust techniques [3]. The methods based on computed-torque such as those of [4]-[11] fall under this heading.

The second category contains methods that exploit the passive nature of the robot [12]. These techniques try to maintain the passivity of the closed-loop robot/controller system despite uncertain knowledge of the robot's parameters. Although not as transparent to linear control techniques as the computed-torque approach is, passivity-based methods can nonetheless guarantee the robust stability of the closed-loop robot/controller system. The works described in [13],[14],[40] fall under this category and will be discussed in this paper.

Next, we group methods that are for the most part Lyapunov-based nonlinear control schemes. These include variable-structure and saturation controllers which attempt to robustly control a rigid robot. Some of these techniques may actually rely on the feedback-linearizability or the passivity of the robot dynamics and may have been included in those approaches.

Finally, we briefly survey approaches that combine robust and adaptive techniques. It should be noted that other classifications of robust controllers in robotics are possible and that this survey reflects our own philosophy rather than a universally accepted division.

Let the rigid robot dynamics be given in joint-space by the Lagrange-Euler equations [19] where q is an n vector of generalized coordinates representing the joints positions, and τ is the generalized n torque input vector. The matrix $D(q)$ is an $n \times n$ symmetric positive-definite inertia matrix and $h(q,\dot{q})$ is an n vector containing the Coriolis, centrifugal, and gravity terms.

$$D(q)\ddot{q} + h(q,\dot{q}) = \tau. \tag{1}$$

In general, (1) arises as a solution to the Lagrange equations of motion for natural systems [20]. In this paper, we survey methods which deal primarily with designing controllers that will make q and $\dot{q}$ track some desired q_d and $\dot{q}_d$ when some entries of $D(q)$ and $h(q,\dot{q})$ are uncertain. This survey is by no means exhaustive and will also exclude the important case when the robot comes in contact with the environment.

Presented at the 1990 American Control Conference, San Diego, CA, May 23-25, 1990. C. Abdallah, P. Dorato, and M. Jamshidi are with the CAD Laboratory for Systems and Robotics, Electrical and Computer Engineering Department, University of New Mexico, Albuquerque, NM 87131. D. Dawson is with the Department of Electrical and Computer Engineering, Clemson University, Clemson, SC 29634-0915.

Linear-Multivariable Approach

In this section we review different designs which use linear multivariable techniques to obtain robust robot controllers. In the early days of robot control, the idea of linearizing the nonlinear robot dynamics about their desired trajectory (using a Taylor series expansion for example) was popular, and many controllers were designed that way [21],[23],[42],[43]. Later however, the physics and special structure of equation (1), coupled with the fact that the control τ provides an independent input for each degree of freedom [2],[12], led to the "global" linearization of the nonlinear robotic system. It is this later approach that is stressed in this section. For an excellent description of the exact linearization of robots see [2]. By defining the trajectory error vector, $e_1 = q - q_d$, $e_2 = \dot{e}_1$, one is able to globally linearize the nonlinear error system, to the following:

$$\begin{aligned} \dot{e} &= A\,e + B\,v \\ A &= \begin{bmatrix} 0 & I \\ 0 & 0 \end{bmatrix} \\ B &= \begin{bmatrix} 0 \\ I \end{bmatrix} \\ e &= \begin{bmatrix} e_1 \\ e_2 \end{bmatrix} \\ v &= D(q)^{-1}[\tau - h(q,\dot{q})] - \ddot{q}_d. \end{aligned} \tag{2}$$

The problem is then reduced to finding a linear control v which will achieve a desired closed-loop performance, i.e. find F,G,H and J in

$$\begin{aligned} \dot{z} &= F\,z + G\,e, \\ v &= H\,z + J\,e, \end{aligned}$$

or

Reprinted from *IEEE Contr. Syst. Mag.*, vol. 11, no. 2, pp. 24–30, Feb. 1991.

$$v(t) = [H(sI-F)^{-1}G + J\,]\,e(t) \equiv C(s)\,e(t) \quad (3)$$

Note that the above notation indicates that $v(t)$ is the output of a system $C(s)$ when an input $e(t)$ is applied. The following static stat-feedback controller is often used:

$$\begin{aligned} &F = G = H = 0, \\ &J = -K \\ &v = -K_1\,e_1 - K_2\,e_2 \equiv -K\,e \end{aligned} \quad (4)$$

leading to the nonlinear controller

$$\tau = D(q)\,[\ddot{q}_d + v\,] + h(q, \dot{q}) \quad (5)$$

which, due to the invertibility of $D(q)$ gives the following closed-loop system:

$$\ddot{e}_1 + K_2\,\dot{e}_1 + K_1\,e_1 = 0. \quad (6)$$

Unfortunately, the control law (5) can not usually be implemented due to its complexity or to uncertainties present in $D(q)$ and $h(q,\dot{q})$. Instead, one applies τ in (7) where $\hat{D}$ and $\hat{h}$ are estimates of D and h:

$$\tau = \hat{D}\;[\,\ddot{q}_d + v\,] + \hat{h}\;. \quad (7)$$

This in turn leads to (Fig. 1):

$$\begin{aligned} &\dot{e} = A\,e + B\,(v + \eta) \\ &\eta = E\,(v + \ddot{q}_d) + D^{-1}\,\Delta h \\ &E = D^{-1}\,\hat{D} - I_n\,, \\ &\Delta h = \hat{h} - h\;. \end{aligned} \quad (8)$$

The vector η is a nonlinear function of both e and v and can not be treated as an external disturbance. It represents a disturbance of the globally linearized error dynamics which is caused by modeling uncertainties, parameter variations, external disturbances and maybe even noisy measurements [4]. The linear multivariable approaches then revolve around the design of linear controllers $C(s)$ (which may be dynamical), such that the complete closed-loop system (Fig. 1) is stable in some suitable sense, e.g. uniformly ultimately bounded [38], globally asymptotically stable, etc. for a given class of nonlinear perturbation η. In other words, choose $C(s)$ in (3) such that the error $e(t)$ in (9) is stable,

$$\dot{e} = A\,e + B\,(v + \eta) \quad (9)$$

$$v(t) = C(s)\,e(t)\,. \quad (10)$$

The reasonable assumptions (11)-(13) below are often made for revolute-joint robots when using this approach [24]. In the following, d_1, d_2, α, β_0, β_1 and β_2 are nonnegative finite constants which depend on the size of the uncertainties:

$$(d_2)^{-1}\,I_n \le \|\,D^{-1}\| \le (d_1)^{-1}\,I_n \quad (11)$$

$$\|\,E\,\| \le \alpha \quad (12)$$

$$\|\,\Delta h\,\| \le \beta_0 + \beta_1\,\|\,e\,\| + \beta_2\,\|\,e\,\|^{\,2} \quad (13)$$

Note that assumption (13) must be modified for robots with prismatic joints.

In general, the small-gain theorem [25], the passivity theorem [25], or the total stability theorem [26] are invoked to find $C(s)$. The most general of these controllers have been designed using Youla parametrization and H^∞ control theory [27],[28] and will be discussed first.

Spong and Vidyasagar [4] used the factorization approach [27] to design a class of linear compensators $C(s)$, parametrized by a stable transfer matrix $R(s)$, which guarantee that the solution $e(t)$ to the linear system (8) has a bounded L_∞ norm. The authors actually assumed that the bound on Δh is linear, i.e. $\beta_2=0$ in (13) and found the family of all L_∞ stabilizing compensators of the nominal plant. A particular compensator may then be obtained by choosing the parameter $R(s)$ to satisfy other design criteria such as surpressing the effects of η. As was discussed in [24], including the more reasonable quadratic bound will not destroy the L_∞ stability result, but will exclude any L_2 results unless the problem is reformulated and more assumptions are made. In particular, noisy measurements are no longer tolerated. Craig [29] discussed the L_2 problem in a similar setting, and under certain conditions, was able to show the boundedness of the error signals.

Static feedback compensators such as the ones given in (4) have also been used extensively starting with the works of Freund [30], and Tarn *et al.* [6], where

$$v = C(s)\,e = -K\,e \quad (14)$$

such that

$$\begin{aligned} \dot{e} &= A\,e + B\,(v + \eta) \\ &= (A - B\,K)\,e + B\,\eta = A_c\,e + B\,\eta\,. \end{aligned} \quad (15)$$

In these papers, the authors use state feedback to either place the poles sufficiently far in the left-half-plane [9], therefore guaranteeing stability in the presence of η (by the total stability theorem for example), or an extra control loop [6] to correct for the effects of η. In [40], the state-feedback controller was used to define an appropriate output Ke such that the input-output closed-loop linear systems $K(sI-A+BK)^{-1}B$ is Strictly-Positive-Real (SPR). The closed-loop stability was then assured for all η resulting from a passive nonlinear system by using the passivity theorem [25]. In Kuo and Wang [31], the internal model principle developed by Francis and Wonham [32] is used to design a linear controller which minimizes the effects of the disturbance term η. However, since η is a nonlinear function of e and v, minimizing its effects does not necessarily guarantee closed-loop stability. In Gilbert and Ha [10], Proportional-Integral-Derivative control is applied in order to obtain some sensitivity improvements. Cai and Goldenberg [33] use Proportional-Integral control to improve the robustness properties of the controller. Arimoto and Miyazaki [34] use Proportional-Integral-Derivative feedback control to robustly stabilize robot manipulators.

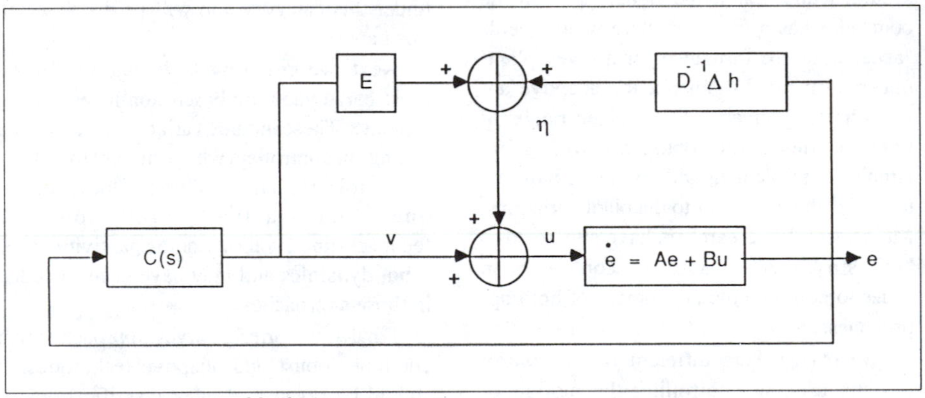

Fig. 1. Linear multivariable design.

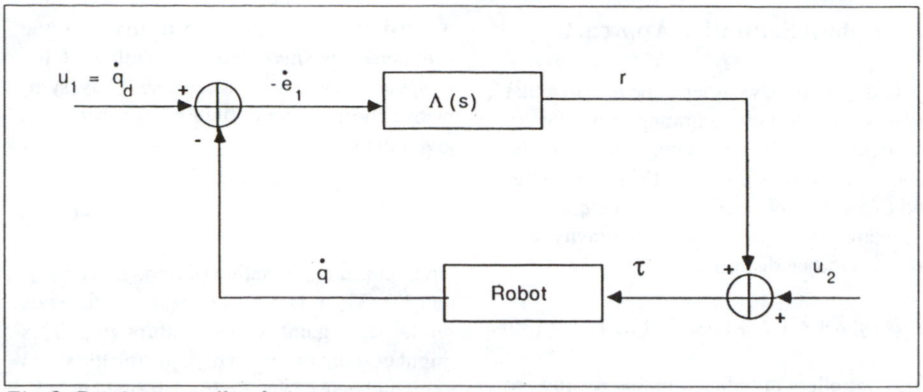

Fig. 2. Passive controller design.

The feedback-linearization approach has been popular (under different names) in the robotics field. Its main advantage is obviously the wealth of linear techniques which may be used in the linear outer loop. In the presence of contact forces however, this approach becomes much more involved as was discussed in [14]. In addition, many controllers designed using this approach are not practical because they require a large control effort.

In some cases, the previously mentioned local linearization approach was combined with other techniques in order to guarantee robust stability [21],[23],[42]. In particular, Desa and Roth [23] used the internal model principle to minimize the effects of disturbances for a robot model linearized over segments of the total operating time. Here also, closed-loop stability is not guaranteed.

Passivity-Based Approach

In this section, we review approaches which rely on the passive structure of rigid robots as described in equation (16) where $h(q,\dot{q}) = C(q,\dot{q})\dot{q} + g(q)$, and $\dot{D}(q) - 2C(q,\dot{q})$ is skew-symmetric by an appropriate choice of $C(q,\dot{q})$ [12]:

$$D(q)\ddot{q} + C(q,\dot{q})\dot{q} + g(q) = \tau . \quad (16)$$

As a result of the skew-symmetry of $\dot{D}$–2C, the following theorem is obtained.

Theorem [1]: The Lagrange-Euler dynamical equations of a rigid robot (1) define a passive mapping from τ to $\dot{q}$, i.e. for some $\beta>0$ and all T finite, the following inequality holds:

$$\langle \dot{q}, \tau \rangle_T \equiv \int_0^T \dot{q}^T \tau \, dt \geq -\beta . \quad (17)$$

Based on this theorem, if one can close the loop from $\dot{q}$ to τ with a passive system (along with l_2 bounded inputs) as in Fig. 2, the closed-loop system will be asymptotically stable using the passivity theorem [25]. This however, will only show the asymptotic stability of $\dot{e}_1$ and not of e_1. On the other hand, if one can show the passivity of the system which maps τ to a new vector r which is a filtered version of e_1, then a controller which closes the loop between $-r$ and τ will guarantee the asymptotic stability of both e_1 and $\dot{e}_1$. This indirect use of the passivity property was illustrated in [1],[55] and will be discussed next. Let the controller be given by (18)-(21) where $F(s)$ is a strictly proper, stable, rational function and K_r is a positive definite matrix,

$$\tau = D(q)a + C(q,\dot{q})v + g(q) - K_r(\dot{q} - v) \quad (18)$$

$$v = \dot{q} - r \quad (19)$$

$$r = -\left[sI + \frac{K(s)}{s} \right] e_1 \quad (20)$$

$$= -F(s)^{-1} e_1$$

$$a = \dot{v} . \quad (21)$$

Then it may be shown that both e_1 and $\dot{e}_1$, are asymptotically stable. This approach was used in the adaptive control literature to design passive controllers [1] but its modification in the design of robust controllers when D, C and G are not exactly known is not obvious.

On the other hand, consider the control law (22) where $\Lambda(s)$ is an SPR transfer function,

$$\tau = -\Lambda(s)\dot{e}_1 + u_2 . \quad (22)$$

The external input u_2 has to be bounded in the L_2 norm. Unfortunately, the inclusion of an integrator which reconstructs the error e_1 will destroy the SPR condition. Substituting the above control law into equation (16), one gets from Fig. 2:

$$r = -\Lambda(s)\dot{e}_1 . \quad (23)$$

By an appropriate choice of $\Lambda(s)$ and u_2, one can apply the passivity theorem and deduce that $\dot{e}_1$ and r are bounded in the L_2 norm, and since $\Lambda(s)^{-1}$ is SPR (being the inverse of an SPR function), one deduces that $\dot{e}_1$, is asymptotically stable because

$$\dot{e}_1 = -\Lambda(s)^{-1} r . \quad (24)$$

Unfortunately, as discussed above, this will only imply that the position error e_1 is bounded but not its asymptotic stability in the case of time-varying trajectories $[\, q_d^T \;\; \dot{q}_d^T \,]\, T$. In the set-point tracking case however, and with gravity precompensation, the asymptotic stability of e_1 may be deduced using LaSalle's theorem [19]. The robustness of the controller (22) is guaranteed as long as $\Lambda(s)$ is SPR and that u_2 is L_2 bounded, regardless of the exact values of the robot's parameters. Note that the controller (22) may be deduced from (5) by choosing the nonlinear controller

$$\tau = D[\ddot{q} - v] + C\dot{q} + g$$
$$v = -D^{-1}[\Lambda(s)\dot{e}_1 + C\dot{q} + g]$$
$$u_2 = D\ddot{q}_d . \quad (25)$$

The passivity approach in (22) is then a modified version of the feedback-linearization approaches. In [13],[14], however, Anderson demonstrated using network-theoretic concepts, that even in the absence of contact forces, a feedback-linearization-based controller is not passive and may therefore cause instabilities in the presence of uncertainties. His solution to the problem consisted of using Proportional-Derivative (PD) controllers with variable gains $K_1(q)$ and $K_2(q)$ which depend on the inertia matrix $D(q)$, i.e.

$$\tau = -K_1(q)e_1 - K_2(q)e_2 + g . \quad (26)$$

Even though $D(q)$ is not exactly known, the stability of the closed-loop error is guaranteed by the passivity of the robot and the feedback law. The advantage of this approach is that contact forces and larger uncertainties may now be accommodated. Its main disadvantage is that although robust stability is guaranteed, the closed-loop performance depends on the knowledge of $D(q)$ whose singular values are needed in order to find K_1 and K_2.

Variable-Structure Controllers

In this section, we group designs that use variable-structure controllers [15]. The VSS theory has been applied to the control of many nonlinear processes [63]. One of the main features of this approach is that one only needs to drive the error to a "switching surface," after which the system is in "sliding mode" and will not be affected by any modeling uncertainties and/or disturbances [15],[16]. The first application of this theory to robot control seems to be in the work of Young [16] where the set point regulation problem ($\dot{q}_d=0$) was solved using the following controller:

$$\tau_i = \begin{cases} \tau_i^+ , & \text{if } s_i(e_{1i}, \dot{q}_i) > 0 \\ \tau_i^- , & \text{if } s_i(e_{1i}, a_i) < 0 \end{cases} \tag{27}$$

where i=1,...,n for an n - link robot, and s_i are the switching planes,

$$s_i(e_{1i}, \dot{q}_i) = c_i e_{1i} + \dot{q}_i , \quad c_i > 0. \tag{28}$$

It is then shown using the hierarchy of the sliding surfaces $s_1, s_2, \ldots, s_n$ and given bounds on the uncertainties in the manipulators model, that one can find τ^+ and τ^- in order to drive the error signal to the intersection of the sliding surfaces after which the error will "slide" to zero. This controller eliminates the nonlinear coupling of the joints by forcing the system into the sliding mode. In [58], a modification of the Young controller was presented. Other VSS robot controllers may be found in [53],[59],[60]. Unfortunately, for most of these schemes, the control effort as seen from (26) is discontinuous along $s_i = 0$ and will therefore create "chattering" which may excite unmodeled high-frequency dynamics.

To address this problem, Slotine modified the original VSS controllers using the so-called "suction control" [17],[41]. In this approach, the sliding surface s is allowed to be time-varying and the control procedure consists of two steps. In the first, the control law forces the trajectory towards the sliding surface while in the second step, the controller is smoothed inside a possibly time-varying boundary layer. This will achieve optimal trade-off between control bandwidth and tracking precision, therefore eliminating chattering and the sensitivity of the controller to high-frequency unmodeled dynamics. The controller structure in this case is given by (29) where Λ is a diagonal matrix of positive elements λ_i (which may be time-varying) and $\phi(.)$ is a nonlinear term determined by the extent of the parametric uncertainties and the suction control modifications [17],

$$\begin{aligned} \tau &= \hat{D}\,[\,\ddot{q}_d - K_2 \dot{e}_1 - K_1^2 e_1 - \phi(q, \dot{q}, t)\,] + \hat{h} \\ K_1 &= \Lambda^2 \\ K_2 &= 2\Lambda \end{aligned} \tag{29}$$

More recently, in [61],[62], VSS controllers which avoided the inversion of the inertia matrix were introduced. The VSS approach although theoretically appealing, does not fully exploit the physics of the robots. In addition, in practice and to avoid chattering, the asymptotic stability of the error is sacrificed.

Robust Saturation Approach

In this section, we review the research that utilizes an auxiliary saturating controller to compensate for the uncertainty present in the robot dynamics as given by (30) where $D(q)$ and $C(q,\dot{q})$ are defined in (16), and $Z(q,\dot{q})$ is an n- vector representing friction, gravity and bounded torque disturbances:

$$D(q)\,\ddot{q} + C(q,\dot{q})\,\dot{q} + Z(q,\dot{q}) = \tau . \tag{30}$$

The controllers introduced in this section are robust due to the fact they only depend on uncertainty bounds rather than on the actual values of the parameters. The following bounds are needed and may be physically justified. The d_i's and ζ_i's in (31) and (32) are positive scalar constants and the trajectory error e is defined before:

$$d_1 I_n = D(q)\, d_2 I_n \tag{31}$$

$$\begin{aligned} &\| C(q,\dot{q})\,\dot{q} + Z(q,\dot{q}) \| \\ &\quad \le \zeta_0 + \zeta_1 \| e \| + \zeta_2 \| e \|^2 . \end{aligned} \tag{32}$$

Note the similarity between (11)-(13) and (31) and (32).

Based on (31),(32), Spong [8] used Lyapunov stability theory to guarantee the ultimate boundedness of e, a concept defined in [38] for example. The control strategy is actually based on the works of Cvetkovic [35] and the linear high-gain theory of Barmish [36], Gutman [37] and Corless [38]. Spong's controller is representative of this class and is given as follows:

$$\begin{aligned} \tau = &(2\, d_1 d_2)(d_1 + d_2)^{-1} \\ &\cdot [\ddot{q}_d - K_2 e_2 - K_1 e_1 - v_r] \\ &+ \hat{C}(q,\dot{q}) + \hat{Z}(q,\dot{q}) \end{aligned} \tag{33}$$

where

$$v_r = \begin{cases} (B^T P e)\,(\| B^T P e \|)^{-1}\,\rho ; \\ \qquad \text{if } \| B^T P e \| > \varepsilon \\ (B^T P e)\,\varepsilon^{-1} \rho ; \\ \qquad \text{if } \| B^T P e \| \le \varepsilon \end{cases} \tag{34}$$

and

$$\begin{aligned} \rho = &(1-\alpha)^{-1}\,[\,\alpha \| \ddot{q}_d \| + \| K_1 \| \\ &\cdot \| e_1 \| + \| K_2 \| \cdot \| e_2 \| + (d_1)^{-1} \phi \end{aligned} \tag{35}$$

$$\phi = \beta_0 + \beta_1 \| e \| + \beta_2 \| e \|^2 \tag{36}$$

$$\alpha = (d_2 - d_1)\,(d_2 + d_1)^{-1} . \tag{37}$$

Note that in the equations above, the matrix B is defined as in (2), the β_i's are defined as in (13), and the matrix P is the symmetric, positive-definite solution of the Lyapunov equation (38), where Q is symmetric and positive-definite matrix and A_c is given in (15):

$$A_c^T P + P A_c = -Q . \tag{38}$$

Upon closer examination of Spong's controller (33)-(37), it becomes clear that v_r depends on the servo gains K_1 and K_2 through ρ. This might obscure the effect of adjusting the servo gains and may be avoided as described in [45]. In fact, let the controller be given by

$$\tau = -K_2 e_2 - K_1 e_1 - v_r(\rho, e_1, e_2, \varepsilon) \tag{39}$$

where v_r is given as in (34) and

$$\rho = \delta_0 + \delta_1 \| e \| + \delta_2 \| e \|^2 \tag{40}$$

where δ_i's are positive scalars. Note that ρ no longer contains the servo gains and as such, one may adjust K_1 and K_2 without tampering with the auxiliary control v_r. As was also shown in [45], if the initial error $e(0)$=0 and by choosing $K_2 = 2K_1 = k_v I_n$, the tracking error may be bounded by the following which shows the direct effect of the control parameters on the tracking error,

$$\| e \| \le \left[\, 4\left(2 k_v + \frac{3 d_2}{2}\right) \varepsilon\, (k_v d_1)^{-1} \right]^{1/2} . \tag{41}$$

In [44], Corless presented a simulation of a similar controller using a Manutec R3 robot. A similar control scheme was given by Chen in [39]. Chen's controller however, requires acceleration measurements. In [5], Gilbert and Ha used a saturating-type feedback derived from Lyapunov-stability theory in order to guarantee the ultimate boundedness of the tracking error. Similarly in [11], Samson derived a "high-gain" controller which guarantees the ultimate boundedness of the error.

Robust Adaptive Approach

In this section, we briefly review some approaches that combine adaptive and robust control concepts. Since so much work has been done in the field of adaptive control of robotic manipulators [1], we only concentrate on schemes that are robust in addition to being adaptive. Let us first review one of the most commonly used robot adaptive controllers. This scheme was derived by Slotine [46] and a simplified version is given by the following where $\hat{\phi}$ is an r vector of the estimated parameters, and $Y(.)$ is an $n \times r$ regression matrix of known time functions:

$$\tau = \tau_a = Y(.)\hat{\phi} - K_2 e_2 - K_1 e_1 \quad (42)$$

$$\frac{d\hat{\phi}}{dt} = -Y^T(.)[e_1 + e_2]\ . \quad (43)$$

If there are no disturbances in the model (16), the tracking error is shown to be asymptotically stable with the above controller. However, the parameter estimate $\hat{\phi}$ in (42) may become unbounded in the presence of a bounded disturbance T_d, or unmodeled dynamics [29],[54]. Robust-Adaptive controllers have attempted to robustify adaptive schemes against such uncertainties.

In [47], Slotine showed that the parameter estimates remain bounded if one uses

$$\tau = \tau_a + k_d \ \mathrm{sgn}\ (e_1 + e_2) \quad (44)$$

where τ_a is given in (42) and k_d is a positive scalar constant satisfying

$$k_d > \| T_d \|\ . \quad (45)$$

More recently [48], Reed introduced the σ-modification method originated by Ioannou [49] in order to compensate for both unmodeled dynamics and bounded disturbances. The control law is now given by

$$\tau = \tau_a = Y(.)\hat{\phi} - K_2 e_2 - K_1 e_1 \quad (46)$$

$$\frac{d\hat{\phi}}{dt} = -Y^T(.)\ [e_1 + e_2] - \sigma\hat{\phi} \quad (47)$$

where

$$\sigma = \begin{cases} 0, & \text{if } \|\hat{\phi}\| < \phi_0 \\ \|\hat{\phi}\|(\phi_0)^{-1} - 1, & \text{if } \phi_0 < \|\hat{\phi}\| < 2\phi_0 \\ 1, & \text{if } \|\hat{\phi}\| > 2\phi_0 \end{cases} \quad (48)$$

and

$$\phi_0 > \| \phi \| \quad (49)$$

Using this controller, Reed was able to show that the tracking error and all closed-loop signals are bounded.

Another approach in this section is that of Singh [50] which combines Spong's controller in (33) with adaptive techniques to estimate the uncertainty terms β_0, β_1, and β_2 in (36). Therefore, no prior knowledge about the exact size of the uncertainties is needed.

In [64], Spong and Ghorbel addressed certain instability mechanisms in the adaptive control of robots. A composite control law was used to damp out the fast dynamics, then a slow adaptive control law based on the algorithm of Slotine and Li [47] was robustified using the σ– modification [65]. Unfortunately, asymptotic stability is then lost if tracking a time-varying trajectory is desired. The algorithm is modified again using the switching σ–modification to ensure the asymptotic stability to a class of time-varying trajectories.

Conclusions

The robust motion control of rigid robot was reviewed. Five main areas were identified and explained. All controllers were robust with respect to a range of uncertain parameters although some of them could only guarantee the boundedness of the position-tracking error rather than its asymptotic convergence. In the last section, we also included adaptive controllers that are also robust. The question of which robust control method to choose is difficult to answer analytically but the following guidelines are suggested. The linear-multivariable approach is useful when linear performance specifications (Percent overshoot, Damping ratio, etc.) are available. This approach may however result in high-gain control laws in the attempt to achieve robustness. The passive controllers are easy to implement but do not provide easily quantifiable performance measures. The robust version of these controllers does not exploit the physics of the robot as their adaptive versions do. The variable-structure controllers should not be used when the flexibilities of the links are considerable for fear of exciting their high frequency dynamics. The saturation controllers, are most useful when a short transient error can be tolerated but ultimately, the error will have to be bounded. The robust adaptive controllers require more computing power and an adaptation time. On the other hand, they are most useful when repetitive or long duration tasks are performed. Their performance actually improves with time and they should be used when a high degree of performance is required. It is useful to note, that although the robot's dynamics are highly nonlinear, most successful controllers have exploited their physics and their very special structure [55],[56]. This observation should be useful as we try to include force control, and flexibility effects in the current and future robotics research.

References

[1] R. Ortega and M.W. Spong, "Adaptive motion control of rigid robots: A tutorial," *Proc. IEEE Conf. Dec. & Contr.*, pp. 1575-1584, Austin, TX, Dec. 1988.

[2] K. Kreutz, "On manipulator control by exact linearization," *IEEE Trans. Auto. Cont.*, Vol. 34, pp. 763-767, July 1989.

[3] P. Dorato, Ed., *Robust Control.* New York: IEEE Press, 1987.

[4] M.W. Spong and M. Vidyasagar, "Robust linear compensator design for nonlinear robotic control," *IEEE J. Rob. Autom.*, Vol. RA-3, pp. 345-351, Aug. 1987.

[5] E.G. Gilbert and I.J. Ha, "An approach to nonlinear feedback control with applications to robotics," *IEEE Trans. Sys., Man, and Cyber.*, Vol. SMC-14, pp. 879-884, Nov./Dec. 1984.

[6] T.J. Tarn, A.K. Bejczy, A. Isidori, and Y. Chen, "Nonlinear feedback in robot arm control," in *Proc. IEEE Conf. Dec. and Contr.*, Las Vegas, NV, Dec. 1984.

[7] E.J. Davison and A. Goldenberg, "The robust control of a general servomechanism problem: The servo compensator," *Automatica*, Vol. 11, pp. 461-471, 1975.

[8] M.W. Spong, J.S. Thorp, and J.M. Kleinwaks, "Robust microprocessor control of robot manipulators, *Automatica*, Vol. 23, pp. 373-379, 1987.

[9] M.S. Fadali, M. Zohdy, and B. Adamczyk, "Robust pole assignment for computed torque robotic manipulators control," in *Proc. IEEE Amer. Contr. Conf.*, June 1989, pp. 37-41.

[10] E.G. Gilbert and I.J. Ha, "Robust tracking in nonlinear systems," *IEEE Trans. Autom. Contr.*, Vol. AC-32, pp. 763-771, 1987.

[11] C. Samson, "Robust nonlinear control of robotic manipulators," in *Proc. 22nd IEEE Conf. Dec. Contr.*, San Antonio, TX, Dec. 1983, pp. 1211-1216.

[12] M.W. Spong, "Control of flexible joint robots: A survey," Rep. UILU-ENG-90-2203, DC-116, Coordinated Science Laboratory, University of Illinois at Urbana-Champaign, Feb. 1990.

[13] R.J. Anderson, "Passive Computed Torque Algorithms For Robots," *Proc. IEEE Conf. Dec. & Contr.*, Tampa, FL, Dec. 1989, pp. 1638-1644.

[14] R.J. Anderson, "A network approach to force control in robotics and teleoperation," Ph.D. thesis, Dept. Electrical and Computer Engineering, Univ. Illinois at Urbana-Champaign, 1989.

[15] V.I. Utkin, "Variable structure systems with sliding modes," *IEEE Trans. Auto. Contr.*, Vol. AC-22, pp. 212-222, Apr. 1977.

[16] K-K.D. Young, "Controller design for a manipulator using theory of variable structure systems," *IEEE Trans. Syst., Man, and Cyber.*, Vol. SMC-8, pp. 210-218, Feb. 1978.

[17] J-J.E. Slotine, "The robust control of robot manipulators," *Int. J. Rob. Res.*, Vol. 4, pp. 49-64, Summer 1985.

[18] P. Dorato and R.K. Yedavali, *Recent Advances in Robust Control.* New York: IEEE Press, 1990.

[19] M.W. Spong and M. Vidyasagar, *Robot Dynamics and Control.* New York: Wiley, 1989.

[20] L. Meirovitch, *Methods of Analytical Dynamics.* New York: McGraw-Hill, 1970.

[21] P. Misra, R.V. Patel, and C.A. Balafoutis, "A robust servomechanism control scheme for tracking cartesian space trajectories," Preprint.

[22] L.R. Hunt, R. Su, and G. Meyer, "Global transformations of nonlinear systems," *IEEE Trans. Autom. Contr.,* Vol. AC-28, pp. 24-31, 1983.

[23] S. Desa and B. Roth, "Synthesis of control systems for manipulators using multivariable robust servo mechanism theory," *Int. J. Robotic Res.,* Vol. 4, pp. 18-34, Fall 1985.

[24] N. Becker and W.M. Grimm, "On L_2 and L_∞ stability approaches for the robust control of robot manipulators," *IEEE Trans. Automa. Contr.* , Vol. 33, pp. 118-122, Jan. 1988.

[25] C. Desoer and M. Vidyasagar, *Feedback Systems: Input-Output Properties.* New York: Academic, 1975.

[26] B.D.O. Anderson *et al.*, *Stability of Adaptive Systems: Passivity and Averaging Techniques.* Englewood Cliffs, NJ: Prentice-Hall, 1989.

[27] M. Vidyasagar, *Control Systems Synthesis: A Factorization Approach.* Cambridge, MA, M.I.T. Press, 1985.

[28] B.A. Francis, *A Course in H∞ Control Theory.* Berlin: Springer-Verlag, 1987.

[29] J.J. Craig, *Adaptive Control of Mechanical Manipulators.* Reading, MA: Addison-Wesley, 1988.

[30] E. Freund, "Fast nonlinear control with arbitrary pole-placement for industrial robots and manipulators," *Int. J. Rob. Res.* , Vol. 1, pp. 65-78, Spring 1982.

[31] C.Y. Kuo and S.P.T. Wang, "Nonlinear robust industrial robot control," *Trans. ASME J. Dyn. Syst. Meas. Control,* Vol. 111, pp. 24-30, Mar. 1989.

[32] B.A. Francis and W.M. Wonham, "The internal model principle of control theory," *Automatica,* Vol. 12, pp. 457-465, 1976.

[33] L. Cai and A.A. Goldenberg, "Robust control of position and force for a robot manipulator in non-contact and contact tasks," in *Proc. Amer. Contr. Conf.,* Pittsburgh, PA, June 1989, pp. 1905-1911.

[34] S. Arimoto and F. Miyazaki, "Stability and robustness of pid feedback control for robot manipulators of sensory capability," in *Proc. 1st Int. Symp. Robotics Res.,* pp.783-799, 1983.

[35] V. Cvetkovic and M. Vukobratovic, "One robust, dynamic control algorithm for manipulation systems," *Int. J. Robotics Res.,* Vol. 1, pp. 15-28, 1982.

[36] B.R. Barmish, I.R. Petersen, and A. Fever, "Linear ultimate boundedness control of uncertain dynamical systems," *Automatics,* Vol. 19, pp. 523-532, 1983.

[37] S. Gutman, "Uncertain dynamic systems - A Lyapunov min-max approach," *IEEE Trans. Autom. Contr.,* Vol. AC-24, pp. 437-443, 1979.

[38] M.J. Corless and G. Leitman, "Continuous state feedback guaranteeing uniform ultimate boundedness for uncertain dynamic systems," *IEEE Trans. Autom. Contr.,* Vol. AC-26, pp. 1139-1144, 1981.

[39] Y.H. Chen, "On the deterministic performance of uncertain dynamical systems," *Int. J. Contr.,* Vol. 43, pp. 1557-1579, 1986.

[40] C. Abdallah and R. Jordan, "A positive-real design for robotic manipulators," *Proc. IEEE Amer. Contr. Conf.,* San Diego, CA, May 1990.

[41] J.J. Slotine and S.S. Sastry, "Tracking control of nonlinear systems using sliding surfaces with applications to robot manipulators," *Int. J. Contr.,* Vol. 38, pp. 465-492, 1983.

[42] M. Whitehead and E.W. Kamen, "Control of serial manipulators with unknown variable loading," in *Proc. IEEE Conf. Dec. & Contr.,* Ft. Lauderdale, FL, Dec. 1985.

[43] J.Y.S. Luh, "Conventional controller design for industrial robots: A tutorial," *IEEE Trans. Sys., Man, & Cyber.,* Vol. SMC-13, pp. 298-316, May/June 1983.

[44] M. Corless, "Tracking controllers for uncertain systems: Application to a Manutec R3 robot," *J. Dyn. Sys., Meas., & Contr.,* Vol. 111, pp. 609-618, Dec. 1989.

[45] D. Dawson and F. Lewis, "Robust and adaptive control of robot manipulators without acceleration measurement," in *Proc. IEEE Conf. Dec. & Contr.,* Tampa, FL, 1989.

[46] J.J. Slotine and W. Li, "Theoretical issues in adaptive control," presented at Fifth Yale Workshop on Applic. of Adaptive Systems Theory, New Haven, CT, 1987.

[47] J.J. Slotine and W. Li, "On the adaptive control of robot manipulators," *Int. J. Rob. Res.,* Vol. 3, 1987.

[48] J. Reed and P. Ioannou, "Instability analysis and robust adaptive control of robot manipulators," in *Proc. IEEE Conf. Dec. & Contr.,* 1988, pp. 1607-1612.

[49] P. Ioannou and K. Tsakalis, "A robust direct adaptive controller," *IEEE Trans. Auto. Contr.,* Vol. AC-31, pp. 1033-1043, Nov. 1986.

[50] S.N. Singh, "Adaptive model following control of nonlinear robotic systems," *IEEE Trans. Auto. Contr.,* Vol. AC-30, pp. 1099-1100, Nov. 1985.

[51] M. Corless and G. Leitmann, "Adaptive control of systems containing uncertain functions and unknown functions with uncertain bounds," *J. Optim. Theory Applic.,* Jan. 1983.

[52] K.Y. Lim and M. Eslami, "Robust adaptive controller designs for robot manipulator systems," *IEEE Conf. Rob. & Auto.,* San Francisco, CA, Apr. 1986.

[53] S. Nicosia and P. Tomei, "Model reference adaptive control algorithms for industrial robots," *Automatica,* pp. 635-644, May 1984.

[54] S. Sastry and M. Bodson, *Adaptive Control: Stability, Convergence and Robustness.* Englewood Cliffs, NJ, Prentice-Hall, 1989.

[55] J-J.E. Slotine, "Putting physics in control - The example of robotics," *IEEE Contr. Syst. Mag.,* Vol. 8, pp. 12-17, Dec. 1988.

[56] P.E. Crouch and A.J. van der Schaft, *Variational and Hamiltonian Control Systems.* Thoma and Wyner Ed. Berlin: Springer-Verlag, 1987.

[57] B.E. Paden and S.S. Sastry, "A calculus for computing Filipov's differential inclusion with application to the variable structure control of robot manipulators," *IEEE Trans. Circ. Syst.,* Vol. CAS-34, pp. 73-82, Jan. 1987.

[58] R.G. Morgan and U. Ozgunner, "A decentralized variable structure control algorithm for robotic manipulators," *IEEE J. Rob. & Auto.,* Vol. RA-1, pp. 57-65, Mar. 1985.

[59] E. Bailey and A. Arapostathis, "Simple sliding mode control scheme applied to robot manipulators," *Int. J. Contr.,* Vol. 45, pp. 1197-1209, 1987.

[60] G. Bartolini and T. Zolezzi, "Variable structure nonlinear in the control law," *IEEE Trans. Auto. Contr.,* Vol. AC-30, pp. 681-684, July 1985.

[61] K.S. Yeung and Y.P. Chen, "A new controller design for manipulators using the theory of variable structure systems," *IEEE Trans. Auto. Contr.,* Vol. 33, pp. 200-206, Feb. 1988.

[62] Y-F Chen, T. Mita, and S. Wakui, "A new and simple algorithm for sliding mode trajectory control of the robot arm," *IEEE Trans. Auto. Contr.,* Vol. 35, pp. 828-829, July 1990.

[63] R.A. DeCarlo, S.H. Żak, and G.P. Matthews, "Variable structure control of nonlinear multivariable systems," *IEEE Proc.,* Vol. 76, pp. 212-232, Mar. 1988.

[64] M.W. Spong and F. Ghorbel, "Robustness of adaptive control of robots," in *Proc. Symp. Control of Robots and Manufacturing Systems,* Arlington, TX, Nov. 1990.

[65] P.A. Ioannou and P.V. Kokotovic, "Instability analysis and improvement of robustaness of adaptive control," *Automatica,* Vol. 20, pp. 583-594, 1984.

Editor's Note: The authors of the paper have taken the opportunity in this reprinting to make corrections to some equations and equation numbers.

Jean-Jacques E. Slotine
Department of Mechanical Engineering
Massachusetts Institute of Technology
Cambridge, Massachusetts 02139

The Robust Control of Robot Manipulators

Abstract

A new scheme is presented for the accurate tracking control of robot manipulators. Based on the more general suction control methodology, the scheme addresses the following problem: Given the extent of parametric uncertainty (such as imprecisions or inertias, geometry, loads) and the frequency range of unmodeled dynamics (such as unmodeled structural modes, neglected time delays), design a nonlinear feedback controller to achieve optimal tracking performance, in a suitable sense. The methodology is compared with standard algorithms such as the computed torque method and is shown to combine in practice improved performance with simpler and more tractable controller designs.

1. Introduction

This paper presents a new scheme to achieve accurate tracking control of robot manipulators in the presence of model uncertainty and disturbances.

The development of efficient feedback control algorithms for robot manipulators has recently been the object of considerable interest (Brady et al. 1982). The complexity of the manipulator control problem largely reflects that of manipulator dynamics itself: Beyond three degrees of freedom, the derivation of manipulator dynamics becomes cumbersome to manage analytically and requires the use of sophisticated on-line computational algorithms (Luh, Waker, and Paul 1980*a*; Hollerbach 1980; Silver 1982; Renaud 1983). A major drawback of such algorithms is that they lack physical tractability, that is, they do not allow one to effectively exploit engineering insight during the design process (Bejczy and Lee 1983; Luh and Gu 1984). Further, the performance of standard control schemes based on these algorithms (for example, the computed torque or inverse method) is very sensitive to parametric uncertainty, that is, to imprecision on manipulator inertias, geometry, loads, and so on (Gilbert and Ha 1983).

In this paper, the *suction control* methodology (Slotine 1984) is proposed as a remedy to these drawbacks. For a class of nonlinear systems, suction control addresses the following problem: Given the extent of parametric uncertainty (such as imprecisions on inertias, geometry, loads) and of disturbances (such as Coulomb or viscous friction) and the frequency range of unmodeled dynamics (such as unmodeled structural modes, neglected time delays), design a nonlinear feedback controller to achieve optimal tracking precision, in a suitable sense. The explicit robustness guarantees provided by the methodology are shown to allow the use of simpler, more tractable manipulator models (such as "decoupled" arm and hand in a six-degree-of-freedom robot) while preserving stability in the face of model imprecision.

Section 2 summarizes the major features of suction controller design. The specific application of the methodology to the accurate tracking control of robot manipulators is described in Section 3. Section 4 concludes that the gains in engineering insight and controller simplicity offered by the methodology generally come for free because suction control schemes based on simplified models outperform standard algorithms based on higher-order models at most practical levels of parametric uncertainty.

2. Suction Control of Nonlinear Systems

In this section we review the basic results of Slotine and Sastry (1983) and Slotine (1984). For notational simplicity, the development is presented for systems with a single control input, although the extension to a large class of multi-input nonlinear systems is straightforward, as illustrated in an example. The specific design of suction controllers for robot manipulators is detailed in Section 3.

The author formerly was with the Robotics Systems Research Department, AT&T Bell Laboratories, Holmdel, New Jersey.

The International Journal of Robotics Research,
Vol. 4, No. 2, Summer 1985.
0278-3649/85/020049-16 $05.00/0,

2.1. Sliding Surfaces

Consider the dynamic system

$$x^{(n)}(t) = f(\mathbf{X};t) + b(\mathbf{X};t)u(t) + d(t), \qquad (1)$$

where $u(t)$ is the control input (say the applied torque at a manipulator joint) and $\mathbf{X} = [x, \dot{x}, \ . \ . \ . \ , \ x^{(n-1)}]^T$ is the state. In Eq. (1), the function $f(\mathbf{X};t)$, in general nonlinear, is not exactly known, but the extent of the imprecision $|\Delta f|$ on $f(\mathbf{X};t)$ is upper-bounded by a known continuous function of $\mathbf{X}$ and t; similarly, control gain $b(\mathbf{X};t)$ is not exactly known but is of constant sign and is bounded by known, continuous functions of $\mathbf{X}$ and t. Disturbance $d(t)$ is unknown but upper-bounded by a known continuous function of $\mathbf{X}$ and t. The control problem is to get the state $\mathbf{X}$ to track a specific state $\mathbf{X}_d = [x_d, \dot{x}_d, \ . \ . \ . \ , x_d^{(n-1)}]^T$ in the presence of model imprecision on $f(\mathbf{X};t)$ and $b(\mathbf{X};t)$ and of disturbances $d(t)$. For this to be achievable using a finite control u, we must assume

$$\tilde{\mathbf{X}}|_{t=0} = 0, \qquad (2)$$

where

$$\tilde{\mathbf{X}} := \mathbf{X} - \mathbf{X}_d = [\tilde{x}, \dot{\tilde{x}}, \ . \ . \ . \ , \tilde{x}^{(n-1)}]^T$$

is the tracking error vector.

We define a *time-varying sliding surface* $S(t)$ in the state-space $\mathbf{R}^n$ as

$$S(t) : s(\mathbf{X};t) = 0,$$

with

$$s(\mathbf{X};t) := \left(\frac{d}{dt} + \lambda\right)^{n-1} \tilde{x}, \quad \lambda > 0, \qquad (3)$$

where λ is a positive constant (design parameter λ will later be interpreted as the desired control bandwidth). Given initial condition (2), the problem of tracking $\mathbf{X} \equiv \mathbf{X}_d$ is equivalent to that of remaining on the surface $S(t)$ for all $t > 0$. Now a sufficient condition for such positive invariance of $S(t)$ is to choose the control law u of Eq. (1) such that outside of sliding condition $S(t)$,

Fig. 1. The sliding condition.

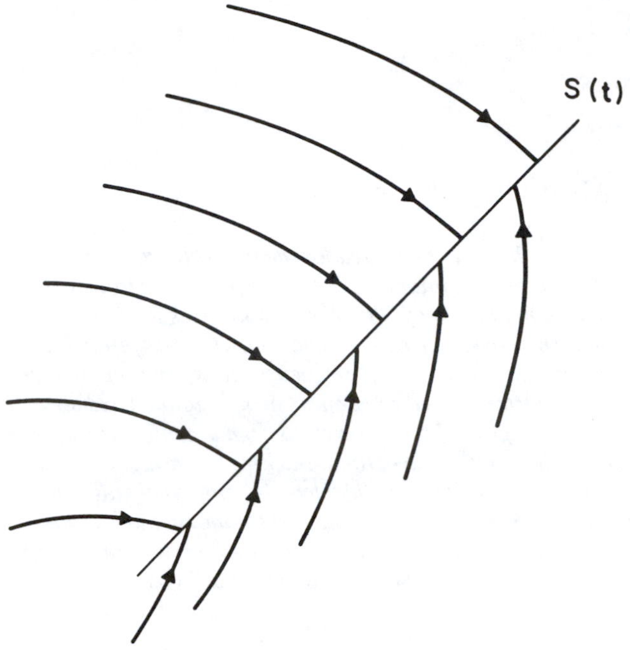

$$\frac{1}{2}\frac{d}{dt} s^2(\mathbf{X};t) \leq -\eta|s|, \qquad (4)$$

where η is a positive constant. Sliding condition (4) constrains state trajectories to point toward $S(t)$, as illustrated in Fig. 1.

The idea behind Eqs. (3) and (4) is to pick up a well-behaved function of the tracking error, s, according to Eq. (3) and then select the feedback control law u in Eq. (1) such that s^2 remains a Lyapunov function (Vidyasagar 1978) of the closed-loop system, in a suitable sense, despite the presence of model imprecision and disturbances. Further, satisfying Eq. (4) guarantees that if condition (2) is not exactly verified, that is, if $\mathbf{X}|_{t=0}$ is actually off $\mathbf{X}_d|_{t=0}$, the surface $S(t)$ will nonetheless be reached in a finite time (inferior to $s(\mathbf{X}(0); 0)/\eta$), while definition (3) then guarantees that $\tilde{\mathbf{X}} \rightarrow 0$ as $t \rightarrow \infty$.

The controller design procedure consists of two steps. First, a feedback control law u is selected so as to verify sliding condition (4). To account for the presence of modeling imprecision and of disturbances, however, such a control law is discontinuous across $S(t)$, which leads to control chattering. Chattering is undesirable in practice because it involves high control activity and further may excite high-frequency dynamics neglected in the course of modeling (such as unmodeled structural modes, neglected time delays, and the like). Thus, in a second step, discontinuous control law u is suitably smoothed to achieve an optimal trade-off between control bandwidth and tracking precision. While the first step accounts for parametric uncertainty, the second steps achieves robustness to high-frequency unmodeled dynamics.

Remark: The notion of sliding surface and the associated theory of variable structure systems (VSS) have been studied in great detail in the Soviet literature (see Utkin 1977 for a review). As discussed extensively in Slotine and Sastry (1983) and Slotine (1983), the VSS methodology has several important drawbacks, particularly high control authority and control chattering. An application of classical VSS theory to manipulator control is described in Young (1978).

2.2. Perfect Tracking Using Switched Control Laws

Given the bounds on uncertainties on $f(\mathbf{X};t)$ and $b(\mathbf{X};t)$ and on disturbances $d(t)$, constructing a control law to verify sliding condition (4) is straightforward, as we now illustrate on a simple multivariable example (numerous examples can be found in Slotine 1983).

Example: Consider the coupled multi-input system

$$\left[\begin{array}{ll} \ddot{\theta}_1 = a(t)(\theta_1)^3 + 2(\dot{\theta}_2)^2 + u_1 + d_1(t), & (5) \\ \ddot{\theta}_2 = f_2(\theta_1, \dot{\theta}_1, \theta_2, \dot{\theta}_2; t) + u_2 + d_2(t), & (6) \end{array}\right.$$

where parameter $a(t)$ is estimated as $\hat{a}(t)$, with

$$|a(t) - \hat{a}(t)| \leqslant \alpha.$$

Consider the problem of getting θ_1 to track a desired trajectory θ_{d1}, specified in real time, such that

$$|\ddot{\theta}_{d1}(t)| \leqslant v, \quad \forall t \geqslant 0$$

in the presence of disturbances $d_1(t)$ such that

$$|d_1(t)| \leqslant \bar{d}, \quad \forall t \geqslant 0,$$

where v and $\bar{d}$ are known positive constants. We define the sliding surface $S_1(t)$ according to Eq. (3), namely

$$S_1(t) : s_1 = 0, \quad \text{with} \quad s_1 = \dot{\tilde{\theta}}_1 + \lambda\tilde{\theta}_1,$$

where $\tilde{\theta}_1 = \theta_1 - \theta_{d1}$ is the tracking error. Note that s_1 depends on θ_1 only. To satisfy a sliding condition of the form (4), we define control law u_1 as

$$u_1 = -\hat{a}(t)\theta_1^3 - 2\dot{\theta}_2^2 - \lambda\dot{\tilde{\theta}}_1 - (\alpha|\theta_1|^3 + \gamma)\,\text{sgn}\, s_1, \quad \gamma > \bar{d} + v, \tag{7}$$

where $\text{sgn}(s) = -1$ for $s < 0$, and $\text{sgn}(s) = +1$ for $s > 0$. Control law (7) is composed of a term $(-\hat{a}(t)\theta_1^3 - 2\dot{\theta}_2^2 - \lambda\dot{\tilde{\theta}}_1)$ that merely compensates for the known part of the dynamics of the variable s and of a term $(-[\alpha|\theta_1|^3 + \gamma]\,\text{sgn}\, s_1)$, discontinuous across $S_1(t)$, that allows one to keep sliding condition (4) verified despite the presence of disturbances and parametric uncertainty. We have indeed

$$\frac{1}{2}\frac{d}{dt}(s_1)^2 = s_1 \cdot \dot{s}_1 = s_1 \cdot (\ddot{\theta}_1 - \ddot{\theta}_{d1} + \lambda\dot{\tilde{\theta}}_1),$$

so control law (7) applied to system dynamics (5) leads to

$$\frac{1}{2}\frac{d}{dt}(s_1)^2 = s_1 \cdot [(a(t) - \hat{a}(t))\theta_1^3 - \ddot{\theta}_{d1}(t) + d_1(t)] - |s_1| \cdot [\alpha|\theta_1|^3 + \gamma],$$

where

$$\alpha \geqslant |a(t) - \hat{a}(t)|, \quad \gamma > v + \bar{d} \geqslant |-\ddot{\theta}_{d1} + d_1(t)|.$$

Hence this satisfies sliding condition (4) for surface $S_1(t)$:

$$\frac{1}{2}\frac{d}{dt}(s_1)^2 \leqslant -[\gamma - (v + \bar{d})] \cdot |s_1|.$$

Note that control discontinuity at $S_1(t)$ increases with parametric imprecision and strength of disturbances to be compensated for. Further, if desired acceleration $\ddot{\theta}_{d1}(t)$ is explicitly available, we may rather use

$$u_1 = \ddot{\theta}_{d1}(t) - \hat{a}(t)\theta_1^3 - 2\dot{\theta}_2^2 - \lambda\dot{\tilde{\theta}}_1 - (\alpha|\theta_1|^3 + \gamma)\text{sgn}\, s_1, \quad \gamma > \bar{d}$$

instead of Eq. (7), thus reducing control discontinuity at $S_1(t)$.

Assume now that Eq. (5) is replaced by

$$\ddot{\theta}_1 = a(t)\theta_1^3 + 2\dot{\theta}_2^2 + b_1(t)u_1 + d_1(t), \quad (8)$$

where gain $b_1(t)$ is estimated by $\hat{b}_1(t)$ such that

$$\frac{1}{\beta_1} \leq \frac{\hat{b}_1(t)}{b_1(t)} \leq \beta_1, \quad \beta_1 \geq 1. \quad (9)$$

Parameter β_1 in Eq. (9) can be interpreted as the *gain margin* on control law u_1 relative to (possibly time-varying or state-dependent) control gain estimate $\hat{b}_1(t)$. Letting

$$u_{10} := \hat{a}(t)\theta_1^3 + 2\dot{\theta}_2^2 + \lambda\dot{\tilde{\theta}}_1,$$

we can satisfy condition (4) by now choosing

$$u_1 = -\frac{1}{\hat{b}_1(t)}\left[\frac{2u_{10}}{\beta_1 + (1/\beta_1)} + \beta_1\left(\frac{|u_{10}|[\beta_1 - (1/\beta_1)]}{\beta_1 + (1/\beta_1)} + \alpha|\theta_1|^3 + \gamma\right)\operatorname{sgn} s_1\right], \quad \gamma > \bar{d} + v \quad (10)$$

instead of (7). We thus have further increased control discontinuity at $S_1(t)$ to compensate for uncertainty on gain $b_1(t)$ (note that the choice of u_1 to satisfy sliding condition 4 is not unique; expression 10 minimizes the continuous part of $\dot{s}_1$).

Notice the decoupling in the design: If θ_2 were also prescribed to track a given trajectory, one would define $s_2 := \dot{\tilde{\theta}}_2 + \lambda_2\tilde{\theta}_2$ and repeat the procedure independently.

As illustrated in the preceding example, and shown in further detail in Slotine and Sastry (1983), the methodology can easily be extended to multivariable systems of the form

$$x_j^{(n_j)} = f_j(\mathbf{X}_1, \ldots, \mathbf{X}_m; t) + b_j(\mathbf{X}_1, \ldots, \mathbf{X}_m; t)u_j + d_j(t), \quad j = 1, \ldots, m, \quad (11)$$

where $\mathbf{X}_j = [x_j, \dot{x}_j, \ldots, x_j^{(n_j-1)}]^T$. We assume again that disturbances d_j and imprecisions on f_j and b_j are bounded by known continuous functions of $\mathbf{X}_j$ and of time. Dynamics of the form (11) describe a large number of nonlinear systems encountered in practice, including a vast class of mechanical systems (from Lagrange's equations). Further, Hunt, Su, and Meyer (1983) have shown that a wide class of controllable nonlinear systems could be put in the form (11) by using appropriate "global" transformations: Being able to deal explicitly with imprecision on the system model allows for numerical conditioning problems that may affect such transformations to be easily accounted for.

2.3. Suction Control

2.3.1. *Continuous Control Laws to Approximate Switched Control*

As seen in the example of Section 2.2, control laws that satisfy sliding condition (4) are discontinuous across the surface $S(t)$, thus leading to control chattering. Chattering is in general highly undesirable in practice because it involves extremely high control activity and further may excite high-frequency dynamics neglected in the course of modeling. We can remedy this situation by smoothing out the control discontinuity in a thin boundary layer neighboring the switching surface (see Fig. 2):

$$B(t) = \{\mathbf{X}, |s(\mathbf{X};t)| \leq \Phi\}, \quad \Phi > 0, \quad (12)$$

where Φ is the boundary layer thickness and $\epsilon := \Phi/\lambda^{n-1}$ is the boundary layer width. This is achieved by choosing outside of $B(t)$ control law u as before—that is, satisfying sliding condition (4), which guarantees boundary layer attractiveness, and hence positive invariance; all trajectories starting inside $B(t = 0)$ remain inside $B(t)$ for all $t \geq 0$—and then interpolating u inside $B(t)$—for instance, replacing in the expression of u the term sgn s by s/Φ, inside $B(t)$, as illustrated in Figs. 2 and 3. As proved in Slotine (1983), this leads to tracking to within a guaranteed precision ϵ and more generally guarantees that for all trajectories starting inside $B(t = 0)$,

$$|\tilde{x}^{(i)}(t)| \leq (2\lambda)^i\epsilon, \quad i = 0, \ldots, n-1.$$

These bounds are understood asymptotically, with a time constant $(n-1)/\lambda$; they hold for all $t \geq 0$ if $\tilde{\mathbf{X}}|_{t=0} = 0$. We now show that the smoothing of control

Fig. 2. Construction of the boundary layer in the case that n = 2.

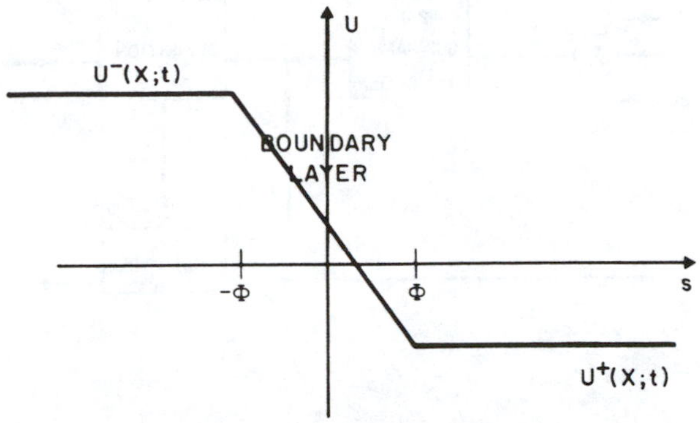

Fig. 3. Sample interpolation of the control law in the boundary layer.

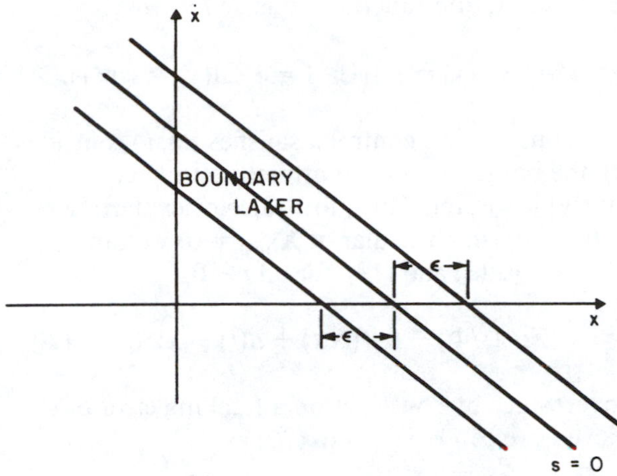

discontinuity inside $B(t)$ essentially assigns a lowpass filter structure to the local dynamics of the variable s, thus eliminating chattering. Recognizing this filter structure allows us to tune up the control law so as to achieve a trade-off between tracking precision and robustness to unmodeled dynamics.

For clarity, we first consider the case of no gain margin ($\beta = 1$) and then generalize. Further details on the following development can be found in Slotine (1984).

Consider the system

$$X^{(n)} = f(\mathbf{X};t) + u + d(t), \tag{13}$$

where

$$f(\mathbf{X};t) = \hat{f}(\mathbf{X};t) + \Delta f(\mathbf{X};t). \tag{14}$$

In Eq. (14), $\hat{f}(\mathbf{X};t)$ is the available model of $f(\mathbf{X};t)$. Further, as in Section 2, we assume

$$|\Delta f(\mathbf{X};t)| \leq F(\mathbf{X};t), \tag{15}$$

$$|d(t)| \leq D(\mathbf{X};t), \tag{16}$$

where F, D, and $\hat{f}$ are known continuous functions of $\mathbf{X}$ and t; uncertainty $\Delta f(\mathbf{X};t)$ on dynamics $f(\mathbf{X};t)$ is assumed to be continuous in $\mathbf{X}$.

The control problem is to get the state $\mathbf{X}$ to track a desired state $\mathbf{X}_d$, specified in real time, such that a priori:

$$|x_d^{(n)}(t)| \leq v(t),$$

although $x_d^{(n)}(t)$ itself is not explicitly available. Defining $s(\mathbf{X};t)$ according to Eq. (3) and letting

$$k(\mathbf{X};t) := F(\mathbf{X};t) + D(\mathbf{X};t) + v(t) + \eta \geqslant \eta > 0, \tag{17}$$

where η is a positive constant, the control law

$$u = -\hat{f}(\mathbf{X};t) - \sum_{p=1}^{n-1} \binom{n-1}{p} \lambda^p \tilde{x}^{(n-p)} - k(\mathbf{X};t)\,\mathrm{sgn}(s) \tag{18}$$

satisfies sliding condition (4).[1] As in the example of Section 2.2, control law (18) is composed of a term,

$$-\hat{f}(\mathbf{X};t) - \sum_{p=1}^{n-1} \binom{n-1}{p} \lambda^p \tilde{x}^{(n-p)},$$

that compensates for the known part of the dynamics of the variable s and of a term $-k(\mathbf{X};t)\,\mathrm{sgn}(s)$, discontinuous across $S(t)$, that keeps sliding condition (4) verified in spite of parametric uncertainty and disturbances. Let us now smooth out the control discontinuity inside the boundary layer $B(t)$ of thickness Φ, defined by Eq. (12). Control law (18) then becomes

$$u = -\hat{f}(\mathbf{X};t) - \sum_{p=1}^{n-1} \binom{n-1}{p} \lambda^p \tilde{x}^{(n-p)} - k(\mathbf{X};t)\,\mathrm{sat}(s/\Phi), \tag{19}$$

1. $\begin{bmatrix} n \\ m \end{bmatrix} = \dfrac{n!}{m!(n-m)!}$ for $m \leqslant n$.

where the saturation function sat is defined by

$$|y| \leq 1 \Rightarrow \text{sat}(y) = y; \; |y| > 1 \Rightarrow \text{sat}(y) = \text{sgn}(y).$$

Since by construction control u satisfies Eq. (4) outside of $B(t)$, the boundary layer is attractive and hence (positively) invariant. Thus, for trajectories starting inside $B(t = 0)$ (in particular, if $\tilde{\mathbf{X}}|_{t=0} = 0$) we can write from Eqs. (3) and (19), for all $t \geq 0$,

$$\dot{s} = -k(\mathbf{X};t)[s/\Phi] + (\Delta f(\mathbf{X};t) + d(t) - x_d^{(n)}(t)). \quad (20)$$

Moreover, since by construction a tracking error of ϵ is achieved, we can rewrite Eq. (20) as

$$\dot{s} = -k(\mathbf{X}_d;t)[s/\Phi] + (\Delta f(\mathbf{X}_d;t) + d(t) - x_d^{(n)}(t) + 0(\epsilon)), \quad (21)$$

since $\Delta f(\mathbf{X};t)$ and $k(\mathbf{X};t)$ are continuous in $\mathbf{X}$. We see from Eq. (21) that the variable s, which is a measure of the algebraic distance to the surface $S(t)$, is the output of a stable first-order filter whose dynamics depend only on the desired state $\mathbf{X}_d(t)$ and perhaps explicitly on time and whose inputs are (to first order) perturbations: disturbance $d(t)$, uncertainty $\Delta f(\mathbf{X}_d;t)$ on the dynamics, and nth derivative $x_d^{(n)}$ of the desired trajectory.

Equation (21) shows that chattering is indeed eliminated, as long as high-frequency unmodeled dynamics are not excited. The dynamic structure of the closed-loop system is summarized in Fig. 4: perturbations are filtered according to Eq. (21) to give s, which in turn provides tracking error $\tilde{x}$ by further lowpass filtering, according to definition (3); control u is a function of s, $\mathbf{X}$ and $\mathbf{X}_d$ as specified by Eq. (19). Since λ is the break frequency of the filter described by Eq. (3), it must be chosen to be "small" with respect to high-frequency unmodeled dynamics (such as unmodeled structural modes or neglected time delays). Assume now that $F(\mathbf{X}_d;t)$, $D(\mathbf{X}_d;t)$, and $v(t)$ can be a priori upper-bounded so that $k(\mathbf{X}_d;t)$ can be upper-bounded, say by k_{max}. The constant k_{max}/Φ may be thought of as the "break frequency" of Eq. (21); as λ, it must also be chosen to be small with respect to high-frequency unmodeled dynamics. Thus, if λ is set to be the largest acceptable break frequency of Eq. (3), we must have

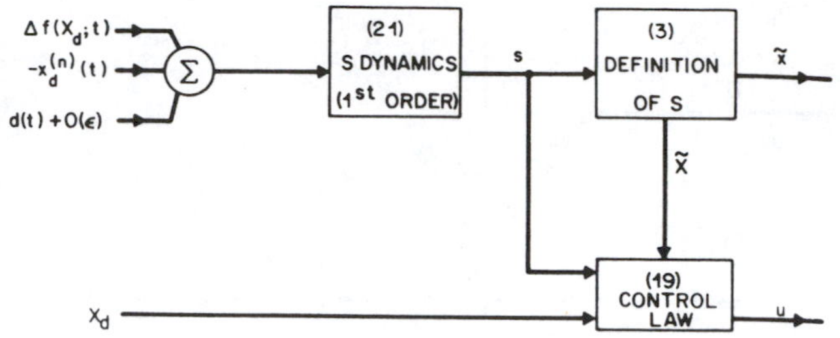

Fig. 4. Dynamic structure of the closed-loop system.

$$k_{\text{max}}/\Phi \leq \lambda,$$

which fixes the best attainable tracking precision ϵ:

$$\lambda\Phi = k_{\text{max}}, \quad \text{that is,} \quad \lambda^n\epsilon = k_{\text{max}}. \quad (22)$$

We will refer to Eq. (22) as the *balance condition* for the control system. Intuitively, it amounts to tuning up the control law so that the closed-loop system will be as close as possible to "critical damping" (if $k(\mathbf{X}_d;t)$ were constant, Eq. 22 would exactly correspond to critical damping). Of course, desired trajectory $x_d(t)$ itself must be slow with respect to unmodeled dynamics.

The preceding development was for $\beta = 1$ (no gain margin). Assume now that Eq. (13) is replaced by

$$x^{(n)} = f(\mathbf{X};t) + b(\mathbf{X};t)u + d(t),$$

where gain $b(\mathbf{X};t)$ is estimated by $\hat{b}(\mathbf{X};t)$, with a gain margin β:

$$1/\beta \leq \hat{b}(\mathbf{X};t)/b(\mathbf{X};t) \leq \beta.$$

If, instead, $\beta_{\text{min}} \leq \hat{b}(\mathbf{X};t)/b(\mathbf{X};t) \leq \beta_{\text{max}}$, set $\beta := \sqrt{\beta_{\text{max}}/\beta_{\text{min}}}$ and use $\hat{b}_{\text{new}}(\mathbf{X};t) := \hat{b}(\mathbf{X};t)/\sqrt{\beta_{\text{min}}\beta_{\text{max}}}$ as the new estimate of $b(\mathbf{X};t)$; note that β_{min}, β_{max}, and β can be time-dependent. One can then show that the balance condition is now of the form

$$\lambda\Phi = \beta k_{\text{max}}, \quad \text{that is,} \quad \lambda^n\epsilon = \beta k_{\text{max}}, \quad (23)$$

instead of Eq. (22). Thus, having chosen λ to be small with respect to high-frequency unmodeled dynamics, the balance condition (23) fixes the best attainable tracking precision ϵ, given bandwidth requirements and a choice of $k(\mathbf{X};t)$ accounting for modeling uncertainties, disturbances to be rejected, and range of trajectories to be tracked. Further, as shown in Slotine (1983), extension of the balance condition to the class of multivariable systems in Eq. (10) is straightforward: for each degree of freedom $j (j = 1, \ldots, m)$ we obtain a condition of the form

$$\lambda_j \Phi_j = \beta_j k_{j\max}, \quad \text{that is} \quad (\lambda_j)^{n_j} \epsilon_j = \beta_j k_{j\max}, \qquad (24)$$

where in general $k_{j\max}$ depends on all $\mathbf{X}_{d1}, \ldots, \mathbf{X}_{dm}$.

Example: Consider again the system (8) of Section 2.2, with (discontinuous) control law u_1 defined as in Eq. (10). The corresponding gain $k_1(\mathbf{X};t)$ is

$$k_1(\mathbf{X};t) = \beta_1(\zeta|u_{10}| + \alpha|\theta_1|^3 + \gamma),$$

where $\zeta := (\beta_1 - 1/\beta_1)/(\beta_1 + 1/\beta_1)$, so that

$$k_1(\mathbf{X}_d;t) = \beta_1(\zeta|\hat{a}(t)\theta_{d1}^3 + 2\dot{\theta}_{d2}^2| + \alpha|\theta_{d1}|^3 + \gamma).$$

Assume now that

$$|\theta_{d1}| \le \theta_{d1}^{\max}, \quad |\dot{\theta}_{d2}| \le \dot{\theta}_{d2}^{\max}, \quad |\hat{a}(t)| \le a^{\max}, \quad \forall t \ge 0.$$

Then

$$k_1(\mathbf{X}_d;t) \le k_1^{\max} := \beta_1[\zeta \cdot (a^{\max}(\theta_{d1}^{\max})^3 + 2(\dot{\theta}_{d2}^{\max})^2) + \alpha|\theta_{d1}^{\max}|^3 + \gamma].$$

Assume further that the system exhibits an unmodeled structural mode at $\nu_{\text{structure}} \approx 5$ Hz. A reasonable choice for λ may then be, for instance,

$$\lambda := 10 \text{ rad/s} < (2\pi/3) \cdot \nu_{\text{structure}},$$

which then fixes tracking precision ϵ_1:

$$\epsilon_1 := \beta_1 k_{1\max}/\lambda^2 = \beta_1 k_{1\max}/100.$$

Continuous control law u_1 is then obtained by replacing the term sgn s_1 by sat$(s_1/\lambda\epsilon_1)$ in expression (10).

Remarks:

1. Effect of data sampling can also be interpreted as part of high-frequency unmodeled dynamics. It can be shown that the corresponding "soft" upper bound that sampling rate ν_{sampling} imposes on λ is

$$2(1 + \tau_{\text{process}} \cdot \nu_{\text{sampling}})\lambda < \nu_{\text{sampling}},$$

where $(\tau_{\text{process}} \cdot \nu_{\text{sampling}})$ is the ratio of processing delay over sampling period. For instance, in the case of a full-period processing delay we must choose

$$\lambda < \nu_{\text{sampling}}/4.$$

2. By an argument similar to that of the above discussion, it can be shown that the specific choice of the dynamics of Eq. (3) used to define sliding surfaces is the best conditioned among linear dynamics, in the sense that it guarantees the best tracking precision ϵ given the desired control bandwidth and the extent of parametric uncertainty and disturbances. Conversely, the definition of the sliding surface can be shaped to address more general applications than tracking control. For instance, in the context of compliance control (Hogan 1984), $s = 0$ can be chosen to describe a desired dynamic response to measured forces exerted by the environment. The methodology developed in this paper largely extends to such problems and in particular represents an effective and systematic procedure to reduce sensitivity to modal uncertainty in compliance control schemes for robot manipulators.

2.3.2. The Dynamic Balance Conditions

The above trade-off between robustness and tracking precision, quantified by balance conditions (24), can be further improved by allowing boundary layer widths to be time-dependent.

In Section 2.3.1, by seeking the smallest constant ϵ such that

$$\beta k(\mathbf{X}_d;t)/(\lambda^{n-1}\epsilon) \le \lambda, \quad \forall k(\mathbf{X}_d;t),$$

we obtained the static balance condition (22):

$$\lambda^n \epsilon = \beta k_{\max}.$$

If we now allow boundary layer width ϵ to be time-varying, we expect to be able to refine the balance condition, that is, to define a boundary layer width history $\epsilon(t)$ such that

$$\epsilon(t) \leq \beta k_{\max}/\lambda^n, \quad \forall t \geq 0, \tag{25}$$

while preserving system bandwidth λ However, to maintain boundary attractiveness while allowing for variations of boundary layer thickness $\Phi = \lambda^{n-1}\epsilon$, we must now choose the control law u such that outside of $B(t)$

$$\frac{1}{2}\frac{d}{dt}s^2 \leq (\dot{\Phi} - \eta)|s|, \quad \eta > 0, \tag{26}$$

instead of Eq. (4). The additional term $\dot{\Phi}|s|$ in Eq. (26) reflects the fact that the boundary layer attractiveness condition is more stringent during boundary layer contraction ($\dot{\Phi} < 0$) and less stringent during boundary layer expansion ($\dot{\Phi} > 0$).

It is shown in Slotine (1984) that, to satisfy condition (26) while preserving control bandwidth λ, one can simply replace gain $k(\mathbf{X};t)$ of Eq. (19) by a new gain $\bar{k}(\mathbf{X};t)$, where

$$\bar{k}(\mathbf{X};t) := [k(\mathbf{X};t) - k(\mathbf{X}_d;t)] + \lambda\Phi/\beta, \tag{27}$$

with static balance condition (23) replaced by the *dynamic balance conditions*

$$\left[\begin{array}{ll} k(\mathbf{X}_d;t) \geq \lambda\Phi/\beta => \dot{\Phi} + \lambda\Phi = \beta k(\mathbf{X}_d;t), & (28) \\ k(\mathbf{X}_d;t) \leq \lambda\Phi/\beta => \dot{\Phi} + \lambda\Phi/\beta^2 = k(\mathbf{X}_d;t)/\beta, & (29) \end{array}\right.$$

with, initially,

$$\Phi(0) := \beta k(\mathbf{X}_d(0); 0)/\lambda. \tag{30}$$

This particular type of sliding control, which uses time-varying boundary layer widths to account for time dependence of parametric uncertainty, will be referred to as *suction control.* It is often the case in practice that $k(\mathbf{X}_d;t)$ shows large variations along a desired trajectory. In these instances, the use of time-varying boundary layer widths, as specified by the dynamic balance conditions, greatly improve tracking performance while introducing only modest additional complexity. In robot manipulator control, for instance, $k(\mathbf{X};t)$ involves centripetal terms $\dot{\theta}_i^2$ and Coriolis terms $\dot{\theta}_i\dot{\theta}_j$ (where the θ_i are joint angles). The dynamic balance conditions then allow one to efficiently trade off speed against tracking precision while preserving system robustness to unmodeled dynamics. Also, recalling that $k(\mathbf{X};t)$ reflects uncertainty on system dynamics, $k(\mathbf{X}_d;t)$ may be decreased as the result of a parameter estimation process, for instance. The dynamic balance conditions allow one to easily account for such on-line improvement on modeling precision. The suction control methodology is thus likely to provide robust "adaptive" schemes because it guarantees stability and fixes control system bandwidth while achieving best tracking precision given current modeling uncertainties.

Further, in some practical instances, desired bandwidth itself may vary with time. In robot manipulator control, for example, structural resonant frequencies decrease as the load mass at the tip of the arm gets larger (Paul 1981). The control law, initially tuned not to excite the lowest expectable mode (that is, to handle maximum load), can thus exploit on-line load estimation by increasing control bandwidth in addition to decreasing $k(\mathbf{X}_d;t)$. Further, structural resonant frequencies actually vary with manipulator configuration (although because of its complexity this effect is rarely modeled). Similarly, it is desirable to monitor mechanical compliance when performing automatic assembly of close-tolerance parts (Hogan 1981; Asakawa, Akiya, and Tabata 1982). Let us call λ_0 the constant value of λ that we previously used, that is, the desired control bandwidth based on a uniform lower bound on frequencies of unmodeled dynamics. It can be shown that Eqs. (27)–(29) remain valid for time-varying $\lambda = \lambda(t)$, provided control u is now defined as

$$u := -\hat{f}(\mathbf{X};t) - \sum_{p=1}^{n-1} \binom{n-1}{p} \lambda^{p-1}(\lambda\tilde{x}^{(n-p)} + p\dot{\lambda}\tilde{x}^{(n-p-1)}) - \bar{k}(\mathbf{X};t)\,\mathrm{sat}(s/\Phi), \tag{31}$$

instead of Eq. (19). Time derivative $\dot{\lambda}$ in Eq. (31) must be chosen smooth enough not to excite high-frequency unmodeled dynamics. A practical way to do so is to generate $\lambda(t)$ by filtering desired bandwidth value λ_{desired} through a second-order critically damped filter of break frequency λ_0. Of course, initial condition (30) is now replaced by

$$\Phi(0) := \beta k(\mathbf{X}_d(0){:}0)/\lambda(0). \tag{32}$$

Remarks:

1. Since ϵ essentially varies as λ^{-n} for given parametric uncertainty, bandwidth variations are strongly reflected in tracking performance.
2. In the case that $\beta = 1$ (no gain margin), Eqs. (28)–(29) reduce to

$$\dot{\Phi} + \lambda\Phi = k(\mathbf{X}_d;t).$$

3. When the dynamics of $k(\mathbf{X}_d;t)$ is slow with respect to frequency λ_0/β^2, one can simply use

$$\bar{k}(\mathbf{X};t) := k(\mathbf{X};t),$$
$$\Phi := \beta k(\mathbf{X}_d;t)/\lambda(t),$$

instead of Eqs. (27)–(29).
4. The preceding development remains valid in the case that gain margin β is time-varying.
5. The s-trajectory, that is, the variation of s/Φ with time, is a compact descriptor of the closed-loop system behavior: Control activity directly depends on s/Φ, while by definition (3) tracking error $\tilde{x}$ is merely a filtered version of s. Further, the s-trajectory represents a time-varying measure of the validity of the assumptions on model uncertainty. Similarly, boundary layer thickness Φ describes the evolution of dynamic model uncertainty with time. The use of s-trajectories for multivariable control system design and testing is demonstrated in Slotine, Yoerger, and Sheridan, in press).

2.4. Summary

By controlling the system along time-varying sliding surfaces in the state-space, we achieved perfect tracking of desired trajectories for a class of multivariable nonlinear time-varying systems.

By substituting smooth transitions across a boundary layer to control switching at the sliding surface, we eliminated chattering and obtained a trade-off between tracking precision and robustness to unmodeled dynamics.

By allowing boundary layer width to be time-varying, we refined the above trade-off to account for time-dependence of parameter uncertainties, thus improving tracking precision while still maintaining robustness to high-frequency unmodeled dynamics.

Finally, we monitored the orientation of the boundary layer in the state-space (defined by λ) to account for possible time dependence of desired bandwidth (whether due to actual changes in the plant or to on-line modeling improvements), thus further improving tracking performance.

We now describe more specifically the application of the methodology to the feedback control of robot manipulators. The scheme is compared to standard algorithms such as the computed torque method and is shown to significantly simplify controller design, thus permitting us to retain and exploit engineering insight all along the design and implementation process.

3. The Robust Control of Robot Manipulators

3.1. Robustness Issues in Manipulator Control

There has recently been a considerable interest in developing efficient control algorithms for robot manipulators (Brady et al. 1982). The complexity of the control problem for manipulators arises mainly from that of manipulator dynamics itself: The dynamics of articulated mechanisms in general, and of robot manipulators in particular, involves strong coupling effects between joints as well as centripetal and Coriolis forces (particularly significant at high speeds). The equations of motion for an n-link manipulator may be expressed as follows (Paul 1981):

$$\mathbf{R}(\theta)\ddot{\theta} + \mathbf{h} = \mathbf{T}, \tag{33}$$

where θ is the n-vector of joint angles (or more generally of joint displacements), $\mathbf{T}$ is the vector of applied torques (or generalized forces), $\mathbf{R}(\theta)$ is the inertia matrix (symmetric, positive definite) that reflects coupling effects between joints, and $\mathbf{h} = \mathbf{h}(\theta, \dot{\theta}; t)$ contains centripetal and Coriolis forces as well as friction and gravitational terms. For an articulated manipulator, the dynamics is reasonably manageable with pencil and paper only up to three degrees of freedom (Horn, Hirokawa, and Varizani 1977). Thus, much effort has been devoted to developing efficient procedures for real-time computation of the dynamics. A substantial improvement in computational efficiency was obtained by using recursive algorithms (Hollerbach 1980; Luh, Walker, and Paul 1980*a*; Silver 1982; Renaud 1983) to generate the torques required to support a desired motion (inverse dynamics), achieving linear variation of the computation complexity with the number of links.

While the inverse dynamics computes the torques theoretically needed to compensate for nonlinearities and to follow a specific trajectory, assuming an exact model explicit values of inertia matrix $\mathbf{R}(\theta)$ and nonlinear terms $\mathbf{h}$ are required to analyze how joint accelerations (hence subsequent motion) are affected not only by control torques and known dynamics but also by disturbances (such as Coulomb and viscous friction) and modeling errors: parametric uncertainties such as inaccuracies on inertias, geometry, loads; and high-frequency unmodeled dynamics, such as unmodeled structural modes or neglected time delays. Currently, the most efficient algorithm to compute $\mathbf{R}(\theta)$ and $\mathbf{h}$ (in fact, to estimate these terms, precisely because of parametric uncertainties) is due to Walker and Orin (1982) and uses the inverse recursive dynamics formulation of Luh, Walker, and Paul (1980*a*) as an important component. The algorithm could be further simplified by exploiting geometric features of "well-structured" manipulators, using the results of Hollerbach and Sahar (1983), and even further by customizing the computations for a specific robot, along the lines of Kanade, Khosla, and Tanaka (1984).

Now there are several reasons for insisting on explicit robustness guarantees for the robot control system. The first is obvious: Control instabilities are unpleasant, especially at high speeds (an elegant stability analysis of computed-torque-like schemes can be found in Gilbert and Ha 1983). Conversely, guaranteed robustness properties allow one to design simple controllers. Consider, for instance, a six-degree-of-freedom manipulator composed of a three-degree-of-freedom arm and a three-degree-of-freedom hand. Kinematic decoupling between hand and arm can be achieved by having the three rotational axe of the wrist intersect at a point (spherical wrist). Physically it seems natural to seek a similar decoupling at the dynamic level, in other words to be able to consider motions of arm and hand as "disturbances" to one another (each of these disturbances being possibly further decomposed into an average, quickly estimated term to be directly compensated for and a genuine perturbation term to be accounted for by the controller robustness). Further, desired bandwidth is likely to be much larger for the hand than for the arm itself (Salisbury 1984). Robust controller design does allow such natural reduction of the original problem into two lower-order control problems: In the case of a six-degree-of-freedom manipulator, both of these problems are amenable to closed-form, pencil-and-paper treatment, thus allowing one to maintain clear physical insight and exploit engineering judgment all along the design and implementation process.[2] Finally, by easily accounting for large model imprecision, robust controllers simplify higher-level programming.

The relevance of suction controllers to robot manipulator control is precisely that they feature such explicit, built-in robustness properties. Depending on the structure of model uncertainty and on the type of user space considered (configuration space, task-oriented or hybrid (force/position) coordinates, and so on), the suction control ideas of Section 2 may be applied in various ways. The following scheme assumes for simplicity that desired trajectories are specified in joint space. Also, as discussed above, if N is the total number of degrees of freedom of the manipulator and N_H is the number of degrees of freedom of the wrist and hand, the first $N - N_H$ and the last N_H degrees of freedom of the arm can be treated separately (for a wrist-partitioned robot). Thus, in the sequel, system size n refers to either $N - N_H$ or N_H.

2. The importance of model tractability has been stressed by several authors in a different context; see, for example, Bejczy and Lee (1983) and Luh and Gu (1984).

3.2. Suction Controller Structure

Let λ be the desired control bandwidth, chosen small enough not to excite unmodeled structural modes or interefere with neglected time delays (see the example of Section 2.3.1), and let

$$\hat{\mathbf{R}}(\theta)\ddot{\theta} + \hat{\mathbf{h}} = \mathbf{T} \tag{34}$$

be the available model of the manipulator described by Eq. (33). Discrepancies between Eqs. (33) and (34) may arise from several factors: imprecisions on the manipulator geometry or inertias, uncertainties on the friction terms or the loads, on-line computation limitations, or purposeful model simplification, as discussed earlier. The suction controller for the manipulator of Eq. (33) takes the form

$$\mathbf{T} = \hat{\mathbf{R}}\mathbf{u} + \hat{\mathbf{h}}, \tag{35}$$

where the components u_i ($i = 1, \ldots, n$) of the vector $\mathbf{u}$ are defined as

$$u_i := \ddot{\theta}_{di} - 2\lambda\dot{\tilde{\theta}}_i - \lambda^2\tilde{\theta}_i - G_i(\theta)\bar{k}_i(\theta, \dot{\theta}; t)\ \mathrm{sat}(s_i/\Phi_i). \tag{36}$$

In Eq. (36), $\ddot{\theta}_{di}$ is the acceleration of the desired trajectory at joint i, $\tilde{\theta}_i$ is the tracking error at joint i, and $\bar{k}_i(\theta, \dot{\theta}; t)$ and Φ_i are defined according to the dynamic balance conditions (27)–(30) of Section 2 (the scaling factor $G_i(\theta)$ of Eq. (36) and the bounds on model uncertainty used in the computation of $\bar{k}_i(\theta, \dot{\theta}; t)$ will be discussed next). The surfaces s_i in Eq. (36) are set to be

$$s_i := \dot{\tilde{\theta}}_i + 2\lambda\tilde{\theta}_i + \lambda^2 \int_0^t \tilde{\theta}_i(\tau)d\tau = \left(\frac{d}{dt} + \lambda\right)^2 \left(\int_0^t \tilde{\theta}_i(\tau)d\tau\right) \tag{37}$$

so as to use integral control and reduce the effects of friction. It is important to remark that the major difference with the computed-torque method is the presence of the robustifying terms in $G_i(\theta)\ \bar{k}_i(\theta, \dot{\theta}; t)\ \mathrm{sat}(s_i/\Phi_i)$ in Eq. (36), which allow one to maintain stability and optimize performance in the face of model uncertainty.

3.3. Practical Evaluation of Parametric Disturbances

There remains only to evaluate explicit bounds on parametric uncertainty so as to generate a set of control discontinuity gains $k_i(\theta, \dot{\theta}; t)$, as in Section 2; these gains will then be fed into the dynamic balance conditions (27)–(30) to compute modified gains $\bar{k}_i(\theta, \dot{\theta}; t)$ and boundary layer thicknesses Φ_i of Eq. (36). A simplified, easily implementable approach to such evaluation is as follows: Define two sets of n-vectors, $\{\mathbf{L}_j; j = 1, \ldots, n\}$ and $\{\Delta\mathbf{R}_j; j = 1, \ldots, n\}$ by

$$\mathbf{R}^{-1} =: [\mathbf{L}_1 \cdots \mathbf{L}_n]$$
$$\hat{\mathbf{R}} - \mathbf{R} =: \Delta\mathbf{R} =: [\Delta\mathbf{R}_1 \cdots \Delta\mathbf{R}_n]$$

The vectors $\mathbf{L}_j$ and $\Delta\mathbf{R}_j$ are functions of configuration θ only. Substituting control law (35) into robot dynamics (33), we can write the dynamics of the closed-loop system as

$$\ddot{\theta} = (\mathbf{I} + \mathbf{R}^{-}\Delta\mathbf{R})\mathbf{u} + \mathbf{R}^{-1}\Delta\mathbf{h}, \tag{38}$$

where $\mathbf{I}$ is the $n \times n$ identity matrix and $\Delta\mathbf{h} = \hat{\mathbf{h}} - \mathbf{h}$ reflects the effects of unmodeled gravitational loads, friction, or model simplification. Define then a set of scalars $\Delta_i = \Delta_i(\theta)$ such that for all θ,

$$\Delta_i \geqslant (\mathbf{L}_i^+)^T\Delta\mathbf{R}_i^+, \quad i = 1, \ldots, n, \tag{39}$$

where the components of $\Delta\mathbf{R}_i^+$ and $\mathbf{L}_i^+$ are the absolute values of the components of $\Delta\mathbf{R}_i$ and $\mathbf{L}_i$. We assume that

$$\Delta_i < \mathbf{L}_i^T\mathbf{R}_i = 1, \quad i = 1, \ldots, n, \tag{40}$$

for all configurations θ. This condition will later be relaxed. The function of multiplier $G_i = G_i(\theta)$ in Eq. (36) is then simply to center the estimated geometric mean gain, according to Section 2.3.1:

$$G_i := ((1 - \Delta_i)(1 + \Delta_i))^{-1/2}, \quad i = 1, \ldots, n. \tag{41}$$

Corresponding gain margins $\beta_i = \beta_i(\theta)$ are

$$\beta_i := \left(\frac{1 + \Delta_i}{1 - \Delta_i}\right)^{1/2}, \quad i = 1, \ldots, n. \tag{42}$$

Gains $k_i(\theta, \dot{\theta}; t)$ are computed so as to meet the simplified conditions

$$k_i(\theta, \dot{\theta}; t) \geq \beta_i((\mathbf{L}_i^+)^T \cdot \sigma(t) + \eta_i), \quad i = 1, \ldots, n, \tag{43}$$

with

$$\sigma(t) := \mathbf{\Delta h}^+ + \sum_{j=1}^{n} |\ddot{\theta}_{dj}| \cdot \mathbf{\Delta R}_j^+, \tag{44}$$

where the η_i in Eq. (43) are the positive constants used in Eq. (4) of each sliding condition and, by definition, the components of $\mathbf{\Delta h}^+$ in Eq. (44) are the absolute values of the components of $\mathbf{\Delta h}$. Conditions (39) and (43) are fairly straightforward to satisfy for $n = 3$ (hence in particular for a six-degree-of-freedom wrist-partitioned robot) as closed-form dynamics remain manageable analytically; in particular, an off-line worst-case analysis may be sufficient in practice (an extreme example of such analysis can be found in Slotine and Sastry 1983). In the general case, an adequate first-order approximation is to evaluate bounds (39) and (43) by using estimates $\hat{\mathbf{L}}_i^+$ (obtained by inverting $\hat{\mathbf{R}}$) in place of the actual $\mathbf{L}_i^+$. Similarly, approximate upper bounds on the components of $\mathbf{\Delta R}_i^+$ or $\mathbf{\Delta h}^+$ may be easily expressed in terms of $\hat{\mathbf{R}}_i^+$ or $\hat{\mathbf{h}}^+$ (with obvious notations); for instance, if the only source of uncertainty on estimated inertia matrix $\hat{\mathbf{R}}$ is that manipulator inertias are known to within a 10 percent precision, one may use in practice the approximation $\mathbf{\Delta R}_i^+ \leq 10\% \ \hat{\mathbf{R}}_i^+$, where the inequality is understood componentwise.

Remarks:

1. Complete expressions of the $k_i(\theta, \dot{\theta}; t)$ are derived in the Appendix.
2. In the case that λ is time-varying (as discussed in Section 2.3.2), the term

$$u_i' := -2\dot{\lambda}(\dot{\tilde{\theta}}_i + \lambda\tilde{\theta}_i)$$

should be added to control law u_i in Eq. (36) while augmenting gain $k_i(\theta, \dot{\theta}; t)$ of Eq. (43) accordingly by the quantity $\beta_i|u_i'|(\beta_i - 1/\beta_i)/(\beta_i + 1/\beta_i)$.

3. Assumption (40) can be relaxed to

$$\mathbf{L}_i^T \cdot \hat{\mathbf{R}}_i > 0, \quad i = 1, \ldots, n. \tag{45}$$

This condition is quite mild: It simply means that u_i contributes to $\ddot{\theta}_i$ with a predictable sign—actually, inequality (45) can always be satisfied by letting $\hat{\mathbf{R}}$ be a (possibly time-varying) positive definite diagonal matrix. In the general case, with

$$0 < \beta_i^{\min} \leq \mathbf{L}_i^T \cdot \hat{\mathbf{R}}_i \leq \beta_i^{\max}, \quad i = 1, \ldots, n, \tag{46}$$

where $\beta_i^{\min} = \beta_i^{\min}(\theta)$ and $\beta_i^{\max} = \beta_i^{\max}(\theta)$, Eqs. (41), (42), and (36) are generalized as

$$G_i := (\beta_i^{\min} \cdot \beta_i^{\max})^{-1/2}, \tag{47}$$

$$\beta_i := (\beta_i^{\max}/\beta_i^{\min})^{+1/2}, \tag{48}$$

$$u_i = -G_i(\theta)[(2/(\beta_i + 1/\beta_i))(-\ddot{\theta}_{di} + 2\lambda\dot{\tilde{\theta}}_i + \lambda^2\tilde{\theta}_i) + k_i(\theta, \dot{\theta}; t)\ \mathrm{sat}(s_i/\Phi_i)], \tag{49}$$

with conditions (43) and (44) replaced by

$$k_i(\theta, \dot{\theta}; t) \geq \beta_i\left((\mathbf{L}_i^+)^T \mathbf{\Delta h}^+ + \sum_{j=1}^{n} |\ddot{\theta}_{dj}| \cdot |\mathbf{L}_i^T \tilde{\mathbf{R}}_j| \cdot [2G_j|(\beta_j + 1/\beta_j)] + \eta_i\right). \tag{50}$$

The degree of simplification in the system model may be varied according to the on-line computing power available: The balance conditions clearly quantify the trade-off between model precision and tracking accuracy. Further, the s-trajectories provide a measure of the validity of the assumptions on model uncertainty and of the adequacy of bound simplifications.

4. Concluding Remarks

Conceptually, the preceding development can be illustrated by Fig. 5. Consider a fast manipulator motion, say a 1/2-s stop-to-stop trajectory across the workspace, including a full flipping of the wrist, and plot

Fig. 5. Effect of parametric uncertainty on controller performance.

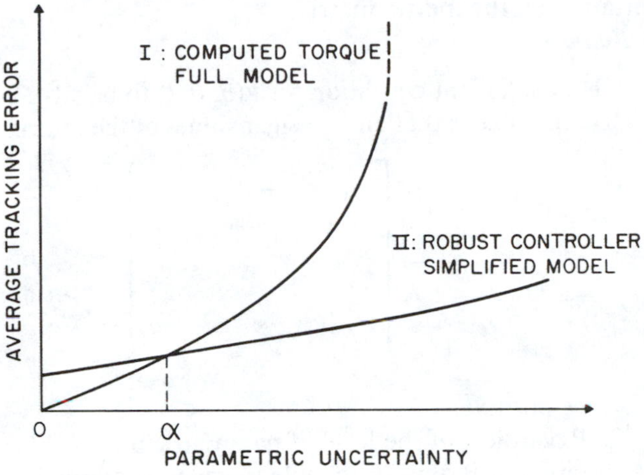

the average tracking precision against parametric uncertainty (say average imprecision on manipulator inertias). Using the computed-torque method based on a full six-degree-of-freedom model, we obtain curve *I*: Tracking is perfect in the absence of model uncertainty and then quickly degrades as uncertainty increases, with the system eventually becoming unstable. Using a robust suction controller based on a simplified model (two "decoupled" three-degree-of-freedom systems), the result is curve II. Because of model simplification, performance is not perfect in the absence of parametric uncertainty, as when using the full six-degree-of-freedom model; but the robust controller quickly outperforms the computed-torque scheme as uncertainty increases. The crossing point α of Fig. 5 typically corresponds to a very low level of parametric uncertainty; further, $\alpha \rightarrow 0$ as allowable bandwidth λ increases, as could be expected from the dynamic balance conditions of Section 2. For direct-drive arms, for instance, the crossing point α of Fig. 5 corresponds to a few percent imprecision on manipulator inertias—a value fairly smaller than typical uncertainty. This effect would likely be further enhanced for geared manipulators involving significant friction terms. The suction control methodology thus combines in practice improved performance with simpler and more tractable controller designs.

Appendix: Uncertainty on the Manipulator Mass Properties

Although simplified conditions (43) or (50) of Section 3.3 were found to be generally adequate in practice, exact expressions for the gains $k_i(\theta, \dot{\theta}; t)$ are presented here for completeness. The derivation is of interest in itself since it provides insight into the effect of uncertainty about the manipulator mass properties. The development is presented directly for the general case of Eq. (46).

Let us define a set of scalars $\Delta_{ij} = \Delta_{ij}(\theta)$ such that, for all θ,

$$\Delta_{ij} \geqslant |\mathbf{L}_i^T \hat{\mathbf{R}}_j|, \quad i = 1, \ldots, n, j = 1, \ldots, n. \quad \text{(A1)}$$

Given the expression of the closed-loop dynamics,

$$\ddot{\theta} = \mathbf{R}^{-1}\hat{\mathbf{R}}\mathbf{u} + \mathbf{R}^{-1}\Delta\mathbf{h},$$

and definitions (47), (48), and (49), gains $k_i(\theta, \dot{\theta}; t)$ must verify

$$k_i(\theta, \dot{\theta}; t) \geqslant \beta_i \left(|u_{i0}|(\beta_i - 1/\beta_i)/(\beta_i + 1/\beta_i) + (\mathbf{L}_i^+)^T \Delta\mathbf{h}^+ + \sum_{j \neq i} \Delta_{ij}|u_j| + \eta_i \right), \quad \text{(A2)}$$

where

$$u_{i0} := -\ddot{\theta}_{di} + 2\lambda\dot{\tilde{\theta}}_i + \lambda^2\tilde{\theta}_i.$$

From the expression (49) of the u_j and the definition (27) of the $\bar{k}_i(\theta, \dot{\theta}; t)$, a sufficient condition for (A2) to be satisfied is that

$$k_i(\theta, \dot{\theta}; t) = k_i'(\theta, \dot{\theta}; t) + \beta_i \sum_{j \neq i} (G_j \Delta_{ij} k_j(\theta, \dot{\theta}; t)|\text{sat}(s_j/\Phi_j)|), \quad \text{(A3)}$$

where $k_i'(\theta, \dot{\theta}; t)$ is defined so that

$$k_i'(\theta, \dot{\theta}; t) \geqslant \beta_i \Bigg(|u_{i0}|(\beta_i - 1/\beta_i)/(\beta_i + 1/\beta_i) + (\mathbf{L}_i^+)^T \Delta\mathbf{h}^+ + \sum_{j \neq i} G_j \Delta_{ij} \Big|2u_{j0}/(\beta_j + 1/\beta_j) + \left(\frac{\lambda\Phi_j}{\beta_j} - k_j(\theta_d, \dot{\theta}_d; t)\right) \text{sat}(s_j/\Phi_j)\Big| + \eta_i \Bigg). \quad \text{(A4)}$$

Now since all terms in $\text{sat}(s_j/\Phi_j)$ vanish for $\theta \equiv \theta_d$, $\dot{\theta} \equiv \dot{\theta}_d$, expressions (A3) and (A4) allow us to compute $k_i(\theta_d, \dot{\theta}_d; t) = k_i'(\theta_d, \dot{\theta}_d; t)$ directly. Further, from the dynamic balance conditions (28)–(30), boundary layer thicknesses Φ_j depend only on the $k_j(\theta_d, \dot{\theta}_d; t)$, so that in turn the knowledge of the $k_j(\theta_d, \dot{\theta}_d; t)$ allows expressions (A3) and (A4) to completely define the $k_i(\theta, \dot{\theta}; t)$: The vector $\mathbf{k}(\theta, \dot{\theta}; t)$ of components $k_i(\theta, \dot{\theta}; t)$ is the solution of the linear system

$$\mathbf{A}(\theta, \dot{\theta}; t)\mathbf{k}(\theta, \dot{\theta}; t) = \mathbf{k}'(\theta, \dot{\theta}; t), \qquad \text{(A5)}$$

where $\mathbf{k}'(\theta, \dot{\theta}; t)$ is the vector of components $k_i'(\theta, \dot{\theta}; t)$, and the matrix $\mathbf{A}(\theta, \dot{\theta}; t)$ is defined as

$$\mathbf{A}(\theta, \dot{\theta}; t) := \begin{bmatrix} 1 & -\beta_1 G_2 \Delta_{12} |\text{sat}(s_2/\Phi_2)| & \cdots \\ -\beta_2 G_1 \Delta_{21} |\text{sat}(s_1/\Phi_1)| & 1 & \\ \cdot & & \cdot \\ \cdot & & \quad \cdot \\ \cdot & & \qquad \cdot \end{bmatrix}.$$

Further, all components $k_i(\theta, \dot{\theta}; t)$ of the solution $\mathbf{k}(\theta, \dot{\theta}; t)$ of Eq. (A5) must be strictly positive for this solution to be admissible: Given the Metzler structure (Siljak 1978) of the matrix $-\mathbf{A}$, this is the case as long as all real eigenvalues of $\mathbf{A}$ remain positive. Equivalently, defining $\mathbf{A}_0 = \mathbf{A}_0(\theta, \dot{\theta}; t)$ as

$$\mathbf{A}_0 := \mathbf{diag}(\beta_j^{-1})\mathbf{A}\ \mathbf{diag}(G_j^{-1})$$

$$= \begin{bmatrix} \beta_1^{\min} & -\Delta_{12}|\text{sat}(s_2/\Phi_2)| & \cdots \\ -\Delta_{21}|\text{sat}(s_1/\Phi_1)| & \beta_2^{\min} & \\ \cdot & & \cdot \\ \cdot & & \quad \cdot \\ \cdot & & \qquad \cdot \end{bmatrix}, \qquad \text{(A6)}$$

where the second equality stems from definitions (47) and (48) of the G_j and β_j, then $\mathbf{k}(\theta, \dot{\theta}; t)$ of Eq. (A5) is admissible as long as all real eigenvalues of $\mathbf{A}_0$ remain positive. Note that $\mathbf{A} \equiv \mathbf{A}_0 \equiv \mathbf{I}$ in the absence of uncertainty on the inertia matrix.

Remarks:

1. A sufficient condition for $\mathbf{k}(\theta, \dot{\theta}; t)$ to be admissible is that all real eigenvalues of the matrix

$$\mathbf{A}_0' := \begin{bmatrix} \beta_1^{\min} & -\Delta_{12} & \cdots \\ -\Delta_{21} & \beta_2^{\min} & \\ \cdot & & \cdot \\ \cdot & & \quad \cdot \\ \cdot & & \qquad \cdot \end{bmatrix}$$

be positive.

2. Regardless of the level of parametric uncertainty, it is always possible to generate admissible $\mathbf{k}(\theta, \dot{\theta}; t)$ by selecting large enough $\mathbf{k}'(\theta, \dot{\theta}; t)$. This can be achieved by multiplying the right-hand side of inequality (A4) by an appropriately large scaling factor ρ, such as a constant upper bound on the Frobenius-Perron roof[3] ρ_1 of the matrix

$$\mathbf{A}_1 := \begin{bmatrix} 0 & \beta_1 G_2 \Delta_{12} & \cdots \\ \beta_2 G_1 \Delta_{21} & 0 & \\ \cdot & & \cdot \\ \cdot & & \quad \cdot \\ \cdot & & \qquad \cdot \end{bmatrix}.$$

We assume that $\rho_1 > 1$, or else gain vector $\mathbf{k}(\theta, \dot{\theta}; t)$ is already guaranteed to be admissible. This scaling has the effect of artificially increasing boundary layer thicknesses Φ_j by the same factor ρ without actually modifying the control action inside the (original and still effective) boundary layers B_j^0 since gains $\bar{k}_j(\theta, \dot{\theta}; t)$ also scale up by the same factor. The values of the s_j are thus left unchanged (as is the tracking performance, which allows us to use constant ρ's), so that the off-diagonal terms of the matrix $\mathbf{A}$ are divided by the scaling fac-

3. See, for example, Siljak (1978). The Frobenius-Perron roof ρ_1 of a matrix $\mathbf{A}_1$ with nonnegative elements is the largest (automatically nonnegative) real eigenvalue of $\mathbf{A}_1$. The equation $(\mathbf{I} - \rho^{-1}\mathbf{A}_1)\mathbf{y} = \mathbf{y}'$ admits positive solutions $\mathbf{y}$ for positive $\mathbf{y}'$ if $\rho > \rho_1$.

tor, which in turn guarantees that the solution $\mathbf{k}(\theta, \dot{\theta}; t)$ of Eq. (A5) is admissible. Further, it is desirable that (A5) be guaranteed to have admissible solutions regardless of the values of the s_j. This can be achieved by substituting $\rho_I^{-1}\text{sat}(\rho_I s_j/\Phi_j)$ to $\text{sat}(s_j/\Phi_j)$ in the expression of $\mathbf{A}$, with $\rho_1 \leq \rho_I \leq \rho$, which further guarantees that the B_j^0 are attractive for all trajectories starting inside boundary layers of thicknesses $(\rho/\rho_I)\Phi_j^0$ (where Φ_j^0 is the thickness of B_j^0). Note that a convenient upper bound of ρ_1 is

$$\rho_1 \leq \min\left(\max_i(G_i \sum_{j\neq i} \Delta_{ji}\beta_j),\ \max_i(\beta_i \sum_{j\neq i} \Delta_{ij}G_j)\right).$$

3. The solution $\mathbf{k}(\theta, \dot{\theta}; t)$ of Eq. (A5) is bounded for bounded θ_d, provided all real eigenvalues of $\mathbf{A}_0$ (or $\mathbf{A}$) are uniformly bounded away from zero (that is, remain larger than some strictly positive constant). Indeed, $\mathbf{A}^{-1}(\theta, \dot{\theta}; t)$ is then bounded for bounded θ, $\dot{\theta}$, and, further, $\mathbf{k}(\theta, \dot{\theta}; t)$ is admissible. This in turn implies that tracking error is indeed limited by boundary layer thicknesses Φ_j, which depend only on the $k_j(\theta_d, \dot{\theta}_d; t)$, so that θ, $\dot{\theta}$ and thus both $\mathbf{A}^{-1}(\theta, \dot{\theta}; t)$ and $\mathbf{k}'(\theta, \dot{\theta}; t)$ are bounded.
4. The condition that $k(\theta, \dot{\theta}; t)$ of Eq. (A5) be admissible can be given a slightly different interpretation (based again on the Metzler structure of $-\mathbf{A}_0$). It is satisfied if

$$\det \begin{bmatrix} \beta_1^{\min} & -\Delta_{12}\chi_2 & \cdots \\ -\Delta_{21}\chi_1 & \beta_2^{\min} & \\ \cdot & & \cdot \\ \cdot & & \cdot \\ \cdot & & \cdot \end{bmatrix} > 0, \quad \text{(A7)}$$

for all χ_j such that $0 \leq \chi_j \leq |\text{sat}(s_j/\Phi_j)|$. Let us now assume that

$$\det \hat{\mathbf{R}} > 0. \quad \text{(A8)}$$

Condition (A8) is automatically satisfied for any physically motivated choice of $\hat{\mathbf{R}}$, since then $\hat{\mathbf{R}} = \hat{\mathbf{R}}^T > 0$. Now from $\mathbf{R} = \mathbf{R}^T > 0$, condition (A8) implies that

$$\det(\mathbf{R}^{-1}\hat{\mathbf{R}}) > 0, \quad \text{(A9)}$$

which can be written

$$\det \begin{bmatrix} \mathbf{L}_1^T\hat{\mathbf{R}}_1 & \mathbf{L}_1^T\hat{\mathbf{R}}_2 & \cdots \\ \mathbf{L}_2^T\hat{\mathbf{R}}_1 & \mathbf{L}_2^T\hat{\mathbf{R}}_2 & \\ \cdot & & \cdot \\ \cdot & & \cdot \\ \cdot & & \cdot \end{bmatrix} > 0. \quad \text{(A10)}$$

The comparison of expressions (A7) and (A10) shows that it is desirable that the $\beta_i^{\min}$ of Eq. (46) and the Δ_{ij} of Eq. (A1) be compatible, that is, be generated in a way that preserves the natural structure of Eq. (A10).

References

Arimoto, S., and Miyazaki, F. 1983 (Bretton Woods, N.H.). Stability and robustness of PID feedback control for robot manipulators of sensory capability. *Proc. 1st Int. Symp. Robotics Res.* Cambridge: MIT Press.

Asakawa, K., Akiya, F., and Tabata, F. 1982. *A variable compliance device and its application for automatic assembly.* Kawasaki, Japan: Fujitsu Laboratories.

Åström, K. J. 1983. Theory and applications of adaptive control: A survey. *Automatica* 19(5):471–486.

Bejczy, A. K., and Lee, S. 1983 (San Antonio). Robot arm dynamic model reduction for control. *Proc. IEEE Conf. Dec. Contr.*, vol. 3, pp. 1466–1471.

Brady, M., et al. 1982. *Robot motion.* Cambridge: MIT Press.

Freund, E. 1982. Fast nonlinear control with arbitrary pole-placement for industrial robots and manipulators. *Int. J. Robotics Res.* 1(1):65–78.

Gilbert, E. G., and Ha, I. J. 1983 (San Antonio). An approach to nonlinear feedback control with applications to robotics. *Proc. IEEE Conf. Dec. Contr.*, vol. 1, pp. 134–140.

Hogan, N. 1981. Impedance control of a robotic manipulator. Paper delivered at the winter annual meeting of the ASME, Washington, D.C.

Hogan, N. 1984. Impedance control of industrial robots. *Robotics and Computer Integrated Manufacturing* 1(1):97–114.

Hollerbach, J. M. 1980. A recursive formulation of lagrangian manipulator dynamics. *IEEE Trans. Sys., Man., Cyber.* SMC-10(11):730–736.

Hollerbach, J. M., and Sahar, G. 1983. Wrist-partitioned inverse kinematic accelerations and manipulator dynamics. *Int. J. Robotics Res.* 2(4):61–76.

Horn, B. K. P., Hirokawa, K. and Varizani, V. V. 1977. Dynamics of a three degree-of-freedom kinematic chain. Memo. 478. Cambridge: MIT Artificial Intelligence Laboratory.

Hunt, L. R., Su, R. and Meyer, G. 1983. Design for multi-input nonlinear systems. *Differential Geom. & Contr. Theory Conf.* Boston, Mass.: Birkhaüser, pp. 268–298.

Kanade T., Khosla, P. K. and Tanaka, N. 1984. Real-time control of the CMU direct-drive arm II using customized inverse dynamics. *Proc. IEEE Conf. Dec. Contr.*, vol. 3, pp. 1345–1352.

Lehtomaki, N. A. 1981. Practical robustness measures in multivariable control systems analysis. Ph.D. thesis, Massachusetts Institute of Technology.

Luh, J. Y. S., and Gu, Y. L. 1984 (Princeton). Lagrangian formulation of robot dynamics with dual-number transformation for computational simplification. *Proc. 18th Conf. Info. Sci. Sys.*

Luh, J. Y. S., Walker, M. W. and Paul, R. P. C. 1980*a*. On-line computational scheme for mechanical manipulators. *J. Dyn. Sys., Meas., Contr.*, vol. 102, pp. 69–76.

Luh, J. Y. S., Walker, M. W. and Paul, R. P. C. 1980*b*. Resolved acceleration control for mechanical manipulators. *IEEE Trans. Automatic Contr.* 25(3):468–474.

Paul, R. P. C. 1981. *Robot manipulators: Mathematics, programming, and control.* Cambridge: MIT Press.

Renaud, M. 1983. An efficient iterative analytical procedure for obtaining a robot manipulator dynamic model. *Proc. Int. Symp. Robotics Res.*

Salisbury, J. K. 1984 (Tokyo). Design and control of an articulated hand. *Proc. Int. Symp. Des. Synth.*, pp. 459–466.

Siljak, D. D. 1978. *Large scale dynamic systems.* New York: North-Holland.

Silver, W. 1982. On the equivalence of Lagrangian and Newton-Euler dynamics for manipulators. *Int. J. Robotics Res.* 1(2):60–70.

Slotine, J.-J. E. 1983. Tracking control of nonlinear systems using sliding surfaces, Ph.D. thesis, Massachusetts Institute of Technology.

Slotine, J.-J. E. 1984. Sliding controller design for nonlinear systems. *Int. J. Cont.* 40(2):421–434.

Slotine, J.-J. E., and Sastry, S. S. 1983. Tracking control of nonlinear systems using sliding surfaces, with applications to robot manipulators. *Int. J. Contr.* 38(2):465–492.

Slotine, J.-J. E., Yoerger, D. R. and Sheridan, T. B. In press. Suction control of a large hydraulic manipulator.

Utkin, V. I. 1977. Variable structure systems with sliding mode: A survey. *IEEE Trans. Automatic Contr.*, vol. 22, pp. 212–222.

Vidyasagar, M. 1978. *Nonlinear systems analysis.* Englewood Cliffs, N.J.: Prentice-Hall.

Young, K. K. D. 1978. Controller design for a manipulator using theory of variable structure systems. *IEEE Trans. Sys., Man, Cyber.* SMC-8(2):101–109.

Walker, M. W., and Orin, D. E. 1982. Efficient dynamic computer simulation of robotic mechanisms. *J. Dyn. Sys., Meas., Contr.*, vol. 104, pp. 205–211.

On L_2- and L_∞-Stability Approaches for the Robust Control of Robot Manipulators

NORBERT BECKER AND WOLFGANG M. GRIMM

Abstract—**Recently, interesting robustness results for the control of robot manipulators were obtained [1], [8] using the L_2- and L_∞-stability approaches, respectively. This note presents critical remarks on these approaches and further investigations. In order to investigate L_2-stability, it is shown that additional assumptions have to be made and a reformulation of the control problem has to be carried out.**

I. Introduction

References [1], [8] deal with the promising attempt to investigate the robustness of robot manipulator feedback systems, where robustness means stability of the feedback loop in the face of model uncertainties and changes. The controller is put up in a general fashion and includes nonlinear decoupling [4], [5] as well as pure linear control. The sufficient stability condition derived in [1], [8] is closely related to the small gain theorem [2], [3]. Its application is not restricted to robot manipulators and can be applied to control systems consisting of a stable linear and a memoryless nonlinear part.

The main objective of this note is to point out difficulties which arise investigating the stability of the closed-loop manipulator system by methods of functional analysis. In particular, this includes the L_2-stability approach in [1] as well as the L_∞ approach in [8]. In order to apply the L_2-stability approach in [1], it is shown that the control problem has to be reformulated and additional assumptions have to be made.

This note is organized as follows. Section II deals with notational arrangements and Section III presents the general formulation of the sufficient stability conditions which is provided in addition to [1], [8]. In Section IV the application to robot manipulator control is described.

II. Notation and Preliminaries

R^n will denote the n-dimensional vector space over the field R of real numbers.

Let $x \in R^n$. Then

$$\|x\| = (x^Tx)^{1/2}\ [1]$$

is the Euclidean norm. The corresponding induced norm of a matrix $A \in C^{n\times n}$ is [2, p. 13]

$$\|A\| = (\max_i \lambda_i(A^*A))^{1/2}\ [2]$$

where $\lambda_i(A^*A)$ is an eigenvalue of A^*A.

Let $f(t) \in R^n$ and $f(t) = 0$ for $t < 0$; then the L_2- and L_∞-norms are defined by [6, p. 226]

$$\|f\|_2 = \left(\int_0^\infty \|f(t)\|^2\, dt\right)^{1/2} \tag{2.1}$$

$$\|f\|_\infty = \operatorname*{ess\,sup}_{t\in[0,\infty)} \|f(t)\|. \tag{2.2}$$

Let $P_i(s) = L\{p_i(t)\}$ be an asymptotically stable and proper transfer function matrix. Then the induced operator norm β_i[3] is defined by [2, p. 23]

$$\beta_i = \sup_{\|f\|_p\neq 0} \frac{\|p_i(t) * f(t)\|_p}{\|f(t)\|_p}, \qquad p = 2, \infty. \tag{2.3}$$

In the sequel $f_T(t)$ denotes the truncated function

$$f_T(t) = \begin{cases} f(t), & 0 \le t \le T \\ 0, & t > T. \end{cases}$$

Further,

$$\|f\|_{Tp} := \|f_T\|_p;\ p = 2, \infty.$$

Definition 2.1: A causal system with inputs $f_i(t)(i = 0, 1, \cdots, m)$, $f_i(t) \in R^{n_i}$, and a corresponding output $e \in R^n$ is weak L_p-stable, if there exist nonnegative finite constants c_W, c_I, and $c_i(i = 0, 1, \cdots, m)$ such that

$$\|e\|_p \le \sum_{i=0}^m c_i \|f_i\|_p^{m_i} + c_I + c_W \tag{2.4}$$

where m_i represents the m_i-fold dependence of the nonlinearity on the time functions $f_i(t)$. The constant c_I vanishes in the case of zero initial conditions. If $c_W = 0$ the system is defined to be L_p-stable.

III. Robust Stability for a Class of Nonlinear Control Systems

In the following the class of nonlinear control systems shown in Fig. 1 is considered, where A, B, C, D, F, H, G, K, and L are constant matrices of appropriate dimensions. $\eta(e, v, f_1(t), \cdots, f_m(t))$ represents a memoryless nonlinearity, where $f_i(t)(i = 0, 1, \cdots, m)$ are known time functions.

The input-output relations of the closed loop are represented by

$$e(t) = p_1(t) * \eta(t) + p_2(t) * f_0(t) + i_e(t)$$
$$v(t) = p_3(t) * \eta(t) + p_4(t) * f_0(t) + i_v(t) \tag{3.1}$$

where i_e and i_v are the responses of the linear feedback system due to the initial conditions of the plant and the controller, respectively, and $p_i(t)$, $i = 1, \cdots, 4$ can be obtained easily from Fig. 1.

Assumption 3.1: The eigenvalues of the linear feedback system ($\eta = 0$) in Fig. 1 are all located in the left-half plane.

[1] x^T denotes the transposed vector.
[2] A^* denotes the conjugate complex transposed matrix.
[3] $*$ in (2.3) and (3.1) denotes the convolution operator.

Manuscript received February 25, 1986; revised March 2, 1987 and June 15, 1987. This work was supported in part by the Stiftung Volkswagenwerk under Contract I/61 394.

The authors are with the Institut für Meß- und Regelungstechnik, FB 9, University of Duisburg, D-4100 Duisburg 1, West Germany.

IEEE Log Number 8716714.

Reprinted from *IEEE Trans. Automat. Contr.*, vol. 33, no. 1, pp. 118–122, Jan. 1988.

Assumptions 3.2:

$$\|\eta\| \le k_0\|e\|^2 + k_1\|e\| + k_2\|v\| + \sum_{i=1}^{m} k_{fi}\|f_i\|^{m_i} + \rho \tag{3.2}$$

for some nonnegative finite constants k_0, k_1, k_2, $k_{fi}(i = 1, 2, \cdots, m)$ and ρ.

Assumption 3.3:

$$\|f_i\|_p < \infty, \quad i = 0, 1, \cdots, m, \quad p = 2, \infty.$$

In the following theorems are derived separately for L_2- and L_∞-stability of the nonlinear closed-loop system of Fig. 1 which are based on Definition 2.1. They are considered to be extensions of the small gain theorem [2], [3].

A. The L_∞-Stability Approach

With Assumption 3.1 the induced L_∞-norm (2.3) of $p_i(t)$ in (3.1) is given by [2, p. 23]

$$\beta_i = \|p_{0i}\| + \int_0^\infty \|p_{si}(t)\|\, dt; \qquad i = 1, \cdots, 4$$

where

$$p_i(t) = p_{0i}\delta(t) + p_{si}(t)$$

and $\delta(t)$ denotes the unit impulse function.

Theorem 1: If $\|e(t = 0)\|$ is sufficiently small and if Assumptions 3.1–3.3 hold, then the closed-loop system in Fig. 1 is:

i) weak L_∞-stable if

$$1 - \beta_1 k_1 - \beta_3 k_2 > 2(\beta_1 k_0 d)^{1/2} \tag{3.3}$$

where

$$d = \beta_1 \left(\sum_{i=1}^{m} k_{fi}\|f_i\|_\infty^{m_i} + \rho \right) + (\beta_1\beta_4 k_2 + \beta_2(1 - \beta_3 k_2))\|f_0\|_\infty + (1 - \beta_3 k_2)\|i_e\|_\infty + \beta_1 k_2 \|i_v\|_\infty \tag{3.4}$$

ii) L_∞-stable if (3.3) holds and $\rho = 0$ in (3.2) and (3.4).

Proof: Refer to Appendix I.

Remark 3.1: If the nonlinearity does not depend in a quadratic manner from $\|e\|$, i.e., $k_0 = 0$ in (3.2), then (3.3) reduces to the well-known small gain condition $1 - \beta_1 k_1 - \beta_3 k_2 > 0$ which is a *global* result. In the case of $k_0 \ne 0$, however, the result is *local* since boundedness of $\|e\|_\infty$ can be shown only for sufficiently small $\|e(t = 0)\|$ (see Appendix I). Moreover, d in (3.4) depends on the L_∞-norms of the inputs and the free responses which, therefore, influence the stability condition (3.3). It should be further noted that in the case of weak L_∞-stability a limit cycle is considered to be stable!

B. The L_2-Stability Approach

With Assumption 3.1 the induced L_2-norm (2.3) of $p_i(t)$ in (3.1) is given by [2, p. 25]

$$\beta_i = \sup_{\omega \ge 0} \|P_i(j\omega)\|; \quad i = 1, \cdots, 4.$$

Theorem 2: If Assumptions 3.1–3.3 hold *and* $k_0 = \rho = 0$ in (3.2), then the closed-loop system in Fig. 1 is L_2-stable if

$$1 - \beta_1 k_1 - \beta_3 k_2 > 0. \tag{3.5}$$

Proof: Refer to Appendix I.

Remark 3.2: In contrast to Theorem 1, L_2-stability cannot be shown if ρ or k_0 is nonzero. As it can be seen from the proof in Appendix I, an additive positive constant ρ in (3.2) would lead to the trivial estimate $\|e\|_2 \le \infty$ which does not prove boundedness. If $k_0 \ne 0$, an equivalent of

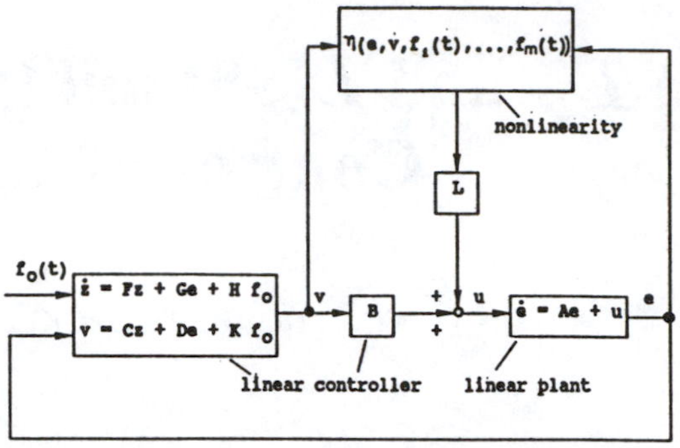

Fig. 1. Feedback system under consideration.

Theorem 1 cannot be derived because $\|e^Te\|_{T2} \le c\|e\|_{T2}^2$ does not hold in general. Thus, the L_2-approach is less generally applicable than the L_∞-approach which seems to make sense because the former proves $e(t \to \infty) = 0$, which is a stronger result than the boundedness of $e(t)$.

IV. Application to Robot Manipulator Control

Let the manipulator dynamics be described by [7]

$$M(q)\ddot{q} + h(q, \dot{q}) = u \tag{4.1}$$

where $q \in R^n$ denotes the vector of generalized coordinates, and $u \in R^n$ the vector of generalized input forces. $h(q, \dot{q})$ represents Coriolis, centripetal, and gravity forces and moments as well as frictional forces. $M(q)$ is the positive definite $n \times n$ generalized interia matrix.

Assumption 4.1: All quantities in (4.1) are dimensionless by suitable normalization.

Let $x = [q^T, \dot{q}^T]^T$ and $x_d = [q_d^T, \dot{q}_d^T]^T$, where $q_d(t)$ is the vector of desired generalized coordinates. Define the error

$$e := x - x_d \tag{4.2}$$

and implement the input forces according to

$$u = \hat{M}(\hat{q})(\ddot{q}_d(t) + v) + \hat{h}(\hat{q}, \dot{\hat{q}}) \tag{4.3}$$

where $\hat{M}$ and $\hat{h}$ are the available models of M and h. $\hat{q}$ and $\dot{\hat{q}}$ are the measured quantities of q and $\dot{q}$. v is considered to be the output of a linear controller. If $M = \hat{M}$ and $h = \hat{h}$, then (4.3) represents complete nonlinear decoupling [4], [5], whereas pure linear control is obtained if $\hat{M}$ is a constant matrix and $\hat{h} = 0$.

Remark 4.1: In contrast to [1], [8] measurement noise is not considered in the sequel ($\hat{q} \doteq q$) since the L_2-approach fails. The L_∞-approach can easily treat measurement noise which results only in a modification of the constants α and ρ_0 in (4.8)–(4.10).

Equations (4.1), (4.2), and (4.3) yield [1]

$$\dot{e} = Ae + Bv + L\eta \tag{4.4}$$

as shown in Fig. 1, where

$$A = \begin{bmatrix} 0 & I \\ 0 & 0 \end{bmatrix}, \quad B = L = \begin{bmatrix} 0 \\ I \end{bmatrix},$$

$$\eta = Ev + M^{-1}\Delta h + E\ddot{q}_d(t), \tag{4.5}$$

$$E = M^{-1}\hat{M} - I, \quad \Delta h = \hat{h} - h. \tag{4.6}$$

The following assumptions are made on the manipulator and its model.

Assumption 4.2:

$$\|M^{-1}\| \le a, \tag{4.7}$$

$$\|E\| \le \alpha, \tag{4.8}$$

$$\|\Delta h\| \le \delta_0\|x\|^2 + \delta\|x\| + \rho_0 \tag{4.9}$$

where a, α, δ_0, δ, and ρ_0 are nonnegative finite constants.

For $\delta_0 = 0$ the assumptions above are identical to those in [1], [8]. $\delta_0 \neq 0$ provides an estimate of the cross-product velocity terms of the Coriolis and centripetal forces.

From (4.2) and (4.5) the nonlinearity is obtained in the form $\eta = \eta(e, v, x_d(t), \ddot{q}_d(t))$ as shown in Fig. 1. Thus, using (4.5), (4.7)-(4.9) and the triangle inequality leads to[4]

$$\|\eta\| \leq 2a\delta_0\|e\|^2 + a\delta\|e\| + \alpha\|v\| + 2a\delta_0\|x_d\|^2 + a\delta\|x_d\| + \alpha\|\ddot{q}_d\| + a\rho_0. \quad (4.10)$$

Comparison of (3.2) in Assumption 3.2 and (4.10) yields $k_0 = 2a\delta_0$, $k_1 = a\delta$, $k_2 = \alpha$, $\rho = a\rho_0$,

$$\sum_{i=1}^{m} k_{fi}\|f_i\|^{m_i} = a\delta\|x_d\| + 2a\delta_0\|x_d\|^2 + \alpha\|\ddot{q}_d\|.$$

Assumption 4.3: Let $q_{d\infty}$ denote a bounded constant vector. Then

$$x_{d\infty} := \lim_{t\to\infty} x_d(t) = [q_{d\infty}^T,\ 0^T]^T.$$

A. The L_∞-Stability Approach

Clearly, Theorem 1 can be applied to the control of a robot manipulator under the Assumptions 4.1-4.3; however, if the control system is described by the deviation from the desired stationary vector $x_{d\infty}$, the bounds on $\|\eta\|$ in (4.10) can be improved as shown in paragraph C for the L_2-approach. This might lead to a less conservative stability condition.

B. The L_2-Stability Approach

Recalling Remark 3.2, the L_2-approach cannot be applied if ρ_0 or δ_0 in (4.9) are nonzero.

Requiring $\rho_0 = 0$ excludes Coulomb frictional forces from consideration. Further, $\rho_0 = 0$ implicates $\Delta h(x = 0) = 0$ which can be fulfilled choosing an appropriate reference position of the robot manipulator; however, in (4.10) $\|x_d\|_2$ is only bounded for the special reference vector $q_{d\infty} = 0$ which rules out general applications. This can be avoided by the reformulation of the control problem as discussed below.

In [1] there is a mistake in the derivation of the upper bound of $\|e\|_2$ [1, (5.19)], i.e., the constant ρ in term b in (5.3) in [1] must be replaced by $\rho\cdot\sqrt{T}$, which is unbounded as $T \to \infty$ and $\rho \neq 0$.

For $\delta_0 = 0$ the estimate (4.9) holds only in a finite domain of the state space, because the Coriolis and centripetal forces contain cross products of joint velocities. Therefore, from a mathematical point of view, applying the L_2-stability approach leads to doubtful stability results; however, from an engineering point of view, for many applications Coriolis and centripetal forces may be neglected.

The following reformulation of the control problem is needed in order to apply the L_2-stability approach when neglecting Coulomb frictional, centripetal, and Coriolis forces. As mentioned above this might also be useful for the L_∞-stability approach.

C. Reformulation of the Control Problem

Assumption 4.4: The linear controller in Fig. 1 has integral actions.

In the case of stability, this assumption is necessary in order to guarantee $e(t \to \infty) = 0$ if η has a constant stationary value which is the case for the robot manipulator. The following reformulation simply consists of describing plant and controller quantities by the deviations from their desired stationary values.

In the sequel the desired stationary quantities are indicated by a lower index ∞, and deviation quantities by an upper dash.

[4] Use $\|x\|^2 \leq (\|e\| + \|x_d\|)^2 \leq 2(\|e\|^2 + \|x_d\|^2)$.

Definition 4.1:

$$h_\infty := h(q_{d\infty}, 0),\ \hat{h}_\infty := \hat{h}(q_{d\infty}, 0),\ \Delta h_\infty := h_\infty - \hat{h}_\infty,$$
$$\hat{M}_\infty := \hat{M}(q_{d\infty}),\ u_\infty := h_\infty,$$
$$v_\infty := \hat{M}_\infty^{-1}\Delta h_\infty, \quad (4.11)$$

$$\bar{x} := x - x_{d\infty},\ \bar{x}_d := x_d - x_{d\infty},\ \bar{u} := u - u_\infty,\ \bar{v} := v - v_\infty,$$
$$\Delta\bar{h} := h - h_\infty,\ \Delta\bar{\hat{h}} := \hat{h} - \hat{h}_\infty,\ \Delta\bar{\hat{M}} := \hat{M} - \hat{M}_\infty. \quad (4.12)$$

Note that e is not affected by this reformulation. Substituting (4.11), (4.12), (4.5), (4.6) in (4.4) yields

$$\dot{e} = Ae + B\bar{v} + L\bar{\eta}, \quad (4.13)$$

where[5]

$$\bar{\eta} = E(\bar{v} + \ddot{q}_d) + M^{-1}(\Delta\bar{\hat{M}}v_\infty + \Delta\bar{\hat{h}} - \Delta\bar{h}). \quad (4.14)$$

Assumption 4.5

$$\|\Delta\bar{h}\| \leq \bar{\delta}_1\|\bar{x}\| \quad (4.15)$$

$$\|\Delta\bar{\hat{h}}\| \leq \bar{\delta}_2\|\bar{x}\| \quad (4.16)$$

$$\|\Delta\bar{\hat{M}}\| \leq \bar{\delta}_3\|\bar{x}\| \quad (4.17)$$

where $\bar{\delta}_i = (i = 1, 2, 3)$ are nonnegative finite constants. Note that due to the reformulation there is no need to include a positive constant on the right-hand side of (4.15) anymore.

With (4.7), (4.8) of Assumption 4.2 and Assumption 4.5 the estimate

$$\|\bar{\eta}\| \leq a\bar{\delta}\|e\| + \alpha\|\bar{v}\| + a\bar{\delta}\|\bar{x}_d\| + \alpha\|\ddot{q}_d\| \quad (4.18)$$

is obtained, where

$$\bar{\delta} = \bar{\delta}_1 + \bar{\delta}_2 + \bar{\delta}_3\|v_\infty\|.$$

In contrast to (4.10) there is no constant on the right-hand side of (4.18). Furthermore, the L_2-norm of $\bar{x}_d$ exists.

Now consider the original feedback system shown in Fig. 1. Let, in accordance to [1], e be the only input of the linear controller ($f_0 = 0$). In the desired stationary case $e = 0$ the desired controller output v_∞ in (4.11) does not vanish in general since the gravitational forces will not be cancelled exactly by $\hat{h}$. Thus, it is evident that the control must include integrators (see Assumption 4.4). Since the controller with output v is a linear device, it can also be described by deviations from its stationary values without changing its structure. Hence, the reformulated control problem can be identified with Fig. 1 while Assumptions 3.1-3.3 are satisfied for $k_0 = \rho = 0$ and $m_i = 1$. Therefore, Theorem 2 yields the stability condition

$$\beta_1 a\bar{\delta} + \beta_3\alpha < 1. \quad (4.19)$$

Thus, this reformulation allows us to take into account the gravitational forces and general reference trajectories for the L_2-stability approach. It can also be applied for the L_∞-stability approach, where the general case $\rho_0 \neq 0$ and $\delta_0 \neq 0$ can be included. The derivation is straightforward and omitted here.

V. Conclusions

In order to apply the L_∞-stability approach to the control of robot manipulators the small gain theorem was extended in Theorem 1 for nonlinearities which do not satisfy sector conditions. It was shown that an equivalent cannot be derived for the L_2-stability approach. Therefore, the L_∞-approach can generally be applied to the investigation of robust stability of robot manipulator control loops, whereas the L_2-approach requires the neglection of Coulomb frictional forces, Coriolis, and centripetal forces as well as a reformulation of the control problem. In this

[5] The quantities E, M, $\hat{M}$, h, $\hat{h}$ are considered to be functions of the corresponding deviation variables which is not indicated explicitly.

note the question of the existence of a suitable controller such that the stability conditions are satisfied has not been answered. In [9] a discussion of the existence of such controllers is provided which would not be adequate for the length of this contribution. There it is shown using simple mathematics that the stability conditions (3.3) and (3.5) can always be satisfied by a set of classical P-PI cascade controllers or PI-P state feedback controllers if $\alpha < 1$ in (4.8).

The described methodology will be investigated for elastic robots in the future.

APPENDIX I

Equation (3.1) yields the following estimates in the time domain:

$$\|e\|_{Tp} \leq \beta_1 \|\eta\|_{Tp} + \beta_2 \|f_0\|_{Tp} + \|i_e\|_{Tp}$$

$$\|v\|_{Tp} \leq \beta_3 \|\eta\|_{Tp} + \beta_4 \|f_0\|_{Tp} + \|i_v\|_{Tp} \quad \text{(A1)}$$

where $p = 2, \infty$.

From (3.2) we obtain using the triangle inequality

$$\|\eta\|_{Tp} \leqslant k_0 \|e^T e\|_{Tp} + k_1 \|e\|_{Tp} + k_2 \|v\|_{Tp} + \sum_{i=1}^{m} k_{fi} \| \, \|f_i\|^{m_i} \|_{Tp} + \|\rho\|_{Tp}. \quad \text{(A2)}$$

Substituting (A2) in (A1) yields with the assumption

$$1 - \beta_3 k_2 > 0 \quad \text{(A3)}$$

the estimate

$$\beta_1 k_0 \|e^T e\|_{Tp} - (1 - \beta_1 k_1 - \beta_3 k_2) \|e\|_{Tp} + d_p \geq 0, \quad \text{(A4)}$$

where

$$d_p = \beta_1 \left(\sum_{i=1}^{m} k_{fi} \| \, \|f_i\|^{m_i} \|_{Tp} + \|\rho\|_{Tp} \right) + (\beta_1 \beta_4 k_2 + \beta_2 (1 - \beta_3 k_2)) \|f_0\|_{Tp} + (1 - \beta_3 k_2) \|i_e\|_{Tp} + \beta_1 k_2 \|i_v\|_{Tp}, \quad p = 2, \infty. \quad \text{(A5)}$$

Proof of Theorem 1: For $p = \infty$ we obtain from (A5) and with Assumption 3.3

$$\lim_{T \to \infty} d_\infty = d$$

[see (3.4)]. Thus, (A4) and

$$\|e^T e\|_{T\infty} = \|e\|^2_{T\infty}$$

lead to

$$f(\|e\|_{T\infty}) := (\|e\|_{T\infty} - b)^2 + c - b^2 \geq 0 \quad \text{(A6)}$$

where

$$b = (1 - \beta_1 k_1 - \beta_3 k_2)/2\beta_1 k_0$$

$$c = d/\beta_1 k_0.$$

In order to prove boundedness from (A6) the parabola $f(\cdot)$ must have a positive real zero which is equivalent to

$$c - b^2 < 0. \quad \text{(A7)}$$

Equation (3.3) follows directly from (A7). Then, if $\|e(t = 0)\|$ is sufficiently small, the boundedness of $\|e\|_{T\infty}$ for $T \in [0, \infty)$ follows by continuity, and proves weak L_∞-stability. It is straightforward to show that $\rho = 0$ yields L_∞-stability. Q.E.D.

Proof of Theorem 2: In the case of $p = 2$ one has to require $k_0 = \rho = 0$ since $\|e^T e\|_{T2}$ in (A4) cannot be estimated by

$$\|e^T e\|_{T2} \leq c \|e\|^2_{T2}$$

for $T \to \infty$, and $\|\rho\|_{T2} \to \infty$ for $T \to \infty$. With (3.5) and Assumption 3.3 equation (A4) leads to

$$\|e\|_{T2} \leq (1 - \beta_1 k_1 - \beta_3 k_2)^{-1} d_2$$

which proves Theorem 2 for $T \to \infty$. Q.E.D.

REFERENCES

[1] M. W. Spong and M. Vidyasagar, "Robust linear compensator design for nonlinear robotic control," in *Proc. IEEE Conf. on Robotics and Automat.*, St. Louis, MO, Mar. 1985, pp. 954-959.
[2] C. A. Desoer and M. Vidyasagar, *Feedback Systems: Input-Output Properties.* New York: Academic, 1975.
[3] G. Zames, "On the input-output stability of time-varying nonlinear feedback systems—Part I," *IEEE Trans. Automat. Contr.*, vol. AC-11, pp. 228-238, 1966.
[4] E. Freund, "The structure of decoupled non-linear systems," *Int. J. Contr.*, vol. 21, pp. 443-450, 1975.
[5] E. Freund, "Fast nonlinear control with arbitrary pole-placement for industrial robots and manipulators," *Int. J. Robotics Res.*, vol. 1, pp. 65-78, 1982.
[6] M. Vidyasagar, *Nonlinear Systems Analysis.* Englewood Cliffs, NJ: Prentice-Hall, 1978.
[7] R. P. Paul, *Robot Manipulators: Mathematics, Programming and Control.* Cambridge, MA: M.I.T. Press, 1982.
[8] M. W. Spong and M. Vidyasagar, "Robust nonlinear control of robot manipulators," in *Proc. 24th Conf. on Decision and Contr.*, Ft. Lauderdale, FL, Dec. 1985, pp. 1767-1772.
[9] N. Becker and W. M. Grimm, "Design of robust controllers for robots. Part 1: Theoretical fundamentals. Part 2: Application to the design of P-PI cascade and PI-P state feedback controllers" (in German), *Automatisierungstechnik*, submitted for publication.

Adaptive Model Following Control of Nonlinear Robotic Systems

SAHJENDRA N. SINGH

Abstract—**An adaptive model following control law for nonlinear robotic systems with rotational joints is presented. The derivation of the controller does not require any knowledge of nonlinear system matrics and the uncertainty in the system. In the closed-loop system the joint angles asymptotically converge to the reference trajectories.**

I. Introduction

The equations of motion of robotic systems are highly nonlinear and coupled and derivation of such equations is not easy for systems of large degrees of freedom. Moreover, the parameters of the system are not known in advance since it must handle variable payloads. Modeling errors also exist due to unknown static and dynamic friction forces at the joints and uncertainty in the link parameters.

Recently, some attempt has been made in designing control systems for uncertain robotic systems [1]–[7]. The results of [1]–]5] are based on linearized time-varying models about the reference trajectories and assume that during the adaptation process the elements of the linearized system remain constant. However, for fast motion of robots, this assumption is not valid. Using Lyapunov's method [6]–[8], hyperstability theory [9], sliding mode theory [10], and stable factorization [11], controllers have been designed for nonlinear uncertain robotic systems. The controllers of [7], [8] do not yield zero tracking error and [6] must use unbounded feedback gains for the convergence of the tracking error to zero. The discontinuous control law of [10] causes the trajectory to chatter along the sliding surface. Although it is possible to avoid the chattering phenomenon by smoothing out the control law in the neighborhood of the sliding surface, nonzero tracking error exists. Moreover, the results of [6], [11] are based on the rather restrictive assumption that the bounds on the parameter uncertainty are known. Although one may argue that in the absence of the precise bounds on the uncertainty, it is possible to design controllers following [6], [10] based on certain overestimated values of these bounds, the resulting controller will require excessive control magnitudes. Of course, some involved calculations are required to obtain the feedback gains using these bounds on the uncertainty and the system matrices. However, such computations are difficult to carry out as the number of links increase. Furthermore, the controllers of [7] and [11] allow only limited variation in the parameters.

In this paper, we present an adaptive control law for accomplishing trajectory tracking in nonlinear uncertain robotic systems. The controller includes a dynamic compensator in the feedback path. Interestingly, unlike [6], [11] we assume here that the system matrices are completely unknown and no information on the possible size of the uncertainty is given. This has an important implication. The involved computations of [6], [11] to obtain the feedback gains are not required here. Instead, the feedback elements are generated as functions of the state of the dynamic compensator and the trajectory error. We also note that unlike [6], [8], in the closed-loop system, the tracking error asymptotically converges to zero, and the restriction on the magnitude of the uncertainty imposed in [7] and [11] is avoided here. The derivation of the controller is based on a recent contribution of Corless and Leitmann [12]–[14] on adaptive control of uncertain systems.

Manuscript received September 17, 1984; revised January 11, 1985 and June 5, 1985.
The author is with Vigyan Research Associates, Inc., Hampton, VA 23666.

II. Robotic Systems

The dynamics of an n-degree of freedom manipulator is given by

$$D(q)\ddot{q}+F(q,\ \dot{q})\dot{q}+g(q,\ \dot{q})=u(t) \tag{1}$$

where q is the vector of n joint coordinates, $D(q)$ is the positive definite symmetric inertia matrix, $F(q, \dot{q})\dot{q}$ is the vector grouping the Coriolis and centrifugal forces, and $g(q, \dot{q})$ represents the "parasitic" forces such as the force of gravity and friction.

In the following we shall consider robotic systems with rotational joints. Then the system matrices in (1) satisfy for all q, $\dot{q} \in R^n$ [6]

$$\begin{aligned} a_1 &\leqq \|D^{-1}(q)\| \leqq a_2 \\ \|F(q,\ \dot{q})\dot{q}\| &\leqq \|\dot{q}\|^2 b_1 \\ \|g(q,\ \dot{q})\| &\leqq c_1+c_2\|\dot{q}\| \end{aligned} \tag{2}$$

where $a_i > 0$, $b_1 > 0$, $c_i > 0$ are some positive numbers. Note that if no velocity-dependent friction forces are present in the system, then $c_2 = 0$ in (2).

Defining $x = (q^T, \dot{q}^T)^T$ (T denotes transposition), (1) can be written in a state variable form

$$\dot{x}=\begin{bmatrix} \dot{q} \\ D^{-1}(q)(-F(q,\ \dot{q})\dot{q}-g(q,\ \dot{q})) \end{bmatrix}+\begin{bmatrix} 0 \\ I \end{bmatrix} D^{-1}(q)u. \tag{3}$$

It is desired that x follow a desired trajectory $\hat{x}$ of the reference model

$$\frac{d}{dt}\begin{pmatrix} \hat{q} \\ \dot{\hat{q}} \end{pmatrix}=\begin{bmatrix} 0 & I \\ P & Q \end{bmatrix}\begin{bmatrix} \hat{q} \\ \dot{\hat{q}} \end{bmatrix}+\begin{bmatrix} 0 \\ B_1 \end{bmatrix} r$$

$$\triangleq A_M\hat{x}+B_M r \tag{4}$$

where $\hat{x} = (\hat{q}^T, \dot{\hat{q}}^T)^T$, $P = \text{diag}\,(P_i)$, $Q = \text{diag}\,(Q_i)$, $r \in R^n$ is an external input, and A_M and B_M are given constant matrices. We assume that A_M is a stable matrix.

Defining $e = (q - \hat{q})$, $z = (e^T, \dot{e}^T)^T$, using (3) and (4) we obtain

$$\dot{z}=A_M z+Bh(q,\ \dot{q},\ u,\ r),\ z(t_o)=z_o \tag{5}$$

where $B = (0, I)^T$, and

$$h(q,\ \dot{q},\ u,\ r)=D^{-1}(q)u+\bar{h}(q,\ \dot{q},\ r)$$

$$\bar{h}(q,\ \dot{q},\ r)=-Pq-Q\dot{q}-B_1 r+D^{-1}(q)[-F(q,\ \dot{q})\dot{q}-g(q,\ \dot{q})]. \tag{6}$$

We are interested in deriving a control law u such that for any uncertainty in the parameters of the robotic system

$$e(t)\to 0, \qquad \text{as } t\to\infty.$$

III. Control Law

In this section we shall derive an adaptive control law such that $z(t) \to 0$, as $t \to \infty$ for any uncertainty in the system parameters using a result of [9].

Reprinted from *IEEE Trans. Automat. Contr.*, vol. AC-30, no. 11, pp. 1099–1100, Nov. 1985.

First, let us evaluate

$$u^T h(q, \dot{q}, u, r) \geqq a_1\|u\|^2 - \|u\|\|\hat{h}(q, \dot{q}, r)\|$$
$$\geqq a_1\|u\|^2 - \|u\|[\|P\|\|q\| + \|Q\|\|\dot{q}\|$$
$$+ \|B_1 r\| + a_2\{\|\dot{q}\|^2 b_1 + c_1 + c_2\|\dot{q}\|\}]. \quad (7)$$

Assuming that $||\hat{x}(0)|| \leq d_0$, and $||r(t)|| \leq d_1 < \infty$, there exists a $d_2 > 0$ such that the solution of (4) satisfies

$$\|\hat{x}(t)\| \leqq d_2 \quad \text{for all } t.$$

Noting that $\|q(t)\| \leq \|e(t)\| + d_2$, $\|\dot{q}(t)\| \leq \|\dot{e}(t)\| + d_2$ and, in view of (7), it follows that there exist positive constant β_i such that

$$u^T h(q, \dot{q}, u, r) \geqq \beta_o\|u\|[\|u\| - \beta_1\|e\| - \beta_2\|\dot{e}\| - \beta_3\|\dot{e}\|^2 - \beta_4]$$
$$= \beta_o\|u\|[\|u\| - \Pi(e, \dot{e}, \beta)] \quad (8)$$

where $\beta = (\beta_1, \cdots, \beta_4)$ *and*

$$\Pi(e, \dot{e}, \beta) = \beta_1\|e\| + \beta_2\|\dot{e}\| + \beta_3\|\dot{e}\|^2 + \beta_4.$$

We note that β_i is unknown, since the parameters of the system are unknown.

Since A_M is a stable matrix, for any given S, a symmetric positive definite matrix (denoted as $S > 0$), there exists a unique $R > 0$ satisfying

$$RA_M + A_M^T R = -S. \quad (9)$$

The system (5) satisfying (8) belongs to a class of systems for which an adaptive control law has been obtained in [9]. Following [9], one obtains the adaptive control law of the form

$$u(t) = -\Pi(z, \hat{\beta})s(z, \hat{\beta}, \epsilon)$$

$$\dot{\hat{\beta}}(t) = L\frac{\partial \Pi^T}{\partial \beta}(z, \hat{\beta})\|\alpha(z)\|$$

$$\dot{\epsilon}(t) = -l\epsilon(t)$$

$$\hat{\beta}(t_o) \in (0, \infty)^4, \ \epsilon(t_o) \in (0, \infty), \quad l > 0 \quad (10)$$

where $L > 0$ is a diagonal matrix,

$$\alpha(z) = B^T Rz, \quad (11)$$

the function s: $R^{2n}x(0, \infty)^5 \rightarrow R^n$ is any strongly Carathéodory function which satisfies

$$s(z, \hat{\beta}, \epsilon)\|B^T Rz\| = \|s(z, \hat{\beta}, \epsilon)\|B^T Rz \quad (12)$$

and for all $(z, \hat{\beta}, \epsilon) \in R^{2n}x(0, \infty)^5$

$$\|\Pi(z, \hat{\beta})B^T Rz\| > \epsilon \Rightarrow s(z, \hat{\beta}, \epsilon) = \alpha(z)/\|\alpha(z)\|. \quad (13)$$

A particular example of such a function s is given by [13]

$$s(z, \hat{\beta}, \epsilon) = \text{sat } [\Pi(z, \hat{\beta})B^T Rz/\epsilon] \quad (14)$$

where

$$\text{sat } \eta = \begin{cases} \eta, & \|\eta\| \leqq 1 \\ \eta/\|\eta\|, & \|\eta\| > 1. \end{cases} \quad (15)$$

Define the parameter "estimate" vector $\hat{p}(t) = (\hat{\beta}^T(t), \epsilon(t))^T$, $\hat{p}(t_0) = \hat{p}_0$, and the parameter vector $p = (\beta^T, 0)^T$. Now the following results can be stated.

Theorem 1: Consider the closed-loop robotic system (1) and (10) with rotational joints. Then the closed-loop system has the following properties.

P1: For each initial condition $(t_0, z_0 \hat{p}_0)$ there exists a solution $(z(\cdot), \hat{p}(\cdot))$ of the system (1) and (10) defined on $[t_0, \infty)$.

P2: For each $\eta > 0$ there exists $\delta > 0$ such that if $(z(\cdot), \hat{p}(\cdot))$ is any solution of (1) and (10) with $\|z(t_0)\|$, $\|\hat{p}(t_0) - p\| < \delta$ then $\|z(t)\|$, $\|\hat{p}(t) - p\| < \eta$ for all $t \geqq t_0$, and the solutions of (1) and (10) are uniformly bounded.

P3: If $(z(\cdot), \hat{p}(\cdot))$: $[t_0, \infty) \rightarrow R^{2n} \times (0, \infty)^5$ is a solution of (1) and (10) then

$$\lim_{t\rightarrow\infty} z(t) = 0$$

and, therefore,

$$(q(t), \dot{q}(t)) \rightarrow (\hat{q}(t), \dot{\hat{q}}(t)) \quad \text{as } t \rightarrow \infty.$$

Proof: The proof can be completed following [13].

We notice that the control law in (10) is the product of the scalar function $\Pi(z, \hat{\beta})$ and the vector $s(z, \hat{\beta}, \epsilon)$ which depend on $\hat{\beta}(t)$ and $\epsilon(t)$. The state variables $\hat{\beta}(t)$ *and* $\epsilon(t)$ of the dynamic compensator evolve according to (10). We note that the matrix R is precomputed using (9). The function $s(z, \hat{\beta}, \epsilon)$ is saturation type and is generated according to (15). Since the structure of the controller is simple, it can be implemented on a microprocessor.

IV. Conclusion

For an uncertain robotic system, an adaptive control law for following the trajectory of a reference model was derived. In this approach, the system matrices of the robotic system were assumed to be completely unknown. The controller of this paper includes a dynamic system which can be easily synthesized.

References

[1] S. Dubowsky and D. T. DesForges, "The application of model referenced adaptive control of robotic manipulators," *J. Dynam. Syst. Meas. Contr.*, vol. 101, pp. 193-200, Sept. 1979.

[2] R. Horowitz and M. Tomizuka, "An adaptive control scheme for mechanical manipulators: compensation of nonlinearity and decoupling control," in *Proc. ASME*, 1980, paper 80, Wa/DSC-6.

[3] A. J. Koivo and T. H. Guo, "Adaptive linear controller for robotic manipulators," *IEEE Trans. Automat. Contr.*, vol. AC-28, pp. 162-171, Feb. 1983.

[4] C. S. G. Lee and M. J. Chung, "An adaptive control strategy for mechanical manipulators," *IEEE Trans. Automat. Contr.*, vol. AC-29, pp. 837-841, Sept. 1984.

[5] B. K. Kim and K. G. Shin, "An adaptive model following control of industrial manipulators," *IEEE Trans. Aerosp. Electron. Syst.*, vol. AES-19, pp. 805-814, Nov. 1983.

[6] C. Samson, "Robust nonlinear control of robotic manipulators," in *Proc. IEEE Conf. Decision Contr.*, 1983, pp. 1211-1216.

[7] M. W. Spong, J. S. Thorp, and J. W. Kleinwaks, "The control of robot manipulators with bounded control, Part II: Robustness and disturbance rejection," in *Proc. 23rd IEEE Conf. Deision Contr.*, Las Vegas, NV, Dec. 1984.

[8] S. N. Singh, "Nonlinear control of uncertain robotic systems," in *Proc. Int. Conf. Robotics and Factories of the Future*, Charlotte, NC, Dec. 1984.

[9] S. Nicosia, and P. Tomei, "Model reference adaptive control algorithms for industrial robots," *Automatica*, pp. 635-644, 1984.

[10] S. S. Sastry and J. J. Slotine, "Tracking control of nonlinear systems using sliding surfaces and application to robot manipulators," *Int. J. Contr.*, vol. 38, no. 2, pp. 465-492, 1983.

[11] M. W. Spong and M. Vidyasagar, "Robust linear compensator design for nonlinear robotic control," in *Proc. IEEE Int. Conf. Robotics Automation*, St. Louis, MO, Mar. 1985.

[12] M. Corless and G. Leitmann, "Adaptive control of systems containing uncertain functions and unknown functions with uncertain bounds," *J. Opt. Theory Appl.*, vol. 41, pp. 155-168, Sept. 1983.

[13] ——, "Adaptive control for uncertain dynamic systems," in *Dynamical Systems and Microphysics, Control Theory and Mechanics*, A. Blaquière and G. Leitmann, Eds. New York: Academic, 1984.

[14] ——, "Adaptive controllers for a class of uncertain systems," *Fondation Louis de Broglie*, vol. 9, pp. 65-95, 1984.

Part 4
Adaptive Control of Robot Manipulators

JEAN-JACQUES E. SLOTINE
Massachusetts Institute of Technology

EFFECTIVE adaptive controllers represent an important step toward developing high-speed and high-precision robots. Robots have to face uncertainty in many dynamic parameters, in particular the parameters describing the dynamic properties of grasped loads. Sensitivity to such parameter uncertainty is especially severe in high-speed operations or when controlling direct-drive robots, for which no gear reduction is available to mask effective inertia variations. More generally, if advanced robots are designed to be capable of precisely affecting their environment (e.g., providing accurate force or impedance control), they are likely to exhibit high sensitivity to external forces and load variations. Two classes of approach have been actively studied to maintain performance in the presence of parameter uncertainties: robust control, as discussed in Part 3, and adaptive control. Because the adaptation mechanism keeps extracting parameter information in the course of operation, adaptive controllers potentially can provide consistent performance in the face of even very large load variations.

Two phases may be discerned in the history of adaptive robot control research, an *approximation* phase (1979–1985) and a *linear-parameterization* phase (1986–present). In the approximation phase, researchers relied on restrictive assumptions or approximations for adaptive-control design and analysis, for example, linearization of robot dynamics, decoupling assumption for join motions, or slow variation of the inertia matrix. Pioneering papers of this first phase included [1], [2], [3], [4]. A reasonably complete review of these early methods is presented in [5].

The explicit introduction in adaptive robotic control research of the linear parameterization of robot dynamics represents a turning point inspired by earlier results developed in the context of parameter estimation [6], [7], [8]. Research on adaptive robot control could fully account for the nonlinear, time-varying and coupled nature of robot dynamics, based on the possibility of selecting a proper set of equivalent parameters such that the manipulator dynamics depend linearly on these parameters. The resulting adaptive controllers can be classified into three categories: direct, indirect, and composite. The direct adaptive controllers (e.g., [9], [10], [11], [12]) use tracking errors of the joint motion to drive parameter adaptation. The indirect adaptive controllers (e.g., [13], [14], [15]), on the other hand, use prediction errors on the filtered joint torques to generate parameter estimates to be used in the control law. The composite adaptive controllers ([16], [17]), use both tracking errors in the joint motion and prediction errors on the filtered torques to drive parameter adaptation.

1. Direct Adaptive Controllers

The direct adaptive controllers use tracking errors of the joint motion to drive the parameter adaptation. In this class of adaptive controllers, the predominant concern of the adaptation laws is to reduce the tracking errors.

An adaptive controller based on computed-torque control is proposed by [9] and [18], which show its global convergence. The authors' desire to maintain the structure of the computed-torque control scheme brings about the need to use acceleration measurements, and to invert the estimated inertia matrix as part of the algorithm. Related results can be found in [19].

In [10], [20], [21], the global tracking convergence of a new adaptive feedforward-plus-PD controller is established and its good performance is demonstrated experimentally. The algorithm only requires the system's state (joint positions and velocities) to be measured, and avoids inversion of the estimated inertia matrix by taking advantage of the inherent positive definiteness of the actual inertia matrix. Related work appears in [11], [12], [22], [23], [24], [25], [26], [27], [28], [29], [30].

2. Indirect Adaptive Controllers

The indirect adaptive-manipulator controllers, pioneered by Middleton and Goodwin [13], use prediction errors on the filtered joint torques to generate parameter estimates to be used in the control law. They show the global tracking convergence of their adaptive controller ([13], [31]), which is composed of a modified computed-torque controller and a modified least-square estimator. The computation of their adaptive controller again requires inversion of the estimated inertia matrix. The indirect adaptive controller in [15] avoids this requirement by using a different modification of the computed-torque controller. Most of the indirect algorithms (as well as [9], [18]) have to assume (or to develop procedures to guarantee) that the estimated inertia matrix remains positive definite in the course of adaptation. If only the load is to be estimated, a projection approach can be used to maintain this positive definiteness while preserving convergence properties, as the convexity result of [15] shows. Indirect controllers allow the vast parameter-estimation literature to be used to select time-variations of the adaptation gains.

3. Composite Adaptive Controllers

The composite adaptive controllers, studied in [16], [17], use both tracking errors in the joint motions and prediction errors on the filtered torques to drive the parameter adaptation. They also allow standard parameter-estimation techniques to be exploited.

4. Concluding Remarks

Adaptive manipulator control has been evolving very rapidly in the past five years, and this brief introduction by no means represents an exhaustive survey of the many research papers and results it has yielded. Trajectory control algorithms can be extended to compliant motion control. Current research directions include issues of computational efficiency [27], [32], transmission dynamics [33], design of exciting desired trajectories [34], and experimental implementations [21], [35], [32].

References

[1] S. Dubowsky and D. DesForges, "The application of model-referenced adaptive control to robotic manipulators," *J. Dyn. Sys. Meas., and Cont.*, vol. 101, pp. 193–200, 1979.

[2] R. Horowitz and M. Tomizuka, "An adaptive control scheme for mechanical manipulators—compensation of nonlinearity and decoupling control," ASME paper 80 WA/DSC-6, 1980. (Also in *J. Dyn. Sys., Meas., and Cont.*, 33-7, pp. 659–668, 1986.)

[3] A. Koivo, "Control of robotic manipulator with adaptive controller," *IEEE Conf. Dec. and Cont.*, San Diego, CA, 1981.

[4] A. Balestrino, G. DeMaria, and L. Sciavicco, "Adaptive control of robotic manipulators," *AFCET Congres Automatique*, Nantes, France, 1981.

[5] T. C. Hsia, 1986, "Adaptive control of robot manipulators—a review," *IEEE Int. Conf. Robotics and Automation*, San Francisco.

[6] F. Nicolo and Katende. *IASTED Conf. Robotics and Automation*, Lugano, 1982.

[7] C. H. An, C. G. Atkeson, and J. M. Hollerbach, "Estimation of inertial parameters of rigid body links of manipulators," *IEEE Conf. Dec. and Cont.*, Fort Lauderdale, 1985.

[8] P. Khosla and T. Kanade, "Parameter identification of robot dynamics," *IEEE Conf. Dec. and Cont.*, Fort Lauderdale, 1985.

[9] J. J. Craig, P. Hsu, and S. Sastry, "Adaptive control of mechanical manipulators," *IEEE Int. Conf. Robotics and Automation*, San Francisco, 1986.

[10] J-J. E. Slotine and W. Li, "On the adaptive control of robot manipulators," *ASME Winter Annual Meeting*, Anaheim, CA, 1986.

[11] D. S. Bayard and J. T. Wen, "Simple adaptive control laws for robotic manipulators," *Proc. Fifth Yale Workshop Applications of Adapt. Sys. Theory*, 1987.

[12] N. Sadegh and R. Horowitz, "Stability analysis of an adaptive controller for robotic manipulators," *IEEE Int. Conf. Robotics and Automation*, Raleigh, NC, 1987.

[13] R. H. Middleton and G. C. Goodwin, "Adaptive computed-torque control for rigid-link manipulators," *25th IEEE Conf. Dec. and Cont.*, Athens, Greece, 1986.

[14] P. Hsu, S. Sastry, M. Bodson, and B. Paden, "Adaptive identification and control of manipulators with joint acceleration measurements," *IEEE Int. Conf. Robotics and Automation*, Raleigh, NC, 1987.

[15] W. Li and J-J. E. Slotine, Indirect adaptive robot control, *5th IEEE Int. Conf. Robotics and Automation*, Philadelphia, 1988.

[16] J-J. E. Slotine and W. Li, "Adaptive manipulator control: a new perspective," *IEEE Conf. Dec. and Cont.*, Los Angeles, 1987.

[17] J-J. E. Slotine and W. Li, Composite adaptive manipulator control, *Automatica*, vol. 25-4, pp. 509–519, 1989.

[18] J. J. Craig, P. Hsu, and S. Sastry, "Adaptive control of mechanical manipulators," *Int. J. Robotics Res.*, vol. 6-2, pp. 10–20, 1987.

[19] G. Bastin and G. Campion, "Adaptive control of manipulators," University of Louvain, Dept. of Engineering, 1986.

[20] J-J. E. Slotine and W. Li, "On the adaptive control of robot manipulators," *Int. J. Robotics Res.*, vol. 6-3, pp. 49–59, 1987.

[21] J-J. E. Slotine and W. Li, "Adaptive manipulator control: a case study," *IEEE Int. Conf. Rob. and Autom.*, Raleigh, NC, 1987.

[22] N. Sadegh and R. Horowitz, "Stability and robustness analysis of a class of adaptive controllers for robot manipulators," *Int. J. Robotics Res.*, vol. 9-3, pp. 74–92, 1990.

[23] D. E. Koditschek, "Adaptive techniques for mechanical systems," *Proc. Fifth Yale Workshop Applications of Adapt. Sys. Theory*, 1987.

[24] B. Paden and R. Panja, "A globally asymptotically stable 'PD + ' controller for robot manipulators," *Int. J. Contr.*, vol. 47, no. 6, 1988.

[25] R. Kelly and R. Carelli, "Unified approach to adaptive control of robotic manipulators," *Proc. IEEE Conf. Decision and Control*, vol. 2, pp. 1598–1603, 1988.

[26] I. Landau and R. Horowitz, "Synthesis of adaptive controllers for robot manipulators using a passive feedback systems approach," *Proc. IEEE Int. Conf. Robot. Automat.*, pp. 1028–1033, 1988.

[27] M. Walker, "An efficient algorithm for the adaptive control of a manipulator," *IEEE Conf. Robotics and Autom.*, Philadelphia, 1988.

[28] J. S. Reed and P. A. Ioannou, "Instability analysis and robust adaptive control of robotic manipulators," *IEEE Trans. Robot. Automat.*, vol. 5, no. 3, pp. 381–386, 1989.

[29] R. Ortega and M. Spong, "Adaptive motion control of rigid robots: A tutorial," *Proc. IEEE Conf. Decision and Control*, vol. 2, pp. 1575–1584, 1988.

[30] R. H. Middleton, "Hybrid adaptive control of robot manipulators," *Proc. IEEE Conf. Decision and Control*, vol. 2, pp. 1592–1597, 1988.

[31] R. H. Middleton and G. C. Goodwin, "Adaptive computed-torque control for rigid-link manipulators," *Sys. Cont. Letters*, vol. 10, pp. 9–16, 1988.

[32] G. Niemeyer and J-J. E. Slotine, "Performance in adaptive manipulator control," *IEEE Conf. Dec. and Cont.*, Austin, TX, 1988.

[33] F. Ghorbel, J. Y. Yung, and M. W. Spong, "Adaptive control of flexible joint manipulators," *IEEE Cont. Sys. Magazine*, vol. 9-7, pp. 9–14, 1989.

[34] B. Armstrong, "On finding 'exciting' trajectories for identification experiments involving systems with nonlinear dynamics," *IEEE Conf. Robotics and Autom.*, 1987.

[35] J-J. E. Slotine and W. Li, "Adaptive manipulator control: a case study," *IEEE Trans. Autom. Cont.*, vol. 33-11, pp. 995–1003, 1988.

Jean-Jacques E. Slotine
Weiping Li
Nonlinear Systems Laboratory
Massachusetts Institute of Technology
Cambridge, Massachusetts 02139

On the Adaptive Control of Robot Manipulators

Abstract

A new adaptive robot control algorithm is derived, which consists of a PD feedback part and a full dynamics feedforward compensation part, with the unknown manipulator and payload parameters being estimated online. The algorithm is computationally simple, because of an effective exploitation of the structure of manipulator dynamics. In particular, it requires neither feedback of joint accelerations nor inversion of the estimated inertia matrix. The algorithm can also be applied directly in Cartesian space.

1. Introduction

Adaptive control, as a branch of systems theory, is not yet quite mature (see, for instance, Åström 1983; 1984). Yet, the practically motivated drive to make robot manipulators capable of handling large loads in the presence of uncertainty on the mass properties of the load or its exact position in the end-effector, as well as the old "cybernetic" ideal of developing learning capabilities in machines, has spurred much research on adaptive control of robot manipulators (see, e.g., Hsia 1986, for a recent review). The nonlinearity of robot dynamics, however, makes them even more complex to analyze than the linear dynamic systems on which most of the existing adaptive control theory has been traditionally focused.

Several approaches have been considered. Some choose to ignore the dynamic complexity and fit the measured data to a second-order, linear, time-varying model, using for instance a recursive least-squares approach (see, e.g., Koivo 1986). Others do exploit the known structure of the system dynamics (e.g., Khosla and Kanade 1985; Atkeson et al. 1985; Craig et al. 1986), although they generally require estimation of joint accelerations. Another class of algorithms considers the "learning" of specific tasks through the use of feedforward signals (Arimoto et al. 1985; Atkeson et al. 1986), without explicitly updating the manipulator model itself.

In this paper a new adaptive robot control algorithm is derived, which consists of a PD feedback part and a full dynamics feedforward compensation part, with the unknown manipulator and payload parameters being estimated online. The algorithm is computationally simple, because of an effective exploitation of the particular structure of manipulator dynamics. As in Khosla and Kanade (1985) and Atkeson et al. (1985), we use the remark that the dependence of the system dynamics on the unknown parameters can be made linear in terms of a suitably selected set of robot and load parameters. However, contrary to most algorithms in the literature, there is no need to measure the joint accelerations or to invert the estimated inertia matrix.

The layout of the paper is as follows: Section 2 presents our basic adaptive structure in joint space, and in Section 3 we discuss its extension to Cartesian space control. Simulation results are presented in Section 4. Section 5 offers brief concluding remarks.

Extensive experimental results are presented in Slotine and Li (1987).

2. Adaptive Robot Controller in Joint Space

2.1. Dynamic Model of Robot Manipulators

In the absence of friction or other disturbances the dynamics of an n-link rigid manipulator can be written as

$$\mathbf{H}(\mathbf{q})\ddot{\mathbf{q}} + \mathbf{C}(\mathbf{q}, \dot{\mathbf{q}})\dot{\mathbf{q}} + \mathbf{G}(\mathbf{q}) = \boldsymbol{\tau}, \tag{1}$$

This research was supported in part by a grant from the Sloan Fund.

The International Journal of Robotics Research,
Vol. 6, No. 3, Fall 1987,

where $\mathbf{q}$ is the $n \times 1$ vector of joint displacements, τ is the $n \times 1$ vector of applied joint torques (or forces), $\mathbf{H(q)}$ is the $n \times n$ symmetric positive definite manipulator inertia matrix, $\mathbf{C(q, \dot{q})\dot{q}}$ is the $n \times 1$ vector of centripetal and Coriolis torques, and $\mathbf{G(q)}$ is the $n \times 1$ vector of gravitational torques.

Two simplifying properties should be noted about this dynamic structure. First, as remarked by several authors (e.g., Arimoto and Miyazaki 1984; Kodistcheck 1984), the matrices $\mathbf{H}$ and $\mathbf{C}$ are not independent. Specifically, given a proper definition of $\mathbf{C}$, the matrix $\dot{\mathbf{H}} - 2\mathbf{C}$ is *skew-symmetric*, as shown in Appendix II. Physically, this property can be easily understood: The derivative of the manipulator's kinetic energy $\dot{\mathbf{q}}^T\mathbf{H}\dot{\mathbf{q}}$ must equal the power input provided by the actuators and the gravitational torques:

$$\frac{1}{2}\frac{d}{dt}[\dot{\mathbf{q}}^T\mathbf{H}\dot{\mathbf{q}}] = \dot{\mathbf{q}}^T[\tau - \mathbf{G(q)}],$$

which implies that at all times

$$\dot{\mathbf{q}}^T(\tfrac{1}{2}\dot{\mathbf{H}} - \mathbf{C})\dot{\mathbf{q}} \equiv 0.$$

Another important property is that the dynamic structure is *linear* in terms of a suitably selected set of robot and load parameters (Khosla and Kanade 1985; Atkeson et al. 1985), as illustrated in Appendix I for a two-link manipulator.

2.2. Controller Design

The controller design problem is as follows: Given the desired trajectory $\mathbf{q}_d(t)$, and with some or all the manipulator parameters being unknown, derive a control law for the actuator torques and an estimation law for the unknown parameters such that the manipulator output $\mathbf{q}(t)$ tracks the desired trajectories after an initial adaptation process.

We derive our adaptive controller in two steps. First, in Section 2.2.1 a simple globally stable adaptive controller is obtained from a Lyapunov stability analysis. The controller strongly exploits the structure of the manipulator dynamics pointed out in the previous section. After the initial transients, however, although the adaptive controller does yield zero velocity errors, it may present nonzero position errors. We solve this problem in Section 2.2.2 by restricting the residual tracking errors to lie on a sliding surface (see Slotine 1985), thus guaranteeing asymptotic convergence of the tracking.

2.2.1. A Globally Stable Adaptive Controller

To derive the control algorithm and adaptation law, we consider the Lyapunov function candidate

$$V(t) = \tfrac{1}{2}(\dot{\tilde{\mathbf{q}}}^T\mathbf{H(q)}\dot{\tilde{\mathbf{q}}} + \tilde{\mathbf{a}}^T\Gamma\tilde{\mathbf{a}} + \tilde{\mathbf{q}}^T\mathbf{K}_p\tilde{\mathbf{q}}), \tag{2}$$

where $\mathbf{a}$ is an m-dimensional vector containing the unknown manipulator and load parameters, and $\hat{\mathbf{a}}$ is its estimate; $\mathbf{K}_P$ and Γ are symmetric positive definite matrices, usually diagonal; $\tilde{\mathbf{q}}(t) = \mathbf{q}(t) - \mathbf{q}_d(t)$ is the tracking error; and $\tilde{\mathbf{a}} = \hat{\mathbf{a}}(t) - \mathbf{a}$ denotes the parameter estimation error vector. Differentiating V yields

$$\begin{aligned}\dot{V}(t) &= \dot{\tilde{\mathbf{q}}}^T\mathbf{H}\ddot{\tilde{\mathbf{q}}} + \tfrac{1}{2}\dot{\tilde{\mathbf{q}}}^T\dot{\mathbf{H}}\dot{\tilde{\mathbf{q}}} + \tilde{\mathbf{a}}^T\Gamma\dot{\tilde{\mathbf{a}}} + \tilde{\mathbf{q}}^T\mathbf{K}_P\dot{\tilde{\mathbf{q}}}\\ &= \dot{\tilde{\mathbf{q}}}^T(\tau - \mathbf{C(q, \dot{q})\dot{q}} - \mathbf{G(q)} - \mathbf{H}\ddot{\mathbf{q}}_d)\\ &\quad + \dot{\tilde{\mathbf{q}}}^T[\tfrac{1}{2}(\dot{\mathbf{H}} - 2\mathbf{C}) + \mathbf{C}]\dot{\tilde{\mathbf{q}}} + \tilde{\mathbf{a}}^T\Gamma\dot{\tilde{\mathbf{a}}} + \dot{\tilde{\mathbf{q}}}^T\mathbf{K}_P\tilde{\mathbf{q}}\\ &= \dot{\tilde{\mathbf{q}}}^T[\tau - \mathbf{H(q)}\ddot{\mathbf{q}}_d - \mathbf{C(q, \dot{q})}\dot{\mathbf{q}}_d - \mathbf{G(q)} + \mathbf{K}_p\tilde{\mathbf{q}}]\\ &\quad + \tilde{\mathbf{a}}^T\Gamma\dot{\tilde{\mathbf{a}}},\end{aligned}$$

where we have used the property of skew-symmetry to eliminate the term $\frac{1}{2}\dot{\tilde{\mathbf{q}}}^T(\dot{\mathbf{H}} - 2\mathbf{C})\dot{\tilde{\mathbf{q}}}$. Let us define the control law as

$$\tau = \hat{\mathbf{H}}\ddot{\mathbf{q}}_d + \hat{\mathbf{C}}(\mathbf{q}, \dot{\mathbf{q}})\dot{\mathbf{q}}_d + \hat{\mathbf{G}}(\mathbf{q}) - \mathbf{K}_P\tilde{\mathbf{q}} - \mathbf{K}_D\dot{\tilde{\mathbf{q}}}, \tag{3}$$

where the positive definite matrix $\mathbf{K}_D$ may be chosen to be time varying. Then

$$\dot{V}(t) = \dot{\tilde{\mathbf{q}}}^T[\tilde{\mathbf{H}}(\mathbf{q})\ddot{\mathbf{q}}_d + \tilde{\mathbf{C}}(\mathbf{q}, \dot{\mathbf{q}})\dot{\mathbf{q}}_d + \tilde{\mathbf{G}}(\mathbf{q}) - \mathbf{K}_D\dot{\tilde{\mathbf{q}}}] + \tilde{\mathbf{a}}^T\Gamma\dot{\tilde{\mathbf{a}}},$$

where

$$\begin{aligned}\tilde{\mathbf{H}}(\mathbf{q}) &= \hat{\mathbf{H}}(\mathbf{q}) - \mathbf{H}(\mathbf{q}),\\ \tilde{\mathbf{C}}(\mathbf{q}, \dot{\mathbf{q}}) &= \hat{\mathbf{C}}(\mathbf{q}, \dot{\mathbf{q}}) - \mathbf{C}(\mathbf{q}, \dot{\mathbf{q}}),\\ \tilde{\mathbf{G}}(\mathbf{q}) &= \hat{\mathbf{G}}(\mathbf{q}) - \mathbf{G}(\mathbf{q}).\end{aligned}$$

Choice (3) cancels the terms associated with the known

manipulator parameters, so only the unknown manipulator parameters have to be retained and estimated in $\hat{\mathbf{a}}$. Further, since the matrices **H**, **C**, and **G** are linear in terms of the manipulator parameters, we can write

$$\tilde{\mathbf{H}}(\mathbf{q})\ddot{\mathbf{q}}_d + \tilde{\mathbf{C}}(\mathbf{q}, \dot{\mathbf{q}})\dot{\mathbf{q}}_d + \tilde{\mathbf{G}}(\mathbf{q}) = \mathbf{Y}\tilde{\mathbf{a}}, \tag{4}$$

where $\mathbf{Y} = \mathbf{Y}(\mathbf{q}, \dot{\mathbf{q}}, \dot{\mathbf{q}}_d, \ddot{\mathbf{q}}_d)$ is an $n \times m$ matrix, and therefore

$$\dot{V}(t) = -\dot{\tilde{\mathbf{q}}}^T \mathbf{K}_D \dot{\tilde{\mathbf{q}}} + \tilde{\mathbf{a}}^T [\Gamma \dot{\tilde{\mathbf{a}}} + \mathbf{Y}^T \dot{\tilde{\mathbf{q}}}].$$

This suggests choosing the adaptation law such that

$$\Gamma \dot{\tilde{\mathbf{a}}} + \mathbf{Y}^T \dot{\tilde{\mathbf{q}}} = \mathbf{0};$$

that is

$$\dot{\hat{\mathbf{a}}} = -\Gamma^{-1}\mathbf{Y}^T(\mathbf{q}, \dot{\mathbf{q}}, \dot{\mathbf{q}}_d, \ddot{\mathbf{q}}_d)\dot{\tilde{\mathbf{q}}}. \tag{5}$$

Note that $\dot{\hat{\mathbf{a}}} = \dot{\tilde{\mathbf{a}}}$, since the unknown parameters **a** are constants. The resulting expression of $\dot{V}$ is

$$\dot{V}(t) = -\dot{\tilde{\mathbf{q}}}^T \mathbf{K}_D \dot{\tilde{\mathbf{q}}} \leq 0. \tag{6}$$

Therefore the control law (3) and the adaptation law (5) yield a globally stable adaptive controller.

Expression (6) implies that the steady-state joint velocity error is zero. However, it does not necessarily guarantee that the steady-state position error is also zero. We now modify the previous adaptive scheme in order to solve this potential problem.

2.2.2. Elimination of Steady-State Position Errors

Undesirable steady-state position errors can be eliminated if we restrict them to lie on a sliding surface

$$\dot{\tilde{\mathbf{q}}} + \Lambda\tilde{\mathbf{q}} = \mathbf{0},$$

where Λ is a constant matrix whose eigenvalues are strictly in the right-half complex plane. Formally, we achieve this by replacing the desired trajectory $\mathbf{q}_d(t)$ in the above derivation by the virtual "reference trajectory"

$$\mathbf{q}_r = \mathbf{q}_d - \Lambda \int_0^t \tilde{\mathbf{q}}\, dt. \tag{7a}$$

Accordingly, $\dot{\mathbf{q}}_d$ and $\ddot{\mathbf{q}}_d$ are replaced by

$$\dot{\mathbf{q}}_r = \dot{\mathbf{q}}_d - \Lambda\tilde{\mathbf{q}}, \tag{7b}$$

$$\ddot{\mathbf{q}}_r = \ddot{\mathbf{q}}_d - \Lambda\dot{\tilde{\mathbf{q}}}. \tag{7c}$$

If we define

$$\mathbf{s} = \dot{\tilde{\mathbf{q}}}_r = \dot{\mathbf{q}} - \dot{\mathbf{q}}_r = \dot{\tilde{\mathbf{q}}} + \Lambda\tilde{\mathbf{q}},$$

the control law and adaptation law become

$$\tau = \hat{\mathbf{H}}(\mathbf{q})\ddot{\mathbf{q}}_r + \hat{\mathbf{C}}(\mathbf{q}, \dot{\mathbf{q}})\dot{\mathbf{q}}_r + \hat{\mathbf{G}}(\mathbf{q}) - \mathbf{K}_D\, \mathbf{s}, \tag{8}$$

$$\dot{\hat{\mathbf{a}}} = -\Gamma^{-1}\mathbf{Y}^T(\mathbf{q}, \dot{\mathbf{q}}, \dot{\mathbf{q}}_r, \ddot{\mathbf{q}}_r)\, \mathbf{s}. \tag{9}$$

Note that the matrix **Y** is now a function of $\dot{\mathbf{q}}_r$ and $\ddot{\mathbf{q}}_r$ rather than $\dot{\mathbf{q}}_d$ and $\ddot{\mathbf{q}}_d$. We can again demonstrate global convergence of the tracking by now using the Lyapunov function

$$V(t) = \tfrac{1}{2}\mathbf{s}^T\mathbf{H}\mathbf{s} + \tfrac{1}{2}\tilde{\mathbf{a}}^T\Gamma\,\tilde{\mathbf{a}}, \tag{10}$$

instead of (2), which yields

$$\dot{V}(t) = -\mathbf{s}^T\mathbf{K}_D\mathbf{s} \leq 0, \tag{11}$$

instead of (6). Note that control law (8) does not contain a term in $\mathbf{K}_P$, since the position error $\tilde{\mathbf{q}}$ is already included in $\dot{\tilde{\mathbf{q}}}_r$. Expression (11) shows that the output errors converge to the sliding surface

$$\mathbf{s} = \dot{\tilde{\mathbf{q}}} + \Lambda\tilde{\mathbf{q}} = \mathbf{0}. \tag{12}$$

This in turn implies that $\tilde{\mathbf{q}} \rightarrow 0$ as $t \rightarrow \infty$. Thus, the adaptive controller defined by (8) and (9) is globally asymptotically stable and guarantees zero steady-state error for joint positions.

The previous proof of tracking convergence may seem somewhat unorthodox to readers not familiar with sliding control theory. Let us detail the basic features. First, the vector **s** conveys information about boundedness and convergence of **q** and $\dot{\mathbf{q}}$, since *the definition of* **s** *can also be viewed as a stable first-order differential equation in* $\tilde{\mathbf{q}}$, with **s** as an input. Thus, for bounded initial conditions, boundedness of **s** implies boundedness of $\tilde{\mathbf{q}}$ and $\dot{\tilde{\mathbf{q}}}$ and, therefore, of **q** and $\dot{\mathbf{q}}$;

Fig. 1. Structure of the joint space adaptive controller.

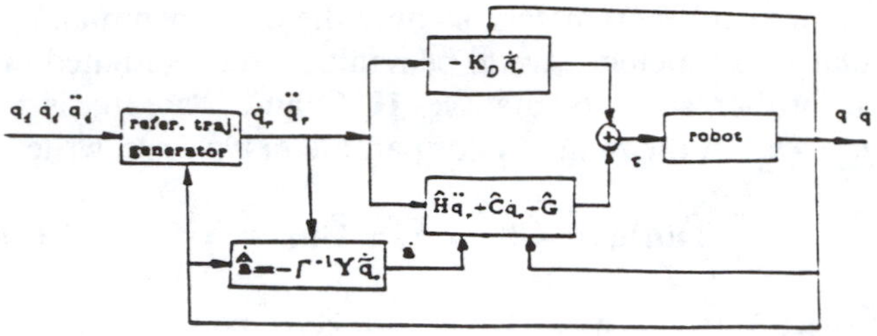

similarly, one can easily show that if $\mathbf{s}$ tends to $\mathbf{0}$ as $t \rightarrow \infty$, so do $\tilde{\mathbf{q}}$ and $\dot{\tilde{\mathbf{q}}}$. Second, the function V is actually a quasi-Lyapunov function, in our case simply a positive continuous function of time. Let us now detail the proof itself. Since $\dot{V}$ is negative or zero and V is lower bounded (by zero), V tends to a constant as $t \rightarrow \infty$ and therefore remains bounded for $t \in [0, \infty]$. Given the definition (10) of V, this in turn implies, since $\mathbf{H}$ is uniformly positive definite (i.e., $\mathbf{H} \geqslant h\mathbf{I}$ for some strictly positive h), that $\mathbf{s}$ is bounded and, therefore, that $\mathbf{q}$ and $\dot{\mathbf{q}}$ are bounded; it also implies that $\tilde{\mathbf{a}}$ is bounded and, therefore, that $\hat{\mathbf{a}}$ is bounded. From the system dynamics this then makes $\dot{\mathbf{s}}$ bounded, and thus $\mathbf{s}$ is *uniformly continuous* on $t \in [0, \infty]$. Assuming that the (perhaps time-varying) matrix $\mathbf{K}_D$ is chosen to be uniformly continuous (as is typically the case, for instance, with $\mathbf{K}_D$ constant, or with $\mathbf{K}_D = \lambda\hat{\mathbf{H}}$), $\dot{V}$ is then uniformly continuous on $t \in [0, \infty]$; therefore, since V is bounded on that time interval and $\dot{V}$ is of constant sign ($\dot{V} \leqslant 0$), *$\dot{V}$ tends to zero as $t \rightarrow \infty$*. Assuming that $\mathbf{K}_D$ is uniformly positive definite (as is again the case if $\mathbf{K}_D$ is chosen to be constant, or if $\mathbf{K}_D = \lambda\hat{\mathbf{H}}$), this implies from (11) that $\mathbf{s} \rightarrow \mathbf{0}$ as $t \rightarrow \infty$, and therefore that $\tilde{\mathbf{q}} \rightarrow \mathbf{0}$ as $t \rightarrow \infty$.

The structure of the adaptive controller given by (8) and (9) is sketched in Fig. 1. The controller consists of two parts. The first part consists of three feedforward terms corresponding to inertial, centripetal and Coriolis, and gravitational torques. The second part contains two terms representing PD feedback. The required inputs to the controller are the desired joint position $\mathbf{q}_d$, velocity $\dot{\mathbf{q}}_d$, and acceleration $\ddot{\mathbf{q}}_d$ from the trajectory planner, and the required measurements are the joint position $\mathbf{q}$ and velocity $\dot{\mathbf{q}}$. Contrary to several algorithms in the literature (e.g., Craig et al. 1986), there is no need for measuring the joint accelerations $\ddot{\mathbf{q}}$ or for inverting the estimated inertia matrix. Note that if measurements of joint accelerations were indeed explicitly available online, one could easily show (Slotine 1986) that the effect of parametric uncertainty on performance could in principle be made arbitrarily small by simply increasing the value of the acceleration gain, without using adaptation; however, this procedure would be extremely sensitive to imprecision on the joint acceleration measurement, which then essentially would enter as a pure disturbance added to $\ddot{\mathbf{q}}$.

Note from Fig. 1 that the integral term $\int_0^t \tilde{\mathbf{q}}\, dt$ of (7) need not be actually computed, since only $\dot{\mathbf{q}}_r$ and $\ddot{\mathbf{q}}_r$ (not $\mathbf{q}_r$) are *explicitly* used in the control law. Therefore, the formal definition of $\mathbf{q}_r$ is, in effect, equivalent to adding a feedback loop.

2.3. Discussion

In this section we discuss implementation aspects, computational efficiency, and strategies that combine adaptation on certain parameters with robustness to uncertainty on others and to disturbances.

2.3.1. *Implementation Aspects*

Since the load is usually fixed with respect to the last link, it can be regarded as part of that link. In practice, the parameters of the robot itself can be measured or estimated beforehand (Khosla and Kanade 1985; Atkeson et al. 1985), so only the parameters of the load are unknown. Models of Coulomb and viscous friction may also be included in (1), and the corresponding coefficients can be identified similarly.

Although convergence of the trajectory tracking is guaranteed in the previous derivation, the parameter estimates themselves do not necessarily converge to their exact values. Intuitively, to guarantee parameter convergence, the desired trajectory must be "sufficiently rich" so that only the true set of parameters can yield exact tracking. A formalization of this concept in the context of robot control and the generation of trajectories that speed up parameter convergence constitute interesting research topics in themselves (Morgan and Narendra 1977; Craig et al. 1986).

We stop updating a given parameter when it reaches

its known bounds, and we resume updating as soon as the corresponding derivative changes signs. This intuitively motivated procedure can easily be shown to preserve convergence of the tracking.

2.3.2. Computational Efficiency

In the practical implementation of the previous adaptive controller, the matrices $\hat{\mathbf{H}}$, $\hat{\mathbf{C}}$, and $\hat{\mathbf{G}}$ may be updated at a low rate, whereas a high update rate is used for $\dot{\mathbf{q}}_r$, $\ddot{\mathbf{q}}_r$, and $\mathbf{s}$, since typically the error terms vary much faster than the dynamic coefficient matrices (see, e.g., Khatib 1986). Further, the matrix $\mathbf{Y}$, whose calculation is naturally coupled to the dynamics computation, can also be updated at the slow rate, since the choice of the adaptation gain matrix Γ is generally such that the adaptation process is slower than the control bandwidth.

Because of the presence of $\dot{\mathbf{q}}_r$ in the second term of control law (8), however, the controller cannot be implemented directly with fast recursive formulations, such as the Newton-Euler method, and, therefore, requires explicit computations of $\hat{\mathbf{H}}$, $\hat{\mathbf{C}}$, and $\hat{\mathbf{G}}$. The same is true of adaptation law (9). We now introduce a recursive Newton-Euler method as an alternative way of implementing the control and adaptation laws. This Newton-Euler formulation can be seen as an approximation of the previous development, for which new stability conditions are derived.

Assume that the second term $\hat{\mathbf{C}}\dot{\mathbf{q}}_r$ in (8) is approximated by $\hat{\mathbf{C}}\dot{\mathbf{q}}$. Then we can compute the first three terms in (8) by a recursive Newton-Euler method, based on the parameters obtained from the adaptation law. The resulting control torque is

$$\tau = \hat{\mathbf{H}}(\mathbf{q})(\ddot{\mathbf{q}}_d + \Lambda\dot{\tilde{\mathbf{q}}}) + \hat{\mathbf{C}}(\mathbf{q}, \dot{\mathbf{q}})\dot{\mathbf{q}} + \hat{\mathbf{G}}(\mathbf{q}) - \mathbf{K}_D\mathbf{s}, \quad (13)$$

which is computed through a number of operations proportional to the number of links. Accordingly, the same approximation is made in the calculation of the matrix $\mathbf{Y}$, namely,

$$\mathbf{Y}(\mathbf{q}, \dot{\mathbf{q}}, \dot{\mathbf{q}}, \ddot{\mathbf{q}}_r)\hat{\mathbf{a}} = \hat{\mathbf{H}}\ddot{\mathbf{q}} + \hat{\mathbf{C}}\dot{\mathbf{q}} + \hat{\mathbf{G}}. \quad (14)$$

Let us examine the effects of these approximations. We have

$$\dot{\hat{\mathbf{a}}} = -\Gamma^{-1}\mathbf{Y}^T(\mathbf{q}, \dot{\mathbf{q}}, \dot{\mathbf{q}}, \ddot{\mathbf{q}}_r)\mathbf{s}, \quad (15)$$

with now

$$\mathbf{Y}(\mathbf{q}, \dot{\mathbf{q}}, \dot{\mathbf{q}}, \ddot{\mathbf{q}}_r)\tilde{\mathbf{a}} = \tilde{\mathbf{H}}\ddot{\mathbf{q}} + \tilde{\mathbf{C}}\dot{\mathbf{q}} + \tilde{\mathbf{G}}. \quad (16)$$

From (10),

$$\dot{V}(t) = \mathbf{s}^T[\tau - \mathbf{H}\ddot{\mathbf{q}}_r - \mathbf{C}\dot{\mathbf{q}}_r - \mathbf{G}] + \tilde{\mathbf{a}}^T\Gamma\dot{\hat{\mathbf{a}}}.$$

Thus from (13), (15), and (16), we obtain

$$\begin{aligned}\dot{V}(t) &= \mathbf{s}^T[\hat{\mathbf{H}}\ddot{\mathbf{q}}_r + \hat{\mathbf{C}}\dot{\mathbf{q}} + \hat{\mathbf{G}} - \mathbf{K}_D\mathbf{s} - \mathbf{H}\ddot{\mathbf{q}}_r - \mathbf{C}\dot{\mathbf{q}}_r - \mathbf{G}] \\ &\quad - \tilde{\mathbf{a}}^T\mathbf{Y}^T(\mathbf{q}, \dot{\mathbf{q}}, \dot{\mathbf{q}}, \ddot{\mathbf{q}}_r)\mathbf{s} \\ &= \mathbf{s}^T[\tilde{\mathbf{H}}\ddot{\mathbf{q}}_r + \tilde{\mathbf{C}}\dot{\mathbf{q}}_r + \tilde{\mathbf{G}} - \mathbf{K}_D\mathbf{s} + \mathbf{C}\dot{\mathbf{q}} - \mathbf{C}\dot{\mathbf{q}}_r] \\ &\quad - \mathbf{s}^T[\tilde{\mathbf{H}}\ddot{\mathbf{q}}_r + \tilde{\mathbf{C}}\dot{\mathbf{q}} + \tilde{\mathbf{G}}] \\ &= -\mathbf{s}^T[\mathbf{K}_D - \mathbf{C}]\mathbf{s} = -\mathbf{s}^T[\mathbf{K}_D - \tfrac{1}{2}\dot{\mathbf{H}}]\mathbf{s},\end{aligned}$$

using the skew-symmetry of the matrix $(\dot{\mathbf{H}} - 2\mathbf{C})$. Therefore, the stability of this recursive formulation of the adaptive controller is guaranteed as long as $\mathbf{K}_D$ is chosen large enough (perhaps time varying) to satisfy $\mathbf{K}_D > \frac{1}{2}\dot{\mathbf{H}}$.

2.3.3. Combining Adaptation with Robustness

In practice, we may simplify the algorithm by not explicitly estimating all unknown parameters. Some parameters may have relatively minor importance in the dynamics, in which case we may choose to make the controller robust to the uncertainty on these parameters rather than explicitly estimating them online. Similarly, some geometric parameters may already be known with reasonable precision or may have been estimated through sorting devices or visual information. Further, the controller must be robust to residual time-varying disturbances, such as stiction or torque ripple.

We categorize the unknown parameters $\mathbf{a}$ into two groups: group $\mathbf{a}_E$ contains the parameters estimated online; group $\mathbf{a}_R$ contains the parameters not estimated online. A sliding control term is then incorporated into the torque input (8) to account for the effects of uncertainties on the parameters in $\mathbf{a}_R$ and of disturbances.

Assume, without loss of generality, that only the first α unknown parameters are to be actually estimated:

$$\mathbf{a} = [\mathbf{a}_E^T \quad \mathbf{a}_R^T]^T,$$

$$\text{with } \mathbf{a}_E = \{a_j\}^T_{j=1,\ldots,\alpha}, \quad \mathbf{a}_R = \{a_j\}^T_{j=\alpha+1,\ldots,m},$$

and let, correspondingly, $\mathbf{Y} = [\mathbf{Y}_E \quad \mathbf{Y}_R]$. Assume that the uncertainties on $\mathbf{a}_R$, as well as the disturbance torques d_i reflected to the manipulator joints, are bounded:

$$|\tilde{a}_j| \leq A_j, \quad j = \alpha + 1, \ldots, m; \qquad |d_i(t)| \leq D_i(t), \quad i = 1, \ldots, n.$$

Add a sliding control term to torque input (8):

$$\tau = \hat{\mathbf{H}}\ddot{\mathbf{q}}_r + \hat{\mathbf{C}}\dot{\mathbf{q}}_r + \hat{\mathbf{G}} - \mathbf{K}_D\mathbf{s} - \mathbf{k}\,\mathrm{sgn}\,(\mathbf{s}), \qquad (17)$$

where the notation $\mathbf{k}$ sgn ($\mathbf{s}$) stands for the $n \times 1$ vector of components k_i sgn (s_i), with the k_i yet to be specified. With $\mathbf{a}_E$ and Γ_E in place of $\mathbf{a}$ and Γ in the Lyapunov function (10), we obtain

$$\dot{V}(t) = -\mathbf{s}^{\mathrm{T}}[\mathbf{K}_D\mathbf{s} + \mathbf{Y}_R\tilde{\mathbf{a}}_R - \mathbf{k}\,\mathrm{sgn}\,(\mathbf{s})] + \tilde{\mathbf{a}}_E^{\mathrm{T}}[\Gamma_E\dot{\hat{\mathbf{a}}}_E + \mathbf{Y}_E^{\mathrm{T}}\mathbf{s}].$$

Since

$$\mathbf{Y}_R\tilde{\mathbf{a}}_R - \mathbf{k}\,\mathrm{sgn}\,(\mathbf{s}) = \left\{\sum_{j=\alpha+1}^{m} Y_{ij}\tilde{a}_j + d_i(t) - k_i\,\mathrm{sgn}\,(s_i)\right\}^{\mathrm{T}}_{i=1,\ldots,n},$$

we let

$$k_i = \sum_{j=\alpha+1}^{m} |Y_{ij}|A_j + D_i + \eta_i, \qquad i = 1, \ldots, n,$$

$$\dot{\hat{\mathbf{a}}}_E = -\Gamma_E^{-1}\mathbf{Y}_E^{\mathrm{T}}\mathbf{s},$$

where the η_i are positive constants. This yields

$$\dot{V}(t) \leq -\sum_{i=1}^{n} \eta_i|s_i| - \mathbf{s}^{\mathrm{T}}\mathbf{K}_D\mathbf{s} \leq 0. \qquad (18)$$

The system trajectories are thus guaranteed to reach sliding surface $\mathbf{s} = \mathbf{0}$, and therefore convergence of the tracking is achieved.

Further, to avoid undesirable control chattering, we can use saturation functions sat (s_i/ϕ_i) in place of the switching function sgn (s_i), with the ϕ_i representing the thicknesses of the corresponding "boundary layers." Similarly to Slotine (1984), $\mathbf{s}$ is then guaranteed to converge to the boundary layers, with corresponding small tracking errors; further, the ϕ_i can be modulated based on bandwidth considerations. Similarly to Slotine and Coetsee (1986), parameter adaptation must then be stopped when the system trajectories are inside the boundary layers; indeed, by definition, disturbances and errors on $\mathbf{a}_R$ can drive the trajectories anywhere in the boundary layers without this providing any information about the estimation error on $\mathbf{a}_E$. This procedure also has the advantage of avoiding long-term drift of the estimated parameters.

Note from (18) that $\mathbf{K}_D\mathbf{s}$ can be eliminated from control input (17), since the sliding control action makes it unnecessary; however, this term must be kept in a Newton–Euler implementation of the algorithm to compensate for the approximation of $\hat{\mathbf{C}}\dot{\mathbf{q}}_r$ by $\hat{\mathbf{C}}\dot{\mathbf{q}}$, as discussed earlier. It may also be retained in order to accelerate convergence. Note that fixed-parameter sliding control is obtained if none of the unknown parameters is explicitly estimated ($\alpha = m$).

3. Extension to Cartesian Space Control

In this section we extend the previous joint space adaptive controllers to task space. To this effect, for a nonredundant manipulator, we simply replace the reference trajectories in (7b) and (7c) by

$$\dot{\mathbf{q}}_r = \mathbf{J}^{-1}[\dot{\mathbf{x}}_d + \Lambda(\mathbf{x}_d - \mathbf{x})], \qquad (19a)$$

and, accordingly,

$$\ddot{\mathbf{q}}_r = \mathbf{J}^{-1}\{[\ddot{\mathbf{x}}_d + \Lambda(\dot{\mathbf{x}}_d - \dot{\mathbf{x}})] - \dot{\mathbf{J}}\dot{\mathbf{q}}_r\}, \qquad (19b)$$

so that

$$\mathbf{s} = \dot{\tilde{\mathbf{q}}}_r = \dot{\mathbf{q}} - \dot{\mathbf{q}}_r = \mathbf{J}^{-1}[\mathbf{J}\dot{\mathbf{q}} - \dot{\mathbf{x}}_d + \Lambda\tilde{\mathbf{x}}].$$

The same control and adaptation laws (8) and (9) are then used, again with (10) as the Lyapunov function. Following the same derivation as before, we obtain

$$\dot{V}(t) = -\mathbf{s}^{\mathrm{T}}\mathbf{K}_D\mathbf{s} = -[\mathbf{J}\dot{\mathbf{q}} - \dot{\mathbf{x}}_d + \Lambda\tilde{\mathbf{x}}]^{\mathrm{T}}\mathbf{J}^{-\mathrm{T}}\mathbf{K}_D\mathbf{J}^{-1}[\mathbf{J}\dot{\mathbf{q}} - \dot{\mathbf{x}}_d + \Lambda\tilde{\mathbf{x}}] \leq 0,$$

Fig. 2. Two-link manipulator carrying a large unknown load.

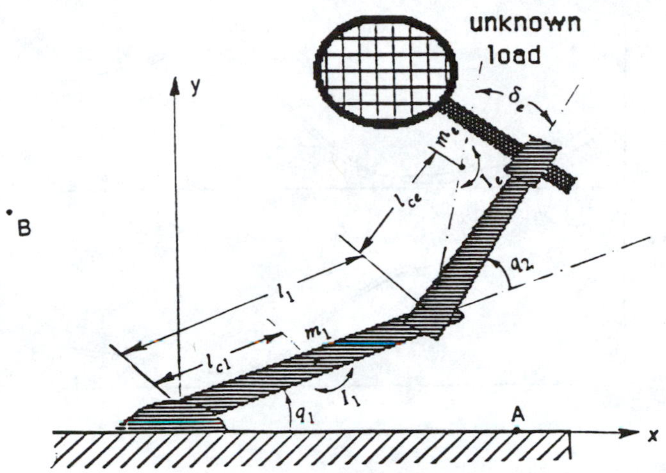

which implies convergence to

$$\mathbf{J}\dot{\mathbf{q}} - \dot{\mathbf{x}}_d + \Lambda\tilde{\mathbf{x}} = \mathbf{0}. \tag{20}$$

Using the kinematic relation $\dot{\mathbf{x}} = \mathbf{J}\dot{\mathbf{q}}$, we recognize expression (20) as the equation of the sliding surface $\dot{\tilde{\mathbf{x}}} + \Lambda\tilde{\mathbf{x}} = \mathbf{0}$, which in turn guarantees that $\tilde{\mathbf{x}} \to \mathbf{0}$ as $t \to \infty$. Therefore, the previous adaptive controller is globally stable and guarantees zero steady-state, Cartesian space, position error.

Note from (19a) and (19b) that only the desired trajectories in Cartesian space $\mathbf{x}_d$, $\dot{\mathbf{x}}_d$, and $\ddot{\mathbf{x}}_d$ have to be given (i.e., explicit inverse kinematics is not necessary). The quantities to be measured are joint positions $\mathbf{q}$ and joint velocities $\dot{\mathbf{q}}$. End-effector position $\mathbf{x}$ and velocity $\dot{\mathbf{x}}$ can be obtained from the direct kinematics, and therefore do not need to be explicitly measured. Also, note that the inverse Jacobian $\mathbf{J}^{-1}$ appears in (19a) and (19b), and therefore singularity points should be avoided (see Khatib 1986 for a relaxation of this condition).

4. Simulation Results

We present computer simulations using the two-link planar manipulator considered in Appendix I, carrying a large load of unknown mass properties (Fig. 2). The

Fig. 3. Desired joint trajectories for Examples 1 and 2.

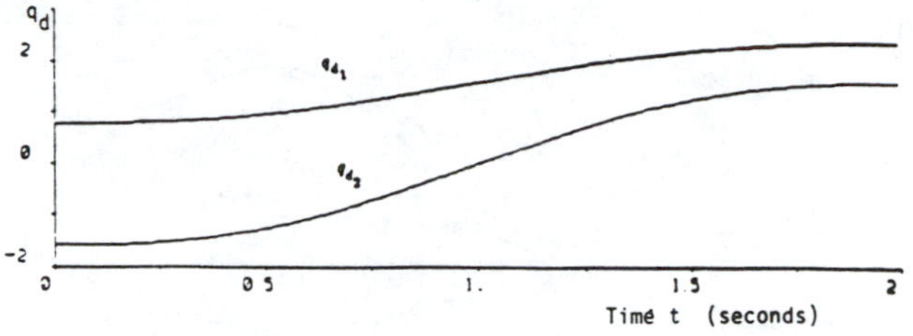

two links are identical uniform beams, with actuators mounted at the joints. In the simulations the unknown load actually has the same geometry as the links but is twice as heavy. For simplicity, the parameters of the robot itself are assumed to be exactly known. The parameters to be adapted are α, β, ϵ, and η, whose true values are $\alpha = 6.7$, $\beta = 3.4$, $\epsilon = 3.0$, and $\eta = 0$. The initial estimates of the load mass properties assume that the load is identical to the second link. The corresponding initial parameter estimates are $\hat{\alpha} = 4.1$, $\hat{\beta} = 1.9$, $\hat{\epsilon} = 1.7$, and $\hat{\eta} = 0$. In the simulation plots the estimates of the first three parameters are normalized by the true values, and $\hat{\eta}$ is normalized by 3 (the true value of ϵ), since η is itself zero.

Example 1: Comparison with conventional controllers

The task is to move the load from position A to position C, as indicated in Fig. 2. Three controllers are used: (1) PD controller, (2) PD + full dynamics feedforward compensation, and (3) adaptive controller given by (3) and (5). The desired joint trajectories are chosen to be fifth-order polynomials and are shown in Fig. 3. The matrices $\mathbf{K}_P$ and $\mathbf{K}_D$ are chosen to be identical for all three controllers, with $\mathbf{K}_P = 800I$ and $\mathbf{K}_D = 160I$. The results are plotted in Fig. 4 for controller a, Fig. 5 for controller b, and Fig. 6 for controller c. The maximum joint position errors are about 7.5° for controller a, 3° for controller b, and only about 0.5° for the adaptive controller. The maximum actuator torques are smaller for the adaptive controller than for controllers a and b. The parameter estimates do not converge to their exact values, since the desired trajectory is not persistently exciting. Also, as anticipated in Section 2.2.1, the joint position errors do not exactly converge to zero, a problem that we now remedy using the development of Section 2.2.2.

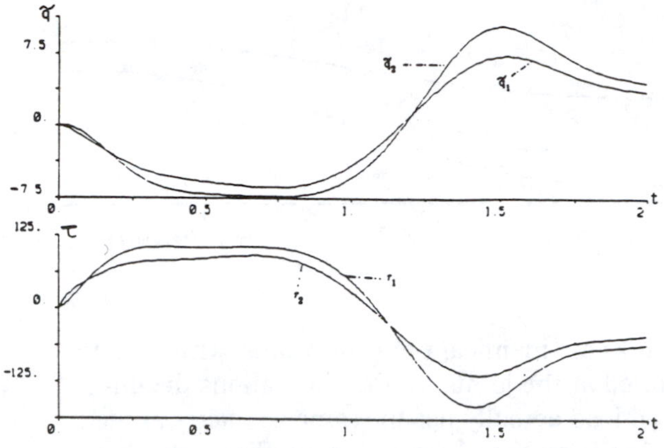

Fig. 4. PD controller in Example 1.

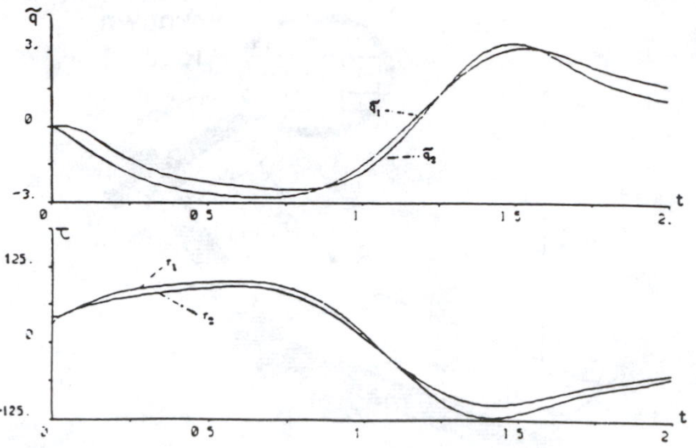

Fig. 5. PD + full dynamics feedforward controller in Example 1.

Example 2: Elimination of steady-state position error

The adaptive controller given by (7) and (8) is simulated with the same parameters as in Example 1, and $\Lambda = 30I$. The joint position errors now converge to zero (Fig. 7). We also note that the maximum joint position errors have been reduced to only 0.08° without significant increase in actuator torques.

A smaller value of Λ is also simulated. With $\Lambda = 5I$, the product of $\mathbf{K}_D$ and Λ is the same as $\mathbf{K}_P$ of controller c in Example 1; however, the resulting maximum position errors are only 0.12°, and convergence to zero is observed.

Example 3: Parameter convergence

In this example the desired trajectory is chosen to be

$$\theta_{d1} = a_1 \sin(t) + a_2 \cos(t) + a_3 \sin(3t) + a_4 \cos(3t) + a_5 \sin(5t) + a_6 \cos(5t),$$
$$\theta_{d2} = b_1 \sin(2t) + b_2 \cos(2t) + b_3 \sin(4t) + b_4 \cos(4t) + b_5 \sin(6t) + b_6 \cos(6t).$$

The coefficients a_i and b_i are chosen to make the desired trajectory satisfy the initial and final conditions on position, velocity, and acceleration. The same adaptive controller as in Example 2 is used. Although it may not be necessary to have six frequency components for the desired trajectory to be persistently exciting, this example demonstrates that sufficiently rich desired trajectories do yield convergence of the parameter estimation (Fig. 8).

Example 4: Cartesian space adaptive controller

The same task as that in previous examples is performed by the adaptive Cartesian space controller of Section 3. The desired path is now a straight line from A to B in Fig. 2. A fifth-order polynomial is constructed for the desired displacement along the path, which has zero velocities and accelerations at the start and the end of the path. The feedback gains and all other parameters are the same as before. The performance of this controller (Fig. 9) is similar to that of the joint space adaptive controller. The steady-state Cartesian position errors are zero, and the maximum Cartesian path errors in the x- and y-directions are about 8×10^{-4} m.

Extensive experimental results (Slotine and Li 1987) confirm these simulations.

5. Concluding Remarks

It is of interest to further investigate specific choices of the adaptation gain matrix Γ that yield optimal convergence rates while still avoiding the excitation of high-frequency unmodeled dynamics (such as structural resonant modes, actuator dynamics, or sampling effects). This may involve employing a time-varying Γ, based, e.g., on a Gauss–Newton algorithm. Although

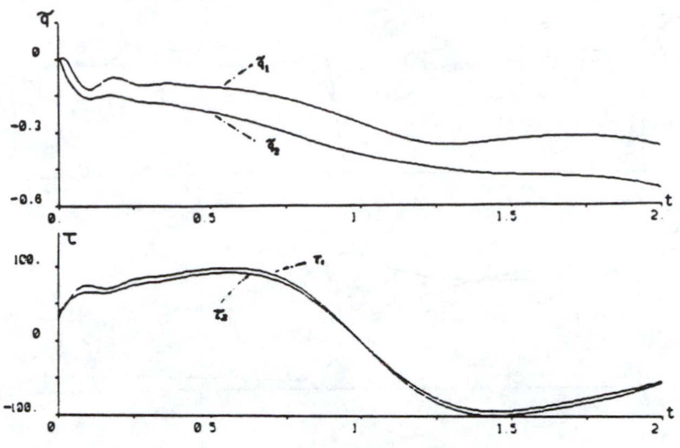

Fig. 6. Adaptive controller (3), (5).

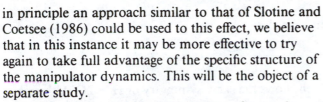

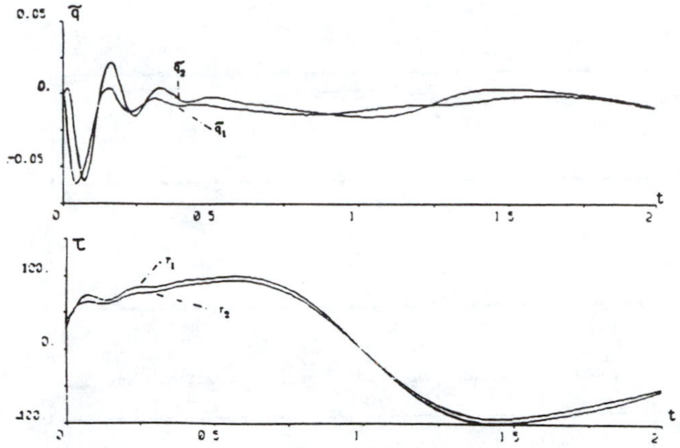

Fig. 7. Adaptive controller with steady-state position error eliminated.

in principle an approach similar to that of Slotine and Coetsee (1986) could be used to this effect, we believe that in this instance it may be more effective to try again to take full advantage of the specific structure of the manipulator dynamics. This will be the object of a separate study.

Further, in the more general context of control system design for physical nonlinear systems, we believe that the approach that consists of modifying, through feedback, the system's natural energy function rather than its explicit expanded dynamics is worthy of further investigation in its own right.

Appendix I: Two-Link Manipulator with Large Unknown Load

A two-link planar manipulator carrying an unknown payload is shown in Fig. 2. The second link, with the payload attached, can be regarded as an augmented link with four unknown parameters, namely, mass m_a, moment of inertia I_e, the distance l_{ce} of its mass center to the second joint, and the angle δ_e relative to the original second link. The dynamics of the manipulator with payload can then be written as

$$\begin{bmatrix} \alpha + 2\epsilon \cos(q_2) + 2\eta \sin(q_2) & \beta + \epsilon \cos(q_2) + \eta \sin(q_2) \\ \beta + \epsilon \cos(q_2) + \eta \sin(q_2) & \beta \end{bmatrix} \begin{bmatrix} \ddot{q}_1 \\ \ddot{q}_2 \end{bmatrix} + \begin{bmatrix} \epsilon Y_1 + \eta Y_2 + (\alpha - \beta + e_1) e_2 \cos(q_1) \\ \epsilon Y_3 + \eta Y_4 \end{bmatrix} = \begin{bmatrix} \tau_1 \\ \tau_2 \end{bmatrix},$$

where

$$Y_1 = -2 \sin(q_2)\dot{q}_1\dot{q}_2 - \sin(q_2)\dot{q}_2^2 + e_2 \cos(q_1 + q_2),$$
$$Y_2 = 2 \cos(q_2)\dot{q}_1\dot{q}_2 + \cos(q_2)\dot{q}_2^2 + e_2 \sin(q_1 + q_2),$$
$$Y_3 = \sin(q_2)\dot{q}_1^2 + e_2 \cos(q_1 + q_2),$$
$$Y_4 = -\cos(q_2)\dot{q}_1^2 + e_2 \sin(q_1 + q_2),$$
$$e_1 = m_1 l_1 l_{c1} - I_1 - m_1 l_{m1}^2,$$
$$e_2 = g/l_1,$$

where g is the acceleration of gravity, and the four unknown parameters α, β, ϵ, and η are functions of the unknown physical parameters:

$$\alpha = I_1 + m_1 l_{c1}^2 + I_e + m_e l_{ce}^2 + m_e l_1^2,$$
$$\beta = I_e + m_e l_{ce}^2,$$
$$\epsilon = m_e l_1 l_{ce} \cos(\delta_e),$$
$$\eta = m_e l_1 l_{ce} \sin(\delta_e).$$

Conversely, the four unknown physical parameters are uniquely determined by α, β, ϵ, and η.

Appendix II: The Matrix $\dot{\mathbf{H}} - 2\mathbf{C}$

We show here that, with a proper definition of the matrix $\mathbf{C}$, the matrix $\dot{\mathbf{H}} - 2\mathbf{C}$ is skew-symmetric, thus making more precise the result obtained earlier from conservation of energy.

Fig. 8. Showing the convergence of the estimates for persistently exciting trajectories: (a) normalized $\hat{\alpha}$ and $\hat{\beta}$; (b) normalized $\hat{\epsilon}$ and $\hat{\eta}$.

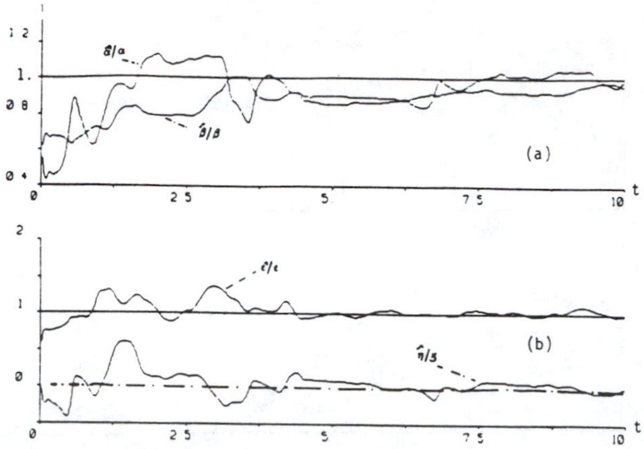

Fig. 9. Adaptive controller in Cartesian space.

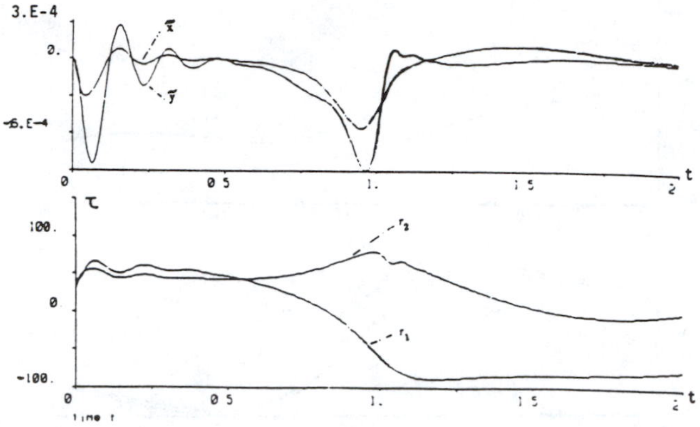

The ith element of the vector $\mathbf{C}\dot{\mathbf{q}}$ is (see, e.g., Asada and Slotine 1986)

$$\sum_{j=1}^{n} C_{ij}\dot{q}_j = \sum_{j=1}^{n}\sum_{k=1}^{n} h_{ijk}\dot{q}_j\dot{q}_k, \tag{A1}$$

where the Christoffel coefficients h_{ijk} verify

$$h_{ijk} = \frac{\partial H_{ij}}{\partial q_k} - \frac{1}{2}\frac{\partial H_{jk}}{\partial q_i}.$$

Thus, (A1) can be written

$$\sum_{j=1}^{n} C_{ij}\dot{q}_j = \frac{1}{2}\sum_{j=1}^{n}\sum_{k=1}^{n}\frac{\partial H_{ij}}{\partial q_k}\dot{q}_j\dot{q}_k + \frac{1}{2}\sum_{k=1}^{n}\sum_{j=1}^{n}\left(\frac{\partial H_{ik}}{\partial q_j} - \frac{\partial H_{jk}}{\partial q_i}\right)\dot{q}_k\dot{q}_j,$$

where we used reindexing to obtain the second term on the right side. Now take

$$C_{ij} = \frac{1}{2}\sum_{k=1}^{n}\frac{\partial H_{ij}}{\partial q_k}\dot{q}_k + \frac{1}{2}\sum_{k=1}^{n}\left(\frac{\partial H_{ik}}{\partial q_j} - \frac{\partial H_{jk}}{\partial q_i}\right)\dot{q}_k$$
$$= \frac{1}{2}\dot{H}_{ij} + \frac{1}{2}\sum_{k=1}^{n}\left(\frac{\partial H_{ik}}{\partial q_j} - \frac{\partial H_{jk}}{\partial q_i}\right)\dot{q}_k,$$

and let $\mathbf{W} = \dot{\mathbf{H}} - 2\mathbf{C}$. Then

$$W_{ij} = \sum_{k=1}^{n}\left(\frac{\partial H_{jk}}{\partial q_i} - \frac{\partial H_{ik}}{\partial q_j}\right)\dot{q}_k.$$

Thus for all i, j

$$W_{ij} = -W_{ji},$$

which shows the skew-symmetry of $\dot{\mathbf{H}} - 2\mathbf{C}$. Although other choices of C_{ij} could satisfy (A1), they usually do not possess this skew-symmetry property.

References

An, C. G., Atkeson, C. G., and Hollerbach, J. M. 1986 (San Francisco, Calif.). Experimental determination of the effect of feedforward control on trajectory tracking errors. *IEEE Int. Conf. Robotics and Automation.*

Arimoto, S., and Miyazaki, F. 1984 (Bretton Woods, N.H.). On the stability of P.I.D. feedback with sensory information. *Robotics Research,* eds. M. Brady and R. P. Paul. Cambridge: MIT Press.

Arimoto, S., Kawamura, S., Miyazaki, F., and Tamaki, S. 1985 (Fort Lauderdale, Fla.). Learning control theory for dynamical systems. *IEEE Conf. Decision and Control.*

Asada, H., and Slotine, J-J. E. 1986. *Robot Analysis and Control.* New York: John Wiley and Sons.

Åström, K. J. 1983. Theory and applications of adaptive control: A survey. *Automatica* 19:471.

Åström, K. J. 1984 (Las Vegas, Nev.). Interaction between excitation and unmodeled dynamics in adaptive control. *IEEE Conf. Decision and Control.*

Atkeson, C. G., An, C. G., and Hollerbach, J. M. 1985 (Gouvieux, France). Estimation of inertial parameters of manipulator loads and links. *Robotics Research,* eds. O. Faugeras and G. Giralt. Cambridge: MIT Press.

Craig, J. J., Hsu, P., and Sastry, S. 1986 (San Francisco, Calif.). Adaptive control of mechanical manipulators. *IEEE Int. Conf. Robotics and Automation.*

Dubowsky, S., and DesForges, D. T. 1979. The application of model-reference adaptive control to robotic manipulators. *J. Dyn. Syst. Meas. Contr.* 101:193–200.

Hsia, T. C. 1986 (San Francisco, Calif.). Adaptive control of robot manipulators—A review. *IEEE Int. Conf. Robotics and Automation.*

Khatib, O. 1986 (Osaka, Japan). *U.S.–Japan Symp. Flexible Automation.*

Khosla, P., and Kanade, T. 1985 (Fort Lauderdale, Fla.). Parameter identification of robot dynamics. *IEEE Conf. Decision and Control.*

Koditschek, D. 1984 (Las Vegas, Nev.). Natural motion of robot arms. *IEEE Conf. Decision and Control.*

Koivo, A. J. 1986 (San Francisco, Calif.). Force-position-velocity control with self-tuning for robotic manipulators. *IEEE Int. Conf. Robotics and Automation.*

Morgan, A. P., and Narendra, K. S. 1977. On the uniform asymptotic stability of certain linear nonautonomous differential equations. *SIAM J. Control Optim.* 15.

Slotine, J. J. E. 1984. Sliding controller design for nonlinear systems. *Int. J. Control* 40(2).

Slotine, J. J. E. 1985. The robust control of robot manipulators. *Int. J. Robotics Research* 4(2).

Slotine, J. J. E. 1986 (San Francisco, Calif.). On robustness and adaptation in robot control. *IEEE Int. Conf. Robotics and Automation.*

Slotine, J. J. E., and Coetsee, J. A. 1986. Adaptive sliding controller synthesis for nonlinear systems. *Int. J. Control* 42:6.

Slotine, J. J. E., and Li, W. 1987 (Raleigh, N.C.). Adaptive manipulator control: A case study. *IEEE Int. Conf. Robotics and Automation.*

Slotine, J. J. E., and Sastry, S. S. 1983. Tracking control of nonlinear systems using sliding surfaces, with application to robot manipulators. *Int. J. Control* 38(2).

Adaptive computed torque control for rigid link manipulations

R.H. MIDDLETON and G.C. GOODWIN

Department of Electrical and Computer Engineering, University of Newcastle, New South Wales, 2308, Australia

Received 26 June 1987
Revised 15 September 1987

Abstract: In this paper we shall examine the adaptive control of rigid link manipulator systems. Linear estimation techniques together with a computed torque control law are shown to give a globally convergent adaptive system which does not require measurements of accelerations.

Keywords: Adaptive control, Robotics, Non-linear systems, Computed torque control.

1. Introduction

The desire for improved dynamic accuracy in mechanical manipulators has led to research into improved control systems. Early works include the computed torque method of control (e.g. Paul [13] or Markewicz [11]) and adaptive controllers which treat the robot as a time varying linear system (e.g. Koivo and Guo [9], Dubowsky and DesForges [4], Tomizuka and Horowitz [18] or Koivo and Sorvari [10]). A complete analysis for this latter case is not available at present, however.

Hyperstability theory (e.g. Nicosia and Tomei [12]), sliding mode theory (e.g. Sastry and Slotine [15]), and stable factorization controllers (e.g. Spong and Vidyasagar [17]) have been used to propose adaptive control laws. However, these approaches suffer from one or more of the following difficulties: (i) non-zero tracking errors, (ii) unbounded feedback gains or (iii) 'chattering' effects. Recent work by Singh [16] has overcome the above difficulties, in the case of a revolute joint robot.

In this paper we use a robot model which is linear in the unknown parameters. Parameter estimation techniques for robotic systems which require measurement of additional torques or accelerations, have been proposed in many references, for example Khasla and Kanade [7] or Chae, Atkeson and Hollerback [2]. This paper has its origins in the work of Elliott et al. [5], where a discrete time algorithm is considered. Craig, Hsu and Sastry [3] have proposed a linear parameterization, based on acceleration measurement, together with an adaptive computed torque control law and have established global convergence for their scheme. Bastin, Campion and Guillaume [1] have examined a similar scheme, and its robustness to bounded disturbances. Here we extend these works in the following areas: (i) we give a decoupled estimation and control procedure, (ii) measurements of extra torques or accelerations is not required and (iii) least squares estimation is given.

2. Linear parameter estimation for robots

The dynamic equations of motion for a general rigid link manipulator having n degrees of freedom can be described (using Lagrangian mechanics, eg. Kibble [8]) as follows:

$$\frac{\mathrm{d}}{\mathrm{d}t}\{I(\theta)\dot{\theta}\} = \tau_G(\theta) + \tau_M + \frac{1}{2}\frac{\partial}{\partial\theta}\left[\dot{\theta}^{\mathrm{T}}I(\theta)\dot{\theta}\right] \tag{1}$$

where θ, $\dot{\theta} \in \mathbb{R}^n$ are the robot angles, and angular velocities.

$I(\theta) = I^{\mathrm{T}}(\theta) > 0$, $I(\theta) \in \mathbb{R}^{n\times n}$ is the inertia matrix and τ_M, $\tau_G(\theta) \in \mathbb{R}^n$ are the gravity torques, and actuator torques. We wish to perform on-line identification, since some parameters, particularly load related parameters, can undergo changes as the robot performs different tasks. It is well known (e.g. [2]) that the dynamic equations (1) may be written in a form that is linear in the unknown parameters such as mass and moments of interia. This can also be extended to include estimation of effects such as non-linear friction (see [3]). We now define:

$$I(\theta) = J(\theta) + \sum_{k=1}^{N} K_k(\theta) m_k \tag{2}$$

where $J(\theta)$ represents the known portion of the inertia matrix, $K_k(\theta)$ is a known function of θ, and m_k, $k = 1, \ldots, N$, are the unknown parameters. Clearly we can also write

$$\tau_G(\theta) = G(\theta) + \sum_{k=1}^{N} H_k(\theta) m_k \tag{3}$$

and in view of (1) we shall write

$$\frac{1}{2} \frac{\partial}{\partial \theta} \left[\dot{\theta}^{\mathrm{T}} I(\theta) \dot{\theta} \right] = D(\theta, \dot{\theta}) + \sum_{k=1}^{N} E_k(\theta, \dot{\theta}) m_k. \tag{4}$$

Using (2)–(4) we can rewrite (1) as

$$\phi^{\mathrm{T}} M = Y + \tau_M \tag{5}$$

where

$$Y = -\frac{\mathrm{d}}{\mathrm{d}t} \{ J(\theta) \dot{\theta} \} + G(\theta) + D(\theta, \dot{\theta}) \}, \tag{6}$$

$$M^{\mathrm{T}} = [m_1, m_2, \ldots, m_N], \tag{7}$$

$\phi \in \mathbb{R}^{N \times n}$, and the k-th row of ϕ is

$$\phi_k = \left[\frac{\mathrm{d}}{\mathrm{d}t} \{ K_k(\theta) \dot{\theta} \} - H_k(\theta) - E_k(\theta, \dot{\theta}) \right]^{\mathrm{T}}. \tag{8}$$

Note that equation (5) is linear in the parameter vector M and so linear estimation can be used provided ϕ, Y and τ_m are available. Note, however, that both ϕ and Y involve $\ddot{\theta}$. In this paper we shall show how estimation may be performed based on state variable measurements only. To achieve this, we operate on the left of (5) by $w/(D+w)$ where $D = \mathrm{d}/\mathrm{d}t$ and w is some constant:

$$\Psi^{\mathrm{T}} M = Y_F + \tau_F \tag{9}$$

where

$$\Psi = \left(\frac{w}{D+w} \right) \phi, \tag{10}$$

$$Y_F = \left(\frac{w}{D+w} \right) Y, \tag{11}$$

and

$$\tau_F = \left(\frac{w}{D+w} \right) \tau_m. \tag{12}$$

Ψ and Y_F are functions of θ and $\dot{\theta}$, but not $\ddot{\theta}$. Given an estimate $\hat{M}$ of M, define the vector prediction error $e(\hat{M})$ by

$$e(\hat{M}) = Y_F + \tau_F - \Psi^{\mathrm{T}} \hat{M} = -\psi^{\mathrm{T}} \tilde{M} \tag{13}$$

where

$$\tilde{M} \triangleq \hat{M} - M. \tag{14}$$

Parameter estimation procedures can be derived based on the above equations by simple extensions to the estimators in (for example) Goodwin and Mayne [6]. We propose the following unnormalized least squares estimation algorithm:

$$\dot{\hat{M}} = P \Psi e, \tag{15}$$

$$\dot{P} = -P \Psi \Psi^{\mathrm{T}} P, \tag{16}$$

where $P \in \mathbb{R}^{N \times N}$, $P(0) = P(0)^{\mathrm{T}} > 0$. The following lemma establishes some of the properties of the above estimator.

Lemma 1. *The estimator*, (15), (16), *applies to the system* (9) *yields the following properties*, *regardless of the control law*:

(i) $\hat{M}$ *is bounded*, *and*

(ii) e *belongs to* L_2.

Proof. Consider the function

$$V = \tilde{M}^{\mathrm{T}} P^{-1} \tilde{M}. \tag{17}$$

Then using (13)–(16) we can show that

$$\dot{V} = -e^{\mathrm{T}} e. \tag{18}$$

The desired results then follow. □

Note also that any positive semi-definite modification to (16) can be included provided P remains bounded for all time.

3. Computed torque servo control

Several authors (see for example Sahba and Mayne [14], Markewicz [11], Paul [13] and Craig, Hsu and Sastry [3], have proposed the use of a computed torque controller for robots. This method, which involves computation and cancellation of all non-linearities and cross-coupling terms, shall now be described. Given some reference signal $\boldsymbol{r}(t)$ our objective shall be to make $\boldsymbol{\theta}$ track $\boldsymbol{\theta}^*$, where

$$(D^2 + \beta_1 D + \beta_2)\boldsymbol{\theta}^* = \beta_2 \boldsymbol{r} \tag{19}$$

and β_1 and β_2 are positive scalars. In the case where all the parameters are known, the following control achieves this objective (for any positive constants, γ_1 and γ_2):

$$\tau_M = -\tau_G(\theta) - \frac{1}{2}\frac{\partial}{\partial \theta}\left[\dot{\theta}^{\mathrm{T}} I(\theta)\dot{\theta}\right] + \dot{I}(\theta)\dot{\theta} - I(\theta)\left[\gamma_1\dot{\theta} + \gamma_2\theta - \ddot{\theta}^* - \gamma_1\dot{\theta}^* - \gamma_2\theta^*\right], \tag{20}$$

since upon substitution of (20) into (1) we obtain

$$I(\theta)[\ddot{\varepsilon} + \gamma_1\dot{\varepsilon} + \gamma_2\varepsilon] = 0 \tag{21}$$

where

$$\varepsilon \triangleq \theta - \theta^*. \tag{22}$$

In the case where the parameters are not known but are estimated as in Section 3, we propose the following adaptive computed torque control law:

$$\tau_M = T_1 + T_2 + T_3 + T_4 \tag{23}$$

where

$$T_1 = -G(\theta) - D(\theta, \dot{\theta}) + \dot{J}(\theta)\dot{\theta} + \sum_{k=1}^{N}\left(\dot{K}_k(\theta)\dot{\theta} - E_k(\theta, \dot{\theta}) - H_k(\theta)\right)\hat{m}_k, \tag{24}$$

$$T_2 = -\hat{I}(\theta)\left[\gamma_1(\dot{\theta} - \dot{\theta}^*) + \gamma_2(\theta - \theta^*) - \ddot{\theta}^*\right], \tag{25}$$

$$T_3 = \frac{1}{w}\Psi^{\mathrm{T}}\dot{\hat{M}}, \tag{26}$$

$$T_4 = -\frac{1}{w}\hat{I}(\theta)\left[\frac{\mathrm{d}}{\mathrm{d}t}\{\hat{I}^{-1}(\theta)\}\right]e, \tag{27}$$

and

$$\hat{I}(\theta) = J(\theta) + \sum_{k=1}^{N} K_k(\theta)\hat{m}_k. \tag{28}$$

The terms T_1 and T_2 in (23) are the certainty equivalence adaptive form of the computed torque controller (20). T_3 and T_4 have been included to deal with terms which arise in the adaptive case due to the commuting of time varying operators.

4. Global convergence for adaptive computed torque control

In this section we shall establish global convergence for the adaptive computed torque controller. To achieve this, we shall make use of the following result.

Lemma 2. *Suppose H is a linear, strictly proper, exponentially stable transfer function, and f_1 belongs to L_1, f_2 belongs to L_2 and f_∞ is bounded. Then $g = H*(f_1 + f_2 + f_\infty)$ is bounded.*

The proof is given in the appendix.

Technical difficulties arise in the global convergence proof if $\hat{I}(\theta)$ is not uniformly positive definite. For this reason we shall assume the knowledge of a convex region, $C \subset \mathbb{R}^N$ such that $M \in C$ and $\forall\ \hat{M} \in C$, $\forall\ \theta \in \mathbb{R}^n$, $\hat{I}(\theta)$ is uniformly positive definite, i.e. $\hat{I}^{-1}$ is bounded. In this case it is a simple task to alter the parameter estimator (15) and (16) such that $\hat{M}$ belongs to C for all time, and properties (i) and (ii) of Lemma 1 hold. (See [6] for details of this modification.) The following lemma will also be used in the global convergence proof.

Lemma 3. *Suppose H is a linear, strictly proper exponentially stable transfer function, and $V(t)$ satisfies*

$$V(t) = H*(V(t)[f_1 + f_2] + g_1 + g_2 + g_\infty) \tag{29}$$

and

$$V(t) = V_\infty + V_2, \tag{30}$$

where f_1, f_2, g_1, g_2, g_∞, V_2 and V_∞ are functions of time and f_1, g_1 belong to L_1; f_2, g_2 and V_2 belong to L_2 and g_∞, V_∞ are bounded. Then $V(t)$ is bounded.

For the proof, see the appendix. We now proceed to our main result.

Theorem 1. *Consider the adaptive computed torque control law (23) applied to the system (1) with parameter estimator (15) and (16), modified to ensure $\hat{I}^{-1}(\theta)$ is bounded. Then provided $r(t)$ is bounded, this adaptive controller is globally convergent, i.e.*

(i) *θ, $\dot{\theta}$ and $\ddot{\theta}$ are bounded.*

(ii) *e, ε and $\dot{\varepsilon}$ tend to zero (where ε is defined in (22)).*

Proof. (i) Operating on the left of equation (13) by $(D + w)/w$ we obtain

$$\left(\frac{D+w}{w}\right)e = Y + \tau_M - \phi^{\mathrm{T}}(\hat{M}) - \frac{1}{w}\Psi^{\mathrm{T}}\dot{\hat{M}}. \tag{31}$$

(The reason for T_3 in (23) thus becomes clear.) Also from (24) and (6) we can show that

$$Y + T_1 - \phi^{\mathrm{T}}(\hat{M}) = -\hat{I}(\theta)\ddot{\theta}. \tag{32}$$

Thus using (32), (23) to (27) in (31) we see that

$$-\left(\frac{D+w}{w}\right)e = \hat{I}(\theta)(\ddot{\varepsilon} + \gamma_1\dot{\varepsilon} + \gamma_2\varepsilon) + \frac{1}{w}\hat{I}(\theta)\left[\frac{\mathrm{d}}{\mathrm{d}t}\{\hat{I}^{-1}(\theta)\}\right]e \tag{33}$$

or

$$-\left(\frac{D+w}{w}\right)[\hat{I}^{-1}(\theta)e] = \ddot{\varepsilon} + \gamma_1\dot{\varepsilon} + \gamma_2\varepsilon \tag{34}$$

and thus we see the reason for introducing T_2 in (23). We define $f = -(1/w)\hat{I}^{-1}e$ and note that since $\hat{I}^{-1}$ is bounded and e belongs to L_2, then f belongs to L_2. We then have

$$[D^2 + \gamma_1 D + \gamma_2]\varepsilon = [D + w]f \tag{35}$$

and so ε is bounded, ε and $\dot{\varepsilon}$ belong to L_2. The remainder of the proof are the technicalities required to ensure that we have boundedness (i.e. L_∞) as well as L_2 properties, and that we have uniform continuity of ϵ and $\dot{\epsilon}$, so that convergence can be assured. To establish boundedness of $\dot{\theta}$ and $\ddot{\theta}$, we shall use the torque expression, (23), and the robot dynamical equation (1). T_3 of (26) can be rewritten as

$$T_3 = -\frac{\alpha}{w}(\Psi^{\mathrm{T}} P \Psi)e. \tag{36}$$

P and w are bounded, e belongs to L_2 and so we shall consider Ψ. Using (8), we can show that the kth row of Ψ is

$$\Psi_k = \left[wK_k(\theta)\dot{\theta} - \left(\frac{w}{D+w}\right)\{wK_k(\theta)\dot{\theta} + H_k(\theta) + E_k(\theta, \dot{\theta})\}\right]. \tag{37}$$

Since θ is bounded, we have that $K_k(\theta)$ and $H_k(\theta)$ are both bounded. Also, from the definition of $E_k(\theta, \dot{\theta})$, we note that since θ is bounded, E_k is bounded by $\dot{\theta}^2$, i.e. $\exists\ c_E < \infty$, such that $\| E_k(\theta, \dot{\theta}) \| \leqslant c_E \| \dot{\theta} \|^2$ for all $\dot{\theta}$, θ.

Since $\dot{\varepsilon} \in L_2$,

$$\dot{\theta} = (L_2) + (L_\infty) \tag{38}$$

(where '(L_p)' denotes some function belonging to the space L_p). We then have from (37) and Lemma 2 that

$$\Psi = (L_\infty)\dot{\theta} + (L_\infty) = (L_\infty) + (L_2). \tag{39}$$

Now considering (36), we have

$$T_3 = [(L_2) + (L_1)]\dot{\theta} + (L_2) + (L_1) \tag{40}$$

(since P is bounded and e belongs to L_2).

We shall now consider T_4 in more detail:

$$T_4 = \frac{1}{w}\left[J(\dot{\theta}) + \sum_{k=1}^{N}(\dot{K}_k\hat{m}_k + K_k\dot{\hat{m}}_k)\right]\hat{I}^{-1}(\theta)e. \tag{41}$$

Since θ is bounded,

$$\dot{J}(\theta) = \frac{\partial J(\theta)}{\partial\theta}\dot{\theta} = (L_\infty) + (L_2) \tag{42}$$

and

$$\dot{K}_k(\theta) = (L_\infty) + (L_2), \tag{43}$$

$$\begin{aligned}\dot{\hat{M}} &= P\Psi e = [(L_\infty)\dot{\theta} + (L_\infty)](L_2) \quad \text{(using (39))}\\ &= (L_2)\dot{\theta} + (L_2).\end{aligned} \tag{44}$$

Using (42)–(44) in (41), $e \in L_2$ and $\hat{I}^{-1}$ is bounded,

$$T_4 = (L_1)\dot{\theta} + (L_2) + (L_1). \tag{45}$$

We then have

$$\tau_m = [(L_1) + (L_2)]\dot{\theta} + (L_1) + (L_2) + (L_\infty). \tag{46}$$

From (1), (38), and, using the fact that θ is bounded,

$$\ddot{\theta} = [(L_1) + (L_2)]\dot{\theta} + (L_1) + (L_2) + (L_\infty). \tag{47}$$

Adding (38) to (47) and rearranging gives

$$\dot{\theta}=\frac{1}{1+D}\{[(L_1)+(L_2)]\dot{\theta}+(L_1)+(L_2)+(L_\infty). \tag{48}$$

Equations (48) and (38) in Lemma 3 now give that $\dot{\theta}$ is bounded. Then arguing as before (i.e. via the torque (i.e. via the torque expression (23) and then the robot dynamic equation (1)), τ_m is bounded and $\ddot{\theta}$ is bounded.

(ii) From (14) it is clear that

$$\dot{e}=\dot{\Psi}^{\mathrm{T}}\tilde{M}+\Psi^{\mathrm{T}}\dot{\hat{M}}. \tag{49}$$

In view of (i), Ψ and $\dot{\Psi}$ are bounded. Then using (13), e is bounded and so $\dot{\tilde{M}}=\dot{\hat{M}}$ is bounded using (15). Thus it is clear that e is uniformly continuous. Since e belongs to L_2, we then have $e\to 0$. From the definition of f and (35), it is clear that f, ε and $\dot{\varepsilon}\to 0$. □

5. Conclusion

In this paper, we have considered the control of rigid link mechanical manipulators. We have then demonstrated global convergence for an adaptive computed torque control law which requires measurements of state variables only. Further work on robustness, computational aspects and performance considerations warrants further research.

Appendix

Proof of Lemma 2. Let $h(t)$ denote the impulse response of H. Then,

$$g(t)=\int_0^t h(t-\tau)[f_1(\tau)+f_2(\tau)+f_\infty(\tau)]\,\mathrm{d}\tau, \tag{A.1}$$

$$\|g(t)\|\leqslant\int_0^t\|h(t-\tau)\|[\|f_1(\tau)\|+\|f_2(\tau)\|+\|f_\infty(\tau)\|]\,\mathrm{d}\tau. \tag{A.2}$$

In view of the assumptions regarding H, ∞ c, $\sigma>0$ such that

$$\|h(t-\tau)\|\leqslant c\,\mathrm{e}^{-\sigma(t-\tau)}. \tag{A.3}$$

Using (A.3) and Schwarz's inequality in (A.2) gives

$$\|g(t)\|\leqslant\left[\int_0^t c^2\,\mathrm{e}^{-2\sigma(t-\tau)}\,\mathrm{d}\tau\right]^{1/2}\left[\int_0^t\|f_2(\tau)\|^2\,\mathrm{d}\tau\right]^{1/2}$$
$$+c\int_0^t\|f_1(\tau)\|\,\mathrm{d}\tau+\left[\sup_{0\leqslant\tau\leqslant t}\{\|f_\infty(\tau)\|\}\right]\int_0^t c\,\mathrm{e}^{-\sigma(t-\tau)}\,\mathrm{d}\tau. \tag{A.4}$$

From (A.4), and using the assumptions concerning f_1, f_2 and f_∞ it is clear that g is bounded. □

Proof of Lemma 3. From Lemma 2, $H*(g_1+g_2+g)$ is bounded. Therefore we need only consider $H*(V(t)[f_1+f_2])$. If we let $h(t)$ denote the impulse response of H, then the response due to $V(t)[f_1+f_2]$ is

$$V(t)=\int_0^t h(t-\tau)V(\tau)(f_1(\tau)+f_2(\tau))\,\mathrm{d}\tau. \tag{A.5}$$

Because of the assumptions on H, it follows that

$$\|V(t)\|\leqslant\int_0^t m\,\mathrm{e}^{-\lambda(t-\tau)}\|V(\tau)\|[|f_1(\tau)|+|f_2(\tau)|]\,\mathrm{d}\tau. \tag{A.6}$$

Using Schwarz's inequality we have

$$\|V(t)\|\leqslant\int_0^t m\|V(\tau)\||f_1(\tau)|\,\mathrm{d}\tau$$
$$+m\left[\int_0^t\mathrm{e}^{-2\lambda(t-\tau)}\|V(\tau)\|\,\mathrm{d}\tau\right]^{1/2}\left[\int_0^t\|V(\tau)\|f_2^2(\tau)\,\mathrm{d}\tau\right]^{1/2}. \tag{A.7}$$

From Lemma 2 it follows that there exists a constant c such that

$$\| V(t) \| \leq \int_0^t m \| V(\tau) \| \| f_1(\tau) \| \, \mathrm{d}\tau + c \left[\int_0^t \| V(\tau) \| \| f_2(\tau) \|^2 \, \mathrm{d}\tau \right]^{1/2} \tag{A.8}$$

$$\leq 1 + \int_0^t \| V(\tau) \| \| \bar{f}_1(\tau) \| \, \mathrm{d}\tau \tag{A.9}$$

where $\bar{f}_1(\tau) = m \| f_1(\tau) \| + c \| f_2(\tau) \|^2$ and thus $\bar{f}_1$ belongs to L_1. It can now be shown using Gronwall's lemma, that $V(t)$ is bounded. □

References

[1] G. Bastin, G. Campion and A.-M. Guillaume, Adaptive external linearization feedback control for manipulators, Technical Report, Université Catholique de Louvain, Louvain-la-Neuve, Belgium (1986).

[2] H. An Chae, C.G. Atkeson and J.M. Hollerback, Estimation of inertial parameters of rigid body links of manipulators, *24th CDC* (1985).

[3] J.J. Craig, P. Hsu and S.s. Sastry, Adaptive control of mechanical manipulators, Tech. Report, *Dept. of Elec. Eng. and Comp. Science,* University of California, Berkeley, CA (1985).

[4] S. Dubowsky and D.T. DesForges, The application of model reference adaptive control to robotic manipulators, *J. Dynamic Systems Measurement and Control* **101** (1979).

[5] H. Elliot, T. Depkovich, J. Kelly and B. Draper, Non-linear adaptive control of mechanical linkage systems with application to robotics, *Proc. of the American Control Conference*, San Diego, CA (1984).

[6] G.C. Goodwin and D.Q. Mayne, A parameter estimation perspective of continuous time adaptive control, *Automatica* **23** (1987) 57–70.

[7] P.K. Khasla and T. Kanade, Parameter identification of robot arms, *24th CDC* (1985).

[8] T.W.B. Kibble, *Classical Mechanics* (McGraw Hill, London, 1966).

[9] A.J. Koivo and T.H. Guo, Adaptive linear controller for robotic manipulators, *IEEE Trans. Automat. Control* **28** (2) 162–170.

[10] H.N. Koivo and J. Sorvari, On line tuning of a multivariable PID Controller for Robot Manipulators, *Proc. 24th CDC* (1985) 1502–1504.

[11] B. Markewicz, Analysis of a computed torque drive method and comparison with conventional position servo for a computer controlled manipulator, Tech. Memorandum 33-601, Jet Propulsion Lab., Pasadena, CA (1973).

[12] S. Nicosia and P. Tomei, Model reference adaptive control algorithms for industrial robots, *Automatica* **20** (5) (1984) 635–644.

[13] R.P. Paul, *Robot Manipulators: Mathematics, Programming and Control* (MIT Press, Cambridge, MA, 1981).

[14] M. Sahba and D.Q. Mayne, Computer aided design of non-linear controllers for torque controlled robot arms, *Proc. IEE* **131** (1) (1984) 8–14.

[15] S. Sastry and J.J. Slotine, Tracking control of non-linear systems using sliding surfaces and applications to robotics, *Internat. J. Comput.* (1983) 465–492.

[16] S.H. Singh, Adaptive model following control of nonlinear robotic systems, *IEEE Trans Automat. Control* **30** (11) 1099–1100.

[17] M.W. Spong and M. Vidyasagar, Robust linear compensator design for nonlinear robotic control, *Proc. IEEE Internat. Conf. Robotic Automation*, St. Louis, MO (March 1985).

[18] M. Tomizuka and R. Horowitz, Model reference adaptive control of mechanical manipulators, *IFAC Adaptive Systems in Control and Signal Processing*, San Francisco, CA (1983).

Stability Analysis of an Adaptive Controller for Robotic Manipulators

Nader Sadegh
Graduate Student

Roberto Horowitz
Assistant Professor

Department of Mechanical Engineering,
University of California, Berkeley, CA 94720

ABSTRACT

The stability analysis of an adaptive control scheme for robotic manipulators, originally introduced by Horowitz and Tomizuka (1980), is presented in this paper. In the previous stability proof it was assumed that the manipulator parameter variation is negligible compared with the speed of adaptation. It is shown that this key assumption can be removed by introducing two modifications in the adaptive control scheme: 1. Reparametrizing the nonlinear terms in dynamic equations as linear functions of unknown but constant terms. 2. Defining the Coriolis compensation term in the control law as a bilinear function of the manipulator and model reference joint velocities, instead of a quadratic function of the manipulator joint velocities. The modified adaptive control scheme is shown to be globally asymptotically stable.

1. Introduction

Robotic manipulators are mechanical systems with nonlinear dynamic characteristics. Furthermore, their inertia characteristics and disturbance characteristics, such as the one coming from gravity, vary during operation and they are not necessarily predictable in advance. Adaptive control techniques have been suggested as an effective method for decoupling and linearizing the manipulator dynamic characteristics.

The earliest work on model reference adaptive control for manipulators (Dubowsky and DesForges, 1979) is based on a linear decoupled model. The steepest descent technique is utilized for parameter adaptation. The first work on adaptive controls of mechanical manipulators based on stability theories was reported by Horowitz and Tomizuka (1980). They utilized the hyperstability theory for adaptively linearizing and decoupling the nonlinear manipulator dynamics. Preliminary experimental evaluations of this approach have been reported by Anex and Hubbard (1984) and Tomizuka et al. (1985b), (1986). A Lyapunov stability based adaptive control approach for trajectory following has been proposed by Takegaki and Arimoto (1981). In these works the unknown parameter in the manipulator dynamic equations, which are position dependent quantities, are treated as constants in the stability analysis, in order to show the asymptotic stability of the control scheme. Therefore, the underlying assumption is that the parameter adaptation law is much faster than the manipulator dynamics.

Recent works in this field includes those by Balestino et al. (1983), Nicosia and Tomei (1984) and Craig et al. (1986). One trend in these recent works is to assure the stability of the overall system in spite of the nonlinear nature of the parameters in the manipulator dynamic equations. This is achieved by either of the following two methods: 1) Decomposing the nonlinear parameters in the manipulator dynamic equations into the product of two quantities: one constant unknown quantity, which includes the numerical values of the masses and moments of inertia of the links and the payload, and link dimensions, the other a known nonlinear function of the manipulator structural dynamics. The nonlinear functions are then assumed to be known and calculable. The parameter adaptation law is only used to estimate the unknown constant quantities (e.g. Craig et al. (1986)). 2) Utilizing nonlinear switching parameter adaptation laws, making use of the knowledge of upper bounds of the parameters in the manipulator dynamic equations. This parameter adaptation law belongs to the class of variable structure control schemes. (e.g. Balestino et al. (1983)). These approaches are all based on the continuous time control theory.

In this paper it is demonstrated that, by modifying the parameter adaptation law using method 1) outlined above, the adaptive control scheme introduced by Horowitz and Tomizuka (1980) is globally asymptotically stable. Moreover in this control scheme only joint position and velocity feedback information is required (no joint acceleration is used, and no matrix inversion is used as in Craig et al. (1986)).

Adaptive control approaches have also been proposed based on the discrete time control theory. Koivo and Guo (1983) proposes a self tuning regulator approach based on a functional linear difference equation model. Dubowsky (1981) has investigated the discrete time implementation of the model reference adaptive control approach (Dubowsky and DesForges, 1979). Horowitz and Tomizuka (1982) proposed a discrete time adaptive control scheme for mechanical manipulators based on the adaptive control approach by Landau and Lozano (1981). The Landau and Lozano method is also the basis for a more recent decentralized adaptive control scheme (Sundareshan and Koenig, 1985). *

This paper is organized as follows: In Section 1., the manipulator's dynamic equations of motion are presented. Section 2. contains the non-adaptive nonlinearity compensation decoupling control law. In sec-

*The authors have recently learned of a similar stability analysis by Slotine(1986), after the submission of this paper.

Reprinted from *Proc. IEEE ROBOT '87*, Raleigh, NC, pp. 1223–1229, 1987.

tion 4., the adaptive control law is presented and a global asymptotic stability theorem is proven. Section 5. contains the conclusions and the references are in section 6.

2. Dynamic Model of a Robotic Manipulator

In this paper we consider robotic manipulator composed of a serial open chain of rigid links connected with either revolute or prismatic joints. The dynamic equations of motion for the manipulator can be expressed in the following form:

$$\frac{d}{dt}\mathbf{x}_p(t) = \mathbf{x}_v(t) \tag{1}$$

$$\mathbf{M}(\mathbf{x}_p)\frac{d}{dt}\mathbf{x}_v(t) = \mathbf{q}(t) - \mathbf{v}(\mathbf{x}_p,\mathbf{x}_v) - \mathbf{g}(\mathbf{x}_p) - \mathbf{c}(\mathbf{x}_p,\mathbf{x}_v)$$

where

- $\mathbf{x}_p$ is the $n\times 1$ vector of joint positions,
- $\mathbf{x}_v$ is the $n\times 1$ vector of joint velocities,
- $\mathbf{M}(\mathbf{x}_p)$ is the $n\times n$ symmetric and positive definite matrix. Also called the generalized inertia matrix,
- $\mathbf{q}(t)$ is the $n\times 1$ vector of joint torques or forces supplied by the actuators,
- $\mathbf{v}(\mathbf{x}_p,\mathbf{x}_v)$ is the $n\times 1$ vector due to Coriolis and centripetal accelerations,
- $\mathbf{g}(\mathbf{x}_p)$ is the $n\times 1$ vector due to gravitational forces and
- $\mathbf{c}(\mathbf{x}_p,\mathbf{x}_v)$ is the $n\times 1$ vector due to friction forces.

$\mathbf{v}(\mathbf{x}_p,\mathbf{x}_v)$ can be expressed in the following form

$$\mathbf{v}(\mathbf{x}_p,\mathbf{x}_v) = \begin{bmatrix} \mathbf{x}_v{}^T\mathbf{N}^1(\mathbf{x}_p)\mathbf{x}_v \\ \mathbf{x}_v{}^T\mathbf{N}^2(\mathbf{x}_p)\mathbf{x}_v \\ \cdot \\ \cdot \\ \cdot \\ \mathbf{x}_v{}^T\mathbf{N}^n(\mathbf{x}_p)\mathbf{x}_v \end{bmatrix} \tag{2}$$

where the matrices $\mathbf{N}^i$'s are symmetric.

The following relation is satisfied between the matrices $\mathbf{N}^i$'s and the generalized inertia matrix $\mathbf{M}$.

$$\mathbf{N}^i(\mathbf{x}_p) = \left[\frac{\partial \mathbf{m}_i}{\partial \mathbf{x}_p} - \frac{1}{2}\frac{\partial \mathbf{M}}{\partial x_{pi}}\right] \tag{3}$$

where $\mathbf{m}_i$ is the *ith* row (or column) of $\mathbf{M}$ and x_{pi} is the *ith* element of $\mathbf{x}_p$. The proof is found in Appendix 1.

The *ith* element of the friction force vector $\mathbf{c}(\mathbf{x}_p,\mathbf{x}_v)$ can be expressed as

$$c_i(x_{vi},q_i) = c_{ci}(x_{vi},q_i) + c_{li}\, x_{vi}(t) \tag{4}$$

where $c_{ci}(x_{vi},q_i)$ represents the Coulomb friction component and $c_{li}\, x_{vi}(t)$ represents the linear friction component. x_{vi} and q_i are the *ith* components of $\mathbf{x}_v$ and $\mathbf{q}$ respectively.

The Coulomb friction term has a significant effect on performance of indirect robot arms. It is described by

$$c_{ci}(x_{vi},q_i) = \begin{cases} c_{cmi}\, sign[x_{vi}(t)] & \text{if } |x_{vi}(t)| > 0 \\ c_{cmi}\, sign[q_i(t)] & \text{if } |x_{vi}(t)| = 0 \text{ and } |q_i(t)| > c_{cmi} \\ q_i(t) & \text{if } |x_{vi}(t)| = 0 \text{ and } |q_i(t)| \le c_{cmi} \end{cases} \tag{5}$$

where c_{cmi} is the magnitude of the friction force.

The values of friction force magnitudes c_{cmi} and c_{li} in general do not vary substantially with manipulator motion or payload variation. Thus, they can be considered as constant and can be estimated, if desired, by off-line experiments, as shown by Tomizuka et al. (1985a), Kubo et al. (1986) and Anwar et al. (1986).

The elements of the matrices $\mathbf{M}(\mathbf{x}_p)$, $\mathbf{N}(\mathbf{x}_p)$'s and of the vector $\mathbf{g}(\mathbf{x}_p)$ and are in general highly nonlinear functions of the the position vector $\mathbf{x}_p$. They are also a function of the link and payload masses and moments of inertia, which may not be precisely known or may vary during the manipulator task. The variation of these parameters is especially significant in the dynamic response of direct drive robotic arms.

2.1. Dynamic Equations Reparametrization

As discussed in Craig et al. (1986) the nonlinear parameters in $\mathbf{M}(\mathbf{x}_p)$, $\mathbf{N}(\mathbf{x}_p)$'s and $\mathbf{g}(\mathbf{x}_p)$ can be decomposed into products of constant terms, which are functions of the inertia characteristics of the links and the payload, and known nonlinear functions of the joint positions.

One method of reparametrizing the manipulator's dynamic equations consists in decomposing each element of the matrices $\mathbf{M}(\mathbf{x}_p)$, $\mathbf{N}(\mathbf{x}_p)$'s and the vector $\mathbf{g}(\mathbf{x}_p)$ into products of unknown constant terms and known functions of the joint displacement vector $\mathbf{x}_p$ i.e.

$$m_{ij}(\mathbf{x}_p) = \sum_{r=1}^{r=rm_{ij}} m_{rij}\, fm_{rij}(\mathbf{x}_p) \tag{6}$$

where m_{ij} is the *ijth* element of $\mathbf{M}(\mathbf{x}_p)$, m_{rij}'s are constant parameters, which are assumed unknown, and $fm_{rij}(\mathbf{x}_p)$'s are known functions of $\mathbf{x}_p$.
Similarly,

$$n^k{}_{ij}(\mathbf{x}_p) = \sum_{r=1}^{r=rn^k{}_{ij}} n^k{}_{rij}\, fn^k{}_{rij}(\mathbf{x}_p) \tag{7}$$

$$g_i(\mathbf{x}_p) = \sum_{r=1}^{r=rg_i} g_{ri}\, fgr_{ri}(\mathbf{x}_p) \tag{8}$$

where $n^k{}_{ij}$ is the *ijth* element of $\mathbf{N}^k$ and g_i is the *ith* element of $\mathbf{g}$, $n^k{}_{rij}$'s and g_{ri}'s are constant unknown quantities and $fn^k{}_{rij}(\mathbf{x}_p)$'s and $fgr_{ri}(\mathbf{x}_p)$'s are known functions.

A second method, which is the one proposed in Craig et. al. (1986), consists in the reparametrization of Eq. (1) into the product of unknown constant vector, which is function of the unknown masses and moments of inertia of the links and the payload, and a matrix formed by known functions of $\mathbf{x}_p$, $\mathbf{x}_v$ and $\dot{\mathbf{x}}_v$, where $\dot{\mathbf{x}}_v = \frac{d}{dt}\mathbf{x}_v$ are the joint accelerations.

$$\mathbf{M}(\mathbf{x}_p)\dot{\mathbf{x}}_v + \mathbf{v}(\mathbf{x}_p,\mathbf{x}_v) + \mathbf{g}(\mathbf{x}_p) = \mathbf{W}(\mathbf{x}_p,\mathbf{x}_v,\dot{\mathbf{x}}_v)\,\boldsymbol{\theta} \tag{9}$$

where $\mathbf{W}(\mathbf{x}_p,\mathbf{x}_v,\dot{\mathbf{x}}_v)$ is an $n\times r$ matrix and $\boldsymbol{\theta}$ is an $r\times 1$ vector of the unknown constant parameters. Notice that $\mathbf{W}(\mathbf{x}_p,\mathbf{x}_v,\dot{\mathbf{x}}_v)$ is a linear function of $\dot{\mathbf{x}}_v$.

By using this method, a substantial reduction in the number of parameters that need to be estimated is achieved.

3. Nonlinearity Compensation and Decoupling Control

In order to dynamically decouple and linearize the manipulator equations of motion (1), the following controller is proposed (Horowitz and Tomizuka (1980):

$$q(t) = \hat{M}(\cdot)u(t) + \hat{v}(\cdot) + \hat{g}(\cdot) + \hat{c}(\cdot) + F_p[\hat{x}_p(t) - x_p(t)] + F_v[\hat{x}_v(t) - x_v(t)] \quad (10)$$

where $u(t)$, is the acceleration input, and $\hat{x}_p(t)$ and $\hat{x}_v(t)$ are the reference velocities and position vectors:

$$\frac{d}{dt}\hat{x}_p(t) = \hat{x}_v(t) \quad (11)$$

$$\frac{d}{dt}\hat{x}_v(t) = u(t). \quad (12)$$

F_p and F_v are constant diagonal positive definite gain matrices and the functions $\hat{M}$, $\hat{v}$, $\hat{g}$ and $\hat{c}$ are the estimates of $M(x_p)$, $v(x_p,x_v)$, $g(x_p)$ and $c(x_p,x_v)$ respectively.

Note that under the assumption that

$$\hat{M}(\cdot) = M(x_p)\,,\quad \hat{v}(\cdot) = v(x_p,x_v)\,,$$

$$\hat{g}(\cdot) = g(x_p) \text{ and } \hat{c}(\cdot) = c(x_p,x_v)\,,$$

the control law in Eq. (10) achieves

$$\frac{d}{dt}e_p(t) = e_v \text{ and } \frac{d}{dt}e_v(t) = -F_p e_p(t) - F_v e_v(t) \quad (13)$$

where

$$e_p(t) = \hat{x}_p(t) - x_p(t) \text{ and} \quad (14)$$

$$e_v(t) = \hat{x}_v(t) - x_v(t) \quad (15)$$

Note that the control law in Eq. (10) forces the manipulator to follow the reference model in Eqs. (11) and (12) which is an nth order system of decoupled double integrators. Eq. (13) assures that $\lim_{t\to\infty} e_p(t) = 0$ $\lim_{t\to\infty} e_v(t) = 0$ for any arbitrary initial condition since F_v and F_p are positive definite matrices.

In order to complete the controller design, an outer loop consisting of a proportional, integral and derivative (PID) action is used for trajectory tracking purposes:

$$u(t) = \ddot{x}_d(t)+u_I(t)+K_p[x_d(t)-x_p(t)]+K_d[\dot{x}_d(t)-x_v(t)] \quad (16)$$

$$\frac{d}{dt}u_I(t) = K_i[x_d(t) - x_p(t)]$$

where $x_d(t)$, $\dot{x}_d(t)$ and $\ddot{x}_d(t)$ are respectively the position, velocity and acceleration of the desired trajectory, and the matrices K_p, K_d and K_i are the PID gains which are selected such that the characteristic equation:

$$s^3 I + K_p s^2 + K_d s + K_i = 0 \quad (17)$$

has all its root in the left hand side of the complex plane.

Eq. (13) is satisfied under the assumption of perfect knowledge of the manipulator parameters. In the next section we will discuss parameter adaptation laws for estimating $\hat{M}$, $\hat{v}$, and $\hat{g}$.

The friction force estimate $\hat{c}(\cdot)$ can be calculated by Eqs. (4) and (5), replacing the magnitudes c_{li}'s and c_{cmi}'s by their estimates. Since these quantities can be successfully estimated off-line, we will assumed that adequate friction compensation is implemented and neglect the effect of the friction forces altogether. Successful implementation of friction compensation in industrial arms has been shown by Kubo et al. (1986). If required, the estimation of c_{li}'s and c_{cmi}'s can be performed on-line using the adaptive techniques that will be presented in the next section.

4. Adaptive Controller

In this section we discuss an adaptive control scheme for updating the control parameters $\hat{M}(\cdot)$, $\hat{v}(\cdot)$ and $\hat{g}(\cdot)$ in the control law given by Eq. (10). The scheme presented here is a modification of the parameter adaptation law originally introduced by Horowitz and Tomizuka (1980). Thus, we first introduce the original scheme which we call the "Constant Plant Parameter Adaptive Control (CPPAC).

4.1. Constant Plant Parameter Adaptive Control (CPPAC)

$$q(t) = \hat{M}(t)u(t) + \hat{v}(x_v,t) + \hat{g}(t) + F_p e_p(t) + F_v e_v(t) + c(x_p,x_v) \quad (18)$$

where $\hat{v}$ is defined by

$$\hat{v}(x_v,t) = \begin{bmatrix} x_v^T\hat{N}^1(t)x_v \\ x_v^T\hat{N}^2(t)x_v \\ \cdot \\ \cdot \\ x_v^T\hat{N}^n(t)x_v \end{bmatrix} \quad (19)$$

and the elements of the parameters $\hat{M}(t)$, $\hat{N}^k(t)$'s and $\hat{g}(t)$ are updated as follows

$$\frac{d}{dt}\hat{m}_{ij} = k_{m_{ij}}[y_i u_j] \quad (20)$$

$$\frac{d}{dt}\hat{n}^k{}_{ij} = k^k{}_{nij}[y_k x_{vi} x_{vj}] \quad (21)$$

$$\frac{d}{dt}\hat{g}_i = k_{gi} y_i \quad (22)$$

where $\hat{m}_{ij}$ is the $ijth$ element of $\hat{M}$, $\hat{n}^k{}_{ij}$ is the $ij\theta$ element of $\hat{N}^k$, g_i is the ith element of $\hat{g}$ and the vector $y(t)$ is defined as

$$y(t) = [y_1 \cdots y_n]^T = C_v x_v(t) + C_p x_p(t) \quad (23)$$

where $k_{m_{ij}} > 0$, $k^k{}_{nij} > 0$, $k_{gi} > 0$, (24)

$$F_p = \rho_p I\,,\ F_v = \rho_v I\,,\ C_p = \sigma_p I\,,\ C_v = \sigma_v I$$

I is the identity matrix and ρ_p, ρ_v, σ_p, σ_v satisfy

$$\sigma_v > \sigma_p\,,\ \sigma_v \rho_v I - \sigma_p M_{max} > 0 \quad (25)$$

$$\text{and } [\sigma_v \rho_p + \sigma_p \rho_v] - \sigma_p M_{max} > 0$$

where $M_{max} = \max_{x_p} [M(x_p)]$.

Note that conditions (25) are satisfied when

$$y(t) = x_v(t) \quad (22a)$$

As shown in Horowitz and Tomizuka (1980), the control law in Eqs. (16) and (18), and the parameter adaptation laws in Eqs. (19) - (25) guarantee the asymptotic tracking objective:

$$\lim_{t\to\infty} [\hat{x}_p(t)-x_p(t)]=0 \text{ and } \lim_{t\to\infty} [\hat{x}_v(t)-x_v(t)]=0$$

under the assumption that the parameters M, N^k's and g remain constant during the adaptation.

Notice that in this adaptive control scheme no joint acceleration feedback or matrix inversion is required as in Craig et al. (1986).

* Although the gravitational term was not included in the paper. The inclusion of this term in the adaptation law is a trivial extension.

4.2. Modified Adaptive Control Scheme

In order to remove the assumption that the parameters $\mathbf{M}$, $\mathbf{N}^k$'s and $\mathbf{g}$ remain constant during adaptation, the following modifications to the control scheme in Eqs. (19) - (25) should be made:

4.2.1. Modification in the Control law:

$$g(t) = \hat{\mathbf{M}}(\mathbf{x}_p,t)\mathbf{u}(t) + \hat{\mathbf{v}}(\mathbf{x}_v,\hat{\mathbf{x}}_v,\mathbf{x}_p,t) + \hat{\mathbf{g}}(\mathbf{x}_p,t) + \mathbf{F}_v\mathbf{e}_v(t) + \mathbf{c}(\mathbf{x}_p,\mathbf{x}_v) \tag{26}$$

where $\hat{\mathbf{v}}$ is now defined by

$$\hat{\mathbf{v}}(\mathbf{x}_v,\hat{\mathbf{x}}_v,\mathbf{x}_p,t) = \begin{bmatrix} \mathbf{x}_v{}^T\hat{\mathbf{N}}^1(\mathbf{x}_p,t)\hat{\mathbf{x}}_v \\ \mathbf{x}_v{}^T\hat{\mathbf{N}}^2(\mathbf{x}_p,t)\hat{\mathbf{x}}_v \\ \cdot \\ \cdot \\ \cdot \\ \mathbf{x}_v{}^T\hat{\mathbf{N}}^n(\mathbf{x}_p,t)\hat{\mathbf{x}}_v \end{bmatrix} \tag{27}$$

Note that there are only two differences between Eqs. (18) and (26): 1) In Eq. (18) $\hat{\mathbf{v}}$ is a quadratic function of the joint velocity vector $\mathbf{x}_v(t)$, while in Eq. (27) $\hat{\mathbf{v}}$ is a bilinear function of the velocity vector $\mathbf{x}_v(t)$ and the reference model velocity vector $\hat{\mathbf{x}}_v(t)$. 2) In Eq. (26), only the error between the manipulator velocity vector and the reference model velocity vector, $\mathbf{e}_v(t)$ is used. The position error, $\mathbf{e}_p(t)$, between the manipulator and the reference model is not used at all. Thus, Eq. (26) can be viewed as a minor loop velocity feedback compensation control. The use of an adaptive minor loop velocity feedback control and a non-adaptive outer loop in the control system structure has been extensively utilized in experimental studies. (Anwar et. al. (1986) and Tomizuka et. al. (1986)).

4.2.2. Modification in the Parameter Adaptation Law

It is apparent that, in order to guarantee asymptotic convergence of the adaptive control scheme, the parameter adaptation law should track a constant parameter instead of a function of time or the state $\mathbf{x}_p$. The choice of dynamic equation reparametrization determines the choice of parameter adaptation law.

The reparametrization of $\mathbf{M}(\mathbf{x}_p)$, $\mathbf{N}^k$'s and $\mathbf{g}$ in Eqs. (6) - (8) leads to the following parameter adaptation algorithm (PAA):

Define the parameter estimates $\hat{\mathbf{M}}$, $\hat{\mathbf{N}}^k$'s and $\hat{\mathbf{g}}$ by

$$\hat{m}_{ij}(\mathbf{x}_p,t) = \sum_{r=1}^{r=rm_{ij}} \hat{m}_{rij}(t)\, fm_{rij}(\mathbf{x}_p) \tag{28}$$

$$\hat{n}^k{}_{ij}(\mathbf{x}_p,t) = \sum_{r=1}^{r=rn^k{}_{ij}} \hat{n}^k{}_{rij}(t)\, fn^k{}_{rij}(\mathbf{x}_p) \tag{29}$$

$$\hat{g}_i(\mathbf{x}_p,t) = \sum_{r=1}^{r=rg_i} \hat{g}_{ri}(t)\, fgr_{ri}(\mathbf{x}_p) \tag{30}$$

where $fm_{rij}(\mathbf{x}_p)$'s, $fn^k{}_{rij}(\mathbf{x}_p)$'s and $fgr_{ri}(\mathbf{x}_p)$'s are known functions, and the parameters $\hat{m}_{rij}$'s(t), $\hat{n}^k{}_{rij}$'s(t) and $\hat{g}_{ri}$'s(t) are updated by

$$\frac{d}{dt}\hat{m}_{rij} = km_{rij}\, fm_{rij}(\mathbf{x}_p)\,[e_{vi}u_j]\,,\ r = 1, \cdots rm_{ij} \tag{31}$$

$$\frac{d}{dt}\hat{n}^k{}_{rij} = kn^k{}_{rij}\, fn^k{}_{rij}(\mathbf{x}_p)\,[e_{vk}x_{vi}\hat{x}_{vj}]\,,\ r=1, \cdots rn^k{}_{ij} \tag{32}$$

$$\frac{d}{dt}\hat{g}_{ri} = kg_{ri}\, fgr_{ri}(\mathbf{x}_p)\,[e_{vi}]\,,\ r = 1, \cdots rg_i \tag{33}$$

where $km_{rij} > 0$, $kn^k{}_{rij} > 0$, $kg_{ri} > 0$,

Where e_{vk} is the kth element of the velocity error vector, $\mathbf{e}_v$, defined in Eq. (15).

The time varying parameters $\hat{m}_{rij}$'s, $\hat{n}^k{}_{rij}$'s and $\hat{g}_{ri}$'s are the estimates of the constant parameters m_{rij}'s, $n^k{}_{rij}$'s and g_{ri}'s respectively.

If the dynamic equations are reparametrized in the form of Eq. (9), the following PAA should be employed:

From Eq. (26)

$$\hat{\mathbf{M}}(\mathbf{x}_p,t)\mathbf{u}(t)+\hat{\mathbf{v}}(\mathbf{x}_v,\hat{\mathbf{x}}_v,\mathbf{x}_p,t)+\hat{\mathbf{g}}(\mathbf{x}_p,t)=\mathbf{W}(\mathbf{x}_p,\mathbf{x}_v,\hat{\mathbf{x}}_v,\mathbf{u})\hat{\boldsymbol{\theta}}(t) \tag{34}$$

$$\frac{d}{dt}\hat{\boldsymbol{\theta}}(t) = \mathbf{K}\,\mathbf{W}^T(\mathbf{x}_p,\mathbf{x}_v,\hat{\mathbf{x}}_v,\mathbf{u})\,\mathbf{e}_v(t)\ ,\quad \mathbf{K} > 0 \tag{35}$$

where the $r\times1$ vector $\hat{\boldsymbol{\theta}}(t)$ is the estimate of the unknown vector $\boldsymbol{\theta}$ in Eq. (9), and the matrix $\mathbf{W}(\mathbf{x}_p,\mathbf{x}_v,\hat{\mathbf{x}}_v,\mathbf{u})$ is similar to $\mathbf{W}(\mathbf{x}_p,\mathbf{x}_v,\dot{\mathbf{x}}_v)$ in Eq. (9), only that $\dot{\mathbf{x}}_v$ is replaced by $\mathbf{u}(t)$ and $x_{vi}x_{vj}$ is replaced by $x_{vi}\hat{x}_{vj}$, $i,j = 1 \cdots n$.

Notice that in the modified parameter adaptation law only the velocity error signal $\mathbf{e}_v(t)$ is used in Eqs. (31) - (33) and (35). (i.e. $\sigma_p = 0$ in Eqs. (21) and (22)). Also, comparing with the method presented by Craig et. al. (1986), the acceleration input $\mathbf{u}(t)$ is used in the PAA's instead of the joint accelerations $\dot{\mathbf{x}}_v$ which are not measurable in most realistic applications, and no matrix inversion is required in the control algorithm.

Theorem 1

For a mechanical manipulator governed by Eq. (1), given a bounded desired trajectory $\mathbf{x}_d(t)$ with bounded first and second derivatives $\dot{\mathbf{x}}_d(t)$ and $\ddot{\mathbf{x}}_d(t)$, if the adaptive control law given by Eqs. (12),(15),(16),(26) - (33) or by Eqs. (12),(15),(16),(26), (27), (34) and (35) is used, the error between the desired and actual trajectory converges asymptotically to the zero, i.e.

$$\lim_{t\to\infty}[\mathbf{x}_d(t)-\mathbf{x}_p(t)]=0 \text{ and } \lim_{t\to\infty}[\dot{\mathbf{x}}_d(t)-\mathbf{x}_v(t)]=0$$

Proof

In this section we will present the stability proof of the adaptive control system when the adaptive control law given by Eqs. (26),(27),(34) and (35) is utilized. The proof for the case when the adaptive control law given by Eqs. (26)-(33) is employed is almost identical and will be omitted.

The proof of Theorem 1 will be carried out in two stages. In the first stage it will be shown that, regardless of the acceleration input function $\mathbf{u}(t)$, the velocity error signal $\mathbf{e}_v(t)$ between the reference model $\hat{\mathbf{x}}_v(t)$ and the manipulator velocity $\mathbf{x}_v(t)$ remains bounded. In the second stage of the proof it is shown that, given a bounded desired trajectory signal $\mathbf{x}_d(t)$, all signals in the closed loop system remain bounded, and, as a consequence, the tracking error between the desired trajectory $\mathbf{x}_d(t)$ and the manipulator trajectory $\mathbf{x}_p(t)$ converges asymptotically to zero.

We begin by obtaining an expression for the velocity error between the reference model and the manipulator, $\dot{\mathbf{e}}_v(t)$, as a function of the parameter error. Multiplying Eq. (12) by $\mathbf{M}(\mathbf{x}_p)$, subtracting Eq. (1) and utilizing the control law (26) we obtain

$$\mathbf{M}(\mathbf{x}_p)\frac{d}{dt}\mathbf{e}_v(t) = -\mathbf{F}_v\mathbf{e}_v(t) + [\,\mathbf{M}(\mathbf{x}_p) - \hat{\mathbf{M}}(\mathbf{x}_p,t)\,]\,\mathbf{u}(t) + \mathbf{v}(\mathbf{x}_p,\mathbf{x}_v) - \hat{\mathbf{v}}(\mathbf{x}_p,\mathbf{x}_v,\hat{\mathbf{x}}_v,t) + \mathbf{g}(\mathbf{x}_p) - \hat{\mathbf{g}}(\mathbf{x}_p,t) \tag{36}$$

where we have used the definition of $\mathbf{e}_v$ given in Eq. (15).

By Eqs. (2) and (27), Eq. (36) can be rearranged as follows

$$\mathbf{M}(\mathbf{x}_p)\,\frac{d}{dt}\mathbf{e}_v(t) = -\mathbf{F}_v\mathbf{e}_v(t) + [\,\mathbf{M}(\mathbf{x}_p) - \hat{\mathbf{M}}(\mathbf{x}_p,t)\,]\,\mathbf{u}(t)$$

$$+\begin{bmatrix} \mathbf{x}_v^T\left\{\mathrm{N}^1(\mathbf{x}_p) - \hat{\mathrm{N}}^1(\mathbf{x}_p,t)\right\}\hat{\mathbf{x}}_v \\ \mathbf{x}_v^T\left\{\mathrm{N}^2(\mathbf{x}_p) - \hat{\mathrm{N}}^2(\mathbf{x}_p,t)\right\}\hat{\mathbf{x}}_v \\ \vdots \\ \mathbf{x}_v^T\left\{\mathrm{N}^n(\mathbf{x}_p) - \hat{\mathrm{N}}^n(\mathbf{x}_p,t)\right\}\hat{\mathbf{x}}_v \end{bmatrix} - \begin{bmatrix} \mathbf{x}_v^T\mathrm{N}^1(\mathbf{x}_p)\mathbf{e}_v \\ \mathbf{x}_v^T\mathrm{N}^2(\mathbf{x}_p)\mathbf{e}_v \\ \vdots \\ \mathbf{x}_v^T\mathrm{N}^n(\mathbf{x}_p)\mathbf{e}_v \end{bmatrix} + \hat{\mathbf{g}}(\mathbf{x}_p,t) - \mathbf{g}(\mathbf{x}_p)$$

$$\mathbf{M}(\mathbf{x}_p)\,\frac{d}{dt}\mathbf{e}_v(t) = -\mathbf{F}_v\mathbf{e}_v(t) - \mathbf{v}(\mathbf{x}_p,\mathbf{x}_v,\mathbf{e}_v) - \mathbf{W}(\mathbf{x}_p,\mathbf{x}_v,\hat{\mathbf{x}}_v,\mathbf{u})\,\tilde{\boldsymbol{\theta}}(t) \tag{37}$$

where

$$\mathbf{v}(\mathbf{x}_p,\mathbf{x}_v,\mathbf{e}_v) = \begin{bmatrix} \mathbf{x}_v^T\mathrm{N}^1(\mathbf{x}_p)\mathbf{e}_v \\ \mathbf{x}_v^T\mathrm{N}^2(\mathbf{x}_p)\mathbf{e}_v \\ \vdots \\ \mathbf{x}_v^T\mathrm{N}^n(\mathbf{x}_p)\mathbf{e}_v \end{bmatrix} \tag{38}$$

and Eqs. (9) and (34) have been utilized . The parameter error vector $\tilde{\boldsymbol{\theta}}(t)$ is defined by

$$\tilde{\boldsymbol{\theta}}(t) = \hat{\boldsymbol{\theta}}(t) - \boldsymbol{\theta}\,. \tag{39}$$

Lemma 1

Consider the system described by the error equation Eq. (38), and the PAA Eq. (35). Then $\mathbf{e}_v(t)$ and $\tilde{\boldsymbol{\theta}}(t)$ are bounded for all $t \ge 0$.

a.1)

If, in addition $\|\mathbf{W}(\mathbf{x}_p,\mathbf{x}_v,\hat{\mathbf{x}}_v,\mathbf{u})\| \le \mathbf{W}_o < \infty$ then

$$\lim_{t\to\infty} \mathbf{e}_v(t) = \mathbf{0} \tag{40}$$

Proof

Define the Lyapunov function candidate

$$V(t) = \frac{1}{2}\,\mathbf{e}_v^T(t)\,\mathbf{M}(\mathbf{x}_p(t))\,\mathbf{e}_v(t) + \frac{1}{2}\,\tilde{\boldsymbol{\theta}}^T(t)\,\mathrm{K}^{-1}\,\tilde{\boldsymbol{\theta}}(t)\,. \tag{41}$$

Notice that $V(t)$ is a legitimate Lyapunov function candidate since (Horowitz and Tomizuka (1980)) $\mathbf{M}(\mathbf{x}_p)$ is symmetric and

$$0 < \|\mathbf{M}_{\min}\| \le \|\mathbf{M}(\mathbf{x}_p)\| \le \|\mathbf{M}_{\max}\| < \infty\,. \tag{42}$$

From (41) and (38) we obtain

$$\frac{d}{dt}V(t) = -\mathbf{e}_v^T\,\mathbf{F}_v\,\mathbf{e}_v + \mathbf{e}_v^T\left[-\mathbf{v}(\mathbf{x}_p,\mathbf{x}_v,\mathbf{e}_v) - \mathbf{W}(\mathbf{x}_p,\mathbf{x}_v,\hat{\mathbf{x}}_v,\mathbf{u})\,\tilde{\boldsymbol{\theta}}\right] + \frac{1}{2}\mathbf{e}_v^T\,\dot{\mathbf{M}}(\mathbf{x}_p)\,\mathbf{e}_v + \tilde{\boldsymbol{\theta}}^T\,\mathrm{K}^{-1}\,\dot{\tilde{\boldsymbol{\theta}}}$$

$$\frac{d}{dt}V(t) = -\mathbf{e}_v^T\,\mathbf{F}_v\,\mathbf{e}_v + \mathbf{e}_v^T\left[\frac{1}{2}\,\dot{\mathbf{M}}(\mathbf{x}_p)\,\mathbf{e}_v - \mathbf{v}(\mathbf{x}_p,\mathbf{x}_v,\mathbf{e}_v)\right] + \left[\tilde{\boldsymbol{\theta}}^T\,\mathrm{K}^{-1}\,\dot{\tilde{\boldsymbol{\theta}}} - \mathbf{e}_v^T\,\mathbf{W}(\mathbf{x}_p,\mathbf{x}_v,\hat{\mathbf{x}}_v,\mathbf{u})\,\tilde{\boldsymbol{\theta}}\right]. \tag{43}$$

Notice that, from Eq. (A1.4) in appendix 1 and Eq. (3), the second term in Eq. (43),

$$\mathbf{e}_v^T\left[\frac{1}{2}\,\dot{\mathbf{M}}(\mathbf{x}_p)\,\mathbf{e}_v - \mathbf{v}(\mathbf{x}_p,\mathbf{x}_v,\mathbf{e}_v)\right] = -\mathbf{e}_v^T\mathrm{R}(\mathbf{x}_p,\mathbf{x}_v)\mathbf{e}_v = 0\,, \tag{44}$$

since the matrix

$$\mathrm{R}(\mathbf{x}_p,\mathbf{x}_v) = \begin{bmatrix} \mathbf{x}_v^T\left\{\dfrac{\partial \mathbf{m}_1}{\partial \mathbf{x}_p} - \dfrac{\partial \mathbf{M}}{\partial x_{p1}}\right\} \\ \mathbf{x}_v^T\left\{\dfrac{\partial \mathbf{m}_2}{\partial \mathbf{x}_p} - \dfrac{\partial \mathbf{M}}{\partial x_{p2}}\right\} \\ \vdots \\ \mathbf{x}_v^T\left\{\dfrac{\partial \mathbf{m}_n}{\partial \mathbf{x}_p} - \dfrac{\partial \mathbf{M}}{\partial x_{pn}}\right\} \end{bmatrix} \tag{45}$$

is skew symmetric.

The third term in Eq. (43),

$$\tilde{\boldsymbol{\theta}}^T\,\mathrm{K}^{-1}\,\dot{\tilde{\boldsymbol{\theta}}} - \mathbf{e}_v^T\,\mathbf{W}(\mathbf{x}_p,\mathbf{x}_v,\hat{\mathbf{x}}_v,\mathbf{u})\,\tilde{\boldsymbol{\theta}} = 0 \tag{46}$$

by Eq. (35), the PAA.

Thus,

$$\frac{d}{dt}V(t) = -\mathbf{e}_v^T\,\mathbf{F}_v\,\mathbf{e}_v \le 0 \tag{47}$$

which proves the boundness of $\|\mathbf{e}_v\|$ and $\|\tilde{\boldsymbol{\theta}}\|$.

If condition a.1) is satisfied, then, using standard adaptive control arguments (Narendra and Valvani (1980)), Eq. (40) is satisfied.

Notice that $\|\mathbf{e}_v\|$ and $\|\tilde{\boldsymbol{\theta}}\|$ are bounded regardless of the boundness of $\|\mathbf{W}(\mathbf{x}_p,\mathbf{x}_v,\hat{\mathbf{x}}_v,\mathbf{u})\|$.

We will now prove the boundness of $\|\mathbf{W}(\mathbf{x}_p,\mathbf{x}_v,\hat{\mathbf{x}}_v,\mathbf{u})\|$.

Lemma 2

Consider the system described by Eqs. (1), (38), the PAA, Eq. (35), and the control law, Eq. (16). U.t.c.

a.2)

If $\mathbf{x}_d(t)$, $\dot{\mathbf{x}}_d(t)$ and $\ddot{\mathbf{x}}_d(t)$ are all bounded signals, and

a.3)

the characteristic equation $s^3\mathrm{I} + s^2\mathrm{K}_d + s\,\mathrm{K}_p + \mathrm{K}_i = \mathbf{0}$ has all of its roots in LHS of the complex plain,

then condition a.1) is satisfied.

Proof:

Define the tracking error, $\mathbf{e}(t)$, by

$$\mathbf{e}(t) = \mathbf{x}_d(t) - \mathbf{x}_p(t)\,. \tag{48}$$

From Eqs. (12) and (15) we obtain

$$\dot{\mathbf{e}}_v(t) = \mathbf{u}(t) - \ddot{\mathbf{x}}_p(t)\,. \tag{49}$$

Differentiating Eq. (49) and utilizing the control law Eq. (16), we obtain

$$\mathbf{E}(s) = [s^3\mathrm{I} + s^2\mathrm{K}_d + s\,\mathrm{K}_p + \mathrm{K}_i]^{-1}\,s^2\,\mathbf{E}_v(s) \tag{50}$$

and

$$s\,\mathbf{E}(s) = [s^3\mathrm{I} + s^2\mathrm{K}_d + s\,\mathrm{K}_p + \mathrm{K}_i]^{-1}\,s^3\,\mathbf{E}_v(s) \tag{51}$$

where $\mathbf{E}(s)$ and $\mathbf{E}_v(s)$ are the Laplace transforms of $\mathbf{e}(t)$ and $\mathbf{e}_v(t)$ respectively.

By a.3) and Lemma 1, Eqs. (50) and (51) imply the boundness of $\|\mathbf{c}(t)\|$ and $\|\dot{\mathbf{c}}(t)\|$. By a.2) and Eq. (48), the boundness of $\|\mathbf{x}_p(t)\|$ and $\|\mathbf{x}_v(t)\|$ follows. By Lemma 1, $\|\hat{\mathbf{x}}_v(t)\|$ is also bounded.

By Eqs. (16) and (12), since $\mathbf{K}_p$, $\mathbf{K}_d$ and $\mathbf{K}_i$ are constant matrices, and $\|\ddot{\mathbf{x}}_d(t)\|$, $\|\mathbf{e}(t)\|$ and $\|\dot{\mathbf{e}}(t)\|$ are bounded, $\|\mathbf{u}(t)\|$ unbounded would imply that $\|\hat{\mathbf{x}}_v(t)\|$ is also unbounded, which is a contradiction. Since $\mathbf{W}(\mathbf{x}_p,\mathbf{x}_v,\hat{\mathbf{x}}_v,\mathbf{u})$ is an algebraic function of bounded signals, condition a.1) is verified, and the lemma is proven.

The asymptotic convergence of of $\|\mathbf{e}_v(t)\|$, $\|\mathbf{c}(t)\|$ and $\|\dot{\mathbf{e}}(t)\|$ follows from Lemma 1, Lemma 2 and Eqs. (50) and (51).

5. Illustrative Example

Consider a two-link planar manipulator on a horizontal plane(i.e. $g(\mathbf{x}_p)=0$) with rotary joints. The links have length l_1 and l_2. The masses and moment of inertias of the links are $\mathbf{m}_1$, $\mathbf{m}_2$, $\mathbf{I}_1$, and $\mathbf{I}_2$ respectively (figure 1). It can be shown that the elements of inertia matrix and Coriolis vector are as follows:

$$\mathbf{M}_{11} = \mathbf{I}_1+\mathbf{I}_2+(.25\mathbf{m}_1+\mathbf{m}_2)l_1^2+\mathbf{m}_2 l_1 l_2\cos(x_{p2}) \tag{52}$$

$$\mathbf{M}_{12} = \mathbf{M}_{21} = \mathbf{I}_2 + .25\mathbf{m}_2 l_2^2 + .5\mathbf{m}_2 l_1 l_2 \cos(x_{p2})$$

$$\mathbf{M}_{22} = \mathbf{I}_2 + .25\mathbf{m}_2 l_2^2$$

$$\mathbf{v}_1(\mathbf{x}_v,\hat{\mathbf{x}}_v,\mathbf{x}_p) = .5\mathbf{m}_2 l_1 l_2\left[-\hat{\mathbf{x}}_{v1}x_{v2} - x_{v1}\hat{\mathbf{x}}_{v2} + x_{v2}\hat{\mathbf{x}}_{v2}\sin(x_{p2})\right]$$

$$\mathbf{v}_2(\mathbf{x}_v,\hat{\mathbf{x}}_v,\mathbf{x}_p) = .5\mathbf{m}_2 l_1 l_2 x_{v1}\hat{\mathbf{x}}_{v1}\sin(x_{p2})$$

Let's define:

$$\mathbf{\theta}_1 = \mathbf{I}_1 + \mathbf{I}_2 + (.25\mathbf{m}_1 + \mathbf{m}_2)l_1^2 \tag{53}$$

$$\mathbf{\theta}_2 = \mathbf{I}_2 + .25\mathbf{m}_2 l_2^2$$

$$\mathbf{\theta}_3 = .5\mathbf{m}_2 l_1 l_2$$

Then by some algebraic manipulation one obtains the following expression for elements of $\mathbf{W}(\mathbf{x}_p,\mathbf{x}_v,\hat{\mathbf{x}}_v,\mathbf{u})$ matrix.

$$W_{11} = u_1 \tag{54}$$

$$W_{12} = u_2$$

$$W_{13} = 2\mathbf{u}_1\cos(x_{p2}) + \mathbf{u}_2\cos(xp\,2) + x_{v2}\hat{\mathbf{x}}_{v2}\sin(x_{p2}) - (\hat{\mathbf{x}}_{v1}x_{v2} + x_{v1}\hat{\mathbf{x}}_{v2})\sin(x_{p2})$$

$$W_{21} = 0\ , \quad W_{22} = u_1 + u_2$$

$$W_{23} = u_1\cos(x_{p2}) + x_{v1}\hat{\mathbf{x}}_{v1}\sin(x_{p2})$$

6. Computational Efficiency

In order to implement the adaptive controller described above one needs to calculate the elements of $\mathbf{W}(\mathbf{x}_p,\mathbf{x}_v,\hat{\mathbf{x}}_v,\mathbf{u})$ on-line. As can be seen in the example above this procedure may be excessively time consuming since it involves computations of highly nonlinear functions of joint position and velocities. Consequently the real time implementation of such a scheme is rather difficult. To overcome this difficulty we suggest to replace $\mathbf{x}_v$ and $\mathbf{x}_p$ with their desired counterparts, namely $\mathbf{x}_d(t)$ and $\dot{\mathbf{x}}_d(t)$, in $\mathbf{W}(\mathbf{x}_p,\mathbf{x}_v,\hat{\mathbf{x}}_v,\mathbf{u})$ matrix.

The desired quantities are known in advance and therefore all their corresponding calculations can be performed off-line. In a paper, soon to be released, we have proven that in spite of such modifications the asymptotic stability of the adaptive system is still preserved provided:

1) Presence of sufficiently large Velocity and position feedback gains.
2) Addition of an explicitly defined auxiliary nonlinear feedback to the existing controller .

7. Conclusion

A stable continuous time model reference adaptive controller for robotic manipulators was presented. This controller is a modified version of the one originally introduced by Horowitz and Tomizuka (1980). The advantages of the present work are:

1) No slowly time varying assumption about the system parameters is required to prove asymptotic stability. This was the main deficiency of the original scheme.
2) The control scheme requires only joint position and velocity feedback, no acceleration feedback is required.
3) The scheme does not involve any matrix inversion and, therefore, all control parameters can be updated in parallel.

8. Acknowledgment

This work was supported by the National Science Foundation under grant MSM-8511955.

9. References

Anex, R. P. and Hubbard, M. (1984). Modeling and Adaptive Control of Mechanical Manipulator. Proceedings of the 1984 American Control Conference, San Diego, pp. 1237-1245.

Anwar, G., Tomizuka, M., Horowitz, R. and Kubo, T., (1986). Experimental Study on Discrete Time Adaptive Control of an Industrial Arm, Proceedings of the 2nd IFAC Workshop on Adaptive Systems in Signal Processing and Control, Lund, Sweden, July.

Balestino, A., DeMaria, G. and Sciavicco, L. (1983). An Adaptive Model Following Control for Robotic Manipulators. *ASME Journal of Dynamic Systems, Measurement and Control*, Vol. 105, pp. 143-151.

Chew, K-K., (1986). Decentralized Decoupled Adaptive Control for Robotic Manipulators on Multiprocessor Systems, MS report, Mechanical Engineering Department, University of California, Berkeley

Craig, J. J., Hsu, P. and Sastry, S. S. (1986). Adaptive Control of Mechanical Manipulators. Proceedings of the 1986 IEEE International Conference on Robotics and Automation, San Francisco, April.

Dubowsky, S. and DesForges, D. T. (1979). The Application of Model-Reference Adaptive Control to Robotic Manipulators. *ASME Journal of Dynamic Systems, Measurement and Control*, Vol. 101, pp. 193-200.

Dubowsky, S. (1981). On the Adaptive Control of Robotic Manipulators: The Discrete Time Case. Proc. of the 1981 JACC, Vol. 1, TA-2B.

Horowitz, R. and Tomizuka, M. (1980). An Adaptive Control Scheme for Mechanical Manipulators---Compensation of Nonlinearity and Decoupling Control. ASME Paper #80-WA/DSC-6 (also in *the ASME Journal of Dynamic Systems, Measurement and Control*, June 1986).

Horowitz, R. and Tomizuka, M. (1982). Discrete Time Model Reference Adaptive Control of Mechanical Manipulators. Computers in Engineering, 1982, Vol. 2, Robots and Robotics, ASME, pp. 107-112.

Koivo, A. J. and Guo, T. (1983). Adaptive Linear Controller for Robotic Manipulators. *IEEE Trans. on Automatic Control*, AC- 28-2, pp. 162-171.

Kubo, T., Anwar, G. and Tomizuka, M. (1986). Application of Nonlinear Friction Compensation to Robot Arm Control. Proceedings of the 1986 IEEE International Conference on Robotics and Automation, San Francisco, April 1986.

Landau, I. D. and Lozano, R. (1981). Unification and Evaluation of Discrete Time Explicit Model Reference Adaptive Design. *Automatica*, Vol. 17, No. 4, pp. 593-611.

Narendra, K.S. and Valavani, L.S., "A Comparison of Lyapunov and Hyperstability Approaches to Adaptive Control of Continuous Systems," *IEEE Trans. on Automatic Control*, Vol. AC-25, No. 2, pp. 243-247, 1980.

Nicosia, S. and Tomei, P. (1984). Model Reference Adaptive Control Algorithms for Industrial Robots. *Automatica*, Vol. 20, No. 5, pp. 635-644.

Rosemberg, R.M.(1977). *Analytical Dynamics of Discrete Systems*, Plenum Press, 1977.

Slotine, J. and Li, W.(1986). On the Adaptive Control of Robot Manipulators, in Robots: Theory and Applications, edited by Paul, F. and Youcef-Tomi, K.. Proc. of ASME Winter Annual Meeting , 1986.

Sundareshan, M. K. and Koenig, M. A. (1985). Decentralized Model Reference Adaptive Control of Robotic Manipulators. Proc. of 1985 American Control Conference, pp. 44-49.

Takegaki, M. and Arimoto, S. (1981). An Adaptive Trajectory Control of Manipulators. *Int. Journal of Control*, 34, pp. 219-230.

Tomizuka, M. and Horowitz, R. (1983). Model Reference Adaptive Control of Mechanical Manipulators. *Adaptive Systems in Control and Signal Processing 1983*, Pergamon Press, pp. 27-32

Tomizuka, M. et al. (1985a). Modelling and Identification of Mechanical Systems with Nonlinearities. Proc. of the 7th IFAC Symp. on Identification and System Parameter Estimation, pp. 845-851.

Tomizuka, M., Horowitz, R. and Teo, C. L. (1985b). Model Reference Adaptive Controller and Robust Controller for Positioning of Varying Inertia. Applied Motion Control 1985, University of Minnesota, pp. 191-196.

Tomizuka, M. (1985c). Zero-Phase Error Tracking Algorithm for Digital Control. Dynamic Systems: Modelling and Control, ASME Publication, 85-73235, pp. 87-91.

Tomizuka, M., Horowitz, R. and Anwar, G. (1986). Adaptive Techniques for Motion Controls of Robotic Manipulators. Proc. Japan-USA Symposium on Flexible Automation, Osaka, Japan, 1986, pp. 217-224.

Figure (1)

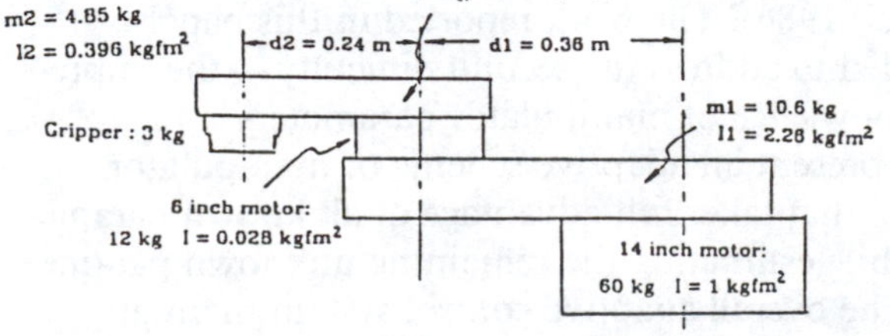

Appendix 1

Proof of Equation (3)

Consider a n degree of freedom robotic manipulator composed by revolute or prismatic joints. The equations of motion for the arm are given by Eq. (1). All the constraints in this system are holonomic scleronomic (Rosemberg (1977)), and the joint position coordinates, x_{pi}'s constitute a set of generalized coordinates. Assume that there are no gravitational or friction forces acting on the system i.e. $g(\mathbf{x}_p) = 0$, $c(\mathbf{x}_p, \mathbf{x}_v) = 0$ in Eq. (1).

Defining the kinetic energy of the system by

$$T = \frac{1}{2}\,\mathbf{x}_v{}^T \mathbf{M}(\mathbf{x}_p)\mathbf{x}_v \quad . \tag{A1.1}$$

Eq. (1) can be obtained by using Lagrange's Equations.

$$\frac{d}{dt}\left[\frac{\partial T}{\partial \mathbf{x}_v}\right] - \frac{\partial T}{\partial \mathbf{x}_p} = \mathbf{q}(t) \tag{A1.2}$$

Expanding EQ. (A1.2),

$$\dot{\mathbf{M}}(\mathbf{x}_p)\mathbf{x}_v(t) + \mathbf{M}(\mathbf{x}_p)\,\dot{\mathbf{x}}_v(t) - \frac{1}{2}\begin{bmatrix} \mathbf{x}_v{}^T \dfrac{\partial \mathbf{M}(\mathbf{x}_p)}{\partial x_{p1}}\mathbf{x}_v \\ \cdot \\ \cdot \\ \cdot \\ \mathbf{x}_v{}^T \dfrac{\partial \mathbf{M}(\mathbf{x}_p)}{\partial x_{pn}}\mathbf{x}_v \end{bmatrix} = \mathbf{q}(t) \tag{A1.3}$$

where we are using the notation $\dot{f}(t) = \frac{d}{dt}f(t)$.

Expanding the first term in Eq. (A1.3), and noticing that

$$\mathbf{M} = \left[\mathbf{m}_1\ \mathbf{m}_2\ \cdots\ \mathbf{m}_n\right] = \begin{bmatrix} \mathbf{m}_1{}^T \\ \mathbf{m}_2{}^T \\ \cdot \\ \cdot \\ \mathbf{m}_n{}^T \end{bmatrix} .$$

$$\dot{\mathbf{M}}(\mathbf{x}_p)\mathbf{x}_v = \left[\frac{\partial \mathbf{M}(\mathbf{x}_p)}{\partial x_{p1}}\mathbf{x}_v\ \cdots\ \frac{\partial \mathbf{M}(\mathbf{x}_p)}{\partial x_{pn}}\mathbf{x}_v\right]\mathbf{x}_v \tag{A1.4}$$

$$= \left[\mathbf{x}_v{}^T \frac{\partial \mathbf{m}_1(\mathbf{x}_p)}{\partial x_{p1}}\mathbf{x}_v\ \cdots\ \mathbf{x}_v{}^T \frac{\partial \mathbf{m}_n(\mathbf{x}_p)}{\partial x_{pn}}\mathbf{x}_v\right]^T$$

Combining Eqs. (A1.3) and (A1.4) we obtain the desired result:

$$\mathbf{M}(\mathbf{x}_p)\,\dot{\mathbf{x}}_v(t) + \begin{bmatrix} \mathbf{x}_v{}^T\left[\dfrac{\partial \mathbf{m}_1(\mathbf{x}_p)}{\partial \mathbf{x}_p} - \dfrac{1}{2}\dfrac{\partial \mathbf{M}(\mathbf{x}_p)}{\partial x_{p1}}\right]\mathbf{x}_v \\ \vdots \\ \mathbf{x}_v{}^T\left[\dfrac{\partial \mathbf{m}_n(\mathbf{x}_p)}{\partial \mathbf{x}_p} - \dfrac{1}{2}\dfrac{\partial \mathbf{M}(\mathbf{x}_p)}{\partial x_{pn}}\right]\mathbf{x}_v \end{bmatrix} = \mathbf{q}(t) \tag{A1.5}$$

John J. Craig
Department of Electrical Engineering
Stanford University
Stanford, California 94305

Ping Hsu
S. Shankar Sastry
Department of Electrical Engineering and Computer Sciences
University of California, Berkeley
Berkeley, California 94720

Adaptive Control of Mechanical Manipulators

Abstract

When an accurate dynamic model of a mechanical manipulator is available, it may be used in a nonlinear, model-based scheme to control the manipulator. Such a control formulation yields a controller that suppresses disturbances and tracks desired trajectories uniformly in all configurations of the manipulator. Use of a poor dynamic model with this kind of model-based decoupling and linearizing scheme, however, may result in performance that is inferior to a much simpler, fixed-gain scheme.

In this paper, we develop a parameter-adaptive control scheme in a set of adaptive laws that can be added to the nonlinear, model-based controller. The scheme is unique because it is designed specifically for the nonlinear, model-based controller and has been proven stable in a full, nonlinear setting. After adaptation, the error dynamics of the joints are decoupled with uniform disturbance rejection in all manipulator configurations. The issues of sufficient excitation and the effect of disturbances are also discussed.

The theory is demonstrated with simulation results and also with data from an implementation for an industrial robot, the Adept One.

J. Craig was supported by a grant from the Systems Development Foundation. P. Hsu and S. Sastry were supported by the Army Research Office under grant No. DAAG 29-85-K-0072 and IBM under faculty development grant 1983. Adept Technology Inc. supplied facilities for the experiments of Section 10.

The International Journal of Robotics Research,
Vol. 6, No. 2, Summer 1987,

1. Introduction

When an accurate dynamic model of a mechanical manipulator is available, it may be used in a nonlinear, model-based scheme to control the manipulator (Craig 1986*a*). This scheme is sometimes called the *computed torque method*. This control formulation yields a controller that suppresses disturbances and tracks desired trajectories uniformly in all configurations of the manipulator. Desirable performance, however, is contingent on two assumptions that have made some implementations less than ideal. First, the dynamic model of the manipulator must be computed quickly enough so that discretization effects do not degrade performance relative to the continuous-time, zero-delay ideal. Second, the values of parameters appearing in the dynamic model in the control law must match the parameters of the actual system if the beneficial decoupling and linearizing effects of the model-based controller are to be realized.

Some recent work in formulating efficient computational algorithms for manipulator dynamics, along with the increase in the performance-price ratio of computing hardware, have decreased the first difficulty of employing this control methodology (Luh, Walker, and Paul 1980; Kanade, Khosla, and Tanaka 1984; Burdick 1986). The work reported in this paper is intended to address the second difficulty—the imprecise knowledge of manipulator parameters.

We present an adaptive scheme of manipulator control that takes full advantage of all known parameters while estimating the remaining unknown parameters. The overall adaptive control system maintains the structure of the nonlinear, model-based controller, but it also has an adaptive element. After sufficient on-line learning, the control algorithm decouples and linearizes the manipulator so that each joint behaves like an independent, second-order system with fixed dynamics.

2. Previous Research

The work of Dubowsky and DesForges (1979), Horowitz and Tomizuka (1980), Koivo and Guo (1983), Leininger (1983), and Lee and Chung (1984) is representative of previous research in parametric adaptive control of manipulators. For a more complete bibliography, see Craig (1986*b*), where 68 publications in this area are cited. We will not attempt a detailed review of the contributions of these researchers here.

Almost all previous work is based on standard MRAC (Sastry 1984) or STR (Astrom et al. 1977) theory for linear, time-invariant plants. Thus, such stability proofs are only valid to the extent that coefficients of the linearized manipulator system vary sufficiently slowly. Modern manipulators move so fast, on the other hand, that the effective joint inertia at a given joint may change by 300% in a fraction of a second. Therefore, most previously published schemes for adaptive control of mechanical manipulators are on questionable theoretical footing. Additionally, previous work ignores the crucial issues of persistent excitation and robustness to disturbances, and published results of experiments with an actual manipulator are almost nonexistent.

Many researchers have tried to use adaptation as a means to simplify or otherwise avoid performing the nonlinear model computations in the control computer. As computing power becomes more readily available and schemes to compute the dynamic model are improved, this motivation diminishes. A more interesting use of adaptive control is to attempt to outperform the simpler, nonadaptive controllers.

3. The Dynamic Model of a Manipulator

The manipulator is modeled as a set of n rigid bodies connected in a serial chain with friction acting at the joints. The vector equation of motion of such a device can be written in the compact form

$$\mathbf{T} = \mathbf{M}(\mathbf{\Theta})\ddot{\mathbf{\Theta}} + \mathbf{Q}(\mathbf{\Theta}, \dot{\mathbf{\Theta}}), \tag{1}$$

where $\mathbf{T}$ is the $n \times 1$ vector of joint forces or torques supplied by the actuators, and $\mathbf{\Theta}$ is the $n \times 1$ vector of joint positions, with $\mathbf{\Theta} = [\theta_1, \theta_2, \ldots, \theta_n]^T$. The matrix, $\mathbf{M}(\mathbf{\Theta})$, is an $n \times n$ matrix sometimes called the *manipulator mass matrix*. The vector $\mathbf{Q}(\mathbf{\Theta}, \dot{\mathbf{\Theta}})$ represents forces or torques arising from centrifugal, Coriolis, gravitational, and frictional forces.

The jth element of Eq. (1) can be written in a sum-of-products form as

$$\tau_j = \sum_{i=1}^{a_j} m_{ji} f_{ji}(\mathbf{\Theta}, \ddot{\mathbf{\Theta}}) + \sum_{i=1}^{b_j} q_{ji} g_{ji}(\mathbf{\Theta}, \dot{\mathbf{\Theta}}), \tag{2}$$

where the m_{ji} and q_{ji} are parameters formed by products of such quantities as link masses, link inertia tensor elements, lengths, friction coefficients, and the gravitational acceleration constant. The $f_{ji}(\mathbf{\Theta}, \ddot{\mathbf{\Theta}})$ and the $g_{ji}(\mathbf{\Theta}, \dot{\mathbf{\Theta}})$ are functions that embody the dynamic structure of the manipulator's motion geometry. In this paper, we assume that the *structure* of these parameters and functions is known, but the numerical values of some or all of the parameters m_{ji} and q_{ji} are unknown. We will assume that bounds on the parameter values are known, although these bounds may sometimes be quite loose.* This is equivalent to knowing the kinematic structure of a manipulator and having parametric models of joint friction effects, but knowing only some, or perhaps none, of the dynamic parameters, such as mass distribution of the links and friction coefficients.

4. Nonlinear Model-Based Control

To control the manipulator, we propose the control law

$$\mathbf{T} = \hat{\mathbf{M}}(\mathbf{\Theta})\ddot{\mathbf{\Theta}}^* + \hat{\mathbf{Q}}(\mathbf{\Theta}, \dot{\mathbf{\Theta}}), \tag{3}$$

* In fact, such bounds are only needed for the parameters appearing in the manipulator's mass matrix. For generality, however, we will assume that bounds on all parameters are known.

where $\hat{\mathbf{M}}(\Theta)$ and $\hat{\mathbf{Q}}(\Theta, \dot{\Theta})$ are estimates of $\mathbf{M}(\Theta)$ and $\mathbf{Q}(\Theta, \dot{\Theta})$, and

$$\ddot{\Theta}^* = \ddot{\Theta}_d + \mathbf{K}_v\dot{\mathbf{E}} + \mathbf{K}_p\mathbf{E}. \tag{4}$$

In Eq. (4), the servo error, $\mathbf{E} = [e_1 e_2 \ . \ . \ . \ e_n]^T$ is defined as

$$\mathbf{E} = \Theta_d - \Theta, \tag{5}$$

and $\mathbf{K}_v$ and $\mathbf{K}_p$ are $n \times n$ constant, diagonal gain matrices with k_{vj} and k_{pj} on the diagonals. Equation (3) is the nonlinear, model-based controller sometimes referred to as the computed torque method of manipulator control. The desired trajectory of the manipulator is assumed known as time functions of joint positions, velocities, and accelerations, $\Theta_d(t)$, $\dot{\Theta}_d(t)$, and $\ddot{\Theta}_d(t)$.

The control law, Eq. (3), is chosen because in the favorable situation of perfect knowledge of parameter values and no disturbances, the jth joint has error dynamics given by the equation

$$\ddot{e}_j + k_{vj}\dot{e}_j + k_{pj}e_j = 0. \tag{6}$$

In this ideal situation, k_{vj} and k_{pj} may be chosen to place the closed-loop poles associated with each joint, and disturbance rejection will be uniform over the entire workspace of the manipulator.

Figure 1 is a block diagram indicating the structure of this controller, which makes use of a dynamic model of the manipulator. An adaptive element is also indicated. This adaptive element observes servo errors and adjusts the parameter estimates, $\hat{m}_{ji}$ and $\hat{q}_{ji}$, which appear in the control law (Eq. 3). The remainder of this paper is concerned with the design of this adaptive element, proof of global stability of the design, and other related issues.

We first consider what we will call the "ideal case," in which we have a perfect structural model of the manipulator dynamics. In this case, parameter errors are the sole source of imperfect decoupling and linearization of the plant. That is, tuning of the parameters would cause the model in the computer to match exactly the dynamics of the actual mechanical manipulator. In Section 8 of this paper, we will address the problem of robustness to disturbances.

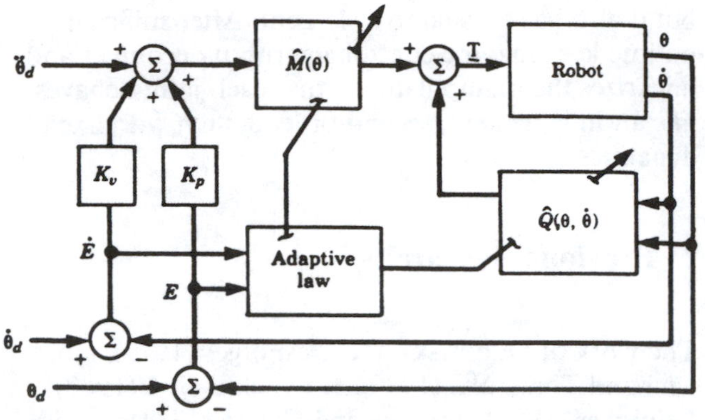

Fig. 1. The controller with an adaptive element.

5. The Error Equation

When estimates of parameters do not match the true parameter values, the closed-loop system will not perform as indicated by Eq. (6). By equating Eqs. (1) and (3), we obtain

$$\ddot{\mathbf{E}} + \mathbf{K}_v\dot{\mathbf{E}} + \mathbf{K}_p\mathbf{E} = \hat{\mathbf{M}}^{-1}(\Theta)[\tilde{\mathbf{M}}(\Theta)\ddot{\Theta} + \tilde{\mathbf{Q}}(\Theta, \dot{\Theta})], \tag{7}$$

where $\tilde{\mathbf{M}}(\Theta) = \mathbf{M}(\Theta) - \hat{\mathbf{M}}(\Theta)$ and $\tilde{\mathbf{Q}}(\Theta, \dot{\Theta}) = \mathbf{Q}(\Theta, \dot{\Theta}) - \hat{\mathbf{Q}}(\Theta, \dot{\Theta})$ represent errors in the dynamic model used in the controller arising from errors in the parameters of the model.

In a given application, we may know some of the parameters m_{ji} and q_{ji}. Of the a_j parameters m_{ji}, and b_j parameters q_{ji} appearing in the dynamic Eq. (2) of the jth joint, let respectively r_j and s_j of them be *unknown*, with $r_j \leq a_j$ and $s_j \leq b_j$ for all j. Re-index the unknown parameters (if necessary) and note that the jth component of the expression in brackets in Eq. (7) can be written

$$\tilde{\tau}_j = \sum_{i=1}^{r_j} \tilde{m}_{ji} f_{ji}(\Theta, \ddot{\Theta}) + \sum_{i=1}^{s_j} \tilde{q}_{ji} g_{ji}(\Theta, \dot{\Theta}), \tag{8}$$

where $\tilde{m}_{ji} = m_{ji} - \hat{m}_{ji}$ and $\tilde{q}_{ji} = q_{ji} - \hat{q}_{ji}$ are parameter errors.

The number of system parameters is

$$r \leq \sum_{j=1}^{n} (r_j + s_j). \tag{9}$$

These r system parameters, which are m_{ji} and q_{ji} either alone or in combination, will now be called $\mathbf{P} = [p_1 \ \ p_2 \ . \ . \ . \ \ p_n]^T$, and their estimates are $\hat{\mathbf{P}} = [\hat{p}_1 \ \ \hat{p}_2 \ . \ . \ . \ \ \hat{p}_r]^T$, so that

$$\mathbf{\Phi} = \mathbf{P} - \hat{\mathbf{P}}. \tag{10}$$

These definitions allow us to construct an adaptive scheme that makes full use of known parameters and only adjusts the estimates of the unknown parameters. For example, we may know the inertial properties of the manipulator, but not the friction coefficients, or we may know the parameters of some links but not others.

The error equation (Eq. 7) may be written in the form

$$\ddot{\mathbf{E}} + \mathbf{K}_v\dot{\mathbf{E}} + \mathbf{K}_p\mathbf{E} = \hat{\mathbf{M}}^{-1}(\mathbf{\Theta})\mathbf{W}(\mathbf{\Theta}, \dot{\mathbf{\Theta}}, \ddot{\mathbf{\Theta}})\mathbf{\Phi}, \tag{11}$$

where $\mathbf{\Phi}$ is the $r \times 1$ vector of parameter errors, and $\mathbf{W}(\mathbf{\Theta}, \dot{\mathbf{\Theta}}, \ddot{\mathbf{\Theta}})$ is an $n \times r$ matrix of functions. For brevity, the arguments of $\hat{\mathbf{M}}^{-1}$ and $\mathbf{W}$ will be dropped in the rest of this paper.

In the following analysis, it is important that the product $\hat{\mathbf{M}}^{-1}\mathbf{W}$ remain bounded at all times. Since $\mathbf{W}$ is composed of bounded functions of a manipulator trajectory, $\mathbf{W}$ will remain bounded if the trajectory of the manipulator remains bounded. The matrix $\hat{\mathbf{M}}(\mathbf{\Theta})$ will remain positive definite and invertible if we insure that all parameters m_{ji} remain within a sufficiently small range near the actual parameter value. With this motivation, we will restrict our estimates of the parameters to lie within bounds, such that

$$l_i - \delta < \hat{p}_i < h_i + \delta, \tag{12}$$

where we know that the actual value, p_i, lies between l_i and h_i and where δ is small, positive, and chosen such that $\hat{\mathbf{M}}^{-1}$ remains bounded as long as Eq. (12) holds. We will see that any choice of δ that satisfies $p_i - l_i > \frac{\delta}{2}$ or $h_i - p_i > \frac{\delta}{2}$ may be used. This formulation will allow our scheme to take advantage of partial knowledge of parameter values in the form of bounds on their true values.

6. The Adaptation Algorithm

The adaptive law will compute how to change parameter estimates as a function of a filtered servo error signal. The filtered servo error, $\mathbf{E}_1$, is

$$\mathbf{E}_1 = \dot{\mathbf{E}} + \mathbf{\Psi}\mathbf{E}, \tag{13}$$

where $\mathbf{\Psi} = \text{diag}(\psi_1 \ \ \psi_2 \ . \ . \ . \ \ \psi_n)$ with $\psi_i > 0$. Note that for manipulators with position and velocity sensors, the value $\mathbf{E}_1$ can be computed simply from sensor readings, and the filter does not need to be implemented.

The ψ_j are chosen so that transfer function

$$\frac{s + \psi_j}{s^2 + k_{vj}s + k_{pj}} \tag{14}$$

is strictly positive real (SPR).* Then, by the positive real lemma (Anderson and Vongpanitlerd 1973) we are assured of the existence of the positive definite matrices $\mathcal{P}_j$ and $\mathcal{Q}_i$, such that

$$\begin{aligned} \mathbf{A}_j^T\mathcal{P}_j + \mathcal{P}_j\mathbf{A}_j &= -\mathcal{Q}_j, \\ \mathcal{P}_j\mathbf{B}_j &= \mathbf{C}_j^T, \end{aligned} \tag{15}$$

where the matrices $\mathbf{A}_j$, $\mathbf{B}_j$, and $\mathbf{C}_j$ are the matrices of a minimal state-space realization of the filtered error equation of the jth joint

$$\begin{aligned} \dot{\mathbf{x}}_j &= \mathbf{A}_j\mathbf{x}_j + \mathbf{B}_j(\hat{\mathbf{M}}^{-1}\mathbf{W}\mathbf{\Phi})_j, \\ e_{1j} &= \mathbf{C}_j\mathbf{x}_j, \end{aligned} \tag{16}$$

where the state vector, $\mathbf{x}_j$, is a linear transformation of $[e_j, \dot{e}_j]^T$.

The filtered error equation of the entire system in state-space form is given by

$$\begin{aligned} \dot{\mathbf{X}} &= \mathbf{A}\mathbf{X} + \mathbf{B}\hat{\mathbf{M}}^{-1}\mathbf{W}\mathbf{\Phi}, \\ \mathbf{E}_1 &= \mathbf{C}\mathbf{X}, \end{aligned} \tag{17}$$

where $\mathbf{A}$, $\mathbf{B}$, and $\mathbf{C}$ are all block diagonal (with $\mathbf{A}_j$, $\mathbf{B}_j$,

* A rational SPR function, $T(s)$, is analytic in the closed right half plane and has $Re(T(j\omega)) < 0 \forall \omega$.

and $\mathbf{C}_j$ on the diagonals, respectively) and $\mathbf{X} = [x_1 \quad x_2 \ . \ . \ . \ x_n]^T$. Forming the $2n \times 2n$ matrices $\mathcal{P} = \text{diag}(\mathcal{P}_1 \quad \mathcal{P}_2 \ . \ . \ . \ \mathcal{P}_n)$ and $\mathcal{Q} = \text{diag}(\mathcal{Q}_1 \quad \mathcal{Q}_2 \ . \ . \ . \ \mathcal{Q}_n)$ we have $\mathcal{P} > 0$, $\mathcal{Q} > 0$, and

$$\begin{aligned} \mathbf{A}^T\mathcal{P} + \mathcal{P}\mathbf{A} &= -\mathcal{Q}, \\ \mathcal{P}\mathbf{B} &= \mathbf{C}^T. \end{aligned} \tag{18}$$

We now use Lyapunov theory to derive an adaptation law (Parks 1966). The Lyapunov function candidate

$$v(\mathbf{X}, \mathbf{\Phi}) = \mathbf{X}^T\mathcal{P}\mathbf{X} + \mathbf{\Phi}^T\mathbf{\Gamma}^{-1}\mathbf{\Phi}, \tag{19}$$

with $\mathbf{\Gamma} = \text{diag}(\gamma_1 \quad \gamma_2 \ . \ . \ . \ \gamma_r)$ and $\gamma_i > 0$ is non-negative in both servo and parameter errors. Differentiation with respect to time makes use of Eq. (18) and leads to

$$\begin{aligned} \dot{v}(\mathbf{X}, \mathbf{\Phi}) = &-\mathbf{X}^T\mathcal{Q}\mathbf{X} \\ &+ 2\mathbf{\Phi}^T(\mathbf{W}^T\hat{\mathbf{M}}^{-1}\mathbf{E}_1 + \mathbf{\Gamma}^{-1}\dot{\mathbf{\Phi}}). \end{aligned} \tag{20}$$

If we choose

$$\dot{\mathbf{\Phi}} = -\mathbf{\Gamma}\mathbf{W}^T\hat{\mathbf{M}}^{-1}\mathbf{E}_1, \tag{21}$$

we have

$$\dot{v}(\mathbf{X}, \mathbf{\Phi}) = -\mathbf{X}^T\mathcal{Q}\mathbf{X}, \tag{22}$$

which is non-positive because $\mathcal{Q}$ is positive definite. Since $\mathbf{\Phi} = \mathbf{P} - \hat{\mathbf{P}}$, we have $\dot{\mathbf{\Phi}} = -\dot{\hat{\mathbf{P}}}$, and from Eq. (21) we have the adaptation law

$$\dot{\hat{\mathbf{P}}} = \mathbf{\Gamma}\mathbf{W}^T\hat{\mathbf{M}}^{-1}\mathbf{E}_1. \tag{23}$$

Equations (19) and (22) imply that $\mathbf{X}$ and $\mathbf{\Phi}$ are bounded. The basic update law is given by Eq. (23). However, in order to restrict the parameter estimates to lie within the bounds given in Eq. (12), we augment the update law for parameter p_i with the reset conditions

$$\begin{aligned} \hat{p}_i(t^+) &= l_i, \qquad \text{if } \hat{p}_i(t) \leq l_i - \delta, \\ \hat{p}_i(t^+) &= h_i, \qquad \text{if } \hat{p}_i(t) \geq h_i + \delta. \end{aligned} \tag{24}$$

Thus, if an estimate moves outside its known bound by an amount δ, it is reset to its bound. This parameter *resetting* causes a step change in $\mathbf{\Phi}$ in Eq. (17). This cannot cause an instantaneous change in $\mathbf{X}$, so we can write the value of the Lyapunov function before and after the reset of p_i to its lower bound at time t_j as

$$\begin{aligned} v(t_j) &= \mathbf{X}^T\mathcal{P}\mathbf{X} + \sum_{\substack{k=1 \\ k \neq i}}^{r} \frac{1}{\gamma_k}\phi_k^2 + \frac{1}{\gamma_i}(p_i - l_i + \delta)^2, \\ v(t_j^+) &= \mathbf{X}^T\mathcal{P}\mathbf{X} + \sum_{\substack{k=1 \\ k \neq i}}^{r} \frac{1}{\gamma_k}\phi_k^2 + \frac{1}{\gamma_i}(p_i - l_i)^2. \end{aligned} \tag{25}$$

The change in v due to the resetting of $\hat{p}_i$ at time t_j is

$$-\epsilon_j = v(t_j^+) - v(t_j) = -(2(p_i - l_i) - \delta)\left(\frac{\delta}{\gamma_i}\right), \tag{26}$$

where ϵ_j is positive and has a nonzero lower bound. For example, if δ is chosen to insure $p_i - l_i \geq \delta$, then ϵ is lower bounded by $\frac{\delta}{\gamma_i}$. Similarly, by resetting p_i to its upper bound at time t_j, we have

$$-\epsilon_j = v(t_j^+) - v(t_j) = (2(p_i - h_i) - \delta)\left(\frac{\delta}{\gamma_i}\right), \tag{27}$$

where ϵ_j is positive and has a nonzero lower bound. With this addition of parameter resetting, Eq. (22) becomes

$$\dot{v}(\mathbf{X}, \mathbf{\Phi}) = -\mathbf{X}^T\mathcal{Q}\mathbf{X} - \sum_{j}^{q} \delta(t - t_j)\epsilon_j, \tag{28}$$

where q resets take place, and $\delta(\cdot)$ here refers to the unit impulse function. The addition of parameter resetting maintains the non-positiveness of $\dot{v}(\mathbf{X}, \mathbf{\Phi})$, and thus the system is stable in the sense of Lyapunov with $\mathbf{X}$ and $\mathbf{\Phi}$ bounded.

Since $\mathbf{X}$, $\mathbf{\Phi}$, $\hat{\mathbf{M}}^{-1}$, and $\mathbf{W}$ are bounded, we see from Eq. (17) that $\dot{\mathbf{X}}$ is bounded as well. Thus, $\mathbf{X}$ is uniformly continuous and so is $\dot{v}(\mathbf{X}, \mathbf{\Phi})$. From Eqs. (19) and (22)

$$\lim_{t \to \infty} v(\mathbf{X}, \mathbf{\Phi}) = v^* \tag{29}$$

with

$$v^* - v(\mathbf{X}_0, \mathbf{\Phi}_0) = -\int_0^\infty \mathbf{X}^T \mathcal{Q} \mathbf{X}\, dt - \sum_{j=1}^{q} \epsilon_j, \quad (30)$$

where q parameter resettings take place. Since the left-hand side is known to be finite, and both terms on the right-hand side have the same sign, we know that each term on the right-hand side must be finite. At most, a finite number, q, of parameter resets take place.

We know (Rudin 1976) that since $\mathbf{X}^T\mathcal{Q}\mathbf{X}$ is non-negative everywhere, uniformly continuous, and has a finite integral that

$$\lim_{t\to\infty} \mathbf{X}^T\mathcal{Q}\mathbf{X} = 0, \quad (31)$$

and thus

$$\lim_{t\to\infty} \mathbf{E} = 0, \qquad \lim_{t\to\infty} \dot{\mathbf{E}} = 0. \quad (32)$$

The adaptive scheme is stable (in the sense that all signals remain bounded), and trajectory tracking errors, $\mathbf{E}$ and $\dot{\mathbf{E}}$, converge to zero. Concerning convergence of the parameter errors, so far we can only say that

$$\lim_{t\to\infty} |\Gamma^{-1/2}\mathbf{\Phi}| = \sqrt{v^*}. \quad (33)$$

Note that $\ddot{\mathbf{\Theta}}$, the actual acceleration of the manipulator, appears in the adaptation law of any parameter representing an inertia. Manipulators do not usually have acceleration sensors. However, the integrating action of the parameter update law reduces the necessity for good acceleration information. This has been verified in simulation and in experiments with an actual manipulator (see Section 10).

7. Parameter Error Convergence

After a finite amount of time, all parameter resets have occurred, and we may write the equations describing the complete system, i.e. Eqs. (17) and (21), as

$$\begin{bmatrix} \dot{\mathbf{X}} \\ \dot{\mathbf{\Phi}} \end{bmatrix} = \begin{bmatrix} \mathbf{A} & \mathbf{BU}^T \\ -\Gamma\mathbf{UC} & 0 \end{bmatrix} \begin{bmatrix} \mathbf{X} \\ \mathbf{\Phi} \end{bmatrix}, \quad (34)$$

where $\mathbf{U} = (\hat{\mathbf{M}}^{-1}\mathbf{W})^T$. Several researchers have studied the asymptotic stability of Eq. (34). In Anderson (1977), Kreisselmeier (1977) and Morgan and Narendra (1977), it is shown that Eq. (34) is uniformly asymptotically stable if the linear system (**A, B, C**) meets the earlier SPR condition and if **U** satisfies the persistent excitation condition

$$\alpha' \mathbf{I}_r \leq \int_{t_0}^{t_0+\rho} \mathbf{U}\mathbf{U}^T\, dt \leq \beta' \mathbf{I}_r \quad (35)$$

for all t_0, where α', β', and ρ are all positive. Equation (35) shows that the integral of $\mathbf{UU}^T$ must be positive definite and bounded over all intervals of length ρ. Note that a matrix of the form $\mathbf{UU}^T$ has dimensions $r \times r$ but can have a rank no greater than n (and usually, $r > n$). Hence, Eq. (35) means that **U** must vary sufficiently over the interval ρ so that the entire r dimensional space is spanned. Note that by restricting the ranges of our estimates we have insured that $\hat{\mathbf{M}}$ remains invertible, and hence **U** is bounded, so that the right-hand inequality in Eq. (35) is already met.

Next we claim that because $\hat{\mathbf{M}}$ is a bounded, positive, definite symmetric matrix, the left-hand inequality of Eq. (35) will be satisfied if for some $\alpha > 0$,

$$\alpha \mathbf{I}_r \leq \int_{t_0}^{t_0+\rho} \mathbf{W}^T\mathbf{W}\, dt \quad (36)$$

is satisfied. A proof of this is given in Craig (1986*b*).

Finally, since we have shown (independent of persistent excitation) that the servo error converges to zero under this control scheme, the persistent excitation condition of Eq. (36) will be met if the *desired trajectory* satisfies

$$\alpha \mathbf{I}_r \leq \int_{t_0}^{t_0+\rho} \mathbf{W}_d^T\mathbf{W}_d\, dt, \quad (37)$$

where $\mathbf{W}_d$ is the **W** function evaluated along the desired rather than the actual trajectory of the manipulator. We have derived a condition on the desired trajectory so that all parameters will be identified after a sufficient learning interval.

8. Robustness to Bounded Disturbances

For the adaptive scheme we have presented in this paper, we will show that the presence of bounded disturbances will not result in loss of stability or unbounded estimates. This property is due to the fact that we have already assumed a priori bounds on the parameter values and have implemented parameter "resetting" rules as part of the adaptive algorithm. However, there are two expected adverse effects of disturbances on the adaptive scheme:

1. Servo errors do not converge asymptotically to zero, but converge to a bounded region near zero. The size of this region is given in a straightforward manner by the choice of servo gains and by the upper bound on the magnitude of the disturbances.
2. Parameter estimates may not converge, but in the worst case, our resetting laws will maintain the estimates in their a priori bounds. If a certain persistent excitation condition is met, then the parameter errors will converge to a bounded region near zero.

To analyze the effect of disturbances, we refer to Eq. (7). We now assume that the modeling error has two components: one due to parameter mismatch, $\mathbf{\Phi}$, and one due to disturbances, $\boldsymbol{\nu}'(t)$. That is:

$$\tilde{\mathbf{M}}(\Theta)\ddot{\Theta} + \tilde{\mathbf{Q}}(\Theta, \dot{\Theta}) = \mathbf{W}(\Theta, \dot{\Theta}, \ddot{\Theta})\mathbf{\Phi} + \boldsymbol{\nu}'. \quad (38)$$

Where the $n \times 1$ vector function, $\boldsymbol{\nu}'(t)$, is completely unknown, but it is known to be upper bounded by

$$\|\boldsymbol{\nu}'(t)\| < \nu'_{\max}. \quad (39)$$

The scope of this analysis lies in what kinds of signals may be in $\boldsymbol{\nu}'$. First, $\boldsymbol{\nu}'$ can certainly contain external, bounded disturbances (assumed uncorrelated with the state of the control system), but $\boldsymbol{\nu}'$ may also contain bounded functions of state; for example, terms like $p_i \cos(\theta_j)$ or $p_i \operatorname{sgn}(\dot{\theta}_j)$. Note, for example, that if $\boldsymbol{\nu}'$ were known to be linear in one or more state variables, then it is generally impossible to know a priori that it is bounded. This is the problem of robustness to unmodeled dynamics, for which a general solution is not yet known. Given Eq. (38), our system from Eq. (17) becomes

$$\begin{aligned} \dot{\mathbf{X}} &= \mathbf{A}\mathbf{X} + \mathbf{B}\hat{\mathbf{M}}^{-1}\mathbf{W}\mathbf{\Phi} + \boldsymbol{\nu}, \\ \mathbf{E}_1 &= \mathbf{C}\mathbf{X}, \end{aligned} \quad (40)$$

where $\boldsymbol{\nu} = \mathbf{B}\hat{\mathbf{M}}^{-1}\boldsymbol{\nu}'$ is $2n \times 1$ with upper bound

$$\|\boldsymbol{\nu}(t)\| < \nu_{\max}. \quad (41)$$

Note that $\nu_{\max}$ is computable from $\nu'_{\max}$. Choosing the same Lyapunov candidate function as before,

$$v(\mathbf{X}, \mathbf{\Phi}) = \mathbf{X}^T \mathcal{P} \mathbf{X} + \mathbf{\Phi}^T \Gamma^{-1} \mathbf{\Phi}. \quad (42)$$

Using the same adaptation law, we obtain

$$\dot{v}(\mathbf{X}, \mathbf{\Phi}) = -\mathbf{X}^T \mathcal{Q} \mathbf{X} + 2\mathbf{X}^T \mathcal{P} \boldsymbol{\nu}. \quad (43)$$

Equation (43) shows that for sufficiently large $\mathbf{X}$, $\dot{v}(\mathbf{X}, \mathbf{\Phi})$ is non-positive, and so $\mathbf{X}$ is bounded. In particular, the region in which $v(\mathbf{X}, \mathbf{\Phi})$ may increase is given by

$$\mathbf{X} < 2\mathcal{Q}^{-1}\mathcal{P}\boldsymbol{\nu}, \quad (44)$$

which is upper bounded by the hypersphere

$$\|\mathbf{X}\|_2 < 2\frac{\lambda_{p\max}}{\lambda_{q\min}}\nu_{\max}, \quad (45)$$

where $\lambda_{p\max}$ is the maximum eigenvalue of $\mathcal{P}$, and $\lambda_{q\min}$ is the minimum eigenvalue of $\mathcal{Q}$. Equation (45) gives an upper bound on the eventual size of servo errors in the presence of bounded disturbances. Note that $\lambda_{p\max}$ and $\lambda_{q\min}$ are a function of the chosen servo gains, and the ψ_j from Eq. (14). Note that Eq. (45) implies that constants $e_{\max}$ and $\dot{e}_{\max}$ exist, such that

$$\begin{aligned} \|\mathbf{E}\| &< e_{\max}, \\ \|\dot{\mathbf{E}}\| &< \dot{e}_{\max}. \end{aligned} \quad (46)$$

From Eq. (7) with Eq. (38) substituted for the bracketed term, $\ddot{e}_{\max}$ exists such that

$$\|\ddot{\mathbf{E}}\| < \ddot{e}_{\max}. \quad (47)$$

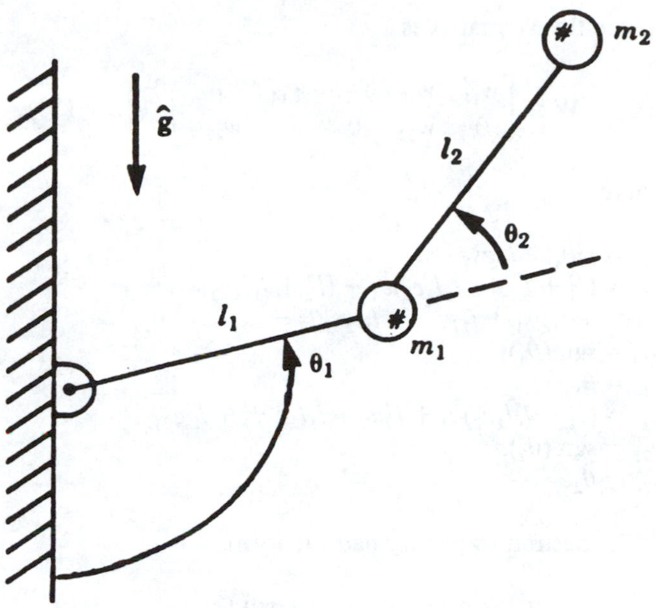

Fig. 2. Manipulator considered for simulation study.

Note that the above analysis does not guarantee that $\mathbf{\Phi}$ goes to zero, or even that it remains small. In fact, without saying anything about excitation of the trajectory, $\mathbf{\Phi}$ may diverge. Because we have implemented resetting rules to insure fixed upper and lower bounds for the elements of $\mathbf{\Phi}$, however, we are assured that our estimates will remain bounded at all times.

To consider persistent excitation in the presence of disturbances, we see that Eq. (34) becomes

$$\begin{bmatrix} \dot{\mathbf{X}} \\ \dot{\mathbf{\Phi}} \end{bmatrix} = \begin{bmatrix} \mathbf{A} & \mathbf{BU}^T \\ -\mathbf{\Gamma UC} & 0 \end{bmatrix} \begin{bmatrix} \mathbf{X} \\ \mathbf{\Phi} \end{bmatrix} + \begin{bmatrix} v \\ 0 \end{bmatrix}. \quad (48)$$

We note (from our previous result) that if

$$\alpha \mathbf{I} \leqslant \int_{t_0}^{t_0+\rho} \mathbf{W}^T\mathbf{W}\, dt, \quad (49)$$

then Eq. (48) represents an asymptotically stable system driven with a bounded input. The state vector $[\mathbf{X}^T\mathbf{\Phi}^T]^T$ converges to a region that can be upper bounded by a ball centered at the origin of this $2n + r$ dimensional space. Hence, $\phi_{\max}$ exists such that

$$\|\mathbf{\Phi}\| < \phi_{\max}. \quad (50)$$

At this point, we would like to give a condition on the *desired* trajectory so that Eq. (49) is true. Since servo errors no longer converge asymptotically to zero, this is no longer trivial. For the case of a linear plant (i.e., $\mathbf{W}$ linear) it has been shown that if a certain measure of the excitation of the desired trajectory exceeds a measure of the disturbance, the actual trajectory will also be sufficiently exciting (Narendra and Annaswamy 1986). For the nonlinear case of interest here, it is not yet apparent how to simply state this requirement.

We have shown that the adaptive control system designed for the ideal case of only structured uncertainty will exhibit robustness to bounded external disturbances and to a certain class of a priori bounded, unmodeled dynamics. Servo errors will be upper bounded by a value computable from the chosen servo gains, and parameter errors will remain small if a certain persistent excitation condition is met. In the worst case, parameters will remain within their a priori known bounds.

9. Simulation Results

A simple, two-degree-of-freedom manipulator (as shown in Fig. 2) was simulated to test the adaptive algorithm. The manipulator was modeled as two rigid links (of lengths l_1, l_2) with point masses at the distal ends of the links (m_1, m_2). It moves in a vertical plane with gravity acting. Both viscous (v_i coefficients) and Coulomb friction (k_i coefficients) are simulated at the joints.

The equations of motion for this device are (Craig 1986*a*):

$$\begin{aligned} \tau_1 &= m_2 l_2^2(\ddot{\theta}_1 + \ddot{\theta}_2) + m_2 l_1 l_2 c_2(2\ddot{\theta}_1 + \ddot{\theta}_2) \\ &\quad + (m_1 + m_2) l_1^2 \ddot{\theta}_1 - m_2 l_1 l_2 s_2 \dot{\theta}_2^2 - 2 m_2 l_1 l_2 s_2 \dot{\theta}_1 \dot{\theta}_2 \\ &\quad + m_2 l_2 g s_{12} + (m_1 + m_2) l_1 g s_1 + v_1 \dot{\theta}_1 \\ &\quad + k_1 \operatorname{sgn}(\dot{\theta}_1), \\ \tau_2 &= m_2 l_1 l_2 c_2 \ddot{\theta}_1 + m_2 l_1 l_2 s_2 \dot{\theta}_1^2 + m_2 l_2 g s_{12} \\ &\quad + m_2 l_2^2(\ddot{\theta}_1 + \ddot{\theta}_2) + v_2 \dot{\theta}_2 + k_2 \operatorname{sgn}(\dot{\theta}_2), \end{aligned} \quad (51)$$

where $c_1 = \cos(\theta_1)$, $s_{12} = \sin(\theta_1 + \theta_2)$, etc.

The unknown parameters were considered to be:

$$P = [m_1 \quad m_2 \quad k_1 \quad v_1 \quad k_2 \quad v_2]^T. \quad (52)$$

Fig. 3. Simulation of the adaptive scheme.

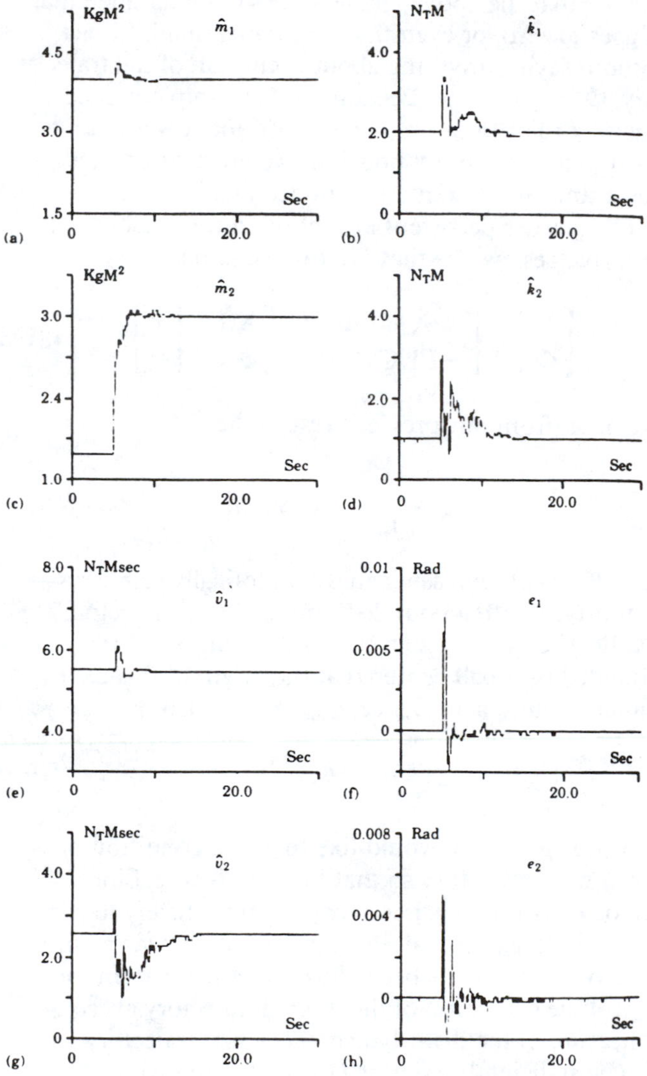

Hence the **W** matrix is

$$\mathbf{W} = \begin{bmatrix} w_{11} & w_{12} & w_{13} & w_{14} & 0 & 0 \\ 0 & w_{22} & 0 & 0 & w_{25} & w_{26} \end{bmatrix}, \tag{53}$$

where

$$\begin{aligned} w_{11} &= l_1^2\ddot{\theta}_1 + l_1 g s_1, \\ w_{12} &= (l_1^2 + l_2^2 + 2l_1l_2c_2)\ddot{\theta}_1 + (l_2^2 + l_1l_2c_2)\ddot{\theta}_2 \\ &\quad + l_1gs_1 + l_2gs_{12} - l_1l_2s_2\dot{\theta}_2^2 - 2l_1l_2s_2\dot{\theta}_1\dot{\theta}_2, \\ w_{13} &= \text{sgn}\,(\dot{\theta}_1), \\ w_{14} &= \dot{\theta}_1, \\ w_{22} &= (l_2^2 + l_1l_2c_2)\ddot{\theta}_1 + l_2^2\ddot{\theta}_2 + l_1l_2s_2\dot{\theta}_1^2 + l_2gs_{12}, \\ w_{25} &= \text{sgn}\,(\dot{\theta}_2), \\ w_{26} &= \dot{\theta}_2. \end{aligned} \tag{54}$$

The desired trajectory had the form

$$\begin{aligned} \theta_{1d} &= a_1 + b_1(\sin(t) + \sin(2t)), \\ \theta_{2d} &= a_2 + b_2(\cos(4t) + \cos(6t)). \end{aligned} \tag{55}$$

Figure 3 shows some results from a typical simulation. In this case, all parameters were initially tuned to their true values, but the value of mass at the end of the second link, m_2, undergoes a step change from 2 kg to 3 kg at $t = 5$ s. These plots indicate how the system can track a step parameter change. Note how all parameters, as well as the servo errors, are temporarily disturbed by the change. Further simulations, including situations in which all parameters are initially mistuned as well as studies of the effects of process noise, unmodeled dynamics, and insufficient excitation, are given in Craig (1986*b*).

The time scale on simulations such as these may be misleading. It would have been possible to adjust the γ_i so that the adaptation was much more rapid, but an attempt was made to use numbers that were felt to be reasonable.

10. Experimental Results

This section describes the results of an experimental implementation of the nonlinear, model-based, adaptive controller. The implementation was done for the two major links of the Adept One, which is a four-axis, "Scara"-style manipulator.

10.1. Dynamics of the Adept One

Figure 4 shows a schematic drawing of links 1 and 2 of the Adept One. Both actuators are located at the base of the robot. Joint 1's actuator applies torque τ_1 to the link 1 structure, which has rotational inertia, I_1.

Fig. 4. Top view of the Adept One.

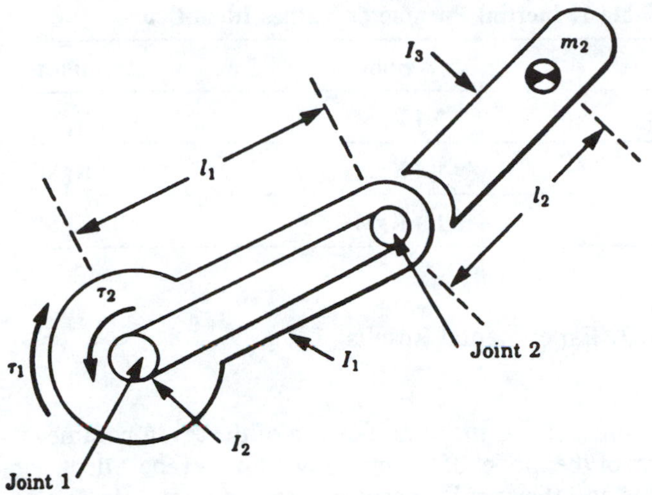

Joint 2's actuator applies a torque τ_2 to the inner column (with inertia I_2), which drives joint 2 through a steel band. Link 2 is of mass m_2 and has its center of mass located at distance l_2 from the axis of joint 2. Link 2 has a rotational inertia about this mass center of I_3. The distance between joint axes 1 and 2 is l_1.

Due to the placement of the actuators and the transmission system, the actuator angles are not equivalent to the joint angles. We will use $\mathbf{\Theta} = [\theta_1 \quad \theta_2]^T$ to indicate the joint angles, and $\mathbf{A} = [A_1 \quad A_2]^T$ to indicate the actuator angles. The transmission coupling is given by:

$$\begin{bmatrix} \theta_1 \\ \theta_2 \end{bmatrix} = \begin{bmatrix} -k & 0 \\ k & k \end{bmatrix} \begin{bmatrix} A_1 \\ A_2 \end{bmatrix}, \tag{56}$$

where k is a constant relating encoder counts to radians.

The dynamics will be written in *actuator space.* That is, we will control (and decouple) in the two-dimensional space defined by the two actuator rotations. Hence, the dynamics were developed that relate actuator accelerations and velocities to torque required at the actuators. These equations are:

$$\begin{aligned} \tau_1 &= p_1 \ddot{A}_1 + p_3(-c\theta_2 \ddot{A}_2 + ks\theta_2 \dot{A}_2^2) \\ &\quad + p_4 \dot{A}_1 + p_6 \operatorname{sgn}(\dot{A}_1), \\ \tau_2 &= p_2 \ddot{A}_2 + p_3(-c\theta_2 \ddot{A}_1 + ks\theta_2 \dot{A}_1^2) \\ &\quad + p_5 \dot{A}_2 + p_7 \operatorname{sgn}(\dot{A}_2), \end{aligned} \tag{57}$$

where for convenience, a dependence on θ_2 has been shown (this could obviously be written in terms of A_1 and A_2, if so desired). Terms representing viscous and Coulomb friction have been added to each actuator. The parameters p_1–p_3 are given in terms of the physical parameters as

$$\begin{aligned} p_1 &= I_1 + m_2 l_1^2, \\ p_2 &= I_2 + I_3 + m_2 l_2^2, \\ p_3 &= m_2 l_1 l_2. \end{aligned} \tag{58}$$

10.2. Experimental Implementation

Prior to implementing the adaptive controller, rough estimates of the inertial parameters were available as

$$\begin{aligned} p_1 &= 3.24, \\ p_2 &= 1.4, \\ p_3 &= 1.4, \end{aligned} \tag{59}$$

in KgM^2.

We need to restrict $\hat{p}_1$, $\hat{p}_2$, and $\hat{p}_3$ to lie on ranges chosen so that $\hat{\mathbf{M}}(\Theta)$ remains invertible. Based on our a priori knowledge, we set parameter bounds as

$$\begin{aligned} 2.9 &< \hat{p}_1 < 3.58, \\ 1.06 &< \hat{p}_2 < 1.74, \\ 1.06 &< \hat{p}_3 < 1.74. \end{aligned} \tag{60}$$

Servo gains were $k_p = 1000\ \text{s}^{-2}$ and $k_v = 65\ \text{s}^{-1}$ for both joints.

Trajectories produced by the Adept controller are used directly. These trajectories are created by smoothly interpolating a set of *via points* specified by the user. The nature of the interpolation used is characterized by short periods of acceleration and deceleration when path segments initiate and terminate, connected in between by constant velocity slews. The device is generally experiencing accelerations only near the start or end of each individual motion, and therefore inertial parameters are adjusted only during these brief periods. The implementation in fixed point representation automatically causes a dead zone so that small amounts of noise do not cause parameter estimates to drift.

Fig. 5. Identification of the Adept parameters.

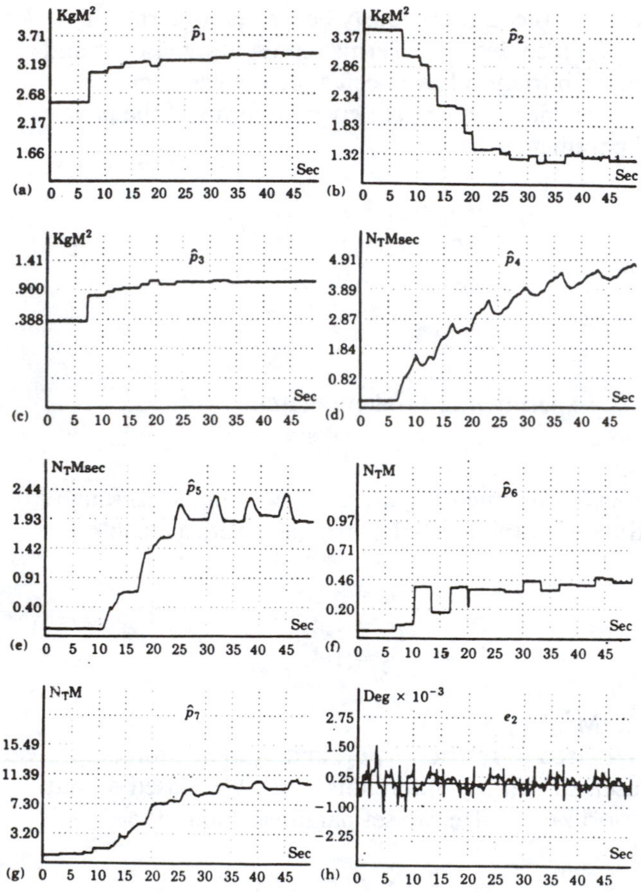

Table 1. Inertial Parameter Values Identified

	A priori	Identified
p_1	3.4 KgM²	3.34 KgM²
p_2	1.4 KgM²	1.32 KgM²
p_3	1.4 KgM²	1.05 KgM²

The algorithm is written in C using 32-bit integer arithmetic running on two Motorola 68000 processors. One processor gets the desired trajectory from the usual Adept controller and implements the computed torque servo for links 1 and 2. The second processor runs the adaptive algorithm and writes the newly updated parameter values into the first processor's memory on each servo cycle. The servo runs at a rate of 250 hz, and each processor was only utilized about 60% of the time. It appears to be possible to control all four joints with the same hardware. If, for example, Motorola 68020 processors were available, or if the coding was carefully checked for efficiency, it would be possible to achieve a higher servo rate or use only one processor. Hence, any fears that this scheme is too complex to be economically implemented for an industrial robot seem unfounded.

10.3. Experimental Results

Figure 5 shows the identification of the seven parameters of the model of the major two links of the Adept One and the angular position error for joint 2. In this experiment, the parameters were initially detuned, and adaptation was enabled at $t = 7.5$ s. The values identified for the inertial parameters came reasonably close to those approximately known a priori by Adept (see Table 1). It should be noted that these results can be made to appear quite different depending on how the γ_i are chosen. Namely, the adaptation can be more rapid, at the expense of more "noise" in the final value identified. In the experiment reported in Fig. 5, since the purpose was to do identification, the γ_i were set quite low, with the effect that low-pass, filtered estimates were obtained.

Figure 6 shows the estimate of the inertia seen by actuator 2, $\hat{p}_2$, when a load is suddenly picked up by the manipulator. The mass of the load was 2.9 kg, and was located at a distance of 0.375 m from joint 2, and so should cause an increase in $\hat{p}_2$ of 0.409 kgm². The change in the estimate is seen to be about 0.38 kgm², an error of 7.1%.

In Fig. 7A, the error on joint 2 is shown for the test of picking up a load at time $t = 20$ s using the adaptive controller with high adaptation rate. Figure 7B shows the same test (here mass is grasped at $t = 25$ s) with the standard Adept controller in action. These results indicate that the adaptive controller is still not outperforming Adept's fixed controller, but is quite close. However, the Adept controller is running with a servo rate of 500 hz, as opposed to the adaptive controller's 250 hz. We feel that with more engineering effort spent on careful implementation, the adaptive controller will outperform the best fixed controllers

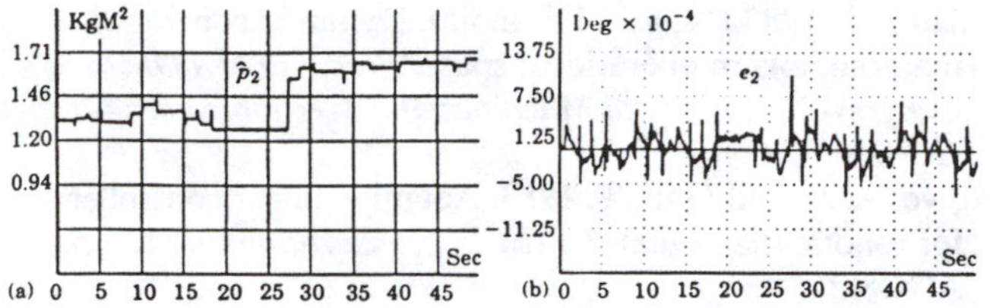

Fig. 6. Adaptation when a mass is grasped by the hand.

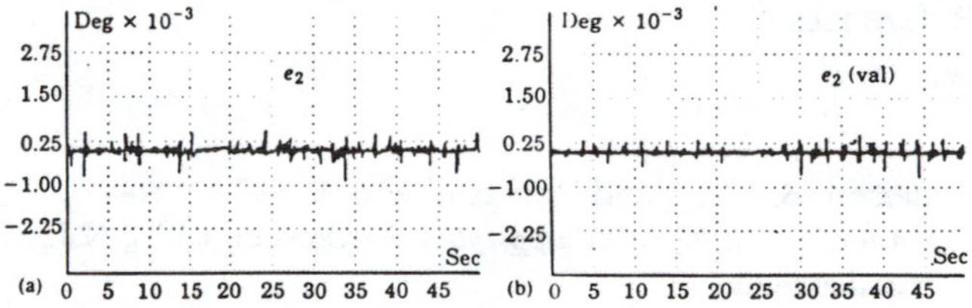

Fig. 7. Tracking error for adaptive controller and Adept controller.

found with present industrial controllers in certain tests. Tracking trajectories with low error is the area in which the adaptive controller will improve performance relative to fixed-gain controllers. However, in another important application of manipulators—fast pick-and-place moves (where exact trajectory shape is not important)—actuators may be saturating, and then, clearly the adaptive controller will make no improvement, and in fact, may not function as well.

11. Conclusions

A globally stable, adaptive, control scheme for a complex, nonlinear system has been designed. The adaptation process can be added to the model-based, control formulation for robot manipulators (sometimes called the computed torque method) without otherwise altering the controller structure.

Using the results in Section 7, trajectories especially well suited to identification of parameters might be preplanned; however, most real-world trajectories carried out by industrial robots are sufficiently exciting, and so our scheme can be used as an on-line controller.

This method could be directly applied to a Cartesian-based (or *operational space*) control scheme, such as the one reported in Khatib (1983). This would be essentially straightforward with minor additions to insure that the product of **W** and the Cartesian mass matrix remain bounded at all times. An interesting application could be made to redundant manipulators controlled in Cartesian space. For such manipulators, internal motions of the manipulator that do not effect the end point might be designed to be well suited to identifying dynamic properties of the manipulator. The method might also be extended to include an adaptive force control servo, where some identified parameters are associated with properties of the task surfaces rather than strictly with the manipulator itself. On the theoretical side, further research and analysis of coping with a more general class of unmodeled dynamics is required, as is presently the case in all adaptive control work.

It is important to realize that although a significant amount of computation is required to implement the controller, it doesn't make the method impractical. This has been shown in Section 10 (albeit in a simple case) for a two-link industrial robot. There have been implementations of the computed torque servo running in several laboratories for some time, and the adaptive controller does not require significantly more computation (Khatib 1983; Kanade, Khosla, and Tanaka 1984). To make a rough estimate, we feel that N (the number of joints) processors with approximately the same power as the Motorola 68000 would be more than sufficient for implementing the complete control algorithm for a general manipulator. Given an industrial robot controller that already must provide power supplies, a backplane, etc., the incremental cost of adding one of these processors is probably presently about $500 and still dropping. On the other hand, the entire system of manipulator and controller could cost $50,000 or more. Schemes like the computed torque servo, together with an adaptive scheme such as the one presented in this paper, are becoming economically justifiable.

Acknowledgments

A special thanks to Dr. Jim Maples, who provided a great deal of help with the experiments at Adept.

References

Anderson, B. D. O., and Vongpanitlerd, S. 1973. *Network synthesis: a state space approach.* Englewood Cliffs, N.J.: Prentice Hall.

Anderson, B. D. O. 1977. Exponential stability of linear equations arising in adaptive identification. *IEEE Trans. Automatic Contr.* AC-22:83–88.

Astrom, K., et al. 1977. Theory and applications of self-tuning regulators. *Automatica* 19:457–476.

Burdick, J. W. 1986 (San Francisco). An algorithm for generation of efficient manipulator dynamic equations. *Proc. 1986 IEEE Int. Conf. Robotics and Automation.* pp. 212–218.

Craig, J. J. 1986*a*. *Introduction to robotics: mechanics and control.* Reading, Mass.: Addison-Wesley.

Craig, J. J. 1986*b*. Adaptive control of mechanical manipulators. Ph.D. thesis, Stanford University, Department of Electrical Engineering.

Dubowsky, S., and DesForges, D. T. 1979. The application of model-referenced adaptive control to robotic manipulators. *ASME J. Dyn. Sys., Meas. Contr.* 101.

Horowitz, R., and Tomizuka, M. 1980. An adaptive control scheme for mechanical manipulators—compensation of nonlinearity and decoupling control. ASME Paper No. 80-WA/DSC-6.

Kanade, T. K., Khosla, P. K., and Tanaka, N. 1984 (Las Vegas, Nevada). Real-time control of the CMU direct drive arm II using customized inverse dynamics. *Proc. 23rd IEEE Conf. Decision and Contr.*

Khatib, O. 1983 (New Delhi, India). Dynamic control of manipulators in operational space. *Proc. 6th IFTOMM Congress on Theory of Machines and Mechanisms, IFTOMM.*

Koivo, A. J., and Guo, T. 1983. Adaptive linear controller for robotic manipulators. *IEEE Trans. Automatic Contr.* AC-28(2).

Kreisselmeier, G. 1977. Adaptive observers with exponential rate of convergence. *IEEE Trans. Automatic Contr.* AC-22.

Lee, C. S. G., and Chung, M. J. 1984. An adaptive control strategy for mechanical manipulators. *IEEE Trans. Automatic Contr.* AC-29(9).

Leininger, G. 1983 (Bretton Woods, N.H.). Adaptive control of manipulators using self-tuning methods. *Proc. 1st ISRR.*

Luh, J. Y. S., Walker, M. W., and Paul R. P. 1980. On-line computational scheme for mechanical manipulators. *Trans. ASME J. Dy. Sys., Meas. Contr.*

Morgan, A. P., and Narendra, K. S. 1977. On the uniform asymptotic stability of certain linear nonautonomous differential equations. *SIAM J. Contr. and Optimization* 15.

Narendra, K. S., and Annaswamy, A. M. 1986. Robust adaptive control in the presence of bounded disturbances. *IEEE Trans. Automatic Contr.* AC-31.

Parks, W. 1966. Lyapunov redesign of model reference adaptive control systems. *IEEE Trans. Automatic Contr.* AC-11(3).

Rudin, W. 1976. *Principles of mathematical analysis.* Menlo Park, Ca.: McGraw-Hill Book Company.

Sastry, S. S. 1984. Model-reference adaptive control—stability, parameter convergence, and robustness. *IMA J. Math. Contr. and Information,* pp. 27–66.

Part 5
Learning Control

S. ARIMOTO
University of Tokyo

SEVERAL researchers struggling in the same scientific frontiers simultaneously and independently came to a common idea. It originated with the observation that modern robot manipulators, being subject to "playback control mode," repeat their motions over and over in cycles. The first to be favored by this fortune were Uchiyama [1] and, second but independently, J. J. Craig (paper 2), G. Casalino and G. Bartolini [2], and S. Arimoto et al. (paper 1). All these people expected that the performance of repetitive tasks such as painting (tracking a preassigned path) and pick-and-place must be improved by using the information (measurement data) gathered in the previous cycles. More specifically, they predicted that if tracking errors appear to remain the same over several repeated trials then the controller must compensate by using a control input that can be properly modified by certain signal components extracted from the previous tracking-error signals. In the earliest paper, by Uchiyama, [1], which was written only in Japanese, use of the phase-gain of the position-error signal was proposed, although the argument was unrefined and the analysis was restricted to one-degree-of-freedom linear servo-systems.

Rigorous and refined exploration of this type of learning control is first discussed in the independent papers reprinted here by Arimoto et al., Craig, and Casalino and Bartolini, all published in the early half of 1984. In Arimoto et al., a simpler algorithm than Uchiyama's is devised for updating the torque input in such a way that

$$u_{k+1}(t) = u_k(t) - \Gamma \frac{d}{dt} \Delta q_k'(t) \tag{1}$$

where

$$\Delta q_k'(t) = q_k'(t) - q_d'(t) \tag{2}$$

and $u_k(t)$ and $q_k'(t)$ denote the input-torque signal and the joint-velocity signal respectively at the kth trial, and $q_d(t)$ denotes the preassigned joint velocity vector. In both Craig and Casalino and Bartolini, the PID law

$$u_{k+1}(t) = u_k(t) + \left(\Gamma \frac{d}{dt} + \Phi + \int \psi \, d\tau\right) \Delta q_k'(t) \tag{3}$$

is proposed. In all three papers the problem of convergence of $q_k'(t)$ to $q_d'(t)$ as $k \to \infty$ was discussed on the basis of robot models linearized in the vicinity of the desired joint trajectory. There is, however, a decisive difference between two types of control concept presented. Craig implicitly assumes that the desired position signal $q_d(t) = (\int_0^t q_d'(\tau)\, d\tau + q_d(0))$ is continuous and periodic with finite period T. Instead of this, Arimoto et al. and Casalino and Bartolini suppose that the manipulator movement in every cycle starts from the same rest position, in effect, it satisfies the reinitialization condition

$$q_k(0) = q_d(0), \qquad q_k'(0) = q_d'(0) \tag{4}$$

for all $k = 1, 2, \cdots$. In Craig an additional assumption of higher gains of PD-closed (local position and velocity) feedback loops is required to attain the convergence, because of lack of the reinitialization. In the paper by Arimoto et al. reprinted here, the convergence is established for a class of linearized robot models. In the latter half of 1984, convergence was established for full nonlinear robot dynamics [3], [4]. The result can be summarized as

THEOREM [3] *Suppose that a desired motion $q_d'(t)$ is preassigned over a finite interval $t \in [0, T]$ and the manipulator is subject to the equation of motion*

$$(H_0 + H(q_k))q_k'' + h(q_k', q_k) = u_k. \tag{5}$$

If an appropriate choice of gain matrix Γ in (1) satisfies the inequality

$$\|I - (H_0 + H(q))^{-1}\Gamma\| < 1 \tag{6}$$

for all q with an appropriate natural matrix norm $\|\cdot\|$, then the D-type learning law described by (1) with the reinitialization of (4) implies the convergence of trajectories $q_k(t)$ to $q_d(t)$ as $k \to \infty$.

The proof goes on quite similarly to Picard's recursive construction of the unique solution to an ordinary differential equation. In the case of the PID learning law described by (3) the convergence is assured under the same condition as in (6) because the existence of Γ (D-gain in PID) is crucial in that proof. The sufficient condition of convergence (inequality (6)) suggests that there is a good margin in choice of Γ because for a wide class of serial-link manipulators. In fact, the inertia matrix $H(q)$ satisfies

$$\alpha I \leqq H(q) \leqq \beta I \tag{7}$$

with appropriate positive constants α and β, where diagonal components of diagonal matrix H_0 refer to a load distribution inside actuators and hence are as large as β. For example, the choice $\Gamma = \beta I + H_0$ satisfies (6), which means that to design the learning law there is no need to know the precise dynamics of the manipulator. Furthermore, it is shown by the use of an adequate function norm that the speed of convergence is exponential.

Based upon these observations, the term *learning control* has been used since 1984. The principles that underlie learning control can now be summarized in the following set of postulates:

P1. Every trial ends in a fixed time of duration $T > 0$.

P2. A desired output $y_d(t)$ is given a priori over that time with duration $t \in [0, T]$.

P3. Repetition of the initial setting is satisfied, that is, the initial state $x_k(0)$ of the objective system can be set the same at the beginning of each trial:

$$x_k(0) = x^0 \qquad \text{for } k = 1, 2, \cdots . \tag{8}$$

P4. Invariance of the system dynamics is ensured throughout these repeated trials.

P5. Every output $y_k(t)$ can be measured and therefore the residual error signal Δy_k

$$\Delta y_k(t) = y_k(t) - y_d(t) \tag{9}$$

can be utilized in construction of the next input $u_{k+1}(t)$.

P6. The system dynamics are invertible, that is, for a given desired output $y_d(t)$ with a piecewise continuous derivative, there is a unique input $u_d(t)$ that excites the system and yields the output $y_d(t)$.

Then the problem is to find a recursive control law

$$u_{k+1}(t) = F(u_k(t), \Delta y_k(t)) \tag{10}$$

and a function norm $\|\cdot\|$ such that $\|\Delta y_k\|$ vanishes as k tends to infinity. The simpler the recursive form $F(u, \Delta y)$ in (10), the better it is for practical implementation of the learning control law, as long as convergence is assured and its speed is satisfactory. Thus, such control laws as (1) and (3) are potential candidates satisfying these conditions. It is implicitly assumed that the initialization condition in postulate P3 can be achieved by a local PD feedback stabilization. In fact, it is well known that the rest state (the equilibrium) is asymptotically stable for a class of robot dynamics with a local PD feedback (see Part 2).

The set of postulates above also reflects program learning and generation for the acquisition of various kinds of fast but skilled motion. Physiology suggests that ideal motion must be acquired through a succession of trainings. As in music and sports, once an idealized form of motion is pointed out by teachers, one must repeat physical exercises to make his/her motion approach the ideal form. Through a sufficient number of repeated trials, a program is formed in the central nervous system that can generate a succession of input command signals that excites a certain system of muscles and tendons concerned with that motion and realizes the desired motion form.

Now let us return to the discussion of recursive learning algorithms. C. G. Atkeson and J. McIntyre (paper 3) discuss problems in the practical implementation of the learning algorithm in equation (1); in particular, the effect of unmodeled terms on the performance of learning. In addition, they propose the use of a digital differentiating filter with an 8 Hz cutoff in numerical differentiation of measured velocity and position signals contaminated by noise. The novelty of Casalino and Bartolini's paper, which is revised from their original work [2] is that it points out the importance of proving the uniform boundedness of trajectories throughout repeated trials. The proof in this paper is also based on a linearized model of robot dynamics. It was just in 1990 [5]–[8] that the uniform boundedness of trajectories was finally and completely proved on the basis of full nonlinear robot dynamics.

In conclusion, we briefly comment on other significant papers listed in chronological order in [9] through [26]. In December 1985 the first special session on learning control was organized by M. Togai and S. Arimoto in the International Conference on Decision and Control, which took place in Fort Lauderdale, Florida. Five papers were presented at this conference. Three of these papers, [9], [10], [11], dealt with learning control based on (1) and the other two, [12], [13], were concerned with repetitive control as it is called in Craig's treatment. In the latter two papers there was an obscure passage that could not eliminate the possibility of ripples excited by small but higher-frequency components of impulsive noises, because signals with a finite time duration are not band-limited in the frequency domain. Repetitive control was also the focus of [14], but it first presented the rigorous treatment by pointing out the importance of passivity of linearized robot models if PD feedback gains are large enough. Experiments [10] and further extensions of learning control to the cases of hybrid control [15], flexible joints [16], and the problem of selecting a better initial input based on the learned knowledge base [17] were carried out by Arimoto's group. Through these practices, Kawamura and Arimoto noticed that a P-type learning law without differentiation of velocity signals, that is, the recursive law

$$u_{k+1}(t) = u_k(t) - \Phi \Delta y_k(t) \tag{11}$$

does work well and leads experimentally to convergence after several repeated trials. It seemed quite difficult, however, to prove it theoretically. Two papers [18] and [19] presented a weak result on convergence of the P-type algorithm. The argument in the proof was based on a linearized model of robot dynamics around the desired trajectory and the ignorance of higher terms. This argument still leaves an important problem concerning robustness of learning control algorithms with respect to the existence of initialization errors (to some extent), small but nonrepeatable fluctuations of dynamics, and measurement noises. Thus, postulates P3 to P5 should be replaced by the following more practical ones:

P3a. The system is initialized at the beginning of every trial within a limited magnitude of errors, in effect,

$$\|x_k(0) - x^0\| \leqq \epsilon_1, \qquad k = 1, 2, \cdots \tag{12}$$

for some small constant $\epsilon_1 > 0$.

P4a. The magnitude of nonrepeatable fluctuations of the system dynamics satisfies

$$\sup_{t \in [0,T]} \|\eta_k(t)\| \leqq \epsilon_2 \tag{13}$$

for some $\epsilon_2 > 0$.

P5a. The magnitude of measurement noise $\xi_k(t)$, which arises in such a way that

$$\Delta y_k(t) + \xi_k(t) = \{y_k(t) + \xi_k(t)\} - y_d(t),$$

satisfies

$$\sup_{t \in [0,T]} \|\xi_k(t)\| \leqq \epsilon_3 \tag{14}$$

for some $\epsilon_3 > 0$.

The robust problem under such relaxed postulates was first discussed by the Arimoto et al. [20] in the case of PID-type learning control. The argument, however, was based on an assumption that the initial trajectory (and thereby all subsequent ones) lies in a neighborhood of the desired trajectory and hence the robot dynamics can be considered to be represented by a linear time-varying system. Recently Heinzinger et al. [21] attacked the same robustness problem for a class of D-type learning control and proved without use of any linearization that learned output trajectories converge to a neighborhood of the desired one. Their proof was indebted to Hauser's [22] complete treatment of the convergence proof. One more important comment was given in [21] by illustrating a counter-example that claims that such a robustness property is not valid for a class of PI-type learning. However, recently a series of papers, [23], [5], [6], [7], [8], has shown that a P-type learning law also becomes robust under P3a through P5a, provided that a forgetting factor $\alpha > 0$ is introduced in the recursive law in the following form:

$$u_{k+1}(t) = (1 - \alpha)u_k(t) + \alpha u_0(t) - \Phi\{\Delta y_k(t) + \xi_k(t)\}. \tag{15}$$

The original idea of using a forgetting factor originated with Heinzinger et al. [21], but it was introduced only into D-type learning. It has been rigorously proved, by exploring the exponential passivity of residual-error dynamics, that the P-type algorithm with a forgetting factor gives rise to the convergence to a neighborhood of the desired one with size $O(\alpha)$. Moreover, if the content of a long-term memory, in effect, $u_0(t)$ in (15), is refreshed after every repeated K trials where K is of $O(1/\alpha)$, then the trajectories converge to an ϵ-neighborhood of the desired one [7], [8]. The size of ϵ can be small, dependent on magnitudes ϵ_1 through ϵ_3 in P3a through P5a. It is also proved on the basis of full nonlinear dynamics that trajectory errors are uniformly bounded in the sense of uniform norm throughout repeated trials. These results were ascertained by experiments using an industrial robot (PUMA-260) [24].

Before finishing this introduction to learning control, we explain how an extended notion of passivity plays a crucial role in the proof of convergence of trajectories. To gain an insight, we consider the problem under the set of idealized postulates P1 through P6. First note that there exists a unique desired input $u_d(t)$ according to P6. Subtracting u_d from both sides of (11), we obtain

$$\Delta u_{k+1} = \Delta u_k - \Phi \Delta y_k \tag{16}$$

where $\Delta u_k = u_k - u_d$. Equation (16) leads to

$$\Delta u_{k+1}^T \Phi^{-1} \Delta u_{k+1} = \Delta u_k^T \Phi^{-1} \Delta u_k + \Delta y_k \Phi \Delta y_k - 2\Delta y_k^T \Delta u_k. \tag{17}$$

Multiplying a function $e^{-\lambda t}$ and integrating over $[0, T]$ yields

$$\|\Delta u_{k+1}\|^2 = \|\Delta u_k\|^2 + \|\Phi \Delta y_k\|^2 - 2\int_0^T e^{-\lambda t} \Delta y_k^T(t) \Delta u_k(t)\, dt \tag{18}$$

where

$$\|\Delta u_k\|^2 = \int_0^T e^{-\lambda t} \Delta u_k(t) \Phi^{-1} \Delta u_k(t)\, dt. \tag{19}$$

Hence, if there are two positive constants λ and β such that

$$\int_0^T e^{-\lambda t} \Delta y_k^T(t) \Delta u_k(t)\, dt \geqq \frac{1+\beta}{2} \|\Phi \Delta y_k\|^2 \tag{20}$$

for all k, then (18) is reduced to

$$\|\Delta u_{k+1}\|^2 \leqq \|\Delta u_k\|^2 - \beta \|\Phi \Delta y_k\|^2. \tag{21}$$

It follows immediately from this inequality that $\Delta y_k \to 0$ as $k \to \infty$ in the sense of L^2-norm. The basic inequality (21) defines the exponentially weighted passivity with a quadratic margin concerning the residual error dynamics. A large class of serial-link manipulators satisfies this generalized passivity condition even if the endpoint of the manipulator is geometrically constrained, as shown in [25].

Current research on learning control seems to be directed at not only simple trajectory-following tasks but also a variety of other robot tasks [25], [26].

References

[1] M. Uchiyama, "Formation of high-speed motion pattern of a mechanical arm by trial," *Trans. SICE* (Society for Instrumentation and Control Engineers), vol. 14, pp. 706–712, 1978, (in Japanese).

[2] G. Casalino and G. Bartolini, "A learning procedure for the control of movements of robotic manipulators," *IASTED Symp. on Robotics and Automation*, pp. 108–111, 1984.

[3] S. Arimoto, S. Kawamura, and F. Miyazaki, "Can mechanical robots learn by themselves?" in *Robotics Research: The Second International Symposium*, H. Hanafusa and H. Inoue, Eds., Cambridge, MA: MIT Press, pp. 127–134, 1984.

[4] S. Arimoto, S. Kawamura, and F. Miyazaki, "Bettering operation of dynamic systems by learning: a new control theory for servomechanism and mechatronics systems," *Proc. 23rd Conf. Dec. and Contr.*, pp. 1064–1069, 1984.

[5] S. Arimoto, "Learning control theory for robotic motion," *Int. J. Adaptive Contr. and Signal Processing*, vol. 4, no. 6, pp. 543–564, 1990.

[6] S. Arimoto, T. Naniwa, and H. Suzuki, "Robustness of p-type learning control with a forgetting factor for robotic motions," *Proc. 29th Conf. Dec. and Contr.*, pp. 2640–2645, 1990.

[7] S. Arimoto, "Learning for skill refinement in robotic systems," *IEICE (Institute of Electronics, Information, and Communications Engineers) Trans.*, vol. E74, no. 2, pp. 235–243, 1991.

[8] S. Arimoto, T. Naniwa, and H. Suzuki, "Selective learning with a forgetting factor for robotic motion control," *Proc. IEEE Int. Conf. Robotics and Automation,* pp. 728–733, 1991.

[9] S. Arimoto, S. Kawamura, F. Miyazaki, and S. Tamaki, "Learning control theory for dynamical systems," *Proc. 24th Conf. Dec. and Contr.*, pp. 1375–1380, 1985.

[10] S. Kawamura, F. Miyazaki, and S. Arimoto, "Applications of learning method for dynamic control of robot manipulators," *Proc. 24th Conf. Dec. and Contr.*, pp. 1381–1386, 1985.

[11] M. Togai and O. Yamano, "Analysis and design of an optimal learning control scheme for industrial robots: a discrete time approach," *Proc. 24th Conf. Dec. and Contr.*, pp. 1399–1404, 1985.

[12] S. Hara, T. Omata, and M. Nakano, "Synthesis of repetitive control systems and its applications," *Proc. 24th Conf. Dec. and Contr.*, pp. 1387–1392, 1985.

[13] T. Mita and E. Kato, "Iterative control and its application to motion of robot arm—a direct approach to servo problems," *Proc. 24th Conf. Dec. and Contr.*, pp. 1393–1398, 1985.

[14] S. Kawamura, F. Miyazaki, and S. Arimoto, "Iterative learning control for robotic systems," *Proc. IECON'84,* pp. 393–398, 1984.

[15] S. Kawamura, F. Miyazaki, and S. Arimoto, "Hybrid position/force control of robot manipulators based on learning method," *Proc. Int. Conf. on Advanced Robotics,* pp. 235–242, 1985.

[16] F. Miyazaki, S. Kawamura, M. Matsumori, and S. Arimoto, "Learning control scheme for a class of robot systems with elasticity," *Proc. 25th Conf. Dec. and Contr.*, pp. 74–79, 1986.

[17] S. Kawamura, F. Miyazaki, and S. Arimoto, "Intelligent control of robot motion based on learning method," *Proc. IEEE Int. Symp. on Intelligent Contr.,* pp. 365–370, 1987.

[18] S. Arimoto, "Mathematical theory of learning with applications to robot control," in *Adaptive and Learning Systems,* K. S. Narendra, Ed., New York: Plenum, pp. 379–388, 1986.

[19] S. Kawamura, F. Miyazaki, and S. Arimoto, "Realization of robot motion based on a learning method," *IEEE Trans. Syst. Man, and Cyber.,* vol. SMC-18, pp. 126–134, 1988.

[20] S. Arimoto, S. Kawamura, and F. Miyazaki, "Convergence, stability, and robustness of learning control schemes for robot manipulators," in *Recent Trends in Robotics: Modeling, Control, and Education*, M. J. Jamshidi, L. Y. Luh, and M. Shahinpoor, Eds. New York: Elsevier, pp. 307–316, 1986.

[21] G. Heinzinger, D. Fenwick, B. Paden, and F. Miyazaki, "Robust learning control," *Proc. 28th Conf. Dec. and Contr.,* 1989.

[22] J. Hauser, "Learning control for a class of nonlinear systems," *Proc. 26th Conf. Dec. and Contr.,* pp. 859–860, 1987.

[23] S. Arimoto, "Robustness of learning control for robot manipulators," *Proc. IEEE Int. Conf. Robotics and Automation*, pp. 1523–1528, 1990.

[24] Y. Nanjo and S. Arimoto, "Experimental studies on robustness of a learning method with a forgetting factor," *Proc. 5th Int. Conf. on Advanced Robotics,* pp. 699–704, 1991.

[25] S. Arimoto and T. Naniwa, "Learning control for motions under geometric endpoint constraints," *Proc. IEEE Int. Conf. Robot. Automat.*, pp. 1914–1919, 1992.

[26] M. Aicardi, G. Cannata, and G. Casalino, "Hybrid learning control for constrained manipulators," to be published in *Advanced Robotics,* vol. 6, no. 1, 1992.

Bettering Operation of Robots by Learning

Suguru Arimoto, Sadao Kawamura, and Fumio Miyazaki
Faculty of Engineering Science, Osaka University, Toyonaka, Osaka, 560 Japan
Received January 26, 1984; accepted March 12, 1984

This article proposes a betterment process for the operation of a mechanical robot in a sense that it betters the next operation of a robot by using the previous operation's data. The process has an iterative learning structure such that the $(k + 1)$th input to joint actuators consists of the kth input plus an error increment composed of the derivative difference between the kth motion trajectory and the given desired motion trajectory. The convergence of the process to the desired motion trajectory is assured under some reasonable conditions. Numerical results by computer simulation are presented to show the effectiveness of the proposed learning scheme.

前回の作動データを用い次回の作動を改善する方式を用いた、ロボット機械系の作動の改善について述べる。この方式は反復学習構造を持ち、関節アクチュエーターへのＫ＋１番目の入力は、Ｋ番目の入力及び軌跡作動の要求値と実値との差の両者でもって決定される。ある妥当なる条件の下でのこの方式の収束性についての確認が出来た。ここで提案された方式の有効性を実証するため、シュミレーションの結果を示す。

I. INTRODUCTION

It is human to make mistakes, but it is also human to learn much from experience. Athletes have improved their form of body motion by learning through repeated training, and skilled hands have mastered the operation of machines or plants by acquiring skill in practice and gaining knowledge from experience. Upon reflection, can machines or robots learn autonomously (without the help of human beings) from measurement data of previous operations and make better their performance of future operations? Is it possible to think of a way to implement such a learning ability in the automatic operation of dynamic systems? If there is a way, it must be applicable to affording autonomy and intelligence to industrial robots.

Motivated by this consideration, we propose a practical approach to the problem of bettering the present operation of mechanical robots by using the data of previous operations. We construct an iterative betterment process for the dynamics of robots so that the trajectory of their motion approaches asymptotically a given desired trajectory as the number of operation trials increases.

In relation to this problem, we should point out that a learning machine called the "perception"[1] was investigated almost twenty years ago in the field of pattern recognition and some related methods of learning were proposed in the field of control engineering,[2] stimulated by the introduction of the perceptron. However, most of the past literature made nothing of the underlying dynamical structures of treated systems, or proposed learning processes had nothing to do with dynamics, since the "perceptron" was a static machine. As opposed to those static

Reprinted from *J. Robot. Syst.*, vol. 1, no. 2, pp. 123–140.

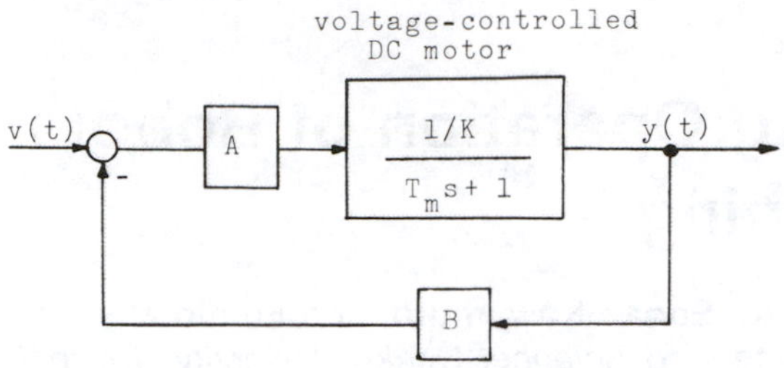

Figure 1. Feedback control system for speed control of a dc motor.

learning processes, our proposed learning scheme uses directly the underlying dynamics of object systems.

In the next section we first introduce a betterment process with iterative learning structure for a linear servomechanism system with single degree of freedom. To see the convergence of the iterative process and its speed, a computer simulation result is given. In Section III the proposed scheme of the betterment process is extended to a class of general linear time-varying dynamical systems, in order to cope with nonlinear dynamics of many degrees of freedom. A mathematically rigorous proof of the convergence for this general class of the iterative betterment process will be given in the Appendix. In Section IV we discuss the dynamics of serial-link-type robot manipulators and formulate a linear time-varying differential equation that governs the motion of manipulators around a given desired motion trajectory. In Section V we discuss the problem of applying the betterment process to the learning control of manipulators and use a few numerical examples that show the possibility of the application.

Some partial results on practical applications of the proposed learning scheme for actual robot manipulators, together with some numerical simulation results, will be given in a subsequent paper.[3]

II. PROPOSAL OF BETTERMENT PROCESS

First we consider a linear servomechanism system with a single degree of freedom, which consists of a feedback control (as shown in Fig. 1) for the speed control of a voltage-controlled dc servomotor (as shown in Fig. 2). It is well known that if the armature inductance is sufficiently small and mechanical friction is ignored, the dynamics of the motor in Figure 2 is expressed by a linear differential equation

$$T_m \frac{d}{dt} y(t) + y(t) = v(t)/K, \tag{1}$$

where y and v denote the angular velocity of the motor and the input voltage,

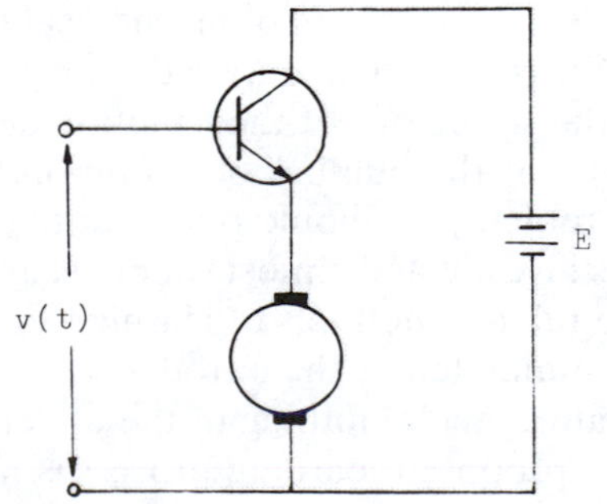

Figure 2. Voltage-controlled dc servomotor.

respectively, and K and T_m denote the torque constant and the time constant of the motor, respectively. Then, the closed-loop system shown in Figure 1 is subject to the following differential equation:

$$T_m \dot{y} + y = A(v - By)/K, \tag{2}$$

where $\dot{y} = dy/dt$. This can be represented in general form

$$\dot{y} + ay = bv, \tag{3}$$

where we set

$$a = (1 + AB/K)/T_m, \qquad b = A/KT_m.$$

It should be noted that the solution of Eq. (3) is expressed in integral form:

$$y(t) = e^{-at}y(0) + \int_0^t be^{-a(t-\tau)}v(\tau)d\tau. \tag{4}$$

Now we suppose that a desired time evolution $y_d(t)$ of angular velocity is given for a fixed finite time interval $t \in [0, T]$. Throughout this article, we implicitly assume that the description of the system dynamics is unknown, that is, exact values of a and b in Eq. (3) or (4) need not be known. It is also assumed implicitly that $y_d(t)$ is continuously differentiable and sufficiently smooth so that every value of $y_d(t)$ can be reproduced approximately with high accuracy by sampled and digitized data $\{y_d(k\Delta t), k = 0, 1, \ldots, N = T/\Delta t\}$, which could be stored on a VLSI RAM memory. We consider the problem of finding an input voltage $v(t)$ whose response is coincident with given $y_d(t)$ over the interval $t \in [0, T]$. To find such an ideal input, we first choose an arbitrary voltage function $v_0(t)$ and try to supply the motor with this input voltage. An error btween the desired output $y_d(t)$ and the first response $y_0(t)$ to supplied input $v_0(t)$ may arise. This error is expressed as

$$e_0(t) = y_d(t) - y_0(t), \tag{5}$$

where

$$\dot{y}_0(t) + ay_0(t) = bv_0(t). \tag{6}$$

If sufficiently densely sampled data on $e_0(t)$ over $t \in [0, T]$ are stored on another RAM memory, it is possible to produce a next input voltage of the form

$$v_1(t) = v_0(t) + b^{-1}\dot{e}_0(t). \tag{7}$$

Next we apply this input to the motor in the second trial of operation, store the second resultant error $e_1(t)$, and so forth, which results in the following iterative process:

$$\begin{aligned} v_{k+1}(t) &= v_k(t) + b^{-1}\dot{e}_k(t), \\ e_k(t) &= y_d(t) - y_k(t), \\ y_k(t) &= e^{-at}y_k(0) + \int_0^t be^{-a(t-\tau)}v_k(\tau)d\tau. \end{aligned} \tag{8}$$

The process of this scheme is depicted in Figure 3. We call this iterative pocess a betterment process because the error $e_k(t)$ vanishes rapidly as the number of trials increases and, moreover, for a norm $\|e_k\|$ it becomes less than for the previous value $\|e_{k-1}\|$. In fact, if $y_0(0) = y_d(0) = y_k(0)$ for any k, then, in view of

Eq. (8), the error is described by

$$\begin{aligned}
\dot{e}_k(t) &= \dot{y}_d(t) - \dot{y}_k(t) \\
&= \dot{y}_d(t) - \left[-ae^{-at} y_k(0) + bv_k(t) - \int_0^t abe^{-a(t-\tau)} v_k(\tau)d\tau \right] \\
&= \dot{y}_d(t) - \left[-ae^{-at} y_{k-1}(0) + bv_{k-1}(t) - \int_0^t abe^{-a(t-\tau)} v_{k-1}(\tau)d\tau \right] \\
&\qquad - \left[\dot{e}_{k-1}(t) - \int_0^t ae^{-a(t-\tau)} \dot{e}_{k-1}(\tau)d\tau \right] \\
&= \dot{y}_d(t) - \dot{y}_{k-1}(t) - \dot{e}_{k-1}(t) + \int_0^t ae^{-a(t-\tau)} \dot{e}_{k-1}(\tau)d\tau \\
&= a\int_0^t e^{-a(t-\tau)} \dot{e}_{k-1}(\tau)d\tau
\end{aligned} \tag{9}$$

which, in general, implies

$$\begin{aligned}
|\dot{e}_k(t)| &\le |a|^2 \int_0^t dt_1 \int_0^{t_1} e^{-a(t-t_2)} |\dot{e}_{k-2}(t_2)| \, dt_2 \\
&\le |a|^k \int_0^t \int_0^{t_1} \cdots \int_0^{t_{k-1}} |\dot{e}_0(t_k)| \, dt_k \, dt_{k-1} \cdots dt_1 \\
&\le \frac{K|a|^k T^k}{k!} \to 0 \qquad \text{as } k \to \infty,
\end{aligned} \tag{10}$$

where we set $K = \max_{0 \le t \le T} |\dot{e}_0(t)|$.

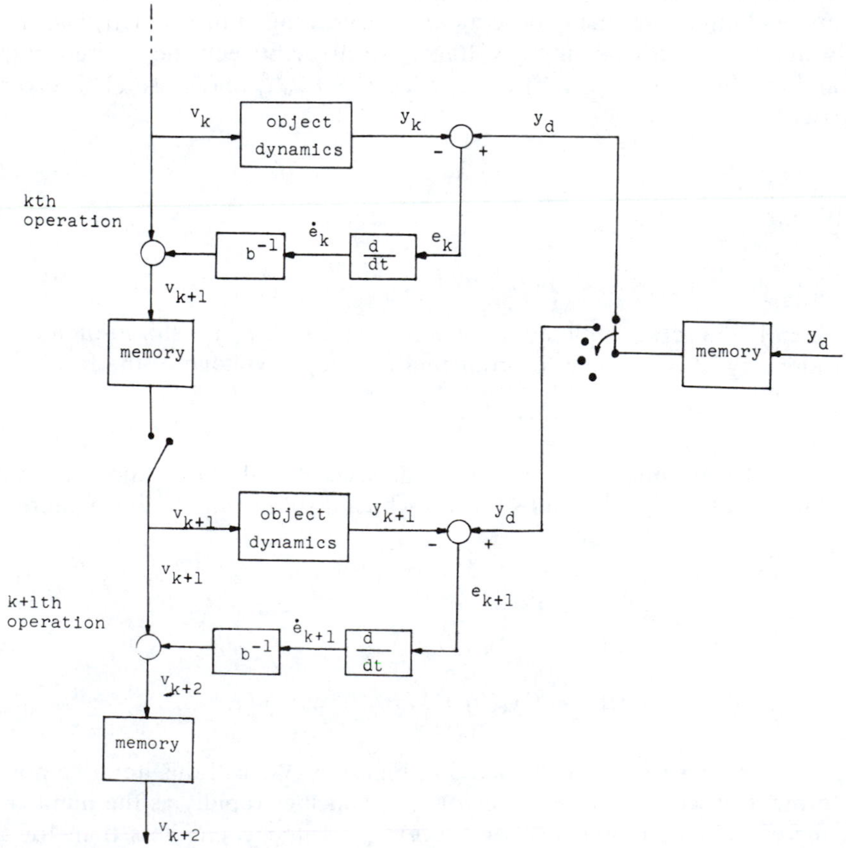

Figure 3. Iterative betterment process.

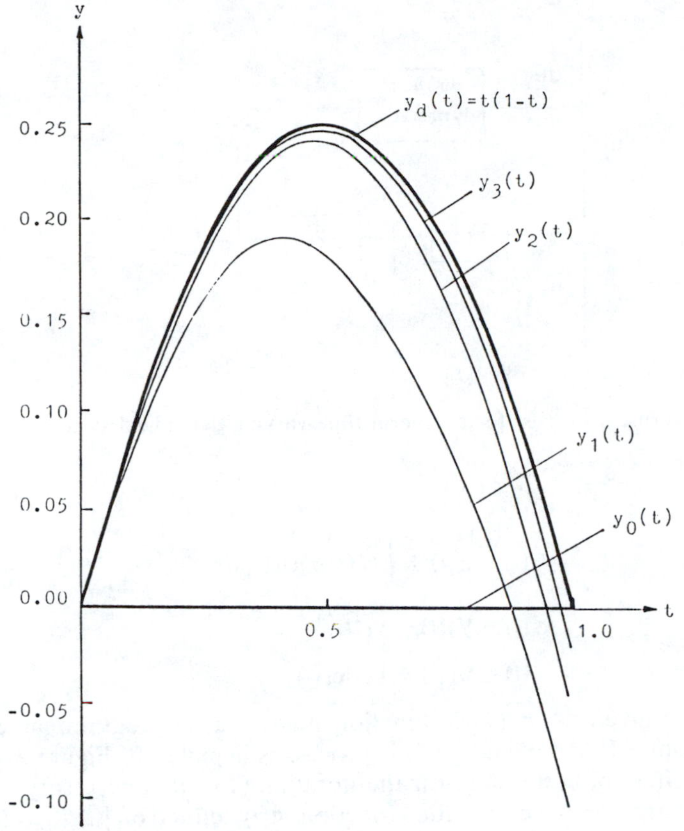

Figure 4. Convergence of betterment process.

To see the speed of convergence for this betterment process a numerical example is shown in Figure 4, in which $a = b = 1$, $y_d(t) \equiv t(1-t)$ over $t \in [0, 1]$, $u_0(t) \equiv 0$, and $y_d(0) = y_k(0)$.

It should be noted that, instead of the exact value of b used in the betterment process defined by Eq. (8), its arbitrary approximate value γ can be used, provided the following condition is satisfied:

$$|1 - b\gamma^{-1}| < 1. \tag{11}$$

This generalization and further extended treatment will be discussed in the next section.

III. BETTERMENT PROCESS FOR GENERAL TIME-VARYING SYSTEMS

Consider a general linear time-varying dynamical system whose input–output relation is described by

$$\mathbf{y}(t) = \mathbf{g}(t) + \int_0^t H(t, \tau)\mathbf{u}(\tau)d\tau, \tag{12}$$

where $\mathbf{u}$ denotes the input vector, $\mathbf{y}$ the output vector, and $\mathbf{g}(t)$ a given fixed vector such that $\mathbf{u}, \mathbf{y}, \mathbf{g} \in R^r$ and hence $H(t, \tau) \in R^{r \times r}$.

Suppose that a desired trajectory of system output $\mathbf{y}_d(t)$ is given over a fixed finite interval $t \in [0, T]$ and an initial input function $\mathbf{u}_0(t)$ is arbitrarily chosen. We assume that $\mathbf{y}_d(t)$, $\mathbf{u}(t)$, $\mathbf{g}(t)$, and $H(t, \tau)$ are continuously differentiable in t and τ.

Now we construct an iterative betterment process for system Eq. (12) as

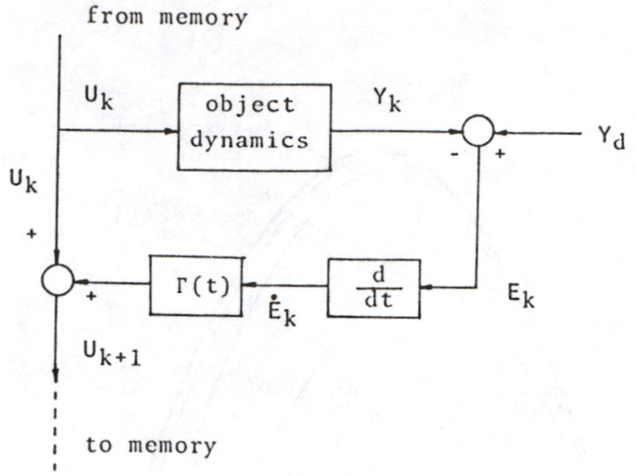

Figure 5. Betterment process for a general time-varying dynamical system.

follows:

$$\mathbf{y}_k(t) = \mathbf{g}(t) + \int_0^t H(t, \tau)\mathbf{u}_k(\tau)d\tau,$$
$$\mathbf{e}_k(t) = \mathbf{y}_d(t) - \mathbf{y}_k(t), \tag{13}$$
$$\mathbf{u}_{k+1}(t) = \mathbf{u}_k(t) + \Gamma(t)\dot{\mathbf{e}}_k(t),$$

where $\Gamma(t)$ is a given $r \times r$ matrix function that is easily programmable or reproducible from a RAM memory. This process is depicted in Figure 5.

To see the effect of betterment for the iteration (13), it is necessary to introduce a vector norm for r-vector-valued functions $\mathbf{e}(t)$ defined on $[0, T]$ as follows:

$$\|\mathbf{e}\|_\lambda = \sup_{0\le t\le T} \{e^{-\lambda t} \max_{1\le i\le r} |e_i(t)|\}, \tag{14}$$

where λ is a positive constant. Then, for the betterment process defined by Eq. (13), the following theorem holds:

THEOREM 1: If $\mathbf{y}_d(0) = \mathbf{g}(0)$, $\|I_r - H(t, t)\Gamma(t)\|_\infty < 1$ for all $t \in [0, T]$, and a given initial input $\mathbf{u}_0(t)$ is continuous on $[0, T]$, then there exist positive constants λ and ρ_0 such that

$$\|\dot{\mathbf{e}}_{k+1}\|_\lambda \le \rho_0 \|\dot{\mathbf{e}}_k\|_\lambda \quad \text{and} \quad 0 \le \rho_0 < 1 \tag{15}$$

for $k = 0, 1, 2, \ldots$, where the matrix norm $\|G\|_\infty$ for $r \times r$ matrix $G = (g_{ij})$ stands for

$$\|G\|_\infty = \max_{1\le i\le r} \left\{ \sum_{j=1}^{r} |g_{ij}| \right\}.$$

The proof will be given in the Appendix. It follows from this theorem that

$$\|\dot{\mathbf{e}}_k(t)\|_\lambda \to 0 \quad \text{as } k \to \infty, \tag{16}$$

and hence

$$\dot{\mathbf{y}}_k(t) \to \dot{\mathbf{y}}_d(t) \quad \text{as } k \to \infty. \tag{17}$$

It is clear by definition of the norm $\|\cdot\|_\lambda$ that these convergences are uniform in $t \in [0, T]$. Moreover, on account of the fact that

$$\mathbf{y}_k(0) = \mathbf{g}(0) = \mathbf{y}_d(0),$$

Eq. (17) implies that

$$\mathbf{y}_k(t) \to \mathbf{y}_d(t) \quad \text{as } k \to \infty \tag{18}$$

uniformly in $t \in [0, T]$.

Finally, we remark that linear time-varying systems described by

$$\dot{\mathbf{x}}(t) = A(t)\mathbf{x}(t) + B(t)\mathbf{u}(t), \tag{19a}$$

$$\mathbf{y}(t) = C(t)\mathbf{x}(t), \qquad \mathbf{x} \in R^n,\ \mathbf{u}, \mathbf{y} \in R^r \tag{19b}$$

are included in the class of dynamical systems described by Eq. (12). In fact, it is well known that Eq. (19) yields

$$\mathbf{y}(t) = C(t)X(t)\mathbf{x}(0) + \int_0^t C(t)X(t)X^{-1}(\tau)B(\tau)\mathbf{u}(\tau)d\tau, \tag{20}$$

where $X(t)$ is a unique matrix solution to the homogeneous matrix differential equation

$$\dot{X}(t) = A(t)X(t), \qquad X(0) = I_n. \tag{21}$$

Evidently, Eq. (20) is reduced to Eq. (12) by setting $\mathbf{g}(t) = C(t)X(t)\mathbf{x}(0)$ and $H(t, \tau) = C(t)X(t)X^{-1}(\tau)B(\tau)$. Hence, if

$$B(t) \equiv B, \qquad C(t) \equiv C,$$

and CB is nonsingular, then there exists a constant matrix such that

$$\| I_r - H(t, t)\Gamma \|_\infty = \| I_r - BC\Gamma \|_\infty < 1. \tag{22}$$

In conclusion, the betterment process for the system (19) with above conditions is convergent in a sense that $\mathbf{y}_k(t) \to \mathbf{y}_d(t)$ uniformly in $t \in [0, T]$ as $k \to \infty$.

IV. DYNAMICS OF ROBOT MANIPULATORS

Now we consider the dynamics of a serial-link manipulator with n degrees of freedom like the one shown in Figure 6. It is well known (see Ref. 4 or 5) that its dynamics can be expressed generally by the form

$$R(\boldsymbol{\theta})\ddot{\boldsymbol{\theta}} + \mathbf{f}(\dot{\boldsymbol{\theta}}, \boldsymbol{\theta}) + \mathbf{g}(\boldsymbol{\theta}) = \boldsymbol{\tau}, \tag{23}$$

where $\boldsymbol{\theta} = (\theta_1, \ldots, \theta_n)$ denotes the joint angle coordinates as shown in Figure 6, and $\boldsymbol{\tau} = (\tau_1, \ldots, \tau_n)$ denotes a generalized force vector. $R(\boldsymbol{\theta})$ is called an inertia matrix and is usually positive definite. The term $\mathbf{g}(\boldsymbol{\theta})$ comes from the potential energy of the manipulator system and has a simpler form than the inertia matrix. The term $\mathbf{f}(\dot{\boldsymbol{\theta}}, \boldsymbol{\theta})$ consists of centrifugal and Coriolis forces and other nonlinear characteristics such as frictional forces.

The task is usually described in terms of Cartesian coordinates or other task-oriented coordinates. However, we assume in this article that a desired motion of the manipulator is determined from a description of the task as a time function $\boldsymbol{\theta}_d(t)$ over $t \in [0, T]$ for joint angle coordinates $\boldsymbol{\theta}$. An extension to the treatment based on the task-oriented coordinates will be discussed in our subsequent paper.[3]

Now we apply a linear PD feedback control law to the manipulator, which we proposed in our previous papers[6–8] where we proved its effectiveness as a positioning control of manipulators. The control input is given by

$$\boldsymbol{\tau} = \mathbf{g}(\boldsymbol{\theta}) + A(\boldsymbol{\theta}_d - \boldsymbol{\theta}) + B(\dot{\boldsymbol{\theta}}_d - \dot{\boldsymbol{\theta}}), \tag{24}$$

where A and B are constant matrices. By using this control method with appropriate gain matrices, the motion of the manipulator usually follows in the neighborhood of a given desired trajectory $\boldsymbol{\theta}_d(t)$ in $\boldsymbol{\theta}$ space. However, there is a small difference between these two trajectories, which we denote

$$\mathbf{x}(t) = \boldsymbol{\theta}(t) - \boldsymbol{\theta}_d(t). \tag{25}$$

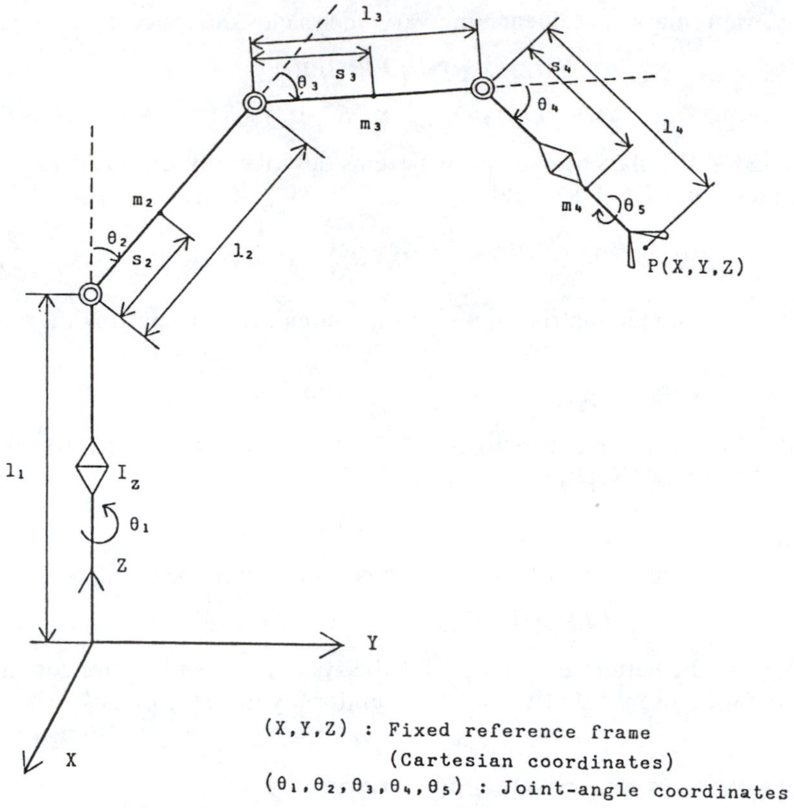

Figure 6. Serial-link manipulator with 5 degrees of freedom.

Since $\mathbf{x}(t)$ is small, it is possible to rewrite Eq. (23) in $\mathbf{x}$ as follows:

$$R[\boldsymbol{\theta}_d(t)]\ddot{\mathbf{x}}(t) + [B + C(t)]\dot{\mathbf{x}}(t) + [A + D(t)]\mathbf{x}(t) = \mathbf{h}(t) + \mathbf{u}(t), \tag{26}$$

where $\mathbf{u}(t)$ is an additional control input to further improve the motion of the manipulator. Since $R[\boldsymbol{\theta}_d(t)]$ is always positive definite, it is possible to multiply both sides of Eq. (26) from the left by $R^{-1}[\boldsymbol{\theta}_d(t)]$, which results in

$$\ddot{\mathbf{x}}(t) + C_1(t)\dot{\mathbf{x}}(t) + C_2(t)\mathbf{x}(t) = \mathbf{g}_1(t) + B_1(t)\mathbf{u}(t). \tag{27}$$

This can be rewritten to yield the following state equation:

$$\frac{d}{dt}\begin{bmatrix}\mathbf{x}(t)\\ \dot{\mathbf{x}}(t)\end{bmatrix} = \begin{bmatrix}0 & \mathrm{I}\\ -C_2(t) & -C_1(t)\end{bmatrix}\begin{bmatrix}\mathbf{x}(t)\\ \dot{\mathbf{x}}(t)\end{bmatrix} + \begin{bmatrix}0\\ \mathbf{g}_1(t)\end{bmatrix} + \begin{bmatrix}0\\ B_1(t)\end{bmatrix}\mathbf{u}(t). \tag{28}$$

At this stage, we must bear in mind that Eq. (28) itself is a linear time-varying dynamical system, like Eq. (19a) [except for the forcing term $\mathbf{g}_1(t)$ in Eq. (28)], though all coefficient time-varying functions together with $\mathbf{g}_1(t)$ cannot be evaluated exactly. However, if we choose an adequate output vector $\mathbf{y}$ for system (28) and a gain matrix $\Gamma(t)$ in order to satisfy the assumption of Theorem 1, then it is possible to apply the betterment process proposed in the previous section. It should be noted again that in the application of such a betterment process all time-varying functions of the coefficients in the linear dynamical system [Eqs. (19a) and (19b)] need not be known in principle, but only the output $\mathbf{y}_k(t)$ for each operation trial must be known.

V. APPLICATION OF THE BETTERMENT PROCESS TO THE CONTROL OF MANIPULATORS

To construct a betterment process for system (28), we choose the output vector as

$$\mathbf{y}(t) = [0 \; I_n]\begin{bmatrix}\mathbf{x}(t)\\ \dot{\mathbf{x}}(t)\end{bmatrix} = \dot{\mathbf{x}}(t) \tag{29}$$

and introduce an iterative process

$$\mathbf{u}_{k+1}(t) = \mathbf{u}_k(t) + \Gamma(t)[\dot{\mathbf{y}}_d(t) - \dot{\mathbf{y}}_k(t)], \tag{30}$$

where $\Gamma(t)$ is an $n \times n$ time-varying or constant matrix and $\mathbf{y}_k(t)$ is the output response to the control input $\mathbf{u}_k(t)$. Next we note that the solution of Eq. (28) is expressed by

$$\begin{aligned}\mathbf{x}(t) &= \bar{\mathbf{g}}_1(t) + \int_0^t H_{12}(t, \tau)B_1(\tau)\mathbf{u}(\tau)d\tau,\\ \dot{\mathbf{x}}(t) &= \bar{\mathbf{g}}_2(t) + \int_0^t H_{22}(t, \tau)B_1(\tau)\mathbf{u}(\tau)d\tau,\end{aligned} \tag{31}$$

where we put

$$X(t)X^{-1}(\tau) = \overset{\qquad\quad n \qquad\qquad\quad n}{\begin{bmatrix} H_{11}(t, \tau) & H_{12}(t, \tau)\\ H_{21}(t, \tau) & H_{22}(t, \tau)\end{bmatrix}}\begin{matrix} n\\ n\end{matrix}.$$

Hence, $X(t)$ is a fundamental solution to a homogeneous $2n \times 2n$ matrix differential equation

$$\frac{d}{dt}X(t) = \begin{bmatrix} 0 & I_n\\ -C_2(t) & -C_1(t)\end{bmatrix}X(t),$$

where

$$X(t) = \begin{bmatrix} X_{11}(t) & X_{12}(t)\\ X_{21}(t) & X_{22}(t)\end{bmatrix}, \qquad X(0) = \begin{bmatrix} I_n & 0\\ 0 & I_n\end{bmatrix}.$$

It should be noted that

$$\begin{aligned}\bar{\mathbf{g}}_1(t) &= X_{11}(t)\mathbf{x}(0) + X_{12}(t)\dot{\mathbf{x}}(0) + \int_0^t H_{12}(t, \tau)\mathbf{g}_1(\tau)d\tau,\\ \bar{\mathbf{g}}_2(t) &= X_{21}(t)\mathbf{x}(0) + X_{22}(t)\dot{\mathbf{x}}(0) + \int_0^t H_{22}(t, \tau)\mathbf{g}_1(\tau)d\tau.\end{aligned} \tag{32}$$

This means that Eq. (28) reduces to a form of Eq. (12) even if a forcing term $g_1(t)$ exists in Eq. (28).

Now we check the validity of the most important assumption in Theorem 1, which is described in this case by

$$\begin{aligned}\| I_r - H(t, t)\Gamma(t) \|_\infty &= \| I_n - [0 \; I_n]X(t)X^{-1}(t)\begin{bmatrix}0\\ B_1(t)\end{bmatrix}\Gamma(t) \|_\infty\\ &= \| I_n - B_1(t)\Gamma(t) \|_\infty .\end{aligned} \tag{33}$$

As seen in the derivation of Eq. (27) from Eq. (26), $B_1(t) = R^{-1}[\boldsymbol{\theta}_d(t)]$. Hence, if we set $\Gamma(t) = R[\boldsymbol{\theta}_d(t)]$, the assumption of Theorem 1 is clearly satisfied with the condition

$$\| I_n - H(t,t)\Gamma(t) \|_\infty = \| I_n - R^{-1}[\boldsymbol{\theta}_d(t)] R[\boldsymbol{\theta}_d(t)] \|_\infty$$
$$= \| I_n - I_n \|_\infty = 0 < 1.$$

In general, most entries of the inertia matrix $R(\boldsymbol{\theta})$ are complicated functions of $\boldsymbol{\theta}$ and therefore it is difficult to compute exact values of $R[\boldsymbol{\theta}_d(t)]$ in real time. However, it is true that $R(\boldsymbol{\theta})$ remains positive definite for any $\boldsymbol{\theta}$. This suggests that in most configurations of the manipulator there is the possibility of selecting an appropriate constant matrix Γ such that

$$\| I_n - B_1(t)\Gamma \|_\infty = \| I_n - R^{-1}(\boldsymbol{\theta}_d)\Gamma \|_\infty < 1.$$

Finally, we remark that, from the original problem of controlling the manipulator posed in the previous section, a desired trajectory $\mathbf{y}_d(t)$ for the variational system should be given as $\mathbf{y}_d(t) \equiv 0$ because $\mathbf{x}(t)$ represents the small difference of $\boldsymbol{\theta}(t)$ with the original desired motion $\boldsymbol{\theta}_d(t)$, that is,

$$\mathbf{x}(t) = \boldsymbol{\theta}(t) - \boldsymbol{\theta}_d(t).$$

Hence, if it holds that $\dot{\mathbf{x}}_k(0) = \mathbf{y}_d(0) = 0$ for all k, then all assumptions of Theorem 1 are satisfied and thereby the betterment process converges. Fortunately, the initial condition $\dot{\mathbf{x}}(0) = \mathbf{0}$ is usually satisfied if the manipulator is still at the beginning of every operation trial, that is, $\dot{\boldsymbol{\theta}}_k(0) = \mathbf{0}$, and $\boldsymbol{\theta}_d(t)$ is chosen to satisfy the initial condition $\dot{\boldsymbol{\theta}}_d(0) = \mathbf{0}$.

In conclusion, it is shown that the betterment process of Eq. (30) for the system described by Eq. (28) and (29) becomes convergent if an appropriate constant $n \times n$ matrix Γ is chosen and the initial condition $\dot{\boldsymbol{\theta}}_d(0) = \dot{\boldsymbol{\theta}}_k(0) = \mathbf{0}$ is satisfied in every operation trial.

To see the applicability of this scheme, we show numerical simulation results in Figures 7(a), 7(b), and 7(c), and Figures 8(a), 8(b), and 8(c), which are

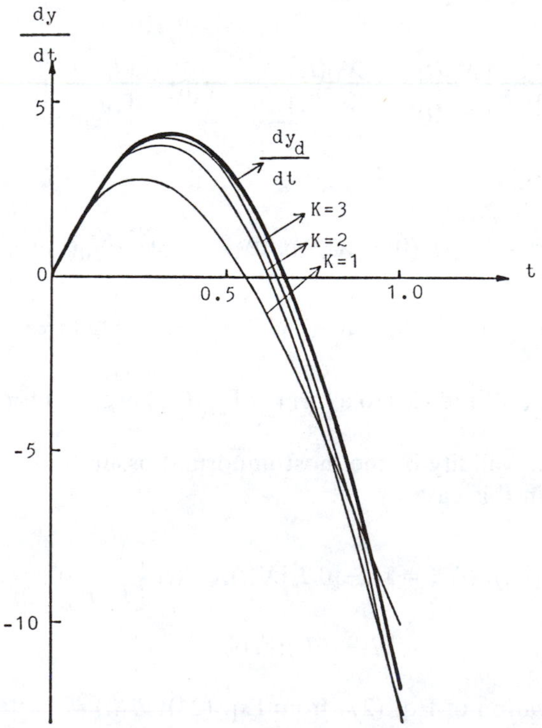

Figure 7(a). Plot of $\dot{y}_k(t) = \ddot{x}_k(t)$ (acceleration) for betterment process with $\Gamma = 1.0$.

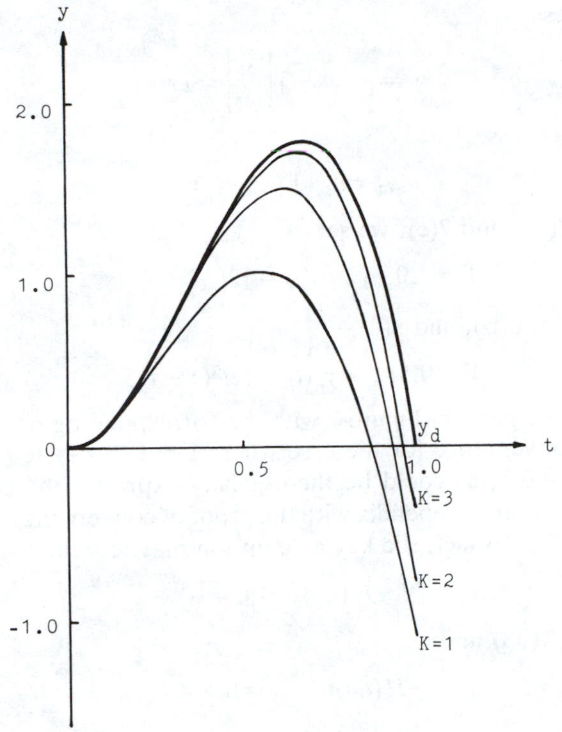

Figure 7(b). Plot of $y_k(t) = \dot{x}_k(t)$ (velocity) for betterment process with $\Gamma = 1.0$.

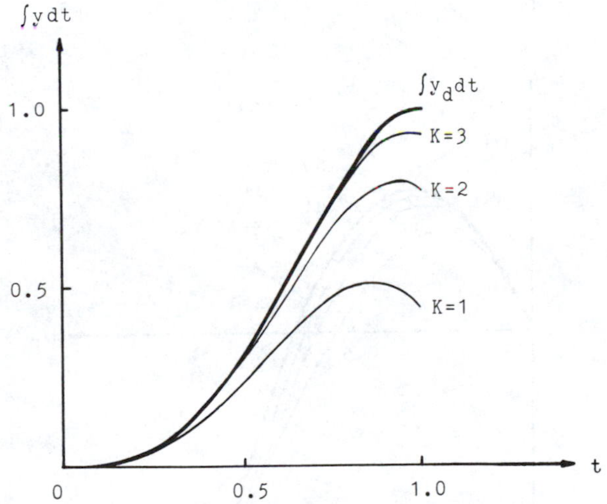

Figure 7(c). Plot of $\int_0^t y_k(\tau)d\tau = x_k(t)$ (position) for betterment process with $\Gamma = 1.0$.

obtained on the basis of the following system:

$$\ddot{x} + 2\dot{x} + x = u \tag{34}$$

or

$$\frac{d}{dt}\begin{bmatrix} x \\ \dot{x} \end{bmatrix} = \begin{bmatrix} 0 & 1 \\ -1 & -2 \end{bmatrix}\begin{bmatrix} x \\ \dot{x} \end{bmatrix} + \begin{bmatrix} 0 \\ 1 \end{bmatrix} u, \tag{35}$$

with output process

$$y = [0 \quad 1]\begin{bmatrix} x \\ \dot{x} \end{bmatrix} = \dot{x} \tag{36}$$

and betterment process

$$u_{k+1} = u_k + \Gamma(\dot{y}_d - \dot{y}_k). \tag{37}$$

In Figures 7(a), 7(b), and 7(c), we set

$$\Gamma = 1.0, \qquad y_d(t) = 12t^2(1-t), \tag{38}$$

and in Figures 8(a), 8(b), and 8(c),

$$\Gamma = 0.5, \qquad y_d(t) = 12t^2(1-t). \tag{39}$$

Comparing the three parts of Figure 7 with the corresponding parts of Figure 8, we see that the convergence for the case where $\Gamma = 1.0$ is faster than the convergence when $\Gamma = 0.5$, as could be theoretically expressed by comparing the proof of Theorem 1 in the Appendix with the proof of convergency for the system described by Eq. (8). In fact, the key assumption of Theorem 1 turns out to be

$$\| I_r - H(t, t)\Gamma(t) \|_\infty = 0$$

for the case of Eq. (38) and

$$\| I_r - H(t, t)\Gamma(t) \|_\infty = 0.5 < 1$$

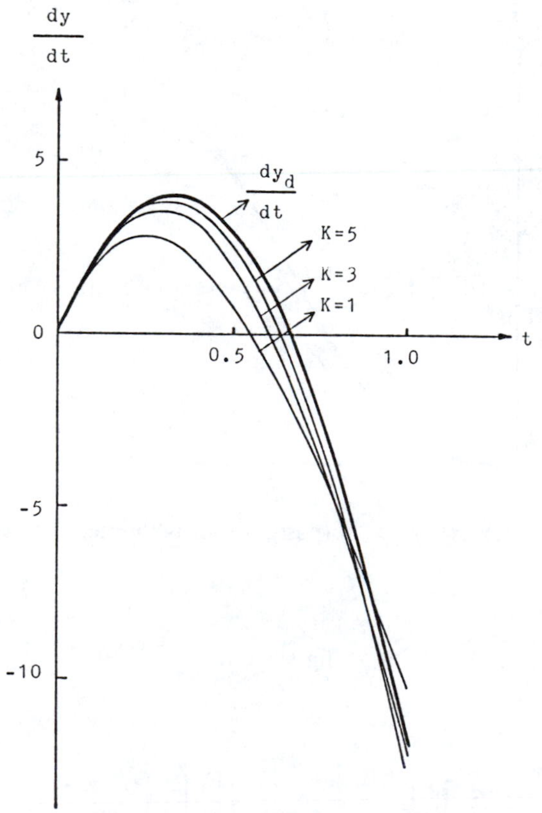

Figure 8(a). Plot of $\dot{y}_k(t) = \ddot{x}_k(t)$ (acceleration) for betterment process with $\Gamma = 0.5$.

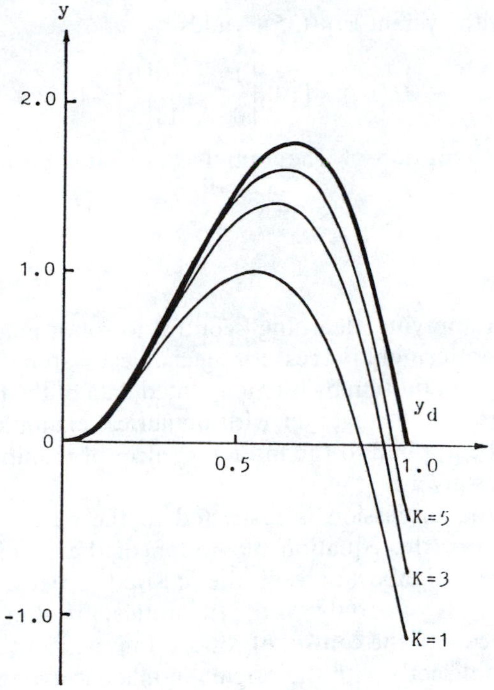

Figure 8(b). Plot of $y_k(t) = \dot{x}_k(t)$ (velocity) for betterment process with $\Gamma = 0.5$.

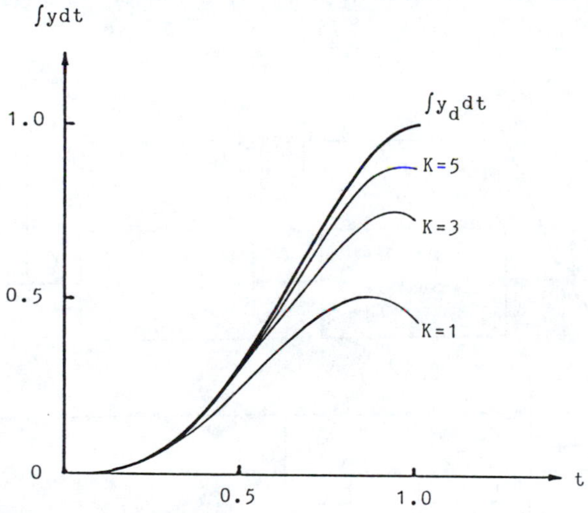

Figure 8(c). Plot of $\int_0^t y_k(\tau)d\tau = x_k(t)$ (position) for betterment process with $\Gamma = 0.5$.

for the case of Eq. (39). In either case, however, the speed of convergence is relatively high and the output approaches the desired trajectory with sufficient accuracy within several trials.

Finally, we must remark that it is impossible to choose the positional variable $x(t)$ as an output. In this case, we write

$$y = [1\ 0]\begin{bmatrix} x \\ \dot{x} \end{bmatrix} = x$$

which, together with system Eq. (35), yields

$$H(t, t) = [1\ 0]\begin{bmatrix} 1 & 0 \\ 0 & 1 \end{bmatrix}\begin{bmatrix} 0 \\ 1 \end{bmatrix} = 0.$$

Hence, the key assumption of Theorem 1 is not satisfied for any choice of Γ.

VI. CONCLUSION

With the goal of applying "learning" control to robot manipulators, we proposed an iterative betterment process for mechanical systems that improves their present performance of motion by use of updated data of the previous operation. It was shown theoretically, together with numerical examples, that the betterment process can be applied to the motion control of manipulators if a desired motion trajectory is given.

In this article, the discussion is restricted to the case of linear dynamical systems and therefore the equation of motion of the manipulator had to be represented in the neighborhood of the desired trajectory. However, this representation was necessary only for the theoretical proof of the applicability of the betterment process to the control of robots. In practice, this iterative scheme could be associated directly with the original nonlinear dynamics of robots, that is, the actual betterment process is constructed as shown in Figure 9.

Finally, we point out that the betterment process for linear dynamical systems with constant coefficients has many interesting aspects in relation to linear system theory. It should also be pointed out that it is possible to prove the convergence of the iterative betterment process constructed for a class of nonlinear

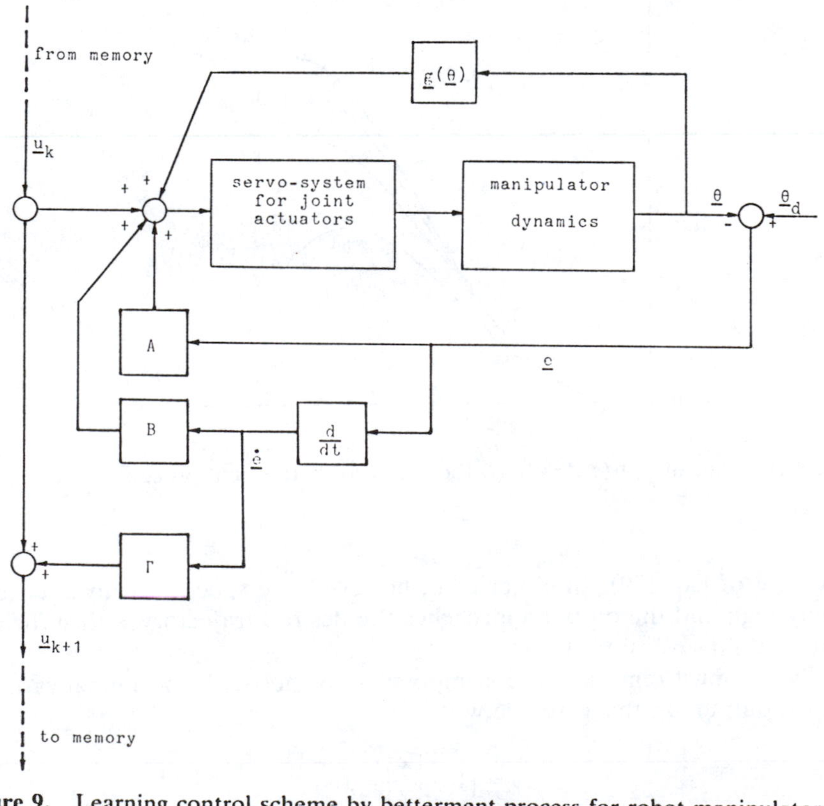

Figure 9. Learning control scheme by betterment process for robot manipulators.

dynamical systems like robot manipulators. These discussions are given in subsequent papers.[9, 10]

APPENDIX

Proof of Theorem 1

First we note that the norm $\| G \|_\infty$ is the natural matrix norm induced by the vector norm

$$\|\mathbf{f}\|_\infty = \max_{1\le i\le r} |f_i|$$

for vector $\mathbf{f} = (f_1, \ldots, f_r)$. Hence, it is evident that

$$\|G\mathbf{f}\|_\infty \le \| G \|_\infty \cdot \| \mathbf{f} \|_\infty . \tag{A1}$$

Now it follows from Eq. (13) that

$$\begin{aligned}
\dot{\mathbf{e}}_k(t) &= \dot{\mathbf{y}}_d(t) - \dot{\mathbf{y}}_k(t)\\
&= \dot{\mathbf{y}}_d(t) - [\dot{\mathbf{g}}(t) + H(t,t)\mathbf{u}_k(t) + \int_0^t \frac{\partial}{\partial t} H(t,\tau)\mathbf{u}_k(\tau)d\tau]\\
&= \dot{\mathbf{y}}_d(t) - [\dot{\mathbf{g}}(t) + H(t,t)\mathbf{u}_{k-1}(t) + \int_0^t \frac{\partial}{\partial t} H(t,\tau)\mathbf{u}_{k-1}(\tau)d\tau]\\
&\quad - [H(t,t)\Gamma(t)\dot{\mathbf{e}}_{k-1}(t) + \int_0^t \frac{\partial}{\partial t} H(t,\tau)\Gamma(\tau)\dot{\mathbf{e}}_{k-1}(\tau)d\tau]\\
&= [I_r - H(t,t)\Gamma(t)]\dot{\mathbf{e}}_{k-1}(t) - \int_0^t \frac{\partial}{\partial t} H(t,\tau)\Gamma(\tau)\dot{\mathbf{e}}_{k-1}(\tau)d\tau .
\end{aligned} \tag{A2}$$

Multiplying both sides of this equation by $e^{-\lambda t}$, taking the norm $\| \cdot \|_\lambda$, and using the property of Eq. (A1), we have

$$\begin{aligned}
\| \dot{\mathbf{e}}_k \|_\lambda &= \max_{0\le t\le T} \{e^{-\lambda t} \| \dot{\mathbf{e}}_k(t) \|_\infty\}\\
&\le \max_{0\le t\le T} \| I_r - H(t,t)\Gamma(t) \|_\infty \cdot \max_{0\le t\le T} \{e^{-\lambda t} \|\dot{\mathbf{e}}_{k-1}(t) \|_\infty\}\\
&\quad + h_0 \cdot \max_{0\le t\le T} \int_0^t e^{-\lambda(t-\tau)} \max_{0\le \tau\le T} \{e^{-\lambda\tau} \| \dot{\mathbf{e}}_{k-1}(\tau) \|_\infty\} d\tau\\
&= \rho \| \dot{\mathbf{e}}_{k-1} \|_\lambda + h_0 \| \dot{\mathbf{e}}_{k-1} \|_\lambda \cdot \max_{0\le t\le T} \int_0^t e^{-\lambda(t-\tau)} d\tau\\
&= \left(\rho + \frac{h_0(1-e^{-\lambda T})}{\lambda}\right) \|\dot{\mathbf{e}}_{k-1} \|_\lambda ,
\end{aligned} \tag{A3}$$

where

$$\rho = \max_{0\le t\le T} \| I_r - H(t,t)\Gamma(t) \|_\infty ,$$

$$h_0 = \max_{0\le t,\tau\le T} \| \frac{\partial}{\partial t} H(t,\tau)\Gamma(\tau) \|_\infty .$$

Since $0 \le \rho < 1$ by assumption, it is possible to choose λ sufficiently large so that

$$\rho_0 = \rho + \frac{h_0(1-e^{-\lambda T})}{\lambda} < 1. \tag{A4}$$

Substituting this into Eq. (A3), we obtain Eq. (15). Thus, Theorem 1 has been proved.

References

1. F. Rosenblatt, *Principles of Neurodynamics, Perceptrons and the Theory of Brain Mechanisms,* Spartan Books Co., Washington, DC, 1961.
2. Ya. Z. Tsypkin, *Adaption and Learning in Automatic Systems,* Academic, New York, 1971.
3. S. Arimoto, S. Kawamura, and F. Miyazaki, "Can mechanical robots learn by themselves?" *Proc. 2nd Int. Symp. Robotics Res.,* Kyoto, Japan, August 1984.
4. R. P. Paul, *Robot Manipulators,* The MIT Press, Cambridge, 1981.
5. P. Vukobratovic, *Scientific Fundamentals of Robotics 1 and 2,* Springer-Verlag, Berlin, 1982.
6. M. Takegaki and S. Arimoto, "A new feedback method for dynamic control of manipulators," *Trans. ASME J Dyn. Syst. Meas. Control,* **103**, 119–125 (1982).
7. S. Arimoto and F. Miyazaki, "Stability and robustness of PID feedback control for robot manipulators of sensory capability," *Proc. 1st Int. Symp. Robotics Res.,* The MIT Press, Cambridge, MA, 1983.
8. F. Miyazaki, S. Arimoto, M. Takegaki, and Y. Maeda, "Sensory feedback based on the artificial potential for robot manipulators," *Proc. 9th IFAC,* Budapest, Hungary, July 1984.
9. S. Kawamura, F. Miyazaki, and S. Arimoto, "Iterative learning control for robotic systems," paper presented at IECON '84, Tokyo, Japan, 1984.
10. S. Arimoto, S. Kawamura, F. Miyazaki, "Bettering operation of dynamic systems by learning; A new control theory for servomechanism or mechatronics systems," paper presented at 23rd IEEE Conf. Decision and Control, Las Vegas, NV, 1984.

Correction to "Bettering Operation of Robots by Learning"

Suguru Arimoto, Sadao Kawamura, and Fumio Miyazaki
Faculty of Engineering Science, Osaka University, Toyonaka, Osaka, 560 Japan

In our previous article,[1] we recently found a mistake in Figure 9, which was presented to explain implementation of the proposed betterment process for learning control of robot manipulators. Hence, we replace Figure 9 by Figure 1 here. In this new figure, the output error $\mathbf{e}_k$ at the kth operation is defined as a difference between the given desired angular velocity vector $\dot{\theta}$ and the measured one $\dot{\theta}$ obtained through tachogenerators. A more detailed discussion related to actual implementation of the betterment process for robot manipulators is found in our more recent paper.[2]

References

1. S. Arimoto, S. Kawamura, and F. Miyazaki, "Bettering operation of robots by learning," *J. Robotic Syst.,* **1**(2), 124–140 (1984).
2. S. Arimoto, S. Kawamura, and F. Miyazaki, "Can mechanical robots learn by themselves?" *Proc. 2nd Int. Symp. Robotics Res.,* Kyoto, Japan, August 1984.

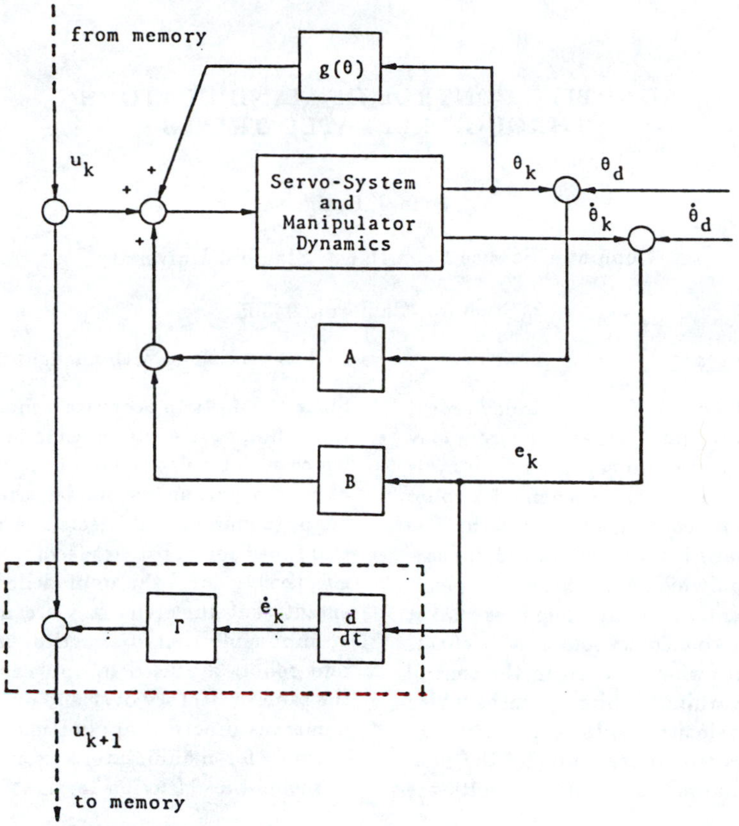

Figure 1. Learning control scheme by betterment process for robot manipulators.

ADAPTIVE CONTROL OF MANIPULATORS THROUGH REPEATED TRIALS

John J. Craig

Computer Science Department, Stanford University

Stanford, California 94305

Abstract

Virtually all manipulators at work in factories today repeat their motions in cycles. Unfortunatly, whatever errors may exist in repeating a trajectory are also repeated from cycle to cycle. This paper presents an adaptive scheme for robot manipulators for the case of repeated trials of a path. The primary benefit of the approach is that it allows compensation for effects such as friction which are otherwise difficult to model. A secondary advantage of the scheme is that it can also adapt to torques due to velocity and gravity effects in order to reduce the computations needed in the control computer. After sufficient learning, arbitrary paths which are within the torque and velocity limits of the arm can be executed with errors lower bounded only by the non-repeatability of the manipulator. Simulation results are included.

1. Introduction

Virtually all manipulators at work in factories today repeat their motions over and over in cycles. Unfortunatly, whatever errors may exist in trajectory following are also repeated from cycle to cycle. If a manipulator posseses largely repeatable dynamics, a control algorithm may be designed so that performance improves from trial to trial.

To a large extent, such trajectory following errors are due to the simple error-driven control systems used in most present day manipulators. The apparent solution is to increase the completeness of the dynamic model of the manipulator used in the control law synthesis. If a perfect model of the dynamics exists, it may be used in a control law which moves the manipulator along arbitrary paths with zero error. Recently there has been much interest in forming dynamic models of manipulators and in specifying algorithms with which to compute the dynamics efficiently [1-3]. These models include all the dynamic terms arising from a rigid body model of the arm. Such models are of course never exact, with the most problematic area being that of adding terms to model friction.

Unfortunately, the effects of friction acting at the joints of present manipulators are significant. The friction has a non-linear dependence on joint velocity, and may depend on joint position as well. Attempts at modelling the friction are further complicated by changes as the device ages, as well as variations with changes in temperature.

Since attempts to accurately model friction will almost certainly fail, we are interested in proposing control methods in which such terms need not be modelled at all. With the lack of even a parametric model, the only plausible scheme to adapt to unmodelled effects is to repeat similar motions several times and construct feedforward torque histories which effectively cancel the unmodelled terms. Obviously, to the extent that the dynamics are not repeatable, adaptation is impossible. As a practical matter, because most current robots are used in applications in which they repeat the same trajectory over and over, such schemes might find numerous practical application. Other published adaptive schemes for manipulators explicitly or implicitly attempt to model these friction terms [4-10].

An important goal of adaptive controller design should be to insure the complete use of all available knowledge of the system dynamics. Intuitively, ignoring what knowledge we may have of system parameters in favor of some general adaptation scheme seems almost certainly to be a bad idea. We wish to model what can be modelled, and use adpatation for what cannot be modelled. We make direct use of what structure is known by partitioning the dynamics of the manipulator into two distinct portions: modelled and unmodelled.

Our initial intent was that the complete rigid body model [1-3] be the modelled portion, with the friction effects representing the unmodelled portion. However, the method allows all terms that do not include acceleration (i.e. all velocity terms and gravity terms) to be included in the unmodelled partition, in the interest of avoiding computation. Additionally, simulations verify that significant error in knowledge of the inertial parameters can be tolerated and compensated for in the unmodelled partition.

The scheme is based upon adaptively constructing feedforward torque histories for each actuator of the mechanism. Since the construction of these feedforward functions is not based on a model of any kind, the learned functions may reflect arbitrarily complex dynamics originating from sources that are unknown to the designer. Compensation may even be learned for forces and torques encountered due to interaction with the environment, to the extent that it is repeatable from trial to trial.

The closer the known part of the dynamics is to the actual dynamics, the more uniform will be the stability margin (or other measure) of the disturbance suppression as configuration varies. In general, our method will cause a repeatable device to move with near zero errors, but disturbances will be suppressed with a varying stability margin in different regions of the workspace. Servo terms in the control law can be viewed as making up a 'robust' controller so that in extreme cases, performance is acceptable. This is, after all, how virtually all present day industrial robots are controlled.

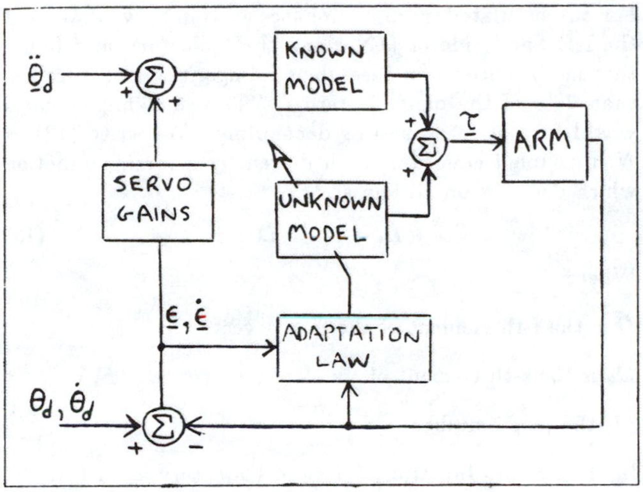

Figure 1 *Adaptive Control Scheme*

2. Outline of the Method

The controller consists of two parts: a control law, and an adaptation law. The control law consists of error driven servo terms, arbitrary amounts of dynamic compensation based on known structure and parameters, and an adaptively learned feedforward torque function. The adaptive law specifies how to construct the next feedforward torque history based on the error history from the last trial. If the mechanism encounters perfectly repeatable dynamics as it moves, the adaptation law will cause a torque history to be learned which forces the device to follow the desired trajectory with zero error. Non-repeatability, which could be viewed as a disturbance in this system, causes errors which are suppressed by the error driven servo terms. The nature of disturbance suppression depends on the partitioning of the known and unknown portions of the model in the control law.

The form of adaptation law proposed here is that of a linear filter. That is, the entire function of servo errors observed on the previous trial is filtered. The output of this filtering operation is a function which is added to the previous torque history to generate the next torque history. Although the filter is linear, through the interactions with the non-linear equations of motion of the mechanism, complex non-linear torque functions are learned after several trials.

Analysis of the entire closed loop system yields a criterion to be met for convergence. This criterion may be met by proper choice of the filter coefficients in the adaptation law. If the criterion is met, the torque history will approach the particular history which causes the device to move along the trajectory with zero error. In the presence of non-repeatability, or of constant disturbances, the system lowers the error to some non-zero limit imposed by the non-repeatability.

3. Manipulator Dynamics

In order to perform stability and convergence analysis on the proposed adaptive controller, a model of the manipulator is required. Modelling the manipulator as a chain of rigid bodies and applying any of several methods of deriving the system's equations of motion [1-3] results in equations which could be written in the form:

$$\tau = \tau(\theta, \dot{\theta}, \ddot{\theta}) = M(\theta)\ddot{\theta} + V(\theta, \dot{\theta}) + F \tag{1}$$

Where:

τ is the $N \times 1$ vector of torques applied at the joints

$M(\theta)$ is the $N \times N$ mass matrix of the manipulator

$V(\theta, \dot{\theta})$ is an $N \times 1$ vector of velocity and gravity effects

F is an $N \times 1$ vector of Friction effects

N is the number of joints

In the case of small errors a reasonable approximation of the dynamics can be made by linearizing about the desired trajectory [11]. Thus, the dynamics might be written:

$$\tau = \tau_d(\theta_d, \dot{\theta}_d, \ddot{\theta}_d) - \frac{\partial \tau}{\partial \ddot{\theta}}\ddot{\epsilon} - \frac{\partial \tau}{\partial \dot{\theta}}\dot{\epsilon} - \frac{\partial \tau}{\partial \theta}\epsilon \tag{2}$$

Where:

τ_d is a vector of torques along the desired path

ϵ is a vector of servo errors, defined as $\epsilon = \theta_d - \theta$

4. Stability

The Control law is proposed as:

$$\tau = \hat{M}(\theta)\left[\ddot{\theta}_d + K_v\dot{\epsilon} + K_p\epsilon\right] + \hat{V}(\theta, \dot{\theta}) + \hat{F}_k \tag{3}$$

Where:

K_v, K_p are positive constant diagonal gain matrices

$\hat{M}(\theta)$ is an approximation of the manipulator mass matrix

$\hat{V}(\theta, \dot{\theta})$ is an approximation of the velocity and gravity terms

$\hat{F}_k$ is the feedforward torque function after trial k

On the first cycle, the feedforward torque term is zero. After each trial, it is adjusted according to an adaptation law.

By combining equations (2) and (3) we write the system dynamics in error space. For brevity, the arguments of the functions are dropped:

$$\hat{M}^{-1}M\ddot{\epsilon}+\left[K_v+\hat{M}^{-1}\frac{\partial\tau}{\partial\dot{\theta}}\right]\dot{\epsilon}+\left[K_p+\hat{M}^{-1}\frac{\partial\tau}{\partial\theta}\right]\epsilon$$
$$=\hat{M}^{-1}\left[(M-\hat{M})\ddot{\theta}_d+(V-\hat{V})+(F-\hat{F}_k)\right] \quad (4)$$

In (4) quantities M, V, and F are evaluated along the desired trajectory as a result of writing the dynamics as in (2). Lyapunov stability analysis can be used to show that control gains exist such that (4) is stable. The mechanism is modelled as:

$$\tau=M(\theta)\ddot{\theta}+V(\theta,\dot{\theta})+F \quad (5)$$

The control law is:

$$\tau=\hat{M}(\theta)\left[\ddot{\theta}_d+K_v\dot{\epsilon}+K_p\epsilon\right]+\hat{V}(\theta,\dot{\theta})+\hat{F}_k \quad (6)$$

The system equation may be written:

$$K_v\dot{\epsilon}+K_p\epsilon=\hat{M}^{-1}\left[(\hat{M}\ddot{\theta}_d-M\ddot{\theta})+(V-\hat{V})+(F-\hat{F}_k)\right] \quad (7)$$

We choose a positive definite function as:

$$L=\frac{1}{2}\epsilon^T K_v\epsilon \quad (8)$$

Therefore:

$$\dot{L}=\epsilon^T K_v\dot{\epsilon}$$
$$\dot{L}=\epsilon^T\hat{M}^{-1}\left[(\hat{M}\ddot{\theta}_d-M\ddot{\theta})+(V-\hat{V})+(F-\hat{F}_k)-K_p\epsilon\right] \quad (9)$$

Which for sufficiently large K_p becomes:

$$\dot{L}=-\epsilon^T K_p\epsilon \quad (10)$$

Which is negative definite. Thus, in the sense of Lyapunov stability, there exists sufficiently large gains for which the system is stable regardless of the dynamics. In some sense, we have shown that sufficiently high gains completely wash out the rest of the dynamics. However, such a proof does not carry strong practical weight since we have no guarantee that gains can indeed be set sufficiently high - we have not modelled sensor noise, discrete time effects, and other limiting factors. However, note that if:

$$M\ddot{\theta}-\hat{M}\ddot{\theta}_d+V-\hat{V}+F-\hat{F}_k\approx 0 \quad (11)$$

Then the gains need not be high to insure stability. Present industrial robots with fixed gain servos are proof that gains can be set such as to guarantee stability in the presence of uncompensated dynamics. In fact, even in simple control systems, fixed gains can often be found which yield a reasonably constant error response over the entire workspace of the device. In the adaptive scheme presented here, a feedforward torque function is learned so that (11) is true (in the absence of noise).

Assuming we have an accurate rigid body model, that is, $\hat{M}=M$ and $\hat{V}=V$, equation (4) becomes:

$$\ddot{\epsilon}+\left[K_v+\hat{M}^{-1}\frac{\partial F}{\partial\dot{\theta}}\right]\dot{\epsilon}+\left[K_p+\hat{M}^{-1}\frac{\partial F}{\partial\theta}\right]\epsilon=\hat{M}^{-1}(F-\hat{F}_k) \quad (12)$$

K_p and K_v are diagonal and generally large, so that (12) is decoupled on the left side to a very good approximation. For our simulated manipulator (see section 8) we evaluated the left hand side of (12) along the trajectory and found that in absolute worst case, the off diagonal terms were less than 20% of the diagonal terms. The use of higher gains would further increase this decoupling. We write (12) as N uncoupled equations each driven by a forcing function which depends on all joints:

$$\ddot{\epsilon}+B\dot{\epsilon}+K\epsilon=D-\hat{D}_k \quad (13)$$

Where:

D is the i-th element of the $N\times 1$ vector $\hat{M}^{-1}F$

$\hat{D}_k$ is the i-th element of the $N\times 1$ vector $\hat{M}^{-1}\hat{F}_k$

i is the joint number

In (12), F is a function of desired trajectory as in (4). Because the mass matrix is composed of smooth functions and errors are expected to be small, $M(\theta_d)=M(\theta)$. So D can be viewed as an unknown but fixed function for each joint.

5. Convergence

The previous stability result is also contingent on the feedforward torque function remaining bounded. More than remaining bounded, we wish to design a system in which the feedforward torque function converges to the function which causes the trajectory to be followed with small error.

The form of adaptation law for each joint is proposed as:

$$\hat{D}_{k+1}=\hat{D}_k+P\star\epsilon_k \quad (14)$$

Where:

P is the impulse response of a linear filter

$\star$ is the convolution operator

k subscript refers to the trial number

Laplace transforming equation (14) yields:

$$\hat{D}_{k+1}(s)=D_k(s)+P(s)\epsilon_k(s) \quad (15)$$

Using the Laplace transform of (13) and substituting into (15) yields:

$$\hat{D}_{k+1}(s)=\hat{D}_k(s)+P(s)H(s)\left[D(s)-\hat{D}_k(s)\right] \quad (16)$$

Where:

$$H(s) = \frac{1}{s^2 + Bs + K} \tag{17}$$

Collecting terms:

$$\hat{D}_{k+1}(s) = [1 - P(s)H(s)]\,\hat{D}_k(s) + P(s)H(s)D(s) \tag{18}$$

Which is a recursion in the transform of the feedforward torque function. The solution to this recursion is seen to be:

$$\hat{D}_k(s) = D(s) + CG^k(s) \tag{19}$$

Where:

$G(s) = 1 - P(s)H(s)$

C is a constant

The two questions we ask are: Under what conditions does (19) converge? and, If it converges, what does it converge to?

It is clear that if the second term in (19) goes to zero with increasing trial number, then the recursion converges, and indeed converges to the transform of the unknown dynamic function along the desired trajectory.

The recursive nature of (18) might be thought of instead as an infinite sequence of filtering operations. In an infinite sequence of identical filters, the frequency response of each filter must be less than one for all frequencies. If it is greater than one for any particular frequency, that frequency component, if excited, will grow without bound. Thus, G(s) must have frequency response less than one for all frequencies. A somewhat more formal proof follows:

Let:

$$g_k(t) = \mathcal{L}^{-1}\left[G^k(s)\right] \tag{20}$$

Where:

$\mathcal{L}^{-1}$ is the inverse Laplace operator

Then, by Parseval's relation:

$$\int_{-\infty}^{+\infty} g_k^2(t)dt = \frac{1}{2\pi}\int_{-\infty}^{+\infty} \left|G^k(j\omega)\right|^2 d\omega \tag{21}$$

And:

$$\left|G^k(j\omega)\right| = \left|G(j\omega)\right|^k \tag{22}$$

Then:

$$\int_{-\infty}^{+\infty} g_k^2(t)dt = \frac{1}{2\pi}\int_{-\infty}^{+\infty} \left|G(j\omega)\right|^{2k} d\omega \tag{23}$$

If

$$|G(j\omega)| < 1 \quad \forall\omega \qquad and \qquad \lim_{\omega\to\infty} |G(j\omega)| = 0 \tag{24}$$

Then:

$$\lim_{k\to\infty} |g_k(t)| = 0 \tag{25}$$

Given (24), the right side of (23) equals zero due to the monotone convergence theorem [12]. The second condition in (24) insures that the convergence theorem can be extended to the infinite interval case. Therefore, given the condition stated in (24) we must have:

$$\lim_{k\to\infty} G^k(s) = 0 \tag{26}$$

Note that the filter which is used in the adaptation law appears as part of this critical transfer function:

$$G(s) = 1 - P(s)H(s) = \frac{s^2 + Bs + K - P(s)}{s^2 + Bs + K} \tag{27}$$

We choose the form of the adaptation filter to be:

$$P(s) = s^2 + (B - \mu)s + (K - \mu) \tag{28}$$

Where:

μ is a constant.

Any reasonable implementation of this filter will be non-causal. However, because we have an entire history of errors from the previous trial to filter, a non-causal filter may be implemented easily.

From (27) and (28):

$$G(s) = \frac{\mu(s+1)}{s^2 + Bs + K} \tag{29}$$

Assuming gains were chosen such that B and K form an approximately critically damped system, the magnitude of (29) along the imaginary axis is:

$$|G(j\omega)| = \mu\frac{\sqrt{\omega^2 + 1}}{\omega^2 + \frac{B^2}{4}} \tag{30}$$

The maximum of (30) occurs at:

$$\omega_{max} = \sqrt{\frac{B^2}{4} - 2} \tag{31}$$

Where the magnitude is:

$$|G(j\omega_{max})| = \mu\frac{\sqrt{\frac{B^2}{4} - 1}}{\frac{B^2}{2} - 2} \tag{32}$$

In order to keep (30) less than one over all frequency, we see we must choose:

$$0 < \mu < \frac{\frac{B^2}{2} - 2}{\sqrt{\frac{B^2}{4} - 1}} \tag{33}$$

This result is based on several approximations. First, that B and K in (13) are known constants. In reality, they may be slightly varying. Secondly, we have assumed the ability to implement a continuous time filter whereas our implementation will be in discrete time. Thus, the adaptation rate, μ, may have to be lowered closer to zero than the bound in (33) would indicate.

In chosing the filter, we must be sure that the frequency response of (30) is less than one at all frequencies. However, the closer the magnitude of the response is to unity, the faster will be the adaptation. We might also choose a more complicated filter than (28) which would allow us to meet the frequency response criteria while also allowing more design freedom. This freedom might be used to insure that a filter with a sharp cutoff frequency is implemented. Such a low pass characteristic may in fact be necessary due to unmodelled vibrations in the mechanism.

6. Extension to Velocity and Gravity Terms

Writing (4) for the case $\hat{M} = M$ and $\hat{V} = 0$ we have:

$$\ddot{\epsilon} + \left[K_v + \hat{M}^{-1}\frac{\partial(V+F)}{\partial\dot{\theta}}\right]\dot{\epsilon} + \left[K_p + \hat{M}^{-1}\frac{\partial(V+F)}{\partial\theta}\right]\epsilon = \hat{M}^{-1}\left[V + F - \hat{F}_k)\right] \tag{34}$$

Which is still quite well approximated by uncoupled equations in the form of (13), so the same adaptive law may be used. Again, for the case of the manipulator simulated (see section 8), the off diagonal terms in (34) are less than 20% of the diagonal terms in worst case.

7. Implementation Detail

The implementation of the system is in discrete time. Although a discrete analysis has been done for the system, it lends no particular insight to the problem, and so the implementation may be considered simply as discrete equivalents of the continous design.

The discrete version of the control law is:

$$\tau(n) = \hat{M}(\theta(n))\left[\ddot{\theta}_d(n) + \alpha\epsilon(n) + \beta\epsilon(n-1)\right] + \hat{V}(\theta(n), \dot{\theta}(n)) + \hat{F}_k(n) \tag{35}$$

The form of the adaptation filter was found by simply forming a discrete equivalvent of the continuous filter of (28). In discretizing we used the relations:

$$s = \frac{1}{2T}(z - z^{-1}) \tag{36a}$$

$$s^2 = \frac{1}{T^2}(z - 2 - z^{-1}) \tag{36b}$$

Where:

T is the sample period

The torque histories are stored in arrays in memory. Memory requirements are not large by today's standards. For example, a six jointed arm being servoed at 100 Hz. over a cyclic path of 5 Seconds in duration would require 3000 stored values.

8. Simulations

A two degree of freedom arm is sufficient to observe any coupling effects which might affect the algorithm. A planar arm with two rotational joints was simulated. The gravity vector lies in the plane of the arm so that gravity forces contribute to the dynamics. Figure 2 shows the manipulator.

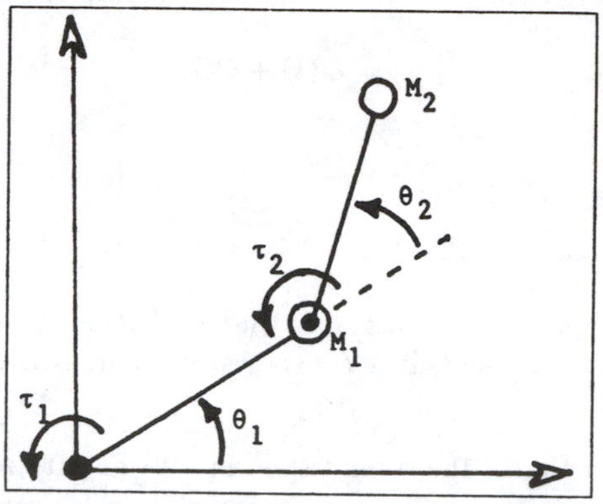

Figure 2 *Two link arm*

Using the simplifying assumption that all mass is located at a point at the distal end of each link, the equations of motion of this device are:

$$\tau_1 = m_2 l_2^2(\ddot{\theta}_1 + \ddot{\theta}_2) + m_2 l_1 l_2 C_2(2\ddot{\theta}_1 + \ddot{\theta}_2) + (m_1 + m_2)l_1^2\ddot{\theta}_1 - m_2 l_1 l_2 S_2 \dot{\theta}_2^2 - 2m_2 l_1 l_2 S_2 \dot{\theta}_1 \dot{\theta}_2 + m_2 l_2 g C_{12} + (m_1 + m_2) l_1 g C_1 \tag{37}$$

$$\tau_2 = m_2 l_1 l_2 \ddot{\theta}_1 C_2 + m_2 l_1 l_2 \dot{\theta}_1^2 S_2 + m_2 g l_2 C_{12} + m_2 l_2^2(\ddot{\theta}_1 + \ddot{\theta}_2) \tag{38}$$

Where:

l_1, l_2 are the lengths of the links (0.5 M, 0.5 M)

m_1, m_2 are the masses of the links (4.6 Kg, 2.3 Kg)

g is the gravity constant (9.8 M/Sec)

C_2 is $\cos(\theta_2)$

C_{12} is $\cos(\theta_1 + \theta_2)$

etc.

In addition to these rigid body terms, friction was simulated at each joint with the model:

$$F = C sgn(\dot{\theta}) + V\dot{\theta} \tag{39}$$

Where:

C is a Coulomb friction constant (2.0 Nt M)

V is a Viscous friction constant (4.0 Nt M Sec)

The control law (35) was implemented with a sample period of 0.01 seconds and gains computed to yield approximate critical damping with a stiffness of 200 Nt M/ Rad:

$$\alpha = \begin{bmatrix} 3028 & 0 \\ 0 & 3028 \end{bmatrix}$$
$$\beta = \begin{bmatrix} -2828 & 0 \\ 0 & -2828 \end{bmatrix} \qquad (40)$$

The desired trajectory can be arbitrarily complex as long as the limits on the actuators are not exceeded. For the experiments reported here, a simple cyclic path with continuous position, velocity, and acceleration was used. This path is defined by specifying 3 points in space, and connecting them with 3 cubic splines. A quintic spline segment is used to move from rest onto the cyclic path, and another is used to decelerate to rest after any number of cycles are performed. The entire path has a parameter of time which can be scaled to change the speed of the path. In these experiments a cycle lasted 2.4 seconds and the cycle was defined by the 3 path points specified in figure 3.

Point	X	Y	θ_1	θ_2
1	0.258 M	-0.023 M	70°	-150°
2	0.633 M	0.112 M	60°	-100°
3	0.953 M	0.211 M	25°	-25°

Figure 3 *Trajectory Via Points*

Figure 4 shows one cycle of the desired trajectory of the end-effector in the X-Y plane.

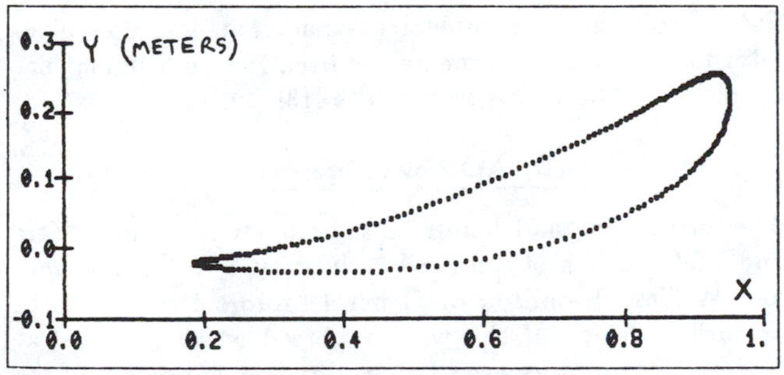

Figure 4 *Desired Spatial Trajectory*

In all the simulations, only knowledge of the mass matrix of the manipulator was assumed. However, the mass parameters were only assumed to be known within 25%, and so values for m_1 and m_2 were used which had a 25% error. Thus, the torque function learned represents the effects of velocity terms, gravity terms, friction terms, plus approximately 25% of the inertial torques as well. When knowledge of the mass constants are better, and/or when velocity and gravity terms are computed, performance is even better than the results shown below.

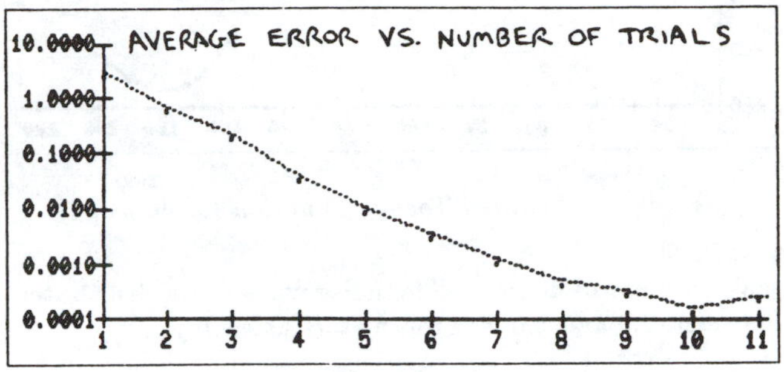

Figure 5 *Learning Curve for Joint 1*

Figures 5 and 6 show how the error is reduced with the number of trials. Here, the error value is calculated as the average error magnitude over an entire cycle. Note the use of a log scale and units of degrees for the average error values.

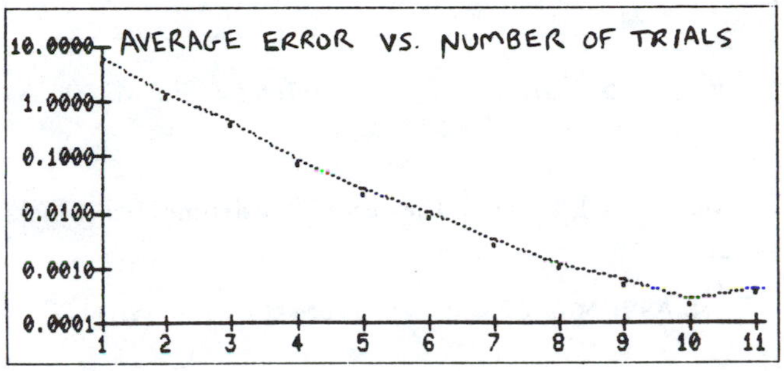

Figure 6 *Learning Curve for Joint 2*

Figures 7 and 8 show the feedforward torque functions that were learned after 10 trials. Because the simulation is perfectly repeatable, the results are extremely good.

Figures 9 and 10 show the learning curves for the case when non-repeatability is simulated through the addition of random torque noise at the joints. The errors are on a log

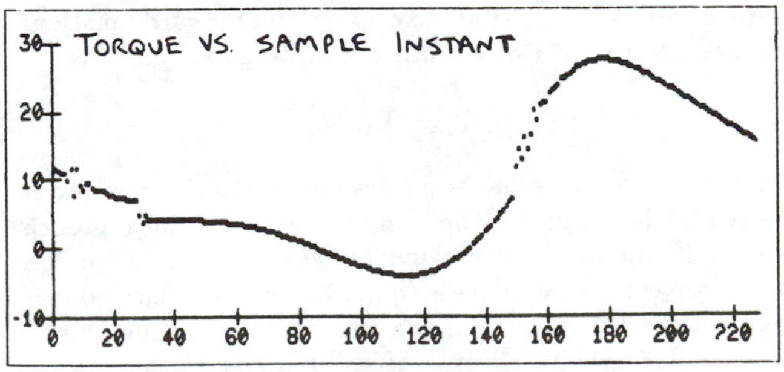

Figure 7 *Learned Torque Function for Joint 1*

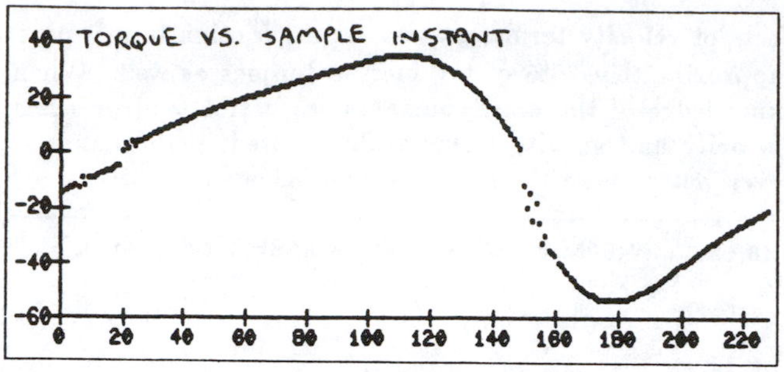

Figure 8 *Learned Torque Function for Joint 2*

scale in units of degrees. The noise was evenly distributed between -2.0 and 2.0 Newton meters at each joint.

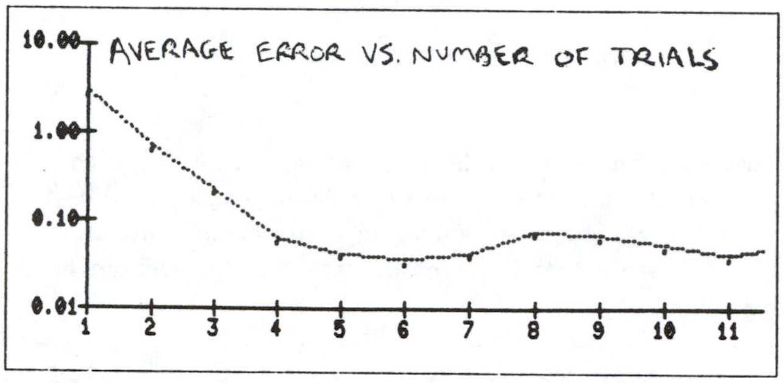

Figure 9 *Learning Curve for Joint 1 (with Noise)*

Figures 11 and 12 show the feedforward functions which

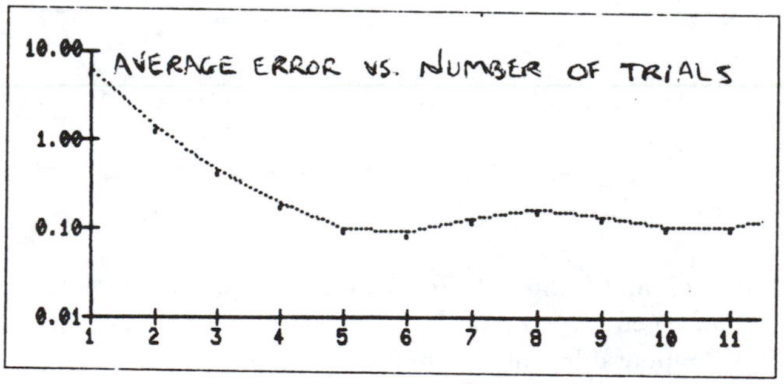

Figure 10 *Learning Curve for Joint 2 (with Noise)*

were learned after 10 trials. Actually, feedforward functions learned after only 1 or 2 trials are quite good.

9. Conclusion

An adaptive scheme has been developed for the case of repeated trials of a path. The primary benefit of the approach is that it allows compensation for effects such as friction which are otherwise difficult to model. A secondary advantage of the scheme is that it can also adapt to torques due to velocity and gravity effects. After sufficient learning, arbitrary paths which are within the torque and velocity limits

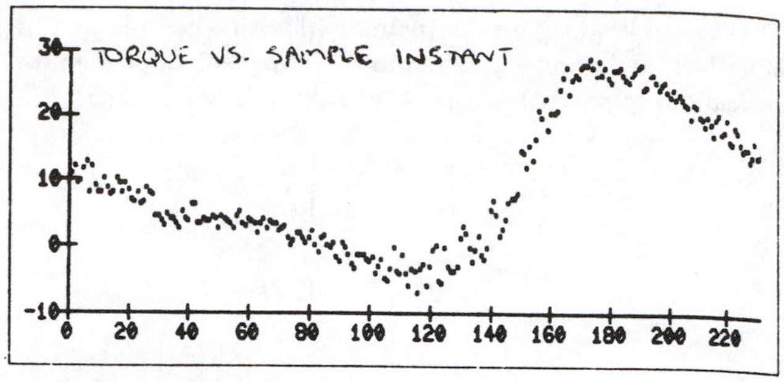

Figure 11 *Learned Torque Function for Joint 1 (Noise)*

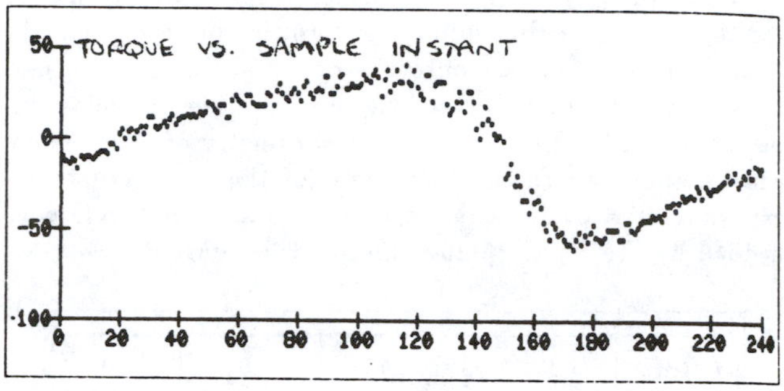

Figure 12 *Learned Torque Function for Joint 2 (Noise)*

of the arm can be executed with errors lower bounded only by the non-repeatability of the manipulator.

Another interesting application of our scheme would be as a method of verification of dynamic models. If a "complete" dynamic model of the manipulator is used in the modelled portion of the control, the feedforward torque learned should be zero. Thus, presence of a non-zero learned torque function indicates that the model is not exact. If the torque function is small or zero over most of the trajectory, the proposed model is proven accurate.

Future work includes implementation on an actual manipulator for testing. The scheme has been run on a single joint of a PUMA 560 with good results [13].

10. Acknowledgements

The author has had helpful discussions with Jeffrey Kerr and Joel Burdick of Stanford University, and Robert Goor and William Hamilton of General Motors Research Laboratories. Much of the work reported was performed at General Motors Research Laboratory. Some portion of the work done at Stanford was supported by NSF grant number MEA 80-19628.

11. References

1. J.M. Hollerbach, "A Recursive Lagrangian Formulation of Manipulator Dynamics and a Comparitive Study of Dynamics Formulation Complexity", IEEE Transactions on Systems, Man, and Cybernetics, SMC-10, November, 1980.

2. J.Y.S. Luh, M.W. Walker, and R. Paul, "On-line Computational Scheme for Mechanical Manipulators", ASME Journal of Dynamic Systems, Measurement, and Control, Vol. 102, 1980.

3. J.M. Hollerbach and G. Sahar, "Wrist-Partitioned Inverse Accelerations and Manipulator Dynamics", MIT A.I. Memo No. 717, April, 1983.

4. R. Horowitz and M. Tomizuka, "An Adaptive Control Scheme for Mechanical Manipulators - Compensation of Nonlinearity and Decoupling Control", Proc. Winter Annual Meeting of the Dynamics System and Control Division of the ASME, Chicago, Nov. 16-21, 1980.

5. S. Dubowsky, "On the Adaptive Control of Robotic Manipulators: The Discrete Time Case", Proc. of the Joint Automatic Control Conference, TA-2B, Virginia, 1981.

6. A.J. Koivo, and T.H. Guo, "Control of Robotic Manipulator with Adaptive Controller", Proc. IEEE Conference on Decision and Control, pp.271-276, San Diego, 1981.

7. M.J. Chung and C.S.G. Lee, "An Adaptive Control Strategy for Computer-Based Manipulators", University of Michigan, RDS-TR-10-82, August, 1982.

8. C. Leininger and S. Wang, "Pole Placement Self-Tuning Control of Manipulators", IFAC Symposium on Computer Aided Design of Multivariable Technological Systems, West Lafayette, Indiana, September, 1982.

9. S. Arimoto and M. Takegaki, "An Adaptive Method for Trajectory Control of Manipulators", IFAC 8th Triennial World Congress, Kyoto, Japan, 1981.

10. B.K. Kim and K.G. Shin, "An Adaptive Model Following Control of Manipulators to Changes in their Payloads and Spatial Configuration", University of Michigan, 1983.

11. M. Vukobratovic and K. Nenad, "Computer-Oriented Method for Linearization of Dynamic Models of Active Spatial Mechanisms", Mechanism and Machine Theory, Vol. 17, No.1, 1982.

12. H.L. Royden, Real Analysis, MacMillan Co., N.Y., 1963.

13. J.J. Craig, "Adaptive Control of Manipulators", Phd. Thesis Proposal, Computer Science Department, Stanford University, April, 1983.

ROBOT TRAJECTORY LEARNING THROUGH PRACTICE

Christopher G. Atkeson and Joseph McIntyre

Artificial Intelligence Laboratory
Massachusetts Institute of Technology
545 Technology Square
Cambridge, Massachusetts 02139

Abstract: We present an algorithm that uses trajectory following errors to improve a feedforward command to a robot. This approach to robot learning is based on explicit modeling of the robot; and uses an inverse of the robot model as part of a learning operator which processes the trajectory errors. Results are presented from a successful implementation of this procedure on the MIT Serial Link Direct Drive Arm. The major point of this paper is that more accurate robot models improve trajectory learning performance, and learning algorithms do not reduce the need for good models in robot control.

1 Introduction

This paper presents an algorithm that enables a robot that is repeating the same trajectory to reduce its trajectory following error on repetition of the motion. The algorithm details how to use the trajectory following error (the time history of the differences between the desired trajectory and the actually executed trajectory) on any particular trial to update an actuator feedforward command that is used for the subsequent attempt.

We assume the desired trajectory has been specified and does not change, and our goal is to follow that trajectory with zero error. This is different from another potentially useful robot learning module which would find a "good" trajectory for a specified task.

Motivation: This algorithm usefully complements standard feedback controllers and also complements the use of a model of the robot to feed forward predicted forces and torques. We illustrate the use of the algorithm with a simple controller structure, although the ideas behind this algorithm can be used to improve a wide variety of controllers. Such a learning algorithm is beneficial even with the use of full dynamic models and high feedback gains, as there will still be trajectory following errors that can be reduced. Furthermore, many proposals for implementing compliant motion require low feedback gains. To combine compliant motion with accurate trajectory following necessitates accurate feedforward commands. Although good models of the robot can initially generate such commands, this algorithm can be used to refine the commands.

Previous Work: The first paper on repeated trajectory learning seems to have been (Uchiyama, 1978). Subsequent work includes (Arimoto, 1985; Arimoto et. al., 1984a, 1984b, 1984c, 1985; Hara, Omata, and Nakano, 1985; Kawamura et. al., 1984, 1985; Craig, 1984, 1983; Mita and Kato, 1985; Togai and Yamano, 1985; Wang, 1984; Wang and Horowitz, 1985)

These papers discuss using linear learning operators and have emphasized stability of the proposed algorithms. There has been little work emphasizing performance, i.e. the convergence rate of the algorithm. Simulations of several of these algorithms have revealed very slow convergence and large sensitivity to disturbances and sensor and actuator noise.

1.1 Features Of This Work

What distinguishes this trajectory learning algorithm from previous robot learning schemes is the following combination:

Makes Optimal Correction Scheme Explicit: This algorithm makes clear the optimal transformation from trajectory error to feedforward command update. If there are no disturbances or sensor or actuator noise and we have perfect models of the robot and its feedback controller, the algorithm generates the correct feedforward command after one motion.

No Unnecessary Approximations: The algorithm makes use of the full robot dynamics in the model used to generate feedforward command corrections. It makes no unnecessary approximations.

Generality: Works With Many Controllers: The algorithm applies to a wide variety of feedback controller structures.

Generality: Works With Many Plants: The algorithm applies to a wide range of plants with (for example): nonlinear dynamics, complex multivariable linear dynamics, actuator dynamics, joint compliance dynamics, or flexible link dynamics. This is an important feature of the algorithm, as it is becoming clear that a typical robot exhibits quite complex dynamics (Sweet and Good, 1985).

Analysis Relevant To Other Schemes: The analysis of this algorithm makes clear why other trajectory improvement schemes that intuitively seem correct often perform badly or actually degrade performance in practice.

2 The Trajectory Learning Algorithm

Assumed Controller Structure: The assumed controller structure is shown in Figure 1. R represents the robot dynamics, while $\mathbf{x}$ represents a vector containing the actual joint positions, velocities, and accelerations of the robot. C is a feedback controller. In robotics an independent PD controller for each joint is often used. We will illustrate the algorithm assuming the feedback controller output is zero when the desired trajectory is achieved exactly. More complex controllers such as a "resolved acceleration" controller (Luh, Walker, and Paul, 1980) may also be used. τ is the actual controller output (joint torques) to the robot.

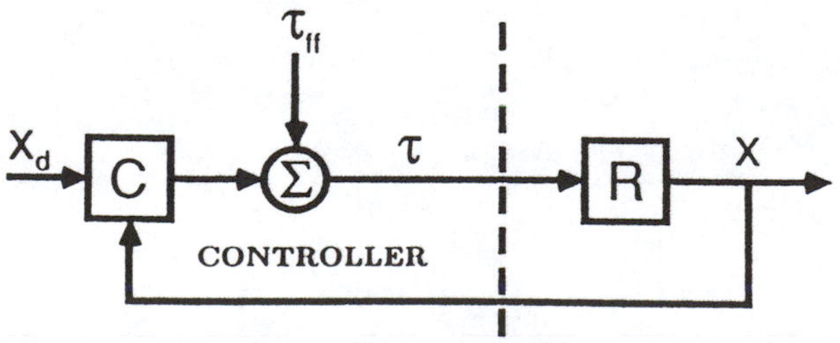

Figure 1: Control system structure.

Reprinted from *Proc. IEEE ROBOT '86*, San Francisco, CA, pp. 1737–1742, 1986.

The commands generated by the feedback controller are augmented by feedforward commands, $\tau_{ff}(t)$. The function of this trajectory learning algorithm is to modify the feedforward command after each attempt at the desired trajectory so as to reduce the trajectory following error on the next attempt. The feedback controller is not modified, although this adaptive feedforward control algorithm could be easily combined with adaptive feedback control algorithms.

Modeling Errors: We assume there is some feedforward control τ^* that will drive the robot exactly along the desired trajectory. Trajectory following errors are caused by errors in modeling the robot, disturbances, and sensor and actuator noise. We model the repeatable components of these error sources as an error in the current feedforward command:

$$\delta\tau_{ff_k}(t) = \tau_{ff_k}(t) - \tau^*(t) \tag{1}$$

The subscript k indicates that the above equation applies to the kth movement.

The Learning Algorithm: The main idea behind this algorithm is to estimate the feedforward command error, $\delta\tau_{ff}(t)$, on each trial and cancel it. This occurs in two steps:

1: The feedback controller reduces the effect of $\delta\tau_{ff_k}$ on the controller output τ_k by measuring the actual state of the robot and generating corrective torques to compensate for any errors. For a linear system the amount of error reduction is given by

$$\delta\tau_k(s) = \tau_k(s) - \tau^*(s) = \frac{1}{1+C(s)R(s)}\delta\tau_{ff_k}(s) \tag{2}$$

in the frequency domain.

2: An inverse of the dynamic model of the robot is now used to estimate the controller output error, $\widehat{\delta\tau}_k(t)$, from the trajectory errors. In general:

$$\widehat{\delta\tau}_k(t) = \tau_k(t) - \tau^*(t) = R^{-1}(\mathbf{x}(t)) - R^{-1}(\mathbf{x}_d(t)) \tag{3}$$

However, the above equation assumes perfect models, in which case $\widehat{\delta\tau}$ would be estimated perfectly after one movement. In reality, an imperfect model, $\hat{R}^{-1}()$, must be used and we can therefore expect trajectory improvement to be an iterative process. For a linear system

$$\widehat{\delta\tau}_k(s) = \frac{1}{\hat{R}(s)}\delta\theta(s) = \frac{1}{\hat{R}(s)}\frac{R(s)}{1+C(s)R(s)}\delta\tau_{ff_k}(s) \tag{4}$$

The estimate of the controller output error, $\widehat{\delta\tau}_k$, is now subtracted from the controller output, τ_k, to find the next feedforward command:

$$\tau_{ff_{k+1}}(t) = \tau_k(t) - \widehat{\delta\tau}_k(t) \tag{5}$$

If our models are perfect and the robot is exactly repeatable we should now be applying the correct input, τ^*. For a linear system we can see how the feedforward command errors are processed. Subtracting Equation 4 from Equation 2.

$$\delta\tau_{ff_{k+1}}(s) = \frac{\hat{R}(s) - R(s)}{\hat{R}(s)(1+C(s)R(s))}\delta\tau_{ff_k}(s) = L(s)\delta\tau_{ff_k}(s) \tag{6}$$

With perfect models, $(\hat{R} = R)$, the learning operator, $L(s)$, is zero, implying convergence in one iteration. The learning algorithm is unstable if $|L(j\omega)| \geq 1$ for any frequency ω.

Initialization Of The Feedforward Command: For the first movement an initial $\tau_{ff}(t)$ must be chosen. If the feedback controller already incorporates a model of the robot and generates the appropriate command signals to drive the robot along the desired trajectory with no errors (given a perfect model of the robot) then the initial $\tau_{ff}(t)$ is simply set to zero. An example of such a controller is the "resolved acceleration" controller.

In simpler feedback controllers that rely on trajectory errors to generate drive signals the feedforward command should be initialized to the best estimate of the actuator commands necessary to drive the robot along the desired trajectory. These commands are generated using:

$$\tau_{ff}(t) = \hat{R}^{-1}(\mathbf{x}_d(t)) \tag{7}$$

Inverse Models of Robots: For robot arms a rigid body dynamics model is often used to predict the forces and torques necessary to achieve a particular motion, and thus serves as a model of the inverted robot dynamics. The rigid body dynamics equations for a robot can be written as

$$\hat{R}^{-1}(\mathbf{x}) = Torques = \mathbf{I}(\boldsymbol{\theta}) \cdot \ddot{\boldsymbol{\theta}} + \dot{\boldsymbol{\theta}} \cdot \mathbf{C}(\boldsymbol{\theta}) \cdot \dot{\boldsymbol{\theta}} + \mathbf{g}(\boldsymbol{\theta}) \tag{8}$$

where $\boldsymbol{\theta}(t)$ is the desired trajectory of the joint angles (we have omitted the time arguments in the above equation), $\mathbf{I}(\boldsymbol{\theta})$ is the inertia matrix of the arm, $\mathbf{C}(\boldsymbol{\theta})$ is the coriolis and centripetal force tensor, and $\mathbf{g}(\boldsymbol{\theta})$ is the gravitational force vector (Hollerbach, 1984).

For some robots it is argued that additional sources of dynamics are important (Goor, 1985; Sweet and Good, 1985; Good, Sweet, and Stroebel, 1985). In this case we can still model these dynamics and invert the model.

3 An Implementation Of The Algorithm

We have implemented the trajectory learning algorithm on the MIT Serial Link Direct Drive Arm. This three joint arm is described in (An, Atkeson, and Hollerbach, 1986). To demonstrate the effectiveness of our learning algorithm we will present results on learning a particular trajectory.

The Test Trajectory: All three joints of the Direct Drive Arm were commanded to follow a fifth order polynomial trajectory with zero initial and final velocities and accelerations and a 1.5 second duration. Figure 2 shows the shape of the trajectory for each joint, and Table 1 gives the initial and final joint positions, the peak joint velocities, and the peak joint accelerations.

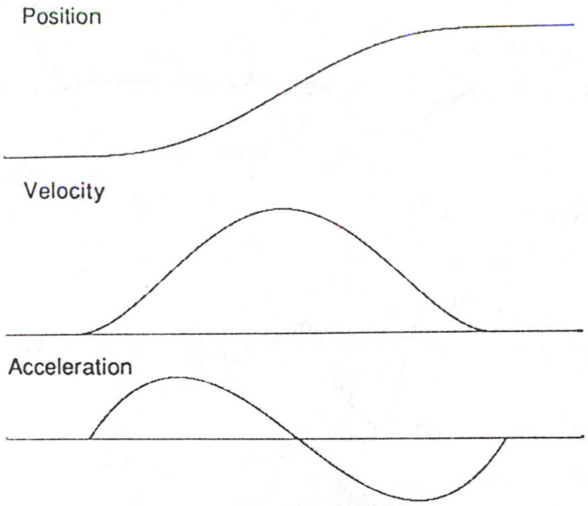

Figure 2. Test Trajectory

Joint	Initial Position *radians*	Final Position *radians*	Peak Velocity *radians/s*	Peak Acceleration *radians/s²*
1	0.5	4.5	5.0	±10.3
2	5.0	1.0	-5.0	±10.3
3	4.0	-0.5	-5.6	±11.5

Table 1: Test trajectory parameters.

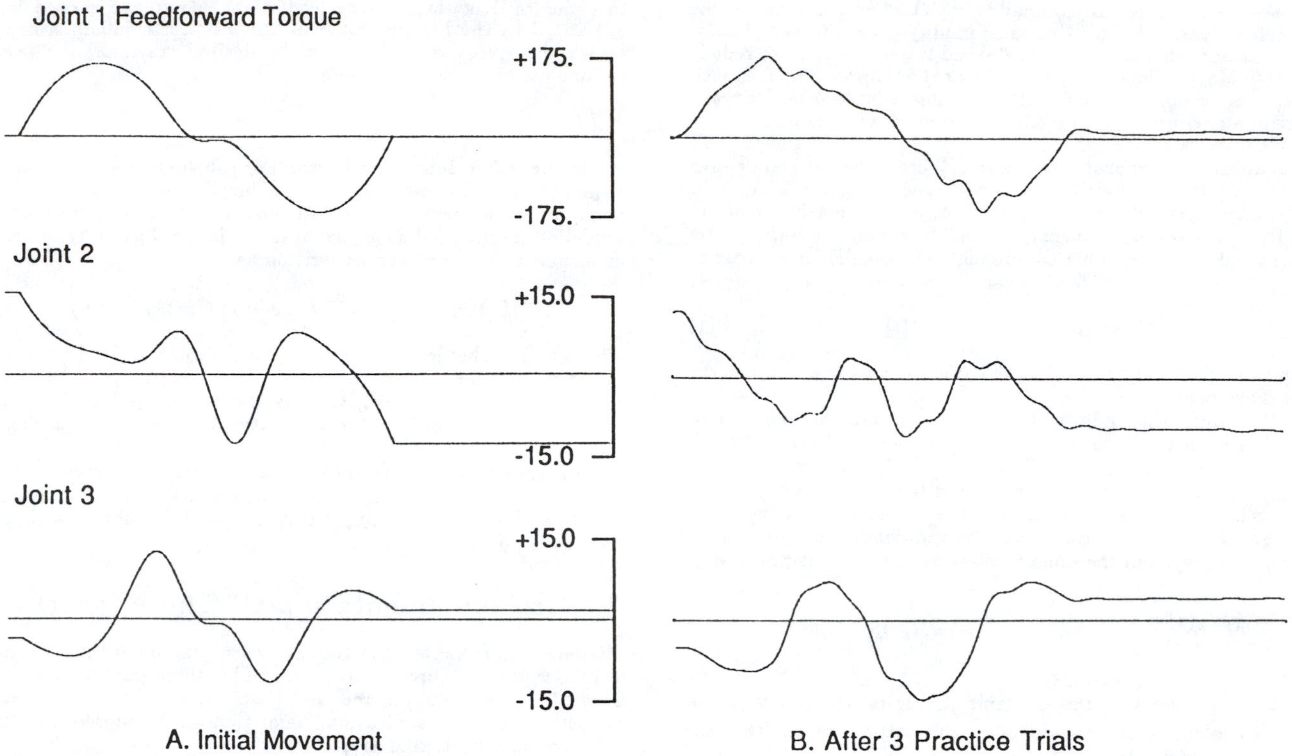

Figure 3. Feedforward Torques (*newton – meters*)

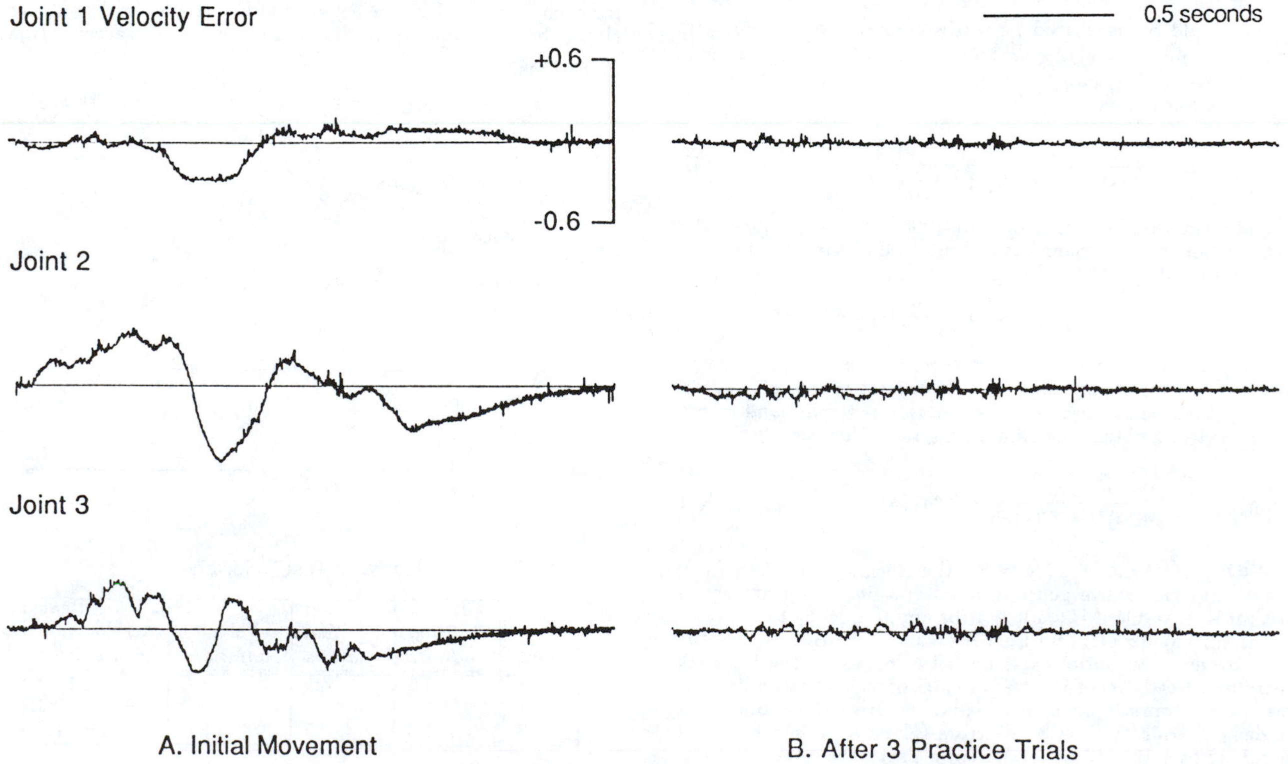

Figure 4. Velocity Errors (*radians/second*)

Figure 5. Simulated Two Joint Learning With Decoupled Model

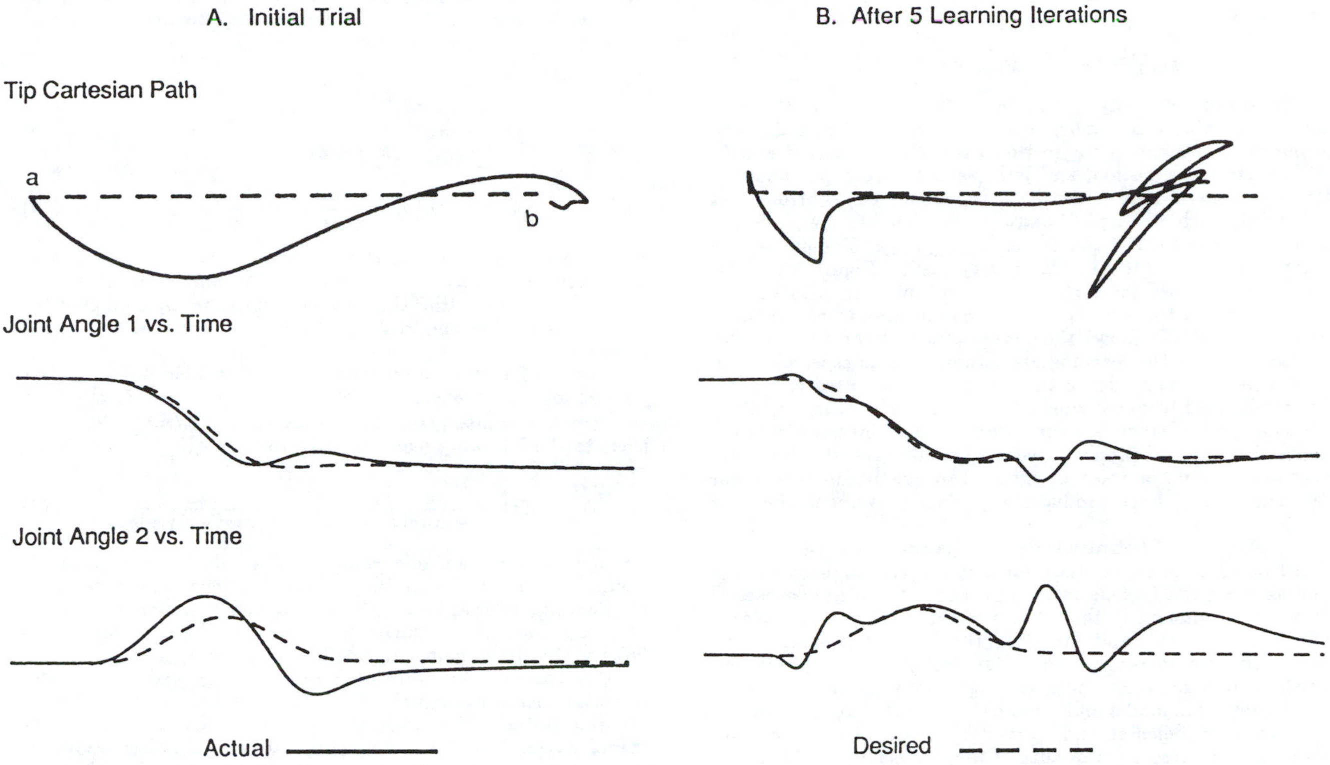

The Feedback Controller: An independent PD (negative position and velocity feedback) digital controller was implemented for each joint and was not modified during learning.

Initialization Of The Feedforward Command: The initial feedforward torques were generated from a rigid body dynamics model. The model and the estimation of its parameters are described in (An, Atkeson, and Hollerbach, 1985). The calculated feedforward torques are shown in Figure 3A.

Initial Trajectory Performance: As an index of trajectory following performance the velocity errors (the difference between the actual joint velocity and the desired joint velocity) for the first movement are shown in Figure 4A. We have plotted the raw velocity error data to give an idea of the relative size of the trajectory errors and sensor noise.

Calculating Acceleration and Filtering: In order to use the rigid body dynamics model for the Direct Drive Arm, the joint accelerations had to be estimated. Joint positions and velocities were measured directly. A digital differentiating filter combined with an 8Hz. low pass filter was applied to the velocity data to estimate accelerations.

For reasons discussed later it is necessary to filter the trajectory errors and controller output to improve the convergence of the learning process. In this implementation we applied digital filters with an 8Hz. cutoff to the data used in the learning process. We filtered the references used by the learning operator with the same filter used on the data. It was necessary to correct for inconsistencies between the velocity sensors and the position measurements. This was done by calibrating the position reference to the feedback controller so that the position error would be consistent with the velocity error.

Final Trajectory Performance: The robot executed two additional training movements which are not shown, and its performance on the fourth attempt of the test trajectory was assessed. Figure 3B shows the modified feedforward commands used on the fourth movement, and should be compared with the predicted torques shown in Figure 3A. Figure 4B shows the velocity errors for the fourth movement, and should be compared with the initial movement velocity errors in Figure 4A. There has been a substantial reduction in trajectory following error after only three practice movements.

4 Using Simplified Models

It may seem unnecessary to use the full rigid body dynamics model of the robot in the learning algorithm. One might think that learning algorithms of this type allow one to avoid modeling the robot in full detail. This is not the case. Simplifying the robot model necessarily introduces additional modeling error. Without careful analysis such modeling errors may cause the learning algorithm to have poor performance, or even to degrade performance. As an illustration of the possible effects of modeling error due to the use of simplified models, we will now present a seemingly reasonable simplified model of a two joint robot arm that when used as the learning operator fails to learn. The robot arm is a planar two link mechanism with rotary joints and is described in the introduction to (Brady, et al., 1982.)

The simplified dynamic model that we have chosen for this example is that of two independent rotary joints with constant moments of inertia. That is, we ignore the centripetal and coriolis torques of the complete rigid body robot dynamics, and we assume constant moments of inertia around each joint. The moment of inertia of link 2 is a constant with respect to θ_2. The moment of inertia around joint 1 depends on θ_2. We approximate the moment around joint 1 as the average of the maximum and minimum moments around that joint, over all possible configurations of the robot. The equations of motion for such a simplified system are:

$$Torques = \mathbf{I} \cdot \ddot{\theta} \quad (9)$$

where I is a constant diagonal 2×2 matrix. This gives a learning operation of:

$$\tau_{ff_{k+1}} = \tau_k - \mathbf{I}_{\mathbf{estimated}} \cdot (\ddot{\theta} - \ddot{\theta}_d) \quad (10)$$

The results of applying this learning algorithm to the simulated two joint robot movement are show in Figure 5. The movement is from point a to point b with zero initial and final velocity, acceleration, and jerk (seventh order polynomial). Feedback control is provided by independent PD controllers at each joint, each having a bandwidth of 1.0 hertz and damping coefficient of 0.707 (based on $\mathbf{I}_{\mathbf{estimated}}$). The simulation is described more fully in (McIntyre, 1986). Figure 5A shows the performance of the system on the initial trial, with the initial feedforward torques, τ_{ff_0}, based on the simplified dynamics model. Figure 5B shows the performance of the system after five iterations of the learning algorithm. In this case, where an inaccurate inverse model of the robot has been used to update the feedforward torque command, no improvement in trajectory following performance is seen. Had the full inverse dynamics model been utilized (under these ideal conditions of no measurement noise, actuator noise, or external torque disturbances), our algorithm would have produced a perfect movement after one iteration.

Appropriate Robot Models: It has been argued that simplified models are appropriate for a robot with high gear ratios such as the PUMA. One must still model the other sources of dynamics prominent in these types of robots (Sweet and Good, 1985; Good, Sweet, and Stroebel, 1985). Higher order actuator dynamics may play an important role, for example (Goor, 1985). Our point is not that the rigid body dynamics are the only appropriate model and must be used, but that *all* dynamics that can be modeled should be modeled. Learning performance can be used to assess the quality of the models used to drive the learning.

5 Convergence

The previous example raises the question of when the algorithm will converge to zero trajectory error. There are many factors affecting convergence including parametric modeling error, unmodeled dynamics, excitation due to process disturbances and sensor and actuator noise, errors in the initial position of the robot, and inconsistency between redundant sensors. Examining these issues in the general case is beyond the scope of this paper. Instead we will explore simple linear models, which still illustrate most of the important considerations. We will assume that the trajectory to be learned has an infinite extent in time so that we can describe the learning operator as a transfer function. This is an important consideration since many previous proofs of convergence of these types of algorithms rely on the finite time interval over which learning is allowed to occur.

The learning operator for linear systems has the transfer function:

$$L(s) = \frac{\hat{R}(s) - R(s)}{\hat{R}(s)(1 + C(s)R(s))} \quad (11)$$

This leads to several observations on learning performance:

Effect of feedback controller performance: As the feedback controller is improved ($1/(1+CR)$ is decreased) learning will converge faster. In the linear case this is due to the feedback controller doing a better job rejecting disturbances. In nonlinear systems there is an additional effect that a better feedback controller generates a trajectory closer to the final trajectory, and the disturbances and modeling errors on that trajectory will be closer to the repeatable disturbances and modeling errors of the final trajectory.

Effect of parametric modeling error: Errors in model parameters typically lead to modeling errors in the frequencies we are interested in controlling. We will use a simple one joint robot with mass m and a PD controller with position gain k and velocity gain b as an example:

$$R(s) = \frac{1}{ms^2} \quad (12)$$

$$C(s) = bs + k \quad (13)$$

$$L(s) = \frac{(m - \hat{m})s^2}{ms^2 + bs + k} \quad (14)$$

$L(j\omega)$ can be plotted on the complex plane as a function of frequency. If it leaves the interior of the unit circle it will be unstable. Manipulating the feedback controller by changing the gains k and b allow one to shape this curve for particular filtering properties.

Effect of unmodeled dynamics: Consider the above one joint robot example with a perfect model ($\hat{m} = m$) except for a very small absolute error Δ, so that $R(s) = 1/ms^2 - \Delta$. This leads to the following learning operator:

$$L(s) = \frac{\Delta m^2 s^4}{-\Delta bms^3 - \Delta kms^2 + ms^2 + bs + k} \quad (15)$$

At high frequencies the learning error is unbounded and clearly unstable. This is inevitable, as the frequency response of all physical plants goes to zero as the frequency increases. We are inverting this small frequency response and therefore are amplifying with a large gain any small unmodeled dynamics.

Excitation: An additional and often ignored problem for trajectory learning algorithms is the increase in errors due to non-repeatable disturbances and sensor and actuator noise. The stability bounds on the learning operator may actually be smaller than ± 1 due to this excitation of the error.

For the above reasons the learning operator must also include some filtering. The filter decreases performance in the ideal case, but for practical use it makes learning more robust against modeling errors and excitation. The filter should have low gain where there are large model errors and/or large excitation. We are still exploring optimal design of filters for learning.

Other robustness techniques: Posture control: Much of the time a robot is at rest or making fine adjustments in its position. The control system can explicitly recognize these essentially static situations and use specialized learning operators which incorporate only static models of the robot. Iterative "trajectory" learning and integrators in the feedback controller play similar roles under these conditions and should be designed to complement each other.

Time weighted updates: As the trajectory execution error decreases, with learning the signal to noise ratio that the learning operator has to deal with decreases. A time varying filter should be applied to handle this problem, with learning gains decreasing with practice. Instead of picking an arbitrary weighting scheme, however, we are developing a filter design procedure based on an optimal estimation/control framework.

Restrict time interval of learning: One can also restrict the amount of time over which learning occurs. Reducing the amount of time during the trajectory when the learning algorithm is active reduces the the amount of time any errors in the feedforward system can have effect. As the actual robot is a causal system, errors can only be propagated forward in time, and since learning occurs only over a finite time interval, eventually errors move out of the learning interval and effectively disappear. This is why many proposed schemes that claim not to need accurate plant models seem to converge. However, such convergence is only a mathematical illusion, as in practice this type of convergence is very sensitive to disturbances and sensor and actuator noise.

Friction: These robustness techniques also can help handle nonlinear systems that are difficult to control such as systems with large amounts of friction. To assess the role of learning in such systems we had to resort to simulation. We have successfully used our learning algorithm on simulated robots with larger amounts of friction than found in the Direct Drive Arm (McIntyre, 1986).

Non-minimum phase systems: A question that often arises is how one inverts a non-minimum phase plant, for which a causal inverse is unstable. Since we process the trajectory data after the trajectory is executed our learning operator may be acausal, and thus stably inverting a non-minimum phase system is possible. This issue is addressed more fully in (Atkeson, 1986).

6 Conclusion

The main message of this paper is that the incorporation of learning in a control system is not a license to do a poor modeling job of the controlled system. The benefits of accurate modeling are better performance in all aspects of control, while the risks of inadequate modeling are poor learning performance or even degradation of performance with practice.

Acknowledgments

This paper describes research done at the Whitaker College, Department of Psychology, and the Artificial Intelligence Laboratory of the Massachusetts Institute of Technology. Support was provided in the Whitaker College by grants AM26710 and NS09343 from the NIH. Support for the AI laboratory's artificial intelligence research is provided in part by the Defense Advanced Research Projects Agency under Office of Naval Research contracts N00014-80-C-050 and N00014-82-K-0334 and by the Systems Development Foundation. Support for JM was provided by NIH training grant NIH-GM07484. Support for CGA was provided by a Whitaker Fund Graduate Fellowship.

References

An, C. H., C. G. Atkeson, and Hollerbach, J. M., "Experimental Determination of the Effect of Feedforward Control on Trajectory Tracking Errors", *IEEE Int. Conf. on Robotics and Automation*, 1986.

An, C. H., C. G. Atkeson, and Hollerbach, J. M., "Estimation of Inertial Parameters of Rigid Body Links of Manipulators", *Proc. 24th Conf. on Decision and Control*, Fort Lauderdale, Florida, Dec. 11-13, 1985.

Arimoto, S. "Mathematical Theory of Learning With Applications to Robot Control", *Proc. of 4th Yale Workshop on Applications of Adaptive Systems Theory*, pp. 215-220, Center for Systems Science, Yale University, May, 1985.

Arimoto, S., S. Kawamura, and F. Miyazaki, "Bettering Operation of Robots by Learning", *J. of Robotic Systems*, 1-2, pp. 123-140, 1984a.

Arimoto, S., S. Kawamura, and F. Miyazaki, "Can Mechanical Robots Learn by Themselves?", *Proc. of 2nd Inter. Symp. Robotics Research*, Kyoto, Japan, August 1984b.

Arimoto, S., S. Kawamura, and F. Miyazaki, "Bettering Operation of Dynamic Systems By Learning: A New Control Theory For Servomechanisms of Mechatronics Systems", *Proc. 23rd IEEE CDC*, Las Vegas, Nevada, Dec. 1984c.

Arimoto, S., S. Kawamura, F. Miyazaki, and S. Tamaki, "Learning Control Theory for Dynamical Systems", *Proc. 24th Conf. on Decision and Control*, Fort Lauderdale, Florida, Dec. 11-13, 1985.

Atkeson, C. G., Thesis Proposal, Dept. of Psychology, MIT, 1986.

Brady, Michael, Hollerbach, John M., Johnson, Timothy L., Lozano-Pérez, Tomás, and Mason, Matthew T., *Robot Motion: Planning and Control*, The MIT Press, Cambridge, MA., 1982

Craig, J.J., "Adaptive Control of Manipulators Through Repeated Trials", *Proc. American Control Conference*, San Diego: June 6-8, 1984, 1566-1574.

Craig, J.J., "Adaptive Control of Manipulators", *PhD. Thesis Proposal*, Computer Science Department, Stanford University, April, 1983.

Good, M.C., Sweet, L.M., and Strobel, K.L., "Dynamic models for control system design of integrated robot and drive systems", *ASME J. Dynamic Systems, Meas., Control,* Vol. 107, pp. 53-59, 1985.

Goor, R.M., "A new approach to robot control" *Proc. American Control Conf.*, pp. 385-389, Boston, June 19-21, 1985.

Hara, S., T. Omata, and M. Nakano "Synthesis of Repetitive Control Systems and its Application", *Proc. 24th Conf. on Decision and Control*, Fort Lauderdale, Florida, Dec. 11-13, 1985.

Hollerbach, J. M. "Dynamic Scaling of Manipulator Trajectories", *Journal of Dynamics Systems, Measurement, and Control* Vol. 106, pp. 102-106, March 1984.

Kawamura, S., F. Miyazaki, and S. Arimoto, "Applications of Learning Method for Dynamic Control of Robot Manipulators", *Proc. 24th Conf. on Decision and Control*, Fort Lauderdale, Florida, Dec. 11-13, 1985.

Kawamura, S., F. Miyazaki, and S. Arimoto, "Iterative Learning Control For Robotic Systems", *Proc. of IECON 84*, Tokyo, Japan, Oct. 1984.

Luh, J.Y.S., M. W. Walker, and R. P. Paul "Resolved Acceleration Control of Mechanical Manipulators", *IEEE Trans. Automatic Control*, **25**, 3, pp. 468-474, 1980.

McIntyre, J. First Year Paper, Dept. of Psychology, MIT, 1986.

Mita, T., and E. Kato "Iterative Control and its Application to Motion Control of Robot Arm – A Direct Approach to Servo-Problems", *Proc. 24th Conf. on Decision and Control*, Fort Lauderdale, Florida, Dec. 11-13, 1985.

Sweet, L. M., and M. C. Good "Redefinition of the Robot Motion-Control Problem", *IEEE Control Systems Magazine*, Vol. 5, No. 3, pp. 18-25, August, 1985.

Togai, M., and O. Yamano "Analysis and Design of an Optimal Learning Control Scheme for Industrial Robots: A Discrete Time Approach", *Proc. 24th Conf. on Decision and Control*, Fort Lauderdale, Florida, Dec. 11-13, 1985.

Uchiyama, M. "Formation of High Speed Motion Pattern of Mechanical Arm by Trial", *Transactions, Society of Instrument and Control Engineers* Vol. 19, #5, pp. 706-712, 1978.

Wang, S. H. "Computed Reference Error Adjustment Technique (CREATE) For The Control of Robot Manipulators", *22nd Annual Allerton Conf. on Communication, Control, and Computing* October, 1984.

Wang, S. H., and I. Horowitz "CREATE - A New Adaptive Technique", *Proc. of the Nineteenth Annual Conf. on Information Sciences and Systems*, March, 1985.

On the Iterative Learning Control Theory for Robotic Manipulators

PAOLA BONDI, GIUSEPPE CASALINO, AND LUCIA GAMBARDELLA

Abstract—**A "high-gain feedback" point of view is considered within the iterative learning control theory for robotic manipulators. Basic results concerning the uniform boundedness of the trajectory errors are established, and a proof of convergence of the algorithm is given.**

I. Introduction

THE SET of proposed methodologies for the control of movements of robotic manipulators can be substantially subdivided into three main categories. The first refers to the use of control systems designed on the basis of classical servomechanism concepts [1]–[3], where the fact that a manipulator is actually a highly nonlinear system is almost completely neglected. Consequently, although the regulators turn out to be very simple and commonly used in normal applications, they cannot face the increasing demand in control performances.

The second category includes control methodologies such as "inverse control" [4], and "decoupling control" [5], [6]. Since in this case the nonlinearities of the system are explicitly taken into account, very good performance can be obtained. The disadvantage is represented by the fact that very accurate mathematical models are required, together with a precise forecasting of the load. Moreover, they give rise to costly and very complicated control systems.

The third category is represented by all the methodologies whose goal is to encompass the previously cited drawbacks by using adaptive control concepts. They generally refer to the so-called "adaptive model following techniques" [7], which do not require a precise description of the mathematical model of the system and can face parameters variations, but they still require a complicated structure for the regulator due to the complexity of the adaptation mechanism. Instead, this paper refers to the so-called "iterative learning technique" applied to robotic manipulators, a technique which has been recently, and independently, proposed by different researchers, including Arimoto *et al.* [12]–[14], Craig [16], and the authors [8]–[11].

As it is now well known, such techniques are based on the use of repeated trials of tracking of a preassigned trajectory. At each trial, the position, velocity, and acceleration errors, with respect to the reference trajectory, are recorded to be exploited by a suitable algorithm in the next trial, with the aim of progressively reducing the trajectory errors. Provided it works, the proposed methodology turns out to be very simple and requires only a very rough description of the manipulator to be controlled, but its main disadvantage is obviously represented by the necessity to perform different trials for any new trajectory which must be tracked. At the present moment, research activities are devoted to investigations about the possibilities of decomposing the problem of following a complex trajectory into a set of subproblems, each one referring to a simpler (and already learned) trajectory.

The paper is partly based on previous work of the authors [8]–[11] and presents an inherently nonlinear analysis of the learning procedure. In particular, a "high-gain feedback" point of view is instrumentally used to prove the possibility of setting up uniform upper bounds to the trajectory errors occurring at each trial. Since such a bounding condition is essential for the proof of convergence of the procedure, it has always been assumed, even if the possibility of its achievement has never been proved completely.

The proof of such possibility actually represents the main contribution of the present work. More particularly, the subsequent analysis of convergence allows us to show that, apart from minor conditions, the existence of a finite (but not necessarily narrow) bound on the trajectory deviations may substantially suffice for guaranteeing the zeroing of the errors after a sufficient number of trials. This fact in turn leaves open the possibility of obtaining the exact tracking of the desired motion, even in presence of moderate values assigned to the feedback gains.

The paper is organized as follows: in Section II the learning methodology is described and the underlying rationale explained. In Section III the results concerning the conditions for the uniform boundedness of the trajectory errors are presented while in Section IV the convergence analysis is performed. Further, in Section V some comments concerning the practical implementability of the procedure are given, and, finally, simulation experiments are presented in Section VI.

II. Description of the Learning Methodology

Let us consider the dynamic equation of a manipulator, relating the vector $q \in R^n$ of the chosen (generalized) coordinates with the vector $M \in R^n$ of the corresponding (generalized) torques, is given by

$$A(a)\ddot{q}+B(q, \dot{q})\dot{q}+C(q)=M \tag{1}$$

Manuscript received June 2, 1986; revised June 18, 1987. This paper was presented in part at the 1986 IEEE Conference on Robotics and Automation, San Francisco, CA, April 7–10, 1986.

P. Bondi and L. Gambardella are with the Dipartimento di Matematica e Applicazioni "R. Caccioppoli," Università di Napoli, Via Mezzo-Cannone 8, 80134 Naples, Italy.

G. Casalino is with the Dipartimento di Informatica, Sistemistica, e Telematica, Università de Genova, Via Opera 11, 16145 Genoa, Italy.

IEEE Log Number 8718069.

Reprinted from *IEEE J. Robot. Automat.*, vol. 4, no. 1, pp. 14–22, Feb. 1988.

where matrices $A(q)$, $B(q, \dot{q})$, and vector $C(q)$ are continuous and locally Lipshitzian functions of their arguments. Matrices $A(q)$ (positive definite) and $B(q, \dot{q})$ are supposed to be completely unknown, and the same assumption substantially holds also with respect to vector $C(q)$, with the exception that the knowledge of its value $C(q(0))$ at the initial time is allowed (more will be said about this point in this section).

Further, it is assumed that the manipulator is operating in closed loop with a linear feedback law of the type

$$M = m + Q - K\ddot{q} - L\dot{q} - Pq \tag{2}$$

where m represents the vector of the effective driving signals, Q another constant vector of external signals, while matrices P, L, and K are the position, velocity, and acceleration gains, respectively.

Remark 1: Note that the acceleration feedback introduces an "algebraic loop" in the closed-loop system. This actually implies that control law (2) cannot be exactly implemented, but only approximated by a suitable choice of the operating conditions. The presence of this drawback will, however, be temporarily neglected and reconsidered in greater detail at the end of the paper.

By substituting (2) in (1), the closed-loop equation is obtained of the form

$$[K + A(q)]\ddot{q} + [L + B(q, \dot{q})]\dot{q} + Pq + [C(q) - Q] = m \tag{3}$$

whose structure allows the interpretation of the terms $A(q)$, $B(q, \dot{q})$, and $[C(q) - Q]$ as (unknown) perturbations occurring in the linear and known model

$$K\ddot{q} + L\dot{q} + Pq = m. \tag{4}$$

For such known model the computation of the signal $m_0(t)$, allowing the output to follow a specified trajectory $q^*(t)$ (hereafter assumed bounded with its first and second derivative and such that $\dot{q}(0) = 0$; $\dot{q}(t) = 0$ for $t \geqslant T > 0$, where T is *a priori* assigned), is straightforward and given by

$$m_0 = K\ddot{q}^* + L\dot{q}^* + Pq^* \tag{5}$$

which also results in a bounded signal. Obviously, if the computed signal $m_0(t)$ is applied to the closed-loop system (3), the corresponding trajectory $q_0(t)$ will be different from the desired one $q^*(t)$.

Nevertheless, assume that a first trial of movement has been carried out using $m_0(t)$ starting from the rest position $q(0) = q^*(0)$ (this requires $Q = C[q^*(0)]$ and explains the assumption about the knowledge of $C[q(0)]$), that the corresponding trajectory errors $\delta q_0(t) \triangleq q_0(t) - q^*(t)$, $\delta\dot{q}_0(t)$, $\delta\ddot{q}_0(t)$, for $t \in [0, T]$, have resulted to be bounded, and that they have been recorded by a suitable device. Then the proposed learning procedure suggests repeating a trial, starting from the same initial rest position $q^*(0)$ and using the new driving signal

$$m_1 = m_0 - \delta m_1 \tag{6}$$

where the correcting term δm_1 is given by

$$\delta m_1 = K\delta\ddot{q}_0 + L\delta\dot{q}_0 + P\delta q_0. \tag{7}$$

The same procedure is in turn applied for all successive trials (provided the errors remain bounded), thus implying that the driving signal applied at the generic ith ($i \geqslant 1$) trial takes on the form

$$m_i = m_{i-1} - \delta m_i = m_0 - \sum_{k=1}^{i} \delta m_k = m_0 - \sum_{k=1}^{i} (K\delta\ddot{q}_{k-1} + L\delta\dot{q}_{k-1} + P\delta q_{k-1}). \tag{8}$$

Remark 2: Obviously, the adoption of the proposed learning procedure requires the presence of a "repositioning system" to be used at the end of each trial, after time T. The problem of devising such a system, however, represents a further step in the development of the learning theory and will not be considered here (see [8] for details of the argument).

In the remainder of this section we shall restrict ourselves to considering some other aspects of the learning process which represent the basis for the discussion of Section III and the convergence analysis of Section IV. To this aim, let us consider a generic driving signal $m_{i+1}(t)$, corresponding to a trial successive to the first, and rewrite it in the following form:

$$m_{i+1} = m^* + \delta m_{i+1}, \qquad i = 0, 1, \cdots \tag{9}$$

where $m^*(t)$ represents the (unknown) signal allowing the closed-loop system to follow the desired trajectory $q^*(t)$, and $\delta m^*_{i+1}(t)$ the corresponding driving signal error.

Then by substituting into $m_{i+1}(t)$ its expression given by (8) (first part) and representing $m_i(t)$ in a form analogous to (9), we immediately get

$$\delta m^*_i = \delta m_{i+1} + \delta m^*_{i+1}, \qquad i = 0, 1, \cdots. \tag{10}$$

Such relation allows interpreting the signal $\delta m_{i+1}(t)$ as an *estimate* for the driving signal error occurring while performing the present ith movement, thus explaining the rationale underlying the use of $-\delta m_{i+1}(t)$ as the compensating term for the next trial. Moreover, the same relation (10) also shows that the *estimation error* at the ith trial actually coincides with the driving signal error that will occur during the next trial, $i + 1$.

Analytic expressions for $\delta m^*_i(t)$ can be obtained as described in the following. Consider the overall closed-loop equation (3) when the generic ith trial is performed ($m = m_i(t)$), and rewrite it in the following form:

$$[K + A(q^* + \delta q_i)](\ddot{q}^* + \delta\ddot{q}_i) + [L + B(q^* + \delta q_i, \dot{q}^* + \delta\dot{q}_i)](\dot{q} + \delta q_i) + P(q^* + \delta q_i) + [C(q^* + \delta q_i) - Q] = m^* + \delta m^*_i, \qquad i = 0, 1, \cdots. \tag{11}$$

Then by substituting into $m^*(t)$ its expression given by

$$m^* = [K + A(q^*)]\ddot{q}^* + [L + B(q^*, \dot{q}^*)]\dot{q}^* + Pq^* + [C(q^*) - Q] \tag{12}$$

with straightforward manipulations, we immediately get

$$\delta m_i^* = K\delta\ddot{q}_i + L\delta\dot{q}_i + P\delta q_i + A_i\delta\ddot{q}_i + B_i\delta\dot{q}_i + (A_i - A_*)\ddot{q}^* + (B_i - B_*)\dot{q}^* + (C_i - C_*), \quad i = 0, 1, \cdots \tag{13}$$

where, for the case of notations, we have defined

$$\begin{aligned} A_i &\triangleq A(q^* + \delta q_i) & A_* &\triangleq A(q^*) \\ B_i &\triangleq B(q^* + \delta q_i, \dot{q}^* + \delta\dot{q}_i) & B_* &\triangleq B(q^*, \dot{q}^*) \\ C_i &\triangleq C(q^* + \delta q_i) & C_* &\triangleq C(q^*). \end{aligned} \tag{14}$$

In (13) the first three terms represent $\delta m_{i+1}(t)$, thus implying (see (10)) that the last five terms necessarily coincide with $\delta m_{i+1}^*(t)$. This is in turn equivalent to saying, for $i \geqslant 1$ only, that $\delta m_i^*(t)$ can also be expressed as

$$\delta m_i^* = A_{i-1}\delta\ddot{q}_{i-1} + B_{i-1}\delta\dot{q}_{i-1} + (A_{i-1} - A_*)\ddot{q}^* + (B_{i-1} - B_*)\dot{q}^* + (C_{i-1} - C_*), \quad i = 1, 2, \cdots. \tag{15}$$

Finally, observe that by equating the right side of (13) and (15) with very simple algebra we obtain

$$(K + A_i)\delta\ddot{q}_i + (L + B_i)\delta\dot{q}_i + P\delta q_i = A_{i-1}\delta\ddot{q}_{i-1} + B_{i-1}\delta\dot{q}_{i-1} - (A_i - A_{i-1})\ddot{q}^* - (B_i - B_{i-1})\dot{q}^* - (C_i - C_{i-1}), \quad i = 1, 2, \cdots \tag{16}$$

$$\text{for all } \epsilon > 0 \quad \exists \begin{cases} \alpha_\epsilon > 0 \\ \rho_\epsilon > 0 \text{ such that} \\ \vartheta_\epsilon > 0 \end{cases} \begin{cases} \alpha \geqslant \alpha_\epsilon \\ \|(\delta q(0), \delta\dot{q}(0))\| \leq \rho_\epsilon \Rightarrow \|(\delta q_\alpha, \delta\dot{q}_\alpha, \delta\ddot{q}_\alpha)\| \leq \epsilon, \text{ for all } t. \\ \|\delta_u\| \leqslant \vartheta_\epsilon, \text{ for all } t \end{cases} \tag{21}$$

which represents a set of differential equations relating, for $i \geqslant 1$, the evolution of the present trajectory errors with those which occurred during the previous trial of movement.

III. Trajectory Errors Boundedness

In this section it is shown that suitable values for the feedback gains K, L, and P can always be devised guaranteeing the uniform boundedness of the trajectory errors $\delta q_i(t)$, $\delta\dot{q}_i(t)$, and $\delta\ddot{q}_i(t)$ for any time t and any trial i. This actually corresponds to proving the possibility of setting up a domain of attraction around the desired trajectory, and, consequently, shows that the correcting signals used at each trial never push the system out of the attraction domain.

This result, though not representing a convergence one, will be fully exploited in Section IV, where the convergence analysis will be effectively performed. To this aim, let us first express the feedback matrices K, L, and P as functions of a common scalar factor $\alpha > 0$, i.e.,

$$K = \alpha\bar{K} \qquad L = \alpha\bar{L} \qquad P = \alpha\bar{P}, \qquad \alpha > 0 \tag{17}$$

and, consequently, represent both the closed-loop system (3) and the reference model (4) in the following equivalent ("normalized") form:

$$[\bar{K} + \bar{A}(q)]\ddot{q} + [\bar{L} + \bar{B}(q, \dot{q})]\dot{q} + \bar{P}q + [\bar{C}(q) - \bar{Q}] = u \tag{18}$$

$$\bar{K}q + \bar{L}q + \bar{P}q = u \tag{19}$$

where

$$\{u, \bar{A}(q), \bar{B}(q, \dot{q}), [\bar{C}(q) - \bar{Q}]\} \triangleq \frac{1}{\alpha}\{m, A(q), B(q, \dot{q}), C(q) - Q\}. \tag{20}$$

For the foregoing normalized representation, we have that $\bar{A}(q)$, $\bar{B}(q, \dot{q})$, and $\bar{C}(q) - \bar{Q}$ all converge to zero for $\alpha \to \infty$ and any finite q, $\dot{q}$, thus implying, contrary to (3), that (18) can also be interpreted as a class of systems parametrized by α and admitting (19) as limit equation for $\alpha \to \infty$. The following lemmas further characterize the system class (18).

Lemma 1: Let $u_\alpha^*(t)$ represent the normalized signal allowing the corresponding system of class (18) to follow the desired trajectory $q^*(t)$. Denote as $\delta u(t)$ an input perturbation superimposed to $u_\alpha^*(t)$, and denote as $(\delta q(0), \delta\dot{q}(0))$ an initial state perturbation. Let $(\delta q_\alpha(t), \delta\dot{q}_\alpha(t), \delta\ddot{q}_\alpha(t))$ represent the vector of the corresponding trajectory errors with respect to $q^*(t)$, $\dot{q}^*(t)$, $\ddot{q}^*(t)$ for a given choice of $\alpha > 0$. Then provided the limit system (19) is asymptotically stable, the following property holds true for class (18):

Proof: For this lemma the main part of the proof is actually quite general, as will be apparent in the following discussion. Consider a family of autonomous systems, parametrized by $\alpha > 0$, of the form

$$\dot{x} = f_\alpha[x, y(x, t)] \tag{22}$$

where $x \in R^n$, $\|y(x, t)\| \leqslant h(x)$ for all t, with $h(\cdot)$ a scalar continuous function, and $f(\cdot, \cdot)$ continuous in its arguments, such that for any $\alpha > 0$ the following conditions are satisfied:

1) $\exists \lim_{\alpha\to\infty} f_\alpha(x, y) = f(x)$ uniformly in any compact set $W \subset R^{2n}$;
2) $f_\alpha(0, y) = 0$ for all $y \in R^n$;
3) the null solution of the "limit system,"

$$\dot{x} = f(x) \tag{23}$$

which exists by 1) and 2), is uniformly asymptotically stable.

Then, we want first to show that the following property holds true for the class (22):

$$\text{for all } \epsilon > 0 \quad \exists \begin{cases} \alpha_\epsilon > 0 \\ \rho_\epsilon > 0 \text{ such that} \\ \vartheta_\epsilon > 0 \end{cases} \begin{cases} \alpha \geqslant \alpha_\epsilon \\ \|x_0\| \leqslant \rho_\epsilon \\ \|g(x, t)\| \leqslant \vartheta_\epsilon \text{ for } \begin{cases} \|x\| \leqslant \epsilon \\ \text{for all } t \end{cases} \end{cases} \Rightarrow \|x_\alpha(t, x_0)\| \leqslant \epsilon \text{ for all } t \tag{24}$$

where $x_\alpha(t, x_0)$ denotes the solution of the perturbed system

$$\dot{x}=f_\alpha(x, y)+g(x, t) \qquad x_\alpha(0, x_0)=x_0. \tag{25}$$

To prove the foregoing property, first observe that assumption 3) implies the total stability of the limit system (see, for instance, [15]), that is, for the "perturbed limit system,"

$$\dot{x}=f(x)+g(x, t) \qquad x(0, x_0)=x_0, \tag{26}$$

the following property holds true:

$$\text{for all } \epsilon>0 \quad \exists \begin{cases}\rho_\epsilon>0 \\ \bar{\vartheta}_\epsilon>0\end{cases} \text{such that} \begin{cases}\|x_0\| \leqslant \rho_\epsilon \\ \|g(x, t)\| \leqslant \bar{\vartheta}_\epsilon \text{ for } \begin{cases}\|x\| \leqslant \epsilon \\ \text{for all } t\end{cases}\end{cases} \Rightarrow \|x(t, x_0)\| \leqslant \epsilon \text{ for all } t. \tag{27}$$

Then due to the foregoing property and taking into account that (25) can be equivalently rewritten as

$$\dot{x}=f(x)+[g(x, t)+f_\alpha(x, y)-f(x)] \qquad x(0, x_0)=x_0 \tag{28}$$

(i.e., in a form analogous to (26) where the perturbation is now $[g(x, t) + f_\alpha(x, y) - f(x)]$), it immediately follows that $\|x_\alpha(t, x_0)\| \leqslant \epsilon$, provided

$$\|x_0\| \leqslant \rho_\epsilon \qquad \|g(x, t)+f_\alpha(x, y)-f(x)\| \leqslant \bar{\vartheta}_\epsilon, \quad \text{for} \begin{cases}\|x\| \leqslant \epsilon \\ \text{for all } t\end{cases} \tag{29}$$

where the second of (29) is in turn certainly satisfied if the perturbation $g(x, t)$ is such that

$$\|g(x, t)\| \leqslant \bar{\vartheta}_\epsilon - \|f_\alpha(x, y)-f(x)\|, \qquad \text{for} \begin{cases}\|x\| \leqslant \epsilon \\ \text{for all } t\end{cases}. \tag{30}$$

Then by keeping in mind that due to assumption 1) we have in the compact set $\{x \in R^n: \|x\| \leqslant \epsilon\}$ the property

$$\text{for all } \delta>0 \qquad \exists \alpha_{\delta,\epsilon}>0 \text{ such that} \begin{cases}\|x\| \leqslant \epsilon \\ \alpha \geqslant \alpha_{\delta,\epsilon}\end{cases} \Rightarrow \|f_\alpha(x, y)-f(x)\| \leqslant \delta \text{ for all } t. \tag{31}$$

It follows that by letting (for instance)

$$\delta=\lambda\bar{\vartheta}_\epsilon, \qquad 0<\lambda<1 \text{ and } \alpha_\epsilon \triangleq \alpha_{\lambda\bar{\vartheta}_\epsilon,\epsilon} \tag{32}$$

condition (30) (and then the second of (29)) turns out to be satisfied for all $\alpha \geqslant \alpha_\epsilon$ by any perturbation such that

$$\|g(x, t)\| \leqslant (1-\lambda)\bar{\vartheta}_\epsilon \triangleq \vartheta_\epsilon, \qquad \text{for} \begin{cases}\|x\| \leqslant \epsilon \\ \text{for all } t\end{cases}, \tag{33}$$

and this completes the proof of property (24).

At this point let us turn to consider system class (18). By expressing the normalized signal $u(t)$ as

$$u=u_\alpha^*+\delta u=[\bar{K}+\bar{A}(q^*)]\ddot{q}^*+[\bar{L}+\bar{B}(q^*, \dot{q}^*)]\dot{q}^* +\bar{P}q^*+[\bar{C}(q^*)-\bar{Q}]+\delta u \tag{34}$$

and substituting in (18), we obtain a normalized form analogous to (13), that is,

$$(\bar{K}+\bar{A})\delta\ddot{q}+(\bar{L}+\bar{B})\delta\dot{q}+\bar{P}\delta q+(\bar{A}-\bar{A}_*)\ddot{q}^* +(\bar{B}-\bar{B}_*)\dot{q}^*+(\bar{C}-\bar{C}_*)=\delta u \tag{35}$$

where $\bar{A} \triangleq A(q^* + \delta q)/\alpha$, $\bar{A}_* = A(q^*)/\alpha$, and the same holds true for the other matrices $\bar{B}$, $\bar{B}_*$ and vectors $\bar{C}$, $\bar{C}_*$.

From (35), by defining the vector $x \triangleq (x_1, x_2) \triangleq (\delta q, \delta\dot{q})$, we immediately get the state-space representation

$$\begin{cases}\dot{x}_1=x_2 \\ \dot{x}_2=-(\bar{K}+\bar{A})^{-1}[(\bar{L}+\bar{B})x_2+\bar{P}x_1+(\bar{A}-\bar{A}_*)\ddot{q}^* \\ \qquad +(\bar{B}-\bar{B}_*)\dot{q}^*+(\bar{C}-\bar{C}_*)+\delta u].\end{cases} \tag{36}$$

Then since $-(\bar{K} + \bar{A})^{-1}\delta u$ plays the same role as $g(x, t)$ in (25), and by noting that the autonomous part of (36) actually satisfies assumptions 1)–3) (with $\dot{x} = f(x) \Leftrightarrow \bar{K}\delta\ddot{q} + \bar{L}\delta\dot{q} + \bar{P}\delta q = 0$), it follows that property (24) directly applies to (36), meaning that Lemma 1 actually holds with respect to vector $(\delta q_\alpha(t), \delta\dot{q}_\alpha(t))$. However, by taking into account the uniform boundedness of $q^*(t)$, $\dot{q}^*(t)$, and $\ddot{q}^*(t)$ and, consequently (since functions $A(\cdot)$, $B(\cdot, \cdot)$, $C(\cdot)$ are locally Lipshitzian), of the Lipshitz condition uniformly satisfied with respect to $\alpha \geqslant \alpha_\epsilon$ by the right side of (36) in the compact sets $\{x:\|x\| \leqslant \epsilon\}$, $\{\delta u:\|\delta u\| \leqslant \vartheta_\epsilon\}$, it easily follows that the same norm limitations can also be made to hold for the whole vector $(\delta q_\alpha(t), \delta\dot{q}_\alpha(t), \delta\ddot{q}_\alpha(t))$.

Lemma 2: Let $u_0(t)$ represent the normalized signal allowing the limit system (19) to follow the desired trajectory $q^*(t)$. Consider $u_0(t)$ applied to a generic system of class (18) and denote as $\delta u_{0,\alpha}^*(t)$ the error with respect to $u_\alpha^*(t)$. Then

$$\exists \lim_{\alpha\to\infty} \delta u_{0,\alpha}^*=0 \text{ uniform with respect to } t. \tag{37}$$

Proof: To prove the lemma, first observe that the input signal error $\delta u_{0,\alpha}^*$ takes on the following form (see (34), and let $u = u_0 = \bar{K}\ddot{q}^* + \bar{L}\dot{q}^* + \bar{P}q^*$):

$$\delta u_{0,\alpha}^*=-\bar{A}(q^*)\ddot{q}^*-\bar{B}(q^*, \dot{q}^*)\dot{q}^*-[\bar{C}(q^*)-\bar{Q}]. \tag{38}$$

Then noting that for $\alpha \to \infty$, $\bar{A}(q^*)$, $\bar{B}(q^*, \dot{q}^*)$, and $\bar{C}(q^*) - \bar{Q}$ all converge to zero uniformly with respect to t (since $q^*(t)$, $\dot{q}^*(t)$, $\ddot{q}^*(t)$ are bounded signals by assumption), the lemma directly follows.

At this point, by reconsidering the proposed learning

procedure and exploiting the results of the previous lemmas, the following main theorem can be stated.

Theorem 1: Let $(\delta q_{i,\alpha}(t), \delta \dot{q}_{i,\alpha}(t), \delta \ddot{q}_{i,\alpha}(t))$ denote the vector of the trajectory errors corresponding to a generic trial, with initial condition $(\delta q_i(0), \delta \dot{q}_i(0))$, and relevant to a specific choice for the feedback parameter α. Then provided the normalized reference model (19) is an asymptotically stable one, the following property holds true:

$$\text{for all } \epsilon>0 \quad \exists \begin{cases} \rho_\epsilon>0 \\ \bar{\vartheta}_\epsilon>0 \end{cases} \text{such that} \begin{cases} \|\delta q_i(0), \delta \dot{q}_i(0)\| \leqslant \rho_\epsilon \\ \alpha \geqslant \bar{\alpha}_\epsilon \end{cases} \Rightarrow \|\delta q_{i,\alpha}(t), \delta \dot{q}_{i,\alpha}(t), \delta \ddot{q}_{i,\alpha}(t)\| \leqslant \epsilon \text{ for all } t, \text{ for all } i. \quad (39)$$

Proof: Let us start by considering the first trial of the procedure. Due to (37) of Lemma 2 we have that α can be sufficiently increased to satisfy the following two conditions:

$$\alpha \geqslant \alpha_\epsilon \qquad \|\delta u^*_{0,\alpha}(t)\| \leqslant \vartheta_\epsilon \text{ for all } t \tag{40}$$

where α_ϵ, ϑ_ϵ are the positive numbers appearing in Lemma 1 for a fixed $\epsilon > 0$, thus implying that $\delta q_{0,\alpha}(t)$, $\delta \dot{q}_{0,\alpha}(t)$, $\delta \ddot{q}_{0,\alpha}(t)$ are all uniformly norm bounded by ϵ, provided that $\|(\delta q_i(0), \delta \dot{q}_i(0)\| \leqslant \rho_\epsilon$.

Now refer to the second trial and consider the expression of the new normalized input error $\delta u^*_{1,\alpha}(t)$, given by (15) after division by α; i.e.,

$$\delta u^*_{1,\alpha} = -\frac{1}{\alpha}[A_0 \delta q_{0,\alpha} + B_0 \delta \dot{q}_{0,\alpha} + (A_0 - A_*)\ddot{q}^* + (B_0 - B_*)\dot{q}^* + (C_0 - C_*)]. \tag{41}$$

Then since $q^*(t)$, $\dot{q}^*(t)$, and $\ddot{q}^*(t)$ are uniformly bounded, functions $A(\cdot)$, $B(\cdot, \cdot)$, $C(\cdot)$ are locally Lipshitzian, and due the fact that $(\delta q_{0,\alpha}(t), \delta \dot{q}_{0,\alpha}(t), \delta \ddot{q}_{0,\alpha}(t))$ range inside the compact set $P_\epsilon \triangleq \{(\delta q, \delta \dot{q}, \delta \ddot{q}) : \|(\delta q, \delta \dot{q}, \delta \ddot{q})\| \leqslant \epsilon\}$, it follows that the whole part in brackets appearing in (41) is Lipshitzian in P_ϵ, meaning that

$$\|\delta u^*_{1,\alpha}(t)\| \leqslant \frac{1}{\alpha}\lambda_\epsilon \epsilon \text{ for all } t, \qquad \text{for all } \alpha \geqslant \alpha_\epsilon \tag{42}$$

where λ_ϵ is the Lipshitz constant in P_ϵ.

From the previous inequality it follows that if α is chosen *a priori* such that

$$\alpha \geqslant \bar{\alpha}_\epsilon \triangleq \max\left(\alpha_\epsilon, \frac{\lambda_\epsilon}{\vartheta_\epsilon}\epsilon\right), \tag{43}$$

we certainly have both of the following conditions satisfied:

$$\|\delta u^*_{0,\alpha}(t)\| \leqslant \vartheta_\epsilon \qquad \|\delta u^*_{1,\alpha}(t)\| \leqslant \vartheta_\epsilon, \qquad \text{for all } t \tag{44}$$

which guarantees ϵ to be a norm upper bound for both the trajectory errors occurring during the first and the second trial (obviously, provided that the same bound ρ_ϵ is maintained for initial conditions). At this point, by repeating the same reasoning, it is easy to see that the same choice for α, as in (43), guarantees condition (44) to hold for all successive normalized input errors, which, consequently, implies $\|(\delta q_{i,\alpha}(t), \delta \dot{q}_{i,\alpha}(t), \delta \ddot{q}_{i,\alpha}(t))\| \leqslant \epsilon$ for all i, for all t. This completes the proof of the theorem.

IV. Convergence Analysis

In this section it is shown that, in the case of null perturbations in the initial states, suitable values for the feedback gains K, L, and P can always be devised guaranteeing not only the uniform boundedness of the trajectory errors $\delta q_i(t)$, $\delta \dot{q}_i(t)$, $\delta \ddot{q}_i(t)$, but also their uniform convergence to zero with the increasing number of trials. To introduce the argument, first assume that an asymptotically stable normalized reference model (19) has been chosen, and the feedback parameter α has been fixed to a sufficiently high value to assure the trajectory errors vector to be uniformly norm bounded by some $\epsilon > 0$ at any trial (see Theorem 1). Moreover, consider the differential equation (16) rewritten as

$$\delta \ddot{q}_i = [\bar{K} + \bar{A}_i]^{-1} \frac{1}{\alpha}[-L\delta \dot{q}_i - P\delta q_i - B_i \delta \dot{q}_i + A_{i-1}\delta \ddot{q}_{i-1} + B_{i-1}\delta \dot{q}_{i-1} - (A_i - A_{i-1})\ddot{q}^* - (B_i - B_{i-1})\dot{q}^* - (C_i - C_{i-1})]. \tag{45}$$

The hypothesis about the uniform boundedness of $q^*(t)$, $\dot{q}^*(t)$, $\ddot{q}^*(t)$, and the local Lipshitzianity of functions $A(\cdot)$, $B(\cdot, \cdot)$, $C(\cdot)$ assure that the vector $f(\delta q_i, \delta \dot{q}_i, \delta q_{i-1}, \delta \dot{q}_{i-1}, \delta \ddot{q}_{i-1}) \triangleq (\delta \dot{q}_i, \bar{f}(\delta q_i, \delta \dot{q}_i, \delta q_{i-1}, \delta \dot{q}_{i-1}, \delta \ddot{q}_{i-1}))$, with $\bar{f}$ equal to the right member of (45), is Lipshitzian in the vectors $(\delta q_i, \delta \dot{q}_i)$, $(\delta q_{i-1}, \delta \dot{q}_{i-1}, \delta \ddot{q}_{i-1})$, with constants G and R, respectively, in the compact set $S_\epsilon \triangleq \{(\delta q_i, \delta \dot{q}_i, \delta \ddot{q}_i), (\delta q_{i-1}, \delta \dot{q}_{i-1}, \delta \ddot{q}_{i-1})$ norm bounded by $\epsilon\}$. Moreover, since R is monotonically decreasing toward zero for increasing α (as can easily be verified from (45)) we suppose that the feedback parameter α has been further increased to assure that R is smaller than 1.

Let $\tilde{C}^2([0, T], R^n)$ be the set of functions $r \in C^2([0, T], R^n)$ such that $r(t)$, $\dot{r}(t)$, $\ddot{r}(t)$ range inside the compact set P_ϵ, and let $\tilde{T}$ be the operator mapping $\tilde{C}^2([0, T], R^n)$ in $\tilde{C}^2([0, T], R^n)$ such that $\tilde{T}(\delta q_{i-1}) = \delta q_i$, with the δq_i solution of (45) corresponding to zero initial data. $\tilde{T}$ admits as fixed point the identically vanishing function in $[0, T]$. Finally, for every $r \in \tilde{C}^2([0, T], R^n)$ we set $\|r\|_\mu \triangleq \sup_{t\in[0,T]} e^{-\mu R t}\|r(t), \dot{r}(t), \ddot{r}(t)\|$, where $\mu > 0$.

Now we shall prove that $\tilde{T}$ is a contractive operator with respect the norm $\|\cdot\|_\mu$, that is, for all $\delta q^1_{i-1}, \delta q^2_{i-1} \in \tilde{C}^2([0, T], R^n)$, we have $\|\delta q^1_i - \delta q^2_i\|_\mu \leqslant h\|\delta q^1_{i-1} - \delta q^2_{i-1}\|_\mu$, where $\delta q^j_i = \tilde{T}(\delta q^j_{i-1})$, $j = 1, 2$, $0 < h < 1$. At first we observe that for all $t \in [0, T]$,

$$\|(\delta q^1_i(t), \delta \dot{q}^1_i(t)) - (\delta q^2_i(t), \delta \dot{q}^2_i(t))\| \leqslant \int_0^t \|f(\delta q^1_i(\tau), \delta \dot{q}^1_i(\tau), \delta q^1_{i-1}(\tau), \delta \dot{q}^1_{i-1}(\tau), \delta \ddot{q}^1_{i-1}(\tau)) - f(\delta q^2_i(\tau),$$

$$\delta\dot{q}_i^2(\tau),\ \delta q_{i-1}^2(\tau),\ \delta\dot{q}_{i-1}^2(\tau),\ \delta\ddot{q}_{i-1}^2(\tau))\|\ d\tau$$

$$\leqslant \int_0^t G\|(\delta q_i^1(\tau),\ \delta\dot{q}_i^1(\tau)) - (\delta q_i^2(\tau),\ \delta\dot{q}_i^2(\tau))\|\ d\tau$$

$$+ \int_0^t Re^{-\mu R\tau} e^{\mu R\tau}\|(\delta q_{i-1}^1(\tau),\ \delta\dot{q}_{i-1}^1(\tau),\ \delta\ddot{q}_{i-1}^1(\tau))$$

$$-(\delta q_{i-1}^2(\tau),\ \delta\dot{q}_{i-1}^2(\tau),\ \delta\ddot{q}_{i-1}^2(\tau))\|\ d\tau \leqslant G\int_0^t \|(\delta q_i^1(\tau),$$

$$\delta\dot{q}_i^1(\tau)) - (\delta q_i^2(\tau),\ \delta\dot{q}_i^2(\tau))\|\ d\tau + \|\delta q_{i-1}^1$$

$$-\delta q_{i-1}^2\|_\mu \frac{e^{\mu Rt}-1}{\mu}$$

where μ is a positive constant such that $\mu R - G > 0$, which we shall specify in the following.

By the generalized Granwall's lemma we have

$$\|(\delta q_i^1(t),\ \delta\dot{q}_i^1(t)) - (\delta q_i^2(t),\ \delta\dot{q}_i^2(t))\|$$
$$\leqslant \|\delta q_{i-1}^1 - \delta q_{i-1}^2\|_\mu \frac{Re^{\mu Rt} - Re^{Gt}}{\mu R - G}. \quad (46)$$

Therefore, for every $t \in [0, T]$,

$$e^{-\mu Rt}\|(\delta q_i^1(t),\ \delta\dot{q}_i^1(t)) - (\delta q_i^2(t),\ \delta\dot{q}_i^2(t))\|$$
$$\leqslant \|\delta q_{i-1}^1 - \delta q_{i-1}^2\|_\mu \frac{R}{\mu R - G}. \quad (47)$$

On the other hand,

$$\|\delta\ddot{q}_i^1(t) - \delta\ddot{q}_i^2(t)\| \leqslant G\|(\delta q_i^1(t),\ \delta\dot{q}_i^1(t)) - (\delta q_i^2(t),$$
$$\delta\dot{q}_i^2(t))\| + R\|(\delta q_{i-1}^1(t),\ \delta\dot{q}_{i-1}^1(t),$$
$$\delta\ddot{q}_{i-1}^1(t)) - (\delta q_{i-1}^2(t),\ \delta\dot{q}_{i-1}^2(t),\ \delta\ddot{q}_{i-1}^2(t))\|,$$

hence by (47)

$$e^{-\mu Rt}\|(\delta\ddot{q}_i^1(t) - \delta\ddot{q}_i^2(t)\| \leqslant G\|\delta q_{i-1}^1 - \delta q_{i-1}^2\|_\mu$$
$$\cdot \frac{R}{\mu R - G} + R\|\delta q_{i-1}^1 - \delta q_{i-1}^2\|_\mu. \quad (48)$$

Then we have

$$\|\delta q_i^1 - \delta q_i^2\|_\mu \leqslant \left(\frac{R}{\mu R - G} + \frac{RG}{\mu R - G} + R\right) \|\delta q_{i-1}^1 - \delta q_{i-1}^2\|_\mu \quad (49)$$

and by choosing μ in order that $R/(\mu R - G) + RG/(\mu R - G) + R < 1$, we have that $\tilde{T}$ is a contractive operator.

Then the Banach contraction principle assures that $\lim_{i\to\infty} \|\delta q_i\|_\mu = 0$, which implies

$$\lim_{i\to\infty} \|(\delta q_i(t),\ \delta\dot{q}_i(t),\ \delta\ddot{q}_i(t)\| = 0, \qquad \text{for all } t \in [0, T]. \quad (50)$$

At this point, it turns out possible to formalize all the considerations previously done via the following conclusion.

Theorem 2: For any triple of nonsingular matrices $\bar{K}$, $\bar{L}$, $\bar{P}$ satisfying the condition

$$\det\,(\bar{K}s^2 - \bar{L}s + \bar{P}) \quad \text{(Hourwitz)}, \quad (51)$$

a number $\bar{\alpha} > 0$ exists such that the convergence of the learning procedure is guaranteed for all the feedback gains of the form

$$K = \alpha\bar{K} \qquad L = \alpha\bar{L} \qquad P = \alpha\bar{P}, \qquad \text{with } \alpha \geqslant \bar{\alpha} \quad (52)$$

provided null position and velocity errors occur at the beginning of any trial.

V. The Acceleration Feedback Problem

In the previous sections a complete solution to the problem of convergence of the learning procedure was presented by assuming the presence of ideal acceleration sensors (with adjustable gains) in the feedback loop. However, since it was noted (see Remark 1) that this assumption makes the presence of "algebraic loops" unavoidable, it is quite clear that the cases for which the theory has been developed actually represent ideal situations that the use of "real" acceleration sensors can only approximate. More specifically, the approximation introduced in the reality is due to the intrinsic time delay of the acceleration sensors which (though preserving causality in the loops) always give rise to a mismatch between the actual acceleration $\ddot{q}(t)$ and the measured $\hat{\ddot{q}}(t)$.

This consideration, however, leads to the natural conjecture that, in practice, convergence to "good approximations" of the desired trajectory $q^*(t)$ can be obtained, provided the mismatches between $\ddot{q}_i(t)$ and $\hat{\ddot{q}}i(t)$ are kept "sufficiently small" at each trial, a condition that, as it can be argued, sets an upper bound on the rate of change of $q^*(t)$, which in turn depends on the acceleration sensors used.

Even if a quantitative and more rigorous analysis would be required, it is important to keep in mind that the simulation experiments described in the next section actually confirm the previous conjecture, thus maintaining the value of the developed theory for practical applications.

However, in this section it seems important to focus attention on the fact that, in practice, the possibility actually exists of devising cases where the results of the learning theory directly apply in exact form, without using any acceleration sensors. To clarify this idea, reconsider the dynamic equation (1) of a manipulator and rewrite matrix $A(q)$ as

$$A(q) = \bar{K} + \tilde{A}(q) \quad (53)$$

where $\bar{K}$ now represents a constant (even rough) "estimate" of the whole (coordinate-dependent) matrix $A(q)$.

Then rewrite (1) as

$$[\bar{K} + \tilde{A}(q)]\ddot{q} + B(q,\ \dot{q})\dot{q} + C(q) = M, \quad (54)$$

and consider a feedback control law (with no acceleration sensors) of the form

$$M = m + Q - \bar{L}q - \bar{P}q \quad (55)$$

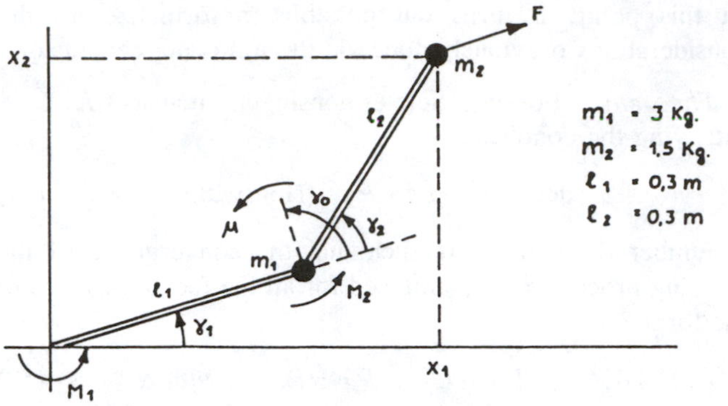

Fig. 1. Structure of simple kinematic chain used for simulation experiments.

giving rise to the closed-loop equation

$$[\bar{K}+\tilde{A}(q)]\ddot{q}+[\bar{L}+B(q, \dot{q})]q+\bar{P}q+[C(q)-Q]=m. \tag{56}$$

Then since (56) is actually the same as that which would have been obtained for the "fictitious" manipulator

$$\tilde{A}(q)\ddot{q}+B(q, \dot{q})\dot{q}+C(q)=M$$

controlled by the "noncausal" control law

$$M=m+Q-\bar{K}\ddot{q}-\bar{L}\dot{q}-\bar{P}q,$$

it directly follows that the same reasoning of Section II can be repeated here, thus allowing the same learning procedure to be devised.

The obvious advantages, in this case, derive directly from the absence of the acceleration sensors in the feedback loop and, moreover, from the opportunity of evaluating the acceleration errors via an off-line (and even noncausal) differentiation of the errors relevant to the velocity. With regard to the asymptotic properties, it is straightforward from Theorem 2 that (provided $\bar{L}$, $\bar{P}$ are fixed in such a way that condition (51) is satisfied) the convergence is assured if the same theorem applies with $\bar{\alpha} < 1$.

VI. Simulation Experiments

The proposed learning methodology has been preliminarily tested via simple simulation examples. They are briefly described and discussed in the present section.

The manipulator considered for the experiments is the simple one depicted in Fig. 1, represented as a two-link inverted pendulum with masses concentrated at the end of each link and driven by the torque vector $M = [M_1, M_2]^T$ acting in correspondence with each joint. A natural choice for the generalized coordinates to be associated with M is represented by the angle vector $q = [\gamma_1, \gamma_2]^T$, which is related to the vector $x = [x_1, x_2]^T$ (the Cartesian coordinates of the end effector) by the following easily devisable purely geometric relations:

$$x=f(q) \qquad \dot{x}=J(q)\dot{q}.$$

As is well-known, an other interesting geometric relation is represented by the following:

$$M=J^T(q)F \tag{57}$$

which allows one to interpret the action of any force vector F applied to the end effector as equivalent to one produced by the torque vector M.

For the chosen example, it is easy to see that all the foregoing relations turn out to be of one-to-one type, provided the invariance of the sign of the angle γ_2 can be somehow assured for all the movements of concern (in this section more will be said about this point). Then provided that this is actually the case, it follows that vector x can also be considered as a set of generalized coordinates for our example, thus, in turn, implying that a dynamic equation of the form

$$\mathcal{A}(x)\ddot{x}+\mathcal{B}(x, \dot{x})\dot{x}+\mathcal{C}(x)=F \tag{58}$$

actually exists, allowing a complete description of the behavior of the manipulator under the assumed conditions.

Since this fact suggests the possibility of implementing the learning procedure by considering desired trajectories and related errors, expressed in terms of Cartesian coordinates of the end effector, the learning-control scheme of Fig. 2 has been devised and used for all simulation experiments to be described.

In such a scheme the computed force F is one line translated into the equivalent torque vector M via relation (57), x and $\dot{x}$ are deduced from the measurements of q, $\dot{q}$ as given by (56), and $\ddot{x}$ is obtained from $\dot{x}$ via the use of a high-pass filter. Moreover, to maintain the condition about the invariance of the sign of the angle γ_2, and elastic torque feedback of the form

$$\mu=k(\gamma_0-\gamma_2)$$

has been inserted *a priori* (see Fig. 1), which is, however, embedded within the term $\mathcal{C}(x)$ of (55).

In our simulation examples, the adopted values for the feedback gains K, L, P correspond to

$$K=\begin{bmatrix}2.5 & 0\\ 0 & 2.5\end{bmatrix} \quad L=\begin{bmatrix}65 & 0\\ 0 & 65\end{bmatrix} \quad P=\begin{bmatrix}50 & 0\\ 0 & 50\end{bmatrix} \tag{59}$$

while the high-pass filters employed for measuring the accelerations each amount to a transfer function of the form

$$T(s)=10s/(s+10). \tag{60}$$

All the considered reference trajectories $x^*(t)$ have always been assigned as a couple of components, each one obtained as the output of a critically dumped second-order system of the form

$$(1/\omega_0^2)\ddot{x}_j+(2/\omega_0)\dot{x}_j+x_j=f_j, \qquad j=1, 2, \cdots \tag{61}$$

driven by a step function with settable amplitude.

By varying the settling time (i.e., ω_0) and the amplitude of the step functions relevant to each component of $x^*(t)$, different forms for the reference trajectories, in the phase plane, can be obtained, all characterized by a bell-shaped velocity profile. Some of the obtained results are reported in

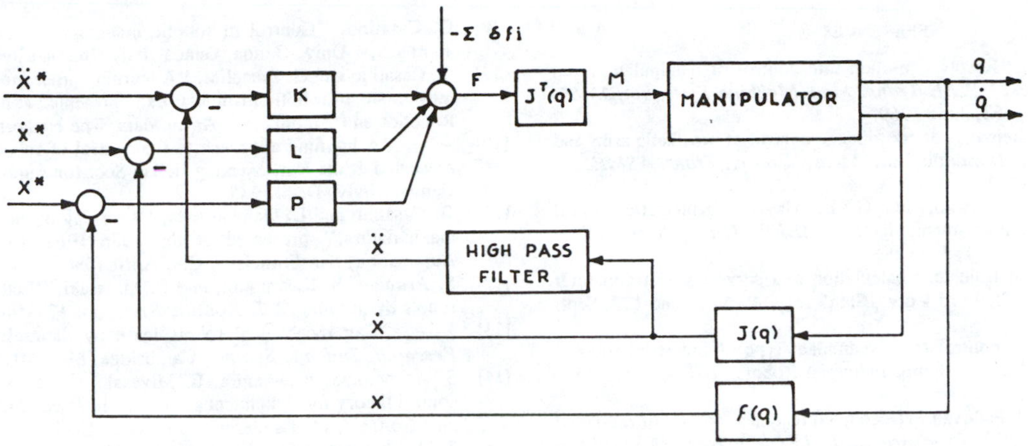

Fig. 2. Simulated learning control scheme.

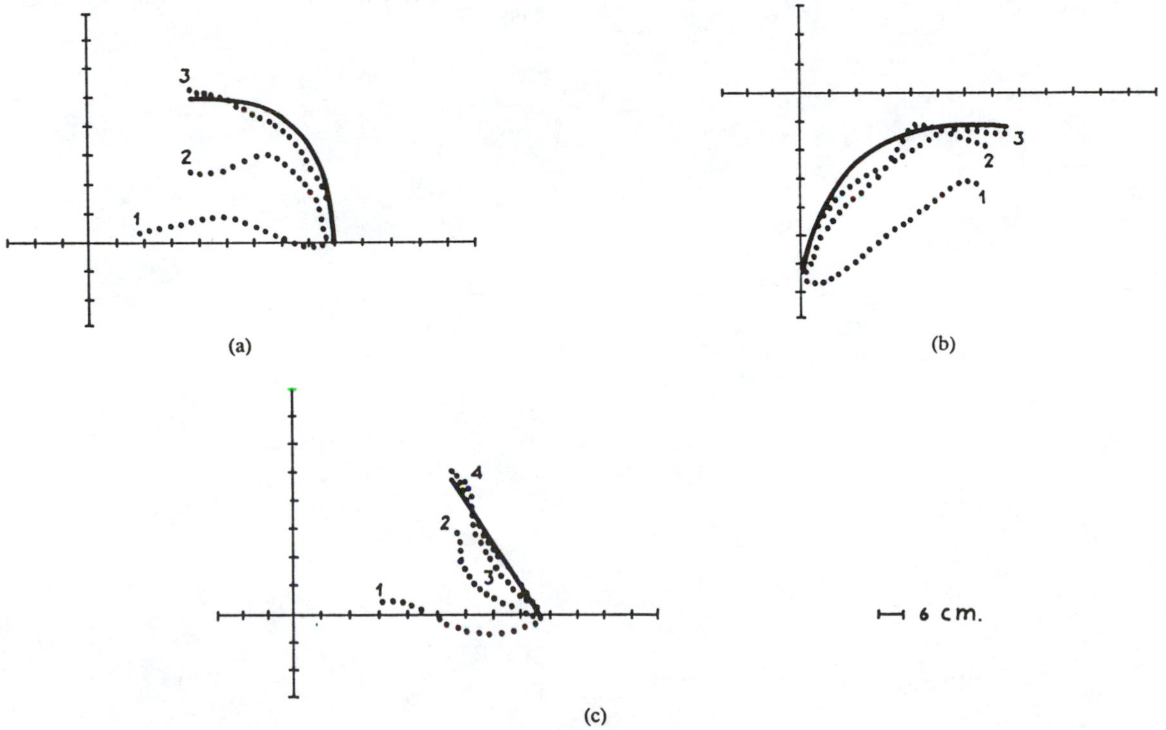

Fig. 3. Some of resulting simulation experiments.

Fig. 3(a)–(c), where the continuous lines represent the reference trajectories (each one having a settling time of the order of 2–4 s), while the dotted ones correspond to the various trials. In these conditions an almost complete convergence to the reference trajectories has always been obtained with the fifth or the seventh trial.

As it was expected, by augmenting the values of the feedback gains, an increase of the convergence rate occurs. For example, by simply doubling the values of the feedback gains and referring to the same desired trajectories, convergence is obtained within the second and the third trial.

VII. Conclusion

This paper has presented an inherently nonlinear analysis of the so-called "iterative learning control procedure" applied to robotic manipulators, by adopting a high-gain feedback point of view. In particular, it has been shown that by simply setting a sufficient (but not necessarily high) amount of linear feedback in the system, it is always possible to achieve the following conditions:

a) the closed-loop system has a domain of attraction around the desired trajectory;
b) the correcting signals used at each trial never push the system out of the attraction domain.

Such conditions are actually essential for the subsequent proof of convergence of the procedure, which has also been reported in the paper. Investigations aiming to give a quantitative measure of the minimal amount of linear feedback required for guaranteeing the learning of a given trajectory are currently under development.

References

[1] D. H. Whitney, "Resolved motion rate control of manipulators and human prostheses," *IEEE Trans. Man-Machine Syst.*, vol. MMS-10, no. 2, pp. 47–53, June 1969.

[2] ——, "The mathematics of coordinated control of prosthetic arms and manipulators," *J. Dynamic Syst., Meas., Contr., Trans. ASME*, pp. 303–309, Dec. 1972.

[3] R. P. Paul, B. Shimano, and G. E. Mayer, "Kinematic control equations for simple manipulators," *IEEE Trans. Syst., Man, Cybern.*, vol. SMC-11, June 1981.

[4] ——, "Modeling trajectory calculation and servoing of a computer controlled arm," Stanford Univ., Stanford, CA, AI Memo 177, Sept. 1972.

[5] E. Freund, "Path control for a redundant type of industrial robot," presented at the 7th Int. Symp. Industrial Robots, Tokyo, Japan, Oct. 1977.

[6] J. R. Hewit and J. Padovan, "Decoupled feedback control of robot and manipulator arms," in *Proc. 3rd CISM-IFT MM Symp. Theory and Practice of Robots and Manipulators*, Udine, Sept. 1978.

[7] S. Dubowski and D. T. Des Forges, "The application of model referenced adaptive control to robotic manipulators," *ASME J. Dynamic Syst., Meas., Contr.*, p. 193, Sept. 1979.

[8] G. Casalino, "Control of robotic manipulators via a trial and error approach," Univ. Genoa, Genoa, Italy, Internal Rep. DIST, 1984.

[9] G. Casalino and G. Bartolini, "A learning procedure for the control of movements of robotic manipulators," presented at the IASTED Symp. Robotics and Automation, Amsterdam, The Netherlands, June 1984.

[10] ——, "A learning approach to the control of robotic manipulators," presented at the Nat. Meeting Italian Soc. for Automation (ANIPLA), Genova, Italy, Dec. 1985.

[11] G. Casalino and L. Gambardella, "Learning of movements in robotic manipulators," presented at the 1986 IEEE Conf. Robotics and Automation, San Francisco, CA, April 1986.

[12] S. Arimoto, S. Kawamura, and F. Miyasaki, "Bettering operation of robots by learning," *J. Robotic Syst.*, pp. 123–140, 1984.

[13] ——, "Can mechanical robots learn by themselves?" in *Robotic Research, 2nd Int. Symp.* Cambridge, MA: MIT Press, 1985.

[14] S. Arimoto, S. Kawamura, F. Miyasaki, and S. Tamaki, "Learning control theory for dynamical systems," in *Proc. 24th Conf. Decision and Control*, Ft. Lauderdale, FL, Dec. 1985.

[15] T. Yoshizawa, "Stability theory by Lyapunov's second method," *Math. Soc. Japan*, 1966.

[16] J. J. Craig, "Adaptive control of manipulators through repeated trials," in *Proc. 1984 Amer. Control Conf.*, San Diego, CA, June 1984.

Part 6
Time-Optimal Control of Robots

KANG G. SHIN
The University of Michigan

DURING the past two decades a great deal of attention has been paid to the use of general-purpose robots for industrial automation. Those industries that in the past have constructed special-purpose devices for manufacturing their products are now looking at the possibility of using robots instead. Unlike special-purpose tools, robots' behavior can easily be modified, so that retooling is kept to a minimum. In some cases they can also make feasible the manufacture of a product in lots that would be too small to justify the creation of a special-purpose machine for their manufacture. Mechanical maintenance is also simpler, because there presumably would be only a few types of robots performing many different tasks. (It should be noted, however, that maintenance of special-purpose machines is replaced with maintenance of special-purpose programs.)

Because robots may be controlled in virtually any manner, one may legitimately ask how a robot should best be controlled. The obvious answer to this question is that the robot should produce as large a profit as possible per unit of time. The usual assumption is that material costs and fixed costs dominate the cost per item produced, so that it is desirable to produce as many units as possible in a given time. Therefore it is important to solve the problem of minimum-time control of robots.

There are a variety of algorithms available for manipulator control. These algorithms usually assume that the control structure of the robot has been divided into three levels [1]. The first level is called *path planning*, the second level is called *trajectory planning,* and the third level is called *trajectory control* or *trajectory tracking*. Trajectory tracking is the process of making the robot's actual position and velocity match some desired values of position and velocity; the desired values are provided to the controller by the trajectory planner. A trajectory planner receives as input some sort of special path descriptor from which it calculates a time history of the desired positions and velocities. A path planner determines geometric path information, without timing information, based on collision avoidance and other task requirements.

The reason for dividing the control scheme in this way is that the process of robot control, if considered in its entirety, is very complicated, because the dynamics of all but the simplest robots are highly nonlinear and coupled. Dividing the controller into the three parts makes the whole process simpler. The trajectory tracker is frequently a linear controller (e.g., a proportional integral derivative (PID) controller). While the nonlinearities of manipulator dynamics frequently are not taken into account at this level, such trackers can generally keep the manipulator fairly close to the desired trajectory. More sophisticated methods can be used, though, such as resolved motion rate control [2], resolved acceleration control [3], and various adaptive techniques [4]–[11].

Unfortunately, the simplicity obtained from the division into trajectory planning and tracking often comes at the expense of efficiency. The source of the inefficiency is the trajectory planner. In order to use the robot efficiently, the trajectory planner must be aware of the robot's dynamic properties, and the more accurate the dynamic model is, the better the robot's capabilities can be used. However, most of the early trajectory planning algorithms assume very little about the robot's dynamics. The usual assumption is that there are constant or piecewise constant bounds on the robot's velocity and acceleration [3], [12], [13]. In fact, these bounds vary with position, payload mass, and even with payload shape. Thus in order to make the constant-upper-bound scheme work, the upper bounds must be chosen to be global greatest lower bounds of the velocity and acceleration limits; in other words, the worst-case limits have to be used. Because the moments of inertia seen at the joints of the robot, and hence the acceleration limits, may vary by an order of magnitude as the robot moves from one position to another, such bounds can result in considerable inefficiency or underutilization of the robot.

One of the early minimum-time trajectory planning systems was developed by Luh and Walker [14]. It describes the desired manipulator path in terms of its initial and final points and a set of intermediate points. Each branch of the path has a maximum velocity assigned to it, and each intermediate point is assigned a maximum acceleration and a maximum position error. The time taken to go from the initial to the final point is, then, the sum of the times taken to traverse each branch of the path plus the sum of the times required to make the transitions from one branch to the next. The minimum possible sum of these times can then be found using linear programming. In this case, one must still choose appropriate maximum velocities and accelerations, and this cannot be done properly without either knowing the dynamic properties and actuator characteristics of the robot or having some experimental results that give maximum velocities and accelerations for given robot configurations. Also, because maximum accelerations and velocities are assumed to be constant over some interval, it is necessary to choose them to be lower bounds of the maximum values over the given interval, in effect, worst-case bounds on acceleration and velocity. Because these bounds will in general depend on position and velocity, this could result in underutilization of the robot's capabilities.

Luh and Lin [13] present a modification of the scheme described above that uses nonlinear programming to generate the minimum-time trajectory. The major difference between the method of Luh and Lin and that of Luh and Walker is in a more careful treatment of the calculation of the times required for the transitions from one path segment to the next. Also, an efficient technique for solving the nonlinear programming problem is presented along with a convergence proof.

Lin et al. [12] present a third variation on this trajectory planning method. Instead of using path segments that are straight lines in Cartesian space, they convert points in Cartesian space into the equivalent joint coordinates, and pass a cubic spline through these points. The maximum velocities and accelerations are, then, joint velocities and accelerations, which are easier to compute from the robot dynamics and actuator characteristics. There is, however, still some calculation (or measurement) required in order to determine these quantities. This work, incidentally, also allows for limits on *jerk* (time-derivative of acceleration). Placing limits on jerk helps prevent mechanism wear.

Kim and Shin [15], [16] present a method similar in some respects to the linear programming methods presented previously, but using the robot dynamic equations to obtain approximate acceleration bounds at each corner point. They also point out a set of conditions under which the linear programming problem reduces to a set of local optimization problems, one for each corner point. This represents a change in computational complexity of from $O(n^3)$ to $O(n)$.

Another type of trajectory planner has been developed by Bobrow et al. [17], Shin and McKay [18], [19], [20], Pfeiffer and Johanni [21], and Slotine and Yang [22]. In these minimum-time trajectory planners, it is assumed that the desired path is given a parameterized form. The parametric equations of the path can then be substituted into the manipulator dynamic equations, giving a set of second-order differential equations in the (scalar) path parameter. Given these dynamic equations, bounds on individual joint torques can be converted into bounds on parametric accelerations (second time-derivatives of the path parameter); the allowable sets of second derivatives (one set per joint) are intersected, giving a single allowable set. Bounds on velocities (first derivatives of the path parameter) can also be found from these equations, since at some velocities there are no admissible accelerations. Then, using the fact that the minimum-time solution will be bang-bang in the acceleration, it is possible to construct phase plane plots that give the optimal trajectory in terms of the parameter and its derivatives.

This last approach to the minimum-time trajectory planning is the main focus of the current part. The coverage begins with the paper by K. G. Shin and N. McKay [19].* This is then followed by the paper by J. Bobrow et al. [23]. The papers differ primarily in two respects: First, in the method of Bobrow et al. the required search for trajectories is carried out by actually constructing trajectories and seeing where they go; in Shin and McKay, the search has been reduced to the problem of finding a sign change in an easily computable function. Second, some complications may arise with respect to computation of the admissible velocities for a given manipulator position. It is possible that there may be several distinct allowable velocity ranges for a given position. In [19] no mention is made about this possibility, and their search technique may fail if there are distinct regions. Shin and McKay present a second algorithm to take care of this possibility.

In the third article, J-J. E. Slotine and H. S. Yang improve the computational efficiency of deriving minimum-time trajectories based on the same problem formulation as in the first two articles. First, instead of the maximum-velocity curves used in [19] and [23], the authors define the *characteristic switching points* that can be uniquely determined from the robot dynamics, the actuator bounds, and the given geometric path. Based on the characteristic switching points, they derive *limit curves* that set sharper admissible regions in the phase plane and shape the minimum-time trajectory accordingly.

The fourth article is somewhat different from the first three, in that it is not specific to minimum-time robot control, and addresses more fundamental issues. Its main results are related to the singular structure of the generic optimal (not necessarily time-optimal) control problem. This is important to robot control, because the nonlinearity of robot dynamics is usually handled in two stages: (1) find an open-loop control that achieves the desired state transfer, and (2) linearize along the resulting trajectory, and use a linear controller for trajectory tracking. Note that a trajectory is singular precisely when this linearization is uncontrollable.

The first three articles assume that the geometric/spatial path planner has generated a parameterized curve. The trajectory planning problem can then be reduced to a problem of small dimension by converting all the dynamic and actuator constraints to constraints on the single parameter used to describe the path, and the parameter's time derivatives. Within this framework, a variety of opti-

*In 1987 it was selected as an Outstanding Paper in the *IEEE Transactions on Automatic Control*.

mization techniques can be applied. The optimization methods described here apply to both minimum-time problems and to more general minimum-cost control problems. It is also possible to modify the constraints to take uncertain dynamics into account at the trajectory planning stage [24].

While most people have taken the separate trajectory planning/trajectory tracking approach, several authors have made attempts at unified approaches to robot control. One early attempt was the near-minimum time control of Kahn and Roth [25], who linearized the dynamic equations, transformed the equations to eliminate coupling terms, and generated switching curves for this linear approximation. The result is a suboptimal control that seems to give fairly good results if the initial and final states of the robot are fairly close. It does, however, have some problems with overshoot, as one would expect from such an approximation.

Another controller using an approximate dynamic model to generate optimal controls is the near-minimum time-fuel method of Kim and Shin [7]. They use a model that is linear over one sample period, and use coefficients in their linear model that result from averaging the coefficients at the current point on the trajectory and at the final point. The controls that result from this approximation depend on whether time or fuel is the predominant term in the objective function, and on the sampling rate. The trajectories for minimum fuel were slower than those for minimum time but had less overshoot. Increasing sampling rate also had the effect of reducing overshoot; thus two parameters could be varied to find a good compromise between manipulator speed and overshoot.

References

[1] K. G. Shin and N. D. McKay, "Selection of near-minimum-time geometric paths for robotic manipulators," *IEEE Trans. Automatic Contr.,* vol. AC-31, no. 6, pp. 501–511, June 1986.

[2] D. E. Whitney, "Resolved motion rate control of manipulators and the human prostheses," *IEEE Trans. Man–Machine Syst.*, vol. MMS-10, no. 2, pp. 47–53.

[3] J. Luh, M. Walker, and R. Paul, "Resolved-acceleration control of mechanical manipulators," *IEEE Trans. Automatic Contr.,* vol. AC-25, no. 3, pp. 468–474, June 1980.

[4] M. Chung and C. Lee, "An adaptive control strategy for computer-based manipulators," *Technical Report RDS-TR-10-82,* Ann Arbor MI: University of Michigan Center for Robotics and Integrated Manfucturing, Aug. 1982.

[5] S. Dubowsky, "On the adaptive control of robotic manipulators: the discrete-time case," *Proc. Joint Automatic Contr. Conf.*, pp. section TA–2B, June 1981.

[6] S. Dubowsky and D. T. DesForges, "The application of model-referenced adaptive control to robotic manipulators," *J. Dyn. Syst., Meas., and Contr.,* pp. 193–200, Sept. 1979.

[7] B. Kim and K. G. Shin, "An adaptive model following control of industrial manipulators," *IEEE Trans. Aerospace and Electronic Syst.,* vol. AES-19, no. 6, pp. 805–814, Nov. 1983.

[8] A. Koivo and T. Guo, "Control of a robotic manipulator with adaptive controller," *Proc. 20th IEEE Conf. Dec. and Contr.*, pp. 271–276, Dec. 1981.

[9] A. Koivo and T. Guo, "Adaptive linear controller for robotic manipulators," *IEEE Trans. Automatic Contr.,* vol. AC-28, no. 2, pp. 162–170, Feb. 1983.

[10] R. Goor, "Continuous time adaptive feedforward control: stability and simulations," Technical Report GMR-4105, Warren, MI: General Motors Research Laboratories, July 1982.

[11] R. Goor, "A new approach to robot control," *Proc. American Contr. Conf.,* pp. 387–389, June 1985.

[12] C.-S. Lin, P.-R. Chang, and J. Luh, "Formulation and optimization of cubic polynomial joint trajectories for mechanical manipulators," *Proc. 21st CDC,* pp. 330–335, Dec. 1982.

[13] J. Luh and C. Lin, "Optimimum path planning for mechanical manipulators," *Dyn. Syst., Meas., and Contr.*, vol. 102, pp. 142–151, June 1981.

[14] J. Luh and M. Walker, "Minimum-time along the path for a mechanical arm," in *Proc. IEEE Conf. Dec. and Contr.,* pp. 755–759, Dec. 1977.

[15] B. Kim and K. G. Shin, "An efficient minimum-time robot path planning under realistic constraints," *Proc. American Contr. Conf.,* pp. 296–303, June 1984.

[16] B. Kim and K. G. Shin, "Minimum-time path planning for robot arms and their dynamics," *IEEE Trans. Syst., Man, and Cyber.,* vol. SMC-15, no. 2, pp. 213–223, March/April 1985.

[17] J. Bobrow, S. Dubowsky, and J. Gibson, "On the optimal control of robotic manipulators with actuator constraints," *Proc. American Contr. Conf.,* pp. 782–787, June 1983.

[18] K. G. Shin and N. D. McKay, "Minimum-time control of a robotic manipulator with geometric path constraints," *Proc. 22nd CDC,* pp. 1449–1457, Dec. 1983.

[19] K. G. Shin and N. D. McKay, "Minimum-time control of a robotic manipulator with geometric path constraints," *IEEE Trans. Automatic Contr.,* vol. AC-30, no. 6, pp. 531–541, June 1985.

[20] K. G. Shin and N. D. McKay, "A dynamic programming approach to trajectory planning of robotic manipulators," *IEEE Trans. Automatic Contr.,* vol. AC-31, no. 6, pp. 491–500, June 1986.

[21] F. Pfeiffer and R. Johanni, "A concept for manipulator trajectory planning," *IEEE J. Robotics and Automation,* vol. RA-3, no. 2, pp. 115–123, April 1987.

[22] J-J. Slotine and H. S. Yang, "Improving the efficiency of time-optimal path-following algorithms," *IEEE J. Robotics and Automation,* vol. 5, no. 1, pp. 118–124, Feb. 1989.

[23] J. Bobrow, S. Dubowsky, and J. Gibson, "Time-optimal control of robotic manipulators along specified paths," *The Int. J. Robotics Res.,* vol. 4, no. 3, pp. 3–17, 1985.

[24] K. G. Shin and N. D. McKay, "Robust trajectory planning for robotic manipulators under payload uncertainties," *IEEE Trans. Automatic Contr.,* vol. AC-32, no. 12, pp. 1044–1054, Dec. 1987.

[25] M. Kahn and B. Roth, "The near-minimum-time control of open-loop articulated kinematic chains," *J. Dyn. Syst., Meas., and Contr.,* pp. 164–172, Sept. 1971.

Minimum-Time Control of Robotic Manipulators with Geometric Path Constraints

KANG. G. SHIN, SENIOR MEMBER, IEEE, AND NEIL D. McKAY

***Abstract*—Conventionally, robot control algorithms are divided into two stages, namely, *path* or *trajectory planning* and *path tracking* (or *path control*). This division has been adopted mainly as a means of alleviating difficulties in dealing with complex, coupled manipulator dynamics. Trajectory planning usually determines the timing of manipulator position and velocity *without* considering its dynamics. Consequently, the simplicity obtained from the division comes at the expense of efficiency in utilizing robot's capabilities.**

To remove at least partially this inefficiency, this paper considers a solution to the problem of moving a manipulator in minimum time along a specified geometric path subject to input torque/force constraints. We first describe the manipulator dynamics using parametric functions which represent geometric path constraints to be honored for collision avoidance as well as task requirements. Second, constraints on input torques/forces are converted to those on the parameters. Third, the minimum-time solution is deduced in an algorithm form using phase-plane techniques. Finally, numerical examples are presented to demonstrate utility of the trajectory planning method developed.

I. INTRODUCTION

INDUSTRIAL robots have emerged as a primary means of contemporary automation due to their potential for productivity increase and product quality improvement. Obviously, a robot should be controlled so as to produce as many units as possible per dollar invested. This in turn naturally leads to the need for minimum-time control of robots.

There are a variety of algorithms available for robot control. These algorithms usually assume that the control structure of the robot has been divided into two levels. The lower level is called *control* or *path tracking,* and the upper level is called *path* or *trajectory planning*. The path tracker attempts to make the robot's actual position and velocity match some desired values of position and velocity; the desired values are provided to the controller by the trajectory planner. The trajectory planner receives as input some sort of geometric path descriptor from which it calculates a time history of the desired positions and velocities. The path tracker then tries to minimize the deviation of the actual position and velocity from the desired values.

The control scheme is divided in this way because the process of robot control, if considered in its entirety, is very complicated, since the dynamics of all but the simplest robots are highly nonlinear and coupled. Dividing the controller into the two parts makes the whole process simpler. The path tracker is frequently a linear controller (e.g., a PID controller). While the nonlinearities of manipulator dynamics frequently are not taken into account at this level, such trackers can generally keep the manipulator fairly close to the desired trajectory.

Unfortunately, the simplicity obtained from the division into trajectory planning and path tracking comes at the expense of efficiency. The source of the inefficiency is the trajectory planner. In order to use the robot at maximum efficiency, the trajectory planner must be aware of the robot's dynamic properties, and the more accurate the dynamic model is, the better the robot's capabilities can be used. However, most of the trajectory planning algorithms presented to date assume very little about the robot's dynamics. The usual assumption is that there are constant or piecewise constant bounds on the robots velocity and acceleration [9], [10]. In fact, these bounds vary with position, payload mass, and even with payload shape. Thus, in order to make the constant-upper-bound scheme work, the upper bounds must be chosen to be global greatest lower bounds of the velocity and acceleration values; in other words, the worst case limits have to be used. Since the moments of inertia seen at the joints of the robot, and hence the acceleration limits, may vary by a factor of three or more, such bounds can result in considerable inefficiency or underutilization of the robot.

To alleviate the inefficiency, this paper presents a solution to the minimum-time manipulator control problem subject to constraints on its geometric path and input torques/forces. The solution will be in the form of a trajectory planning algorithm, and will take into account the details of the dynamics of the manipulator. The output of the trajectory planner will be the true minimum-time solution, and so will be useful as a standard against which the performance of other trajectory planning algorithms may be measured. Note that the problem and its solution considered in this paper are different from the near minimum-time control methods in [4], [5].

Bobrow *et al.* [1], [2] have independently come to conclusions similar to our own. Although their formulation of the problem is similar to ours, their solution algorithm and motivations are different from ours in several respects. In their work, no specific form is assumed for the actuator torque bounds except that they be functions only of the robot's current position and velocity. We have assumed that the velocity dependence is at most quadratic, which allows a more concrete treatment of some aspects of the problem while still allowing treatment of most of the actuators found in practice. The assumption of quadratic velocity bounds combined with the assumption that the parametric equations of the curve to be followed are piecewise analytic (again, an assumption which presents no practical difficulties) permits construction of a proof that our algorithm terminates in a finite number of steps. By contrast, Bobrow's work relies on the hypothesis that there are a finite number of switching points. Finally, both [1], [2], this paper make use of phase plane techniques, and both use the idea of a set of admissible velocities for a given position. This naturally leads to the idea of an "admissible region" in the phase plane. Bobrow's solution implicitly assumes that this region is simply connected. However, we show in this paper that 1) the admissible region may *not* be simply connected, possibly making some steps of Bobrow's algorithm impossible, 2) our algorithm terminates within a finite number of steps even for nonsimply connected admissible regions, and 3) the resulting algorithm is optimal for the general case, i.e., for nonsimply connected admissible regions.

The remainder of this paper is divided into five sections.

Manuscript received August 4, 1983; revised November 26, 1983 and April 20, 1984. Paper recommended by Past Associate Editor, J. Y. S. Luh. This work was supported in part by the National Science Foundation under Grant ECS 8409938, the U.S. Air Force Office of Scientific Research under Contract F49620-82-C-0089 and the Robot Systems Division, Center for Research and Integrated Manufacturing (CRIM), The University of Michigan, Ann Arbor, MI.

The authors are with the Department of Electrical Engineering and Computer Science, The University of Michigan, Ann Arbor, MI 48109.

Reprinted from *IEEE Trans. Automat. Contr.*, vol. AC-30, no. 6, pp. 531–541, June 1985.

Section II describes a method for making the manipulator dynamic equations more tractable and a method for handling input torque constraints. Section III contains a detailed formulation of the minimum-time control problem. Also, the form of the optimal solution is deduced using phase-plane techniques. Section IV presents an algorithm for generating minimum-time trajectories, and proofs of the convergence of the algorithm and optimality of the generated trajectories. In Section V, we discuss some simple examples to demonstrate the utility of our trajectory planning algorithm. The final section discusses the significance of the results.

II. Parameterized Robot Dynamics with Input Torque Constraints

Before delving headlong into the problem of minimum-time control, a dynamic model of the manipulator is required. There are a number of ways of obtaining the dynamic equations of a robot arm, i.e., the equations which relate joint forces and torques to positions, velocities, and accelerations. The Lagrange formulation of mechanism dynamics yields a set of differential equations which are easy to manipulate for robot control problems, and so will be used here [3], [11]. The dynamic equations take the form

$$u_i = J_{ij}(q)\ddot{q}^j + R_{ij}\dot{q}^j + C_{ijk}(q)\dot{q}^j\dot{q}^k + G_i(q) \tag{1}$$

where u_i is the ith generalized force, q^i is the ith generalized coordinate, J_{ij} the inertia matrix, G_i the gravitational force on the ith joint, C_{ijk} the Coriolis force array, and R_{ij} is the viscous friction matrix. The Einstein summation convention has been used, and all indexes run from one to n inclusive for an n-degree-of-freedom robot.

The motion of the robot arm will not, of course, be completely unconstrained. In fact, it will later be assumed that the manipulator must be constrained to a fixed path in joint space, and that the path is given as a *parameterized curve.* The curve is assumed to be given by a set of n functions of a single parameter λ, so that we are given

$$q^i = f^i(\lambda), \qquad 0 \le \lambda \le \lambda_{\max} \tag{2}$$

where λ is a parameter for describing the desired path, and it is assumed that the coordinates q^i vary continuously with λ and that the path never retraces itself as λ goes from 0 to $\lambda_{\max}$.

It should be noted that in practice the spatial paths are given in Cartesian coordinates. While it is in general difficult to convert a curve in Cartesian coordinates to that in joint coordinates, it is relatively easy to perform the conversion for individual points. One can then pick a sufficiently large number of points on the Cartesian path, convert to joint coordinates, and use some sort of interpolation technique (e.g., cubic splines) to obtain a similar path in joint space (see [8] for an example).

Returning to the problem at hand, we may use the parameterization of the q^i and differentiate with respect to time, giving

$$\dot{q}^i = \frac{df^i}{d\lambda}\frac{d\lambda}{dt} = \frac{df^i}{d\lambda}\dot{\lambda} = \frac{df^i}{d\lambda}\mu \tag{3}$$

where $\mu \equiv \dot{\lambda}$. The equations of motion along the curve (i.e., the geometric path) then become

$$\dot{\lambda} = \mu \tag{4a}$$

$$u_i = J_{ij}(\lambda)\frac{df^j}{d\lambda}\dot{\mu} + J_{ij}(\lambda)\frac{d^2f^j}{d\lambda^2}\mu^2 + G_i(\lambda) + R_{ij}\frac{df^j}{d\lambda}\mu + C_{ijk}(\lambda)\frac{df^j}{d\lambda}\frac{df^k}{d\lambda}\mu^2. \tag{4b}$$

Note that if λ is used to represent arc length along the path, then μ and $\dot{\mu}$ are the velocity and the acceleration along the path, respectively.

With this parameterization, there are two state variables, i.e., λ and μ, but $(n + 1)$ equations. One way to look at the system is to choose the equation $\dot{\lambda} = \mu$ and one of the remaining equations as state equations, regarding the other equations as constraints on the inputs and on $\dot{\mu}$. However, the problem has a more appealing symmetry if a single differential equation is obtained from the n equations given by multiplying the ith equation by $df^i/d\lambda$ and sum over i, giving

$$u_i\frac{df^i}{d\lambda} = J_{ij}(\lambda)\frac{df^i}{d\lambda}\frac{df^j}{d\lambda}\dot{\mu} + J_{ij}(\lambda)\frac{df^i}{d\lambda}\frac{d^2f^j}{d\lambda^2}\mu^2 + G_i(\lambda)\frac{df^i}{d\lambda} + R_{ij}\frac{df^i}{d\lambda}\frac{df^j}{d\lambda}\mu + C_{ijk}(\lambda)\frac{df^i}{d\lambda}\frac{df^j}{d\lambda}\frac{df^k}{d\lambda}\mu^2. \tag{5}$$

This formulation has a distinct advantage. Note that the coefficient of $\dot{\mu}$ is quadratic in the vector of derivatives of the constraint functions. Since a smooth curve[1] can always be parameterized in such a way that the first derivatives never all disappear simultaneously, and since the inertia matrix is positive definite, the whole equation can be divided by the nonzero positive coefficient of $\dot{\mu}$, providing a solution for $\dot{\mu}$ in terms of λ and μ. Now there are only two state equations, and the original n equations can be regarded as constraints on the inputs and on $\dot{\mu}$ (more on this will be discussed later).

With this formulation, the state equations become

$$\dot{\lambda} = \mu \tag{6a}$$

$$\dot{\mu} = \frac{1}{J_{ij}(\lambda)\frac{df^i}{d\lambda}\frac{df^j}{d\lambda}}\left[u_i\frac{df^i}{d\lambda} - J_{ij}(\lambda)\frac{df^i}{d\lambda}\frac{d^2f^j}{d\lambda^2}\mu^2 - G_i(\lambda)\frac{df^i}{d\lambda} - R_{ij}\frac{df^i}{d\lambda}\frac{df^j}{d\lambda}\mu - C_{ijk}(\lambda)\frac{df^i}{d\lambda}\frac{df^j}{d\lambda}\frac{df^k}{d\lambda}\mu^2\right]. \tag{6b}$$

Consider now the constraints on the inputs, namely, $u^i_{\min} \le u_i \le u^i_{\max}$ and (4b). The dynamic equations (4b) can be viewed as having the following form: $u_i = g_i(\lambda)\dot{\mu} + h_i(\lambda, \mu)$. For a given state, i.e., given λ and μ, this is just a set of parametric equations for a line, where the parameter is $\dot{\mu}$. The admissible controls, then, are those which are on this line in the input space and also are inside the rectangular prism formed by the input magnitude constraints. Thus, the rectangular prism puts bounds on $\dot{\mu}$. The reason for converting from bounds on the input torques/forces to bounds on the *pseudoacceleration* $\dot{\mu}$ is that all the positions, velocities, and accelerations of the various joints are related to one another through the parameterization of the path. Given the current state (λ, μ), the quantity $\dot{\mu}$, if known, determines the input torques/forces for *all* of the joints of the robot, so that manipulation of this one scalar quantity can replace the manipulation of n scalars (the input torques) and a set of constraints (the path parameterization equations).

For evaluating the bounds on $\dot{\mu}$ explicitly, (4b) can be plugged into the inequalities $u^i_{\min} \le u_i \le u^i_{\max}$ so that

$$u^i_{\min} \le J_{ij}\frac{df^j}{d\lambda}\dot{\mu} + \left(J_{ij}\frac{d^2f^j}{d\lambda^2} + C_{ijk}\frac{df^j}{d\lambda}\frac{df^k}{d\lambda}\right)\mu^2 + R_{ij}\frac{df^j}{d\lambda}\mu + G_i \le u^i_{\max}. \tag{7a}$$

[1] We assumed to have a smooth curve for describing the given path between the starting and destination points. If it is not, still we can divide the path into smooth subpaths. Then, the above assumption becomes valid.

Introducing some shorthand notation, let

$$M_i \equiv J_{ij}\frac{df^j}{d\lambda}, \quad Q_i \equiv J_{ij}\frac{d^2f^j}{d\lambda^2} + C_{ijk}\frac{df^j}{d\lambda}\frac{df^k}{d\lambda}, \quad R_i \equiv R_{ij}\frac{df^j}{d\lambda}, \quad S_i \equiv G_i.$$

We then have

$$u^i_{\min} \le M_i\dot{\mu} + Q_i\mu^2 + R_i\mu + S_i \le u^i_{\max}. \tag{7b}$$

Note that the quantities listed above are functions of λ. For the sake of brevity, the functional dependence is not indicated in what follows.

Manipulation of these inequalities gives (assuming that $M_i \neq 0$)

$$\frac{u^i_{\min} - Q_i\mu^2 - R_i\mu - S_i}{|M_i|} \le \operatorname{sgn}(M_i)\dot{\mu} \le \frac{u^i_{\max} - Q_i\mu^2 - R_i\mu - S_i}{|M_i|}$$

$$\text{or } LB_i \le \dot{\mu} \le UB_i \tag{8}$$

where

$$LB_i \equiv \frac{u^i_{\min}(M_i>0) + u^i_{\max}(M_i<0) - (Q_i\mu^2 + R_i\mu + S_i)}{M_i}$$

and

$$UB_i \equiv \frac{u^i_{\max}(M_i>0) + u^i_{\min}(M_i<0) - (Q_i\mu^2 + R_i\mu + S_i)}{M_i}.$$

The expression $(M_i > 0)$ evaluates to one if $M_i > 0$, zero otherwise. Since these constraints must hold for all n joints, $\dot{\mu}$ must satisfy $\max_i LB_i \le \dot{\mu} \le \min_i UB_i$, or $GLB(\lambda, \mu) \le \dot{\mu} \le LUB(\lambda, \mu)$.

Note that it has *not* been assumed here that $u^i_{\min}$ and $u^i_{\max}$ are constants; they may indeed be arbitrary functions of λ and μ. Later these quantities will be assumed to have specific, relatively simple forms, but these forms should be adequate to describe most of the actuators used in practice.

The difference between the trajectory planning algorithm to be presented and those which are conventionally used can be seen in terms of the equation above. Assume that the parameter λ is arc length in Cartesian space. Then μ is the speed and $\dot{\mu}$ the acceleration along the geometric path. Since most conventional trajectory planners put *constant* bounds on the acceleration over some particular (frequently the entire) interval, one would have $GLB(\lambda, \mu) \le \dot{\mu}_{\min} \le \dot{\mu} \le \dot{\mu}_{\max} \le LUB(\lambda, \mu)$, where $\dot{\mu}_{\min}$ and $\dot{\mu}_{\max}$ are constants. The conventional techniques, then, restrict the acceleration more than is really necessary. Likewise, constant bounds on the velocity will also be more restrictive than necessary.

III. Formulation of Optimal Control Problem

With the manipulator dynamic equations and joint torque/force constraints in suitable form, we can address the actual control problem. Problems which require the minimization of cost functions subject to differential equation constraints can be expressed very naturally in the language of optimal control theory. The usual method of solving such a problem is to employ Pontryagin's maximum principle [6]. The maximum principle yields a two-point boundary value problem which is, except in some simple cases, impossible to solve in closed form, and may be difficult to solve numerically as well. We will therefore not use the maximum principle, but will use some simpler reasoning, taking advantage of the specific form of the cost function and of the controlled system.

In the case considered here, minimum cost is equated with minimum time, thus maximizing the operating speed of the robot. The cost function can then be expressed as $T = \int_0^{t_f} 1 \cdot dt$ where the final time t_f is left free. It is assumed here that the desired geometric path of the manipulator has been preplanned,[2] and is provided to the minimum-time controller in *parametric form*, as described earlier [i.e., (3)]. Further, assume that the q^i are parameterized in such a way that the initial point corresponds to $\lambda = 0$, the final point corresponds to $\lambda = \lambda_{\max}$, and that the $df^i/d\lambda$ never all become zero simultaneously. This guarantees that the state equations (6a) and (6b) exist, and also guarantees that as λ increases from 0 to $\lambda_{\max}$ the path never retraces itself.

Given this form for the dynamic equations, we have the minimum time path planning (MTPP) problem as follows.

Problem MTPP: Find $x^* = (\lambda^*, \mu^*)$ and u_i^* by minimizing T subject to (6a), (6b), $u^i_{\min} \le u_i \le u^i_{\max}$, $0 \le \lambda \le \lambda_{\max}$, and the boundary conditions $\mu(0) = \mu_0$, $\mu(t_f) = \mu_f$, $\lambda(0) = 0$, and $\lambda(t_f) = \lambda_{\max}$.

Before doing any further manipulations on the state equations, define the functions.

$$M(\lambda) \equiv J_{ij}(\lambda)\frac{df^i}{d\lambda}\frac{df^j}{d\lambda},$$

$$Q(\lambda) \equiv J_{ij}(\lambda)\frac{df^i}{d\lambda}\frac{d^2f^j}{d\lambda^2} + C_{ijk}(\lambda)\frac{df^i}{d\lambda}\frac{df^j}{d\lambda}\frac{df^k}{d\lambda},$$

$$R(\lambda) \equiv R_{ij}\frac{df^i}{d\lambda}\frac{df^j}{d\lambda}, \quad S(\lambda) \equiv G_i(\lambda)\frac{df^i}{d\lambda}, \text{ and } U(\lambda) \equiv u_i\frac{df^i}{d\lambda}.$$

Again, for convenience the dependence of the above coefficients on λ will be omitted in the sequel. Now rewrite the state equations in the following form:

$$\dot{\lambda} = \mu \tag{9a}$$

$$\dot{\mu} = \frac{1}{M}[U - Q\mu^2 - R\mu - S]. \tag{9b}$$

The M term is a quadratic form reminiscent of the expression for the manipulator's kinetic energy. In fact, if the parametric expressions for the $\dot{q}_i$ are plugged into the formula for kinetic energy, one obtains the expression $2K = M\mu^2$. The Q term represents the components of the Coriolis and centrifugal forces which act along the path plus the fictitious forces generated by the restriction that the robot stay on the parameterized path. The R term represents frictional components, and S gives the gravitational force along the path. U is the projection of the input vector onto the velocity vector.

At this point, it is instructive to look at the system's behavior in the phase plane. The equations of the phase-plane trajectories can be obtained by dividing (9b) by (9a). This gives

$$\frac{d\mu}{d\lambda} = \frac{\frac{d\mu}{dt}}{\frac{d\lambda}{dt}} = \frac{\dot{\mu}}{\mu} = \frac{1}{\mu M}[U - Q\mu^2 - R\mu - S]. \tag{10}$$

It is interesting to note that the total time T it takes to go from initial to final states is

$$T = \int_0^{t_f} dt = \int_0^{\lambda_{\max}} \frac{dt}{d\lambda}\,d\lambda = \int_0^{\lambda_{\max}} \frac{1}{\mu}\,d\lambda. \tag{11}$$

The idea, then, is to minimize this integral subject to the given

[2] This is done at the stage of task planning to avoid collision as well as to meet task requirements.

constraints. We therefore want to make the pseudovelocity μ as large as possible, a result which would be expected intuitively.

The constraints on $\dot{\mu}$ have two effects. One effect is to place limits on the slope of the phase trajectory. The other is to place limits on the value of μ. To obtain the limits on $d\mu/d\lambda$, one simply divides the limits on $\dot{\mu}$ by μ, since $d\mu/d\lambda = \dot{\mu}/\mu$.

To get the constraints on μ, it is necessary to consider the bounds on $\dot{\mu}$. If, for particular values of λ and μ, we have $LUB(\lambda, \mu) < GLB(\lambda, \mu)$ then there are no permissible values of $\dot{\mu}$. Therefore, for each value of λ we can assign a set of values of μ as determined by the inequality $LUB(\lambda, \mu) - GLB(\lambda, \mu) \geq 0$. This inequality holds if and only if $UB_i(\lambda, \mu) - LB_j(\lambda, \mu) \geq 0$ for all i and j. The intersection of the regions determined by these inequalities produces a region of the phase outside of which the phase trajectory must not stray. This region will hereafter be referred to as the *admissible region* of the phase plane. Using the equations for the lower and upper bounds for all i and j,

$$\frac{u^i_{\max}(M_i>0)+u^i_{\min}(M_i<0)-(Q_i\mu^2+R_i\mu+S_i)}{M_i} - \frac{u^j_{\min}(M_j>0)+u^j_{\max}(M_j<0)-(Q_j\mu^2+R_j\mu+S_j)}{M_j} \geq 0.$$

Rearranging this inequality,

$$\left[\frac{Q_i}{M_i}-\frac{Q_j}{M_j}\right]\mu^2+\left[\frac{R_i}{M_i}-\frac{R_j}{M_j}\right]\mu\left[\frac{S_i}{M_i}-\frac{S_j}{M_j}\right] + \left[\frac{u^i_{\max}(M_i<0)-u^i_{\min}(M_i>0)}{|M_i|} - \frac{u^j_{\max}(M_j<0)-u^j_{\min}(M_j>0)}{|M_j|}\right] \geq 0. \quad (12a)$$

It will prove convenient to "symmetrize" the input torque bounds in the discussion which follows. Each joint has a mean torque u^i_M and a maximum deviation Δ^i given by

$$u^i_M \equiv \frac{u^i_{\max}+u^i_{\min}}{2}, \quad \Delta^i \equiv \frac{u^i_{\max}-u^i_{\min}}{2}.$$

The inequality (12a) can then be rewritten as

$$\left[\frac{Q_i}{M_i}-\frac{Q_j}{M_j}\right]\mu^2+\left[\frac{R_i}{M_i}-\frac{R_j}{M_j}\right]\mu+\left[\frac{S_i}{M_i}-\frac{S_j}{M_j}\right] - \left[\frac{u^i_M}{M_i}-\frac{u^j_M}{M_j}\right]+\left[\frac{\Delta^i}{|M_i|}+\frac{\Delta^j}{|M_j|}\right] \geq 0. \quad (12b)$$

At this point, a specific form for the torque bounds will be assumed. If the maximum and minimum torques for each joint are functions only of the states q^i and $\dot{q}^i$ (i.e., the actuator torques are all independent of one another) and are at most *quadratic* in the velocities $\dot{q}^i$, then the inequality yields a simple quadratic in μ. This allows one to solve for the velocity bounds using the quadratic formula. A particulary simple and useful special case is that encountered when the actuator is a fixed-field DC motor with a bounded voltage input. In this case, the torque constraints take the form $u^i_{\max} = V^i_{\max} + k^i_m\dot{q}^i$ and $\dot{u}^i_{\min} = V^i_{\min} + k^i_m\dot{q}^i$ where $V^i_{\min}$ and $V^i_{\max}$ are proportional to the voltage limits and k^i_m is a constant which depends upon the motor winding resistance, voltage source resistance, and back E.M.F. generated by the motor. Let $V_{\text{ave}} = (V^i_{\max} + V^i_{\min})/2$ and $\Delta^i = V^i_{\max} - V^i_{\min}$. Then we get

$$u^i_M = V_{\text{ave}} + k^i_m\dot{q}^i = V_{\text{ave}} + k^i_m\frac{df^i}{d\lambda}\mu.$$

From here on, the case outlined above will be used for the sake of simplicity. The only changes required for the more general case of quadratic velocity dependence of the torque bounds is a redefinition of the coefficients in some the equations which follow.

Introducing yet more shorthand notation, let

$$A_{ij} \equiv \left[\frac{Q_i}{M_i}-\frac{Q_j}{M_j}\right] \quad B_{ij} \equiv \left[\frac{R_i}{M_i}-\frac{R_j}{M_j}\right] - \left[\frac{V_{\text{ave}}+k^i_m\dfrac{df^i}{d\lambda}\mu}{M_i} - \frac{V_{\text{ave}}+k^j_m\dfrac{df^j}{d\lambda}\mu}{M_j}\right]$$

$$C_{ij} \equiv \left[\frac{\Delta^i}{|M_i|}+\frac{\Delta^j}{|M_j|}\right] \quad D_{ij} \equiv \left[\frac{S_j}{M_j}-\frac{S_i}{M_i}\right].$$

Noting that (at least in this case), $A_{ij} = -A_{ji}$, $B_{ij} = -B_{ji}$, $C_{ij} = C_{ji}$, and $D_{ij} = D_{ji}$, we have the inequalities

$$A_{ij}\mu^2+B_{ij}\mu+C_{ij}+D_{ij}\geq 0 \text{ and } -A_{ij}\mu^2-B_{ij}\mu+C_{ij}-D_{ij}\geq 0. \quad (12c)$$

The second inequality is obtained by interchanging i and j and using the symmetry or antisymmetry of the coefficients. Only the cases where $i \neq j$ need be considered, so there are $n(n-1)/2$ such pairs of equations, where n is the number of degrees of freedom of the robot.

If $A_{ij} = B_{ij} = 0$, we have $C_{ij} - D_{ij} \geq 0$ and $C_{ij} + D_{ij} \geq 0$, which are always true if the robot is "strong" enough so that it can stop and hold its position at all points on the desired path. If $A_{ij} = 0$ and $B_{ij} \neq 0$, then we have a pair of linear inequalities which determine a closed interval for μ. If $A_{ij} \neq 0$, then, without loss of generality, we can assume that $A_{ij} > 0$. Then the left-hand side of the first of the inequalities (12c) is a parabola which is concave upward, whereas for the second, it is concave downward. When the parabola is concave downward, then the inequality holds when μ is between the two roots of the quadratic. If the parabola is concave upward, then the inequality holds outside of the region between the roots (Fig. 1). Thus, in one case μ must lie within a closed interval and in the other it must lie outside an open interval, unless of course the open interval is of length zero. In that case, the inequality is always satisfied and the roots of the quadratic will be complex.

Since the admissible values of μ are those which satisfy all of the inequalities, the admissible values must lie in the intersection of all the regions determined by the inequalities. There are $n(n-1)/2$ inequalities which give closed intervals, so the intersection of these regions is also a closed interval. The other $n(n-1)/2$ inequalities, when intersected with this closed interval, each may have the effect of "punching a hole" in the interval (Fig. 2). It is thus possible to have, for any particular value of λ, a set of admissible values for μ which consists of as many as $n(n-1)/2 + 1$ distinct intervals. When the phase portrait of the optimal path is drawn, it may be necessary to have the optimal trajectory dodge the little "islands" which can occur in the admissible region of the phase plane. (Hereafter, these inadmissible regions will be referred to as *islands of inadmissibility*, or just *islands*.) It should be noted, though, that if there is no friction, then $B_{ij} = 0$, which means that in the concave upward case the inequality is satisfied for all values of μ. Thus, in this case there will be no islands in the admissible region.

In addition to the constraints on μ described above, we must also have $\mu \geq 0$. This can be shown as follows: if $\mu < 0$, then the trajectory has passed below the line $\mu = 0$. Below this line, the trajectories always move to the left, since $\mu \equiv d\lambda/dt < 0$. Since the optimal trajectory must approach the desired final state through positive values of μ, the trajectory would then have to pass through $\mu = 0$ again, and would pass from $\mu < 0$ to $\mu > 0$ at a point to the left of where it had passed from $\mu > 0$ to $\mu < 0$.

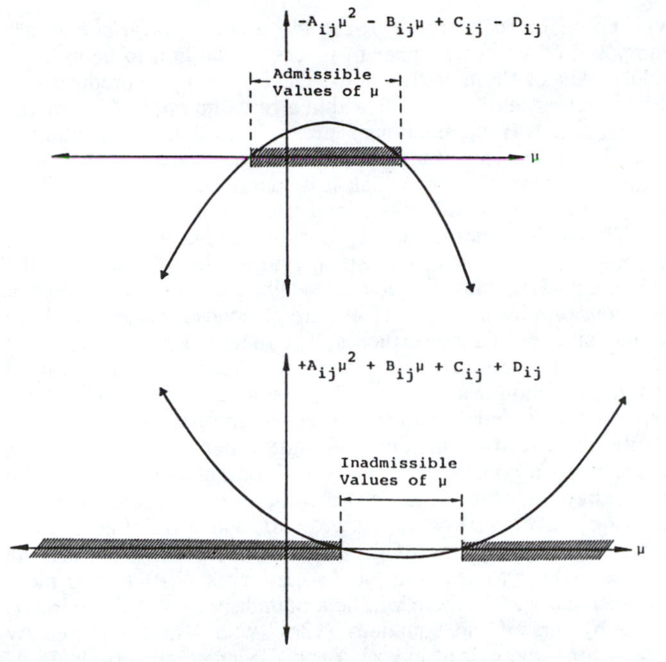

Fig. 1. Admissible regions of μ determined by a pair of parabolic constraints.

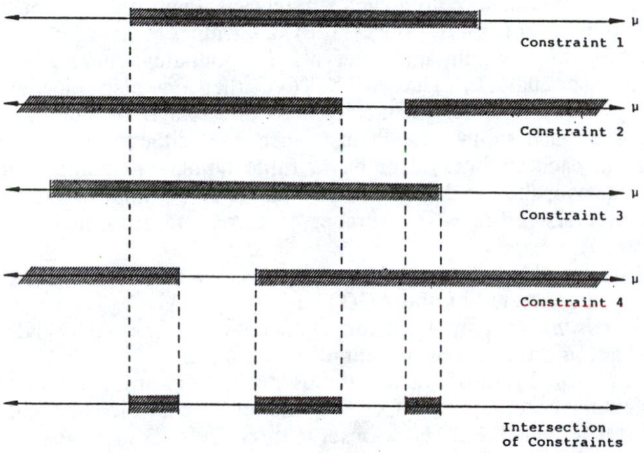

Fig. 2. Intersection of admissible regions of μ.

Thus, in order to get to the desired final state, the trajectory would have to cross itself, forming a loop. But, then, there is no sense in traversing the loop; it would take less time to just use the crossing point as a switching point. Thus, we need consider only those points of the phase plane for which $\mu \geq 0$.

Another way of thinking about the system phase portrait is to assign a pair of vectors to each point in the phase plane. One vector represents the slope when the system is accelerating (i.e., $\dot{\mu}$ is maximized) and the other represents the slope for deceleration (i.e., $\dot{\mu}$ is minimized). This pair of vectors looks like a pair of scissors, and as the position in the phase plane changes, the angles of both the upper and lower jaws of the pair of scissors change. In particular, the angle between the two vectors varies with position. The phase trajectories must, at every point of the phase plane, point in a direction which lies between the jaws of the scissors. At particular points of the phase plane, though, the jaws of the scissors close completely, allowing only a single value for the slope. At other points the scissors may try to go past the closed position, allowing no trajectory at all. This phenomenon, and the condition $\mu \geq 0$, determine the admissible region of the phase plane. This is illustrated in Fig. 3. Note that the boundary of the admissible region passes through those points which have only a single vector associated with them, corresponding to those states where only a single acceleration value is permitted.

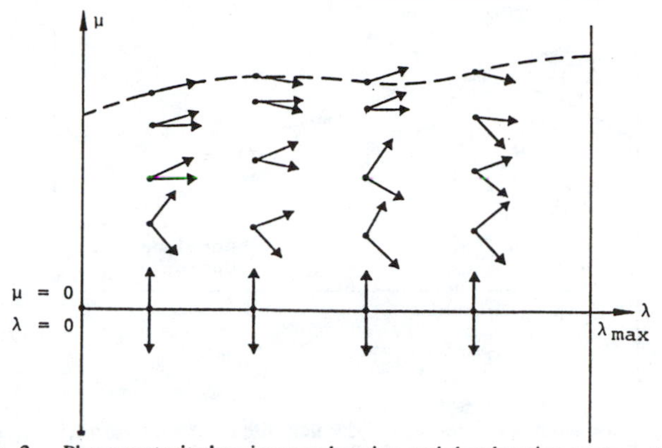

Fig. 3. Phase portrait showing acceleration and deceleration vectors at each state with $\mu_0 = 0$.

IV. Determination of Optimal Trajectories

For illustrative puposes, we first present an algorithm for finding the optimal trajectories for which there are no islands in the phase plane which need to be dodged. The only restrictions, then, will be that μ must lie between a pair of values which are easily calculable, given λ. The optimal trajectory can be constructed by the following steps called the algorithm for constructing optimal trajectories, no islands (ACOTNI).

Step 1: Start at $\lambda = 0$, $\mu = \mu_0$ and construct a trajectory that has the maximum acceleration value. Continue this curve until it either leaves the admissible region of the phase plane or goes past $\lambda = \lambda_{max}$. Note that "leaves the admissible region" implies that if part of the trajectory happens to coincide with a section of the admissible region's boundary, then the trajectory should be extended along the boundary. It is not sufficient in this case to continue the trajectory only until it touches the edge of the admissible region.

Step 2: Construct a second trajectory that starts at $\lambda = \lambda_{max}$, $\mu = \mu_f$ and proceeds *backwards*, so that it is a decelerating curve. This curve should be extended until it either leaves the admissible region or extends past $\lambda = 0$.

Step 3: If the two trajectories intersect, then stop. The point at which the trajectories intersect is the (single) switching point, and the optimal trajectory consists of the first (accelerating) curve from $\lambda = 0$ to the switching point, and the second (decelerating) curve from the switching point to $\lambda = \lambda_{max}$ (Fig. 4).

Step 4: If the two curves under consideration do not intersect, then they must both leave the admissible region. Call the point where the accelerating curve leaves the admissible region λ_1. This is a point on the boundary curve of the admissible region (Fig. 5). If the boundary curve is given by $\mu = g(\lambda)$, then search along the curve, starting at λ_1, until a point is found at which the quantity $\phi(\lambda) \equiv d\mu/d\lambda - dg/d\lambda$ changes sign. (Note that since $g(\lambda)$ determines the boundary of the admissible region, there is only one allowable value of $d\mu/d\lambda$. Also note that if $g(\lambda)$ has a discontinuity, $dg/d\lambda$ must be treated as $+\infty$ or $-\infty$ depending upon the direction of the jump.) This point is the next switching point. Call it λ_d.

Step 5: Construct a decelerating trajectory backwards from λ_d until it intersects an accelerating trajectory. This gives another switching point (see point A in Fig. 6).

Step 6: Construct an accelerating trajectory starting from λ_d. Continue the trajectory until it either intersects the final decelerating trajectory or it leaves the admissible region. If it intersects the decelerating trajectory, then the intersection gives another switching point (see point C in Fig. 6), and the procedure terminates. If the trajectory leaves the admissible region, then go to Step 4.

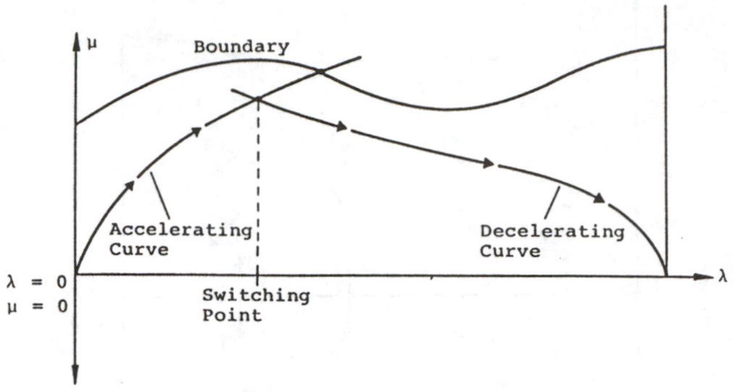

Fig. 4. Case when accelerating and decelerating curves intersect with $\mu_0 = 0$.

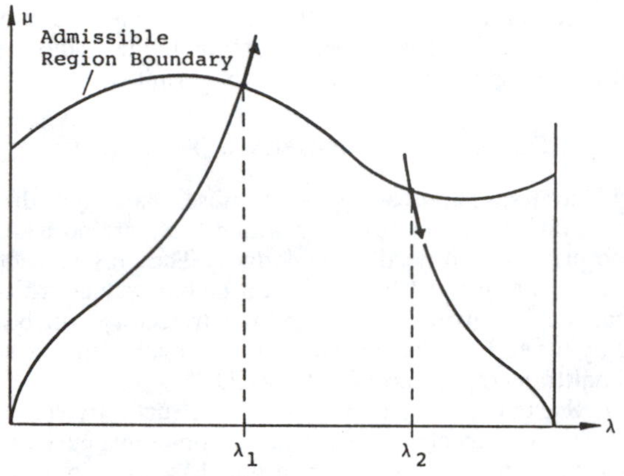

Fig. 5. Case when accelerating and decelerating curves do not intersect.

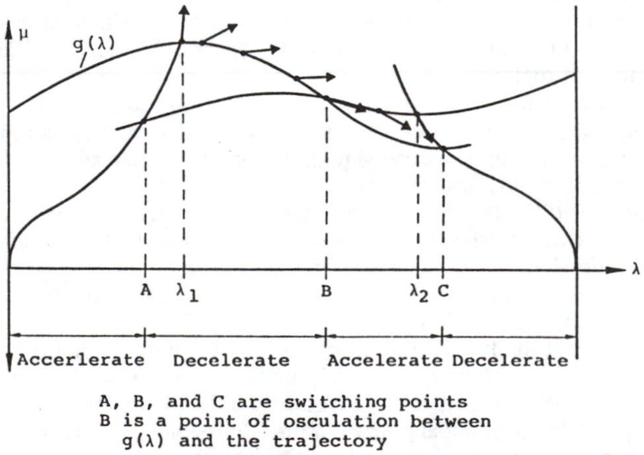

Fig. 6. Complete optimal trajectory formed by ACOTNI with three switching points.

This algorithm yields a sequence of alternately accelerating and decelerating curves which gives the optimal trajectory. Before discussing the optimality of the trajectory, one has to show that all steps of the ACOTNI are possible and that the ACOTNI will terminate.

Addressing the first question, Steps 1–3, 5, and 6 are clearly possible. Step 4 requires finding a sign change of the function $\phi(\lambda)$. Since $\phi(\lambda)$ must be greater than zero where the accelerating trajectory leaves the admissible region and less than zero where the decelerating trajectory leaves, there must be a sign change. Therefore, all steps are possible.

In order to prove that ACOTNI terminates, we must make some assumptions about the form of the functions $f(\lambda)$. In particular, it will be assumed that the f^i-are *piecewise analytic* and are composed of a finite number of pieces in addition to being real-valued. Under these assumptions, the following theorem proves the convergence of ACOTNI within a finite number of iterations.

Theorem 1: If the functions f^i are composed of a finite number of analytic, real-valued pieces, then the function $\phi(\lambda)$ has a finite number of intervals over which it is identically zero and a finite number of zeros outside those intervals.

Proof: The inertia matrix, Coriolis array, and gravitational loading vector are all piecewise analytic in the q^i, and since the $f(\lambda)$ are analytic in λ, the inertia matrix, etc., when expressed as functions of λ [as in (4a) and (4b)] are piecewise analytic and have a finite number of analytic pieces. The functions M_i, Q_i, R_i, S_i of (7b) are, therefore, also piecewise analytic. Since a real-valued analytic function with no singularities in a finite interval must either have a finite number of zeros in that interval or be identically zero, the quantities M_i must either be identically zero in the interval considered or have a finite number of zeros. We cannot have all of the M_i zero, for if that were the case we would have $J_{ij}df^i/d\lambda df^j/d\lambda = M_i df^i/d\lambda = 0$, which is not allowed by hypothesis. If only one of the M_i is nonzero, then there is no boundary curve to deal with, and so no zeros. With two or more not identically zero, there will be a boundary curve. The curve is given by one of the equations (12c) (with "$\geq$" replaced by "$=$") for some pair of indexes i and j. Since the coefficients A, B, C, and D in (12c) are analytic except at the zeros of the M_i, and because the M_i have a finite number of zeros, we can divide the interval under consideration further, using the zeros of the M_i as division points. Within each subinterval, then, only one of the equations (12c) holds. Since (12c) determines μ as an analytic function of λ within this interval, the bounding curve $g(\lambda)$ is piecewise analytic. The curve $\phi(\lambda)$, then, is also piecewise analytic and is either identically zero or has a finite number of zeros in each subinterval. Thus, since $\phi(\lambda)$ either is identically zero in each subinterval or has a finite number of zeros in the subinterval, the number of subintervals is finite, and the number of intervals is finite, the number of zeros and zero-intervals is finite. Q.E.D.

Finally, the following theorem proves the optimality of the solution generated by the ACOTNI.

Theorem 4: Any trajectory generated by the ACOTNI is optimal in the sense of minimum-time control.

Proof: Proof of this theorem is straightforward. Let Γ be the trajectory generated by ACONTI, and let Γ be a trajectory with a shorter traversal time. Now observe three facts. 1) From the form (11) of the cost T, there must be a point (λ_0, μ') on Γ' which is higher than the point (λ_0, μ) on Γ, i.e., $\mu' > \mu$. Otherwise, we would not have a trajectory with a smaller travel time. 2) The trajectory Γ consists of alternately accelerating and decelerating segments, and can therefore be divided into sections which consist of one accelerating and one decelerating segment. 3) The admissible portions of these sections which lie above Γ are bounded on the left and right by either the line $\lambda = 0$, the line $\lambda = \lambda_{max}$, the boundary of the admissible region, or the ACOTNI trajectory itself. Now consider the point (λ, μ') and the trajectory Γ'. This trajectory, if extended backward and forward from (λ_0, μ') must intersect a single section of the ACOTNI trajectory at two or more points, since otherwise it would either leave the admissible region or not meet the initial or final boundary conditions. One such point must occur for $\lambda < \lambda_0$ and one must occur for $\lambda > \lambda_0$. But since the accelerating segment of the trajectory precedes the decelerating segment, the new trajectory must either intersect the accelerating part of the ACOTNI trajectory twice, intersect the decelerating part twice, or first intersect the accelerating part then the decelerating part. But since the torques were chosen so as to minimize or maximize U in (10), any of these situations leads to a contradiction of a theorem on differential inequalities presented in [7]. Q.E.D.

The whole idea of the algorithm is to generate trajectories which come as close as possible to the edge of the admissible region without actually passing outside it. Thus, the trajectories just barely touch the inadmissible region. In practice this would, of course, be highly dangerous, since minute errors in the control inputs or measured system parameters would very likely make the robot stray from the desired path. Theoretically, however, this trajectory is the minimum-time optimum.

We are now in a position to consider the general case, i.e., the case in which friction, copper losses in the drive motor, etc., are sufficient to cause islands in the phase plane. In this case, the algorithm is most easily presented in a slightly different form. Since there may be several boundary curves instead of one, it is not possible to search a single function for zeros, as was done in ACOTNI. Thus, instead of looking for zeros as the algorithm progresses, we look for them all at once instead, and then construct the trajectories which "just miss" the boundaries, whether the boundaries be the edges of the admissible region or the edges of islands. The appropriate trajectories can then be found by searching the resulting directed graph, always taking the highest trajectory possible, and backtracking when necessary. More formally, the algorithm for construction of optimal trajectories (ACOT) is as follows.

Step 1: Construct the initial accelerating trajectory (same as ACOTNI).

Step 2: Construct the final decelerating trajectory (same as ACOTNI).

Step 3: Calculate the function $\phi(\lambda)$ for the edge of the admissible region and for the edges of all the islands. At each of the sign changes of $\phi(\lambda)$, construct a trajectory for which the sign change is a switching point, as in ACOTNI Steps 5 and 6. The switching direction (acceleration-to-deceleration or vice-versa) should be chosen so that the trajectory does not leave the admissible region. Extend each trajectory until it either leaves the admissible region, or goes past λ_{max}.

Step 4: Find all the intersections of the trajectories. These are potential switching points.

Step 5: Starting at $\lambda = 0$, $\mu = \mu_0$, traverse the grid formed by the various trajectories in such a way that the highest trajectory from the initial to the final points is followed. This is described below in the *grid traversal algorithm* (GTA). Traversing the grid formed by the trajectories generated in Steps 3 and 4 above is a search of a directed graph, where the goal to be searched for is the final decelerating trajectory. If one imagines searching the grid by walking along the trajectories, then one would try to keep making left turns, if possible. If a particular turn led to a dead end, then it would be necessary to backtrack, and take a right turn instead. The whole procedure can best be expressed recursively, in much the same manner as tree traversal procedures.

The algorithm consists of two procedures, one which searches accelerating curves and one which searches decelerating curves. The algorithm is as follows.

AccSearch

On the current (accelerating) trajectory, find the last switching point. At this point, the current trajectory meets a decelerating curve. If that curve is the final decelerating trajectory, then the switching point under consideration is a switching point of the final optimal trajectory. Otherwise, call DecSearch, starting at the current switching point. If DecSearch is successful, then the current point is a switching point of the optimal trajectory. Otherwise, move back along the current accelerating curve to the previous switching point and repeat the process.

DecSearch

On the current (decelerating) trajectory, find the first switching point. Apply AccSearch, starting on this point. If successful, then the current point is a switching point of the optimal trajectory. Otherwise, move forward to the next switching point and repeat the process.

These two algorithms always look first for the curves with the highest velocity, since AccSearch always starts at the end of an accelerating curve and DecSearch always starts at the beginning of a decelerating curve. Therefore, the algorithm finds (if possible) the trajectory with the highest velocity, and hence the smallest traversal time.

The proofs of optimality and convergence of this algorithm are virtually identical to those of ACOTNI, and will not be repeated here. Note that in the convergence proof for ACOTNI the fact that there is only a single boundary curve in the zero-friction case is never used; the same proof therefore applies in the high-friction case.

V. Application Examples

To show how the minimum-time algorithm works, a numerical example follows. The robot used in the example is a simple two-degree-of-freedom robot with one revolute and one prismatic joint, i.e., a robot which moves in polar coordinates. Despite its simplicity, the example robot is sufficient to show the most important aspects of our trajectory planning method. The path chosen is a straight line. Before applying the minimum-time algorithm, we must derive the dynamic equations for the robot. This requires calculation of the inertia matrix, so masses and moments of inertia of the robot must be given.

A drawing of our hypothetical robot is shown in Fig. 7. The robot consists of a rotating fixture with moment of inertia J_θ through which slides a uniformly dense rod of length L_r and mass M_r. The payload has mass M_p and moment of inertia J_p, and its center of mass is at the point (x, y) which is L_p units of length from the end of the sliding rod.

In the examples presented here, the robot will be moved from the point (1, 1) to the point (1, -1). The equation of the curve can be expressed as $r = \sec\theta$, where θ ranges from $+\pi/4$ to $-\pi/4$. Introducing the parameter λ, one possible parameterization is

$$\theta = \frac{\pi}{4} - \lambda \quad r = \sec\left(-\frac{\pi}{4} - \lambda\right). \tag{13}$$

Now introduce the shorthand expressions $M_t \equiv M_r + M_p$, $K \equiv M_r(L_r + 2L_p)$ and $J_t \equiv J_\theta + J_p + M_r(L_p^2 + L_rL_p + L_r^2/3)$. Plugging these expressions and the expressions for the derivatives of r and θ into the dynamic equations gives (see [12] for a detailed derivation)

$$\begin{aligned} u_r = &-M_t \sec\left(\frac{\pi}{4}-\lambda\right)\tan\left(\frac{\pi}{4}-\lambda\right)\dot{\mu} \\ &-k_r \sec\left(\frac{\pi}{4}-\lambda\right)\tan\left(\frac{\pi}{4}-\lambda\right)\mu \\ &+\left[M_t \sec\left(\frac{\pi}{4}-\lambda\right)\right. \\ &\cdot\left(\sec^2\left(\frac{\pi}{4}-\lambda\right)+\tan^2\left(\frac{\pi}{4}-\lambda\right)\right) \\ &\left.+\frac{K}{2} - M_t \sec\left(\frac{\pi}{4}-\lambda\right)\right]\mu^2 \end{aligned} \tag{14a}$$

$$\begin{aligned} u_\theta = &-\left[J_t - K\sec\left(\frac{\pi}{4}-\lambda\right) + M_t\sec^2\left(\frac{\pi}{4}-\lambda\right)\right]\dot{\mu} - k_\theta\mu \\ &+\mu^2\left(2M_t\sec\left(\frac{\pi}{4}-\lambda\right) - K\right)\sec\left(\frac{\pi}{4}-\lambda\right) \\ &\cdot\tan\left(\frac{\pi}{4}-\lambda\right). \end{aligned} \tag{14b}$$

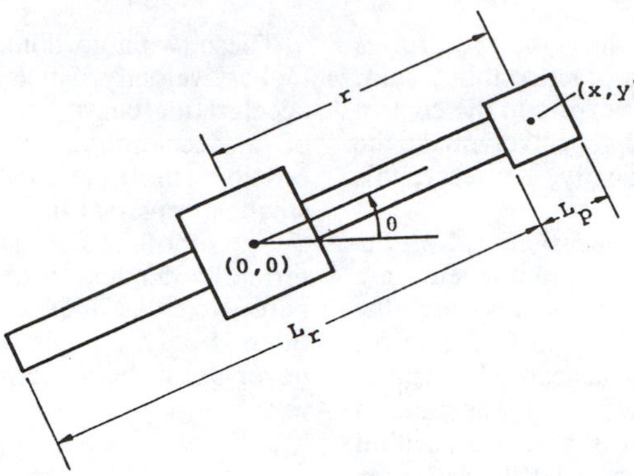

Fig. 7. An example manipulator with extensory and rotational joints.

Solving for $\dot{\mu}$, we have

$$\dot{\mu}=\frac{-1}{M_t \sec\left(\frac{\pi}{4}-\lambda\right)\tan\left(\frac{\pi}{4}-\lambda\right)}\left[u_r+k_r\mu \sec\left(\frac{\pi}{4}-\lambda\right)\tan\left(\frac{\pi}{4}-\lambda\right)-\mu^2\left\{M_t \sec\left(\frac{\pi}{4}-\lambda\right)\left(\sec^2\left(\frac{\pi}{4}-\lambda\right)+\tan^2\left(\frac{\pi}{4}-\lambda\right)\right)+\frac{K}{2}-M_t \sec\left(\frac{\pi}{4}-\lambda\right)\right\}\right] \tag{15a}$$

and

$$\dot{\mu}=\frac{u_\theta+k_\theta\mu-\mu^2\left\{2M_t \sec\left(\frac{\pi}{4}-\lambda\right)-K\right\}\sec\left(\frac{\pi}{4}-\lambda\right)\tan\left(\frac{\pi}{4}-\lambda\right)}{J_t+K\sec\left(\frac{\pi}{4}-\lambda\right)+M_t\sec^2\left(\frac{\pi}{4}-\lambda\right)}. \tag{15b}$$

The signs of the coefficients of u_r and u_θ are

$$\operatorname{sgn}(u_r)=\begin{cases}-1 & 0<\lambda<\frac{\pi}{4}\\ +1 & \frac{\pi}{4}<\lambda<\frac{\pi}{2}\end{cases} \quad \text{and } \operatorname{sgn}(u_\theta)=1.$$

The limits on $\dot{\mu}$ imposed by the θ joint are the same over the whole interval. For simplicity, let $u^i_{\max} = -u^i_{\max}$ for $i = r, \theta$, then the limits are

$$\dot{\mu}\le\frac{u^\theta_{\max}+\left[\left\{2M_t \sec\left(\frac{\pi}{4}-\lambda\right)-K\right\}\sec\left(\frac{\pi}{4}-\lambda\right)\tan\left(\frac{\pi}{4}-\lambda\right)\right]\mu^2-k_\theta\mu}{J_t-K\sec\left(\frac{\pi}{4}-\lambda\right)+M_t\sec^2\left(\frac{\pi}{4}-\lambda\right)} \tag{16a}$$

and

$$\dot{\mu}\le\frac{-u^\theta_{\max}+\left[\left\{2M_t \sec\left(\frac{\pi}{4}-\lambda\right)-K\right\}\sec\left(\frac{\pi}{4}-\lambda\right)\tan\left(\frac{\pi}{4}-\lambda\right)\right]\mu^2-k_\theta\mu}{J_t-K\sec\left(\frac{\pi}{4}-\lambda\right)+M_t\cdot\sec^2\left(\frac{\pi}{4}-\lambda\right)}. \tag{16b}$$

For the r joint, consider the case when $\lambda < \pi/4$. Then we also have

$$\dot{\mu}\le\frac{u^r_{\max}+\left[2M_t \sec\left(\frac{\pi}{4}-\lambda\right)\tan^2\left(\frac{\pi}{4}-\lambda\right)+\frac{K}{2}\right]\mu^2-k_r\mu\sec\left(\frac{\pi}{4}-\lambda\right)\tan\left(\frac{\pi}{4}-\lambda\right)}{M_t\sec\left(\frac{\pi}{4}-\lambda\right)\tan\left(\frac{\pi}{4}-\lambda\right)} \tag{16c}$$

and

$$\dot{\mu} \le \frac{-u^r_{\max} + \left[2M_t \sec\left(\frac{\pi}{4}-\lambda\right) \tan^2\left(\frac{\pi}{4}-\lambda\right) + \frac{K}{2}\right]\mu^2 - k_r\mu \sec\left(\frac{\pi}{4}-\lambda\right) \tan\left(\frac{\pi}{4}-\lambda\right)}{M_t \sec\left(\frac{\pi}{4}-\lambda\right) \tan\left(\frac{\pi}{4}-\lambda\right)}. \tag{16d}$$

For $\lambda > \pi/4$ the limits have the signs of $u^r_{\max}$ reversed.

Equating upper and lower limits on $\dot{\mu}$ gives the boundary of the admissible region. For $\lambda < \pi/4$, (16b) and (16c) give

$$A\mu^2 + B\mu + C \ge 0$$

where

$$A = -KM_t \sec^4\left(\frac{\pi}{4}-\lambda\right) + 2M_t \sec^3\left(\frac{\pi}{4}-\lambda\right) + \frac{3}{2}KM_t \sec^2\left(\frac{\pi}{4}-\lambda\right) - \left(2M_tJ_t + \frac{K^2}{2}\right) \sec\left(\frac{\pi}{2}-\lambda\right) + \frac{KJ_t}{2}$$

$$B = (J_tk_r - M_tk_\theta) \sec\left(\frac{\pi}{4}-\lambda\right) \tan\left(\frac{\pi}{4}-\lambda\right) - Kk_r \sec^2\left(\frac{\pi}{4}-\lambda\right) \cdot \tan\left(\frac{\pi}{4}-\lambda\right) + M_tk_r \tan\left(\frac{\pi}{4}-\lambda\right) \sec^3\left(\frac{\pi}{4}-\lambda\right)$$

$$C = u^r_{\max}\left(J_t - K \sec\left(\frac{\pi}{4}-\lambda\right) + M_t \sec^2\left(\frac{\pi}{4}-\lambda\right)\right) + u^\theta_{\max} M_t \tan\left(\frac{\pi}{4}-\lambda\right) \sec\left(\frac{\pi}{4}-\lambda\right).$$

Likewise, (16a) and (16d) give

$$-A\mu^2 - B\mu + C \ge 0.$$

The same inequalities, with $u^r_{\max}$ negated, work when $\lambda \ge \pi/4$.

Finally, we need to determine the differential equations to be solved. These equations are

$$\dot{\mu} = \frac{1}{J_t - K \sec\left(\frac{\pi}{4}-\lambda\right) + M_t \sec^4\left(\frac{\pi}{4}-\lambda\right)} \cdot \left[-u_\theta - u_r \sec\left(\frac{\pi}{4}-\lambda\right) \tan\left(\frac{\pi}{4}-\lambda\right) + 2\mu^2 M_t \sec^4\left(\frac{\pi}{4}-\lambda\right) \tan\left(\frac{\pi}{4}-\lambda\right) - \left\{k_r \tan^2\left(\frac{\pi}{4}-\lambda\right) \sec^2\left(\frac{\pi}{4}-\lambda\right) + k_\theta\right\}\mu\right] \tag{17a}$$

$$\dot{\lambda} = \mu. \tag{17b}$$

The numerical values of the various constants which describe the robot are given in Table I. Using these data, the differential equations were solved numerically using the fourth-order Runge–Kutta method, the program being written in C and run under the UNIX[3] operating system on a VAX-11/780.[4] The derivative of the boundary curve $g(\lambda)$ [needed to compute the function $\phi(\lambda)$] was calculated numerically, and the sign changes of $\phi(\lambda)$ found by bisection. The graphs of the resulting trajectories and of the boundary of the admissible region are given in Fig. 8 for the zero-friction case and in Figs. 9 and 10 for the high-friction case.

Note in particular the shape of the admissible region boundary in Fig. 9. For values of λ less than about 0.42 there is not a single range of admissible velocities, but two ranges. Thus, there is an "island" in the phase plane, although the island is chopped off by the constraint that λ be positive. While the existence of such islands may at first seem to defy intuition, the example shows that they do indeed exist. In this case, the island does not really come into play in the calculation of the optimal trajectory. Nevertheless, the example does demonstrate that there may be situations where the admissible region has a fairly complicated shape. Since most practical manipulators have more than two joints and have more complicated dynamic equations than those of the simple robot used here, it is conceivable that the admissible region of the phase plane for a practical robot arm could have quite a complicated shape.

As a final example, to demonstrate clearly the existence of islands in the phase plane, we include a sketch of the admissible region of the phase plane for a two-dimensional Cartesian robot moving along a circular path. In this case, the dynamic equations are a simple pair of uncoupled, linear differential equations with constant coefficients, i.e., $u_x = m\ddot{x} + k_x\dot{x}$, $u_y = m\ddot{y} + k_y\dot{y}$ where $m \equiv$ mass of x and y joints, $k_x \equiv$ coefficient of friction of x joint, and $k_y \equiv$ coefficient of friction of y joint.

Moving this manipulator in a unit circle, say in the first quadrant, requires that

$$x = \cos\lambda, \; y = \sin\lambda, \; 0 \le \lambda \le \frac{\pi}{2}.$$

Plugging these expressions and their derivatives into the dynamic equations gives

$$u_x = -m\dot{\mu}\sin\lambda - m\mu^2\cos\lambda - k_x\mu\sin\lambda$$
$$u_y = -m\dot{\mu}\cos\lambda - m\mu^2\sin\lambda - k_y\mu\cos\lambda.$$

Now let the torque bounds be $-T \le u_x, u_y \le +T$. Then the bounds on $\dot{\mu}$ are

$$\frac{-T - m\mu^2\cos\lambda - k_x\mu\sin\lambda}{m\sin\lambda} \le \dot{\mu} \le \frac{+T - m\mu^2\cos\lambda - k_x\mu\sin\lambda}{m\sin\lambda}$$

and

$$\frac{-T + m\mu^2\sin\lambda - k_y\mu\cos\lambda}{m\cos\lambda} \le \dot{\mu} \le \frac{+T + m\mu^2\sin\lambda - k_y\mu\cos\lambda}{m\cos\lambda}.$$

[3] UNIX is a trademark of Bell Laboratories.

[4] VAX is a trademark of the Digital Equipment Corporation.

TABLE I
DATA FOR THE EXAMPLE ROBOT

Constant	Description	Value
J_t	Moment of inertia of θ joint	10^{-3} Kg–M
M_r	Mass of sliding rod	4.0 $Kg.$
L_r	Length of rod	2.0 $M.$
J_p	Moment of inertia of payload	10^{-5} $Kg.$–M
M_p	Payload mass	1.0 $Kg.$
L_p	Length of payload	0.1
u^r_{max}	Maximum force on r joint	1.0 $Kg.$–M/sec^2
u^t_{max}	Maximum torque on θ joint	1.0 $Kg.$–M^2/sec^2
k_r	Friction coefficient of r joint	0.0 (low friction)
	Friction coefficient of r joint	15.0 (high friction)
k_t	Friction coefficient of θ joint	0.0

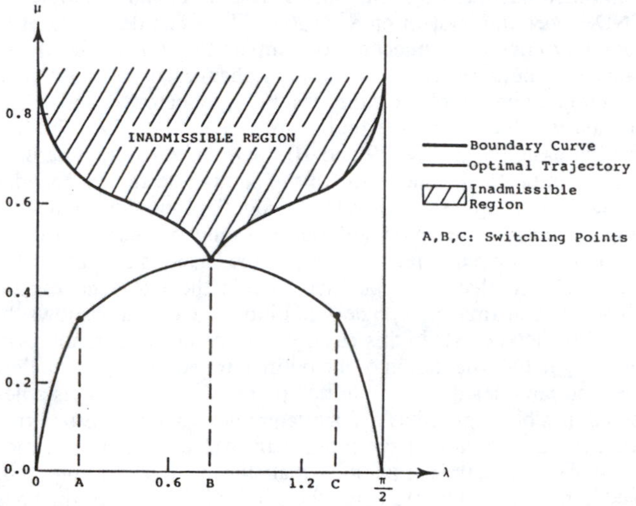

Fig. 8. Optimal trajectory and inadmissible region in case of no island.

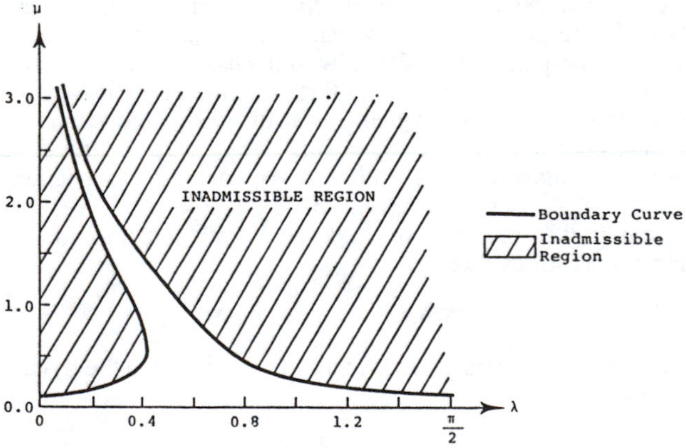

Fig. 9. Inadmissible region in high-friction case.

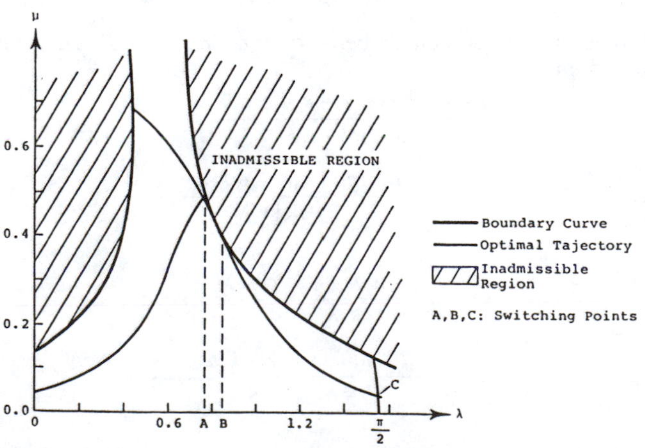

Fig. 10. Optimal trajectory in high-friction case with expanded view of Fig. 9.

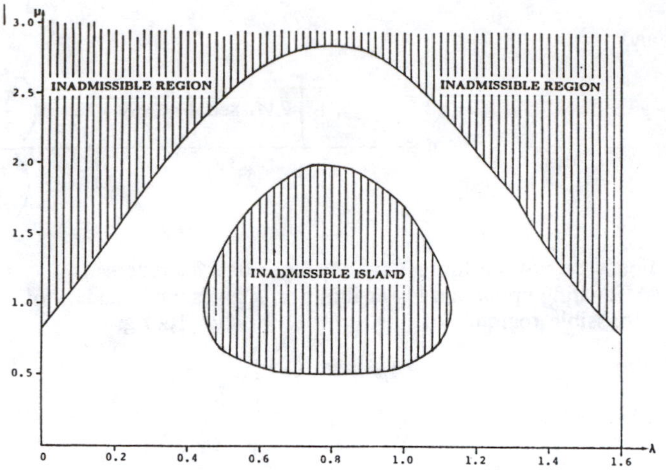

Fig. 11. Example of feasible region with an inadmissible island.

The admissible region consists of the region where the inequalities given above allow some value of the acceleration $\dot{\mu}$, as previously described. Simplifying the resulting inequalities gives the admissible region as that area of the phase plane where

$$m\mu^2+(k_x-k_y)\mu\ \sin\ \lambda\ \cos\ \lambda+T(\sin\ \lambda+\ \cos\ L)\geq 0$$

and

$$-m\mu^2+(k_x-k_y)\mu\ \sin\ \lambda\ \cos\ \lambda+T(\sin\ \lambda+\ \cos\ L)\geq 0.$$

Using the values $m = 2$, $k_x = 0$, $k_y = 10$, and $T = \sqrt{2}$ gives the region plotted in Fig. 11 and clearly shows the island.

VI. Discussion and Conclusion

In this paper we have presented a method for obtaining trajectories for minimum-time control of a mechanical arm given the desired geometric path and input torque constraints.

As was already pointed out, the optimal trajectory may actually touch the boundary of the admissible region, generating a rather dangerous case. However, if slightly conservative torque bounds are used in the calculations, then the actual admissible region will be slightly larger than the calculated admissible region, giving some margin for error.

The algorithm has been presented for both the case in which there are no islands in the phase plane and that in which islands do occur. In both cases, the algorithms produce trajectories which "just miss" the inadmissible region, whether the portion of the inadmissible region missed is an island or the region determined by the upper velocity limit. Since the algorithm generates the true minimum-time solution, rather than an approximation to it, the results from the algorithm can provide an absolute reference against which other trajectory planning algorithms can be measured.

Acknowledgment

The authors would like to thank anonymous reviewers for constructive comments and pointing out the missing references [1], [2] in the original version of this paper.

References

[1] J. E. Bobrow, "Optimal control of robotic manipulators," Ph.D. dissertation, Univ. Calif., Los Angeles, CA, Dec. 1982.
[2] J. E. Bobrow, S. Dubowsky, and J. S. Gibson, "On the optimal control robotic manipulators with actuator constraints," in *Proc. 1983 Amer. Contr. Conf.*, San Francisco, CA, June 1983, pp. 782–787.
[3] D. Ter Haar, *Elements of Hamiltonian Mechanics*, 2nd ed. New York: 1971, pp. 35–49.

[4] M. E. Kahn and B. E. Roth, "The near minimum-time control of open-loop articulated kinematic chains," *ASME J. DSMC,* vol. 93, pp. 164–172, Sept. 1971.
[5] B. K. Kim and K. G. Shin, "Suboptimal control of industrial manipulators with a weighted minimum time fuel criterion," *IEEE Trans. Automat. Contr.,* vol. AC-30, pp. 1-10, Jan. 1985.
[6] D. E. Kirk, *Optimal Control Theory: An Introduction.* Englewood Cliffs, NJ: Prentice-Hall, 1971, pp. 227–238.
[7] V. Lakshmikantham and S. Leela, *Differential and Integral Inequalities.* New York: Academic, 1969, pp. 41–43.
[8] C.—S. Lin, P.-R. Chang, and J. Y. S. Luh, "Formulation and optimization of cubic polynomial joint trajectories for mechanical manipulators," *IEEE Trans. Automat. Contr.,* vol. AC-28, pp. 1066–1074, Dec. 1983.
[9] J. Y. S. Luh and M. W. Walker, "Minimum-time along the path for a mechanical arm," in *Proc. 16th Conf. Decision Contr.,* Dec. 1977, pp. 755–759.
[10] J. Y. S. Luh, and C. S. Lin, "Optimum path planning for mechanical manipulators," *ASME J. Dynam. Syst., Measurement, Contr.,* vol. 2, pp. 330–335, June 1981.
[11] R. P. C. Paul, *Robot Manipulators: Mathematics, Programming, and Control.* Cambridge, MA: M.I.T. Press, 1981, pp. 157–195.
[12] K. G. Shin and N. D. McKay, "Open-loop minimum-time control of mechanical manipulators and its application," in *Proc. 1984 Amer. Contr. Conf.,* San Diego, CA, June 1984, pp. 1231–1236.

J. E. Bobrow
Department of Mechanical Engineering
University of California, Irvine
Irvine, California 92717

S. Dubowsky
Department of Mechanical Engineering
Massachusetts Institute of Technology
Cambridge, Massachusetts 02139

J. S. Gibson
Mechanical, Aerospace, and Nuclear Engineering
University of California, Los Angeles
Los Angeles, California 90024

Time-Optimal Control of Robotic Manipulators Along Specified Paths

Abstract

The minimum-time manipulator control problem is solved for the case when the path is specified and the actuator torque limitations are known. The optimal open-loop torques are found, and a method is given for implementing these torques with a conventional linear feedback control system. The algorithm allows bounds on the torques that may be arbitrary functions of the joint angles and angular velocities. This method is valid for any path and orientation of the end-effector that is specified. The algorithm can be used for any manipulator that has rigid links, known dynamic equations of motion, and joint angles that can be determined at a given position on the path.

1. Introduction

For many industrial applications, present robotic manipulators are too slow to justify their use economically. Their speed and hence their productivity are limited by the capability of their actuators. Increasing actuator size and power is not the best solution; it is largely self-defeating because of the increased inertia of the actuators themselves and because of the increased cost and power consumption of the larger actuators. A more successful approach is to minimize the time needed to perform a given task, subject to the constraints imposed by the actuators. The subject of this paper is the minimum-time control problem for applications where the path of the manipulator is specified.

Work on minimum-time control problems for manipulators began as early as the late 1960s (Kahn 1970; Kahn and Roth 1971). The limits on the actuator torques were assumed to be constant, and the path was not constrained (only the endpoints were specified). Although this approach is suitable for some applications, it is often necessary to specify the manipulator trajectory in order to avoid obstacles. This additional collision-avoidance constraint may be added to the unconstrained path minimum-time problem. The result is a highly nonlinear, difficult-to-solve optimal control problem.

Niv and Auslander (1984) show some progress toward solving this problem using a parameter optimization scheme on the joint actuator switching times. During the motion, each actuator exerts maximum control torque (bang-bang) while enabling the manipulator to avoid all obstacles and reach its final destination. This method involves considerable computation and may be difficult to implement for general manipulators. Another approach (Dubowsky and Shiller 1984) is to minimize the time along any known path using the algorithm described in this paper and to vary the path to find the one that avoids all obstacles and gives the shortest time.

The research reported here was supported by the National Science Foundation under grants ECS 78-04753 and CME 80-08926.

The International Journal of Robotics Research,
Vol. 4, No. 3, Fall 1985,

Several other researchers have addressed the time-optimal control problem. Lynch (1981) developed a specialized minimum-time algorithm for the Stanford-type manipulator, assuming that the actuator torque bounds were constant. Another method (Hollerbach 1984) has been developed that scales any known path velocity profile to make full use of the actuators. While this method produces shorter traveling times than conventional techniques, it does not produce the minimum-time solution. A technique also was developed (Luh and Walker 1977; Luh and Lin 1981) to minimize the time required to move along a specified path consisting of straight lines and circular arcs. In this work, piecewise constant acceleration and maximum velocity constraints were assumed. Although these assumptions are common in manipulator control, the maximum achievable accelerations and velocities actually can vary substantially with manipulator configuration and angular velocities, both because of the nonlinear manipulator dynamics and because the maximum torques that electric motors and hydraulic actuators can produce depend on the angular velocities of the joints (Herrick 1982; Kollmorgen Corporation 1983).

This paper presents in detail the solution derived by Bobrow (1982) and presented in Bobrow, Dubowsky, and Gibson (1983) of the minimum-time control problem with specified manipulator path and state-dependent constraints on the actuators. The actuator torque and force constraints can be arbitrary functions of the joint positions and velocities. Rigid links are assumed, and the full manipulator dynamics are modeled. The solution given here is valid for any smooth path along which the joint angles, including end-effector orientation, can be determined uniquely at each point.

In the solution presented here, the distance and velocity of the end-effector along the specified path are taken as the state vector, and the nonlinear manipulator dynamics and actuator constraints are transformed into state-dependent constraints on the acceleration along the path. The problem thus becomes a time-optimal problem for a second-order linear system with nonlinear state-dependent constraints on the control, which is the acceleration along the path. The basic idea of the solution is to select the acceleration profile that produces the largest velocity profile such that, at each point on the path, the velocity is no greater than the maximum velocity at which the actuators can hold the manipulator on the path.

This solution is given in terms of a switching curve in the phase plane for the tip motion along the path. When the velocity lies below the switching curve, maximum acceleration is optimal; when the velocity lies on the switching curve, either maximum acceleration or maximum deceleration is optimal, depending on the location on the switching curve. Because the actuator efforts can be determined from the position, velocity, and acceleration along the path, the switching curve provides a graphical representation of the feedback law for the time-optimal control. The switching curve can be computed efficiently by iteration on one switching point at a time while solving a first-order nonlinear differential equation.

We should note that Shin and McKay (1984) have subsequently derived a similar algorithm. With the assumptions that the bound on each actuator torque is a quadratic function of the joint velocity and that each joint position is written as a polynomial in the path parameter, they have obtained added insight into the case of high joint friction.

Because the mathematical optimal control problem involves a second-order system with constraints on the control, we first tried standard optimal control methods — in particular, Pontryagin's maximum principle — for the solution. However, in even the simplest cases, the numerical algorithms did not converge to a solution. The solution finally came from the nonstandard, though conceptually straightforward, approach in this paper, and the resulting numerical algorithm has performed well on numerous examples.

The control problem is formulated first in Section 2 for a three-degree-of-freedom elbow-type manipulator. After the mathematical optimal control problem is derived, the solution is given in the form of an algorithm for constructing the switching curve. While the motivation for the algorithm is discussed in Section 2, the complete mathematical proof that the algorithm indeed gives the optimal control is deferred until the Appendix. At the end of Section 2 an example with a curved three-dimensional path is given.

Section 3 shows how the dynamics and orientation of the end-effector can be included in the problem. In fact, any number of degrees of freedom can be han-

Fig. 1. The manipulator model used in this study.

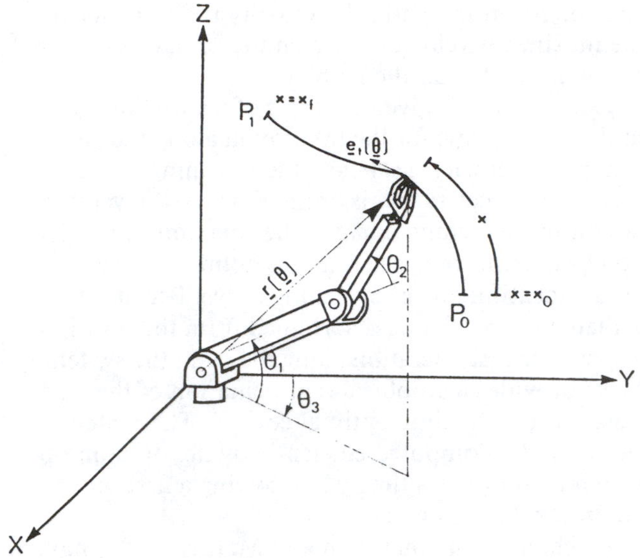

dled as long as all of the joint angles can be determined by an inverse arm solution at each point on the path. In this case, the algorithm in Section 2 still applies. Also, it is pointed out in Section 3 that, while it is easiest initially to think of the state variable in the mathematical optimal control problem as the distance along the path, the results of the paper apply for almost any parameterization of the path.

The development presented in Section 2 uses Lagrange's equations with the joint positions as the generalized coordinates. Section 4 extends this development to cases where more than the minimum required number of coordinates are used to describe the system. Lagrange's equations are used along with Lagrange multipliers to formulate the equations that must be solved. A detailed example is presented that demonstrates that this approach often reduces the complexity of the equations.

2. Formulation and Solution of the Optimal Control Problem

2.1. Problem Formulation

To illustrate the minimum-time control algorithm, we first consider the relatively simple case where the tip of the three-degree-of-freedom manipulator shown in Fig. 1 is required to move along a specified path $\mathbf{P}_0$ to $\mathbf{P}_1$, starting and finishing at rest. This case does not include the motion of the end-effector relative to the second link. In this case, the dynamics of the end-effector are assumed to have negligible influence on the dynamics of the manipulator, as in many industrial robotic systems. The method can handle such motion if the orientation of the end-effector is specified at each point on the path, as illustrated by the third example in this paper, and in Dubowsky and Shiller (1984).

The equations of motion for this system can be derived using Lagrange's equations (see Bobrow 1982), which have the form

$$\mathbf{M}(\theta)\ddot{\theta} + \mathbf{h}(\theta, \dot{\theta}) = \mathbf{T}. \tag{1}$$

The vectors $\theta = (\theta_1, \theta_2, \theta_3)^T$ and $\mathbf{T} = (T_1, T_2, T_3)^T$ are the joint angles and the applied actuator torques. The torque T_i acts about the θ_i-axis, and T_2 acts between links 1 and 2. The detailed definitions of the mass matrix $\mathbf{M}$ and vector $\mathbf{h}$ are given in Bobrow (1982).

The optimal control problem is as follows: Given the manipulator equations of motion (Eq. 1), a path through which the tip must move (Fig. 1), and actuator torque constraints of the form

$$T_{i_{\min}}(\theta, \dot{\theta}) \leq T_i \leq T_{i_{\max}}(\theta, \dot{\theta}), \tag{2}$$

find the torques $\mathbf{T}(t)$ that will drive the manipulator from the initial position at $\mathbf{P}_0$ to the final position at $\mathbf{P}_1$ in minimum time.

In our solution, the time-optimal control problem is transformed into an equivalent mathematical optimal control problem in which the single control variable is $\ddot{x}$, the tip acceleration along the path; that is, $\ddot{x} = d^2x/dt^2$ *where* x *is measured along the path from* $\mathbf{P}_0$. For the transformed problem, we must determine at each position and velocity on the path the constraints on the linear acceleration $\ddot{x}$ corresponding to the actuator torques. This requires that the joint angles be computed as functions of x and that the angular velocities and accelerations be computed as functions of x, $\dot{x}$, and $\ddot{x}$; that is, we must have

$$\theta = \theta(x), \tag{3A}$$

$$\dot{\theta} = \dot{\theta}(x, \dot{x}), \tag{3B}$$

$$\ddot{\theta} = \ddot{\theta}(x, \dot{x}, \ddot{x}). \quad (3C)$$

With the path of the tip specified, the function $\theta(x)$ for Eq. (3A) is determined at least implicitly. We emphasize that θ need not be written as an explicit function of x; the ability to compute θ numerically for each value of x is sufficient. Similarly, instead of explicit expressions for Eqs. (3B) and (3C), numerical evaluation of $\dot{\theta}$ and $\ddot{\theta}$ according to the following kinematic development is sufficient.

To obtain Eqs. (3B) and (3C), we note that the position vector **r** of the tip can be thought of as either a function of the joint angles or a function of the distance along the path. Hence we write

$$\mathbf{r} = \mathbf{r}(\theta) = \tilde{\mathbf{r}}(x). \quad (4)$$

Differentiating Eq. (4) with respect to time yields

$$[r_\theta]\dot{\theta} = \tilde{\mathbf{r}}_x \dot{x}, \quad (5)$$

where $[r_\theta]$ is the Jacobian matrix of partial derivatives of the position vector components with respect to the joint angles and $\tilde{\mathbf{r}}_x$ is the unit vector tangent to the path. Then, when the manipulator is not at a singular point, the Jacobian $[r_\theta]$ is invertible and we can solve Eq. (5) for the expression in Eq. (3B):

$$\dot{\theta}(x, \dot{x}) = [r_\theta]^{-1}\tilde{\mathbf{r}}_x \dot{x}. \quad (6)$$

We should note that, in practice, manipulator trajectories with singularities are avoided.

For Eq. (3C), differentiating Eq. (5) with respect to time yields

$$[r_\theta]\ddot{\theta} + [\dot{r}_\theta]\dot{\theta} = \tilde{\mathbf{r}}_x \ddot{x} + \tilde{\mathbf{r}}_{xx}\dot{x}^2, \quad (7)$$

where $[\dot{r}_\theta]$ is the time derivative of the Jacobian matrix and $\tilde{\mathbf{r}}_{xx}$ is the second derivative of $\tilde{\mathbf{r}}$ with respect to x. (The expression $[\dot{r}_\theta]\dot{\theta}$ contains the term $(\partial^2 r_i/\partial\theta_j\partial\theta_k)\dot{\theta}_j\dot{\theta}_k$, familiar in rigid-body dynamics; see Bobrow 1982.)

Note that the first term on the right-hand side of Eq. (7) is the tangential acceleration of the tip along the path and the second term is the normal acceleration. Still assuming that the manipulator is not in a singular configuration, we can solve Eq. (7) for Eq. (3C):

$$\ddot{\theta} = [r_\theta]^{-1}(\tilde{\mathbf{r}}_x \ddot{x} + \tilde{\mathbf{r}}_{xx}\dot{x}^2 - [\dot{r}_\theta]\dot{\theta}). \quad (8)$$

Now we can derive expressions for the maximum acceleration and deceleration the actuators can produce, at any distance x along the path and velocity $\dot{x}$. Substituting Eq. (8) into Eq. (1) yields

$$\ddot{x}\mathbf{c}_1(x) + \mathbf{c}_2(x, \dot{x}) = \mathbf{T}, \quad (9)$$

where

$$\mathbf{c}_1(x) = \mathbf{M}(\theta)[r_\theta]^{-1}\tilde{\mathbf{r}}_x \quad (10)$$

and

$$\mathbf{c}_2(x, \dot{x}) = \mathbf{M}(\theta)[r_\theta]^{-1}(\tilde{\mathbf{r}}_{xx}\dot{x}^2 - [\dot{r}_\theta]\dot{\theta}) + \mathbf{h}(\theta, \dot{\theta}). \quad (11)$$

Given the distance x along the path, the tip velocity $\dot{x}$, and the tangential acceleration $\ddot{x}$, Eq. (9) shows the unique values of the three actuator torques to be

$$T_i = c_{1i}(x)\ddot{x} + c_{2i}(x, \dot{x}), \qquad i = 1, 2, 3. \quad (12)$$

With Eq. (12), the torque constraints in Eq. (2) yield the following constraints on acceleration:

$$T_{i\min} - c_{2i}(x, \dot{x}) \leq c_{1i}(x)\ddot{x} \leq T_{i\max} - c_{2i}(x, \dot{x}), \qquad i = 1, 2, 3. \quad (13)$$

Since, for any x and $\dot{x}$, $\theta(x)$ and $\dot{\theta}(x, \dot{x})$ can be computed from the inverse arm solution and Eq. (6), $T_{i\min}$ and $T_{i\max}$ can be written as functions of x and $\dot{x}$. Then, if $c_{1i}(x) \neq 0$, Eq. (13) can be written as

$$f_i(x, \dot{x}) \leq \ddot{x} \leq g_i(x, \dot{x}), \quad (14)$$

where

$$f_i(x, \dot{x}) = \begin{cases} (T_{i\min} - c_{2i})/c_{1i}, & c_{1i} > 0, \\ (T_{i\max} - c_{2i})/c_{1i}, & c_{1i} < 0, \end{cases} \quad (15)$$

and

$$g_i(x, \dot{x}) = \begin{cases} (T_{i\max} - c_{2i})/c_{1i}, & c_{1i} > 0, \\ (T_{i\min} - c_{2i})/c_{1i}, & c_{1i} < 0. \end{cases} \quad (16)$$

When $c_{1i}(x) \neq 0$, Eqs. (14)–(16) give the range of tip acceleration $\ddot{x}$ for which the actuators can hold the tip on the path without violating the ith constraint in Eq. (2). For the tip to remain on the path, $\ddot{x}$ must lie in the intersection of the intervals $[f_i(x, \dot{x}), g_i(x, \dot{x})]$, with the intersection taken over the values of i for which $c_{1i}(x) \neq 0$. It can happen, usually when the velocity is too great, that the intervals $[f_i(x, \dot{x}), g_i(x, \dot{x})]$ do not intersect, in which case the manipulator tip will leave the path immediately. On the other hand, if $c_{1i}(x) \neq 0$ for all i and the three intervals $[f_i(x, \dot{x}), g_i(x, \dot{x})]$ have nonempty intersection, then the necessary and sufficient condition for the tip to stay on the path is that Eq. (14) hold for all i.

Although the constraints in Eq. (13) must hold for all i, if $c_{1i}(x) = 0$ for some i, then the selection of $\ddot{x}$ cannot affect whether Eq. (13) holds for that i. In this case, $\ddot{x}$ must be chosen to satisfy the two remaining constraints. As long as the manipulator is not in a singular configuration, there will be at least one nonzero $c_{1i}(x)$. This follows from the right-hand side of Eq. (10) since the mass matrix $\mathbf{M}(\theta)$ is always positive definite and $\tilde{\mathbf{r}}_x$ is a unit vector.

For given x and $\dot{x}$, an *admissible acceleration* is any tangential acceleration $\ddot{x}$ at which the actuators can hold the tip on the prescribed path without violating the constraints. The foregoing discussion shows that, if any admissible acceleration exists, then the range of admissible accelerations is given by

$$f(x, \dot{x}) \leq \ddot{x} \leq g(x, \dot{x}), \tag{17}$$

where

$$f(x, \dot{x}) = \max_i f_i(x, \dot{x}) \tag{18}$$

and

$$g(x, \dot{x}) = \min_i g_i(x, \dot{x}), \tag{19}$$

with the maximum and minimum taken over those i for which $c_{1i}(x) \neq 0$. See Fig. 2. Note that if two of the intervals $[f_i(x, \dot{x}), g_i(x, \dot{x})]$ do not intersect, then $f(x, \dot{x}) > g(x, \dot{x})$. Maintaining $f(x, \dot{x}) \leq g(x, \dot{x})$, so that an admissible acceleration exists, is a key idea in solving the time-optimal problem. We now have the mathematical time-optimal control problem:

Fig. 2. Ranges of acceleration possible for each actuator at some instant.

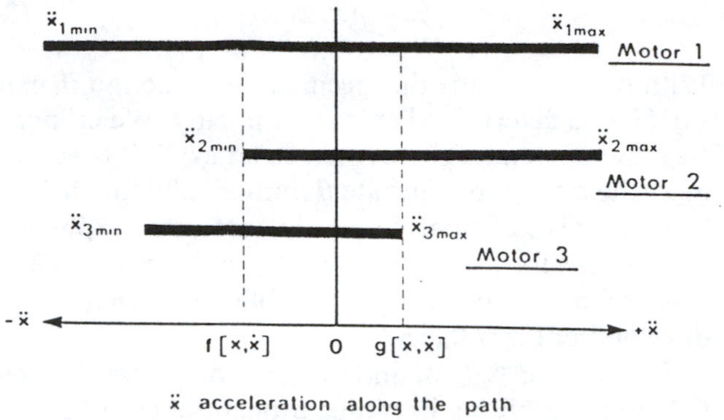

> Given $x(0)$ and $\dot{x}(0)$, choose $\ddot{x}(t)$ to minimize the final time t_f for which $x(t_f) = x_f$ and $\dot{x}(t_f) = \dot{x}_f$, subject to Eq. (17) at each t.

Although we have transformed the manipulator control problem so that the tangential acceleration becomes the control variable for the mathematical time-optimal problem, we have not lost sight of the torques, which are the physical controls. Recall that Eq. (12) gives the actuator torques/forces in terms of x, $\dot{x}$, and $\ddot{x}$.

2.2. Solution of the Time-Optimal Problem

The basic idea of the time-optimal solution is to choose the acceleration $\ddot{x}$ to make the velocity $\dot{x}$ as large as possible at every point without violating the condition $f(x, \dot{x}) \leq g(x, \dot{x})$. This is suggested by the identity

$$t_f = \int_{x_0}^{x_f} (dx)/\dot{x} \tag{20}$$

It is proved in the Appendix that to minimize t_f, $\ddot{x}$ always takes either its largest or its smallest possible value; that is, either $\ddot{x} = g(x, \dot{x})$ or $\ddot{x} = f(x, \dot{x})$. Therefore, finding the optimal control law amounts to finding the times, or positions, at which $\ddot{x}$ switches between maximum acceleration and maximum deceleration.

Fig. 3. A typical minimum-time trajectory with one switching time.

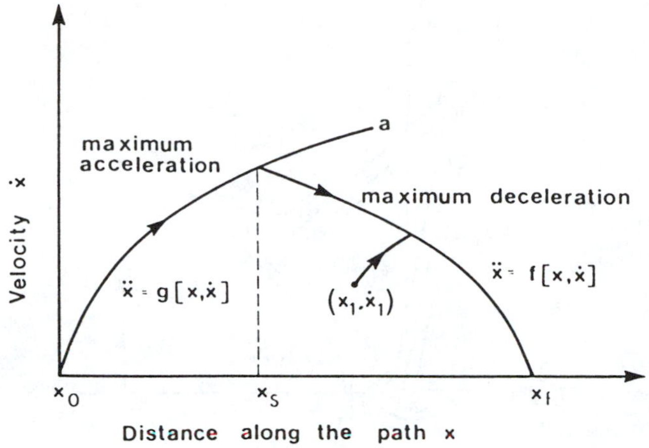

Fig. 4. An example showing the minimum-time trajectory construction method.

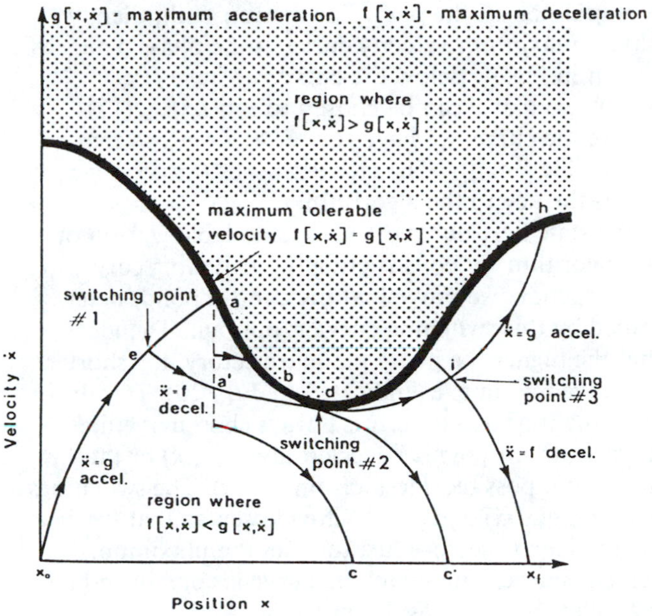

2.2.1. *Problems with One Switching*

The best approach for finding the switching position is to construct the switching curves in the $x - \dot{x}$ phase plane. We will give an algorithm for the construction for problems with multiple switchings, but first it should help to consider the case of one switching. A typical minimum-time trajectory with one switching is shown in Fig. 3. The manipulator tip starts accelerating from the initial position x_0 with $\ddot{x} = g(x, \dot{x})$. At the switching position x_s, the acceleration switches to $\ddot{x} = f(x, \dot{x})$ and continues this deceleration until coming to rest at the final position x_f.

To find x_s, we solve $\ddot{x} = g(x, \dot{x})$ forward in time from the point $x = x_0$, $\dot{x} = 0$ to some point *a* as shown in Fig. 3. Then we solve $\ddot{x} = f(x, \dot{x})$ backward in time from $x = x_f$, $\dot{x} = 0$ until the two trajectories intersect at x_s. The phase plane trajectory that results from solving $\ddot{x} = f(x, \dot{x})$ backward from $x = x_f$, $\dot{x} = 0$ is the switching curve for this case. If the manipulator starts at some position x_1 with tip velocity $\dot{x}_1$, as shown in the figure, the optimal control policy is to use the maximum acceleration $\ddot{x} = g(x, \dot{x})$ until the phase plane trajectory intersects the switching curve and then switch to maximum deceleration $\ddot{x} = f(x, \dot{x})$.

2.2.2. *Multiple Switchings*

The minimum-time problem becomes considerably more difficult when the maximum acceleration curve $\ddot{x} = g(x, \dot{x})$ proceeding from x_0 and the maximum deceleration curve $\ddot{x} = f(x, \dot{x})$ proceeding backward from x_f do not intersect before the velocity becomes too large and the condition $f(x, \dot{x}) \leqslant g(x, \dot{x})$ is violated. In this case, before the final switch to the deceleration curve that brings the tip to rest at x_f, the optimal control policy requires earlier switches between acceleration and deceleration to avoid building up velocities, and hence large inertial forces, at which the actuators can no longer hold the tip on the specified path.

Finding the multiple switching points is the most difficult part of the minimum-time problem. Several approaches were tried, including a conjugate gradient optimization algorithm that treated the switching times as parameters and the numerical solution of the two-point boundary value problem that results from the maximum principle with state-dependent constraints on the control (Leitmann 1966). In neither of these approaches did the numerical algorithm converge to a solution in even the simplest cases. We finally developed a nonstandard approach that is actually more straightforward and that has been successful on numerous examples. This method yields a simple numerical algorithm for constructing the switching curve.

The method is motivated by noting that, for most points on the path, there is a tip velocity above which no combination of admissible torques will hold the manipulator on the path. For a given path, then, we have a maximum velocity curve in the phase plane, as shown in Fig. 4. Below the curve, we have $f(x, \dot{x}) \leq g(x, \dot{x})$, so that $\ddot{x}$ can be chosen to satisfy Eq. (17). Above the curve, we have $f(x, \dot{x}) > g(x, \dot{x})$, so that no admissible $\ddot{x}$ exists. The curve satisfies $f(x, \dot{x}) = g(x, \dot{x})$ and is the key to the algorithm.

We state here our most understandable version of the algorithm for constructing the switching curve. In the Appendix we prove that the control policy determined by this switching curve is optimal. The idea is that the higher the phase plane trajectory, the shorter the traveling time, as indicated by Eq. (20). To minimize the traveling time, $\ddot{x}$ is always chosen as either the maximum possible acceleration $g(x, \dot{x})$ or the maximum possible deceleration $f(x, \dot{x})$. The switchings between $g(x, \dot{x})$ and $f(x, \dot{x})$ are chosen so that the phase plane trajectory just touches the maximum velocity curve. The switching curve is constructed by the following six steps. Refer to Fig. 4.

Step 1. Integrate the equation $\ddot{x} = g(x, \dot{x})$ from the initial state $(x_0, \dot{x}_0)$ until the maximum velocity curve is reached at some point a.

Step 2. From point a, drop to some lower velocity on the dotted vertical line and then integrate the equation $\ddot{x} = f(x, \dot{x})$ forward in time. One of two things will happen: Either the trajectory will intersect the maximum velocity curve again at some point b or the trajectory will intersect the x-axis at some point c. The object is to find, by iteration, the point a' such that the deceleration ($\ddot{x} = f$) trajectory emanating from a' just touches the maximum velocity curve at a single point d and then continues downward, intersecting the x-axis at c'. If $c' \geq x_f$, then there is only one switching point.

Step 3. From a', integrate $\ddot{x} = f(x, \dot{x})$ backward in time until the acceleration trajectory from x_0 to a is intersected at some point e.

Step 4. Integrate the equation $\ddot{x} = g(x, \dot{x})$ forward in time from point d until either x_f is passed or the trajectory again intersects the maximum velocity curve, as at point h. It is proved in the Appendix that it is possible to resume maximum acceleration ($\ddot{x} = g$) at d without immediately violating $f(x, \dot{x}) \leq g(x, \dot{x})$.

Fig. 5. A three-dimensional path. Dimensions are in feet.

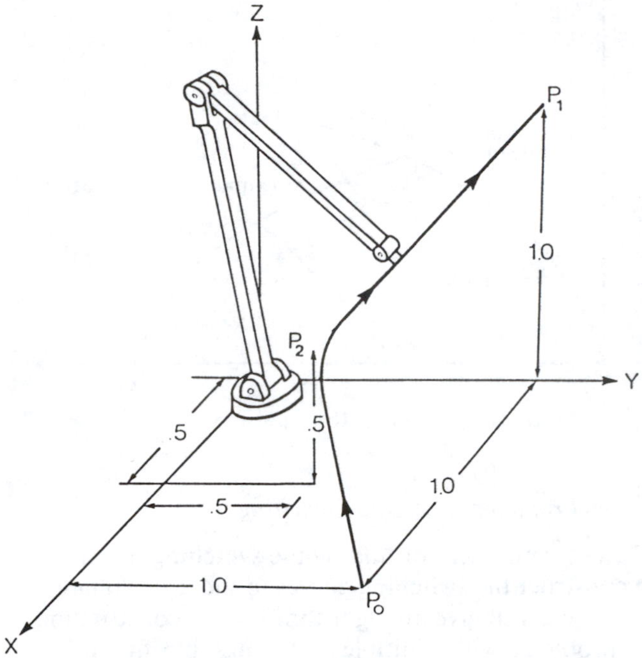

Step 5. Finally, integrate the equation $\ddot{x} = f(x, \dot{x})$ backward in time from the final position until the trajectory from d to h is crossed at point i.

Step 6. If the deceleration trajectory proceeding backward from x_f does not intersect the acceleration trajectory from d to h, then there are more than three switchings. In this case the switching point between d and h is determined as point e is determined in steps 2 and 3, and the algorithm is continued until x_f is reached.

The curve x_0edix_f is the switching curve. For any initial conditions $(x(0), \dot{x}(0))$ that lie beneath this switching curve, the optimal control policy is to use maximum acceleration until the switching curve is reached and then switch to deceleration and follow the switching curve to x_f.

Some computation can be saved by solving the first-order equations $d\dot{x}/dx = g/\dot{x}$ and $d\dot{x}/dx = f/\dot{x}$ instead of $\ddot{x} = g$ and $\ddot{x} = f$. Also, these first-order equations are more convenient for the more rigorous statement of the algorithm in the Appendix.

Finally, note that the initial and final velocities actually need not be zero. By simply letting the x-axis in Fig. 4 intersect the $\dot{x}$-axis at $\dot{x} = \dot{x}_f$, the desired final

Fig. 6. The actuator torque constraints.

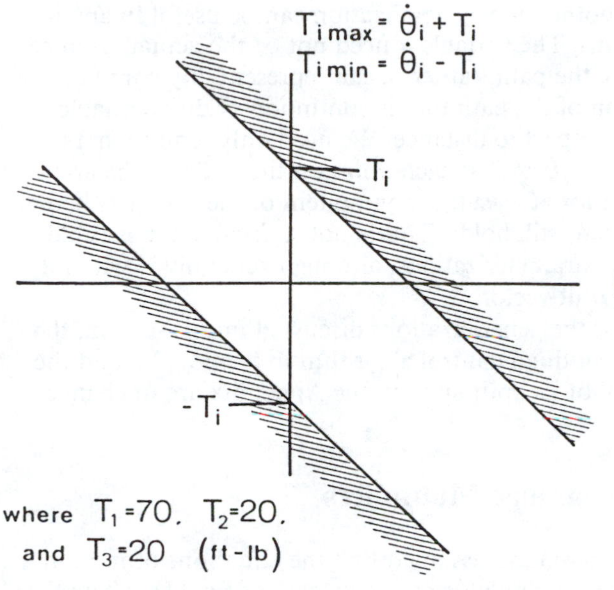

Fig. 7. The time-optimal trajectory for the path in Fig. 5.

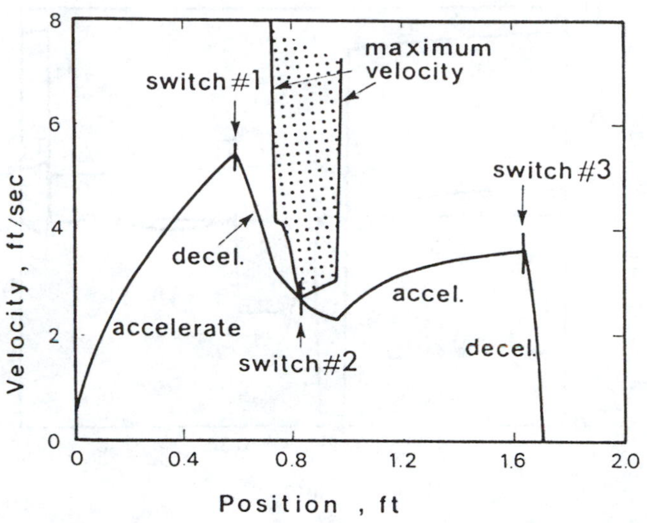

velocity along the path, we can construct the switching curve just as above. Therefore, the method can be applied directly, for example, to a problem where an object is to be moved from one moving conveyor to another in minimum time.

2.3. Example 1

For each link of the manipular in Fig. 1, we take the length to be 1 ft and the weight to be 32.2 lb. We require the tip to move along the three-dimensional path in Fig. 5, which consists of two straight lines connected by a circular arc of radius .2 ft. The arc smooths out the corner formed by the intersecting lines P_0P_2 and P_1P_2, where $P_0 = (1', 0, 1')$, $P_1 = (1', 1', 0)$ and $P_2 = (.5', .5', .5')$. The torque constraints are the linear functions of the motor angular velocities shown in Fig. 6.

The minimum-time algorithm given in steps 1–6 yields the maximum velocity and switching curves shown in Fig. 7. Note that the segment of the switching curve just past the second switch indicates that the maximum acceleration $g(x, \dot{x})$ is actually negative for a short time. Figure 8 shows the optimal actuator torques as functions of time.

3. Modeling the Dynamics and Orientation of the End-Effector

Most industrial manipulators have two or three degrees of freedom for the end-effector in addition to the three degrees of freedom for the links considered so far. In many applications, the dynamics and orientation of the end-effector are significant and must be modeled. The method of Section 2 can include such applications if the orientation of the end-effector is prescribed for each point on the trajectory, as usually is the case with programmed trajectories. We now show how the method of Section 2 applies to this case.

We assume that there are k(1, 2, or 3) additional degrees of freedom for the end-effector, each represented by a joint variable θ_i, and that there is an additional actuator for each degree of freedom, producing a joint torque/force T_i about the θ_i-axis. The θ and T vectors then are expanded to $\theta = (\theta_1, \theta_2, \ldots, \theta_{3+k})^T$ and $T = (T_1, T_2, \ldots, T_{3+k})^T$.

We assume that on the end-effector there is a point P whose path is specified as in Fig. 1. The 3-vector $\mathbf{r}(\theta)$ contains the coordinates of P as before. The orientation of the end-effector is represented by a k-vector $\mathbf{s}(\theta)$. Since the position of P can be written as a function of θ and of x, we have

$$\mathbf{r}(\theta) = \tilde{\mathbf{r}}(x). \quad (21)$$

Fig. 8. Optimal actuator torque versus time.

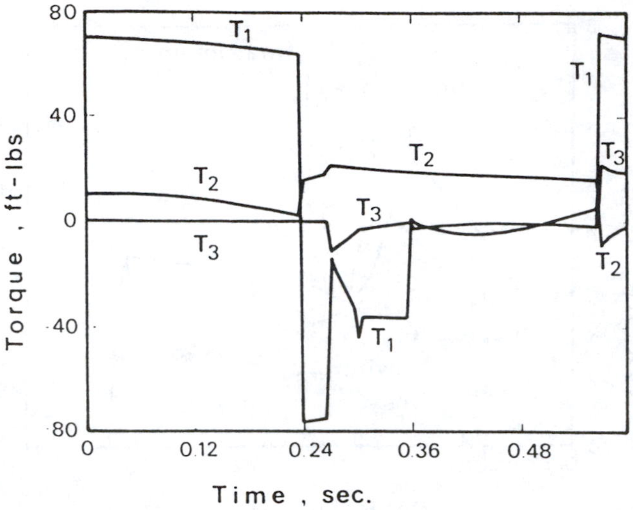

Also, the orientation of the end-effector will be prescribed along the path by k scalar equations having the vector form

$$\mathbf{s}(\theta) = \tilde{\mathbf{s}}(x). \qquad (22)$$

To generalize the method of Section 2 to include the present case, we expand the vector $\mathbf{r}(\theta)$ by writing

$$\mathbf{R}(\theta) = \begin{pmatrix} \mathbf{p}(\theta) \\ \mathbf{s}(\theta) \end{pmatrix} = \tilde{\mathbf{R}}(x). \qquad (23)$$

Then Eq. (23) replaces Eq. (4) and, from here, the development in Section 2 can be retraced easily to see that, after Eq. (4), all the equations hold with the more general $(3 + k)$-vectors θ, $\mathbf{R}(\theta)$, and $\mathbf{T}$.

Recall that when the dynamics and orientation of the end-effector were neglected, the vector $\tilde{\mathbf{r}}_x$ was the unit vector tangent to the path. We now have the more general

$$\tilde{\mathbf{R}}_x = \frac{d\tilde{\mathbf{R}}}{dx} = \begin{pmatrix} \tilde{\mathbf{r}}_x \\ \tilde{\mathbf{s}}_x \end{pmatrix}, \qquad (24)$$

where $\tilde{\mathbf{r}}_x$ is the unit vector tangent to the path of P and $\tilde{\mathbf{s}}_x = d\tilde{\mathbf{s}}/dx$. The $\tilde{\mathbf{R}}_x$ of Eq. (24) is now used in Eqs. (5)–(11). The $(3 + k) \times (3 + k)$ Jacobian $[r_\theta]$ will be singular at most at isolated points in θ-space, which, again, are avoided in practice.

Another easy generalization can be useful in applications. The variable x need not be the actual distance along the path; rather, x can represent any parameterization of the path that is continuously differentiable with respect to distance. We need only require that $\tilde{\mathbf{r}}_x = d\tilde{\mathbf{r}}/dx \neq 0$ at each point on the path, so the argument for at least one component of the vector $\mathbf{c}_1$ being nonzero still holds. This is not restrictive for sensible path parameterizations, although generally $\tilde{\mathbf{r}}_x$ will not be a unit vector.

For the generalizations discussed in this section, the time-optimal control algorithm in Section 2.2 and the proof of its optimality in the Appendix are unchanged.

4. Lagrange Multipliers

So far we have assumed that the equations of motion of the manipulator have the form of Eq. (1), where the joint angles θ_i are independent generalized coordinates. In applying our optimal control algorithm, we have found it convenient to use additional coordinates along with Lagrange multipliers. With Lagrange multipliers, the coordinates of the center of mass of the end-effector can be included among the generalized coordinates to make the chore of writing Lagrange's equations more tractable.

4.1. Including the Extra Degrees of Freedom and Constraints

If $\mathbf{z}$ is the vector containing the Cartesian coordinates of the center of mass of the end-effector, we have the holonomic constraints

$$\mathbf{z} = \mathbf{z}(\theta), \qquad (25)$$

which result from writing the position vector $\mathbf{z}$ in terms of the joint angles. In general, $\mathbf{z}$ is a 3-vector; however, if (as in Example 3 below) the end-effector moves only in the plane, $\mathbf{z}$ can be a 2-vector. Differentiating Eq. (25) with respect to time, we obtain the nonholonomic form

$$[z_\theta]\dot{\theta} - \dot{\mathbf{z}} = 0, \qquad (26)$$

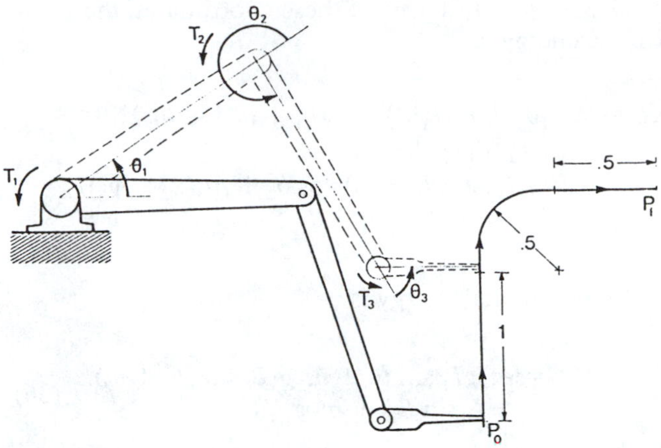

Fig. 9. A three-degree-of-freedom planar manipulator.

where the matrix $[z_\theta]$ contains the partial derivatives $\partial z_i/\partial\theta_j$. Then, selecting the generalized coordinate vector

$$\mathbf{q} = \begin{pmatrix}\theta \\ \mathbf{z}\end{pmatrix}, \tag{27}$$

we obtain the Lagrange equations

$$\begin{aligned}\mathbf{M}(\mathbf{q})\ddot{\mathbf{q}} + h(\mathbf{q}, \dot{\mathbf{q}}) &= \begin{bmatrix}\mathbf{B}_1 & [z_\theta]^T \\ 0 & -\mathbf{I}\end{bmatrix}\begin{pmatrix}\mathbf{T} \\ \lambda\end{pmatrix} \\ &= B(\mathbf{q})\begin{pmatrix}\mathbf{T} \\ \lambda\end{pmatrix},\end{aligned} \tag{28}$$

where $\mathbf{M}(\mathbf{q})$ is the mass matrix; $h(\mathbf{q}, \dot{\mathbf{q}})$ is a vector that arises from the rotation of the links and from gravity; $\mathbf{T}$ is the vector of actuator torques; and λ is the Lagrange multiplier vector. The term $B_1\mathbf{T}$ is the generalized force corresponding to the torques and forces on the joints.

Since we are assuming an independent actuator for each degree of freedom (i.e., for each θ_i), the matrix B_1 is invertible, and therefore

$$B(\mathbf{q})^{-1} = \begin{bmatrix}B_1^{-1} & B_1^{-1}[z_\theta]^T \\ 0 & -I\end{bmatrix}. \tag{29}$$

This allows us to solve Eq. (28) for $\mathbf{T}$ as a function of q, $\dot{q}$, and $\ddot{q}$. Now, just as before, after using the inverse arm solution to obtain $\theta(x)$, we can use Eqs. (6) and (8) to obtain $\dot{\theta}(x, \dot{x})$ and $\ddot{\theta}(x, \dot{x}, \ddot{x})$. Then Eqs. (25)–(27) show how to compute

$$\mathbf{q} = \mathbf{q}(x), \tag{30A}$$

$$\dot{\mathbf{q}} = \dot{\mathbf{q}}(x, \dot{x}), \tag{30B}$$

and

$$\ddot{\mathbf{q}} = \ddot{\mathbf{q}}(x, \dot{x}, \ddot{x}). \tag{30C}$$

In particular, $\ddot{\mathbf{q}}$ has the form

$$\ddot{\mathbf{q}} = v(x, \dot{x}) + \ddot{x}\mathbf{w}(x). \tag{31}$$

With these expressions, Eq. (28) can be solved for $\mathbf{T}$ and λ to obtain

$$\mathbf{T} = \ddot{x}\mathbf{c}_1(x) + \mathbf{c}_2(x, \dot{x}), \tag{32}$$

which is the same as Eq. (9) except that

$$c_1(x) = [B_1^{-1}B_1^{-1}[z_\theta]^T]\mathbf{M}(q(x))w(x) \tag{33}$$

and

$$\begin{aligned}c_2(x, \dot{x}) = [B_1^{-1}B_1^{-1}[z_\theta]^T](&\mathbf{M}(\mathbf{q}(x))\mathbf{v}(x, \dot{x}) \\ &+ \mathbf{h}(\mathbf{q}(x), \dot{\mathbf{q}}(x, \dot{x}))).\end{aligned} \tag{34}$$

Recall also that $[z_\theta] = [z_\theta(\theta(x))]$.

With Eqs. (32)–(34), the functions $f_i(x, \dot{x})$ and $g_i(x, \dot{x})$ for the time-optimal algorithm are defined as in Eqs. (15) and (16), and the algorithm proceeds just as in Section 2.

4.2. Example 2

This example is a planar problem in which we have added a third link to represent the end-effector. The third link must traverse the path P_0P_1 shown in Fig. 9 while remaining horizontal, perhaps to avoid spilling a payload.

Introducing the Cartesian coordinates of the center of mass of the third link reduces the effort required to write the kinetic energy and derive Lagrange's equations. We will denote these extra coordinates by z_1 and z_2. The generalized coordinate vector then is $q = (\theta_1,$

$\theta_2, \theta_3, z_1, z_2)^T$. In terms of these coordinates, the kinetic energy is

$$KE = \tfrac{1}{2}[J_1\dot{\theta}_1^2 + m_1 I_1'^2\dot{\theta}_1^2 + J_2(\dot{\theta}_1 + \dot{\theta}_2)^2 + m_2 I_1^2\dot{\theta}_1^2 + m_2 I_2'^2(\dot{\theta}_1 + \dot{\theta}_2)^2 + 2m_2 I_1 I_2'\dot{\theta}_1(\dot{\theta}_1 + \dot{\theta}_2)\cos\theta_2 + m_3(\dot{x}_3^2 + \dot{y}_3^2) + J_3(\dot{\theta}_1 + \dot{\theta}_2 + \dot{\theta}_3)^2], \quad (35)$$

and the potential energy is

$$PE = m_1 g I_1' \sin\theta_1 + m_2 g(I_2' \sin(\theta_1 + \theta_2) + I_1 \sin\theta_1) + m_3 g y_3. \quad (36)$$

The constraints in Eq. (25) are

$$\begin{aligned} z_1 &= I_1 \cos\theta_1 + I_2 \cos(\theta_1 + \theta_2) + I_3 \cos(\theta_1 + \theta_2 + \theta_3), \\ z_2 &= I_1 \sin\theta_1 + I_2 \sin(\theta_1 + \theta_2) + I_3 \sin(\theta_1 + \theta_2 + \theta_3). \end{aligned} \quad (37)$$

By applying Lagrange's equations to the kinetic and potential energies, we obtain the following nonzero components of the mass matrix $\mathbf{M}(q)$:

$$\begin{aligned} m_{11} &= J_1 + J_2 + J_3 + m_2(I_1^2 + I_2'^2 + 2I_1 I_2 \cos\theta_2) + m_1 I_1'^2, \\ m_{12} &= m_{21} = J_2 + J_3 + m_2(I_2'^2 + I_1 I_2' \cos\theta_2), \\ m_{13} &= m_{31} = J_3, \\ m_{22} &= J_2 + J_3 + m_2 I_2'^2, \\ m_{23} &= m_{32} = J_3, \\ m_{33} &= J_3, \\ m_{44} &= m_3, \\ m_{55} &= m_3. \end{aligned} \quad (38)$$

The vector $\mathbf{h}(q, \dot{q})$ in Eq. (28) has components

$$\begin{aligned} h_1 &= -m_2 I_1 I_2'\dot{\theta}_2(2\dot{\theta}_1 + \dot{\theta}_2)\sin\theta_2 + m_2 g(I_2' \cos(\theta_1 + \theta_2) + I_1 \cos\theta_1), \\ h_2 &= m_2 I_1 I_2'\dot{\theta}_1^2 \sin\theta_2 + m_2 g I_2' \cos(\theta_1 + \theta_2), \\ h_3 &= h_4 = 0, \\ h_5 &= m_3 g. \end{aligned} \quad (39)$$

Finally, the 5×5 $\mathbf{B}$ matrix is

$$\mathbf{B} = \left[\begin{array}{c|cc} I & b_{14} & b_{15} \\ & b_{24} & b_{25} \\ (3 \times 3) & b_{34} & b_{35} \\ \hline O & \multicolumn{2}{c}{-I} \\ (2 \times 3) & \multicolumn{2}{c}{(2 \times 2)} \end{array}\right], \quad (40)$$

where

$$\begin{aligned} b_{14} &= -I_1 \sin\theta_1 - I_2 \sin(\theta_1 + \theta_2) - I_3 \sin(\theta_1 + \theta_2 + \theta_3), \\ b_{15} &= I_1 \cos\theta_1 + I_2 \cos(\theta_1 + \theta_2) + I_3 \cos(\theta_1 + \theta_2 + \theta_3), \\ b_{24} &= -I_2 \sin(\theta_1 + \theta_2) - I_3 \sin(\theta_1 + \theta_2 + \theta_3), \\ b_{25} &= I_2 \cos(\theta_1 + \theta_2) + I_3 \cos(\theta_1 + \theta_2 + \theta_3), \\ b_{34} &= -I_3 \sin(\theta_1 + \theta_2 + \theta_3), \\ b_{35} &= I_3 \cos(\theta_1 + \theta_2 + \theta_3). \end{aligned} \quad (41)$$

For this case, we have the very nice property that $B^{-1} = B$ (this can be seen by computing BB to obtain the identity matrix).

We now have all the terms in Eqs. (32)–(34), so we can apply the time-optimal control algorithm of Section 2.2 to obtain the switching curve.

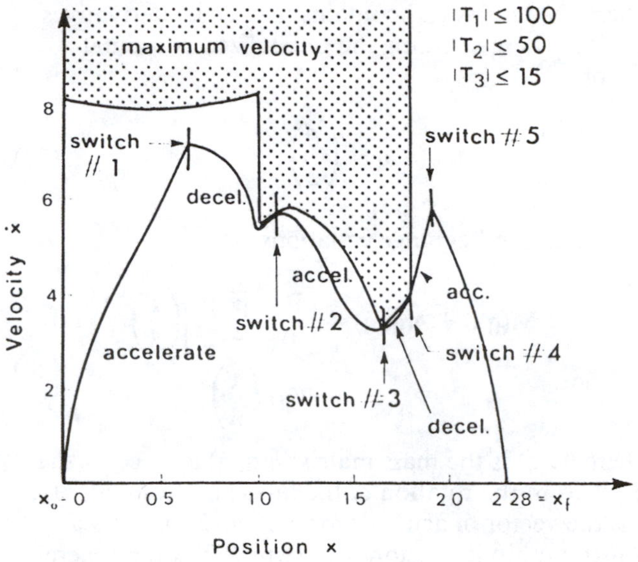

Fig. 10. The minimum-time trajectory for the path in Fig. 9.

For this problem, we used constant torque constraints $|T_i| \leq T_{i\max}$, as indicated in Fig. 10, with torque in lb·ft. Figure 10 shows the maximum velocity and switching curves. Note that the maximum velocity curve is discontinuous where the tip passes into and out of the circular arc. This results from the normal acceleration on the arc suddenly appearing and then disappearing. Also note that there are five switchings between maximum acceleration and maximum deceleration in this example. The second switch occurs tangent to the maximum velocity curve after the first discontinuity. It would probably be impossible to find these switching points using maximum principle-based algorithms.

5. Discussion

The preceding results show that the actuator torques required for a typical optimal trajectory are discontinuous functions of time. In practice, the dynamic properties of the actuators make it impossible to produce these torques exactly. For typical direct-current motors used in manipulators, the time constants relating the input voltage to output torques range from 0.0001 sec to 0.025 sec (Kollmorgen Corporation 1983), which is short enough that for most robots these dynamic effects can be neglected (Bobrow 1982). The open-loop input voltage that produces the optimal open-loop torque found with our algorithm can then be computed algebraically. For cases when the actuator time constant is not small, the actuator input voltage can be found by solving an optimal linear quadratic tracking problem (Kirk 1970). The actuator input voltage thus calculated will produce an output torque that is as close as possible to the desired open-loop torque.

Once the open-loop actuator input voltage has been determined, it should drive the manipulator along the desired path if the dynamic model is accurate. However, in most cases this is not a realistic assumption; there is always some uncertainty about the values of the system parameters and unmodeled disturbances. For this reason it is necessary to apply the open-loop voltage in conjunction with feedback. Bobrow (1982) presents a straightforward method for accomplishing this. The method was tested with a complete dynamic simulation of the system, including the motor dynamics. The results showed that in spite of significant errors in the manipulator model, the feedback control did a remarkable job of keeping the manipulator on the required trajectory.

6. Conclusion

We have presented an algorithm for computing the actuator torques that will move the manipulator along a specified path in minimum time, subject to constraints on the torques.

For a specific manipulator, the algorithm requires that

1. the path of the end-effector, including its orientation, is specified;
2. the joint angles can be calculated in terms of the position on the path;
3. the dynamic equations of motion for the manipulator are known;
4. the maximum and minimum torques that can be produced by each actuator are known as functions of the joint angles and angular velocities.

The algorithm is computationally efficient since it involves the numerical integration of only a second-order differential equation, and it requires iteration on only one variable to find the switching curve in the phase plane. While the method can be used to generate the optimal open-loop torques, it also gives the optimal feedback control law for motions along the path in terms of a switching curve in the phase plane.

Appendix

Rules for Constructing a Switching Curve

Before proving rigorously that the algorithm presented in Section 2 does in fact produce the minimum-time control, we must state more precisely the rules for constructing the switching curve. First note that we can write the second-order differential equation $\ddot{x} = g$ in first-order form by the relation $d\dot{x}/dt = (d\dot{x}/dx)(dx/dt) = g$. Thus, the equivalent equations for maximum acceleration and deceleration are

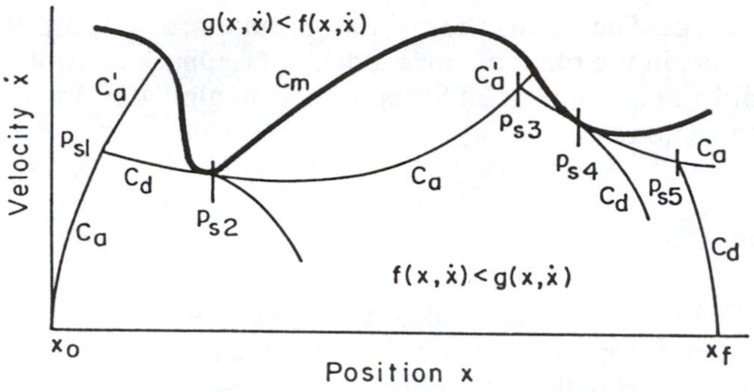

Fig. A1. An optimal trajectory with sets of points labeled.

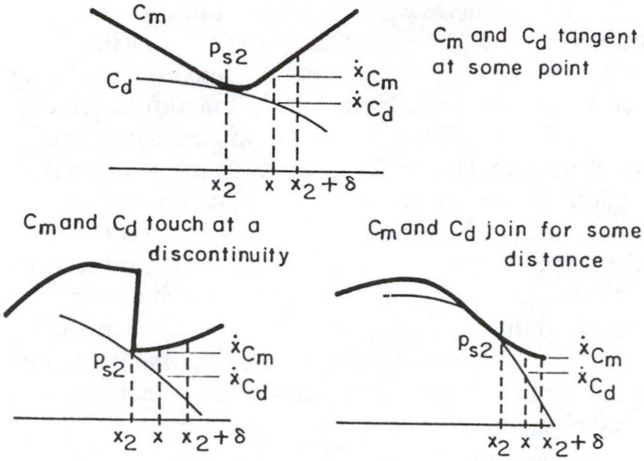

Fig. A2. Three possible occurrences of the second switch.

$$d\dot{x}/dx = g(x, \dot{x})/\dot{x} \qquad \text{acceleration,} \qquad \text{(A1)}$$

$$d\dot{x}/dx = f(x, \dot{x})/\dot{x} \qquad \text{deceleration} \qquad \text{(A2)}$$

When $\dot{x} = 0$, we can obtain the solution to these differential equations by solving $\ddot{x} = g$ and $\ddot{x} = f$ in place of Eqs. (A1) and (A2), respectively. Also, as stated earlier, points on the maximum velocity curve satisfy the relation $f(x, \dot{x}) = g(x, \dot{x})$. Let C_m be the set of these points p:

$$C_m = \{p = (x, \dot{x}): f(x, \dot{x}) = g(x, \dot{x})\}.$$

To construct the optimal trajectory, we will make the following assumptions:

Hypothesis 1: The time required for the manipulator to move from initial condition x_0 to final condition x_f is finite.

Hypothesis 2: There are a finite number of switches between maximum acceleration and maximum deceleration.

We then use the following rules for the construction:

Step A1. Generate the solution to the acceleration equation (A1) from the initial condition $(x_0, \dot{x}_0)$ until the constraint C_m is violated. Let C_a be the set of points on this trajectory (see Fig. A1); i.e.,

$$C_a = \{p = (x, \dot{x}): (x, \dot{x}) \text{ are the solution to Eq. A1) starting from } (x_0, x_0)\}.$$

Step A2. Let

$$C_a' = \{p \in C_a: \text{the solution of Eq. (A2) emanating from } p \text{ intersects } C_m\}.$$

Let $P_{s1} = \min C_a'$ that is, the point in C_a' that has the smallest value for x. [*Note:* The method for choosing P_{s1} requires that the solution C_a touches but does not violate the constraint C_m.]

Step A3. Generate the solution to the deceleration equation A2 starting at P_{s1} and ending some distance beyond the intersection with C_m. Call the set of points p on this trajectory C_d:

$$C_d = \{p = (x, \dot{x}): (x, \dot{x}) \text{ lies on the solution to Eq. (A2) starting from } P_{s1}\}.$$

Step A4. The second switch occurs at P_{s2}, the last (usually only) point that C_m has in common with C_a. That is, p_{s2} satisfies the following two conditions:

(a) $\quad p_{s2} \in C_d \cap C_m$,

(b) $\quad$ for any $\delta > 0$,
$\quad$ if $x_2 < x < x_2 + \delta$, $\quad$ then $\dot{x}_{C_m} > \dot{x}_{C_a}$,

where x_2 is the position of p_{s2} and $\dot{x}_{C_m}$ and $\dot{x}_{C_d}$ are the velocities in C_m and C_d corresponding to position x (see Fig. A2).

Step A5. Generate the solution to the acceleration equation (A1) as in step A1, but start from p_{s2}. Redefine C_a as the set of these points:

$$C_a = \{p = (x, \dot{x}): (x, \dot{x}) \text{ are the solution to Eq. (A1) starting from } p_{s2}\}.$$

Fig. A3. Switch positions for the minimum-time proof.

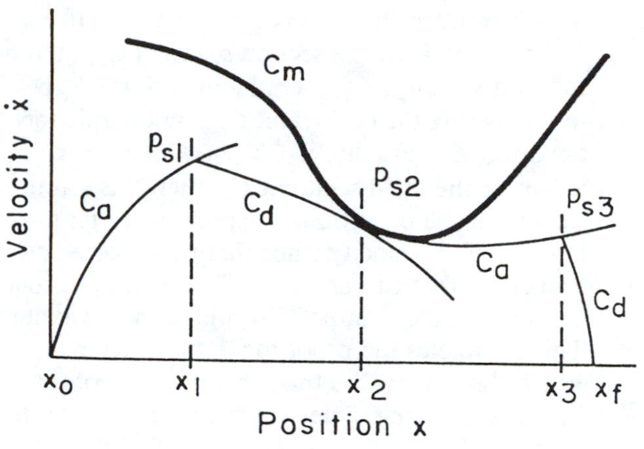

Step A6. Repeat the entire process to find more switch points until the solution goes beyond x_f (see Fig. A1). This will ensure that when (A2) is solved backward from the final condition, the solution will cross the trajectory constructed.

Proof That the Construction Gives the Minimum-Time Trajectory

To prove that it is possible to switch from deceleration to acceleration at the p_{s2} found in step A4 without immediately violating the constraint, we need the following standard result (Birkhoff and Rota 1969, pp. 24–26).

Theorem 1 Comparison Theorem. Let y and z be solutions of the differential equations

$$y' = F(x, y), \qquad z' = G(x, z),$$

respectively, where $F(x, y) \leq G(x, z)$ in the strip $a \leq x \leq b$ and F or G satisfies a Lipschitz condition. Also let $y(a) = z(a)$. Then $y(x) \leq z(x)$ for all $x \in [a, b]$.

For our problem, the theorem says that for any position x, the velocity corresponding to the solution of the deceleration equation (A2) is less than or equal to the velocity corresponding to the solution of the acceleration equation (A1) for the same initial condition. The theorem is valid in our case because under the maximum velocity curve, $f/\dot{x} < g/\dot{x}$. Also, f and g are piecewise continuously differentiable functions of x, so they satisfy a Lipschitz condition (Birkhoff and Rota 1969). This can be seen by noting how f and g were obtained (see Eqs. 15–19) and observing that it is possible to differentiate these functions except when the Jacobian $[r_\theta]$ is singular (these positions are normally avoided).

We are now able to prove that it is possible to switch from deceleration to acceleration at p_{s2} (Fig. A2).

Theorem 2. The solution to $d\dot{x}/dx = g/\dot{x}$ starting at p_{s2} remains below C_m for some $\delta > 0$.
Proof. First observe that C_d was constructed so that it continues below C_m. Also, below C_m, $f/\dot{x} < g/\dot{x}$, and above C_m, $f/\dot{x} > g/\dot{x}$. If the solution to Eq. (A1) starting from p_{s2} went immediately above C_m, then by the comparison theorem the solution to Eq. (A2) would be above that, since $f/\dot{x} > g/\dot{x}$ above C_m. This cannot happen since C_d (the solution to Eq. A2) is below C_m. Hence the solution to the acceleration equation (A1) must be below C_m for some $\delta > 0$.

Finally, we will prove that the trajectory constructed is the minimum-time solution.

Theorem 3. The solution that moves the manipulator from initial condition x_0, $\dot{x}_0$ to final condition x_f, $\dot{x}_f$ in minimum time is constructed by steps A1–A6.
Proof. Assume first that there are three switches between acceleration and deceleration, at the positions x_1, x_2, and x_3 (see Fig. A3). In the arguments given below, we have implicitly used the result that the solutions to Eqs. (A1) and (A2) are unique. This is true because the functions $f(x, \dot{x})$ and $g(x, \dot{x})$ satisfy a Lipschitz condition.

The time for the maneuver is given by $t = \int_{x_0}^{x_f} dx/\dot{x}$. This integral exists under our hypothesis that it is possible to move from x_0 to x_f in finite time. Assume there is another trajectory that gives a shorter time $t = \int_{x_0}^{x_f} dx/\dot{\tilde{x}}$. Then $\dot{\tilde{x}} > \dot{x}$ for some $x < x_f$. In the first segment, $x_0 < x < x_1$. This cannot happen because $d\dot{x}/dx < g/\dot{x}$, and, since g is as large as possible, the comparison theorem guarantees no higher solution. We can make the same argument in the last segment by making the change of variable $\tilde{x} = x_f - x$ and integrating backward from the final condition x_f, $\dot{x}_f$. If $\dot{\hat{x}} > \dot{x}$ for $x_1 < x < x_2$ in the second segment C_d, then no solution starting at $\dot{x}$, $\dot{\hat{x}}$ could go beneath C_d before violating the constraint C_m, by the rule for obtaining C_d. Finally, if $\dot{\hat{x}} > \dot{x}$ for some $x_2 < x < x_3$ in the third segment C_a, then this solution starting at the position corresponding to p_{s2} must have a higher velocity since there is no acceleration curve that can cross C_a. This is not possible because the constraint C_m would then be violated. This completes the proof for three switches. When there are more than three, the proof that the intermediate segments are optimal is the same.

References

Birkhoff, G., and Rota, G. 1969. *Ordinary differential equations.* New York: Blaisdell.

Bobrow, J. E. 1982. Optimal control of robotic manipulators. Ph.D. dissertation, University of California, Los Angeles.

Bobrow, J. E., Dubowsky, S., and Gibson, J. S. 1983 (San Francisco). On the optimal control of robotic manipulators with actuator constraints. *Proc. American Control Conference* 2:782–787.

Dubowsky, S., and Desforges, D. T. 1979. The application of model-referenced adaptive control to robotic manipulators. *ASME J. Dyn. Sys., Meas., Contr.* 101:193–200.

Dubowsky, S., and Shiller, Z. 1984 (Udine, Italy). Optimal dynamic trajectories for robotic manipulators. *Proc. V CISM-IFToMM Symp. Theory and Practice of Robots and Manipulators.*

Herrick, S. L. 1982. Model referenced adaptive control of an industrial robot. Master's thesis, University of California, Los Angeles.

Hollerbach, J. M., 1984. Dynamic scaling of manipulator trajectories. *ASME J. Dyn. Sys., Meas., Contr.* 106:102–106.

Kahn, M. E. 1970. The near-minimum time control of open-loop articulated kinematic chains. Ph.D. dissertation, Stanford University.

Kahn, M. E., and Roth, B. 1971. The near-minimum-time-control of open-loop articulated kinematic chains. *ASME J. Dyn. Sys., Meas., Contr.*, Sept., pp. 164–171.

Kirk, D. E. 1970. *Optimal control theory.* Englewood Cliffs, N.J.: Prentice-Hall, pp. 219–227.

Kollmorgen Corporation, Inland Motor Specialty Products Division, 1983. *Direct drive dc motors.* Radford, Va.

Leitmann, G. 1966. *An introduction to optimal control.* New York: McGraw-Hill, pp. 88–97.

Luh, J. Y. S., and Lin C. S. 1981. Optimal path planning for mechanical manipulators. *ASME J. Dyn. Sys., Meas., Contr.* 12:142–151.

Luh, J. Y. S., and Walker, W. M. 1977 (New Orleans). Minimum-time along the path for a mechanical arm. *Proc. 1977 IEEE Conf. Dec. Contr.* 2:755–759.

Lynch, P. M. 1981 (Charlottesville, Va.). Minimum-time, sequential axis operation of a cylindrical, two axis manipulator. *Proc. Joint Automatic Contr. Conf.*

Niv, M., and Auslander, D. M. 1984 (San Diego). Optimal control of a robot with obstacles. *Proc. American Contr. Conf.* 1:280–287.

Ogata, K. 1970. *Modern control engineering.* Englewood Cliffs, N.J.: Prentice-Hall, pp. 98–101.

Paul, R. P. 1981. *Robotic manipulators.* Cambridge: MIT Press, pp. 59–62.

Shin, K. G., and McKay, N. D. 1984 (San Diego). Open-loop minimum-time control of mechanical manipulators and its application. *Proc. American Contr. Conf.*, pp. 296–303.

Improving the Efficiency of Time-Optimal Path-Following Algorithms

JEAN-JACQUES E. SLOTINE AND HYUN S. YANG

Abstract—**This communication presents a new method which significantly improves the computational efficiency of time-optimal path-following planning algorithms for robot manipulators with limited actuator torques. *Characteristic switching points* are identified and characterized analytically as functions of the single parameter defining the position along the path. *Limit curves* are then constructed from the characteristic switching points, in contrast with the so-called *maximum velocity curve* approach of existing methods. A new efficient algorithm is proposed which exploits the characteristics of the newly defined concepts. The algorithm can also account for viscous friction effects and smooth state-dependent actuator bounds. A numerical example, while showing the consistence of the algorithm with the existing techniques, demonstrates its potential for increasing computational efficiency by several orders of magnitude.**

I. Introduction

Finding minimum-time planning strategies for robot manipulators, given actuator constraints, has been a long-standing concern in the robotics literature. This interest is largely motivated by the obvious relationship between execution time of specific tasks and productivity. Because of the nonlinear multi-input dynamics of robot manipulators, however, finding true minimum-time solutions is difficult. In early methods, various assumptions or simplifications on the manipulator dynamics were used to obtain near-minimum time solutions. Kahn and Roth [4] used linearization techniques and studied the application of linear optimal control theory. Purely kinematic approaches were proposed by Lin, Chang, and Luh [5] while Scheinman and Roth, [11], Wen and Desrochers [16], and Singh and Leu [15] used other kinds of simplifying assumptions. In particular, the problem is considerably simplified if the robot arm is assumed to be statically balanced [6]; such an assumption, however, precludes the robot from manipulating various loads of weights and sizes similar to its own, as the human arm routinely does.

The true minimum-time solution *along a prescribed path* was derived by Bobrow, Dubowsky, and Gibson, [1], [2] and Shin and McKay, [13], [14], based on the possibility of parametrizing the path with a single scalar variable. Their methods consider the full nonlinear manipulator dynamics and the torque constraints, and thus provide true minimum-time solutions. Pfeiffer and Johanni [7], [8] express the parameterized dynamic equations slightly differently from the former works, while using in essence the same procedure as in Shin and McKay, [13], [14]. These algorithms, however, require a search over the whole range of the scalar parameter and that of its time derivative, in order to find the maximum possible manipulator tip velocity limit for each value of the parameter.

In this communication, new characteristics of the time-optimal planning problem for robot manipulators are identified and applied to the existing prescribed-path algorithms. A modification is proposed, which greatly simplifies the original procedures and provides a new interpretation of the time-optimal problem. *Characteristic switching points* are defined, which are uniquely determined given the manipulator dynamics, the actuator bounds, and the prescribed path. Necessary conditions describing the characteristic switching points are derived as functions of the scalar parameter defining the position along the path. Based on the characteristic switching points, *limit curves* are drawn in the phase plane, in contrast with the maximum velocity curves of existing methods. The limit curves set sharper admissible regions for the phase-plane trajectory, and shape the solution for the time-optimal trajectory accordingly.

After a review, in Section II, of the methods of Bobrow *et al.* [1], [2] and Shin and McKay, [13], [14], the new minimum-time algorithm is described in Section III. A numerical example, in Section IV, demonstrates the potential for computational improvements of several orders of magnitude over existing methods, making quasi-real-time trajectory generation feasible. Section V offers brief concluding remarks.

Manuscript received December 16, 1987; revised May 18, 1988.

The authors are with the Nonlinear Systems Laboratory, Massachusetts Institute of Technology, Cambridge, MA 02139.

IEEE Log Number 8823367.

II. Existing Solutions to the Optimal Planning Problem

In the absence of friction or other disturbances, the dynamics of a rigid manipulator can be written as

$$H(q)\ddot{q} + \dot{q}^T Q(q)\dot{q} + g(q) = \tau \quad (1)$$

where q is the $n \times 1$ vector of joint displacements, τ is the $n \times 1$ vector of applied joint torques (or forces), $H(q)$ is the $n \times n$ symmetric positive-definite manipulator inertia matrix, $\dot{q}^T(Q(q)\dot{q}$ is the $n \times 1$ vector of centripetal and Coriolis torques (with $Q(q)$ an $n \times n \times n$ array), and $g(q)$ is the $n \times 1$ vector of gravitational torques. When the path for the manipulator tip to follow is specified, the vector q of joint displacements can be written as a function of a single parameter s, either in task space [1], [2] or in joint space [13], [14]. Therefore, the manipulator dynamics can be expressed as n equations in the parameter s. Namely, in task space, the position and orientation of the end-effector can be represented by a 6×1 vector p as

$$p = (x; a) = (x, y, z; \alpha_1, \alpha_2, \alpha_3) \quad (2)$$

where x is the end-effector position vector, and a is composed of Euler angles representing the orientation of the end-effector with respect to a fixed frame. The vector p is a known smooth function of the displacement s along the path, and can be expressed in terms of the joint variables q as

$$p(s) = r(q) \quad (3)$$

where $r(q)$ defines the direct kinematics of the manipulator. Differentiating (3) with respect to time twice, and solving for $\dot{q}$ and $\ddot{q}$, yields

$$\dot{q} = r_q^{-1} p_s \dot{s} \quad (4)$$

$$\ddot{q} = r_q^{-1}[p_s \ddot{s} + p_{ss}\dot{s}^2 - (r_q^{-1}p_s)^T r_{qq}(r_q^{-1}p_s)\dot{s}^2] \quad (5)$$

where the subscripts s and q denote derivatives with respect to the scalar s and the vector q, with r_q being the manipulator Jacobian matrix and r_{qq} the Hessian of the vector function r. Similarly, in joint space, the path is given as a curve smoothly parameterized by

Reprinted from *IEEE Trans. Robot. Automat.*, vol. 5, no. 1, pp. 118–124, Feb. 1989.

the single scalar s as

$$q = f(s). \tag{6}$$

Differentiating (6) with respect to time twice, the joint velocity and acceleration can be written as

$$\dot{q} = f_s \dot{s} \tag{7}$$

$$\ddot{q} = f_s \ddot{s} + f_{ss} \dot{s}^2. \tag{8}$$

Using (4) and (5), or (7) and (8), with (1), we get n equations of motion parametrized by the single parameter s. In task space, these are written

$$a_{TS}(s)\ddot{s} + b_{TS}(s)\dot{s}^2 + c_{TS}(s) = \tau \tag{9}$$

where the $n \times 1$ vectors $a_{TS}(s)$, $b_{TS}(s)$, and $c_{TS}(s)$, are defined as (with the subscript TS denoting task-space quantities)

$$a_{TS}(s) = Hr_q^{-1} p_s \tag{10}$$

$$b_{TS}(s) = Hr_q^{-1}\{p_{ss} - (r_q^{-1}p_s)^T r_{qq}(r_q^{-1}p_s)\} + (r_q^{-1}p_s)^T Q (r_q^{-1}p_s) \tag{11}$$

$$c_{TS}(s) = g(s). \tag{12}$$

Similarly, in joint space, we obtain (with the subscript JS denoting joint-space quantities)

$$a_{JS}(s)\ddot{s} + b_{JS}(s)\dot{s}^2 + c_{JS}(s) = \tau \tag{13}$$

where the $n \times 1$ vectors, $a_{JS}(s)$, $b_{JS}(s)$, and $c_{JS}(s)$, are defined as

$$a_{JS}(s) = Hf_s \tag{14}$$

$$b_{JS}(s) = Hf_{ss} + f_s^T Q f_s \tag{15}$$

$$c_{JS}(s) = g(s). \tag{16}$$

Note that (9) and (13) have the same form, and that only the coefficient vectors a, b, and c, are different. We shall write both equations in the form

$$a_i(s)\ddot{s} + b_i(s)\dot{s}^2 + c_i(s) = \tau_i, \qquad i = 1, 2, \cdots, n. \tag{17}$$

The joint actuator torques (or forces) τ_i are assumed to be bounded by constants

$$\tau_i^{\min} \le \tau_i \le \tau_i^{\max}, \qquad i = 1, 2, \cdots, n. \tag{18}$$

Viscous damping effects, and especially the back EMF of the joint actuator motors, can also be included in the dynamic equations (1); such case shall be detailed later in Section III-D.

Given the actuator bounds (18), the maximum and minimum possible values of the parameter $\ddot{s}$ can be determined as functions of s and $\dot{s}$ as

$$\tau_i^{\min} \le a_i(s)\ddot{s} + b_i(s)\dot{s}^2 + c_i(s) \le \tau_i^{\max}, \qquad \text{for all } i = 1, 2, \cdots, n \tag{19}$$

which can be written formally as n sets of constraints on $\ddot{s}$

$$\alpha_i(s, \dot{s}) \le (\ddot{s})_i \le \beta_i(s, \dot{s}) \tag{20}$$

where

$$\alpha_i = (\tau_i^\alpha - b_i\dot{s}^2 - c_i)/a_i \tag{21}$$

$$\beta_i = (\tau_i^\beta - b_i\dot{s}^2 - c_i)/a_i \tag{22}$$

and the values of τ_i^α and τ_i^β depend on the sign of the "inertia" term a_i

if $a_i > 0$, $\quad \tau_i^\alpha = \tau_i^{\min}$ and $\tau_i^\beta = \tau_i^{\max}$

if $a_i < 0$, $\quad \tau_i^\alpha = \tau_i^{\max}$ and $\tau_i^\beta = \tau_i^{\min}$

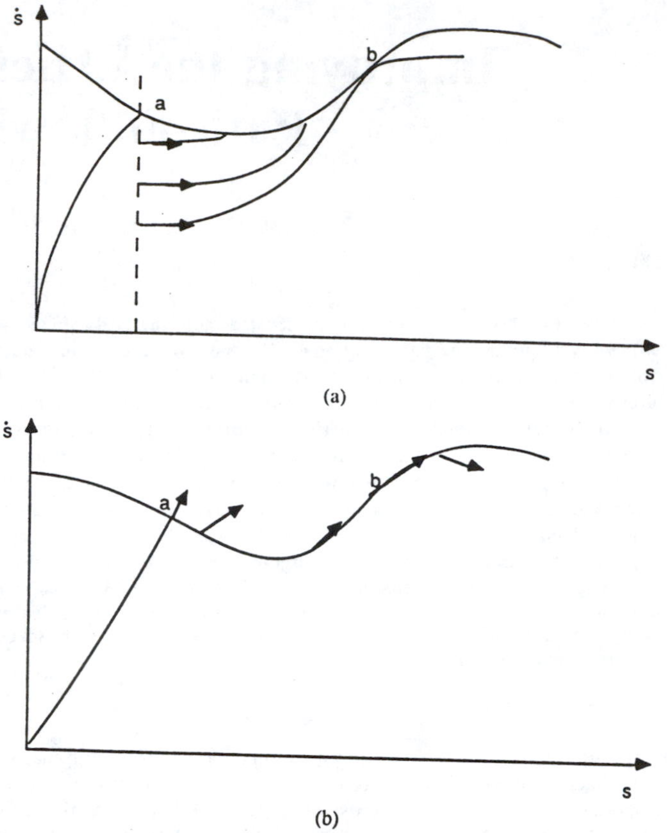

Fig. 1. Existing methods to find switching points (the arrows indicate $d\dot{s}/ds$). (a) The Bobrow algorithm. (b) The Shin and McKay algorithm.

(the case $a_i = 0$ shall be discussed later, in Section III-A). Thus the bounds on $\ddot{s}$ are determined as

$$\alpha(s, \dot{s}) \le \ddot{s} \le \beta(s, \dot{s}) \tag{23}$$

where

$$\alpha(s, \dot{s}) = \max\{\alpha_i(s, \dot{s})\} \qquad \beta(s, \dot{s}) = \min\{\beta_i(s, \dot{s})\}.$$

Bobrow *et al.* [1], [2] and Shin and McKay [13], [14] suggest to find for each value of s the maximum possible value of $\dot{s}$, and draw accordingly the *maximum velocity curve* in the s–$\dot{s}$ phase plane (which corresponds to $\alpha(s, \dot{s}) = \beta(s, \dot{s})$). To find the point where the phase plane trajectory meets the maximum velocity curve, Bobrow *et al.* [1], [2] integrate the equation $\ddot{s} = \beta(s, \dot{s})$ from the initial state until the maximum velocity curve is reached, at some point a (Fig. 1(a)), then drop to some lower velocity on the dotted vertical line, and then integrate the equation $\ddot{s} = \alpha(s, \dot{s})$ forward in time, choosing various initial values of $\dot{s}$ by trial and error until the maximum velocity curve just touches at a single point b, as shown in the figure. Shin and McKay [13], [14] offer a different method. When the phase plane trajectory hits the maximum velocity curve $\dot{s} = g(s)$, their approach is to search along the maximum velocity curve to find a point where

$$\kappa(s) = \frac{d\dot{s}}{ds} - \frac{dg}{ds}$$

changes signs. The term $d\dot{s}/ds$ represents the slope of the phase-plane trajectory at the maximum velocity curve, and is obtained from the values $\ddot{s}$ and $\dot{s}$ at the maximum velocity curve ($d\dot{s}/ds = \ddot{s}/\dot{s}$). This is where the phase plane trajectory meets the maximum velocity curve at a point without violating the actuator constraints, as shown in Fig. 1(b). Pfeiffer and Johanni [7], [8] use the same method with a minor difference in the parametrization of the path.

III. Improving the Efficiency of the Optimal Planning Algorithm

In this section, we show that the points, which we shall refer to as *characteristic switching points,* where the phase plane trajectory just meets the maximum velocity curve without violating the actuator constraints, can be exhaustively classified into three possible types. We call these types the *zero-inertia point,* the *discontinuity point,* and the *tangent point.* We describe how the characteristic switching points can be directly obtained without computing the maximum velocity curve explicitly, and how this property can be exploited to simplify the derivation of the time-optimal solution.

Section III-A finds the switching points where the maximum velocity curve is continuous but not differentiable (zero-inertia points), which correspond to having one of the a_i change signs. Section III-B covers the case where the maximum velocity curve is discontinuous (discontinuity points). Section III-C, which represents the main result of this communication, derives a simple procedure to find the switching points where the maximum velocity curve is continuous and differentiable (tangent points). Inclusion of viscous friction effects and state-dependent actuator bounds is detailed in Section III-D. The resulting algorithm is summarized in Section III-E.

A. Case 1: The Zero-Inertia Point

If, in (17), $a_i(s) = 0$ for some i, then the corresponding terms α_i and β_i of (21) and (22) cannot be defined. In this case the acceleration $\ddot{s}$ at the maximum velocity $\dot{s}_{\max}$ is not uniquely determined. The time-optimal phase-plane trajectory may include this singular point, which is, therefore, a candidate characteristic switching point, as noticed in [7]. We call this case the *zero-inertia point,* since $a_i(s)$ represents an inertia-like term in the parameterized dynamic equation. Zero-inertia points can be found directly from the expression of $a_i(s)$.

B. Case 2: The Discontinuity Point

In this subsection and the next, we shall assume that none of the $a_i(s)$ is zero in the vicinity of the characteristic switching point considered (this case having been treated in the previous subsection). We can further assume, without loss of generality, that the first derivative of the parameterized path function, namely p_s of [1] and [2] or f_s of [13] and [14], is *continuous.* Indeed, if this is not the case at a particular point, then at this point the velocity along the path is necessarily zero (since, physically, the velocity vector cannot be discontinuous), so that the task can be partitioned into two independent optimal control problems.

The *second* derivative p_{ss} (or f_{ss}), however, may be discontinuous. Assume that for a given value of s

$$\alpha_m(s, \dot{s}) \leq \ddot{s} \leq \beta_k(s, \dot{s})$$

which means that joint m gives the maximum of decelerations among α_i's and joint k gives the minimum of accelerations among β_i's. At the maximum velocity limit $\dot{s} = \dot{s}_{\max}$ the condition $\alpha_k = \beta_m$ should be satisfied. Furthermore, for infinitesimally smaller and larger values of s in the vicinity of given s, if any component in the expressions of α_k and β_m changes discontinuously, then the maximum velocity $\dot{s}_{\max}$ is discontinuous in that vicinity. Assuming as we do that none of the $a_i(s)$ is zero, the only component which may be discontinuous in α_k or β_m is p_{ss} (or f_{ss}). Therefore, the maximum velocity curve is discontinuous (Fig. 2) if and only if p_{ss} (or f_{ss}) is discontinuous.

From the analysis just described, the corresponding *discontinuity points* are easy to find without constructing the maximum velocity curve. One must compute the values of s where p_{ss} (or f_{ss}) is discontinuous, then calculate the associated maximum velocity limits. Two values of $\dot{s}$ are obtained for each value of s, the smaller of which defines the possible point where the phase-plane trajectory can touch. For example, if the path is parameterized in task space, and the parameter s is the displacement along the path from the starting point, then the position components x_{ss} of p_{ss} are normal to the path, with the magnitude of x_{ss} being the reciprocal of the radius of the path. In this case, the discontinuity point is where the curvature of the path changes discontinuously.

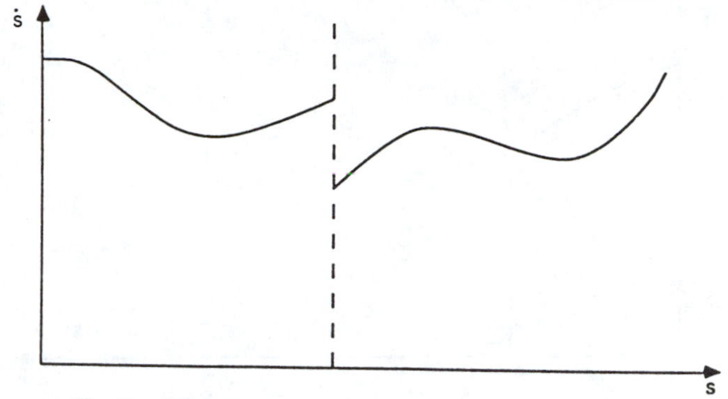

Fig. 2. Maximum velocity curve with a discontinuity point.

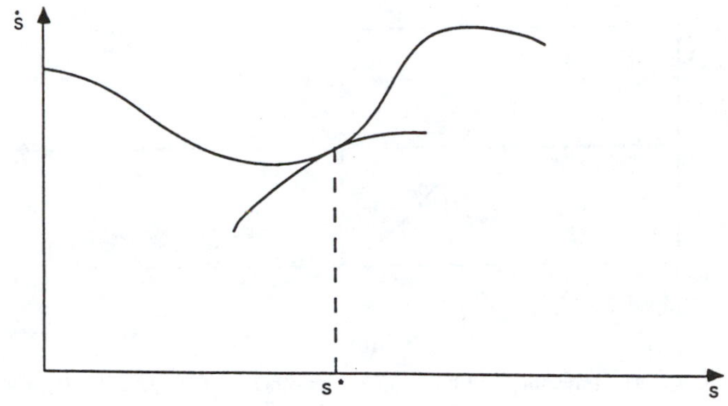

(a)

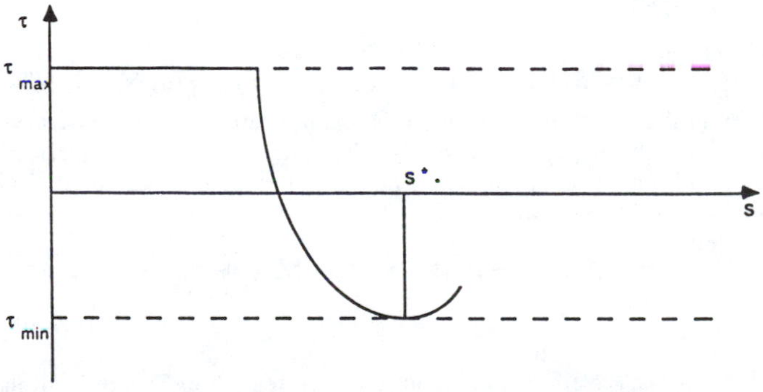

(b)

Fig. 3. The trajectory and the torque meet tangentially. (a) The trajectory meets tangentially. (b) The torque meets tangentially.

C. Case 3: The Tangent Point

Subsection III-A determined the switching points where the maximum velocity curve is continuous but not differentiable (which corresponds to having one of the a_i be zero). Subsection III-B covered the case where the maximum velocity curve is discontinuous. This subsection completes the analysis, by deriving a simple procedure to find the switching points where the maximum velocity curve is both continuous and differentiable.

The smoothness of the maximum velocity curve at the switching point implies that $\ddot{s}$ is continuous in the vicinity of that point, and thus that the phase-plane trajectory must be locally continuous and differentiable. Therefore, if the trajectory hit the maximum velocity curve other than tangentially, then it would enter the region above the maximum velocity curve (Fig. 4(a)). Hence, the trajectory must meet the maximum velocity curve tangentially (Fig. 3(a)). At any time, at least one of the actuators must be saturated [1], [2], and at the switching point there must exist *another* actuator which is also saturated. Specifically, assume that deceleration continues just before and after the switching point, with the kth actuator saturated,

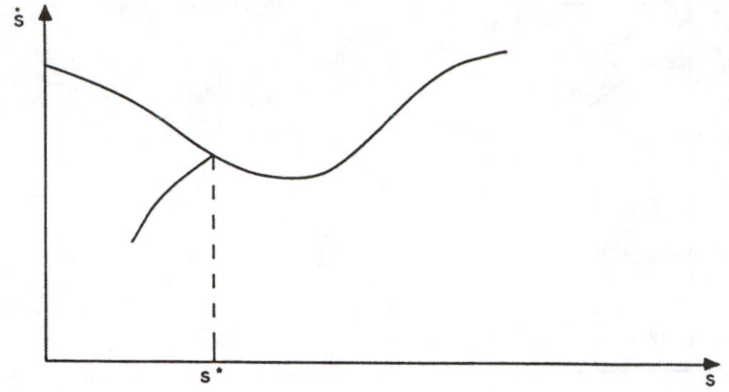

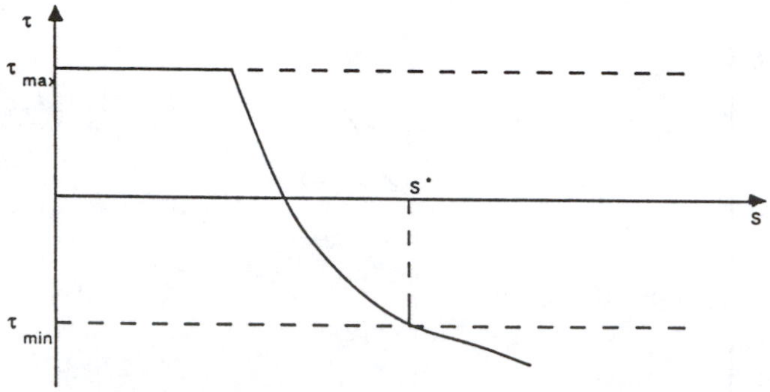

Fig. 4. The trajectory and the torque meet other than tangentially. (a) The trajectory meets other than tangentially. (b) The torque meets other than tangentially.

i.e.,

$$\ddot{s} = \alpha_k(s, \dot{s}) = \{\tau_k^\alpha - b_k(s)\dot{s}^2 - c_k(s)\}/a_k(s) \tag{24}$$

then at the switching point there is another satured torque τ_m ($m \neq k$). Since by using (17) and (24) the torque τ_m can be expressed smoothly in terms of s and $\dot{s}$, which are both continuous and differentiable at the switching point

$$\tau_m = a_m(s)\{\tau_k^\alpha - b_k(s)\dot{s}^2 - c_k(s)\}/a_k + b_m(s)\dot{s}^2 + c_m(s) \tag{25}$$

thus τ_m must itself be continuous and differentiable. Therefore, if its torque trajectory hit the corresponding bound other than tangentially, then τ_m would have to violate its constraint (Fig. 4(b)). Hence the torque τ_m must meet its constraint bound tangentially (Fig. 3(b)).

Based on this analysis, a necessary condition for the point $(s^*, \dot{s}^*)$ in the phase plane to be a tangent point can be derived as follows. Since τ_m must meet the actuator constraint tangentially

$$d\tau_m = 0. \tag{26}$$

Since, from (25), τ_m can be expressed as a function of only s and $\dot{s}$, condition (26) is equivalent to

$$\frac{\partial \tau_m}{\partial s}\dot{s} + \frac{\partial \tau_m}{\partial \dot{s}}\ddot{s} = 0. \tag{27}$$

Using (24), (25), and (27), yields

$$\phi_1(s)\dot{s}^2 + \phi_3(s) = 0 \tag{28}$$

where

$$\phi_1(s) = \frac{d\eta_1}{ds} - 2\eta_1\frac{b_k}{a_k}$$

$$\phi_3(s) = \frac{d\eta_2}{ds} - 2\eta_1\frac{c_k}{a_k} \tag{29}$$

and

$$\eta_1(s) = b_m - b_k a_m/a_k$$

$$\eta_2(s) = c_m - c_k a_m/a_k. \tag{30}$$

Therefore, for each value of s, we can solve (28) for $\dot{s}^*$ (recalling that by definition $\dot{s} \geq 0$), and then check for the value $(s^*, \dot{s}^*)$ whether

$$\tau_m = \tau_m^{\max} \quad (\text{if } a_m > 0) \qquad \text{or} \qquad \tau_m = \tau_m^{\min} \quad (\text{if } a_m < 0)$$

while the other actuator torques remain within their admissible bounds. If this is the case, then $(s^*, \dot{s}^*)$ is a possible tangent point. Note that in (29) the derivative terms d/ds need not be computed explicitly, but rather can be approximated with appropriate accuracy using, e.g., a central differentiation method, since the actuator torque is assumed to be a smooth function of s and $\dot{s}$ in the vicinity of the tangent point. Also, since, by definition, k and m cannot be equal, there are only $n(n-1)/2$ possible combinations of joints for (28). These combinations can be computed in a parallel and independent fashion, so that the computation time may only increase modestly with the number of links.

D. Viscous Friction Effects and State-Dependent Actuator Bounds

When viscous friction effects (such as the back EMF of the motors) are included, the parametrized equations can be expressed as

$$a_i(s)\ddot{s} + b_i(s)\dot{s}^2 + c_i(s) + d_i(s)\dot{s} = 0, \qquad i = 1, \cdots, n$$

instead of (17). More generally, we may include any smooth state-dependent actuator bounds. Assume for instance that the actuator bounds have the following form, which corresponds to a fixed-field dc motor with a bounded input voltage:

$$\tau_i^{\max} = V_i^{\max} + K_i\dot{q}_i = V_i^{\max} + K_i\sigma(s)\dot{s}$$

$$\tau_i^{\min} = V_i^{\min} + K_i\dot{q}_i = V_i^{\min} + K_i\sigma(s)\dot{s}$$

where $V_i^{\max}$ and $V_i^{\min}$ are (scaled) input voltage bounds, K_i is a constant coefficient, and

$$\sigma(s) = r_q^{-1}p_s \qquad (\text{in task space})$$

$$= f_s \qquad (\text{in joint space}).$$

The tangentiality condition (26) is then modified as

$$d\tau_m = d\tau_m^\alpha$$

and the necessary condition (28) becomes accordingly

$$\phi_1(s)\dot{s}^3 + \phi_2(s)\dot{s}^2 + \phi_3(s)\dot{s} + \phi_4(s) = 0 \tag{31}$$

where

$$\phi_1(s) = \frac{d\eta_1}{ds} - 2\eta_1\frac{b_k}{a_k}$$

$$\phi_2(s) = \frac{d\eta_3}{ds} - K_m\frac{d\sigma}{ds} + 2\eta_1\left(K_k\frac{\sigma}{a_k} - \frac{d_k}{a_k}\right) - \frac{b_k}{a_k}(\eta_3 - K_m\sigma)$$

$$\phi_3(s) = \frac{d\eta_2}{ds} + 2\eta_1\left(\frac{V_k^\alpha}{a_k} - \frac{c_k}{a_k}\right) + (\eta_3 - K_m\sigma)\left(\frac{K_k\sigma - d_k}{a_k}\right)$$

$$\phi_4(s) = (\eta_3 - K_m\sigma)\left(\frac{V_k^\alpha - c_k}{a_k}\right)$$

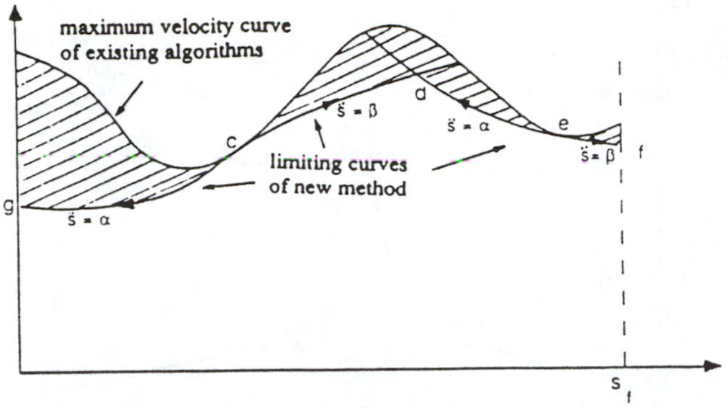

Fig. 5. The procedure to get the limit curve and superposition of the maximum velocity curve.

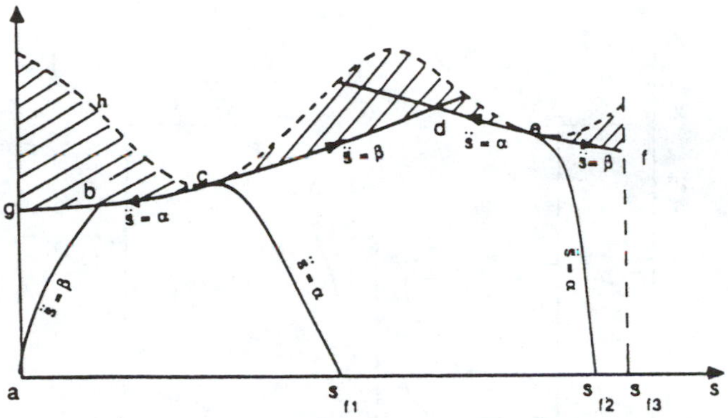

Fig. 6. The complete optimal trajectory with superposition of the maximum velocity curve.

and

$$\eta_1(s) = b_m - b_k a_m / a_k$$

$$\eta_2(s) = V_k^\alpha a_m / a_k + c_m - c_k a_m / a_k$$

$$\eta_3(s) = K_k \sigma a_m / a_k + d_m - d_k a_m / a_k.$$

In this case, as remarked in [14], an exclusion "island" may occur in the phase plane, which in our formalism simply corresponds to the fact that for each s, (31) may have multiple positive solutions in $\dot{s}$. The necessary condition developed above must hold at the boundary of the island.

E. Summary

We have described how to find all the possible points where the phase-plane trajectory can meet the maximum velocity curve (or exclusion islands) in the time-optimal solution. These three types of points are uniquely determined from the given path, the manipulator dynamics, and the actuator constraints, and are independent of the initial and final positions and velocities along the path. The conditions determining the characteristic switching points are expressed as functions of only the single parameter s. This allows us to avoid the exhaustive searches of the earlier methods over the whole ranges of s and $\dot{s}$. The new algorithm can be summarized as follows:

1) Using the methods explained in Sections III-A, -B, and -C, obtain all the candidate characteristic switching points by searching once over the values of s. The maximum velocity curve need not be found explicitly.

2) From those points, integrate $\ddot{s} = \alpha(s, \dot{s})$ backward in time, and $\ddot{s} = \beta(s, \dot{s})$ forward in time, until the phase-plane trajectory hits the s axis, the $\dot{s}$ axis, or the $s = s_f$ line, or until one of the actuator constraints is violated. If, from a candidate characteristic switching point, one cannot integrate both forward and backward without violating the actuator constraints, then such point should be discarded. Then, the admissible region for the phase-plane trajectory is under resulting *limit curves*. The procedure is described graphically in Fig. 5. The maximum velocity curve is also shown, in order to help understand how the new algorithm differs from the earlier methods. Note that the shaded region in Fig. 5 is inadmissible. This means that once the phase-plane trajectory gets into the shaded region, then later in the trajectory it cannot get out of the region without hitting the maximum velocity curve other than tangentially (in other words, without violating at least one of the actuator bounds). Therefore, the admissible region is under these *limit curves*, and not merely under the *maximum velocity curve* used in the earlier methods.

3) The characteristic switching points (points c and e in Fig. 6) have been found in steps 1) and 2). In step 3), we get the rest of the switching points as follows (Fig. 6). If the final point is between 0 and s_{f1}, then there is only one switching point between a and b, which is the point where the deceleration trajectory, integrated backward in time from the final point, meets the acceleration trajectory, integrated forward in time from the starting point. If the final point is between s_{f1} and s_{f2}, then there are three switching points, namely, b, c, and some point between c and d, where the deceleration trajectory from the final point hits. Similarly, if the final point is between s_{f2} and s_{f3}, then there are five switching points, namely, b, c, d, e, and some point between e and f. By repeating the above reasoning, all switching points can be obtained. Note that this implies that the number of switching points must be odd.

4) If the condition for a tangent point is satisfied over a some finite interval, then the phase-plane trajectory coincides with the maximum velocity curve along a finite arc.

IV. Numerical Example

The efficiency of the new algorithm is demonstrated on a hypothetical planar two-link manipulator. Each link has unit mass and unit length, and the actuator torques τ_1 and τ_2 are limited in absolute value to 30 and 10 N · m. The parametrized equations for these two paths are

$$x = s + 0.5 \qquad y = 4(s - 0.5)^3.$$

The result of the algorithm is shown in Fig. 7. Fig. 7(a) is the limit curve which is obtained by the above method. The maximum velocity curve from the earlier method is also drawn with the limit curve in Fig. 7(b), in order to show the consistency between the two methods. Fig. 7(c) represents the complete time-optimal phase plane trajectory. If s_{max} is between 0 and 0.737, then there is a single switching point. If s_{max} is between 0.737 and 0.944, then there are three switching points. If s_{max} is between 0.944 and 1, then there are five switching points. Fig. 7(d) gives the corresponding actuator histories.

This algorithm owes it efficiency to the fact that it does not need to compute the maximum velocity curve, and that all the characteristic switching points can be obtained by searching just once over the value of s using a simple method, while earlier approaches need systematic searches to find a switching point with the help of the maximum velocity curve, and should repeat them as long as there are more switching points ahead. An exact efficiency comparison between the earlier methods and this new algorithm is difficult, since, for instance, the step sizes in s and $\dot{s}$ which are used to get the maximum velocity curve largely affect the computation time of Bobrow's algorithm, as do other implementation choices in all methods (such as dichotomic searches in Bobrow's algorithm). In the implementation of the above numerical examples, however, we found that for this simple two-link manipulator case, our new algorithm is more efficient than Bobrow's method roughly by a factor of 50 or more. Furthermore, we expect the relative efficiency of our algorithm to increase with the number of links, since the algorithm does not require iterative searches.

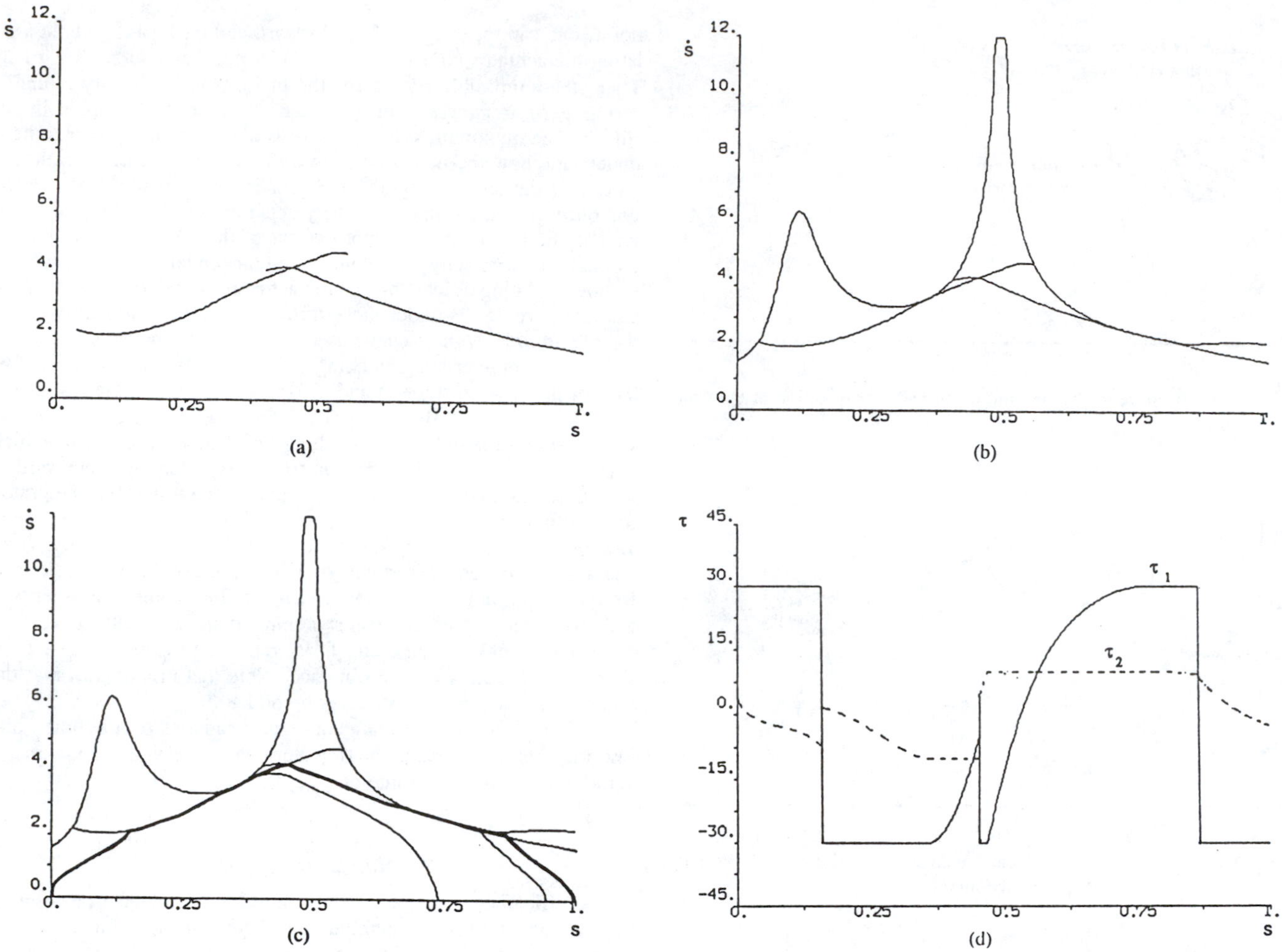

Fig. 7. Simulation results. (a) The limiting curves. (b) The limiting curves with superposition of the maximum velocity curve. (c) The complete optimal trajectory. (d) The corresponding optimal torque history.

V. Concluding Remarks

In this communication, an improved solution concept to address the time-optimal path-following problem for robot manipulators is presented and a new efficient trajectory planning algorithm is proposed. Characteristic switching points are defined and are exhaustively classified into three types. A simple method is developed to find such points. Limiting curves are introduced according to the characteristic switching points, and they set a new admissible region for the phase-plane trajectory, which differs from the admissible region defined by the maximum velocity curve of the existing methods. Numerical examples show the consistency of the new method with the existing methods, while demonstrating significant improvements in computational efficiency.

Solutions of more complex time-optimal problems, such as finding a minimum-time path for given end states or taking obstacles into account, and issues of robustness to model uncertainty, are important further research topics to be addressed. The analytic tools presented here are likely to contribute to research on those complex problems as well. Sahar and Hollerbach [10] simplify the global time-optimal path search by using dynamic time scaling and joint tesselation, but their approach is still very intensive computationally. Other global time-optimal studies such as Dubowsky, Norris, and Shiller [3], Shiller [12], and Rajan [9] use the prescribed path algorithm as a major component. Besides greatly simplifying the prescribed path algorithm itself, the necessary conditions determining the characteristic switching points, developed in Section III, must be satisfied for any globally time-optimal path as well. Thus they may allow us to analytically investigate the behavior of the characteristic switching points as the path changes, and therefore provide stronger tools to address the global time-optimal problem.

References

[1] E. Bobrow, S. Dubowsky, and J. S. Gibson, "On the optimal control of robotic manipulators with actuator constraints," presented at the American Control Conf., San Francisco, CA, 1983.

[2] ——, "Time-optimal control of robotic manipulators along specified paths," *Int. J. Robotics Res.*, vol. 4, no. 3, 1985.

[3] S. Dubowsky, M. A. Norris, and Z. Shiller, "Time optimal trajectory planning for robotic manipulators with obstacle avoidance: A CAD approach," presented at the IEEE Conf. on Robotics and Automation, San Francisco, CA, 1986.

[4] M. E. Kahn and B. Roth, "The near-minimum-time-control of open-loop articulated kinematic chains," *J. Dyn. Syst., Meas., Contr.*, vol. 93, 1971.

[5] C. S. Lin, P. R. Chang, and J. Y. S. Luh, "Formulation and optimization of cubic polynomial trajectories for industrial robots," *IEEE Trans. Automat. Contr.*, vol. AC-28, 1983.

[6] W. S. Newman, "High-speed robot control in complex environments," Ph.D. dissertation, M.I.T., 1987.

[7] F. Pfeiffer and R. Johanni, "A concept for manipulator trajectory planning," presented at the IEEE Conf. on Robotics and Automation, San Francisco, CA, 1986.

[8] ——, "A concept for manipulator trajectory planning," *IEEE J. Robotics Automat.*, vol. RA-3, no. 2, 1987.

[9] V. T. Rajan, "Minimum time trajectory planning," presented at the IEEE Conf. on Robotics and Automation, St. Louis, MO, 1985.

[10] G. Sahar and J. M. Hollerbach, "Planning of minimum-time trajectories for robot arms," *Int. J. Robotics Res.*, vol. 5, no. 3, 1986.

[11] V. Scheinman and B. Roth, "On the optimal selection and placement of manipulators," in *Theory and Practice of Robots and Manipulators, RoManSy '84* (Proc. 5th CISM-IFToMM Symp.), 1985.

[12] Z. Shiller, "Time-optimal motion of robotic manipulators," Ph.D. dissertation, M.I.T., 1987.

[13] K. G. Shin and N. D. McKay, "An efficient robot arm control under geometric path constraints," presented at the IEEE Conf. on Decision and Control, San Francisco, CA, 1983.

[14] ——, "Minimum-time control of robotic manipulators with geometric path constraints," *IEEE Trans. Automat. Contr.*, vol. AC- 25, 1985.

[15] S. Singh and M. C. Leu, "Optimal trajectory planning and control of robotic manipulators," in *Proc. Japan-U.S.A. Symp. on Flexible Automation*, 1986.

[16] J. T. Wen and A. Desrochers, "Sub-time-optimal control strategies for robotic manipulators," presented at the IEEE Conf. on Robotics and Automation, San Francisco, CA, 1986.

TIME-OPTIMAL CONTROL OF MANIPULATORS

Eduardo D. Sontag[*] and *Hector J. Sussmann*[*]

Department of Mathematics
Rutgers, The State University
New Brunswick, NJ 09803

ABSTRACT

This paper studies time-optimal control questions for a certain class of nonlinear systems. This class includes a large number of mechanical systems, in particular, rigid robotic manipulators with torque constraints. As nonlinear systems, these systems have many properties that are false for generic systems of the same dimensions.

1. Introduction.

A recent paper by the authors ([SS]) established a number of results about the time optimal control problem for the two link rigid rotational manipulator model described in Paul's book ([PA], equations 6.16 and 6.20). It became evident later that many of the results could in fact be derived just from the general form of the equations of a manipulator, and in fact that a great deal of these results apply to (rigid) manipulators with more than two links as well as to a rather general class of nonlinear systems. This class of systems, which we shall call *mechanical systems with full control*, is characterized by the fact that, as with many Lagràngian formulations of mechanical systems, the evolution equations arise from a nonsingular set of second order differential equations (an Euler-Lagrange equation, typically) for a set of variables ("positions"), and are such that derivatives of these variables ("velocities") appear only quadratically. (Alternatively, a Hamiltonian formulation is also possible.) This note will establish a number of basic optimal control results for mechanical systems, and then apply them to a two-link manipulator model. We view this work as only a (small) first step towards the understanding of this class of systems.

Of course, we do not mean to imply that our "mechanical systems" encompass all possible models of control systems in mechanics. For instance, certain types of frictional effects cannot be included in such models.

Mechanical systems -in the sense of this paper- constitute a very restrictive class of systems when viewed in the context of general nonlinear systems. In particular, the Lie algebra of vector fields associated to such a system must satisfy a large number of nongeneric relations. Since the structure of this Lie algebra characterizes most interesting optimal control properties, one can expect, and indeed one finds, many properties of the time optimal problem for these systems which are false in general for nonlinear systems of the same order. For instance, it is almost trivial to establish that if all controls except possibly one are singular along an extremal, then the remaining control cannot be singular, and if fact must be bang-bang.

Most of the results that we obtain are about the singular structure of the optimal control problem, rather than about optimal controls themselves. The study of the singular structure of the problem is of great interest in itself, for the following reason. One of the main techniques used in practical robotic control consists in dividing the design effort into two stages: (1) find an open-loop control which achieves the desired state transfer, and (2) linearize along the resulting trajectory, and use a linear controller to regulate deviations from this trajectory. The essential point is that this last step will typically depend on controllability of the obtained linearization (as a time-varying linear system), and *a trajectory is singular precisely when this linearization is uncontrollable.* Thus, our characterizations of singular trajectories should help in determining if a trajectory suggested by step (1) is suitable for step (2).

The literature in (numerical) optimal control of manipulators is rather extensive; see for instance the papers [RA], [SD], and [SH], as well as the references there and other papers in the conference volume in which they appear. As far as we are aware, a systematic study of singularities as the one started here has not been attempted in previous work. We intend to direct further research both to theoretical topics and to the understanding of what implications our results have for the algorithms given in the literature. For instance, they may help in the "pruning" of possibilities in dynamic programing numerical methods.

[*]Research supported in part by US Air Force Grant 85-0247 (Sontag) and by NSF Grant MCS78-02442-03 (Sussmann)

Reprinted from *Proc. IEEE ROBOT '86*, San Francisco, CA, pp. 1692–1697, April 1986.

The only other theoretical work on this problem that we are aware of is that in [AL]. That paper is devoted mainly to the proof of existence of optimal controls (basically, one needs to establish that there are no finite escape times), but the last section proves that for a two link manipulator both controls cannot be simultaneously singular. Note that the result proved here, mentioned earlier, will imply a much stronger fact, namely, that one of the controls has only finitely many switchings. (The proof given in [AL] does not rule out a phenomenon like "Fuller's problem", which occurs in many systems, and in which optimal controls switch infinitely often in a bounded interval, nor for that matter even more pathological behaviors, like Cantor sets of discontinuities for the switching function.)

We shall not repeat most of the material in [SS], which should be consulted for further and more detailed discussions. The model in that reference was, as mentioned earlier, that in Paul's textbook. We have since noticed that there seems to be an error in the Lagrangian derivations in the book, which results in an extra term which cannot be disregarded if both links are moving at high speed. This means that a few of the statements in [SS], while true for the system studied in [PA], are not necessarily true for a two-link manipulator, -specifically, lemmas 4.5 and 5.4, and formulas 4.2-4.5,- though they can be in most cases modified in trivial ways. The theorems in [SS] are still valid (with signs interchanged in 6.1), using the corrected formulas, and in fact will be established in more generality in this note.

2. Mechanical systems with full control.

All vectors will be column vectors, but for printing purposes we shall often display them as rows.

A *(finite dimensional) mechanical system (with full control)* will be, for the purposes of this paper, a system defined by equations (omitting the time arguments for simplicity):

$$u = M(\theta)\ddot{\theta} + N(\theta,\dot{\theta}) ,$$

where θ is a vector (of positions) in $\Re^n$, u is a vector (of controls) in $\Re^n$, and where M is an n×n matrix of functions of θ, symmetric positive definite for each $\theta\in\Re^n$, and N is an n-vector of functions of θ and $\dot{\theta}$ with the property that, as functions of $\dot{\theta}$, each of its entries is quadratic, i.e. is a polynomial of degree at most 2. (In the robotics literature, N is usually displayed as a sum of two terms, N+Q, the first homogeneous of degree 2 in $\dot{\theta}$ and the second independent of $\dot{\theta}$; we are allowing also for linear terms in the velocities.) The entries u_i are bounded in magnitude:

$$L_i \le u_i \le M_i \ , \ i=1,\cdots,n,$$

where the $L_i < M_i$ are given constants. As a function of t, each u_i is measurable essentially bounded. We assume that all the functions of θ and $\dot{\theta}$ that appear are real-analytic (in most applications, functions belong to finitely generated algebras spanned by trigonometric functions and polynomials).

This model includes mechanical manipulators with rigid rotational links as well as many other systems of interest. (The "full control" qualifier refers to the fact that every degree of freedom can be independently controlled; certain lumped models for flexible arms, as well as models that include actuator dynamics, result in very similar equations but without this latter property.) We take the positions θ as belonging to $\Re^n$ rather than to a subset -or even a manifold like S^1, as is natural for some robotics problems- for notational simplicity; in any case, all of the results obtained depend on local methods.

The state space model associated to the above is given by equations

$$\dot{x} = f(x) + G(x)u \ , \qquad (2.1)$$

where x is a vector in $\Re^{2n}$; denoting the first n entries of x as θ and last n entries as $\dot{\theta}$, f(x) is the 2n-vector

$$\begin{pmatrix} \dot{\theta} \\ -M(\theta)^{-1}N(\theta,\dot{\theta}) \end{pmatrix}$$

(thus as functions of $\dot{\theta}$ the last n coordinates of f are polynomials of degree at most 2) and

$$G(x) = \begin{pmatrix} 0 \\ L(\theta) \end{pmatrix},$$

with $L(\theta)$ being the inverse of $M(\theta)$ -hence also symmetric positive definite for each x.

Such systems are "linearizable under feedback", in the sense that the the transformation

$$u = N(\theta,\dot{\theta}) + M(\theta)v \ ,$$

where v is a new control, results in a set of decoupled double integrators. This transformation is typically used in control ("computed torque method", etc.); however in optimal control it does not seem to be useful, since the torque constraints get transformed into state-dependent constraints for the linear problem. A discussion is given in [SS] showing that full-control mechanical systems may have very different behavior than double integrators, in terms of certain degeneracies that may appear in optimal problems.

For each i=1,⋯,n, we shall let g_i denote the ith column of G; its entries are all functions of the first n coordinates θ of x, and are nonzero for every x (positive definiteness of L).

2.1. Basic Lie theoretic properties.

We shall identify functions $\Re^{2n}\to\Re^{2n}$ with vector fields on $\Re^{2n}$, and apply differential geometric notation. In particular, $ad_a(b) = [a,b]$ denotes the Lie bracket of the vector fields a and b, i.e. b'a-a'b, where prime indicates Jacobian. When we refer to the "ith coordinate" of a vector field, we mean coordinates with respect to above identification.

We denote by $\mathcal{G}$ the module over $C^\omega(\Re^{2n})$ generated by the vector fields $g_1,\cdots,g_n$, and by $\mathcal{L}$ the Lie algebra of vector fields generated by $\{f,g_1,\cdots,g_n\}$, , the vector fields appearing in the model (2.1); we also use the shorthand notation:

$$X = X_1X_2\cdots X_k := [X_1,[X_2,[\cdots,[X_{k-1},X_k]\cdots]]] \tag{2.2}$$

for iterated Lie compositions of vector fields. Note that $X^kY = \mathrm{ad}_X^k Y$ with this notation. By an *iterated bracket* we shall mean an expression as above for which each X_i is one of f, g_1, ..., g_n. (We also say, abusing terminology, that the vector field X is an iterated bracket.) Given one such $X = X_1X_2\cdots X_k$, the *index* $\iota(X)$ of X is defined as the difference $n_f - n_g$, where n_f is the number of times that the vector field f appears in the expression, and n_g is the total number of g_i's, counting multiplicities.

With these notations, $fg_i = [f,g_i]$, $i=1,\cdots,n$. Because of the form of f and of the vector fields g_i, it follows that the matrix $(fg_1,\cdots,fg_n)$ has the form

$$-\binom{L(\theta)}{\cdot},$$

from which it follows that the set of vector fields

$$\{g_1,\cdots,g_n,fg_1,\cdots,fg_n\} \text{ constitutes a frame,} \tag{2.3}$$

i.e. is everywhere linearly independent, and hence generates (as a module over analytic functions) the set of all analytic vector fields on the state-space $\Re^{2n}$.

Let Z_i denote the set of all functions on R^{2n} which, as functions of θ_1, ..., θ_n, are polynomials of degree at most i. (With the convention that $Z_i = \{0\}$ if $i<0$.)

Let S_i be the set of vector fields X with the property that the first n coordinates of X are in Z_i, and the last n coordinates are in Z_{i+1}. (So, for instance, vectors in S_{-1} have their first n coordinates zero and last n coordinates dependent only on θ, and hence belong to $\mathcal{G}$.) If X is in S_i and Y is in S_j, then [X,Y] belongs to S_{i+j}. The sets S_i provide a filtration of $\mathcal{L}$. In fact, since f is in S_1 and all g_i are in S_{-1}, by induction on the formation of iterated brackets it follows that:

Lemma 2.1: For any iterated bracket X, $X \in S_{\iota(X)}$■

In particular, for all i,j,

$$[g_i,g_j] = 0$$
$$g_ifg_j \in \mathcal{G}. \tag{2.4}$$

3. Singular extremals.

An *extremal* on an interval $I = [0,T]$ is given by functions (x,λ,u) on I which satisfy (2.1) together with

$$\dot\lambda = D[f(x)+G(x)u]\lambda \tag{3.1}$$

(where D[...] denotes transpose Jacobian), $\lambda\neq0$ everywhere, u is measurable essentially bounded, and, for each $i=1,\cdots,n$ and for almost every t where the *i-th switching function*

$$\phi_i(t) := <\lambda(t),g_i(x((t))>$$

is nonzero,

$$u_i(t) = \begin{cases} M_i & if\ \phi_i(t)>0 \\ L_i & if\ \phi_i(t)<0 \end{cases}.$$

An extremal is *admissible* if $L_i\le u_i(t)\le M_i$ for each i and almost all t. A *u_i-singular extremal* is one for which ϕ_i is identically zero; a singular extremal is one that is u_i-singular for some i.

From the maximum principle, it follows that, if $\{(x(t),u(t)),\ t\in[0,T]\}$ is a time-optimal trajectory then there exists a λ such that (x,λ,u) is an extremal. If the set of zeroes of ϕ_i is *finite* along such an extremal, then necessarily u_i is a *bang-bang control*, i.e., it equals a.e. a function that is piecewise constant, with values equal to either L_i or M_i in each of finitely many subintervals.

We fix an extremal and an $i=1,\cdots,n$. Consider the switching function ϕ_i; its derivative is

$$\begin{aligned}\phi_i' &= <\lambda,[f,g_i]> + \textstyle\sum_{j=1}^{n} u_j<\lambda,[g_j,g_i]> \\ &= <\lambda,[f,g_i]>.\end{aligned} \tag{3.2}$$

In particular, the derivative exists everywhere, and is itself absolutely continuous. (Note that we are abusing notation, in that the precise meaning of the above is as follows: $\phi_i'(t) = <\lambda(t),[f,g_i](x(t))>$. We shall often omit the arguments t, or x(t), for notational simplicity.)

Let J_i [resp., J_i'] be the set of limit points of zeroes of ϕ_i [resp., of ϕ_i']. Since ϕ_i is continuously differentiable, J_i is contained in J_i'. Thus, at points in J_i the equations

$$<\lambda(t),g_i(x(t))> = <\lambda(t),[f,g_i](x(t))> = 0 \tag{3.3}$$

hold. If t is a limit point of all ϕ_i, then these equations hold for all i, contradicting the facts that λ is always nonzero and that (2.3) holds. We have then established:

Lemma 3.1: $J_1 \cap J_2 \cap \cdots \cap J_n$ is empty.■

Thus, if n-1 of these sets equal all of I, the remaining one, say J_i, must be empty, i.e. the set of zeroes of ϕ_i is finite. This says that not all controls may be simultaneously singular, and more precisely:

Corollary 3.2: If the extremal (x,λ,u) is u_j-singular for all $j\neq i$, then u_i is bang-bang.■

This result suggests the study of extremals that are singular for n-1 of the controls. We now consider this question in some more detail, and then specialize to the case n=2. There are of course many other cases of interest (e.g., n-2 controls are singular and the rest are bang-bang), but they will not be studied in this note.

3.1. A generic situation.

By the second formula in (2.4), there exist analytic functions $\{\alpha_{ijk},\ i,j,k = 1,\cdots,n\}$ such that, for each $i,j = 1,\cdots,n$:

$$g_ifg_j = \textstyle\sum_{k=1}^{n} \alpha_{ijk}g_k .$$

These coefficients can be computed explicitely, as follows. Let v be any vector field whose first n coordinates vanish identically, and let w be the vector function obtained from the last n coordinates. Then, v can be expressed as $\sum_{k=1}^{n} \alpha_k g_k$, where α_k is the k-th entry of the n-vector $M(\theta)w$. When $v = g_i f g_j$, which is in S_{-1}, the obtained coefficients are in fact functions of θ alone.

Assume an extremal (x,λ,u) has been fixed. Taking a further derivative in (3.2) results for each i in:

$$\phi_i'' = \langle\lambda, ffg_i\rangle + \sum_{k=1}^{n} \beta_{ik}\phi_k , \tag{3.4}$$

where for each i,k:

$$\beta_{ik}(t) = \sum_{j=1}^{n} \alpha_{ijk}(x(t))u_j(t) .$$

This formula holds under no assumptions of singularity whatsoever. Assume now, however, that the extremal in question is u_i-singular for all $i \neq k$, for some given k. This means that ϕ_i vanishes identically for $i \neq k$, and hence (3.4) reduces to:

$$\phi_i'' = \langle\lambda, ffg_i\rangle + \beta_{ik}\phi_k . \tag{3.5}$$

Thus, for every $i \neq k$, the equations in (3.3) hold at all t, as well as the almost everywhere vanishing of ϕ_i'' for such i. Assume that, at some t_o in the interval of definition of the extremal, $\phi_k(t_o) = 0$. In that case,

$$\langle\lambda(t_o), ffg_i(x(t_o))\rangle = 0$$

for all $i \neq k$. (The precise argument is as follows: the coefficient $\beta_{ik}(t)$ is essentially bounded, and $\phi_k \to 0$ as t approaches t_o. Thus the last term in (3.5) approaches 0 except at most along a set of measure zero. Since the first term is continuous, it must be zero at t_o.) Consider, for the given k, the set of vector fields

$$\{g_i, i=1,\cdots,n\} \cup \{fg_i, i=1,\cdots,n, i\neq k\} \cup \{ffg_i, i=1,\cdots,n, i\neq k\} ,$$

and let S_k be the set of states x at which these span the entire (2n-dimensional) tangent space. This set S_k is an open set, and since the vector fields are all analytic it is in fact open dense, provided only that it be nonempty. Since there are 3n-2 vectors in $\Re^{2n}$, one may expect that it is indeed nonempty (assuming $n \geq 2$), and this does happen in the 2-link manipulator example discussed later. The above arguments establish, by contradiction, the following fact:

Theorem 3.1: If (x,λ,u) is an u_i-singular extremal for all $i \neq k$ and $x(t)$ remains in S_k for all t, then u_k is constant (equal to L_k or M_k).∎

(Of course, as with all statements in optimal control, "constant" here means equal to a constant almost everywhere.) Later, we shall see how one may sometimes determine whether u_k is equal to L_k or M_k, based on higher order conditions.

For each k=1,⋯,n, let A_k be the $(n-1)\times(n-1)$ matrix (α_{ijk}), where i and j each take the values $1,\cdots,k-1,k+1,\cdots,n$. (That is, delete the k-th row and column of (α_{ijk}), seen for fixed k as an $n\times n$ matrix.) Let

$$\Delta_k := \det(A_k) .$$

This is again an analytic function of x. Finally, let

$$R_k := S_k \cap \{x \mid \Delta_k \neq 0\} .$$

This set is either empty or open dense. If an extremal as in the previous theorem is such that $x(t)$ in fact remains in R_k, then not only is u_k constant, but we may determine the remaining controls. Fix one such extremal, and assume $u \equiv c$ constant (one of the above two values). The set of simultaneous equations $\{\phi_i''=0, i\neq k\}$, is by (3.5) equivalent to the following matrix equation:

$$A_k(x(t))\omega_k(t) = \psi_k(t) , \tag{3.6}$$

where ω_k is the column (n-1)-vector $(u_1,\cdots,\hat{u}_k,\cdots,u_n)'$ (we use the "^" to indicate a missing element), and where

$$\psi = (\psi_{k1},\cdots,\hat{\psi}_{kk},\cdots,\psi_{kn})'$$

and for each $i \neq k$,

$$\psi_{ki} := \langle\lambda(t), ffg_i(x(t))\rangle/\phi_k(t) - \alpha_{ikk}c .$$

(Recall that ϕ_k is always nonzero along this type of extremal.) If $x(t)$ remains in R_k, then we can solve (3.6) for ω_k as an analytic function of $\lambda(t)$ and $x(t)$. We may substitute the obtained expressions for $u_i(t)$, $i\neq k$, as well as $u_k \equiv c$, into the system equation (2.1) and the adjoint equation (3.1). These two together become a system of 2n ordinary differential equations with analytic right-hand side (there are no controls u left), and the solutions, that is both $\lambda(t)$ and $x(t)$, are analytic functions of time. Since ω_k was expressed as an analytic function of them, we also conclude:

Theorem 3.2: If (x,λ,u) is an u_i-singular extremal for all $i\neq k$ and $x(t)$ remains in R_k for all t, then all controls u_i are analytic as functions of time. (And can in fact be computed in the above way.)∎

Note that the conditions $\langle\lambda, g_i\rangle \equiv \langle\lambda, fg_i\rangle \equiv 0$ for $i\neq k$ result in 2n-2 independent constraints, by (2.3). So the costate λ depends globally on only two parameters, and it is possible to give thus a 2-dimensional equation for these parameters. We discuss this in more detail below when treating the 2-link manipulator case.

3.2. A degenerate case.

It may happen that Δ_k is identically zero, so that R_k is empty and the above theorem doesn't provide any information. On sets where the rank of A_k is constant, it is possible to provide some results, using pseudoinverses instead of inverses. In particular, assume that there is a row, say the i-th, of some A_k, which is identically zero. For this particular i, then, (3.5) says that the equation

$$\phi_i'' = \langle\lambda, ffg_i\rangle \equiv 0$$

must also hold. There are now 2n-1 conditions for λ, and if independent these determine λ up to a constant multiple (and hence essentially uniquely as a multiplier). If the rank of A_k is constantly n-2, one also can determine by an argument as above the controls u_j, $j \neq i,k$ as feedback functions of the state alone. However, also the condition $\phi_i''' \equiv 0$ must then hold, and this may result in yet another equation for λ, inconsistent with the previous 2n-1. This case appears in the two-link manipulator discussed next.

4. A manipulator example.

We computed explicitely with the 2-link (n=2) model given in [SH], with the numerical parameters provided in their Figure 1. With the use of MACSYMA, we deduced the following facts:

$$\alpha_{ij1} \equiv 0 \quad \text{for all i,j (so } A_1 \equiv 0), \qquad (4.1)$$
$$\Delta_2 = A_2 = \alpha_{112} = \gamma(\theta_2)\sin 2\theta_2,$$
$$S_2 = \{x | \ \beta(\theta_2)(\dot{\theta}_1+\dot{\theta}_2)\sin\theta_2 = 0\},$$

where γ and β are functions which are nonzero, (in fact, always negative,) for all θ_2. Note also that

$$g_2 ffg_2 = fg_2 fg_2 = L_f(\alpha_{222})g_2 + \alpha_{222} fg_2 \ . \qquad (4.2)$$

4.1. Second control singular.

We first consider the case corresponding to k=1 in the discussion in the previous section. Thus fix any extremal which is u_2-singular. We calculate the third derivative of ϕ_2, and obtain:

$$\phi_2''' = \langle\lambda, fffg_2\rangle + u_1\langle\lambda, g_1 ffg_2\rangle + u_2\langle\lambda, g_2 ffg_2\rangle \ .$$

By (4.2), the last term is a linear combination of ϕ_2 and ϕ_2'', and hence vanishes by singularity. The control u_1 must be constant. Let B be the set in which the vectors

$$\{g_2, fg_2, ffg_2, fffg_2 + cg_1 ffg_2\}$$

are linearly independent, for $c=L_i$ and $c=M_i$. This is open, and a calculation shows that it is nonempty; thus:

Theorem 4.1: There are no v-singular extremals for which x(t) intersects the open dense set B.■

States with $\theta=0$ are especially interesting. The intersection of B with the set of such states is still nonempty (hence, open dense in that subset).

4.2. First control singular.

Consider now the case k=2. From the calculations in (4.1) it follows that there is an easy geometric characterization of R_2:

$$R_2 = \{x \mid \theta_2 \neq k\pi/2 \text{ and } \dot{\theta}_1+\dot{\theta}_2 \neq 0\} \ .$$

There, u_2 is constant=c (=M_2 or L_2), and u_1 is analytic, as discussed earlier. We now provide some details of the way u_1 is computed.

Note that g_1 has the form $(0,0,\mu,\nu)'$, where -by positive definiteness of M- μ is everywhere nonzero (positive). Correspondingly, fg_1 has the form $(-\mu,-\nu,0,0)'$. The vectors

$$a := (-\nu/\mu, 1, 0, 0)' \quad \text{and}$$
$$b := (0, 0, -\nu/\mu, 1)'$$

are orthogonal to fg_1 and g_1 respectively. Since λ is also orthogonal to these vectors along a u_1-singular trajectory, it follows that λ is a combination of the two independent vectors a,b, i.e.:

$$\lambda(t) = \lambda_2(t)a(x(t)) + \lambda_4(t)b(x(t)) \ .$$

Because of the fact that the last 2 entries of a vanish, a is also ortogonal to g_1 and g_2 along this trajectory. It follows from the definition of S_2 that $\langle a, ffg_1\rangle$ can never vanish. Further,

$$\phi_2 = \langle\lambda, g_2\rangle = \lambda_4\langle b, g_2\rangle \ ,$$

so λ_4 cannot vanish at any point of the interval (otherwise, this would give $\phi_2=0$, a contradiction). It follows that

$$q(t) := \lambda_2(t)/\lambda_4(t)$$

is well-defined. We compute the derivative of q using the adjoint equations for λ; this results in a Riccati differential equation

$$q(t) = q^2(t) + \psi(x(t))q(t) + \chi(x(t),u_1(t)) \ , \qquad (4.3)$$

where $\chi(x,u_1)$ and $\psi(x)$ are explicitely computed functions, the former linear in u_1 (and dependent on the constant value c of the control u_2). If x(t) remains in R_2, then we may as before solve $\phi_1''=0$ for u_1, there resulting the control law

$$u_1(t) = r(x(t))q(t) + s(x(t)) \ , \qquad (4.4)$$

where s(x) is easily computed (and depends on c) and where

$$r(x) := -\langle a, ffg_1\rangle \qquad (4.5)$$

is always nonzero as remarked earlier. If we substitute the control law (4.4) into (4.3), we get a similar equation but with the function χ now independent of u. Alternatively, we may solve for q in (4.4), and substitute into (4.3) in order to obtain a similar differential equation for u.

The construction can be reversed, in the following sense. Given any x_o in R_2, solving (4.3) for any given initial q(0) results, via the rule (4.4), in a singular extremal (for either fixed value of u_2), defined at least for small time. Moreover, *since* r(x) *is never zero*, we may always find q(0) so that $L_1 < u_1(0) < M_1$, and hence so that the extremal is admissible (for small enough time). We have then recovered theorem 5.1 of [SS], for the present system (and with a somewhat simpler proof, based on considerably less computations):

Theorem 4.2: Assume that (x,p,u) is a u_1-singular extremal such that x(t) is in the open set R_2 for all t in I = [0,T]. Then there is a solution q(t) of the Riccati equation (4.3) on I such that the control law (4.4) holds, while u_2 equals one of the constant values c = L_2 or M_2.

Conversely, for each $x_o \in R_2$, each c = L_2 or M_2, and each real q_o, there is a u_1-singular extremal (x,p,u), and a solution of equation (4.3), both defined on an interval I which contains 0 in its interior, such that $x(0)=x_o$, $q(0)=q_o$, and equation (4.4) holds. Moreover, there is for each x_o in R_2 a nonempty open interval $Q(x_o) \subseteq \Re$ with the following property: If $q_0 \in Q$ then the singular extremal so constructed, for either of the two values of c, is *admissible.*■

One can apply a higher order test for optimality in order to determine the exact value of c for the above extremals, just as done in [SS]. Various authors (see e.g. [KR], [HE], [MO], and references in these papers,) have found stronger constraints than those implied by the maximum principle. The simplest of these generalizes the classical *Legendre-Clebsch* condition from variational calculus. We apply these conditions to the single-control system that results when u_2 is set identically equal to c. The necessary condition is then that, along the singular extremal,

$$\langle \lambda, g_1 f g_1 \rangle \geq 0 \ . \qquad (4.6)$$

Note that, by definition of α_{112},

$$\langle \lambda, g_1 f g_1 \rangle = \alpha_{112} \langle \lambda, g_2 \rangle = \alpha_{112} \phi_2 \ .$$

Since α_{112} here equals Δ_2, we know that it is never zero along an extremal for which x(t) is in R_2. Thus, since ϕ_2 never vanishes either, it follows that the inequality (4.6) *is strict*, and hence:

$$\text{sign of } \phi_2 = \text{sign of } \Delta_2 \ .$$

Equivalently, we obtain the following precise characterization of the value of the constant control:

$$u_2(t) = \begin{cases} M_i \ \textit{if} \ \sin 2\theta_2 < 0 \\ L_i \ \textit{if} \ \sin 2\theta_2 > 0 \end{cases} .$$

5. References.

[AL] Ailon,A. and G.Langholz, "On the existence of time-optimal control of mechanical manipulators," *J.Opt.Theory & Appls.* **46**(1985): 1-21.

[HE] Hermes,H., "Lie algebras of vector fields and local approximation of attainable sets," *SIAM J.Cntr. and Opt.* **16**(1978): 715-727.

[KR] Krener,A.J., "The high order maximal principle and its application to singular extremals," *SIAM J.Contr. and Opt.*, **15**(1977): 256-293.

[MO] Moyer,H.G., "Sufficient conditions for a strong minimum in singular control problems," *SIAM J.Control* **11**(1973): 620-636.

[PA] Paul, Richard P., *Robot Manipulators: Mathematics, Programming, and Control*, MIT Press, 1982.

[RA] Rajan, V.T., "Minimum time trajectory planning," *IEEE 1985 International Conf.on Robotics and Automation*, IEEE Computer Society, St.Louis, MO 1985, pp.759-764.

[SH] Sahar,G. and J.M.Hollerbach, "Planning of minimum-time trajectories for robot arms," *IEEE 1985 International Conf.on Robotics and Automation*, IEEE Computer Society, St.Louis, MO 1985, pp. 751-758.

[SD] Shiller,Z. and S.Dubowsky, "On the optimal control of robotic manipulators with actuator and end-effector constraints," *IEEE 1985 International Conf.on Robotics and Automation*, IEEE Computer Society, St.Louis, MO 1985, pp.614-626

[SS] Sontag, E.D., and H.J. Sussmann, "Remarks on the time-optimal control of two-link manipulators," *Proc. IEEE Conf. Dec. and Control, 1985, pp.1643-1652.*

Part 7
Force and Impedance Control

N. HARRIS McCLAMROCH
The University of Michigan

THE MOST successful applications of industrial robots have involved tasks, such as pick and place operations, that require little or no interaction of the robot with its environment. Much is known about the dynamics, planning, and control of robots where external forces from the environment are viewed as force disturbances and the desired task can be described solely in terms of motion objectives. The most challenging robot tasks often involve some specified interaction with the robot environment: Examples of such challenging tasks include parts insertions and assembly, crank turning, polishing, deburring, grinding, contour following; these tasks may involve single or multiple robots working together. Our current knowledge base, both in terms of theory and practical experience, is inadequate to deal with these tasks and others that involve complex interactions of a robot with its environment. Although there has been considerable research on these topics in recent years, there is still much research to be carried out. A common framework for the study of such problems is not yet clear; a variety of mathematical models have been proposed as suitable and a variety of planning and control methodologies have been proposed; there is no current consensus about how such tasks should be studied. In this part, an overview is given of the problem components that need to be addressed; a summary of some of the main developments in the literature is given; and the roles of the four reprinted papers in this context are described.

1. Modeling Issues

The common nature of the tasks mentioned above require careful description. Each of those tasks is characterized by three features: the robot, the environment, and the contact conditions between the robot and the environment. Fundamental for such tasks is the presence of a contact force between some part of the robot and the environment; in particular, the task is defined in terms of the motion of the contact and the contact force. In fact, the key variables in the study of these tasks must include the contact forces as primary. In particular two important issues are: (1) How can the contact force be controlled in a desired way? and (2) How can the contact force be used for feedback to achieve some desired objective? There is ample experimental evidence indicating that control issues associated with contact force variables present substantial practical difficulty [1], [2]. As we have indicated, a variety of assumptions can be made that characterize a robot, its environment, and its contact configuration. Some of the possible assumptions are now identified.

1.1 Robot Assumptions

Standard robot models, modified only to take into account the effects of contact force that arise from contact with the environment, have been commonly used. The typical mathematical model used to describe the robot dynamics is given as

$$M(q)q'' + F(q,q') = \tau + u$$

where q denotes the vector of generalized joint coordinates, u denotes the control torques applied at the joints, and τ denotes the torque on the robot, reflected to the joint coordinates, caused by contact with the environment. The notation in the above equation is standard. In addition, it is often important to express the coordinates of a point on the robot that is in contact with the environment, denoted by x, in terms of the joint coordinates q through the algebraic relation

$$x = X(q);$$

in addition, the velocity of the point of contact and the contact force f satisfy

$$x' = J(q)q', \qquad \tau = J'(q)f$$

where $J(q)$ is the Jacobian matrix defined from the above algebraic relation.

1.2 Environment Assumptions

There are numerous assumptions that can be made about the environment that characterize its reaction to an external force or torque. The environment can be viewed as a fixed body, as a rigid body, or as a deformable body. If the environment is deformable, its local stiffness and damping properties need to be specified and they may be spatially uniform or not; if the environment is movable, its inertial properties need to be specified and they may be spatially uniform or not. In addition, the geometry of the surface of the environment must be specified; the surface may be planar or curved, it might be smooth, or it might have creases or discontinuities in its curvature. It is clear that there are an enormous number of possible assumptions that could be made about the environment; it is by no means clear that a general theory can be developed that would include all or most of the important assumptions.

Most of the relevant literature can be classified on the basis of the environment assumptions according to one of three categories. These categories are: complete avoidance of explicit assumptions about the environment, assumptions that the environment is characterized by a geometrically simple surface with an impedance (often described by a simple mass-spring-damper model constrained to move in one dimension) [2]–[7]; assumptions that the environment is characterized by a geometrically complex surface that is fixed and nondeformable [8]–[13]. It seems certain that the performance of a robot task depends, in an important way, on the environment assumptions. It can be seen that only a few cases have yet been studied in any detail.

A typical model of the environment, of the second category, assumes that the deformation of the environment is characterized by a scalar variable, say $\boldsymbol{B}'x$, and the contact force is given by

$$f = -\boldsymbol{B}(c\boldsymbol{B}'\dot{x} + k\boldsymbol{B}'x)$$

which defines the environment impedance. In the above c and k are the local damping and elastic coefficients of the environment at the point of contact and the vector $\boldsymbol{B}$ characterizes the direction of the contact force.

A typical model of the environment, from the third category, is that the environment is characterized by a fixed and rigid surface defined by the algebraic equation

$$\phi(x) = 0$$

If the contact occurs at a single point and if there is no friction at the contact point then the contact force is given by

$$f = \boldsymbol{J}(x)\lambda$$

where λ is a multiplier that is implicitly defined by the condition that the constraint equation be satisfied. Such models allow for a complicated surface geometry, but they necessarily ignore deformation of the surface at the point of contact.

It is clear that the second set of assumptions is most suitable for contact with a soft environment where impedance effects are dominant while the third set of assumptions is most suitable for contact with a hard environment where surface geometry effects are dominant. One approach to combining these two classes of models, using singular perturbation theory, has been described in [14].

If contact between the robot and the environment occurs over an area, both forces and torques can be generated by the contact; this is a complicated situation that has not received much attention. Even if there is point contact, the contact can be characterized as sliding contact or rolling contact. The case of rolling contact can be quite complicated and is only beginning to receive attention. It has been assumed, in the previous discussion, that contact between the robot and its environment is maintained. The more general case is to assume that contact can occur intermittently. Examples of such tasks might include hammering, digging, and sweeping. In such cases, important issues are the description of when contact is initiated, where contact occurs, the dynamics of contact including possible impact effects, and when contact is terminated. Satisfactory models for such complicated tasks have not yet been developed, and issues of planning and control have not been carefully studied. The mathematical relations described in the special cases mentioned previously are both characterized by the simplest assumption of point contact with sliding.

2. Dynamics and Control Problems

As indicated above, there are a variety of assumptions that can be made as a basis for obtaining a mathematical model that describes a particular task. There are many modeling issues that need to be addressed in particular instances; some of these issues have been mentioned. Associated modeling issues are model simplification and order reduction, linearization and local approximations, and parameter identification. Such mathematical models can be used for theoretical purposes involving task planning and feedback control. In addition the models can be used as a basis for computer simulation using numerical methods.

A fundamental problem is the translation of the specific robot task into desired motion of the contact and contact force specifications. The desired motion of the contact and the desired contact force must be selected to be consistent with the mathematical models of the robot, the environment, and the contact assumptions. Where there are many possible ways in which a specific task can be accomplished, it may be possible to determine motions and contact forces that satisfy certain optimality properties. A few example problems of this type have been studied, for example in [15].

It is usually the case that feedback control must be used to control the motion and contact force to achieve desired closed-loop properties that can arise from the appropriate use of feedback. There are a variety of possible feedback loops that can be introduced. These include the use of feedback of joint positions and joint velocities and, possibly, the use of feedback of the motion of the point of contact and feedback of the contact force. There is a substantial literature on the design of feedback controllers that achieve some of the benefits of feedback. Design procedures have been proposed using several of the classes of assumptions mentioned previously and one of several possible control design approaches. There have been few results that demonstrate, in an explicit way, the value of feedback in achieving robustness and other desired closed-loop properties.

It is possible to classify most of the control design procedures that have been suggested as being based on impedance-control or hybrid-control concepts. The concept of *impedance control* recognizes that both motion of the contact point and the contact force are to be controlled simultaneously. The specific approach is to close feedback loops on the robot so that a specified impedance, relating the motion of the contact point and the contact force, is achieved. In other words, feedback loops at the robot joints are closed so that the robot, from the perspective of the environment, appears as a specified impedance. A variety of control design procedures have been developed using this approach [4], [5], [16], [17], with two of the reprinted papers selected from this group. One of the disadvantages of this approach is that it is not necessarily clear how the specified impedance at the contact point should be chosen in order to complete a specific task; another problem is that there is no guarantee that the closed loop has the desired properties when the robot is in contact with an environment with a particular set of physical characteristics.

The concept of *hybrid control* also recognizes that both motion of the contact point and the contact force are to be controlled simultaneously in a way that is consistent with the constraints imposed by the models of the robot, the environment, and the contact assumptions. For example, if it is assumed that there is point contact of the robot with a fixed rigid surface, as described previously, then the motion of the contact point must necessarily lie within the surface and the contact force must be in a direction that is normal to the surface. In other words, feedback loops at the robot joints are closed so that the motion of the point of contact is in a certain set of possible directions and the direction of the contact force is in another set of possible directions. A variety of control design procedures have been developed using this approach [9], [10], [12], [13], [18]–[23]. Two of the reprinted papers have been selected from this group; in these two papers the environment assumptions are explicitly stated. One of the disadvantages of this approach is that it requires specific knowledge of the allowable directions for the motion vector and for the contact force vector; such information depends on the specific model of the environment.

3. Conclusions

An overview of research on force and impedance control and related topics is given, in which we have placed the research within a general context that is motivated by the automation of a class of robotic tasks defined in terms of contact relations. It is clear that much of this research has been concentrated on relatively few topics while many important topics have been little studied. Future research in robot task applications should be directed toward consideration of new contact assumptions that have not been previously studied; future theoretical research is needed to develop a deeper and more unified approach to modeling, planning, and control of robots carrying out tasks involving contact.

References

[1] D. E. Whitney, "Historical perspective and state of the art in robot force control," *Int. J. Robotics Res.*, pp. 3–14, 1987.

[2] S. D. Eppinger and W P. Seering, "Introduction to dynamic models for robot force control," *IEEE Contr. Syst. Magazine,* vol. 7, pp. 48–52, 1987.

[3] M. T. Mason, "Compliance and force control for computer-controlled manipulators," *IEEE Trans. Syst., Man Cyber.*, vol. SMC-11, pp. 418–432, 1981.

[4] O. Khatib, "A unified approach for motion and force control of robot manipulators: the operational space formulation," *IEEE J. Robotics Automation,* vol. RA-3, pp. 43–53, 1987.

[5] R. J. Anderson and M. W. Spong, "Hybrid impedance control of robotic manipulators," *IEEE J. Robotics Automation,* vol. RA-4, pp. 549–556, 1988.

[6] A. A. Goldenberg, "Implementation of force and impedance control in robot manipulators," *Proc. Conf. Robotics Automation,* pp. 1626–1632, 1988.

[7] T. Yabuta, A. J. Chona, and G. Beni, "On the asymptotic stability of the hybrid position/force control scheme for robot manipulators," *Proc. Conf. Robotics and Automation,* pp. 338–343, 1988.

[8] N. H. McClamroch, "Singular systems of differential equations as dynamic models for constrained robot systems," *Proc. Conf. Robotics and Automation,* pp. 21–28, 1986.

[9] T. Yoshikawa, "Dynamic hybrid position/force control of robot manipulators—description of hand constraints and calculation of joint driving force," *IEEE J. Robotics and Automation,* vol. RA-3, pp. 386–392, 1987.

[10] T. Yoshikawa, T. Sugie, and M. Tanaka, "Dynamic hybrid position/force control of robot manipulators—controller design and experiment," *Proc. Conf. Robotics and Automation,* pp. 2005–2010, 1987.

[11] R. K. Kankaanranta and H. N. Koivo, "Dynamic and simulation of compliant motion of a manipulator," *IEEE J. Robotics and Automation,* vol. RA-4, pp. 163–173, 1988.

[12] N. H. McClamroch and D. Wang, "Feedback stabilization and tracking of constrained robots," *IEEE Trans. Automatic Contr.,* vol. AC-33, pp. 419–426, 1988.

[13] J. K. Mills and A. A. Goldenberg, "Force and position control of manipulators during constrained motion tasks," *IEEE Trans. Robotics and Automation,* vol. RA-5, pp. 30–46, 1989.

[14] N. H. McClamroch, "A singular perturbation approach to modeling and control of manipulators constrained by a stiff environment," *Proc. 28th Conf. Dec. and Contr.,* pp. 2407–2411, 1989.

[15] H. P. Huang and N. H. McClamroch, "Time-optimal control for a robotic contour-following problem," *IEEE J. Robotics and Automation,* vol. RA-4, pp. 419–426, 1988.

[16] N. Hogan, "Impedance control: an approach to manipulation: part I–theory, part II–implementations, part III–applications," *J. Dyn. Syst., Meas., and Contr.,* pp. 1–24, 1985.

[17] H. Kazerooni, T. B. Sheridan, and P. K. Houpt, "Robust compliant motion for manipulators, part I: the fundamental concepts of compliant motion," *IEEE J. Robotics and Automation,* vol. RA-2, pp. 83–92, 1986.

[18] M. H. Raibert and J. J. Craig, "Hybrid position/force control of manipulators," *J. Dyn. Syst., Meas., and Contr.,* pp. 126–133, 1981.

[19] X. Yun, "Dynamic-state feedback control of constrained robot manipulators," *Proc. 27th Conf. Dec. and Contr.,* pp. 622–626, 1988.

[20] R. K. Kankaanranta and H. N. Koivo, "Stability analysis of position-force control using linearized cartesian space model," *Proc. IFAC Symp. Robot Contr.,* 1988.

[21] A. A. Cole, "Control of robot manipulators with constrained motion," *Proc. 28th Conf. Dec. and Contr.,* pp. 1657–1658, 1989.

[22] A. DeLuca, C. Manes, and G. Ulivi, "Robust hybrid dynamic control of robot arms," *Proc. 28th Conf. Dec. and Contr.,* pp. 2641–2646, 1989.

[23] W. Li and J-J. E. Slotine, "A unified approach to compliant motion control," *Proc. American Contr. Conf.,* pp. 1944–1949, 1989.

A Unified Approach for Motion and Force Control of Robot Manipulators: The Operational Space Formulation

OUSSAMA KHATIB, MEMBER, IEEE

Abstract—**A framework for the analysis and control of manipulator systems with respect to the dynamic behavior of their end-effectors is developed. First, issues related to the description of end-effector tasks that involve constrained motion and active force control are discussed. The fundamentals of the operational space formulation are then presented, and the unified approach for motion and force control is developed. The extension of this formulation to redundant manipulator systems is also presented, constructing the end-effector equations of motion and describing their behavior with respect to joint forces. These results are used in the development of a new and systematic approach for dealing with the problems arising at kinematic singularities. At a singular configuration, the manipulator is treated as a mechanism that is redundant with respect to the motion of the end-effector in the subspace of operational space orthogonal to the singular direction.**

I. Introduction

RESEARCH in dynamics of robot mechanisms has largely focused on developing the equations of joint motions. These joint space dynamic models have been the basis for various approaches to dynamic control of manipulators. However, task specification for motion and contact forces, dynamics, and force sensing feedback are closely linked to the end-effector. The dynamic behavior of the end-effector is one of the most significant characteristics in evaluating the performance of robot manipulator systems. The problem of end-effector motion control has been investigated, and algorithms resolving end-effector accelerations have been developed [7], [11], [22], [30], [33].

The issue of end-effector dynamic modeling and control is yet more acute for tasks that involve combined motion and contact forces of the end-effector. Precise control of applied end-effector forces is crucial to accomplishing advanced robot assembly tasks. This is reflected by the research effort that has been devoted to the study of manipulator force control. Accommodation [35], joint compliance [26], active stiffness [31], impedance control [9], and hybrid position/force control [28] are among the various methods that have been proposed.

Force control has been generally based on kinematic and static considerations. While in motion, however, a manipulator end-effector is subject to inertial, centrifugal, and Coriolis forces. The magnitude of these dynamic forces cannot be ignored when large accelerations and fast motions are considered. Controlling the end-effector contact forces in some direction can be strongly affected by the forces of coupling created by the end-effector motion that can take place in the subspace orthogonal to that direction. The description of the dynamic interaction between end-effector motions and the effects of these motions on the end-effector's behavior in the direction of force control are basic requirements for the analysis and design of high-performance manipulator control systems. Obviously, these characteristics cannot be found in the manipulator joint space dynamic model, which only provides a description of the interaction between joint motions. High-performance control of end-effector motion and contact forces requires the description of how motions along different axes are interacting, and how the apparent or equivalent inertia or mass of the end-effector varies with configurations and directions.

The description, analysis, and control of manipulator systems with respect to the dynamic characteristics of their end-effectors has been the basic motivation in the research and development of the operational space formulation. The end-effector equations of motion [13], [14] are a fundamental tool for the analysis, control, and dynamic characterization [18] of manipulator systems. In this paper, we will discuss, from the perspective of end-effector control, the issue of task description, where constrained motions and contact forces are involved. The fundamentals of the operational space formulation are presented, and the unified approach for the control of end-effector motion and contact forces is developed.

Treated within the framework of joint space control systems, redundancy of manipulator mechanisms has been generally viewed as a problem of resolving the end-effector desired motion into joint motions with respect to some criteria. Manipulator redundancy has been aimed at achieving goals such as the minimization of a quadratic criterion [29], [34], the avoidance of joint limits [5], [21] the avoidance of obstacles [4], [6], [20], kinematic singularities [23], or the minimization of actuator joint forces [8]. The end-effector equations of motion for a redundant manipulator are established and its behavior with respect to generalized joint forces is described. The unified approach for motion and active force control is then extended to these systems.

Kinematic singularities is another area that has been considered within the framework of joint space control and

Manuscript received August 6, 1985; revised May 29, 1986. This work was supported in part by the National Science Foundation and in part by the Systems Development Foundation.

The author is with the Artificial Intelligence Laboratory, Computer Science Department, Cedar Hall, Stanford University, Stanford, CA 94305.

IEEE Log Number 8610323.

Reprinted from *IEEE J. Robot. Automat.*, vol. RA-3, no. 1, pp. 43–53, Feb. 1987.

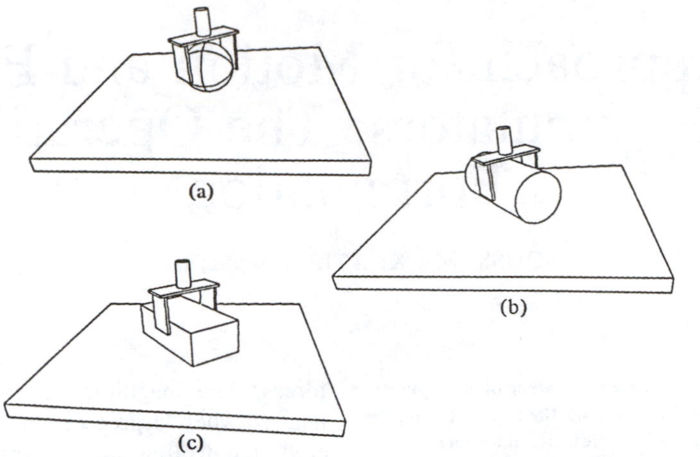

Fig. 1. Constrained end-effector freedom of motion. (a) Five degrees of freedom. (b) Four degrees of freedom. (c) Three degrees of freedom.

formulated in terms of resolution of the task specifications into joint motions. Generalized inverses and pseudo-inverses have been used, and recently an interesting solution based on the singularity robust inverse has been proposed [25]. In this paper, a new approach for dealing with the problem of kinematic singularities within the operational space framework is presented. In the neighborhood of a singular configuration the manipulator is treated as a redundant mechanism with respect to the motion of the end-effector in the subspace of operational space orthogonal to its singular direction. Control of the end-effector for motions along the singular direction is based on the use of the kinematic characteristic of the Jacobian matrix.

II. Generalized Task Specification Matrices

The end-effector motion and contact forces are among the most important components in the planning, description, and control of assembly operations of robot manipulators. The end-effector configuration is represented by a set of m parameters, $x_1, x_2, \cdots, x_m$, specifying its position and orientation in some reference frame. In free motion operations, the number of end-effector degrees of freedom m_0 is defined [13] as the number of independent parameters required to specify completely, in a frame of reference $\mathcal{R}_0$, its position and orientation. A set of such independent configuration parameters forms a system of operational coordinates.

In constrained motion operations, the displacement and rotations of the end-effector are subjected to a set of geometric constraints. These constraints restrict the freedom of motion (displacements and rotations) of the end-effector. Clearly, geometric constraints will affect only the freedom of motion of the end-effector, since static forces and moments at these constraints can still be applied. The number of degrees of freedom of the constrained end-effector is given by the difference between m_0 and the number of independent equations specifying the geometric constraints, assumed to be holonomic. Examples of five-, four-, and three-degree-of-freedom constrained end-effectors are shown in Fig. 1.

An interesting description of the characteristics of end-effectors and their constraints uses a mechanical linkage representation [5], [24]. The end-effector, tool, or manipulated object forms, with the fixture or constrained object, a pair of two rigid bodies linked through a joint. A constrained motion task can be described, for instance, by a spherical, planar, cylindrical, prismatic, or revolute joint.

However, when viewed from the perspective of end-effector control, two elements of information are required for a complete description of the task. These are the vectors of total force and moment that are to be applied to maintain the imposed constraints, and the specification of the end-effector motion degrees of freedom and their directions.

Let f_d be a unite vector, in the frame of reference $\mathcal{R}_0(\mathcal{O}, x_0, y_0, z_0)$, along the direction of the force that is to be applied by the end-effector. The positional freedom, if any, of the constrained end-effector will therefore lie in the subspace orthogonal to f_d.

A convenient coordinate frame for the description of tasks involving constrained motion operations is a coordinate frame $\mathcal{R}_f(\mathcal{O}, x_f, y_f, z_f)$ obtained from $\mathcal{R}_0$ by a rotation transformation described by S_f such that z_f is aligned with f_d. For tasks where the freedom of motion (displacement) is restricted to a single direction orthogonal to f_d, one of the axes $\mathcal{O}x_f$ or $\mathcal{O}y_f$ will be selected in alignment with that direction, as shown for the task in Fig. 2.

Let us define, in the coordinate $\mathcal{R}_f$, the position specification matrix

$$\Sigma_f = \begin{pmatrix} \sigma_x & 0 & 0 \\ 0 & \sigma_y & 0 \\ 0 & 0 & \sigma_z \end{pmatrix} \tag{1}$$

where σ_x, σ_y, and σ_z are binary numbers assigned the value 1 when a free motion is specified along the axes $\mathcal{O}x_f$, $\mathcal{O}y_f$, and $\mathcal{O}z_f$, respectively, and zero otherwise. A nonzero value of σ_z implies a full freedom of the end-effector position. This case of unconstrained end-effector position is integrated here for completeness. The coordinate frame $\mathcal{R}_f$ in this case is assumed to be identical to $\mathcal{R}_0$, and the matrix S_f is the identity matrix.

The directions of force control are described by the force specification matrix $\bar{\Sigma}_f$ associated with Σ_f and defined by

$$\bar{\Sigma}_f = I - \Sigma_f \tag{2}$$

where I designates the 3×3 identity matrix.

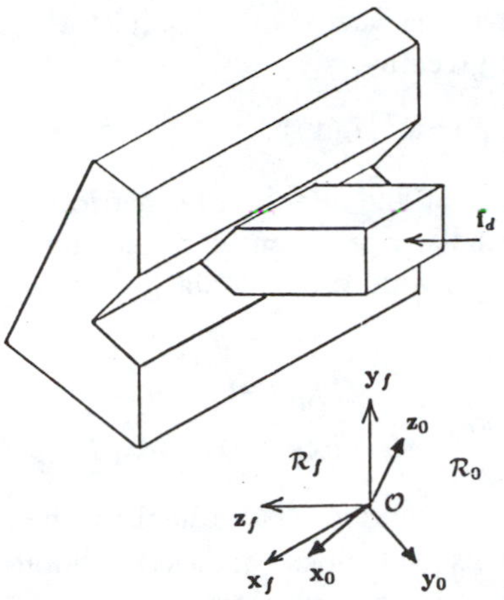

Fig. 2. One-degree-of-freedom motion.

Let us now consider the case where the end-effector task involves constrained rotations and applied moments. Let τ_d be the vector, in the frame of reference $\mathcal{R}_0(\mathcal{O}, x_0, y_0, z_0)$, of moments that are to be applied by the end-effector, and $\mathcal{R}_\tau(\mathcal{O}, x_\tau, y_\tau, z_\tau)$ be a coordinate frame obtained from $\mathcal{R}_0(\mathcal{O}, x_0, y_0, z_0)$ by a rotation S_τ that brings z_τ into alignment with the moment vector τ_d. In $\mathcal{R}_\tau$, the rotation freedom of the end-effector lies in the subspace spanned by $\{x_\tau, y_\tau\}$. To a task specified in terms of end-effector rotations and applied moments in the coordinate frame $\mathcal{R}_\tau$, we associate the rotation and moment specification matrices Σ_τ and $\tilde{\Sigma}_\tau$, defined similarly to Σ_f and $\tilde{\Sigma}_f$.

For general tasks that involve end-effector motion (both position and orientation) and contact forces (forces and moments) described in the frame of reference $\mathcal{R}_0$, we define the *generalized task specification matrices*

$$\Omega = \begin{pmatrix} S_f^T \Sigma_f S_f & 0 \\ 0 & S_\tau^T \Sigma_\tau S_\tau \end{pmatrix} \tag{3}$$

and

$$\tilde{\Omega} = \begin{pmatrix} S_f^T \tilde{\Sigma}_f S_f & 0 \\ 0 & S_\tau^T \tilde{\Sigma}_\tau S_\tau \end{pmatrix} \tag{4}$$

associated with specifications of motion and contact forces, respectively.

Ω and $\tilde{\Omega}$ act on vectors described in the reference frame $\mathcal{R}_0$. A position command vector, for instance, initially expressed in $\mathcal{R}_0$ is transformed by the rotation matrix S_f to the task coordinate frame $\mathcal{R}_f$. The motion directions are then selected in this frame by the application of Σ_f. Finally, the resulting vector is transformed back in $\mathcal{R}_0$ by S_f^T.

The construction of the generalized task specification matrices is motivated by the aim of formulating the selection process in the same coordinate frame (reference frame $\mathcal{R}_0$) where the manipulator geometric, kinematic, and dynamic models are formulated. This allows a more efficient implementation of the control system for real-time operations. Control systems using specifications based only on the matrices Σ_f and Σ_τ will require costly geometric, kinematic, and dynamic transformations between the reference frame and the task coordinate frames.

The task specification matrices Ω and $\tilde{\Omega}$ can be constant, configuration-varying, or time-varying matrices. Nonconstant generalized task specification matrices correspond to specifications that involve changes in the direction of the applied force vector and/or moment vector, e.g., moving the end-effector while maintaining a normal force to a nonplanar surface. Ω and $\tilde{\Omega}$ have been expressed here with respect to the frame of reference $\mathcal{R}_0$. For control systems implemented for tasks specified with respect to the end-effector coordinate frame, these matrices will be specified with respect to that coordinate frame as well.

III. End-Effector Equations of Motion

Joint space dynamic models, which establish the equations of manipulator joint motions, provide means for the analysis and control of these motions, and for the description of the configuration dependency and interactive nature of these mechanisms. However, the control of end-effector motion and contact forces, or the analysis and characterization of end-effector dynamic performance requires the construction of the model describing the dynamic behavior of this specific part of the manipulator system.

The end-effector motion is the result of those combined joint forces that are able to act along or about the axes of displacement or rotation of the end-effector. These are, indeed, the forces associated with the system of operational coordinates selected to describe the position and orientation of the end-effector. The construction of the end-effector dynamic model is achieved by expressing the relationships between its operational positions, velocities, accelerations, and the virtual operational forces acting on it.

First, let us consider the case of nonredundant manipulators, where a set of operational coordinates can be selected as a system of generalized coordinates for the manipulator. The manipulator configuration is represented by the column matrix q of n joint coordinates, and the end-effector position and orientation is described, in a frame of reference $\mathcal{R}_0$, by the $m_0 \times 1$ column matrix x of independent configuration parameters, i.e., operational coordinates. With the manipulator nonredundancy assumption we have the equality $n = m_0$.

Now let us examine the conditions under which a set of independent end-effector configuration parameters can be used as a generalized coordinate system for a nonredundant manipulator. In the reference frame $\mathcal{R}_0$, the system of m_0 equations expressing the components of x as functions of joint coordinates, i.e., the geometric model, is given by

$$x = G(q). \tag{5}$$

Let $\underline{q}_i$ and $\bar{q}_i$ be, respectively, the minimal and maximal bounds of the ith joint coordinate q_i. The manipulator configuration represented by the point q in joint space is confined to the hyperparallelepiped

$$\mathcal{D}_q = \prod_{i=1}^{n} [\underline{q}_i, \bar{q}_i]. \tag{6}$$

Obviously, for arbitrary kinematic linkages, and general joint boundaries, the set of functions G defined from $\mathcal{D}_q$ to the domain $\mathfrak{D}_x$ of the operational space given by

$$\mathfrak{D}_x = G(\mathfrak{D}_q) \tag{7}$$

is not one-to-one.

Different configurations of the manipulator links can, in fact, be found for a given configuration of the end-effector. The restriction to a domain where G is one-to-one is therefore necessary to construct, with the operational coordinates, a system of generalized coordinates for the manipulator mechanism.

In addition, for some configurations of the manipulator, the end-effector motion is restricted by the linkage constraints and its freedom of motion locally decreases. These are the singular configurations, which can be found by considering the differentiability characteristics of the geometric model G. Singular configurations $q \in \mathfrak{D}_q$ are those where the Jacobian matrix $J(q)$ involved in the variational or kinematic model associated with G,

$$\delta x = J(q)\delta q, \tag{8}$$

is singular. The end-effector behavior at singular configurations is treated in Section VIII.

Let $\tilde{\mathfrak{D}}_q$ be the domain obtained from $\mathfrak{D}_q$ by excluding the manipulator singular configurations and such that the vector function G of (5) is one-to-one. Let $\tilde{\mathfrak{D}}_x$ designate the domain

$$\tilde{\mathfrak{D}}_x = G(\tilde{\mathfrak{D}}_q). \tag{9}$$

The independent parameters $x_1, x_2, \cdots, x_{m_0}$ form a complete set of configuration parameters for a nonredundant manipulator, in the domain $\tilde{\mathfrak{D}}_x$ of operational space and thus constitute a system of generalized coordinates for the manipulator system.

The kinetic energy of the holonomic articulated mechanism is a quadratic form of the generalized operational velocities

$$T(x, \dot{x}) = \frac{1}{2}\dot{x}^T\Lambda(x)\dot{x} \tag{10}$$

where $\Lambda(x)$ designates the $m_0 \times m_0$ symmetric matrix of the quadratic form, i.e., the kinetic energy matrix. Using the Lagrangian formalism, the end-effector equations of motion are given by

$$\frac{d}{dt}\left(\frac{\partial L}{\partial \dot{x}}\right) - \frac{\partial L}{\partial x} = F \tag{11}$$

where the Lagrangian $L(x, \dot{x})$ is

$$L(x, \dot{x}) = T(x, \dot{x}) - U(x) \tag{12}$$

and $U(x)$ represents the potential energy due to gravity. F is the operational force vector. Let $p(x)$ be the vector of gravity forces

$$p(x) = \nabla U(x). \tag{13}$$

The end-effector equations of motion in operational space can be written [13], [14] in the form

$$\Lambda(x)\ddot{x} + \mu(x, \dot{x}) + p(x) = F \tag{14}$$

where $\mu(x, \dot{x})$ is the vector of end-effector centrifugal and Coriolis forces given by

$$\mu_i(x, \dot{x}) = \dot{x}^T\Pi_i(x)\dot{x}, \qquad i = 1, \cdots, m_0. \tag{15}$$

The components of the $m_0 \times m_0$ matrices $\Pi_i(x)$ are the Christoffel symbols $\pi_{i,jk}$ given as a function of the partial derivatives of $\Lambda(x)$ with respect to the generalized coordinates x by

$$\pi_{i,jk} = \frac{1}{2}\left(\frac{\partial \lambda_{ij}}{\partial x_k} + \frac{\partial \lambda_{ik}}{\partial x_j} - \frac{\partial \lambda_{jk}}{\partial x_i}\right). \tag{16}$$

The equations of motion (14) establish the relationships between positions, velocities, and accelerations of the end-effector and the generalized operational forces acting on it. The dynamic parameters in these equations are related to the parameters involved in the manipulator joint space dynamic model. The manipulator equations of motion in joint space are given by

$$A(q)\ddot{q} + b(q, \dot{q}) + g(q) = \Gamma \tag{17}$$

where $b(q, \dot{q})$, $g(q)$, and Γ represent, respectively, the Coriolis and centrifugal, gravity, and generalized forces in joint space. $A(q)$ is the $n \times n$ joint space kinetic energy matrix. The relationship between the kinetic energy matrices $A(q)$ and $\Lambda(x)$ corresponding, respectively, to the joint space and operational space dynamic models can be established [13], [14] by exploiting the identity between the expressions of the quadratic forms of the mechanism kinetic energy with respect to the generalized joint and operational velocities,

$$\Lambda(x) = J^{-T}(q)A(q)J^{-1}(q). \tag{18}$$

The relationship between the centrifugal and Coriolis forces $b(q, \dot{q})$ and $\mu(x, \dot{x})$ can be established by the expansion of the expression of $\mu(x, \dot{x})$ that results from (11),

$$\mu(x, \dot{x}) = \dot{\Lambda}(x)\dot{x} - \nabla T(x, \dot{x}). \tag{19}$$

Using the expression of $\Lambda(x)$ in (18), the components of $\mu(x, \dot{x})$ in (19) can be written as

$$\dot{\Lambda}(x)\dot{x} = J^{-T}(q)\dot{A}(q)\dot{q} - \Lambda(q)h(q, \dot{q}) + \dot{J}^{-T}(q)A(q)\dot{q}$$

$$\nabla T(x, \dot{x}) = J^{-T}(q)l(q, \dot{q}) + \dot{J}^{-T}(q)A(q)\dot{q} \tag{20}$$

where

$$h(q, \dot{q}) = \dot{J}(q)\dot{q} \tag{21}$$

and

$$l_i(q, \dot{q}) = \frac{1}{2}\dot{q}^TA_{q_i}(q)\dot{q}, \qquad i = 1, \cdots, n. \tag{22}$$

The subscript q_i indicates the partial derivative with respect to the ith joint coordinate. Observing from the definition of $b(q, \dot{q})$ that

$$b(q, \dot{q}) = \dot{A}(q)\dot{q} - l(q, \dot{q}) \tag{23}$$

yields

$$\mu(x, \dot{x}) = J^{-T}(q)b(q, \dot{q}) - \Lambda(q)h(q, \dot{q}). \tag{24}$$

The relationship between the expressions of gravity forces can be obtained using the identity between the functions expressing the gravity potential energy in the two systems of generalized coordinates and the relationships between the partial derivatives with respect to these coordinates. Using the definition of the Jacobian matrix (8) yields

$$p(x) = J^{-T}(q)g(q). \tag{25}$$

In the foregoing relations, the components involved in the end-effector equations of motion (14), i.e., Λ, μ, p, are expressed in terms of joint coordinates. This resolves the ambiguity in defining the configuration of the manipulator corresponding to a configuration of the end-effector in the domain $\mathfrak{D}_x$ of (7). With these expressions, the restriction to the domain $\tilde{\mathfrak{D}}_x$, where G is one-to-one, then becomes unnecessary. Indeed, the domain of definition of the end-effector dynamic model of a nonredundant manipulator can be extended to the domain $\bar{\mathfrak{D}}_x$ defined by

$$\bar{\mathfrak{D}}_x = G(\bar{\mathfrak{D}}_q) \tag{26}$$

where $\bar{\mathfrak{D}}_q$ is the domain resulting from $\mathfrak{D}_q$ of (6) by excluding the kinematic singular configurations.

Finally, let us establish the relationship between generalized forces, i.e., F and Γ. Using (18), (24), and (25) the end-effector equations of motion (14) can be rewritten as

$$J^{-T}(q)[A(q)\ddot{q} + b(q, \dot{q}) + g(q)] = F. \tag{27}$$

Substituting (17) yields

$$\Gamma = J^T(q)F \tag{28}$$

which represents the fundamental relationship between operational and joint forces consistent with the end-effector and manipulator dynamic equations. This relationship is the basis for the actual control of manipulators in operational space.

IV. End-Effector Motion Control

The control of a manipulator in operational space is based on the selection of the generalized operational forces F as a command vector. These forces are produced by submitting the manipulator to the corresponding joint forces Γ obtained from (28).

As with joint space control systems, the control in operational space can be developed using a variety of control techniques. In operational space control systems, however, errors, performance, dynamics, simplifications, characterizations, and controlled variables are directly related to manipulator tasks.

One of the most effective techniques for dealing with these highly nonlinear and strongly coupled systems is the *nonlinear dynamic decoupling approach* [36], [37], which fully exploits the knowledge of the dynamic model structure and parameters. Within this framework of control and at the level of the uncoupled system linear, nonlinear, robust [32], and adaptive [3] control structures can be implemented.

Nonlinear dynamic decoupling in operational space is obtained by the selection of the following control structure,

$$F = F_m + F_{ccg} \tag{29}$$

with

$$F_m = \hat{\Lambda}(x)F_m^*$$
$$F_{ccg} = \hat{\mu}(x, \dot{x}) + \hat{p}(x) \tag{30}$$

where $\hat{\Lambda}(x)$, $\hat{\mu}(x, \dot{x})$, and $\hat{p}(x)$ represent the estimates of $\Lambda(x)$, $\mu(x, \dot{x})$, and $p(x)$. F_m^* is the command vector of the decoupled end-effector. With a perfect nonlinear dynamic decoupling, the end-effector becomes equivalent to a *single unit mass* I_{m_0}, moving in the m_0-dimensional space. To simplify the notations, the symbol ^ will be dropped in the following development.

At the level of the decoupled end-effector, F_m^*, various control structures can be selected. For tasks where the desired motion of the end-effector is specified, a linear dynamic behavior can be obtained by selecting

$$F_m^* = I_{m_0}\ddot{x}_d - k_p(x - x_d) - k_v(\dot{x} - \dot{x}_d) \tag{31}$$

where x_d, $\dot{x}_d$, and $\ddot{x}_d$ are the desired position, velocity, and acceleration, respectively, of the end-effector. I_{m_0} is the $m_0 \times m_0$ identity matrix. k_p and k_v are the position and velocity gain matrices.

An interesting approach for tasks that involve large motion to a goal position, where a particular path is not required, is based on the selection of the decoupled end-effector command vector F_m^* as

$$F_m^* = -k_v(\dot{x} - \nu\dot{x}_d) \tag{32}$$

where

$$\dot{x}_d = \frac{k_p}{k_v}(x_d - x)$$
$$\nu = \min\left(1, \frac{V_{\max}}{\sqrt{\dot{x}_d^T\dot{x}_d}}\right). \tag{33}$$

This allows a straight line motion of the end-effector at a given speed $V_{\max}$. The velocity vector $\dot{x}$ is in fact controlled to be pointed toward the goal position while its magnitude is limited to $V_{\max}$. The end-effector will then travel at $V_{\max}$ in a straight line, except during the acceleration and deceleration segments. This command vector is particularly useful when used in conjunction with the gradient of an artificial potential field for collision avoidance [15].

Using the relationship between generalized forces given in (28), the joint forces corresponding to the operational command vector F, in (29) and (30), for the end-effector dynamic decoupling and control, can be written as

$$\Gamma = J^T(q)\Lambda(q)F_m^* + \tilde{b}(q, \dot{q}) + g(q) \tag{34}$$

where $\tilde{b}(q, \dot{q})$ is the vector of joint forces under the mapping into joint space of the end-effector Coriolis and centrifugal force vector $\mu(x, \dot{x})$. To simplify the notation, Λ has also been used here to designate the kinetic energy matrix when expressed as a function of the joint coordinate vector q. $\tilde{b}(q, \dot{q})$ is distinct from the vector of centrifugal and Coriolis forces $b(q, \dot{q})$ in (17) that arises when viewing the manipulator motion in joint space. These vectors are related by

$$\tilde{b}(q, \dot{q}) = b(q, \dot{q}) - J^T(q)\Lambda(q)h(q, \dot{q}). \quad (35)$$

A useful form of $\tilde{b}(q, \dot{q})$ for real-time control and dynamic analysis can be obtained by a separation of its dependency on position and velocity.

The joint space centrifugal and Coriolis force vector $b(q, \dot{q})$ of (17) can, in fact, be developed in the form

$$b(q, \dot{q}) = B(q)[\dot{q}\dot{q}] + C(q)[\dot{q}^2] \quad (36)$$

where $B(q)$ and $C(q)$ are, respectively, the $n \times n(n-1)/2$ and $n \times n$ matrices of the joint space Coriolis and centrifugal forces associated with $b(q, \dot{q})$. $[\dot{q}\dot{q}]$ and $[\dot{q}^2]$ are the symbolic notations for the $n(n-1)/2 \times 1$ and $n \times 1$ column matrices

$$[\dot{q}\dot{q}] = [\dot{q}_1\dot{q}_2 \quad \dot{q}_1\dot{q}_3 \cdots \dot{q}_{n-1}\dot{q}_n]^T$$
$$[\dot{q}^2] = [\dot{q}_1^2 \quad \dot{q}_2^2 \cdots \dot{q}_n^2]^T. \quad (37)$$

With $[\dot{q}\dot{q}]$ and $[\dot{q}^2]$, the vector $h(q, \dot{q})$ can be developed in the form

$$h(q, \dot{q}) = H_1(q)[\dot{q}\dot{q}] + H_2(q)[\dot{q}^2] \quad (38)$$

where the matrices $H_1(q)$ and $H_2(q)$ have, respectively, the dimensions $n \times n(-1)/2$ and $n \times n$. Finally, the vector $\tilde{b}(q, \dot{q})$ can be written as

$$\tilde{b}(q, \dot{q}) = \tilde{B}(q)[\dot{q}\dot{q}] + \tilde{C}(q)[\dot{q}^2] \quad (39)$$

where $\tilde{B}(q)$ and $\tilde{C}(q)$ are the $n \times n(n-1)/2$ and $n \times n$ matrices of the joint forces under the mapping into joint space of the end-effector Coriolis and centrifugal forces. These matrices are

$$\tilde{B}(q) = B(q) - J^T(q)\Lambda(q)H_1(q)$$
$$\tilde{C}(q) = C(q) - J^T(q)\Lambda(q)H_2(q) \quad (40)$$

With the relation (39), the dynamic decoupling of the end-effector can be obtained using the configuration dependent dynamic coefficients $\Lambda(q)$, $\tilde{B}(q)$, $\tilde{C}(q)$, and $g(q)$. The joint force control vector (34) becomes

$$\Gamma = J^T(q)\Lambda(q)F_m^* + \tilde{B}(q)[\dot{q}\dot{q}] + \tilde{C}(q)[\dot{q}^2] + g(q). \quad (41)$$

By isolating these coefficients, end-effector dynamic decoupling and control can be achieved in a two-level control system architecture [15]. The real-time computation of these coefficients can then be paced by the rate of configuration changes, which is much lower than that of the mechanism dynamics. This leads to the following architecture for the control system

- a low rate *dynamic parameter evaluation level:* updating the end-effector dynamic parameters;
- a high rate *servo control level:* computing the command vector (41) using the updated dynamic coefficients.

This approach has also been proposed [10] for real-time dynamic control of manipulators in joint space.

V. Constrained Motion Operations

The matrix Ω defined earlier specifies, with respect to the frame of reference $\mathcal{R}_0$, the directions of motion (displacement and rotations) of the end-effector. Forces and moments are to be applied in or about directions that are orthogonal to these motion directions. These are specified by the matrix $\tilde{\Omega}$.

An important issue related to the specification of axes of rotation and applied moments is concerned with the compatibility between these specifications and the type of representation used for the description of the end-effector orientation. In fact, the specification of axes of rotations and applied moments in the matrices Σ_r and $\tilde{\Sigma}_r$ are only compatible with descriptions of the orientation using instantaneous angular rotations. However, instantaneous angular rotations are not quantities that can be used as a set of configuration parameters for the orientation. Representations of the end-effector orientation such as Euler angles, direction cosines, or Euler parameters, are indeed incompatible with specifications provided by Σ_r and $\tilde{\Sigma}_r$.

Instantaneous angular rotations have been used for the description of orientation error of the end-effector. An angular rotation error vector $\delta\phi$ that corresponds to the error between the actual orientation of the end-effector and its desired orientation can be formed from the orientation description given by the selected representation [13], [22].

The time derivatives of the parameters corresponding to a representation of the orientation are related simply to the angular velocity vector. With linear and angular velocities is associated the matrix $J_0(q)$, termed the *basic Jacobian,* defined independently of the particular set of parameters used to describe the end-effector configuration

$$\begin{pmatrix} v \\ \omega \end{pmatrix} = J_0(q)\dot{q}. \quad (42)$$

The Jacobian matrix $J(q)$ associated with a given representation of the end-effector orientation x_r can then be expressed in the form [13]

$$J(q) = E_{x_r}J_0(q) \quad (43)$$

where the matrix E_{x_r} is simply given as a function of x_r.

For end-effector motions specified in terms of Cartesian coordinates and instantaneous angular rotations, the dynamic decoupling and motion control of the end-effector can be achieved [13] by

$$\Gamma = J_0^T(q)\Lambda_0(x)F_m^* + \tilde{b}_0(q, \dot{q}) + g(q) \quad (44)$$

where $\Lambda_0(q)$ and $\tilde{b}_0(q, \dot{q})$ are defined similarly to $\Lambda(q)$ and $\tilde{b}(q, \dot{q})$ with $J(q)$ being replaced by $J_0(q)$.

Using the relationship (43) similar control structures can be designed to achieve dynamic decoupling and motion control with respect to descriptions using other representations for the orientation of the end-effector.

The unified operational command vector for end-effector dynamic decoupling, motion, and active force control can be

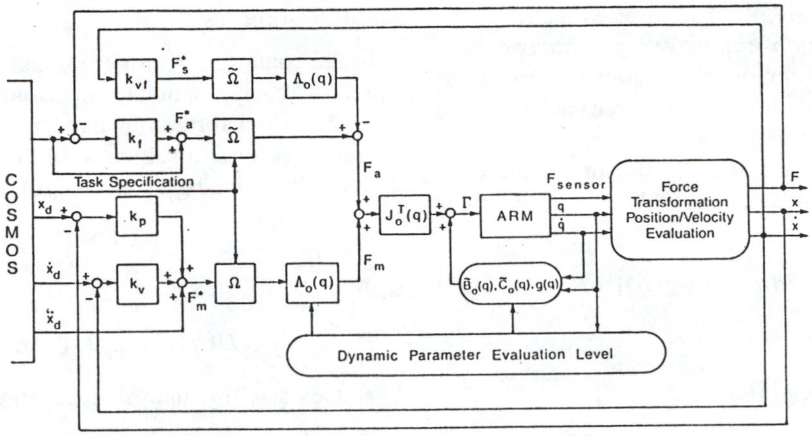

Fig. 3. Operational space control system architecture.

written as

$$F = F_m + F_a + F_{ccg} \tag{45}$$

where F_m, F_a, and F_{ccg} are the operational command vectors of motion, active force control, and centrifugal, Coriolis, and gravity forces given by

$$F_m = \Lambda_0(q)\Omega F^*_m$$
$$F_a = \tilde{\Omega} F^*_a + \Lambda_0(q)\tilde{\Omega} F^*_a$$
$$F_{ccg} = \tilde{b}_0(q, \dot{q}) + g(q) \tag{46}$$

where F^*_s represents the vector of end-effector velocity damping that acts in the direction of force control. The joint force vector corresponding to F in (45) is

$$\Gamma = J_0^T(q)[\Lambda_0(q)(\Omega F^*_m + \tilde{\Omega} F^*_s) + \tilde{\Omega} F^*_a] + \tilde{b}_0(q, \dot{q}) + g(q). \tag{47}$$

The control system architecture is shown in Fig. 3, where k_f represents the force error gain and k_{vf} denotes the velocity gain in F^*_s. An effective strategy for the control of the end-effector during the transition from free to constrained motions is based on a pure dissipation of the energy at the impact. The operational command vector F_a during the *impact transition control* stage is

$$F_a = \Lambda_0(q)\tilde{\Omega} F^*_s. \tag{48}$$

The duration of the impact transition control is a function of the impact velocity and the limitations on damping gains and actuator torques (this duration is typically on the order of tens of milliseconds). Force rate feedback has also been used in F^*_s. A more detailed description of the components involved in this control system, real-time implementation issues, and experimental results can be found in [19].

VI. Redundant Manipulators

The configuration of a redundant manipulator cannot be specified by a set of parameters that only describes the end-effector position and orientation. An independent set of end-effector configuration parameters, therefore, does not constitute a generalized coordinate system for a redundant manipulator, and the dynamic behavior of the entire redundant system cannot be represented by a dynamic model in coordinates only of the end-effector configuration. The dynamic behavior of the end-effector itself, nevertheless, can still be described, and its equations of motion in operational space can still be established.

The end-effector is affected by forces acting along or about the axes of its freedom of motion. These are the operational forces associated with the operational coordinates selected to describe its position and orientation. Let us consider the end-effector dynamic response to the application, on the end-effector, or an operational force vector F. In this case of redundant manipulator systems, the joint forces that can be used to produce a given operational force vector are not unique. The joint force vector

$$\Gamma = J^T(q)F \tag{28}$$

represents, in fact, one of these solution.

The application of the joint forces (28) to the manipulator (17), and the use of the relation

$$\ddot{x} = J(q)\ddot{q} + h(q, \dot{q}) \tag{49}$$

allow us to establish [13] the equations of motion of the end-effector

$$\Lambda_r(q)\ddot{x} + \mu_r(q, \dot{q}) + p_r(q) = F \tag{50}$$

where

$$\Lambda_r(q) = [J(q)A^{-1}(q)J^T(q)]^{-1}$$
$$\mu_r(q, \dot{q}) = \bar{J}^T(q)b(q, \dot{q}) - \Lambda_r(q)h(q, \dot{q})$$
$$p_r(q) = \bar{J}^T(q)g(q) \tag{51}$$

and

$$\bar{J}(q) = A^{-1}(q)J^T(q)\Lambda_r(q). \tag{52}$$

$\bar{J}(q)$ is actually a generalized inverse of the Jacobian matrix corresponding to the solution that minimizes the manipulator's instantaneous kinetic energy.

Equation (50) describes the dynamic behavior of the end-effector when the manipulator is submitted to a generalized joint force vector of the form (28). The $m \times m$ matrix $\Lambda_r(q)$ can be interpreted as a *pseudo-kinetic energy matrix* corresponding to the end-effector motion in operational space. $\mu_r(q, \dot{q})$ represents the centrifugal and Coriolis forces acting on the end-effector and $p_r(q)$ the gravity force vector.

The effect on the end-effector of the application of arbitrary joint forces, can be determined by (50) which can be rewritten as

$$\bar{J}^T(q)[A(q)\ddot{q}+b(q, \dot{q})+g(q)]=F. \quad (53)$$

Substituting (17) yields

$$F=\bar{J}^T(q)\Gamma. \quad (54)$$

This relationship determines how the joint space dynamic forces are reflected at the level of the end-effector.

Lemma: The unconstrained end-effector (50) is subjected to the operational force F if and only if the manipulator (17) is submitted to the generalized joint force vector

$$\Gamma=J^T(q)F+[I_n-J^T(q)\bar{J}^T(q)]\Gamma_0 \quad (55)$$

where I_n is the $n \times n$ identity matrix, $\bar{J}(q)$ is the matrix given in (52) and Γ_0 is an arbitrary joint force vector.

When the applied forces Γ are of the form (55), it is straightforward from (54) to verify that the only forces acting on the end-effector are the operational forces F produced by the first term in the expression of Γ. Joint forces of the form $[I_n - J^T(q)\bar{J}^T(q)]\Gamma_0$ correspond in fact to a null operational force vector.

The uniqueness of (55) is essentially linked to the use of a generalized inverse $\bar{J}(q)$ that is consistent with the dynamic equations of the manipulator and end-effector. The form of the decomposition (55) itself is general. A joint force vector Γ can always be expressed in the form of (55).

Let $P(q)$ be a generalized inverse of $J(q)$, and let us submit the manipulator to the joint force vector

$$\Gamma=J^T(q)F+[I_n-J^T(q)P^T(q)]\Gamma_0. \quad (56)$$

If, for any Γ_0, the end-effector is only subjected to F, (56) yields

$$J(q)A^{-1}(q)=[J(q)A^{-1}(q)J^T(q)]P^T(q) \quad (57)$$

which implies the equivalence of $P(q)$ and $\bar{J}(q)$.

VII. Control of Redundant Manipulators

As in the case of nonredundant manipulators, the dynamic decoupling and control of the end-effector can be achieved by selecting an operational command vector of the form of (29), (30). The corresponding joint forces are

$$\Gamma=J^T(q)\Lambda_r(q)F_m^*+\tilde{b}_r(q, \dot{q})+g(q) \quad (58)$$

where $\tilde{b}_r(q, \dot{q})$ is defined similarly to $\tilde{b}(q, \dot{q})$.

The manipulator joint motions produced by this command vector are those that minimize the instantaneous kinetic energy of the mechanism.

Stability Analysis

In the command vector (58), and with the assumption of a "perfect" compensation (or noncompensation) of the centrifugal and Coriolis forces, the manipulator can be considered as a conservative system subjected to the dissipative forces due to the velocity damping term $(-k_v\dot{x})$ in F_m^*. These forces are

$$\Gamma_{\text{dis}}=D(q)\dot{q} \quad (59)$$

with

$$D(q)=-k_vJ^T(q)\Lambda_r(q)J(q). \quad (60)$$

Lyapunov stability analysis leads to the condition

$$\dot{q}^TD(q)\dot{q}\leq 0 \quad (61)$$

which is satisfied, since $D(q)$ is an $n \times n$ negative semidefinite matrix of rank m. However, the redundant mechanism can still describe movements that are solutions of the equation

$$\dot{q}^TD(q)\dot{q}=0. \quad (62)$$

An example of such behavior is shown in Fig. 4(a). The end-effector of a simulated three-degree-of-freedom planar manipulator is controlled under (58). The end-effector goal position coincides with its current position, while the three joints are assumed to have initially nonzero velocities (0.5 rad/s has been used).

Asymptotic stabilization of the system can be achieved by the addition of dissipative joint forces [13]. These forces can be selected to act in the null space of the Jacobian matrix [16]. This precludes any effect of the additional forces on the end-effector and maintains its dynamic decoupling. Using (55) these additional stabilizing joint forces are of the form

$$\Gamma_{ns}=[I_n-J^T(q)\bar{J}^T(q)]\Gamma_s. \quad (63)$$

By selecting

$$\Gamma_s=-k_{vq}A(q)\dot{q}, \quad (64)$$

the vector Γ_{ns} becomes

$$\Gamma_{ns}=\Gamma_s+J^T(q)\Lambda_r(q)F_{rs} \quad (65)$$

with

$$F_{rs}=k_{vq}\dot{x}. \quad (66)$$

Finally, the joint force command vector can be written as

$$\Gamma=J^T(q)\Lambda_r(q)(F_m^*+F_{rs})+\Gamma_s+\tilde{b}_r(q, \dot{q})+g(q). \quad (67)$$

Under this form, the evaluation of the generalized inverse of the Jacobian matrix is avoided. The matrix $D(q)$ corresponding to the new expression for the dissipative joint forces Γ_{dis} in the command vector (67) becomes

$$D(q)=-[(k_v-k_{vq})J^T(q)\Lambda_r(q)J(q)+k_{vq}A(q)]. \quad (68)$$

Now, the matrix $D(q)$ is negative definite and the system is asymptotically stable. Fig. 4(b) shows the effects of this

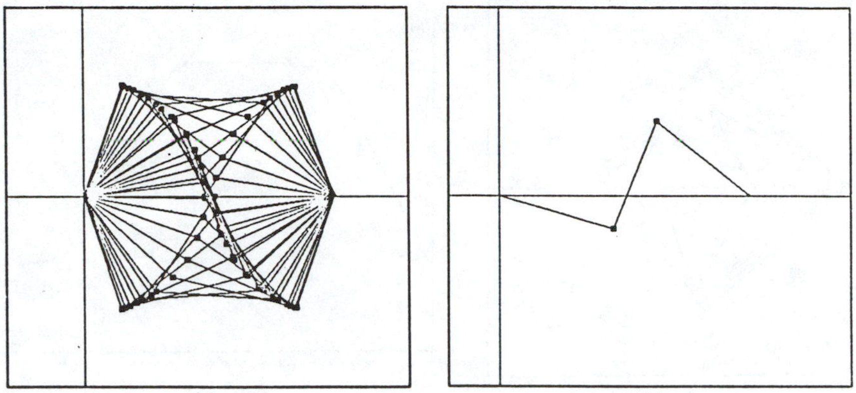

(a) (b)

Fig. 4. Stabilization of redundant manipulator.

stabilization on the previous example of a simulated three-degree-of-freedom manipulator.

Constrained Motion Control

The extension to redundant manipulators of the results obtained in the case of nonredundancy is straightforward. The generalized joint forces command vector becomes

$$\Gamma = J_0^T(q)[\Lambda_{r0}(q)(\Omega F_m^* + \tilde{\Omega} F_s^* + F_{rs}) + \tilde{\Omega} F_a^*] + \Gamma_s + \tilde{b}_{r0}(q, \dot{q}) + g(q) \quad (69)$$

where $\Lambda_{r0}(a)$ and $\tilde{b}_{r0}(q, \dot{q})$ are defined with respect to the basic Jacobian matrix $J_0(q)$.

VII. Singular Configurations

A *singular configuration* is a configuration q at which some column vectors of the Jacobian matrix become linearly dependent. The mobility of the end-effector can be defined as the rank of this matrix [5]. In the case of nonredundant manipulators considered here, the end-effector at a singular configuration loses the ability to move along or rotate about some direction of the Cartesian space; its mobility locally decreases. Singularity and mobility can be characterized, in this case, by the determinant of the Jacobian matrix.

Singularities can be further specified by the posture of the mechanism at which they occur. Different types of singularities can be observed for a given mechanical linkage. These can be directly identified from the expression of the determinant of the Jacobian matrix. The expression of this determinant can, in fact, be developed into a product of terms, each of which corresponds to a type of singularity related to the kinematic configuration of the mechanism, e.g., alignment of two links or alignment of two joint axes.

To each singular configuration there corresponds a singular "direction." It is in this direction that the end-effector presents infinite inertial mass for displacements or infinite inertia for rotations. Its movements remain free in the subspace orthogonal to this direction. This behavior extends, in reality, to a neighborhood of the singular configuration. The extent of this neighborhood can be characterized by the particular expression $s(q)$ in the determinant of the Jacobian matrix that vanishes at this specific singularity. The neighborhood of a given type of singularity $\mathfrak{D}_s$ can be defined as

$$\mathfrak{D}_s = \{ q \| s(q) | \leq s_0 \} \quad (70)$$

where s_0 is positive.

The basic concept in our approach to the problem of kinematic singularities can be formulated as follows. In the neighborhood $\mathfrak{D}_s$ of a singular configuration q, the manipulator is treated as a mechanism that is redundant with respect to the motion of the end-effector in the subspace of operational space orthogonal to the singular direction. For end-effector motion in that subspace, the manipulator is controlled as a redundant mechanism. Joint forces selected from the associated null space are used for the control of the end-effector motion along the singular direction. When moving out of the singularity, this is achieved by controlling the rate of change of $s(q)$ according to the value of the desired velocity for this motion at the configuration when $|s(q)| = s_0$. Selecting the sign of the desired rate of change of $s(q)$ allows the control of the manipulator posture among the two configurations that it can generally take when moving out of a singularity. A position error term on $s(q)$ is used in the control vector for tasks that involve a motion toward goal positions located at or in the neighborhood of the singular configuration.

Using polar or singular value decomposition, this approach can be easily extended to redundant manipulator systems. The extension to configurations where more than one singularity is involved can be also simply achieved. An example of a simulated two-degree-of-freedom manipulator is shown in Fig. 5(a). The manipulator has been controlled to move into and out of the singular configuration while displaying two different postures. The time-response of the motion in the singular direction $x(t)$ is shown in Fig. 5(b).

IX. Summary and Discussion

A methodology for the description of end-effector constrained motion tasks based on the construction of generalized task specification matrices has been proposed. For such tasks where both motion and active force control are involved, a unified approach for end-effector dynamic control within the operational space framework has been presented. The use of the generalized task specification matrix has provided a more

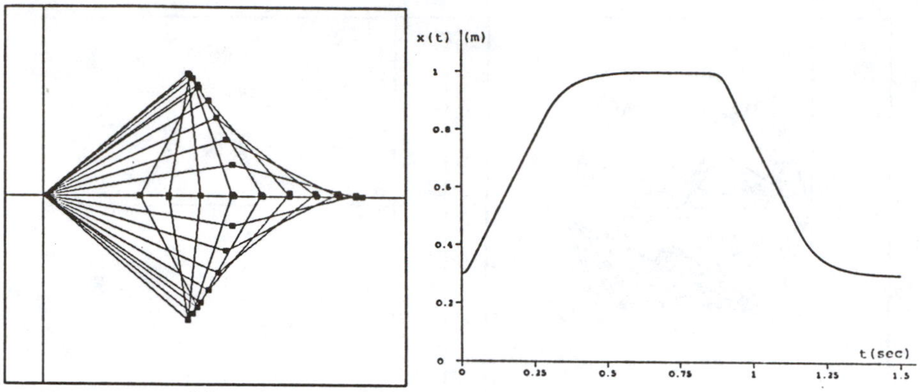

Fig. 5. Control at singular configuration.

efficient control structure for real-time implementations, further enhanced by a two-level control architecture.

Results of the implementation of this approach have shown the operational space formulation to be an effective means of achieving high dynamic performance in real-time motion control and active force control of robot manipulator systems. This approach has been implemented in an experimental manipulator programming system COSMOS (Control in Operational Space of a Manipulator-with-Obstacles System). Using a PUMA 560 and wrist and finger sensing, basic assembly operations have been performed. These include contact, slide, insertion, and compliance operations [17]. With the recent implementation of COSMOS in the multiprocessor computer system NYMPH [2], where four National Semiconductor 32016 microprocessors have been used, a low-level servo rate of 200 Hz and a high-level dynamics rate of 100 Hz have been achieved.

The impact transition control strategy was effective in the elimination of bounces at contact with a highly stiff surface. The end-effector normal velocities at impact were up to 4.0 in/s. Experiments with square wave force input have also been conducted, and responses with rise times of less than 0.02 s and steady force errors of less than 12 percent have been observed. This performance has been obtained despite the limitations in controlling the manipulator joint torques [27]. Accurate identification of the PUMA 560 dynamic parameters [1] has contributed to a nearly perfect dynamic decoupling of the end-effector.

For redundant manipulator systems, the end-effector equations of motion have been established, and an operational space control system for end-effector dynamic decoupling and control has been designed. The expression of joint forces of the null space of the Jacobian matrix consistent with the end-effector dynamic behavior has been identified and used for the asymptotic stabilization of the redundant mechanism. The resulting control system avoids the explicit evaluation of any generalized inverse or pseudo-inverse of the Jacobian matrix. Joint constraints, collision avoidance [12], [15], and control of manipulators' postures can be naturally integrated in this framework of operational space control systems. Also, a new systematic solution to the problem of kinematic singularities has been presented. This solution constitutes an effective alternative to resolving end-effector motions into joint motions generally used in joint space based control systems.

Acknowledgments

The author would like to thank Brian Armstrong, Harlyn Baker, Joel Burdick, Bradley Chen, John Craig, Ron Fearing, Jean Ponce, and Shashank Shekhar for the discussions, comments, and help during the preparation of the manuscript.

References

[1] B. Armstrong, O. Khatib, and J. Burdick, "The explicit dynamic model and inertial parameters of the PUMA 560 arm," in *Proc. 1986 IEEE Int. Conf. Robotics and Automation*, pp. 510–518.

[2] B. Chen, R. Fearing, B. Armstrong, and J. Burdick, "NYMPH: A multiprocessor for manipulation applications," in *Proc. 1986 IEEE Int. Conf. Robotics and Automation*, pp. 1731–1736.

[3] J. J. Craig, P. Hsu, S. S. Sastry, "Adaptive control of mechanical manipulators," in *Proc. 1986 IEEE Int. Conf. Robotics and Automation*, pp. 190–195.

[4] B. Espiau and R. Boulic, "Collision avoidance for redundant robots with proximity sensors," in *Preprints 3rd Int. Symp. Robotics Research*, 1985, pp. 94–102.

[5] A. Fournier, "Génération de mouvements en robotique—application des inverses généralisées et des pseudo inverses," Thèse d'etat, Mention Science, Univ. des Sciences et Techniques du Languedoc, Montpellier, France, Apr. 1980.

[6] H. Hanafusa, T. Yoshikawa, and Y. Nakamura, "Analysis and control of articulated robot arms with redundancy," in *Proc. 8th IFAC World Congress*, vol. XIV, 1981, pp. 38–83.

[7] J. R. Hewit and J. Padovan, "Decoupled feedback control of a robot and manipulator arms," in *Proc. 3rd CISM-IFToMM Symp. Theory and Practice of Robots and Manipulators*. New York: Elsevier, 1979, pp. 251–266.

[8] J. M. Hollerbach and K. C. Suh, "Redundancy resolution of manipulators through torque optimization," in *Proc. 1985 Int. Conf. Robotics and Automation*, pp. 1016–1021.

[9] N. Hogan, "Impedance control of a robotic manipulator," presented at the Winter Annual Meeting of the ASME, Washington, DC, 1981.

[10] A. Izaguirre, and R. P. Paul, "Computation of the inertial and gravitational coefficients of the dynamic equations for a robot manipulator with a load," in *Proc. 1985 Int. Conf. Robotics and Automation*, pp. 1024–1032.

[11] O. Khatib, M. Llibre, and R. Mampey, "Fonction décision-commande d'un robot manipulateur," Rapport Scientifique 2/7156, DERA-CERT, Toulouse, France, July 1978.

[12] O. Khatib and J. F. Le Maitre, "Dynamic control of manipulators operating in a complex environment," in *Proc. 3rd CISM-IFToMM Symp. Theory and Practice of Robots and Manipulators*. New York: Elsevier, 1979, pp. 267–282.

[13] O. Khatib, "Commande dynamique dans l'espace opérationnel des robots manipulateurs en présence d'obstacles," thèse de docteur-ingénieur, École National Supérieure de l'Aéronautique et de l'Espace (ENSAE). Toulouse, France, 1980.

[14] ——, "Dynamic control of manipulators in operational space," in *Proc. 6th CISM-IFToMM Congress on Theory of Machines and Mechanisms.* New York: Wiley, 1983, pp. 1128–1131.

[15] ——, "Real-time obstacle avoidance for manipulators and mobile robots," in *Proc. IEEE Int. Conf. Robotics and Automation,* pp. 500–505.

[16] ——, "Operational space formulation in the analysis, design, and control of manipulators," in *Preprints 3rd Int. Symp. Robotics Research,* 1985, pp. 103–110.

[17] O. Khatib, J. Burdick, and B. Armstrong, "Robotics in three acts—Part II" (film), Artificial Intell. Lab., Stanford Univ., Stanford, CA, 1985.

[18] O. Khatib and J. Burdick, "Dynamic optimization in manipulator design: The operational space formulation," presented at the 1985 ASME Winter Annual Meeting, Miami, FL.

[19] ——, "Motion and force control of robot manipulators," in *Proc. 1986 IEEE Int. Conf. Robotics and Automation,* pp. 1381–1386.

[20] M. Kircanski and M. Vukobratovic, "Trajectory planning for redundant manipulators in presence of obstacles," in *Preprints 5th CISM-IFToMM Symp. Theory and Practice of Robots and Manipulators,* 1984, pp. 43–58.

[21] A. Liegois, "Automation supervisory control of the configuration and behavior of multibody mechanisms," *IEEE Trans. Syst., Man, Cybern.,* vol. SMC-7, Dec. 1977.

[22] J. Y. S. Luh, M. W. Walker, and P. Paul, "Resolved acceleration control of mechanical manipulators," *IEEE Trans. Automat. Contr.,* pp. 468–474, June 1980.

[23] J. Y. S. Luh and Y. L. Gu, "Industrial robots with seven joints," in *Proc. 1985 IEEE Int. Conf. Robotics and Automation,* pp. 1010–1015.

[24] M. T. Mason, "Compliance and force control for computer controlled manipulators," *IEEE Trans. Syst., Man, Cybern.,* vol. SMC-11, pp. 418–432, 1981.

[25] Y. Nakamura, "Kinematical studies on the trajectory control of robot manipulators," Ph.D. dissertation, Kyoto, Japan, June 1985.

[26] R. P. Paul and B. Shimano, "Compliance and control," in *Proc. Joint Automatic Control Conf.,* pp. 694–699, July 1976.

[27] L. Pfeffer, O. Khatib, and J. Hake, "Joint torque sensory feedback in the control of a PUMA manipulator," presented at the 1986 American Control Conf., Seattle, WA, June 1986.

[28] M. Raibert and J. Craig, "Hybrid position/force control of manipulators," *ASME J. Dynamic Syst., Meas., Contr.,* June 1981.

[29] M. Renaud, "Contribution à l'étude de la modélisation et de la commmande des systèmes mécaniques articulés," thèse de docteur ingénieur, Univ. Paul Sabatier, Toulouse, France, Dec. 1975.

[30] M. Renaud and J. Zabala-Iturralde, "Robot manipulator control," in *Proc. 9th Int. Symp. Industrial Robots,* 1979, pp. 463–475.

[31] J. K. Salisbury, "Active stiffness control of a manipulator in Cartesian coordinates," presented at the 19th IEEE Conf. Decision and Control, Albuquerque, NM, Dec. 1980.

[32] J. J. Slotine and O. Khatib, "Robust control in operational space for goal positioned manipulator tasks," presented at the 1986 American Control Conf., Seattle, WA, June 1986.

[33] K. Takase, "Task-oriented variable control of manipulator and its software servoing system," in *Proc. IFAC Int. Symp.,* 1977, pp. 139–145.

[34] D. E. Whitney, "Resolved motion rate control of manipulators and human prostheses," *IEEE Trans. Man-Machine Syst.,* vol. MMS-2, pp. 47–53, June 1969.

[35] D. E. Whitney, "Force feedback control of manipulator fine motions," *ASME J. Dynamic Syst., Meas., Contr.,* pp. 91–97, June 1977.

[36] J. Zabala-Iturralde, "Commande des robots manipulateurs à partir de la modélisation de leur dynamique," thèse de 3eme cycle, Univ. Paul Sabatier, Toulouse, France, July 1978.

[37] E. Freund, "The structure of decoupled nonlinear systems," *Int. J. Contr.,* vol. 21, no. 3, pp. 443–450, 1975.

Hybrid Impedance Control of Robotic Manipulators

ROBERT J. ANDERSON AND MARK W. SPONG

Abstract—**In order for robots to improve their adaptability, it is necessary for the underlying control strategies to become more sophisticated. This is especially true in contact tasks such as grinding or deburring where it is desired not only for the robot to apply a steady force, but also to reject undesirable high-frequency disturbances. This work presents a foundation for such force control strategies. Realizing that the type of control strategy that is employed depends fundamentally on the characteristics of the environment, this correspondence classifies the types of environments by use of a duality condition, and demonstrates which control is appropriate for each type of environment. A general control approach is introduced, called hybrid impedance control (HIC), which in its simplest forms reduces to Khatib's operational space control [3], or to Hogan's impedance control [14]. The control law is formulated in a general enough fashion, however, to allow for higher order controllers.**

I. Introduction

Force control of robotic manipulators has been an active area of research for many reasons. The inclusion of force information in the control of robots increases their adaptability to uncertain environments, such as are found in deburring, grinding, and assembly tasks, and provides safety against breakage due to excessive contact forces. Typically, forces are measured by a force sensor connected to the wrist of the robot [1]. It is assumed that this force sensor can accurately delineate forces along six degrees of freedom (DOF).

A. Definitions and Terms

Force control implicitly involves contact with the "environment," which we define to be any element connected to or contacting the robot anywhere past the wrist force sensor. With this definition the end-effector's mass and compliance are both considered to be part of the environment.

Two descriptions of the end-effector position are of interest. The first coincides with the Denavit–Hartenberg [2] representation where the generalized variable q_i represents joint angles, and generalized joint torques are given by τ_i. This representation of the manipulator's position is called joint space. The second description is called task space or constraint space. The six DOF vector x consists of the three orthogonal displacements, which represent the distance between the end-effector and a fixed Cartesian coordinate frame, and the three respective angles, which represent the orientation of the end-effector with respect to the fixed Cartesian coordinate frame. Similarly, the six DOF vector F consists of the three orthogonal force components and the three orthogonal torque components operating at the end-effector with respect to the fixed Cartesian coordinate system.

When the task space is not fixed (e.g., when following a contour) a third, "world" space is necessary. This space is a fixed Cartesian coordinate frame, which serves as a reference point for the task space. For simplicity we shall consider world space and task space to be identical.

Manuscript received April 7, 1987; revised January 5, 1988. This research was partially supported by the National Science Foundation under Grant DMC-8516091. Part of the material in this correspondence was presented at the IEEE International Conference on Robotics and Automation, Raleigh, NC, March 1987.

The authors are with the Coordinated Science Laboratory, University of Illinois, Urbana, IL 61801.

IEEE Log Number 8820705.

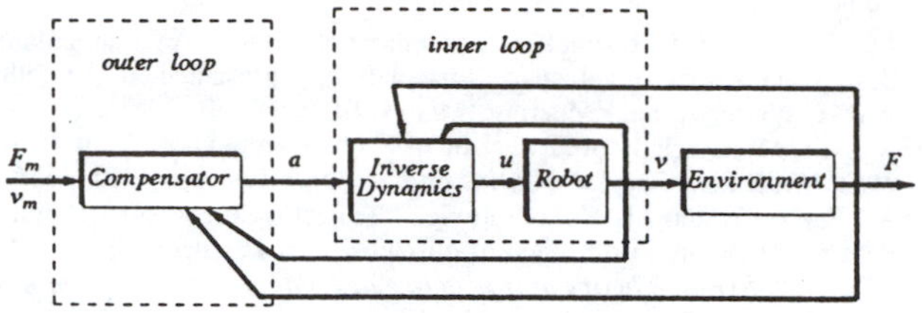

Fig. 1. Inner/outer loop control.

The equations of motion of an n-degree-of-freedom robot can be written using task space coordinates [3] as

$$D(x)\ddot{x}+h(x,\dot{x})-F=J^{-T}u \tag{1}$$

where $D(x)$ is the $n \times n$ inertia matrix in task space, $h(x, \dot{x})$ contains the Coriolis, centripetal, and gravitational force terms, F represents contact forces, J represents the Jacobian, and u is the input torque.

B. Inner/Outer Loop Control Strategy

The control problem for the system (1) is to choose the input $u(t)$ to cause the end-effector to execute a desired motion in task space while regulating the forces of interaction of the end-effector with the environment. Depending on the task, this can be specified either as a trajectory tracking problem, that is, as the problem of tracking a given motion and/or force trajectory in Cartesian space, or as the problem of obtaining a desired impedance, for example, controlling the manipulator to respond as a second-order system with stiffness K, damping B, and mass M as

$$M\ddot{x}+B\dot{x}+Kx=Kx_m. \tag{2}$$

We will use the concept of inner/outer loop control shown in Fig. 1, where the inner loop is a nonlinear feedback linearization or inverse dynamics control [4] and the outer loop is an additional control to achieve the more classical control theoretic goals such as tracking, disturbance rejection, robustness, etc. Given the system (1), the inner loop control law is assumed to be of the form

$$u=J^T(D(x)a+h(x,\dot{x})-F) \tag{3}$$

and cancels the nonlinearities present in (1). Since the inertia matrix is invertible and assuming that we are in a region free of kinematic singularities, the control law (3) applied to (1) results in the familiar double integrator system

$$\ddot{x}=a=\dot{v} \tag{4}$$

where a represents the outer loop control whose design constitutes the main thrust of this correspondence.

Two comments are in order at this point. First, the control law (3) differs from the standard inverse dynamics control considered by many authors in the sense that the end-effector forces, represented by F, are also canceled. We are assuming that these forces are directly measurable and hence F in (3) is a measured rather than a computed term. Second, we do not consider the robustness issue in this

Reprinted from *IEEE J. Robot. Automat.*, vol. 4, no. 5, pp. 549–556, Oct. 1988.

correspondence that arises from inexact cancellation of the nonlinearities. We are concerned here only with the outer loop design in the absence of inner loop uncertainty. Numerous robust algorithms [5]–[8] have been applied to inverse dynamics controllers and may also be applied to HIC. The initial presentation [9] of the HIC algorithm demonstrated how a Model Reference Adaptive Controller (MRAC) might be implemented within the inner loop of the HIC controller.

II. Research in Force Control

A number of force control algorithms have been proposed with such names as damping control, stiffness control, and explicit force control. Recently, Whitney surveyed the bulk of these approaches, and gave a historical perspective to the field [10]. Despite the diversity of approaches, however, most of these approaches can be separated into two classes, *hybrid control* and *impedance control*.

A. Hybrid Control

Hybrid position/force control, or hybrid control for short, was first proposed by Raibert and Craig [11], based on an orthogonal decomposition of task space [12]. Improvements have been suggested since then [13], but the central concept remains the same. The position of the end-effector and the contact force with the environment along one DOF cannot be controlled independently. The task space is split into two subspaces, called the position-controlled and force-controlled subspaces. As the name would imply, positions are commanded and controlled along one subspace, and forces are commanded and controlled along the other. A selection matrix S is used to determine which DOF are to be position-controlled, and which are to be force-controlled.

B. Impedance Control

Impedance control is more an approach to force control than an established algorithm. Hogan realized that although both position and force cannot be controlled simultaneously, by controlling the amount of compliance, or "impedance" in the manipulator, the contact forces can be regulated [14].

Impedance control has been implemented in many forms. In its simplest form it can be considered a generalization of damping and stiffness control schemes [10]. In this form, it is essentially a PD position controller, with position and velocity feedback gains adjusted to obtain different apparent impedances. Unfortunately, given constant feedback gains, the robot impedance changes with configuration, due to the high nonlinearity of the robot dynamics, making it difficult to determine exactly what impedance the environment will see.

In order to achieve a constant, known impedance, independent of manipulator configuration, some sort of inverse dynamics is necessary. Hogan [15] suggests one way to implement this.

C. Problems with Earlier Approaches

Hybrid control is a highly intuitive approach to force control, which properly recognizes the distinction between force-controlled and position-controlled subspaces. The problem with hybrid control is its failure to recognize the importance of manipulator impedance.

Impedance control considers the effects of impedance on robot/environment interactions. When performed in task space, a known impedance can be maintained for all configurations. It is considered, however, to be solely a position-control scheme, with small adjustments made to react to contact forces. Positions are commanded, and impedances are adjusted to obtain the proper force response. No attempt is made to follow a commanded force trajectory and any distinction between force-controlled subspaces and position-controlled subspaces is ignored. Furthermore, the impedance schemes published to date consider only second-order impedances for the manipulator, where in some cases higher order impedances are desirable.

III. Hybrid Impedance Control

The control approach introduced in this section, namely, hybrid impedance control (HIC), combines impedance control and hybrid position/force control into one strategy, while allowing for more sophisticated impedances.

For the remainder of this correspondence we shall assume that an inverse dynamics controller has been implemented in task space, and has effectively decoupled the manipulator into single-DOF linear subsystems. When only linear systems are considered, systems concepts, such as Norton equivalence, Thèvenin equivalence, and impedance, may be readily applied to manipulator/environment interaction. Our interest will now be directed to these subsystems.

A. The System Approach

From a systems point of view, the input/output behavior of a linear continuous system of the type considered here is described by the ratio of two variables, effort (F) and flow ($v = \dot{x}$). For a mechanical system, effort is represented by force and torque, and flow is represented by linear and angular velocity. Motors and batteries are considered equivalent, in a system sense, both being effort sources. Similarly, a current generator or a rotating cam shaft are both flow sources.

Passive elements are characterized by resistance (B), capacitance (K), and inertia (M). Resistance represents the proportional relationship between effort and flow, $B = F/v$, capacitance represents the integral relationship between effort and flow, $K = F/\int v\, dt$, and inertia represents the differential relationship between effort and flow, $M = F/\dot{v}$. For linear, time-invariant continuous systems, the impedance Z may be defined as the ratio of the Laplace transform of the effort $F(s)$ to the Laplace transform of the flow $v(s)$. For nonlinear systems, the term impedance can still be used to describe the relationship between effort and flow. In this case, the impedance is operating point dependent. That is, the impedance of the nonlinear system is defined as the equivalent linear impedance for the system linearized about a particular operating point.

As any electrical engineer knows, the determination of how two complex subcircuits interact is greatly simplified when the subcircuits are represented by their Norton or Thèvenin equivalent circuits. The same is true for force control. We are concerned with how two complex systems, the robot and the environment, will react when connected together. In the spirit of systems theory, the robot and the environment can both be represented by Norton and Thèvenin equivalent circuit models.

B. Modeling the Environment

The environment is central to any force control strategy. For instance, no force algorithm, no matter how sophisticated, can command a force with the end-effector in free space, nor command a motion while the end-effector is held fast. The environment is usually modeled as a linear spring K_e which is sometimes in parallel with a dashpot B_e. Both are considered to be known and constant, and a force law is chosen accordingly. This may be valid when a rigid tool meets a compliant workpiece, but is far too simplistic to represent the entire spectrum of environments.

We will describe the robot and environment systems based on the low-frequency behavior of the impedance. For a linear environment, the impedance is defined as the ratio of the Laplace transforms of effort and flow. For any given frequency ω, this is a complex number with real part $R(\omega)$ and imaginary part $X(\omega)$

$$Z(\omega) = R(\omega) + jX(\omega). \tag{5}$$

As ω approaches zero, one of three things can happen to the magnitude of the environment's impedance. It can approach infinity, it can approach a nonzero finite number, or it can approach zero. We introduce the following definitions:

Definition 3.1: A system with impedance given by (5) is *inertial* iff $|Z(0)| = 0$.

Definition 3.2: A system with impedance given by (5) is *resistive* iff $|Z(0)| = c$, where $0 < c < \infty$.

Definition 3.3: A system with impedance given by (5) is *capacitive* iff $|Z(0)| = \infty$.

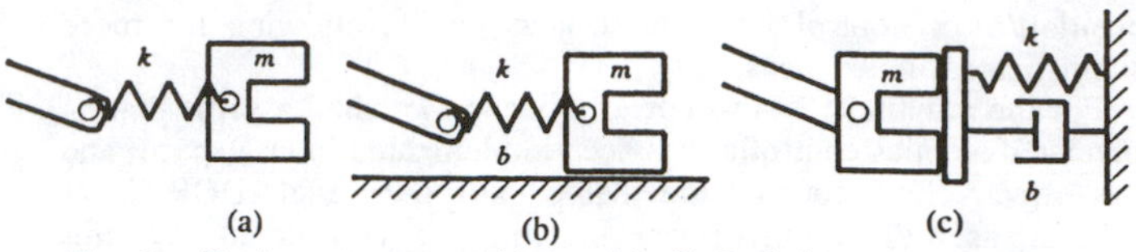

Fig. 2. Environment types. (a) Inertial. (b) Resistive. (c) Capacitive.

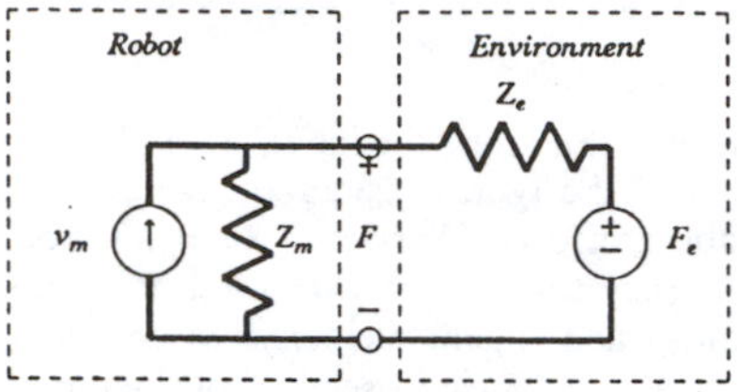

Fig. 3. Position-control model.

Fig. 2 shows examples of three environments (recall that the environment, by definition, includes everything past the wrist force sensor). The first environment is a mass-spring system with impedance $Z_e = mks/(ms + k)$ and, by application of the defintion, is inertial. The second environment consists of a compliant mass sliding across a viscous surface. It has an impedance of $Z_e = b + mks/(ms + k)$ and is resistive. The final environment consists of a damped, compliant surface with impedance $Z_e = ms + b + k/s$ and is, by definition, capacitive.

Capacitive and inertial environments represent dual impedances in the sense that the inverse of a capacitive system is inertial, and the inverse of an inertial system is capacitive. A resistive environment is self-dual. To represent this duality we shall use Norton and Thèvenin equivalents.

Recall that a Norton equivalent consists of an impedance in parallel with a flow source, and a Thèvenin equivalent consists of an impedance in series with an effort source. We shall use a Norton equivalent to represent a capacitive system, and conversely, we shall use a Thèvenin equivalent to represent an inertial system. Either representation will suffice for representing a resistive environment.

C. Duality

Once the environment has been properly modeled, the desired manipulator response may be determined. A fundamental goal for designing a controller is zero steady-state error to a step input. This will be obtained if we adhere to the following duality principle.

Duality Principle:

The manipulator should be controlled to respond as the dual of the environment.

This statement is most easily described in terms of Norton and Thèvenin equivalents. When the environment is capacitive we represent it as an impedance in parallel with a flow source, and the corresponding manipulator dual is an effort source in series with a noncapacitive impedance. When the environment is inertial we represent it as an impedance in series with an effort source, and the corresponding manipulator dual is a flow source in parallel with a noninertial impedance. When the environment is resistive, either equivalent may be used but the dual manipulator impedance must be nonresistive. Simply stated, capacitive environments require a force-controlled manipulator, inertial environments require a position-controlled manipulator, and resistive environments allow either position or force control. Once the type of servo control is known, the manipulator impedances should be chosen accordingly.

To show that this condition insures zero steady-state error to a step (assuming no environmental inputs) is straightforward. First assume that the environment is found to be inertial so that $Z_e(0) = 0$. Fig. 3 shows the environment, and the corresponding manipulator, where Z_e is the environment's impedance, Z_m is the manipulator's impedance, and v_m is the input velocity. The input/output transfer function

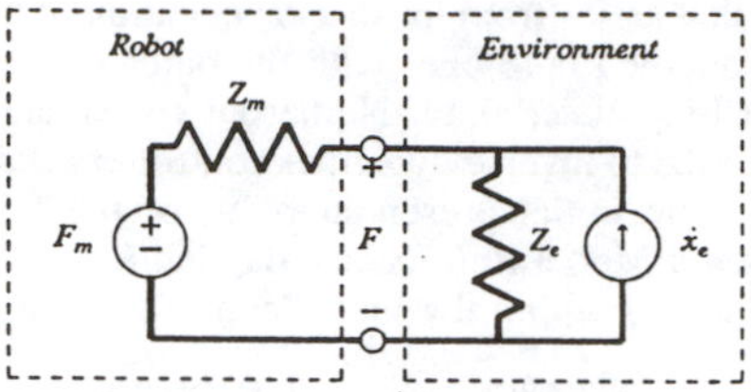

Fig. 4. Force-control model.

for velocity is given as

$$\frac{v}{v_m} = \frac{Z_m(s)}{Z_m(s) + Z_e(s)}. \tag{6}$$

Assuming stability, the steady-state error to a step input $1/s$ is given by the final value theorem as

$$e_{ss} = \lim_{t \to \infty} (v - v_m) = \frac{-Z_e(0)}{Z_m(0) + Z_e(0)} = 0 \tag{7}$$

as long as $Z_m(0) \neq 0$, i.e., as long as the manipulator impedance is noninertial. Next assume that the environment is capacitive so that $Z_e(0) = \infty$. Fig. 4 shows the environment and the corresponding manipulator. The input/output transfer function for force is given by

$$\frac{F}{F_m} = \frac{Z_e(s)}{Z_m(s) + Z_e(s)} \tag{8}$$

and the steady-state error to a step is likewise given by

$$e_{ss} = \lim_{t \to \infty} (F - F_m) = \frac{-Z_m(0)}{Z_m(0) + Z_e(0)} = 0 \tag{9}$$

as long as $Z_m(0)$ is finite, i.e., as long as the manipulator impedance is noncapacitive. It is simple to show that zero steady-state error is also obtained for resistive environments, as long as either $Z_m(0) = 0$ and the manipulator is force-controlled, or $Z_m(0) = \infty$ and the manipulator is position-controlled.

The duality condition encompasses the notion that neither two different flows nor two different efforts can be maintained simultaneously at the junction of a one-port. An environment following a position trajectory and a position-controlled robot attempting to track this environment are inconsistent. The dual combination of one Norton equivalent flow source and a Thèvenin equivalent effort source, however, can exist simultaneously. Although we have only discussed linear environments, duality is also a desirable property for nonlinear manipulators [14].

HIC requires that duality be maintained along every degree of freedom of the manipulator. This duality principle will allow us to determine, based on the character of the environment, when to use position control or force control in a given task. To demonstrate the application of the duality principle in a force-control problem we give two examples.

First, consider the setup shown in Fig. 5. The environmental impedance is given by $M_e s + B_e$. Because the environment is resistive, the system can be either force-controlled or position-controlled. Duality requires that if we choose to position-control the manipulator then the manipulator impedance should be capacitive, such as $Ms + B + K/s$. The resulting velocity through the environment port is given by

$$\frac{v}{v_m} = \frac{Bs + K}{(M_e + M)s^2 + (B + B_e)s + K}. \tag{10}$$

As a second example consider the system shown in Fig. 6. The environmental impedance is given by $B_e + K_e/s$. Furthermore, we shall assume that the environmental velocity v_e has a high noise content over a narrow band. Because the environment is capacitive it should be modeled by its Norton equivalent. Duality demands that the manipulator appears as a Thèvenin equivalent system, a force source

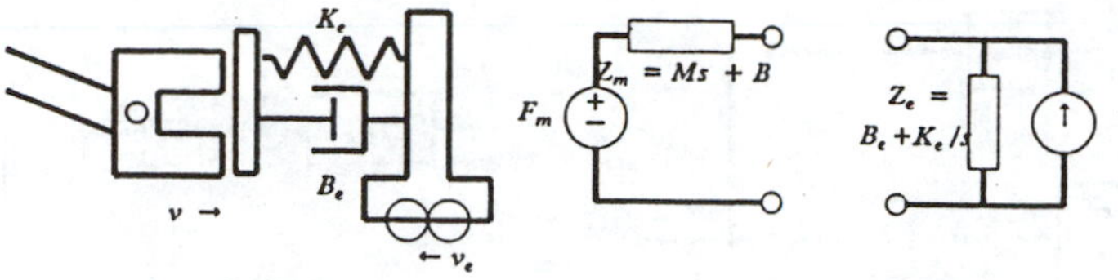

(a) (b) (c)

Fig. 5. Inertial environment example. (a) Mechanical representation. (b) Circuit model for manipulator. (c) Circuit model for environment.

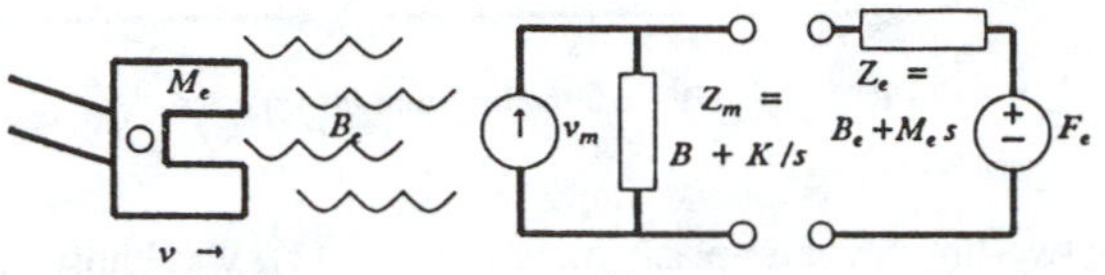

(a) (b) (c)

Fig. 6. Capacitive environment example. (a) Mechanical representation. (b) Circuit model for manipulator. (c) Circuit model for environment.

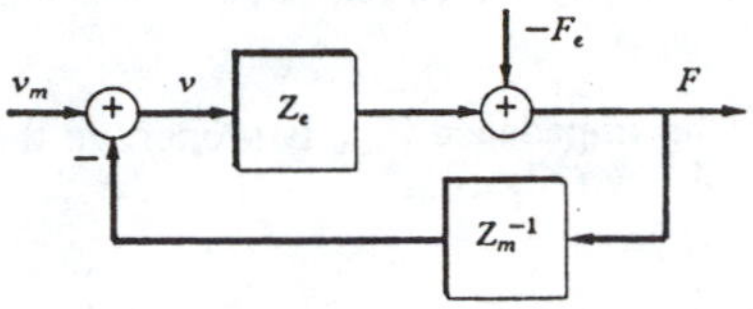

Fig. 7. Position-control block diagram.

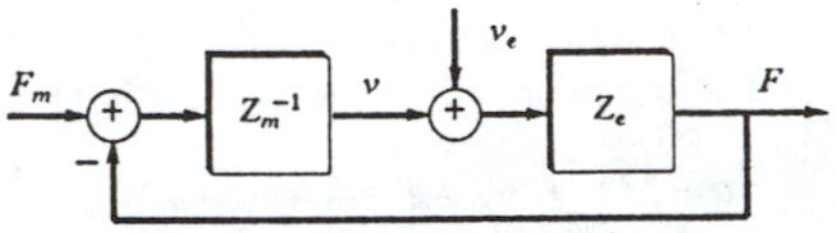

Fig. 8. Force-control block diagram.

in series with a noncapacitive impedance. In order to filter out the environmental noise (which might occur due to mechanical vibrations in a grinding wheel, for instance) we choose

$$Z_m = Ms + B\frac{(s+\omega)^2}{s^2+\omega^2} \tag{11}$$

where the second term is tuned to the environmental noise frequency. Upon connecting the manipulator port with the environment port the resulting force is given by

$$F = \frac{(s^2+\omega^2)(B_e s+K_e)}{(Ms^2+B_e s+K_e)(s^2+\omega^2)+Bs(s+\omega)^2} F_m - \frac{Ms^2(s^2+\omega^2)(B_e s+K_e)+B(s+\omega)^2(B_e s+K_e)}{(Ms^2+B_e s+K_e)(s^2+\omega^2)+B(s+\omega)^2} v_e. \tag{12}$$

For $F_m = 0$ we see that

$$v = -\frac{Z_e}{Z_m+Z_e} v_e = \frac{(B_e s+K_e)(s^2+\omega^2)}{(Ms^2+B_e s+K_e)(s^2+\omega^2)+B(s+\omega)^2} v_e \tag{13}$$

and the manipulator is responsive to all environmental motions, except for those motions in the narrow band of disturbances.

Position-controlled and force-controlled subsystems are shown in Figs. 3 and 4, respectively. The control engineer has no influence over either the environment's impedance Z_e or the environment sources v_e and F_e. The control engineer can, however, determine Z_m by properly designing a controller and feeding back state variables. How this may be done, without interfering with the inner loop controller, is the subject of the next two sections.

D. Position-Controlled Subsystem

The transfer function for the position-controlled circuit shown in Fig. 3 is

$$v = \frac{Z_m}{Z_m+Z_e} v_m + \frac{1}{Z_m+Z_e} F_e. \tag{14}$$

This response can be realized be feedback of the contact force, combined with information about the desired acceleration for the manipulator. Fig. 7 shows a block diagram of the position-control implementation.

The commanded acceleration a is given implicitly by

$$a = \dot{v} = \frac{d}{dt}\left(v_m - \frac{F}{Z_m}\right). \tag{15}$$

In practice, we would like to obtain the control a explicitly, without differentiators and using only measurements of F, x, and v. This is possible if the impedance can be written as

$$Z_m = Ms + Z_{\text{rem}} \tag{16}$$

where the remaining impedance Z_{rem} is proper. It should be mentioned here that Z_m must still satisfy the duality condition, and will do so only if Z_{rem} contains the appropriate capacitive elements. The position control law (15) can now be rewritten as

$$a = \frac{d}{dt}\left(v_m - \frac{F}{Ms+Z_{\text{rem}}}\right) = \dot{v}_m + \frac{1}{M} Z_{\text{rem}}(v_m - v) - \frac{1}{M}F. \tag{17}$$

As an example consider the system in Fig. 5, where the manipulator impedance was given as $Z_m = Ms + B + K/s$. Here, $Z_{\text{rem}} = B + K/s$, and the commanded acceleration a is,

$$a = \dot{v}_m + \frac{B}{M}(v_m - v) + \frac{K}{M}(x_m - x) - \frac{1}{M}F. \tag{18}$$

The position control law (18) is seen to be equivalent to Hogan's impedance control law [15].

E. Force-Controlled Subsystem

The transfer function for the force-controlled subsystem shown in Fig. 4 is

$$F = \frac{Z_e}{Z_m+Z_e} F_m + \frac{Z_e Z_m}{Z_m+Z_e} v_e. \tag{19}$$

This response can be realized by feedback of the force signal, in combination with a controller with transfer function Z_m^{-1}. Fig. 8 shows a block diagram of the resulting system. The outer loop control a is defined implicitly by

$$a = \frac{d}{dt}\frac{(F_m - F)}{Z_m} \tag{20}$$

where the force F is a measured quantity given by

$$F = Z_e(v - v_e) \tag{21}$$

and F_m is the input command corresponding to a desired Thèvenin equivalent force. Note that if the inputs v_m and F_m are set to zero then the force control law (20) and the position control law (15) are identical.

As before, we would like to obtain a without differentiators, and using only measurements of F, x, and v. Again this is possible if the impedance can be written as

$$Z_m = Ms + Z_{\text{rem}} \tag{22}$$

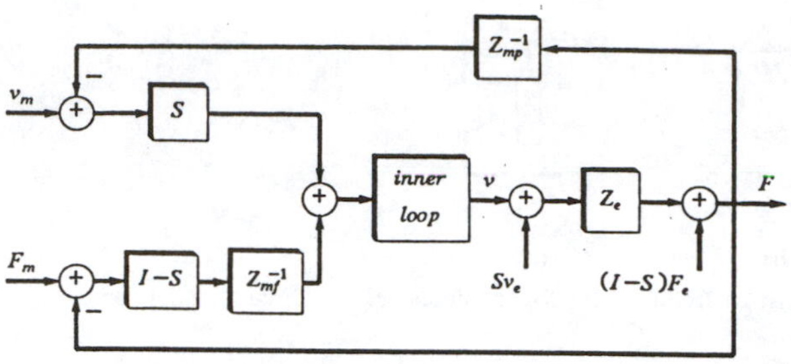

Fig. 9. Hybrid impedance block diagram.

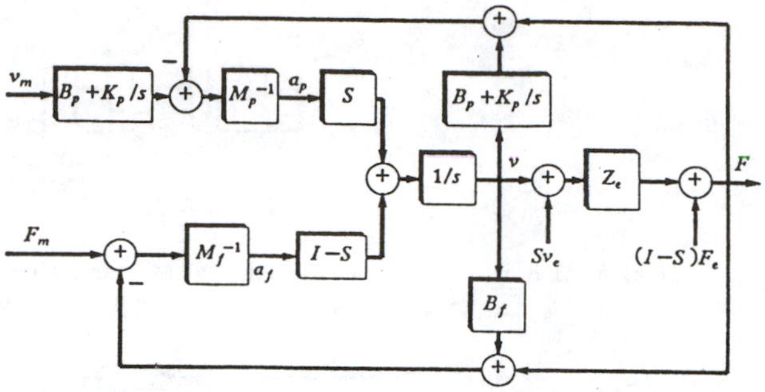

Fig. 10. HIC controller for two-link arm.

where the remaining impedance Z_{rem} is proper. In this case we can rewrite our control law (20) as

$$a=\frac{d}{dt}\frac{(F_m-F)}{Ms+Z_{\text{rem}}}=-\frac{1}{M}Z_{\text{rem}}v+\frac{1}{M}(F_m-F). \tag{23}$$

As an illustration, consider the system shown previously in Fig. 6, in which the robot impedance was chosen to be $Z_m = Ms + B(s+\omega)^2/(s^2+\omega^2)$. The commanded acceleration a is given by

$$a=M^{-1}((F_m-F)-B(v-v_f)) \tag{24}$$

where v_f is an additional filtering compensation term obtained from the filter $v_f = 2\omega s/(s^2+\omega^2)v$.

F. Combining the Subsystems

Now that the force-controlled and position-controlled subsystems have been described in detail for a single DOF, the entire system may be put together. The hybrid system is similar to Raibert's and Craig's hybrid controller, but with control loops operating in task space rather than in joint space. This has been achieved by using an inverse dynamics inner loop controller with additional end-effector force cancellation. A selection matrix S, which consists of ones and zeros down the diagonal, is used to separate the force-controlled and position-controlled subspaces.

It should be reiterated that as the manipulator encounters different environments, different controller gains are necessary to maintain a desired response. A robot packaging eggs needs far more internal compliance than a robot packaging cold cuts. An intelligent controller would not just determine the elements of the selection matrix, but should also be actively adjusting system parameters. This might be done by adaptive control, table look-up, or other means. The details of such an implementation, however, are beyond the intended scope of this work.

The hybrid impedance control algorithm for a six-DOF manipulator is given in block diagram form in Fig. 9. Here, the robot impedance terms Z_{mp} and Z_{mf} represent diagonal matrices with terms equal to the impedance along each degree of freedom. The control signal a is now a vector and may be defined implicitly as

$$\begin{aligned}a&=Sa_p+(I-S)a_f\\&=S\frac{d}{dt}(v_m-Z_{mp}^{-1}F)+(I-S)\frac{d}{dt}Z_{mf}^{-1}(F_m-F)\end{aligned} \tag{25}$$

which to be implemented should be of the form

$$\begin{aligned}a=&S(\dot v_m+M_p^{-1}Z_{\text{rem}\,p}(v_m-v)-M_p^{-1}F)\\&-(I-S)(M_f^{-1}Z_{\text{rem}\,f}v+M_f^{-1}(F_m-F)).\end{aligned} \tag{26}$$

IV. Simulations

In order to demonstrate the implementation of HIC, an assembly task was simulated. The robot arm is modeled in continuous time, using a second-order Runge–Kutta integration scheme, while the control is implemented in discrete time. The simulations were run in Fortran.

The two-link revolute-prismatic arm (Fig. 11), was chosen since it was the simplest nonlinear manipulator which could be simultaneously force- and position-controlled. The first subsection develops the HIC equations for this manipulator. The second subsection shows the results of a simulation involving the two-link manipulator and an assembly task.

A. The Two-Link Manipulator

The dynamics and kinematics for the circular manipulator have been derived from the Euler–Lagrange equation. The actuators are assumed to be ideal, and the effects of friction have been ignored. The dynamics equation for this system is

$$\begin{bmatrix}u_1\\u_2\end{bmatrix}=\begin{bmatrix}I_1&0\\0&m_2\end{bmatrix}\begin{bmatrix}\ddot q_1\\\ddot q_2\end{bmatrix}+\begin{bmatrix}2m_2q_2\dot q_2\dot q_1+m_2gq_2\cos(q_1)\\-m_2q_2\dot q_1^2-m_2g\sin(q_1)\end{bmatrix}. \tag{27}$$

The inverse dynamics equation, (3), can be applied with $D(x)$ and $h(x, v)$ given by

$$D(x)=\begin{bmatrix}m_2+\dfrac{\sin^2(q_1)}{q_2^2}I_1 & \dfrac{-I_1\sin(q_1)\cos(q_1)}{q_2^2}\\ \dfrac{-I_1\sin(q_1)\cos(q_1)}{q_2^2} & m_2+\dfrac{\cos^2(q_1)}{q_2^2}I_1\end{bmatrix} \tag{28}$$

$$h(x,v)=\begin{bmatrix}2\sin(q_1)m_2\dot q_2\dot q_1-2\cos(q_1)m_2q_2\dot q_1^2\\-2\cos(q_1)m_2\dot q_2\dot q_1-2\sin(q_1)m_2q_2\dot q_1^2-m_2g\end{bmatrix}. \tag{29}$$

In free space, the end-effector impedance is given by $Z_e = mK_es(mIs^2+K_e)^{-1}$, and the second-order manipulator impedance was chosen accordingly as $Z_{mp} = M_ps + B_p + K_p/s$, where M_p, B_p, and K_p are all diagonal matrices.

When the end-effector comes in contact with a stiff surface, the end-effector impedance becomes $Z_e = K_e/s$ and the corresponding manipulator impedance is chosen as $Z_{mf} = M_fs + B_f(M_fs^2 + B_fs + K_f)^{-1}$. The inverse dynamics inputs a are obtained as in Section I-B. In two dimensions, the position subspace inputs are

$$\begin{aligned}a_p=\begin{bmatrix}\dot v_{m1}\\\dot v_{m2}\end{bmatrix}&+M_p^{-1}\begin{bmatrix}B_p&0\\0&B_p\end{bmatrix}\begin{bmatrix}v_{m1}-v_1\\v_{m2}-v_2\end{bmatrix}\\&+M_p^{-1}\begin{bmatrix}K_p&0\\0&K_p\end{bmatrix}\begin{bmatrix}x_{m1}-x_1\\x_{m2}-x_2\end{bmatrix}-\begin{bmatrix}F_1\\F_2\end{bmatrix}\end{aligned} \tag{30}$$

and the force subspace inputs are

$$a_f=M_f^{-1}\begin{bmatrix}F_{m1}-F_1\\F_{m2}-F_2\end{bmatrix}-B_fM_f^{-1}\begin{bmatrix}v_1\\v_2\end{bmatrix}. \tag{31}$$

The resulting control signal a is obtained from a_f and a_p as in (26). A block diagram of the HIC controller with the inner loop suppressed is shown in Fig. 10.

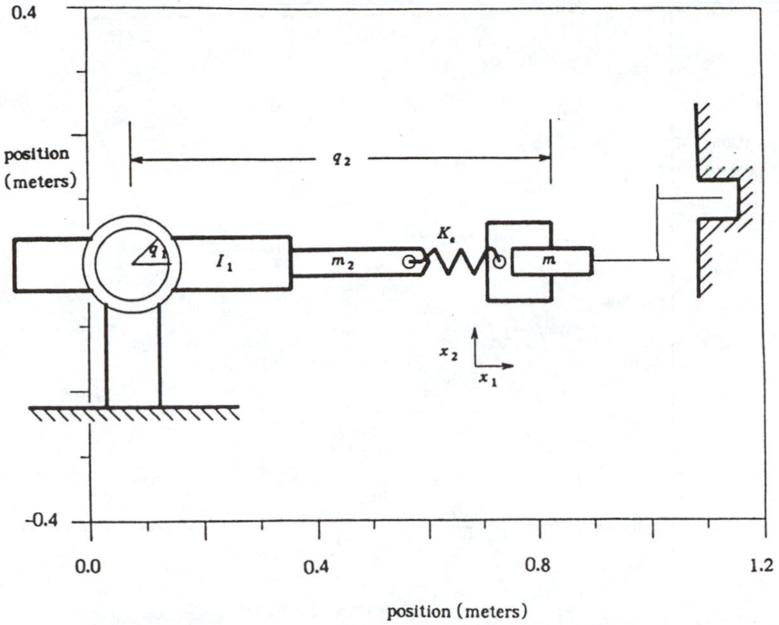

Fig. 11. Manipulator and assembly task.

B. Assembly Problem

The peg-in-the-hole problem is a classic problem, which involves moving a peg to a surface, moving along the surface until a hole is discovered, centering over the hole, and finally driving the peg into the hole. What is easily done by a human is not so easily implemented in a robot. Very subtle mechanisms are used to monitor forces. Slight adjustments are made to prevent jamming, and excessive forces. Determining algorithms is difficult because the task is done on such a low level. We do not think out our reflexes. Nevertheless, a robust systematic approach to the peg-in-the-hole problem may be implemented using hybrid impedance control.

The task may be broken into six stages:

1) Move to the surface along the x_1 direction, beneath the peg-hole with constant impedance along both DOF. When measured forces exceed preset thresholds, the surface has been contacted.
2) With the x_1-axis force controlled and the x_2-axis position controlled, maintain constant position, and let the force along the x_1 direction settle to within a reasonable bound.
3) Proceed in the x_2 direction, with constant force in the x_1 direction. Maintain a reasonable impedance along x_2 so that friction does not cause skipping. Continue until the x_1 position jumps sharply. The hole has been found.
4) Switch the x_2 direction to force control, and the x_1 position to position control. Back away from the far wall of the hole by commanding the force in the x_2 direction to zero. Keep the x_1 position constant.
5) Once the forces along x_2 have settled, proceed into the hole in the positive x_1 position. Maintain sufficient flexibility so that the peg does not jam. Continue until the force in the x_1 direction exceeds preset bounds.
6) Finally, let the forces felt from exerting pressure on the bottom of the hole settle to within a prescribed minimum. The peg is now securely in the hole.

This algorithm was implemented in Fortran using the HIC strategy. Instead of changing the impedance to best match each stage of the task, however, a constant force impedance, and a constant position impedance were used.

The results of the simulations are shown in Figs. 11–13. The six stages begin at t = 0.0, t = 0.95, t = 1.50, t = 2.15, t = 2.9, and t = 3.5 s, respectively. Fig. 11 shows the manipulator operating in the plane. The environment with peg-hole is also shown. The solid black line represents the movement of the wrist. Fig. 12 shows both the wrist position along x_1 and the desired wrist position along x_1 as a function of time. It is shown that when x_1 is position-controlled, x_1 tracks the desired position. When movement along the x-axis is force-controlled, however, actual and desired positions deviate. Fig. 13 shows the force and position in the x_1 direction both plotted against time. When movement is force-controlled, the force approaches the desired force exponentially. Small oscillations in force are due to vibrations of the passive end-effector spring.

V. Conclusion

Hybrid impedance control represents a unification of many previous approaches to force and position control of robotics. By combining recent advances in inverse dynamics with already existing force control strategies, such as impedance control and hybrid control, a cohesive general plan has been developed.

The careful modeling of the environment has been shown to be instrumental in determining the proper control strategy. Proper models with appropriate controls make it possible for output responses to match input commands in the steady state. Duality between Thèvenin and Norton equivalents insures that a robot is being controlled consistently with respect to the environment.

The implementation of HIC was demonstrated for a simple two-link manipulator, applied to the peg-in-the-hole task. Inverse dynamics in task space were calculated, and the general impedance notation used in the development of HIC was made specific in terms of spring and damping constants.

Despite the generality of hybrid impedance control, a number of issues remain. In the simulations, a constant impedance was maintained, and only force and position commands were adjusted. Human impedances change constantly, however, and such an approach might prove advantageous in robotics. Programming is also significantly more difficult since it is hard to understand a task in terms of applied forces. Perhaps an AI system could be developed for determining appropriate force and position commands for a certain task. Furthermore, numerous simplifications were made in the model. Gear backlash, flexibility, static friction, torque constraints, and noise all detract from the ideal system. It was assumed that a devoted inverse dynamics controller existed which not only computed inverse dynamics in task space, but also included measured force signals as part of the process.

Hybrid impedance control is just one step towards developing more

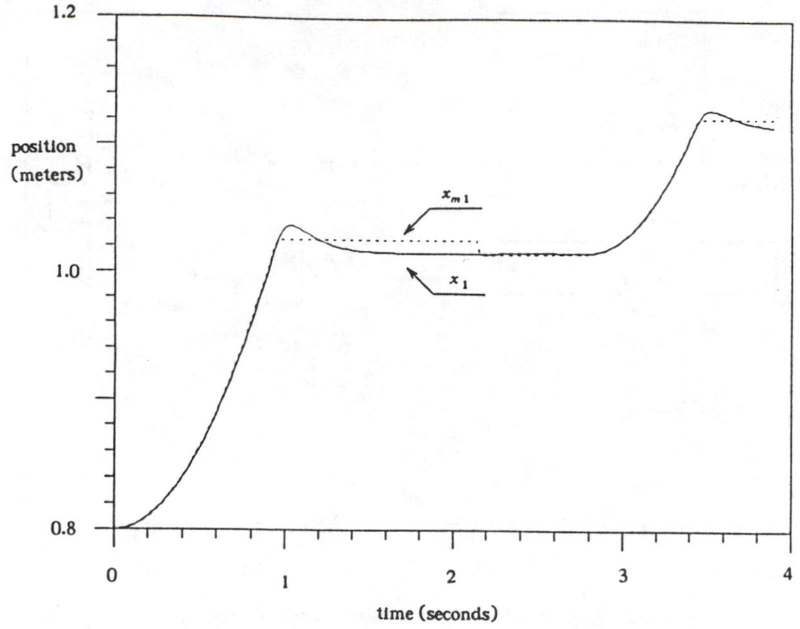

Fig. 12. Wrist position during assembly.

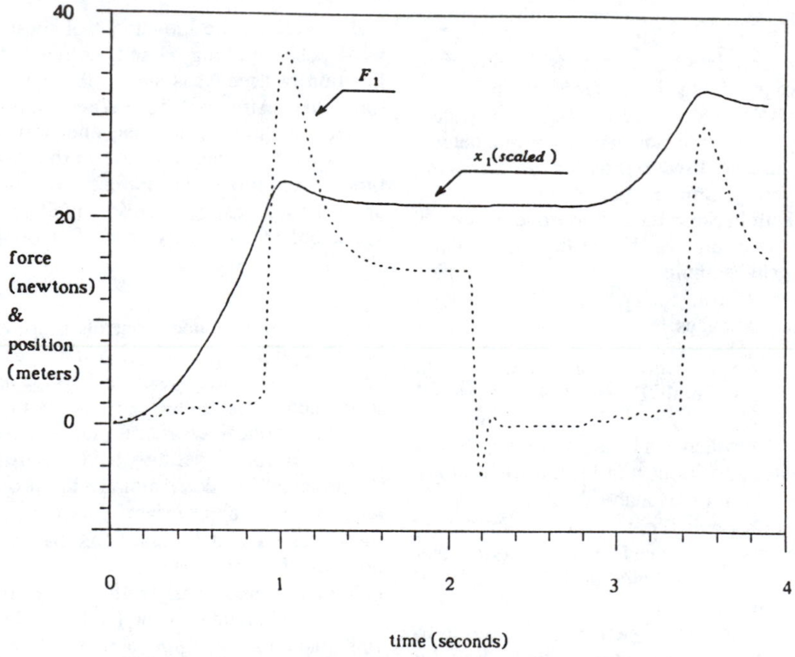

Fig. 13. Force and position in x_1 direction during assembly.

general and more sophisticated robots. Further reduction in the cost of computing power and the increased availability of sensors should insure, however, the future of such control strategies.

References

[1] B. Shimano and B. Roth, "On force sensing information and its use in controlling manipulators," in *Proc. 8th Int. Symp. on Industrial Robots* (Washington, DC), pp. 119-126.

[2] A. K. Bejczy, "Robot arm dynamics and control," NASA-JPL Tech. Memo., 33-69, 1974.

[3] O. Khatib and J. Burdick, "Motion and force control of robot manipulators," in *Proc. 1986 Int. Conf. on Robotics and Automation*, pp. 1381-1386.

[4] J. Y. S. Luh, M. W. Walker, and R. P. Paul, "Resolved-acceleration control of mechanical manipulators," *IEEE Trans. Automat. Contr.*, vol. AC-25, no. 3, pp. 195-200, June 1980.

[5] S. Dubowsky and D. T. DesForges, "The application of model-reference adaptive control to robotic manipulators," *Trans. ASME, J. Dyn. Syst., Meas. Contr.*, vol. 101, pp. 193-200, Sept. 1979.

[6] J. Slotine, "Robustness issues in robot control," in *Proc. 1985 Int. Conf. on Robotics and Automation* (St. Louis, MO), pp. 656-661.

[7] M. Spong, J. Thorp, and J. Kleinwaks, "Robust microprocessor control of robot manipulators," *Automatica*, vol. 23, no. 3, pp. 373-379, 1987.

[8] M. Spong, "Robust stabilization for a class of nonlinear systems," in *Theory and Applications of Nonlinear Control Systems*. Amsterdam, The Netherlands: North-Holland, 1986, pp. 155-165.

[9] R. J. Anderson, "Hybrid admittance/impedance force control of robotic manipulators," Masters thesis submitted to the University of Illinois at Urbana-Champaign, Oct. 1986.

[10] D. E. Whitney, "Historical perspective and state of the art in robot force control," in *Proc. 1985 IEEE Int. Conf. on Robotics and Automation* (St. Louis, MO), pp. 262-268.

[11] M. H. Raibert and J. J. Craig, "Hybrid Position/Force Control of Manipulators," *J. Dyn. Syst. Contr.*, vol. 102, June 1981.
[12] M. T. Mason, "Compliance and force control for computed controlled manipulators," *IEEE Trans. Syst., Man Cybern.*, vol. SMC-11, no. 6, pp. 418–432, June 1981.
[13] H. Zhang and R. P. Paul, "Hybrid control of robot manipulators," in *Proc. 1985 IEEE Int. Conf. on Robotics and Automation* (St. Louis, MO), pp. 602–607.
[14] N. Hogan, "Impedance control: An approach to manipulation," *J. Dyn. Syst., Meas., Contr.*, vol. 107, pp. 1–7, Mar. 1985.
[15] ——, "Beyond regulators: Modeling control systems as physical systems," preprint. Nov. 1986.

Feedback Stabilization and Tracking of Constrained Robots

N. HARRIS McCLAMROCH, FELLOW, IEEE, AND DANWEI WANG

Abstract—**Mathematical models for constrained robot dynamics, incorporating the effects of constraint forces required to maintain satisfaction of the constraints, are used to develop explicit conditions for stabilization and tracking using feedback. The control structure allows feedback of generalized robot displacements, velocities, and the constraint forces. Global conditions for tracking based on a modified computed torque controller and local conditions for feedback stabilization using a linear controller are presented. The framework is also used to investigate the closed-loop properties if there are force disturbances, dynamics in the force feedback loops, or uncertainty in the constraint functions.**

I. Introduction and Motivation

Current industrial robots are characterized by a wide diversity of physical design configurations. However, the basic task capabilities of most current robots are quite limited. The most common tasks involve so-called "pick and place" operations, which characterize the vast majority of current robot applications. If robot technology is to have a more wide-ranging impact on industrial practice, it is essential that the task capabilities of robots be substantially expanded.

A robot can be viewed as a physical mechanism for performing work; the mechanism is often defined as a connection of articulated links constructed so that the end of the last link (which may include a gripper holding an object or a tool holder containing a tool) is the location at which the work is performed. It is natural to focus on the so-called end effector of the robot and to define particular robot tasks in terms of desired motions of the end effector; this is the common view in dealing with "pick and place" operations. However, there are many industrial tasks (most of which cannot be automated using current robots) that are defined in a fundamentally different way. In particular, there are numerous tasks which cannot be defined solely in terms of motion of the end effector. Of specific interest in this paper are tasks which are characterized by physical contact between the end effector and a constraint surface. A long list of such tasks can be given, including scribing, writing, deburring, grinding, and others [1]–[7].

There have been numerous research publications which have dealt with such applications. Although the primary focus of such research has not often been on the role of constraints in defining the tasks, research on compliant control [8]–[11] and force feedback control [12]–[19] are closely related. Several formal control design approaches have been proposed and there have been descriptions of related robot experiments. However, there has been no theoretical framework established which can serve as a basis for the study of robot performance of constrained tasks.

It is the premise of this paper that there is a need for a carefully developed theoretical framework for investigation into application of robots to such tasks. Furthermore, it is our premise that such a theoretical framework should explicitly incorporate the effects of the forces of constraint. Note that the constraint forces are not exogenous but are implicitly defined as the forces required to maintain satisfaction of the constraints. Specifically, we present a dynamic model of a robot, incorporating constraint effects; the form of the model follows from classical results in dynamics [20]–[21] and has recently been recognized as the proper theoretical model for constrained robot problems [22]–[28].

This model is used in this paper to develop theoretical conditions for closed-loop stabilization and tracking using a class of feedback controllers. The constraints are assumed to be nonlinear and the development is based on a coordinate transformation for which the constraints are expressed in a simple form. Global tracking conditions are developed for the case of nonlinear dynamics using a modification of the computed torque method. Local stabilization conditions are also developed using a linear controller. The results are also presented for the case of linear constraints, for simplicity. Such stabilization conditions are new; they serve to provide a theoretical basis for the use of force feedback in constrained systems such as has been suggested in [16]. The constrained model studied in this paper can also form the basis for a rigorous analysis of the closed-loop properties that arise from the use of other control approaches described in the literature [8]–[19].

We further investigate the properties of the closed loop using the proposed controller structure; we show that "high gain displacement feedback loops" reduce the steady-state displacement regulation error for constant force disturbances and uncertainty in the constraint function. We also show that "high gain force feedback loops" reduce the steady-state constraint force regulation error for constant force disturbances and uncertainty in the constraint function. Further, closed-loop stabilization is shown not to be affected by a certain class of dynamics in the constraint force feedback loops.

II. Formulation of Constrained Dynamic Equations

Our development is based on a Lagrangian formulation of robot dynamics, in a coordinate system convenient for characterizing the robot motion. Let $q \in R^n$ denote the vector of generalized displacements, in robot coordinates. If $q:R^1 \rightarrow R^n$ is differentiable, then $\dot{q}$ denotes its time derivative; $q \in R^n$ is viewed as a column vector so that its transpose q' is a row vector. We assume the existence of a symmetric, positive definite matrix valued inertia function $M:R^n \rightarrow R^{n\times n}$, and a scalar valued potential function $V:R^n \rightarrow R^1$, such that the equations of motion of the robot are defined in terms of the Lagrangian function $L(q, \dot{q}) = 0.5\, \dot{q}'M(q)\, \dot{q} - V(q)$ as

$$\frac{d\partial L}{dt\partial \dot{q}}(q, \dot{q}) - \frac{\partial L}{\partial q}(q, \dot{q}) = f + u$$

Manuscript received July 6, 1987; revised October 26, 1987. This paper is based on a prior submission of September 17, 1986. Paper recommended by Associate Editor, S. S. Sastry. This work was supported in part by the Center for Research on Integrated Manufacturing (CRIM) at The University of Michigan.

N. H. McClamroch is with the Department of Aerospace Engineering and the Department of Electrical Engineering and Computer Science, The University of Michigan, Ann Arbor, MI 48109.

D. Wang is with the Program in Computer, Information and Control Engineering, The University of Michigan, Ann Arbor, MI 48109.

IEEE Log Number 8819551.

Reprinted from *IEEE Trans. Automat. Contr.*, vol. 33, no. 5, pp. 419–426, May 1988.

where $f \in R^n$ denotes the vector of generalized constraint forces in robot coordinates, and $u \in R^n$ denotes the vector of other generalized force inputs, which includes the control input, disturbance inputs, and any dissipative effects, in robot coordinates. Thus, the equations of motion for the constrained robot, in robot coordinates, are given by

$$M(q)\ddot{q}+F(q, \dot{q})=f+u \tag{1}$$

where

$$F(q, \dot{q})=\frac{d}{dt}[M(q)]\dot{q}-0.5\frac{\partial \dot{q}M(q)\dot{q}}{\partial q}+\frac{\partial V(q)}{\partial q}.$$

Let $p \in R^n$ denote the generalized position vector of the robot end effector, in coordinates in which constraints on the end effector are defined. These (environmental) constraints are such that the generalized position vector of the robot end effector is assumed to satisfy the algebraic equation $\theta(p) = 0$ where the constraint function $\theta: R^n \rightarrow R^m$. We assume that the generalized position vector of the robot end effector, in constraint coordinates, can be expressed in terms of the generalized displacement in robot coordinates according to the algebraic equation $p = H(q)$, where the mapping $H: R^n \rightarrow R^n$ is invertible. Our subsequent development does not assume that the inverse kinematic relations be expressible in closed form. Thus, the constraint function defined by $\phi(q) = \theta(H(q))$, in robot coordinates, satisfies the constraint equation

$$\phi(q)=0. \tag{2}$$

Since the constraints are holonomic it follows, as shown in [20], [21], that the generalized constraint forces, in robot coordinates, are given by the relation

$$f=J'(q)\lambda \tag{3}$$

where $\lambda \in R^m$ is a vector of generalized multipliers associated with the constraints, and the Jacobian matrix $J(q) = \partial\phi(q)/\partial q$. The constraint force on the end effector, i.e., the contact force between the end effector and the constraint surface, can be expressed, in the constraint coordinates, in terms of the constraint multiplier vector λ.

Note that if $\phi(q) = 0$ is identically satisfied, then also $J(q)\dot{q} = 0$. Thus, the constraint manifold S in R^{2n} defined by $S = \{(q, \dot{q}): \phi(q) = 0, J(q)\dot{q} = 0\}$ is fundamental in the development. In particular, if conditions given in [25]–[26] are satisfied, then there is a unique solution of (1)–(3), denoted by $q(t)$ satisfying $(q(t), \dot{q}(t)) \in S$, for each $(q(0), \dot{q}(0)) \in S$. That is, S is an invariant manifold. Thus, the constraints on the robot can be viewed as restricting the dynamics to the manifold S only rather than to the space R^{2n}; in other words the model described by (1)–(3) is singular on R^{2n} as described in [25]. It is this feature that is critical in defining the constrained robot dynamics; this is also the fundamental source of difficulty in analysis and control of the robot dynamics.

We again emphasize the important role of the constraints in the constrained dynamics, especially with relation to the stabilization problem. In particular, it is easy to see that a closed-loop system that is asymptotically stable if constraints are ignored may, in fact, not be asymptotically stable if the constraints are imposed. A simple theoretical example has been presented in [29] which demonstrates this possibility. In addition, robot experiments have also indicated that a closed loop may be destabilized when a hard constraint (infinitely stiff environment) is imposed on the end effector [30].

III. Transformation of Constrained Dynamic Equations

In order to carry out our subsequent development, we assume that the constraint function satisfies the following.

Assumption 1: There is an open set $\Theta \subset R^{n-m}$ and a function $\Omega: \Theta \lesssim R^m$ such that

$$\phi(\Omega(q_2), q_2)=0 \quad \text{for all } q_2 \in \Theta.$$

Suppose a constant vector $\bar{q} \in R^n$ satisfies $\phi(q) = 0$; Assumption 1 holds in some neighborhood of $\bar{q}$ if rank $J(\bar{q}) = m$, according to the implicit function theorem, although a reordering of the variables may be required. Let Assumption 1 hold with $\Theta = R^{n-m}$. A transformation is now made so that the constraint equations are written in a simple form. First define the vector partition $q' = (q_1', q_2')$, where $q_1 \in R^m$, $q_2 \in R^{n-m}$. Define the nonlinear transformation $X: R^n \rightarrow R^n$ by

$$x=X(q)=\begin{vmatrix} q_1-\Omega(q_2) \\ q_2 \end{vmatrix}$$

which is differentiable and has a differentiable inverse transformation $Q: R^n \rightarrow R^n$ given by

$$q=Q(x)=\begin{vmatrix} x_1+\Omega(x_2) \\ x_2 \end{vmatrix}$$

where the vector partition $x' = (x_1', x_2')$, $x_1 \in R^m$, $x_2 \in R^{n-m}$, is used. We also define the Jacobian matrix of the inverse transformation

$$T(x)=\frac{\partial Q}{\partial x}(x)=\begin{vmatrix} I_m & \frac{\partial \Omega}{\partial x_2}(x_2) \\ 0 & I_{n-m} \end{vmatrix}$$

which is necessarily nonsingular. By abuse of notation we often write $T(x_2)$ or $T(q)$ in place of $T(x)$. The differential equations (1) can be expressed in terms of the variables x as

$$T'(x)M(Q(x))T(x)\ddot{x}+T'(x)\{F(Q(x), T(x)\dot{x}) +M(Q(x))\dot{T}(x)\dot{x}\}=T'(x)u+T'(x)f.$$

For simplicity, introduce the function definitions $\bar{M}(x) = T'(x)M(Q(x))T(x)$ and $\bar{F}(x, \dot{x}) = T'(x)\{F(Q(x), T(x)\dot{x}) + M(Q(x))\dot{T}(x)\dot{x}\}$ and introduce the partitioning of the identity matrix $I_n = [E_1' \mid E_2']$ where E_1 is an $m \times n$ matrix and E_2 is an $(n-m) \times n$ matrix. These equations can be written in a so-called reduced form as

$$E_1\bar{M}(x_2)E_2'\ddot{x}_2+E_1\bar{F}(x_2, \dot{x}_2)=E_1T'(x_2)u+E_1T'(x_2)f \tag{4}$$

$$E_2\bar{M}(x_2)E_2'\ddot{x}_2+E_2\bar{F}(x_2, \dot{x}_2)=E_2T'(x_2)u \tag{5}$$

where we note that in (5), $E_2T'(x_2)f = 0$ follows from (3) and Assumption 1. The notation $F(x_2, \dot{x}_2)$ denotes $F(x, \dot{x})$ evaluated at $x' = (0, x_2')$, $\dot{x} = (0, \dot{x}_2')$, etc. In the transformed coordinates the constraint equation is

$$x_1=0 \tag{6}$$

and the constraint force satisfies

$$f=J'(x_2)\lambda. \tag{7}$$

This reduced form has a useful interpretation that forms the basis for our subsequent development. *The ordinary differential equation (5) characterizes the motion of the robot on the constraint manifold; equation (4) can be viewed as an algebraic equation for the constraint force expressed in terms of the motion on the constraint manifold. It is this structure that is crucial for our approach.*

An important special case occurs if the constraints are linear; it can then be shown that the above development can be carried out explicitly in terms of a singular value decomposition. We make the following assumption.

Assumption 2: Assume that $\phi(q) = Jq$ where J is a constant $m \times n$ matrix with rank of J being m.

Then it follows that there is a singular value decomposition of J such that $J = U\Sigma V'$ where U and V are $m \times m$ and $n \times n$ orthogonal matrices, respectively, and $\Sigma = [D \vdots 0]$, $D = \text{diag}\,[d_1, \cdots, d_m]$ with $d_i > 0$, $i = 1, \cdots, m$ being the singular values of J.

Now the previous development holds with $Q(x) = Vx$, $X(q) = V'q$, and with $T(x) = V$. Our development is considerably simplified in this case; consequently specific results are subsequently presented for the case of linear constraints.

IV. Global Tracking Using Nonlinear Feedback

In this section, we consider a general tracking problem for constrained robots. For simplicity, we formulate the tracking problem in terms of robot coordinates. The desired motion and desired constraint forces, in robot coordinates, are defined by vector functions $q_d: R^1 \to R^n$, $f_d: R^1 \to R^n$. For consistency with the imposed constraints, it is necessary that $\phi(q_d) = 0$ and $f_d = J'(q_d)\lambda_d$, identically, for some multiplier function $\lambda_d: R^1 \to R^m$. Note that λ_d also defines the desired contact force on the end effector in the constraint coordinates.

Our objective is to determine a nonlinear feedback controller to solve the following tracking problem. A feedback control u, depending on the tracking functions q_d, $\dot q_d$, $\ddot q_d$, f_d, and feedback of the generalized displacement q, the generalized velocity $\dot q$, and the generalized constraint force f, is to be selected so that for all $(q(0), \dot q(0)) \in S$ it follows that the closed-loop responses satisfy

$$q(t) \to q_d(t) \quad \text{as } t \to \infty$$

$$f(t) \to f_d(t) \quad \text{as } t \to \infty.$$

We consider a version of the computed torque controller [12], [31], modified to accommodate the presence of the constraints, feedback of the constrained motion and the constraint forces, and the simultaneous motion and force tracking objectives. As indicated previously, our development is based on application of the computed torque controller concept to the reduced equations (4)–(7). The controller is chosen so that

$$\begin{aligned} E_1T'(x_2)u = {} & E_1\bar M(x_2)E_2'\ddot x_{2d} - E_1T'(x_2)J'(x_2)\lambda_d + E_1\bar F(x_2, \dot x_2) \\ & + E_1\bar M(x_2)E_2'[G_v(\dot x_{2d} - \dot x_2) + G_d(x_{2d} - x_2)] \\ & + E_1E_1'G_fE_1T'(x_2)J'(x_2)(\lambda - \lambda_d) \end{aligned}$$

$$\begin{aligned} E_2T'(x_2)u = {} & E_2\bar M(x_2)E_2'\ddot x_{2d} - E_2T'(x_2)J'(x_2)\lambda_d + E_2\bar F(x_2, \dot x_2) \\ & + E_2\bar M(x_2)E_2'[G_v(\dot x_{2d} - \dot x_2) + G_d(x_{2d} - x_2)] \\ & + E_2E_1'G_fE_1T'(x_2)J'(x_2)(\lambda - \lambda_d) \end{aligned}$$

where G_v and G_d are $(n-m) \times (n-m)$ constant feedback gain matrices and G_f is an $m \times m$ constant feedback gain matrix. The controller can be expressed in terms of the variables in robot coordinates as

$$\begin{aligned} u = {} & M(q)T(q)T(q_d)^{-1}\ddot q_d - J'(q)\lambda_d + F(q, \dot q) \\ & + M(q)\{\dot T(q)T(q)^{-1}\dot q - T(q)T(q_d)^{-1}\dot T(q_d)^{-1}\dot q_d\} \\ & + M(q)\{T(q)E_2'G_vE_2[T(q_d)^{-1}\dot q_d - T(q)^{-1}\dot q] \\ & + T(q)E_2'G_dE_2[X(q_d) - X(q)]\} \\ & + T'(q)^{-1}E_1'G_fE_1T'(q)J'(q)(\lambda - \lambda_d). \end{aligned} \tag{8}$$

The controller can also be expressed in terms of the constraint force and the desired constraint force as

$$\begin{aligned} u = {} & M(q)T(q)T(q_d)^{-1}\ddot q_d - J'(q)[J(q_d)J'(q_d)]^{-1}J(q_d)f_d \\ & + F(q, \dot q) + M(q)\{\dot T(q)T(q)^{-1}\dot q - T(q)T(q_d)^{-1}\dot T(q_d) \\ & \cdot T(q_d)^{-1}\dot q_d\} + M(q)\{T(q)E_2'G_vE_2[T(q_d)^{-1}\dot q_d - T(q)^{-1}\dot q] \\ & + T(q)E_2'G_dE_2[X(q_d) - X(q)]\} + T'(q)^{-1}E_1'G_fE_1T'(q) \\ & \cdot \{f - J'(q)[J(q_d)J'(q_d)]^{-1}J(q_d)f_d\}. \end{aligned} \tag{9}$$

Using the relations $E_2T'(x_2)J'(x_2) = 0$, $E_1E_1' = I_m$, and $E_2E_1' = 0$, the closed-loop equations, in reduced form, can be shown to be given by the linear equations

$$E_1\bar M(x_2)E_2'\{\ddot e_2 + G_v\dot e_2 + G_de_2\} = (I_m + G_f)E_1T'(x_2)J'(x_2)(\lambda - \lambda_d) \tag{10}$$

$$E_2\bar M(x_2)E_2'\{\ddot e_2 + G_v\dot e_2 + G_de_2\} = 0 \tag{11}$$

$$e_1 = 0 \tag{12}$$

where $e_2 = x_2 - E_2X(q_d)$, $e_1 = x_1$.

Conditions on the gain matrices so that the tracking problem is solved are readily obtained from these equations, since (11) is an ordinary differential equation which characterizes the motion on the constraint manifold. The matrix gains G_v and G_d can be selected so that, according to (11), $e_2 \to 0$, and hence $q \to q_d$ as $t \to \infty$. Then from (10) it follows that $\lambda \to \lambda_d$, and hence $f \to f_d$ as $t \to \infty$.

Using this framework, the following result is obtained.

Theorem 1: Suppose that Assumption 1 is satisfied with $\Theta = R^{n-m}$. The closed-loop system defined by the plant equations (1)–(3) and the controller (9) is globally asymptotically stable in the sense that

$$q(t) \to q_d(t) \quad \text{as } t \to \infty$$

$$f(t) \to f_d(t) \quad \text{as } t \to \infty$$

for any $(q(0), \dot q(0)) \in S$ if G_v and G_d are symmetric and positive definite and G_f is symmetric and nonnegative definite.

We now consider the simpler case where the constraints are linear. In such case the feedback controller (9) can be written as

$$\begin{aligned} u = {} & M(q)\ddot q_d - f_d + F(q, \dot q) + M(q)[VE_2'G_vE_2V'(\dot q_d - \dot q) \\ & + VE_2'G_dE_2V'(q_d - q)] + VE_1'G_fE_1V'(f - f_d) \end{aligned} \tag{13}$$

where G_v and G_d are $(n - m) \times (n - m)$ constant feedback gain matrices and G_f is an $m \times m$ constant feedback gain matrix. Conditions so that the closed-loop system solves the tracking problem in this case are as follows.

Corollary 2: Suppose that Assumption 2 is satisfied. The closed-loop system defined by (1)–(3) and (13) is globally asymptotically stable in the sense that

$$q(t) \to q_d(t) \quad \text{as } t \to \infty$$

$$f(t) \to f_d(t) \quad \text{as } t \to \infty$$

for any $(q(0), \dot q(0)) \in S$ if G_v and G_d are symmetric and positive definite and G_f is symmetric and nonnegative definite.

It is important to note that the dependence of the controller on the constraint force in expressions (9) and (13) is crucial; in particular, the second term in the control expression in (13), namely the desired tracking constraint force $-f_d$, cannot be replaced by the feedback constraint force $-f$, because the closed-loop system in such case is ill-posed. Such an incorrect approach has, in fact, been proposed in the literature. One proper form for

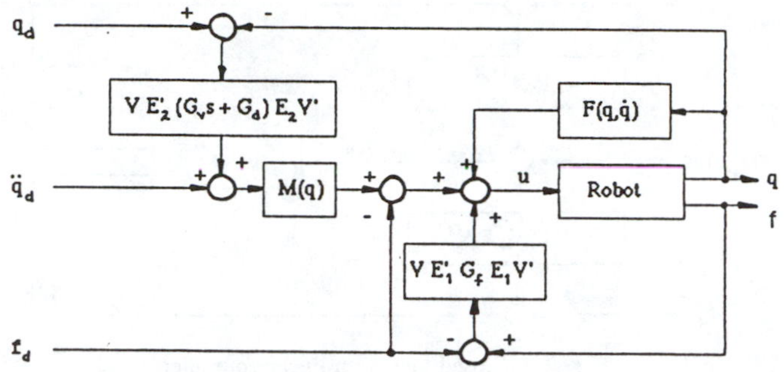

Fig. 1. Closed loop with modified computed torque controller.

introducing feedback of the constraint force is through the controller expressions given by (9) or (13).

Note that the closed loop can be stabilized even if $G_f = 0$; feedback of the constraint force is not required for stabilization. But, as we subsequently indicate, there are potential robustness advantages in using feedback of the constraint force.

The control relations (9) and (13) suggest that feedback of the generalized displacement q, the generalized velocity $\dot q$, and the generalized constraint force f, in robot coordinates, is required; but (9) and (13) could be modified to depend on feedback of displacement and velocity of the end effector and feedback of the contact force on the end effector, in the constraint coordinates. Of course, the control expressions (9) and (13) could also be modified to depend on feedback of the motion on the constraint manifold, e.g., through feedback of q_2 and $\dot q_2$.

A schematic diagram of the closed-loop system, indicating the control structure (13), is shown in Fig. 1, for the case where the constraints are linear. The viewpoint of the controller as a computed torque controller, modified to conform to the constraints, and depending on feedback of q, $\dot q$, and f is clarified from that figure. It should be noted that the control structure (13) is a generalization of the hybrid control architecture presented by Raibert and Craig in [16]; the specific motion and force selection matrices introduced in [16] correspond to the specific case that $\phi(q) = q_1$ in our notation. In addition, we have presented conditions guaranteeing that the closed loop, subject to the imposed constraints, is asymptotically stable, for the general case.

V. Local Stabilization Using Linear Feedback

In this section, we consider a regulation problem where desired *constant* regulation vectors $q_d \in R^n$, $f_d \in R^n$ are given. For consistency with the imposed constraints, it is necessary that $\phi(q_d) = 0$ and $f_d = J'(q_d)\lambda_d$, identically, for some constant multiplier vector λ_d. Our objective is to determine a linear feedback controller to solve the following regulation problem. A feedback control u, depending on the constant regulation vectors q_d and f_d, and feedback of the generalized displacement q, the generalized velocity $\dot q$, and the generalized constraint force f, is to be selected so that there is a neighborhood N of $(q_d, 0)$ in R^{2n} such that for all $(q(0), \dot q(0)) \in S \cap N$ it follows that the closed-loop responses satisfy

$$q(t) \to q_d \quad \text{as } t \to \infty$$

$$f(t) \to f_d \quad \text{as } t \to \infty.$$

We now consider a linear controller, with feedback of the constrained motion and the constraint forces, that accommodates the presence of the constraints and the simultaneous motion and force regulation objectives. As indicated previously, our development is based on application of linear control concepts to the reduced equations (4)–(7). The controller is chosen to be a linear feedback function of the indicated form and to satisfy the following:

$$E_1T'(x_{2d})[u-F(q_d,0)+f_d] = E_1E_2'G_v(\dot x_{2d}-\dot x_2)+E_1E_2'G_d(x_{2d}-x_2) + E_1E_1'G_fE_1T'(x_{2d})J'(x_{2d})(\lambda-\lambda_d)$$

$$E_2T'(x_{2d})[u-F(q_d,0)+f_d] = E_2E_2'G_v(\dot x_{2d}-\dot x_2)+E_2E_2'G_d(x_{2d}-x_2) + E_2E_1'G_fE_1T'(x_{2d})J'(x_{2d})(\lambda-\lambda_d)$$

where G_v and G_d are $(n-m)\times(n-m)$ constant feedback gain matrices and G_f is an $m\times m$ constant feedback gain matrix. This controller can be expressed in terms of the variables in robot coordinates as

$$u=F(q_d,0)-f_d+T'(q_d)^{-1}E_2'G_vE_2T(q_d)^{-1}(\dot q_d-\dot q) + T'(q_d)^{-1}E_2'G_dE_2T(q_d)^{-1}(q_d-q) + T'(q_d)^{-1}E_1'G_fE_1T'(q_d)J'(q_d)(\lambda-\lambda_d). \tag{14}$$

The controller can also be expressed in terms of the difference of the constraint force and the desired constraint force; the difference $(\lambda-\lambda_d)$ can be expressed in terms of (q_d-q) and $(f-f_d)$ using (3), to first order. The result can be substituted into (14) to obtain

$$u=F(q_d,0)-f_d+T'(q_d)^{-1}E_2'G_vE_2T(q_d)^{-1}(\dot q_d-\dot q) + T'(q_d)^{-1}E_2'G_dE_2T(q_d)^{-1}(q_d-q) + T'(q_d)^{-1}E_1'G_fE_1T'(q_d)[(f-f_d) - \left.\frac{\partial(J'(q)\lambda_d)}{\partial q}\right|_{q=q_d} T(q_d)E_2'E_2T(q_d)^{-1}(q-q_d)]. \tag{15}$$

Using the relations $E_1E_2' = 0$, $E_1E_1' = I_m$, $E_2E_2' = I_{n-m}$, and $E_2E_1' = 0$, the linearized closed-loop equations, in reduced form, can be shown to be given by

$$E_1\bar M(x_2)E_2'\ddot e_2+E_1\bar K(x_{2d})E_2'e_2 = (I_m+G_f)E_1T'(x_{2d})J'(x_{2d})(\lambda-\lambda_d) \tag{16}$$

$$E_2\bar M(x_{2d})E_2'\ddot e_2+G_v\dot e_2+[G_d+E_2\bar K(x_{2d})E_2']e_2=0 \tag{17}$$

$$e_1=0 \tag{18}$$

where

$$\bar K(x_{2d})=\frac{\partial}{\partial x}\{T'(x)[F(Q(x),T(x)\dot x)-F(Q(0,x_{2d}),0) + J'(x_{2d})\lambda_d-J'(x)\lambda_d]\}\Big|_{x=(0,x_{2d}),\,\dot x=(0,0)}.$$

Conditions on the gain matrices so that the regulation problem is solved are readily obtained using the ordinary differential equation (17) which characterizes the linearized motion on the constraint manifold. The matrix gains G_v and G_d can be selected so that, according to (17), $e_2 \to 0$, and hence $q \to q_d$ as $t \to \infty$ hold locally. Then from (16) it follows that $\lambda \to \lambda_d$, and hence $f \to f_d$ as $t \to \infty$ hold locally.

The following result is obtained.

Theorem 3: Suppose that Assumption 1 is satisfied in some neighborhood of q_d. The closed-loop system defined by the plant equations (1)–(3) and controller (15) is locally asymptotically stable in the sense that there is a neighborhood N of $(q_d, 0)$ such that

$$q(t) \to q_d \quad \text{as } t \to \infty$$

$$f(t) \to f_d \quad \text{as } t \to \infty$$

for any $(q(0), \dot{q}(0)) \in S \cap N$ if G_v and G_d are symmetric and positive definite such that all $2(n - m)$ zeros of

$$\det \{E_2 T'(q_d)[M(q_d)s^2 + K(q_d)]T(q_d)E_2' + G_v s + G_d\}$$

have negative real parts and G_f is symmetric and nonnegative definite where

$$K(q_d) = T'(q_d)^{-1} \frac{\partial}{\partial q}\{T'(q)[F(q, \dot{q}) - F(q_d, 0) + J'(q_d)\lambda_d - J'(q)\lambda_d]\}|_{q=q_d, \dot{q}=0}.$$

We now consider the simpler case where the constraints are linear. In such case the linear feedback controller (15) can be written as

$$u = F(q_d, 0) - f_d - VE_2' G_v E_2 V'\dot{q} + VE_2' G_d E_2 V'(q_d - q) + VE_1' G_f E_1 V'(f - f_d) \quad (19)$$

where G_v and G_d are $(n - m) \times (n - m)$ constant feedback gain matrices and G_f is an $m \times m$ constant feedback gain matrix. Conditions so that the closed-loop system solves the regulation problem are as follows.

Corollary 4: Suppose that Assumption 2 is satisfied. The closed-loop system defined by the plant equations (1)–(3) and controller (19) is locally asymptotically stable in the sense that there is a neighborhood N of $(q_d, 0)$ such that

$$q(t) \to q_d \quad \text{as } t \to \infty$$

$$f(t) \to f_d \quad \text{as } t \to \infty$$

for any $(q(0), \dot{q}(0)) \in S \cap N$ if G_v and G_d are symmetric and positive definite such that all $2(n - m)$ zeros of

$$\det \{E_2 V'[M(q_d)s^2 + K(q_d)] VE_2' + G_v s + G_d\}$$

have negative real parts and G_f is symmetric and nonnegative definite and

$$K(q_d) = \frac{\partial F(q, \dot{q})}{\partial q}\bigg|_{q=q_d, \dot{q}=0}.$$

Again we mention that the control relations (15) and (19) have been written in terms of feedback of the motion and constraint forces in robot coordinates, but they could be modified to depend on the motion of the end effector and the contact force on the end effector in constraint coordinates.

A schematic diagram of the closed-loop system, indicating the control structure (19), is shown in Fig. 2, for the case where the constraints are linear. The controller consists of a constant bias input plus linear terms proportional to the generalized displacement error, the generalized velocity error, and the generalized constraint force error, all in robot coordinates.

VI. Consequences of Model Imperfections

We have developed conditions which indicate how feedback controllers can be developed so that the closed-loop is asymptotically stable. In this case stability is defined as robustness to changes in the initial data that are consistent with satisfaction of the imposed constraints. As has been demonstrated, feedback can be used to improve the closed-loop properties with respect to such uncertainties.

But feedback can be expected to play a further important role in possibly reducing the effects of disturbances and model imperfections. In this section we briefly consider the implications of using feedback in the case that there are external force disturbances, additional dynamics in the constraint force feedback loops, and uncertainties in the constraints. To avoid unnecessary complications, our developments are based on the linearized equations, assuming that the constraints are linear, i.e., that Assumption 2 holds. Thus, our conclusions are approximations only valid near an equilibrium. Nevertheless, the qualitative features of our conclusions are of substantial importance as they suggest the general implications of the feedback control structure studied. Our conclusions about the effects of model uncertainties are limited; but the indicated linearized equations could form the basis for a more detailed study.

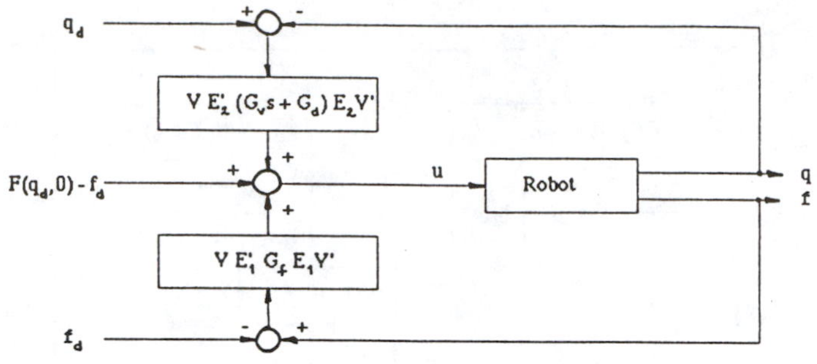

Fig. 2. Closed loop with linear controller.

Effects of Force Disturbances

Suppose there is an external force disturbance so that the constrained system is described by

$$M(q)\ddot{q} + F(q, \dot{q}) = u + f + \tau$$

$$Jq = 0$$

$$f = J'\lambda$$

where τ represents the n-vector force disturbance. Recall that constant vectors q_d and f_d are assumed to be consistent with the constraints and they define a constrained equilibrium, corresponding to $\tau = 0$. Assume that the controller is given by (19). Then, following the development indicated previously, the linearized equations, in the reduced form, for the closed loop can be shown to be given by

$$E_1\bar{M}(x_{2d})E_2'\ddot{e}_2 + E_1\bar{K}(x_{2d})E_2'e_2 = (I_m + G_f)E_1 V'(f - f_d) + E_1 V'\tau$$

$$E_2\bar{M}(x_{2d})E_2'\ddot{e}_2 + G_v\dot{e}_2 + \{G_d + E_2\bar{K}(x_{2d})E_2'\}e_2 = E_2 V'\tau$$

$$e_1 = 0.$$

The effects of the force disturbance on the regulation accuracy, at least locally, are characterized by these equations.

Suppose that the feedback gain matrices satisfy the conditions of Corollary 4. If τ is a constant force disturbance, then there are steady-state position errors and constraint force errors such that, at least locally, the closed-loop responses satisfy

$$q - q_d \to VE_2'[G_d + E_2 V'K(q_d) VE_2']^{-1}E_2 V'\tau \quad \text{as } t \to \infty$$

$$f - f_d \to J(E_1 V'J')^{-1}[I_m + G_f]^{-1}\{E_1 V'K(q_d) VE_2'[G_d + E_2 V'K(q_d) VE_2']^{-1}E_2 - E_1\} V'\tau \quad \text{as } t \to \infty.$$

Thus, the steady-state displacement error does not depend on the force feedback gain matrix G_f and is inversely proportional to the displacement feedback gain matrix G_d. The steady-state constraint force error does depend on the force feedback gain matrix G_f and is inversely proportional to it. Thus, "high gain" in the displacement feedback loops results in improved steady-state displacement accuracy for additive force disturbances. And "high gain'' in the force feedback loops results in improved steady-state constraint force accuracy for additive force disturbances.

Effects of Dynamics in Force Feedback Loops

Suppose that there are dynamics in the force feedback loops such as might be due to force sensor dynamics. We make a simple assumption about the nature of these dynamics so results are easily obtained; we do not examine the effects of dynamics in the displacement and velocity feedback loops although that might also be of importance.

Recall that q_d and f_d are assumed to be consistent with the constraints and they define a constrained equilibrium if there are no sensor dynamics. Assume that the controller is given by

$$u=F(q_d, 0)-f_d-VE_2'G_vE_2V'\dot{q}+VE_2G_dE_2V'(q_d-q)+VE_1'G_fz$$

$$\dot{z}=-\mu z+\mu E_1V'(f-f_d)$$

where z represents the m vector state of the force feedback loops. This assumes first-order feedback dynamics as a consequence of measurement of the vector component of the constraint force, $E_1V'f$, normal to the constraint surface. For simplicity, we take μ as a positive scalar.

Then, following the development indicated previously, the linearized equations, in the reduced form, for the closed loop given can be shown to be given by

$$E_1\bar{M}(x_{2d})E_2'\ddot{e}_2+E_1\bar{K}(x_{2d})E_2'e_2=G_fz+E_1V'(f-f_d)$$

$$E_2\bar{M}(x_{2d})E_2'\ddot{e}_2+G_v\dot{e}_2+\{G_d+E_2\bar{K}(x_{2d})E_2'\}e_2=0$$

$$\dot{z}=-\mu z+\mu E_1V'(f-f_d)$$

$$e_1=0.$$

The effects of the dynamics in the force feedback loops on the regulation accuracy, at least locally, are characterized by the above equations.

Suppose that the feedback gain matrices satisfy the conditions of Corollary 4. It is easy to show that if $\mu > 0$, the closed-loop equations are locally asymptotically stable in the sense that for initial data near the equilibrium

$$q(t)\to q_d \quad \text{as } t\to\infty$$

$$f(t)\to f_d \quad \text{as } t\to\infty.$$

That is, the dynamics in the force feedback loops do not destabilize the closed-loop system as long as the feedback dynamics are of the assumed simple form and the gains satisfy the conditions of Corollary 4. Note that the assumption that the force feedback gain G_f is nonnegative definite is critical here.

Effects of Constraint Uncertainties

Suppose that there are uncertainties in the constraint function; it is of interest to determine the effect of such uncertainties on the regulation accuracy of the closed loop. Although there are various assumptions that could be made, for simplicity we assume that the constraint is linear and given by

$$Jq=\Delta$$

where Δ represents the constant m vector of constraint uncertainty. Recall that constant regulation vectors q_d, f_d are assumed to be consistent with the constraints and they define a constrained equilibrium corresponding to $\Delta = 0$. Assume that the controller is given by (19). The linearized equations, in the reduced form, for the closed loop can be shown to be given by

$$E_1\bar{M}(x_{2d})E_2'\ddot{e}_2+E_1\bar{K}(x_{2d})E_2'e_2=(I_m+G_f)E_1V'(f-f_d)-E_1K(x_{2d})E_1'(E_1V'J')^{-1}\Delta$$

$$E_2\bar{M}(x_{2d})E_2'\ddot{e}_2+G_v\dot{e}_2+\{G_d+E_2\bar{K}(x_{2d})E_2'\}e_2=-E_2\bar{K}(x_{2d})E_1'(E_1V'J')^{-1}\Delta$$

$$e_1=(E_1V'J')^{-1}\Delta.$$

The effects of the constraint uncertainty on the regulation accuracy, at least locally, are characterized by these equations.

Suppose that the feedback gain matrices satisfy the conditions of Corollary 4. Then there are steady-state position errors and constraint force errors such that, at least locally, the closed-loop responses satisfy

$$q-q_d\to -VE_2'[G_d+E_2V'K(q_d)VE_2']^{-1}E_2V'K(q_d)VE_1'(E_1V'J')^{-1}\Delta \quad \text{as } t\to\infty$$

$$f-f_d\to -J(E_1V'J')^{-1}[I_m+G_f]^{-1}\{E_1V'K(q_d)VE_2'[G_d+E_2V'K(q_d)VE_2']^{-1}E_2V'K(q_d)VE_1'-E_1V'K(q_d)VE_1'\}(E_1V'J')^{-1}\Delta \quad \text{as } t\to\infty.$$

Again we see that the steady-state displacement errors do not depend on the force feedback gain matrix G_f and are inversely proportional to the displacement feedback gain matrix G_d. The steady-state constraint force errors do depend on the force feedback gain matrix G_f and are inversely proportional to it. Thus, ''high gain'' in the displacement feedback loops results in improved steady-state displacement accuracy for uncertainty in the constraint function; and ''high gain'' in the force feedback loops results in improved steady-state constraint force accuracy for uncertainty in the constraint function.

VII. An Example

Consider a simple example of a planar Cartesian manipulator constrained so that the end effector follows an elliptic arc as shown in Fig. 3. We take the equations of motion to be given by

$$\ddot{q}_1=u_1+f_1$$

$$\ddot{q}_2=u_2+f_2$$

with scalar constraint equation given by

$$4(q_1)^2+(q_2)^2-1=0.$$

Thus, the forces of constraint are

$$f_1=8q_1\lambda$$

$$f_2=2q_2\lambda$$

where λ is the constraint multiplier.

The control objective is to choose a feedback controller so that the closed loop is stable and the motion and constraint forces are regulated about the constant values $q_{1d}, q_{2d}, f_{1d}, f_{2d}$, assumed to be consistent with the imposed constraints. Our approach is based on the developments in Sections III–V.

As in the previous notation, let $q' = (q_1, q_2)$ and let $x' = (x_1, x_2)$ be defined by

$$x=X(q)=\begin{vmatrix} q_1-0.5[1-(q_2)^2]^{1/2} \\ q_2 \end{vmatrix}.$$

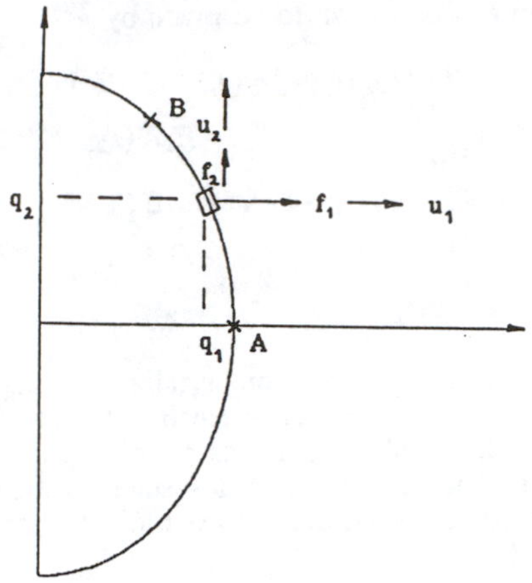

Fig. 3. Cartesian manipulator constrained to ellipse.

Thus

$$T(x)=\begin{vmatrix}1 & -0.5q_2[1-(q_2)^2]^{1/2}\\ 0 & 1\end{vmatrix}.$$

The open-loop nonlinear equations, in the reduced form, are

$$-0.5x_2(1-x_2^2)^{-1/2}\ddot{x}_2-0.5(1-x_2^2)^{-3/2}\dot{x}_2^2=u_1+f_1$$

$$[1+0.25(1-x_2^2)^{-1}x_2^2]\ddot{x}_2+0.25(1-x_2^2)^{-2}x_2\dot{x}_2^2$$
$$=-0.5(1-x_2^2)^{-1/2}x_2u_1+u_2$$

$$x_1=0.$$

Results are now presented for two different desired regulation objectives. *We first consider the regulation objective defined by* q_d' = (0.5, 0) *and* f_d' = (1, 0), *corresponding to stable regulation about point A in Fig. 3 with desired normal contact force of 1.*

Corresponding to the development in Section IV, the modified computed torque controller can be written in the complicated form

$$u_1=-2q_1-0.25(q_1\dot{q}_2-q_2\dot{q}_1)\dot{q}_2q_1^{-2}$$
$$+(0.25q_2q_1^{-1})(g_v\dot{q}_2+g_dq_2)+g_f(f_1-2q_1)$$

$$u_2=-0.5q_2-g_v\dot{q}_2-g_dq_2+(0.25q_2q_1^{-1})g_f(f_1-2q_1).$$

The closed-loop equations, in reduced form, are

$$(-0.25q_2q_1^{-1})[\ddot{e}_2+g_v\dot{e}_2+g_de_2]=(1+g_f)(f_1-2q_1)$$

$$\ddot{e}_2+g_v\dot{e}_2+g_de_2=0$$

$$e_1=0.$$

Thus, if $g_v>0$, $g_d>0$, $g_f>0$, it follows that $e_2\to 0$ as $t\to\infty$, and therefore

$$q_1\to 0.5,\ q_2\to 0 \qquad \text{as } t\to\infty,$$

$$f_1\to 1,\ f_2\to 0 \qquad \text{as } t\to\infty.$$

Also, a linear feedback controller can be developed using the procedure indicated in Section V, leading to the linear controller

$$u_1=-1+g_f(f_1-1)$$

$$u_2=-g_v\dot{q}_2-g_dq_2.$$

The resulting linearized equations, in the reduced form, for the closed loop are given by

$$0=(1+g_f)(f_1-1)$$

$$\ddot{e}_2+g_v\dot{e}_2+(g_d-0.5)e_2=0$$

$$e_1=0.$$

Thus, if $g_v>0$, $g_d>0.5$, $g_f>0$, it follows that, locally, $e_2\to$ 0 as $t\to\infty$, and therefore

$$q_1\to 0.5,\ q_2\to 0 \qquad \text{as } t\to\infty$$

$$f_1\to 1,\ f_2\to 0 \qquad \text{as } t\to\infty.$$

We next consider the regulation objective defined by q_d' = (0.25, 0.866) *and* f_d' = (0, 0), *corresponding to stable regulation about point B in Fig. 3 with zero desired normal contact force.*

Corresponding to the development in Section IV, the modified computed torque controller can be written in the complicated form

$$u_1=-0.25(q_1\dot{q}_2-q_2\dot{q}_1)\dot{q}_2q_1^{-2}$$
$$+(0.25q_2q_1^{-1})[g_v\dot{q}_2+g_d(q_2-0.866)]+g_ff_1$$

$$u_2=-g_v\dot{q}_2-g_d(q_2-0.866)+(0.25q_2q_1^{-1})g_ff_1.$$

The closed-loop equations, in reduced form, are

$$(-0.25q_2q_1^{-1})[\ddot{e}_2+g_v\dot{e}_2+g_de_2]=(1+g_f)f_1$$

$$\ddot{e}_2+g_v\dot{e}_2+g_de_2=0$$

$$e_1=0.$$

Thus, if $g_v>0$, $g_d>0$, $g_f>0$, it follows that $e_2\to 0$ as $t\to\infty$, and therefore

$$q_1\to 0.25,\ q_2\to 0.866 \qquad \text{as } t\to\infty$$

$$f_1\to 0,\ f_2\to 0 \qquad \text{as } t\to\infty.$$

Also, a linear feedback controller can be developed using the procedure indicated in Section V, leading to the linear controller

$$u_1=g_ff_1$$

$$u_2=-g_v\dot{q}_2-g_d(q_2-0.866)+0.866g_ff_1.$$

The resulting linearized equations, in the reduced form, for the closed loop are given by

$$-0.866\ddot{e}_2=(1+g_f)f_1$$

$$1.75\ddot{e}_2+g_v\dot{e}_2+g_de_2=0$$

$$e_1=0.$$

Thus, if $g_v>0$, $g_d>0$, $g_f>0$, it follows that, locally, $e_2\to 0$ as $t\to\infty$, and therefore

$$q_1\to 0.25,\ q_2\to 0.866 \qquad \text{as } t\to\infty$$

$$f_1\to 0,\ f_2\to 0 \qquad \text{as } t\to\infty.$$

VIII. Conclusions

Conditions for stabilization of a closed-loop constrained robot have been developed using mathematical models which explicitly include the constraint functions. Global stabilization conditions have been developed using a nonlinear controller which is based on a modification of the computed torque method. Local stabilization conditions have also been developed using a linear controller with a specified feedback structure.

We have also investigated the properties of closed-loop systems using the proposed controller structure; we have shown that "high gain displacement feedback loops" reduce the steady-state displacement regulation errors for constant force disturbances and uncertainty in the constraint function. We also have shown that "high gain constraint force feedback loops" reduce the steady-state constraint force regulation errors for constant force disturbances and uncertainty in the constraint function. Further, closed-loop stabilization has been shown not to be affected by a certain class of dynamics in the constraint force feedback loops.

The suggested approach to investigation of applications of robots to tasks defined by constraints is new. We have shown that, although the underlying mathematical issues are complicated, a formal mathematical approach is tractable. Our emphasis here has been on feedback stabilization. But, as indicated in [23], [25] we believe that this approach provides a theoretical basis for the investigation of a variety of problems that involve the use of force feedback in constrained robot systems.

References

[1] H. Asada and N. Goldfine, "Optimal compliance design for grinding robot tool holders," in *Proc. IEEE Conf. Robotics Automation*, St. Louis, MO, 1985, pp. 316–322.

[2] ——, "Compliance analysis and tool holder design for grinding with robots," in *Proc. Amer. Contr. Conf.*, Boston, MA, 1985, pp. 65–68.

[3] L. Gustafsson, "Deburring with industrial robots," presented at the SME Deburring and Surface Conditioning Conf., 1983.

[4] N. Hogan and S. R. Moore, "Part referenced manipulation—A strategy applied to robotic drilling," in *Proc. ASME Winter Annual Meet.*, Boston, MA, 1983, pp. 183–191.

[5] G. Plant and G. Hirzinger, "Controlling a robot's motion speed by a force-torque sensor for deburring problems," in *Proc. IFIC/IFAP Symp. Inform. Contr. Problems in Manufacturing Technol.*, 1982, pp. 97–102.

[6] T. M. Stepien, *et al.*, "Control of tool/workpiece contact force with application to robotic deburring," in *Proc. IEEE Conf. Robotics Automation*, St. Louis, MO, 1985.

[7] G. P. Starr, "Edge following with a Puma 560 manipulator using VAL-II," in *Proc. IEEE Conf. Robotics Automation*, San Francisco, CA, 1986, pp. 379–383.

[8] N. Hogan, "Impedance control: An approach to manipulation: Part I—Theory; Part II—Implementation; Part III—Applications," *ASME J. Dynam. Syst., Measurement Contr.*, vol. 107, 1985.

[9] H. Kazerooni, *et al.*, "Robust compliant motion for manipulators; Part I: The fundamental concepts of compliant motion; Part II: Design methods," in *Proc. IEEE J. Robotics Automation*, vol. RA-2, pp. 83–92, June 1986.

[10] R. P. Paul and B. Shimano, "Compliance and control," in *Proc. Joint Automat. Contr. Conf.*, 1976, pp. 694–699.

[11] J. K. Salisbury, "Active stiffness control of a manipulator in Cartesian coordinates," in *Proc. IEEE Conf. Decision Contr.*, Albuquerque, NM, 1980, pp. 95–100.

[12] J. J. Craig, *Introduction to Robotics: Mechanics and Control.* Reading, MA: Addison-Wesley, 1986.

[13] G. Hirzinger, "Force feedback problems in robotics," presented at the IASTED Symp. on Modelling, Identification and Contr., Davos, 1982.

[14] ——, "Direct digital robot control using a force torque sensor," in *Proc. IFAC Symp. Real Time Digital Contr. Appl.*, 1983, pp. 243–255.

[15] M. T. Mason, "Compliance and force control for computer-controlled manipulators," *IEEE Trans. Syst., Man., Cybern.*, vol. SMC-11, 1981.

[16] M. H. Raibert and J. J. Craig, "Hybrid position/force control of manipulators," *ASME J. Dynam. Syst., Measurement Contr.*, vol. 102, pp. 126–133, 1981.

[17] H. West and H. Asada, "A method for the design of hybrid position/force controllers for manipulators constrained by contact with the environment," in *Proc. IEEE Conf. Robotics Automation*, St. Louis, MO, 1985.

[18] D. Whitney, "Force feedback control of manipulator fine motions," *ASME J. Dynam. Syst., Measurement Contr.*, pp. 91–97, 1971.

[19] ——, "Historical perspective and state of the art in robot force control," in *Proc. IEEE Conf. Robotics Automation*, St. Louis, MO, 1985, pp. 262–268.

[20] V. I. Arnold, *Mathematical Methods of Classical Mechanics.* New York: Springer-Verlag, 1978.

[21] D. T. Greenwood, *Principles of Dynamics.* Englewood Cliffs, NJ: Prentice-Hall, 1965.

[22] H. Hemami and B. F. Wyman, "Modelling and control of constrained dynamic systems with application to biped locomotion in the frontal plane," *IEEE Trans. Automat. Contr.*, vol. AC-24, pp. 526–535, 1979.

[23] H. P. Huang, "Constrained manipulators and contact force control of contour following problems," Ph.D. dissertation, Dep. Elect. Eng. Comput. Sci., Univ. Michigan, Ann Arbor, July 1986.

[24] R. Kankaanranta and H. N. Koivo, "A model for constrained motion of a serial link manipulator," in *Proc. IEEE Conf. Robotics Automation*, San Francisco, CA, 1986.

[25] N. H. McClamroch, "Singular systems of differential equations as dynamic models for constrained robot systems," in *Proc. IEEE Conf. Robotics Automation*, San Francisco, CA, 1986.

[26] N. H. McClamroch and H. P. Huang, "Dynamics of a closed chain manipulator," in *Proc. Amer. Contr. Conf.*, Boston, MA, 1985.

[27] T. Yoshikawa, "Dynamic hybrid position/force control of robot manipulators-description of hand constraints and calculation of joint driving forces," in *Proc. IEEE Conf. Robotics Automation*, San Francisco, CA, 1986, pp. 1393–1398.

[28] M. Takegaki and S. Arimoto, "A new feedback method for dynamic control of manipulators," *J. Dynam. Syst., Measurement Contr.*, vol. 102, pp. 119–125, 1981.

[29] N. H. McClamroch and D. Wang, "Feedback stabilization and tracking of constrained robots," Center for Research in Integrated Manufacturing, Robot Syst. Division, Univ. Michigan, RSD-25, Oct. 1986.

[30] S. D. Eppinger and W. P. Seering, "Introduction to dynamic models for robot force control," *IEEE Contr. Syst. Mag.*, vol. 7, no. 2, pp. 48–52, Apr. 1987.

[31] J. Y. S. Luh, "Conventional controller design for industrial robots—A tutorial," *IEEE Trans. Syst., Man, Cybern.*, vol. SMC-11, pp. 298–316, 1983.

Force and Position Control of Manipulators During Constrained Motion Tasks

JAMES K. MILLS, MEMBER, IEEE, AND ANDREW A. GOLDENBERG, SENIOR MEMBER, IEEE

Abstract—**Trajectory control of a manipulator constrained by the contact of the end-effector with the environment represents an important class of control problems. In this paper, a method is proposed whereby both contact force exerted by the manipulator, and the position of the end-effector while in contact with the surface are controlled. The controller parameters are derived based on a linearized dynamic model of the manipulator during constrained motion. Hence the method is valid only in a neighborhood about the point of linearization. Additionally, a perfect kinematic model of the contact surface is assumed. The proposed method exploits the fundamental structure of the dynamic formulation of the manipulator's constrained motion. With this formulation, the trajectory control problem is naturally expressed in terms of the state vector variables of the model of the constrained dynamic system. A detailed numerical example illustrates the proposed method.**

I. Introduction

CONSIDERABLE EFFORT has been focussed on the problem of control of manipulators during the execution of tasks in which the manipulator does not come in contact with the environment. Such tasks include paint spraying, and arc welding. Other tasks require contact to be made with the environment by the manipulator end-effector. There are many tasks of this nature such as assembly operations, parts deburring, and cutting to name a few. Control of robotic manipulators during contact with the environment has recently been the subject of much interest in the robotics control literature. Certain published works study this control problem with the environment modeled dynamically as a mechanical impedance, i.e., spring, mass, dashpot [3], [19]–[22], [25], [26]. This is a very general approach to the control problem associated with control of contact tasks. Generally, the objective of control during contact with the environment is to regulate the force/torque that the manipulator end-effector exerts on the environment, while simultaneously regulating the free position/orientation coordinates of the end-effector.

The problem addressed in this paper is control of a robotic manipulator while the end-effector is in contact with an environment with infinite stiffness. Specifically, control during contact with rigid surfaces is examined. While seemingly restrictive in nature, the problem of control during contact with such a surface admits many common types of contact tasks that might be performed by a robot. Many objects a manipulator comes in contact with in its environment are, for practical purposes, rigid, i.e., the environmental stiffness coefficients are extremely large. As a result, although the environment is not treated in a general manner by modeling it as a mechanical impedance, for example, the practical importance of control during contact with rigid structures must not be underestimated.

Manuscript received May 12, 1987; revised April 22, 1988. Part of the material in this paper was presented at the 1st IEEE International Symposium on Intelligent Control, Philadelphia, PA, January 23–25, 1987.

The authors are with the Robotics and Automation Laboratory, Department of Mechanical Engineering, University of Toronto, Toronto., Ont., Canada M5S 1S5.

IEEE Log Number 8823370.

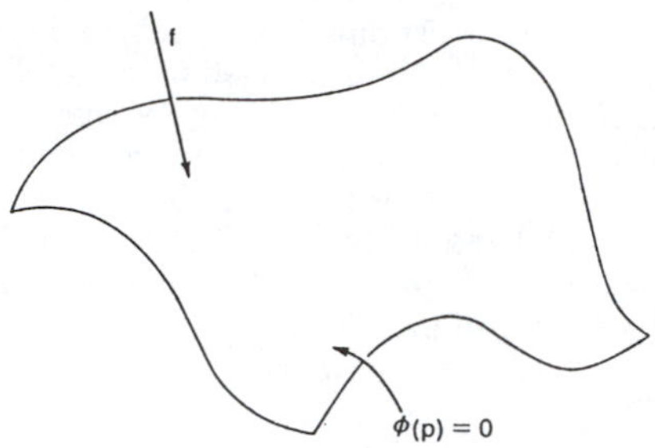

Fig. 1. Constraint surface with normal force.

As a result of the kinematic constraint imposed on the motion of the manipulator through contact with a rigid surface, tasks of this nature will be referred to as constrained motion tasks. An extremely important aspect of constrained motion while in contact with a rigid surface is the control of forces applied by the manipulator end-effector on the environment. It is essential that control be exercised over these forces to ensure that i) neither the manipulator nor the environment is damaged due to contact, ii) contact is maintained during the task, and iii) the required forces are applied to successfully complete the task.

Two types of forces exist during contact that must be controlled. The first, termed *workless* force, results from the normal (constraint) force that arises due to contact with a surface. This is illustrated in Fig. 1. This force generates no virtual work, hence is called *workless force.*

The second type of forces arises due to the application of forces such that work is performed on the environment. These forces are termed *applied* forces. It is clear, with this definition, that the reaction due to surface friction during motion of the manipulator on a surface is an *applied* force. The consideration of applied forces is necessary for many tasks such as mechanical deburring, cutting, and grinding to name a few. Often it is desirable to simultaneously exercise control over both *workless* and *applied* forces. Such situations arise when the manipulator must move while in contact with a surface.

The control of both workless force and position while the manipulator is in contact with a rigid frictionless surface is the

Reprinted from *IEEE Trans. Robot. Automat.*, vol. 5, no. 1, pp. 30–46, Feb. 1989.

subject of this paper. The objective of the controller is as follows: to regulate the contact force and position to some predetermined values in the presence of manipulator dynamic model parameters uncertainties and unknown payloads. Control in the presence of kinematic parameter uncertainty is not considered. It is assumed that a perfect kinematic model of both the manipulator and the environmental surface exists. The controller sought must be "robust" to dynamic parameter perturbations in the following sense. As long as closed-loop stability of the system is maintained, asymptotic trajectory tracking of contact force and position will occur, even in the presence of dynamic parameter perturbations caused by unknown payloads and/or manipulator dynamic parameter uncertainties.

The proposed approach to control of workless force and position, while the manipulator end-effector is in contact with a rigid surface, is through the application of results obtained for the control of linear descriptor variable systems [7], [8]. While the dynamic behavior of a manipulator during unconstrained motion is correctly described by a set of nonlinear ordinary differential equations, often a force must be applied at a particular position on a surface, for example. Under this situation, a linearized dynamic model derived from a Taylor series expansion of the nonlinear manipulator equations of motion is an accurate model of the manipulator dynamics. This approach has been taken by others in the study of control during contact, i.e., Hogan [19] and Kazerooni *et al.* [25], [26]. Controller synthesis then, is based on a linearized dynamic model of the nonlinear manipulator dynamics.

During constrained motion, the manipulator dynamics are described by a set of nonlinear differential algebraic equations with holonomic constraints, referred to as nonlinear descriptor variable systems in the relevant literature. In a manner identical to that of the authors cited above, a local dynamic model derived from a Taylor series expansion of the nonlinear descriptor variable system is derived, again valid in a neighborhood about the point of linearization. Controller synthesis for constrained motion is then based on a linearized dynamic model of the constrained manipulator. The resultant control law, valid in a neighborhood about the point of linearization, is then applied to the nonlinear system. Thus control of a constrained manipulator is treated naturally as a problem of control of a descriptor variable system.

Descriptor variable systems are known to exhibit a dynamic behavior which is notably different from regular state variable systems, as discussed in Rosenbrock [34]. Basically, impulsive behavior of a descriptor variable system response can result from arbitrary, but finite initial conditions. This situation arises because in general, arbitrary initial conditions violate the algebraic constraint imposed in the descriptor variable system formulation. As a result, a step change in some state variables occurs so that the constraint equation is not violated. Further, impulsive behavior is exhibited by those states which are functions of derivatives of states that have undergone a step change. One of the accomplishments of the theory associated with control of descriptor variables systems is the elimination of impulsive modes due to arbitrary initial conditions [7]. However, in this paper, it is shown that under the assumptions made, impulsive modes cannot be excited during motion of the manipulator, hence are not considered in the controller design.

Section II reviews existing methods of force control reported in the literature. Manipulator dynamics are formulated in Section III, followed by a review of descriptor variable systems in Section IV. Section V deals with synthesis of a feedback controller to achieve robust *workless* force control. Section VI gives a detailed treatment of a three-degree-of-freedom robot manipulator chosen to illustrate the design method. Numerical simulation results are included which verify the proposed approach to control of constrained manipulators.

II. Review of the Literature

In the published literature, much attention has been given to the control of forces and torques during contact of the manipulator with the environment. For a general review of this literature, the reader is referred to Whitney [39]. In many of these publications, i.e., Hogan [19]–[22], Kazerooni *et al.* [29], [26], and Asada and Slotine [3], the environment is modeled as a mechanical impedance with corresponding stiffness, damping, and inertia terms. This represents a general mathematical description of the environmental dynamics, as encountered by the manipulator end-effector. Generally, control of manipulators during contact with environments described above is termed impedance control. Control algorithms have appeared addressing this problem in published papers by Anderson and Spong [1], Hogan [23], in addition to those given above, to cite a few.

The above papers have dealt with control during contact with an environment that deforms due to an applied force or torque. This paper deals with contact which results in a kinematic constraint imposed on the motion of the manipulator. Perhaps the first paper to address the issue of robotic constrained motion control is that of Hemami and Wyman [17]. It deals with the problem of control of biped locomotion. Their approach to control of the constrained motion is completely general. Holonomic constrains are included in the dynamic model formulation. The control solution is derived by stabilization of the closed-loop system through feedback of states which correspond to the submanifold of the system states described by the holonomic constraint equation. Linearization of the nonlinear system results in a linear feedback which is then applied to the nonlinear system dynamics. Further papers on constrained motion control by Goddard, Hemami, and Weimer [14] discuss control of the same physical system. Recent work by Hemami, Wongchaisuwat, and Brinker [18] proposes a solution to the constrained robot control problem through relegation. In this work, through the introduction of kinematic redundancy, the task of trajectory design and control is shown to be simplified.

Hybrid control is the subject of a number of published papers, i.e., Raibert and Craig [33] and Zhang and Paul [44]. Originally, Mason [27] discussed kinematic constraints imposed on manipulator motion due to a particular task geometry. The discussion in this paper is quite general and includes many types of constraints that can occur during a variety of

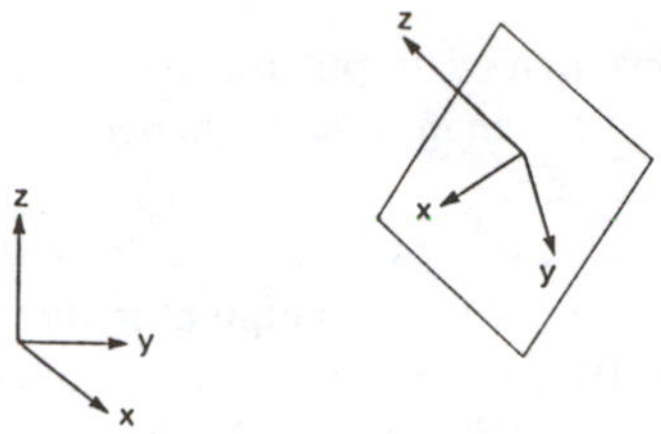

Fig. 2. Constraint frame, dependent on task geometry.

tasks. Hybrid control is proposed to address the issue of control in the presence of natural constraints imposed by the task geometry and artificial constraints imposed by the performance of the task itself. The key to hybrid control is the specification of a task constraint frame, shown in Fig. 2, for example. With respect to this coordinate frame, position and orientation is constrained in certain directions by task geometry. In these directions, constraint forces and torques can be controlled, while in other coordinate directions, position and orientation is controlled.

Control of constrained motion is implicit in the hybrid control formulation, but is never explicitly discussed in the hybrid control literature. Interestingly, the constrained dynamic formulation, introduced by McClamroch [28], which accurately models motion of the manipulator when in contact with a rigid environment, has been largely ignored in the hybrid control literature.

Recent work on a time-optimal control problem during constrained motion is presented by Huang and McClamroch [24]. This work uses kinematic inequality constraints or unilateral constraints as discussed by Mason [27]. A different approach to the problem is found in a paper by West and Asada [37]. Here a specialized mechanical device was constructed to increase the robot's stiffness in the direction normal to a rigid contact surface. Control of this specialized device is achieved through the use of a hybrid position/force controller applied to the device.

In summary, relatively few papers on the subject of constrained motion control exist. This is probably due to the somewhat specialized nature of constrained motion. However, as pointed out earlier, although a very special case of manipulator motion, it is a good mathematical approximation to many tasks in which the end-effector comes in contact with a very stiff environment.

III. Manipulator Dynamics

When the end-effector is in contact with a rigid surface, a kinematic constraint is imposed on the manipulator motion, which corresponds to an algebraic constraint among the manipulator state variables. This algebraic constraint gives rise to a system of differential algebraic equations which are characterized by a singular matrix premultiplying the vector of state derivatives. McClamroch and Huang [29] and McClamroch [28] present singular dynamic models which represent manipulator dynamics under several different circumstances when subject to kinematic constraints. As a result of the singularity of the premultiplying matrix, these dynamic systems are often referred to as singular or descriptor variable systems, or simply descriptor systems.

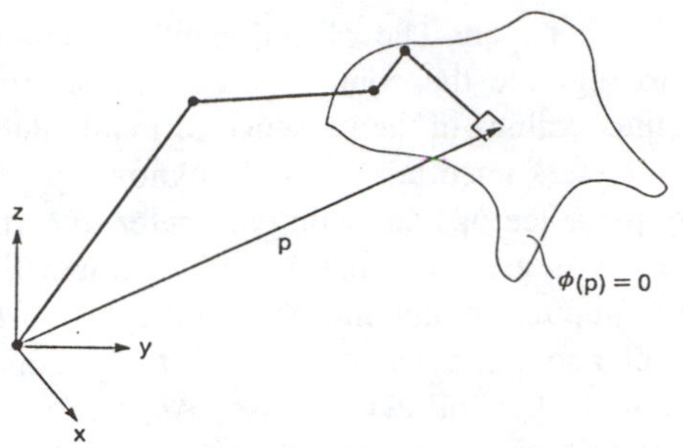

Fig. 3. Task geometry with constraint surface.

Singular systems of differential equations are of a fundamentally different nature than nonsingular or state variable systems, which describe unconstrained manipulator dynamics. A significant difference is the introduction of impulsive behavior of the system, which depends on the system initial conditions. Section IV reviews the properties of descriptor systems as they pertain to the manipulator force control problem discussed in this paper.

Many robot tasks are of the following form: a period of unconstrained motion, followed by a collision with the environment, and finally a period of constrained motion, while the manipulator is in contact with the environment [24]. In this paper, only constrained motion is considered, in which the workless contact force and position of the manipulator end-effector are to be regulated. We do not consider the regulation of applied forces or the collision phase of the manipulator motion. The latter is a complex problem and a research area in itself, see Zheng and Hemami [45]. Further results on dynamic modeling of manipulators with stiction, friction, and constraint addition–deletion can be found in Haug *et al.* [16] and Wu *et al.* [41], [42].

Here, the dynamic equations of motion of a rigid, n-degree-of-freedom manipulator in contact with a rigid, frictionless constraint surface are considered. It is assumed that contact with the environment is made by the end-effector and occurs at a point. Let $p \in R^3$ denote a position vector from a fixed reference frame to the constraint surface. The constraint surface is assumed to satisfy the following scalar relation:

$$\phi(p)=0 \tag{1}$$

where $\phi(\cdot) \triangleq \phi{:}R^3 \rightarrow R$, a given scalar function with continuous gradient. Fig. 3 illustrates this geometry.

Let the manipulator forward kinematics be defined as follows:

$$z=g(q) \tag{2}$$

where

$g(\cdot) \triangleq$ map from joint space to task space,
$q \triangleq$ n-tuple of manipulator generalized coordinates.

The definitions of joint and task spaces are given here.

Definition: An n-tuple of generalized joint displacements $q \in D$ where

$$D \triangleq \{q : q_i \in [q_{i_{\min}}, q_{i_{\max}}] i=1, \cdots, n\}, \qquad D \subset R^n$$

an open set, determines the configuration of the robotic manipulator. D is known as the joint space of the manipulator.

Definition: The position and orientation of the manipulator end-effector is specified as follows. The three-tuple $p = (p_x, p_y, p_z)^T$ denotes the Cartesian position of the end-effector with respect to a fixed reference frame. Given an orthogonal rotation matrix $R \in SO(3)$ which specifies the orientation of the manipulator end-effector, an invariant vector $u_i \in R^3$ parallel to the axis of rotation of R is defined by vect $(R) = u_i \sin(\theta)$ where θ is the angle of rotation about u_i, and vect (R) is a three-dimensional vector invariant of R, its ith component given by $1/2\epsilon_{ijk}r_{kj}$, r_{kj} an element of R, where ϵ_{ijk} is a tensor of rank three and summation notation is used in the above with

$$\epsilon_{ijk}=\begin{cases}+1, & \text{even permutation of } ijk\\ -1, & \text{odd permutation of } ijk\\ 0, & \text{any two of } ijk \text{ identical}\end{cases}$$

Angeles [2]. Note that u_i is the eigenvector of R corresponding to the eigenvalue $\lambda = 1$, and θ can be obtained from the trace of the matrix R. As u_i is a unit vector, only two of its coordinates are specified independently. To uniquely define a real rotation, we define a three-tuple $v = (u_{i1}, u_{i2}, \theta)$ where u_{i1} and u_{i2} are the first two components of u_i and θ is the angle of rotation about u_i. Together, the three tuples, p, and v specify the position and orientation of the manipulator. $z = (p^T, v^T)T \in T$ where $T \subset SO(3) \times R^3$, an open set [13] is referred to as the task space configuration of the manipulator.

As noted earlier, the manipulator is in contact with a rigid constraint surface hence the position vector p is defined as a function of the joint coordinates as follows:

$$p=L(q). \tag{3}$$

A contact force exists and the contact or *workless* force is given by the following (Goldstein [15] and McClamroch [28]):

$$f=D^T(p)\lambda \tag{4}$$

where

$$D(p)=\frac{\partial\phi(p)}{\partial p} \tag{5}$$

$\lambda \triangleq$ Lagrange multiplier.

Thus based on the Lagrange formulation, the equations of motion of a manipulator, constrained by one point in contact with a rigid frictionless surface are given by

$$\Phi(q)\ddot{q}+H(\dot{q},q)+G(q)=\tau+J^T(q)D^T(p)\lambda \tag{6}$$

where

$\Phi(q)$ $\triangleq$ manipulator inertia matrix, $\Phi(q) \in R^{n\times n}$

$H(\dot{q}, q)$ $\triangleq$ vector of Coriolis and centripetal forces, $H(\cdot, \cdot) \in R^{n\times 1}$

$G(q)$ $\triangleq$ vector of gravity forces, $G(\cdot) \in R^{n\times 1}$

τ $\triangleq$ vector of generalized forces applied at each joint, $\tau \in R^{n\times 1}$

$\dot{q}, \ddot{q}$ $\triangleq$ velocity and acceleration of generalized joint coordinates

$J(q)$ $\triangleq$ $\partial L(q)/\partial q$ manipulator Jacobian, $J(i) \in R^{3\times n}$.

Equation (6) can be rewritten as in [28] to give

$$\begin{bmatrix} I & 0 & 0\\ 0 & \Phi(q) & 0\\ 0 & 0 & 0\end{bmatrix}\begin{pmatrix}\dot{q}\\ \ddot{q}\\ \dot{\lambda}\end{pmatrix}=\begin{bmatrix}0 & I & 0\\ 0 & 0 & J^TD^T\\ 0 & 0 & 0\end{bmatrix}\begin{pmatrix} q\\ \dot{q}\\ \lambda\end{pmatrix} + \begin{bmatrix}0\\ -H(\dot{q},q)-G(q)\\ \phi(p)\end{bmatrix}+\begin{bmatrix}0\\ I\\ 0\end{bmatrix}\tau \tag{7}$$

where all terms have been previously defined.

Inspection of (7) reveals that the constrained dynamic equations of motion are equivalent to a singular system of differential algebraic equations, which is the result of the singularity of the matrix on the left-hand side of (7).

We assume that the application of a force at a point on a constraint surface is such that the amplitude of variations in joint position, velocity, and acceleration about some nominal value remains small. This permits the manipulator dynamics to be represented by a perturbation model about a nominal state given by

$$(q_0, \dot{q}_0, \lambda_0)^T \tag{8}$$

where $q_0 \triangleq$ nominal manipulator joint configuration, $\dot{q}_0 \triangleq$ nominal joint velocity; $\lambda_0 \triangleq$ nominal value of the Lagrange multiplier.

We wish to linearize the nonlinear dynamic equations, given by (6), about the nominal state (8). At the nominal operating point, the manipulator is at rest hence $\dot{q}_0 = \mathbf{0}$ and $\ddot{q}_0 = \mathbf{0}$, resulting in simplification of the linearized equations of motion. Writing (6) in the form

$$\tau=\tau(q, \dot{q}, \ddot{q}, \lambda) \tag{9}$$

and expanding about the nominal state (8) using a multivariable Taylor series with the definitions $\delta q \triangleq q - q_0$, $\delta\dot{q} \triangleq \dot{q} - \dot{q}_0$, $\delta\ddot{q} \triangleq \ddot{q} - \ddot{q}_0$, $\delta\tau \triangleq \tau - \tau_0$, and $\delta\lambda \triangleq \lambda - \lambda_0$ yields, where higher order terms have been dropped

$$\delta\tau=\Phi(q_0)\delta\ddot{q}+\frac{\partial}{\partial q}(G-J^TD^T\lambda)|_0\,\delta q-J^TD^T|_0\,\delta\lambda \tag{10}$$

where we have made use of the fact that $(\partial/\partial q)H(\dot{q}, q)|_0 = 0$ due to the fact that $H(\dot{q}, q)$ is homogeneous in $\dot{q}$, as in [26]. Note that τ_0 and λ_0 are as yet undefined. We must also linearize the constraint equation given by (1). Differentiating (1) with respect to q and evaluating the result at q_0 gives

$$\left.\frac{\partial\phi(p)}{\partial p}\frac{\partial p}{\partial q}\right|_0\delta q=0 \tag{11}$$

or

$$D(p)J(q)|_0\delta q=0. \tag{12}$$

Using (10) and (12), the linearized dynamic equations are written in a form similar to (7) as follows:

$$\begin{bmatrix} I & 0 & 0 \\ 0 & \Phi(q_0) & 0 \\ 0 & 0 & 0 \end{bmatrix} \begin{pmatrix} \delta\dot{q} \\ \delta\ddot{q} \\ \delta\dot{\lambda} \end{pmatrix} = \begin{bmatrix} 0 & I & 0 \\ -\frac{\partial}{\partial q}(G - J^T D^T \lambda)|_0 & 0 & J^T D^T|_0 \\ DJ|_0 & 0 & 0 \end{bmatrix} \begin{pmatrix} \delta q \\ \delta\dot{q} \\ \delta\lambda \end{pmatrix} + \begin{bmatrix} 0 \\ I \\ 0 \end{bmatrix} \delta\tau. \tag{13}$$

Once again, (13) represents a singular system of differential algebraic equations, due to the singularity of the matrix on the left-hand side of (13). This is equivalent to a small oscillation model with constraints. All coefficients of (13) are time-invariant due to the linearization of the nonlinear system (7) about a nominal state given by (8).

In order to determine the nominal applied torque τ_0 corresponding to the nominal state (8), it is only necessary to evaluate (7) at $q = q_0$, $\dot{q}_0 = 0$, $\ddot{q}_0 = 0$, $\lambda = \lambda_0$ to yield the following:

$$\tau_0 = G(q_0) - J^T(q_0) D^T(L(q_0))\lambda_0. \tag{14}$$

We note that the *workless* force is to be controlled, thus at equilibrium

$$D^T(L(q_0))\lambda_0 = f^{\text{ref}} \tag{15}$$

and using (14), τ_0 becomes

$$\tau_0 = G(q_0) - J^T(q_0) f^{\text{ref}} \tag{16}$$

where all quantities on the right-hand side of (15) are known. The nominal Lagrange multiplier λ_0 is easily determined from (15) as follows:

$$\lambda_0 = (DD^T)^{-1} D f^{\text{ref}} \tag{17}$$

where the scalar (DD^T) is always invertible.

Linearized equations of motion have been developed for a robotic manipulator constrained to be in contact, at a fixed point, with a rigid frictionless constraint surface. The resultant dynamic equations form a singular set of differential algebraic equations which have properties unlike those of regular state variable systems. In the following, singular systems of equations such as those given by (7) and (13) will be referred to as descriptor variable systems or simply descriptor systems as is customary in the relevant literature.

IV. Descriptor Variable Systems

In this section, a brief overview of linear time-invariant descriptor variable systems is given. Due to space limitations, only the salient features, as they pertain to the problem of manipulator *workless* force control, are discussed. For general background on descriptor variable systems, the reader is referred to the following: Rosenbrock [34], Campbell [5], [6], and Verghese, Levy, and Kailath [35]. These references provide a good introduction to descriptor systems, however, the results of Cobb [7]–[10] are used most extensively here. While not exhaustive, the above references are essential to an understanding of the method by which the manipulator contact force is to be controlled.

Descriptor variable theory is motivated because it is not always possible to express a mathematical model of a dynamic system in state variable form, when the state variables are chosen in a natural way. In many cases, the natural choice of system variables will not be minimal leading to a system model

$$E\dot{x} = Ax + Bu \qquad x(0) = x_0 \tag{18}$$

where

$E \triangleq$ singular matrix, $E \in R^{n \times n}$, $\det(E) = 0$
$A \triangleq$ singular matrix, $A \in R^{n \times n}$
$B \triangleq$ input distribution matrix, $B \in R^{n \times r}$
$u \triangleq$ input vector, $u \in R^{r \times 1}$
$x \triangleq$ state vector, $x \in R^{n \times 1}$.

The fact that E is singular characterizes (18) as a descriptor variable system. Stated differently, E is singular as a result of algebraic constraints which exist among the state variables of a dynamic system. This means that any dynamic system subject to holonomic or nonholonomic constraints can be written in descriptor variable form. It is emphasized here that E is always singular, and not at a possibly finite set of points in the state space X.

Much of the literature which deals with descriptor variable systems focuses on the solvability, e.g., Campbell [5], [6] and numerical integration, e.g., Gear and Petzold [11], Gear, Leimkuhler, and Gupta [12], of these systems. These results are not restricted to the linear system case. Comparatively few published papers deal with the control of descriptor variable systems, and those that do, deal with the linear time-invariant system given by (18). Of these papers, the work of Cobb [7]–[10] proves to be very useful in the study of control of contact force.

Descriptor variable systems exhibit time responses that are markedly different from regular state variable systems. It is possible that for an arbitrary finite initial condition x_0 in (18), the time response $x(t)$, $t > 0$ will exhibit impulsive behavior along with derivatives of these impulses. This behavior is not seen in regular state variable systems with finite initial conditions. In the following, a decomposition of (18) will be given which clearly identifies the source of these impulsive solutions. Later, it is made clear how a feedback can be found to restructure the system so that impulses no longer occur, as well as to shift the poles of the descriptor variable system to obtain a desired closed-loop system response. However, as applied to the robotic control problem stated here, it is shown that the system initial conditions are not arbitrary, hence impulses may not occur in the time response. In this paper, as in the relevant literature, arbitrary initial conditions refer to those initial conditions that generate impulsive behavior.

A. Decomposition of Descriptor System

Most of the material in this section is taken from Cobb [7]. We wish to decompose the system given by (18) into two subsystems referred to as "slow" and "fast" subsystems. The standard assumption is made that $(\lambda E - A)^{-1}$ exists for some $\lambda \in C - \sigma(E, A)$ where C represents the complex plane and $\sigma(E, A)$ is the set of generalized eigenvalues of the pair (E, A), denoted by $\sigma(E, A) = \{\lambda_1, \cdots, \lambda_k\}$. Let

$$\det(\lambda E - A) = \phi_0 \prod_{i=1}^{k} (\lambda - \lambda_i)^{n_i}$$

where ϕ_0 is a constant. As in Cobb, the following subspaces are defined. Note that the decomposition is independent of λ as discussed in Cobb [7].

$$S = \bigoplus_{i=1}^{k} \ker \left((\lambda E - A)^{-1} E - \operatorname{diag}\left(\frac{1}{\lambda - \lambda_i}\right)\right)^{n_i} \tag{19}$$

$$F = \ker\, (\lambda E - A)^{-1} E)^{n-r} \tag{20}$$

where $r = \deg(\det(\lambda E - A)) \leq n$.

$$n = \sum_{i=1}^{k} n_i. \tag{21}$$

Cobb [7, Theorem 2.1] gives a canonical decomposition of the state space X with respect to the pair (E, A). This decomposition results in a "slow" subsystem corresponding to the finite eigenvalues of $sE - A$ and a "fast" subsystem with infinite eigenvalues corresponding to $sE - A$ losing rank at $s = \infty$. The statement of the theorem by Cobb is given here. The notation "$|$" means restriction.

Theorem:

1) $S \oplus F = X$, $\dim(S) = r$.
2) There exists an invertible $M: X \rightarrow X$ such that
   a) S and F are both ME and MA invariant
   b) $ME|S = I$, $MA|F = I$
   c) $ME|F$ is nilpotent
   d) $\det(Is - MA|S) = \prod_{i=1}^{k} (s - \lambda_i)^{n_i}$.

Applying M to (18) gives

$$ME\dot{x} = MAx + MBu. \tag{22}$$

Now, letting P and Q be the skew projection operators on S along F and on F along S, respectively, we have

$$B_s = PMB \tag{23}$$

$$B_f = QMB. \tag{24}$$

Equation (22) may be rewritten as

$$\dot{x}_s = L_s x_s + B_s u \qquad \text{slow} \tag{25}$$

$$L_f \dot{x}_f = x_f + B_f u \qquad \text{fast} \tag{26}$$

where $L_s = MA|S$, $L_f = ME|F$, $x_s = Px$, $x_f = Qx$, and $x_{s0} = Px_0$, $x_{0f} = Qx_0$. Thus two subsystems which operate on independent state spaces with $x = x_s + x_f$ are defined. Note that the slow subsystem is a regular state variable system and L_f is nilpotent, of degree ν, i.e., $L_f^{\nu} = 0$.

Based on the decomposition (25), (26), we now examine the solutions to (18). Equation (25) represents a state variable system and has the solution

$$x_s = e^{L_s t} P x_0 + \int_0^t e^{L_s(t-\tau)} B_s u(\tau)\, d\tau. \tag{27}$$

This is in sharp contrast to the solution of (26) which is written as

$$x_f = -\sum_{i=1}^{\nu-1} \delta^{i-1} L_f^i x_{0f} - \sum_{i=0}^{\nu-1} L_f^i B_f u^i \tag{28}$$

where

$\delta^j \triangleq$ jth derivative of the Dirac delta
$u^j \triangleq$ jth derivative of the input u^i
$L_f^j \triangleq$ L_f raised to the jth power
$x_{0f} \triangleq Qx_0$.

The solution for the fast subsystem may be obtained through the use of formal Laplace transforms as in [35]. Thus it is the response of the fast subsystem that causes the response of (18) to be fundamentally different than a regular state variable system such as (19). Arbitrary initial conditions can result in an impulsive time responses, however, restriction of the initial conditions to the slow subspace will eliminate the impulsive response. As pointed out by Cobb [7], often the initial conditions acting on a system are unknown, thus impulsive modes may be present.

B. Controllability and Observability of Descriptor Variable Systems

The concepts of controllability and observability have been defined by various authors for both continuous and discrete linear descriptor systems. The definitions of these properties as given by various authors are not identical. The definitions found in Yip and Sincovec [43], Rosenbrock [34], Pandolfi [32], and Verghese, Levy, and Kailath [35] are limited by the assumption that the initial condition $x_{0f} = 0$ and hence impulsive solutions to (18) are not admitted. Cobb [10] has defined controllability and observability to include arbitrary initial conditions x_0 in (18), i.e., $x_{0f} \neq 0$.

C. Pole Placement

As described in Cobb [7] placement of the closed-loop poles of the slow subsystem is possible if and only if

$$\operatorname{Im}(\lambda_i E - A) + \operatorname{Im}(B) = X \tag{29}$$

where $\lambda_i \in \sigma(E, A)$. Similarly, the fast subsystem is controllable if and only if

$$\operatorname{Im}(E) + \operatorname{Im}(B) = X. \tag{30}$$

Given the linear feedback

$$u = Kx + v \tag{31}$$

applied to (18) it is assumed that λ is chosen such that $\lambda \in C - \sigma(E, A + BK)$ so that

$$\det(\lambda E - A - BK) \neq 0 \tag{32}$$

ensuring that the closed-loop system behavior

$$E\dot{x} = (A + BK)x + Bv \tag{33}$$

is well-posed. In this paper, as in the relevant literature, the term well-posed refers to the existence of a formal Laplace transform solution to (33), ensured by the condition given by (32) being satisfied. The purpose of this section is to demonstrate that pole placement for both the slow and fast subsystems can be achieved, and to give the conditions under which this is possible.

With the decomposition given in (25), (26), it is possible to consider feedback of the slow states and fast states separately. Slow feedback is considered first. It is demonstrated in Cobb [7] that feedback of the slow state x_s does not alter the structure of the fast subspace F. This permits feedback of the slow states without concern regarding the fast states. A control

$$u = K_s x_s + v \tag{34}$$

with $K_s = K|S$ and $\ker(K) \subset F$ allows arbitrary placement of a pole of (18), λ_i, if and only if λ_i is controllable [7].

Similarly, fast feedback of the form

$$u = K_f x_f + v \tag{35}$$

with $K_f = K|F$ can be applied to (18) which may induce additional eigenvalues into the system. It is shown that feedback of the fast state x_f does not alter the structure of the slow subspace S [7]. The decomposition of X into $S \oplus F$ is considered because of the property that feedback of the slow and fast states separately does not alter the structure of the fast and slow subsystems, respectively.

Impulses in the response of the system for arbitrary initial conditions can only be eliminated if $L_f \equiv 0$ in (28) [7]. L_f can be made identically zero if K_f can be found such that r_k = rank (E) [7], where r_k is given by

$$r_k = \deg(\det(E\lambda - A - BK) \tag{36}$$

and $\lambda \in C - \sigma(E, A + BK)$. Under these conditions, $L_f \equiv \mathbf{0}$, and no delta functions can exist in the system for arbitrary initial conditions.

V. Feedback Control

A. Introduction

In this section, a feedback control law for control of *workless* contact force and end-effector position on a rigid surface is proposed. The control law permits asymptotic trajectory tracking for constant *workless* force and position in the presence of constant disturbances due to manipulator dynamic parameter uncertainty. It is essential that close trajectory control be maintained in the presence of dynamic parameter modeling uncertainties, such as the mass and inertia properties of a payload held by the manipulator end-effector which are assumed to be unknown.

In the following, the meaning of position control on a constraint surface is made precise. This clarification is necessary because due to task geometry modeling errors, the exact position of the constraint surface may be uncertain resulting in an inability to regulate all three position coordinates of the end-effector. Measurement information required to implement the feedback law is also discussed.

The linearized manipulator dynamic model is rederived to include dynamic parameter modeling inaccuracy resulting in a constant disturbance term. With this model, the time response of the constrained system is then discussed. Particular attention is given to the impulsive modes that are known to be possible for arbitrary initial conditions. Finally, the feedback controller which provides robust asymptotic trajectory tracking for input and disturbance signals is presented.

B. Position Control on a Constraint Surface

We now consider control of the position of the manipulator end-effector while constrained by contact with a rigid surface. In this section, end-effector position errors are discussed. Although constraint surface kinematic model errors are not considered in the solution to the control problem posed in this paper, these errors are discussed here for completeness. It is demonstrated that regulation of two of the three coordinates of p is possible in spite of this kinematic error. As it will be seen in the next section, this choice of position error specification permits the form of the linearized dynamic equations of motion, given by (13), to be exploited. As shown in the example of Fig. 2 then, position on the surface could be specified in terms of two coordinate directions, namely x and y. Due to the constraint, only two coordinates can be specified independently. As seen in Fig. 3, the position vector p determines the end-effector position. Just as with the hybrid control approach, since only two of the three coordinates of the position vector p can be specified independently when the manipulator end-effector is constrained by contact with a rigid surface, we wish to regulate only two of the three position coordinates of the vector p. The difference between specification of errors in hybrid control and the method proposed here is the following. In the method proposed in this paper, it is unnecessary to specify an intermediate constraint frame, as is the case in hybrid control [33]. Control of position error is performed directly in the task space.

Usually only approximate kinematic information is known concerning the exact position of the constraint surface the manipulator is to come in contact with. Consider now the case, where due to task kinematic modeling inaccuracy, $\phi(p)$ is known only approximately. We wish to regulate the end-effector position to p_0, but due to errors in the constraint surface kinematic model, p_0 is given such that

$$\phi(p_0) = \Delta \tag{37}$$

$$\phi(p') = 0 \tag{38}$$

where $p' \neq p_0$. A simple formula is developed which relates the error in the unregulated coordinate of p to Δ in (37) in the presence of inaccuracy in $\phi(p)$ as given above.

Expansion of (1) about p_0 gives

$$\phi(p)=\phi(p_0)+\left.\frac{\partial\phi}{\partial p}\right|_{p_0}\delta p \tag{39}$$

where the higher order terms have been dropped. Evaluation of (39) at $p = p'$ gives

$$\phi(p')=\phi(p_0)+\left.\frac{\partial\phi}{\partial p}\right|_{p_0}\delta p. \tag{40}$$

Expanding the partial derivative in (40)

$$\left.\frac{\partial\phi}{\partial p}\right|_{p_0}=\left.\left(\frac{\partial\phi}{\partial p_1}\ \frac{\partial\phi}{\partial p_2}\ \frac{\partial\phi}{\partial p_3}\right)\right|_{p_0}.$$

Under the assumption that the first two coordinates of p are regulated, i.e., $\delta p_1 = 0$, $\delta p_2 = 0$, and substituting the above along with (37) and (38) into (40) gives

$$\left.\frac{\partial\phi}{\partial p_3}\right|_{p_0}\partial p_3=-\Delta. \tag{41}$$

Equation (41) gives a simple formula which relates the error in an unregulated coordinate of p with the uncertainty in the constraint function.

From this discussion, the following is clear. Regulation of three of the position coordinates of the manipulator end-effector is unnecessary, and only two coordinates need to be regulated, as motion is constrained by contact with the constant surface. If two coordinates of the position error of the end-effector, with respect to the manipulator base frame, are regulated to zero, then the error in the third coordinate is related to the constraint kinematic error given by (41).

C. Feedback Measurements

As defined earlier, the state of the constrained manipulator linearized dynamics is given as

$$x \triangleq \begin{pmatrix}\delta q\\ \delta\dot{q}\\ \delta\lambda\end{pmatrix}. \tag{42}$$

We wish to regulate contact force given by

$$f=D^T\lambda \tag{43}$$

and two position coordinates of the position vector p as defined in (3). Regulation of contact force means $f \rightarrow f^{\text{ref}}$ as $t\rightarrow\infty$, where $f^{\text{ref}} = D^T\lambda_0$.

Thus in order that $f \rightarrow f^{\text{ref}}$ as $t \rightarrow \infty$, we require the perturbation in the Lagrange multiplier be regulated to zero. If $\delta\lambda \rightarrow 0$ as $t\rightarrow\infty$ then $\lambda\rightarrow\lambda_0$ as $t\rightarrow\infty$ and $f\rightarrow f^{\text{ref}}$ as $t\rightarrow\infty$ assuming D in (43) is perfectly known. Given a force measurement from a contact force sensor, we must obtain an estimate of the state element λ. From (43), λ is determined from the measured contact force f^m

$$\lambda=(DD^T)^{-1}Df^m \tag{44}$$

using the gradient of the constraint function

$$D \triangleq \frac{\partial\phi(p)}{\partial p}.$$

From (44), the state vector element $\delta\lambda$ is constructed as

$$\delta\lambda=\lambda-\lambda_0 \tag{45}$$

where λ is given by (44).

Similarly, we wish to regulate two of the position coordinates of p. Linearizing p about the nominal position p_0, q_0 we obtain

$$\delta p=J(q_0)\delta q \tag{46}$$

where higher order terms are ignored. Since only two coordinates of p are to be regulated we write

$$\begin{pmatrix}\delta p_i\\ \delta p_j\end{pmatrix}=\begin{bmatrix}j_i\\ j_j\end{bmatrix}\delta q \tag{47}$$

where

$\delta p_k \triangleq$ kth coordinate of p
$j_k \triangleq$ kth row of $J(q_0)$.

Thus as long as $\delta p_k\rightarrow 0$ as $t\rightarrow\infty$, $p_k\rightarrow p_{k0}$ as $t\rightarrow\infty$ for $k = i, j$ in the above, regulating two of the three coordinates of p as desired.

With the foregoing, it is clear that we wish to regulate the following signal:

$$y=Cx \tag{48}$$

where

$y \triangleq$ output of perturbation model

$x \triangleq \begin{pmatrix}\delta q\\ \delta\dot{q}\\ \delta\lambda\end{pmatrix}$ perturbation model state

$C \triangleq \begin{bmatrix} & 0 & \cdots & 0 & 1\\ j_i & 0 & \cdots & & 0\\ j_j & 0 & \cdots & & 0\end{bmatrix}$ output map

$$C \in R^{3\times(2n+1)}$$

$n \triangleq$ number of degrees of freedom of manipulator.

The structure of the constrained dynamic formulation has been exploited in a very natural way such that the problem of force and position control is an output regulation problem. As long as $y\rightarrow 0$ as $t\rightarrow\infty$ in (48), $f\rightarrow f^{\text{ref}}$ and two of the position coordinates p_i, p_j are regulated, as desired.

D. Dynamic Model with Constant Disturbance Term

The linearized equations of motion of a rigid n degree of freedom manipulator in contact with a rigid surface, as derived in Section III, are given by

$$\begin{bmatrix} I & 0 & 0 \\ 0 & \Phi(q_0) & 0 \\ 0 & 0 & 0 \end{bmatrix}\begin{pmatrix} \delta\dot{q} \\ \delta\ddot{q} \\ \delta\dot{\lambda} \end{pmatrix} = \begin{bmatrix} 0 & I & 0 \\ \frac{\partial}{\partial q}(G-J^TD^T\lambda)|_0 & 0 & J^TD^T|_0 \\ DJ|_0 & 0 & 0 \end{bmatrix}\begin{pmatrix} \delta q \\ \delta\dot{q} \\ \delta\lambda \end{pmatrix} + \begin{bmatrix} 0 \\ I \\ 0 \end{bmatrix}\delta\tau. \tag{13}$$

Equation (13) represents a linearization of (7) about (q_0, $\dot{q}_0$, λ_0) expressed in state vector form, as discussed in Section III. Introduction of uncertainty in manipulator dynamic parameters introduces a constant disturbance term in (13) as follows. Linearizing (6) about (q_0, 0, λ_0) gives

$$\tau = \hat{\tau}_0 + \Phi(q_0)\delta\ddot{q} + \frac{\partial}{\partial q}(G-J^TD^T\lambda)|_0 - J^TD^T\delta\lambda. \tag{49}$$

Due to dynamic parameter modeling errors

$$\hat{\tau}_0 \triangleq \tau_0 + \Delta\tau \tag{50}$$

where

$$\tau_0 \triangleq G(q_0) - J^TD^T\lambda_0$$

$$\Delta\tau \triangleq \Delta G(q_0),$$

which is error due to gravity compensation. Thus $\Delta\tau$ represents a constant bias term due to inaccurate gravity compensation caused by unknown payload or inaccurate manipulator dynamic parameters. Errors as a result of kinematic modeling are not considered in this paper. With this uncertainty, (49) and (13) become

$$\begin{bmatrix} I & 0 & 0 \\ 0 & \Phi(q_0) & 0 \\ 0 & 0 & 0 \end{bmatrix}\begin{pmatrix} \delta\dot{q} \\ \delta\ddot{q} \\ \delta\dot{\lambda} \end{pmatrix} = \begin{bmatrix} 0 & I & 0 \\ \frac{\partial}{\partial q}(G-J^TD^T\lambda)|_0 & 0 & J^TD^T|_0 \\ DJ|_0 & 0 & 0 \end{bmatrix}\begin{pmatrix} \delta q \\ \delta\dot{q} \\ \delta\lambda \end{pmatrix} + \begin{bmatrix} 0 \\ I \\ 0 \end{bmatrix}\delta\tau + \begin{bmatrix} 0 \\ \Delta\tau \\ 0 \end{bmatrix}. \tag{51}$$

We now define the following in order to rewrite (51) in a more compact form:

$$E \triangleq \begin{bmatrix} I & 0 & 0 \\ 0 & \Phi(q_0) & 0 \\ 0 & 0 & 0 \end{bmatrix}$$

$$A \triangleq \begin{bmatrix} 0 & I & 0 \\ \frac{\partial}{\partial q}(G-J^TD^T\lambda)|_0 & 0 & J^TD^T|_0 \\ DJ|_0 & 0 & 0 \end{bmatrix}$$

$$B \triangleq \begin{bmatrix} 0 \\ I \\ 0 \end{bmatrix}$$

$$u \triangleq \delta\tau$$

$$d \triangleq \begin{bmatrix} 0 \\ \Delta\tau \\ 0 \end{bmatrix}.$$

Using these definitions as well as the definition of the state x given previously in (42), (51) can be written

$$E\dot{x} = Ax + Bu + d \tag{52}$$

where det (E) = 0.

E. Decomposition of Robotic System

In this section the robotic system given by (52) is decomposed into slow and fast subsystems, as introduced in Section IV-A. The effects of initial conditions and the constant disturbance term introduced in the dynamic formulation in the system given by (52) are examined in detail. A physical interpretation is given to the impulsive modes that may exist for arbitrary initial conditions in (52), but it is demonstrated that impulsive modes do not appear in the time response of the constrained robotic system.

We consider the system (52). It is desired to decompose this dynamic system into Kronecker canonical form using the theory of matrix pencils as introduced in Section IV. This permits the Laplace transform solution to (52) to be written explicitly in terms of the slow and fast states of (52).

Initially, we consider the dimensions of the slow and fast subspaces of the robotic system in descriptor form, given by (52). Robotic systems (7) fall into a broad class of dynamic systems which are characterized by a global nilpotency index equal to three, [28] and [11]. The global nilpotency index of a nonlinear system is identically equal to the index of a singular system in the linear case. The index corresponds to the degree of nilpotency [5] of the matrix L_f in the decomposition of a linear descriptor system into slow and fast subspaces as given by (25) and (26). Gear and Petzold [11] give an algorithm to determine the global nilpotency index of nonlinear or index of linear descriptor systems, respectively. The application of this algorithm to the linearized system dynamics given by (52) indicates that the index of (52) is three. Details of this calculation are given in the Appendix.

An alternate approach [7] to determine the index of (52) and hence the dimension of the slow and fast subspaces is to evaluate

$$\deg(\det(\lambda E - A)). \tag{53}$$

The dimension of the fast and slow subspaces for (7) are

$$\dim(S) = 2n + 1 - 3$$
$$\dim(F) = 3 \tag{54}$$

where $\dim(X) = \dim(S) + \dim(F)$ and the number of degrees of freedom of the manipulator equals n.

With the dimension of the slow and fast subsystems known, the system (52) is now decomposed as in Cobb [7] to yield the following:

$$\dot{x}_s = L_s x_s + B_s u + d_s \tag{55}$$

$$L_f \dot{x}_f = x_f + B_f u + d_f \tag{56}$$

where

$$MA|S = L_s$$
$$MA|F = I$$
$$ME|S = I$$

$ME|F = L_f$, a nilpotent matrix with index = 3, i.e., $L_f^3 = 0$

$$B_s = PMB$$
$$B_f = QMB$$
$$d_s = PMd$$
$$d_f = QMd$$

with

$$x_{0s} = Px_0$$
$$x_{0f} = Qx_0.$$

Transformation of the fast subsystem given by (56) into Jordan form with the nonsingular transformation T gives

$$J\dot{x}_f' = x_f' + B_f' u + d_f' \tag{57}$$

where

$$J \triangleq T^{-1} L_f T = \begin{bmatrix} 0 & 1 & 0 \\ 0 & 0 & 1 \\ 0 & 0 & 0 \end{bmatrix}, \quad \text{Jordan block}$$

$$x_f' \triangleq T^{-1} x_f$$
$$B_f' \triangleq T^{-1} B_f$$
$$d_f' \triangleq T^{-1} d_f.$$

With the fast subsystem in the form given by (57), and $u \equiv 0$, the solution is easily written in terms of Laplace transforms as

$$(Js - I)x_f'(s) = Jx_{0f}'(0) + d_f'(s) \tag{58}$$

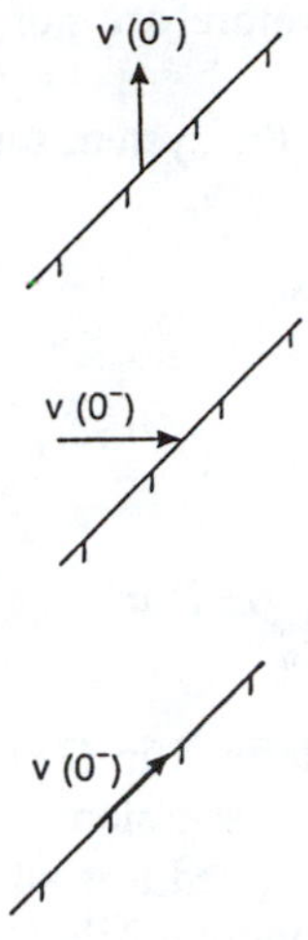

Fig. 4. Initial conditions for constrained system.

where $x_{0f}'(0) = T^{-1}x_{0f}$. Solving for $x_f'(s)$ we have

$$x_f' = -\begin{bmatrix} 1 & s & s^2 \\ 0 & 1 & s \\ 0 & 0 & 1 \end{bmatrix} Jx_{0f}' - \begin{bmatrix} 1 & s & s^2 \\ 0 & 1 & s \\ 0 & 0 & 1 \end{bmatrix} d_f'(s). \tag{59}$$

Equation (59) represents the free system response or the initial condition response of the robotic system. It is clear that initial conditions and disturbance terms can give rise to impulsive behavior of the fast subsystem states. Physically, this impulsive behavior results from initial conditions that do not satisfy the algebraic constraints that are imposed by the system, i.e.

$$x_{0^-} \neq x_{0^+}. \tag{60}$$

In order for the constraint equation to be satisfied, for arbitrary initial conditions, certain states may undergo step changes. Corresponding to these states are those which represent first or higher order derivatives of the step change, i.e., velocity or acceleration components. These states then exhibit impulsive behavior due to the arbitrary nature of the initial condition.

In this paper, it is assumed that the manipulator end-effector is initially in contact with the constraint surface. In addition, it is also assumed that any initial component of end-effector velocity is tangent to the constraint surface, i.e., the component of velocity normal to the surface is zero. Fig. 4 illustrates various initial conditions. The initial condition assumed here is such that the constraint equation, $\phi(p) = 0$, is never violated, hence the initial conditions are not arbitrary (Fig. 4—iii).

Similarly, only disturbance terms $d(s)$ are considered which do not result in impulsive behavior of the fast subsystem. With the foregoing, in order for the fast subsystem to exhibit no impulsive behavior, the following must be true:

$$x_{0^-} \subset S \tag{61}$$

$$d \subset S. \tag{62}$$

Thus arbitrary initial conditions and disturbance terms are not permitted and furthermore are not physically meaningful in this context.

With (61) and (62) the system response of (52) becomes

$$x_s = e^{L_s t} x_{0s} + \int_0^t e^{L_s(t-\tau)} B_s u(\tau)\, d\tau$$

$$+ \int_0^t e^{L_s(t-\tau)} d(\tau)\, d\tau$$

$$x_f = -\sum_{i=0}^{\nu-1} L_f^i B_f u^i \tag{63}$$

where all terms are defined in (25) and (26).

As a result of this discussion of impulsive behavior, it is seen that there is little need to find a fast state feedback such that $L_f \equiv 0$ to eliminate impulsive behavior. We need only be concerned with feedback of the slow subsystem. It is clear that with the restrictions given by (61) and (62), the impulsive modes of the robotic system are never excited.

It is assumed here that the control exercised over the contact force is such that the force will always hold the end-effector on the constraint surface. If a disturbance were encountered such that the end-effector would have left the constraint surface, then the constrained equations of motion would no longer apply and a control for unconstrained motion would be required. In this paper, we only consider the case where contact is maintained, hence the problems that arise due to loss of contact are not considered.

F. Controller Structure

Since the effects of arbitrary initial conditions and the disturbance d have been eliminated through physical argument, it is only necessary to consider feedback of the slow states of the system (52). The problem as posed here is the following. It is desired to regulate asymptotically to zero the output

$$y = Cx \tag{48}$$

of the dynamic system

$$E\dot{x} = Ax + Bu + d \tag{52}$$

i.e., $y_{ss} = 0$. We introduce the integrator states

$$\dot{\eta} = y \tag{64}$$

forming the augmented system as follows:

$$E_a \dot{z} = A_a z + B_a u + d_a, \qquad z \in X_a$$

$$y = C_a z \tag{65}$$

where

$$E_a \triangleq \begin{bmatrix} E & 0 \\ 0 & I \end{bmatrix}$$

$$A_a \triangleq \begin{bmatrix} A & 0 \\ C & 0 \end{bmatrix}$$

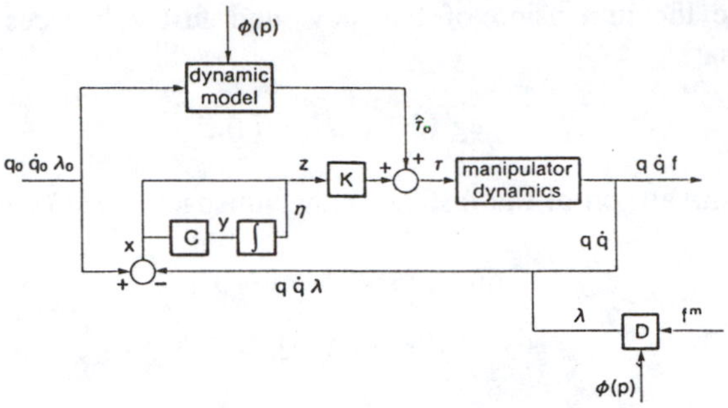

Fig. 5. Closed-loop system.

$$B_a \triangleq \begin{bmatrix} B \\ 0 \end{bmatrix}$$

$$C_a \triangleq [C \quad 0]$$

$$z \triangleq \begin{pmatrix} x \\ \eta \end{pmatrix}$$

$$d_a \triangleq \begin{pmatrix} d \\ 0 \end{pmatrix}.$$

Decomposing (65) into slow and fast subsystems as before we have

$$\dot{z}_s = L_{sa} z_s + B_{sa} u + d_{sa} \tag{66}$$

$$L_{fa} \dot{z}_f = z_f + B_{fa} u \tag{67}$$

where the definitions of L_{sa}, etc., are obvious.

A stabilizing feedback for the slow subsystem

$$u = K_s z_s \tag{68}$$

is found such that the closed-loop system

$$\dot{z}_s = (L_{sa} + B_{sa} K_s) z_s + d_{sa} \tag{69}$$

is asymptotically stable, and well-posed as discussed previously, where K is found by transforming K_s into the original state space X_a, i.e., $K_s = K|S$.

Thus the control input driving the manipulator using the nominal control, given by (14), at the operating point (q_0, $\dot{q}_0$, λ_0) is

$$\tau = \tau_0 + Kz. \tag{70}$$

Equation (70) represents i) a nominal control τ_0 which provides a nominal input to cause the manipulator to exert a contact force f^{ref} at a fixed position on the constraint surface and ii) a feedback Kz which causes a regulation of force and position to occur in spite of inaccuracy in the nominal input due to unavoidable dynamic parameter modeling errors. The control (70) is valid in the neighborhood of (q_0, $\dot{q}_0$, λ_0) in which the linearization of the manipulator dynamics holds. Within this region, asymptotic trajectory tracking will occur for step disturbances and constant reference inputs. Fig. 5 illustrates the closed-loop scheme.

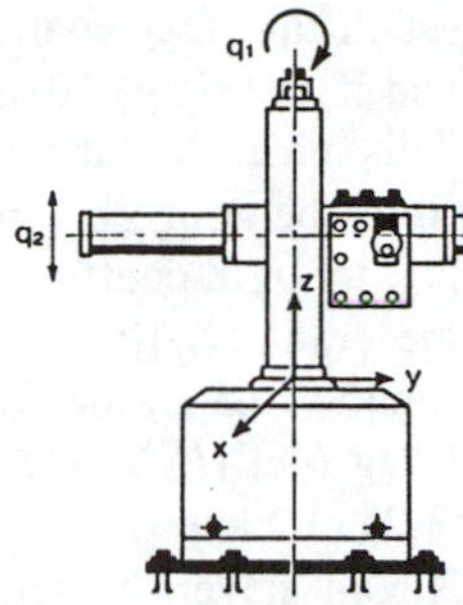

Fig. 6. UMS-2 robot.

VI. Numerical Example

The numerical example chosen to illustrate the design procedure is the manipulator UMS-2 [36], illustrated in Fig. 6. The dynamic equations of motion of this manipulator in unconstrained form are given by

$$\tau_i = \Phi_i \ddot{q}_i + H_i \tag{71}$$

where

$$\Phi_1 = J_{z1} + J_{z2} + J_{z3} + m_3(q_3 + l_3)^2$$
$$H_1 = 2m_3(q_3 + l_3)\dot{q}_1\dot{q}_3$$
$$\Phi_2 = m_2 + m_3$$
$$H_2 = (m_2 + m_3)g$$
$$\Phi_3 = m_3$$
$$H_3 = -m_3(q_3 + l_3)\dot{q}_1^2.$$

The position vector p is given by

$$p = \begin{pmatrix} (q_3 + l_3)\cos q_1 \\ (q_3 + l_3)\sin q_1 \\ q_2 \end{pmatrix}. \tag{72}$$

The constraint function chosen is the following:

$$\phi(p) = x + y + z - c = 0 \tag{73}$$

where

$$c = x_0 + y_0 + z_0.$$
$$x_0 = (q_{30} + l_3)\cos q_{10}$$
$$y_0 = (q_{30} + l_3)\sin q_{10}$$
$$z_0 = q_{20}.$$

The dynamic parameters are given in Table I, as well as the nominal joint values about which the constrained manipulator system is linearized. Fig. 7 illustrates the manipulator task geometry. The constrained dynamics formulation is given by

$$\Phi(q)\ddot{q} + H(\dot{q}, q) + G(q) = \tau + J^T D^T \lambda \tag{74}$$

where the elements of (74) are constructed from the definitions given in (71) and (73). This system is linearized about the nominal state given by $(q_0, \dot{q}_0, \lambda_0)$ and three new states are

TABLE I
NOMINAL DYNAMIC AND KINEMATIC PARAMETERS

Nominal State	Nominal Dynamic Parameters
$q_{10} = 25°$	$l_3 = 0.2$ m
$q_{20} = 0.3$ m	$m_2 = 1$ kg
$q_{30} = 0.2$ m	$m_2 = 2$ kg
$\lambda_0 = \frac{-1}{\sqrt{3}} \frac{\text{kg-m}}{\text{s}^2}$	$J_{z1} = 0.1$ kg-m^2
$\dot{q}_{10} = 0$	$J_{z2} = 0.2$ kg-m^2
$\dot{q}_{20} = 0$	$J_{z3} = 0.1$ kg-m^2
$\dot{q}_{30} = 0$	
$\ddot{q}_{10} = 0$	
$\ddot{q}_{20} = 0$	
$\ddot{q}_{30} = 0$	

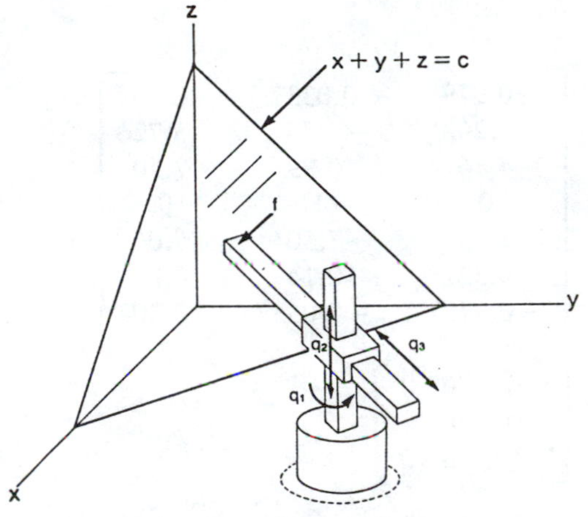

Fig. 7. Manipulator task geometry.

defined as follows:

$$\dot{\eta} = \begin{pmatrix} \delta\lambda \\ \delta p_1 \\ \delta p_2 \end{pmatrix}. \tag{75}$$

The linearized matrices for the system (65) are given by

$$E_a = \text{diag}\,(1\ 1\ 1\ 0.9\ 3\ 2\ 0\ 1\ 1\ 1), \qquad E_a \in R^{10\times 10}$$

$$A_a = \begin{bmatrix} 0 & I_{3,3} & 0 & 0 \\ \frac{\partial}{\partial q}(G - J^T D^T \lambda)|_0 & 0 & J^T D^T|_0 & 0 \\ DJ|_0 & 0 & 0 & 0 \\ C & 0 & 0 & 0 \end{bmatrix},$$

$$A_a \in R^{10\times 10}$$

$$DJ|_0 = (0.242\ 1\ 1.33)^T$$

$$\frac{\partial}{\partial q}(G-J^TD^T\lambda)|_0=\begin{bmatrix} 0.384 & 0 & -0.279 \\ 0 & 0 & 0 \\ -0.279 & 0 & 0 \end{bmatrix}$$

$$C=\begin{bmatrix} -0.211 & 0 & 0.91 & 0 & 0 & 0 & 0 \\ 0 & 1 & 0 & 0 & 0 & 0 & 0 \\ 0 & 0 & 0 & 0 & 0 & 0 & 1 \end{bmatrix}$$

$$B_a=\begin{bmatrix} 0 \\ I_{3,3} \\ 0 \end{bmatrix}, \qquad B_a \in R^{10\times 3}.$$

Decomposition of (E_a, A_a, B_a) gives

$$L_s=\begin{bmatrix} 0.7105 & 0 & 0 & 0 & 0 & 0 & 0 \\ 0 & 0 & 0 & 0 & 0 & 0 & 0 \\ 0 & 0 & 0 & 0 & 0 & 0 & 0 \\ 0 & 0 & 0 & 0 & 0 & 0 & 0 \\ 0 & 0 & 0 & 0 & 0 & 0.149 & 0 \\ 0 & 0 & 0 & 0 & -0.149 & 0 & 0 \\ 0 & 0 & 0 & 0 & 0 & 0 & -0.7105 \end{bmatrix} \tag{76a}$$

$$B_s=\begin{bmatrix} -0.979 & -0.0337 & 0.204 \\ 3.262 & -5.74 & 3.726 \\ -4.76 & 10.99 & -7.40 \\ 0 & -1 & 0 \\ 3.20 & -7.50 & 5.06 \\ 4.327 & -10.14 & 6.844 \\ -0.9797 & -0.0337 & 0.204 \end{bmatrix} \tag{76b}$$

$$L_f=\begin{bmatrix} 0 & 0 & 0 \\ 0 & 0 & 1 \\ 1.505 & 0 & 0 \end{bmatrix} \tag{76c}$$

$$B_f=\begin{bmatrix} 0 & 0 & 0 \\ 0.209 & 0.260 & 0.518 \\ 0 & 0 & 0 \end{bmatrix}. \tag{76d}$$

Before constructing a feedback

$$u=K_sx_s \tag{77}$$

to stabilize the linearized system given by (66), the possibility of eliminating impulsive behavior with feedback of the fast states as outlined by Cobb [7] is examined. Physically, this is impossible for the robotic system considered here, because for any feedback applied to the system (66), initial conditions can always be found such that impulsive behavior exists, i.e., during a collision, velocities change discontinuously resulting in impulsive acceleration behavior. To demonstrate this, the fast subsystem matrices (76c), (76d) are examined in the following to verify that impulsive modes cannot be eliminated by a feedback of the fast states.

To ensure that there are no impulsive modes in the closed-loop response of the descriptor system (52), for arbitrary initial conditions, the following must hold, $L_{fK} \equiv \mathbf{0}$, where L_{fK} is the matrix L_f which results from decomposition of (52) with a feedback K_f applied to the fast subsystem states. We wish to find a feedback K_f such that $L_{fK} \equiv \mathbf{0}$. According to Cobb [7], this can be achieved if and only if deg (det $(E_a\lambda - A_a - B_aK)$) = rank (E_a). This means that the r_kth coefficient of det $(E_a\lambda - A_a - B_aK)$ must be nonzero where r_k = rank (E_a). From matrix algebra [7], the r_kth coefficient of det $(E_a\lambda - A_a - B_aK)$ is equal to the {rank (L_f)]th coefficient of det $(L_f\lambda - I - B_fK_f)$. Examination of L_f and B_f reveals that there exists no K_f such that deg (det $(L_f\lambda - I - B_fK_f)$) = rank (L_f). Otherwise, the (3,2) element of $L_f\lambda - I - B_fK_f$ must be nonzero, and this can never be true due to the structure of B_f. Thus as verified by this numerical example, impulsive modes due to arbitrary initial conditions acting on the example robotic system cannot be eliminated.

At the present time, it is unknown if it is true in general that impulsive modes cannot be eliminated from a constrained robotic system. Further work in this area is required to develop a mathematical proof of such a concept. However, the control problem considered here addresses constrained motion only. If an initial condition violates the constraint equation, then the motion of the manipulator is represented by an unconstrained mathematical model, and a different control strategy may have to be considered.

Examination of the pair (L_s, B_s) reveals that the slow subsystem is controllable, hence the slow subsystem can be stabilized with state feedback, as given by (68). A stabilizing feedback K_s for the slow subsystem was obtained through minimization of the following:

$$\min \int_0^\infty (z_s^TQz_s+u^TRu)\,dt \tag{78}$$

$$\text{subject to: } \dot{z}_s=L_sz_s+B_su \tag{79}$$

with

$Q \triangleq$ diag (0.05 0.05 0.05 0.2 0.2 0.2 0.2)
$R \triangleq$ diag (0.05 0.05 0.05).

This gain is then transformed to the original state space so the feedback control is of the form

$$u=Kz+v \tag{80}$$

where K is given by $K_s = K|S$.

Under the assumption that the state remains close to a nominal state about which the manipulator dynamics were linearized, application of the control given by (70) is valid. The results of a numerical simulation of the control (70) applied to the nonlinear system dynamics given by (7) is discussed in the next section.

VII. Numerical Simulation

In order to verify the control system approach outlined in this paper, the nonlinear equations of motion of the example of Section VI were integrated numerically. The response to a step change in mass of link 3 was investigated to demonstrate the asymptotic trajectory tracking capability of the proposed controller. This step change in link mass is equivalent to a sudden reduction or increase in payload mass that might occur while in contact with a constraint surface during execution of a task. Spreading glue on a surface is an example of such a task.

Since the method proposed is clearly not robust to constraint surface kinematic error, a simulation under these conditions was not considered.

As discussed by Gear and Petzold [11], there are no known methods of integration for singular systems of index three or higher. As a result, it is necessary to remove the singularity that exists in the dynamic formulation given by (7) through a reduction transformation as outlined in Wittenberg [40] and McClamroch [28]. The procedure to carry this out is given below. Differentiation of the constraint equation $\phi(p) = 0$ with respect to time gives

$$DJ\dot{q}=0. \tag{81}$$

Differentiation of (81) again with respect to time gives

$$DJ\ddot{q}+\Gamma(q)=0 \tag{82}$$

where

$$\Gamma(q) \triangleq \frac{d}{dt}(DJ)\dot{q}.$$

Solving (6) for $\ddot{q}$ and substituting this into (82) gives

$$DJ\Phi^{-1}(-H-G+\tau)+DJ\Phi^{-1}J^TD^T\lambda+\Gamma(q)=0. \tag{83}$$

Equation (83) can be solved for λ, where the term $DJ\Phi^{-1}J^TD^T$ is always invertible [17], and results in an expression for λ of the form

$$\lambda=\Lambda(q, \dot{q}, \tau). \tag{84}$$

Substitution of (84) into (6) gives the final form of the equations to be integrated

$$\Phi(q)\ddot{q}+H(\dot{q}, q)+G(q)=\tau+J^TD^T\Lambda(q, \dot{q}, \tau). \tag{85}$$

Equation (85) represents a well-posed initial value problem that can be integrated by standard numerical integration techniques. However, integration of (85) is equivalent to integration of a constrained system [40], subject to a constraint given by

$$\frac{d^2}{dt^2}\phi(p)=0. \tag{86}$$

Equation (86) does not ensure that $\phi(p) = 0$, and in general, due to the unavoidable integration of truncation errors, $\phi(p) = 0$ will be violated.

To avoid this problem, numerical damping is added to the system as follows. Rather than substituting (6) into (82), which is equivalent to $\ddot{\phi}(p) = 0$, (6) is substituted into

$$\ddot{\phi}(p)+2\alpha\dot{\phi}(p)+\beta^2\phi(p)=0 \tag{87}$$

where the coefficients of (87) are chosen so that (87) is asymptotically stable. Thus rather than (82), we obtain

$$DJ\ddot{q}+\Gamma'(q)=0 \tag{88}$$

where $\Gamma'(q)$ includes the damping terms of (87). The idea of this kind of damping for singular systems was first introduced by Baumgarte [4]. Specification of α and β in (87) permits the damping poles of (87) to be placed arbitrarily in the left-hand complex plane. When feedback is applied to the overall system, these damping poles shift. Analysis of the linearized closed-loop system with numerical damping revealed that the damping poles, originally chosen to lie at $s = -50, -50$ have been shifted to $s = -50 \pm j50$. Thus in a closed loop, these poles are still stable, with an acceptable transient response. In all simulations performed, the constraint equations were not violated by more than 10^{-9}, indicating that the numerical damping was successful.

In addition to (85), integrator states were included as follows:

$$\begin{aligned}\dot{\eta}_1&=\lambda-\lambda_0\\ \dot{\eta}_2&=p_1-p_{10}\\ \dot{\eta}_3&=p_2-p_{20}.\end{aligned} \tag{89}$$

The simulation performed was designed to demonstrate the trajectory tracking capabilities of the proposed control scheme. A step change in mass was introduced in the manipulator while in contact with the constraint surface to demonstrate stability and trajectory tracking. Certain physical situations are accurately represented by this simulation such as spreading glue on a surface, where the glue reservoir is held by the manipulator. The mass of the manipulator changes while in contact with the environment, requiring both workless force and position to be regulated during this parameter perturbation.

Exact measurement of the workless force is also represented by the simulation. In addition, an exact model of the gradient of the surface is assumed, permitting the Lagrange multiplier to be calculated exactly, as in (44). Figs. 8 and 9 illustrate the response to a step change in m_2 of the manipulator of 0.1 kg, applied at $t = 10$, with a feedback gain matrix K given by

$$K=\begin{bmatrix} -3.1 & 4.6 & -1.0 & -6.4 & 26.5 & -11.2 & 0.0 & 0.90 & 0.39 & 0.33\\ 1.03 & -10.6 & 4.9 & 7.9 & -67.8 & 28.9 & 0.0 & 0.212 & -0.76 & 1.26\\ 0 & 7.7 & -3.9 & -5.0 & 44.0 & -22 & 0.0 & -0.37 & 0.534 & 1.52 \end{bmatrix}. \tag{94}$$

This represents a 10-percent change in the mass of link three. As a result of this change in mass it is observed in Fig. 8 that the value of the Lagrange multiplier λ, and hence the contact force, changes immediately. As a result of control, it is seen that the value of λ approaches λ_0, the nominal value after approximately 6 s. Fig. 8 also illustrates the joint torques and forces that are applied during the disturbance. It is seen that there is an abrupt change in two of the joint torques when the mass of link three is changed. Due to the manipulator geometry, the steady-state forces τ_1 and τ_3 are unaffected by gravity, hence τ_1 and τ_3 return to their nominal values at $t = 20$ s. In contrast, the steady-state force applied at joint 2 is

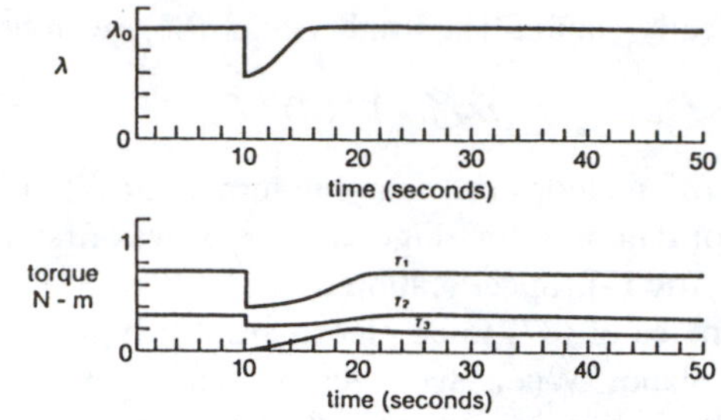

Fig. 8. Force transient response.

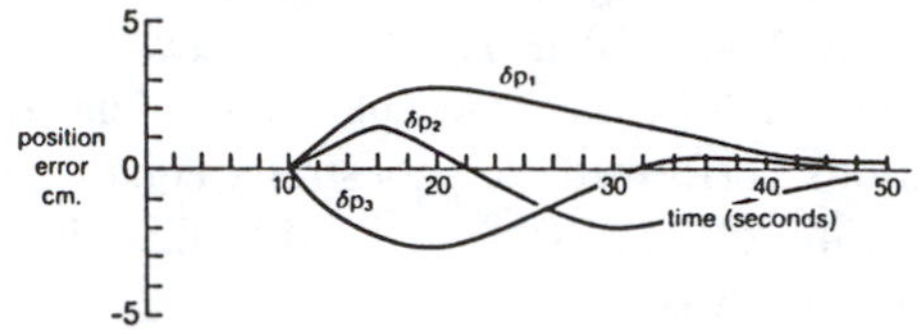

Fig. 9. Position transient response.

affected by the change in mass of link 3, hence an additional compensating force is applied at this joint as illustrated in Fig. 8.

It is observed in Fig. 8 that the Lagrange multiplier λ approaches λ_0 before the torques have reached their steady-state values. This is explained in the following way. The manipulator tip initially slips down the constraint surface, causing errors in λ and position, as seen in Figs. 8 and 9. While position error still exists, regulation of λ is achieved, regulating contact force. However, due to the presence of position errors, the torques continue to change until the position error decreases to zero, without a significant effect on the value of λ. Thus the torques continue to change even after $\lambda = \lambda_0$. Due to the scale of the plot in Fig. 8, the small changes in the torque while position continues to change cannot be seen.

As discussed, Fig. 9 illustrates the position errors generated as a result of the change in mass of link 3. Initially, before the disturbance is applied to link 3, there is zero position error. As a result of the added mass, the manipulator tip slips down the constraint surface generating position errors. Due to the regulation of the coordinates p_1 and p_2, the position error is brought to zero by the controller as is seen in Fig. 9.

The results obtained here compare favorably with results obtained with hybrid control. Experiments performed by Merlet [30] with a hybrid controller regulating normal contact force and position on a constraint surface illustrate that a hybrid controller can perform the same task as the controller proposed here. In both cases, regulation of contact force and constrained position is obtained. Similar experiments with regulation of contact force but with a disturbance in one position coordinate are seen in Raibert and Craig [33]. Here a hybrid controller is used to achieve regulation of the remaining position coordinate and contact force.

VIII. Conclusion

In this paper, a new method is presented to control simultaneously both position and contact force during constrained motion of the manipulator. The specific problem addressed is that of control of force and constrained position in the presence of manipulator dynamic parameter uncertainty with a perfect kinematic model of the constraint surface. The controller proposed here, valid in a neighborhood about the point of linearization, is seen to be a useful concept since in many contact task applications, it is desired to apply a force at a particular point on a constraint surface.

As discussed in this paper, the time response of the descriptor system due to arbitrary initial conditions admits impulsive behavior. Through decomposition of the robot linearized dynamic system into slow and fast subspaces, it is clearly seen how this impulsive behavior arises. Subsequently, through consideration of the physics of constrained motion, it is determined that impulsive behavior cannot occur if the end-effector remains in contact with the constraint surface at all times. As a result, a controller which uses feedback of the slow subspace states of an augmented state space is found to provide asymptotic trajectory tracking of both position and contact force in the presence of constant disturbances.

The control problem formulation, as presented in this paper, exploits the structure of the constrained dynamic system in a natural way leading to an output regulation problem that can be solved using linear time-invariant descriptor variable theory. The method is significant as contact tasks involving very stiff surfaces, which from a theoretical point of view impose a kinematic constraint on the motion of the manipulator, represent an important class of tasks commonly encountered.

Appendix

In this Appendix, the algorithm of Gear and Petzold [11] is used to determine the index of the linearized descriptor system given by

$$\begin{bmatrix} I & 0 & 0 \\ 0 & \Phi(q_0) & 0 \\ 0 & 0 & 0 \end{bmatrix} \begin{pmatrix} \delta\dot{q} \\ \delta\ddot{q} \\ \delta\dot{\lambda} \end{pmatrix} = \begin{bmatrix} 0 & I & 0 \\ \dfrac{\partial}{\partial q}(G - J^T D^T \lambda)|_0 & 0 & J^T D^T|_0 \\ DJ|_0 & 0 & 0 \end{bmatrix} \begin{pmatrix} \delta q \\ \delta\dot{q} \\ \delta\lambda \end{pmatrix} + \begin{bmatrix} 0 \\ I \\ 0 \end{bmatrix} \delta\dot{\tau}. \tag{A1}$$

Equation (A1) is written in a compact form as follows to simplify manipulation:

$$\begin{bmatrix} I & 0 & 0 \\ 0 & \Phi_0 & 0 \\ 0 & 0 & 0 \end{bmatrix} \begin{pmatrix} \delta\dot{q} \\ \delta\ddot{q} \\ \delta\dot{\lambda} \end{pmatrix} = \begin{bmatrix} 0 & I & 0 \\ R_1 & 0 & R_2 \\ R_3 & 0 & 0 \end{bmatrix} \begin{pmatrix} \delta q \\ \delta\dot{q} \\ \delta\lambda \end{pmatrix} + \begin{bmatrix} 0 \\ I \\ 0 \end{bmatrix} \delta\tau \tag{A2}$$

where

$$\Phi_0 \triangleq \Phi(q_0)$$

$$R_1 \triangleq \frac{\partial}{\partial q}(G - J^T D^T \lambda)|_0$$

$$R_2 \triangleq J^T D^T|_0$$

$$R_3 \triangleq DJ|_0.$$

Step 1

Differentiation with respect to time and rearrangement of the last row of (A2) gives

$$\begin{bmatrix} I & 0 & 0 \\ 0 & \Phi_0 & 0 \\ R_3 & 0 & 0 \end{bmatrix} \begin{pmatrix} \delta\dot{q} \\ \delta\ddot{q} \\ \delta\dot{\lambda} \end{pmatrix} = \begin{bmatrix} 0 & I & 0 \\ R_1 & 0 & R_2 \\ 0 & 0 & 0 \end{bmatrix} \begin{pmatrix} \delta q \\ \delta\dot{q} \\ \delta\lambda \end{pmatrix} + \begin{bmatrix} 0 \\ I \\ 0 \end{bmatrix} \delta\tau. \quad \text{(A3)}$$

The system given by (A3) is still singular, thus with a nonsingular transformation given by

$$P_1 \triangleq \begin{bmatrix} I & 0 & 0 \\ 0 & I & 0 \\ -R_3 & 0 & 1 \end{bmatrix} \quad \text{(A4)}$$

applied to (A3), the last row of the matrix on the left-hand side of (A3) is set equal to zero to give

$$\begin{bmatrix} I & 0 & 0 \\ 0 & \Phi_0 & 0 \\ 0 & 0 & 0 \end{bmatrix} \begin{pmatrix} \delta\dot{q} \\ \delta\ddot{q} \\ \delta\dot{\lambda} \end{pmatrix} = \begin{bmatrix} 0 & I & 0 \\ R_1 & 0 & R_2 \\ 0 & -R_3 & 0 \end{bmatrix} \begin{pmatrix} \delta q \\ \delta\dot{q} \\ \delta\lambda \end{pmatrix} + \begin{bmatrix} 0 \\ I \\ 0 \end{bmatrix} \delta\tau. \quad \text{(A5)}$$

Step 2

Since (A5) is still a singular system, the last row of (A5) is differentiated with respect to time to yield

$$\begin{bmatrix} I & 0 & 0 \\ 0 & \Phi_0 & 0 \\ 0 & -R_3 & 0 \end{bmatrix} \begin{pmatrix} \delta\dot{q} \\ \delta\ddot{q} \\ \delta\dot{\lambda} \end{pmatrix} = \begin{bmatrix} 0 & I & 0 \\ R_1 & 0 & R_2 \\ 0 & 0 & 0 \end{bmatrix} \begin{pmatrix} \delta q \\ \delta\dot{q} \\ \delta\lambda \end{pmatrix} + \begin{bmatrix} 0 \\ I \\ 0 \end{bmatrix} \delta\tau. \quad \text{(A6)}$$

This system is singular, thus zeroing the last row of the left-hand side of (A6) with the nonsingular transformation

$$P_2 \triangleq \begin{bmatrix} I & 0 & 0 \\ 0 & I & 0 \\ 0 & R_3\Phi_0^{-1} & 1 \end{bmatrix} \quad \text{(A7)}$$

applied to (A6) gives

$$\begin{bmatrix} I & 0 & 0 \\ 0 & \Phi_0 & 0 \\ 0 & 0 & 0 \end{bmatrix} \begin{pmatrix} \delta\dot{q} \\ \delta\ddot{q} \\ \delta\dot{\lambda} \end{pmatrix} = \begin{bmatrix} 0 & I & 0 \\ R_1 & 0 & R_2 \\ R_3\Phi_0^{-1}R_1 & 0 & R_3\Phi_0^{-1}R_2 \end{bmatrix} \begin{pmatrix} \delta q \\ \delta\dot{q} \\ \delta\lambda \end{pmatrix} + \begin{bmatrix} 0 \\ I \\ R_3\Phi_0^{-1} \end{bmatrix} \delta\tau. \quad \text{(A8)}$$

Step 3

Since (A8) is a singular system, the last row of (A8) is differentiated with respect to time to yield

$$\begin{bmatrix} I & 0 & 0 \\ 0 & \Phi_0 & 0 \\ 0 & 0 & -R_3\Phi_0^{-1}R_2 \end{bmatrix} \begin{pmatrix} \delta\dot{q} \\ \delta\ddot{q} \\ \delta\dot{\lambda} \end{pmatrix}$$

$$= \begin{bmatrix} 0 & I & 0 \\ R_1 & 0 & R_2 \\ 0 & R_3\Phi_0^{-1}R_1 & 0 \end{bmatrix} \begin{pmatrix} \delta q \\ \delta\dot{q} \\ \delta\lambda \end{pmatrix} + \begin{bmatrix} 0 \\ I \\ R_3\Phi_0^{-1} \end{bmatrix} \delta\tau + \begin{bmatrix} 0 \\ I \\ -R_3\Phi_0^{-1} \end{bmatrix} \delta\dot{\tau}. \quad \text{(A9)}$$

Since the system (A9) is nonsingular, due to the invertibility of the matrix on the left-hand side of (A9), the index of the system (A1) is equal to three, the number of differentiations of the last row of (A1) required to bring the system to nonsingular form. Note that

$$R_3\Phi_0^{-1}R_2 = DJ\Phi_0^{-1}J^TD^T \quad \text{(A10)}$$

is always nonsingular [40].

Acknowledgment

The authors wish to acknowledge fruitful discussions with Prof. D. Cobb at the University of Wisconsin and Prof. B. Francis at the University of Toronto.

References

[1] R. J. Anderson and M. Spong, "Hybrid impedance control of robotic manipulators," in *Proc. IEEE Conf. on Robotics and Automation* (Rayleigh, NC, 1987), pp. 1073–1080.

[2] J. Angeles, "On the numerical solution of the inverse kinematic problem," *Int. J. Robotics Res.*, vol. 4, no. 2, pp. 21–37, 1985.

[3] H. Asada and J-J. Slotine, *Robotic Analysis and Control.* New York, NY: Wiley, 1985.

[4] J. Baumgarte, "Stabilization of constraints and integrals of motion in dynamical systems," *Comput. Meth. Appl. Mech., Eng.*, vol. 1, pp. 1–16, 1972.

[5] S. L. Campbell, *Singular Systems of Differential Equations.* London, UK: Pitman, 1980.

[6] ——, *Singular Systems of Differential Equations II.* London, UK: Pitman, 1982.

[7] D. J. Cobb, "Descriptor variable and generalized singularly perturbed systems: A geometric approach," Ph.D. dissertation, University of Illinois, Urbana, IL, 1980.

[8] ——, "On the solutions of linear differential equations with singular coefficients," *J. Differ. Equat.*, vol. 46, no. 3, pp. 310–323, 1982.

[9] ——, "Descriptor variable systems and optimal state regulation," *IEEE Trans. Automat. Contr.*, vol. AC-28, no. 5, pp. 601–611, 1983.

[10] ——, "Controllability, observability and duality in singular systems," *IEEE Trans. Automat. Contr.*, vol. AC-29, pp. 1076–1082, 1984.

[11] C. W. Gear and L. R. Petzold, "ODE methods for the solution of differential algebraic systems," *SIAM J. Numer. Anal.*, vol. 24, no. 4, pp. 716–728, 1984.

[12] C. W. Gear, B. Leimkuhler, and G. K. Gupta, "Automatic integration of Euler-Lagrange equations with constraints," *J. Computat. Appl. Math.*, vols. 12 and 13, pp. 77–90, 1985.

[13] E. G. Gilbert and I. J. Ha, "An approach to nonlinear feedback control with applications to robotics," in *Proc. 22nd IEEE Conf. on Decision and Control* (San Antonio, TX, Dec. 14–16, 1983).

[14] R. E. Goddard, H. Hamami, and F. C. Weimer, "Biped side step in the frontal plane," *IEEE Trans. Automat. Contr.*, vol. AC-28, no. 2, pp. 129–187, 1983.

[15] H. Goldstein, *Classical Mechanics* Cambridge, MA: Addison-Wesley, 1950.

[16] E. J. Haug, S. C. Wu, and S. M. Yang, "Dynamics of mechanical systems with Coulomb friction, stiction, impact and constraint addition-deletion I," *J. Mech. Mach. Theory,* vol. 21, no. 5, pp. 401–406, 1986.

[17] H. Hemami and B. F. Wyman, "Modeling and control of constrained dynamic systems with application to biped locomotion in the frontal plane," *IEEE Trans. Automat. Contr.*, vol. AC-24, no. 4, pp. 526–535, 1979.

[18] H. Hemami, C. Wongchaisuwat, and J. L. Brinker, "A heuristic study of relegation of control in constrained robotic systems," *ASME J. Dyn. Syst. Meas. Contr.*, vol. 109, pp. 224–231, 1987.

[19] N. Hogan, "Impedance control of a robotic manipulator," in *Proc. ASME Winter Annual Meeting,* 1981.

[20] ——, "Impedance control: An approach to manipulation, Part I: Theory," *ASME J. Dyn. Syst. Meas. Contr.*, vol. 107, pp. 1–7, 1985.

[21] ——, "Impedance control: An approach to manipulation, Part II: Implementation," *ASME J. Dyn. Syst. Meas. Contr.*, vol. 107, pp. 8–16, 1985.

[22] ——, "Impedance control: An approach to manipulation, Part III: Applications," *ASME J. Dyn. Syst. Meas. Contr.*, vol. 107, pp. 17–24, 1985.

[23] ——, "Stable execution of contact tasks using impedance control," in *Proc. IEEE Conf. on Robotics and Automation,* 1987.

[24] H.-P. Huang and N. H. McClamroch, "Time-optimal control for a robotic contour following problem," in *Proc. IEEE Conf. on Robotics and Automation,* pp. 1140–1145, 1987.

[25] H. Kazerooni, P. K. Houpt, and T. B. Sheridan, "Robust compliant motion of manipulators, Part I: The Fundamental concepts of compliant motion," *IEEE J. Robotics Automat.*, vol. RA-2, no. 2, pp. 83–92, 1986.

[26] ——, "Robust compliant motion for manipulators, Part II: Design method," *IEEE J. Robotics Automat.*, vol. RA-2, no. 2, pp. 93–105, 1986.

[27] M. T. Mason, "Compliance and force control of computer controlled manipulators," *IEEE Trans. Syst. Man Cybern.*, vol. SMC-11, no. 6, pp. 418–432, 1981.

[28] N. H. McClamroch, "Singular systems of differential equations as dynamic models for constrained robot systems," in *Proc. IEEE Int. Conf. on Robotics and Automation* (San Francisco, CA, Apr. 8–10, 1986), pp. 21–28.

[29] N. H. McClamroch and H. P. Huang, "Dynamics of a closed chain manipulator," in *Proc. American Control Conf.*, pp. 50–54, 1985.

[30] J.-P. Merlet, "C-surface applied to the design of an hybrid force-position robot controller," in *Proc. IEEE Conf. on Robotics and Automation* (Rayleigh, NC, 1987), pp. 1055–1059.

[32] L. Pandolfi, "Controllability and stabilization for linear systems of algebraic and differential equations," *J. Optimiz. Theory Appl.*, vol. 30, no. 4, pp. 601–620, 1980.

[33] M. H. Raibert and J. J. Craig, "Hybrid position/force control of manipulators," *J. Dyn. Syst. Meas., Contr.*, vol. 102, pp. 126–133, 1981.

[34] H. H. Rosenbrock, "Structural properties of linear dynamic systems," *Int. J. Contr.*, vol. 20, no. 2, pp. 191–202, 1974.

[35] G. C. Verghese, B. Levy, and T. Kailath, "A generalized state space for singular systems," *IEEE Trans. Automat. Contr.*, vol. AC-26, no. 4, 1981.

[36] K. M. Vukobratović and V. Potkonjak, *Dynamics of Manipulator Robots: Theory and Application.* Berlin, FRG: Springer Verlag, 1982.

[37] H. West and H. Asada, "A method for the control of robot arms constrained by contact with the environment," in *Proc. American Control Conf.* (Boston, MA, 1985), pp. 383–386.

[38] D. E. Whitney, "Quasi-static assembly of compliantly supported rigid parts,"*ASME J. Dyn. Syst. Meas. Contr.*, vol. 104, pp. 65–77, 1982.

[39] ——, "Historical perspective and state of the art of robot force control," *Int. J. Robotics Res.*, vol. 6, no. 1, pp. 3–14, 1987.

[40] J. Wittenberg, *Dynamics of Systems of Rigid Bodies.* Stuttgart, FRG: B. G. Teubner, 1977.

[41] S. C. Wu, S. M. Yang, and E. J. Haug, "Dynamics of mechanical systems with coulomb friction, stiction, impact and constraint addition-deletion II," *J. Mech. Mach. Theory,* vol. 21, no. 5, pp. 407–416, 1986.

[42] ——, "Dynamics of mechanical systems with Coulomb friction, stiction, impact and constraint addition-deletion III," *J. Mech. Mach. Theory,* vol. 21, no. 5, pp. 417–425, 1986.

[43] E. Yip and R. Sincovec, "Solubility, controllability and observability of continuous descriptor systems," *IEEE Trans. Automat. Contr.*, vol. AC-26, pp. 702–707, 1981.

[44] H. Zhang and R. P. Paul, "Hybrid control of robot manipulators," in *Proc. IEEE Int. Conf. on Robotics and Automation*, pp. 602–607, 1985.

[45] Y.-F. Zheng and H. Hemami, "Mathematical modeling of a robot collision with its environment," *J. Robotic Syst.*, vol. 2, no. 3, pp. 289–307. 1985.

Part 8
Flexibility Effects on Performance and Control

STEPHEN YURKOVICH
The Ohio State University

CONTROL of mechanical manipulators to maintain accurate position in pick-and-place applications is an extremely important problem. The complexity of this problem increases dramatically when the robot possesses distributed flexibility. Efforts in the modeling and control of flexible-link manipulators have been motivated by the demand for lightweight, accurate, high-speed robots. While a good deal of the work to date focuses on space-based robotics, where relatively large endpoint deflections are likely, some researchers have treated the problem of vibration compensation for small deflections, which also has application in earth-based systems. The emphasis of Part 8 is on the effects of link flexibility on performance and control, especially with regard to large, lightweight manipulators. Space limitations do not permit citations of all the important works in these areas; rather, a representative collection is mentioned here, and three articles are reproduced that lead the interested reader through the literature on modeling, identification, and control of flexible-link manipulators.

Numerous works in the last few years both from a theoretical viewpoint and to a lesser extent from an experimental viewpoint have greatly increased our knowledge of flexible-link robots. References [1] and [2] serve as excellent summaries of existing works in flexible manipulator modeling and contain comprehensive modeling formulations. Generally speaking, the model describing the dynamics of a flexible-link manipulator consists of highly coupled nonlinear integro-partial differential equations. Several approximations to this model have appeared in the literature in efforts to reduce complexity. The assumed-modes method produces a set of nonlinear ordinary equations based on an orthonormal series expansion of the flexure variables. One basic assumption is that the multiple input, multiple output (MIMO) system can be approximated by a set of independent single input, single output (SISO) nonlinear systems, where the coupling between flexible modes of the links is neglected while the number of the retained flexible modes in the model is selected on a trial-and-error basis, often driven by the application.

Equally numerous as the schemes for modeling are the various approaches that have appeared in the literature for controller design. The greatest number of these works have dealt in simulation studies only, and some have developed quite elaborate and complex control schemes. A topical area that must be mentioned because of its appeal and promise for applicability is that of two-time scale and singular perturbation approaches, such as in [3], [4], and others. Also appealing is the use of computed-torque and inverse-dynamic techniques [5], [6], and input-shaping techniques [7] to reduce effects of flexibility. A promising approach successful in aerospace and related areas, which has yet to be exploited fully in the flexible-robotics field, is the use of special-purpose polymers for sensing and actuation [8].

System identification techniques in the modeling of flexible manipulators comprise different peculiarities because of the distributed nature of the system and its lightly damped behavior. Measurement noise, limited sampling capabilities, quantifying errors, and, most important, nonlinearities inherent in the system's dynamic behavior complicate the identification of the system's infinite spectrum. One may argue on the other hand that, for the control objective of vibration suppression in endpoint positioning after large angle movements, less complicated models (such as linear auto-regressive, moving average (ARMA) representations) may be appropriate. From a modeling and control viewpoint, on-line updating of such models is equivalent to linearizing the dynamic model about some configuration. The literature for system identification in flexible-link mechanisms is much less abundant than for control-related work; the interested reader may wish to pursue [9], [10], [11]. The problem of controller adaptation to account for changes in system characteristics caused by payload manipulation combines the aspects of system identification and control, and is indeed a particularly important problem; the topic is treated in [11]–[15] and others.

Several successful laboratory setups have demonstrated the effectiveness of algorithms for flexible manipulator control. While most experimental studies have focused on single-link manipulators, or on multilink manipulators with a single flexible link, such setups have served as valuable testbeds for modeling, system identification, and controller design. A case can be made, for example, that research on single-link mechanisms is important because of the envisioned likelihood that a space-based robotic manipulator could be constructed of a long, lightweight link (or boom) with serial connections of rigid mechanisms for articulation. This line of thinking could be applied to any multilink mechanism, space-based or otherwise, in which one of the links exhibits flexible dynamics. Relatively little activity has focused on the controller design for multilink flexible manipulators in experimentation, and at the time of this writing preliminary work has only recently begun to appear [16]–[22].

1. Selected Papers

The three papers selected in this part serve as an introduction to the topics of modeling, identification, and control in flexible-link mechanisms. The first paper, by R. C. Canon and E. Schmitz, and the third paper, by S. Yurkovich and A. P. Tzes, provide both theoretical and experimental studies, while the second paper, by W. J. Book focuses on modeling of flexible-link manipulators.

The paper by Canon and Schmitz, one of the most often cited works in the area, represents some of the earliest work reported from the Stanford group. An analytical model is derived for a single-link apparatus assuming a pinned-free configuration, and a Lagrangian formulation results in the simplified model. Important parameters relative to the unconstrained modes are then determined experimentally. For the experiments described in this paper, a 1-m-long, very flexible arm is confined to motion in the horizontal plane and is driven by a DC motor with position- and velocity-sensing at the hub. Off-structure sensing is accomplished via a camera detecting a light bulb at the arm endpoint; use of this noncolocated sensing and actuation setup leads to interesting problems in the controller design. A linear quadratic Gaussian (LQG) control law is developed, and the results of the experiments indicate that tracking response is limited by (mechanical) wave propagation delay inherent in the structure. Good results are obtained for endpoint position control that relies on an accurate model of the arm. Several results followed this basic study on the same apparatus (some cited above), including treatment of the varying-payload problem.

The paper by Book develops the dynamic equations for flexible manipulator arms consisting of rotary joints. The modeling approach described here is appealing in that the familiar 4×4 transition matrices are utilized to describe joint and deflection motion. Assuming small deflections, the resulting nonlinear equations account for angular rate and deflection interaction. The method relies on the choice of mode shapes and a linear representation of elasticity, and the Lagrangian formulation gives rise to computations reduceable to a recursive form. Implementation of the equations for computation, because of their resulting form, is relatively simple, and comparisons to rigid-link cases are offered in the paper in terms of computational efficiency. As noted above, many researchers have investigated the problem of modeling flexible-link mechanisms; what makes this paper appealing is its conceptually straightforward approach, offering the interested reader a good introduction to the issues associated with the modeling problem.

The paper by Yurkovich and Tzes serves to summarize work done at Ohio State toward system identification and control of a single-link mechanism, while offering an introduction to the topic of controller tuning for the payload variance problem. Approaches in both the time and frequency domains for identification are presented and results of implementation on a 1-m planar arm are discussed. Time-domain auto-tuning results employ an auto-tuning proportional integral derivative (PID) control law, whereas for frequency-domain identification and controller-tuning results, a new recursive algorithm is outlined. The experimental results emphasize the need for changing the controller to adapt to changes in payload conditions. Interesting features of this particular study include the use of relatively large payloads (up to 70% of the arm weight) added to the arm endpoint, and the use of endpoint acceleration in the identification and feedback control algorithms.

2. Future Directions

The extension of ideas for identification and control of single-link flexible mechanisms to multilink systems is an extremely difficult problem, primarily because nonlinearities are not manifested in a single-link system. Indeed, such an extension is currently an active area of research and to a great extent must rely upon results of all the single-link work, as well as on results established for rigid manipulators. For example, while the use of ARMA model representations will undoubtedly have limited appeal for multiple-link mechanisms, progress in self-tuning control for rigid manipulators along these lines gives direction to the work in flexible mechanisms. Nonlinear control schemes, such as those incorporating feedback-linearization or inverse-dynamics approaches, also have appeal for future work. That is, while exact linearization may not be possible for flexible dynamics, focusing on the rigid dynamics portion, perhaps in conjunction with two-time scale approaches where flexible dynamics are accounted for separately, could lead to useful results. In addition to investigation of enumerable schemes for robust control, ideas alluded to above such as distributed actuation and sensing as well as precompensation control with feedback schemes, will also lead the way for research in accounting for effects of flexibility on robot control.

References

[1] W. J. Book, "Recursive Lagrangian formulation of flexible manipulator arms," *Int. J. Robotics Res.*, vol. 3, pp. 87–101, Oct. 1984.

[2] T. J. Tarn, A. K. Bejczy, and X. Ding, "On the modeling of flexible robot arms," *Technical Report SSM-RL*-88-11, Washington University, St. Louis, MO, 1988.

[3] B. Siciliano, A. J. Calise, and V. P. R. Jonnalagadda, "Optimal output fast feedback in two-time scale control of flexible arms," *Proc. IEEE Conf. Dec. and Contr.*, pp. 1400–1403, 1986.

[4] F. Khorrami and U. Ozguner, "Perturbation methods in control of flexible-link manipulators," *Proc. IEEE Int. Conf. Robotics and Automation*, pp. 310–315, April 1988.

[5] E. Bayo, "Computed torque for the position control of open-chain flexible robots," *Proc. IEEE Int. Conf. Robotics and Automation*, pp. 316–321, April 1988.

[6] H. Asada and Z. D. Ma, "Inverse dynamics of flexible robots,'" *Proc. American Contr. Conf.*, pp. 2352–2359, June 1989.

[7] N. C. Singer, "Residual vibration reduction in computer controlled machines," Ph.D. thesis, Massachusetts Institute of Technology, January 1989.

[8] S. E. Burke and J. E. Hubbard, "Distributed actuator control design for flexible beams," *Automatica*, vol. 24, no. 5, pp. 619–627, September 1988.

[9] A. Tzes and S. Yurkovich, "A frequency-domain identification scheme for flexible structure control," *J. Dyn. Syst., Meas., and Cont.*, vol. 11, pp. 427–434, Sept. 1990.

[10] A. Tzes and S. Yurkovich, "Application and comparison of on-line identification methods for flexible manipulator control," *Int. J. Robotics Res.*, vol. 10, pp. 515–527, Oct. 1991.

[11] D. M. Rovner and R. H. Cannon, "Experiments toward on-line identification and control of a very flexible one-link manipulator," *Int. J. Robotics Res.*, vol. 6, no. 4, pp. 3–19, 1987.

[12] D. M. Rovner and G. F. Franklin, "Experiments in load-adaptive control of a very flexible one-link manipulator," *Automatica*, vol. 24, no. 4, pp. 541–548, July 1988.

[13] S. Yurkovich and F. E. Pacheco, "On controller turning for a flexible-link manipulator with varying payload," *J. Robotic Syst.*, vol. 6, no. 3, pp. 233–254, June 1989.

[14] S. Yurkovich, F. E. Pacheco, and A. P. Tzes, "On-line frequency domain information for control of a flexible-link robot with varying payload," *IEEE Trans. Automatic Contr.*, vol. AC-33, no. 12, pp. 1300–1303, Dec. 1989.

[15] V. Feliu, K. S. Rattan, and H. B. Brown, "Adaptive control of a single-link flexible manipulator," *IEEE Contr. Syst. Magazine*, vol. 10, no. 2, pp. 29–33, Feb. 1990.

[16] B. Gebler, "Feed-forward control strategy for an industrial robot with elastic links and joints," *Proc. IEEE Int. Conf. Robotics and Automation*, pp. 923–928, Raleigh, NC, April 1987.

[17] F. Pfeiffer, B. Gebler, and U. Kleeman, "On dynamics and control of elastic robots," *Proc. 2nd IFAC Symp. Robot Contr.*, pp. 41–45, October 1988.

[18] E. Schmitz, "Modeling and control of a planar manipulator with an elastic forearm," in *Proc. IEEE Int. Conf. Robotics and Automation*, pp. 894–899, May 1989.

[19] C. M. Oakley and R. C. Cannon, "End-point control of a two-link manipulator with a very flexible forearm: issues and experiments," *Proc. American Contr. Conf.*, pp. 1381–1388, June 1989.

[20] S. Yurkovich, A. P. Tzes, and K. L. Hillsley, "Controlling coupled flexible links rotating in the horizontal plane," *Proc. American Contr. Conf.*, pp. 362–367, June 1990.

[21] M. F. Ramey and E. Schmitz, "LQR Design for an experimental planar elastic arm with a large tip payload," *Proc. American Contr. Conf.*, pp. 1729–1732, June 1990.

[22] K. L. Hillsley and S. Yurkovich, "Vibration control of a two-link flexible robot arm," *Proc. IEEE Int. Conf. Robotics and Automation*, pp. 2121–2126, April 1991.

Robert H. Cannon, Jr.
Eric Schmitz
Stanford University
Department of Aeronautics and Astronautics
Stanford, California 94305

Initial Experiments on the End-Point Control of a Flexible One-Link Robot

Abstract

It has been known for some time (Gevarter 1970) that if a flexible structure is controlled by locating every sensor exactly at the actuator it will control, then stable operation is easy to achieve. Nearly all commercial robots are controlled in this way, for this reason. *So are most flexible spacecraft.*

Conversely, when one attempts to control a flexible structure by applying control torques at one end that are based on a sensor at the other *end, the problem of achieving stability is severe. Solving it is an essential step for better control in space: the space-shuttle arm is a cogent example. The next generation of industrial robots will also need such control capability, for they will need to be much lighter in weight (to achieve quick response with modest energy), and they will need to achieve greater precision by employing end-point sensing.*

A set of experiments has been constructed to demonstrate control strategies for a single-link, very flexible manipulator, where the position of one end is to be sensed and precisely positioned by torquing at the other end. The objective of this first set of experiments is to uncover and solve problems related to the control of very flexible manipulators where sensors are not colocated with the actuator. The experimental arrangement described here is also a test bed for new designs for flexible-structure controllers, designs that use insensitive, reduced-order control and adaptive control methods, for example. This paper describes the experimental arrangement, model identification, control design, and first experimental results.

Some interesting results are the following. First, good stability can be achieved for such noncolocated systems, and reponse can be achieved that is effectively three times faster than the first natural cantilever period of the system: but a good model of the system dynamics and rather sophisticated control algorithms are essential to doing so. Even then, the system will always be conditionally stable. In addition to the tip sensor, a colocated rate sensor and nearly colocated strain gauges have been found to be very useful for achieving good closed-loop performance, that is, high gain and high bandwidth. Second, there is an ultimate physical limit to achievable response time, namely, the time required for a wave to travel the length of the member. Well-designed controllers can approach this limit. Third, the use of end-point sensing makes less critical the elaborate dynamic conditioning of position-command signals—"model-following" differentiators, feed-forward, and the like—such as are typically needed in present-generation robots that use "dead reckoning" in lieu of end-point sensing. With end-point sensing, feedback alone (suitably conditioned) is sufficient to whip the tip to the commanded position and hold it there precisely. Even more important, a shift in, for example, workpiece with respect to robot base, no longer produces an error.

The research reported upon here was sponsored by the National Aeronautics and Space Administration through the Jet Propulsion Laboratory and the Langley Research Center. Robert H. Cannon, Jr. is Charles Lee Powell Professor and Chairman. Eric Schmitz is a Graduate Research Assistant and Research and Development engineer at Ford Aerospace and Communications Corporation.

The International Journal of Robotics Research,
Vol. 3, No. 3, Fall 1984,
0278-3649/84/030062-14 $05.00/00,

1. Introduction

The object of the research reported here was to develop and carry out initial experiments designed to address, under laboratory conditions, a specific unsolved control problem that seemed to us to be central to advances in the art of robotics: namely, control of a flexible member (one link of a robot system) where the position of the end-effector (called here the *end point*, or *tip*), is controlled by *measuring* that position and using the measurement as a basis for applying control torque to the *other* end of the flexible member (the robot's elbow joint, for example).

Most of today's robots do not use such control. Instead, in their control schemes they employ only sensors that are colocated with actuators. Thus, to position fingertips (or a tool), the desired fingertip loca-

Fig. 1. The experimental arm.

tion is converted by real-time kinematic computation into the equivalent angles that each of the robot's joints must assume. The joints are then driven simultaneously to said angles, each joint using a tight proportional, integral, derivative (PID) servo loop made up of a torquer and a colocated angle sensor (e.g., a digital encoder from which angular rate is derived); and the robot is presumed stiff enough that the end point will thus (by dead reckoning) be in the intended location.

The present-generation procedure for controlling robots has two severe limitations: (1) the inherent flexibility in the structure and drive trains of robots (as well as in the bases on which they are mounted) makes it impossible to achieve truly high precision; and (2) members and drive trains of the robot have to be made very stiff, and must therefore be heavy, in order to get some degree of precision; this typically limits the robot to slow speeds and/or requires high levels of drive power.

For space applications, the use of noncolocated servo control will be essential for any automated satellite-servicing module: once the module gets close to its satellite "target," the module manipulator will have to operate in such a control mode. Note that, in addition, this problem requires that the system be able to track a moving target.

The reason that current robots are controlled by the dead-reckoning approach described above, using only colocated sensors and actuators, is that colocation guarantees stable servo control (a principle noted by Gevarter in 1970). Conversely, the technology of achieving stable control using *noncolocated* sensors and actuators (e.g., tip sensor with elbow torquer) has not been developed: in fact, it is fundamentally a very difficult control problem. It is to contribute experimentally to this technology that the experiments described here have been developed.

In the sections that follow, we first describe the features of the first experimental arm that has been made, and then outline the general strategy for controlling it using its tip sensor and shoulder torquer. An accurate mathematical model is essential to successful noncolocated control, and the longest section of the paper discusses that subject. Finally, the details of designing effective control algorithms are described, and experimental results are presented.

2. The Experimental Flexible Arm

Figure 1 shows the first experimental arm that was used for the tests described here. It is a 1-m-long, very flexible structure that can bend freely in the horizontal plane but not in the vertical plane or in torsion. Torsional stiffness is obtained by connecting two thin aluminum beams with a series of flexible bridges, made of aluminum and beryllium copper, which act as pinned joints for horizontal bending (Fig. 2).

At one end, the arm is clamped on a rigid hub mounted directly on the vertical shaft of a direct-current (dc) motor; a torque applied by the motor rotates the arm in the horizontal plane. Because the motion occurs only in the horizontal plane, we do not have to take into account the effect of gravity. The dc motor (by Magtech) is driven by a current amplifier and is

Fig. 2. A flexible bridge. The bridges provide torsional stiffness while permitting flexibility.

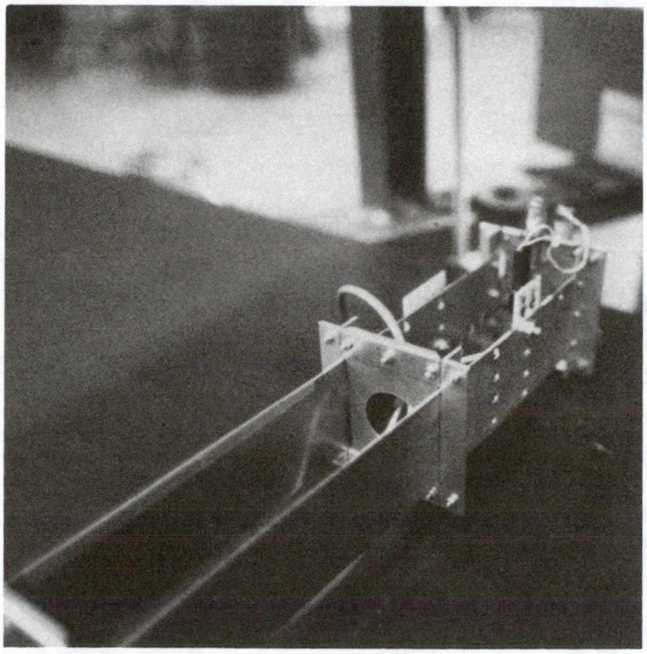

provided with a film wire potentiometer and an analog tachometer (also by Magtech).

The other end of the arm is equipped with a small light bulb for optical sensing of tip position. The tip sensor consists of a focusing lens 1 m above the tip's light bulb and, behind the lens, a linear fast silicon photodetector (by United Detector Technology [U.D.T.]) that provides two analog output voltages specifying the x and y positions of the light spot created (through the lens) by the light bulb. These two output voltages are amplified through a special-purpose amplifier (also made by U.D.T.). The sensor's field of view for our setup is $\pm 20°$ providing for measurement of motions of the arm ± 40 cm about the centerline. The position accuracy is better than ± 0.5 mm. The sensor's performance was greatly improved by the addition of a "hood" around the sensor fixture, which eliminated background "noise" caused by the ambient light (Maples 1982). Three pairs of strain gauges are glued on both sides of one of the two beams to serve as additional sensors for control purposes, if desired. The tip can also be fitted with a payload gripper.

3. Control Strategy

The tip-position controller design is intended to achieve the following:

- Fast tip response to a step command (as fast as the first flexible mode at 2.0 Hz, with as small an overshoot as possible)
- "Good" disturbance rejection for disturbance forces acting on the tip
- High loop gain for the transfer function from actuator disturbance to control torque in order to reduce the effects of friction, stiction, and cogging

If the arm were rigid, these three goals would be lumped into a single one. Alternatively, using end-point feedback enables us to introduce artificial stiffening at the strategic points of the beam, namely the tip end and the actuator end.

As usual in feedback control (Fig. 3), a commanded tip position (versus time) is compared with the sensor's measurement of actual tip position, and the difference (tip-position error) is used, together with other estimated system states, as the basis — via the chosen control algorithm — for applying torque at the hub through the motor. Estimation of the system's states, in turn, is based on dynamic calculation using signals from the tip-position sensor and the hub-rate sensor. The strain-gauge signals were not used in the first experiments reported here.

Both design of the control algorithm and real-time estimation of the system's states depend — crucially, in the case of noncolocated control — on having a suitable mathematical model of the dynamics of the arm. This model is the subject of Section 4. Design of the control algorithm is discussed in Section 5. In a sophisticated system, the parameters of the model may be adapted in real time, but the model's structure must be sufficiently valid to begin with.

4. Mathematical Modeling of the Experimental Arm

An accurate dynamic model for the flexible arm is required in order to design a tip-position controller. A simplified analytic model provides the general form of

Fig. 3. The feedback-control system.

Fig. 4. Geometry of the flexible arm.

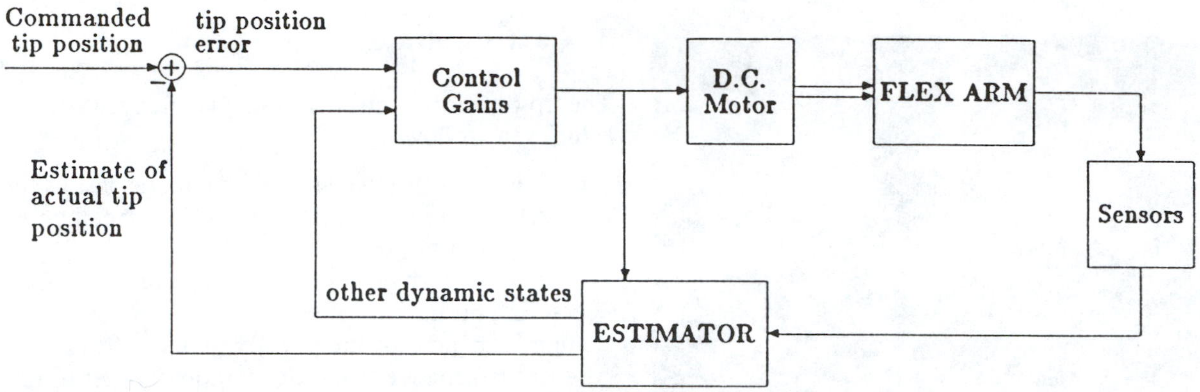

the equations of motion, whose coefficients are then experimentally identified by sine dwell open-loop tests.

4.1. Analytic Model

The arm is modeled as a continuous, pinned-free beam of length L whose moment of inertia about the root is I_B, with an additional lumped inertia I_H at the actuator end (hub). It is the total inertia of the arm. Displacement of any point P, along the beam at a distance x from the hub is given by the hub angle $\theta(t)$ and the deflection $w(x, t)$ measured from the line Ox, as shown in Fig. 4.

The following assumptions are made:

The deflection w is small ($<0.1L$), and any extension is neglected.

All terms involving $\dot{\theta}^2$ are negligible.

For the beam we use the Euler-Bernoulli model, for which rotary inertia and shear deformation effects are ignored.

If we define $y(x, t)$ as

$$y(x, t) = w(x, t) + x\theta(t), \tag{1}$$

then the kinetic energy T_k and potential energy V are

$$2T_k = I_H\dot{\theta}^2 + \int_0^L \left(\frac{\partial y}{\partial t}\right)^2 dm, \tag{2}$$

Fig. 4

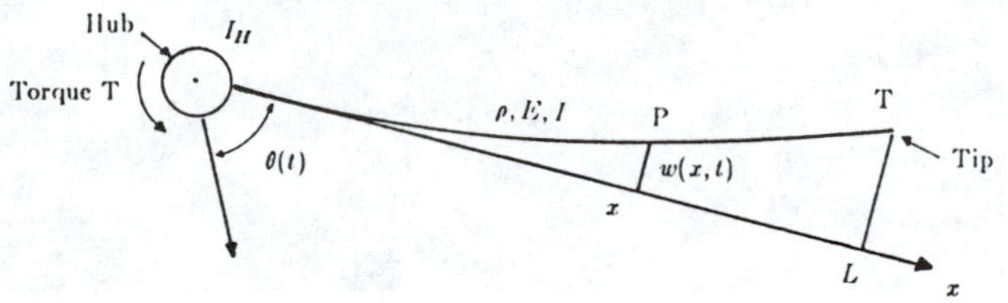

$$2V = \int_0^L EI\left(\frac{\partial^2 y}{\partial x^2}\right)^2 dx - T\theta. \tag{3}$$

From Hamilton's principle, we obtain a fourth-order partial differential equation with four boundary conditions:

$$EI\frac{\partial^4 y}{\partial x^4} + \rho\frac{\partial^2 y}{\partial t^2} = 0, \tag{4}$$

$$\begin{aligned} EI\frac{\partial^2 y}{\partial x^2}\Big|_{x=0} + T - I_H\ddot{\theta} &= 0 \\ w(0) &= 0 \\ EI\frac{\partial^2 y}{\partial x^2}\Big|_{x=L} &= 0 \\ EI\frac{\partial^3 y}{\partial x^3}\Big|_{x=L} &= 0. \end{aligned} \tag{5}$$

We then solve for the normal modes $\phi_i(x)$ and the pinned-free frequencies ω_i's such that $y(x, t) = \phi_i(x)e^{j\omega_i t}$ is a solution of (Eqs. 4 and 5). Given the bending stiffness of the beam EI, its rigid inertia I_B and the ratio I_H/I_B, we obtained numerical values for the ω_i's and the ϕ_i's. For our final design, however, we relied on the experimental data as discussed in Section 4.2.

Note that these modes (sometimes called *unconstrained* modes in the literature on flexible spacecraft) are here the most natural ones to use because they are readily identified in the experimental procedure. Their choice is not easily generalized for the flexible multi-link robot case, however.

The orthogonality relation used here is (Breakwell 1980)

$$\int_0^L \phi_i(x)\phi_j(x)\rho dx + I_H \frac{d\phi}{dx} i(0) \frac{d\phi}{dx} j(0) = (I_H + I_B)\delta_{ij}. \quad (6)$$

The general solution y is then written as

$$y(x, t) = \sum_{i=0}^{\infty} \phi_i(x)q_i(t) \quad (7)$$

where $i = 0$ is the rigid-body mode ($\phi_o(x) = x$). By use of (Eqs. 2, 3, 6, and 7), we get a simple expression for the Lagrangian L of the system:

$$2L = \sum_{i=0}^{\infty} (I_H + I_B)\dot{q}_i^2 - \sum_{i=0}^{\infty} (I_H + I_B)\omega_i^2 q_i^2 + T(q_0 + \sum_{i=0}^{\infty} \frac{d\phi_i}{dx}(0)q_i). \quad (8)$$

We could have obtained the system-dynamics equatins from (Eqs. 4 and 5) alone, but it is easier to use the Lagrange formulation at this point. After applying Lagrange's equations, we obtain an infinite set of ordinary differential equations that are *decoupled* from each other. Retaining the first $n + 1$ of them, we write the final model equations.

$$\frac{d\mathbf{x}}{dt} = \mathbf{F}\mathbf{x} + \mathbf{G}T$$
$$\mathbf{z} = \mathbf{H}\mathbf{x}(\textit{sensors}), \quad (9)$$

where $\mathbf{x}$, $\mathbf{F}$, $\mathbf{G}$, $\mathbf{H}$ are the following matrices:

$$\mathbf{x} = \begin{bmatrix} q_0 \\ \dot{q}_0 \\ q_1 \\ \dot{q}_1 \\ \vdots \\ q_n \\ \dot{q}_n \end{bmatrix}$$

$$\mathbf{F} = \begin{bmatrix} 0 & 1 & & & & & \\ 0 & 0 & & & & & \\ & & & & 0 & & \\ & & 0 & 1 & & & \\ & & -w_1^2 & -2\zeta_1 w_1 & & & \\ & & & & \ddots & & \\ & & 0 & & & 0 & 1 \\ & & & & & -w_n^2 & -2\zeta_n w_n \end{bmatrix}$$

$$\mathbf{G} = \frac{1}{I_T}\begin{bmatrix} 0 \\ 1 \\ 0 \\ \frac{d\phi_1}{dx}(0) \\ \vdots \\ 0 \\ \frac{d\phi_n}{dx}(0) \end{bmatrix}$$

$$\mathbf{H} = \begin{bmatrix} L & 0 & \phi_1(L) & 0 & \cdots & \phi_n(L) & 0 \\ 0 & 1 & 0 & \frac{d\phi_1}{dx}(0) & \cdots & 0 & \frac{d\phi_n}{dx}(0) \end{bmatrix}.$$

The two rows of $\mathbf{H}$ are the tip-sensor and the hub-rate-sensor measurement vectors. A linear model such as (Eq. 9) is often used in the literature on the control of large space structures. The linear model is useful for several reasons:

It provides the general structure of the equations for

an accurate experimental identification. Note that this form does not depend on the assumption of the Euler-Bernoulli model.

The analytic model was useful in some initial control designing and in understanding the nature of the open-loop transfer functions.

The same equation form would be obtained for a general flexible robot model linearized around a given robot configuration (without gravity effects and for small angular velocities).

4.2. Experimental Identification

We need now to identify the numerical values of the elements of the matrices **F**, **G**, **H** given in (Eq. 9). The transfer functions from torque to any sensor are, from (Eq. 9),

$$z = \mathbf{H}(sI - \mathbf{F})^{-1}\mathbf{G}T(s).$$

For the two sensors of interest here, we get

$$\frac{\dot{\theta}}{T}(s) = \frac{1}{I_T s} + \frac{1}{I_T}\sum_{i=1}^{\infty}\left[\frac{d\phi_i}{dx}(0)\right]^2 \frac{s}{s^2 + 2\zeta_i\omega_i s + \omega_i^2}, \quad (10)$$

$$\frac{y_T}{T}(s) = \frac{L}{I_T s^2} + \frac{1}{I_T}\sum_{i=1}^{\infty}\frac{\phi_i(L)\dfrac{d\phi}{dx}i(0)}{s^2 + 2\zeta_i\omega_i s + \omega_i^2}. \quad (11)$$

It has been shown that any colocated transfer function for a linear flexible structure is made up of alternating poles and zeros on the imaginary axis (Gevarter 1970; Martin 1978). Therefore, (Eq. 10) can also be written as

$$\frac{\dot{\theta}(s)}{T(s)} = \frac{1}{I_T s}\prod_{i=1}^{\infty}\frac{\left(\dfrac{s^2}{\Omega_i^2} + 2\zeta_i\dfrac{s}{\Omega_i} + 1\right)}{\left(\dfrac{s^2}{\omega_i^2} + 2\zeta_i\dfrac{s}{\omega_i} + 1\right)}. \quad (12)$$

The experimental procedure for obtaining the values of the constants in (Eqs. 10–12) is as follows. A sinusoidal voltage is input to the dc motor amplifier and its frequency is swept in the range of interest (0–10 Hz). The resonance frequencies' ω_i's (*pinned-free* modes) are easily identified for a lightly damped structure because, for these, the sensor outputs have maximum amplitude and show a 90° change in the phase lag with the input signal. Similarly the zeros' Ω_i's of the transfer function $\dot{\theta}/T(s)$ (*cantilevered* modes) are identified as the frequencies for which the tachometer (or the potentiometer) output amplitude goes through a minimum. The residues $[d\phi_i/dx(0)]^2$ in (Eq. 9) are negligible beyond the third flexible mode (pole/zero cancellation). Therefore, from ω_i and Ω_i for $i =$ 1, 2, 3, using (Eqs. 10 and 12), we can solve for the coefficients $d\phi_i/dx(0)$, $i = 1, 2, 3$.

Next, if the excitation frequency is close to the resonance mode ω_i, then from (Eqs. 10 and 11) we have

$$\dot{\theta}(s = jw) = \frac{1}{I_T}\frac{\left[\dfrac{d\phi_i}{dx}(0)\right]^2}{2\zeta_i\omega_i}T(j\omega)$$
$$y(s = j\omega) = \frac{1}{I_T}\frac{\dfrac{d\phi_i}{dx}(0)\phi_i(L)}{2\zeta_i j\omega_i^2}T(j\omega). \quad (13)$$

This is valid because (1) the structure is lightly damped and (2) the resonance frequencies are well separated for any beamlike structure. From (Eq. 13) we get the tip-sensor modal gains:

$$\phi_i(L) = \left[\frac{d\phi_i}{dx}(0)\right]\omega_i\left[\frac{y}{\dot{\theta}}\right]_{\text{measured}}. \quad (14)$$

The open-loop gains were also measured experimentally (the total moment of inertia I_T is equal to 0.44 kgm^2). Figure 5 shows the s-plane pole-zero pattern for the tachometer and tip-sensor open-loop transfer functions (from torque to sensor output) obtained from an eighth-order model. These transfer functions are valid only for small elastic deformations of the beam. Note that the tip-transfer function is *nonminimum phase:* it is this property that makes the control design difficult (owing to limited gain margin). The location of the first pair of nonminimum-phase zeros does not change more than 15%, whether we include one, two, or three flexible modes in our model.

Fig. 5. The s-plane display of measured transfer functions for the experimental arm. A. Transfer function from actuator to tachometer. B. Transfer function from actuator to tip-sensor output.

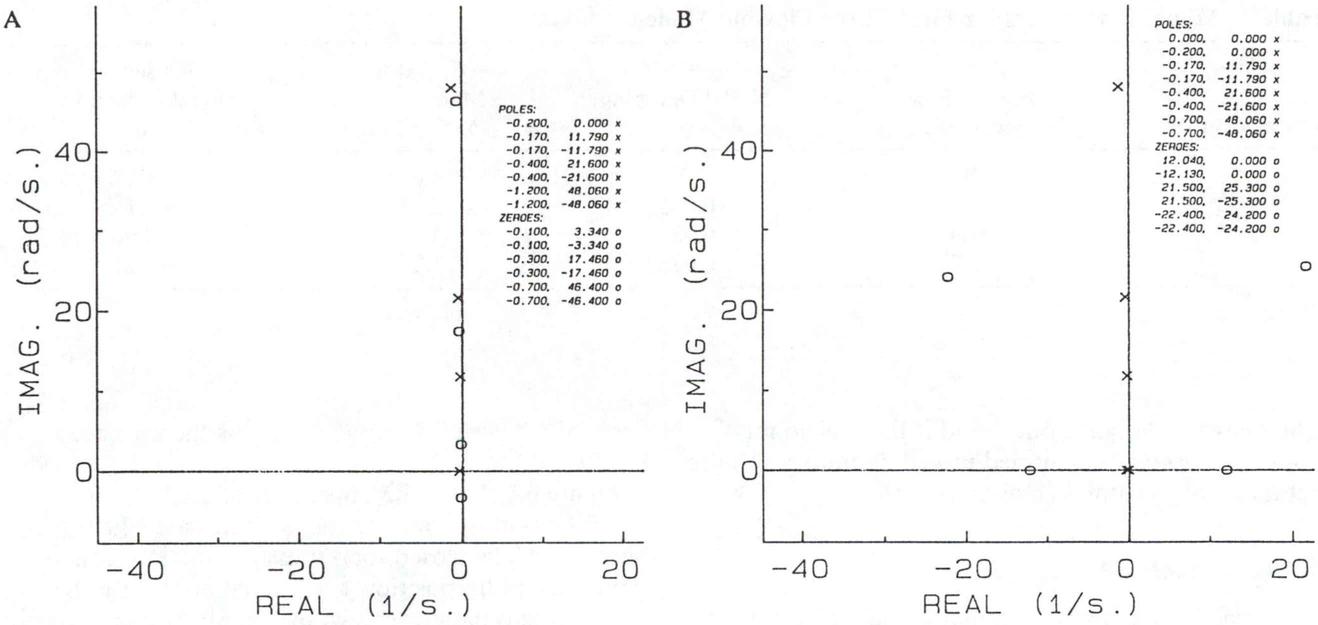

The model parameters for the first three flexible modes are given in Table 1.

5. Tip Controller

Given a perfect model of the system to be controlled, the design of an acceptable controller for the noncolocated, sensor-and-actuator case is straightforward (but not trivial!). Any of a number of design techniques can be used. We demonstrate one of them below.

In practice, of course, the model will be quite *imperfect*, for the plant is a complex, distributed-parameter system whose *parameters* are uncertain and may change with time. High "loop gain" and high robustness to parameter change are important qualities to achieve.

First we design a control algorithm (using the linear quadratic Gaussian [LQG] approach). Then we illustrate the loop-gain robustness problem.

5.1. Design

We chose to design the control algorithm using (primarily) optimal control theory based on the quadratic performance index approach. The main reasons for this choice (for a first design) are the following:

1. It handles easily the use of more than one sensor for feedback; it can also address the multi-input, multi-output control of a linearized dynamic model for a multilink flexible robot.
2. It permits trade-offs between end-point speed of response and damping and available actuator power.

We used the dynamic model identified here to design an eighth-order, LQG compensator based on the Stanford discretized version of the optimal steady-state linear quadratic Gaussian algorithm (DISC program) (Katz and Powell 1974). The sampling period T_s used for the design discussed here was 50 Hz. The DISC algorithm has two steps:

1. It computes a set of constant, full-state feedback gains that "optimally" shift the open-loop plant poles toward desired stable locations.
2. Assuming that the plant is perturbed by a white noise process input, it computes for a given set of measurements, z, contaminated by white noise, the filter gains that optimally reconstruct the dynamic states x.

Table 1. Model Parameters for First Three Flexible Modes

Mode Number	Pinned-Free Frequency F_i(Hz)	Modal Damping ζ_i	Actuator Modal Gain $d\phi_i/dx(0)$(rad)	Tip-sensor Modal Gain $\phi_i(L)$ (m)
0*	0	0	1.0	1.12
1	1.88	0.015	2.97	−1.1
2	3.44	0.02	3.0	0.90
3	7.70	0.02	1.25	−1.21

* Rigid body.

The two-step design is only valid if there is no mismatch between the identified model dynamics and the actual plant dynamics (*separation theorem*).

5.1.1. Regulator

The steady-state, discrete regulator gains C_k, in $T_k = -Cx_k$, are obtained by minimizing the discretized expression of the continuous performance index J_R. As our goal is to control the tip position y_T, we choose J_R to be

$$J_R = \int_0^{\infty} \left(y_T^2 + \tau^2 \dot{y}_T^2 + \frac{T^2}{T_{max}^2} \right) dt. \qquad (15)$$

One essential feature of the DISC program is that the designer inputs the continuous-weighting matrices for the performance index, which is then discretized internally. For sufficiently fast sampling rates, the discrete closed-loop roots will match the one obtained from a continuous s-plane design. From optimal control theory (Bryson 1979; Franklin and Powell 1980), we know that the closed-loop regulator roots for an s-plane design are the stable roots of a symmetric root locus drawn versus the gain $\rho = T_{max}^2$:

$$1 + T_{max}^2 (1 - \tau^2 s^2) \frac{y}{T}(+s) \frac{y}{T}(-s) = 0. \qquad (16)$$

As the gain ρ tends to infinity (i.e., decreasing penalty on actuator control), seven of the eight closed-loop roots are converging toward the six stable open-loop zeros of $G(s) = y/T(+s)y/T(-s)$ plus the zero at $s = -1.0/\tau$.

Figure 6A shows the upper part of the locus for $\tau = 0.1$ s. Let us consider the limiting case where ρ is very large. The closed-loop transfer function from commanded tip position y_c to actual position y_T has the pole-zero pattern shown in Fig. 6B. It is an all-pass filter plus an additional pole at infinity. This all-pass filter can be closely approximated to a third-order Padé approximation for a delay of 200 ms (with 5% error); this shows the limitation for the maximum speed of response to a step for our system. This limitation is inherent in any nonminimum-phase system. Physically, it represents the time required for a bending wave to travel from the motor end of the beam to the tip at the other end. For $T_{max}^2 = 10^3$, for example, we find that we can achieve the following well-damped poles:

$$-6.1 \pm j2.65, -7.7 \pm j11.4,$$
$$-6.4 \pm j23.8, -5.2 \pm j48.5.$$

Note that these are the s-plane roots obtained from the z-plane roots through the relation $z = e^{sT_s}$.

5.1.2. Estimator

The discrete estimator reconstructs the states x_k from the tip sensor y_T and the colocated rate sensor $\dot{\theta}$. The DISC algorithm computes the steady-state discrete filter gains K (current estimator) by trading off process and measurement noise in the least-squares sense (Katz & Powel 1974; Franklin and Powell 1980). The current estimator equations are

Fig. 6. A. Symmetric root locus for designing tip regulator of flexible arm (Eq. 16). B. Closed-loop transfer function from tip command to tip output for infinite gain ρ.

A

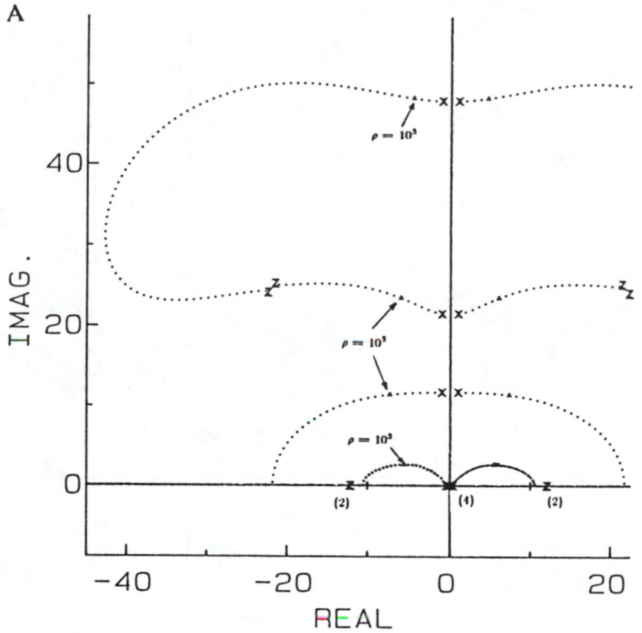

B

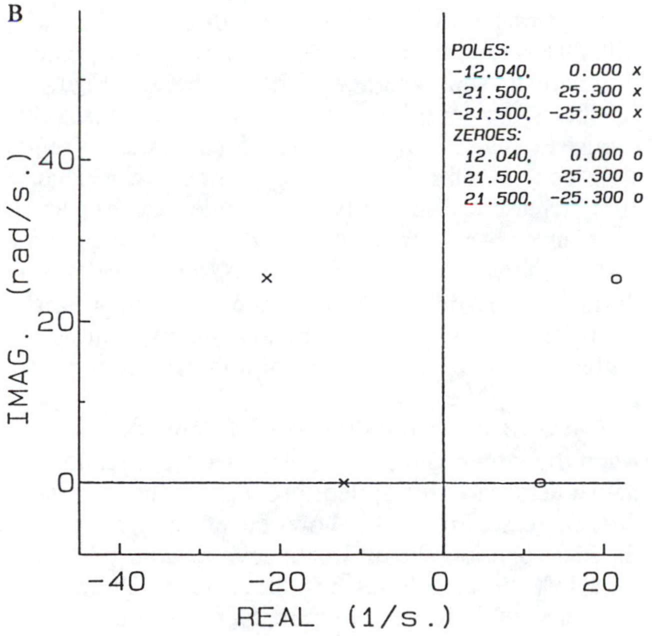

$$\hat{x}_k = \bar{x}_k + K(z_k - H\bar{x}_k),$$
$$\bar{x}_{k+1} = \Phi\hat{x}_k + \Gamma T_k, \tag{17}$$

where Φ and Γ are the discretized F and G matrices for a zero-order hold process, and H is as given in (Eq. 9).

The process noise is modeled here as an actuator disturbance. We use DISC as a pole-placement algorithm. The main difficulty is to choose adequate numerical values for the process noise covariance (Q_d) and the measurement noise covariances (R_{y_T}, R_θ) in order to achieve desired locations for the closed-loop estimator poles.

In a classic design, the feedback gain for each sensor would be chosen by closing one feedback loop after another. This process becomes fairly complicated when more than two sensors are used, however.

In estimator design, common practice is to choose the ratios R_{y_T}/Q_d and $R_{\dot\theta}/Q_d$ by trial and error until one comes up with well-damped and sufficiently fast poles in the bandwidth of interest (which includes the rigid body plus the first bending mode at 2 Hz). It is also helpful to draw a symmetric root locus versus the process-noise covariance, for a given ratio of the tip-sensor over the rate-sensor noise covariances. The left-half plane zeros of this root locus indicate the maximum estimator bandwidth, given the process-noise disturbance model.

With $Q_d/R_{y_T} = 200$ and $Q_d/R_{\dot\theta} = 0.10$, the equivalent s-plane closed-loop estimator poles are

$$-6.6, -5.9 \pm j6.6, -3.1 \pm j18.3,$$
$$-1.94 \pm j46.7, -63.4 \pm j48.6.$$

Note that the rate sensor used for the experimental recording discussed in Section 6 is obtained from the hub potentiometer by analog differentiation (Maples 1982). We added one more state in the filter to model the reconstructed rate: this explains why we have nine estimator roots.

5.1.3. Resulting Compensator

The control torque T is then obtained by feeding back the estimated state $\hat{x}(k)$: $T_k = -C\hat{x}_k$. The combination of the regulator and the estimator can be interpreted as an equivalent classic compensator (Bryson

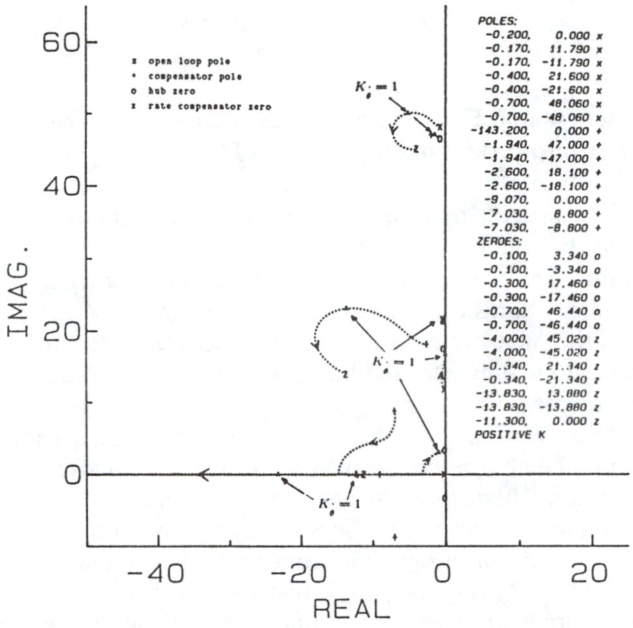

Fig. 7. Control design in the s-plane using successive loop closing: locus of roots versus hub-rate-loop gain.

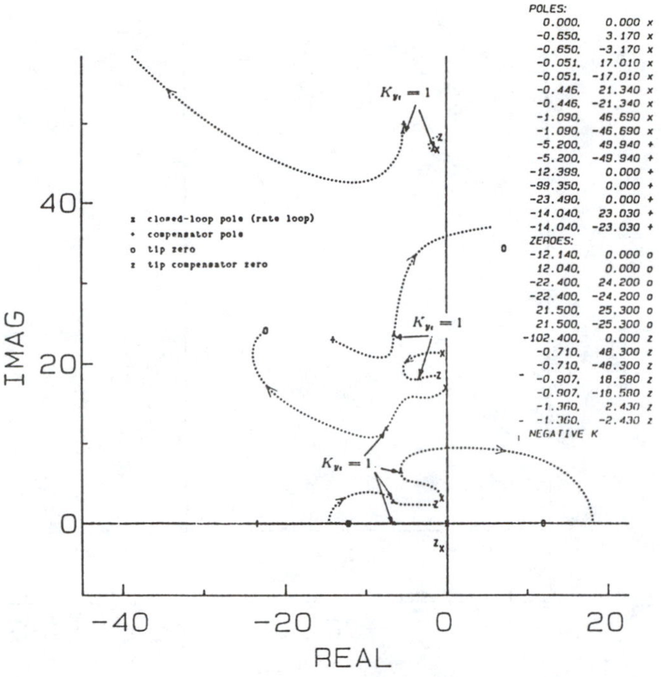

Fig. 8. Control design in the s-plane using successive loop closing: locus of roots versus tip-position gain.

1979). T is rewritten as

$$T(z) = -K_{y_T} G_{y_T}(z) - K_{\dot{\theta}} G_{\dot{\theta}}(z). \qquad (18)$$

In the remainder of the discussion we shall use, for ease of interpretation, a continuous s-plane design similar to the above z-plane discrete-domain one. In order to analyze the closed-loop system, we use the technique of successive loop closure. Figure 7 shows the rate loop versus the gain $K_{\dot{\theta}}$ varied from its nominal value. The basic feature of the rate compensator is that it provides a simple proportional gain, which mainly damps the first flexible mode. Figure 8 shows the tip-position loop root locus versus the gain K_{y_T} after the rate loop has been closed: the two zeros at $-1.3 \pm 2.4j$ provide lead compensation; they nearly cancel two of the rate loop closed-loop poles, and by this action delay the migration of the tip loop closed-loop poles toward the nonminimum-phase zeroes when the gain K_{y_T} is increased. Instability occurs when the tip position gain K_{y_T} is increased by a factor of 1.35.

5.1.4. System-Model Mismatch

When the computer's dynamic model does not agree with the actual system dynamics, the closed-loop roots of the overall tip-control system will move in the s-plane. It was found that the estimator root at 8.8 rad/s will always tend to move toward the right-half plane for relatively small change in the parameters. Figure 9A shows the damping ratio ζ for that root versus the change $(+\Delta\phi_1(L)/\phi_1(L))$, where $\phi_1(L)$ is the tip-sensor modal gain for the first flexible mode. More than 10% increase will seriously degrade the response, and the controller will be unstable for +35% change. Note that a positive change $+\Delta\phi_1(L)/\phi_1(L)$ means that the flexible beam's deformation for the first mode is larger than predicted; it also corresponds directly to an error in the location of the first pair of nonminimum-phase zeros.

Figure 9B shows the same pair of estimator roots when the rate-sensor gain is decreased from its nominal value. Again, the system becomes unstable with a 26% decrease. Figure 9C shows the effect of decreasing the first open-loop beam frequency. Instability occurs for a 45% decrease. These results, although obtained for a particular design, are quite typical of the control problem addressed here.

It is possible to increase the stability gain margins of the closed-loop system by using low-order compensators instead of the full-order LQG design: in fact, some of the LQG compensators' poles and zeros nearly

Fig. 9. Affect of model-parameters mismatch upon stability of the closed-loop system. A. Effect of the tip-sensor modal gain. B. Effect of the rate-sensor gain. C. Effect of the first-mode frequency.

A

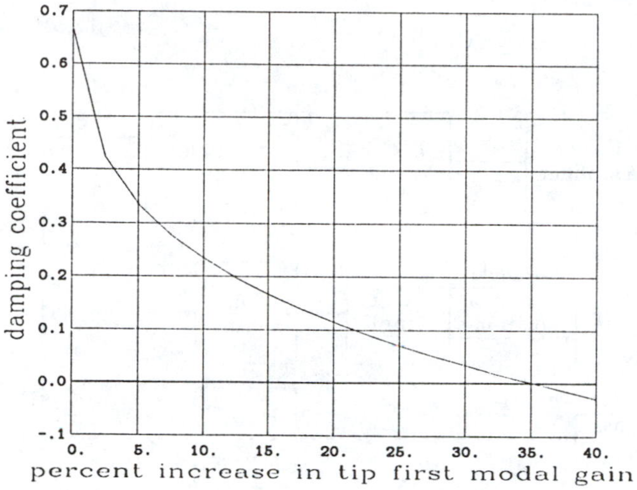

B

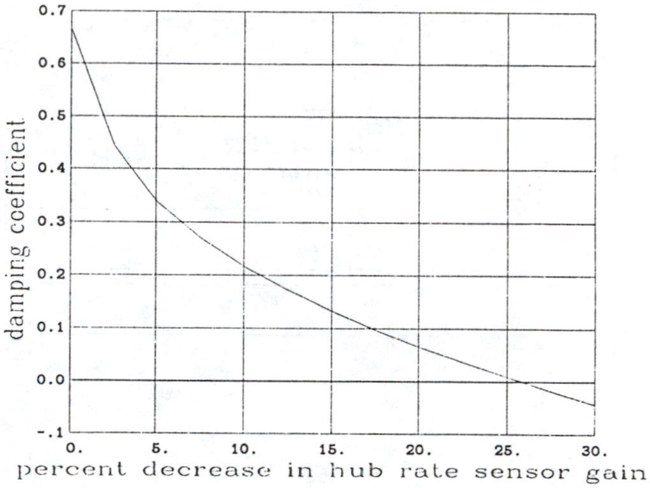

C

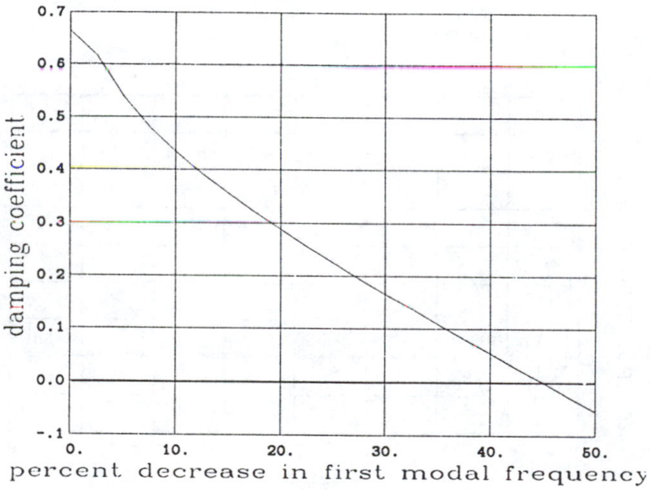

cancel each other! An efficient, optimal, low-order compensator design algorithm has been implemented at Stanford (Uy Loy Lee 1982), and we have used it for this problem.

5.2. Implementation

A MincII computer (a 16-bit LSI 1123 processor) was used to implement the compensator described in Section 5.1. It is equipped with analog-to-digital (eight inputs) and digital-to-analog converters (with 12-bit resolution). All the software is written in Fortran except the real-time routines. A real-time software package is used to interface with the programmable gains, ADC, and DAC, and to drive the programmable clock (Maples 1981). Several sampling rates, ranging from 10 to 75 Hz, have been tried. A block diagram (Fig. 10) shows the discrete overall tip-controller implementation.

6. Experimental Results

6.1. Fundamental Closed-Loop Behavior

A typical response to a step-command tip position of 10 cm for the controller discussed in Section 5 is shown in detail in Fig. 11A, which presents experimental time traces of sensor outputs and control effort. A great deal can be seen from Fig. 11 that is generic and fundamental to the basic problem of controlling the position of one end of a flexible beam by sensing that position and applying appropriate control torque to the *other* end.

Recall that the laboratory experiment described here and depicted in Fig. 11 is a scaled version of a real system, which for industrial applications might be 100 times as stiff; its dynamic time constants would thus be about 10 times as fast. To convert the time traces of Fig. 11A to a real industrial application, one would have to speed things up 10 times: Fig. 11A is a "slow-motion" picture of a real system.

Fig. 10. Overall controller implementation.

Fig. 11. A. Response of the tip-sensor-controlled flexible arm to a step command in tip position at t = 0 (wave propagation time is 0.13). B. Intermediate beam shapes during the step command.

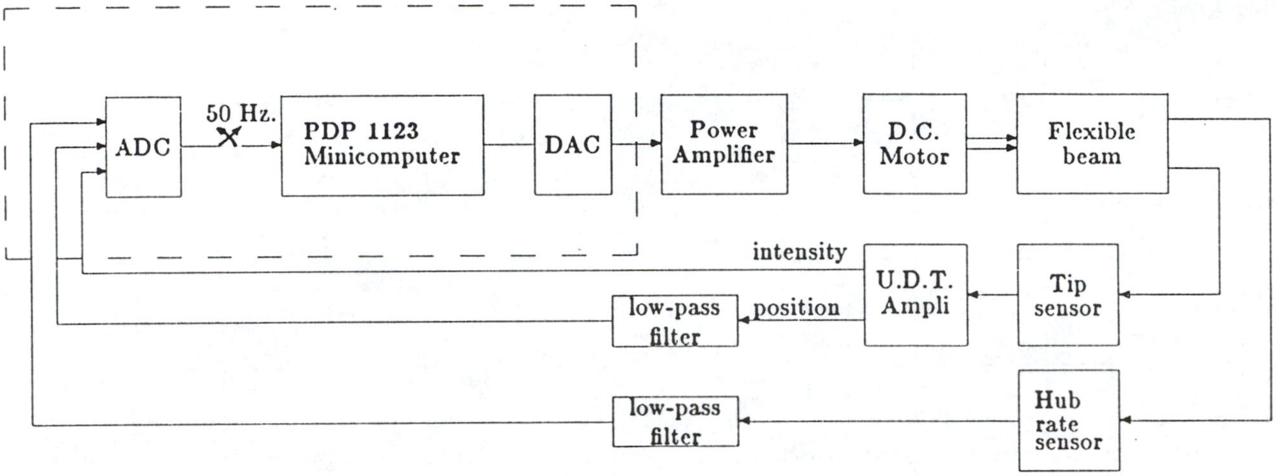

Fig. 11

A

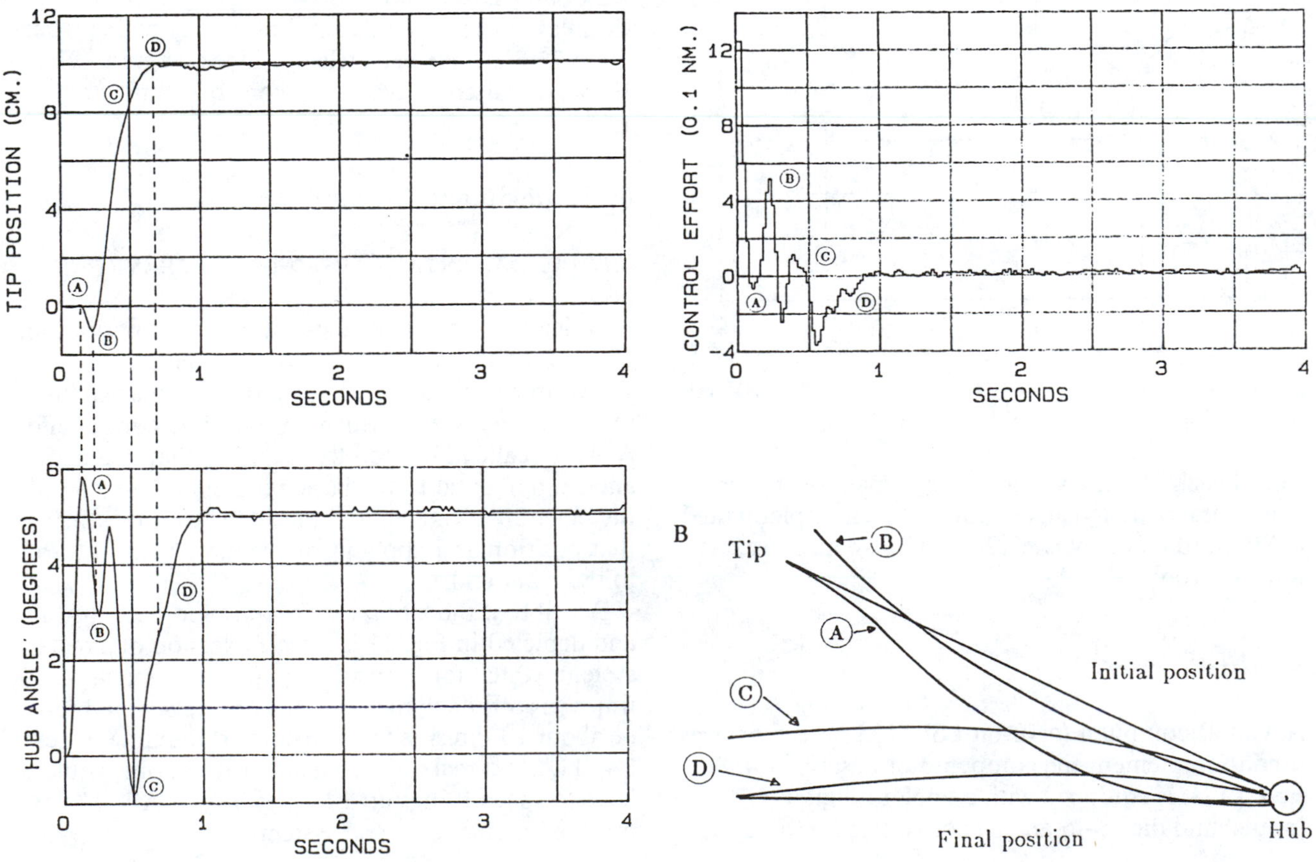

For the space-shuttle manipulator, (a remote manipulator system [RMS]), on the other hand, the times involved are quite comparable to those in Figure 11A; perhaps even slower. With its links all extended and its joints locked, the RMS has a natural cantilever frequency of about 3 s. (The open-loop, or human-operator control currently employed results in motion much slower than this.)

Figure 11A shows that, despite torque applied to the hub at time 0, and despite large rotation of the hub at time 0 plus (say) 40 s, the tip remains perfectly stationary until 130 ms has passed (point A in Fig. 11A). Then (point B) the tip moves initially in the wrong direction! Thereafter, the tip moves rapidly to the commanded position, reaching it less than 450 ms after reaching point B, and with nearly zero overshoot. It is possible to get faster rise time at the expense of overshoot by increasing the regulator gains. Thus, the tip of a beam whose natural cantilever period is 2 s has been effectively repositioned in 0.7 s by control of torque on the hub using a good control algorithm based on measured tip position.

Two interesting features, (1) 130-ms delay before the tip moves at all and (2) the tip's initial motion in the wrong direction, are quickly understood if we look also at the time history of hub angle θ (Fig. 11A), from which the sequence of configurations of the entire beam can be deduced (Fig. 11B). The time delay of 130 ms before the tip moves at all is clearly the inherent time for a bending-type wave to propagate the length of the beam! This will be a property of any physical beam, and represents an absolute, ultimate limit to the time that will be required to move the tip of such a beam (robot link) by torquing on the other end of it. The "wrong-way" start is seen in Fig. 11B to be a flip of the tip as the beam begins its whipping action to the new position. This appears mathematically in transfer function (Eq. 11), which is nonminimum phase.

Subsequent points in the time traces of Fig. 11A reveal the remainder of the control action. Motion from B to D is an effective whipping action. As the tip approaches its commanded position (point C), the controller, responding to eighth-order prediction in its algorithm, has already rotated the hub back nearly to $\theta = 0$ in order to apply a large breaking force to the tip's position. Thereafter, the hub reaches its final position more slowly in order to achieve a smooth "touch-down" of the tip, with no overshoot.

It is noted that good, quick, start-stop control was achieved with straightforward linear control (albeit with high-order prediction) and—notably—*without* any command shaping, model following, feed-forward, or the like: just old-fashioned feedback control based on a verified mathematical model of the system and some advanced versions of optimal control methods for gain selection. (In subsequent tests, it has been found that modest rounding of the corners of the square-wave command of Fig. 11A greatly reduces the amplitude of the θ vibrations at the end of the motion. Notch filtering of a square wave, with notches to take out the natural bending-mode-frequency content, accomplished the same thing in a more sophisticated way, but is not necessary.) With end-point sensing and good feedback-control algorithms, the elaborate command signal processing that must be used in today's operating robots can be largely eliminated, responses will be quicker, and control more precise.

6.2. Closed-Loop Performance Improvements

It is possible to increase significantly the tip-compensator static gain by using, in addition to the rate sensor, a strain gauge placed close to the hub. We added this sensor to deal with the actuator nonlinearities in our first experimental setup, namely, stiction and cogging torque. Such nonlinearities are typical in any real-life robot system and therefore need to be dealt with. The use of a nearly colocated strain gauge enables, in effect, the mechanization of a tight torque loop around the actuator.

7. Conclusions and Future Research

Initial techniques have been developed and demonstrated for stable and precise position control of one end of a very flexible beam by use of direct measurement of that end's position as a basis for torquing at the other end. It has been shown that the speed of response to commands is ultimately limited by the inherent wave-propagation delay for the beam. It has been demonstrated that satisfactory tip-control response can be achieved with a good dynamic model of the flexible arm.

In a future paper, we will discuss the following points:

1. Use of low-order compensators instead of the LQG controllers, in order to gain robustness in modeling uncertainties.
2. Dynamic switching between sensor sets. A typical application of end-point sensing would be in pick-and-place maneuvers. If tip sensors with limited fields of view are placed at different locations in the workspace, end-point information is not available while the robot is moving between these locations, and a different control system must be used. Then, the problem is to avoid undesirable transients when switching controls as the robot tip moves in and out of a sensor's field of view.
3. Adaptation. A first step is to design simple adaptive laws (gain scheduling) to compensate for sudden changes in the arm's payload mass.

Work is in progress in our laboratory to extend this research to multilink-flexible-robot control that will make full use of the concept of end-point sensor feedback.

References

Breakwell, J. A. 1980. Control of flexible spacecraft. Ph.D. thesis, Stanford University. Department of Aeronautics and Astronautics.

Bryson, A. E. Jr. 1979 (June). Some connections between modern and classical control concepts. *ASME J. Dyn. Syst. Measurements Contr.* 101:91–98.

Franklin, G., and Powell, J. D.: 1980. *Digital control of dynamic systems.* Reading, Mass.: Addison-Wesley.

Gevarter, W. B. 1970 (Apr.). Basic relations for control of flexible vehicles. *AIAA J.* 8(4):666–672.

Katz, P., and Powell, J. D. 1974. Selection of sampling rate for digital control of aircrafts. Ph.D. thesis, Stanford University Department of Aeronautics and Astronautics.

Maples, J. 1981. Real-time software manual for control applications (RT-11 operating system). Stanford University Aeronautics and Astronautics Robotics Laboratory.

Maples, J. 1982. Electronics design drawings for the flexible arm experiment facility. Stanford University Aeronautics and Astronautics Robotics Laboratory.

Martin, G. 1978. On the control of flexible mechanical systems. Ph.D. thesis, Stanford University Department of Aeronautics and Astronautics.

Paul, R. P. 1981. *Robot manipulators; Mathematics, programming and control.* Cambridge, Mass.: MIT Press.

Uy Loy Lee. 1982. A design algorithm for robust low order controllers. Ph.D. thesis, Stanford University Department of Aeronautics and Astronautics.

Wayne J. Book
School of Mechanical Engineering
Georgia Institute of Technology
Atlanta, Georgia 30332

Recursive Lagrangian Dynamics of Flexible Manipulator Arms

Abstract

Nonlinear equations of motion are developed for flexible manipulator arms consisting of rotary joints that connect pairs of flexible links. Kinematics of both the rotary-joint motion and the link deformation are described by 4 × 4 transformation matrices. The link deflection is assumed small so that the link transformation can be composed of summations of assumed link shapes. The resulting equations are presented as scalar and 4 × 4 matrix operations ready for programming. The efficiency of this formulation is compared to rigid-link cases reported in the literature.

1. Introduction

Improving the performance of most engineering systems requires the ability to model the system's behavior with improved accuracy. The evolution of the mechanical arm from teleoperator and crane to present day industrial and space robots and large space manipulators is no exception. Initial simple kinematic and dynamic models are no longer adequate to improve performance in the most critical applications. Both the mechanical system and control system require improved models for design simulation. Proposed new control algorithms require dynamic models for control calculation. Planning and programming activities as well as human-in-the-loop simulation also require accurate models of arms.

Accuracy is usually acquired at some cost. The application of mechanical arms to economically sensitive endeavors in industry and space is also an incentive to improve the efficiency of the formulation and simulation of dynamic models. Control algorithms and human-in-the-loop simulation require real-time calculation of dynamic behavior. Formulation of the dynamics in an easy-to-understand conceptual approach is also important if maximum utilization of the results is to be obtained.

The International Journal of Robotics Research,
Vol. 3, No. 3, Fall 1984,
0278-3649/84/030087-15 $05.00/00,

1.1. Sketch of Prior Work

Much work has been done in the formulation of the dynamic equations of motion for mechanical arms with rigid links. Work on the "inverse dynamic formulation" used in control is described elsewhere (Bejczy 1974; Stepanenko and Vukobratović 1976; Luh, Walker, and Paul 1980; Silver 1981), as is work on the dynamic formulation for simulation of rigid-link arms (Sturges 1973; Liegois et al. 1976; Derby 1981; Thomas and Tesow 1981; Walker and Orin 1982). The efficiency of these formulations and alternatives to their real-time calculation are discussed and referred to by Albus (1975) and Raibert and Horn (1978).

The limitation of these works is that rigid links are assumed. With this assumption, the techniques become at some point self-defeating, if their purpose is to improve performance. Maintaining rigidity of the links inhibits improved performance but is necessary if the rigid-link assumption is to be accurate.

Consideration of flexibility and control of the links in arm-type devices was reported in 1972 by Mirro. This early work considered both the modeling and control of a single-link device. Book (1973) considered the linear dynamics of spatial flexible arms represented as lumped mass and spring components via 4 × 4 transformation matrices. Later, this concept was refined (1979). Book and Whitney (Book 1974; Book, Maizza-Neto, and Whitney 1975) considered linear distributed dynamics of planar arms via transfer matrices and the limitations flexibility imposed on control system performance (Book 1976). Maizza-Neto (1974) and Book, Maizza-Neto, and Whitney (1975) utilized a planar, nonlinear model with modal representation of the flexibility and considered modal control as a technique for overcoming the limitations of the flexibility. Whitney, Book, and Lynch (1974) and Book, Maizza-Neto, and Whitney (1975) considered the design implications of flexibility. Distributed frequency-domain analysis of nonplanar arms using transfer-matrix techniques has been used by Book, Majette, and Ma (1979; 1981) in verifying the accuracy of truncated modal models of the nonlinear spatial dynamics of flexible manipulators (the remote manipulator of the space shuttle). The nonlinear modal model appearing here was first presented by the author in 1982. A more classic approach to manipulator dynamics, both rigid (Huston and Passerello 1980) and flexible (Kelly and Huston 1981), has been undertaken by others.

The work in flexible spacecraft has spawned a line of research pertaining to the interaction of articulated structures, which has great relevance to the manipulator modeling problem (Likins 1972; Nguyen and Hughes 1976). This activity produced the spatial, nonlinear, flexible manipulator model reported by Ho et al. (1974) and corresponding computer code for simulation. The simulation required great amounts of computer time and was unsuitable for even off-line simulation. Further work for the purposes of simulating the space shuttle remote manipulator was performed by Hughes. He described both linearized (1977) and more general models (1979). The Hughes model ignores the interaction between structural deformation and angular rate as might be appropriate for the space shuttle arm. This work, and associated work at SPAR Aerospace and the Charles Stark Draper Laboratory, probably represents the most intensive work on the modeling, simulation, and control of flexible arms. Unfortunately, little of this work has been reported in the available literature. Recent examination of experimental results from the operation of the shuttle arm in space has confirmed the validity of these models. More recently, Singh and Likins (1983) reported an efficient flexible-arm simulation program.

Other researchers have addressed the flexible-manipulator dynamics problem through the study of flexible mechanisms. Dubowsky and Gardner (1977) and Winfrey (1972) provide bibliographies on this work. Sunada and Dubowsky (1981) have developed modeling techniques applicable to both spatial closed-loop mechanisms and such open-loop chains as manipulator arms. Such techniques assume a known nominal motion over time about which the flexible-arm equations are linearized. They fall short of true simulation of the flexible, nonlinear equations, but are an interesting compromise for the sake of computational speed. Sunada and Dubowsky's technique is oriented toward finite-element analysis to obtain modal characteristics of the links, which are then combined using a time-varying compatibility matrix. It uses 4×4 matrices to represent the nominal kinematics and derivation of the compatibility matrix.

1.2. Perspective on This Work

The work described in this paper stresses an efficient, complete, and conceptually straightforward modeling approach utilizing the 4×4 transformation matrices that are familiar to workers in the field of robotics. It is unique in several respects. It uses 4×4 matrices to represent both the joint and deflection motion. The deflection transformation is represented in terms of a summation of modal shapes. The computations resulting from the Lagrangian formulation of the dynamics are reduced to recursive form similar to that which has proved so efficient in the rigid-link case. The equations are free from assumptions of a nominal motion and do not ignore the interaction of angular rates and deflections. They do assume small deflections of the links that can be described by a summation of the modal shapes and a linear model of elasticity. Only rotational joints are allowed. The results are quite tractable for automated computer solution for arbitrary rotary joints. Preliminary programs written to evaluate computational efficiency show that this method requires about 2.7 times as many computations as the most efficient rigid formulations with the same number of degrees of freedom. The rigid model could incorporate 21 degrees of freedom compared to 12 degrees of freedom (6 of which are joints) for this flexible model. Thus 15 degrees of freedom in the rigid model could be used to approximate the flexibility that the 6 flexible degrees of freedom of the model presented here approximate. The relative accuracy of the two approximations has not been determined. These issues are discussed in more detail in Section 6.

2. Flexible-Arm Kinematics

The previous works on rigid-arm dynamics make heavy use of the serial nature of manipulator arms, which results in multiplicative terms in the kinematics. The modal representation of flexible-structure dynamics, on the other hand, is a parallel or additive representation of system behavior. One of the purposes of this paper is to resolve this difference in a concise way. As with much of the previous work on rigid dynamics, the 4×4 matrices of Denavit and Hartenberg (1955) are used. Sunada and Dubowsky (1981) used this representation for their flexible-arm simulations but did not produce a complete nonlinear dynamic simulation. Other workers, such as Hughes (1979), relied on the more general formulation provided by a vector-dyadic representation. While Silver (1981), Hollerbach (1980), and others have pointed out the relative inefficiency of the 4×4 formulation, the conceptual framework is most advantageous for tackling the complexity of the flexible dynamics.

Define the position of a point in Cartesian coordinates by an augmented vector:

$$[1 x\text{-component } y\text{-component } z\text{-component}]^T.$$

Define the coordinate system $[x\ y\ z]_i$ on link i with origin O_i at the proximal end (nearest the base, oriented so that the x axis is coincident with the neutral axis of the beam in its undeformed condition. The orientation of the remaining axes will be done so as to allow efficient description of the joint motion. A point on the neutral axis at $x = \eta$ when the beam is undeformed is located at ${}^i h_i(\eta)$ under a general condition of deformation, in terms of system i.

By a homogeneous transformation of coordinates, the position of a point can be described in any other coordinate system j if the transformation ${}^j\mathbf{W}_i$ is known (see Fig. 1). The form of this matrix is

$${}^j\mathbf{W}_i = \left[\begin{array}{c|c} 1 & \mathbf{0}^T \\ x_j \text{ component of } O_i & \\ y_j \text{ component of } O_i & {}^j\mathbf{R}_i \\ z_j \text{ component of } O_i & \end{array}\right], \tag{1}$$

where ${}^j\mathbf{R}_i$ = a 3×3 matrix of direction cosines, and $\mathbf{0}$ = a 1×3 vector of zeros.

Fig. 1. Basic nomenclature for transformation of coordinates.

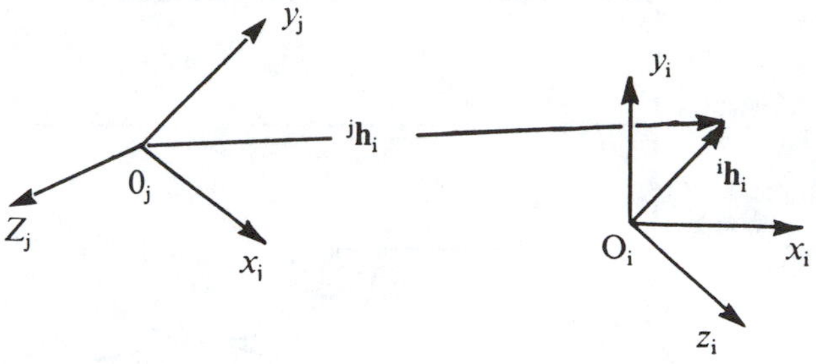

$${}^j\mathbf{h}_i = {}^j\mathbf{W}_i {}^i\mathbf{h}_i$$

Thus, in terms of the fixed inertial coordinates of the base, the position of a point on link i is given as

$$\mathbf{h}_i = {}^0\mathbf{W}_i {}^i\mathbf{h}_i = \mathbf{W}_i {}^i\mathbf{h}_i, \tag{2}$$

where the special case of ${}^0\mathbf{W}_i = \mathbf{W}_i$. It is useful to separate the transformations due to the joint from the transformation due to the flexible link, as shown in Fig. 2:

$$\mathbf{W}_j = \mathbf{W}_{j-1}\mathbf{E}_{j-1}\mathbf{A}_j = \hat{\mathbf{W}}_{j-1}\mathbf{A}_j, \tag{3}$$

where

$\mathbf{A}_j$ = the joint transformation matrix for joint j;
$\mathbf{E}_{j-1}$ = the link transformation matrix for link $j - 1$ between joints $j - 1$ and j; and
$\hat{\mathbf{W}}_{j-1}$ = the cumulative transformation from base coordinates to $\hat{O}_{j-1}$ at the distal end of link j.

$\hat{O}_{j-1}$ is fixed to the link $j - 1$, and with no deflection $[\hat{x}\ \hat{y}\ \hat{z}]_{j-1}$ is parallel to $[x\ y\ z]_{j-1}$, with x_{j-1} coincident with $\hat{x}_{j-1}$.

To incorporate the deflection of the link, the approach of modal analysis is used. This approach is valid for small deflection of the link.

$${}^i\mathbf{h}_i(\eta) = \begin{bmatrix} 1 \\ \eta \\ 0 \\ 0 \end{bmatrix} + \sum_{j=1}^{m_i} \delta_{ij} \begin{bmatrix} 0 \\ x_{ij}(\eta) \\ y_{ij}(\eta) \\ z_{ij}(\eta) \end{bmatrix}, \tag{4}$$

where

x_{ij}, y_{ij}, z_{ij} = the x_i, y_i, and z_i displacement components of mode j of link i's deflection, respectively;

Fig. 2. Nomenclature for joint transformations ($\mathbf{A}_j$) and link transformations ($\mathbf{E}_j$).

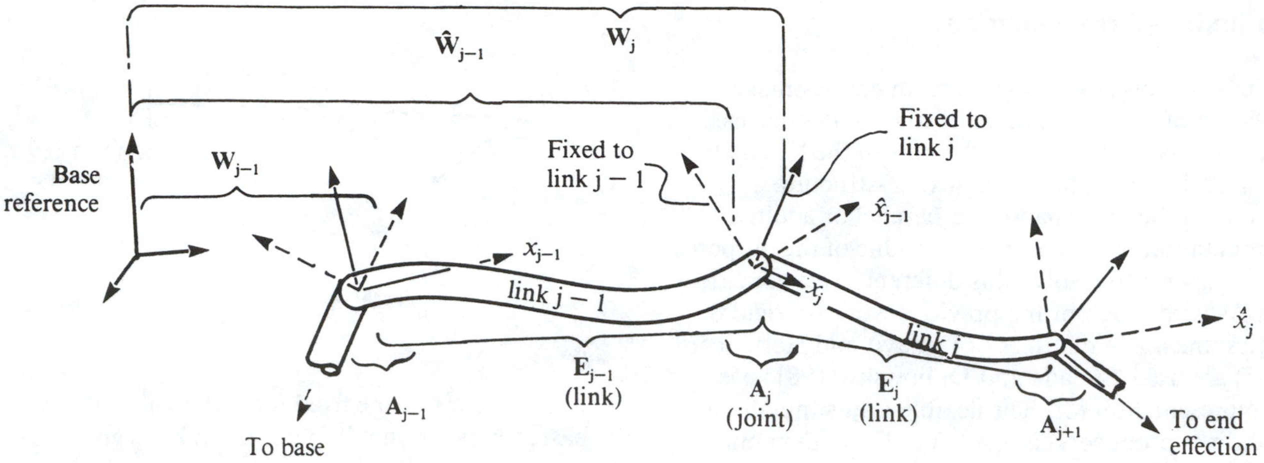

$$W_j = W_{j-1}E_{j-1}A_j = \hat{W}_{j-1}A_j$$

δ_{ij} = the time-varying amplitude of mode j of link i; and

m_i = the number of modes used to describe the deflection of link i.

The link-transformation matrix must also incorporate the deflection of the link. Here the rotations as well as the translations of the deflection must be represented. If one consistently requires small rotations the direction cosine matrix simplifies as noted elsewhere (Book, 1979), and furthermore the small angles can be assumed to add vectorally. This is basic to the approach used here. The link-transformation matrix can then be written as

$$\mathbf{E}_i = [\mathbf{H}_i + \sum_{j=1}^{m_i} \delta_{ij}\mathbf{M}_{ij}], \tag{5}$$

where

$$\mathbf{H}_i = \begin{bmatrix} 1 & 0 & 0 & 0 \\ l_i & 1 & 0 & 0 \\ 0 & 0 & 1 & 0 \\ 0 & 0 & 0 & 1 \end{bmatrix}, \tag{6}$$

and

$$\mathbf{M}_{ij} = \begin{bmatrix} 0 & 0 & 0 & 0 \\ x_{ij} & 0 & -\theta_{zij} & \theta_{yij} \\ y_{ij} & \theta_{zij} & 0 & -\theta_{xij} \\ z_{ij} & -\theta_{yij} & \theta_{xij} & 0 \end{bmatrix}, \tag{7}$$

and where

All variables in brackets are evaluated at l_i;

$\theta_{xij}, \theta_{yij}, \theta_{zij}$ = the x_i, y_i, and z_i rotation components of link i, respectively; and

l_i = the length of link i.

To find the velocity of a point on link i, take the time derivative of the position:

$$\frac{d\mathbf{h}_i}{dt} = \dot{\mathbf{h}}_i = \dot{\mathbf{W}}_i{}^i\mathbf{h}_i + \mathbf{W}_i{}^i\dot{\mathbf{h}}_j. \tag{8}$$

Due to the serial nature of the kinematic chain, it is computationally efficient to relate the position of a point and its derivatives to preceding members in the chain. By differentiating 2, one obtains

$$\dot{\mathbf{W}}_j = \dot{\hat{\mathbf{W}}}_{j-1}\mathbf{A}_j + \hat{\mathbf{W}}_{j-1}\dot{\mathbf{A}}_j, \tag{9}$$

and

$$\ddot{W}_j = \ddot{\hat{W}}_{j-1} A_j + 2\dot{\hat{W}}_{j-1} \dot{A}_j + \hat{W}_{j-1} \ddot{A}_j, \tag{10}$$

where

$$\dot{A}_j = U_j \dot{q}_j; \tag{11}$$

$$\ddot{A}_j = U_{2j} \dot{q}_j^2 + U_j \ddot{q}_j; \tag{12}$$

$U_j = \partial A_j / \partial q_j$;
$U_{2j} = \partial^2 A_j / \partial q_j^2$; and
q_j = the joint variable of joint j.

Thus, $\dot{W}_j$ and $\ddot{W}_j$ can be computed recursively from $\hat{W}_{j-1}$, its derivatives, and the partials with respect to the variables of link j − 1 and joint j. No mixed partials are explicitly present. This computational approach is similar to that proposed by Hollerbach (1980) for rigid-link arms. Here, one additionally needs $\hat{W}_{j-1}$ and its derivatives. These can be computed recursively from W_{j-1} and its derivatives:

$$\hat{W}_j = W_j E_j, \tag{13}$$

$$\dot{\hat{W}}_j = \dot{W}_j E_j + W_j \dot{E}_j, \tag{14}$$

$$\ddot{\hat{W}}_j = \ddot{W}_j E_j + 2\,\dot{W}_j \dot{E}_j + W_j \ddot{E}_j, \tag{15}$$

$$\dot{F}_j = \sum_{k=1}^{m_j} \dot{\delta}_{jk} M_{jk}, \tag{16}$$

and

$$\ddot{E}_j = \sum_{k=1}^{m_j} \ddot{\delta}_{jk} M_{jk}. \tag{17}$$

One can see from (Eqs. 16 and 17) that the deflection transformations enter even more simply into the kinematics on a per-variable basis than do the joint variables. This is due to the small deflection assumption and the form chosen for the transformation. The recursive nature of the velocity and acceleration is preserved from the rigid case. For the simulation equations, the terms involving second derivatives of the joint and deflection variables will be separated from the above expressions and included in the inertia matrix to make up the coefficient matrix of the derivatives of the state variables. The "inverse dynamics" solution that proceeds directly from the Langrangian formulation has little obvious utility.

3. The System's Kinetic Energy

In this section, the expression for the system's kinetic energy is developed for use in Lagrange's equations. First, the kinetic energy for a differential element is written. Then, integration of this differential kinetic energy over the link gives the link's total contribution. This produces terms that are the equivalent of the moment-of-inertia matrices of rigid-link arms. Summation over all the links provides the total kinetic energy.

The kinetic energy of a point on the i-th link is

$$dk_i = \frac{1}{2}\, dm \;\mathrm{Tr}\,\{\dot{h}_i \dot{h}_i^T\}, \tag{18}$$

where dm is the differential mass of the point, and Tr{.} is the trace operator.
Expanding (Eq. 18) and using the fact that $\mathrm{Tr}\{A\,B^T\} = \mathrm{Tr}\{B\,A^T\}\rangle$, the expression for dk_i becomes

$$dk_i = \frac{1}{2}\, dm\; \mathrm{Tr}\{\dot{W}_i {}^i h_i {}^i h_i^T \dot{W}_i^T + 2\,\dot{W}_i {}^i h_i {}^i \dot{h}_i^T W_i^T + W_i {}^i \dot{h}_i {}^i \dot{h}_i^T W_i^T\}, \tag{19}$$

where

$${}^i \dot{h}_i = \sum_{j=1}^{m_j} \dot{\delta}_{ij} [0 \; x_{ij} \; y_{ij} \; y_{ij} \; z_{ij}]^T. \tag{20}$$

By integrating over the link, one can obtain the link's total kinetic energy. In this paper, it is assumed that the links are slender beams, because this assumption makes the central development clearer. Other mass distributions could be used with a slight departure here in the development. For slender beams, $dm = \mu\, d\eta$, and one can integrate over η from 0 to l_i. Only the terms in ${}^i h_i$ and its derivatives are functions of η for this link. Thus, the integration can be performed without knowledge of W_i and its derivative. Summing over all n links, one finds the system's kinetic energy to be

$$K = \sum_{i=1}^{n} \int_0^{l_i} dk_i, \tag{21}$$

$$K = \sum_{i=1}^{n} \mathrm{Tr}\{\dot{\mathbf{W}}_i \mathbf{B}_{3i} \dot{\mathbf{W}}_i^T + 2\dot{\mathbf{W}}_i \mathbf{B}_{2i} \mathbf{W}_i^T + \mathbf{W}_i \mathbf{B}_{1i} \mathbf{W}_i^T\}, \tag{22}$$

where

$$\mathbf{B}_{1i} = \frac{1}{2} \int_0^{l_i} \mu^i \dot{\mathbf{h}}_i {}^j\dot{\mathbf{h}}_i^T \, d\eta. \tag{23}$$

By interchanging the integration in (Eq. 23) and the summations involved in the definition of ${}^j\mathbf{h}_i$ in (Eq. 20), one obtains

$$\mathbf{B}_{1i} = \sum_{j=1}^{m_i} \sum_{k=1}^{m_i} \dot{\delta}_{ij} \dot{\delta}_{ik} \mathbf{C}_{ikj}, \tag{24}$$

where

$$\mathbf{C}_{ikj} = \frac{1}{2} \int_0^{l_i} \mu [0 \; x_{ik} \; y_{ik} \; z_{ik}]^T [0 \; x_{ij} \; y_{ij} \; z_{ij}] d\eta \tag{25}$$

It should be noted that $\mathbf{C}_{ikj}$ has units of an inertia matrix and serves a similar function. While shown here as a 4×4 matrix, it is nonzero only in the 3×3 lower-right-hand corner. It can also be shown that $\mathbf{C}_{ikj} = \mathbf{C}_{ijk}^T$. By choosing the assumed mode shapes in an appropriate manner, it is possible to reduce the number of nonzero terms in (Eq. 24). This matter is discussed in light of computational speed in the conclusions.

The other terms in (Eq. 22) can similarly be found:

$$\mathbf{B}_{2i} = \frac{1}{2} \int_0^{l_i} \mu^i \mathbf{h}_i {}^i\dot{\mathbf{h}}_i^T d\eta, \tag{26}$$

and

$$\mathbf{B}_{2i} = \sum_{j=1}^{m_i} \dot{\delta}_{ij} \mathbf{C}_{ij} + \sum_{k=1}^{m_i} \sum_{j=1}^{m_i} \delta_{ik} \dot{\delta}_{ij} \mathbf{C}_{ikj}, \tag{27}$$

where

$$\mathbf{C}_{ij} = \frac{1}{2} \int_0^{l_i} \mu [1 \; \eta \; 0 \; 0]^T [0 \; x_{ij} \; y_{ij} \; z_{ij}] d\eta. \tag{28}$$

Finally, by a similar approach,

$$\mathbf{B}_{3i} = \frac{1}{2} \int_0^{l_i} \mu^i \mathbf{h}_i {}^i\mathbf{h}_i^T d\eta,$$

$$\mathbf{B}_{3i} = \mathbf{C}_i + \sum_{j=1}^{m_i} \delta_{ij} [\mathbf{C}_{ik} + \mathbf{C}_{ik}^T] + \sum_{k=1}^{m_i} \sum_{j=1}^{m_i} \delta_{ik} \delta_{ij} \mathbf{C}_{ikl}, \tag{29}$$

where

$$\mathbf{C}_i = \frac{1}{2} \int_0^{l_i} \mu [1 \; \eta \; 0 \; 0]^T [1 \; \eta \; 0 \; 0] d\eta. \tag{30}$$

This final term contains the rigid-body-inertia terms.

It should be noted that the terms defined in (Eqs. 26–30) are easily simplified if one link in the system is to be considered rigid, in which $m_i = 0$. Should a link consist of a flexible member with rigid appendages, the above derivation is readily extended to modify the matrices $\mathbf{C}_{ikj}$, $\mathbf{C}_{ik}$, and $\mathbf{C}_i$ with no further modifications to the succeeding development. In fact, these matrices could be obtained by finite-element analysis should the link shape be irregular, as is often the case. Furthermore, the expression for $\mathbf{B}_{3i}$ contains a term of order δ^2, which is by definition small and a candidate for later elimination. Finally, much of the complexity of the integration of the modal shape products can be done off-line one time for a given link structure.

3.1. Derivatives of Kinetic Energy

For construction of Lagrange's equations, one needs

$$\partial K/\partial q_j, \; \partial K/\partial \delta_{jf}, \; \frac{d}{dt}(\partial K/\partial \dot{q}_j), \text{ and } \frac{d}{dt}(\partial K/\partial \dot{\delta}_{jf}).$$

First consider $\partial K/\partial \dot{q}_j$. This will involve the partials of all the terms in (Eq. 22), some of which are zero. In fact, only $\mathbf{W}_i$ for $j \leq i \leq n$ provides nonzero partials with respect to $\dot{q}_j$. The time derivative of the partial is then taken. In this respect, the following equivalences should be noted:

$$\partial \dot{\mathbf{W}}_i / \partial \dot{q}_j = \partial \mathbf{W}_i / \partial q_j, \tag{31}$$

$$\frac{d}{dt}(\partial \dot{W}_i/\partial \dot{q}_j) = \partial \dot{W}_i/\partial q_j, \quad (32)$$

$$\partial \dot{W}_i/\partial \dot{\delta}_{jf} = \partial W_i/\partial \delta_{jf}, \quad (33)$$

$$\frac{d}{dt}(\partial \dot{W}_i/\partial \dot{\delta}_{jf}) = \partial \dot{W}_i/\partial \delta_{jf}. \quad (34)$$

Also helpful in simplifying the result is that $Tr\{A\} = Tr\{A^T\}$ for any square matrix A, and that B_{3i} is symmetric. Considerable cancellation and combination results when the terms in Lagrange's equation involving kinetic energy are combined. The result of this combination is

$$\frac{d}{dt}(\partial K/\partial \dot{q}_j) - \partial K/\partial q_j = 2\sum_{i=j}^{n} Tr\left\{\frac{\partial W_i}{\partial q_j}\left[\left[C_i + \sum_{k=1}^{m_i}\delta_{ik}\left(C_{ik} + C_{ik}^T + \sum_{l=1}^{m_i}\delta_{il}C_{ilk}\right)\right]\ddot{W}_i^T + \left[\sum_{k=1}^{m_i}\ddot{\delta}_{ik}\left(C_{ik} + \sum_{l=1}^{m_i}\delta_{il}C_{ilk}\right)\right]W_i^T + \left[2\sum_{k=1}^{m_i}\dot{\delta}_{ik}\left(C_{ik} + \sum_{l=1}^{m_i}\delta_{il}C_{ilk}\right)\right]\dot{W}_i^T\right]\right\}. \quad (35)$$

Terms of the form $\delta_{ik}\,\delta_{il}$, which are noted above, are second-order. These can be ignored consistent with the assumption that the deflections are small. Because of recurrence of certain terms above, it is convenient to define the following:

$$D_{ik} = C_{ik} + \sum_{l=1}^{m_i}\delta_{il}C_{ilk}, \quad (36)$$

$$G_i = C_i + \sum_{k=1}^{m_i}\delta_{ik}(C_{ik} + C_{ik}^T). \quad (37)$$

When these definitions are substituted into (Eq. 35), one obtains

$$\frac{d}{dt}(\partial K/\partial \dot{q}_j) - \partial K/\partial q_j = 2\sum_{i=j}^{n} Tr\left\{\frac{\partial W_i}{\partial q_j}\left[G_i\ddot{W}_i^T + \sum_{k=1}^{m_i}\ddot{\delta}_{ik}D_{ik}W_i^T + 2\sum_{k=1}^{m_i}\dot{\delta}_{ik}D_{ik}\dot{W}_i^T\right]\right\}. \quad (38)$$

The partials of K with respect to δ_{jf} and $\dot{\delta}_{jf}$ are considerably more complex due to the fact that $\mathbf{B}_{1i}$, $\mathbf{B}_{2i}$, and $\mathbf{B}_{3i}$ are functions of the deflection variables. The techniques of simplification are similar. An additional simplification arises due to the fact that if **A** is any antisymmetric matrix, and if **W** is compatible for mutiplication, then

$$Tr\{\mathbf{W}\,\mathbf{A}\,\mathbf{W}^T\} = 0.$$

An antisymmetric matrix in fact occurs as the difference of a matrix and its transpose.

$$\frac{d}{dt}(\partial K/\partial \dot{\delta}_{jf}) - \partial K/\partial \delta_{jf} = 2\sum_{i=j+1}^{n} Tr\left\{\frac{\partial W_i}{\partial \delta_{jf}}\left[G_i\ddot{W}_i^T + \sum_{k=1}^{m_i}\ddot{\delta}_{ik}D_{ik}W_i^T + 2\sum_{k=1}^{m_i}\dot{\delta}_{ik}D_{ik}\dot{W}_i^T\right]\right\} + Tr\left\{2\left[\ddot{W}_jD_{jk} + 2\dot{W}_j\sum_{k=1}^{m_i}\dot{\delta}_{jk}C_{jkf} + W_j\sum_{k=1}^{m_i}\ddot{\delta}_{jk}C_{jkf}\right]W_j^T\right\}. \quad (39)$$

4. The System's Potential Energy

The potential energy of the system arises from two sources considered here: elastic deformation and gravity. Both are included by first writing the potential energy contribution of a differential element, integrating over the length of the link, and then summing over all links.

4.1. Elastic Potential Energy

Consider a point on the i-th link as undergoing small deflections. First, restrict the link to be of the slender-beam type. The elastic potential is accounted for to a good approximation by bending about the transverse y_i and z_i axes and twisting about the longitudinal x_i axis. Compression is not initially included, as it is generally much smaller. Along an incremental length $d\eta$, the elastic potential is

$$dv_{ei} = \frac{1}{2} d\eta \left\{ E \left[I_z \left(\frac{\partial \theta_{zi}}{\partial \eta} \right)^2 + I_y \left(\frac{\partial \theta_{yi}}{\partial \eta} \right)^2 \right] + G\, I_x \left(\frac{\partial \theta_{xi}}{\partial \eta} \right)^2 \right\}, \tag{40}$$

where

θ_{xi}, θ_{yi}, and θ_{zi} are the rotations of the neutral axis of the beam at the point η in the x_i, y_i, and z_i directions, respectively. Since deflections are small, these directions are essentially parallel or perpendicular to the neutral axis of the beam.
E = Young's modulus of elasticity of the material.
G = the shear modulus of the material.
I_x = the polar area moment of inertia of the link's cross section about the neutral axis.
I_y, I_z = the area moment of inertia of the link's cross section about the y_i and z_i axes respectively.

With a truncated modal approximation for the beam deformation, the angles θ_{xi}, θ_{yi}, and θ_{zi} are represented as summations of modal coefficients times the deflection variables. The x rotation, for example, is

$$\theta_{xi} = \sum_{k=1}^{m_i} \delta_{ik} \theta_{xik}, \tag{41}$$

where θ_{xik} is the angle about the x_i axis corresponding to the k-th mode of link i at the point η. When dv_{ei} is integrated over the link, the integration can be taken inside the modal summations of (Eq. 41) and its corresponding y and z components. The following definitions then prove useful:

$$K_{ikl} = K_{xikl} + K_{yikl} + K_{zikl}, \tag{42}$$

where

$$K_{xikl} = \int_0^{l_i} G\, I_x(\eta) \frac{\partial \theta_{xil}}{\partial \eta} \frac{\partial \theta_{xik}}{\partial \eta} d\eta, \tag{43}$$

$$K_{yikl} = \int_0^{l_i} E\, I_y(\eta) \frac{\partial \theta_{yil}}{\partial \eta} \frac{\partial \theta_{yik}}{\partial \eta} d\eta, \tag{44}$$

$$K_{zikl} = \int_0^{l_i} E\, I_z(\eta) \frac{\partial \theta_{zil}}{\partial \eta} \frac{\partial \theta_{zik}}{\partial \eta} d\eta. \tag{45}$$

Note that $K_{ikl} = K_{ilk}$, and that for certain special cases the orthorgonality of the modal functions can eliminate many of the terms in (Eqs. 43–45). The elastic potential for the total system, V_e can then be written as

$$V_e = \frac{1}{2} \sum_{i=1}^{n} \sum_{k=1}^{m_i} \sum_{l=1}^{m_i} \delta_{ik} \delta_{il} K_{ikl}. \tag{46}$$

Note that the V_e is independent of q_i, the joint variables.

$$\frac{\partial V_e}{\partial q_j} = 0. \tag{47}$$

For deflection variables,

$$\frac{\partial V_e}{\partial \delta_{jf}} = \sum_{k=1}^{m_i} \delta_{jk} K_{jkf}. \tag{48}$$

The form of (Eq. 48) is much more general than the initial assumptions made regarding the contributions to the elastic potential energy would allow. Compression strain energy and link forms other than beams can be represented in this form. The values of the coefficients K_{jkf} can be determined analytically or numerically, for example, by finite-element methods.

4.2. Potential Energy Due to Gravity

For a differential element on the i-th link of length $d\eta$, the potential energy due to gravity is

$$dv_{gi} = -\mu \mathbf{g}^T \mathbf{W}_i {}^i\mathbf{h}_i d\eta, \tag{49}$$

where the gravity vector $\mathbf{g}$ has the form

$$\mathbf{g}^T = [0\ g_x\ g_y\ g_z].$$

When integrated over the length of the beam and summed over all beams, the gravity potential becomes

$$V_g = -\mathbf{g}^T \sum_{i=1}^{n} \mathbf{W}_i \mathbf{r}_i, \tag{50}$$

where

$$\mathbf{r}_i = M_i \mathbf{r}_{ri} + \sum_{k=1}^{m_i} \delta_{ik} \boldsymbol{\epsilon}_{ik}; \qquad (51)$$

M_i = the total mass of link i; $\mathbf{r}_{ri} = [1\ r_{xi}\ 0\ 0]$, a vector to the center of gravity from joint i (undeformed); and

$$\boldsymbol{\epsilon}_{ik} = \int_0^{l_i} \mu[0\ x_{ik}\ y_{ik}\ z_{ik}]^T d\eta. \qquad (52)$$

Note that $\boldsymbol{\epsilon}_{ik}$ is found in the top row of $\mathbf{C}_{ik}$. It is the distance from the undeformed center of gravity to the center of gravity when all δ are zero except δ_{ik}, which is one. The total distance to the center of gravity from O_i (joint i) is multiplied by the mass to give $\mathbf{r}_i$.

Upon taking the partial derivatives required by Lagrange's equations, one finds for the joint variables:

$$\frac{\partial V_g}{\partial q_j} = -\mathbf{g}^T \sum_{i=j}^{n} \frac{\partial \mathbf{W}_i}{\partial q_j} \mathbf{r}_i. \qquad (53)$$

For the deflection variables, for $1 \leq j \leq n-1$,

$$\frac{\partial V_g}{\partial \delta_{jf}} = -\mathbf{g}^T \sum_{i=j+1}^{n} \left(\frac{\partial \mathbf{W}_i}{\partial \delta_{jf}} \mathbf{r}_i \right) - \mathbf{g}^T \mathbf{W}_j \boldsymbol{\epsilon}_{jf}. \qquad (54)$$

For $j = n$,

$$\frac{\partial V_g}{\partial \delta_{nf}} = -\mathbf{g}^T \mathbf{W}_n \boldsymbol{\epsilon}_{nf}. \qquad (55)$$

5. Lagrange's Equations in Simulation Form

The components of the complete equations of motion in Lagrange's formulation, except for the external forcing terms, have been evaluated in (Eqs. 38, 47, and 53) for the joint equations and in (Eqs. 39, 48, 54, and 55) for deflection equations. The *external forcing terms* are the generalized forces corresponding to the generalized coordinates: the joint and deflection variables, in this case. The generalized force corresponding to joint variable q_i is the joint torque F_i. For the deflection variables, the corresponding generalized force will be zero if the corresponding modal deflections or rotations have no displacement at those locations where external forces are applied. Thus it is assumed for the present development that the modal functions are selected so that that is the case. This is convenient for utilizing the results as well. All motion at the joint is described in terms of the joint variable. (This is not true in the approach taken by Sunada and Dubowsky [1981]). The form of Lagrange's equations will then be

1. The joint equation, j:

$$\frac{d}{dt}(\partial K/\partial \dot{q}_j) - \partial K/\partial q_j + \frac{\partial V_e}{\partial q_j} + \frac{\partial V_g}{\partial q_j} = F_j. \qquad (56)$$

2. The deflection equation, j,f:

$$\frac{d}{dt}(\partial K/\partial \dot{\delta}_{jf}) - \partial K/\partial \delta_{jf} + \frac{\partial V_e}{\partial \delta_{jf}} + \frac{\partial V_g}{\partial \delta_{jf}} = 0. \qquad (57)$$

These equations are in the *inverse dynamic* form. To convert them to the simulation (*dynamic*) form, one must extract the coefficients of the second derivatives of the generalized coordinates to compose an inertia matrix for the system. The second and first derivatives together make up the derivative of the state vector, which can be used in one of the available integration schemes (e.g., Runga-Kutta), to solve for the state as a function of time for given initial conditions and inputs F_i.

5.1. Kinematics Revisited

This section will extend the kinematics to separate the second derivatives of the joint variables and deflection variables from the expressions for $\ddot{\mathbf{W}}_i$ and $\ddot{\hat{\mathbf{W}}}_i$. Other occurrences of these derivatives are already explicit in the formulation as it exists.

First, consider the product of transformations that make up $\hat{\mathbf{W}}_i$, and two alternative ways of expressing it:

$$\hat{\mathbf{W}}_i = \mathbf{A}_1\mathbf{E}_1\mathbf{A}_2\mathbf{E}_2 \ldots \mathbf{A}_h\mathbf{F}_h \ldots \mathbf{A}_i\mathbf{E}_i = \hat{\mathbf{W}}_{h-1}\mathbf{A}_h{}^h\mathbf{W}_i \qquad (58)$$

$$= \mathbf{W}_h\mathbf{E}_h{}^h\hat{\mathbf{W}}_i. \qquad (59)$$

Carrying through the derivatives, one obtains

$$\ddot{\hat{\mathbf{W}}}_i = \sum_{h=1}^{i} \left(\hat{\mathbf{W}}_{h-1}\mathbf{U}_h{}^h\bar{\mathbf{W}}_i\ddot{q}_h + \sum_{k=1}^{m_h} \mathbf{W}_h\mathbf{M}_{hk}{}^h\hat{\mathbf{W}}_i\ddot{\delta}_{hk} \right) + \ddot{\hat{\mathbf{W}}}_{vi}. \qquad (60)$$

For the corresponding expression for $\mathbf{W}_i$, write

$$\mathbf{W}_i = \mathbf{A}_i\mathbf{E}_1\mathbf{A}_2\mathbf{E}_2 \ldots \mathbf{A}_h\mathbf{E}_h \ldots \mathbf{E}_{i-1}\mathbf{A}_i = \hat{\mathbf{W}}_{h-1}\mathbf{A}_h{}^h\tilde{\mathbf{W}}_h \tag{61}$$

$$= \mathbf{W}_h\mathbf{E}_h{}^h\mathbf{W}_i. \tag{62}$$

$$\ddot{\mathbf{W}}_i = \sum_{h=1}^{i} \hat{\mathbf{W}}_{h-1}\mathbf{U}_h{}^h\tilde{\mathbf{W}}_i\ddot{q}_h + \sum_{h=1}^{i-1}\sum_{k=1}^{m_h} \mathbf{W}_h\mathbf{M}_{hk}{}^h\hat{\mathbf{W}}_i\ddot{\delta}_{hk} + \ddot{\mathbf{W}}_{vi}. \tag{63}$$

The value of $\ddot{\hat{\mathbf{W}}}_{vi}$ and $\ddot{\mathbf{W}}_{vi}$ can be calculated recursively, as shown in (Eqs. 15 and 10), respectively (for $\ddot{\hat{\mathbf{W}}}_i$ and $\ddot{\mathbf{W}}_i$) by only eliminating terms involving $\ddot{q}_j$ and $\ddot{\delta}_{jk}$. The result is

$$\ddot{\hat{\mathbf{W}}}_{vj} = \ddot{\mathbf{W}}_{v,j-1}\mathbf{A}_j + 2\dot{\mathbf{W}}\,\dot{\mathbf{A}}_j + \hat{\mathbf{W}}_{j-1}\mathbf{U}_{2j}\dot{q}_j^2, \tag{64}$$

$$\ddot{\mathbf{W}}_{vj} = \ddot{\hat{\mathbf{W}}}_{vj}\mathbf{E}_j + 2\dot{\hat{\mathbf{W}}}_j\dot{\mathbf{E}}_j. \tag{65}$$

5.2. Inertia Coefficients

To obtain the inertia coefficients that multiply the second derivatives, substitute (Eqs. 63 and 60) into the relevant parts of the equations of motion, (Eqs. 38 and 39), respectively. Collecting the terms and arranging them for efficient computation requires the steps outlined in this section.

5.2.1. Inertia Coefficients of Joint Variables in the Joint Equations

All occurrences of $\ddot{q}_j$ in (Eq. 38) are in the expression for $\ddot{\mathbf{W}}_i^T$. When these terms are isolated, a double summation over the indices i and h exists. Interchange the order of the summation as follows:

$$\sum_{i=j}^{n}\sum_{h=1}^{i} = \sum_{h=1}^{n}\sum_{i=\max(h,j)}^{n}.$$

The resulting coefficient for joint variable q_h in the joint equation j is

$$J_{jh} = 2\,\mathrm{Tr}\{\hat{\mathbf{W}}_{j-1}\mathbf{U}_j{}^j\tilde{\mathbf{F}}_h\mathbf{U}_h^T\hat{\mathbf{W}}_{h-1}^T\}, \tag{66}$$

where

$${}^j\tilde{\mathbf{F}}_h = \sum_{i=\max(h,j)}^{n} {}^j\tilde{\mathbf{W}}_i\mathbf{G}_i{}^h\tilde{\mathbf{W}}_i^T. \tag{67}$$

Note that if one exchanges j and h and transposes inside the trace operation, an identical expression is obtained. This indicates the symmetry of the inertia matrix that is used to reduce the number of computations required. The expression for ${}^j\tilde{\mathbf{F}}_h$ can be computed recursively, as will be described later to improve the efficiency of calculation further.

5.2.2. Inertia Coefficients of the Deflection Variables in the Joint Equations

The deflection variables appear both in the expression for $\ddot{\mathbf{W}}_i^T$ and explicitly in (Eq. 38). After substituting $\ddot{\mathbf{W}}_i^T$ into Eq. 38), collect terms in $\ddot{\delta}_{jf}$ and exchange the order of summations as follows:

$$\sum_{i=j}^{n}\sum_{h=1}^{i-1} = \sum_{h=1}^{n-1}\sum_{i=\max(h+1,j)}^{n}.$$

The resulting coefficient of $\ddot{\delta}_{hk}$ in joint equation j is J_{jhk}. The terms to be included depend on the relative values of j and h. The following hold for $1 \le k \le m_h$. For $h = n$, $j = 1 \ldots n$,

$$J_{jnk} = 2\,\mathrm{Tr}\{(\hat{\mathbf{W}}_{j-1}\mathbf{U}_j){}^j\tilde{\mathbf{W}}_n\mathbf{D}_{nk}\mathbf{W}_n^T\}; \tag{68}$$

for $h = j \ldots n-1$, $j = 1 \ldots n-1$,

$$J_{jnk} = 2\,\mathrm{Tr}\{(\hat{\mathbf{W}}_{jh1}\mathbf{U}_j)[{}^j\mathbf{F}_h\mathbf{M}_{hk}^T + {}^j\tilde{\mathbf{W}}_h\mathbf{D}_{hk}]\mathbf{W}_h^T\}; \tag{69}$$

for $h = 1 \ldots j-1$, $j = 2 \ldots n$,

$$J_{jnk} = 2\,\mathrm{Tr}\{(\hat{\mathbf{W}}_{j-1}\mathbf{U}_j){}^j\mathbf{F}_h\mathbf{M}_{hk}^T\mathbf{W}_n^T\}; \tag{70}$$

where for $h = 1 \ldots n-1$, $j = 1 \ldots n$,

$${}^j\mathbf{F}_h = \sum_{i=\max(h+1,j)}^{n} {}^j\tilde{\mathbf{W}}_i\mathbf{G}_i{}^h\mathbf{W}_i^T. \tag{71}$$

It can be shown that the inertia coefficient for the deflection variable δ_{hk} in the joint equation j is the same as the coefficient for the joint variable q_j in the deflection equation h,k. This further extends the symmetry of the inertia matrix and reduces the computation necessary.

5.2.3. Inertia Coefficients of the Deflection Variables in the Deflection Equation

In a manner much the same as for the previous two types of coefficients, the inertia coefficients of the deflection variables in the deflection equations are evaluated. Symmetry of the coefficients can be shown such that the coefficient of variable h,k in equation j,f is the same as the coefficient of variable j,f in equation h,k. Substituting (Eq. 63) into (Eq. 39), isolating the second derivatives of the deflection variables, and interchanging the order of summations enables one to identify the inertia coefficients. Further simplification is based on the identity that, for any three square matrices, **A**, **B**, and **C**,

$$\mathrm{Tr}\{\mathbf{A\,B\,C}\} = \mathrm{Tr}\{\mathbf{C\,A\,B}\} = \mathrm{Tr}\{\mathbf{B\,C\,A}\}.$$

Furthermore, the rotation matrices in the transformation matrices are orthogonal, so that $\mathbf{R}_i\,\mathbf{R}_i^T = \mathbf{I}$, a 3×3 identity matrix. This, coupled with the zero first row and column of $\mathbf{C}_{jkf}$, results in an especially simple form for two of the four cases. The following hold for $1 \leq k \leq m_h$ and $1 \leq f \leq m_j$.

For $j = h = n$,

$$I_{nfnk} = 2\,\mathrm{Tr}\{\mathbf{C}_{nkf}\}. \tag{72}$$

For $j = h = 1 \ldots n-1$,

$$I_{jfjk} = 2\,\mathrm{Tr}\{\mathbf{M}_{jf}\,{}^j\mathbf{\Phi}_j\mathbf{M}_{jk}^T + \mathbf{C}_{jkf}\}. \tag{73}$$

For $h = n;\ j = 1 \ldots n-1$,

$$I_{jfnk} = 2\,\mathrm{Tr}\{\mathbf{W}_j\mathbf{M}_{jf}\,{}^j\mathbf{W}_n\mathbf{D}_{nk}\mathbf{W}_n^T\}. \tag{74}$$

For $j = 1 \ldots n-1;\ h = j+1 \ldots n-1$,

$$I_{jfhk} = 2\,\mathrm{Tr}\{\mathbf{M}_{jf}[{}^j\mathbf{\Phi}_h + \mathbf{M}_{hk}^T + {}^j\mathbf{W}_h\mathbf{D}_{hk}]\mathbf{W}_h^T\}. \tag{75}$$

Terms in the above defined for $j = 1 \ldots n-1$; $h = 1 \ldots n-1$ are

$${}^j\mathbf{\Phi}_h = \sum_{i=\max(j+1,h+1)}^{n} {}^{jW_i}\mathbf{G}_i\,{}^h\mathbf{W}_i^T. \tag{76}$$

5.2.4. Recursions in the Calculation of the Inertia Coefficients

Since the inertia matrix is a square matrix, it requires the calculation of n_t^2 terms, where n_t is the total number of variables:

$$n_t = n + \sum_{i=1}^{n} m_i.$$

The fact that the matrix is symmetric reduces the number of distinct terms to $n_t(n_t + 1)/2$, which still has a second-power dependence. Thus, while the inverse dynamics computation complexity can be made linear in n_t simulation requires the inertia matrix with complexity dependent on n_t^2. Since n_t can be quite large for practical arms, it is important to reduce the coefficient of the squared term as much as possible. Note further that due to their short or even zero length it is possible for some links to be essentially rigid. Anthropomorphic arms, for example, have two links that are much longer than the others and tend to dominate the compliance. Thus it is possile that many of the terms derived above will not be needed for these links—four of the six links in the anthropomorphic example. Any recursive scheme for calculating the terms in the equations should not require these calculations as a means to get to needed terms, if possible.

Consider the calculation of (Eq. 67, 71, and 76). Several recursive schemes could be arranged for the efficient calculation of these quantities. Equation (71) is only needed if the link corresponding to the variable, link h, is flexible. That is, if $m_h > 0$. Equation (76) is only needed if both the link of the variable and the link of the equation, link j, is also flexible. Thus, the following recursive scheme for calculating ${}^j\tilde{\mathbf{F}}_h$, ${}^j\mathbf{F}_h$, and ${}^j\mathbf{\Phi}_h$ is proposed. The following hold for $1 \leq k \leq m_h$; $1 \leq f \leq m_j$.

Initialization:

$$^{n}\tilde{\mathbf{F}}_{n} = \mathbf{G}_{n}. \quad (77)$$

For $j > h \leq n$,

$$^{j}\tilde{\mathbf{F}}_{h} = \mathbf{E}_{j}\mathbf{A}_{j}\,^{j+1}\tilde{\mathbf{F}}_{h}. \quad (78)$$

For $j = h$,

$$^{h}\tilde{\mathbf{F}}_{h} = \mathbf{G}_{h} + {}^{h}\tilde{\mathbf{F}}_{h+1}(\mathbf{E}_{j}\mathbf{A}_{j+1})^{T}. \quad (79)$$

If $m_h > 0$, calculate

$$^{j}\mathbf{F}_{h} = {}^{j}\tilde{\mathbf{F}}_{h+1}\mathbf{A}_{h}^{T}. \quad (80)$$

If $m_h > 0$ and $m_j > 0$, calculate

$$^{j}\boldsymbol{\Phi}_{h} = \mathbf{A}_{j+1}\,^{j}\tilde{\mathbf{F}}_{h}. \quad (81)$$

5.3. Assembly of Final Simulation Equations

The complete simulation equations have now been derived. It remains to assemble them in final form and to point out some remaining recursion relations that can be used to reduce the number of calculations. The second derivatives of the joint and deflection variables are desired on the "left-hand side" of the equation as unknowns, and the remaining dynamic effects and the inputs are desired on the "right-hand side." To carry out this process completely, one would take the inverse of the inertia matrix **J** and premultiply the vector of other dynamic effects. This inverse can only be evaluated numerically, because of its complexity. Thus, for the purposes of this paper, the equations will be considered complete in the following form:

$$\mathbf{J}\,\ddot{\mathbf{z}} = \mathbf{R}, \quad (82)$$

where

- $\mathbf{J}$ = inertia matrix consisting of coefficients previously defined in the order for multiplication appropriate for $\mathbf{z}$;
- $\mathbf{z}$ = the vector of generalized coordinates $= [q_1\ \delta_{11}\ \delta_{12}\ \ldots\ \delta_{1m_1}\ q_2\ \delta_{21}\ \ldots\ \delta_{2m_2}\ \ldots\ q_h\ \delta_{h1}\ \ldots\ \delta_{hm_h}\ \ldots\ \delta_{nm_n}\ ]^T$;
- q_h = the joint variable of the h-th joint;
- δ_{hk} = the deflection variable (amplitude) of the k-th mode of link h;
- $\mathbf{R}$ = vector of remaining dynamics and external forcing terms $= [R_1\ R_{11}\ R_{12}\ \ldots\ R_{1m_1}\ R_2\ R_{21}\ \ldots\ R_{2m_2}\ \ldots\ R_j\ R_{j1}\ \ldots\ R_{jf}\ \ldots\ R_{jm_j}\ \ldots\ R_{nm_n}\ ]^T$;
- R_j = dynamics from the joint equation j (Eq. 56), excluding second derivatives of the generalized coordinates; and
- R_{jf} = dynamics from the deflection equation jf (Eq. 57), excluding second derivatives of the generalized coordinates.

The elements of **J** have just been formulated and can be arranged to form the proper equations in the order described above. This order has been selected because it results in the symmetric appearance of **J**. The elements of **R** have not been explicitly given with the second derivatives removed. These are given below, with some recursions to facilitate their computation.

$$R_1 = -2\,\mathrm{Tr}\{\mathbf{U}_1\mathbf{Q}_1\} + \mathbf{g}^T\mathbf{U}_1\mathbf{P}_1 + F_1, \quad (83)$$

$$R_j = -2\,\mathrm{Tr}\{\hat{\mathbf{W}}_{j-1}\mathbf{U}_j\mathbf{Q}_j\} + \mathbf{g}^T\hat{\mathbf{W}}_{j-1}\mathbf{U}_j\mathbf{P}_j + F_j, \quad (84)$$

$$R_{nf} = -2\,\mathrm{Tr}\{[\ddot{\mathbf{W}}_{vn}\mathbf{D}_{nf} + 2\dot{\mathbf{W}}_n \sum_{k=1}^{m_n} \dot{\delta}_{nk}\mathbf{C}_{nkf}]\mathbf{W}_n^T\} - \sum_{k=1}^{m_n} \delta_{nk}K_{nkf} + \mathbf{g}^T\mathbf{W}_n\boldsymbol{\epsilon}_{nf}, \quad (85)$$

$$R_{jf} = -2\,\mathrm{Tr}\{\mathbf{W}_j\mathbf{M}_{jf}\mathbf{A}_{j+1}\mathbf{Q}_{j+1}[\ddot{\mathbf{W}}_{vj}\mathbf{D}_{jf} + 2\dot{\mathbf{W}}_j \sum_{k=1}^{m_j} \dot{\delta}_{jk}\mathbf{C}_{jkf}]\mathbf{W}_j^T - \sum_{k=1}^{m_j} \delta_{jk}K_{jkf} + \mathbf{g}^T\mathbf{W}_j\mathbf{M}_{jf}\mathbf{A}_{j+1}\mathbf{P}_{j+1} + \mathbf{g}^T\mathbf{W}_j\boldsymbol{\epsilon}_{jf}, \quad (86)$$

where

$$\mathbf{Q}_n = \mathbf{G}_n\ddot{\mathbf{W}}_{vn}^T + 2\left(\sum_{k=1}^{m_n} \dot{\delta}_{nk}\mathbf{D}_{nk}\right)\dot{\mathbf{W}}_n^T; \quad (87)$$

$$\mathbf{Q}_j = \mathbf{G}_j \ddot{\mathbf{W}}_{vj}^T + 2\left(\sum_{k=1}^{m_j} \dot{\delta}_{jk} \mathbf{D}_{jk}\right) \dot{\mathbf{W}}_j^T + \mathbf{E}_j \mathbf{A}_{j+1} \mathbf{Q}_{j+1}; \tag{88}$$

$$\mathbf{P}_n = M_n \mathbf{r}_n + \sum_{k=1}^{m_j} \delta_{nk} \epsilon_{nk}; \tag{89}$$

$$\mathbf{P}_j = M_j \mathbf{r}_j + \sum_{k=1}^{m_j} \delta_j \epsilon_{jk} + \mathbf{E}_j \mathbf{A}_{j+1} \mathbf{P}_{j+1}. \tag{90}$$

6. Conclusions

Two measures of the above model's success are its accuracy and its speed. The two are somewhat related in that accuracy of the flexible representation can be improved by increasing the number of modes used to represent the link deflection at the expense of calculation time. The issue is further complicated by the choice of mode shapes, range of motion considered, and the arm configuration. Furthermore, limited information is available in the literature for comparison. A simple comparison has been used in the past and can be performed for calculation complexity. Hollerbach (1980) compares several approaches (by different authors) to the inverse dynamics problem of rigid arms. Walker and Orin (1982) give a similar count for four approaches to the simulation problem. Sunada and Dubowsky (1981) have given computation times for a given manipulator, trajectory, and computer for their flexible simulation. Comparison to the calculation counts of rigid models are given for a rough comparison of speeds in this section. No attempt is made at a quantitative comparison of accuracy.

In determining the number of calculations from the equations one must choose how some matrix products are to be implemented. Hollerbach took the approach that the most straightforward implementation of the equations should be used. The attitude here is quite different. Obvious simplifications in the multiplication of matrices with known constant rows, the top row of a transformation matrix, for example, are assumed in these computations. The 4×4 matrix transformation was chosen for its conceptual convenience, and the calculation count will not be intentionally penalized due to that choice. Furthermore, certain products appear in multiple equations and are assumed to be saved for later use. Special-purpose multiply routines are used whenever they can capitalize on the special structure of a given matrix. Finally, in the simulation form, the calculations needed to invert the inertia matrix are not included, and no consideration is given to the calculations of the integration routine. The general form of the modal parameters is used, however. This results in all combinations of modes h and k in the matrix $\mathbf{C}_{ihk}$ to be computed and used and hence introduces a squared dependence on the number of modes on *each* inertial coefficient of the deflection variables. With these assumptions, the number of calculations is approximately

1. Number of multiplications:

$$6\, n_f^2 m^2 + 17.5\, n_f m^2 + 118\, n_f^2 m + 74\, n\, n_f m + 137.5\, 6 n_f m + 84\, n^2 + 86\, n\, n_f + 279\, n + 126\, n_f - 57$$

2. Number of additions:

$$6.5\, n_f^2 m^2 + 19\, n_f m^2 + 115.5\, n_f^2 m + 68\, n\, n_f m + 123\, n_f m + 85\, n^2 + 80\, n\, n_f + 329\, n + 111\, n_f - 91,$$

where n = total number of joints; n_f = number of flexible links; and m = number of modes describing each flexible link.

The above approximation assumes an "average" joint complexity over two common types of rotary joints, the same number of modes on each flexible link, a rigid last link, and a flexible first link.

If assumed mode shapes are restricted so that the shape functions in the *x*, *y*, and *z* directions are orthogonal, only $\mathbf{C}_{ikk}$ will be nonzero. This is a stronger requirement than the orthogonality of the set of complete mode shapes, but would often be realized with simple mode shapes. It has not been determined whether this would improve the combination of speed and accuracy.

This calculation count can be roughly compared to rigid-link results available in the literature mentioned above. For a 12-degree-of-freedom rigid problem, the inverse 3×3 transformation-matrix formulation requires 2.66 times as many multiples as the Newton-

Euler formulation. Walker's method 3 (his best) for simulation requires 4,491 multiples. For 6 joints and two flexible links with 3 modes each, the method of this paper requires approximately 12,009 multiples. The ratio of these simulation methods is 2.67, almost exactly the same as for the inverse dynamic methods with the same number of degrees of freedom. A modal representation of flexibility would be much more accurate than adding 6 imaginary joints to represent compliance, but one could expect to use 15 imaginary joints and 6 real joints with Walker's method with fewer multiples than with the method of this paper.

Thus it seems that in order to be competitive with possible Newton-Euler, nontransfer-matrix approaches, the assumed mode shapes will have to be simplified. It is not clear that the conceptual convenience of the transformation-matrix approach can be justified relative to vector dyadic approaches of Hughes (1979) and Likins (Singh and Likins 1983). Unfortunately, computation counts are not available for that work.

Acknowledgments

The author gratefully acknowledges the support of this work by The Robotics Institute at Carnegie-Mellon University and the Georgia Institute of Technology while the author was on leave as Visiting Scientist. The encouragement of Marc Raibert of The Robotics Institute was also important.

References

Albus, J. S. 1975. A new approach to manipulator control: The cerebellar model articulation controller (CMAC). *Trans. ASME J. Dyn. Syst. Measurement Cont.* 97:270–277.

Bejczy, A. K. 1974 (Feb.). Robot arm dynamics and control. NASA Tech. Memo 33-669. Pasadena, Calif.: California Institute of Technology Jet Propulsion Laboratory.

Book, W. J. 1974 (Apr.). Modeling, design and control of flexible manipulator arms. Ph.D. thesis, Massachusetts Institute of Technology Department of Mechanical Engineering.

Book, W. J., Maizza-Neto, O., and Whitney, D. E. 1975 (Dec.). Feedback control of two beam, two joint systems with distributed flexibility. *Trans. ASME J. Dyn. Syst. Measurement Contr.* 97G(4):424–431.

Book, W. J. 1973 (Feb.). Study of design and control of remote manipulators. Part II: Vibration considerations in manipulator design. Contract rept. NAS8-28055. Cambridge, Mass.: Massachusetts Institute of Technology Department of Mechanical Engineering.

Book. W. J. 1976 (Sept., Warsaw). Characterization of strength and stiffness constraints on manipulator control. *Proc. Symp. Theory Robots Manipulators.* New York: Elsevier/North-Holland, pp. 28–37.

Book, W. J. 1979 (Sept.). Analysis of massless elastic chains with servo controlled joints. *Trans. ASME J. Dyn. Syst. Measurement Contr.* 101(3):187–192.

Book. W. J. 1982 (Sept.). Recursive Lagrangian dynamics of flexible manipulator arms via transformation matrices. paper delivered at IFAC Symp. Computer Aided Design of Multivariable Technological Systems, Purdue University, West Lafayette, Ind.

Book, W. J., Majette, M., and Ma, K. 1979 (July). The distributed systems analysis package (DSAP) and its application to modeling flexible manipulators. Georgia Institute of Technology, School of Mechanical Engineering, Subcontract no. 551 to Charles Stark Draper Laboratory, Inc., NASA contract no. NAS9-13809.

Book, W. J., Majette, M., and Ma, K. 1981 (Feb.). Frequency domain analysis of the space shuttle manipulator arm and its payloads; Vol. 1—Analysis and conclusions; Vol. II—Computer program description and listings. Georgia Institute of Technology, School of Mechanical Engineering, Subcontract no. 586 to Charles Stark Draper Laboratory, Inc., NASA contract no. NAS9-13809.

Denavit, J., and Hartenberg, R. S. 1955 (June). A kinematic notation for lower-pair mechanisms based on matrices. *ASME J. Applied Mechanics* 22:215–221.

Derby, S. J. 1981 (Aug.). Kinematic elasto-dynamic analysis and computer graphics simulation of general purpose robot manipulators. Ph.D. thesis, Rensselaer Polytechnic Institute.

Dubowsky, S., and Gardner, T. N. 1977 (Feb.). Design and analysis of multi-link flexible mechanisms with multiple clearance connections. *ASME J. Engineering for Industry* 99(1).

Ho, J. Y. L. et al. 1974 (Oct.). *Remote manipulator system (RMS) simulation,* vol. 1. Palo Alto, Calif.: Lockheed Palo Alto Research Laboratory.

Hollerbach, J. M. 1980 (Nov.). A recursive Lagrangian formulation of manipulator dynamics and a comparative study of dynamics formulation complexity. *IEEE Trans. Syst. Man. Cybern.* SMC-10(11):730–736.

Hughes, P. C. 1977 (Sept.). Dynamics of a flexible manipulator arm for the space shuttle. Paper delivered at AAS/AIAA Astrodynamics Conference, American Institute of Aeronautics and Astronautics, Jackson Lake Lodge, Grand Teton National Park, Wyo.

Hughes, P. C., 1979 (Oct.-Dec.). Dynamics of a chain of flexible bodies. *J. Astronautical Sci.* 27(4):359–380.

Huston, R. L., and Passerello, C. E., 1980 (Nov.). Multibody structural dynamics including translation between bodies. *Computers and Structures* 12(5):713–720.

Kelly, F. A., and Huston, R. L., 1981 (June). Modeling of flexibility effects in robot arms. *Proc. 1981 Joint Automatic Contr. Conf.* Green Valley, Ariz.: American Automatic Control Council, paper WP-2C.

Liegois, A., et al. 1976 (Sept., Warsaw). Mathematical and computer models of interconnected mechanical system. *Proc. Symp. Theory Pract. Robots Manipulators.* New York: Elsevier/North-Holland, pp. 5–17.

Likins, P. W. 1972. Finite element appendage equations for hybrid coordinate dynamic analysis. *Int. J. Solids Structures* 8:709–731.

Luh, J. Y. S., Walker, M. W., and Paul, R. P. C., 1980 (June). On-line computational scheme for mechanical manipulators. *Trans. ASME J. Dyn. Syst. Measurement Contr.* 102:69–76.

Maizza-Neto, O. 1974. *Modal analysis and control of flexible manipulator arms.* Ph.D. thesis, Massachusetts Institute of Technology Department of Mechanical Engineering.

Mirro, J., 1972 (Aug.). Automatic feedback control of a vibrating beam. Master's thesis, Massachusetts Institute of Technology, Department of Mechanical Engineering. Rept. T-571. Cambrige, Mass.: C. S. Draper Laboratory.

Nguyen, P. K., and Hughes, P. C., 1976 (June). Finite-element analysis of CTS-like flexible spacecraft. Tech. Rept. 205. Toronto: University of Toronto Institute for Aerospace Studies.

Raibert, M. H., and Horn, B. K. P. 1978 (June). Manipulator control using the configuration space method. *Industrial Robot* 5(2):69–73.

Silver, W. M. 1981 (June). On the equivalence of Lagrangian and Newton-Euler dynamics for manipulators. *Proc. 1981 Joint Automatic Contr. Conf.* Green Valley, Ariz.: American Automatic Control Council, paper no. TA-2A.

Singh, R. P., and Likins, P. W. 1983 (June). Manipulator interactive design with interconnected flexible elements. Paper delivered at the 1983 Automatic Contr. Conf., San Francisco, Calif.

Stepanenko, Y., and Vukobratović, M. 1976. Dynamics of articulated open-chain active mechanisms. *Math. Biosci.* 28:137–170.

Sturges, R. 1973 (Feb.). Teleoperator arm design program (TOAD). Tech. Rept. E-2746. Cambridge, Mass.: C. S. Draper Laboratory.

Sunada, W., and Dubowsky, S. 1981 (July). The application of finite element methods to the dynamic analysis of spatial and coplanar linkage system. *ASME J. Mechanical Design* 103:643–651.

Thomas, M., and Tesar, D. 1981 (June). Dynamic modeling of serial manipulator arms. *Proc. 1981 Joint Automatic Contr. Conf.* Green Valley, Ariz.: American Automatic Control Conference. *1982 Trans. ASME J. Dyn. Syst. Measurement Contr.* 104(3):218–228.

Walker, M. W., and Orin, D. E. 1982 (Sept.). Efficient dynamic computer simulation of robotic mechanisms. *Trans. ASME J. Dyn. Syst. Measurement Contr.* 104(3):205–211.

Whitney, D. E., Book, W. J., and Lynch, P. M. 1974. Design and control considerations for industrial and space manipulators. *Proc. 1974 Jount Automatic Contr. Conf. AIChE,* pp. 591–598.

Winfrey, R. C. 1972 (May). Dynamic analysis of elastic mechanisms by reduction of coordinates. *ASME J. Engineering for Industry* 94(1).

Errata

Several equations have been corrected as follows:

$$\frac{d\mathbf{h}_i}{dt} = \dot{\mathbf{h}}_i = \dot{\mathbf{W}}_i\,{}^i\mathbf{h}_i + \mathbf{W}_i\,{}^i\dot{\mathbf{h}}_i\,. \tag{8}$$

By differentiating (3), one obtains

$$\dot{\mathbf{W}}_j = \dot{\hat{\mathbf{W}}}_{j-1}\,\mathbf{A}_j + \hat{\mathbf{W}}_{j-1}\dot{\mathbf{A}}_j\,, \tag{9}$$

$$\ddot{\mathbf{W}}_j = \ddot{\hat{\mathbf{W}}}_{j-1}\,\mathbf{A}_j + 2\dot{\hat{\mathbf{W}}}_{j-1}\,\dot{\mathbf{A}}_j + \hat{\mathbf{W}}_{j-1}\,\ddot{\mathbf{A}}_j\,, \tag{10}$$

$$\dot{\mathbf{E}}_j = \sum_{k=1}^{m_j} \dot{\delta}_{jk}\mathbf{M}_{jk} \tag{16}$$

$$ {}^{i}\dot{\mathbf{h}}_{i} = \sum_{j=1}^{m_j} \dot{\delta}_{ij} \left[0\ x_{ij}\ y_{ij}\ z_{ij} \right]^{T} \tag{20} $$

$$ \mathbf{B}_{ii} = \frac{1}{2} \int_{0}^{l_i} \mu\ {}^{i}\dot{\mathbf{h}}_{i}\ {}^{i}\dot{\mathbf{h}}_{i}^{T}\, d\eta \tag{23} $$

$$ \mathbf{G}_{i} = \mathbf{C}_{i} + \sum_{k=1}^{m_i} \delta_{ik} (\mathbf{C}_{ik} + \mathbf{C}_{ik}^{T} + \sum_{l=1}^{m_i} \delta_{il} \mathbf{C}_{ilk}) \tag{37} $$

$$ \mathbf{I}_{jfhk} = 2\,\mathrm{Tr}\left\{ \mathbf{M}_{jf} \left[{}^{j}\Phi_{h}\ \mathbf{M}_{hk}^{T} + {}^{j}\mathbf{W}_{h}\ \mathbf{D}_{hk} \right] \mathbf{W}_{h}^{T} \right\}. \tag{75} $$

Experiments in Identification and Control of Flexible-Link Manipulators

Stephen Yurkovich and Anthony P. Tzes

ABSTRACT: This paper reports on an ongoing effort for end-point position control of flexible-link manipulators under realistic conditions in laboratory setups consisting of one- and two-link manipulators. The paper treats modeling, identification, and control of flexible-link manipulators that are required to carry payloads, possibly unknown and varying, while undergoing disturbance effects from the environment and the workspace.

Introduction

Efforts in the modeling and control of flexible-link manipulators have been motivated by the foreseen demand for lightweight, accurate, high-speed robots in space telerobotics and several other applications. Presently, studies in these areas have reached a fairly high level of maturity, due primarily to numerous works in the last few years from both an analytical viewpoint and, to a lesser extent, an experimental viewpoint. Analytical studies in modeling flexible-link robots abound and are, in fact, too numerous to cite here; Refs. [1] and [2] serve as excellent summaries of existing works in flexible manipulator modeling. Equally numerous are the various approaches that have appeared in the literature for controller design schemes. The greatest number of these works have dealt in simulation studies only, and some have developed quite elaborate and complex control schemes.

On the other hand, several successful laboratory setups have demonstrated the effectiveness of relatively simple algorithms for flexible manipulator control. Although most experimental studies have focused on single-link manipulators, or multilink manipulators with a single flexible link, such setups have served as valuable test beds for modeling, system identification, and controller design. Some of the more visible experimental efforts can be found in Refs. [3]–[14]. Because of the particular relevance to the work reported here, we note the contribution of [3], where the use of measurements from a linear accelerometer in vibration compensation of the robot end point were shown to be extremely successful, proving the concept of acceleration feedback for flexible-link manipulator control. The use of acceleration feedback has intuitive appeal from an engineering design viewpoint, due to ease of implementation, relatively low cost, and advantages of structure-mounted sensing.

Despite this recent activity, relatively little has appeared involving laboratory verification of tuning controllers for realistic flexible-link manipulators, which are required to maintain end-point accuracy while manipulating loads that are possibly unknown and varying. This paper discusses several techniques for flexible-link systems and presents experimental results in system identification and control of a one-link flexible manipulator carrying an unknown, varying payload. In doing so, this paper serves to summarize some previous works, such as [4], [5], [15], as well as report on other recent findings.

The authors are with The Ohio State University, Department of Electrical Engineering, Columbus, OH 43210.

Problem Description

Two laboratory setups are currently utilized in the Flexible Structures Facility at Ohio State (Department of Electrical Engineering) [16], [17]. The one-link system is the subject of the experimental results reported in this paper and is described in detail below. The identification and control techniques described are, however, currently being investigated primarily for the second system, which consists of two flexible links situated in the horizontal plane.

One-Link Setup

The flexible-link manipulator arm of this study is a beam made of $\frac{1}{16}$-in. 6061-T6 aluminum, 1 m in length and 10 cm in height. The arm is counterbalanced about the motor axis with a rigid aluminum attachment 38 cm in length. Experimentation with different payloads is made possible with a series of machined aluminum and brass fittings that are attached easily to the manipulator end point. Actuation at the hub is accomplished by a direct-drive DC motor with rated stall torque at 680 oz.-in. The two sensors for use in feedback control are the accelerometer, located at the arm end point, and the optical shaft encoder located at the hub. The encoder is rated at 3600 pulses per revolution, allowing measurement of the shaft angular position with a resolution of 0.05 deg. The piezoelectric accelerometer is rated at $\pm 250g$ with a sensitivity of 1.15 mV/g. A linear array line scan camera is used for data recording, reading a light source at the arm end point, but is not used in feedback control (results of end-point position feedback for this setup were presented in [3]). The computer used in the data acquisition and control is the DEC MicroVax II.

Identification and Control

The laboratory setup has furnished an excellent test bed for investigation of many ideas in the areas of identification and control, several of which are described in the sections to follow. Specifically, methodologies under study, the results of which have been reported elsewhere, include position feedback, fixed controller designs; acceleration feedback, fixed controller designs; eigenstructure realization algorithms for identification; autotuning control designs with time-domain identification; autotuning control designs with frequency-domain identification; and input shaping with acceleration feedback [18].

A characteristic of flexible-link manipulators situated in the horizontal plane is that the modal frequencies are reduced when a payload is added. Motivated by this and the fact that a fixed controller will not perform well over a range of payloads, the idea pursued in many of the techniques listed above is to tune a nominal control configuration according to the changing characteristics of the arm. As an illustration, consider the *nominal* case, i.e., when no load is carried by the arm. This control scheme utilizes end-point acceleration through a static feedback gain, with shaft position in a separate static gain feedback loop. We note that more complex schemes have been investigated, such as linear quadratic regulators, or inclusion of

Reprinted from *IEEE Contr. Mag.*, vol. 10, no. 2, pp. 41–47, Feb. 1990.

dynamics in the compensation network, but the primary objective was to attain good performance with the simplest possible control technique. This acceleration feedback control scheme is very robust to disturbance effects (can maintain end-point position even when the arm is jolted), but, as might be expected, cannot perform nearly as well for significant payload variation. This effect is verified experimentally by having the arm carry a payload weighing 0.67 lb, which is approximately 63 percent of the weight of the arm itself. Figure 1 shows the results of this exercise, where the *same controller gains* utilized in the nominal, no-load case (dashed curve) are employed for the case with payload; the response is to a commanded input, which basically demands that the arm end point follow a square-wave reference. Attempts at designing fixed controllers with more complicated dynamics were only slightly more successful. Indeed, the large overshoot could be avoided, and end-point position accuracy maintained, if the control gains were tuned appropriately. For purposes of comparison, Fig. 2 offers the

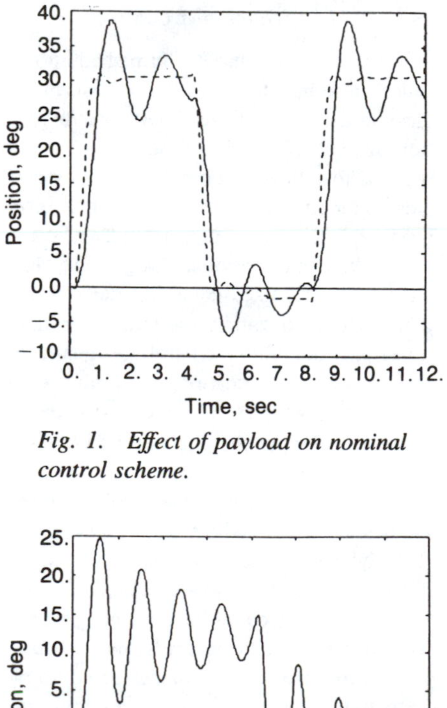

Fig. 1. Effect of payload on nominal control scheme.

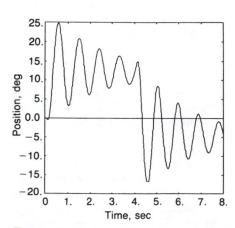

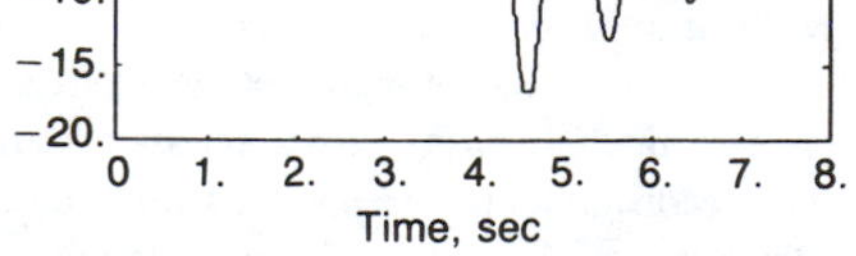

Fig. 2. End-point position with no control applied.

open-loop response, which illustrates the flexibility of the arm, even for a relatively small slew angle (relative to the hub) of only about 15 deg.

Time-Domain Autotuning Control

Identification

Since the control objective considered in this problem involves vibration suppression *after* the manipulator has undergone the nonlinear slew maneuver, a reasonable choice of model structure amenable to control design for the input to shaft angle and the input to end-point acceleration transfer functions (filter) is the Autoregressive Moving Average (ARMA) model. To this end, define θ as the vector of filter coefficients; $w(k)$ is stationary, zero-mean process noise; $y(k)$ and $u(k)$ are the system output and input, respectively; and d is the inherent delay (in sampling time multiples) between the commanded input (for hub torque actuation) and the response seen in the shaft angle or, more noticeably, in the end-point acceleration; the regression vector is then

$$\phi^T(k) = [y(k-1), \ldots, y(k-n), u(k-d-1), \ldots, u(k-d-n)]$$

Within this setting, the model structure takes the form

$$y(k) = \phi^T(k)\theta + w(k) \quad (1)$$

For filter parameter updates we limit our discussion here to the least-squares (LS) and recursive least-squares (RLS) algorithms. Both techniques are based on computation of the optimal value of the parameter vector θ based on minimization of a scalar loss function of the squared equation error. That is, the well-known nonrecursive solution to this procedure for the model (1) is given by the following, where the information matrix $M = [\Phi^T\Phi]^{-1}$ is constructed from the data ϕ, and $\hat{\theta}$ is the estimate of θ:

$$\hat{\theta} = M\Phi^T y \quad (2)$$

Although computationally fast, the amount of data needed for reliable parameter convergence made RLS only slightly faster than a recursive implementation of an *information matrix* form of standard least squares. Therefore, we have chosen to use such a form of LS that tended to give better estimator performance, traded off against computational burden. In the robotic system application we consider here, payload changes are of a discrete nature at a given point in time, implying that there is no requirement for remembering previous load characteristics. Best results then were obtained using a weighted version of the nonrecursive expression

$$M^{-1}(k+1)\,\hat{\theta}(k+1) = \Phi^T(k+1)\,y(k+1) \quad (3)$$

Recursive data information updates can then be made according to

$$M^{-1}(k+1) = \lambda(k)\,M^{-1}(k) + \phi(k+1)\,\phi^T(k+1) \quad (4)$$

$$\Phi^T(k+1)\,y(k+1) = \lambda(k)\,\Phi^T(k)\,y(k) + \phi(k+1) \cdot y(k+1) \quad (5)$$

In the preceding, the weight $\lambda(k)$ is the forgetting factor, and for these applications typically took a value in the range 0.96–0.99.

PID Tuning Controller

Automatic tuning for a proportional-integral-derivative (PID) control law has been given attention in several recent investigations [19]. Motivation for autotuning schemes lies in the fact that oftentimes PID controllers are difficult to tune manually, particularly when a high level of precision is important.

To state the discrete version of the ideal analog PID controller, let T represent the sampling time, $u(k)$ the control input, $e(k)$ the deviation between the controlled signal and a desired reference, and let K_P, K_I, K_D represent the proportional, integral, and derivative gains, respectively. Then the control law takes the form

$$u(k) = K_P e(k) + \frac{T}{K_I}\sum_{i=0}^{k-1} e(i) + \frac{K_D}{T}[e(k) - e(k-1)] \quad (6)$$

A recursive expression for the control input follows from Eq. (6) in the form

$$u(k) - u(k-1) = b_0 e(k) + b_1 e(k-1) + b_2 e(k-2) \quad (7)$$

Choice of the parameters $[b_0, b_1, b_2]$ can follow several design criteria, such as classical pole-placement or pole-cancellation design. However, for this application, it typically is not apparent a priori what the desired closed-loop poles should be, since their choice may depend on the effect of payload variation (with a larger payload, a slower slew maneuver may be required). For this reason, we choose the PID parameters via an optimization of a performance criterion that is *time varying* and weights the control deviation and the end-point acceleration, where $e_s(k)$ represents shaft position error, $\Delta u = u(k) - u(k-1)$, and $\alpha(k)$ is the end-point acceleration:

$$J(k) = \sum_{k=1}^{q} [ke_s^2(k) + 50{,}000(\Delta u)^2 + 6k^2\alpha^2(k)] \quad (8)$$

End-point deflections are weighted more heavily as time increases in this index since, in general, when end-point movement is minimal the shaft position error term is larger than the acceleration term. Moreover, for a given selection of PID parameters, the shaft position error remains virtually the same when a payload is added, due to the fact that a high-gain voltage-to-current amplifier for the hub actuator is employed. However, the end-point acceleration is noticeably reduced with payload and the relative weight of the square of the acceleration drops. For this system, it was observed that end-point oscillations continue for a relatively long period when a payload is added; this accounts for the k^2 factor in Eq. (8). Thus, minimization of Eq. (8) reduces the duration of oscillation. A period of 100 samples (T = 30 msec) was found to be an adequate interval over which to evaluate the performance index.

To illustrate the relationship between the ARMA parameters and the performance criterion [4], represent the z-domain transfer function of Eq. (6) with numerator $N_c(z^{-1})$ (a polynomial in the coefficients b_0, b_1, b_2) and denominator $D_c(z^{-1})$, and introduce the notation for the transformed variables $U(z)$, $Y(z)$, and $E(z)$. Then, for a step reference input,

$$E(z) = Y(z) - \frac{1}{1 - z^{-1}} \quad (9)$$

Using the circumflex notation for estimated variables [estimated coefficients from Eq. (1)], it is easy to show that

$$E(z) = z^{-d}\frac{\hat{N}(z^{-1})}{\hat{D}(z^{-1})}E(z) - \frac{1}{1 - z^{-1}} \quad (10)$$

Moreover,

$$\Delta U(z) = (1 - z^{-1})\,U(z) = N_c(z^{-1})\,E(z) \quad (11)$$

Finally, it follows that

$$E(z) = \frac{\hat{D}(z^{-1})}{D_c(z^{-1})\,\hat{D}(z^{-1}) + N_c(z^{-1})z^{-d}\,\hat{N}(z^{-1})} \quad (12)$$

$$\Delta U(z) = \frac{N_c(z^{-1})\,\hat{D}(z^{-1})}{D_c(z^{-1})\,\hat{D}(z^{-1}) + N_c(z^{-1})z^{-d}\,\hat{N}(z^{-1})} \quad (13)$$

Minimization of Eq. (8) consists of calculating Eqs. (12) and (13) at each step in the optimization scheme. Since the reference acceleration is zero in this application, $\alpha(k)$ may be thought of as an acceleration error or deviation in the sense of $e(k)$ in Eq. (6). Thus, $e_s(k)$ and $\alpha(k)$ in Eq. (8) are compatible with the general formulation given earlier.

The controller design now reduces to computing Eq. (8) and carrying out an optimization over possible PID parameters. It is quite obvious, and easily verifiable via experimental tests, that one way to reduce vibrations at the manipulator end point when a load is added is to reduce the slew rate. This, of course, is viable only to a degree, since our objective is to attempt to slew as fast as possible with the best possible performance.

Several combinations of proportional, PI, PD, or full-PID designs in either or both feedback loops are possible [4]. Here we consider the case for gain adjustment within each feedback path (shaft angle gain and end-point acceleration gain). The effectiveness of this identification/control scheme is illustrated in Fig. 3 for the following profile. First, the arm without payload undergoes a 25-deg slew with no control applied (open loop), where large oscillations are apparent. Tuning takes place at this point and then the arm undergoes a further 20-deg slew in the same direction, then reverses direction for a 45-deg slew. During this phase, the control has been extremely effective in end-point vibration compensation. In the next phase, a 0.415-lb payload is added and the arm undergoes a 20-deg slew; after 5 sec (allowing for damping of deflections), tuning is performed. In the same direction, the arm is then slewed an additional 25 deg, and in 5 sec the payload is removed and the direction is reversed to complete the profile with a 45-deg slew. The points at which controller tuning was performed are marked on the plot; the total time span for identification and tuning, given the limitations of the laboratory computer, is anywhere from 15 to 25 sec depending on the order of the controller (number of parameters tuned in the optimization). Thus, in Fig. 3 the points at which tuning occurs show a break in the position measurement to emphasize the time lapse that actually occurred.

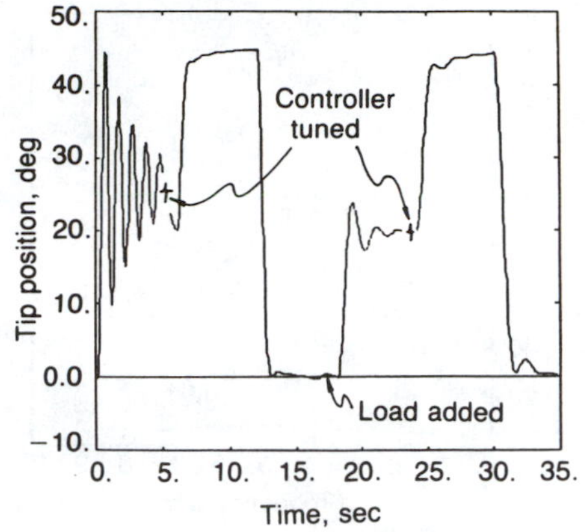

Fig. 3. Time-domain tuning scheme.

Frequency-Domain Autotuning Control

Identification

An alternative to time-domain methods for adaptive filtering is the use of techniques based in the frequency domain [20]–[22]. Frequency-domain adaptive filters enjoy several advantages over their time-domain counterparts, including reduced computation and a fast rate of convergence. A disadvantage of methods that identify system frequency response, however, is that autotuning (on-line) control design is often at best ad hoc.

As an illustration of one of the major shortcomings of the RLS method, consider the case of identifying the end-point acceleration of the one-link apparatus, using zero-mean white noise as input. A typical characteristic of time-domain methods is the requirement for a persistently exciting input during the identification starting process. This was the case, for example, in [7], where after a significant amount of data was gathered the identifier was turned on and the estimated parameters converged "fast" to the actual ones. That is, although the convergence of RLS is superior to most other time-identification methods, many iterations are required for convergence to the actual parameters. Shown in Fig. 4, the estimated transfer-function spectrum of the one-link manipulator is plotted for the cases of 64, 128, 256, and 512 iterations after the start-up of the identification process (30-msec sampling). In all cases, a Butterworth filter of sixth order with a cutoff at 48 rad/sec was used to prefilter the data, the order of the estimated ARMA system was 5 (these values were found to produce the best results [23]), and all the initial estimated parameters were set to zero. RLS can predict the first mode (at approximately 1 Hz) after only 512 iterations, where the corresponding peak begins

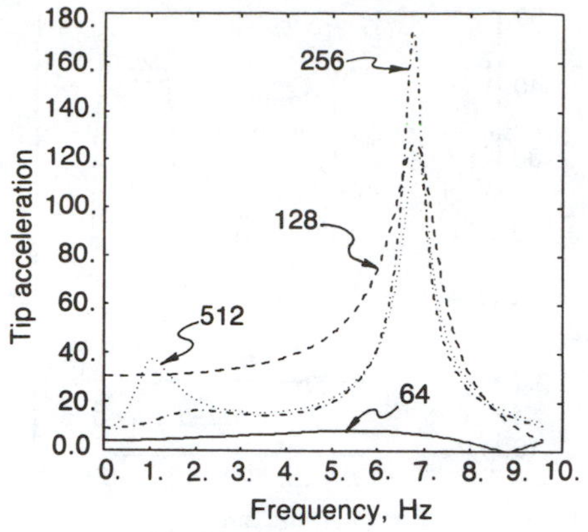

Fig. 4. RLS convergence.

to appear. During convergence, the estimated poles and zeros of the system were far away from the actual ones. In the case of an abrupt change of the carried payload for the manipulator under consideration, the RLS algorithm needed a significant amount of time to converge to the new system parameters. This characteristic was noticed for the identification and control experiments described in the preceding section and may not be satisfactory if the control law update scheme is required to respond quickly.

By contrast, for on-line filter (transfer-function) update, with frequency-domain methods the system input signal is transformed to the frequency domain before adaptive filtering is applied. The simplest frequency-domain adaptive filter is one in which the input signal $u(n)$ and output $y(n)$ are accumulated into buffer memories to form N-point data blocks. These blocks are then transformed by N-point fast Fourier transforms (FFTs) to their equivalent frequency transformed blocks U, Y at the kth time instant.

A simple yet effective representation for transfer-function identification is the Empirical Transfer Function Estimate (ETFE) [24]. A nonrecursive updating scheme for the transfer function $H_i(k)$ [complex conjugate denoted $H_i^*(k)$] at a given time k, for $i \in \{0 \leq i < (N/2),\ U(i) \neq 0\}$, where i corresponds to the ith bin in the frequency domain, is utilized in the following manner:

$$H_i(k) = Y_i(k)/U_i(k) \tag{14a}$$

$$H_{N-i}(k) = H_i^*(k) \tag{14b}$$

Notice that $H_i(k)$ can be updated every l samples, where $1 \leq l < N$. The main properties of this technique are as follows: (1) the variance in the estimation is equal to the signal-to-noise ratio at the frequency under consideration; (2) estimates at different frequencies are uncorrelated (asymptotically as $k \to \infty$).

A recursive implementation of this idea is possible via the Time-Varying Transfer Function Estimation (TTFE) method [15], [25]. The TTFE technique can be used to reduce the variance of the estimated frequency response through two distinguishing characteristics. To illustrate these characteristics, define a modulus Δ_i with a corresponding weight ϵ_j^i (scalar constant) for the frequency point (bin) ω_j. Then the adjacent frequency bins $H_i(k)$, $H_j(k)$ from Eq. (14) can be correlated via the ratio

$$H_i(k) = \left[\sum_{j=(i-\Delta_i)\,\mathrm{mod}\,N}^{(i+\Delta_i)\,\mathrm{mod}\,N} \epsilon_j^i H_j(k)\right] \div \left[\sum_{j=(i-\Delta_i)\,\mathrm{mod}\,N}^{(i+\Delta_i)\,\mathrm{mod}\,N} \epsilon_j^i\right] \tag{15}$$

That is, the estimate $H_i(k)$ is related to all the adjacent frequencies within the modulus Δ_i. The case $\Delta_i = 0$ corresponds to a frequency windowing technique [26] used to reduce the bias and variance of the estimated transfer function. Moreover, the case $\Delta_i = \Delta$ for all i and $\epsilon_j^i = \epsilon^i = \Phi_i(u)$ [where $\Phi_i(u)$ is the input spectral density] corresponds to the Blackman-Tukey Procedure [27] for smoothing the estimated transfer function. Therefore, based on this relation, the estimated transfer function is a "smoothed" version of the one obtained from ETFE.

The second distinguishing feature of TTFE is that the relationship between the frequency bin $H_i(k)$ and the $H_i(k-1), \ldots, H_i(k-\beta_i)$ bins $(i = 1, \ldots, N/2)$, for forgetting factor λ_i $(0 < \lambda_i \leq 1)$ and covariance matrix P_i, is implemented via RLS by

$$H_i(k) = H_i(k-1) + \frac{P_i(k-1)\ U_i(k)}{\lambda_i + P_i(k-1)\ [U_i(k)]^2} \cdot [Y_i(k) - H_i(k-1)\ U_i(k)] \tag{16a}$$

$$P_i(k) = \frac{1}{\lambda_i}\left[P_i(k-1) - \frac{[P_i(k-1)\ U_i(k)]^2}{\lambda_i + P_i(k-1)\ [U_i(k)]^2}\right] \tag{16b}$$

In case of a sudden change of system dynamics, this recursion results in a smooth transient from the old transfer function to the new one, representing a substantial difference when compared with the nonrecursive ETFE technique, which suffers a less smooth transition due to the assumption of orthogonalized input/output data blocks. The computational complexity of TTFE is reasonable and can be decreased by assuming that the frequency bins $H_i(k)$, $H_j(k)$ for the same time instant k are uncorrelated $(\Delta_i = 0)$, over a time period of γ samples, where γ corresponds to the updated interval for the adaptation algorithm.

It would appear that real-time implementation of the TTFE scheme requires frequent recalculation of the discrete Fourier transform (DFT, via the FFT) within the RLS framework. Although this basic operation can be performed quite fast, the computational properties may be enhanced greatly via a recursive relationship in the "moving window" estimation, primarily to allow more time for controller update calculation.

Consider calculation of an N-point FFT. The objective in TTFE is to calculate the input and output FFTs in a time-varying manner as data are collected. Let ω_k denote the kth frequency bin at time $n-1$ and let $W_N^i = e^{-j(2\pi i/N)}$. For data [denoted $x(k)$] up to time $n-1$, the transformed vector variable $X_N^{n-1}(\omega_k)$ has the form (DFT)

$$X_N^{n-1}(\omega_k) = \sum_{l=0}^{N-1} x(l+n-N)\,W_N^{kl} \tag{17}$$

Using this same notation, including the *next data point* and computing the subsequent DFT results in

$$X_N^n(\omega_k) = \sum_{m=0}^{N-1} x(m+n-N+1)\,W_N^{km} \tag{18}$$

However, notice that upon rearrangement and premultiplication by the term W_N^k, it follows that

$$W_N^k X_N^n(\omega_k) + x(n-N)\,W_N^0 - x(n)\,W_N^k W_N^{k(N-1)} = X_N^{n-1}(\omega_k) \tag{19}$$

Intuitively, this operation amounts to adding and subtracting the appropriate term (data point) from the beginning and end of the data sequence. More importantly, expression (19) relates the immediate past DFT, $X_N^{n-1}(\omega_k)$, to the present DFT, $X_N^n(\omega_k)$, allowing for a recursive implementation. Now introduce the notation $\mathcal{R}^n$ and $\mathcal{I}^n$ for the real and imaginary parts of $X_N^n(\omega_k)$ [likewise for $X_N^{n-1}(\omega_k)$]. With this notation, assuming *real data* leads to

$$\mathcal{R}^n = [\mathcal{R}^{n-1} + x(n) - x(n-N)] * \cos\frac{2\pi k}{N} - \mathcal{I}^{n-1} * \sin\frac{2\pi k}{N} \tag{20}$$

$$\mathcal{I}^n = [\mathcal{R}^{n-1} + x(n) - x(n-N)] * \sin\frac{2\pi k}{N} + \mathcal{I}^{n-1} * \cos\frac{2\pi k}{N} \quad (21)$$

From the preceding, it is easy to see that the "current" DFT $X_N^n(\omega_k)$ is formulated in terms of the previous DFT, or, more precisely, $\mathcal{R}^n$ and $\mathcal{I}^n$ are written in terms of $\mathcal{R}^{n-1}$ and $\mathcal{I}^{n-1}$, respectively. This facilitates a simple recursive update in the "sliding window" transfer-function estimation scheme of TTFE. The recursive update requires only four multiplications, explicitly denoted in Eqs. (20) and (21) as "*"; calculation of the sine and cosine terms can be tabulated off-line. As an example, a 1024-point FFT, zero-padded and pruned to 32 output points and 64 input points, would require only 128 real multiplications for the update, given the previous DFT, whereas calculation of the FFT, even with input and output pruning, requires over 3000 real multiplications (over 10,000 without pruning). Note that because this is a recursive scheme, any roundoff and quantization errors will propagate as time progresses. However, the time savings of the scheme allows for occasional recalculation of the entire FFT to "restart" the process.

Controller Tuning

The critical information for control purposes sought by frequency-domain methods is the location of poles and zeros of the transfer function. These locations correspond to the peaks and valleys of the estimated magnitude response. Because of the lightly damped nature of the manipulator, these locations are easily recognizable with the TTFE technique, even with signal-to-noise ratios up to 15 dB [23].

In light of the preceding discussion on convergence of the parameter estimation, the performance of the TTFE approach in estimating the first two modal frequencies of the system is demonstrated via experiment. Consider again the motion of Fig. 3. For such a profile, Fig. 5 depicts the end-point acceleration magnitude plots as identified with the TTFE scheme for various points during the slew motion. The arm initially carries no payload, and the first two modal frequencies are at about 1 Hz (6.3 rad/sec) and 7 Hz (44 rad/sec). At this point, TTFE is able to identify these points accurately, as is RLS due to the high level of excitation. The control is then tuned and the arm is slewed to its maximum angle of 45 deg (second phase), after which time TTFE is still able to place accurately the two modal frequencies; RLS was unable to do so because of the lack of sufficient levels of excitation [23]. After the next phase, the payload is added, resulting in a change in frequencies. Estimates from the TTFE scheme proved to be adequate (for control purposes), in terms of speed of convergence as well as accuracy, when the payload is first added, and after the subsequent 20- and 25-deg slews. Migration of the modal frequencies back up to their position for the unloaded case is evident in the final identification exercise once the payload is removed.

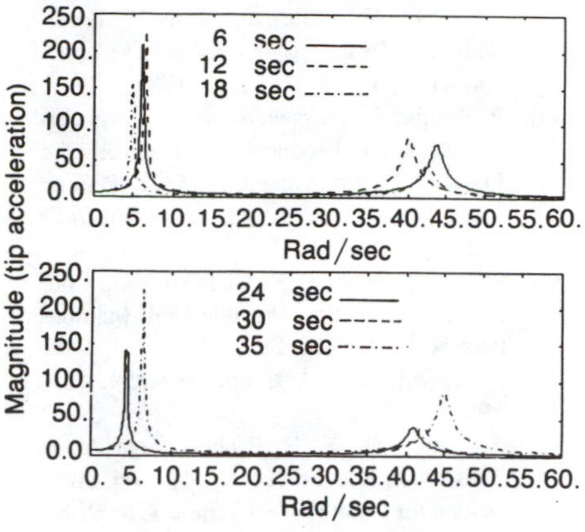

Fig. 5. TTFE estimated spectrum for Fig. 3.

Several algorithms for control design using the identified frequency response directly have been implemented for the experimental setup, including a frequency-weighted quadratic regulator design [15]. Here we describe results of a design in which the control structure is set within a scheduling framework comprised of two feedback loops, one in which the end-point acceleration is used as input to a control law, and the other in which the motor shaft angle is input to a separate control law. These two loops are then summed to give a commanded motor input voltage. Motivated by the desire to achieve end-point position accuracy while maintaining a relatively straightforward implementation structure, simple proportional schemes make up the above-mentioned control laws in the separate loops; it is the individual proportional gains that are scheduled as correlated to frequency-domain information over a wide range of payloads. The scheme discussed in the previous section [PID tuning using Eq. (8)] was used to establish a "look-up" table for various payloads corresponding to the first modal frequency of the arm. That is, the fundamental frequency (first mode) was found using FFT analysis; because these calculations are carried out off-line, an ample amount of data can be accumulated for the best possible accuracy. The motivation for using the fundamental frequency as the "pointer" in a look-up table of scheduled optimal controller parameters is the obvious relationship with the payload. This fact is exploited in the control law by interpolating four such data points (using four different payloads) in construction of a functional relationship, filling out the look-up table, for use in real-time control.

Figure 6 shows results of the scheduling controller using frequency-domain estimation; shown is the arm end-point position as read by the camera. As before, the gross motion control objective is to track a staircase-shaped reference trajectory. At the end of the first and fourth segments, as indicated, the FFT of the end-point acceleration is computed and the controller is tuned according to the estimated frequency of the first mode. The first segment is performed in the open loop for a slew angle of 8 deg; the absence of any control effort is evident from the large overshoot. The controller is then adjusted with the estimated frequency—this operation, including FFT calculation, requires less than 0.3 sec of CPU time on the MicroVax computer. In the next two segments, the arm is slewed another 32 deg, then 40 deg in the opposite direction; the stabilizing effect of the acceleration feedback controller is evident. A payload of 0.74 lb (69 percent of total arm weight) is added at the beginning of the fourth segment, as indicated, and the arm is slewed through a commanded angle of 5 deg. This small level of excitation is sufficient to estimate accurately the first modal frequency, so that the controller is retuned to account for the addition of the payload. In the final two segments, the arm is slewed first another 35 deg, and then 40 deg in the opposite direction. Again, the control has compensated adequately for deflections at the end point. Since the 5-deg slew with payload generates relatively small deflections, to illustrate the effectiveness of the scheduling control the

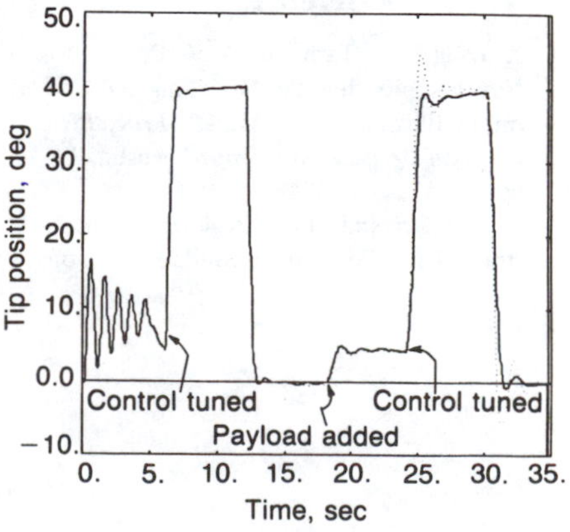

Fig. 6. Scheduling control.

end-point position for the case of *not* retuning the controller after addition of payload is overlayed in the figure (dashed line). Similar results were obtained for a variety of payload conditions.

Conclusion

This paper has presented a summary of the various identification and control techniques being investigated in the laboratory at Ohio State for flexible-link manipulator systems. Primary attention in these techniques focuses on the ability of the controller to adjust to changes in dynamics, payload, and working environment. Time-domain methods offer identified model structures that are readily available for control design, whereas frequency-domain methods, particularly the Time-Varying Transfer Function Estimation approach developed for this application, are more desirable when rapid controller tuning is required. Currently, these techniques are being developed further for multilink flexible manipulators in both theoretical and experimental aspects. As one might imagine, the degree of difficulty is increased greatly over the case of the single rotating link case, primarily due to nonlinearities in the large-angle slews, interaction effects due to the coupled flexible members, and the effects of varying system inertias on fundamental frequencies to be controlled.

Acknowledgments

This work was supported in part under NASA Grant NAG-1-720. The authors wish to acknowledge the contributions of several students in this ongoing effort, particularly Paul T. Kotnik and Fernando E. Pacheco, who were instrumental in generation of the experimental results presented here.

References

[1] X. Ding, T. J. Tarn, and A. K. Bejczy, "A Novel Approach to the Modelling and Control of Flexible Robot Arms," *Proc. IEEE Conf. on Decision and Control*, Austin, TX, pp. 52–57, Dec. 1988.

[2] E. Barbieri and Ümit Özgüner, "Unconstrained and Constrained Mode Expansions for a Flexible Slewing Link," *Trans. ASME, J. Dynam. Syst., Measur., Contr.*, vol. 111, pp. 416-421, Dec. 1988.

[3] P. Kotnik, S. Yurkovich, and Ü. Özgüner, "Acceleration Feedback for Control of a Flexible Manipulator Arm," *J. Robotic Syst.*, vol. 5, no. 3, pp. 181–196, June 1988.

[4] S. Yurkovich and F. E. Pacheco, "On Controller Tuning for a Flexible-Link Manipulator with Varying Payload," *J. Robotic Syst.*, vol. 6, no. 3, pp. 233–254, June 1989.

[5] S. Yurkovich, F. E. Pacheco, and A. P. Tzes, "On-line Frequency Domain Information for Control of a Flexible-Link Robot with Varying Payload," *Proc. IEEE Intl. Conf. on Robotics and Automation*, 1989; also, *IEEE Trans. Automat. Contr.*, vol. AC-33, pp. 876–881, Dec. 1989, Scottsdale, AZ, May 1989.

[6] R. H. Canon and E. Schmitz, "Initial Experiments on the End-Point Control of a Flexible One-Link Robot," *Int. J. Robotics Res.*, vol. 3, no. 3, pp. 62–75, 1984.

[7] D. M. Rovner and R. H. Cannon, "Experiments Toward On-line Identification and Control of a Very Flexible One-Link Manipulator," *Intl. J. Robotics Res.*, vol. 6, no. 4, pp. 3–19, Winter 1987.

[8] D. M. Rovner and G. F. Franklin, "Experiments in Load-Adaptive Control of a Very Flexible One-Link Manipulator," *Automatica*, vol. 24, no. 4, pp. 541–548, July 1988.

[9] E. Schmitz, "Modeling and Control of a Planar Manipulator with an Elastic Forearm," *Proc. IEEE Intl. Conf. on Robotics and Automation*, pp. 894–899, Scottsdale, AZ, May 1989.

[10] C. M. Oakley and R. C. Cannon, "End-Point Control of a Two-Link Manipulator with a Very Flexible Forearm: Issues and Experiments," *Proc. Amer. Contr. Conf.*, pp. 1381-1388, Pittsburgh, PA, June 1989.

[11] G. G. Hastings and W. J. Book, "Experiments in Optimal Control of a Flexible Arm," *Proc. 1985 Amer. Contr. Conf.*, pp. 728–729, Boston, MA, June 1985.

[12] G. G. Hastings and W. J. Book, "Verification of a Linear Dynamic Model for Flexible Robotic Manipulators," *Proc. IEEE Intl. Conf. on Robotics and Automation*, pp. 1024–1029, San Francisco, Apr. 1986.

[13] V. Feliu, K. S. Rattan, and H. B. Brown, "Adaptive Control of a Single-Link Flexible Manipulator in the Presence of Joint Friction and Load Changes," *Proc. IEEE Intl. Conf. on Robotics and Automation*, pp. 1036–1041, Scottsdale, AZ, May 1989.

[14] H. Krishnan and M. Vidyasagar, "Control of a Single-Link Flexible Beam Using a Hankel-Norm-Based Reduced Order Model," *Proc. IEEE Intl. Conf. on Robotics and Automation*, pp. 9–14, Philadelphia, PA, Apr. 1988.

[15] A. Tzes and S. Yurkovich, "A Frequency Domain Identification Scheme for Flexible Structure Control," *Trans. ASME, J. Dynam. Syst., Measur., Contr.*, (to appear).

[16] Ü. Özgüner, S. Yurkovich, J. Martin, and P. Kotnik, "A Laboratory Facility for Flexible Structure Control Experiments," *IEEE Contr. Syst. Mag.*, vol. 8, no. 4, pp. 27–33, Aug. 1988.

[17] S. Yurkovich and Ümit Özgüner, "Recent Developments in the OSU Flexible Structure Control Laboratory," *Proc. 7th VPI & AIAA Symp. on Dynamics and Control of Large Structures*, Blacksburg, VA, May 1989.

[18] A. Tzes and S. Yurkovich, "Adaptive Precompensators for Flexible-Link Manipulator Control," *Proc. IEEE Conf. on Decision and Control*, Tampa, FL, Dec. 1989.

[19] F. Radke and R. Isermann, "A Parameter-Adaptive PID Controller with Stepwise Parameter Optimization," *Automatica*, vol. 23, no. 4, pp. 449–457, 1987.

[20] P. J. Parker and R. R. Bitmead, "Adaptive Frequency Response Identification," *Proc. IEEE Conf. on Decision and Control*, pp. 348-353, Los Angeles, Dec. 1987.

[21] M. Dentino, B. Widrow, and J. McCool, "Adaptive Filtering in the Frequency Domain," *Proc. IEEE*, vol. 66, pp. 1658–1659, Dec. 1978.

[22] R. LaMaire, L. Valavani, M. Athans, and G. Stein, "A Frequency-Domain Estimator for Use in Adaptive Control Systems," *Proc. Amer. Contr. Conf.*, pp. 238–244, Minneapolis, MN, June 1987.

[23] A. Tzes and S. Yurkovich, "Application and Comparison of On-line Identification Methods for Flexible Manipulator Control," *Proc. Intl. Conf. on Advanced Robotics*, Columbus, OH, June 1989.

[24] L. Ljung, *System Identification Theory for the User*, Englewood Cliffs, NJ: Prentice-Hall, 1987.

[25] A. Tzes and S. Yurkovich, "A New Approach to Frequency Domain Identification for Flexible Structure Control," *Proc. IEEE Conf. on Decision and Control*, pp. 1627–1632, Austin, TX, Dec. 1988.

[26] D. R. Brillinger, *Time Series: Data Analysis and Theory*, San Francisco: Holden Day, 1981.

[27] S. M. Kay, *Modern Spectral Estimation*, Englewood Cliffs, NJ: Prentice-Hall, 1988.

Part 9
Dexterous End-Effectors and Grasping

SHANKAR S. SASTRY, University of California, Berkeley and PING HSU, San Jose State University

THE STUDY of dexterous end-effectors for grasping has its roots in prosthetics and indeed a great deal of the inspiration for the study of multifingered robot hands comes from anthropomorphic considerations. In recent years, however, research on the development of multifingered hands has been spurred by the search for intelligent end-effectors and tooling for robotic manipulation and assembly. The literature, however, reflects this dual perspective in that it alternately switches between anthropomorphic and flexible-manufacturing points of view.

Research on the development of a general purpose end-effector for robot manipulators has been active since the 1970s. This trend was motivated because the versatility of a robot is mainly limited by the flexibility of its end-effector. A typical industrial manipulator is equipped with a specially designed grasping tool for performing specific tasks. The cost of special tooling and fixturing is often substantial and this makes a robot manipulator not very different from a conventional machine. Research has, consequently, been focused on the development of a multifingered hand, because such a device is capable of manipulating and securing objects of various shapes. Many multifingered hands have been developed for industrial applications or research (see, for example [1]–[5]). The papers ([6]–[9]) included in this chapter provide a sample of research results in the control and planning area in recent years. Detailed discussions on earlier development and design can be found in, for example, [2] and [10]. The papers included in this survey are by no means exhaustive; rather they are intended to give readers a flavor of the field. We apologize to authors of other important papers that we have been unable to include. Research in tactile sensing, microactuation, and artificial intelligence also contribute to the development of robotic hand systems. These areas will not be discussed here. The following is a brief introduction to issues concerning planning and control of a multifingered hand system.

1. Grasp Planning

The task of securing an object by a multifingered hand involves three key issues: (1) Characterization of contacts and secure grasping of objects, (2) Planning of contact positions and finger configurations, and (3) Determination of internal force and force distribution. The first problem involves fundamental principles in mechanics and kinematics of contacts between rigid bodies. The later two are related to the implementation of a grasp operation.

1.1 Characterization of Contacts and Secure Grasp of Objects

Methods of constraining a rigid body by contacts are well known in the theory of mechanisms for designing mechanical fixtures or jigs [11]. In this theory the ability to secure an object by a set of contacts is referred to as *force-closure condition* (if contact friction is taken into account) or *form-closure condition* (if all contacts are assumed to be frictionless). For example, the typical way of holding a pen is a force-closure grasp but not a form-closure grasp because the pen is free to slide if no friction force is acting between the pen and the fingers. An eyeglass frame holding a noncircular lens is a form-closure grasp because the lens is perfectly immobilized even without contact friction. As the pen-holding example indicates, most assembly tasks can be carried out with a force-closure grasp. In the case that contact friction is relatively small (e.g. holding a wet soap), a form-closure grasp is necessary. Whether a set of contacts is a force closure or not depends on the following factors:

Friction Characteristics. Coulomb's law is commonly used for modeling interaction forces at contacts between fingers and the grasped object. These forces are commonly referred to as *finger forces* or *contact forces*. According to this model, to maintain a contact, an interaction force's tangential component must be less than the product of the force's normal component and a constant (called the *coefficient of static friction*). Because of this proportionality, the set of allowable finger forces form a cone-shaped region (called *friction cone*) in the space of contact forces. In order to maintain the contact, the finger force must be within the interior of this region. In [12], a procedure is proposed to determine the minimum friction coefficient required for a certain task. For example, grasping a basketball in the face-down position with one hand is possible only if the friction coefficient of the basketball surface exceeds a certain critical value.

Contact Geometry. In [2], contacts are classified into a number of types according to their geometry and friction characteristics; for example, a point contact with friction, a line contact without friction, and so on. This classification of contact types is widely used in the research community.

Number of Contacts. It has been shown in [11] that to achieve a form closure of a rigid body, seven contact points are necessary. A force closure of a rigid body requires at least three contact points. These conditions are necessary but not sufficient. For example, a form-closure grasp on a spherical object is impossible since the object is free to rotate regardless of the number of contacts. The minimum number of contacts for securing an object has a direct implication on the design of hands and grasp planning.

Contact Position. The most difficult task in the planning of a grasp is the determination of the contact positions on the object to be grasped. Because of the large number of possible grasp configurations, extensive analysis is required to synthesize an optimal grasp even for a simple object.

1.2 Determination of Contact Positions and Finger Configuration

The task of grasp planning is generally referred to as the determination of the contact positions (on the object to be grasped) and the fingers' postures. Form- and force-closure theory in the theory of mechanisms provides a guideline in the analysis of a grasp. The results of the theory, however, are mostly conceptual and are not easy to use in synthesis. In [13], Nguyen proposes an algorithm for finding a set of force-closure contact points based on the shape of the grasped object. In [14], Mishra et al. propose a procedure for synthesizing a form-closure grasp. From experience we know that a proper grasp configuration should depend not only on the shape of the object but also on the task to be performed. For example, we do not hold a small screwdriver the way that we hold a pen although these two objects are similar in shape. This observation motivates the so-called *task oriented grasp planning* of [9] as explained next.

For rigidly grasping an object, the best grasp will be the one that has all fingers conform and enclose the object (i.e., the 'power grasp' [10]). In such a grasp, the object is completely immobilized, and disturbance force can be easily rejected. If our objective is to manipulate the object, we will grasp the object with our fingertips (these are the so-called *dexterous* or *manipulable* grasps [15]). The disturbance-rejection ability and manipulation ability are directional properties. In a task-oriented grasp planning, high disturbance rejection direction and high manipulation ability are planned as called for by a task. In other words, it synthesizes a grasp that provides the highest mechanical advantage for performing a certain task.

1.3 Determination of Internal Force and Force Distribution

It is intuitive that a hand should slightly squeeze the object it is holding to ensure a firm grasp. Without such a squeezing force, the hand only accommodates the object, rather than grasping it.

As mentioned earlier, finger forces must stay in the friction cone to maintain the contacts. A disturbance action on the object in the tangential direction of a finger contact, for example, cannot be rejected by simply applying an opposing finger force. An additional inward force component must be added to keep the combined finger force in the friction cone. Since the purpose of these inward forces is to cause no object motion, they must result in zero resultant force on the object. This is the reason why a squeezing force is formally referred to as an *internal force*. Many methods have been proposed for internal-force planning.

In [16], the optimal internal force is formulated as the internal force furthest away from the boundary formed by friction cones and finger-actuator torque limits. The distance between the nominal force and these boundaries provides a margin for additional force for accelerating the object or for rejecting disturbances. The magnitude of the optimal force suggested by this approach is proportional to the maximum torque of finger actuators. Such a hard squeezing approach is not desirable if the grasped object is fragile or if there are positional uncertainties as indicated in the following example. If we squeeze too hard, when we grasp a coin by pinching its two opposing edges, the coin collapses to a position in which one's fingers are pressing against the faces. This is because a true internal force is impossible to realize because we always have some degree of uncertainty about the position of the fingers. If the pinching force is low, the net force resulting from it can be compensated by our finger-control mechanism. If a high pinching force is applied, the high net force exceeds our control mechanism's ability to correct it. Consequently, the coin collapses.

To carry out a certain task (e.g., to accelerate an object or to reject a disturbance force), the finger forces must result in an appropriate resultant force on the object. This resultant force is a linear function (i.e., the grasp matrix in [9]) of the finger forces. Typically, the grasp matrix has a nontrivial null space. Consequently, there is a subspace of finger forces corresponding to a given desired resultant force. The procedure of choosing a particular combination of finger forces to produce the desired resultant force is commonly referred to as *force distribution*. Because the null space is precisely the set of internal forces, determining a force distribution and choosing an internal force achieve the same goal, namely, to resolve the finger force at each and every contact.

Without taking into account the friction-cone constraints, the force distribution can be carried out by a generalized inverse of the grasp matrix as suggested in many papers (see, for example, [17], [9]). In [18], a scheme is proposed for finding an optimal finger force in the sense that the angle between each finger force and the inward normal at the corresponding contact is minimized.

In this way, the friction-cone constraint can be satisfied even on a slippery object.

2. Manipulation and Control

Assembly tasks often involve precision but only a small range of motions. Such a maneuver is difficult for a full-size manipulator because positioning accuracy is usually inversely related to the overall dimension of a robot, and large excursions at main joints may be required to produce a relatively small orientation change of the object. With a multifingered hand as end-effector, these maneuvers can be accomplished by the fingers. A general procedure for planning such a maneuver is nearly impossible to derive because of the diversity of the problems that may be involved. Even for a specific task, planning requires extensive analysis, as indicated in [19] and [20].

Typically, a planned parts-matching maneuver is first expressed in the form of a set of desired object trajectories. Based on these trajectories and the kinematics constraints imposed by the contacts, the corresponding finger joint motions can be derived. The derivation procedure is straightforward if all contacts are fixed relative to the object and the fingertips [9]. In a case where fingertips roll or slide on the object, the derivation becomes very complicated. This problem has been investigated by many researchers (see, for example, [6], [16], [21]).

In [16], a set of differential equations describing the relation between the finger-joint motions and the object motion is derived. In [22], a velocity-constraint equation imposed by rolling contacts is developed for formulating a dynamic control law. In [21], both rolling and sliding motions are taken into account in formulating the equation of motion of two rigid bodies in contact. Using concepts from differential geometry, Montana [6] formulated a set of differential equations relating fingers' motions on an object surface to metric properties of the fingers and object. This work is later extended to the case in which the rolling surfaces are not rigid [23]. In [24], the planning procedure takes into account the nonholonomy of kinematic constraints imposed by rolling contacts.

Motion control of each finger can be carried out in exactly the same way as manipulator control. To control a group of fingers manipulating an object is much more difficult than to control a group of manipulators performing separate tasks for the following two reasons: 1) unlike a manipulator, a finger is not rigidly attached to the object and 2) fingers closely interact with each other. Two approaches have been taken: stiffness control and dynamic control. In stiffness control, a desired object stiffness is achieved by controlling the stiffness of each finger joint. The object motion control can be achieved by sending a sequence of intermediate position commands to the stiffness controller. In dynamic control, a complete system dynamic equation is first derived and, based on this equation, a trajectory tracking control law is developed. These two approaches are explained in the following sections.

2.1 Stiffness Control

Stiffness of a mechanical system is a linearized function that relates forces acting on the system and the resulting displacement. Stiffness is usually represented by a 6×6 matrix. Compliance is the inverse of stiffness. Stiffness of a grasped object depends mainly on the finger-joint servo stiffness. A high precision task requires a high stiffness grasp to minimize the unmodeled effects. A high stiffness grasp may, however, result in excessive contact forces (because of collision or jamming) between parts during a parts-matching maneuver. A well-planned object compliance has not only an appropriate magnitude but also a geometrical property that helps the matching of parts and prevents jamming as shown in [25].

In [10], [26], and [27], the grasp stiffness is used in characterizing a stability measure of grasp. Roughly speaking, if the object is perturbed by a small amount and if the changes of finger force (caused by the perturbation) tend to oppose the perturbation, then this grasp is called stable. The coin-pinching grasp described earlier is an unstable grasp [10], if the pinching force is too high and if fingers are not stiff enough. This definition of a stable grasp involves grasp configuration as well as the stiffness of fingers.

2.2 Dynamic Control

In this approach, control laws are developed based on the equation of motion of a hand system. A detailed discussion on this subject can be found in [9] and [28]. The dynamic control of a hand system is similar to the control of a multiple-manipulator system, which has been investigated by many researchers (see, for example, [29], [30]). The difference, however, is that to maintain contacts, the controller of a hand system must keep all finger forces within the friction cones while performing the object-tracking control. In [15], based on a linearized model, a state-dependent gain-scheduling method is used in such a way that the closed-loop system becomes a stable, linear, and state-independent system. In [21], a control scheme is developed for rolling contacts.

3. Summary of the Included Papers

In this collection of papers we have chosen to highlight only conceptual papers rather than those dealing with construction and computer architecture issues for two reasons: (1) the conceptual issues are more in tune with the other chapters of this book and (2) the technology associated with design and fabrication of multifingered hands is changing so rapidly as to render several papers on these topics out of date.

In the paper "The Kinematics of Contact and Grasp" by D. J. Montana, a set of differential equations is formulated for relating fingers' motions on an object surface to metric properties of the fingers and object. The paper uses concepts from differential geometry to rigorize the formulation and yet it is perfectly readable to readers with a background in vector calculus and rigid-body kinematics. The main contribution of the paper is the derivation of the contact equation. The equation relates the velocity of the contact points (expressed in local coordinate systems) to the relative velocity of the two objects. Unlike the equation derived in [21], the equation is valid for sliding contacts, rolling contacts, and contacts with both sliding and rolling motions. Several examples are given in the paper to demonstrate the equation's applications. In one of the examples, the curvature of an object is estimated based on readings taken by rolling a tactile sensor of known curvature on the object.

A procedure for computing the compliance of a grasped object is proposed in the paper "Computing and Controlling the Compliance of a Robotic Hand" by M. R. Cutkosky and I. Kao. In their paper, three types of compliances are considered: (1) compliance of finger-joint servos, (2) finger structural compliance, and (3) compliance caused by change of contact positions. The change of contact positions (e.g., fingers rolling on the object) is caused by the small change of configuration in the compliance analysis. The first two types of compliances are transformed to and combined at the finger tips. Depending on the contact types, some components of these fingertip compliances are reflected to the object. For example, in the case of a point contact, only translational compliance contributes to the object compliance. The overall object compliance is derived by combining all reflected compliances and the third type of compliance. The paper also proposes a procedure (reverse procedure) that gives the compliance of each finger joint so a desired object compliance is achieved.

In this paper "Dynamics and Stability in Coordination of Multiple Robotic Mechanisms" by Y. Nakamura et al., a dynamic control scheme is developed in two steps: 1) a desired resultant force on the grasped object is determined according to a PD-type feedback law and 2) the finger-joint controllers control the finger motion so the kinematic constraints (imposed by the contacts) are satisfied and, at same time, the desired resultant force (as previously determined) is applied to the object.

The paper proposes a procedure for finding the minimal-norm finger force that satisfies the friction-cone constraint and also results in the given resultant force. This force is called *minimal force* in this paper. As indicated in the paper, a minimal force is impractical since it may be close to or right on the edge of a given friction cone. Consequently, any small disturbances may cause the fingers to slip. This observation suggests that a finger force should be placed within the friction cone and with a safety margin of force proportional to the magnitude of the disturbances. This leads to the notion of *contact-stability ellipsoid*. Suppose the object and the fingers are accelerated by a disturbance force acting on the object. The inertial forces experienced by the object when it accelerates the fingers are functions of the acceleration and the fingers' inertia. The set of inertial forces caused by a spherical set of accelerations of a contact point is an ellipsoid and is defined as contact-stability ellipsoid. In other words, the ellipsoid is the set of interaction forces (at the contacts) induced by a bounded set of disturbances. To maintain a contact, a finger force must be such that the sum of the finger force and the contact-stability ellipsoid is contained in the friction cone. The minimum-norm finger force that has such a property and results in the given resultant force is defined to be optimal. A procedure for finding an optimal force is stated in the paper.

In the paper "Grasping and Coordinated Manipulation by a Multifingered Robot Hand" by Z. Li et al., a task-oriented grasp planning and a dynamic control law are proposed. In the paper, a task is characterized in force space and/or velocity space. In force space, a task is characterized by a set of orthogonal vectors representing the expected resistance forces during the execution of the task. For example, in a peg-in-hole task, the resistance force along the positive direction of insertion is expected to be less than that along the remaining directions. Similarly, in velocity space a task is characterized by a set of orthogonal vectors indicating the required mobility for performing the task. For example, the screwing-nut-on-a-bolt task requires a greater rotational mobility along the screw axis. These two sets of vectors are used as principle axes in the definition of *task ellipsoids*. These ellipsoids are reflected to finger-joint space through equivalent transformation matrices. Based on the images of the task ellipsoids and a number of practical considerations (e.g., joint torque and speed limitations), two quality measures of a grasp are defined. A grasp can then be planned optimize these quality measures.

The dynamic control scheme proposed in this paper is developed based on a dynamic equation of a hand-object system in the form of a standard second-order Lagrangian equation with the object's position and orientation as the generalized coordinates. As in a typical Lagrangian system, the input force (i.e., finger-joint torque) is transformed into the generalized force through a state-dependent $6 \times n$ matrix where n is the number of total finger joints. Since the number of total finger joints is generally greater than six, the input matrix has a nontrivial null space. The null space component of the input force controls the internal force (i.e., the grasping force) and its orthogonal component controls the object's motion in the same way as the *computed-torque* control law [31] does.

4. Conclusion

In conclusion, we would like to say that research in dexterous end-effectors and multifingered hands is in its infancy. As of the writing of this introduction, there are a large number of projects aimed at building multifingered robot hands at various scales and for varied reasons. Hands such as the Salisbury hand, the UTAH/MIT Hand, and the Belgrade/USC Hand are some of the older research hands. Other projects are underway at Harvard, University of California at Irvine, New York University, University of Pennsylvania, and at University of California at Berkeley building the next generation of hands with specialized force and tactile sensing in addition to the positioning capability of earlier hands. Newer avenues of application are being sought. For instance, at Berkeley, multifingered hands are being fabricated at the scale of integrated circuit chips of fiber-optic positioning requirements and at the scale of a few millimeters for endoscopic (surgical endoscopy) operating procedures. A lot of the advances needed are those in hardware, computer architecture, motion control languages and also advanced sensors. In the present series of papers we have for the most part chosen not to highlight technological issues but have instead presented those topics we consider to be technology independent, for a conceptual introduction into this extremely promising new area of research in robotics.

References

[1] T. Okada, "Computer control of multi-jointed finger system for precise object handling," *IEEE Trans. on Sys., Man., and Cyber.*, vol. SMC-12, pp. 289–299, May 1982.

[2] M. T. Mason and J. K. Salisbury, *Robot Hand and the Mechanics of Manipulation*, Cambridge, MA: MIT Press, 1985.

[3] S. C. Jacobsen, J. E. Wood, D. F. Knutti, and K. B. Biggers, "The UTAH/MIT dextrous hand: work in progress," *Int. J. Robotics Res.*, vol. 3, no. 4, pp. 21–50, 1984.

[4] J. Demmel, G. Lafferriere, and J. Schwartz, "Theoretical and experimental studies using a multifinger manipulator," *Proc. IEEE Conf. Robotics and Automation*, pp. 390–395, 1988.

[5] R. M. Murray and S. S. Sastry, "Control Experiments in Planar Manipulation and Control," *Proc. IEEE Int. Conf. Robotics and Automation*, pp. 624–629, 1989.

[6] D. J. Montana, "The kinematics of contact and grasp," *Int. J. Robotics Res.*, vol. 7, no. 3, pp. 17–32, June 1988.

[7] M. R. Cutkosky and I. Kao, "Computing and controlling the compliance of a robotic hand," *IEEE Trans. Robotics and Automation*, vol. 5, no. 2, pp. 151–165, April 1989.

[8] Y. Nakamura, K. Nagai, and T. Yoshikawa, "Dynamics and stability in coordination of multiple robotic mechanisms," *Int. J. Robotics Res.*, vol. 8, no. 2, pp. 44–61, Aug. 1989.

[9] Z. Li, P. Hsu, and S. S. Sastry, "Grasping and coordinated manipulation by a multifingered robot hand," *Int. J. Robotics Res.*, vol. 8, no. 4, pp. 33–50, Aug. 1989.

[10] M. R. Cutkosky, *Robotic Grasping and Fine Manipulation*, Boston: Kluwer Academic, 1985.

[11] K. Lakshminarayana, "Mechanics of Form Closure," ASME 78-DET-32, ASME Design Engineering Tech. Conf., 1978.

[12] W. Holzmann and J. M. McCarthy, "Computing the friction forces associated with a three-fingered hand," *IEEE J. Robotics and Automation*, vol. RA-1, no. 4, pp. 206–210, Dec. 1985.

[13] V. Nguyen, "Constructing Force-Closure Grasps," *Int. J. Robotics Res.*, vol. 7, no. 3, pp. 3–16, June 1988.

[14] B. Mishra, J. T. Schwartz, and M. Sharir, "On the existence and synthesis of multifinger positive grips," *Algorithmica*, vol. 2, pp. 541–558, 1987.

[15] H. Kobayashi, "Control and Geometrical Considerations for an Articulated Robot Hand," *Int. J. Robotics Res.*, vol. 4, no. 1, pp. 3–12, 1985.

[16] J. Kerr and B. Roth, "Analysis of Multifingered Hands," *Int. J. Robotics Res.*, vol. 4, no. 4, pp. 3–17, 1986.

[17] V. Kumar and K. Waldron, "Sub-optimal algorithms for force distribution in multifingered grippers," *Proc. IEEE Conf. Robotics and Automation*, pp. 252–257, 1987.

[18] J. Demmel and G. Lafferriere, "Optimal three-finger grasps," *Proc. IEEE Conf. Robotics and Automation*, pp. 936–942, 1989.

[19] R. Fearing, "Implementing a force strategy for object reorientation," *Proc. IEEE Conf. Robotics and Automation*, pp. 96–102, 1986.

[20] D. L. Brock, "Enhancing the dexterity of a robot hand using controlled slip," *Proc. IEEE Conf. Robotics and Automation*, pp. 249–251, 1988.

[21] C. Cai and B. Roth, "On the spatial motion of a rigid body with point contact," *Proc. IEEE Conf. Robotics and Automation*, pp. 686–695, 1987.

[22] A. Cole, J. Hauser, and S. S. Sastry, "Kinematics and control of multifingered hands with rolling contact," *IEEE Trans. Automatic Contr.*, vol. 34, no. 4, pp. 398–404, April 1989.

[23] D. J. Montana, "The kinematics of contact with compliance," *Proc. IEEE Conf. Robotics and Automation*, pp. 770–774, 1989.

[24] Z. Li, J. F. Canny, and S. S. Sastry, "On motion planning for dextrous manipulation part I: the problem formulation," *Proc. IEEE Conf. Robotics and Automation*, pp. 775–780, 1989.

[25] D. E. Whitney, "Quasi-static assembly of compliantly supported rigid parts," *J. Dyn. Sys., Meas., and Cont.*, pp. 65–77, March 1982.

[26] H. Hanafusa and H. Asada, "Stable prehension by a robot hand with elastic fingers," *Proc. 7th Int. Symp. Industrial Robots*, pp. 361–368, 1977.

[27] V. Nguyen, "Constructing stable grasps," *Int. J. Robotics Res.*, vol. 8, no. 1, pp. 26–37, February 1989.

[28] R. Murray and S. S. Sastry "Grasping and manipulation using multifingered robot hands," *Robotics: Proc. Symp. in Appl. Mathematics*, vol. 41, pp. 91–128, 1990, Providence, RI.

[29] T. Tarn, A. Bejczy, and X. Yuan, "Control of two coordinated robots," *Proc. IEEE Conf. Robotics and Automation*, pp. 1193–1202, 1986.

[30] Y. F. Zheng and J. Y. S. Luh, "Optimal load distribution for two industrial robots handing a single object," *Proc. IEEE Conf. Robotics and Automation*, pp. 344–349, 1988.

[31] Mark W. Spong and M. Vidyasagar, *Robot Dynamics and Control*, New York: Wiley, 1989.

David J. Montana
BBN Laboratories, Inc.
70 Fawcett Street
Cambridge, Massachusetts 02138

The Kinematics of Contact and Grasp

Abstract

The kinematics of contact describe the motion of a point of contact over the surfaces of two contacting objects in response to a relative motion of these objects. Using concepts from differential geometry, I derive a set of equations, called the contact equations, that embody this relationship. I employ the contact equations to design the following applications to be executed by an end-effector with tactile sensing capability: (1) determining the curvature form of an unknown object at a point of contact; and (2) following the surface of an unknown object. The contact equations also serve as a basis for an investigation of the kinematics of grasp. I derive the relationship between the relative motion of two fingers grasping an object and the motion of the points of contact over the object surface. Based on this analysis, we explore the following applications: (1) rolling a sphere between two arbitrarily shaped fingers; (2) fine grip adjustment (i.e., having two fingers that grasp an unknown object locally optimize their grip for maximum stability).

1. Introduction

A kinematic relation describes the dependence of one set of motion parameters on another such set due to the geometry and mechanics of the physical world. One prominent example of a kinematic relation is that of the kinematic chain, which is discussed in most texts on robotics, including Craig (1986). A kinematic chain is a coordinate transformation that relates the position and orientation of an end-effector to the joint angles and displacements of the attached manipulator.

The International Journal of Robotics Research,
Vol. 7, No. 3, June 1988,

Another example of a kinematic relation is the grip Jacobian defined in Salisbury (1982). This linear transformation calculates the velocity of an object in the grasp of the fingers of a hand given the velocities of the joints of the fingers.

In this paper I discuss the kinematics of rigid bodies that maintain contact while in relative motion. In particular, I examine the kinematic relation between the relative motion of two objects and the motion of a point of contact over the surfaces of these objects. Investigations of this kinematic relation have previously put simplifying restrictions on the shapes of the objects (e.g., flat, spherical, or two-dimensional) and/or the type of relative motion (pure sliding or pure rolling) (e.g., see Cai and Roth 1986; Kerr and Roth 1986; Bajcsy 1984; Mason 1981). A general description of this kinematic relation has been derived by myself (Montana 1986) and, independently, by Cai and Roth (1987). Using methods from differential geometry, I provide a formulation and solution of the kinematics of contact that is more mathematically rigorous and concise.

The contact equations are the equations that I derive which encapsulate this kinematic relation. Based on the contact equations, I investigate two tasks for a single end-effector with tactile sensing capability. (Tactile sensing is needed because it allows us to measure the position of a point of contact on the end-effector surface [Fearing and Hollerbach 1985].) First, I describe how such an end-effector can determine the curvature form of an unknown object at a point of contact by performing rotational probes and measuring the motion of the point of contact across its own surface. The curvature form of the object is estimated as that which fits these measurements in a least-squares way. Second, I show how to have such an end-effector follow the surface of an unknown object. Tactile data is used to close a loop around the kinematics of contact and steer the point of contact as desired on the end-effector surface. This contour-following algorithm adapts to the unknown and changing curvature of the object. A contour-following scheme based on the kinematics of contact is also presented in Cai and Roth (1987). However, there it is assumed that the curvature of the object is already known, and they are therefore solving a different (and easier) problem.

I also use the contact equations to investigate the kinematics of grasp. This is the problem of manipulating an object with a number of independent end-effectors, usually the fingers of a hand. Most research on mechanical hands has focused on particular hands and/or particular applications (Hanafusa and Asada 1977; Okada 1982). A general theory of manipulation was formulated in Salisbury (1982). Assuming stationary points of contact, Salisbury's grip Jacobian determines the finger joint velocities needed to produce a given velocity of the grasped object relative to the palm. In Kerr and Roth (1986), Salisbury's analysis is extended to allow rolling contact. However, the kinematic relation of interest is still the same. Allowing the points of contact to move just provides extra freedom in how to choose the joint motions to produce a desired object motion. Like Kerr and Roth, I examine grasps with rolling contact, but I derive the kinematic relation between the relative motion of two fingers grasping an object and the motion of the points of contact on the object surface. To do this, I apply the contact equations at each point of contact and perform suitable coordinate transformations to combine the two sets of equations into one.

I use this kinematic relation to investigate a couple of tasks for two fingers. First, I examine the problem of rolling a spherical object between two arbitrarily shaped fingers. This problem reduces to choosing a relative motion of the fingers such that the two points of contact remain diametrically opposed on the object surface. I also investigate the task of fine grip adjustment, showing how two fingers grasping an unknown object can locally optimize their respective points of contact with the object to achieve maximum stability. This is done by iterating on the following two steps: (1) determine the local geometry (position, surface orientation, and curvature) of the object at each point of contact, and (2) move the points of contact to new positions on the object surface so as to improve a certain grip stability criterion.

2. Mathematical Background

In this section I discuss concepts concerning rigid-body motion (Craig 1986) and the geometry of curves and surfaces (Spivak 1979).

Notation 1 Let C_{s_1} and C_{s_2} be two coordinate frames, where s_1 and s_2 are arbitrary subscripts. Then, $\mathbf{p}_{s_2s_1}$ and $R_{s_2s_1}$ denote the position and orientation of C_{s_1} relative to C_{s_2}. Furthermore, $\mathbf{v}_{s_2s_1} = R^{\mathrm{T}}_{s_2s_1}$ and $\Omega_{s_2s_1} = R^{\mathrm{T}}_{s_2s_1}\dot{R}_{s_2s_1}$ are the translational velocity and rotational velocity of C_{s_1} relative to C_{s_2}. The vector form of angular velocity is denoted by $\omega_{s_2s_1}$. For instance, $\mathbf{p}_{21}$, R_{21}, $\mathbf{v}_{21}$, and Ω_{21} describe the motion of a frame named C_1 relative to a frame named C_2. Similarly, $\mathbf{p}_{a_1b}$, R_{a_1b}, $\mathbf{v}_{a_1b}$, and Ω_{a_1b} are the motion parameters of C_b relative to C_{a_1}.

Proposition 1 *Consider three coordinate frames C_1, C_2, and C_3. The following relation exists between their relative velocities:*

$$\begin{aligned} \mathbf{v}_{13} &= R^{\mathrm{T}}_{23}\mathbf{v}_{12} + R^{\mathrm{T}}_{23}\Omega_{12}\mathbf{p}_{23} + \mathbf{v}_{23}, \\ \Omega_{13} &= R^{\mathrm{T}}_{23}\Omega_{12}R_{23} + \Omega_{23}. \end{aligned} \tag{1}$$

Equivalently, in terms of the vector form of angular velocity, we have

$$\begin{aligned} \mathbf{v}_{13} &= R^{\mathrm{T}}_{23}(\mathbf{v}_{12} + \omega_{12} \times \mathbf{p}_{23}) + \mathbf{v}_{23}, \\ \omega_{13} &= R^{\mathrm{T}}_{23}\omega_{12} + \omega_{23}. \end{aligned} \tag{2}$$

Proof: The positions and orientations are composed according to

$$\mathbf{p}_{13} = \mathbf{p}_{12} + R_{12}\mathbf{p}_{23}, \qquad R_{13} = R_{12}R_{23}. \tag{3}$$

Hence, the translational and rotational velocities can be expressed as

$$\begin{aligned} \mathbf{v}_{13} &= R^{\mathrm{T}}_{13}\dot{\mathbf{p}}_{13} = R^{\mathrm{T}}_{23}R^{\mathrm{T}}_{12}(\dot{\mathbf{p}}_{12} + \dot{R}_{12}\mathbf{p}_{23} + R_{12}\dot{\mathbf{p}}_{23}) \\ &= R^{\mathrm{T}}_{23}\mathbf{v}_{12} + R^{\mathrm{T}}_{23}\Omega_{12}\mathbf{p}_{23} + \mathbf{v}_{23}, \end{aligned} \tag{4}$$

$$\begin{aligned} \Omega_{13} &= R^{\mathrm{T}}_{13}\dot{R}_{13} = R^{\mathrm{T}}_{23}R^{\mathrm{T}}_{12}(\dot{R}_{12}R_{23} + R_{12}\dot{R}_{23}) \\ &= R^{\mathrm{T}}_{23}\Omega_{12}R_{23} + \Omega_{23}. \end{aligned} \tag{5}$$

Definition 1.

A *coordinate patch* S_0 for a surface $S \subset \Re^3$ is an open, connected subset of S with the following property: There exists an open subset U of $\Re^2$ and an invertible map $f: U \rightarrow S_0 \subset \Re^3$ such that the partial derivatives $f_u(\mathbf{u})$ and $f_v(\mathbf{u})$ are linearly independent for all $\mathbf{u} = (u, v) \in U$. The pair (f, U) is called a *coordinate system* for S_0. The *coordinates* of a point $s \in S_0$ are $(u, v) = f^{-1}(s)$. A *2-manifold embedded in* $\Re^3$ (which we henceforth call a *manifold*) is a surface $S \subset \Re^3$ that can be written $S = \cup_{i=1}^{n} S_i$, where the S_i's are coordinate patches for S. The set $\{S_i\}_{i=1}^{n}$ is called an *atlas* for S.

Definition 2.

A *Gauss map* (or *normal map*) for a manifold S is a continuous map $g: S \rightarrow S^2 \subset \Re^3$ such that for every $s \in S$, $g(s)$ is perpendicular to S at s. (Recall that S^2 is the unit sphere.) An *orientable* manifold S is one for which a Gauss map exists. When S is the surface of a solid object, we call the Gauss map that points outward the *outward normal map* and the one that points inward the *inward normal map*.

Definition 3.

Consider a manifold S with Gauss map g, a coordinate patch S_0 for S, and a coordinate system (f, U) for S_0. The coordinate system (f, U) is *orthogonal* if $f_u(\mathbf{u}) \cdot f_v(\mathbf{u}) = 0$ for all $\mathbf{u} \in U$. When (f, U) is orthogonal, we can define the *normalized Gauss frame* at a point $\mathbf{u} \in U$ as the coordinate frame with origin at $f(\mathbf{u})$ and coordinate axes

$$\mathbf{x}(\mathbf{u}) = f_u(\mathbf{u})/\|f_u(\mathbf{u})\|, \quad \mathbf{y}(\mathbf{u}) = f_v(\mathbf{u})/\|f_v(\mathbf{u})\|, \quad \mathbf{z}(\mathbf{u}) = g(f(\mathbf{u})). \tag{6}$$

Note that the coordinate axes are functions mapping U to $\Re^3$. We call an orthogonal coordinate system (f, U) *right-handed* if its induced normalized Gauss frame is everywhere right-handed.

NOTE 1 1. For any coordinate patch with an associated Gauss map there exists a right-handed, orthogonal coordinate system.

2. The normalized Gauss frame is an example of what Cartan called a moving frame (Cartan 1946). Cartan used moving frames to define the curvature form and torsion form, and we now adapt his definitions into the present context.

Definition 4.

Consider a manifold S with Gauss map g, coordinate patch S_0, and orthogonal coordinate system (f, U). At a point $s \in S_0$, the *curvature form* K is defined as the 2×2 matrix

$$K = [\mathbf{x}(\mathbf{u}), \mathbf{y}(\mathbf{u})]^{\mathrm{T}}[\mathbf{z}_u(\mathbf{u})/\|f_u(\mathbf{u})\|, \mathbf{z}_v(\mathbf{u})/\|f_v(\mathbf{u})\|], \tag{7}$$

where $\mathbf{u} = f^{-1}(s)$. The *torsion form* T at s is the 1×2 matrix

$$T = \mathbf{y}(\mathbf{u})^{\mathrm{T}}[\mathbf{x}_u(\mathbf{u})/\|f_u(\mathbf{u})\|, \mathbf{x}_v(\mathbf{u})/\|f_v(\mathbf{u})\|]. \tag{8}$$

We define the *metric* M at s as the 2×2 diagonal matrix

$$M = \mathrm{diag}(\|f_u(\mathbf{u})\|, \|f_v(\mathbf{u})\|). \tag{9}$$

Our metric is the square root of the Riemannian metric (Spivak 1979).

EXAMPLE 1 Consider the set

$$U = \{(u, v) | -\pi/2 < u < \pi/2, -\pi < v < \pi\} \tag{10}$$

and the map

$$f: U \rightarrow \Re^3, \quad (u, v) \mapsto (R \cos u \cos v, -R \cos u \sin v, R \sin u) \tag{11}$$

for some $R > 0$. Let $S_0 = f(U)$. The reader can verify that (f, U) is a coordinate system for S_0. Let S be the sphere of radius R. Then S_0 is a coordinate patch for S. The coordinates u and v are known as the latitude and longitude, respectively. We can define another map

$$\tilde{f}: U \rightarrow \Re^3, \quad (u, v) \mapsto (-R \cos u \cos v, R \sin u, R \cos u \sin u). \tag{12}$$

Let $\tilde{S}_0 = \tilde{f}(U)$. Then $\{S_0, \tilde{S}_0\}$ is an atlas for S. Hence,

Fig. 1. The coordinate frames at time t *(with $\tau > 0$).*

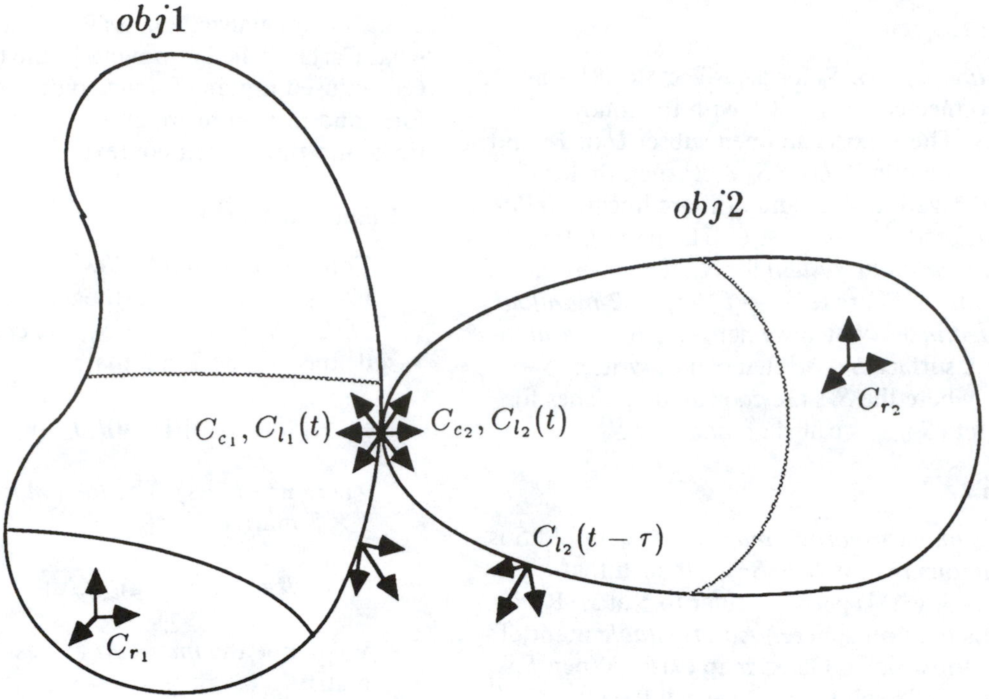

S is a manifold. If we view the sphere as the surface of a ball, then the outward normal map is

$$g: S \rightarrow S^2, \qquad \mathbf{v} \mapsto (1/R)\mathbf{v}. \tag{13}$$

With this normal map, (f, U) is right-handed. It can be shown that (f, U) is an orthogonal coordinate system. Therefore, the normalized Gauss frame exists for all $\mathbf{u} = (u, v) \in U$. Its coordinate vectors are

$$\mathbf{x}(\mathbf{u}) = \begin{bmatrix} -\sin u \cos v \\ \sin u \sin v \\ \cos u \end{bmatrix}, \quad \mathbf{y}(\mathbf{u}) = \begin{bmatrix} -\sin v \\ -\cos v \\ 0 \end{bmatrix}, \tag{14}$$
$$\mathbf{z}(\mathbf{u}) = \begin{bmatrix} \cos u \cos v \\ -\cos u \sin v \\ \sin u \end{bmatrix}.$$

On the spherical surface of the earth, the x-, y-, and z-directions are called north, west, and up, respectively. The curvature form, torsion form, and metric are

$$K = \begin{bmatrix} 1/R & 0 \\ 0 & 1/R \end{bmatrix}, \quad T = \begin{bmatrix} 0 & \frac{-\tan u}{R} \end{bmatrix}, \tag{15}$$
$$M = \begin{bmatrix} R & 0 \\ 0 & R\cos u \end{bmatrix}.$$

3. The Kinematics of Contact

We now consider two rigid objects that move while maintaining contact with each other. Rigid bodies will generally make contact at isolated points rather than over areas of their surfaces. In this section we investigate the motion of one of these points of contact across the surfaces of the objects in response to a relative motion of the objects.

Call the objects obj 1 and obj 2. Choose reference frames C_{r_1} and C_{r_2} fixed relative to obj 1 and obj 2, respectively. Let $S_1 \subset \Re^3$ and $S_2 \subset \Re^3$ be the embeddings of the surfaces of obj 1 and obj 2 relative to C_{r_1} and C_{r_2}, respectively. Surfaces S_1 and S_2 are orientable manifolds.Let g_1 and g_2 be the outward normal maps for S_1 and S_2. Choose atlases $\{S_{1_i}\}_{i=1}^{n_1}$ and $\{S_{2_j}\}_{j=1}^{n_2}$ for S_1 and S_2. Let (f_{1_i}, U_{1_i}) be an orthogonal, right-handed coordinate system for S_{1_i} with normal map g_1. Similarly, let (f_{2_j}, U_{2_j}) be an orthogonal, right-handed coordinate system for S_{2_j} with g_2.

Let $c_1(t) \in S_1$ and $c_2(t) \in S_2$ be the positions at time t of the point of contact relative to C_{r_1} and C_{r_2}, respectively. In general, $c_1(t)$ will not remain in a single coordinate patch of the atlas $\{S_{1_i}\}_{i=1}^{n_1}$ for all time, and likewise for $c_2(t)$ and the atlas $\{S_{2_j}\}_{j=1}^{n_2}$. Therefore, we restrict our attention to an interval I such that $c_1(t) \in S_{1_i}$ and $c_2(t) \in S_{2_j}$ for all $t \in I$ and some i and j. The coordinate systems (f_{1_i}, U_{1_i}) and (f_{2_j}, U_{2_j}) induce a normalized Gauss frame at all points in S_{1_i} and S_{2_j}. We define the contact frames, C_{c_1} and C_{c_2} as the coordinate frames that coincide with the normalized Gauss frames at $c_1(t)$ and $c_2(t)$, respectively, for all $t \in I$. We also define a continuous family of coordinate frames, two for each $t \in I$, as follows. Let the local frames at time t, $C_{l_1}(t)$ and $C_{l_2}(t)$, be the coordinate frames fixed relative to C_{r_1} and C_{r_2}, respectively, that coincide at time t with the normalized Gauss frames at $c_1(t)$ and $c_2(t)$ (see Fig. 1).

We now define the parameters that describe the 5 degrees of freedom for the motion of the point of contact. The coordinates of the point of contact relative to the coordinate systems (f_{1_i}, U_{1_i}) and (f_{2_j}, U_{2_j}) are given by $\mathbf{u}_1(t) = f_{1_i}^{-1}(c_1(t)) \in U_{1_i}$ and $\mathbf{u}_2(t) = f_{2_j}^{-1}(c_2(t)) \in U_{2_j}$. These account for 4 degrees of freedom. The final parameter is the angle of contact $\psi(t)$, which is defined as the angle between the x-axes of C_{c_1} and C_{c_2}. We choose the sign of ψ so that a rotation of C_{c_1} through angle $-\psi$ around its z-axis aligns the x-axes.

We describe the motion of obj 1 relative to obj 2 at time t, using the local coordinate frames $C_{l_1}(t)$ and $C_{l_2}(t)$. Let v_x, v_y, and v_z be the components of translational velocity of $C_{l_1}(t)$ relative to $C_{l_2}(t)$ at time t. Similarly, let ω_x, ω_y, and ω_z be the components of rotational velocity. Then v_x, v_y, v_z, ω_x, ω_y, and ω_z provide the 6 degrees of freedom for the relative motion between the objects (see Fig. 2).

The symbols K_1, T_1, and M_1 represent, respectively, the curvature form, torsion form, and metric at time t at the point $c_1(t)$ relative to the coordinate system (f_{1_i}, U_{1_i}). We can analogously define K_2, T_2, and M_2. We also let

$$R_\psi = \begin{bmatrix} \cos\psi & -\sin\psi \\ -\sin\psi & -\cos\psi \end{bmatrix}, \qquad \tilde{K}_2 = R_\psi K_2 R_\psi. \quad (16)$$

Note that R_ψ is the orientation of the x- and y-axes of C_{c_1} relative to the x- and y-axes of C_{c_2}. Hence, $\tilde{K}_2$ is the curvature of obj 2 at the point of contact relative to the x- and y-axes of C_{c_1}. Call $K_1 + \tilde{K}_2$ the *relative curvature form.*

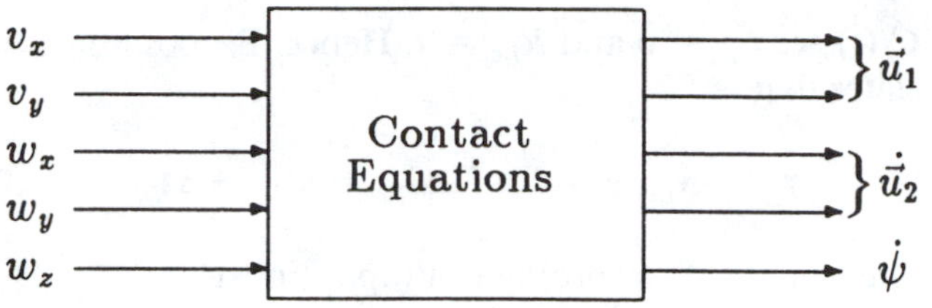

Fig. 2. Sliding contact.

THEOREM 1 *At a point of contact, if the relative curvature form is invertible, then the point of contact and angle of contact evolve according to*

$$\dot{\mathbf{u}}_1 = M_1^{-1}(K_1 + \tilde{K}_2)^{-1}\left(\begin{bmatrix} -\omega_y \\ \omega_x \end{bmatrix} - \tilde{K}_2 \begin{bmatrix} v_x \\ v_y \end{bmatrix}\right), \quad (17)$$

$$\dot{\mathbf{u}}_2 = M_2^{-1} R_\psi (K_1 + \tilde{K}_2)^{-1}\left(\begin{bmatrix} -\omega_y \\ \omega_x \end{bmatrix} + K_1 \begin{bmatrix} v_x \\ v_y \end{bmatrix}\right), \quad (18)$$

$$\dot{\psi} = \omega_z + T_1 M_1 \dot{\mathbf{u}}_1 + T_2 M_2 \dot{\mathbf{u}}_2, \quad (19)$$

$$0 = v_z. \quad (20)$$

Proof: Recall the notation introduced in Notation 1. Since $C_{l_1}(t)$ is fixed relative to C_{r_1}, the velocity at time t of $C_{l_1}(t)$ relative to C_{r_1} is given by $\mathbf{v}_{r_1 l_1} = 0$ and $\Omega_{r_1 l_1} = 0$. Therefore, according to Proposition 1,

$$\mathbf{v}_{r_1 c_1} = \mathbf{v}_{l_1 c_1}, \qquad \Omega_{r_1 c_1} = \Omega_{l_1 c_1}. \quad (21)$$

Similarly, we find that

$$\mathbf{v}_{r_2 c_2} = \mathbf{v}_{l_2 c_2}, \qquad \Omega_{r_2 c_2} = \Omega_{l_2 c_2}. \quad (22)$$

At time t the position and orientation of C_{c_1} relative to

$C_{l_1}(t)$ are $\mathbf{r}_{l_1c_1} = 0$ and $R_{l_1c_1} = I$. Hence, Proposition 1 states that

$$\mathbf{v}_{l_2c_1} = \mathbf{v}_{l_1c_1} + \mathbf{v}_{l_2l_1}, \qquad \Omega_{l_2c_1} = \Omega_{l_1c_1} + \Omega_{l_2l_1}. \tag{23}$$

Since $\mathbf{p}_{c_2c_1} = 0$, according to Proposition 1,

$$\begin{aligned} \mathbf{v}_{l_2c_1} &= \mathbf{v}_{c_2c_1} + R^{\mathrm{T}}_{c_2c_1}\mathbf{v}_{l_2c_2}, \\ \Omega_{l_2c_1} &= \Omega_{c_2c_1} + R^{\mathrm{T}}_{c_2c_1}\Omega_{l_2c_2}R_{c_2c_1}. \end{aligned} \tag{24}$$

Combining Eqs. (21–24) yields

$$\mathbf{v}_{r_1c_1} + \mathbf{v}_{l_2l_1} = \mathbf{v}_{c_2c_1} + R^{\mathrm{T}}_{c_2c_1}\mathbf{v}_{r_2c_2}, \tag{25}$$

$$\Omega_{r_1c_1} + \Omega_{l_2l_1} = \Omega_{c_2c_1} + R^{\mathrm{T}}_{c_2c_1}\Omega_{r_2c_2}R_{c_2c_1}. \tag{26}$$

We now find the values of each of the quantities in Eqs. (25) and (26) in terms of the contact parameters and motion parameters. To start, we observe that

$$\begin{aligned} R_{c_2c_1} &= \begin{bmatrix} R_\psi & 0 \\ 0 & -1 \end{bmatrix}, \quad \mathbf{v}_{c_2c_1} = 0, \\ \Omega_{c_2c_1} &= \begin{bmatrix} 0 & -\dot\psi & 0 \\ \dot\psi & 0 & 0 \\ 0 & 0 & 0 \end{bmatrix}. \end{aligned} \tag{27}$$

By the definition for v_x, v_y, v_z, ω_x, ω_y, and ω_z we gave above,

$$\mathbf{v}_{l_2l_1} = \begin{bmatrix} v_x \\ v_y \\ v_z \end{bmatrix}, \qquad \Omega_{l_2l_1} = \begin{bmatrix} 0 & -\omega_z & \omega_y \\ \omega_z & 0 & -\omega_x \\ -\omega_y & \omega_x & 0 \end{bmatrix}. \tag{28}$$

To examine the motion of C_{c_1} relative to C_{r_1}, let $\mathbf{x}_1(\mathbf{u}_1)$, $\mathbf{y}_1(\mathbf{u}_1)$, and $\mathbf{z}_1(\mathbf{u}_1)$ be the coordinate vectors of the normalized Gauss frame for obj 1 at the point $\mathbf{u}_1 \in U_{1i}$. Then,

$$\begin{aligned} \mathbf{p}_{r_1c_1} &= c_1(t) = f_{1_i}(\mathbf{u}_1(t)), \\ R_{r_1c_1} &= [\mathbf{x}_1(\mathbf{u}_1(t)), \mathbf{y}_1(\mathbf{u}_1(t)), \mathbf{z}_1(\mathbf{u}_1(t))], \end{aligned} \tag{29}$$

$$\begin{aligned} \mathbf{v}_{r_1c_1} &= R^{\mathrm{T}}_{r_1c_1}\dot{\mathbf{p}}_{r_1c_1} = [\mathbf{x}_1(\mathbf{u}_1), \mathbf{y}_1(\mathbf{u}_1), \mathbf{z}_1(\mathbf{u}_1)]^{\mathrm{T}} \\ &\times [(f_{1_i})_u(\mathbf{u}_1), (f_{1_i})_v(\mathbf{u}_1)]\dot{\mathbf{u}}_1 = \begin{bmatrix} M\dot{\mathbf{u}}_1 \\ 0 \end{bmatrix}, \end{aligned} \tag{30}$$

$$\Omega_{c_1r_1} = R^{\mathrm{T}}_{c_1r_1}\dot{A}_{c_1r_1} \tag{31}$$

$$\begin{aligned} &= [\mathbf{x}_1, \mathbf{y}_1, \mathbf{z}_1]^{\mathrm{T}} \\ &\times [[(\mathbf{x}_1)_u, (\mathbf{x}_1)_v]\dot{\mathbf{u}}_1, [(\mathbf{y}_1)_u, (\mathbf{y}_1)_v]\dot{\mathbf{u}}_1, [(\mathbf{z}_1)_u, (\mathbf{z}_1)_v]\dot{\mathbf{u}}_1] \end{aligned} \tag{32}$$

$$= \begin{bmatrix} 0 & -T_{1M_1}\dot{\mathbf{u}}_1 & K_1M_1\dot{\mathbf{u}}_1 \\ T_1M_1\dot{\mathbf{u}}_1 & 0 & \\ -(K_1M_1\dot{\mathbf{u}}_1)^{\mathrm{T}} & & 0 \end{bmatrix}. \tag{33}$$

We similarly find that

$$\begin{aligned} \mathbf{v}_{r_2c_2} &= \begin{bmatrix} M\dot{\mathbf{u}}_2 \\ 0 \end{bmatrix}, \\ \Omega_{r_2c_2} &= \begin{bmatrix} 0 & -T_2M_2\dot{\mathbf{u}}_2 & K_2M_2\dot{\mathbf{u}}_2 \\ T_2M_2\dot{\mathbf{u}}_2 & 0 & \\ -(K_2M_2\dot{\mathbf{u}}_2)^{\mathrm{T}} & & 0 \end{bmatrix}. \end{aligned} \tag{34}$$

Substituting Eqs. (27), (28), (30), (33), and (34) into Eqs. (25) and (26) and equating components, we get

$$M_1\dot{\mathbf{u}}_1 + \begin{bmatrix} v_x \\ v_y \end{bmatrix} = M_2\dot{\mathbf{u}}_2, \tag{35}$$

$$v_z = 0, \tag{36}$$

$$K_1M_1\dot{\mathbf{u}}_1 + \begin{bmatrix} \omega_y \\ -\omega_x \end{bmatrix} = -R_\psi K_2M_2\dot{\mathbf{u}}_2, \tag{37}$$

$$T_1M_1\dot{\mathbf{u}}_1 + \omega_z = \dot\psi - T_2M_2\dot{\mathbf{u}}_2. \tag{38}$$

After some algebraic manipulation, we can write Eqs. (35–38) in the form given in Eqs. (17–20).

We call Eqs. (17)–(19) the first, second, and third *contact equations* respectively. We call Eq. (20) the *kinematic constraint of contact* because it expresses the constraint on the relative motion necessary to maintain contact.

NOTE 2 For some of the applications discussed below, obj 2 will be an object of unknown shape. Hence, we will not be able to choose a coordinate system for it. We therefore now re-express the second contact equation in a form that is independent of the coordinate system chosen for obj 2. (The first contact equation is already in such a form.) Define $\dot{\tilde{s}}_2 = R_\psi M_2\dot{\mathbf{u}}_2$. Then, $\dot{\tilde{s}}_2$ is the rate at which the point of contact traverses arc length across the surface of obj 2 as measured relative to the x- and y-axes of the local

Fig. 3. Rolling without slipping.

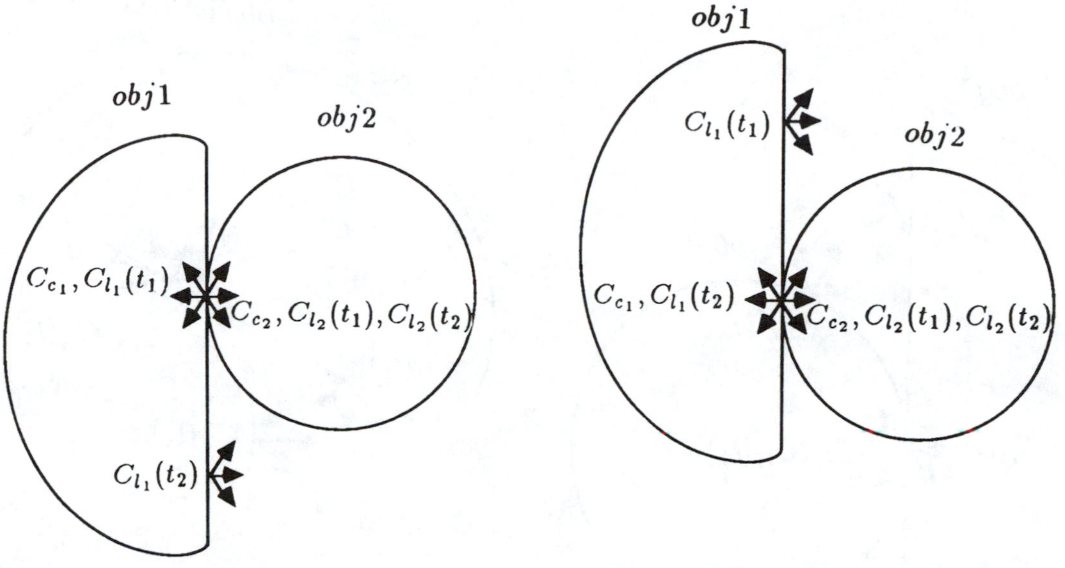

(a) before: $t = t_1$ (b) after: $t = t_2$

coordinate frame of obj 1. This quantity is independent of the coordinate system chosen for obj 2. Substituting into the second contact equation gives

$$\dot{\mathbf{s}}_2 = (K_1 + \tilde{K}_2)^{-1}\left(\begin{bmatrix} -\omega_y \\ \omega_x \end{bmatrix} + K_1 \begin{bmatrix} v_x \\ v_y \end{bmatrix}\right). \qquad (39)$$

EXAMPLE 2 Let obj 1 be an object whose surface has a planar coordinate patch. Choosing a Cartesian coordinate system for this coordinate patch yields $K_1 = 0$, $T_1 = 0$, and $M_1 = I$ at all points. Let obj 2 be a unit ball. Using the coordinate patch investigated in Example 1 gives values for the curvature form, torsion form, and metric of $K_2 = I$, $T_2 = [0, -\tan u]$, and $M_2 = \text{diag}(1, \cos u)$. Let obj 1 and obj 2 be oriented so that at time t_0 the x-axis of $C_{l_1}(t)$ coincides with the x-axis of $C_{l_2}(t)$. Then at time t_0, $R_\psi = \text{diag}(1, -1)$, and the contact equations are

$$\dot{\mathbf{u}}_1 = \begin{bmatrix} -\omega_y - v_x \\ \omega_x - v_y \end{bmatrix}, \quad \dot{\mathbf{u}}_2 = \begin{bmatrix} -\omega_y \\ -\omega_x \sec u_2 \end{bmatrix}, \quad \dot{\psi} = \omega_z - \omega_x \tan u_2, \qquad (40)$$

where $\mathbf{u}_2 = [u_2, v_2]^{\mathrm{T}}$.

When there is sliding contact, $\omega_x = \omega_y = \omega_z = 0$. Therefore, Eq. (40) becomes

$$\dot{\mathbf{u}}_1 = \begin{bmatrix} -v_x \\ -v_y \end{bmatrix}, \quad \dot{\mathbf{u}}_2 = 0, \quad \dot{\psi} = 0. \qquad (41)$$

This motion is pictured in Fig. 3.

When the relative motion is rolling without slipping, $v_x = v_y = \omega_z = 0$. Hence, Equation (40) is

$$\dot{\mathbf{u}}_1 = \begin{bmatrix} -\omega_y \\ \omega_x \end{bmatrix}, \quad \dot{\mathbf{u}}_2 = \begin{bmatrix} -\omega_y \\ -\omega_x \sec u_2 \end{bmatrix}, \quad \dot{\psi} = \omega_x \tan u_2. \qquad (42)$$

This motion is pictured in Fig. 4.

When the relative motion is rotation around the normal, $\omega_x = \omega_y = v_x = v_y = 0$. Then Eq. (40) becomes

$$\dot{\mathbf{u}}_1 = 0, \quad \dot{\mathbf{u}}_2 = 0, \quad \dot{\psi} = \omega_z. \qquad (43)$$

For such motion the point of contact is fixed on both surfaces, and only the angle of contact changes.

Fig. 4. The coordinate frames at time t.

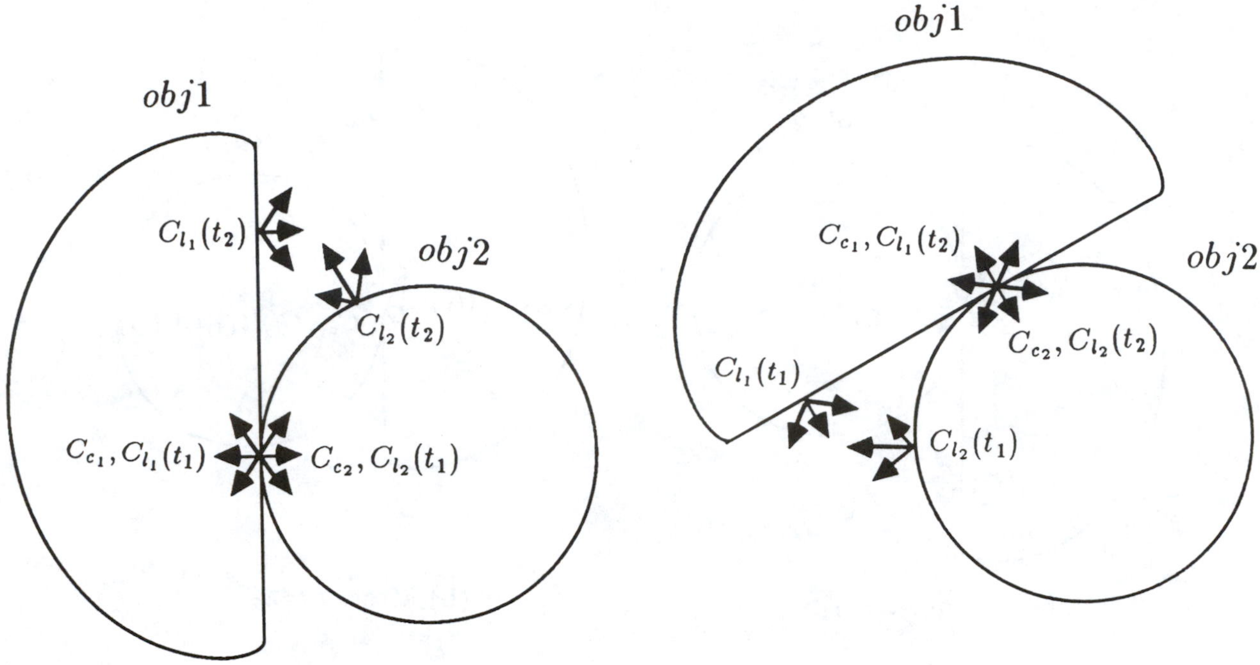

(a) before: $t = t_1$ (b) after: $t = t_2$

4. Application 1: Finding Curvature

Let obj 1 be a tactile sensor attached to a manipulator, and let obj 2 be an object of unknown shape. Assume that there is a single point of contact between them. We now discuss how to determine $\tilde{K}_2$, the curvature form of the unknown object at the point of contact, through a series of experiments. The ith experiment consists of rotating the sensor without slippage relative to the object through a small angle $[\Delta\theta_{x_i}, \Delta\theta_{y_i}, 0]^T$. Assume that the point of contact remains in one coordinate patch for the experiments. Then the tactile sensor can measure the resulting change in the coordinates of the point of contact on its surface $\Delta\mathbf{u}_{1_i}$. Since the inverse of the relative curvature form is symmetric, we can write it as

$$(K_1 + \tilde{K}_2)^{-1} = \begin{bmatrix} k_{r_1} & k_{k_2} \\ k_{r_2} & k_{r_3} \end{bmatrix}.$$

Because the shape of the sensor and the chosen coordinate system are known and the coordinates of the point of contact on the sensor surface can be measured, we can compute M_1 and K_1.

PROPOSITION 2 *Consider n such rotational probes. The values of* k_{r_1}, k_{r_2}, and k_{r_3} *that minimize the sum of the squares of the errors in the measurements of* $M_i\,\Delta\mathbf{u}_{1_i}$ *are given by*

$$\begin{bmatrix} k_{r_1} \\ k_{r_2} \\ k_{r_3} \end{bmatrix} = (A^T A)^{-1} A^T B, \qquad (44)$$

where A and B are defined as

$$A = \begin{bmatrix} -\Delta\theta_{y_1} & \Delta\theta_{x_1} & 0 \\ 0 & -\Delta\theta_{y_1} & \Delta\theta_{x_1} \\ \cdots & \cdots & \cdots \\ -\Delta\theta_{y_n} & \Delta\theta_{x_n} & 0 \\ 0 & -\Delta\theta_{y_n} & \Delta\theta_{x_n} \end{bmatrix}, \quad B = \begin{bmatrix} M_1\Delta\mathbf{u}_{1_1} \\ \cdots \\ M_1\Delta\mathbf{u}_{1_n} \end{bmatrix}. \qquad (45)$$

Proof: According to the first contact equation,

$$M_1 \Delta \mathbf{u}_{1_i} + e_i = (K_1 + \tilde{K}_2)^{-1} \begin{bmatrix} -\Delta\theta_{y_i} \\ \Delta\theta_{x_i} \end{bmatrix}, \quad (46)$$

where e_i is the error in the measurement of $M_1 \Delta \mathbf{u}_{1_i}$. This can be rewritten as

$$\begin{bmatrix} -\Delta\theta_{y_i} & \Delta\theta_{x_i} & 0 \\ 0 & -\Delta\theta_{y_i} & \Delta\theta_{x_i} \end{bmatrix} \begin{bmatrix} k_{r_1} \\ k_{r_2} \\ k_{r_3} \end{bmatrix} = M_1 \Delta \mathbf{u}_{1_i} + e_i. \quad (47)$$

Combining the results of all n experiments gives

$$A \begin{bmatrix} k_{r_1} \\ k_{r_2} \\ k_{r_3} \end{bmatrix} = B + \begin{bmatrix} e_1 \\ \cdots \\ e_n \end{bmatrix}. \quad (48)$$

The value of $[k_{r_1}, k_{r_2}, k_{r_3}]^{\mathrm{T}}$ that minimizes the square of the error term is as shown in Eq. (44) (Campbell and Meyer 1979).

Given the inverse of the relative curvature form, we can solve for the curvature form of the unknown object as

$$\tilde{K}_2 = \begin{bmatrix} k_{r_1} & k_{r_2} \\ k_{r_2} & k_{r_3} \end{bmatrix}^{-1} - K_1. \quad (49)$$

5. Application 2: Contour Following

Take obj 1 to be an end-effector attached to a manipulator. Let obj 2 be some arbitrary object of unknown shape fixed relative to the base of the manipulator. We assume that the two objects meet at a single point of contact. We specify that the end-effector has tactile-sensing capability. With tactile sensing it is possible to measure the position of the point of contact on the surface of the end-effector. We also assume that we have proprioceptive sensors to measure the velocity of the end-effector relative to its base and hence relative to the fixed object.

In this section, we describe a closed-loop servosystem that drives the end-effector to steer the point of contact to some desired location on its own surface while following the surface of the unknown object. The main problem in designing such a servosystem is that the contact equations depend on the curvature form of the object whose shape is unknown. Our servosystem adapts to the changing shape of the unknown object and provides a partial estimate of its curvature form.

We start by choosing one coordinate patch on the surface of the end-effector in which we try to maintain the point of contact. (For human fingers, this coordinate patch would be the fingertip.) This allows us to always specify the position of the point of contact on the end-effector by its coordinates in this coordinate patch.

We assume that we can command the manipulator to produce any desired values for $\dot{v}_x$ $\dot{v}_y$, $\dot{w}_x$, $\dot{w}_y$, and $\dot{w}_z$. Let the velocity parameter C be an arbitrarily chosen two-vector. Choose the set point $\mathbf{u}_s$ to be a two-vector, which is the coordinates of some point in the selected coordinate patch for the end-effector. Define

$$e_1 = \begin{bmatrix} -\omega_y \\ \omega_x \end{bmatrix} + K_1 \begin{bmatrix} v_x \\ v_y \end{bmatrix} - C, \qquad e_2 = \mathbf{u}_1 - \mathbf{u}_s. \quad (50)$$

Let $(e_1)_m$, $(e_2)_m$, and $(\dot{e}_2)_m$ be the measured values of e_1, e_2, and $\dot{e}_2$, respectively.

Proposition 3 *If $\dot{K}_1 \approx 0$ and $\dot{C} \approx 0$ (i.e., K_1 and C are quasi-static), then the control law CL1,*

$$\begin{bmatrix} -\dot{w}_y \\ \dot{w}_x \end{bmatrix} + K_1 \begin{bmatrix} \dot{v}_x \\ \dot{v}_y \end{bmatrix} = -a_1(e_1)_m - a_2 \int (e_1)_m \, dt \quad (51)$$

with a_1 and a_2 positive constants, will steer e_1 to zero.

Proof: Differentiating the expression for e_1 in Eq. (50) gives

$$\begin{aligned} \dot{e}_1 &= \begin{bmatrix} -\dot{w}_y \\ \dot{w}_x \end{bmatrix} + K_1 \begin{bmatrix} \dot{v}_x \\ \dot{v}_y \end{bmatrix} \\ &= -a_1(e_1)_m - a_2 \int (e_1)_m \, dt. \end{aligned} \quad (52)$$

This is a proportional-integral (PI) system, which is known to steer e_1 to zero.

Proposition 4 *If M_1, K_1, $\tilde{K}_2$, $\mathbf{u}_s$, and e_1 are all quasi-static, then the control law CL2,*

Fig. 5. Closed-loop contour following.

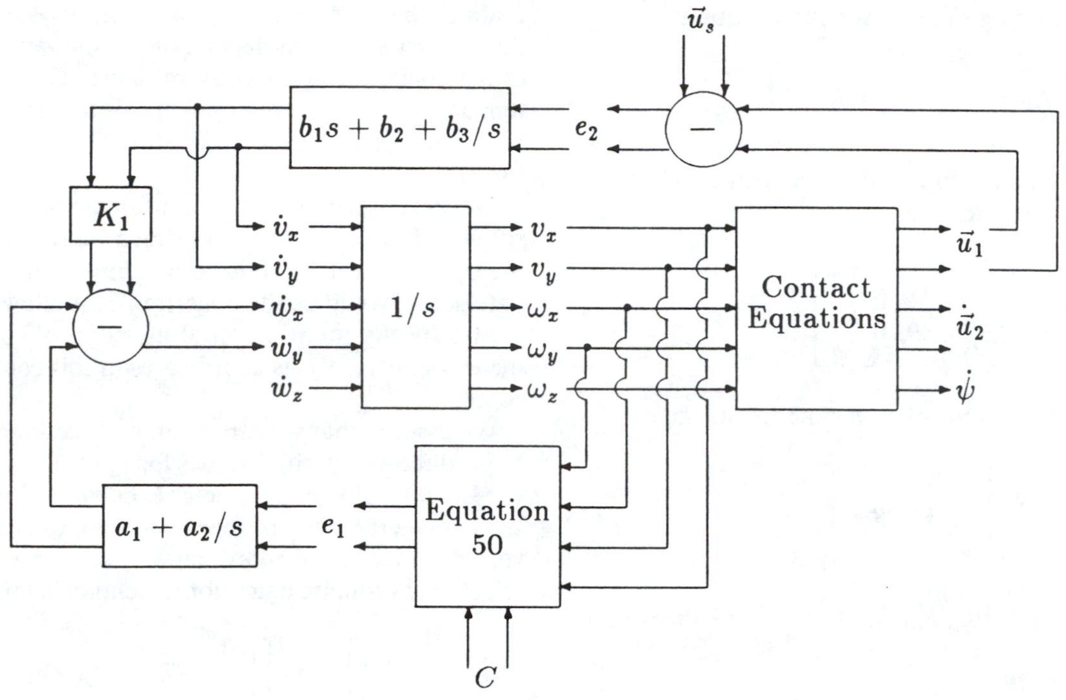

$$\begin{bmatrix} \dot{v}_x \\ \dot{v}_y \end{bmatrix} = M_1 \left(b_1(\dot{e}_2)_m + b_2(e_2)_m + b_3 \int (e_2)_m \, dt \right) \quad (53)$$

with b_1, b_2, and b_3 positive constants, steers e_2 to zero.

Proof: The first contact equation can be written as

$$\dot{e}_2 + \dot{\mathbf{u}}_s = \dot{\mathbf{u}}_1 = M_1^{-1}(K_1 + \tilde{K}_2)^{-1} \times \left(C - (K_1 + \tilde{K}_2) \begin{bmatrix} v_x \\ v_y \end{bmatrix} + e_1 \right). \quad (54)$$

Differentiating Eq. (54) gives

$$\ddot{e}_2 = -M_1^{-1} \begin{bmatrix} \dot{v}_x \\ \dot{v}_y \end{bmatrix} = b_1(\dot{e}_2)_m + b_2(e_2)_m + b_3 \int (e_2)_m \, dt. \quad (55)$$

This is a proportional-integral-derivative (PID) system, which is known to steer e_2 to zero.

Combining Propositions 3 and 4 gives the following theorem.

THEOREM 2 *Assume that M_1, K_1, $\tilde{K}_2$, C, and $\mathbf{u}_s$ are all quasi-static and that the time scale for control law CL1 is small enough compared to that for CL2 so that CL1 appears to always be in steady state from the viewpoint of CL2. Then the control law obtained by combining CL1 and CL2 steers e_1 and e_2* to zero.

The quasi-static assumptions need *not* hold at all times. Any deviation from these assumptions causes a disturbance on the system that, if not too large, is compensated by the closed-loop control.

The control scheme of Theorem 2 is pictured in Fig. 5. The time scale of the lower loop is smaller than that of the upper loop. The free parameters in this system are C, $\mathbf{u}_s$, and $\dot{\omega}_z$. This contour-following algorithm is discussed further in Montana (1986). There it is shown how we can vary these free parameters in order to have the point of contact follow a line of curvature on the object surface. Also described in Montana (1986) is an initial implementation of this control scheme.

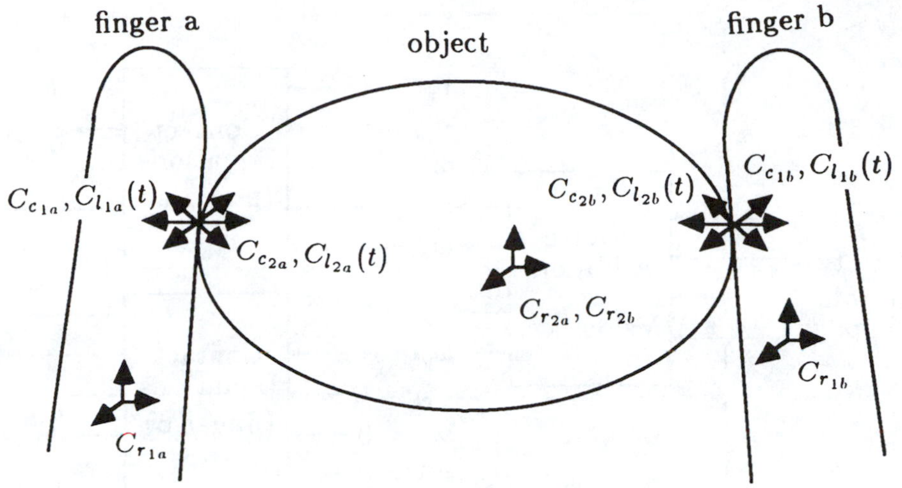

Fig. 6. Manipulation without slippage as an input-output system.

6. The Kinematics of Grasp

In this section we examine the problem of manipulating a rigid object with two end-effectors, which we refer to as fingers. We assume that the object has exactly one point of contact with each finger. We require that the fingers constantly grasp the object so as not to risk dropping it. Therefore, at each point of contact, the finger is constrained to roll without slipping so that static friction can be maintained.

We take the finger to be obj 1 and the object to be obj 2 at both points of contact. We refer to the two fingers as finger *a* and finger *b*. All symbols with subscript *a* refer to the point of contact between the object and finger *a*, and similarly for subscript *b*. The various coordinate frames are pictured in Fig. 6. The constraint that the fingers must roll without slipping can thus be expressed as $v_{xa} = v_{ya} = \omega_{za} = 0$ and $v_{xb} = v_{yb} = \omega_{zb} = 0$. To avoid the long subscripts induced by Notation 1, we let $\mathbf{p}_f$, R_f, $\mathbf{v}_f$, and $\mathbf{w}_f$ be the motion parameters of $C_{l_{1a}}(t)$ relative to $C_{l_{1b}}(t)$ at time t. Then $\mathbf{p}_f$, R_f, v_f, and $\mathbf{w}_f$ describe the relative motion of the two fingers at time t.

Definition 5 We say that the two points of contact form a *grip* if

$$\cos^{-1}([0, 0, 1]\mathbf{p}_f/\|\mathbf{p}_f\|) < \tan^{-1}(\kappa_s),$$
$$\cos^{-1}(-[0, 0, 1]R_f^{\mathrm{T}}\mathbf{p}_f/\|\mathbf{p}_f\|) < \tan^{-1}(\kappa_s) \qquad (56)$$

where κ_s is the static coefficient of friction (Mason 1982). (When the points of contact form a grip, the fingers can exert opposing forces and thus grasp the object.)

Definition 6 We define the *addition of velocities* map $V(\mathbf{p}_f, R_f)$ as

$$V(\mathbf{p}_f, R_f): \Re^4 \longrightarrow \Re^6, \quad \begin{bmatrix} \omega_{xa} \\ \omega_{ya} \\ \omega_{xb} \\ \omega_{yb} \end{bmatrix} \mapsto \begin{bmatrix} -R_f^{\mathrm{T}}\left(\begin{bmatrix} \omega_{xb} \\ \omega_{yb} \\ 0 \end{bmatrix} \times \mathbf{p}_f\right) \\ -R_f^{\mathrm{T}}\begin{bmatrix} \omega_{xb} \\ \omega_{yb} \\ 0 \end{bmatrix} + \begin{bmatrix} \omega_{xa} \\ \omega_{ya} \\ 0 \end{bmatrix} \end{bmatrix}. \qquad (57)$$

Theorem 3 *If the position and orientation of finger a relative to finger b are $\mathbf{p}_f$ and R_f and finger a and finger b roll without slipping relative to the object with angular velocity components ω_{xa}, ω_{ya}, ω_{xb}, and ω_{yb}, then the velocity of finger a relative to finger b is*

$$\begin{bmatrix} \mathbf{v}_f \\ \mathbf{w}_f \end{bmatrix} = V(\mathbf{p}_f, R_f)([\omega_{xa}, \omega_{ya}, \omega_{xb}, \omega_{yb}]^{\mathrm{T}}). \qquad (58)$$

Fig. 7. The contact equations as an input-output system.

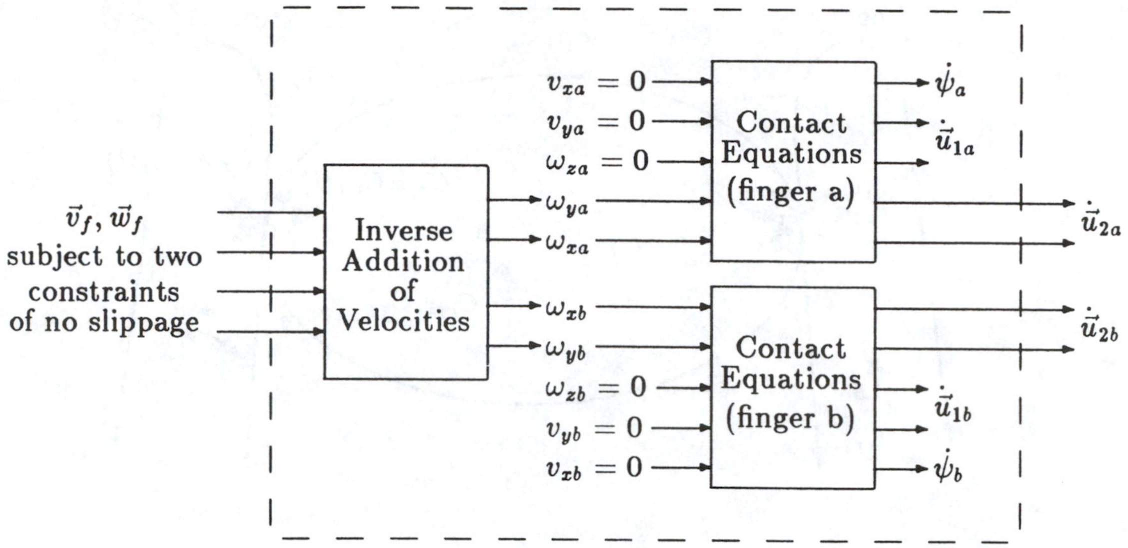

Furthermore, if the points of contact form a grip, then $V(\mathbf{p}_f, R_f)$ is an injective map.

Proof: From Proposition 1 we find that

$$\mathbf{v}_{l_{2b}l_{1a}} = R^{\mathrm{T}}_{l_{1b}l_{1a}}(\mathbf{v}_{l_{2b}l_{1b}} + \mathbf{w}_{l_{2b}l_{1b}} \times \mathbf{p}_{l_{1b}l_{1a}}) + \mathbf{v}_{l_{1b}l_{1a}}, \tag{59}$$

$$\mathbf{w}_{l_{2b}l_{1a}} = R^{\mathrm{T}}_{l_{1b}l_{1a}}\mathbf{w}_{l_{2b}l_{1b}} + \mathbf{w}_{l_{1b}l_{1a}}, \tag{60}$$

Since the object is a rigid body, $\mathbf{v}_{l_{2b}l_{2a}} = \mathbf{w}_{l_{2b}l_{2a}} = 0$. Hence, Proposition 1 states that

$$\mathbf{v}_{l_{2b}l_{1a}} = \mathbf{v}_{l_{2a}l_{1a}}, \qquad \mathbf{w}_{l_{2b}l_{1a}} = \mathbf{w}_{l_{2a}l_{1a}}. \tag{61}$$

According to the statement of the theorem,

$$\begin{aligned} &\mathbf{w}_{l_{2a}l_{1a}} = [\omega_{xa}, \omega_{ya}, 0]^{\mathrm{T}}, \quad w_{l_{2b}l_{1b}} = [\omega_{xb}, \omega_{yb}, 0]^{\mathrm{T}}, \\ &\mathbf{v}_{l_{2a}l_{1a}} = \mathbf{v}_{l_{2b}l_{1b}} = 0. \end{aligned} \tag{62}$$

Equations (59)–(62) can be combined to yield

$$\begin{aligned} 0 &= R_f^{\mathrm{T}}([\omega_{xb}, \omega_{yb}, 0]^{\mathrm{T}} \times \mathbf{p}_f) + \mathbf{v}_f, \\ [\omega_{xa}, \omega_{ya}, 0]^{\mathrm{T}} &= R_f^{\mathrm{T}}[\omega_{xb}, \omega_{yb}, 0]^{\mathrm{T}} + \mathbf{w}_f, \end{aligned} \tag{63}$$

which is equivalent to Eq. (58).

As to the injectivity of the addition of velocities map, let $\mathbf{p}_f = [p_{f_x}, p_{f_y}, p_{f_z}]^{\mathrm{T}}$. Then,

$$\begin{aligned} \cos^{-1}(p_{f_z}/\|\mathbf{p}_f\|) &= \cos^{-1}([0, 0, 1]\mathbf{p}_f/\|\mathbf{p}_f\|) \\ &< \tan^{-1}(\kappa_s) < \pi/2. \end{aligned} \tag{64}$$

We thus deduce that $p_{f_z} > 0$. From Eq. (63) we find that

$$-R_f\mathbf{v}_f = \begin{bmatrix} \omega_{xb} \\ \omega_{yb} \\ 0 \end{bmatrix} \times \begin{bmatrix} p_{f_x} \\ p_{f_y} \\ p_{f_z} \end{bmatrix} = \begin{bmatrix} \omega_{yb}p_{f_z} \\ -\omega_{xb}p_{f_z} \\ \omega_{xb}p_{f_y} - \omega_{yb}p_{f_x} \end{bmatrix}. \tag{65}$$

Therefore,

$$\mathbf{v}_f = 0 \Rightarrow \omega_{xb} = \omega_{yb} = 0. \tag{66}$$

From Eq. (63) we observe that

$$(\mathbf{w}_f = 0 \wedge \omega_{xb} = \omega_{yb} = 0) \Rightarrow \omega_{xa} = \omega_{ya} = 0. \tag{67}$$

Combining the logical implications of Eqs. (66) and (67) gives

$$\mathbf{v}_f = \mathbf{w}_f = 0 \Rightarrow \omega_{xb} = \omega_{yb} = \omega_{xa} = \omega_{ya} = 0. \tag{68}$$

So, the kernel of $V(\mathbf{p}_f, R_f)$ is zero-dimensional.

Since $V(\mathbf{p}_f, R_f)$ is injective, $V(\mathbf{p}_f, R_f)(\Re^4)$ is four-dimensional. Therefore, there exist two independent six-vectors $\mathbf{a}_1$ and $\mathbf{a}_2$ such that

$$\begin{bmatrix} \mathbf{v}_f \\ \mathbf{w}_f \end{bmatrix} \in V(\mathbf{p}_f, R_f)(\Re^4)$$

if and only if

$$\mathbf{a}_1 \cdot \begin{bmatrix} \mathbf{v}_f \\ \mathbf{w}_f \end{bmatrix} = 0, \qquad \mathbf{a}_2 \cdot \begin{bmatrix} \mathbf{v}_f \\ \mathbf{w}_f \end{bmatrix} = 0. \tag{69}$$

We call these conditions on the relative velocity of the sensors the *kinematic constraints of no slippage.* We can further conclude that there exists a linear map

$$\begin{aligned} V^{-1}(\mathbf{p}_f, R_f)&: V(\mathbf{p}_f, R_f)(\Re^4) \to \Re^4, \\ &V(\mathbf{p}_f, R_f)(\mathbf{b}) \mapsto \mathbf{b}. \end{aligned} \tag{70}$$

We call $V^{-1}(\mathbf{p}_f, R_f)$ the *inverse addition of velocities map.* In physical terms, it calculates the motion of each sensor relative to the object in response to a relative motion between the sensors that satisfies the kinematic constraints of no slippage.

The relationship between the relative motion of the fingers and the motion of the points of contact is as shown in Fig. 7. Importantly, this relation can be inverted. Given desired values for the velocities of the points of contact on the surfaces of the objects, $\dot{\mathbf{u}}_{2a}$ and $\dot{\mathbf{u}}_{2b}$, we can find the unique relative velocity $(\mathbf{v}_f, \mathbf{w}_f)$ that produces these velocities for the points of contact and satisfies the slippage constraints.

7. Application 3: Rolling a Sphere

Consider two fingers grasping a sphere of radius R with one point of contact for each finger. Assume that, to start, the points of contact are diametrically opposed. Recall the coordinate system for a subset of the sphere described in Example 1. Embed the sphere in $\Re^3$ so that in this coordinate system the two points of contact have u coordinates (latitudes) equal to zero (i.e., lie on the equator). If the points of contact move on the surface of the sphere according to $\dot{\mathbf{u}}_{2a} = \dot{\mathbf{u}}_{2b} = [0, \dot{\mathbf{v}}]^T$, then they will remain on the equator diametrically opposed. Hence, a grip is maintained. When the points of contact move thus and when both fingers rotate without slipping relative to the sphere, we say that the fingers are rolling the sphere.

PROPOSITION 5 *The unique velocity of finger a relative to finger b that satisfies the kinematic constraints of no slippage and produces velocities for the points of contact of $\dot{\mathbf{u}}_{2a} = \dot{\mathbf{u}}_{2b} = [0, \dot{\mathbf{v}}]^T$ is*

$$\begin{bmatrix} \mathbf{v}_f \\ \mathbf{w}_f \end{bmatrix} = \begin{bmatrix} 2(I + R\tilde{\tilde{K}}_{1b}) \\ 0 \quad 0 \\ J(\tilde{\tilde{K}}_{1b} - K_{1a}) \\ 0 \quad 0 \end{bmatrix} \begin{bmatrix} \dot{v} \sin \psi_a \\ \dot{v} \cos \psi_a \end{bmatrix}, \tag{71}$$

where

$$\begin{aligned} J &= \begin{bmatrix} 0 & 1 \\ -1 & 0 \end{bmatrix}, \\ R_{ab} &= \begin{bmatrix} \cos(\psi_a + \psi_b) & -\sin(\psi_a + \psi_b) \\ -\sin(\psi_a + \psi_b) & -\cos(\psi_a + \psi_b) \end{bmatrix}, \\ \tilde{\tilde{K}}_{1b} &= R_{ab} K_{1b} R_{ab}. \end{aligned} \tag{72}$$

Proof: Since there is no slippage, $v_{xa} = v_{ya} = v_{xb} = v_{yb} = 0$. Hence, the second contact equation yields

$$\begin{aligned} \begin{bmatrix} -\omega_{ya} \\ \omega_{xa} \end{bmatrix} &= (K_{1a} + \tilde{K}_{2a}) R_\psi M_{2a} \dot{\mathbf{u}}_{2a} \\ &= \left(K_{1a} + \frac{1}{R} I\right) \begin{bmatrix} -\dot{v} \sin \psi_a \\ -\dot{v} \cos \psi_a \end{bmatrix}, \end{aligned} \tag{73}$$

$$\begin{aligned} \begin{bmatrix} -\omega_{yb} \\ \omega_{xb} \end{bmatrix} &= (K_{1b} + \tilde{K}_{2b}) R_\psi M_{2b} \dot{\mathbf{u}}_{2b} \\ &= \left(K_{1b} + \frac{1}{R} I\right) \begin{bmatrix} -\dot{v} \sin \psi_b \\ -\dot{v} \cos \psi_b \end{bmatrix}. \end{aligned} \tag{74}$$

Observe that the position and orientation of finger a relative to finger b are given by

$$\mathbf{p}_f = \begin{bmatrix} 0 \\ 0 \\ 2R \end{bmatrix}, \quad R_f = \begin{bmatrix} R_{ab} & 0 \\ 0 & 1 \end{bmatrix}. \tag{75}$$

After substituting Eqs. (73)–(75) into Eq. (58) and performing algebraic simplification, we get Eq. (71).

8. Application Four: Fine Grip Adjustment

In this section we examine the problem of controlling two fingers with tactile-sensing capability to actively adjust their grip so as to locally optimize some criterion rating possible grips. The criterion we choose is as follows. Let $\phi_b = \cos^{-1}([0, 0, 1]\mathbf{p}_f)$ and $\phi_a = \cos^{-1}(-[0, 0, 1]R_f^T\mathbf{p}_f)$. Then the smaller the value of $\max(\phi_a, \phi_b)$ the better is the grip. To see why, recall from Eq. (56) that two-fingered grips are characterized by the condition $\max(\phi_a, \phi_b) < \tan^{-1}(\kappa_s)$. Hence the smaller the value of $\max(\phi_a, \phi_b)$, the larger is the error required for the grip to be lost.

We now investigate how ϕ_a and ϕ_b depend on the motion of the points of contact. Let $\mathbf{n}_a(t)$ and $\mathbf{n}_b(t)$ be the inward normals to the object at the points of contact at time t. Let $\mathbf{d}_{ba}(t)$ be the vector from the point of contact b to the point of contact a. Relative to $C_{l_{1b}}(t)$, the local coordinate frame for finger b, these vectors are

$$\mathbf{n}_a(t) = R_f(t)\begin{bmatrix}0\\0\\1\end{bmatrix}, \quad \mathbf{n}_b(t) = \begin{bmatrix}0\\0\\1\end{bmatrix}, \quad \mathbf{d}_{ba}(t) = \mathbf{p}_f(t). \tag{76}$$

So, $\phi_b = \cos^{-1}(\mathbf{n}_b \cdot \mathbf{d}_{ba})$ and $\phi_a = \cos^{-1}(-\mathbf{n}_a \cdot \mathbf{d}_{ba})$. Over the time interval Δt the points of contact traverse small arc lengths $\Delta\tilde{s}_{2a}$ and $\Delta\tilde{s}_{2b}$ across the surface of the object. To first-order approximation, relative to the coordinate frame $C_{l_{1b}}(t)$,

$$\mathbf{n}_a(t+\Delta t) = R_f(t)\begin{bmatrix}-\tilde{K}_{2a}\,\Delta\tilde{s}_{2a}\\1\end{bmatrix}, \quad \mathbf{n}_b(t+\Delta t) = \begin{bmatrix}-\tilde{K}_{2b}\,\Delta\tilde{s}_{2b}\\1\end{bmatrix}, \tag{77}$$

$$\mathbf{d}_{ba}(t+\Delta t) = \mathbf{p}_f + R_f(t)\begin{bmatrix}\Delta\tilde{s}_{2a}\\0\end{bmatrix} - \begin{bmatrix}\Delta\tilde{s}_{2b}\\0\end{bmatrix}. \tag{78}$$

Since dot products are invariant under coordinate frame transformation,

$$\phi_a(t+\Delta t) = \cos^{-1}(-\mathbf{n}_a(t+\Delta t)\cdot\mathbf{d}_{ba}(t+\Delta t)), \tag{79}$$

$$\phi_b(t+\Delta t) = \cos^{-1}(\mathbf{n}_b(t+\Delta t)\cdot\mathbf{d}_{ba}(t+\Delta t)), \tag{80}$$

where $\mathbf{n}_a(t+\Delta t)$, $\mathbf{n}_b(t+\Delta t)$, and $\mathbf{d}_{ba}(t+\Delta t)$ are as given in Eqs. (77) and (78). We can think of $\phi_a(t+\Delta t)$ and $\phi_b(t+\Delta t)$ as functions of $\Delta\tilde{s}_{2a}$ and $\Delta\tilde{s}_{2b}$. Thus, we define the function

$$f_1(\Delta\tilde{s}_{2a}, \Delta\tilde{s}_{2b}) = \max(\phi_a(\Delta\tilde{s}_{2a}, \Delta\tilde{s}_{2b}), \phi_b(\Delta\tilde{s}_{2a}, \Delta\tilde{s}_{2b})), \tag{81}$$

which is a rating of the grip obtained from the present one by motion of the points of contact across the surface of the object through arc lengths $\Delta\tilde{s}_{2a}$ and $\Delta\tilde{s}_{2b}$.

We further observe that, according to the second contact equation (as given in Eq. (39)), the angles of rotation needed to produce the arc length traversals $\Delta\tilde{s}_{2a}$ and $\Delta\tilde{s}_{2b}$ are

$$\Delta\theta_a = \begin{bmatrix}-\Delta\theta_{ya}\\\Delta\theta_{xa}\end{bmatrix} = (K_{1a} + \tilde{K}_{2a})\Delta\tilde{s}_{2a}, \quad \Delta\theta_b = \begin{bmatrix}-\Delta\theta_{yb}\\\Delta\theta_{xb}\end{bmatrix} = (K_{1b} + \tilde{K}_{2b})\Delta\tilde{s}_{2b}. \tag{82}$$

We can then define the function

$$f_2(\Delta\tilde{s}_{2a}, \Delta\tilde{s}_{2b}) = \|\Delta\theta_a(\Delta\tilde{s}_{2a})\| + \|\Delta\theta_b(\Delta\tilde{s}_{2b})\|, \tag{83}$$

which is a measure of the size of the motion of the fingers.

We can perform a hill-climbing search to locally optimize the grip based on the following iterative step.

1. Use tactile sensing to measure the position of the points of contact on the two fingers. With proprioceptive sensing, determine the positions and orientations of the fingers. Based on these measurements, compute $\mathbf{p}_f$ and R_f, the relative position and orientation of the local coordinate frames, and K_{1a} and K_{1b}, the curvature forms of the fingers.
2. Perform curvature experiments to find K_{2a} and K_{2b}, the curvature forms of the object at each point of contact (recall Section 4). Curvature experiments involve only motions of the finger relative to the object such that the finger is

rolling without slipping. Hence, they can be performed while grasping the object based on the analysis of Section 6.

3. Find the values of $\Delta\tilde{s}_{2a}$ and $\Delta\tilde{s}_{2b}$, such that $f_2(\Delta\tilde{s}_{2a}, \Delta\tilde{s}_{2b}) \leq \delta$, that minimize $f_1(\Delta\tilde{s}_{2a}, \Delta\tilde{s}_{2b})$. The parameter $\delta > 0$ is the maximum step size. If there are multiple sets of $\Delta\tilde{s}_{2a}$ and $\Delta\tilde{s}_{2b}$ that provide a minimum for $f_1(\Delta\tilde{s}_{2a}, \Delta\tilde{s}_{2b})$, choose one that minimizes $f_2(\Delta\tilde{s}_{2a}, \Delta\tilde{s}_{2b})$.
4. (optional) If, for the chosen values of $\Delta\tilde{s}_{2a}$ and $\Delta\tilde{s}_{2b}$, $f_1(0, 0) - f_1(\Delta\tilde{s}_{2a}, \Delta\tilde{s}_{2b}) < \epsilon$, where $\epsilon > 0$ is an appropriately chosen parameter, then stop the iteration and maintain the present grip.
5. Move the points of contact through arc lengths $\Delta\tilde{s}_{2a}$ and $\Delta\tilde{s}_{2b}$ across the surface of the object by rotating the fingers without slipping relative to the object through angles $\Delta\theta_a$ and $\Delta\theta_b$ as given in Eq. (82). Substituting into Eq. (58) gives the unique relative motion of the fingers that accomplishes this.
6. Repeat.

9. Conclusion

Using concepts from differential geometry, I have derived a set of equations, called contact equations, that are a general description of the kinematics of contact between two rigid bodies. Because of their generality, the contact equations are potentially a powerful tool for analyzing any task that involves contact evolving in time. Based on these equations, I have examined the following applications for a single end-effector: (1) determining the curvature form of an unknown object at a point of contact, and (2) following the surface of an unknown object. I have also used the contact equations to examine the kinematics of grasp. Based on this analysis, I have investigated these applications for two end-effectors: (1) rolling a sphere between two arbitrarily shaped fingers, and (2) fine grip adjustment (i.e., having two fingers that grasp an object locally optimize their grip for maximum stability).

Experimental work to corroborate the theory has been hampered by lack of resources, although there have been some preliminary but promising experiments performed. I have implemented a contour-following algorithm similar to that examined in this paper. Its performance is detailed on Montana (1986). Also described is a set of experiments investigating the effect of compliance on the kinematics of contact (the theory of which is discussed in Montana (1986) but not here).

Acknowledgments

The research for this paper was performed at the Division of Applied Sciences of Harvard University under the supervision of Roger W. Brockett. Financial support for this research was provided by NSF grants MEA-83-18972 and ECF-81-21428 and ARO grant DAAG29-83-K-0027.

References

Bajcsy, R. 1984. What can we learn from one finger experiments? *1st Symp. on Robotics Research:* 509–527. Cambridge, Mass.: MIT Press.

Cai, C. and Roth, B. 1986. On the planar motion of rigid bodies with point contact. *Mechanism and Machine Theory.*

Cai, C. and Roth, B. 1987 (Raleigh, N.C.). On the spatial motion of rigid bodies with point contact. *Proc. 1987 IEEE Conf. on Robotics and Automation:* 686–695.

Campbell, S. L. and Meyer, C. D., Jr. 1979. *Generalised inverses of linear transformations.* London: Pitman.

Cartan, E. 1946. *Lecons sur la géometrie de Riemann.* Paris: Gauthier-Villars.

Craig, J. J. 1986. *Introduction to robotics.* Reading, Mass.: Addison-Wesley.

Fearing, R. S. and Hollerbach, J. H. 1985. Basic solid mechanics for tactile sensing. *Int. J. Robotics Res.* 4(3):40–54.

Hanafusa, H. and Asada, H. 1977 (Tokyo). Stable prehension by a robot hand with elastic fingers. *Proc. 7th ISIR:* 361–368.

Kerr, J. and Roth, B. 1986. Analysis of multifingered hands. *Int. J. Robotics Res.* 4(4):3–17.

Mason, M. T. 1981. Compliance and force control for computer controlled manipulators. *IEEE Trans. Systems, Man Cybernet.* SMC-11(6):418–432.

Mason, M. T. 1982. Manipulator grasping and pushing operations. Ph.D. Thesis, Department of Electrical Engineering and Computer Science, Massachusetts Institute of Technology.

Montana, D. J. 1986. Tactile sensing and the kinematics of contact. Ph.D. Thesis, Division of Applied Sciences, Harvard University.

Okada, T. 1982. Computer control of multi-jointed finger systems for precise object handling. *IEEE Trans. Systems, Man Cybernet.* SMC-12(3):289–299.

Salisbury, J. K. 1982. Kinematic and force analysis of articulated hands. Ph.D. Thesis, Department of Mechanical Engineering, Stanford University.

Spivak, M. 1979. *A comprehensive introduction to differential geometry.* Berkeley: Publish or Perish.

Computing and Controlling the Compliance of a Robotic Hand

MARK R. CUTKOSKY AND IMIN KAO, STUDENT MEMBER, IEEE

Abstract—Compliance is one of the most important quantities for characterizing the grasp of a robotic hand on a tool or workpiece. This is particularly true in fine manipulation (as in assembling components) where small motions and low velocities lead to dynamic equations that are dominated by compliance, friction, and contact conditions.

In this paper we express the compliance of a grasp as a function of grasp geometry, contact conditions between the fingers and the grasped object, and mechanical properties of the fingers. We argue that the effects of structural compliance and small changes in the grasp geometry should be included in the computation. We then examine factors that can lead a grasp to become unstable, independently of whether it satisfies force closure. Finally, we examine the reverse problem of how to specify servo gains at the joints of a robotic hand so as to achieve, as nearly as possible, a desired overall grasp compliance. We show that coupling between the joints of different fingers is useful in this context.

NOMENCLATURE

δx_b	Motion in the body coordinates.
δx_p	Motion in the contact coordinates, $\delta x_p = {}^{P}_{B}J\delta x_b$
δx_f	Motion at the fingertip, $\delta x_f = J_\theta \delta\theta$.
δx_{tr}	Motion transmitted through the contact, $H\delta x_p = H\delta x_f = \delta x_{tr}$.
$\delta\theta$	Changes of joint angles.
f_b	Force in the body coordinates.
f_p	Grasp force at point P on the body.
f_f	Grasp force at point F at the fingertip, $f_f = f_p$.
f_{tr}	Force transmitted through the contact, $f_p = f_f = H^T f_{tr}$.
τ	Joint torques.
${}^{P}_{B}J$	Coordinate transformation matrix, ${}^{P}_{B}J\delta x_b = \delta x_p$.
H	Contact constraint matrix.
J_θ	Joint Jacobian matrix, $J_\theta \delta\theta = \delta x_f$.
D	Coordinate transformation matrix relating contact coordinates before and after infinitesimal motion.
${}^{P}_{B}\Delta J_i^T$	${}^{P}_{B}J^T(D_i^T - I)$ that represents a differential Jacobian.
Ω	$\Omega = (HC_s H^T - HC_s H^T H K_p H^T H C_s H^T)$.
B	HJ_θ.
${}^{P}_{B}\mathcal{J}$	Concatenated coordinate transformation matrix.
$\mathcal{H}$	Concatenated contact constraint matrix.
$\mathcal{J}_\theta$	Concatenated joint Jacobian matrix.
K	Stiffness, which is generally defined as $\partial f/\partial x$ as in (1).
K_θ	The joint stiffness matrix, $K_\theta = C_\theta^{-1}$.
K_f	The stiffness matrix in the fingertip coordinates, $K_f = C_f^{-1}$ if it is invertible.
K_b	Stiffness matrix due to servo and structural compliance in the body coordinates.
K_J	Stiffness matrix due to changes in geometry.
K_e	Effective stiffness matrix, $K_e = K_b + K_J$.
$\mathcal{K}_p$	Concatenated stiffness matrix at contact points P.
$\mathcal{K}_\theta$	Concatenated servo stiffness.
$\mathcal{C}_\theta$	Concatenated servo compliance.
$\mathcal{C}_s$	Concatenated structural compliance at the tip coordinates.
$\mathcal{C}_f$	Concatenated fingertip compliance, $\mathcal{C}_f = \mathcal{J}_\theta \mathcal{C}_\theta \mathcal{J}_\theta^T + \mathcal{C}_s$.
m_i	Number of joints for finger i.
n_i	Degrees of freedom at contact i.
nf	Number of fingers.
l	The Lagrange multipliers, $l = [\lambda_1 \lambda_2 \lambda_3 \cdots \lambda_n]$.
M_{ij}	Submatrices of the inverse of $\begin{bmatrix} K_{f_i} & H_i^T \\ H_i & 0 \end{bmatrix}$ as derived in (20).
$B(x, y, z)$	Body coordinates.
$F(a, b, c)$	Contact coordinates at the contact point F on the fingertip.
$P(l, m, n)$	Contact coordinates at the contact point P on the body.
$O(x, y, z)$	World reference frame.
$\theta_1 \cdots \theta_m$	Joint angles of a finger.

I. INTRODUCTION

THE EFFECTIVE compliance or stiffness of an object held in a robotic hand is an important quantity in tasks where forces and small motions are imparted to the object. For example, in assembling components it has been demonstrated that the chances of successful assembly are highest if the compliance matrix associated with the grasped part is diagonal at the point where the part first touches a mating component [15].

Manuscript received October 10, 1987; revised June 2, 1988. This work was supported by the National Science Foundation under Grants DMC8552691 and DMC8602847, with additional support from the Standard Institute for Manufacturing and Automation.

The authors are with the Mechanical Engineering Department, Stanford University, Stanford, CA 94305-3030.

IEEE Log Number 8823936.

Reprinted from *IEEE Trans. Robot. Automat.*, vol. 5, no. 2, pp. 151–165, April 1989.

Similarly, in robotic grinding and deburring, the compliance matrix associated with the tool can be used to simplify the control problem [3]. The importance of compliance is also evident when examining the relative magnitudes of terms in the dynamic equations associated with fine motion tasks [12]. Low velocities and small relative motions lead to small inertia terms and a substantially linear analysis which is dominated by stiffness or its inverse, compliance.

In this paper we build upon recent grasp analyses by Salisbury [9], Kerr [7], Cutkosky [2], Kobayashi [8], and Nguyen [10], in which the instantaneous kinematics of multifingered hands have been defined in terms of Jacobian matrices and in which servo gains associated with the fingers contribute to the overall stiffness of the grip. As in [2], we are most interested in cases for which soft, anthropomorphic fingers are in contact with the object so that deformation and rolling of the fingertips cannot be ignored.

The first problem addressed in this paper is to obtain the effective compliance of an object held by a multifingered hand. In contrast to previous efforts, we assume that the fingers may be constructed from relatively flexible materials (such as molded thermoplastics) so that their structural compliance must be added to the compliance of the fingertips and to the controllable servo compliance of the joints. As Table II shows, even when fingers are made of aluminum, elasticity in the drive cables and in the joints produces significant compliance.

The effective compliance or stiffness of the grasped object depends not only on the compliances of the fingers but also on terms that arise from small changes in the grasp geometry as the object is perturbed by external forces. When grasp forces are large, these geometric terms may make the grasp statically unstable. In Section III we provide a new, systematic method for computing the matrix, K_J, that represents these terms.

The overall grasp thickness matrix, containing structural, servo, and geometric terms, is a useful measure of the grasp. The rank and the eigenvalues of this matrix tell us about the mobility and stability of the grasp.[1] The relative magnitudes of the elements tell us how sensitive the grasp forces are to small motions in various directions. In addition, as discussed in [6], the grasp stiffness is useful in predicting the onset of sliding motion since it provides information about how rapidly the normal and tangential force components will change at each contact when the object is disturbed.

Finding the stiffness of a part grasped in a robotic hand is useful for predicting the grasp behavior and for discriminating among competing grasps when contemplating a task with particular requirements. For controls purposes, however, the reverse problem is of more interest: adjusting the servo gains at the finger joints so as to achieve a desired grasp stiffness. In many cases it is not possible to achieve exactly the required stiffness, but in Section IV we explore the methods that involve coupling the control of joints on different fingers to achieve a best approximation. The results of the forward and reverse compliance analysis are illustrated with several examples.

[1] In this paper we address only quasi-static stability, ignoring drive-train and actuator dynamics and the effects of rapidly varying, task- induced forces.

II. Previous Investigations of Grasp Compliance and Stiffness

Grasp compliance and stiffness have been studied by several investigators. The earliest analysis is that of Asada [1] in which a two-dimensional model of the object and the fingers leads to a potential function for describing stable grasps. The analysis assumes that fingers are essentially springs that move along a given locus with a single degree of freedom. Friction is ignored at the contacts between the fingers and the object and stable grasps are those for which the object will slide back to its initial equilibrium position after being displaced. The analysis gives intuitively satisfying results for handling slippery objects and where the main concern is that the object should not be dropped. However, in everyday manipulation tasks people and robots take advantage of friction. Consequently, subsequent grasp analyses have generally included friction at the contact points.

Salisbury [9] augments the resultant force and torque on the object with internal grasp forces so that an invertible 9×9 stiffness matrix is obtained for a three-fingered hand. The resulting stiffness matrix is useful for grasp control and simplifies programming the hand for certain manipulations. Cutkosky [2] examines the effective stiffness of a grasped object and shows that it is a function not only of the servo parameters but also of fingertip models and of small changes in the grasp geometry as the object is perturbed by external forces. Nguyen [10] addresses the stiffness and stability of planar and three-dimensional objects grasped by "virtual springs" corresponding to force components, or wrenches, at the finger/object contact points. Nguyen also proposes a least squares solution for specifying the stiffnesses of the "virtual springs" to obtain a desired grasp stiffness. However, he does not address the relation of "virtual springs" to joints of an actual hand. In this paper we consider a related problem: specifying the joint servoing in fingers with known structural properties to achieve a desired grasp compliance. The inclusion of the finger kinematics produces somewhat more complexity, but, due to coupling terms, also provides more opportunities for controlling the stiffness matrix.

III. Stiffness in Grasping

In its most general form, the stiffness of a grasp is a linearized expression of the relationships between forces applied to the grasp and the resulting motions

$$K = \frac{\partial f}{\partial x}. \tag{1}$$

The resulting matrix can describe force/motion relationships both internal and external to the grasped object [9], but in this section we concern ourselves only with stiffness of the object with respect to external forces and moments.

As a partial derivative, the grasp stiffness permits us to look beyond instantaneous properties such as connectivity[2] and force-closure[3] and to examine the *sensitivity* of the grasp

[2] The connectivity of a grasp is the number of degrees of freedom between the grasped object and the palm of the hand [9].

[3] A grasp has force-closure if it can resist forces and moments from any direction, assuming the fingertips maintain contact with the object [11].

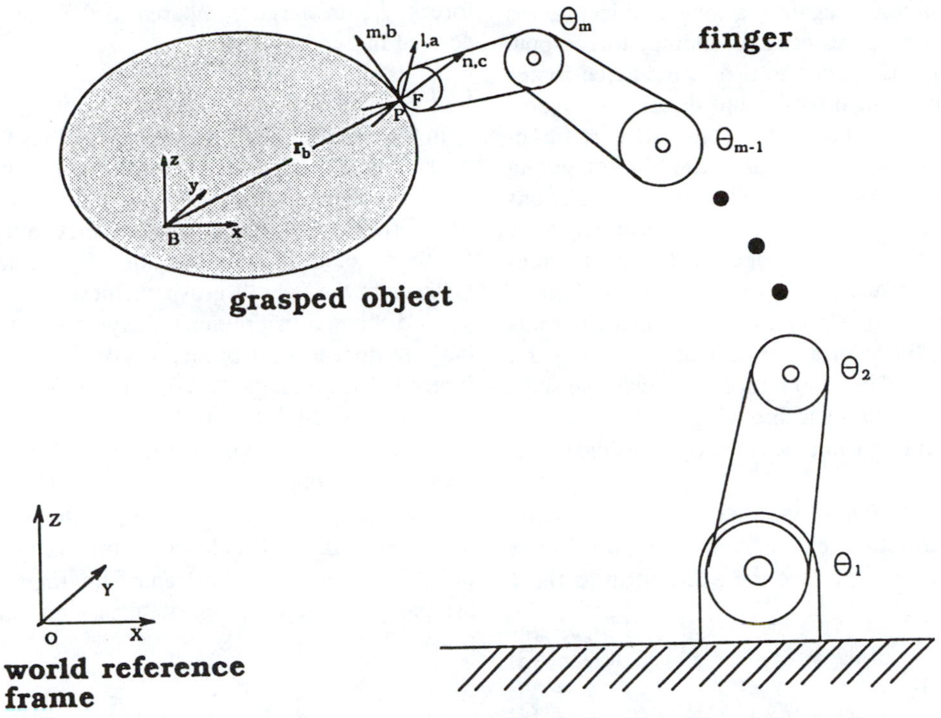

Fig. 1. Coordinate systems for a single finger touching a grasped object.

TABLE I
FORWARD FORCE/MOTION RELATIONSHIPS FOR FINGER AND OBJECT IN FIG. 1

motion	$J_\theta \delta\theta = \delta x_f$ $(6 \times m)(m \times 1)(6 \times 1)$		$H\delta x_f = \delta x_{tr} = H\delta x_p$ $(n \times 6)(6 \times 1)(n \times 1)(n \times 6)(6 \times 1)$		${}^{P}_{B}J\delta x_b = \delta x_p$ $(6 \times 6)(6 \times 1)(6 \times 1)$
	JOINTS	$\rightarrow$ $\leftarrow$	**CONTACT**	$\leftarrow$ $\rightarrow$	**OBJECT**
force	$J_\theta^T f_f = \tau$ $(m \times 6)(6 \times 1)(m \times 1)$		$f_f = H^T f_{tr} = f_p$ $(6 \times 1)(6 \times n)(n \times 1)(6 \times 1)$		${}^{P}_{B}J^T f_p = f_b$ $(6 \times 6)(6 \times 1)(6 \times 1)$

to disturbances. In fact, as Salisbury [9] points out, K is a direct measure of quasi-static grasp stability. As long as K is positive-definite, the changes in the forces on the object will have a stabilizing effect.

In the following subsections, we establish the force/motion relationships for the grasp and develop general expressions for the grasp stiffness in terms of finger compliances, contact types, and changes in the grasp geometry.

A. Forward Force/Motion Relationships

In this section we develop a framework for stiffness computation that includes the effects of

- servo compliance
- structural compliance
- changes in geometry
- coupling among different joints and fingers
- different contact types
- changes in contact location (e.g., due to rolling).

Fig. 1 shows the coordinate systems and Table I summarizes the forward force/motion relationships for a single finger touching an object. For derivations of the force/velocity transformations see [2], [7], [8]. In Table I, J_θ is the Jacobian matrix relating joint velocities to tip velocities in the $F(abc)$ coordinate system and ${}^{P}_{B}J$ is a Cartesian transformation matrix between the $B(xyz)$ coordinate system in the body and the $P(lmn)$ system at the contact point. At each contact, the force and velocity constraints are represented by an $n \times 6$ matrix H. Thus for a soft finger as defined by Salisbury [9]

$$H = \begin{bmatrix} 1 & 0 & 0 & 0 & 0 & 0 \\ 0 & 1 & 0 & 0 & 0 & 0 \\ 0 & 0 & 1 & 0 & 0 & 0 \\ 0 & 0 & 0 & 0 & 0 & 1 \end{bmatrix}.$$

The reciprocal force constraints at the contact can be derived from force equilibrium

$$f_p = f_f = H^T f_{tr}. \tag{2}$$

Thus for a point contact with friction $n = 3$, and the number of nonzero elements in f_p or f_b is three. In the following derivations we will also make use of the fact that H^T is the generalized inverse of H.

Concatenated Force/Motion Relationships: When several fingers like the finger in Fig. 1 grasp an object, the resulting configuration includes both serial and parallel kinematics. Each finger is a serial kinematic chain, while the combination of several fingertips pressing against a single object forms a parallel mechanism. As a result, any attempt to compute the joint forces or motions needed to produce desired forces or motions of the object will involve both direct and inverse kinematic computations [7], and must accommodate static indeterminacy, redundancy, and over-constraint. In computing the grasp stiffness, we are subject to the same complications since we need to propagate motions from the object out to the finger joints and then propagate changes in the joint forces back to the object. However, as we shall see in the following section, the inclusion of structural compliances permits a systematic approach for arbitrary combinations of fingers, joints, and contact types. This is an important advantage since fingers are continually making and breaking contact with the object, and rolling or sliding from faces to edges, as the object is manipulated.

We concatenate ${}^{\mathrm{P}}_{\mathrm{B}}J$, H, and J_θ, so that we can simultaneously express the force/motion relationships for all the fingers of a grasp. The new equations have the same form as those in Table I

$$ {}^{\mathrm{F}}\mathfrak{J}_\theta \delta\theta = \delta x_f \tag{3}$$

$$ ({}^{\mathrm{F}}\mathfrak{J}_\theta)^T f_f = \tau \tag{4}$$

where τ and $\delta\theta$ are concatenated vectors of all the joint torques and infinitesimal joint motions and f_f and δx_f are concatenated vectors of forces and infinitesimal motions at the fingertips. The contact equations over all fingertips now become

$$\mathfrak{H}\delta x_p = \mathfrak{H}\delta x_f \tag{5}$$

$$f_p = f_f = \mathfrak{H}^T f_{\mathrm{tr}}. \tag{6}$$

Finally, between the contact and the object $B(xyz)$ coordinate system we have

$$\underset{(6\cdot nf\times 6)}{{}^{\mathrm{P}}_{\mathrm{B}}\mathfrak{J}} \quad \underset{(6\times 1)}{\delta x_p} = \underset{(6\cdot nf\times 1)}{\delta x_p} \tag{7}$$

and

$$\underset{(6\times 6\cdot nf)}{({}^{\mathrm{P}}_{\mathrm{B}}\mathfrak{J})^T} \quad \underset{(6\cdot nf\times 1)}{f_p} = \underset{(6\times 1)}{f_b} \tag{8}$$

where ${}^{\mathrm{P}}_{\mathrm{B}}\mathfrak{J}$ is a $(6\cdot nf\times 6)$ matrix transforming infinitesimal motions δx_b of the object to equivalent motions at the contact.

B. Forward Stiffness Computation

We are now in a position to expand (1) by the chain rule, using (6) and (8)

$$\frac{\partial f_b}{\partial x_b} = \frac{\partial({}^{\mathrm{P}}_{\mathrm{B}}\mathfrak{J}^T f_f)}{\partial x_b} = \underbrace{({}^{\mathrm{P}}_{\mathrm{B}}\mathfrak{J}^T)\frac{\partial f_f}{\partial x_b}}_{K_b} + \underbrace{\frac{\partial({}^{\mathrm{P}}_{\mathrm{B}}\mathfrak{J}^T)f_f}{\partial x_b}}_{K_J}. \tag{9}$$

The first part of the right-hand side of (9) expresses the restoring forces at the contacts resulting from the structural and servo stiffness of the fingers. The matrix associated with these terms is K_b, and is derived in Section III-C. The second term K_J represents the effects of small changes in the grasp configuration. These terms become important when the grasp forces, f_f, are large compared to the restoring forces. K_J is derived in Section III-D.

C. Computing K_b: The Effects of Restoring Forces

In this subsection we derive expressions for the compliances of the fingers, beginning with the individual joint and structural compliances. We then impose the contact constraints and transform the fingertip stiffness matrix to the body coordinate frame to obtain the contribution of joint and structural compliances to the overall grasp stiffness.

As discussed in previous analyses [9], the joint stiffnesses $\mathfrak{K}_{\theta_{ij}}$ are due, to first order, to position feedback gains in the fingers. Thus a single finger with m joint servos would produce an $m\times m$ joint stiffness matrix, which could be inverted to obtain a joint compliance matrix. Often, however, the fingers will be coupled. For example, the human hand exhibits both active (servo) and passive (structural) coupling among the fingers. If we wiggle our fourth fingers, it is nearly impossible to keep the third and fifth fingers from moving in sympathy. Therefore, we establish a concatenated joint compliance matrix

$$\mathfrak{C}_\theta = \mathfrak{K}_\theta^{-1} = \begin{bmatrix} C_{\theta_{11}} & C_{\theta_{12}} & \cdots & C_{\theta_{1,nf}} \\ C_{\theta_{21}} & C_{\theta_{22}} & \cdots & C_{\theta_{2,nf}} \\ & & \ddots & \\ & \cdots & & C_{\theta_{nf,nf}} \end{bmatrix}. \tag{10}$$

Applying (3) and (4) and using the principal of virtual work, we obtain an equivalent compliance matrix for the fingertips

$$\mathfrak{C}_j = \mathfrak{J}_\theta \mathfrak{C}_\theta \mathfrak{J}_\theta^T. \tag{11}$$

If the fingers are not coupled, $\mathfrak{C}_j$ is block-diagonal. More generally, $\mathfrak{C}_j$ is a symmetric $(6\cdot nf\times 6\cdot nf)$ matrix (where nf is the number of fingers) but will be singular as long as the fingers have less than six joints.

To the controllable joint compliances we must add the uncontrollable structural compliance. In industrial grippers, structural compliance may be negligible but in dexterous hands, the use of actuating cables and soft gripping surfaces leads to significant compliance. In some cases, the compliance is easiest to determine experimentally, applying known loads and recording deflections. For example, in recent tests on the Stanford/JPL hand, we measured the typical compliances shown in Table II. These values were measured with the fingers extended. The combined structural compliance, at approximately 10^{-4} N/m, is roughly 10 percent of the minimum achievable servo compliance. In more anthropomorphic hands, with softer links and fingertips, we expect structural compliance to be correspondingly more important.

The structural compliance matrix varies considerably as a function of finger position, and the problem of computing it is similar to that of computing inertia matrices in robot dynamics. As in dynamics, one possibility is to interpolate among stored values corresponding to different configurations [13]. If the structural compliance is dominated by known flexibilities in the fingertips, links, or cables we can also express the compliance directly in terms of finger orientation. For example, if flexibility in the links is dominant, the structural

TABLE II
TYPICAL STRUCTURAL COMPLIANCES IN THE STANFORD/JPL HAND

Component	Compliance in m/N
Cable compliance	3.1×10^{-5}
Joint and link compliance	1.0×10^{-5}
Fingertip compliance	1.1×10^{-4}

compliance at each fingertip becomes

$$ {}^{F}C_s = \sum_{j=1}^{m} {}^{F}J_{l_j} C_{l_j} {}^{F}J_{l_j}^T + C_{\text{tip}} \tag{12} $$

where ${}^{F}J_{l_j}$ is a 6×6 Jacobian relating a coordinate system in the jth link to the $F(abc)$ coordinate system at the fingertip. An example of C_{tip} is derived in [3] for "very soft" fingertips and C_{l_j} can be estimated from elementary beam theory [6]. In general, ${}^{F}C_s$ is a nonsingular 6×6 symmetric matrix, although the terms in ${}^{F}C_s$ may be small compared to those in ${}^{F}C_j$.

The concatenated fingertip compliance matrix is obtained by summing the joint and structural compliances for each finger

$$ \mathcal{C}_f = \mathcal{C}_j + \mathcal{C}_s. \tag{13} $$

If $\mathcal{C}_f$ is invertible, which it usually will be if $\mathcal{C}_s$ is not negligible, we can immediately obtain the concatenated fingertip stiffness matrix $\mathcal{K}_f$. However, as observed earlier, the contact matrices $\mathcal{H}$ "filter out" some of the motions of the fingertips. Hence not all elements of $\mathcal{K}_f$ will be experienced by the grasped object. Therefore, we use (5) and (6) to define a new stiffness matrix, $(\mathcal{H}\mathcal{C}_f\mathcal{H}^T)^{-1}$ that represents the finger stiffness components seen through the contacts. Then, applying (5) and (6) again, we can expand this stiffness to a (singular) $(nf \cdot n \times nf \cdot n)$ matrix for the contact points on the object

$$ \mathcal{K}_p = \mathcal{H}^T(\mathcal{H}\mathcal{C}_f\mathcal{H}^T)^{-1}\mathcal{H}. \tag{14} $$

Note that if structural compliance is negligible, the product $(\mathcal{H}\mathcal{C}_f\mathcal{H}^T)^{-1}$ will still be invertible as long as the grasp is fully "manipulable," i.e., if the fingers can impart arbitrary motions to the object [9]. As discussed in [6], the expression for $\mathcal{K}_p$ is particularly useful for sliding analyses, as its partitions $\mathcal{K}_{p_{ij}}$ indicate the stiffnesses that the object "sees" at each contact point.

Finally, in terms of the $B(xyz)$ coordinate system of the object, we apply (7) and (8) to obtain

$$ K_b = {}^{P}_{B}\mathcal{J}^T \mathcal{K}_p {}^{P}_{B}\mathcal{J} \tag{15} $$

where K_b is a 6×6 matrix representing the stiffness of the grasped object, due to servo and structural terms.

D. Computing K_J: The Effects of Changes in Geometry

When a grasped object is displaced slightly, two things happen which may affect the overall grasp stiffness and stability: the fingers shift slightly with respect to the object and, if the fingers roll or slide, the contact points move upon the object. To compute the effects of these changes in the grasp geometry we first need to find an expression for the complete motions of the fingertips.

The following results will be derived for a single finger, as in Fig. 1 and Table I, since the overall effect can be obtained simply by summing the contributions of all fingers

$$ K_J = \sum_{i=1}^{nf} K_{J_i}. \tag{16} $$

From the contact velocity constraints in Table I, we can specify some, but not all, of the components of the fingertip motion when the object is displaced slightly. We need to make some assumptions about the fingertip control to determine the remaining components. One possibility is to assume that the fingertip orientation remains essentially fixed with respect to some global coordinate system [6]. For long, multijointed fingers with point contract, this is probably a reasonable assumption. A more general assumption is that like any underconstrained elastic mechanism, the grasp will adopt a configuration that minimizes its potential energy, subject to kinematic constraints. In this case we can solve for the motion of the fingertips using Lagrange multipliers. Since the grasp is initially at equilibrium, displacing the object will increase the potential energy

$$ \Delta\text{p.e.} = \frac{1}{2}(\delta x_f^T \mathcal{K}_f \delta x_f) - \delta x_f^T f_f \tag{17} $$

where $\delta x_f^T(-f_f)$ represents work done against the initial grasp forces. To minimize Δp.e. subject to the constraint $\mathcal{H}(\delta x_f - \delta x_p) = 0$ in (5), we define the cost function c as

$$ c = \Delta\text{p.e.} + l^T\mathcal{H}(\delta x_f - {}^{P}_{B}\mathcal{J}\,\delta x_b) \tag{18} $$

where l is a vector of Lagrange multipliers. To find the optimal solution, we differentiate c with respect to δx_f and obtain

$$ \mathcal{K}_f\,\delta x_f + \mathcal{H}^T l = f_f. \tag{19} $$

Combining (5) and (19), we have (20). Note that the potential energy includes the combined structural and servo stiffness. The matrix equation with Lagrange multipliers can be expressed in fingertip coordinates:

$$ \begin{bmatrix} \mathcal{K}_f & \mathcal{H}^T \\ \mathcal{H} & 0 \end{bmatrix} \begin{bmatrix} \delta x_f \\ l \end{bmatrix} = \begin{bmatrix} f_f \\ \mathcal{H}\delta x_p \end{bmatrix} \tag{20} $$

where $l = [\lambda_1 \lambda_2 \lambda_3 \cdots \lambda_n]$ are the Lagrange multipliers. If $\mathcal{K}_f$ cannot be obtained by inverting $\mathcal{C}_f$, due to negligible structural compliance, the equations can be solved in the finger joint space. Solving for δx_f (where δx_f is the concatenated vector of all fingertip motions), we obtain

$$ \begin{bmatrix} \delta x_f \\ l \end{bmatrix} = \begin{bmatrix} \mathcal{K}_f & \mathcal{H}^T \\ \mathcal{H} & 0 \end{bmatrix}^{-1} \begin{bmatrix} f_f \\ \mathcal{H}\delta x_p \end{bmatrix}. \tag{21} $$

For particular contact types with decoupled servo stiffness, the partitions of the inverse matrix in (21) can be computed symbolically in terms of C_f for each finger. Appendix I shows the solution for δx_{f_i} (where δx_f is the motion of each fingertip) for the case of independent point contacts with friction, where $\mathcal{K}_f$ is block-diagonal.

While δx_{f_i} represents the absolute motion of a fingertip,

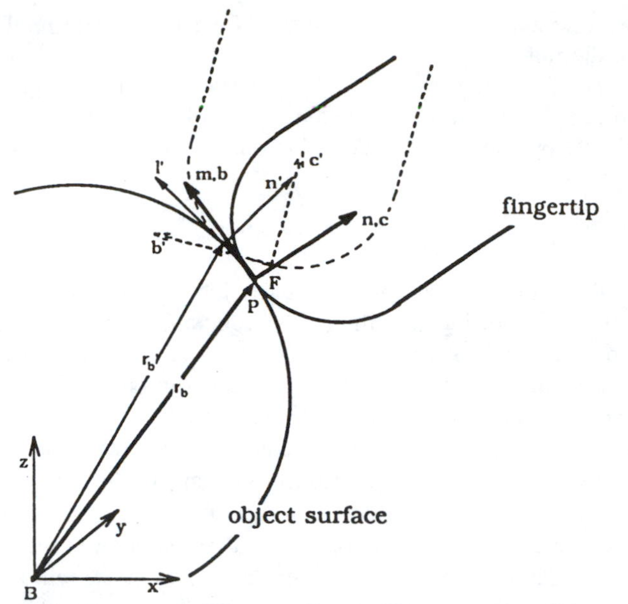

Fig. 2. Coordinate systems for a finger rolling on a grasped object.

$(\delta x_{f_i} - \delta x_{p_i})$ represents the relative motion of the fingertip with respect to the object. Expressed in the body coordinate frame $B(xyz)$, the relative motion becomes

$$ {}^{P}_{B}J_i^{-1}(\delta x_{p_i} - \delta x_{f_i}) = \delta x_b - {}^{P}_{B}J_i^{-1}\delta x_{f_i}. \tag{22}$$

By comparing δx_{f_i} and δx_{p_i}, it is possible for a given contact geometry and contact type (e.g., rolling contact) to determine how the contact point is moving upon the object. Fig. 2 shows the contact coordinate systems before and after a fingertip has rolled slightly. The details of computing the motion of the contact can be complex, but solutions are discussed in [7] and [4] and, for the special case in which the finger does not twist about its own axis while rolling, a first-order approximation is given in [2]. For our purposes, it suffices to find the change in the resultant *force* on the object as the finger rotates with respect to the object and the contact point shifts upon the object. Put another way, from (6) we know that initially $f_{p_i} = f_{f_i}$, but after infinitesimal motion the force acting at the contact becomes

$$f'_{p_i} = D_i^T f_{f_i} \tag{23}$$

where D_i is a matrix describing a small, Cartesian transformation between the new position and orientation of the fingertip force f_{f_i} and the initial $P(lmn)$ contact coordinate system. Then, in body coordinates we have

$$f'_{b_i} = {}^{P}_{B}J_i^T D_i^T f_{f_i} = {}^{P}_{B}J_i^T f_{f_i} + {}^{P}_{B}\Delta J_i^T f_{f_i}. \tag{24}$$

If no rolling or sliding occurs, D_i will contain only rotation terms, but in general both translations and rotations may be present. Solving for ${}^{P}_{B}\Delta J_i^T$, we have

$${}^{P}_{B}\Delta J_i^T = {}^{P}_{B}J_i^T(D_i^T - I) \tag{25}$$

where ${}^{P}_{B}\Delta J_i^T$ represents a differential Jacobian due to the changes in the geometry. Details are given in Appendix II.

Multiplying ${}^{P}_{B}\Delta J_i^T$ by f_{f_i} gives us K_{J_i}, that represents the change of forces due to geometry effects only. Since ${}^{P}_{B}\Delta J^T$ can be expressed as a linear function of δx_b, as shown in Appendix II, it is clear that ${}^{P}_{B}\Delta J^T f_f$ is a *linear* function of f_f and δx_b only. Since f_f is held constant in defining K_J in (9), the term ${}^{P}_{B}\Delta J^T f_f$ can be written as

$${}^{P}_{B}\Delta J^T f_f = \frac{\partial({}^{P}_{B}\Delta J^T f_f)}{\partial \delta x_b}\,\delta x_b. \tag{26}$$

Therefore, we obtain the stiffness matrix due to geometry changes as follows:

$$K_{J_i} = \frac{\partial({}^{P}_{B}\Delta J_i^T f_{f_i})}{\partial \delta x_b}. \tag{27}$$

Finally, the effective grasp stiffness is obtained by summing K_b from Section III-C and K_{J_i} for all fingers

$$K_e = K_b + \sum_{i=1}^{nf} K_{J_i}. \tag{28}$$

E. Forward Procedure Examples

In the following examples we illustrate some of the issues that arise in computing the overall compliance of a grasp. To reduce the algebra, Example 1 illustrates a grasp with structural compliance but without changes in geometry and Example 2 illustrates the effects of changes in grasp geometry, without structural compliance.

Example 1: Inserting a Rivet into a Hole: Fig. 3 shows two three-joint fingers about to insert a small rivet into a chamfered hole. We wish to find K_b due to servo and structural compliance. The grasp is inspired by the thumb-index finger "pinch" that people use in precision tasks with tiny objects. The grasp also approximates a two-fingered grasp between the thumb and one of the fingers of the Stanford/JPL hand, except that the base axis of the thumb has been rotated parallel to the base axes of the fingers to simplify the Jacobian. Note that while the grasp appears planar in Fig. 3, it can manipulate the rivet along the x-axis, out of the plane of the page. For realism, we choose servo and structural compliances comparable to those listed for the Stanford/JPL hand in Table II and assume "very-soft" contacts between the fingertips and the rivet. The Jacobians and the compliance matrices are given in Appendix III-A. Applying (10)–(15), we arrive at the object stiffness matrix at the tip of the rivet

$$K_b = \begin{bmatrix} 2490 & 0 & 0 & 0 & 258 & 0 \\ 0 & 28900 & 0 & 191 & 0 & 0 \\ 0 & 0 & 61610 & 0 & 0 & 0 \\ 0 & 191 & 0 & 22 & 0 & 0 \\ 258 & 0 & 0 & 0 & 37 & 0 \\ 0 & 0 & 0 & 0 & 0 & 35 \end{bmatrix}$$

where the units are in newtons per meter and newton-meters for translational and rotational stiffness, respectively.

Discussion: Looking at K_b, we notice that while the structural compliances are small compared to the servo compliances (about 10 percent), they prevent both the grasp and fingertip compliance matrices from becoming singular. By

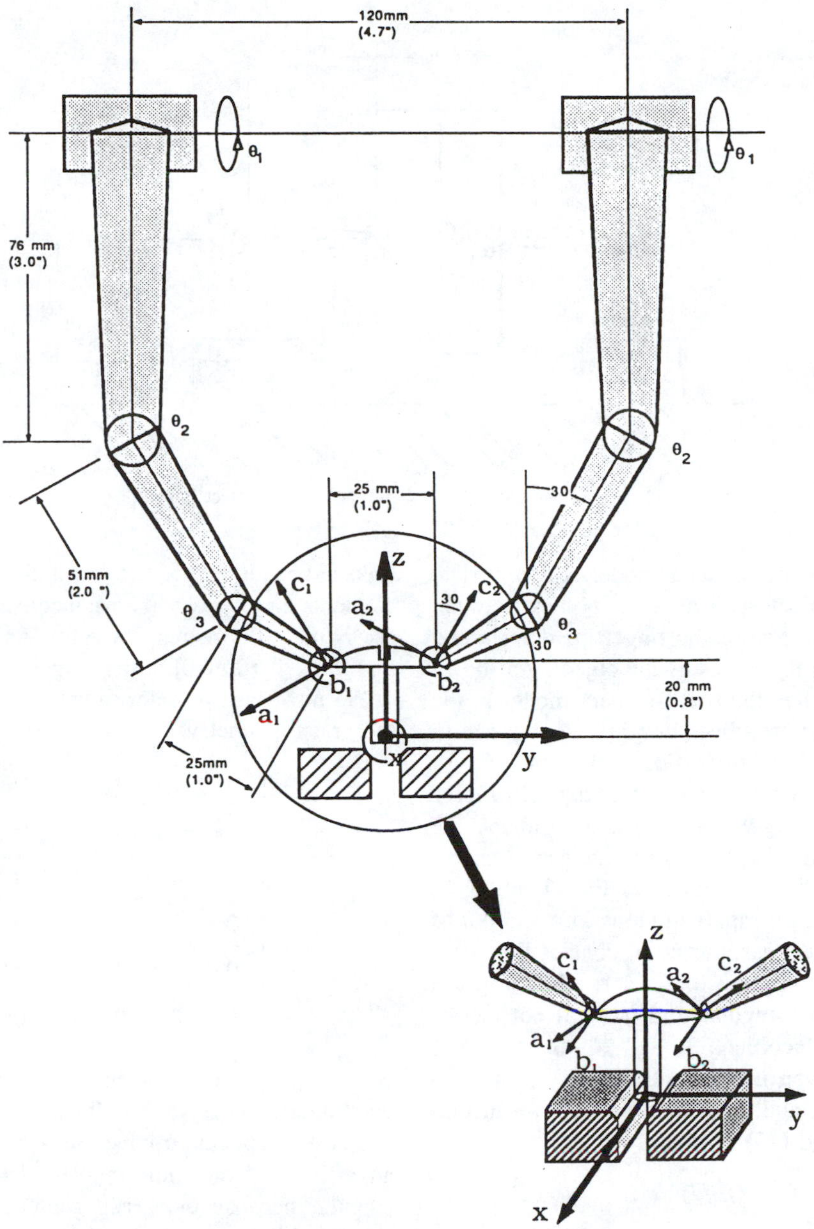

Fig. 3. Assembling a rivet with two soft-tipped fingers.

evaluating the eigenvalues of K_b, we can readily find the stiffest and softest directions. The eigenvalues are all positive, confirming that K_b is stable. Also, since K_b has full rank, the grasp is a force-closure grasp and can resist forces and moments from any direction.[4] Finally, we observe that the stiffness matrix is decoupled in the z-direction since the grasp is symmetric about the z-axis. If K_b were completely diagonal, the center of compliance of the rivet would be at the $B(xyz)$ origin.

Before leaving this example, it is worth recalling that the stability and full rank properties of this grasp stem from our assumption of "very-soft" fingertips. In fact, the choice of contact model for an example such as this requires some care. The contact areas are small since the head of the rivet is thin. However, if we had assumed point contacts with friction, the change in $\mathcal{H}$ would result in a new stiffness matrix with a singular direction: $\vec{d} = [0\,0\,0\,0\,1\,0]$. In other words, the grasp could not resist moments about the y-axis and would not be a force-closure grasp—which seems unrealistic. A soft-finger model (in which the fingertips are free to roll about their a and b axes, but prevented from twisting about their c axes) seem more reasonable, but can also lead to an unrealistic stiffness matrix. Part of the problem is that the surface normal (the direction of the c-axis) is poorly defined at the edge of the rivet. In addition, the combination of a rotational constraint about the c-axis and a lack of constraint for rotations about the a and b axes, tends to produce a grasp stiffness matrix that is unrealistically stiff with respect to rotations about the z and y axes. As a result, the best model is the "very-soft" finger in which a full 6×6 compliance matrix exists for the contact, but for which rotational compliances are much larger than translational compliances, due to the small contact area. The

[4] However, this grasp cannot *impart* independent forces and motions in all directions, due to kinematic coupling.

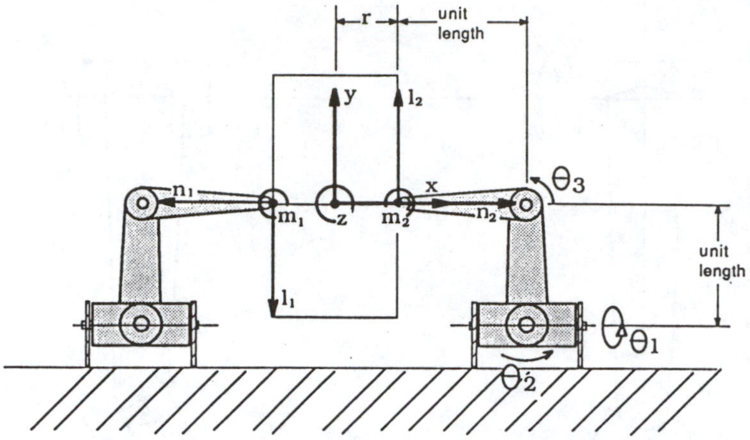

Fig. 4. A possible unstable grasp with point-contact fingers.

reader can also verify that this contact model is appropriate when grasping a small object such as a thumbtack between the thumb and index fingers, since the fingertips deform and may even partially enclose the edges of the object. As shown in [3], this model approaches the point-contact model in the limiting case as the contact area becomes small compared to the characteristic lengths of the fingertips or the object.

Example 2: The Effect of Changes in Grasp Geometry for Two Point-Contact Fingers: We wish to compute the effective stiffness, including the effects of changes in geometry, for the grasp shown in Fig. 4. As in Fig. 3, although the grasp looks planar, it can impart motions into and out of the page. For simplicity, we assume point-contact fingertips and negligible structural compliance. As a result, the grasp stiffness matrix K_b will be singular and we will not have a force-closure grasp. The Jacobians $\mathfrak{J}_\theta$ and ${}^{P}_{B}\mathfrak{J}$ and the joint stiffness matrix K_θ are given in Appendix III-B. For a diagonal joint stiffness matrix, and the dimensions shown in Fig. 4, we apply (11), (14), and (15) to obtain

$$K_b = 2\begin{bmatrix} (k_b+k_c) & 0 & 0 & 0 & 0 & -rk_c \\ 0 & k_c & 0 & 0 & 0 & 0 \\ 0 & 0 & k_a & 0 & 0 & 0 \\ 0 & 0 & 0 & 0 & 0 & 0 \\ 0 & 0 & 0 & 0 & r^2k_a & 0 \\ -rk_c & 0 & 0 & 0 & 0 & r^2k_c \end{bmatrix} \quad (29)$$

where r is one half the width of the object. (Because the links have unit length, the translational and rotational stiffness elements may appear to have inconsistent dimensions, but they are N/m and Nm, respectively.) As expected, K_b is singular and $\vec{y} = [0\,0\,0\,1\,0\,0]$ is the singular direction of the grasp.

We now wish to determine the effects of small changes in the grasp geometry. Following (16)–(27), we obtain K_J

$$K_J = 2\begin{bmatrix} 0 & 0 & 0 & 0 & 0 & 0 \\ 0 & -f_n & 0 & 0 & 0 & 0 \\ 0 & 0 & 0 & 0 & 0 & 0 \\ 0 & 0 & 0 & 0 & 0 & 0 \\ 0 & 0 & 0 & 0 & -rf_n & 0 \\ 0 & 0 & 0 & 0 & 0 & -rf_n(1+r) \end{bmatrix} \quad (30)$$

where f_n represents the initial normal grasp force applied at the fingertips. As in K_b, linear and rotational stiffness elements may appear to have the wrong dimensions, but this is an artifact of the unit link lengths. We observe that K_J is singular because the grasp geometry is unaffected by moving the object along the x-axis or rotating it about the x-axis. In addition, since the fingertips cannot rotate about their b-axes, there is no change in the grasp geometry for small motions in the z-direction. More importantly, the stiffness terms corresponding to rotations about the y and z axes are negative. Thus K_J tends to *destabilize* the grasp.

The effective grasp stiffness is obtained by adding K_b and K_J as in (28). The resulting expression for K_e makes intuitive sense

$$K_e = 2\begin{bmatrix} (k_b+k_c) & 0 & 0 & 0 & 0 & -rk_c \\ 0 & (k_c-f_n) & 0 & 0 & 0 & 0 \\ 0 & 0 & k_a & 0 & 0 & 0 \\ 0 & 0 & 0 & 0 & 0 & 0 \\ 0 & 0 & 0 & 0 & r(rk_a-f_n) & 0 \\ -rk_c & 0 & 0 & 0 & 0 & r(rk_c-f_n-rf_n) \end{bmatrix} \quad (31)$$

Discussion: The effective stiffness matrix of this grasp reveals the important grasp properties. For example, since K_e is singular, we know that the grasp is not a force-closure grasp. Examination of the fourth row and column reveals that this is because the grasp cannot impart or resist moments about the x-axis. In addition, we find that if either

$$f_n \geq rk_a \quad \text{or} \quad f_n \geq \left(\frac{rk_c}{r+1}\right)\left(\frac{k_b}{k_b+k_c}\right)$$

K_e will not be positive-definite and the grasp will be unstable. We also see that for a given object size and finger stiffness, pressing harder without increasing the finger stiffness makes the grip less stable. This effect is easily demonstrated by pressing a small coin on edge between two fingers and gradually increasing the grasp force until the coin snaps over. In fact, since human muscles tend to become stiffer when larger forces are applied, there is some automatic compensation for this instability. If we purposefully tense our muscles, thereby increasing K_b, we can preserve the positive definiteness of K_e for large grasp forces.

IV. Controlling the Grasp Stiffness

Given an initial grasp and a manipulation task, how should we control the finger joints so as to achieve a desired grasp stiffness?

Working backwards from (28), we have

$$K_b = K_e - \sum_{i=1}^{nf} K_{J_i}$$

where K_e is the desired grasp stiffness and K_J is a function of the grasp geometry. There are then two major steps in the reverse procedure: 1) obtain $\mathcal{K}_p$ from the desired K_b, and 2) obtain $\mathcal{C}_\theta$ or $\mathcal{K}_\theta$ from $\mathcal{K}_p$. We see from (14) and (15) that we can expand the contact stiffness as

$$\mathcal{K}_p = \begin{bmatrix} K_{p_{11}} & K_{p_{12}} & \cdots & K_{p_{1,nf}} \\ K_{p_{21}} & K_{p_{22}} & \cdots & K_{p_{2,nf}} \\ & & \ddots & \vdots \\ & \cdots & & K_{p_{nf,nf}} \end{bmatrix} \tag{32}$$

where $K_{p_{ij}}$ are the desired elements of $\mathcal{K}_p$ that we wish to specify. Note that for $i \neq j$, $K_{p_{ij}}$ represent coupling terms among the joints on different fingers. These terms provide us with considerably more control over K_b than we could obtain from independently servoed fingers.

Equation (15) presents an underconstrained problem: there are many possible $\mathcal{K}_p$ that will satisfy K_b. However, there are also numerous constraints on K_p imposed by the kinematics of the fingers, the contact types, and the joint configurations. The reverse compliance procedure is consequently difficult, often involving both overconstraint and underconstraint in a single grasp. The problem is similar to reverse force or velocity computations, for which solutions have been presented by Kerr [7] and Kobayashi [8], but is more complicated due to off-diagonal terms in the stiffness matrix and due to structural compliances. However, as we shall see in the following section, these extra terms also afford us more opportunities to achieve a desired grasp stiffness.

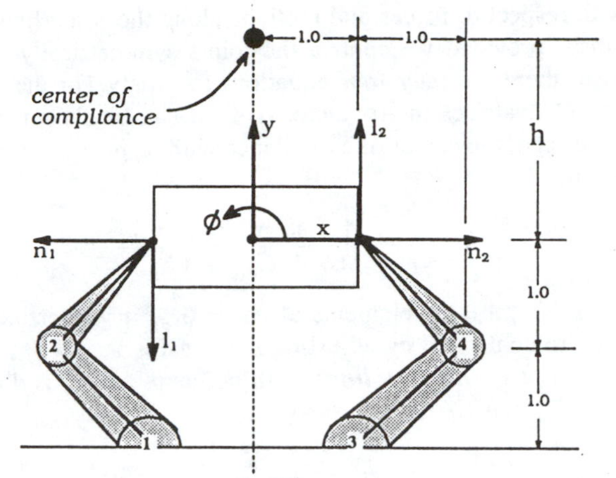

Fig. 5. Coordinates and center of compliance for a planar grasp.

Before launching into an exploration of general, three-dimensional reverse compliance it is worth examining a simple case that provides us with some physical insight for what can happen.

Example 3: Controlling the Compliance of a Planar Grasp: The dimensions and configuration of a simple, two-fingered grasp are shown in Fig. 5. Since that is a planar problem, forces and motions are represented by three-element vectors with two translational components and one rotation: $[x\ y,\ \phi]$. The left finger Jacobian is

$$J_{\theta_1} = \begin{bmatrix} 0 & -1 \\ 2 & 1 \\ 1 & 1 \end{bmatrix} \tag{33}$$

which maps joints to the Cartesian coordinates $[l_1, n_1, \phi_{m_1}]$, and the right finger follows from symmetry. We will assume that structural compliance is confined to the fingertips and that the contact areas are small enough that the contact behaves nearly as a point contact. The symbolic $\mathcal{C}_\theta$, $\mathcal{K}$, and $\mathcal{C}_s$ matrices are given in Appendix III-C. If we keep the grasp symmetric, the forward and reverse compliance computations are simple enough to perform symbolically.

We want to locate the Center of Compliance of the object (the point for which the stiffness or compliance matrix becomes diagonal) at an arbitrary location, while also maintaining a stable (positive-definite) stiffness matrix and controlling the magnitude of the stiffnesses in the x, y, and ϕ directions. We examine four different cases, of increasing complexity:

1) no coupling in the control of the joints diagonal),
2) coupling only between the joints on each finger ($\mathcal{K}_\theta$ or $\mathcal{C}_\theta$ is block-diagonal),
3) full coupling among all joints,
4) an overconstrained case in which the lower right joint is locked.

In general, the planar graphs stiffness matrix K_b is a 3×3 symmetric matrix, resulting in six equations. However, since this grasp is symmetric, its stiffness will always be uncoupled with respect to forces and motions along the centerline of the grasp, provided we control the joints symmetrically. In this case, there are only four equations to satisfy. For the $\mathcal{C}_\theta$, $\mathcal{JC}$, and $\mathcal{C}_s$ matrices in Appendix III-C, it can be shown that the location of the center of compliance with respect to the $B(xyz)$ origin is

$$h = \frac{C_{\theta_{24}} + C_{\theta_{22}} + 2C_{\theta_{14}} + 2C_{\theta_{12}}}{C_s + C_{\theta_{24}} + C_{\theta_{22}}} \tag{34}$$

where $C_{\theta_{ij}}$ are the elements of $\mathcal{C}_\theta = \mathcal{K}_\theta^{-1}$ in Appendix III-C, and are controlled by adjusting servo gains.

Case 1—no coupling: With no coupling, $\mathcal{C}_\theta$ is diagonal and $C_{\theta_{24}}$ and $C_{\theta_{12}}$ are zero so that

$$h = \frac{C_{\theta_{22}}}{C_s + C_{\theta_{22}}}.$$

As we might expect from examining the designs of Remote-Center-Compliance devices [15], [3], $h \to 0$ as the structural compliance becomes very large and $h \to 1$ when structural compliance is negligible. The latter case corresponds to the instantaneous center of rotation of a linkage formed by the fingers and object in Fig. 5. By adjusting the compliance of the second joints of the fingers we could, in theory, vary h between 0 and 1, except that there are practical upper and lower bounds on $C_{\theta_{22}}$. The lower bound is established by the maximum servo stiffness $(1/C_{\theta_{22}})$ that we can reasonably obtain while the upper bound is determined by the need to keep $\mathcal{C}_\theta$ positive-definite so that the grasp is stable.

Case 2—intra-finger coupling: If we allow coupling between the first and second joint of each finger we obtain another independent variable, $C_{\theta_{12}}$, that we can use in controlling K_b and h. The resulting expression for h is

$$h = \frac{C_{\theta_{22}} + 2C_{\theta_{12}}}{C_s + C_{\theta_{22}}}.$$

It would now appear that we could place the Center of Compliance anywhere along the centerline of the grasp except that again, we are limited by achievable upper bounds on the servo stiffnesses, and by the need to keep $\mathcal{C}_\theta$ positive-definite. If we look at the limiting case in which we are trying to maximize h (in which case C_s should approach zero), we arrive at the following bounds on h, in terms of the elements $\mathcal{C}_\theta$ to keep the joint compliance matrix positive-definite:

$$1 - 2\sqrt{\frac{n}{C_{\theta_{22}}} - 1} < h < 1 + 2\sqrt{\frac{n}{C_{\theta_{22}}} - 1} \tag{35}$$

where $n = (C_{\theta_{11}} + C_{\theta_{22}})$ is the trace of C_θ.

Case 3—inter- and intra-finger coupling: In contrast with the last two cases, if we permit coupling among the joints on different fingers, the expression for h is given by (34), from which it is clear that we now have considerably more freedom in controlling h while keeping $\mathcal{C}_\theta$ positive-definite by choosing appropriate off-diagonal terms.

Case 4—overconstraint: Suppose that the lower right joint is locked, so that we have a hand with just three joints. Can we still control K_b? If we use coupling among all the joints, we have six independent variables, which would seem to be enough to control the six independent elements of K_b (the grasp is no longer symmetric). However, our intuition tells us that with only a single link on the right side, we will run into difficulties, since the tip of the link cannot move in the x direction without also moving in the y direction.

The symbolic expression for this nonsymmetric grasp matrix is complicated, but some special cases are worth examining. To begin with, we *must* have structural compliance to achieve any control of K_b. In addition, if we try to diagonalize K_b at the origin we discover that this is only possible if the diagonal terms $k_{b_{11}}$ and $k_{b_{33}}$ (corresponding to stiffnesses in the x and ϕ directions, respectively) are also zero. More generally, it is possible to achieve a variety of K_b matrices, but our options are severely constrained. For example, if we try to place the center of compliance at a location [0 1 0] with the stiffnesses in the x, y, and ϕ directions of 50, 75, and 100, the resulting joint stiffness matrix

$$K_\theta = \begin{bmatrix} -100 & 0 & 100 \\ 0 & 50 & -50 \\ 100 & -50 & 50 \end{bmatrix} \tag{36}$$

is not positive-definite.

This raises an interesting issue: How can we be sure that the reverse compliance procedure will not produce joint stiffness matrices with negative eigenvalues, and more importantly, how can we be sure that a positive-definite joint stiffness matrix will not produce an unstable K_b matrix? Fortunately, it can be shown that while a positive-definite K_b matrix does not always produce a positive-definite $\mathcal{K}_\theta$ matrix, the reverse is always true. The proof is given in Appendix IV. Thus a safe approach is to require that $\mathcal{K}_\theta$ be positive-definite and to use the off-diagonal terms resulting from coupling among the joints to increase the positive definiteness of $\mathcal{K}_\theta$.

A. A General Approach to the Reverse Procedure

The planar example in the last section has given us some idea of what to expect when solving for $\mathcal{K}_\theta$ to achieve a desired K_b. The basic problem is to ensure that the limitations of the contacts and finger joints are reflected in $\mathcal{K}_p$.

We now proceed to elaborate the two steps of the reverse procedure. First, we have to find a matrix $\mathcal{K}_p$ that satisfies (15). As discussed previously, this is an underconstrained problem, and many solutions exist. However, $\mathcal{K}_p$ must also satisfy the following contact and joint constraints:

1) *Contact constraints:* $\mathcal{K}_p$ should reflect the constraints represented by $\mathcal{H}$. In other words, $\mathcal{K}_p$ shoud have the same zero rows and columns as $\mathcal{H}^T\mathcal{H}$. This is because $\mathcal{K}_p = \mathcal{H}^T(\mathcal{H}\mathcal{C}_f\mathcal{H}^T)^{-1}\mathcal{H}$, as shown in (14).

2) *Joint constraints:* To ensure that the fingers can fully control the $\mathcal{K}_p$ matrix, we require that the nullspace of the product

$$\Omega = (\mathcal{C}_s' - \mathcal{C}_s'\mathcal{H}\mathcal{K}_p\mathcal{H}^T\mathcal{C}_s') \tag{37}$$

should be the same as the left nullspace of

$$\mathcal{B} = \mathcal{H}\mathcal{J}_\theta. \tag{38}$$

The derivation is given in Appendix V. As an example, in the overconstrained case from Example 3, the finger tip on the right is constrained to move in the $\pm[1\,1\,0]$ direction, which is orthogonal to the null space of Ω ([1 −1 0]).

The remaining requirement, which surfaced in the overconstrained example in the last section is, that $\mathcal{C}_\theta$ should be kept positive-definite so that K_b will be positive-definite (or positive-semidefinite for a non-force-closure grasp). In addition, it may be useful to impose other restrictions for numerical or controls reasons.

Once we have found a suitable $\mathcal{K}_p$ matrix that satisfies the above constraints, it is straightforward to find $\mathcal{K}_\theta$. The presence of structural compliance provides some complication but, following the derivation in Appendix V, it can be shown that

$$\mathcal{K}_\theta = \mathcal{C}_\theta^{-1} = [(\mathcal{H}\mathcal{J}_\theta)^*(\mathcal{C}_s' - \mathcal{C}_s'\mathcal{H}\mathcal{K}_p\mathcal{H}^T\mathcal{C}_s')(\mathcal{J}_\theta^T\mathcal{H}^T)^*]^{-1} - (\mathcal{J}_\theta^T\mathcal{H}^T\mathcal{C}_s'^{-1}\mathcal{H}\mathcal{J}_\theta). \quad (39)$$

Example 4: Reverse Compliance Computation for an Assembly Problem: We return to the three-dimensional grasp of Example 1, and attempt to determine appropriate joint stiffnesses so as to achieve a desired stiffness for the rivet. As Whitney [15] and others have demonstrated, it is ideal if the Center of Compliance of the rivet is placed at or near its tip. Therefore, we desire a stiffness matrix that will be diagonal at the $B(xyz)$ origin. In this example, we neglect the changes in geometry for small grasp forces, i.e., K_J is negligible. In addition, it is desirable for the stiffness along the x and y axes to be relatively small. We therefore specify the following desired grasp stiffness matrix

$$K_e = \begin{bmatrix} 4000 & 0 & 0 & 0 & 0 & 0 \\ 0 & 10400 & 0 & 0 & 0 & 0 \\ 0 & 0 & 49000 & 0 & 0 & 0 \\ 0 & 0 & 0 & 0.1 & 0 & 0 \\ 0 & 0 & 0 & 0 & 14 & 0 \\ 0 & 0 & 0 & 0 & 0 & 27 \end{bmatrix}$$

where units are in N/m and Nm for linear and rotational stiffnesses.

Referring to (37) and (38), we construct a concatenated grasp stiffness matrix $\mathcal{K}_p$ that satisfies the criteria listed earlier in this section. The result is shown in Appendix III-D. Then, applying (39), we obtain $\mathcal{K}_\theta$

$$\mathcal{K}_\theta = \begin{bmatrix} 37.8 & 0 & 0 & -1.8 & 0 & 0 \\ 0 & 6.6 & -0.0006 & 0 & -3.5 & 3.2 \\ 0 & -0.0006 & 6.7 & 0 & 3.2 & -3 \\ -1.8 & 0 & 0 & 37.8 & 0 & 0 \\ 0 & -3.5 & 3.2 & 0 & 6.6 & -0.0006 \\ 0 & 3.2 & -3 & 0 & -0.0006 & 6.7 \end{bmatrix}$$

which can be checked through the forward procedure. The result is nearly identical to our desired K_b matrix (see Appendix III-D).

Discussion: As expected $\mathcal{K}_\theta$ is a symmetric 6×6 matrix which, like K_b, has rank 6. In addition, although $\mathcal{K}_\theta$ is the concatenated stiffness matrix of two three-joint fingers, it is not block-diagonal. In other words, inter- and intra-finger coupling occurs to achieve the desired grasp stiffness.

Suppose that we do not permit coupling between the fingers? Then we cannot expect to achieve an arbitrary K_b matrix. The problem is that we have only two soft fingers, each with a 4×4 symmetric K_f matrix providing ten independent variables. Applying (15), we have

$$K_b = {}^{\mathrm{P}}_{\mathrm{B}}\mathcal{J}^{T}\mathcal{K}_p\,{}^{\mathrm{P}}_{\mathrm{B}}\mathcal{J}$$

which yields up to 21 equations. Therefore, we have a problem with 21 equations for 20 unknowns. If we rewrite the equations in the form $Ax = b$, where x represents the free variables in matrices K_{p_i} and A and b are the known coefficients, the necessary and sufficient conditions for a solution to exist are

$$\det[A \mid b] \equiv 0 \quad \text{and} \quad i_{21}^T y \neq 0 \quad (40)$$

where i_{21} is a 21×1 unit vector, with the 21st element being 1, and y is a vector in the null space of augmented matrix $[A\ b]$. It can be further shown that there is no solution for this grasp if coupling is not allowed. More generally, Table III shows the *minimum* number of fingers required to fully specify the grasp stiffness for some common contact types.

V. Conclusions

The stiffness or compliance of a robotic grasp represents the rate of change of grasp forces with respect to small motions of the grasped object. The stiffness depends on structural compliances in the fingers and fingertips, on servo gains at the finger joints, and on small changes in the grasp geometry that affect the way in which the grasp forces act upon the object. As such, the grasp stiffness is a useful measure of the grasp. The rank of the stiffness matrix immediately reveals whether the grasp is a force-closure grasp and can resist forces and moments from arbitrary directions. The singular directions $\vec{d}$ are readily found through $K_e\vec{d} = 0$. The eigenvalues of the matrix reveal the stiffest and softest directions. Generally, the stiffest directions will be those in which only structural compliance is present. Finally, the positive definiteness of the grasp matrix is a measrue of the quasi-static grasp stability. As discussed in Section III-D, an unstable grasp is one for which the contributions of changes in the grasp geometry K_J are significant, and negative.

For controlling a hand, we are interested in specifying servo gains at the finger joints for a desired grasp stiffness. For example, we may wish the grasp to have a center of compliance

TABLE III
MINIMUM NUMBER OF FINGERS NEEDED TO CONTROL K_e WITH AND WITHOUT COUPLING

Contact Type	No Coupling	Intra-Finger Coupling	Inter-Finger Coupling
Point contact without friction	21	21	6
Point contact with friction	7	4	2
Soft contact	6	3	2

at a particular point. Achieving the desired stiffness may be either an overconstrained or underconstrained problem, depending on the number of fingers and the number of joints per finger. One way to check this is to look at the stiffness matrix obtained through the forward procedure—it should have full rank and no extremely stiff directions.

Generally, it is useful to couple the servoing of joints on different fingers so that we can better control the off-diagonal terms of K_b. In the fourth example, we show that without inter-finger coupling we are not able to satisfy the compliance requirements of the task.

Appendix I
Symbolic Matrix Inverse for Point Contacts

If K_f is block-diagonal, we can treat each fingertip separately. If we divide the fingertip stiffness matrix in (20) into 3×3 partitions

$$K_{f_i} = \begin{bmatrix} K_{11} & K_{12} \\ K_{21} & K_{22} \end{bmatrix}.$$

Then, for point contacts, the inverse of the matrix in (21) can be expressed as

$$\begin{bmatrix} K_{f_i} & H_i^T \\ H_i & 0 \end{bmatrix}^{-1} = \begin{bmatrix} 0 & 0 & I_3 \\ 0 & K_{22}^{-1} & -K_{22}^{-1}K_{21} \\ I_3 & -K_{12}K_{22}^{-1} & (K_{12}K_{22}^{-1}K_{21} - K_{11}) \end{bmatrix}$$

where the matrix H_i is

$$H_i = \begin{bmatrix} 1 & 0 & 0 & 0 & 0 & 0 \\ 0 & 1 & 0 & 0 & 0 & 0 \\ 0 & 0 & 1 & 0 & 0 & 0 \end{bmatrix}$$

for point contact, and from (21), δx_{f_i} is simply

$$\delta x_{f_i} = \begin{bmatrix} H_i \delta x_{p_i} \\ -K_{22}^{-1} K_{21} H_i \delta x_{p_i} \end{bmatrix}$$

for point contacts. Similar symbolic expressions can be derived for soft contacts and very soft contacts. Thus the computation of δx_{f_i} need not be computationally expensive.

Appendix II
Differential Jacobian

In (23) we define a 6×6 Cartesian transformation matrix between the initial $P(lmn)$ coordinate system and the new position and orientation of the grasp force after a small motion has taken place. If we denote the translation and rotation of the grasp force with respect to the initial $P(lmn)$ coordinate system as $\delta x = [dx, dy, dz, d\theta_x, d\theta_y, d\theta_z]$, we can express the Cartesian transformation as

$$D^T = \left[\begin{array}{c|c} \Delta A + I & 0 \\ \hline \Delta R(\Delta A + I) & \Delta A + I \end{array}\right] \tag{41}$$

where ΔA^T and ΔR^T are small rotation and cross-product matrices, respectively. They can be generated from δx as

$$\Delta A^T = \begin{bmatrix} 0 & -d\theta_z & d\theta_y \\ d\theta_z & 0 & -d\theta_x \\ -d\theta_y & d\theta_x & 0 \end{bmatrix}$$

$$\Delta R^T = \begin{bmatrix} 0 & -dz & dy \\ dz & 0 & -dx \\ -dy & dx & 0 \end{bmatrix}.$$

If the orientation and the point of application of the grasp force do not change, A is simply an identity matrix and R a null matrix.

If we neglect second-order and smaller terms, we can express the differential Jacobian in (25) as

$${}^{P}_{B}\Delta J^T = \left[\begin{array}{c|c} \Delta A & 0 \\ \hline \Delta R & \Delta A \end{array}\right]. \tag{42}$$

Appendix III
Equations and Expressions in Examples

A. Example 1

We assume servo stiffnesses comparable to those achievable in the Stanford/JPL hand and add some coupling among the joints on different fingers to achieve a typical joint stiffness matrix

$$\mathcal{K}_\theta = \mathcal{C}_\theta^{-1} = \begin{bmatrix} 10 & 0 & 0 & -1 & 0 & 0 \\ 0 & 5.65 & 0 & 0 & -1.5 & 0 \\ 0 & 0 & 3.66 & 0 & 0 & -1.1 \\ -1 & 0 & 0 & 10 & 0 & 0 \\ 0 & -1.5 & 0 & 0 & 5.65 & 0 \\ 0 & 0 & -1.1 & 0 & 0 & 3.1 \end{bmatrix}$$

where units are in Nm, for rotational stiffness. The joint Jacobian and the structural compliance matrix for the first finger are

$$J_{\theta_1} = \begin{bmatrix} 0 & -0.0728 & -0.022 \\ -0.1329 & 0 & 0 \\ 0 & 0.0127 & 0.0127 \\ -0.866 & 0 & 0 \\ 0 & 1 & 1 \\ -0.5 & 0 & 0 \end{bmatrix}$$

$$C_{s_1} = \begin{bmatrix} 1.02 & 0 & -0.141 & 0 & -11.1 & 0 \\ 0 & 1.71 & 0 & 9.21 & 0 & 5.32 \\ -0.141 & 0 & 0.184 & 0 & 2.64 & 0 \\ 0 & 9.21 & 0 & 856 & 0 & 34.6 \\ -11.1 & 0 & 2.64 & 0 & 1008 & 0 \\ 0 & 5.32 & 0 & 34.6 & 0 & 520 \end{bmatrix} \times 10^{-4}.$$

The matrices for the second finger follow from symmetry. The structural compliance matrix was obtained by choosing values for the links comparable to those of the Stanford/JPL hand. A fingertip compliance matrix for a "very soft" finger was then added, with large rotational compliances in the l and m directions, corresponding to a small contact area and approaching the limiting case of a soft finger contact. For very soft fingertips, the contact matrix, $\mathcal{K}$, is a 12 × 12 identity matrix.

B. Example 2

We assume a diagonal joint stiffness matrix, with identical stiffnesses for both fingers

$$\mathcal{K}_\theta = \begin{bmatrix} k_a & 0 & 0 & 0 & 0 & 0 \\ 0 & k_b & 0 & 0 & 0 & 0 \\ 0 & 0 & k_c & 0 & 0 & 0 \\ 0 & 0 & 0 & k_a & 0 & 0 \\ 0 & 0 & 0 & 0 & k_b & 0 \\ 0 & 0 & 0 & 0 & 0 & k_c \end{bmatrix}.$$

The joint Jacobian and coordinate transformation matrix for the left finger are

$$J_{\theta_1} = \begin{bmatrix} 0 & -1 & -1 \\ 1 & 0 & 0 \\ 0 & 1 & 0 \\ 0 & 0 & 0 \\ 0 & 1 & 1 \\ -1 & 0 & 0 \end{bmatrix}$$

$${}^{\mathrm{P}}_{\mathrm{B}}J_1 = \begin{bmatrix} 0 & -1 & 1 & 0 & 0 & r \\ 0 & 0 & 1 & 0 & r & 0 \\ -1 & 0 & 0 & 0 & 0 & 0 \\ 0 & 0 & 0 & 0 & -1 & 1 \\ 0 & 0 & 0 & 0 & 0 & 1 \\ 0 & 0 & 0 & -1 & 0 & 0 \end{bmatrix}.$$

The right finger follows from symmetry.

C. Example 3

In Example 3, the general servo stiffness matrix is

$$\mathcal{K}_\theta = \mathcal{C}_\theta^{-1} = \begin{bmatrix} K_{\theta_{11}} & K_{\theta_{12}} & K_{\theta_{13}} & K_{\theta_{14}} \\ K_{\theta_{12}} & K_{\theta_{22}} & K_{\theta_{23}} & K_{\theta_{24}} \\ K_{\theta_{13}} & K_{\theta_{23}} & K_{\theta_{33}} & K_{\theta_{34}} \\ K_{\theta_{14}} & K_{\theta_{24}} & K_{\theta_{34}} & K_{\theta_{44}} \end{bmatrix}$$

where the elements of $\mathcal{K}_\theta$ correspond to individual joint stiffnesses or coupling terms between different joints, as shown in Fig. 6. We also assume symmetry in this example so that $K_{\theta_{23}} = K_{\theta_{14}}$, $K_{\theta_{33}} = K_{\theta_{11}}$, $K_{\theta_{44}} = K_{\theta_{22}}$, and $K_{\theta_{34}} = K_{\theta_{12}}$. The structural compliance matrix is assumed to be diagonal (fingertip compliance only) so that

$$\mathcal{K}\mathcal{C}_s = \begin{bmatrix} C_s & 0 & 0 & 0 \\ 0 & C_s & 0 & 0 \\ 0 & 0 & C_s & 0 \\ 0 & 0 & 0 & C_s \end{bmatrix}.$$

D. Example 4

The null space vectors of Ω in this example are [0 3.76 0 −1 0 0 0 0] and [0 0 0 0 0 3.76 0 1] (in a, b, c, and rotation about c for each contact) which correspond to a single coupled motion in the object space, in which translation along the x-axis is unavoidably coupled with rotation about y.

If we take a K_p matrix that satisfies the constraints and apply the forward procedure, we obtain a K_b matrix that almost matches the desired K_b matrix. The only significant difference is due to the unavoidable coupling between x-axis translation and y-axis rotation:

$$K_b' = \begin{bmatrix} 3530 & 0 & 0 & 0 & 142 & 0 \\ 0 & 10400 & 0 & 0 & 0 & 0 \\ 0 & 0 & 49000 & 0 & 0 & 0 \\ 0 & 0 & 0 & 0.1 & 0 & 0 \\ 142 & 0 & 0 & 0 & 14 & 0 \\ 0 & 0 & 0 & 0 & 0 & 27 \end{bmatrix}$$

Appendix IV

Positive Definiteness of the Forward Procedure

We want to prove that a positive-definite $\mathcal{K}_\theta$ or $\mathcal{C}_\theta$ will result in a positive-semidefinite K_b. We begin by recalling the "Law

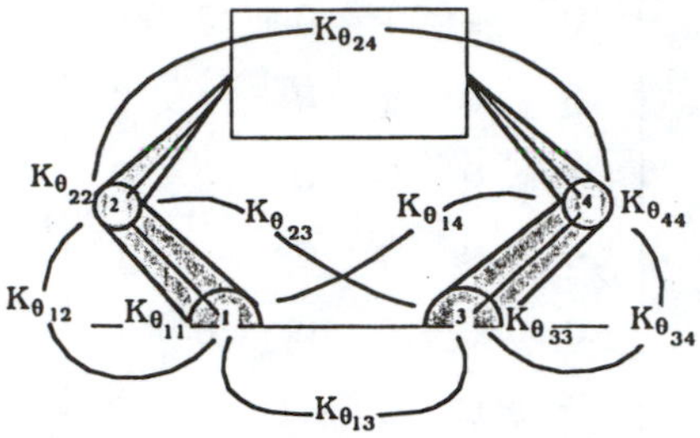

Fig. 6. Correlation between elements of K_θ and coupled joint control.

of Inertia'' for a congruence transformation:

Law of Inertia: If the transformation matrix T in the congreuence transformation T^TQT is nonsingular, the signs of the eigenvalues are preserved [14]. Therefore, we have the following two theorems:

Theorem 1: If Q is positive-definite, its congruence transformation is positive-semidefinite, i.e.,

$$\forall Q > 0 \Rightarrow T^TQT \geq 0, \qquad \text{for any matrix } T$$

also

$$Q > 0 \Rightarrow T^TQT > 0, \qquad \text{when } \mathfrak{N}(T) = \{\ \}$$

where $\mathfrak{N}(T)$ is the null space of matrix T.

Proof: To prove the positive definiteness of T^TQT, we pre- and post-multiply by a vector x^T and x. Therefore

$$x^T(T^TQT)x = y^TQy$$

where $y = Tx$ is a vector. Clearly, from definition of positive definiteness, $y^TQy > 0$ if $y = Tx$ is nontrivial. That is, if $y \notin \mathfrak{N}(T)$ then $T^TQT > 0$; otherwise, $T^TQT \geq 0$. ■

Theorem 2: If $\mathcal{C}_\theta$ or $\mathcal{K}_\theta$ is positive-definite, then $\mathcal{K}_b$ obtained in the forward procedure will be positive-semidefinite. If the grasp is force-closure, then $\mathcal{K}_b$ is positive-definite.

Proof: If we begin with a positive-definite $\mathcal{C}_\theta$, then $\mathcal{C}_j$ in (11) is positive-semidefinite according Theorem 1. Since $\mathcal{C}_s$ is always positive-definite, $\mathcal{C}_f$ in (13) is positive-definite. Therefore, $\mathcal{H}\mathcal{C}_f\mathcal{H}^T$ is positive-definite from Theorem 1, as is its inverse. $\mathcal{K}_p$ and K_b in (15) are therefore positive-semidefinite. The stiffness matrix K_b is thus positive-semidefinite. If the grasp is force-closure then K_b is nonsingular and therefore positive-definite. ■

Appendix V

Obtaining Joint Stiffness in the Reverse Procedure

In this Appendix, we find the expression for the joint stiffness matrix K_θ in terms of the structural compliance and the desired K_p matrix. Note that all matrices discussed here can be the concatenated matrices for generalization. First, we use (11) and (13) to expand the product $(HC_fH^T)^{-1}$

$$(HC_fH^T)^{-1} = (HC_sH^T + HJ_\theta C_\theta J_\theta^T H^T)^{-1}.$$

Then, we use the standard matrix inverse formula [5]

$$(A + BCD)^{-1} = A^{-1} - A^{-1}B(DA^{-1}B + C^{-1})^{-1}DA^{-1}$$

(where A and C are nonsingular square matrices) to expand (14) as

$$K_p = H^TC_s'^{-1}H - H^TC_s'^{-1}B(B^TC_s'^{-1}B + C_\theta^{-1})^{-1}BC_s'^{-1}H \tag{43}$$

where $C_s' = HC_sH^T$ represents structural compliance seen by the object through contact, and $B = HJ_\theta$, such that $\delta x_{tr} = B\delta\theta$.

We wish to invert the relationship in (43) to obtain C_θ^{-1} in terms of K_p and C_s'. As a first step

$$K_p - H^TC_s'^{-1}H = -H^TC_s'^{-1}B(B^TC_s'^{-1}B + C_\theta^{-1})^{-1}B^TC_s'^{-1}H.$$

All the matrix products in the above equation have the same form, with rows and columns of zeros determined by the contact matrix H. Therefore, it makes no difference if we only consider only the nontrivial submatrices, that is, if we pre- and post-multiply by H, and H^T. Then since $HH^T = I$, and $n \times n$ identity matrix

$$(C_s' - C'_sHK_pH^TC'_s) = B(B^TC_s'^{-1}B + C_\theta^{-1})^{-1}B^T. \tag{44}$$

Now, since $B = HJ_\theta$ is usually not square, there are two possibilities.

I: If $m \leq n$, then $B^*B = I_m$ and

$$C_\theta^{-1} = K_\theta = [B^*(C_s' - C'_sHK_pH^TC_s')B^{T*}]^{-1} - (B^TC_s'^{-1}B) \tag{45}$$

where the superscript * stands for generalized inverse and I_m is an $m \times m$ identity matrix.

II: If $m > n$, the finger has redundant joints and the optimal solution can be obtained from generalized inverse of B by minimizing the norm of C_θ in which case (45) for C_θ still holds.

Ω in the Reverse Procedure: In the reverse procedure, we require that the desired matrix K_p be chosen such that product $\Omega = (C_s' - C_s'HK_pH^TC_s')$ has the same null space as B. Ω represents the coupled directions, as seen through the contacts, of the motions of the fingers.

Theorem: The null space of matrix Ω is the same as that of B, which is independent of the structural compliance, $\mathcal{C}_s$ and of $\mathcal{C}_\theta$, as long as $\mathcal{C}_\theta$ is positive-definite.

Proof: Substituting K_p from (14) into Ω and expanding, we obtain:

$$\Omega = C_s' - C_s'HH^T(BC_\theta B^T + C_s')^{-1}HH^TC_s'$$

$$= C_s' - C_s'[C_s'^{-1} - C_s'^{-1}B(B^TC_s'^{-1}B + C_\theta^{-1})^{-1}B^TC_s'^{-1}]C_s'$$

or

$$\Omega = B(B^TC_s'^{-1}B + C_\theta^{-1})^{-1}B^T. \tag{46}$$

Since $\mathcal{C}_s$ is always positive-definite, if we choose a positive-definite C_θ matrix then $(B^TC_s'^{-1}B + C_\theta^{-1})$ will be positive-definite along with its inverse. From the theorem and proof in Appendix IV, we conclude that Ω has the same null space as B^T or the same left null space as B, which is independent of the values of C_s and C_θ. ■

Acknowledgment

The authors wish to thank P. Akella, and R. Howe for their comments and their help in doing the experiments on the Stanford/JPL hand. They also wish to thank M. Nagurka and J. Jourdain at CMU for many useful discussions. Finally, they thank the reviewers for the excellent suggestions on the draft.

References

[1] H. Asada, "Studies on prehension and handling by robot hands with elastic fingers," Ph.D. dissertation, Kyoto University, Kyoto, Japan, Apr. 1979.

[2] M. R. Cutkosky, *Robotic Grasping and Fine Manipulation.* Boston, MA: Kluwer, 1985.

[3] M. R. Cutkosky and P. K. Wright, "Active control of a compliant wrist in manufacturing tasks," *ASME J. Eng. for Ind.*, vol. 108, no. 1, pp. 36–43, 1985.

[4] Z. Ji, "Dexterous hand: Optimizing grasp by design and planning," Ph.D. dissertation, Stanford University, Stanford, CA, 1987.

[5] T. Kailath, *Linear Systems.* Englewood Cliffs, NJ: Prentice-Hall, 1st ed., 1980.

[6] I. Kao and M. Cutkosky, "Effective stiffness and compliance, and their application to sliding analysis," Tech. Rep. SIMA, Stanford University, Stanford, CA, Mar. 1987.

[7] J. Kerr, "Analysis of multifingered hands," Ph.D. dissertation, Stanford University, Stanford, CA, 1986.

[8] H. Kobayashi, "Control and geometrical considerations for an articulated robot hand," *Robotics Res.*, vol. 4, no. 1, pp. 3–12, 1985.

[9] M. T. Mason and J. K. Salisbury, *Robot Hands and the Mechanics of Manipulation* Cambridge, MA: MIT Press, 1985.

[10] V. Nguyen, "Constructuring force-closure grasps in 3-d," in *Proc. 1987 IEEE Conf. on Robotics and Automation*, pp. 240–245, Mar. 1987.

[11] M. Ohwovoriole, "An extension to screw theory and its applications to the automation of industrial assemblies," Ph.D. dissertation, Stanford University, Stanford, CA, Apr. 1980.

[12] M. A. Peshkin, "Planning robotic manipulation strategies for sliding objects," Ph.D. dissertation, Carnegie Mellon University, Pittsburgh, PA, 1986.

[13] M. H. Raibert and B. K. P. Horn, "Manipulator control using the configuration space method," *Industrial Robot (UK)*, vol. 5, no. 2, pp. 69–73, June 1978.

[14] G. Strang, *Lienar Algebra and Its Applications.* New York, NY: Academic Press, 2nd ed., 1980.

[15] D. E. Whitney, "Part mating theory for compliant parts," First. Rep. The Charles Stark Draper Lab., Inc. Aug. 1980, NSF Grant DAR79-10341.

Yoshihiko Nakamura
Mechanical and Environmental Engineering and Center for Robotic Systems in Microelectronics
University of California, Santa Barbara, California 93106

Kiyoshi Nagai
Department of Computer and Systems Science
Ritsumeikan University, Kyoto, 603 Japan

Tsuneo Yoshikawa
Automation Research Laboratory, Kyoto University
Uji, Kyoto, 611 Japan

Dynamics and Stability in Coordination of Multiple Robotic Mechanisms

Abstract

This paper discusses the dynamical coordination of multiple robot manipulators or a multifingered robot hand. The coordination problem is divided into two phases: determining the resultant force by multiple robotic mechanisms; determining the internal force between mechanisms. The resultant force is used for maintaining dynamic equilibrium and for generating the restoring force. A dynamic coordinative control scheme which guarantees the object stability is proposed, associated with the determination of the resultant forces. The internal force is used to satisfy the static frictional constraints and is related to contact stability. The minimizing internal force is defined as the internal force which yields the minimal norm force satisfying static frictional constraints and the minimizing internal force is solved by applying a nonlinear programming method. The concept of the minimizing internal force is fundamental and important because it can theoretically answer whether the object can be grasped and manipulated with the given finger placement. Contact stability means the ability for the end-effector to maintain contact with an object without slipping when the object is subjected to disturbing external forces. A method of evaluating contact stability is developed. The minimal norm force satisfying the specified degree of contact stability can be obtained by using an algorithm similar to that for the minimizing internal force. Numerical examples of computing the minimizing internal forces are given.

The International Journal of Robotics Research,
Vol. 8, No. 2, April 1989,

1. Introduction

Advanced applications sometimes require multiple robotic manipulators to perform a single task coordinatively. Grasping and manipulation by a multifingered robot hand are important technologies for applying robots to fine applications. These problems substantially have the same physical characteristics as the problem of coordination of multiple robotic mechanisms and should be discussed in a unified way.

For two manipulators, Nakano et al. (1974) adopted a scheme to control one as a master (by position control) and the other as a follower (by force control). Kurone (1975) proposed to change compliance depending upon the directions. Mason (1981) sought to determine the forces of two manipulators without specifying either one as a master or a follower. Uchiyama, Hakomori, and Shimizu (1983) proposed a method to obtain the forces of manipulators based on the maintaining force and the restoring force. On the other hand, for three manipulators Takase (1985) suggested controlling two positional variables of each arm by position control, and one positional variable and three orientational variables by force control.

On grasping and manipulation by a robot hand, many papers have been published. Hanafusa and Asada (1977) proposed to grasp an object in such a way that the potential energy in the elastic fingers is minimized. Salisbury (1982) and Salisbury and Craig (1982) discussed the contact condition between fingers and an object, and suggested that the internal forces should be determined so that the net forces have positive magnitude in the inner normal direction at the contact points on the object surface. Hanafusa, Kobayashi, and Terasaki (1983) and Kobayashi (1985) clarified a kinematical necessary and sufficient condition to manipulate an object by considering the degrees of freedom of both fingers and contact points. Nguyen (1986a, b, c) discussed the force-closure grasp of polyhedra and the stability when each finger's stiffness is controlled. Fearing (1986) proposed a method for stable grasping of two-dimensional polygons. The problems of friction and stability of a soft-finger contact model were investigated by Cutkosky (1985) and Cutkosky and Wright (1986). Jameson and Leifer (1986) also studied the stability of a frictional point contact model and the frictional soft-finger contact model. On the other hand, frictionless enveloping graping was studied by Trinkle, Abel, and Paul (1985). Hanafusa et al. (1985) defined the magnitude of the grasping force of three-fingered robot hands as the area of a triangle made by internal force vectors and proposed a performance criterion to determine the internal forces based on it. On the other hand, Kerr and Roth (1986) approximated the frictional constraints by linear constraints, and proposed to compute the optimal internal forces for the constraints and the joint torque constraints by a linear programming method. Kumar and Waldron (1987) discussed suboptimal algorithms for distributing finger forces.

Although stability is one of the main problems discussed in most papers, Jameson and Leifer (1986) have mentioned that "grasping stability" has at least two meanings. One is the ability to return to the static equilibrium position when the object position is perturbed. This should be termed "object stability." Object stability has been discussed as the problem of restoring forces (Hanafusa and Asada 1977; Nguyen 1986c, Jameson and Leifer 1986; Fearing 1986). The other meaning is the ability to maintain contact when the object is subjected to disturbing forces. This should be termed "contact stability." Hanafusa et al. (1985) and Kerr and Roth (1986) believe that hard squeezing of an object with fingers will make the contacts more resistant to slipping due to disturbing forces.

The propriety of the asymmetric control schemes where one manipulator is controlled as a master and the other as a follower seems not to be very clear. Concerning object stability of a robot hand, the static restoring force has often been discussed, but no clear mathematical approach has been made in a dynamic sense. It seems to be because the analysis and synthesis of grasping and manipulation by robot fingers has been done statically. Moreover, the relationship between contact stability and large grasping forces has only been discussed intuitively. These facts indicate the need for studies on dynamical coordination of multiple robotic mechanisms.

In this paper, we discuss the mechanics involved in coordinating multiple robotic mechanisms by considering the dynamics of an object and explicitly describing the resultant force and the internal forces applied to it. The resultant force is the sum of the force vectors applied to an object that contributes directly to motion control of the object. On the other hand, the internal forces represent the elements of force vectors which are canceled within the object and do not contribute to motion control.

First, we discuss the coordinative manipulation by the resultant force, taking the dynamics of the object into consideration, which allows us to attain dynamical object stability. Second, we discuss the resolution of the resultant force into the individual net forces of robotic mechanisms. The main problem here is to determine internal forces. Hanafusa et al. (1985) and Kerr and Roth (1986) proposed to keep contact stability by assigning large internal forces. However, such forces imply low-stability grasping because even small position errors or disturbances may cause a large disturbing force and moment (Cutkosky and Wright 1986).

It is quite obvious that large grasping forces are not appropriate for grasping breakable objects. We define the minimizing internal forces as the internal forces that give the minimal net forces under static frictional constraints and develop its computational algorithm. The minimizing internal forces are obtained by solving, at most, $\Sigma_{j=1}^{m}\binom{m}{j}\Sigma_{i=0}^{j}\binom{j}{i}$ sets of algebraic equations, if they exist (m is the number of robotic mechanisms). This is important because, by checking the existence of the solution, we can answer the fundamental question of whether a robot hand can grasp the object with the given finger placement.

Third, contact stability against unexpected disturbing forces is discussed and the measure of contact stability is defined as the mathematical evaluation method. The optimal forces are defined as the end-

Fig. 1. Mechanical model of coordinative manipulation of a rigid object by m robotic mechanisms.

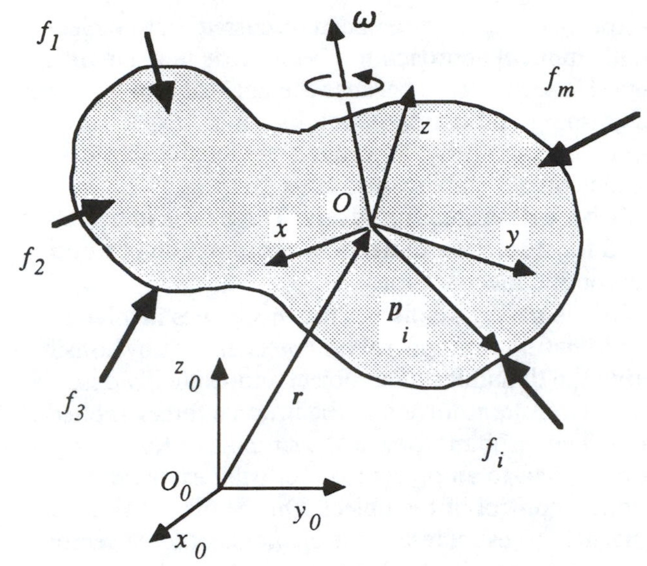

effector forces which generate the given resultant force and have the specified measure of contact stability. It is also shown that the optimal forces and the optimal internal forces are obtained by using algorithms similar to those for the minimal forces and the minimizing internal forces. Finally, numerical examples of the computation of the minimal forces and the minimizing internal forces are given.

2. Coordinative Manipulation by Multiple Robotic Mechanisms

See the appendix for a list of notation. Vectors are assumed to be represented by the object coordinates unless otherwise specified.

2.1. Basic Equations

We derive the basic equations for coordinative manipulation of a rigid object by m robotic mechanisms as shown in Fig. 1. To simplify the problem, we assume that

1. A robotic mechanism makes a frictional point contact with the object.
2. A contact point does not move on the object by changing the relative orientation between the object and the end-effector of the robotic mechanism.

The first assumption allows us to consider only forces at the contact points. This assumption is common and reasonable and enables us to simplify the discussion. The second assumption means that the contact point of the end-effector is modeled not by a point on a smooth curved surface like spherical contact but by a vertex of a polyhedron or a cone. Although this assumption is not necessary for our dynamical problem, without it we would have to consider an additional kinematic problem because the contact points would change in accordance with the motion of the object. This kinematic problem is caused by nonholonomic constraints between the object and the end-effector, which was studied in detail by Kerr and Roth (1986).

The resultant force f_0 and moment n_0 of the external forces applied to the object are represented by

$$\mathbf{f}_0 = \sum_{i=1}^{m} \mathbf{f}_i + m_0\mathbf{g}, \tag{1}$$

$$\mathbf{n}_0 = \sum_{i=1}^{m} \mathbf{p}_i \times \mathbf{f}_i. \tag{2}$$

The motion equations of the object are represented by the following Newton and Euler equations:

$$m_0\ddot{\mathbf{r}} = \mathbf{f}_0, \tag{3}$$

$$\mathbf{I}\dot{\omega} + \omega \times (\mathbf{I}\omega) = \mathbf{n}_0. \tag{4}$$

Equations (1)–(4) are summarized by the following equations:

$$\begin{aligned} &\mathbf{I}_0\ddot{\phi} + \mathbf{Q}_0 = \mathbf{Q}, \\ &\ddot{\phi} = (\ddot{\mathbf{r}}^{\mathrm{T}} \quad \dot{\omega}^{\mathrm{T}})^{\mathrm{T}} \in R^6, \\ &\mathbf{I}_0 = \begin{pmatrix} m_0\mathbf{E}_3 & \mathbf{O} \\ \mathbf{O} & \mathbf{I} \end{pmatrix} \in R^{6\times 6}, \\ &\mathbf{Q}_0 = (-m_0\mathbf{g}^{\mathrm{T}} \quad \{\omega \times (\mathbf{I}\omega)\}^{\mathrm{T}})^{\mathrm{T}}, \end{aligned} \tag{5}$$

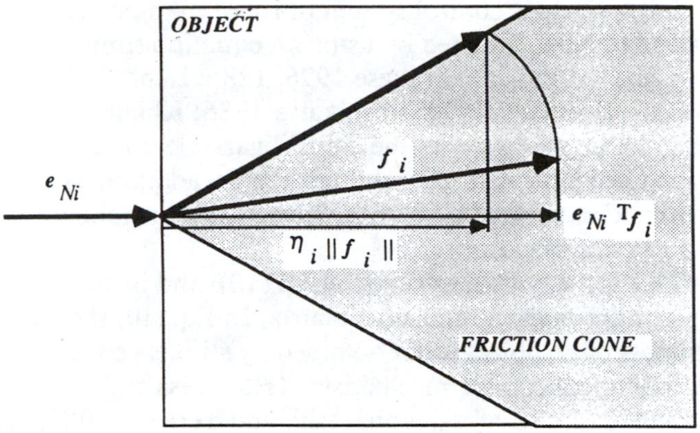

Fig. 2. Static frictional constraints.

where

$$
\begin{aligned}
\mathbf{Q} &= \mathbf{WF} \in R^6, \\
\mathbf{F} &= (\mathbf{f}_1^T \ \ \mathbf{f}_2^T \ \ \cdots \ \ \mathbf{f}_m^T)^T \in R^{3m}, \\
\mathbf{W} &= \begin{pmatrix} \mathbf{E}_3 & \mathbf{E}_3 & \cdots & \mathbf{E}_3 \\ \mathbf{P}_1 & \mathbf{P}_2 & \cdots & \mathbf{P}_m \end{pmatrix} \in R^{6\times 3m}, \\
\mathbf{P}_i &= \begin{pmatrix} 0 & -p_{i3} & p_{i2} \\ p_{i3} & 0 & -p_{i1} \\ -p_{i2} & p_{i1} & 0 \end{pmatrix} \in R^{3\times 3}, \\
\mathbf{p}_i &= (p_{i1} \ \ p_{i2} \ \ p_{i3})^T \in R^3.
\end{aligned}
\tag{6}
$$

2.2. Static Frictional Constraints

Suppose that the shape of the object is represented by

$$S(x, y, z) = 0, \tag{7}$$

where $S(x, y, z) > 0$ means the inside of the object and $S(x, y, z)$ is spatially once differentiable at the contact points. Then the inner unit normal vector at contact point p_i is

$$\mathbf{e}_{Ni} = \frac{\text{grad } S(\mathbf{p}_i)}{\| \text{grad } S(\mathbf{p}_i) \|} \in R^3, \tag{8}$$

where $\| * \|$ means the Euclidean norm of vector $*$. The force applied at the ith contact point is constrainted as shown in Fig. 2 by the maximum static frictional condition

$$\mathbf{e}_{Ni}^T \mathbf{f}_i \geq \eta_i \| \mathbf{f}_i \|, \qquad \eta_i = \frac{1}{\sqrt{1 + \mu_i^2}}. \tag{9}$$

2.3. Dynamic Coordination and Object Stability

The main problem of grasping by robot hands is to stabilize an object within the robot's fingers. This stability should be called object stability. It has been attained by controlling each finger by controlling stiffness (Hanafusa and Asada 1977; Nguyen 1986a, c). The main concern in this approach is to design finger stiffness so that their total stiffness should generate the restoring forces for any positional perturbation of the object. However, dynamic behavior, including dynamic stability, has not been considered so far because grasping and manipulation have been discussed statically. It is also not self-evident whether stiffness control of each finger is the unique solution to object stability. We answer these problems here by discussing the synthesis of the resultant force, considering the object dynamics.

Let the desired trajectory of the object be given by $\boldsymbol{\phi}_d(t) \in R^6$. If the resultant force

$$\mathbf{Q} = \mathbf{Q}_0 + \mathbf{I}_0\{\ddot{\boldsymbol{\phi}}_d + \mathbf{K}_1(\dot{\boldsymbol{\phi}}_d - \dot{\boldsymbol{\phi}}) + \mathbf{K}_2(\boldsymbol{\phi}_d - \boldsymbol{\phi})\}, \tag{10}$$

is applied to the object, then its motion is governed by the equation

$$(\ddot{\boldsymbol{\phi}}_d - \ddot{\boldsymbol{\phi}}) + \mathbf{K}_1(\dot{\boldsymbol{\phi}}_d - \dot{\boldsymbol{\phi}}) + \mathbf{K}_2(\boldsymbol{\phi}_d - \boldsymbol{\phi}) = 0, \tag{11}$$

where $\mathbf{K}_1$ and $\mathbf{K}_2 \in R^{6\times 6}$ are constant matrices which guarantee asymptotic stability. Equation (11) means that $\boldsymbol{\phi}(t)$ converges to $\boldsymbol{\phi}_d(t)$, which means that the object is dynamically stable. From Eq. (5), the orientational element of $\boldsymbol{\phi}$ becomes $\int_{t_0}^{t} \boldsymbol{\omega}\, dt$, whose physical meaning is not very clear. If we adopt the orientational representation by a 3×3 orthogonal matrix, $\boldsymbol{\omega}$ and $\dot{\boldsymbol{\omega}}$ can be obtained by differentiating it, and $\boldsymbol{\phi}_d - \boldsymbol{\phi}$ can be computed from the difference between the desired orientational matrix and the actual one (Luh, Walker, and Paul, 1980).

The trajectory control system of robot manipulators has often been designed by using an equation similar to Eqs. (10) and (11) (Takase 1976, 1985; Luh, Walker, and Paul 1980; Yoshikawa 1986; Khatib 1985). Our method uses the control variable not as the joint torque but as the resultant force. In addition to dynamic manipulation of the object, it has the following physical meaning.

$\mathbf{I}_0\mathbf{K}_2$ is the stiffness matrix in Eq. (10), and its inverse implies the compliance matrix. In Eq. (10) the stiffness of an object is not obtained by stiffness control of individual robotic mechanisms (Hanafusa and Asada 1977; Hanafusa, Kobayashi, and Terasaki 1983; Kobayashi 1985; Nguyen 1986a, c), but is specified explicitly by the resultant forces. We have redundancy in determining the individual forces because we have nine force elements even in the case of three robotic mechanisms, while $\mathbf{Q}$ has only six. This redundancy is reserved as the degrees of freedom in determining the internal forces, which will be discussed in detail in Sections 4 and 5. Note that when each finger is controlled by stiffness, the finger forces are determined uniquely only by the displacement of an object, which means that no degree of freedom is left for adjustment of internal forces. For example, when each finger is stiffness controlled, finger stiffness has to be large enough in order to grasp an object with high object stiffness. This will result in large net finger forces. On the other hand, when Eq. (10) is used to determine finger forces, it is possible in this control scheme to make the net finger forces smaller by choosing appropriate internal forces for the same object stiffness.

2.4. Requirement for Force Controller

To manipulate an object dynamically, we must generate the resultant force as shown in Eq. (10). The distribution of the resultant force to individual end-effector forces will be discussed in the following sections. In this subsection, the basic requirement for the force controller is clarified.

To maintain contact with constraint surfaces, force control should be in the constrained directions (that is, the normal directions of the constraint surfaces); to move the end-effector, position control should be in the unconstrained directions (that is, the tangential directions of the constraint surfaces; Mason 1981). The force controller based on this formulation is usually termed a *hybrid position/force control system* (Raibert and Craig 1981; Yoshikawa 1986). Since the constrained directions and the unconstrained directions span the orthogonal complement spaces to each other in six-dimensional linear space (Mason 1981), it is an important feature of hybrid position/force control that the end-effector not move in the force-controlled directions. Therefore, the role of its force controller is to keep static equilibrium in the constrained directions.

The force control required in dynamic coordinative manipulation should be in all directions of a six-dimensional space, and an object is manipulated by the work done by the end-effector forces.[1] In other words, the force control is in the direction in which the end-effector moves. Therefore, the role of the force controller is completely different from that of hybrid position/force control systems. The force controller must consider static and dynamic equilibrium. Static equilibrium works to generate the desired end-effector force at the contact point. On the other hand, dynamic equilibrium works to maintain contact with the object by accelerating the end-effector to follow the motion of the contact point. The force control is mathematically formulated as follows.

From Fig. 1, the acceleration of the contact point is

$$\ddot{\mathbf{r}}_i = \ddot{\mathbf{r}} + \dot{\boldsymbol{\omega}} \times \mathbf{p}_i + \boldsymbol{\omega} \times (\boldsymbol{\omega} \times \mathbf{p}_i), \tag{12}$$

$\ddot{\mathbf{r}}$ and $\dot{\boldsymbol{\omega}}$ are the accelerations of the object caused by the resultant force of Eq. (10), and are represented from Eqs. (10) and (11) by

$$\begin{aligned}(\ddot{\mathbf{r}}^{\mathrm{T}} \quad \dot{\boldsymbol{\omega}}^{\mathrm{T}}) = \ddot{\boldsymbol{\phi}} &= \ddot{\boldsymbol{\phi}}_d + \mathbf{K}_1(\dot{\boldsymbol{\phi}}_d - \dot{\boldsymbol{\phi}}) + \mathbf{K}_2(\boldsymbol{\phi}_d - \boldsymbol{\phi}) \\ &= \mathbf{I}_0^{-1}(\mathbf{Q} - \mathbf{Q}_0).\end{aligned} \tag{13}$$

Let τ_{ip} be the generalized driving force of the ith robotic mechanism required to generate acceleration $\ddot{\mathbf{r}}_i$ when the mechanism is not making contact with the

1. The internal force elements of end-effector forces do not work totally, which will be discussed in Section 4.1. However, the internal force element, as well as the noninternal force element, of each end-effector force does work at each contact point.

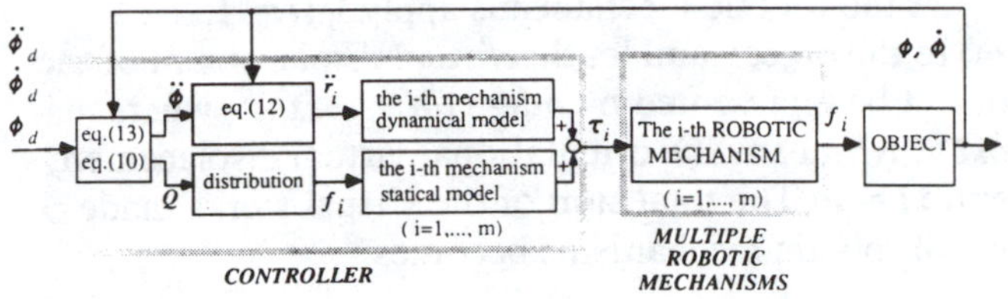

Fig. 3. Requirement for force controller: feedforward passes for position and force control.

object. τ_{ip} is computed considering the dynamic characteristics of the robotic mechanism, such as the inertial tensor and the Colioris, centrifugal, and gravity forces. Also let τ_{if} be the generalized driving force of the ith robotic mechanism required to statically apply force $\mathbf{f}_i$ to the object. τ_{if} is computed from the static equilibrium condition by multiplying the transposed Jacobian matrix of the ith robotic mechanism to $\mathbf{f}_i$. According to D'Alembert's principle, the generalized driving force τ_i to manipulate the object is the sum of τ_{ip} and τ_{if}. The fundamental structure of the control system for coordinative manipulation is synthesized as shown in Fig. 3.

Note that the above discussion does not mean that an arbitrary combination of $\mathbf{f}_i$ and $\ddot{\mathbf{r}}_i$ can be realized. As shown in Eqs. (12) and (13), $\ddot{\mathbf{r}}_i$ is constrained so that its trajectory should be the resultant motion of the force control. Therefore, τ_{if} means the generalized driving force for force control, and τ_{ip} implies the generalized driving force for accompanying position control. The above consideration is summarized by the following proposition.

PROPOSITION 1 If a robotic mechanism is force controlled in the directions in which the end-effector works, it must also be position controlled so that the trajectory should be the resultant motion of the force control.

Figure 3 shows only the feedforward passes of position control and force control. Generally speaking, the control system should include feedback loops for both position control and force control (Raibert and Craig 1981; Yoshikawa 1986). Although position control should depend on the force control, the feedback loop for position control may disturb this dependency. Further study on the synthesis of feedback control systems must be done from the theoretical and practical viewpoints.

Note that the discussion in this subsection does not mean that an object cannot be manipulated without considering the dynamic equilibrium condition. When robotic mechanisms are controlled only by the static equilibrium condition, large internal forces may play a role to maintain contact with the object. However, the driving force computed from the static equilibrium is used to accelerate the end-effector of the robotic mechanism, and the desired resultant force will not be attained accurately, which implies that the desired object manipulation will be disturbed. The neglect of the dynamic equilibrium would practically be possible only when the driving force from the dynamic equilibrium is significantly smaller than that from the static one.

3. Coordinative Manipulability

Can the robotic mechanisms manipulate the object arbitrarily? This question is basic. To answer it, we have to consider the static frictional constraints and the fact that a robotic mechanism can push but not pull the object. In this section, we derive a method to verify if the robotic mechanisms can generate an arbitrary acceleration $\ddot{\phi}$ under the static frictional constraints of Eq. (9).

Since the inertial tensor is positive definite in Eq. (5), generating an arbitrary $\mathbf{f}$ is equivalent to generating an arbitrary $\mathbf{Q}$. The grasp which enabled an arbitrary force to be exerted was called *force-closure grasp* by Ohwovoriole (1980) and discussed in detail in the two-dimensional case (Nguyen 1986c). Our result gives a method for investigating, for any number of fingers, whether a three-dimensional grasp is force-closure.

Suppose that two linearly independent resultant forces $\mathbf{Q}^1$ and $\mathbf{Q}^2$ can be generated under the constraints of Eq. (9); that is,

$$\mathbf{Q}^j = \mathbf{W}\mathbf{F}^j \qquad \text{for } j = 1, 2, \tag{14}$$

$$\begin{aligned} &\mathbf{F}^j = (\mathbf{f}_1^{jT} \quad \mathbf{f}_2^{jT} \quad \cdots \quad \mathbf{f}_m^{jT})^T \in R^{3m}, \\ &\mathbf{e}_{Ni}^T \mathbf{f}_i^j \geq \eta_i \, \|\mathbf{f}_i^j\| \qquad \text{for } i = 1, \ldots, m, \quad j = 1, 2. \end{aligned} \tag{15}$$

Then, for arbitrary nonnegative scalars k_1 and k_2,

$$\mathbf{Q}^0 = k_1\mathbf{Q}^1 + k_2\mathbf{Q}^2 \tag{16}$$

can also be generated as follows. Let

$$\mathbf{F}^0 = k_1\mathbf{F}^1 + k_2\mathbf{F}^2 = (\mathbf{f}_1^{0\mathrm{T}} \quad \mathbf{f}_2^{0\mathrm{T}} \quad \cdots \quad \mathbf{f}_m^{0\mathrm{T}})^\mathrm{T}. \quad (17)$$

The following inequality derived from Eq. (15) implies that $\mathbf{F}^0$ can be generated under static frictional constraints.

$$\begin{aligned}\mathbf{e}_{Ni}^\mathrm{T}\mathbf{f}_i^0 = \mathbf{e}_{Ni}^\mathrm{T}(k_1\mathbf{f}_i^1 + k_2\mathbf{f}_i^2) &\geq \eta_i(k_1 \|\mathbf{f}_i^1\| + k_2 \|\mathbf{f}_i^2\|) \\ &\geq \eta_i(\|k_1\mathbf{f}_i^1 + k_2\mathbf{f}_i^2\|) = \eta_i \|\mathbf{f}_i^0\|.\end{aligned} \quad (18)$$

From Eqs. (14), (16), and (17), $\mathbf{F}^0$ obviously yields the resultant force $\mathbf{Q}^0$. This fact shows that if $\pm\mathbf{Q}^1$ and $\pm\mathbf{Q}^2$ can be generated, any linear combination of $\mathbf{Q}^1$ and $\mathbf{Q}^2$ can be generated. Accordingly, the following proposition is given.

PROPOSITION 2 Let $\mathbf{Q}^j$ ($j = 1, \ldots, 6$) be linearly independent resultant forces. Then the robotic mechanisms can generate an arbitrary acceleration of the object if and only if $\mathbf{Q}^j$ and $-\mathbf{Q}^j$ ($j = 1, \ldots, 6$) can be generated under the static frictional constraints.

Proposition 2 shows that we can judge the coordinative manipulability by checking 12 resultant forces. Since an object should be grasped at the contact points which allow coordinative manipulability, the proposition can be used to determine the contact points. In order to examine the coordinative manipulability by using the proposition, we must discuss whether each resultant force can be generated. The computational algorithm to do it is given in the next section. In addition, a necessary condition for generating six linearly independent resultant forces is that $\mathbf{W}$ should be full rank in Eq. (14).

4. Computation of Minimal Forces

4.1. Internal Force in Coordinative Manipulation

The internal force in grasping and manipulation by robot hands was discussed based on Eq. (6) and specified by the element of the null space of $\mathbf{W}$. In this subsection, we give a physical meaning of the internal force and an alternative definition.

Let the robotic mechanisms apply $\mathbf{f}_i$ ($i = 1, \ldots, m$) to the object, and let the virtual displacement of the object be represented by $\delta\boldsymbol{\phi} = (\delta\mathbf{r}^\mathrm{T} \quad \delta\boldsymbol{\Omega}^\mathrm{T})^\mathrm{T}$, where $\delta\boldsymbol{\Omega} \in R^3$ means the orientational virtual displacement and $\dot{\boldsymbol{\Omega}} = \omega$ The total sum of the virtual works made by all robotic mechanisms becomes

$$\delta\omega = \sum_{i=1}^{m} (\delta\boldsymbol{\Omega} \times \mathbf{p}_i + \delta\mathbf{r})^\mathrm{T}\mathbf{f}_i. \quad (19)$$

Equation (19) is simplified to

$$\delta\omega = \delta\boldsymbol{\phi}^\mathrm{T}\mathbf{W}\mathbf{F}. \quad (20)$$

The following equation is a sufficient and necessary condition to make $\delta\omega$ equal to zero for an arbitrary $\delta\boldsymbol{\phi}$:

$$\mathbf{W}\mathbf{F} = 0. \quad (21)$$

Accordingly, the internal forces are such a set of $\mathbf{f}_i$ that the total sum of the virtual works done by the fingers is zero for any and all virtual displacement of the object.

DEFINITION 1 The internal forces are defined as such a set of $\mathbf{f}_i$ ($i = 1, \ldots, m$) that the total sum of the virtual works done by the fingers results in zero for any and all virtual displacement of the object.

4.2. The Minimizing Internal Forces

Hanafusa et al. (1985) and Kerr and Roth (1986) proposed to determine the optimal internal forces by considering the maximum value of $\mathbf{f}_i$ and the maximum torques of the finger joints, respectively. As acknowledged in the latter paper, their approaches are aimed at determining "how hard to squeeze an object with the fingers in order to ensure that the object is grasped stably." Accordingly, the larger the available forces or torques are for the robotic mechanisms, the larger the grasping forces these approaches choose. However, with the large grasping forces even a small error of the force direction introduced by the measurement error or the positional disturbance may cause a large disturbing force and moment at the mass center of the object. It is often experienced when we grasp a wet cake of soap that hard squeezing results in slipping and misgrasping. Cutkosky and Wright (1986) pointed out that "increasing the gripping force reduced the chance of slipping but also made the grip less stable with respect to disturbances." These facts suggest that it is an appropriate principle for determining internal forces from the viewpoint of stable grasping to reduce the grasping forces as long as they have enough contact stability.

In this section, we discuss such an extreme problem as obtaining the minimal forces under the static frictional constraints without considering the contact stability. Although the solution of this extreme problem seems not to be practical because it may not have enough contact stability, the minimal forces have a significant physical meaning. Since the object can be grasped or manipulated by the specified resultant force only when the minimal forces exist, by checking the existence of the minimal forces we can judge whether a specified resultant force can be generated. This is necessary for investigating the coordinative manipulability. The algorithms to be developed in this section are extended in Section 5 to obtain the minimal forces, taking the contact stability into consideration.

DEFINITION 2 The minimal norm force **F** among all the forces that generate the specified resultant force **Q** under the static frictional constraints of Eq. (9) is called the *minimal force*, and the internal forces that yield the minimal force are called the *minimizing internal forces.*

In order to cope with the uncertainty of the coefficient of maximum static friction, we should regard

$$\mu_{ai} = \frac{1}{C}\mu_{ei} \tag{22}$$

as the effective coefficient of maximum static friction, where μ_{ei} is the critically estimated coefficient of maximum static friction and $C \geq 1$ is the safety factor. Since **Q** can be generated if and only if the minimal internal forces exist, we can investigate the coordinative manipulability of Proposition 2 by verifying the existence of the minimal internal forces for $\pm \mathbf{Q}^j$ $(j = 1, \ldots, 6)$.

4.3. Application of Nonlinear Programming

The minimizing internal force problem is formulated as follows: for the specified resultant force **Q**, the internal forces which satisfy

$$\min \|\mathbf{F}\| \tag{23}$$

and force **F** are to be solved under the constraints

$$\mathbf{Q} = \mathbf{W}\mathbf{F}, \tag{24}$$

$$\mathbf{e}_{Ni}^T \mathbf{f}_i \geq \eta_i \|\mathbf{f}_i\|, \tag{25}$$

where we assume that the fingers have enough joints and can always apply arbitrary forces. If an **F** exists that satisfies Eq. (24),[2] we can write it as

$$\begin{aligned} \mathbf{F} &= \mathbf{F}_0 + \mathbf{A}\mathbf{y}, & \mathbf{y} &\in R^b, \\ \mathbf{F}_0 &= \mathbf{W}^{\#}\mathbf{Q} \\ &= (\mathbf{f}_{01}^T \quad \mathbf{f}_{02}^T \quad \cdots \quad \mathbf{f}_{0m}^T)^T, & \mathbf{f}_{0i} &\in R^3, \\ \mathbf{A} &= (\mathbf{A}_1^T \quad \mathbf{A}_2^T \quad \cdots \quad \mathbf{A}_m^T)^T, & \mathbf{A}_i &\in R^{3\times b}, \end{aligned} \tag{26}$$

where $\mathbf{W}^{\#} \in R^{3m\times 6}$ is the pseudoinverse of **W** (Rao and Mitra 1971), **A** is the matrix of orthonormals of the null space of **W**,[3] and $b = 3m - \text{rank}\,\mathbf{W}$ is the dimension of the null space of **W**. Since the first term of Eq. (26) is an element of the orthogonal complement of the null space of **W** and the second term is an element of the null space of **W**, the internal force corresponds to the second term of Eq. (26). The constraints of Eq. (25) can be represented by the following $2m$ inequalities:

$$\begin{aligned} & g_i(\mathbf{y}) \leq 0 \quad \text{for } i = 1, \ldots, 2m, \\ & g_i(\mathbf{y}) = \begin{cases} (\mathbf{f}_{0i} + \mathbf{A}_i\mathbf{y})^T \mathbf{B}_i (\mathbf{f}_{0i} + \mathbf{A}_i\mathbf{y}) & \text{for } i = 1, \ldots, m, \\ -\mathbf{e}_{Ni-m}^T(\mathbf{f}_{0i-m} + \mathbf{A}_{i-m}\mathbf{y}) & \text{for } i = m+1, \ldots, 2m, \end{cases} \\ & \quad \mathbf{B}_i = \eta_i^2 \mathbf{E}_3 - \mathbf{e}_{Ni}\mathbf{e}_{Ni}^T. \end{aligned} \tag{27}$$

Equation (23) can also be represented by

$$\max_{\mathbf{y}} h(\mathbf{y}), \qquad h(\mathbf{y}) = -(\mathbf{F}_0 + \mathbf{A}\mathbf{y})^T(\mathbf{F}_0 + \mathbf{A}\mathbf{y}). \tag{28}$$

2. The existence of **F** is verified by checking that $(\mathbf{E}_6 - \mathbf{W}\mathbf{W}^{\#})\mathbf{Q} = 0$.

3. Matrix **A** can be produced by collecting and orthonormalizing the linearly independent column vectors of $(\mathbf{E}_{3m} - \mathbf{W}^{\#}\mathbf{W})$.

From Eqs. (27) and (28), the problem of finding the minimizing internal force has been reduced to maximizing a quadratic function with linear and quadratic constraints.

In order to obtain $\mathbf{y}$ that yields the minimizing internal forces, we apply nonlinear programming using Lagrange multipliers (Mine 1966). We construct a function $\epsilon(\mathbf{y}, \boldsymbol{\lambda})$ by using Lagrange multipliers $\boldsymbol{\lambda} = (\lambda_1 \quad \lambda_2 \quad \cdots \quad \lambda_2 m)^{\mathrm{T}} \in R^{2m}$ as follows:

$$\epsilon(\mathbf{y}, \boldsymbol{\lambda}) = h(\mathbf{y}) - \sum_{i=1}^{2m} \lambda_i g_i(\mathbf{y}). \tag{29}$$

A necessary condition for $\mathbf{y}^0$ to yield the local maximum of $h(\mathbf{y})$ under the condition $g_i(\mathbf{y}) \leq 0$ is that $\boldsymbol{\lambda}^0 \leq 0$ $(\boldsymbol{\lambda}^0 = (\lambda_1^0 \quad \lambda_2^0 \quad \cdots \quad \lambda_{2m}^0)^{\mathrm{T}}, \lambda_i^0 \leq 0, i = 1, \ldots, 2m)$ satisfy the following equations:[4]

$$\left.\frac{\partial \epsilon}{\partial \mathbf{y}}\right|_{y=y^0, \lambda=\lambda^0} = 0, \tag{30}$$

$$\left.\frac{\partial \epsilon}{\partial \boldsymbol{\lambda}}\right|_{y=y^0, \lambda=\lambda^0} \geq 0, \tag{31}$$

$$\left.\left(\frac{\partial \epsilon}{\partial \boldsymbol{\lambda}}\right) \boldsymbol{\lambda}\right|_{y=y^0, \lambda=\lambda^0} = 0. \tag{32}$$

It is well known that Eqs. (30) through (32) are necessary and sufficient conditions of the global maximum when $\epsilon(\mathbf{y}, \boldsymbol{\lambda})$ is a concave function with respect to $\mathbf{y}$ (Mine 1966).

4.4. Concavity of $\epsilon(\mathbf{y}, \boldsymbol{\lambda})$

Substituting $g_i(\mathbf{y})$ and $h(\mathbf{y})$ into Eq. (29) and using the fact that $\mathbf{F}_0$ and $\mathbf{Ay}$ are orthogonal to each other, we obtain

$$\begin{aligned}
\epsilon(\mathbf{y}, \boldsymbol{\lambda}) &= -\mathbf{y}^{\mathrm{T}}\mathbf{A}^{\mathrm{T}}(\mathbf{B}_0 + \mathbf{E}_{3m})\mathbf{Ay} - (2\mathbf{F}_0^{\mathrm{T}}\mathbf{B}_0 + \mathbf{e}_{N0}^{\mathrm{T}})\mathbf{Ay} \\
&\quad - \mathbf{F}_0^{\mathrm{T}}(\mathbf{B}_0 + \mathbf{E}_{3m})\mathbf{F}_0 + \mathbf{e}_{N0}^{\mathrm{T}}\mathbf{F}_0, \\
\mathbf{B}_0 &= \begin{pmatrix} \lambda_1 \mathbf{B}_1 & \cdots & \mathbf{O} \\ \vdots & \ddots & \vdots \\ \mathbf{O} & \cdots & \lambda_m \mathbf{B}_m \end{pmatrix} \in R^{3m \times 3m}, \\
\mathbf{e}_{N0} &= (\lambda_{m+1}\mathbf{e}_{N1}^{\mathrm{T}} \quad \cdots \quad \lambda_{2m}\mathbf{e}_{Nm}^{\mathrm{T}})^{\mathrm{T}} \in R^{3m}.
\end{aligned} \tag{33}$$

Equation (33) shows that $\epsilon(\mathbf{y}, \boldsymbol{\lambda})$ is a quadratic function of $\mathbf{y}$. Concavity of quadratic functions is proved by showing the negative definiteness of the second-order term. The second-order term of Eq. (33) is expanded as follows:

$$\begin{aligned}
&-\mathbf{y}^{\mathrm{T}}\mathbf{A}^{\mathrm{T}}(\mathbf{B}_0 + \mathbf{E}_{3m})\mathbf{Ay} \\
&= -\mathbf{y}^{\mathrm{T}} \left[\sum_{i=1}^{m} \mathbf{A}_i^{\mathrm{T}}\{(1 + \lambda_i \eta_i^2)\mathbf{E}_3 - \lambda_i \mathbf{e}_{Ni}\mathbf{e}_{Ni}^{\mathrm{T}}\}\mathbf{A}_i \right] \mathbf{y}.
\end{aligned} \tag{34}$$

The easiest way to show negative definiteness of Eq. (34) is to prove the positive definiteness of $\mathbf{A}_i^{\mathrm{T}}\{(1 + \lambda_i \eta_i^2)\mathbf{E}_3 - \lambda_i \mathbf{e}_{Ni}\mathbf{e}_{Ni}^{\mathrm{T}}\}\mathbf{A}_i$ for all i. However, since λ_i (≥ 0) is variable, the proof should be done for all possible values of λ_i. It turns out by investigating its physical meaning that $\mathbf{A}_i^{\mathrm{T}}\{(1 + \lambda_i \eta_i^2)\mathbf{E}_3 - \lambda_i \mathbf{e}_{Ni}\mathbf{e}_{Ni}^{\mathrm{T}}\}\mathbf{A}_i$ is positive definite for all i and all $\lambda_i \geq 0$ only when all the possible internal forces are outside of the friction cones. This condition is too strong and would probably not be satisfied in most cases.

Consequently, we have to directly verify the positive definiteness of $\mathbf{A}^{\mathrm{T}}(\mathbf{B}_0 + \mathbf{E}_{3m})\mathbf{A}$ for obtained λ_i. That is, if obtained λ_i (≥ 0) makes $\mathbf{A}^{\mathrm{T}}(\mathbf{B}_0 + \mathbf{E}_{3m})\mathbf{A}$ positive definite, it guarantees the concavity of $\epsilon(\mathbf{y}, \boldsymbol{\lambda})$ with respect to $\mathbf{y}$ and the global solution.

4.5. Computation of the Minimizing Internal Forces

We give a procedure for computing the minimizing internal forces according to Eqs. (30)–(32). Substituting Eq. (33) into Eq. (30) yields

$$\mathbf{y}^0 = -\{\mathbf{A}^{\mathrm{T}}(\mathbf{B}_0 + \mathbf{E}_{3m})\mathbf{A}\}^{-1}\mathbf{A}^{\mathrm{T}}(\mathbf{B}_0\mathbf{F}_0 - \tfrac{1}{2}\mathbf{e}_{N0}). \tag{35}$$

Note that from the discussion in Section 4.4, $\mathbf{A}^{\mathrm{T}}(\mathbf{B}_0 +$

4. Equations (30)–(32) are obtained from Theorem 3.7.1 of Mine (1966) by eliminating the condition $\mathbf{y} \geq 0$, which is unnecessary for our problem.

$\mathbf{E}_{3m})\mathbf{A}$ is positive definite and, therefore, nonsingular for λ_i, which gives the global solutions. Equation (31) is equivalent to the inequalities of Eq. (27). Equation (32) is calculated as follows:

$$\sum_{i=1}^{2m} \lambda_i^0 g_i(\mathbf{y}^0) = 0. \quad (36)$$

Since $\lambda_i^0 \geq 0$ and $g_i(\mathbf{y}^0) \leq 0$, Eq. (36) is equivalent to

$$\lambda_i^0 \begin{cases} = 0 & \text{for } g_i(\mathbf{y}^0) < 0, \\ \geq 0 & \text{for } g_i(\mathbf{y}^0) = 0 \quad (i = 1, \ldots, 2m). \end{cases} \quad (37)$$

Equation (37) implies that $\lambda_i^0 = 0$ when the inequality for i is satisfied inside, and $\lambda_i^0 > 0$ is satisfied only when the inequality for i is satisfied on the boundary of the friction cone.

$g_i(\mathbf{y}^0) \leq 0$ $(i = m + 1, \ldots, 2m)$ is satisfied on the boundary only when $\mathbf{f}_{i-m} = 0$. Therefore, when $\mathbf{f}_{i-m} \neq 0$ $(i = m + 1 \ldots, 2m)$, $g_i(\mathbf{y}^0) \leq 0$ must be satisfied inside the inequality constraint. In this case, Eqs. (35) and (37) are simplified by using $\lambda_i^0 = 0$ $(i = m + 1, \ldots, 2m)$ as follows:

$$\mathbf{y}^0 = -\{\mathbf{A}^T(\mathbf{B}_0 + \mathbf{E}_{3m})\mathbf{A}\}^{-1}\mathbf{A}^T\mathbf{B}_0\mathbf{F}_0,$$

$$\lambda_i^0 \begin{cases} = 0 & \text{for } g_i(\mathbf{y}^0) < 0, \\ \geq 0 & \text{for } g_i(\mathbf{y}^0) = 0 \quad (i = 1, \ldots, m). \end{cases} \quad (38)$$

Assuming $\mathbf{f}_i \neq 0$ $(i = 1, \ldots, m)$, the minimizing internal forces and the minimal forces are obtained according to the following procedure:[5]

ALGORITHM 1

1. $k = 0$.
2. $j = 1$.
3. Assume k robotic mechanisms are on the boundaries of the friction cones and the others are inside them. Choose an unchecked combination among $\binom{m}{k}$ combinations. Compute $\mathbf{y}^0$ according to Eq. (38) by setting $\lambda_i^0 = 0$ for $m - k$ robotic mechanisms assumed to be inside the constraints. Obtain the rest of λ_i^0 by solving k equations $g_i(\mathbf{y}^0) = 0$ $(1 \leq i \leq m)$ for the robotic mechanisms assumed to be on the boundary of the constraints.
4. If obtained λ_i^0 satisfies $\lambda_i^0 \geq 0$ and $2m$ inequalities of Eq. (27) and makes $\mathbf{A}^T(\mathbf{B}_0 + \mathbf{E}_{3m})\mathbf{A}$ positive definite, then go to step 7.
5. If $j = \binom{m}{k}$ and $k < m$, then $k + 1 \rightarrow k$ and go to step 2. If $j = \binom{m}{k}$ and $k = m$, then go to step 8.
6. $j + 1 \rightarrow j$ and go to step 3.
7. Compute $\mathbf{y}^0$ by substituting λ_i^0 $(i = 1, \ldots, m)$ into Eq. (38). $\mathbf{A}\mathbf{y}^0$ is the minimizing internal force. The minimal forces are obtained by substituting $\mathbf{y}^0$ into Eq. (26). End.
8. The resultant force $\mathbf{Q}$ cannot be generated under the constraints.
End.

The global solution for the general case where $\mathbf{f}_i = 0$ is allowed can be computed according to the following procedure:

ALGORITHM 2

1. $m' = m$.
2. $j = 1$.
3. Assume $\mathbf{f}_i \neq 0$ for m' robotic mechanisms, and $\mathbf{f}_i = 0$ for the rest. Choose an unchecked combination among $\binom{m}{m'}$ combinations. Investigate whether the robotic mechanisms for which $\mathbf{f}_i = 0$ is assumed satisfy $\mathbf{f}_{0i} \in \mathcal{R}(\mathbf{A}_i)$ ($\mathcal{R}(\mathbf{A}_i)$ is the range space of $\mathbf{A}_i$). If not, then go to step 5.[6]
4. Apply Algorithm 1 for m' robotic mechanisms for which $\mathbf{f}_i \neq 0$ is assumed. If the global solutions are obtained, then go to step 7.
5. If $j = \binom{m}{m'}$ and $m' > 1$, then $m' - 1 \rightarrow m'$ and go to step 2. If $j = \binom{m}{m'}$ and $m' = 1$, then go to step 8.
6. $j + 1 \rightarrow j$ and go to step 3.
7. The minimizing internal forces and the minimal forces have been obtained for m' robotic mechanisms for which $\mathbf{f}_i \neq 0$ is assumed. For $m - m'$ robotic mechanisms for which $\mathbf{f}_i = 0$ is assumed, $-\mathbf{f}_{0i}$ becomes the minimizing internal force and $\mathbf{f}_i = 0$ becomes the minimal force.

5. $\mathbf{f}_i = 0$ may be obtained as the minimal force in some cases, though $\mathbf{f}_i \neq 0$ is assumed, because $\lambda_i^0 = 0$ can be satisfied even when $g_i(\mathbf{y}^0) = 0$. There is no problem in treating $\mathbf{f}_i = 0$ as the solution.

6. Since $\mathbf{f}_i = 0$ holds from Eq. (26) only when $\mathbf{f}_{0i} \in \mathcal{R}(\mathbf{A}_i)$, step 4 can be skipped when $\mathbf{f}_{0i} \in \mathcal{R}(\mathbf{A}_i)$ is not satisfied.

8. The resultant force **Q** cannot be generated under the constraints.
   End.

When resultant force **Q** cannot be generated, step 4 must be repeated $\Sigma_{m'=1}^{m}\binom{m}{m'}$ times until we arrive at step 8. In step 4, Algorithm 1 requires at most $\Sigma_{i=0}^{m'}\binom{m'}{i}$ times computations. Therefore, in the worst cases, we have to solve a set of algebraic equations $\Sigma_{j=1}^{m}\binom{m}{j}\Sigma_{i=0}^{j}\binom{j}{i}$ (26 for $m = 3$) times. However, the case where some robotic mechanisms satisfy $\mathbf{f}_i = 0$ is very rare because the mechanisms are not necessary and $\mathbf{f}_i \neq 0$ of the mechanisms results in the larger norm of **F**. Also the case where some robotic mechanisms are simultaneously on the boundary of the constraints is considered practically rare. Therefore, in most cases, the global solution would be found among the cases of $j = m$ and $i = 0, 1$; that is, $\binom{m}{0} + \binom{m}{1} = m + 1$ cases.

5. Contact Stability and Optimal Forces

5.1. Measure of Contact Stability

Object stability was obtained by generating the resulting force specified by Eq. (10), where it was assumed that the end-effectors maintain contact with an object. Contact stability means the ability for the end-effectors to maintain contact with an object without slipping. The dynamic equilibrium condition consider in Section 2.5 will work for contact stability by accelerating the end-effectors to track the contact points, and it will be enough in the ideal situation where there is neither parameter ambiguity nor any unexpected disturbing force. In this section, we investigate contact stability against unexpected disturbing forces.

Previous works (Hanafusa et al. 1985; Kerr and Roth 1986) were motivated by the idea that strong squeezing will ensure contact stability. We have already mentioned that hard squeezing may make gripping less stable because of measurement error or positional disturbances. Another problem is that hard squeezing may cause large slipping acceleration, because once slipping happens the friction cone suddenly shrinks by the change of friction coefficient from static to kinetic. Therefore, all the force elements inside of the static friction cones but outside of the kinetic friction cones will become the accelerating forces, which are much larger in the case of hard squeezing. The motivation of our approach is that in order to be more resistant to disturbing forces caused by measurement error or positional disturbances, and in order not to generate large slipping acceleration, even if slipping occurs, the end-effectors should apply as weak forces as possible so long as contact stability is maintained for a reasonable class of disturbing forces. As preparation for discussing the optimal forces based on this motivation, we develop an evaluation method of contact stability.

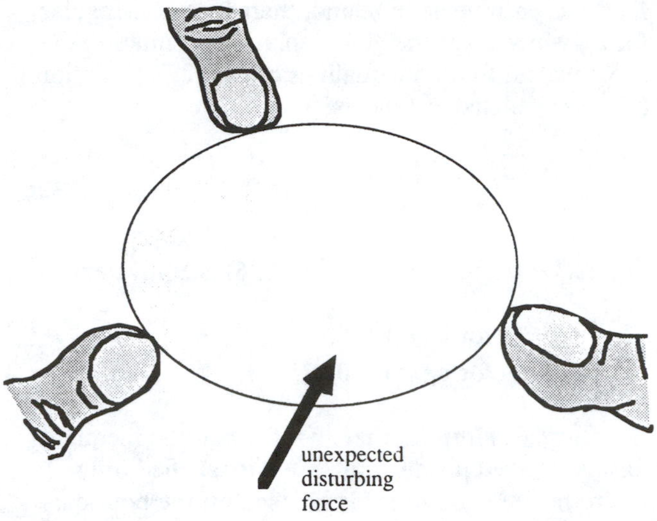

Fig. 4. Can the contacts be maintained for unexpected disturbing forces?

Our problem is schematically represented by Fig. 4. To generate the resultant force specified by Eq. (10), each end-effector applies some force inside the friction cone. With the force controller in Section 2.5, the end-effectors can follow the motion of contact points expected as a result of the resultant force. However, if unexpected external disturbing forces are exerted, the contacts of the end-effectors may not be maintained. We have to clarify the extent of disturbing forces that the contacts can maintain without slipping.

Since dynamical equilibrium is maintained by applying τ_{ip} (Section 2.5), this problem is reduced to the static problem related to τ_{if}, so we can assume without loss of generality that the end-effectors are stationary

Fig. 5. Contact stability and contact point acceleration.

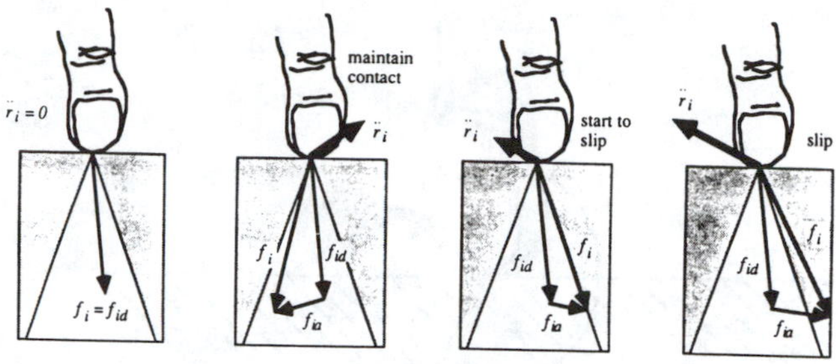

and not subjected to the gravity force. Since $\dot{\theta}_i = 0$ is satisfied and no gravity exists from the above assumption, the dynamics of the ith robotic mechanism are described by

$$\tau_{if} = \mathbf{I}_i \ddot{\theta}_i + \mathbf{J}_i^T \mathbf{f}_i, \quad (39)$$

where $\mathbf{I}_i \in R^{k_i \times k_i}$ is the inertia matrix, $\theta_i \in R^{k_i}$ are the generalized coordinates, and $\mathbf{J}_i = \partial \mathbf{r}_i / \partial \theta_i \in R^{3 \times k_i}$ is the Jacobian matrix of the ith robotic mechanism. We also assume that

$$\tau_{if} = \mathbf{J}_i^T \mathbf{f}_{id} \quad (40)$$

is generated by the joints and at the initial time the end-effector exerts $\mathbf{f}_i = \mathbf{f}_{id}$, which is included in the friction cone.[7]

First, we discuss what will happen to $\ddot{\theta}_i$ and $\mathbf{f}_i$ if the object is suddenly accelerated by the disturbing force. Assume that the acceleration of the object at the contact point caused by the disturbing force is represented by $\ddot{\mathbf{r}}_i$. Now, we suppose that the end-effector can follow the motion of the contact point. Then, from $\dot{\theta}_i = 0$, the following equation is satisfied:

$$\ddot{\mathbf{r}}_i = \mathbf{J}_i \ddot{\theta}_i + \dot{\mathbf{J}}_i \dot{\theta}_i = \mathbf{J}_i \ddot{\theta}_i. \quad (41)$$

Considering the positive definiteness of inertia matrix $\mathbf{I}_i$ and using Eqs. (40) and (41), the following equations are derived from Eq. (39):

$$\mathbf{f}_i = \mathbf{f}_{id} + \mathbf{f}_{ia}, \quad (42)$$

$$\mathbf{f}_{ia} = (\mathbf{J}_i \mathbf{I}_i^{-1} \mathbf{J}_i^T)^{-1} \ddot{\mathbf{r}}_i, \quad (43)$$

where $\text{rank}(\mathbf{J}_i) = 3$ is assumed and $\mathbf{I}_i$ is positive definite; therefore, $\mathbf{J}_i \mathbf{I}_i^{-1} \mathbf{J}_i^T \in R^{3 \times 3}$ becomes positive definite. Equation (42) shows the force exerted on the object when the contact point is accelerated by $\ddot{\mathbf{r}}_i$. Since the end-effector starts to slip when $\mathbf{f}_i$ is on the boundary of the friction cone, it can follow the motion of the contact point so long as $\mathbf{f}_i$ is in the friction cone. The relationship between contact stability and $\ddot{\mathbf{r}}_i$ is schematically shown in Fig. 5.

7. A necessary and sufficient condition to be able to exert $\mathbf{f}_{id}$ by generating τ_{if} of Eq. (40) was discussed in detail by Nakamura (1987).

Now, we consider the following sphere set of contact point accelerations:

$$\{\ddot{\mathbf{r}}_i;\ \|\ddot{\mathbf{r}}_i\| \le a\}, \quad (44)$$

where a is a positive constant scalar. Then the set of $\mathbf{f}_{ia}$ made by all the possible $\ddot{\mathbf{r}}_i$ is

$$\{\mathbf{f}_{ia};\ \mathbf{f}_{ia}^T (\mathbf{J}_i \mathbf{I}_i^{-1} \mathbf{J}_i^T)^T (\mathbf{J}_i \mathbf{I}_i^{-1} \mathbf{J}_i^T) \mathbf{f}_{ia} \le a^2\}. \quad (45)$$

Since $\mathbf{J}_i \mathbf{I}_i^{-1} \mathbf{J}_i^T$ is positive definite, from Eqs. (42) and (45) the set of end points of $\mathbf{f}_i$ becomes an ellipsoid with center at $\mathbf{f}_{id}$ and principal axes $2a/\sigma_{ij}$ ($j = 1, 2, 3$) of length, where $\sigma_{ij} > 0$ is the singular value of $\mathbf{J}_i \mathbf{I}_i^{-1} \mathbf{J}_i^T$ (Golub and Van Loan 1983). Let a_c be the minimal value of a that makes the ellipsoid contact with the friction cone. Then a_c implies the maximum radius of the sphere set of $\ddot{\mathbf{r}}_i$ of which no element causes slipping. This relationship is shown in Fig. 6. Therefore, the following definition is given.

DEFINITION 3 The minimal value of the radius of the sphere set of contact point acceleration that makes the ellipsoid set of $\mathbf{f}_i$ of Eq. (45) contact the friction cone is called the *measure of contact stability,* and the correspondig ellipsoid set of $\mathbf{f}_i$ is called the *contact stability ellipsoid.*

The contact stability ellipsoid equals the generalized inertia ellipsoid (Asada 1983) when $\mathbf{f}_{id} = 0$ and $a_c = 1$.

Next, we discuss the relationship between squeezing and the object acceleration by disturbing forces. Suppose $\omega = 0$, there is no gravity, and the resultant force of $\mathbf{f}_{id}$ is zero. Then, from Eqs. (5), (6), and (42) we obtain

$$\mathbf{I}_0 \ddot{\phi} = \mathbf{W} \mathbf{F}_a + \mathbf{Q}_{dis}, \quad (46)$$

where $\mathbf{F}_a = (\mathbf{f}_{1a}^T \quad \mathbf{f}_{2a}^T \quad \cdots \quad \mathbf{f}_{ma}^T)^T$, and $\mathbf{Q}_{dis}$ is the

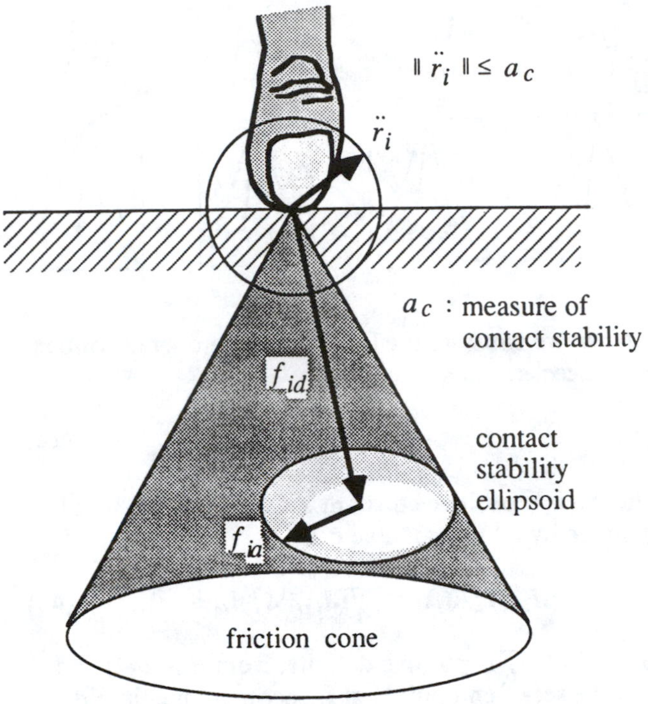

Fig. 6. Measure of contact stability and contact stability ellipsoid.

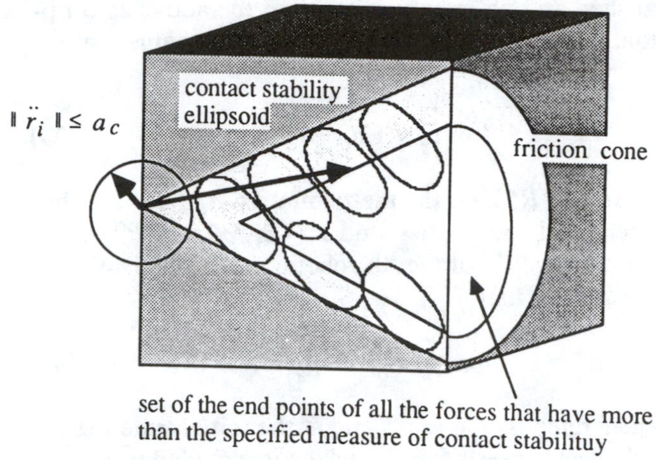

Fig. 7. Set of the end points of all the forces that have more than the specified measure of contact stability.

resultant force of disturbing forces. Using $\omega = 0$ and the definition of $\mathbf{W}$ in Eq. (6), we derive, from Eq. (12),

$$\ddot{\mathbf{r}}_c = \mathbf{W}^T \ddot{\boldsymbol{\phi}}, \tag{47}$$

where $\ddot{\mathbf{r}}_c = (\ddot{\mathbf{r}}_1^T \;\; \ddot{\mathbf{r}}_2^T \;\; \cdots \;\; \ddot{\mathbf{r}}_m^T)^T$. By using Eqs. (42), (43), (46), and (47), we get

$$\ddot{\boldsymbol{\phi}} = (\mathbf{I}_0 + \mathbf{W}\mathbf{I}_r\mathbf{W}^T)^{-1}\mathbf{Q}_{dis},$$
$$\mathbf{I}_r = \begin{pmatrix} (\mathbf{J}_1\mathbf{I}_1^{-1}\mathbf{J}_1^T)^{-1} & \cdots & \mathbf{O} \\ \vdots & \ddots & \vdots \\ \mathbf{O} & \cdots & (\mathbf{J}_m\mathbf{I}_m^{-1}\mathbf{J}_m^T)^{-1} \end{pmatrix} \in R^{3m \times 3m}. \tag{48}$$

In Eq. (48), $\mathbf{I}_0 + \mathbf{W}\mathbf{I}_r\mathbf{W}^T$ is the combined inertia matrix of the object and the robotic mechanisms. From Eq. (48) we see that the object acceleration does not depend on how hard the robotic mechanisms squeeze the object. In other words, hard squeezing is effective for contact stability, but has no advantage for reducing the object acceleration caused by the disturbing forces. Once the acceleration causes the change of the object velocity and position, and if disturbing forces are not exerted thereafter, the object behavior is totally determined by the resultant force control strategy of Eq. (10).

5.2. Optimal Forces

In Section 4, the internal forces were determined so as to minimize the end-effector forces. We mentioned that the minimal force is worth computing to investigate the fundamental characteristics of grasping, but it is not very practical for control because it may not have enough contact stability. In this subsection, the optimal forces are defined as the end-effector forces which generate the specified resultant force and have the specified measure of contact stability.

When the measure of contact stability is specified, the set of end points of all the possible force vectors that have at least the specified measure of contact stability make the elliptic cone as shown in Fig. 7. Since this elliptic cone seems to be too complicated to be considered in real-time force optimization, we propose to approximate it by the following circular cone:

$$\mathbf{e}_{Ni}^T \bar{\mathbf{f}}_i \geq \eta_i \, \| \bar{\mathbf{f}}_i \|, \tag{49}$$

$$\bar{\mathbf{f}}_i = \mathbf{f}_i - \bar{\mathbf{f}}_{0i}, \tag{50}$$

$$\bar{\mathbf{f}}_{0i} = \frac{\sqrt{\mu_i^2 + 1}}{\mu_i} \frac{a}{\sigma_{i3}} \mathbf{e}_{Ni}. \tag{51}$$

Fig. 8. Contact stability cone.

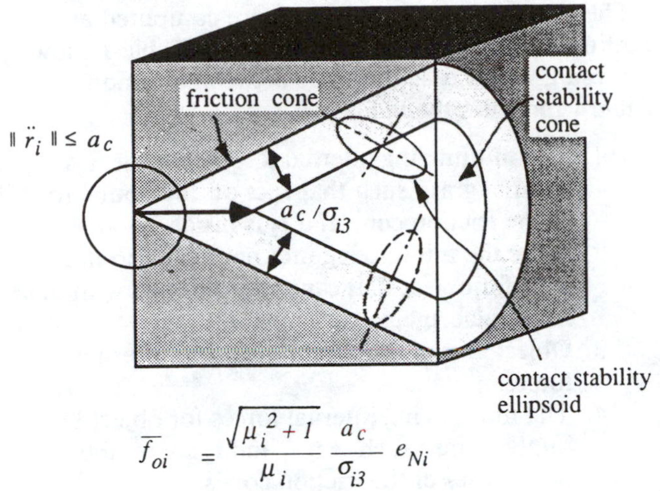

The relationship between this circular cone and the friction cone is shown in Fig. 8. The surface of this cone is parallel to the friction cone. The direction of the constant vector $\bar{\mathbf{f}}_{0i}$ is the same as that of the internal unit vector at the contact point, and its length is determined so that the distance between the surfaces of two cones is equal to a half of the longest principal axis of the contact stability ellipsoid. In Eq. (51), σ_{i3} is the least singular value of $\mathbf{J}_i\mathbf{I}_i^{-1}\mathbf{J}_i^T \in R^{3\times3}$.

We call the circular cone defined by Eqs. (49)–(51) *the contact stability cone,* based on which the optimal forces are defined as follows.

DEFINITION 4 The minimal norm force $\mathbf{F}$ among the forces that generate the specified resultant force $\mathbf{Q}$ under the contact stability constraints of Eqs. (49)–(51) is called the *optimal force,* and the internal forces that yield the optimal force are called the *optimal internal forces.*

5.3. Computation of Optimal Forces

The problem of obtaining the optimal force and the optimal internal force is mathematically formulated as follows: for the given resultant force $\mathbf{Q}$, the internal forces which satisfy

$$\min \|\mathbf{F}\| \tag{52}$$

and its net force $\mathbf{F}$ are solved under the constraints

$$\mathbf{Q} = \mathbf{W}\mathbf{F}, \tag{53}$$

$$\mathbf{e}_{Ni}^T\bar{\mathbf{f}}_i \geq \eta_i \|\bar{\mathbf{f}}_i\|. \tag{54}$$

If there exists an $\mathbf{F}$ that satisfies Eq. (53), then Eq. (53) is equivalently replaced by

$$\begin{aligned}\bar{\mathbf{F}} &= \mathbf{F}_0 - \bar{\mathbf{F}}_0 + \mathbf{A}\mathbf{y},\\ \bar{\mathbf{F}} &= (\bar{\mathbf{f}}_1^T \;\; \bar{\mathbf{f}}_2^T \;\; \cdots \;\; \bar{\mathbf{f}}_m^T)^T,\\ \bar{\mathbf{F}}_0 &= (\bar{\mathbf{f}}_{01}^T \;\; \bar{\mathbf{f}}_{02}^T \;\; \cdots \;\; \bar{\mathbf{f}}_{0m}^T)^T.\end{aligned} \tag{55}$$

This formulation is quite similar to the formulation of the minimal forces and the minimizing internal forces discussed in Sections 4.3–4.5. Therefore, the algorithms to obtain the optimal forces and the optimal internal forces are obtained by slightly modifying Algorithms 1 and 2 and replacing Eqs. (27) and (38) by

$$\bar{g}_i(\mathbf{y}) \leq 0 \qquad \text{for } i = 1, \ldots, 2m, \tag{56}$$

$$\bar{g}_i(\mathbf{y}) = \begin{cases}(\mathbf{f}_{0i} - \bar{\mathbf{f}}_{0i} + \mathbf{A}_i\mathbf{y})^T\mathbf{B}_i(\mathbf{f}_{0i} - \bar{\mathbf{f}}_{0i} + \mathbf{A}_i\mathbf{y}) & \text{for } i = 1, \ldots, m,\\ -\mathbf{e}_{Ni-m}^T(\mathbf{f}_{0i-m} - \bar{\mathbf{f}}_{0i-m} + \mathbf{A}_{i-m}\mathbf{y}) & \text{for } i = m+1, \ldots, 2m,\end{cases} \tag{57}$$

$$\mathbf{y}^0 = -\{\mathbf{A}^T(\mathbf{B}_0 + \mathbf{E}_{3m})\mathbf{A}\}^{-1}\mathbf{A}^T\mathbf{B}_0(\mathbf{F}_0 - \bar{\mathbf{F}}_0).$$

Assuming $\bar{\mathbf{f}}_i \neq 0$ $(i = 1, \ldots, m)$, the optimal internal forces and the optimal forces are obtained according to Algorithm 1 by replacing $g_i(\mathbf{y}^0)$, Eqs. (27), and (38) with $\bar{g}_i(\mathbf{y}^0)$, Eqs. (56), and (57), respectively.

The global solutions in the general case when $\bar{\mathbf{f}}_i = 0$ is allowed are computed using Algorithm 2 by replacing $\mathbf{f}_i$ and $\mathbf{f}_{0i}$ with $\bar{\mathbf{f}}_i$ and $\mathbf{f}_{0i} - \bar{\mathbf{f}}_{0i}$, respectively.

Accordingly, it turns out that the computational procedure and complexity of the optimal forces and the optimal internal forces are almost the same as those of the minimal forces and the minimizing internal forces, except for the computation of σ_{i3} to give $\bar{\mathbf{f}}_{0i}$.

Fig. 9. Numerical examples of minimal force computation: two-dimensional objects.

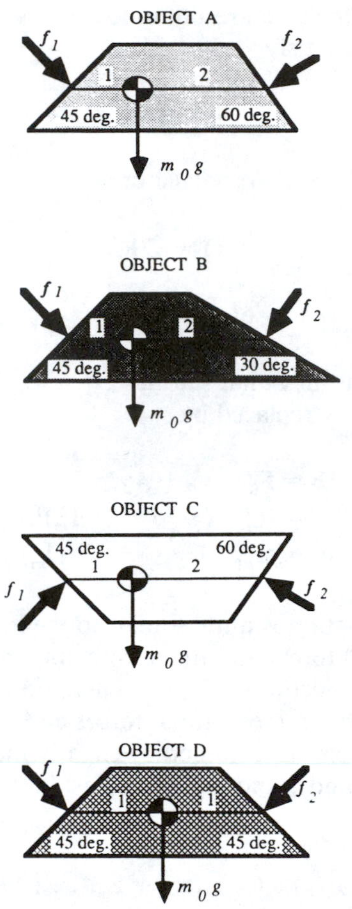

6. Numerical Examples of Minimal Force Computation

In this section, we give numerical examples of the computation of the minimizing internal forces. Since the computation of the optimal internal forces is almost the same, these examples will also show this. The examples are to grasp four kind of objects as shown in Fig. 9, applying the resultant forces to cope with the gravity force. The object coordinates are fixed at the mass center of the objects with the y axis in the upper vertical direction. The x axis is chosen to make a right-hand coordinate system. The contact points are on the x axis. The maximum static frictional coefficient is $\mu_i = \sqrt{3}$ $(i = 1, 2)$, which means that the angle of the friction cone is 60°.

The minimizing internal forces are computed according to the procedure in Section 4.5. Table 1 shows the result, where α_i is the angle between $\mathbf{e}_{Ni}$ and $\mathbf{f}_i^0$. Table 1 indicates the following:

1. The minimizing internal forces for object A mean a grasp such that $\mathbf{f}_1$ is on the boundary of the friction cone and $\mathbf{f}_2$ is inside.
2. Since the minimizing internal forces do not exist for object B, it cannot be grasped with this finger placement.
3. Object C can be grasped with zero internal forces.
4. The minimizing internal forces for object D imply a grasp such that $\mathbf{f}_1$ and $\mathbf{f}_2$ are on the boundaries of the friction cones.

For two-dimensional problems with two robotic mechanisms, $g_i(\mathbf{y}^0) = 0$ $(1 \le i \le m)$ becomes a quadratic equation. For three-dimensional problems, it becomes a $2b$th-order (b is the dimension of the null space of $\mathbf{W}$) algebraic equation, which means a sixth-order equation even for three robotic mechanisms. Therefore, we have to use some numerical method to solve it in the three-dimensional cases.

7. Conclusion

We discussed the dynamics and stability aspects in coordinatively manipulating an object with multiple robot manipulators or a multifingered robot hand. The main results obtained are summarized as follows:

1. Coordinative manipulation is divided into two phases. One is the determination of the resultant force to control the object's trajectory and to specify the stiffness for the external forces. The other is the determination of the internal forces to cope with uncertainty and variety of the static friction.
2. We proposed to determine the resultant force by considering the dynamics of the object.
3. If a robotic mechanism is force controlled in the direction in which the end-effector works, it must also be position controlled so that it

Table 1. Computed Minimal Forces and Minimizing Internal Forces

	Object A	Object B	Object C	Object D
λ_1^0	$8(3+2\sqrt{3})/3$	No solution	0	$\lambda_1^0+\lambda_2^0$
λ_2^0	0	No solution	0	$=8(3+2\sqrt{3})/3$
f_1^0/m_0g	$4(1+\sqrt{3}/2)/3$ $2/3$	No solution	0 $2/3$	$1+\sqrt{3}/2$ $1/2$
f_2^0/m_0g	$-4(1+\sqrt{3}/2)/3$ $1/3$	No solution	0 $1/3$	$-(1+\sqrt{3}/2)$ $1/2$
A_1y^0/m_0g	$4(1+\sqrt{3}/2)/3$ 0	No solution	0 0	$1+\sqrt{3}/2$ 0
A_2y^0/m_0g	$-4(1+\sqrt{3}/2)/3$ 0	No solution	0 0	$-(1+\sqrt{3}/2)$ 0
α_1	60°	No solution	45°	60°
α_2	37.6°	No solution	60°	60°

can follow the resulting object motion of the force control.

4. The robotic mechanisms can generate an arbitrary acceleration of the object if and only if $\mathbf{Q}^j$ and $-\mathbf{Q}^j$ ($j = 1, \ldots, 6$) can be generated under the static frictional constraints, where $\mathbf{Q}^j$ ($j = 1, \ldots, 6$) are linearly independent resultant forces.
5. We defined the minimizing internal forces as the internal forces that yield the minimal forces for the specified resultant force under the static frictional constraints, and established a computational procedure to obtain them. The minimizing internal forces are necessarily obtained by solving, at most, $\Sigma_{j=1}^{m}\binom{m}{j}\Sigma_{i=0}^{j}\binom{j}{i}$ (m means the number of robotic mechanisms) sets of algebraic equations if they exist. The concept of minimal forces is important because it can answer the fundamental question of whether the object can be grasped with the specified contact point placement.
6. The concept of contact stability was defined as the ability to keep contact with an object without slipping for a class of unexpected disturbing forces, and the measure of contact stability was proposed to evaluate the contact stability of the end-effector forces.
7. We defined the optimal internal forces as the internal forces which yield the minimal forces that give the specified resultant force and have the specified measure of contact stability. It was shown that the optimal internal forces can be computed by using algorithms similar to those for the minimizing internal forces.
8. Examples of computing the minimizing internal forces were given to show the effectiveness of the algorithms.

Acknowledgment

We express our sincere gratitude to Dr. Susan Hackwood and Dr. Gerardo Beni for the encouragement and the support of this work at the Center of Robotics Systems in Microelectronics, University of California, Santa Barbara. This research is partially supported by the National Science Foundation under contract number 8421415. Any opinions, findings, conclusions, or recommendations expressed in this publication are those of the authors and do not necessarily reflect the views of the NSF.

Appendix: Nomenclature

$O_0(x_0y_0z_0)$	The absolute coordinates
$O\,(xyz)$	The object coordinates fixed at the mass center of the object
$\mathbf{r} \in R^3$	Vector from the origin of the absolute coordinates to the mass center of the object, m
$\omega \in R^3$	Rotational velocity vector of the object, rad/s
m_0	Mass of the object, kg
$\mathbf{I} \in R^{3\times 3}$	Inertial tensor of the object represented by the object coordinates, $kg \cdot m^2$
m	Number of robotic mechanisms
$\mathbf{p}_i \in R^3$	Position vector of the ith contact point represented by the object coordinates, m
$\mathbf{f}_i \in R^3$	Force applied to the object at the ith contact point, N
$\mathbf{Q} \in R^6$	Resultant force and moment at the center of mass of the object, N, N·m
$\phi \in R^6$	Position and orientation of the object with respect to the absolute coordinates, m, rad
μ_i	Maximum static frictional coefficient between the object and the ith robotic mechanisms
$\mathbf{r}_i \in R^3$	Position vector of the ith contact point represented by the absolute coordinates, m
$\tau_i \in R^{k_i}$	Generalized driving force of the ith robotic mechanism
k_i	Degrees of freedom of the ith robotic mechanism
$\mathbf{E}_i \in R^{i\times i}$	Unit matrix
$\mathbf{g} \in R^3$	Gravity acceleration, m/s^2

Bibliography

Asada, H. 1983. A geometrical representation of manipulator dynamics and its application to arm design. *Trans. ASME J. Dyn. Sys. Meas. Contr.* 104(3):131–136.

Brady, M., Hollerbach, J. M., Johnson, T. L., Lozano-Perez, T., and Mason, M. T. 1982. *Robot motion.* Cambridge: MIT Press.

Cutkosky, M. R. 1985. *Robotic grasping and fine manipulation.* Boston: Kluwer Academic.

Cutkosky, M. R., and Wright, P. K. 1986. Friction, stability and the design of robotic fingers. *Int. J. Robotics Res.* 5(4):20–37.

Fearing, R. S. 1986. Simplified grasping and manipulation with dextrous robot hands. *IEEE J. Robotics Automat.* 2(4):188–195.

Golub, G. H., and Van Loan, C. F. 1983. *Matrix computations.* Baltimore: Johns Hopkins University Press.

Hanafusa, H., and Asada, H. 1977 (Tokyo). Stable prehension by a robot hand with elastic fingers. *Proc. 7th Int. Symp. Industrial Robots:* 311–368.

Hanafusa, H., Kobayashi, H., and Terasaki, N. 1983 (Tokyo). Fine control of the object with articulated multi-finger robot hands. *Proc. 1983 Int. Conf. Advanced Robotics:* 245–251.

Hanafusa, H., Yoshikawa, T., Nakamura, Y., and Nagai, K. 1985 (Tokyo). Structural analysis and robust prehension of robotic hand-arm system. *Proc. 1985 Int. Conf. Advanced Robotics:* 311–318.

Jameson, J. W., and Leifer, L. J. 1986. Quasi-static analysis: A method for predicting grasp stability. *Proc. 1986 IEEE Int. Conf. on Robotics and Automation:* 876–883.

Kerr, J., and Roth, B. 1986. Analysis of multifingered hands. *Int. J. Robotics Res.* 4(4):3–17.

Khatib, O. 1985 (Tokyo). The operational space formulation in robot manipulator control. *Proc. 15th Int. Symp. Industrial Robots:* 165–172.

Kobayashi, H. 1985. Control and geometrical consideration for an articulated robot hand. *Int. J. Robotics Res.* 4(1):3–12.

Kumar, V., and Waldron, K. 1987 (Raleigh). Sub-optimal algorithms for force distribution in multifingered grippers. *Proc. 1987 IEEE Int. Conf. Robotics and Automation:* 252–257.

Kurono, S. 1975. Coordinated control of a pair of artificial arms. In *Biomechanism 3,* pp. 182–193. Tokyo: Tokyo University Press.

Luh, J. Y. S., Walker, M. W., and Paul, R. P. C. 1980. Resolved acceleration control of mechanical manipulators. *IEEE Trans. Automat. Contr.* 25(3):468–474.

Mason, M. T. 1981. Compliance and force control for computer controlled manipulators. *IEEE Trans. Syst. Man Cybernet.* SMC-11:418–432.

Mine, H. 1966. *Operations Research Vol. 1,* pp. 122–130. Tokyo: Asakura.

Nakamura, Y. 1987 (Los Angeles). Force applicability of robotic mechanisms. *Proc. 26th IEEE Conf. Decision and Control:* 570–575.

Nakano, E., Ozaki, S., Ishida, T., and Kato, I. 1974 (Tokyo).

Cooperational control of the anthropomorphous manipulator 'MELARM'. *Proc. 4th Int. Symp. Industrial Robots:* 251–260.

Nguyen, V. 1986a. The synthesis of stable force-closure grasps. Technical report 905, MIT Artificial Intelligence Laboratory.

Nguyen, V. 1986b. Constructing force-closure grasp. *Proc. 1986 IEEE Int. Conf. Robotics and Automation:* 1368–1373.

Nguyen, V. 1986c. The synthesis of stable grasps in the plane. *Proc. 1986 IEEE Int. Conf. Robotics and Automation:* 884–889.

Ohwovoriole, M. S. 1980. An extension of screw theory and its application to the automation of industrial assemblies. Ph.D. thesis, Stanford University, Dept. of Mechanical Engineering.

Raibert, M. H., and Craig, J. J. 1981. Hybrid position/force control of manipulators. *Trans. ASME J. Dyn. Sys. Meas. Contr.* 102:126–133.

Rao, C. R., and Mitra, S. K., 1971. *Generalized inverse of matrices and its applications.* New York: Wiley.

Salisbury, J. K. 1982. Kinematic and force analysis of articulated hands. Ph.D. thesis, Stanford University, Dept. of Mechanical Engineering.

Salisbury, J. K., and Craig, J. J. 1982. Articulated hands: force control and kinematic issues. *Int. J. Robotics Res.* 1(1):4–17.

Takase, K. 1976. Generalized decomposition and control of a motion of a manipulator. *Trans. Soc. Instru. Contr. Eng.* 12:62–68. *(Japanese)*

Takase, K. 1985. Representation of constraint motion and dynamic control of manipulators under constraints. *Trans. Soc. Instr. Contr. Eng.* 21(5):508–513. *(Japanese)*

Trinkle, J. C., Abel, J. M., and Paul, R. P. 1985. An investigation of frictionless, enveloping grasping in the plane. Technical report MS-CIS-86-57 GRASP LAB 70, Department of Computer and Information Science, University of Pennsylvania.

Uchiyama, M., Hakomori, K., and Shimizu, K. 1983 (Tokyo). A servo synthesis method for multi-arm cooperation. *Proc. 1st Annual Conf. Robotics Society of Japan:* 101–102. *(Japanese)*

Yoshikawa, T. 1986. Dynamic hybrid position/force control of robot manipulators—Description of hand constraints and calculation of joint driving force. *Proc. 1986 IEEE Int. Conf. Robotics and Automation:* 1393–1398.

Zexiang Li
Ping Hsu
Shankar Sastry
Department of Electrical Engineering and Computer Sciences and the Electronics Research Laboratory
University of California
Berkeley, California

Grasping and Coordinated Manipulation by a Multifingered Robot Hand

Abstract

A new avenue of progress in the area of robotics is the use of multifingered robot hands for fine motion manipulation. This paper treats two fundamental problems in the study of multifingered robot hands: grasp planning and the determination of coordinated control laws with point contact models. First, we develop the dual notions of grasp stability and grasp manipulability and propose a procedure for task modeling. Using the task model, we define the structured grasp quality measures, and using these measures we then devise a grasp planning algorithm. Second, based on the assumption of point contact models, we develop a computed torque-like control algorithm for the coordinated manipulation of a multifingered robot hand. This control algorithm, which takes into account both the dynamics of the object and the dynamics of the hand, will realize simultaneously both the position trajectory of the object and any desired value of internal grasp force. Moreover, the formulation of the control scheme can be easily extended to allow rolling and sliding motion of the fingers with respect to the object.

1. Introduction

A new avenue of progress in the area of robotics is the use of a multifingered robot hand for fine motion manipulation. The versatility of robot hands accrues from the fact that fine motion manipulation can be accomplished through relatively fast and small motions of the fingers and from the fact that they can be used on a wide variety of different objects (obviating the need for a large stockpile of custom end effectors). Several articulated hands, such as the JPL/Stanford hand (Salisbury 1982) and the Utah/MIT hand (Jacobsen et al. 1985), have recently been developed to explore problems relating to grasping and manipulation of objects. It is of interest to note that the coordinated action of multiple robots in a single manufacturing cell may be treated in the same framework as a multifingered hand.

Grasping and manipulation of objects by a multifingered robot hand is more complicated than the manipulation of an object rigidly attached to the end of a six-axis robotic arm for two reasons: the kinematic relations between the finger joint motion and the object motion are complicated, and the hand has to firmly grasp the object during its motion.

The majority of the literature on multifingered hands has dealt with kinematic design of hands, the automatic generation of stable grasping configurations, and the use of task requirement as a criterion for choosing grasps (Salisbury 1982; Kerr 1984; Kobayashi 1985; Cutkosky 1985; Nguyen 1987; Li and Sastry 1988). Some of these references (Cutkosky 1985; Nguyen 1987; Asada 1987; Li and Sastry 1988) have suggested the use of a task specification as a criterion for choosing a grasp, albeit in a somewhat preliminary form. A few control schemes for the coordination of a multifingered robot hand or a multiple robotic system have been proposed by Nakamura, Nagai, and Yoshikawa (1987), Zheng and Luh (1985), Arimoto (1987) and Hayati (1986). The most developed scheme is the master-slave methodology (Zheng and Luh 1985; Arimoto 1987) for a two-manipulator system. The schemes developed so far all suffer from the drawback

The International Journal of Robotics Research,
Vol. 8, No. 4, August 1989,

that they either assume rigid attachment of the fingertips to the object or are open loop. The schemes do not account for an appropriate contact model between the fingertips and the object.

This paper treats two fundamental problems in the kinematics and control of multifingered hands: grasp planning and the determination of coordinated control laws with point contact models. We develop dual notions of grasp stability (or force closure) and grasp manipulability and propose a simple procedure for task modeling. Using the task model, we then define the structured (or task-oriented) grasp quality measures, which are subsequently used for devising a grasp planning algorithm. Based on point contact models, we give a computed torque-like control law for the coordinated control of a multifingered robot hand manipulating an object. This control algorithm takes into account both the dynamics of the object and the dynamics of the fingers and can be easily extended to the control of the object rolling or sliding with respect to the fingertips.

A brief outline of the paper is as follows. Section 2 defines the grasp map, its associated effective force domain, and the hand Jacobian. We develop dual generalized force and velocity transformation formulas relating the finger joint torques and velocities to the generalized force on and generalized velocity of the object being manipulated. Using these relations, we define stability and manipulability of a grasp. In section 3, we extend the work by Li and Sastry (1988) to define task-oriented measures for grasp stability and manipulability and use these measures for grasp planning. Section 4 uses the machinery of sections 2 and 3 to develop a new computed torque-like control scheme for the dynamic coordination of the multifingered robot hand, along with a proof of its convergence.

2. Kinematics of Multifingered Robot Hands

This section develops the kinematics of multifingered robot hands. These subjects are discussed in greater depth in Salisbury (1982), Kerr (1984), Kobayashi (1985), Li, Hsu, and Sastry (1988), and Montana (1986).

Fig. 1. A k*-fingered hand grasping an object.*

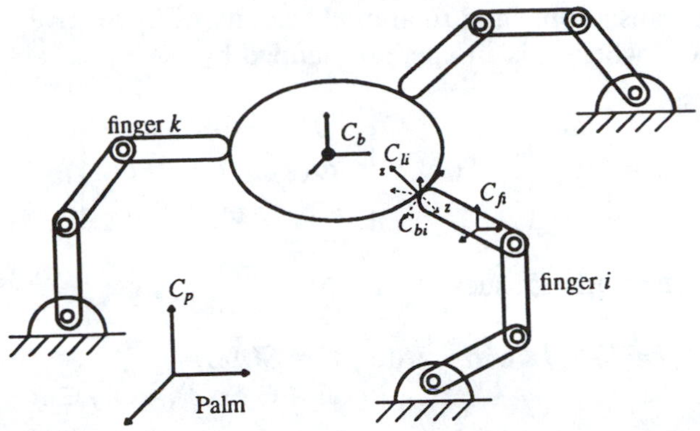

Figure 1 shows a manipulation system with a k-fingered robot hand. We assume that the number of joints of finger i, $i = 1, \ldots, k$, is m_i, and denote by θ_i, $\tau_i \in R^{m_i}$ the joint variable, and the joint torque vector, respectively, of finger i. To describe the contact constraints between the object and the fingers, we define a set of coordinate frames as follows: The reference frame, C_p, is fixed to the hand palm; the body coordinate frame, C_b, is fixed to the mass center of the object; at the ith point of contact with finger i, $i = 1, \ldots, k$, the local frame, C_{bi}, of the object is fixed relative to C_b and the origin is at the point of contact, whereas the z-axis coincides with the outward pointing normal to the object surface. Similarly, the finger frame, C_{fi}, is fixed to the last link of finger i; and the local frame of finger i, C_{li}, is fixed relative to C_{fi} and the origin is at the point of contact, whereas the z-axis coincides with the outward pointing normal to the fingertip. The local frames C_{bi} and C_{li} share a common origin, and their x- and y-axes lie in the common tangent plane. We define the contact angle ϕ_i by the angle between the x-axis of C_{bi} and C_{li}. We choose the sign of ϕ_i so that a rotation of C_{bi} through $-\phi_i$ around its z-axis aligns the x-axis.

We will need the following results concerning relative motion of coordinate frames. For any two coordinate frames C_α, C_β, where α, β are any subscripts, we denote by $r_{\beta,\alpha} \in R^3$ and $A_{\beta,\alpha} \in SO(3)$ the position and orientation of C_β relative to C_α. Furthermore, if $(r_{\beta,\alpha}(t), A_{\beta,\alpha}(t))$ is any curve in $SE(3) \triangleq R^3 \times SO(3)$ representing the trajectory of C_β relative to C_α, we denote by

$$v_{\beta,\alpha} = A^t_{\beta,\alpha}\dot{r}_{\beta,\alpha} \quad \text{and} \quad \omega_{\beta,\alpha} = S^{-1}(A^t_{\beta,\alpha}\dot{A}_{\beta,\alpha})$$

the translational and rotational velocity of C_β relative to C_α, where S is an operator defined by

$$\omega = \begin{bmatrix} \omega_1 \\ \omega_2 \\ \omega_3 \end{bmatrix}, \quad S(\omega) = \begin{bmatrix} 0 & -\omega_3 & \omega_2 \\ \omega_3 & 0 & -\omega_1 \\ -\omega_2 & \omega_1 & 0 \end{bmatrix},$$

which clearly satisfies

$$S(\omega)f = \omega \times f \quad \text{and} \quad AS(\omega)A^t = S(A\omega) \quad \text{for all } A \in So(3), \quad \omega, f \in R^3.$$

The vector $(v^t_{\beta,\alpha}, \omega^t_{\beta,\alpha})^t \in R^6$ is called the generalized velocity (or twist) of C_β relative to C_α.

Consider now three coordinate frames C_α, C_β, and C_ρ. The following relation exists between their relative velocities:

$$\begin{aligned} v_{\rho,\alpha} &= A^t_{\rho,\beta}(v_{\beta,\alpha} + \omega_{\beta,\alpha} \times r_{\rho,\beta}) + v_{\rho,\beta} \\ \omega_{\rho,\alpha} &= A^t_{\rho,\beta}\omega_{\beta,\alpha} + \omega_{\rho,\beta}. \end{aligned} \tag{2-1}$$

In particular, when C_ρ is fixed relative to C_β, the velocity of C_ρ is related to that of C_β by a constant transformation, given by,

$$\begin{aligned} \begin{bmatrix} v_{\rho,\alpha} \\ \omega_{\rho,\alpha} \end{bmatrix} &= \begin{bmatrix} A^t_{\rho,\beta} & -A^t_{\rho,\beta} S(r_{\rho,\beta}) \\ 0 & A^t_{\rho,\beta} \end{bmatrix} \begin{bmatrix} v_{\beta,\alpha} \\ \omega_{\beta,\alpha} \end{bmatrix} \\ &\triangleq T_{\rho,\beta} \begin{bmatrix} v_{\beta,\alpha} \\ \omega_{\beta,\alpha} \end{bmatrix}. \end{aligned} \tag{2-2}$$

In other words, if we let C_α be the reference frame, C_β the body coordinate frame, and C_ρ the object local frame, then the velocity of the object represented by the body frame and by the local frame are related by a constant transformation, which in turn is a function of the contact geometry of the object. A similar relation holds for the finger.

The (generalized) velocity space, or the twist space, of an object is denoted by $se(3)$. Dual to $se(3)$ is the space of generalized forces (or wrenches) exerted to the object, and is denoted by $se^*(3)$. Relative to the body coordinate frame C_b, a generalized force can be written as

$$\eta_b = \begin{bmatrix} f_b \\ m_b \end{bmatrix}$$

where $m_b \in R^3$ is the torque about the origin of C_b, and $f_b \in R^3$ is a linear force. It is well known that a wrench can also be expressed relative to the local frame C_{bi}, and the relation between η_{bi} and η_b is given by the dual transformation of (2-2),

$$\begin{aligned} \begin{bmatrix} f_b \\ m_b \end{bmatrix} &= \begin{bmatrix} A_{bi,b} & 0 \\ S(r_{bi,b})A_{bi,b} & A_{bi,b} \end{bmatrix} \begin{bmatrix} f_{bi} \\ m_{bi} \end{bmatrix} \\ &= T^t_{bi,b} \begin{bmatrix} f_{bi} \\ m_{bi} \end{bmatrix}. \end{aligned} \tag{2-3}$$

We now describe the contact constraints in terms of the relative velocities of the coordinate frames defined earlier. Let (v^i_x, v^i_y, v^i_z) and $(\omega^i_x, \omega^i_y, \omega^i_z)$ denote, respectively, the translational and rotational velocity of C_{bi} relative to C_{li}. These are the velocity of the object relative to finger i expressed in the local frames. Using (2-1), the velocity of C_{bi} can be expressed as (note that $r_{bi,fi} = 0$)

$$\begin{bmatrix} v_{bi,p} \\ \omega_{bi,p} \end{bmatrix} = \begin{bmatrix} A_{\phi_i} & 0 \\ 0 & A_{\phi_i} \end{bmatrix} \begin{bmatrix} v_{li,p} \\ \omega_{li,p} \end{bmatrix} + \begin{bmatrix} v^i_x \\ v^i_y \\ v^i_z \\ \omega^i_x \\ \omega^i_y \\ \omega^i_z \end{bmatrix} \tag{2-4}$$

where

$$\mathbf{A}_{\phi_i} = \begin{bmatrix} \cos\phi_i & -\sin\phi_i & 0 \\ -\sin\phi_i & -\cos\phi_i & 0 \\ 0 & 0 & -1 \end{bmatrix}$$

is the orientation matrix of C_{bi} relative to C_{li}.

On the other hand, the velocity of C_{bi} is related to the velocity of C_b by

$$\begin{aligned} \begin{bmatrix} v_{bi,p} \\ \omega_{bi,p} \end{bmatrix} &= \begin{bmatrix} A^t_{bi,b} & -A^t_{bi,b}S(r_{bi,b}) \\ 0 & A^t_{bi,b} \end{bmatrix} \begin{bmatrix} v_{b,p} \\ \omega_{b,p} \end{bmatrix} \\ &\triangleq T_{bi,b} \begin{bmatrix} v_{b,p} \\ \omega_{b,p} \end{bmatrix} \end{aligned} \tag{2-5a}$$

and similarly one has for finger i that

$$\begin{bmatrix} v_{li,p} \\ \omega_{li,p} \end{bmatrix} = \begin{bmatrix} A^t_{li,fi} & -A^t_{li,fi} S(r_{li,fi}) \\ 0 & A^t_{li,fi} \end{bmatrix} \begin{bmatrix} v_{fi,p} \\ \omega_{fi,p} \end{bmatrix} \triangleq T_{li,fi} \begin{bmatrix} v_{fi,p} \\ \omega_{fi,p} \end{bmatrix}. \quad \text{(2-5b)}$$

Moreover, the velocity of the finger frame, C_{fi}, is related to the velocity of the finger joints, $\dot{\theta}_i$, by the finger Jacobian,

$$\begin{bmatrix} v_{fi,p} \\ \omega_{fi,p} \end{bmatrix} = J_i(\theta_i)\dot{\theta}_i. \quad \text{(2-6)}$$

In this paper, we will consider the following contact models: (a) a point contact without friction, (b) a point contact with friction, (c) a soft finger contact, and (d) a rigid contact. These contact models give rise to contact constraints specified by

(a): $v^i_z = 0$, for a point contact without friction;
(b): $v^i_x = v^i_y = v^i_z = 0$, for a point contact with friction;
(c): $v^i_x = v^i_y = v^i_z = 0$, and $\omega^i_z = 0$, for a soft finger contact;
(d): $v^i_x = v^i_y = v^i_z = 0$, and $\omega^i_x = \omega^i_y = \omega^i_z = 0$, for a rigid contact.

For each of the contact models, substituting the above contact constraints and (2-5) and (2-6) into (2-4) yields an equation that relates the body velocity of the object and the joint velocity of the finger,

$$B^t_i T_{bi,b} \begin{bmatrix} v_{b,p} \\ \omega_{b,p} \end{bmatrix} = B^t_i J_{fi} \dot{\theta}_i, \quad \text{(2-7)}$$

where

$$J_{fi} \triangleq \begin{bmatrix} A_{\phi_i} & 0 \\ 0 & A_{\phi_i} \end{bmatrix} T_{li,fi} J_i(\theta_i)$$

is called the modified finger Jacobian, and B^t_i is the basis matrix defined in (Kerr 1984) representing the contact constraints (a)–(d). For example, for a point contact with friction,

$$B^t_i = \begin{bmatrix} 1 & 0 & 0 & 0 & 0 & 0 \\ 0 & 1 & 0 & 0 & 0 & 0 \\ 0 & 0 & 1 & 0 & 0 & 0 \end{bmatrix}$$

and so on. Equation (2-7) is called the velocity constraint equation between the object and finger i.

We now briefly examine contact constraints in terms of contact wrenches. For a contact model, let n_i denote the total number of independent contact wrenches that finger i can apply to the object. For example, $n_i = 1$ for a point contact without friction (i.e., a force in the normal direction), and $n_i = 3$ for a point contact with friction (i.e., a force in the normal direction plus two components of frictional forces). Note that n_i is just the total number of contact constraints corresponding to the contact model. The resulting body wrench from applied contact wrenches of finger i can be expressed as

$$\begin{bmatrix} f_b \\ m_b \end{bmatrix} = T^t_{bi,b} B_i x_i \quad \text{(2-8)}$$

where $x_i \in R^{n_i}$ is the magnitude vector of applied contact wrenches along the basis directions of B_i. For a frictional point contact, x_i is constrained to lie in the frictional cone

$$K_i = \{x_i \in R^3,\ x_{i,3} \geq 0,\ x^2_{i,1} + x^2_{i,2} < \mu^2 x^2_{i,3}\}$$

where μ is the static Coulomb frictional coefficient.

By the *principle of virtual work*, the joint torque required for maintaining static equilibrium in the presence of contact wrench $x_i \in R^{n_i}$, is given by

$$\tau_i = J^t_{fi} B_i x_i \quad \text{(2-9)}$$

Finally, for the hand manipulation system, we define the total number of joints as

$$m = \sum_{i=1}^{k} m_i,$$

the total number of constraints as

$$n = \sum_{i=1}^{k} n_i,$$

Fig. 2. The force/torque and velocity transformation relations.

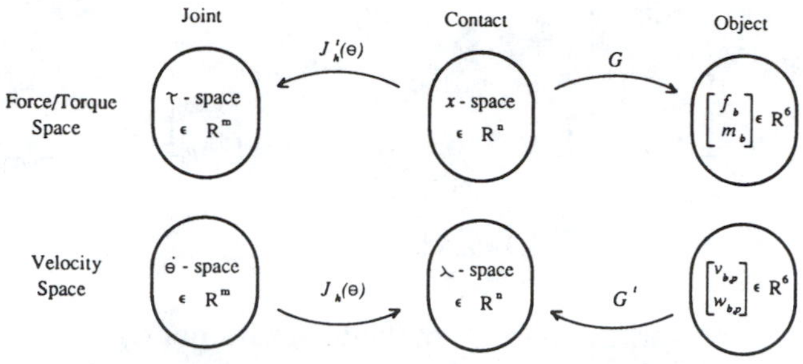

the hand joint variable and the hand joint torque vectors, respectively, as

$$\theta = (\theta_1^t, \cdots, \theta_k^t)^t, \quad \tau = (\tau_1^t, \cdots, \tau_k^t)^t \in R^m,$$

the basis matrix as

$$B = \text{diag}(B_1, \cdots, B_k)$$

the magnitude vector of contact wrenches along the directions of B as

$$x = (x_1^t, \cdots, x_k^t)^t \in R^n,$$

and the force cone as

$$K = K_1 \oplus \cdots \oplus K_k \subset R^n.$$

Then, the contact constraint equation (2-7) can be concatenated for $i = 1, \cdots, k$ to give,

$$G^t \begin{bmatrix} v_{b,p} \\ \omega_{b,p} \end{bmatrix} = J_h \dot{\theta} \tag{2-10}$$

where

$$G = [T_{b1,b} \cdots T_{bk,b}]B \quad \text{and} \quad J_h = B^t \text{ diag } \{J_{f1}, \cdots, J_{fk}\} \tag{2-11}$$

are called, respectively, the grasp matrix and the hand Jacobian. The equation that relates the resulting body wrench and applied contact wrenches is

$$\begin{bmatrix} f_b \\ m_b \end{bmatrix} = Gx \tag{2-13}$$

and the equation that relates contact wrenches to the required joint torques for maintaining static equilibrium is

$$\tau = J_h^t x. \tag{2-14}$$

The relations have been summarized in the table below:

	Force Torque Relations	Velocity Relations
Body to Fingertip	$\begin{bmatrix} f_b \\ m_b \end{bmatrix} = Gx$	$\lambda = G^t \begin{bmatrix} v_{b,p} \\ \omega_{b,p} \end{bmatrix}$
Fingertip to Joints	$\tau = J_h^t(\theta)x$	$J_h(\theta)\dot{\theta} = \lambda$

Figure 2 illustrates transformation of forces and motion in a hand manipulation system. Note that the vector $\boldsymbol{\lambda} \in R^n$ is called the contact velocity. The null space of the grasp map G, denoted as $\eta(G)$, is called the space of internal grasping forces (Kobayashi 1985). Any applied finger forces in $\eta(G)$ do not contribute to the motion of the object. However, during the course of manipulation a set of nonzero internal grasping forces is needed to assure that the grasp is maintained. Kerr and Roth (1986) and Nakamura, Nagai, and Yoshikawa (1987) have presented detailed discussions on the optimal selection of internal grasping forces. When the fingertips roll or slide over the object, the contact parameters $(r_{bi,b}, A_{bi,b})$ of the object and $(r_{li,fi}, A_{li,fi})$ of the finger evolve according to the kinematic equations of contact (Montana 1986) and G becomes time dependent.

The following dual definitions are now intuitive.

Stability and Manipulability of a Grasp

Define the grasp by a multifingered hand by $\Omega \triangleq (G, K, J_h)$ (Fig. 2). Then, for $K = R^n$ we have:

(i) The grasp Ω is said to be stable if, for every body wrench $(f_b^t, m_b^t)^t$, there exists a choice of joint torque τ to balance it.

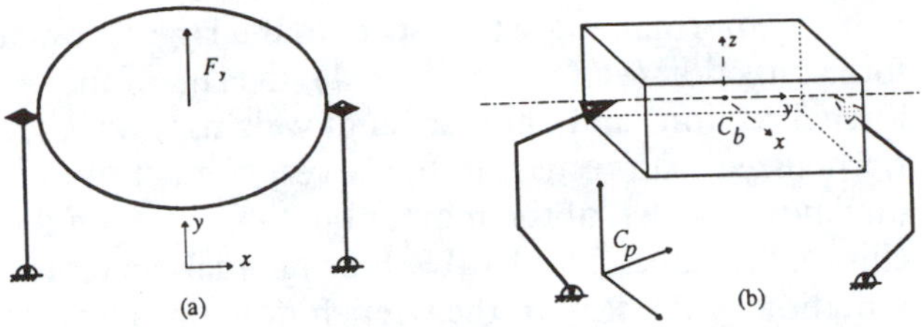

Fig. 3. (A) A stable but not manipulable grasp. (B) A manipulable but not stable grasp.

(ii) The grasp Ω is said to be manipulable if, for every body motion $(v^t_{b,p}, \omega^t_{b,p})^t$, there exists a choice of joint velocity $\dot{\theta}$ to accommodate this motion without breaking contact.

Remark: (1) A stable grasp has been called a force-closure grasp by Salisbury (1982) and Nguyen (1987). It is important to note that stability is not to be understood in the sense of Lyapunov since we are not discussing stability of a differential equation. On the other hand, under appropriate joint level control laws for the fingers, a stable grasp will give rise to a stable hand manipulation system (i.e., stable in the sense of Lyapunov) (Nguyen 1987).

(2) A manipulable grasp is called a grasp with full mobility by Salisbury (1982), or a grasp with full manipulable space by Kobayashi (1985).

Grasp stability and manipulability are now easily characterized for a given position of the fingers by the following proposition.

PROPOSITION: (i) A grasp is stable if and only if G is onto, i.e., the range space of G is the entire R^6. (ii) A grasp is manipulable if and only if $R(J_h(\theta)) \supset R(G^t)$, where $R(\cdot)$ denotes the range space of a map.

Remark: The conditions (i) and (ii) superficially appear to be distinct, but they are related. In particular, a stable grasp that requires zero joint torque to balance a nonzero body wrench will be nonmanipulable. Conversely, a manipulable grasp that requires zero joint motion to accommodate a nonzero body motion will be non-stable. Figure 3A shows a planar two-fingered grasp, where each finger is one-jointed and contacts the object with a point contact with friction. Clearly the grasp is stable, and a force f_y can be resisted with no joint torques. But the grasp is not manipulable, since a y-direction velocity on the body cannot be accommodated. Figure 3B shows a grasp of a body in R^3 by two three-jointed fingers. The contacts are point contacts with friction. The grasp is manipulable, though the object can spin around the y-axis with zero joint velocities $\dot{\theta}$. However, the grasp is not stable since a body torque τ_n about the y-axis cannot be resisted by any combination of joint torques.

In view of the preceding remarks, we will require a grasp to be both manipulable and stable, i.e.,

$$R(G) = R^6 \quad \text{and} \quad R(J_h(\theta)) \supset R(G^t). \qquad (2\text{-}15)$$

Condition (i) suffers from the drawback that the force domain is left completely unconstrained. As we have seen earlier that the forces are constrained to lie in a convex cone K, taking into account the unidirectionality of the contact forces, finite friction, etc., in which case the image of $K \cap R(J_h)$ under G should cover all of R^6. Thus, we have

COROLLARY: A grasp under unisense and finite frictional forces is both stable and manipulable if and only if

$$G(K \cap R(J_h)) = R^6, \quad \text{and} \quad R(J_h) \supset R(G^t). \qquad (2\text{-}16)$$

3. Grasp Planning

Typical tasks associated with multifingered robot hands include scribing, inserting a peg into a hole, and other assembly operations. Common to these tasks is the fact that the robot hand must manipulate an object from one configuration to another while exerting a set of desired contact forces on the environment. Successful execution of such tasks amounts to having the robot hand perform a sequence of operations: (1) selecting a "good" grasp on the object, and (2) using the cooperative action of the fingers to control the object. In this section we study how to generate a good grasp for a given task, and in the next section we study how to manipulate the object by the cooperative action of the fingers.

Fig. 4. Peg-in-hole task.

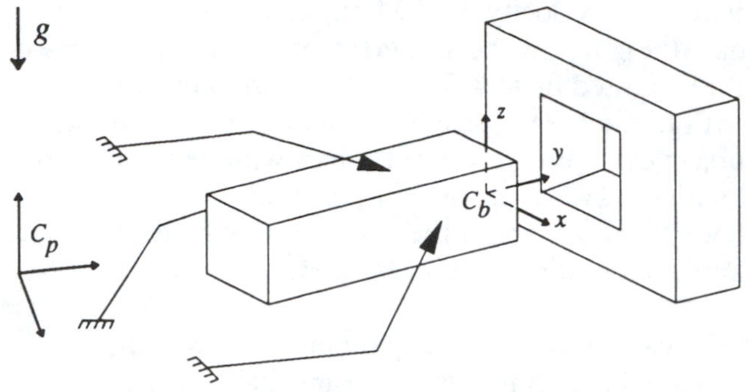

In Li and Sastry (1988) a task was modeled by an ellipsoid, called the task ellipsoid, in the *wrench space* of the object. In this section, we extend the work of Li and Sastry to consider the problem of determining grasps that are both *manipulable* and *stable* with respect to the modeled tasks.

The following two steps constitute our strategy to this problem: (1) proper modeling of the task in both wrench and twist space and (2) using the task models to define grasp quality measures and using these measures for grasp planning.

Consider the following task.

EXAMPLE (PEG-IN-HOLE INSERTION): Consider the peg insertion task depicted in Figure 4 where the robot hand grasps the workpiece and inserts it into the hole. In order to execute the task, a nominal trajectory is planned before grasping. After grasping, the hand follows the planned trajectory until some misalignment of the peg causes the object to deviate from the nominal trajectory and collide with the environment.

Choose the body coordinate frame as shown (the location of the body coordinate frame is arbitrary, since by Eqs. (2-2) and (2-3) twists and wrenches of the body with respect to different body coordinate frames are related by constant transformations). Clearly, contact forces/moments in certain directions will exceed that in the others. For example, contact forces in direction $-f_y$ will dominate contact forces in directions $\pm f_x$ and $\pm f_z$; and contact moments in directions $\pm\tau_z$, $\pm\tau_x$ (caused by $-f_y$ force) will dominate contact moments in the remaining directions. Moreover, contact force in direction $+f_y$ will be much less than $-f_y$, and force in $-f_z$ will be larger than $+f_z$ by the weight of the object. If we assign one set of weights, r_x, r_y, r_z, normalized with respect to the largest contact force direction (i.e., $r_x, r_z \leq r_y = 1$), to each of the force directions and another set of weights, $\gamma_x, \gamma_y, \gamma_z$, normalized with respect to the largest contact moment direction, to each of the moment directions, then the ellipsoid A_α given by (3-1a) will be an analytic representation of the task in the wrench domain. Here, the constant c_1 is the offset of the maximum expected contact force in $+f_y$ from $-f_y$, c_2 the weight of the object, and $\alpha \geq 0$ is a scaling parameter or a normalizing constant.

$$A_\alpha = \left\{ (f_x, \ldots, \tau_z) \in R^6, \frac{f_x^2}{r_x^2} + \frac{(f_y + c_1)^2}{r_y^2} + \frac{(f_z - c_2)^2}{r_z^2} + \frac{\tau_x^2}{\gamma_x^2} + \frac{\tau_y^2}{\gamma_y^2} + \frac{\tau_z^2}{\gamma_z^2} \leq \alpha^2 \right\}. \tag{3-1a}$$

Geometrically, A_α is an ellipsoid in the wrench space, centered at $(0, c_1, c_2, 0, 0, 0)$, with the principal axes given by the generalized force directions, and axis length by the corresponding weights. The size of the ellipsoid is scaled by α. Part of the objective of task modeling is to obtain numerical values for the weights as well as for the constants c_1 and c_2.

While the previous discussion is centered on the *force/moment* property of the task, let us now return to address the *motion* property of the task. For such a task, relatively larger motions in direction v_y and ω_y are required than in the remaining directions. Consequently, we may model the twist space ellipsoid by

$$B_\beta = \left\{ (v_x, \ldots, \omega_z) \in R^6, \frac{v_x^2}{\delta_x^2} + \frac{v_y^2}{\delta_y^2} + \frac{v_z^2}{\delta_z^2} + \frac{\omega_x^2}{\epsilon_x^2} + \frac{\omega_y^2}{\epsilon_y^2} + \frac{\omega_z^2}{\epsilon_z^2} \leq \beta^2 \right\}. \tag{3-1b}$$

Here, the weighting constants are again normalized with respect to the largest motion directions.

Other examples may also be studied and modeled by two task ellipsoids, one in the wrench space that represents the relative force/moment requirement and the other in the twist space that represents the relative motion requirement of the task. The procedure to obtain these models is through experiments or experience with similar tasks.

There are other approaches to develop task ellipsoids. For example, if stiffness control is used for the hand, then the maximum expected positional uncertainties in each of the task directions may be used to scale the axis of B_β. Also, during part mating, jamming can be avoided if certain constraints on the ratios of the contact forces are satisfied (Whitney 1982); the constraints on the force ratios can be used to scale the ellipsoid A_α.

In summary, we will assume that a task is modeled by generalized ellipsoids A_α (wrench space) and B_β (twist space) of the form

$$A_\alpha = \{y \in R^6\colon\ y = \alpha A x + c \text{ where } x, c \in R^6, \|x\| \leq 1, \text{ and } A \in R^{6\times 6}\}, \tag{3-2a}$$

and

$$B_\beta = \{y \in R^6\colon\ y = \beta B x + d \text{ where } x, d \in R^6, \|x\| \leq 1, \text{ and } B \in R^{6\times 6}\}, \tag{3-2b}$$

where the structure matrices A, B are given by

$$A = [U_1, \cdots U_6] \begin{bmatrix} \sigma_1 & \cdots & 0 \\ \cdot & \cdot & \cdot \\ \cdot & \cdot & \cdot \\ \cdot & \cdot & \cdot \\ 0 & \cdots & \sigma_6 \end{bmatrix} \begin{bmatrix} U_1^t \\ \cdot \\ \cdot \\ \cdot \\ U_6^t \end{bmatrix} \tag{3-3a}$$

$$\triangleq U \Sigma U^t, \quad U_i \in R^6, \sigma_i \geq 0, i = 1, \ldots, 6,$$

and

$$B = [U_1, \cdots U_6] \begin{bmatrix} \delta_1 & \cdots & 0 \\ \cdot & \cdot & \cdot \\ \cdot & \cdot & \cdot \\ \cdot & \cdot & \cdot \\ 0 & \cdots & \delta_6 \end{bmatrix} \begin{bmatrix} U_1^t \\ \cdot \\ \cdot \\ \cdot \\ U_6^t \end{bmatrix} \tag{3-3b}$$

$$\triangleq U \Delta U^t, \quad U_i \in R^6, \delta_i \geq 0, i = 1, \ldots, 6.$$

Here, U_i, $(i = 1, \ldots, 6,)$ is the task direction expressed in the body coordinates (in Example 3.1, $U = I$) and σ's (or δ's) are the normalized weights in the wrench space (and the twist space respectively). $c(d) \in R^6$ is the offset constant, which locates the center of the ellipsoid. We may assume in the following development that $c = d = 0$. When the structure matrix, say A, is nonsingular, an alternative expression of A_α may be obtained as

$$A_\alpha = \{y \in R^6\colon \langle y - c, \beta^{-2}(AA^t)^{-1}(y - c)\rangle \leq 1\}, \tag{3-4}$$

where $\langle \cdot, \cdot \rangle$ denotes the inner product in R^6.

When *stiffness control* is used for the hand, σ_i is related to δ_i by $\sigma_i = K_i \cdot \delta_i$, where K_i is the desired stiffness in direction U_i. On the other hand, when hybrid position/force control is used for the hand, $\sigma_i = 0$ if direction U_i is position controlled and $\delta_i = 0$ if direction U_i is force controlled.

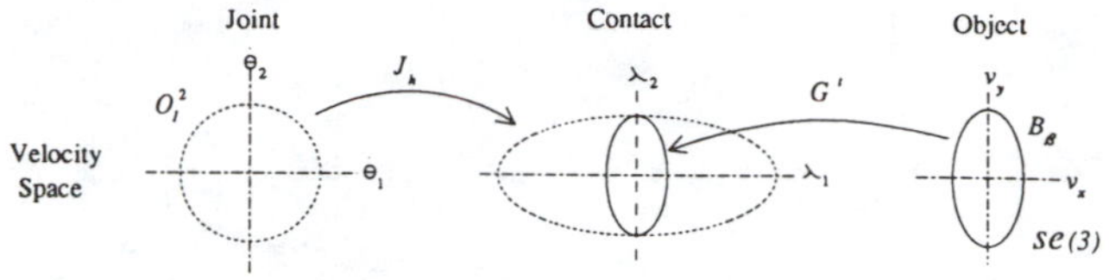

Fig. 5. Geometric interpretation of μ_t.

3.1. Structured Quality Measures for Grasp Planning

We now integrate the definition of a grasp $\Omega = (G, K, J_h)$ and the previously described task models to develop two quality measures for a grasp, one in the twist space and the other in the wrench space.

3.1.1. The Structured Twist Space Quality Measure μ_t

Following the previous notation we let $O_1^m \subset R^m$ denote the unit ball in R^m, the space of finger joint velocities, and define the structured twist space quality measure $\mu_t(\Omega)$ of Ω by

$$\mu_t(\Omega) = \sup_{\beta \in R_+} \{\beta, \text{ such that } J_h(O_1^m) \supset G^t(B_\beta)\}. \tag{3.1-1}$$

The geometric meaning of $\mu_t(\Omega)$ is as follows (Fig. 5): the unit ball O_1^m in the finger joint velocity space is mapped into the space of contact velocity by J_h. On

Fig. 6. Geometric interpretation of μ_w.

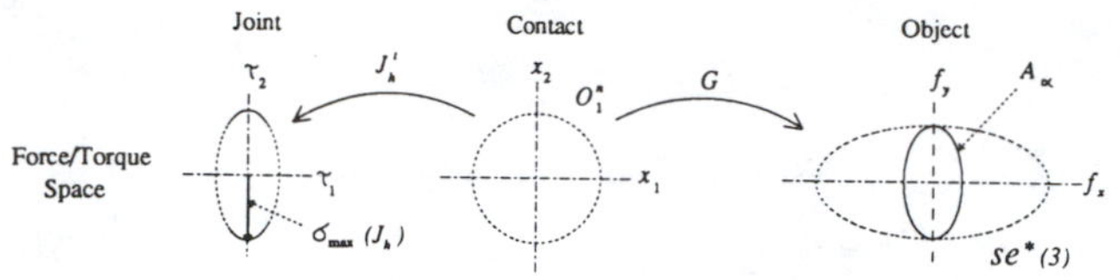

the other hand, a task ellipsoid B_β in the twist space is mapped back into the contact velocity space by G^t; $\mu_t(\Omega)$ is then the largest β such that $G^t(B_\beta)$ is *contained* in $J_h(O_1^m)$. In other words, object velocity of size β can be accommodated by finger joint velocity of unit magnitude. From a theoretical point of view, $\mu_t(\Omega)$ is the ratio of the "structured" output (i.e., the task ellipsoid) over the input (i.e., the finger joint velocity). Also, note from the figure that $\mu_t(\Omega)$ is at its maximum if the inner ellipsoid has the same shape and orientation as the outer ellipsoid.

3.1.2. The Structured Wrench Space Quality Measure μ_w

Let $O_1^n \subset R^n$ be the unit ball in the finger wrench space and $\sigma_{\max}(J_h)$ the maximum singular value of J_h. We define the structured wrench space quality measure μ_w of Ω by

$$\mu_w(\Omega) = \sup_{\alpha \in R_+} \{\alpha, \text{ such that } G(O_1^n \cap K) \supset A_\alpha\} \cdot \sigma_{\max}^{-1}(J_h). \qquad (3.1\text{-}2)$$

Remark: (1) Note from Figure 6 that $\mu_w(\Omega)$ is the largest α such that A_α can be embedded in $G(O_1^n)$ (the output), and the second term is the largest input torque required to generate the finger wrench O_1^n (the input).

(2) Here, we did not consider internal grasping force in $\eta(G)$, which is often a necessary part for securing a grasp. It would be useful to include internal grasping forces in the quality measures by considering how joint torques can be effectively transformed into internal grasp forces.

These quality measures defined in (3.1-1) and (3.1-2) provide useful characterization of a grasp. Clearly, we can say that a grasp Ω is a better grasp than other candidate grasps with respect to a given task modeled by A_α and B_β, if it attains higher structured quality measures μ_t and μ_w. But due to unidirectionality and finite frictional forces as reflected by K, which is usually a proper subset of R^n, it is difficult to evaluate Eq. (3.1-2). To simplify this problem, we will assume that $K = R^n$ and explore the properties of Eqs. (3.1-1) and (3.1-2) further.

PROPOSITION: Under the assumption that $K = R^n$ the structured quality measures (3.1-1) and (3.1-2) are given by

$$\mu_t(\Omega) = \sigma_{\max}^{-1/2}\{B^t G (J_h J_h^t)^{-1} G^t B\}, \qquad (3.1\text{-}3)$$

and

$$\mu_w(\Omega) = \sigma_{\max}^{-1/2}\{A^t (G G^t)^{-1} A\} \cdot \sigma_{\max}^{-1}(J_h). \qquad (3.1\text{-}4)$$

PROOF OF (3.1-3). Using the following expression

$$J_h(O_1^m) = \{y \in R^n, \langle y, (J_h J_h^t)^{-1} y\rangle \leq 1\} \qquad (3.1\text{-}5)$$

and

$$G^t(B_\beta) = \{y \in R^n: \quad y = \beta G^t B x, x \in R^6, \|x\| \leq 1\} \qquad (3.1\text{-}6)$$

in (3.1-1) and notice that $G^t(B_\beta) \subset J_h(O_1^m)$ if and only if

$$\langle \beta G^t B x, (J_h J_h^t)^{-1} \beta G^t B x\rangle \leq 1, \quad \text{for all } \|x\| \leq 1. \qquad (3.1\text{-}7)$$

In particular, (3.1-7) must hold for

$$\beta^2 \sup_{\|x\|=1} \langle G^t B x, (J_h J_h^t)^{-1} G^t B x\rangle = \beta^2 \sup_{\|x\|=1} \langle x, (G^t B)^t (J_h J_h^t)^{-1} G^t B x\rangle \leq 1, \qquad (3.1\text{-}8)$$

which is equivalent to

$$\beta \leq \sigma_{\max}^{-1/2}\{B^t G (J_h J_h^t)^{-1} G^t B\}. \qquad (3.1\text{-}9)$$

But by (3.1-1) we want the largest β such that (3.1-7) is true, hence we have (3.1-3). The proof of (3.1-4) follows similarly.

Remark: The quality measures given here can be easily evaluated using singular value decomposition data of a matrix. If we also want to consider the manipulability of a grasp in a certain direction, say U_i, we may simply apply U_i to (3.1-7) and obtain that

$$\begin{aligned}\beta_i &= \langle G^t B U_i, (J_h J_h^t)^{-1} G^t B U_i \rangle^{-1/2} \\ &= \delta_i^{-1} \cdot \langle G^t U_i, (J_h J_h^t)^{-1} G^t U_i \rangle^{-1/2}.\end{aligned} \quad (3.1\text{-}10)$$

Here, β_i measures the effectiveness of the grasp in imparting motion at direction U_i, and a similar relation holds for the stability measure.

Notice that (3.1-3) and (3.1-4) exhibit an interesting dual relation in the following sense: let the task ellipsoids be the unit balls; if we hold G constant but vary J_h, then $\mu_t(\Omega)$ is directly proportional to $\sigma_{\min}(J_h)$, and $\mu_w(\Omega)$ is inversely proportional to $\sigma_{\max}(J_h)$. On the other hand, if we hold J_h constant but vary G, then $\mu_w(\Omega)$ is directly proportional to $\sigma_{\min}(G)$ and $\mu_t(\Omega)$ is inversely proportional to $\sigma_{\max}(G)$. This observation implies that to a certain point, it is generally not possible to increase the two quality measures simultaneously by varying G and J_h. Namely, increasing one quality measure will sacrifice the other. For instance, if we select a "power grasp" in the scribing task (i.e., a grasp with high quality measure in the wrench space) then the grasp will be very inefficient in imparting motion at the pencil lead. Conversely, if we choose a "precision grasp" with high quality measure in the twist space, the grasp will be very poor in rejecting disturbance forces. The objective of grasp planning, therefore, is to search for a grasp which maximizes some performance measure (PM) defined by

$$\mathrm{PM} = [\mu_t(\Omega)]^{\gamma} \cdot [\mu_w(\Omega)]^{1-\gamma}, \quad \gamma \in [0,1] \quad (3.1\text{-}11)$$

Here, γ is called the selection parameter. $\gamma > 0.5$ indicates that the task is motion oriented, and $\gamma < 0.5$ indicates that the task is force oriented. A grasp that maximizes PM with γ close to 1 will be a "precision grasp," and a grasp that maximizes PM with γ close to 0 will be a "power grasp." More generally, a grasp that maximizes PM with γ close to 0.5 will be both stable and manipulable. Depending on the nature of the task we can set the parameter γ so that the final grasp satisfies the task requirements.

To plan for the final grasp we can, in principle, use

Fig. 7. Planar peg-in-hole task.

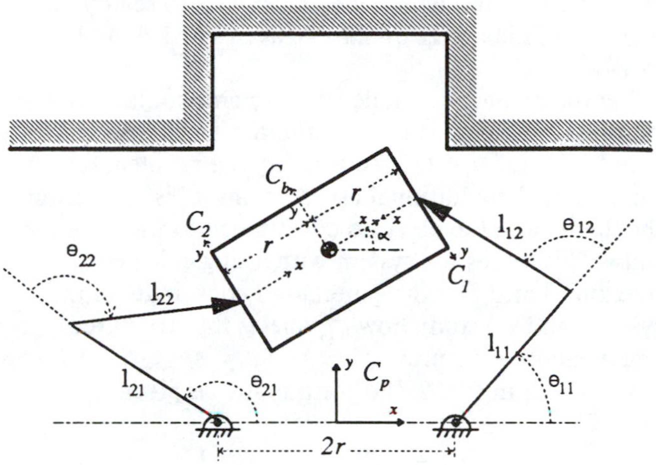

PM, the geometry of the object, and the structure of the hand to formulate the corresponding optimization problem; by solving this optimization problem we obtain the final grasp. However, simultaneously placing all k fingers optimally usually involves considerable computations and is thus impractical.

To reduce the number of computations, we may use the following suboptimal algorithm for grasp planning: *place fingers one after another on the object periphery using the constraint that the objective function PM be maximized.* Here, we present a simple example to illustrate the preceding discussion.

Example: Consider again the planar peg-in-hole task, shown in Figure 7. For simplicity we assume that the body orientation coincides with the orientation of C_p. Hence, the task direction matrix U expressed in C_b is given by $\mathrm{U} = \mathrm{I} \in R^{3\times 3}$. To execute the task, stiffness control is used for the hand (Whitney 1982), and we assume that the desired stiffness matrix is given by

$$K = \begin{bmatrix} K_x & 0 & 0 \\ 0 & K_y & 0 \\ 0 & 0 & K_\theta \end{bmatrix} = \begin{bmatrix} 5 & 0 & 0 \\ 0 & 7 & 0 \\ 0 & 0 & 100 \end{bmatrix}. \quad (3.1\text{-}12)$$

From (Whitney 1982) and the peg-in-hole insertion example we model the task by

$$\begin{aligned} &U = I \in R^{3\times 3}, \qquad \Sigma = \mathrm{diag}\,\{4, 5, 2\}, \\ &\Delta = \mathrm{diag}\,\{0.8, 0.7, 0.02\} = K^{-1} \cdot \Sigma. \end{aligned} \quad (3.1\text{-}13)$$

With this task modeling, the objective is to search for grasps that maximize quality measures (3.1-3) and (3.1-4).

Let the contact be modeled as a point contact with friction, the block width ($2r$) and the finger spacing be 2 and $l_{i,j} = 1$, $i, j = 1,2$. To simplify the problem further, we make additional assumptions: G is fixed as in the figure, and the object is constrained to move vertically. This leaves the system with a single degree of freedom. Let θ_{11} be the generalized coordinate of the system, and we study how θ_{11} affects the structured grasp quality measures.

As shown in Fig. 8, the grasp matrix is given by

$$G = \begin{bmatrix} -1 & 0 & 1 & 0 \\ 0 & -1 & 0 & 1 \\ 0 & -1 & 0 & -1 \end{bmatrix} \tag{3.1-14}$$

and the hand Jacobian J_h is

$$J_h \doteq \begin{bmatrix} J_1 & 0 \\ 0 & J_2 \end{bmatrix}, \tag{3.1-15}$$

where

$$J_1 = \begin{bmatrix} \cos\alpha & -\sin\alpha \\ \sin\alpha & \cos\alpha \end{bmatrix} \begin{bmatrix} -\sin\theta_{11} - \sin(\theta_{11}+\theta_{12}) & -\sin(\theta_{11}+\theta_{12}) \\ \cos\theta_{11} + \cos(\theta_{11}+\theta_{12}) & (\cos\theta_{11}+\theta_{12}) \end{bmatrix}$$

and

$$J_2 = \begin{bmatrix} -\cos\alpha & \sin\alpha \\ -\sin\alpha & -\cos\alpha \end{bmatrix} \times \begin{bmatrix} -\sin\theta_{21} - \sin(\theta_{21}-\theta_{22}) & \sin(\theta_{21}-\theta_{22}) \\ \cos\theta_{21} + \cos(\theta_{21}-\theta_{22}) & -\cos(\theta_{21}+\theta_{22}) \end{bmatrix},$$

where α is the orientation angle of the object. The previous assumptions impose the following constraints:

$$\alpha = 0,\ \theta_{12} = \pi - 2\theta_{11}, \\ \theta_{21} = \pi - \theta_{11},$$

and

$$\theta_{11} - \theta_{22} = \pi - (\theta_{11} + \theta_{12}).$$

Figure 8 shows plots of the quality measures and the performance measure ($\gamma = 0.5$) as a function of θ_{11}. The structured measure $\mu_t(\Omega)$ and PM attain their maximum at $\theta_{11} = 0.475$ radian (27°). Since G is held constant, the task structures have no effect on μ_w, which is inversely propositional to $\sigma_{\max}(J_h)$. Clearly, the optimal posture (or grasp) is $\theta_{11} = 27°$.

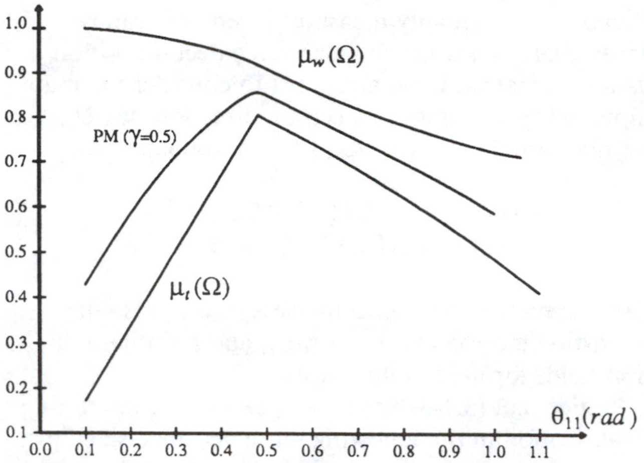

Fig. 8. Quality measures and performance measure versus θ_{11}.

4. Coordinated Control of a Multifingered Robot Hand

In this section, we develop control algorithms for the coordinated control of a multifingered robot hand. The goal of the control scheme is to specify a set of control inputs for the finger motors so that the gripped object undergoes a desired body motion while exerting a set of desired contact forces on the environment.

The control scheme we develop in this section, which is based on a generalization of the computed torque methodology, will have the following desirable features:

1. It assumes a point contact model.
2. It takes both object dynamics and hand dynamics into account.
3. It realizes both the desired position trajectory and the desired internal grasping force trajectory.
4. It can be easily extended to permit rolling and sliding motion of the fingertips on the object.

The last property is especially useful when we study fine manipulation of an object within the hand.

Without loss of generality, we may assume that the desired task is:

1. To manipulate the object along the following prespecified trajectory:
$$(r^d_{b,p}(t), A^d_{b,p}(t)) \in SE(3). \quad (4\text{-}1)$$
2. To maintain a set of desired internal grasping forces during the course of manipulation.

Even though a grasp chosen based on the preceding discussions may significantly benefit the control from an energy stand point, we only need the following simple assumption about the grasp:

(A1): The grasp is both stable and manipulable (see corollary in section 2).

A necessary condition for (A1) to hold is that the grasp map G and the hand Jacobian $J_h(\theta)$ are of full rank. From section 2, we know that in order to maintain the contact during manipulation the finger joint velocity $\dot{\theta}$ and the object velocity $[v^t_{b,p}, \omega^t_{b,p}]^t$ must satisfy the following velocity constraint relation:

$$J_h(\theta)\dot{\theta} = G^t \begin{bmatrix} v_{b,p} \\ \omega_{b,p} \end{bmatrix}. \quad (4\text{-}2)$$

Differentiating (4-2), we obtain the following acceleration constraint equation

$$J_h(\theta)\ddot{\theta} + \dot{J}_h(\theta)\dot{\theta} = G^t \begin{bmatrix} \dot{v}_{b,p} \\ \dot{\omega}_{b,p} \end{bmatrix}. \quad (4\text{-}3)$$

Since $R(J_h(\theta)) \supset R(G^t)$ by assumption (A1), we may express the joint acceleration $\ddot{\theta}$ in terms of the object acceleration $[\dot{v}^t_{b,p}, \dot{\omega}^t_{b,p}]^t$ by

$$\ddot{\theta} = J_h^+ G^t \begin{bmatrix} \dot{v}_{b,p} \\ \dot{\omega}_{b,p} \end{bmatrix} - J_h^+ \dot{J}_h \dot{\theta} + \ddot{\theta}_o. \quad (4\text{-}4)$$

Here $J_h^+ = J_h^t(J_h J_h^t)^{-1}$ is the generalized inverse of J_h, and $\ddot{\theta}_o \in \eta(J_h)$ is the internal motion of redundant joints not affecting the object motion.

Remark: (1) Using (4-4) will lead to a control algorithm in the task space (Khatib 1987). But if we express the object acceleration in terms of $\ddot{\theta}$ by

$$[\dot{v}^t_{b,p}, \dot{\omega}^t_{b,p}]^t = (GG^t)^{-1}(J_h\ddot{\theta} + \dot{J}_h\dot{\theta}),$$

a control algorithm in the joint space of the fingers will be developed. In future work, we will consider this alternative, since it appears to hold some interesting and different possibilities.

(2) When J_h is square, its generalized inverse J_h^+ is just the usual inverse, and $\ddot{\theta}_o$ disappears from (4-4). This also implies that the joint motion is determined uniquely by the motion of the object.

(3) When the fingertips roll on the object, there will be an extra term, $\dot{G}^t[v^t_{b,p}, \omega^t_{b,p}]^t$ on the right of (4-3). By including this term in the development we can study the control of the hand under rolling contact, even though we have not done so here explicitly.

The dynamics of the object are given by the Newton-Euler equations

$$\begin{bmatrix} \hat{m} & 0 \\ 0 & I \end{bmatrix} \begin{bmatrix} \dot{v}_{b,p} \\ \dot{\omega}_{b,p} \end{bmatrix} + \begin{bmatrix} \omega_{b,p} \times \hat{m} v_{b,p} \\ \omega_{b,p} \times I\omega_{b,p} \end{bmatrix} = \begin{bmatrix} f_b \\ m_b \end{bmatrix}, \quad (4\text{-}5)$$

where $\hat{m} \in R^{3\times3}$ is the diagonal matrix with the object mass in the diagonal, $I \in R^{3\times3}$ is the object inertia matrix with respect to the body coordinates, and $[f^t_b, m^t_b]^t$ is the applied body wrench in the body coordinates which is also related to the applied finger wrench $x \in R^n$ through

$$Gx = \begin{bmatrix} f_b \\ m_b \end{bmatrix}. \quad (4\text{-}6)$$

Notice that gravity and interaction forces from the environment can always be added to the right-hand side of (4-5), and corresponding contact wrenches will be generated to counteract them.

Since we have assumed that the grasp is stable (i.e., G is onto), we may solve (4-6) as

$$x = G^+ \begin{bmatrix} f_b \\ m_b \end{bmatrix} + x_o, \quad (4\text{-}7)$$

where $G^+ = G^t(GG^t)^{-1}$ is the left inverse of G, and $x_o \in \eta(G)$ is the internal grasping force. Part of the control objective is to steer the internal grasping force x_o to a certain desired value $x_o^d \in \eta(G)$.

Combining (4-5) and (4-7) yields

$$x = G^+\left\{\begin{bmatrix} \hat{m} & 0 \\ 0 & I \end{bmatrix}\begin{bmatrix} \dot{v}_{b,p} \\ \dot{\omega}_{b,p} \end{bmatrix} + \begin{bmatrix} \omega_{b,p} \times \hat{m} v_{b,p} \\ \omega_{b,p} \times I\omega_{b,p} \end{bmatrix}\right\} + x_o. \quad (4\text{-}8)$$

4.1. The Control Algorithm

The dynamics of the ith finger manipulator is given by

$$M_i(\theta_i)\ddot{\theta}_i + N_i(\theta_i, \dot{\theta}_i) = \tau_i - J_i^t(\theta_i)B_i x_i. \quad (4.1\text{-}1)$$

Here, as is common in the literature, $M_i(\theta_i) \in R^{m_i \times m_i}$ is the moment of inertia matrix of the ith finger manipulator; $N_i(\theta_i, \dot{\theta}_i) \in R^{m_i}$ are the centrifugal, Coriolis, and gravitational force terms; τ_i is the vector of joint torque inputs; and $B_i x_i \in R^6$ the vector of applied finger wrenches. Define

$$M(\theta) = \begin{bmatrix} M_1(\theta_1) & \cdots & 0 \\ \cdot & \cdots & \cdot \\ \cdot & \cdots & \cdot \\ \cdot & \cdots & \cdot \\ 0 & \cdots & M_k(\theta_k) \end{bmatrix},$$

$$N(\theta, \dot{\theta}) = \begin{bmatrix} N_1(\theta_1, \dot{\theta}_1) \\ \cdot \\ \cdot \\ \cdot \\ N_k(\theta_k, \dot{\theta}_k) \end{bmatrix} \quad \text{and} \quad (4.1\text{-}2)$$

$$\tau = \begin{bmatrix} \tau_1 \\ \cdot \\ \cdot \\ \cdot \\ \tau_k \end{bmatrix}.$$

Then, the finger dynamics can be grouped to yield

$$M(\theta)\ddot{\theta} + N(\theta, \dot{\theta}) = \tau - J_h(\theta)^t x. \quad (4.1\text{-}3)$$

The control objective is to specify a set of joint torque inputs τ so that both the desired body motion (4-1) and the desired internal grasping force x_o^d are realized.

Since $SO(3)$ is a compact three dimensional manifold, we may locally parameterize it by either the pitch-roll-yaw variables or the exponential coordinates. Let $\phi_{b,p} = [\phi_1, \phi_2, \phi_3]^t$ be a nonsingular parameterization of $SO(3)$. Then, we can express the body trajectory as

$$(r_{b,p}(t), A_{b,p}(\phi_{b,p}(t))) \in SE(3) \quad (4.1\text{-}4)$$

and the body velocity as

$$\begin{bmatrix} v_{b,p}(t) \\ \omega_{p,b}(t) \end{bmatrix} = U(\phi_{b,p}(t), r_{b,p}(t)) \begin{bmatrix} \dot{r}_{b,p}(t) \\ \dot{\phi}_{b,p}(t) \end{bmatrix}, \quad (4.1\text{-}5)$$

where $U(\phi_{b,p}(t), r_{b,p}(t)) \in R^{6\times 6}$ is a nonsingular parameterization dependent matrix that relates the derivatives of the parameterization to the body velocity. Differentiating (4.1-5) yields

$$\begin{bmatrix} \dot{v}_{b,p}(t) \\ \dot{\omega}_{b,p}(t) \end{bmatrix} = U \begin{bmatrix} \ddot{r}_{b,p}(t) \\ \ddot{\phi}_{b,p}(t) \end{bmatrix} + \dot{U} \begin{bmatrix} \dot{r}_{b,p}(t) \\ \dot{\phi}_{b,p}(t) \end{bmatrix}. \quad (4.1\text{-}6)$$

Theorem: Assume that (A1) holds and that the fingers are nonredundant (i.e., $m_i = n_i$, for $i = 1, \ldots, k$). Define the position error $e_p \in R^6$, and the internal grasping force error $e_f \in R^{n-6}$ to be

$$e_p = \begin{bmatrix} r_{b,p} \\ \phi_{b,p} \end{bmatrix} - \begin{bmatrix} r_{b,p}^d \\ \phi_{b,p}^d \end{bmatrix} \quad \text{and} \quad e_f = x_o - x_o^d, \quad (4.1\text{-}7)$$

where $[(r_{b,p}^d)^t, (\phi_{b,p}^d)^t]^t$ is the desired body trajectory, and x_o^d is the desired internal grasping force. Thus, the control law specified by (4.1-8) realizes not only the desired body trajectory but also the desired internal grasping force.

$$\tau = N(\theta, \dot{\theta}) + J_h^t G^+ \begin{bmatrix} \omega_{b,p} \times \hat{m} v_{b,p} \\ \omega_{b,p} \times I \omega_{b,p} \end{bmatrix} - M(\theta) J_h^{-1} \dot{J}_h \dot{\theta} + M_h \dot{U} \begin{bmatrix} \dot{r}_{b,p} \\ \dot{\phi}_{b,p} \end{bmatrix} + J_h^t (x_o^d - K_I \int e_f) + M_h U \left\{ \begin{bmatrix} \ddot{r}_{b,p}^d \\ \ddot{\phi}_{b,p}^d \end{bmatrix} - K_v \dot{e}_p - K_p e_p \right\}, \quad (4.1\text{-}8a)$$

where

$$M_h = M(\theta) J_h^{-1} G^t + J_h^t G^+ \begin{bmatrix} \hat{m} & 0 \\ 0 & I \end{bmatrix} \quad (4.1\text{-}8b)$$

and K_I is a matrix such that the null space of G is K_I-invariant.

Remarks: (1) (4.1-8) can be generalized to the redundant case and the results are given in the Appendix.

(2) When the fingertips roll on the object, the control law needs to include an extra term that depends on the contact parameters, which involve the kinematic equations of contact (Montana 1986). Combining this development with the work of Montana can lead to a control law with rolling or sliding constraints. An interesting application on rolling is given in Cole, Hauser, and Sastry (1988).

(3) The first four components in (4.1-8a) are used for cancellation of Coriolis, gravitational, and centrifugal forces. These terms behave exactly like the nonlinearity cancellation terms in the computed torque control for a single manipulator; the term $J_h^t(x_o^d - K_I \int e_f)$ is the compensation for the internal grasping force loop, and the last term is the compensation for the position loop. We will see in the proof that the dynamics of the internal grasping force loop and that of the position loop are mutually decoupled. Consequently, we can design the force error integral gain K_I independently from the position feedback gains K_v and K_p.

(4) In the presence of stiction and frictional forces, feedforward compensation can be added to (4.1-8). This has been shown to be very successful in the control of the Berkeley hand by R. Murray (1989).

(5) Nakamura, Nagai, and Yoshikawa (1987) proposed to adjust the internal grasping forces depending on the change of the manipulation force. Using this approach, we can plan the internal grasping force trajectory so that some optimality criterion can be achieved during the course of manipulation.

PROOF: The proof is very procedural and straightforward. First, we substitute (4-4) and (4-8) into (4.1-3) to get

$$M(\theta) \left\{ J_h^{-1} G^t \begin{bmatrix} \dot{v}_{b,p} \\ \dot{\omega}_{b,p} \end{bmatrix} - J_h^{-1} \dot{J}_h \dot{\theta} \right\} + N(\theta, \dot{\theta}) = \tau - J_h^t \left\{ G^+ \begin{bmatrix} \hat{m} & 0 \\ 0 & I \end{bmatrix} \begin{bmatrix} \dot{v}_{b,p} \\ \dot{\omega}_{b,p} \end{bmatrix} + G^+ \begin{bmatrix} \omega_{b,p} \times \hat{m} v_{b,p} \\ \omega_{b,p} \times I \omega_{b,p} \end{bmatrix} \right\} - J_h^t x_o. \quad (4.1\text{-}9)$$

Note that in (4-4) the generalized inverse for nonredundant fingers reduces to the regular inverse J_h^{-1} and $\ddot{\theta}_o = 0$. Linearizing (4.1-9) with the following control

$$\tau = N(\theta, \dot{\theta}) + J_h^t G^t \begin{bmatrix} \omega_{b,p} \times \hat{m} v_{b,p} \\ \omega_{b,p} \times I \omega_{b,p} \end{bmatrix} - M(\theta) J_h^{-1} \dot{J}_h \dot{\theta} + \tau_1, \quad (4.1\text{-}10)$$

where τ_1 is to be determined, we have that

$$\left\{ M(\theta) J_h^{-1} G^t + J_h^t G^+ \begin{bmatrix} \hat{m} & 0 \\ 0 & I \end{bmatrix} \right\} \begin{bmatrix} \dot{v}_{b,p} \\ \dot{\omega}_{b,p} \end{bmatrix} = \tau_1 - J_h^t x_o, \quad \text{or} \quad M_h \begin{bmatrix} \dot{v}_{b,p} \\ \dot{\omega}_{b,p} \end{bmatrix} = \tau_1 - J_h^t x_o. \quad (4.1\text{-}11)$$

Substituting (4.1-6) into the above equation, we have

$$M_h \left\{ U \begin{bmatrix} \ddot{r}_{b,p} \\ \ddot{\phi}_{b,p} \end{bmatrix} + \dot{U} \begin{bmatrix} \dot{r}_{b,p} \\ \dot{\phi}_{b,p} \end{bmatrix} \right\} = \tau_1 - J_h^t x_o. \quad (4.1\text{-}12)$$

Further, let the control input τ_1 be

$$\tau_1 = M_h U \left\{ \begin{bmatrix} \ddot{r}_{b,p}^d \\ \ddot{\phi}_{b,p}^d \end{bmatrix} - K_v \dot{e}_p - K_p e_p \right\} + M_h \dot{U} \begin{bmatrix} \dot{r}_{b,p} \\ \dot{\phi}_{b,p} \end{bmatrix} + J_h^t \left(x_o^d - K_I \int e_f \right) \quad (4.1\text{-}13)$$

and apply it to (4.1-12) to yield:

$$M_h U \{\ddot{e}_p + K_v \dot{e}_p + K_p e_p\} = -J_h^t \left(e_f + K_I \int e_f \right). \quad (4.1\text{-}14)$$

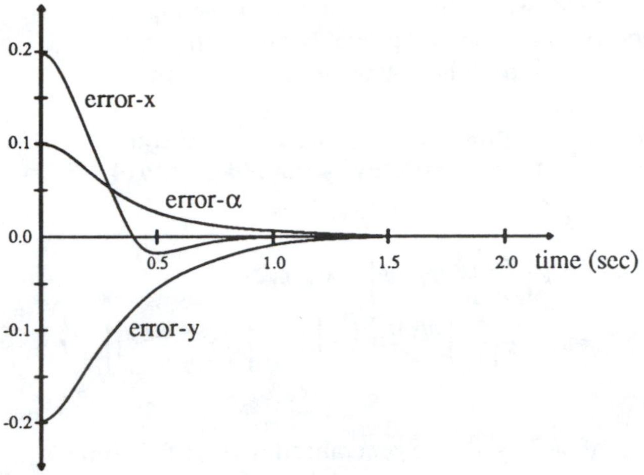

Fig. 9. Position error.

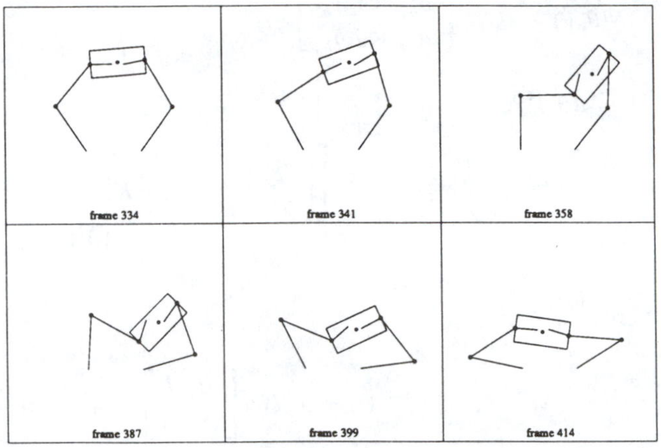

Fig. 10. Simulation with 10 units of internal grasping force.

Multiplying (4.1-14) by GJ_h^{-t}, we obtain the following equation.

$$GJ_h^{-t}M_hU\{\ddot{e}_p + K_v\dot{e}_p + K_pe_p\} = -G\left(e_f + K_I\int e_f\right) = 0, \tag{4.1-15}$$

where we have used the fact that the internal grasping forces lie in the null space of G, i.e.,

$$G\left(e_f + K_I\int e_f\right) = 0. \tag{4.1-16}$$

Since $GJ_h^{-t}M_h = GJ_h^{-t}M(\theta)J_h^{-1}G^t + \begin{bmatrix}\hat{m} & 0\\ 0 & I\end{bmatrix}$ is positive definite and U is nonsingular, (4.1-15) implies that

$$\ddot{e}_p + K_v\dot{e}_p + K_pe_p = 0. \tag{4.1-17}$$

Thus we have shown that the position error e_p can be driven to zero with proper choice of the feedback gain matrices K_v and K_p.

The last step is to show that e_f also goes to zero. If we substitute (4.1-17) into (4.1-14) and notice that J_h is nonsingular, we have the following equation.

$$e_f + K_I\int e_f = 0. \tag{4.1-18}$$

With proper choice of K_I, the above equation implies that the internal grasping force error e_f converges to zero.

Q.E.D.

4.2. Simulation

Consider the two-fingered planar manipulation system shown in Figure 7, where the two fingers are assumed to be identical. We model the contact to be a point contact with friction. The grasp matrix and the hand Jacobian are given in (3.1-14) and (3.1-15). It has been shown in the example in section 3.1.2 that the grasp configuration in the figure is both stable and manipulable. We have simulated the system to follow the following desired trajectory of the body:

$$x(t) = c_1 \sin(t),\quad y(t) = c_1 + c_1 \cos(t),\quad \alpha(t) = c_3 \sin(t). \tag{4.2-1}$$

The simulation used a program designed to integrate differential equations with algebraic constraints. Figure 9 shows that the initial position error (in Cartesian space) diminishes exponentially as predicted by Eq. 4.1-17. The simulation was fed to a movie package that shows the continuous motion. Figure 10 shows a sequence of sampled pictures from a typical simulation. In the figure, the line segment at each contact shows the magnitude and the direction of the total force that is exerted to the object by the finger. The desired internal grasp force is set to 10 unit force. Note that path planning has not been performed to keep the fingers from running into the object. This is a subject of ongoing research.

5. Concluding Remarks

We have studied techniques for the determination of a grasp to an object by a multifingered hand. We have also provided a control algorithm to generate the appropriate motor torques required to manipulate an object in a certain prescribed fashion. The scheme is shown to converge, in the sense that the true body trajectory and the true internal grasp force trajectory converge to their desired values. A simulation of our scheme to a planar manipulation of an object by a two-fingered hand is presented.

In future work we will study more sophisticated models for contact of a body by a multifingered hand and their implications for the schemes of this paper.

Acknowledgments

This research was supported in part by NSF under grant #DMC-8451129. We would like to thank J. Hauser for performing the simulation of Figure 10, R. Fearing for a careful reading of the manuscript, and A. Cole, G. Heinzinger, R. Murray, and K. Pister for their help in prehending a number of ideas. Also, we would like to thank the anonymous reviewers and Y. Wang for their excellent comments and suggestions.

Appendix: Coordinated Control for Redundant Fingers

In this appendix, we supplement the theorem in section 4 with the control algorithm for redundant fingers. For the case the number of joints m_i, $i, \ldots, k$ is greater than the number of constrained directions n_i, $i = 1, \ldots, k$. In many industrial applications, several robots that often have more than three degrees of freedom are integrated to maneuver a massive load. Under the frictional point-contact model, such a system is redundant. The dynamic distinction of a redundant hand from a nonredundant hand is the internal motion given by $\ddot{\theta}_0$ in (4-4). We claim that the following control law will achieve the desired control objective for a hand with redundant degrees of freedom:

$$\begin{aligned}\tau = {} & N(\theta, \dot{\theta}) + J_h^t G^+ \begin{bmatrix} \omega_{b,p} \times \hat{m} v_{b,p} \\ \omega_{b,p} \times I \omega_{b,p} \end{bmatrix} \\ & - M J_h^+ \dot{J}_h \dot{\theta} + M J_h^+ (J_h M^{-1} J_h^t) M_h \dot{U} \begin{bmatrix} \dot{r}_{b,p} \\ \dot{\phi}_{b,p} \end{bmatrix} \\ & + M J_h^+ (J_h M^{-1} J_h^t) \left(x_o^d - K_I \int e_f \right) \\ & + M J_h^+ (J_h M^{-1} J_h^t) M_h U \left\{ \begin{bmatrix} \ddot{r}_{b,p}^d \\ \ddot{\phi}_{b,p}^d \end{bmatrix} - K_v \dot{e}_p - K_p e_p \right\},\end{aligned} \tag{A-1a}$$

where

$$M_h = (J_h M^{-1} J_h^t)^{-1} G^t + G^+ \begin{bmatrix} \hat{m} & 0 \\ 0 & I \end{bmatrix}. \tag{A-1b}$$

Remark: Compare (A-1a,b) with (4.1-8) and observe that M_h is different here.

To see this, we use (4-4) and (4-8) in (4.1-3) and suppress the θ dependence of M to get

$$\begin{aligned} & M \left\{ J_h^+ G^t \begin{bmatrix} \dot{v}_{b,p} \\ \dot{\omega}_{b,p} \end{bmatrix} - J_h^+ \dot{J}_h \dot{\theta} \right\} + M \ddot{\theta}_0 + N(\theta, \dot{\theta}) \\ & = \tau - J_h^t \left\{ G^+ \begin{bmatrix} \hat{m} & 0 \\ 0 & I \end{bmatrix} \begin{bmatrix} \dot{v}_{b,p} \\ \dot{\omega}_{b,p} \end{bmatrix} + G^+ \begin{bmatrix} \omega_{b,p} \times \hat{m} v_{b,p} \\ \omega_{b,p} \times I \omega_{b,p} \end{bmatrix} \right\} \\ & - J_h^t x_0. \end{aligned} \tag{A-2}$$

Comparing to (4.1-9) we see that there is an extra term $M\ddot{\theta}_0$ in (A-2), and J_h^{-1} is replaced by the generalized inverse J_h^+.

Linearizing the equation with the following control input:

$$\begin{aligned}\tau = {} & N(\theta, \dot{\theta}) + J_h^t G^+ \begin{bmatrix} \omega_{b,p} \times \hat{m} v_{b,p} \\ \omega_{b,p} \times I \omega_{b,p} \end{bmatrix} \\ & - M J_h^+ \dot{J}_h \dot{\theta} + \tau_1,\end{aligned} \tag{A-3}$$

where τ_1 is to be determined, we have that

$$\left\{MJ_h^+G^t + J_h^tG^+\begin{bmatrix}\hat{m} & 0\\ 0 & I\end{bmatrix}\right\}\begin{bmatrix}\dot{v}_{b,p}\\ \dot{\omega}_{b,p}\end{bmatrix} + M\ddot{\theta}_0 = \tau_1 - J_h^t x_0. \quad \text{(A-4)}$$

Multiplying the above equation by J_hM^{-1} and using the fact that $J_h\dot{\theta}_0 = 0$, yields

$$\left\{G^t + J_hM^{-1}J_h^tG^+\begin{bmatrix}\hat{m} & 0\\ 0 & I\end{bmatrix}\right\}\begin{bmatrix}\dot{v}_{b,p}\\ \dot{\omega}_{b,p}\end{bmatrix} = J_hM^{-1}\tau_1 - J_hM^{-1}J_h^t x_0. \quad \text{(A-5)}$$

Since J_h is onto, $J_hM^{-1}J_h^t$ is nonsingular, and we can further multiply (A-5) by $(J_hM^{-1}J_h^t)^{-1}$ to obtain

$$\left\{(J_hM^{-1}J_h^t)^{-1}G^t + G^+\begin{bmatrix}\hat{m} & 0\\ 0 & I\end{bmatrix}\right\}\begin{bmatrix}\dot{v}_{b,p}\\ \dot{\omega}_{b,p}\end{bmatrix} = (J_hM^{-1}J_h^t)^{-1}J_hM^{-1}\tau_1 - x_0. \quad \text{(A-6)}$$

Substituting (4.1-6) into (A-6) and choosing the following remaining control input for τ_1 we obtain

$$\tau_1 = MJ_h^+(J_hM^{-1}J_h^t)\left\{M_hU\left(\begin{bmatrix}\ddot{r}_{b,p}^d\\ \ddot{\phi}_{b,p}^d\end{bmatrix} - K_v\dot{e}_p + K_pe_p\right) + M_h\dot{U}\begin{bmatrix}\dot{r}_{b,p}\\ \dot{\phi}_{b,p}\end{bmatrix} + \left(x_0^d - K_I\int e_f\right)\right\} \quad \text{(A-7)}$$

and

$$M_hU\{\ddot{e}_p + K_v\dot{e}_p + K_pe_p\} = -\left(e_f + K_I\int e_f\right). \quad \text{(A-8)}$$

Multiplying (A-8) by G and using the fact that $G(e_f + K_I\int e_f) = 0$ we obtain that

$$GM_hU\{\ddot{e}_p + K_v\dot{e}_p + K_pe_p\} = 0. \quad \text{(A-9)}$$

Since

$$GM_h = G(J_hM^{-1}J_h^t)^{-1}G^t + \begin{bmatrix}\hat{m} & 0\\ 0 & I\end{bmatrix}$$

and U are both nonsingular, (A-9) immediately implies that

$$\ddot{e}_p + K_v\dot{e}_p + K_pe_p = 0, \quad \text{(A-10)}$$

which shows that the position error e_p can be driven to zero by proper choice of the feedback gain matrices K_v and K_p.

On the other hand, using (A-10) in (A-8) we finally obtain that

$$e_f + K_I\int e_f = 0, \quad \text{(A-11)}$$

which shows that the internal grasping force error can also be driven to zero by proper choice of the force integral gain matrix K_I.

References

Arimoto, S. 1987. Cooperative motion control of multi-robot arms or fingers. *Proc. IEEE Int. Conf. on Robotics and Automation,* 1407–1412.

Asada, H. 1987. On the dynamic analysis of a manipulator and its end effector interacting with the environment. *Proc. IEEE Int. Conf. on Robotics and Automation,* 751–756.

Cole, A., Hauser, J. and Sastry, S. 1988. Kinematics and control of multifingered hands with rolling contact. *Proc. IEEE Int. Conf. on Robotics and Automation,* 228–233.

Cutkosky, M. 1985. Grasping and fine manipulation for automated manufacturing. Ph.D. thesis, Department of Mechanical Engineering, Carnegie-Mellon University.

Hayati, S. 1986. Hybrid position/force control of multi-arm cooperating robots. *Proc. IEEE Int. Conf. on Robotics and Automation,* 82–89.

Jacobsen, S., Wood, J., Biggers, K., and Iversen, E. 1985. The Utah/MIT hand: work in progress. *Int. J. of Robotics Res.* 4(3).

Kerr, J. 1984. An analysis of multifingered hands. Ph.D. Dissertation. Mechanical Engineering, Stanford University.

Kerr, J. and Roth, B. 1986. Analysis of multifingered hands. *Int. J. of Robotics Res.* 4(4).

Kobayashi, H. 1985. Geometric considerations for a multi-fingered robot hand. *Int. J. of Robotics Res.* 4(1).

Li, Z. X., and Sastry, S. 1988. Task oriented optimal grasping by multifingered robot hands. *IEEE J. of Robotics and Automation.* RA 2-14(1) 32–44.

Li, Z. X., Hsu, P. and Sastry, S. 1987. On grasping and

dynamic coordination of multifingered robot hands. Memo. No. UCB/ERL M87/63, University of California, Berkeley.

Montana, D. 1986. Tactile sensing and kinematics of contact. Ph.D. thesis, Division of Applied Sciences, Harvard University.

Murray, R. 1989. Control experiments in planar manipulation and grasping. ERL memo. number M89/3. University of California, Berkeley, CA.

Nakamura, Y., Nagai, K., and Yoshikawa, T. 1987. Mechanics of coordinative manipulation by multiple robotic mechanisms. *Proc. IEEE Int. Conf. on Robotics and Automation,* 991–998.

Nguyen, V. 1987. Constructing stable grasps in 3-D. *Proc. IEEE Int. Conf. on Robotics and Automation,* 234–245.

Salisbury, J. 1982. Kinematic and force analysis of articulated hands. Ph.D. Thesis, Department of Mechanical Engineering, Stanford University.

Whitney, D. E. 1982. Quasi-static assembly of compliantly supported rigid parts. *J. Dyn. Sys. Meas. and Contr.* 104:65–77.

Zheng, Y. F., and Luh, J. Y. S. 1985. Control of two coordinated robots in motion. *Proc. 24th Contr. and Dec. Conf.,* 1761–1766.

Part 10
Redundant Robots

ALESSANDRO DE LUCA
Universitá degli Studi di Roma "La Sapienza"

ROBOT MANIPULATORS are said to be *kinematically redundant* when the number n of degrees of freedom (dof) owned by the mechanism is larger than the number m of variables strictly needed for accomplishing a given task. The difference $n - m$ characterizes the *degree of redundancy*. Redundancy is therefore a relative concept for a robot, depending on the particular type of task to be executed. Typically, six joints (e.g., all revolute) are necessary in a robot arm for arbitrary positioning and orienting the end-effector within the workspace. If only the positioning task is of concern, the same arm becomes a redundant one with degree of redundancy equal to three. Similarly, tasks like welding or pointing do not require the full capabilities of six-dof robots, because the final rolling motion around the approach vector is not specified.

Therefore, when the task is of reduced dimensions, exploiting the available degrees of freedom of a conventional arm is already a matter of redundancy utilization. As a result, robots with redundant capabilities already exist in both industrial and research environments. On the other hand, the presence of unavoidable *kinematic singularities*, internal to the robot workspace, restricts feasible motions of a conventional six-dof arm. Thus, its overall functionality and ease of use, for instance in programming realizable end-effector trajectories, is considerably reduced.

This drawback, and in general the need for dexterous manipulation with increased skills, motivated the introduction of a new generation of manipulators with seven or more degrees of freedom. Actual production of redundant robots (i.e., with definitely more than six dof) is in fact growing quite rapidly. Examples are the Cybotech P-15, the K/B-1207 of Robotics Research Corporation [1], or the Cesar manipulator developed at Oak Ridge National Labs [2]. Indeed, efforts in the mechanical design of redundant arms with optimal kinematics [3] have to be complemented by the realization of advanced robot controllers—intended in a broad sense in their supervisory, trajectory planning, and servoing functions—that can make best use of the increased dexterity.

General methodologies for planning and controlling motion of redundant robots are presented in Part 10, rather than specific approaches or solutions. It is easy to find out that most of the reported material applies in a direct way to a variety of robotic systems:

- Single robot arms with seven or more dof, and in particular with a very large degree of redundancy—as in spine-type robots having in practice all continuous configurations;
- Two-robot systems, including also dual-arm robots, or multirobot systems performing strictly cooperative tasks, such as holding together a heavy object;
- Dexterous multifingered hands needed for fine manipulation of complex objects that common grippers cannot grasp firmly;
- Multilegged locomotion systems, where a redundant number of supporting legs is used to improve stability and to allow different possible gaits.

In these illustrative situations, one or more of the following advantages are obtained through redundancy:

- Collision with obstacles in a crowded workspace can be avoided while keeping the motion of selected points (mostly, the end-effector) along prespecified paths;
- Tasks are executed with full utilization of the available joint range, in particular without reaching the geometric limits in the robot configuration space;
- The robotic system is able to assume the best posture for the given task, exerting compatible forces in selected directions and enhancing Cartesian velocity in others;
- Automatic singularity avoidance can be performed so that the feasible workspace, in which the robotic system has full manipulability, may coincide with the reachable one (except for boundaries);
- The same task effort can be distributed over the degrees of freedom, to reduce global joint motion, and/or over the actuators, to minimize total torque demand.

All achieved features can be summarized by the fact that a redundant robot acquires *self-motion* capabilities: A motion can be performed in the joint space, changing the internal arm configuration, without affecting task-space coordinates. This may happen also for conventional arms in correspondence to singular points, but the effective task space would be restricted in that case.

In the face of the above benefits, some factors which have limited the introduction of redundancy should be taken into account:

- The mechanical construction of a redundant arm is usually more complex and requires a larger number of actuating elements, so a more expensive design results;
- Kinematic control algorithms are certainly more sophisticated, because of the one-to-many nature of the inverse kinematic mapping.

Pros and cons have to be traded off, and the success of using redundant robots will depend on the value added to specific applications. On the other hand, many advanced tasks assume high dexterity, self-organized flexibility, and autonomous motion planning as robotic prerequisities. Tasks performed in hazardous environments, currently in a slow fashion and with an expert human operator in the loop (e.g., space servicing or nuclear-plant maintenance), may take advantage of redundant robotic systems capable of autonomous low-level motion adjustment in response to high-level commands. Within the remote manipulation field, an interesting application of redundancy is the use of a simple "master" arm to drive a redundant "slave" manipulator.

The starting point for the formal analysis of redundant robots is in the kinematic relationship between joint-space variables and task-space coordinates:

$$\boldsymbol{p} = \boldsymbol{f}(\boldsymbol{q}), \qquad \boldsymbol{p} \in \boldsymbol{R}^m, \boldsymbol{q} \in \boldsymbol{R}^n, n > m. \tag{1}$$

The inverse kinematic problem consists in finding one configuration $\boldsymbol{q}$ among the set of ∞^{n-m} solutions to (1), for a given instant value of the task-space vector $\boldsymbol{p}$. This is a highly nonlinear problem, with no general closed-form solution available. Therefore, kinematics is usually rewritten at a differential level as

$$\dot{\boldsymbol{p}} = \frac{d\boldsymbol{f}(\boldsymbol{q})}{d\boldsymbol{q}}\dot{\boldsymbol{q}} = \boldsymbol{J}(\boldsymbol{q})\dot{\boldsymbol{q}}, \qquad \boldsymbol{J}: n \times m \text{ robot Jacobian.} \tag{2}$$

For each configuration $\boldsymbol{q}$ and assigned task velocity $\dot{\boldsymbol{p}}$, an underdetermined linear system results, with m equations in the n unknowns $\dot{\boldsymbol{q}}$. The choice of one possible joint velocity $\dot{\boldsymbol{q}}$ is simpler, owing to the linearity of this relation. It can be shown that all solutions to (2) can be set in the form

$$\dot{\boldsymbol{q}} = \boldsymbol{J}^{\dagger}(\boldsymbol{q})\dot{\boldsymbol{p}} + [\boldsymbol{I} - \boldsymbol{J}^{\dagger}(\boldsymbol{q})\boldsymbol{J}(\boldsymbol{q})]\boldsymbol{v}, \tag{3}$$

where $\boldsymbol{J}^{\dagger}$ is the (unique) pseudoinverse of $\boldsymbol{J}$, and $\boldsymbol{v}$ is an arbitrary joint-velocity vector. The second term on the right-hand side is the homogeneous contribution, where the projection matrix into the null space of $\boldsymbol{J}$ appears.

In order to select a specific solution, that is, a $\boldsymbol{v}$ in (3), either an *optimization* or a *task-augmentation* approach is followed. In the first case, several performance criteria have been proposed, reflecting some measure of the desired arm behavior. The most common are the weighted distance from the center of the available joint range [4], the distance from the closest obstacle [5], various kinematic manipulability indices [6], [7], [8] and their extensions to include inertial properties [9]. In particular, the definition of a suitable manipulability index to correctly measure distance from structural singularities is an important issue. The smallest singular value of the Jacobian matrix $\boldsymbol{J}$ has been indicated as a reliable one, but it is interesting to note that most definitions can be given a coordinate-free expression [10]. Once a proper criterion $H(\boldsymbol{q})$ is selected, the additional degrees of freedom are used for its local maximization or minimization, mainly based on the Projected Gradient method, yielding $\boldsymbol{v} = \pm\nabla_q H$. Here, local refers to an instantaneous solution that could be derived on-line, without knowledge of the future reference trajectory. This approach is indeed very convenient from the implementation point of view. Note that the optimization process is in principle a constrained one, because (1) has to be satisfied anyway. When using (3), however, the choice of the n-vector $\boldsymbol{v}$ is completely unconstrained.

The philosophy behind task augmentation is that the additional degrees of freedom of a redundant robot may be used for executing other parallel tasks, implicitly defined through a mechanism of constraint satisfaction. When the overall dimension m of the augmented task reaches n, no more redundancy will be left over. There is a strict connection between the two approaches, and usually one may be translated into the other and vice versa. For example, additional constraints could be introduced as the necessary conditions of optimality for an auxiliary problem [11]. Since optimization and task augmentation can also be combined, it is a matter of convenience whether to define a subtask as a hard constraint or to include it into an objective function. Finally, the above two approaches were developed initially at a kinematic (first-order) level, but reformulations at the dynamic level, including robot dynamics, have been considered later.

The four papers in this part discuss in detail the above issues. These contributions have been selected based also on the influence they have had on robotics research in this field.

The first paper (C. A. Klein and C. H. Huang, "Review of Pseudoinverse Control for Use with Kinematically Redundant Manipulators," is a pioneering, short, and handy review on the use of the Jacobian pseudoinverse $\boldsymbol{J}^{\dagger}$ for velocity control of redundant robots. The authors list the properties of pseudoinverses and the basic techniques for their computation. The inclusion of homogeneous solutions to improve performance is illustrated using the simple but ubiquitous 3R planar robot. For this arm, it is proven for the first time that noncyclic joint paths are obtained from cyclic Cartesian paths when redundancy is resolved by simple pseudoinversion (i.e., using the resolved motion method of [12]). The same technique is shown to generate different arm behaviors if a closed Cartesian path is executed clockwise or counterclockwise. Furthermore, pseudoinverse control, though simple and appealing, does not guarantee avoidance of singularities.

In the second paper, by Y. Nakamura et al., "Task-Priority Based Redundancy Control of Robot Manipulators," the very useful concept of priority of tasks is introduced. Arm redundancy is exploited for augmenting the number of tasks to be performed together with the primary one—usually a trajectory to be followed by the robot end-effector. The designer orders the additional list of tasks, each requiring m_i degrees of freedom, by their relative importance. In general, the sum m of dimensions of all tasks can be greater than, equal to or less than n. When their simultaneous satisfaction is impossible, execution of low-priority tasks will be relaxed without affecting high-priority ones. As opposed to the *Extended Jacobian* method [13] defined for $m = n$, computations are always organized to carefully handle the so-called *algorithmic singularities*. A proper use of projection operators avoids a global deterioration of performance, restraining it to the less-revelant tasks. The idea amounts to automatically solving an inconsistent set of linear equations, giving privilege to some of them and least-squaring on the others. Numerical simulations and experiments are presented, including also preliminary consideration of dynamic issues. An obstacle-avoidance task is successfully executed using the task-priority method. For this problem, a similar technique was presented in [5].

The third paper (J. M. Hollerbach and K. C. Suh, "Redundancy Resolution of Manipulators Through Torque Optimization,") focuses on the inclusion of dynamics in the redundancy resolution algorithm. Some of the results in this paper are outgrowths of initial research reported in [14]. Generalized inverses formulated at the acceleration level and used in conjunction with the robot dynamic model were first considered in [15]. Hollerbach and Suh, however, consider torque requirements in a more direct way: In particular, local minimization of weighted and unweighted norms of joint torques is pursued. These two methods are compared with the inertia-weighted and the unweighted minimum-norm acceleration solutions through extensive simulations. When norm weighting is chosen according to the allowable torque range, the obtained solution tends to stay within actuator limits. On the other hand, a rather unexpected instability problem arises for long end-effector trajectories. In fact, although torque is minimized at each instant, the actual torque demand may suddenly "explode" after a smooth initial behavior. The authors interpret this effect physically as a whiplash action, needed to keep the end-effector on the desired path against the induced high velocity of arm motion. This issue is further examined by Maciejewski [16]. In any case, this stability limitation is intrinsic to the local nature of the problem formulation, providing a strong motivation for exploring globally optimal, or, at least, overall-stable dynamic-resolution schemes. It should be noted that inclusion of dynamic aspects in the redundancy-resolution process does not necessarily imply the use of a dynamic (model-based) control law. For this, the reader may refer to the papers of Hsu et al. [17] and of Egeland [18].

The final paper (D. R. Baker and C. W. Wampler, "On the Inverse Kinematics of Redundant Manipulators,") investigates fundamental properties shared by most redundancy-resolution algorithms. The major concern is on local schemes, because these are the only ones executable on line in a sensor-driven motion. The basic concepts of *inverse kinematic function* and, accordingly, of *tracking algorithm* are formally stated. The problem of existence of such functions and algorithms and their characterization are then addressed. Using topological arguments, it is shown that no continuous inverse kinematic function exists for tasks like pointing over the whole sphere or orienting the end-effector in an arbitrary way (e.g., with a spherical wrist), no matter how large the arm-redundancy degree is. Moreover, the authors prove that a tracking algorithm corresponds naturally to an inverse kinematic function if and only if it maps cyclic end-effector paths into cyclic joint paths. The proof of sufficiency is constructive, showing how to obtain an inverse kinematic function from a cyclic tracking algorithm. It is also pointed out how the Extended Jacobian method fits into this general analysis. These results, although of some negative flavor, set precise limitations on what can be done and what cannot.

Because of the tutorial nature of this collection of papers, some topics have been left out. A list of problems is given next, reflecting current trends of investigation in the area of redundant robots and deserving the attention of active researchers.

Global Optimization. As already mentioned, most schemes resolving redundancy via optimization are defined locally. If the whole end-effector trajectory is known in advance, one may also look for global solutions, optimal along the path. The problem becomes much more difficult, requiring the minimization of an objective functional (an integral criterion) subject to two-point boundary-value conditions. Variational techniques or Pontryagin's principle apply to this formulation. Although the associated optimality conditions are easily stated, the bottleneck stands mainly in the numerical computations involved, for instance, with the multiple-shooting solution algorithm. Minimization of joint velocities or kinetic energy along the whole trajectory are two typical objectives. The problem of global torque optimization, which is roughly twice as hard to solve, has been addressed with two different but similar approaches in [19] and [20]. Boundary conditions, which are specified at the initial and final trajectory points, play a major role in all cases. Interestingly enough, cyclic solutions can be obtained for a closed Cartesian path by imposing equal arm configurations at the initial and final instants. Moreover, contrary to the local situation, no explosion of joint torque is reported in the globally minimizing solution. A different research direction is pursued in [21] and [22]: Cases are found where the solution to a global optimization problem coincides, under suitable conditions, with the solution to a related local problem. Then, numerical complexity is dramatically reduced to the simple integration of differential equations.

Computational Aspects. One common disadvantage in handling redundancy is the need for computing pseudoinverses. This is already apparent in (3), and occurs repeatedly in the task-priority method as well as in dynamic optimization schemes. In general, this matrix operation involves a singular value decomposition (SVD) of the robot Jacobian. Advances in the efficient computation of SVD in the robotic case are reported in [23]. In the full rank case for $\boldsymbol{J}$, ways to rearrange pseudoinverse derivation have been suggested in [24] and [2]. Still, the presence of $\boldsymbol{J}^{\dagger}$ and of the projection operator $\boldsymbol{I} - \boldsymbol{J}^{\dagger}\boldsymbol{J}$ complicates unnecessarily the solution, often obscuring its significance. The use of the simple Jacobian transpose $\boldsymbol{J}^T$ in a closed-loop scheme has been proposed in [25]. This method is also robust w.r.t. singularities, similar to [26] and [27]. On the other hand, full exploitation of the idea of decomposing joint variables to reduce the optimization task to the smaller space of the $n - m$ extra degrees of freedom leads to the fast and efficient Reduced Gradient method of [28].

Cyclicity. By letting the robot configuration be defined on a smooth manifold, the powerful tools of differential geometry can be used to investigate the problem of cyclic, that is, repeatable, motion. Shamir and Yomdin [29] have given one significant outcome of this connection, showing that cyclicity with pseudoinverse control is achieved if and only if the columns of $\boldsymbol{J}^{\dagger}$ are *involutive* vector fields. As a consequence, it is found for the 3R planar robot that cyclic Cartesian paths can still be mapped into cyclic joint paths via $\boldsymbol{J}^{\dagger}$, provided the arm starts from certain configurations. This involutivity result, which is in fact an integrability condition related to Frobenius Theorem, is general and applies to any other inverse mapping $\boldsymbol{K}$ in place of $\boldsymbol{J}^{\dagger}$. Its extension to second-order inverse functions (i.e., in terms of accelerations) would be interesting, together with the definition of a strategy for choosing $\boldsymbol{v}$ in (3) to force repeatability.

Time-optimality. The following planning task can be posed. For a redundant robot, let the desired end-effector path be specified in parametric form from source to destination. We would like to find the optimal timing along this path and the optimal sequence of arm configurations, to minimize the total traveling time under joint torque limits. A feasible initial arm configuration may or may not be assigned. This problem is a variant of the time-optimal motion on a given Cartesian geometric path: In the case of conventional robots, very efficient phase-plane solution techniques have been found (see Part 5). At present, the extended problem for redundant manipulators is still open.

As a final remark, it should be stressed that interest has been focused here on robots with redundant kinematics, not on redundant systems in robots. In the latter category, actuational redundancy [30] as well as sensor redundancy and fusion are very important topics, aimed at achieving a more robust robot behavior in the presence of failures or uncertainties.

A list of papers that give a rather complete picture of the area of redundant robots is included for further reading. Surveys on parts of these investigations have appeared recently, [31], [32], wherein more extensive references can be found.

References

[1] J. P. Karlen, J. M. Thompson, H. I. Vold, J. D. Farrell, and P. H. Eismann, "A dual-arm dexterous manipulator system with anthropomorphic kinematics," *Proc. IEEE Int. Conf. Robotics and Automation*, pp. 368–373, May 1990.

[2] R. V. Dubey, J. A. Euler, and S. M. Babcock, "An efficient gradient projection optimization scheme for a seven-degree-of-freedom redundant robot with spherical wrist," *Proc. IEEE Int. Conf. Robotics and Automation*, pp. 28–36, Apr. 1988.

[3] J. M. Hollerbach, "Optimum kinematic design for a seven-degree-of-freedom manipulator," *Proc. 2nd Int. Symp. Robotics Res.*, pp. 215–222, 1985.

[4] A. Liègeois, "Automatic supervisory control of the configuration and behavior of multibody mechanisms," *IEEE Trans. Syst., Man, and Cyber.*, vol. SMC-7, no. 12, pp. 868–871, 1977.

[5] A. A. Maciejewski and C. A. Klein, "Obstacle avoidance for kinematically redundant manipulators in dynamically varying environments," *Int. J. Robotics Res.*, vol. 4, no. 3, pp. 109–117, 1985.

[6] J. Angeles, "Isotropy criteria in the kinematic design and control of redundant manipulators," *NATO Adv. Research Workshop on Robots with Redundancy*, Salò, Italy, Jun. 1988.

[7] S. Chiu, "Task compatibility of manipulator postures," *Int. J. Robotics Res.*, vol. 7, no. 5, pp. 13–21, 1988.

[8] T. Yoshikawa, "Manipulability of robotic mechanisms," *Int. J. Robotics Res.*, vol. 4, no. 2, pp. 3–9, 1985.

[9] T. Yoshikawa, "Dynamic manipulability of robot manipulators," *J. Robotic Syst.*, vol. 2, no. 1, pp. 113–124, 1985.

[10] J. Baillieul, "A constraint oriented approach to inverse problems for kinematically redundant manipulators," *Proc. IEEE Int. Conf. Robotics and Automation*, pp. 1827–1833, March 1987.

[11] P. H. Chang, "A closed-form solution for inverse kinematics of robot manipulators with redundancy," *IEEE J. Robotics and Automation*, vol. RA-3, no. 5, pp. 393–403, 1987.

[12] D. E. Whitney, "Resolved motion rate control of manipulators and human prostheses," *IEEE Trans. on Man-Machine Syst.*, vol. MMS-10, no. 2, pp. 47–53, 1969.

[13] J. Baillieul, "Kinematic programming alternatives for redundant manipulators," *Proc. IEEE Conf. Robotics and Automation*, pp. 722–728, March 1985.

[14] J. Baillieul, J. M. Hollerbach, and R. Brockett, "Programming and control of kinematically redundant manipulators," *Proc. 23rd IEEE Conf. Dec. and Cont.*, pp. 768–774, Dec. 1984.

[15] O. Khatib, "Dynamic control of manipulators in operational space,"

Proc. 6th IFToMM Congr. on Theory of Machines and Mechanisms, pp. 1123–1131, 1983.

[16] A. A. Maciejewski, "Kinetic limitations on the use of redundancy in robotic manipulators," *Proc. IEEE Int. Conf. Robotics and Automation*, pp. 113–118, May 1989.

[17] P. Hsu, J. Hauser, and S. S. Sastry, "Dynamic control of redundant manipulators," *J. Robotic Syst.*, vol. 6, no. 2, pp. 133–148, 1989.

[18] O. Egeland, "Task-space tracking with redundant manipulators," *IEEE J. Robotics and Automation*, vol. RA-3, no. 5, pp. 471–475, 1987.

[19] Y. Nakamura and H. Hanafusa, "Optimal redundancy resolution control of robot manipulators," *Int. J. Robotics Res.*, vol. 6, no. 1, pp. 32–42, 1987.

[20] K. C. Suh and J. M. Hollerbach, "Local versus global torque optimization of redundant manipulators," *Proc. IEEE Int. Conf. Robotics and Automation*, pp. 619–624, March 1987.

[21] K. Kazerounian and Z. Wang, "Global versus local optimization in redundancy resolution of robotic manipulators," *Int. J. Robotics Res.*, vol. 7, no. 5, pp. 3–12, 1988.

[22] A. Nedungadi and K. Kazerounian, "A local solution with global characteristics for the joint torque optimization of a redundant manipulator," *J. Robotic Syst.*, vol. 6, no. 5, pp. 631–654, 1989.

[23] B. Siciliano and L. Sciavicco, "A solution algorithm to the inverse kinematic problem for redundant manipulators," *IEEE J. Robotics and Automation*, vol. RA-4, no. 4, pp. 403–410, 1988.

[24] C. Chevallereau and W. Khalil, "A new method for the solution of the inverse kinematics of redundant robots," *Proc. IEEE Int. Conf. Robotics and Automation*, pp. 37–42, April 1988.

[25] A. A. Maciejewski and C. A. Klein, "The singular value decomposition: computation and application to robotics," *Int. J. Robotics Res.*, vol. 8, no. 6, pp. 63–79, 1989.

[26] Y. Nakamura and H. Hanafusa, "Inverse kinematic solutions with singularity robustness for robot manipulator control," *J. Dyn. Syst., Meas., and Cont.*, vol. 108, no. 3, pp. 163–171, 1986.

[27] C. W. Wampler, "Manipulator inverse kinematic solution based on vector formulations and damped least-squares methods," *IEEE Trans. Syst., Man, and Cyber.*, vol. SMC-16, no. 1, pp. 93–101, 1986.

[28] A. De Luca and G. Oriolo, "The reduced gradient technique for solving robot redundancy," *Robotersysteme*, vol. 7, no. 2, pp. 117–122, 1991.

[29] T. Shamir and Y. Yomdin, "Repeatability of redundant manipulators: mathematical solution of the problem," *IEEE Trans. Automatic Contr.*, vol. AC-33, pp. 1004–1009, 1988.

[30] Y. Nakamura and M. Ghodoussi, "Dynamics computation of closed-link robot mechanisms with nonredundant and redundant actuators," *IEEE Trans. Robotics and Automation*, vol. RA-5, no. 3, pp. 294–302, 1989.

[31] D. N. Nenchev, "Redundancy resolution through local optimization: A review," *J. Robotic Syst.*, vol. 6, no. 6, pp. 769–798, 1989.

[32] B. Siciliano, "Kinematic control of redundant robot manipulators: A tutorial," *J. Intelligent and Robotic Syst.*, vol. 3, no. 3, pp. 201–212, 1990.

Review of Pseudoinverse Control for Use with Kinematically Redundant Manipulators

CHARLES A. KLEIN MEMBER, IEEE, AND CHING-HSIANG HUANG

Abstract—Kinematically redundant manipulators have a number of potential advantages over current manipulator designs. For this type of arm, velocity control through pseudoinverses has been suggested. Questions associated with pseudoinverse control are examined in detail and show that in some cases this control leads to undesired arm configurations. A method for distributing joint angles of a redundant arm in a good approximation to a true minimax criterion is described. In addition several numerical considerations are discussed.

I. Introduction

In complex systems where there are a large number of degrees of freedom to be controlled, a successful philosophy for designing control software has been to use the concept of supervisory remote control [1], [2]. In order to fully specify the operation of many actuators from a few goal dimensions, it is necessary to either add a number of constraints or to optimize over the number of unspecified dimensions. Use of the pseudoinverse, a method with an implied optimality criteria, has been proposed for this purpose by a number of individuals [2]–[5].

Here the pseudoinverse will be applied to the supervisory control of articulated mechanisms such as manipulator arms, vehicle legs, and prosthetic limbs. For manipulator control, it is usually most convenient to specify goal directions in rectilinear coordinates. Through the method of Jacobian control [6] an external description of the end effector motion in rectilinear coordinates can be related to the internal actuator motion by the Jacobian matrix:

$$\dot{x} = J\dot{\theta}. \tag{1}$$

In this equation $\dot{x}$ is an m element vector of the translational and rotational velocities of the end effector; $\dot{\theta}$ is an n element vector of the actuator velocities and J is an $m \times n$ matrix. For arms with all revolute joints, $\dot{\theta}$ is the joint velocity vector.

An arm can be termed redundant if it contains more degrees of freedom than the number required for a class of tasks. Since six degrees of freedom are required for the usual task of arbitrarily locating and orienting a load in space, an arm will be redundant if it has more than that number of actuators, not counting gripper actuation.

If J is square and nonsingular, $\dot{\theta}$ can be conceptually computed by

$$\dot{\theta} = J^{-1}\dot{x}. \tag{2}$$

When this is not the case, J^{-1} does not exist and mathematically there is not a solution for $\dot{\theta}$; however, there does exist a generalized inverse of J, G, which yields a "useful" answer for $\dot{\theta}$ in the same form as (2):

$$\dot{\theta} = G\dot{x}. \tag{3}$$

Manuscript received September 4, 1981; revised May 17, 1982. This work was supported by NASA Langley Grant NAGI-30, and in part by the National Science Foundation through Grant ECS78-18957.

C. A. Klein is with the Department of Electrical Engineering, The Ohio State University, 2015 Neil Avenue, Columbus, OH 43210.

C. H. Huang was with the Department of Electrical Engineering, The Ohio State University. He is now at General Automation, 1055 S. East Street, Anaheim, CA 92805.

The sense of the usefulness of the answer depends on the choice of generalized inverse used.

The most commonly used generalized inverse is the pseudoinverse. Although the use of this method has been described [2]–[5], there are a number of unappreciated properties of this method that can yield paradoxical results if not understood. One of the purposes here is to clarify these effects.

For the organization of the rest of this correspondence, Section II summarizes the properties of the pseudoinverse, and Section III describes effects of pseudoinverse control for a simpler manipulator model. Section IV examines uses of the homogeneous solution of the Jacobian equation to improve pseudoinverse control, Section V describes improvements to computational methods, and Section VI presents conclusions.

II. Review of Pseudoinverse Properties

While the motivation for a generalized inverse is best explained by the need for a matrix which can multiply the data vector to give an answer even when a true matrix inverse does not exist, most texts define G as a generalized inverse of matrix A if

$$AGA = G. \tag{4}$$

The Moore–Penrose pseudoinverse [7], in addition satisfies

$$GAG = A \tag{5}$$

$$(GA)^* = GA \tag{6}$$

and

$$(AG)^* = AG, \tag{7}$$

where the asterisk superscript indicates the complex conjugate transpose. Most papers using the pseudoinverse define it explicitly as the matrix satisfying (4), (5), (6), and (7) because the terminology for different types of generalized inverses is so confused in the literature. In this correspondence only the pseudoinverse will be used, which will be symbolized by A^+.

The best way to understand the pseudoinverse is through matrix representations by the singular value decomposition theorem (SVD) [8], [9]. For the linear system

$$Ax = y, \tag{8}$$

if A is an $m \times n$ complex matrix with rank k, then by the SVD theorem, A can be represented by

$$A = \sum_{i=1}^{k} \mu_i e_i f_i^*, \tag{9}$$

where μ_i are positive real constants, e_i are a set of mutually orthonormal m-element output vectors, and f_i are a set of orthonormal n-element input vectors. By Gram–Schmidt orthogonalization, e_i for $i = k+1, m$; and f_i for $i = k+1, n$ can be defined so that these vectors span m-space and n-space, respectively. If a matrix is singular or there are fewer rows than columns, then there will be components f_i, $i = k+1, n$ of the vector x which will be mapped into the null vector. For this reason, this subspace can be termed the *null space* of A [10].

Reprinted from *IEEE Trans. Syst., Man, Cybern.*, vol. SMC-13, no. 3, pp. 245–250, March/April 1983.

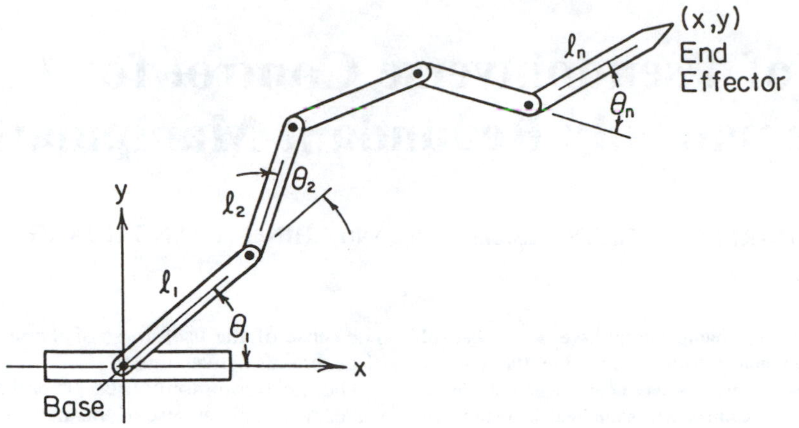

Fig. 1. Two-dimensional revolute arm characterized by relative link angles.

The SVD representation (9) suggests that pseudoinverse is

$$A^{+} = \sum_{i=1}^{k} 1/\mu_i f_i e_i^{*}. \tag{10}$$

Multiplying y by A^+ maps back all components of the output space which do not correspond to the null space of A; components of x in the null space of A have no trace in y and, therefore, cannot be recovered and are assigned zero coefficients. Substituting (10) into (4)–(7) shows that it satisfies the requirements of the pseudoinverse.

From the SVD representation of the pseudoinverse, the optimal properties of this matrix for the solution of the linear systems (8) can be easily seen geometrically. For the underdetermined case, where there are fewer rows than columns and the rank is the number of rows, there will be an infinite number of solutions, and it is usually desired to find the solution with the minimum Euclidean norm. The pseudoinverse solution can be represented by

$$\tilde{x} = A^{+}y = \sum_{i=1}^{k} \alpha_i f_i \tag{11}$$

for some set of coefficients α_i. The general solution for x would be

$$x = \sum_{i=1}^{k} \alpha_i f_i + \sum_{i=k+1}^{n} \alpha_i f_i \tag{12}$$

for any value of α_i, $i = k+1, n$, since this second term has no contribution to the output space. The pseudoinverse solution clearly has the minimum norm since adding orthogonal components to a vector always increases its magnitude.

For the overdetermined case, it can be shown in a similar way that the pseudoinverse yields an answer which best fits the data vector in a least-squares sense [7], [10]. In the fully general case (8) may have no solution, but an infinite number of x may yield the same least-squares value for y. Again, through the SVD analysis, overdetermined and underdetermined subspaces of A can be identified and, in this case, the pseudoinverse provides the minimum norm, least-squares solution. The pseudoinverse can be easily modified to select the least x^*Vx value for all x which minimize r^*Wr, where r is the residual of (8) and V and W are Hermitian positive definite weighting matrices [10]. This modification simply involves transforming spaces to those in which the pseudoinverse yields the usual minimum Euclidean norms.

Thus the pseudoinverse, through its projection property of ignoring unmatchable components, provides a useful answer for (8) independent of the dimension or rank. Part of the reason this is not widely appreciated is that the terminology "minimum norm solution" for the underdetermined and "least-squares solution" for the overdetermined cases [11] obscures the fact that in both cases a minimum norm vector is produced.

For underdetermined cases with full rank ($m < n$, $k = m$) and the overdetermined case with full rank ($m > n$, $k = n$), explicit formulas for the pseudoinverse can be provided. For the undetermined case,

$$A^{+} = A^{*}(AA^{*})^{-1}, \qquad m < n, k = m, \tag{13}$$

and for the overdetermined case,

$$A^{+} = (A^{*}A)^{-1}A^{*}, \qquad m > n, k = n. \tag{14}$$

These results can be derived in terms of projection arguments [12], or (13) can be derived with Lagrangian multipliers with $Ax = y$ as an equality constraint [4]. However these results can be trivially verified by substituting the SVD representation for A in (13) or (14) to obtain (10). Often (13) is believed to define the pseudoinverse, while it is only a formula for the pseudoinverse that is valid in a particular solution; the pseudoinverse, as represented in (10), still exists when neither (13) nor (14) is valid.

III. Use of the Pseudoinverse to Control a Simple Redundant Arm

As mentioned in the introduction, through Jacobian control, the pseudoinverse has been proposed for the control of joint velocities of redundant manipulators. Since energy consumption can be related to the norm of joint velocities, and since the pseudoinverse finds the minimum norm solution, instantaneous power is minimized. Since nearness to singularities is characterized by high joint velocities, the pseudoinverse causes the system to avoid singularities [4].

However pseudoinverse control can provide some paradoxical and unwanted results which will be shown in this section through a simple two-dimensional example. Fig. 1 shows a two-dimensional revolute arm where the internal variables θ_i are the relative angles between adjacent links. The end effector position is given by

$$\begin{pmatrix} x \\ y \end{pmatrix} = \left(\sum_{i=1}^{n} l_i \cos\psi_i, \; \sum_{i=1}^{n} l_i \sin\psi_i \right)^{T}, \tag{15}$$

where ψ_i is the absolute link angle given by

$$\psi_i = \sum_{j=1}^{i} \theta_j. \tag{16}$$

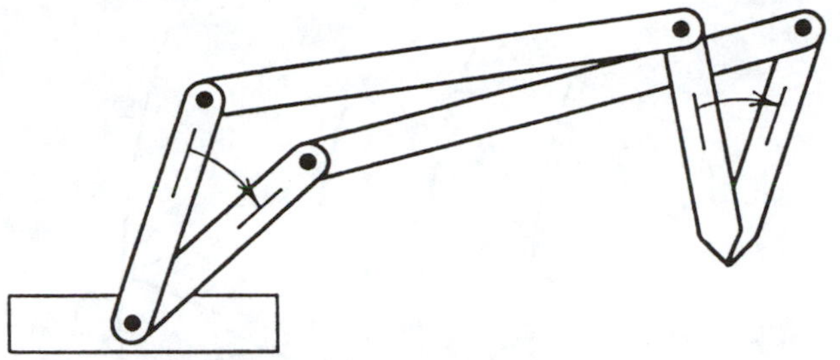

Fig. 2. Homogeneous solution to Jacobian equation is set of joint velocities which cause no end effector motion. No such components are present in the pseudoinverse solution.

Differentiating yields the Jacobian relation (1) where

$$J = \begin{bmatrix} -\sum_{i=1}^{n} l_i \sin\psi_i & -\sum_{i=2}^{n} l_i \sin\psi_i & \cdots & -\sum_{i=n}^{n} l_i \sin\psi_i \\ \sum_{i=1}^{n} l_i \cos\psi_i & \sum_{i=2}^{n} l_i \cos\psi_i & \cdots & \sum_{i=n}^{n} l_i \cos\psi_i \end{bmatrix}. \tag{17}$$

Except when the arm is fully extended, J has rank 2, and (13) can be used to solve for $\dot{\theta}$:

$$\dot{\theta} = J^T(JJ^T)^{-1}\dot{x}. \tag{18}$$

For some interpretations, time dependence of the motion does not need to be explicitly considered at all; by the chain rule,

$$d\theta = J^T(JJ^T)^{-1}\,dx. \tag{19}$$

For a trial task, the arm tip has been specified to trace a square. In addition to the ideal velocity, a correctional velocity is added so that any drift from the path will be corrected in the next time interval. If x_1 and x_2 are the displacement vectors for two adjacent corners, Δt is the time allowed for travel between the two, and m iterations occur on a side, then velocity should be calculated as

$$\dot{x} = (x_2 - x_1)/\Delta t + [(x_2 - x_1)i/m - x]m/\Delta t \tag{20}$$

at the end of the ith iteration where x is the current position.

For this problem, the null space of J, the homogeneous solutions of (1), which are rejected by the pseudoinverse, have a simple physical meaning. As shown in Fig. 2, the homogeneous solutions are the joint velocities which produce no end effector motion. Therefore the joint velocities produced by the pseudoinverse have no component which does not contribute to the motion of the end effector.

After viewing a simulation of the end effector tracing a square, it becomes clear that even though the pseudoinverse finds an n component vector $\dot{\theta}$ (or $d\theta$) from the two-component vector $\dot{x}$, the angular position θ is not made a function of x, even when the arm is started at the same point in x and θ space. Thus, in general, using the pseudoinverse for the derivatives does not yield an inverse function for the variables. The Appendix analytically proves the lack of an inverse for the revolute arm and shows that this is not an effect of roundoff error in the computer simulation. The implication of this result is that after a number of cycles around a closed path, the θ state for a given x will not generally be known and may be undesirable.

A number of special cases are worth mentioning. If one variable is a function of p variables ($p > 1$), then the pseudoinverse solution for moving in the p-variable space for a specified velocity of the single variable always yields an inverse function for p variables in terms of the single variable, except for points where the gradient is zero. For the revolute arm, this situation would exist if the y component were of interest, and the x value could be ignored. A true inverse function would exist because all motion

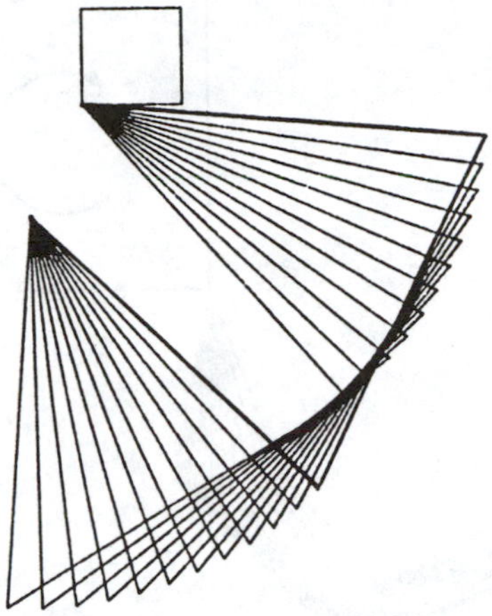

Fig. 3. Three-link two-dimensional arm tracing a square under pseudoinverse control in CW direction shown passing through the same point during successive cycles. For this configuration, no limit is reached.

would be along the path produced by the gradient. In another case, if two xyz tables were stacked, the pseudoinverse would demand equal contributions from each corresponding actuator, and again an inverse function would exist. A general test for whether a pseudoinverse for the derivatives yields an inverse function for the corresponding variables, would involve examining the rows of the pseudoinverse. Since the rows would be gradients of the inverse functions if they exist, they would have zero curl.[1] Unfortunately, in most cases, either the pseudoinverse is calculated numerically, or it is expressed in terms of the wrong set of variables for partial derivatives to be easily calculated.

Not only does pseudoinverse control have a lack of an implied inverse function for θ in terms of x which implies a drift in θ space after traversing a closed path, but also there may not be a simple limit for arm motion after repeatedly covering this same path. Numerical experimentation shows a variety of effects depending on the configuration of the arm and the size of the square path. When the square is close to the arm's base, the configuration drifts a nearly constant amount each cycle, and there will be no limit (see Fig. 3). When there is a limit, there will be different limits depending on the direction of the arm's motion over the path (see Figs. 4 and 5). The effects seen in Figs. 3, 4, and 5 have been seen when very large numbers of path segments have been used. In some cases there will be a limit when an extremely large number of path segments is used, but when a more practical number is applied, then the system drifts indefinitely. Since discrete time control is intrinsic to digital computer control of arms, such discretization effects must be considered an integral property of a control method, not a side effect.

Suppose one integrates $d\theta$ over the arm's travel around the square path to get a path length in θ space. One might suppose that since the pseudoinverse chooses a minimum $d\theta$ at each point that the θ path length would decrease every time around. Paradoxically, this is not the case. Since, if the x motion is reversed, the path in θ space is reversed, it follows that if for one sense of rotation the θ path length is decreasing, it must be increasing for the other sense of rotation. For the case of three links, the paradox can be resolved by visualing the paths in θ space. Fig. 6 shows a conceptual diagram of a four-surfaced object which corresponds to θ motion for the tip to be on the square. The line l_A is the locus of all θ such that the tip is at position A, and all points in the surface containing l_A and l_B have the same y value as A and B. Pseudoinverse control specifies that when the tip is at

[1]Although curl is not usually defined for n space, testing whether the ith partial derivative of the jth component equals the jth partial derivative of the ith component, for all i and j, is a more general test whether an n-element vector is a gradient.

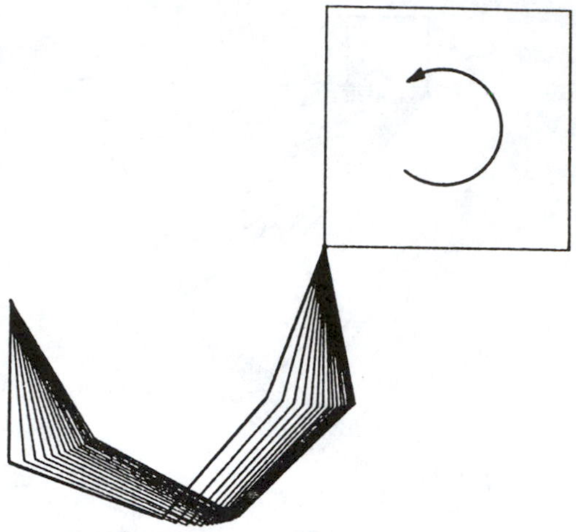

Fig. 4. Four-link two-dimensional arm tracing a square under pseudoinverse control in CCW direction shown passing through the same point during successive cycles. Limit is reached.

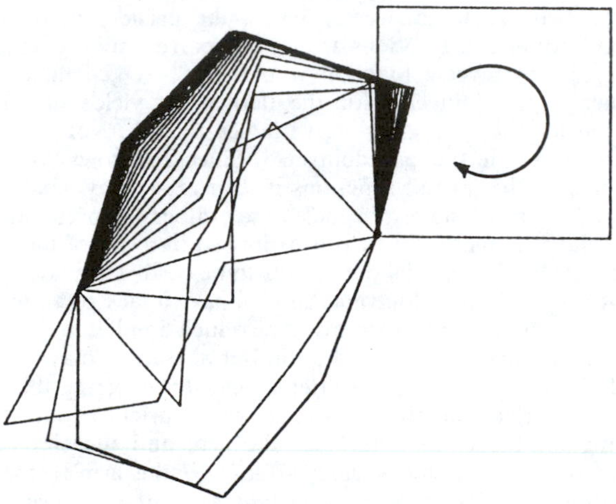

Fig. 5. Four-link, two-dimensional arm tracing a square under pseudoinverse control in CW direction shown passing through the same point during successive cycles. Although starting point is the same as for Fig. 4, a different limit is reached.

A, there will be no component in the direction of l_A, the direction of the homogeneous solution. As the tip traces the path, θ follows a screw-like motion over the surface. Because of the minimum norm property, the path will be a geodesic. However it will be a minimum length path between two different points in θ space, not the same point in x space. Since the points in θ space change each cycle, there is no reason why the θ path should be getting smaller. In the case of a limit, the path closes on itself making a screw of zero pitch.

IV. Use of Homogeneous Solutions to Improve Performance

If one is willing to give up the minimum norm solution, components of the homogeneous solutions to (1) can be included in the solution to optimize additional criterion. Liegeois [2] has shown how components of the gradient of such a criterion can be blended with the pseudoinverse solution. In terms of the variables used here, this modified control becomes

$$\dot{\theta} = J^{+}\dot{x} + \alpha(J^{+}J - I_n)\nabla H(\theta) \tag{21}$$

where $H(\theta)$ is an optimization criterion to be minimized subject to the required rectilinear velocity, I_n is the nth order identity matrix, and α is a gain constant. A realistic criterion for a

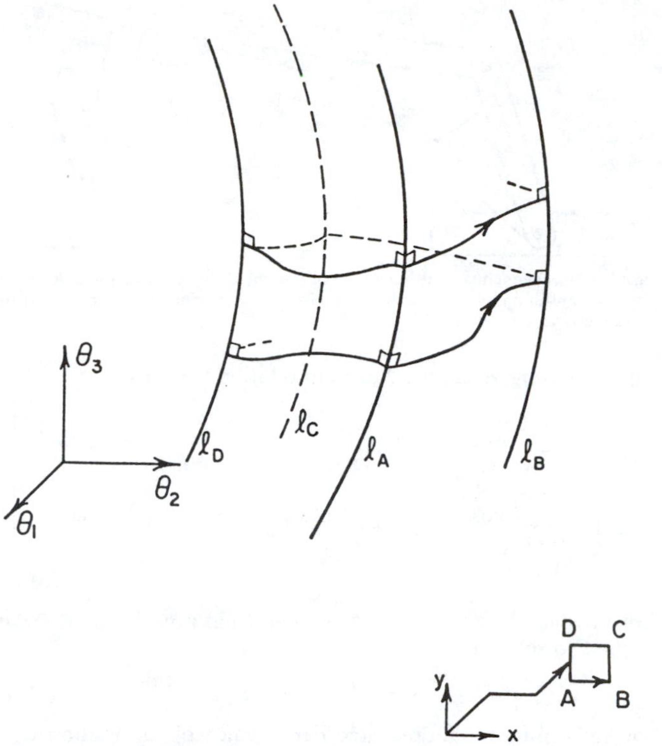

Fig. 6. Conceptual diagram of θ space corresponding to the tip tracing a square in x space under pseudoinverse control.

redundant manipulator might be to solve for joint angles such that each angle is within the physical limits caused by mechanical stops. Liegeois [2] programmed a six-degree-of-freedom manipulator to draw a circle. By specifying only the position of the end effector, the manipulator is redundant and the extra degrees of freedom are used to optimize joint range availability in a least-squares sense. If θ_{ci} is the center position for the ith joint and the maximum one-sided excursion is $\Delta\theta_i$, then the criterion function H is

$$H_2 = \sum_{i=1}^{n} ((\theta_i - \theta_{ci})/\Delta\theta_i)^2. \tag{22}$$

The least-squares norm is used in many problems, not because it is truly desired, but because it is tractable. The ideal criterion to distribute joint angles equitably would be to minimize

$$H_\infty = \max_i |\theta_i - \theta_{ci}|/\Delta\theta_i. \tag{23}$$

Since the least-squares and maximum (box) norms are within a factor of $\sqrt{n}$ of each other [8], the least-squares norm is often used to obtain a reasonable and suboptimal solution in place of the desired norm. However, since the angles that generate the norm are of primary interest, not the value of the norm itself, the closeness of the norm values can be misleading. Fig. 7 defines a simple configuration that illustrates how different the least-squares and minimax solutions can be. For this three-degree-of-freedom arm, the first angle can be used to parametrically describe all configurations with the tip at a fixed point. With zero being the center of travel and assuming all joints have equal ranges, one can solve for θ for the minimum H_2. While the minimax solution is $\theta_1 = \theta_2 = 90°$, the least-squares solution is $\theta_1 = 60°$, $\theta_2 = 120°$, which yields a factor of 2 between maximum and minimum angles.

A natural way to approach the maximum norm with a tractable norm is to use the family of p-norms [13]. These norms are

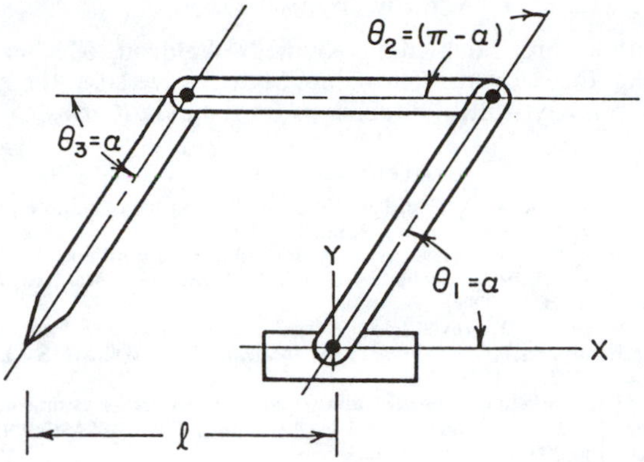

Fig. 7. Two-dimensional arm configuration used to compare optimal joint distribution under different norms. Each link has length l.

defined as

$$\|x\|_p = \left(\sum_{i=1}^{n} |x_i|^p\right)^{1/p} \tag{24}$$

As shown in Fig. 8, as p approaches infinity, the norm becomes the maximum norm. For the configuration in Fig. 7, the minimum p-norms yield the following ratio of maximum-to-minimum angles

$$\theta_{max}/\theta_{min} = 2^{1/(p-1)}. \tag{25}$$

A computer simulation of this control method has been performed for an n-link, two-dimensional linkage where the tip either traces a square or can be directly controlled by a joystick. Behavior of the system depends on the size of the weighting constant α. When an operator uses the joystick to move the tip to a new position and α is small, then there is a slow reorientation of the links without the tip moving. If α is large, though, the arm appears to be instantaneously in the optimal configuration and no reorientation motion is seen when the joystick is released. For a constant rectilinear velocity, when the tip is moved near the base, the joint velocities are relatively large and the reorientation motion is again visible. For best performance, either α should be a function of rectilinear displacement or rectilinear velocity should be automatically reduced near the base.

If the power p in the angular position norm is chosen to be too high, the system can become unstable. For the infinite norm, the system is trying to reduce the maximum angle; if the system crosses a corner of the norm diagram, then the angle to be reduced is discontinuously changed. For p being large, the motion can change quickly, and if the gain α is too high, the system will become unstable. The value $p = 6$ allows an α to produce a reasonable time constant and still gives a joint ratio of 1.15 from (25) for the example in Fig. 7.

Using (21) to include homogeneous solutions in the velocity vector involves a trade-off between minimum energy and optimization of an auxiliary condition. This trade-off can be quantified by first noting that the two terms of (21) are orthogonal. Therefore, one can define an efficiency showing the square of the joint velocities with and without the homogeneous solution:

$$\text{efficiency} = \frac{\|J^+\dot{x}\|^2}{\|J^+\dot{x}\|^2 + \alpha^2\|(J^+J - I)\nabla H\|^2}. \tag{26}$$

Because the system reduces the gradient when it moves to reduce the criterion, the system tends to raise its own efficiency defined in this way.

V. Numerical Considerations

This section will address itself to several numerical questions that must be considered while using the pseudoinverse. The first is that, like regular inverses, it is usually possible to avoid explicit calculation of the pseudoinverse. While this is immediately obvious for the overdetermined case (14), the pseudoinverse solution for the underdetermined case (13) can also be obtained through Gaussian elimination. By solving for w such that

$$AA^*w = y \tag{27}$$

the pseudoinverse solution is

$$x = A^*w. \tag{28}$$

Even when the pseudoinverse is used in the projection operator of (21), the equation can be regrouped so that Gaussian elimination can be used in place of inverting square matrices. Solve for w in

$$(JJ^*)w = \dot{x} + \alpha J\nabla H \tag{29}$$

then

$$\dot{\theta} = -\alpha\nabla H + J^*w. \tag{30}$$

The second point is that the formulas (13) and (14) are not usable at singularities even though the pseudoinverse still exists and is ideal for solving for the components that are still realizable. The above formulas can be reapplied if a linearly dependent row is removed from the Jacobian. In some situations, however, it may be desirable to have a single algorithm that works in all cases. For example, if one processor in a multiprocessor is calculating pseudoinverses [14], it may be necessary for the timing to be constant over all data conditions. In some applications, such solutions are known as uniform solutions. A number of algorithms for pseudoinverses require no explicit estimate of rank and would appear to be such solutions.

Because of a property of pseudoinverses, however, it appears there is no such general formula. Unlike regular inverses, as one matrix approaches another, their pseudoinverses do not always converge [15]. Divergence occurs when the matrices have different ranks. In this application, the difficulty occurs not when a matrix is exactly singular, where the homogeneous solutions are entirely removed, but instead near singularities where the system is still ill-conditioned and very large joint velocities occur. By determining an effective rank from the singular values or from pivot ratios [16], these near-singularities could be recognized. However, because singularities can be most easily characterized in $\boldsymbol{\theta}$ space [17], it is still probably easier and faster to remove dependent rows and apply (13).

VI. Conclusion and Recommendations for Future Work

This paper has reviewed the pseudoinverse control method in terms of the singular value decomposition representation. Although pseudoinverse control of kinematically redundant manipulators avoids singularities, and yields a minimum norm of joint velocities, closed path, cyclical motion either has no joint angle limit, or else has limits in configurations which may not be desirable. However pseudoinverse control can be supplemented with an additional criterion function to approximate a minimax solution for joint angle distribution.

The use of the pseudoinverse in robotic systems is not limited to kinematics. In closed kinematic chains, internal forces and torques may exist which cause no useful work [18]. These forces

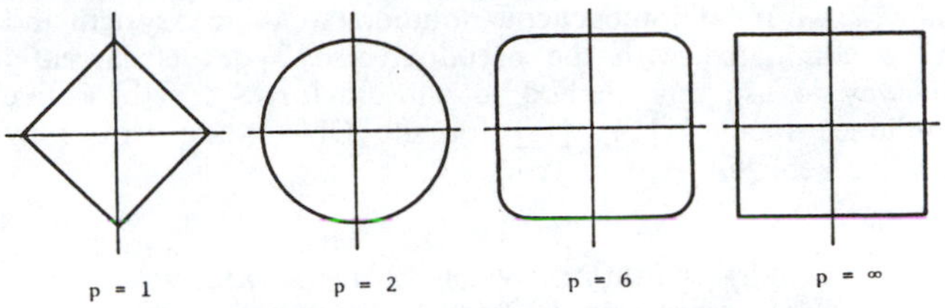

Fig. 8. Family of p-norms in two-dimensional space.

can be shown to be homogeneous solutions to a linear system and can be eliminated with the pseudoinverse. Work is currently underway to use this method to allocate forces for the active compliance method [14], [19] for the Ohio State University Hexapod Vehicle.

Appendix
Proof that Pseudoinverse Control does not Implicitly Yield an Inverse Function from Rectilinear-to-Joint Spaces for a Revolute Arm for the Case $n = 3$

From (24), the rectilinear and joint velocities of a three-link planar arm can be related by

$$\boldsymbol{dx} = \begin{bmatrix} dx \\ dy \end{bmatrix} = \begin{bmatrix} \nabla f^T \\ \nabla g^T \end{bmatrix} \begin{bmatrix} d\theta_1 \\ d\theta_2 \\ d\theta_3 \end{bmatrix}, \tag{31}$$

where

$$\nabla f^T = [-l_1 \sin\psi_1 - l_2 \sin\psi_2 - l_3 \sin\psi_3, \; -l_2 \sin\psi_2 - l_3 \sin\psi_3, \; -l_3 \sin\psi_3] \tag{32}$$

and

$$\nabla g^T = [l_1 \cos\psi_1 + l_2 \cos\psi_2 + l_3 \cos\psi_3, \; l_2 \cos\psi_2 + l_3 \cos\psi_3, \; l_3 \cos\psi_3], \tag{33}$$

where

$$\psi_i = \sum_{j=1}^{i} \theta_j. \tag{34}$$

The pseudoinverse solution for $\boldsymbol{d\theta}$ (call it $\boldsymbol{d\theta}_p$), will have no component of the homogeneous solution to (31). For $n = 3$, the direction of the homogeneous solution is easily generated by a cross-product since it must be orthogonal to both rows of the matrix.

$$d\boldsymbol{\theta}_h \propto \nabla f \times \nabla g, \tag{35}$$

where $d\boldsymbol{\theta}_h$ is the homogeneous solution. Therefore,

$$(\nabla f \times \nabla g) \cdot \boldsymbol{d\theta}_p = 0. \tag{36}$$

This equation can be recognized to be a Pfaffian differential equation and the question of the conservation of pseudoinverse control becomes equivalent to the integrability of (36). A concrete test is provided by a theorem that a necessary and sufficient condition for the Pfaffian differential equation $X \cdot dr = 0$ to be integrable is that [20]

$$X \cdot \text{curl } X = 0. \tag{37}$$

Calculating, $X \cdot \text{curl } X$, where $X = \nabla f \times \nabla g$ yields

$$l_3 l_2 \{ l_1 l_2 \cos\theta_2 - l_2 l_3 \cos\theta_3 \sin\theta_3 + l_1 l_3 (2 \sin\theta_3 \cos(\theta_2 + \theta_3) - \cos\theta_3 \sin(\theta_2 + \theta_3)) \} \neq 0. \tag{38}$$

Since this quantity does not vanish, pseudoinverse control is not conservative and a closed path in x space will not, in general, imply a closed path in $\boldsymbol{\theta}$ space.

Acknowledgment

The authors are indebted to Kenneth Waldron, Mechanical Engineering Department, The Ohio State University, for participation in many fruitful discussions on redundant arms.

References

[1] W. R. Ferrell and T. B. Sheridan, "Supervisory control of remote manipulation," *IEEE Spectrum*, vol. 4, no. 10, Oct. 1967.

[2] A. Liegeois, "Automatic supervisory control of the configuration and behavior of multibody mechanisms," *IEEE Trans. Syst., Man, Cybern.*, vol. SMC-7, no. 12, Dec. 1977.

[3] D. E. Whitney, "Resolved motion rate control of manipulators and human prostheses," *IEEE Trans. Man–Machine Syst.*, vol. MMS-10, no. 2, pp. 47–53, June 1969.

[4] ——, "The mathematics of coordinated control of prostheses and manipulators," *J. Dynamic Syst., Measurement, Contr., Trans. ASME*, vol. 94, series G, pp. 303–309, Dec. 1972.

[5] M. L. Moe, "Kinematics and rate control of the rancho arm," presented at First CISM-IFToMM Symposium, Sept. 1974.

[6] R. Paul, "Manipulator Cartesian path control," *IEEE Trans. Syst., Man, Cybern.*, vol. SMC-9, no. 11, Nov. 1979.

[7] T. L. Boullion and P. L. Odell, *Generalized Inverse Matrices*. New York: Wiley-Interscience, 1971.

[8] G. Forsythe and C. B. Moler, *Computer Solution of Linear Algebraic Systems*. Englewood Cliffs, NJ: Prentice-Hall, 1967.

[9] V. C. Klema and A. J. Laub, "The singular value decomposition: Its computation and some applications," *IEEE Trans. Autom. Contr.*, vol. AC-25, no. 2, Apr. 1980.

[10] A. Ben-Israel and T. N. E. Greville, *Generalized Inverses: Theory and Applications*. New York: Wiley, 1974.

[11] W. L. Brogan, *Modern Control Theory*. New York: Quantum, 1974.

[12] C. R. Rao and S. K. Mitra, *Generalized Inverse of Matrices and Its Applications*. New York: Wiley, 1971.

[13] E. Isaacson and H. B. Keller, *Analysis of Numerical Methods*. New York: Wiley, 1966.

[14] C. A. Klein and W. Wahawisan, "Use of a multiprocessor for control of a robotic system," *Robotics Res.*, vol. 1, no. 2, pp. 45–59, Summer 1982.

[15] B. Noble, "Methods for computing the Moore–Penrose generalized inverses, and related matters," in *Generalized Inverses and Applications*, by M. Z. Nashed, Ed. New York: Academic, 1975, pp. 245–301.

[16] R. Mittra and C. A. Klein, "The use of pivot ratios as a guide to stability of matrix equations arising in the method of moments," *IEEE Trans. Antennas Propagat.*, vol. AP-23, pp. 448–450, May 1975.

[17] M. Renaud, "Coordination control of robot-manipulators: Determination of the singularities of the Jacobian matrix," in *Proc. of First Yugoslav National Symp. Industrial Robots*, Dubrovnik, Yugoslavia, pp. 153–165, Sept. 1979.

[18] R. B. McGhee, C. A. Klein, and R. L. Briggs, "On the role of regulated compliance in manipulation and locomotion tasks involving closed kinematic chains," in *Proc. of First Yugoslav Symp. Industrial Robots and Artificial Intelligence*, Dubrovnik, Yugoslavia, Sept. 1979.

[19] C. A. Klein and R. L. Briggs, "Use of active compliance in the control of legged vehicles," *IEEE Trans. Syst., Man, Cybern.*, vol. SMC-10, no. 7, July 1980.

[20] I. N. Sneddon, *Elements of Partial Differential Equations*. New York: McGraw-Hill, 1957.

Errata:

Equations (4) and (5) should read:

$$AGA = A \tag{4}$$

$$GAG = G \tag{5}$$

Equation (38) should read:

$$l_1 l_2 l_3 \{ -l_3 \sin\theta_2 + l_2 \cos\theta_2 \sin\theta_3 + l_3 \sin\theta_3 \cos(\theta_2 + \theta_3) \} \neq 0. \tag{38}$$

Yoshihiko Nakamura
Hideo Hanafusa
Tsuneo Yoshikawa
Automation Research Laboratory
Kyoto University
Uji, Kyoto
611 Japan

Task-Priority Based Redundancy Control of Robot Manipulators

Abstract

In this paper, we describe a new scheme for redundancy control of robot manipulators. We introduce the concept of task priority in relation to the inverse kinematic problem of redundant robot manipulators. A required task is divided into subtasks according to the order of priority. We propose to determine the joint motions of robot manipulators so that subtasks with lower priority can be performed utilizing redundancy on subtasks with higher priority. This procedure is formulated using the pseudoinverses of Jacobian matrices. Most problems of redundancy utilization can be formulated in the framework of tasks with the order of priority. The results of numerical simulations and experiments show the effectiveness of the proposed redundancy control scheme.

1. Introduction

Robot manipulators have usually been designed to have no more than six degrees of freedom, which is the least degrees of freedom needed to perform 3-D tasks. Although six degrees of freedom are sufficient for conventional mechanical design, robot manipulators are expected to be more flexible and adaptive, like human arms. When a robot manipulator is required to trace a given trajectory of the end effector while avoiding obstacles in the workspace, more degrees of freedom are needed than in a free working space.

Whitney (1972) and Uchiyama (1979) pointed out that redundancy of a robot manipulator is effective for overcoming its kinematical singularities. Redundancy is also effective for enabling a robot manipulator to approach a workpiece from all directions or to trace a given spatial trajectory avoiding obstacles in a workspace (Freund 1977). Whitney (1969, 1972) discussed the redundancy of prosthetic arms. Although he gave a criterion to minimize the integrated value of kinematic energy, he simplified it because of the large amount of computations and proposed to minimize the quadratic form of joint velocity instantaneously. This method implies the motion resolution by means of the weighted pseudoinverse of the full-rank Jacobian matrix. Nakano and Ozaki (1974), on the other hand, suggested *minimum potential energy criterion,* which forces a restriction on the inverse kinematics of an anthropomorphic manipulator so that the position of the elbow should be located in the vertical direction as low as possible. The criteria by Whitney and Nakano and Ozaki seem to be focused not on the active utilization of redundancy but on the unique determination of the solutions.

Liegeois (1977) discussed the active utilization of redundancy of robot manipulators. He expressed the general solution of joint velocity by means of the generalized inverse of the Jacobian matrix and proposed to determine the arbitrary vector as a gradient vector of a scalar function. He also showed the numerical simulation of utilizing redundancy for keeping the joint angles within their physical limitation. Hanafusa, Yoshikawa, and Nakamura (1978) proposed a numerical algorithm to plan a joint trajectory of a redundant manipulator using the general solution expressed by the pseudoinverse of the Jacobian matrix, where the limitation of joint angles and torques and the restriction imposed by obstacles were considered. Hanafusa, Yoshikawa, and Nakamura (1981) analyzed redundancy of robot manipulators in light of the matrix theory and discussed its utilization. Klein and Huang (1983) reviewed the control of redundant manipulators

The International Journal of Robotics Research,
Vol. 6, No. 2, Summer 1987,

by means of pseudoinverses, where they suggested a computational algorithm to reduce the computational amount. Moreover, Klein (1984) showed by numerical simulation that a manipulator can avoid an obstacle by paying attention to the point on the manipulator that is the shortest distance from the obstacle. Konstantinov, Markov, and Nenchev (1981) also discussed the utilization of redundancy for the physical limitation of joint angles by means of the pseudoinverse of the Jacobian matrix. Benati, Morasso, and Tagliasco (1982), on the other hand, proposed a recursive algorithm for inverse kinematics of an anthropomorphic redundant manipulator.

There are two possible approaches to the redundancy utilization problem: (1) instantaneously optimal control of redundancy and (2) globally optimal control of redundancy. Although the first approach requires a small amount of computation, it lacks a guarantee of global optimality. The second approach guarantees global optimality, but it requires a large amount of computation. Thus, the instantaneously optimal control approach is suitable for real-time redundancy control, such as sensor-based obstacle avoidance problems. The globally optimal control approach is better for off-line trajectory planning for tasks requiring strict optimality, such as obstacle avoidance problems in more complicated working spaces and energy minimizing problems. Therefore, these two approaches should be used properly according to the circumstances in which redundancy control is necessary.

In this paper, we will discuss the *instantaneously optimal control of redundancy.* To formulate the problem, the concept of task priority is introduced into kinematic inverse problems. A required task is divided into subtasks according to the order of priority. Then, the joint motions of robot manipulators are resolved in such a way that subtasks with lower priority are realized using redundancy or extra degrees of freedom not committed to satisfying subtask requirements of higher priority. This procedure is formulated based on the Jacobian matrix and its pseudoinverse. Although the basic idea was already published by the authors in Japanese (Hanafusa, Yoshikawa, and Nakamura 1983), this paper describes more complete formulations, including the dynamics of manipulators and the use of potential functions for obstacle avoidance and also shows the new result of their simulations. In Section 3, we formulate the problem in the framework of *the resolved motion rate control* (Whitney 1969). Numerical simulations are carried out in Section 4 to show the effectiveness of the formulation, where the formulation is extended to the framework of *the resolved acceleration control* in order to consider the dynamics of manipulators (Luh, Walker, and Paul 1980). Two examples are simulated in Section 4: in one case, the position of the end effector is prior to the orientation; in the other case, the obstacle avoidance problem, where artificial potential function (Khatib and Le Maitre 1978), is used to describe the lower priority motion. In Section 5, experiments of obstacle avoidance are carried out on a robot manipulator with seven degrees of freedom to discuss the implementation and the actual effectiveness of the formulation. The information on obstacle is taught as a reference joint angle by an operator and used as the desired value of the lower priority task. Operator intervention of this kind would be a simple but effective method to utilize the global judgment of human operators in obstacle avoidance problems.

2. Tasks with the Order of Priority

We sometimes find tasks where the position or the orientation of the end effector is more important than the other. For example, in welding, cutting, and shape measurement, the position of the end effector is more important than the orientation. On the other hand, with spray painting, or directing a camera to objects, orientation is more important. We consider these tasks to be composed of subtasks with different levels of significance and call them *tasks with the order of priority.*

The usual problems of redundancy utilization can be formulated in the framework of tasks with the order of priority. When a redundant manipulator is required to trace a given trajectory of the end effector while avoiding obstacles in the workspace, trajectory tracing is given the first priority and obstacle avoidance is given the second priority. When a redundant manipulator is needed to trace a given trajectory of the end effector while avoiding singular points, trajectory tracing is given the first priority and singularity avoidance is given the second priority.

For the tasks with the order of priority, if it is impossible to perform all of the subtasks completely because of the degeneracy or the shortage of degrees of freedom, it seems reasonable, then, to perform the most significant subtask preferentially and the less important subtasks (as well as possible) using the remaining degrees of freedom. In Section 3, the problem of redundancy utilization will be formulated based on this idea.

3. Inverse Kinematic Solutions Considering the Order of Priority

Here, to simplify the discussion, we formulate the problem for a case with two subtasks. This formulation can easily be extended to a case with more than two subtasks. The subtask with the first priority will be specified using the first manipulation variable, $\mathbf{r}_1 \in R^{m_1}$, and the subtask with the second priority will be specified using the second manipulation variable, $\mathbf{r}_2 \in R^{m_2}$. The kinematic relationships between the joint variable $\theta \in R^n$ and the manipulation variables are expressed as follows:

$$\mathbf{r}_i = \mathbf{f}_i(\theta). \qquad (i = 1, 2) \tag{1}$$

Their differential relationships are expressed as follows:

$$\dot{\mathbf{r}}_i = \mathbf{J}_i(\theta)\dot{\theta}, \qquad (i = 1, 2) \tag{2}$$

where $\mathbf{J}_i(\theta) \triangleq \partial \mathbf{f}_i / \partial \theta \in R^{m_i \times n}$ is the Jacobian matrix for the ith manipulation variable.

The general solution of Eq. (2) for $i = 1$ is obtained using pseudoinverses, as follows (Boullion and Odell 1971):

$$\dot{\theta} = \mathbf{J}_1^{\#}(\theta)\dot{\mathbf{r}}_1 + \{\mathbf{I} - \mathbf{J}_1^{\#}(\theta)\mathbf{J}_1(\theta)\}\mathbf{y}, \tag{3}$$

where $\mathbf{J}_1^{\#}(\theta) \in R^{n \times m_1}$ is the pseudoinverse of $\mathbf{J}_1(\theta)$ and $\mathbf{y} \in R^n$ is an arbitrary vector. If the exact solution does not exist, Eq. (3) represents the least-squares solution, minimizing $\|\dot{\mathbf{r}}_1 - \mathbf{J}_1(\theta)\dot{\theta}\|$.

Now, substituting Eq. (3) into Eq. (2) for $i = 2$, we get the following equation:

$$\mathbf{J}_2(\mathbf{I} - \mathbf{J}_1^{\#}\mathbf{J}_1)\mathbf{y} = \dot{\mathbf{r}}_2 - \mathbf{J}_2\mathbf{J}_1^{\#}\dot{\mathbf{r}}_1. \tag{4}$$

If the exact solution of $\mathbf{y}$ exists for Eq. (4), it means that the second manipulation variable can be realized. However, the exact solution does not generally exist. We get $\mathbf{y}$, which minimizes $\|\dot{\mathbf{r}}_2 - \mathbf{J}_2(\theta)\dot{\theta}\|$, in the same way as Eq. (3). That is,

$$\mathbf{y} = \tilde{\mathbf{J}}_2^{\#}(\dot{\mathbf{r}}_2 - \mathbf{J}_2\mathbf{J}_1^{\#}\dot{\mathbf{r}}_1) + (\mathbf{I} - \tilde{\mathbf{J}}_2^{\#}\tilde{\mathbf{J}}_2)\mathbf{z}, \tag{5}$$

where $\tilde{\mathbf{J}}_2 \triangleq \mathbf{J}_2(\mathbf{I} - \mathbf{J}_1^{\#}\mathbf{J}_1)$, and $\mathbf{z} \in R^n$ is an arbitrary vector.

The solution $\dot{\theta}$ is obtained from Eqs. (3) and (5) as follows:

$$\begin{aligned}\dot{\theta} = \mathbf{J}_1^{\#}\dot{\mathbf{r}}_1 &+ (\mathbf{I} - \mathbf{J}_1^{\#}\mathbf{J}_1)\tilde{\mathbf{J}}_2^{\#}(\dot{\mathbf{r}}_2 - \mathbf{J}_2\mathbf{J}_1^{\#}\dot{\mathbf{r}}_1) \\ &+ (\mathbf{I} - \mathbf{J}_1^{\#}\mathbf{J}_1)(\mathbf{I} - \tilde{\mathbf{J}}_2^{\#}\tilde{\mathbf{J}}_2)\mathbf{z}.\end{aligned} \tag{6}$$

Equation (6) represents the inverse kinematic solution considering task priority.*

We defined the range space of the Jacobian matrix, $\mathcal{R}(\mathbf{J})$, as the manipulable space, and the null space of the Jacobian matrix, $\mathcal{N}(\mathbf{J})$, as the redundant space (Hanafusa, Yoshikawa, and Nakamura 1981). Figure 1 shows the general relationship between the manipulable spaces and the redundant spaces for the first and the second manipulation variables. In Fig. 1, Subspace A, Subspace B, and Subspace C are $S_{R1}^{\perp}$, $S_{R1} \cap (S_{R1} \cap S_{R2})^{\perp}$, and $S_{R1} \cap S_{R2}$ respectively, where $S_{R1} = \mathcal{N}(\mathbf{J}_1)$, $S_{R2} = \mathcal{N}(\mathbf{J}_2)$, and $S_*^{\perp}$ means the orthogonal complement of subspace S_*. Subspace A means the contribution of $\dot{\theta}$ to the first manipulation variable. Subspace B can contribute to the second manipulation variable

* After submission of this paper, a paper by Maciejewski and Klein (1985) was published, which proved that the second term of Eq. (6) can be reduced to $\tilde{\mathbf{J}}_2^{\#}(\dot{\mathbf{r}}_2 - \mathbf{J}_2\mathbf{J}_1^{\#}\dot{\mathbf{r}}_1)$.

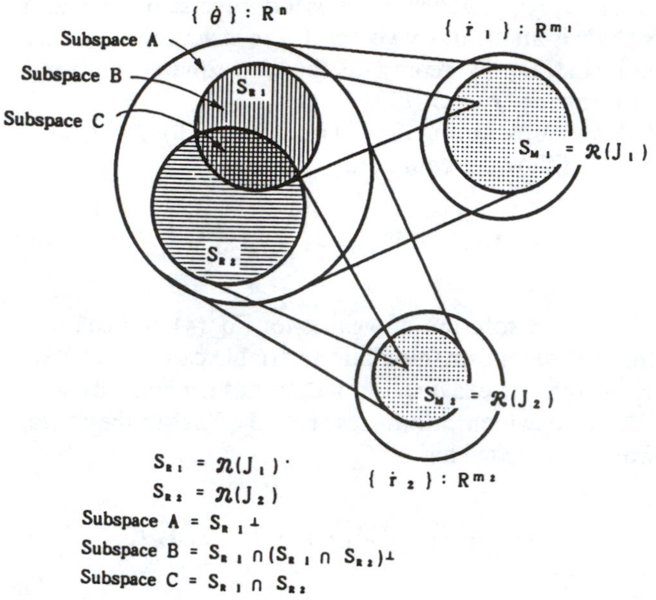

Fig. 1. Manipulable spaces and redundant spaces for the first and the second manipulation variables.

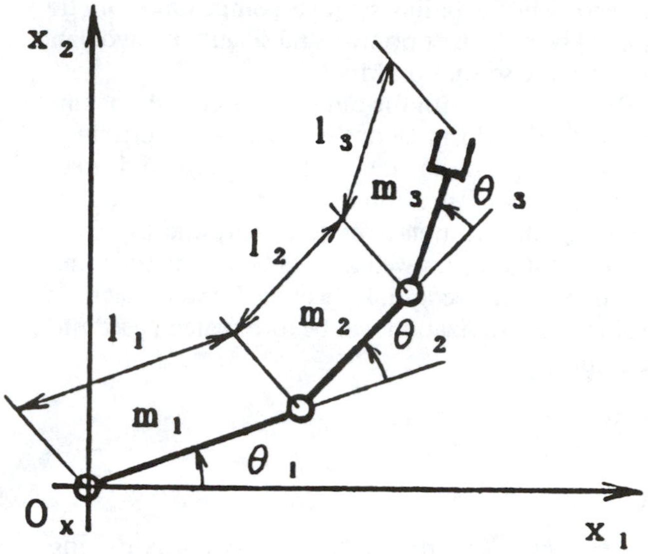

Fig. 2. 3 d.o.f. manipulator.

without disturbing the first manipulation variable. Subspace C means the remaining degrees of freedom and can be used for performing the third manipulation variable, if necessary. The first term in the right-hand side of Eq. (6) is the mapping of $\dot{\mathbf{r}}_1$ onto Subspace A. The second term means the mapping onto Subspace B of $\dot{\mathbf{r}}_2 - \mathbf{J}_2\mathbf{J}_1{}^{\#}\dot{\mathbf{r}}_1$, which is the desired value of the second manipulation variable modified considering the effect of the first term on the second manipulation variable. The third term is the orthogonal projection of the arbitrary vector **z** onto Subspace C. If there is a third manipulation variable, an arbitrary vector **z** is determined in the same way as **y**.

In the case of $\mathbf{r}_2 = \boldsymbol{\theta}$, Eq. (6) can be reduced to a simpler form using $\mathbf{J}_2 = \mathbf{I}$ as follows:

$$\dot{\boldsymbol{\theta}} = \mathbf{J}_1{}^{\#}\dot{\mathbf{r}}_1 + (\mathbf{I} - \mathbf{J}_1{}^{\#}\mathbf{J}_1)\dot{\mathbf{r}}_2, \tag{7}$$

where the relationships of $(\mathbf{I} - \mathbf{M}^{\#}\mathbf{M})^{\#} = \mathbf{I} - \mathbf{M}^{\#}\mathbf{M}$ and $(\mathbf{I} - \mathbf{M}^{\#}\mathbf{M}) \cdot (\mathbf{I} - \mathbf{M}^{\#}\mathbf{M}) = \mathbf{I} - \mathbf{M}^{\#}\mathbf{M}$ are used. In Eq. (7), the term corresponding to the third term in Eq. (6) is equal to zero, which means that no degree of freedom remains for the third manipulation variable because the second manipulation variable $\mathbf{r}_2 = \boldsymbol{\theta}$ requires all of the remaining degrees of freedom.

Table 1. Parameters of 3 d.o.f. Manipulator

i	l_i (cm)	m_i (kg)
1	50.0	30.0
2	43.3	25.0
3	35.0	20.0

4. Numerical Simulations

4.1. A Case in Which the Position of the End Effector is Prior to the Orientation

Here we perform a numerical simulation where the position of the end effector is prior to the orientation. Suppose that a robot manipulator has three degrees of freedom, as shown in Fig. 2. (The parameters of the manipulator are summarized in Table 1.) The mass of each link is distributed uniformly, and the manipulator is constrained within a horizontal plane. Therefore, the dynamics of the manipulator is represented as follows:

$$\mathbf{T} = \mathbf{A}(\boldsymbol{\theta})\ddot{\boldsymbol{\theta}} + \mathbf{B}(\boldsymbol{\theta}, \dot{\boldsymbol{\theta}}), \tag{8}$$

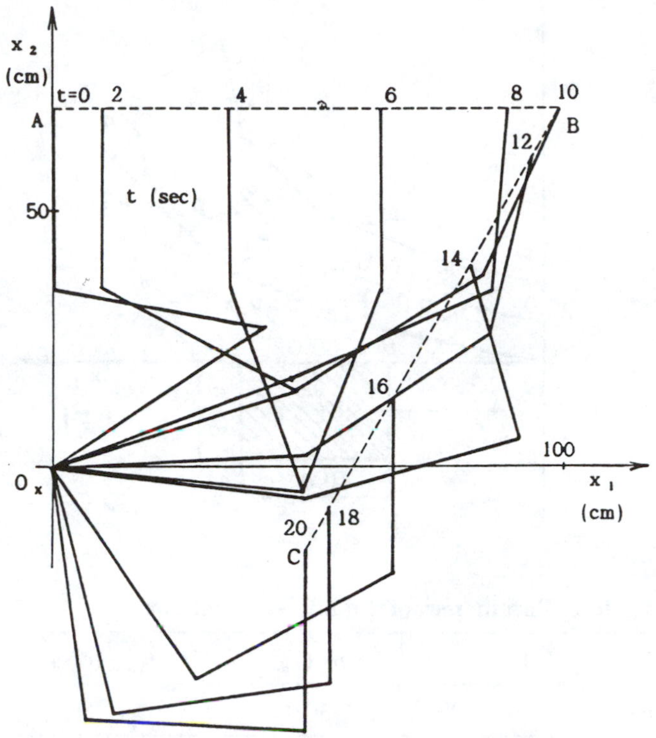

Fig. 3. Simulation results where the position of the end effector is prior to the orientation.

where $\mathbf{T}$, $\boldsymbol{\theta}$, $\dot{\boldsymbol{\theta}}$ and $\ddot{\boldsymbol{\theta}} \in R^3$, $\mathbf{A}(\boldsymbol{\theta}) \in R^{3\times3}$ is an inertia matrix, and $\mathbf{B}(\boldsymbol{\theta}, \dot{\boldsymbol{\theta}}) \in R^3$ implies the effect of Coriolis and centrifugal forces.

The first and second manipulation variables are described as follows:

$$\mathbf{r}_1 = (x_1 \quad x_2)^T, \tag{9}$$

$$r_2 = \cos(\theta_1 + \theta_2 + \theta_3), \tag{10}$$

where x_1 and x_2 are the position of the end effector in Cartesian coordinates and r_2 is the orientation of the end effector.

By differentiating Eq. (2) with respect to time, the following equation is derived:

$$\ddot{\mathbf{r}}_i = \mathbf{J}_i(\boldsymbol{\theta})\ddot{\boldsymbol{\theta}} + \dot{\mathbf{J}}_i(\boldsymbol{\theta})\dot{\boldsymbol{\theta}}. \tag{11}$$

A feedback control scheme is chosen so that the following equation may represent the closed-loop characteristics:

$$(\ddot{\mathbf{r}}_i^o(t) - \ddot{\mathbf{r}}_i) + G_{1i}(\dot{\mathbf{r}}_i^o(t) - \dot{\mathbf{r}}_i) + G_{2i}(\mathbf{r}_i^o(t) - \mathbf{r}_i) = 0, \tag{12}$$

where $\mathbf{r}_i^o(t)$ is the desired trajectory of the ith manipulation variable and G_{1i} and G_{2i} are the scalar feedback coefficients for the ith manipulation variable. If G_{1i} and G_{2i} are so chosen that Eq. (12) is stable, $\mathbf{r}_i(t)$ will converge to $\mathbf{r}_i^o(t)$. From Eqs. (11) and (12), the joint acceleration necessary to realize the feedback control scheme should satisfy the following equation:

$$\mathbf{J}_i(\boldsymbol{\theta})\ddot{\boldsymbol{\theta}} = -\dot{\mathbf{J}}_i(\boldsymbol{\theta})\dot{\boldsymbol{\theta}} + \ddot{\mathbf{r}}_i^o(t) + G_{1i}(\dot{\mathbf{r}}_i^o(t) - \dot{\mathbf{r}}_i) + G_{2i}(\mathbf{r}_i^o(t) - \mathbf{r}_i) \triangleq \mathbf{h}_i(\boldsymbol{\theta}, \dot{\boldsymbol{\theta}}, t). \tag{13}$$

Since Eq. (13) is similar to Eq. (2), the approach discussed in Section 3 can be applied here. To simplify the computation, Eq. (7) was applied in place of Eq. (6) by solving Eq. (13) for $i = 2$ and regarding the solution $\ddot{\boldsymbol{\theta}} = \mathbf{J}_2^{\#}\mathbf{h}_2$ as the desired acceleration of the second manipulation variable. That is,

$$\ddot{\boldsymbol{\theta}} = \mathbf{J}_1^{\#}\mathbf{h}_1 + (\mathbf{I} - \mathbf{J}_1^{\#}\mathbf{J}_1)\mathbf{J}_2^{\#}\mathbf{h}_2. \tag{14}$$

The joint torque to be applied was calculated by substituting Eq. (14) into Eq. (8).

Figure 3 shows the results. The desired trajectory $\mathbf{r}_1^o(t)$ traces the lines connecting points A, B, and C. The translational motion along the trajectory is described by the third-order polynomials with respect to time. The desired trajectory $\mathbf{r}_2^o(t)$ is equal to zero. It is impossible to realize both $\mathbf{r}_1^o(t)$ and $\mathbf{r}_2^o(t)$ for $t = 8$–12 s because the second joint is stretched out and the degrees of freedom degenerate. Since pseudoinverse solutions often cause unstable motions near singularities (Nakamura and Hanafusa 1986), the robot manipulator will stop or oscillate before $t = 8$ s if we apply the inverse or pseudoinverse without considering the order of priority. Figure 3 shows that $\mathbf{r}_1^o(t)$ is always satisfied and only $\mathbf{r}_2^o(t)$ is disturbed if both of them are not realizable, which means that the inverse kinematics considering the order of priority is effective for these problems.

$\mathbf{r}_2(t)$ is disturbed at $t = 14$ s because we applied Eq. (7) for the sake of the computational reduction, and it would be reduced if Eq. (6) were applied. In the simulation, the feedback coefficients are chosen as $G_{11} = 0$ 1/s, $G_{21} = 0$ 1/s^2, $G_{12} = 20$ 1/s and $G_{22} = 100$ 1/s^2.

Fig. 4. 4 d.o.f. manipulator and an obstacle.

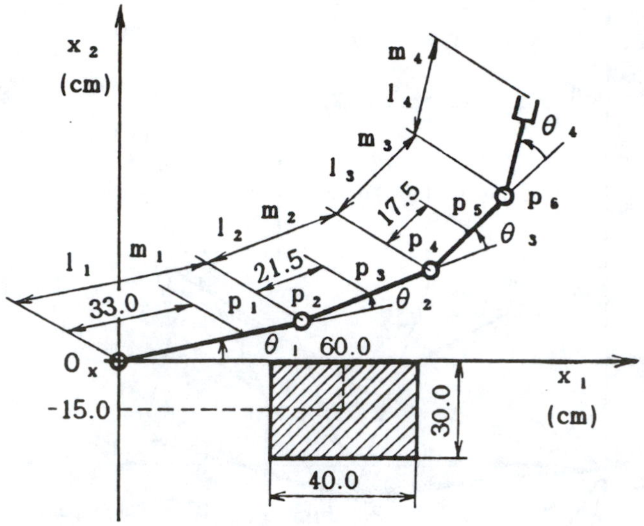

Table 2. Parameters of 4 d.o.f. Manipulator

i	l_i (cm)	m_i (kg)	$\theta_{i\max}$ (degree)
1	50.0	30.0	180
2	43.0	25.0	120
3	35.0	20.0	120
4	22.0	10.0	150

4.2. Obstacle Avoidance Using Potential Functions

Now, inverse kinematics based on the order of priority is applied to an obstacle-avoidance problem. Consider a situation where there is a robot manipulator with four degrees of freedom and an obstacle (as shown in Fig. 4). The parameters of the manipulator are summarized in Table 2, where $\theta_{i\max}$ is the limit of the *i*th joint angle, namely, $|\theta_i| \leqq \theta_{i\max}$. The mass of each link is supposed to be uniformly distributed. The manipulator is constrained within a horizontal plane. Therefore, the dynamics is represented by

$$\mathbf{T} = \mathbf{A}(\boldsymbol{\theta})\ddot{\boldsymbol{\theta}} + \mathbf{B}(\boldsymbol{\theta}, \dot{\boldsymbol{\theta}}), \tag{15}$$

where $\mathbf{T}, \boldsymbol{\theta}, \dot{\boldsymbol{\theta}}, \ddot{\boldsymbol{\theta}} \in R^4$, $\mathbf{A}(\boldsymbol{\theta}) \in R^{4\times4}$ and $\mathbf{B}(\boldsymbol{\theta}, \dot{\boldsymbol{\theta}}) \in R^4$.

Fig. 5. Simulation results for an obstacle avoidance problem. A. Without obstacle. B. With obstacle.

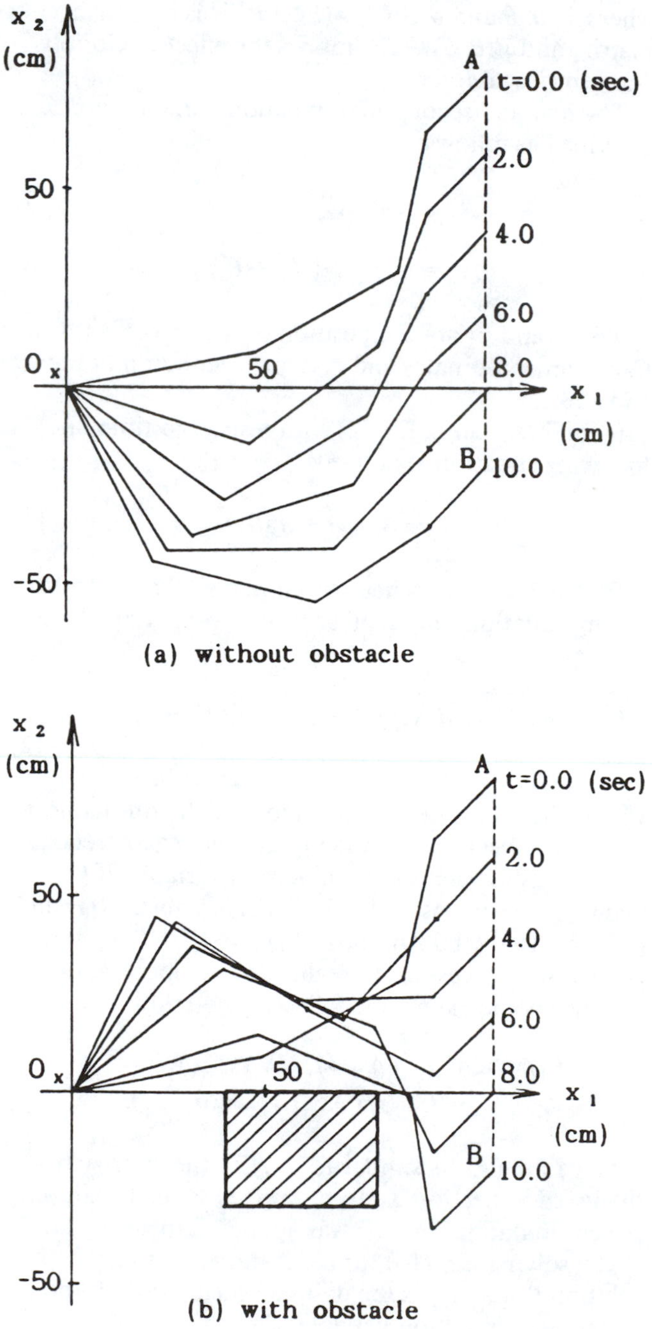

The first manipulation variable $\mathbf{r}_1 \in R^3$ is described as follows:

$$\mathbf{r}_1 = (x_1 \quad x_2 \quad \cos(\theta_1 + \theta_2 + \theta_3 + \theta_4))^T. \tag{16}$$

Khatib and Le Maitre (1978) proposed to use artificial potential and dissipative functions in determining the joint torque for obstacle avoidance. According to this method, the joint torque is calculated as follows:

$$\mathbf{T} = -(\mathrm{d}P/\mathrm{d}\boldsymbol{\theta} + \mathrm{d}D/\mathrm{d}\dot{\boldsymbol{\theta}})^T, \tag{17}$$

where P and D are artificial potential and dissipative functions, respectively. If the joint torque of Eq. (17) is applied, the following joint acceleration is generated from Eq. (15):

$$\ddot{\boldsymbol{\theta}} = -\mathbf{A}^{-1}(\boldsymbol{\theta})\{(\mathrm{d}P/\mathrm{d}\boldsymbol{\theta} + \mathrm{d}D/\mathrm{d}\dot{\boldsymbol{\theta}})^T + \mathbf{B}(\boldsymbol{\theta}, \dot{\boldsymbol{\theta}})\}. \tag{18}$$

Now, let us consider the joint acceleration of Eq. (18) as the acceleration of the second manipulation variable. Then, from Eqs. (7), (13), and (18), the joint acceleration considering the first and the second manipulation variables is calculated as follows:

$$\ddot{\boldsymbol{\theta}} = \mathbf{J}_1^{\#}\mathbf{h}_1 - (\mathbf{I} - \mathbf{J}_1^{\#}\mathbf{J}_1)\mathbf{A}^{-1}\{(\mathrm{d}P/\mathrm{d}\boldsymbol{\theta} + \mathrm{d}D/\mathrm{d}\dot{\boldsymbol{\theta}})^T + \mathbf{B}\}. \tag{19}$$

The joint torque to be applied is calculated by substituting $\boldsymbol{\theta}$, $\dot{\boldsymbol{\theta}}$, and Eq. (19) into Eq. (15).

The artificial potential and dissipative functions are defined by

$$P \triangleq P_O + P_J, \tag{20}$$

$$P_O \triangleq k_O \sum_{i=1}^{6} 1/\{C_O(\mathbf{p}_i) - 1\}, \tag{21}$$

$$P_J \triangleq k_J \sum_{i=1}^{4} 1/(\theta_{i\,\max}^2 - \theta_i^2), \tag{22}$$

$$D \triangleq k_D \sum_{i=1}^{4} \frac{1}{2}\dot{\theta}_i^2, \tag{23}$$

where $\mathbf{p}_i \triangleq (x_{1i} \quad x_{2i})^T$ (cm) are the positions of points on the manipulator (as shown in Fig. 4) and are used in order to evaluate the distance between the manipulator and the obstacle. $C_O(\mathbf{p}_i)$ is defined by the following equation (Khatib and Le Maitre 1978):

$$C_O(\mathbf{p}_i) \triangleq \left(\frac{x_{1i} - 60}{20}\right)^8 + \left(\frac{x_{2i} + 15}{15}\right)^8, \tag{24}$$

Fig. 6. Distribution of degrees of freedom of UJIBOT, a robot manipulator with 7 d.o.f.

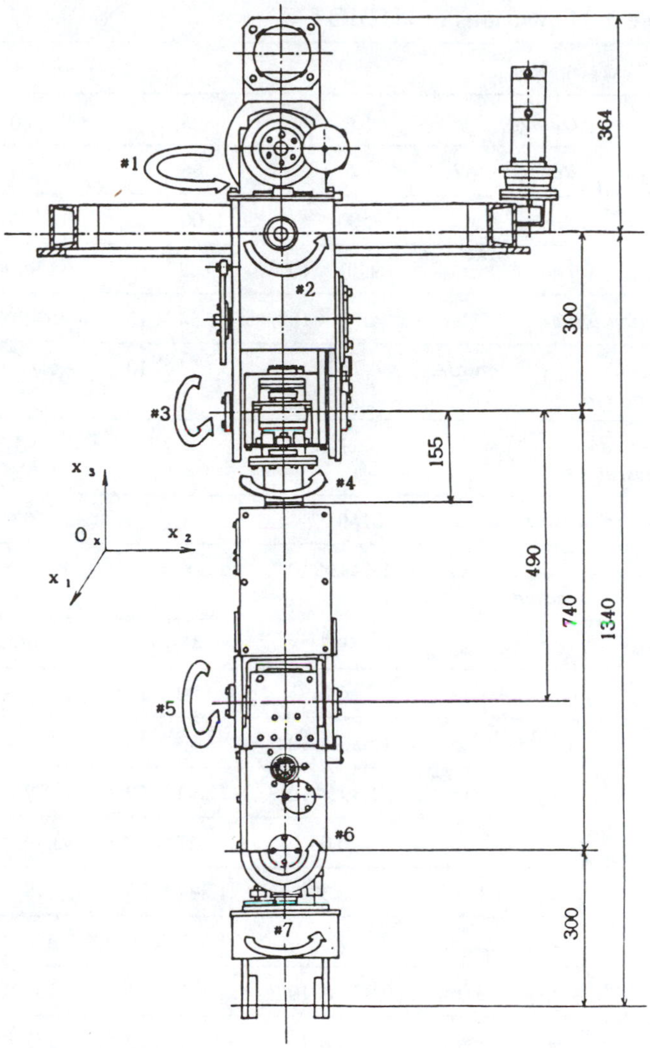

where $C_O(\mathbf{p}_i) = 1$ approximates the contour of the obstacle.

Figure 5 illustrates the results of the numerical simulation. In Fig. 5A, the obstacle does not exist. Figure 5A was computed by neglecting the second term in the right-hand side of Eq. (19). In Fig. 5B, the obstacle is present. The desired trajectory of the first manipulation variable $\mathbf{r}_1(t)$ is the constant velocity motion along the line connecting points A and B. Figure 5B clearly shows that inverse kinematics based on the order of priority is applicable and effective for obstacle avoidance problems. k_O, k_J, and k_D in Eqs. (20)–(23) were

Table 3. Dimensions of UJIBOT

		i		1	2	3	4	5	6	7
Motor	Rated	*Output*	W	35	95	80	64	64	35	10
		Voltage	V	24	55	31.3	30.8	30.8	24	17
		R.p.m.		3500	1100	4000	4000	4000	3500	3200
		Torque	nm	9.80×10^{-2}	9.02×10^{-1}	1.96×10^{-1}	1.57×10^{-1}	1.57×10^{-1}	9.80×10^{-2}	2.94×10^{-2}
	Inductance		vsec	4.74×10^{-2}	4.54×10^{-1}	6.49×10^{-2}	6.44×10^{-2}	6.44×10^{-2}	4.74×10^{-2}	3.72×10^{-2}
	Torque constant		nm/A	4.74×10^{-2}	4.51×10^{-1}	6.47×10^{-2}	6.43×10^{-2}	6.43×10^{-2}	4.70×10^{-2}	3.72×10^{-2}
	Armature resistance		Ω	1.68	4.7	1.3	1.7	1.7	1.68	4.0
Reduction ratio				3160	2321.4	3703.7	2222.2	2133.3	853.3	110
Inertia of motor and transmission			kgm^2	1.444×10^{-4}	1.800×10^{-1}	1.627×10^{-4}	1.098×10^{-4}	1.018×10^{-4}	1.607×10^{-4}	2.040×10^{-5}
Mass of link			kg	49.75	43.16	10.79	14.16	13.18	2.29	0.60
Length of link			m	0.0	3.00×10^{-1}	2.781×10^{-1}	2.119×10^{-1}	2.50×10^{-1}	6.853×10^{-2}	2.315×10^{-1}
Gravity center $\mathbf{l}_i$		l_{i1}	m	-1.856×10^{-2}	3.33×10^{-3}	7.75×10^{-3}	-1.80×10^{-3}	9.82×10^{-3}	0.0	0.0
		l_{i2}	m	-3.39×10^{-2}	-2.62×10^{-3}	-1.977×10^{-2}	1.432×10^{-1}	-1.84×10^{-3}	-1.189×10^{-2}	3.20×10^{-6}
		l_{i3}	m	0.0	-1.427×10^{-2}	3.90×10^{-3}	0.0	8.97×10^{-2}	1.580×10^{-2}	0.0
Inertia matrix $\mathbf{I}_i$		I_{i11}	kgm^2	8.374	1.898×10^{-1}	3.119×10^{-2}	2.01×10^{-2}	3.861×10^{-2}	1.03×10^{-3}	1.368×10^{-3}
		I_{i22}	kgm^2	7.873	1.877×10^{-1}	2.456×10^{-2}	4.089×10^{-2}	1.465×10^{-2}	2.32×10^{-3}	1.377×10^{-3}
		I_{i33}	kgm^2	9.022×10^{-1}	1.6167	4.303×10^{-2}	2.466×10^{-1}	1.118×10^{-1}	4.40×10^{-3}	1.136×10^{-3}
		I_{i12}	kgm^2	-1.036×10^{-1}	-3.71×10^{-4}	2.952×10^{-3}	-5.23×10^{-4}	8.79×10^{-4}	0.0	0.0
		I_{i23}	kgm^2	-9.79×10^{-2}	-5.284×10^{-3}	5.540×10^{-3}	1.437×10^{-2}	5.476×10^{-3}	3.84×10^{-4}	2.15×10^{-5}
		I_{i31}	kgm^2	2.007×10^{-1}	-1.622×10^{-2}	4.981×10^{-3}	-4.591×10^{-3}	-5.696×10^{-3}	0.0	0.0

chosen as $k_O = 5.0 \times 10^5$ $kgcm^2/s^2$, $k_J = 6.0 \times 10^4$ $kgcm^2/s^2$ and $k_D = 1.0 \times 10^4$ $kgcm^2/s^2$. The feedback coefficients for the first manipulation variable were chosen as $G_{11} = G_{21} = 0$. Concerning the synthesis of the artificial potential field, Khatib (1985) showed that the residual perturbation, which is caused by the potential field to the lower priority task, can be avoided by selecting a contour or a distance behind which the potential field is zero.

5. Experiments

5.1. UJIBOT, a Robot Manipulator with 7 Degrees of Freedom

Figure 6 shows the distribution of degrees of freedom of UJIBOT, a robot manipulator with 7 degrees of

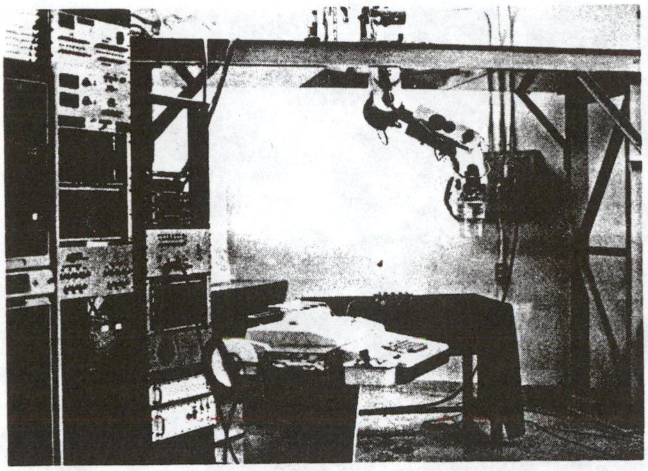

Fig. 7. Overall view of the experimental setup.

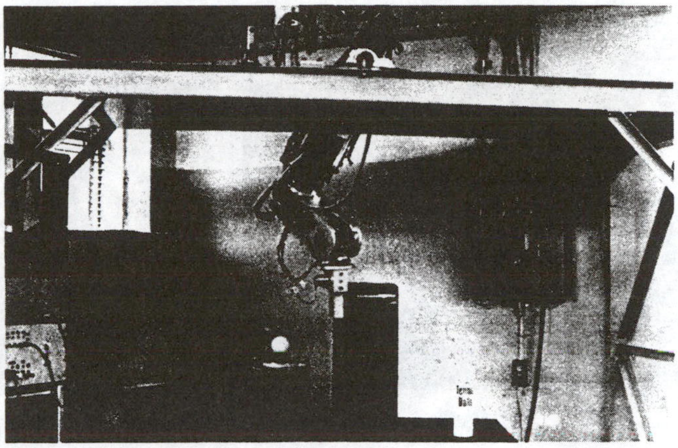

Fig. 8. Reference arm posture $\boldsymbol{\theta}_r$.

Table 4. Identified Dynamic Characteristics of UJIBOT

i	P_{Ui} ($Vsec^2$)	Q_{Ui} ($Vsec$)	R_{Ui} (V)	S_{Ui} (V/Nm)
1	3.095	7.034×10^{1}	6.010×10^{-1}	0
2	6.800	1.812×10^{2}	3.446×10^{-1}	7.306×10^{-4}
3	2.326	7.060×10^{1}	2.881×10^{-1}	3.386×10^{-3}
4	1.107	3.905×10^{1}	2.789×10^{-1}	6.264×10^{-3}
5	1.058	5.506×10^{1}	3.293×10^{-1}	1.007×10^{-2}
6	4.548	1.962×10^{1}	5.309×10^{-1}	1.188×10^{-1}
7	3.568×10^{-1}	2.843	5.065×10^{-1}	0

freedom driven by D.C. servo motors. The dimensions of UJIBOT are summarized in Table 3. A gravity center vector $\mathbf{l}_i$, a vector from the gravity center of the ith link to the ith joint, and an inertia matrix $\mathbf{I}_i$ are described based on the orthogonal coordinates fixed at the gravity center of the ith link, of which three axes are parallel with the corresponding axes of the base Cartesian coordinates when UJIBOT looks as shown in Fig. 6, which means $\boldsymbol{\theta} = 0$.

Since the reduction ratios are large (as shown in Table 3), the inertial, centrifugal, and Coriolis forces of links are negligible in the dynamic characteristics of UJIBOT. Therefore, the approximate dynamics of UJIBOT is represented by

$$\mathbf{V} = \mathbf{P}_U\ddot{\boldsymbol{\theta}} + \mathbf{Q}_U\dot{\boldsymbol{\theta}} + \mathbf{R}_U \text{ sign } \dot{\boldsymbol{\theta}} + \mathbf{S}_U\mathbf{C}(\boldsymbol{\theta}), \tag{25}$$

where $\mathbf{V} \in R^7$ is the input voltage vector of motors, $\boldsymbol{\theta} \in R^7$ is the joint variable vector, $\mathbf{C}(\boldsymbol{\theta}) \in R^7$ is the gravity torque vector, sign $\dot{\boldsymbol{\theta}} \triangleq$ col. (sign $\dot{\theta}_i$) and $\mathbf{P}_U \triangleq$ diag. (P_{Ui}), $\mathbf{Q}_U \triangleq$ diag. (Q_{Ui}), $\mathbf{R}_U \triangleq$ diag. (R_{Ui}), $\mathbf{S}_U \triangleq$ diag. (S_{Ui}) $\in R^{7\times 7}$ are the constant matrices whose values are shown in Table 4. Figure 7 shows the overall view of the experimental setup.

5.2. Task Description and Control Scheme

We describe the task as follows: to reach for a tennis ball on a box, grasp it, return it to its initial position, transport the ball along the x_2-axis at the constant rate of 0.1 m/s for 6.58 s, stop and release the ball into a can below, while avoiding another box that can prevent the motion.

The first and the second manipulation variables are chosen as follows:

$$\mathbf{r}_1 = (x_1 \quad x_2 \quad x_3)^T, \tag{26}$$

$$\mathbf{r}_2 = \boldsymbol{\theta}, \tag{27}$$

where x_1, x_2 and x_3 are the positions of the end effector. Since the acceleration term is much less than the velocity term in Eq. (25) (see Table 4), velocity control is adopted as the control scheme. The velocity command for the manipulation variable is calculated by

Fig. 9. Motion of UJIBOT without provisions for the obstacle ($G_1 = 3.0$ 1/s, $G_2 = 0.0$ 1/s).

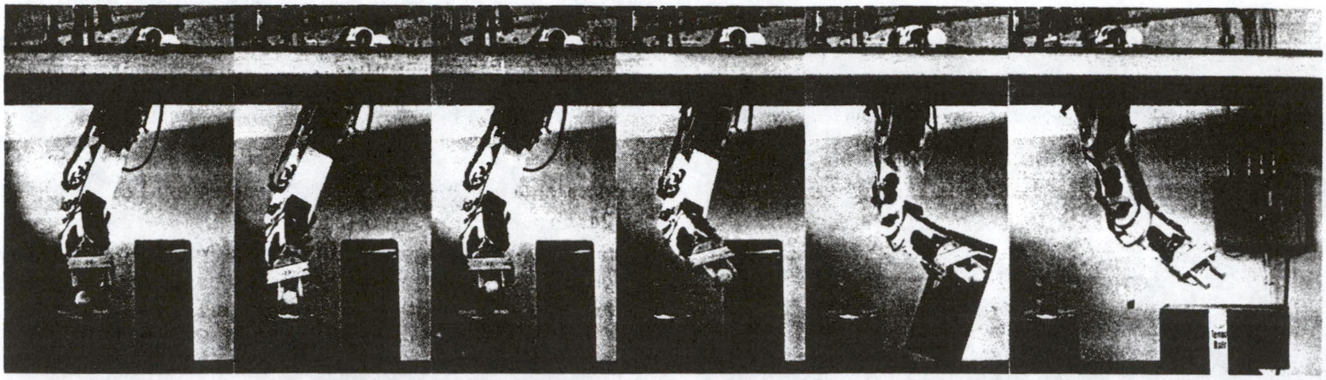

Fig. 10. Motion of UJIBOT with provisions for the obstacle ($G_1 = 3.0$ 1/s, $G_2 = 0.3$ 1/s).

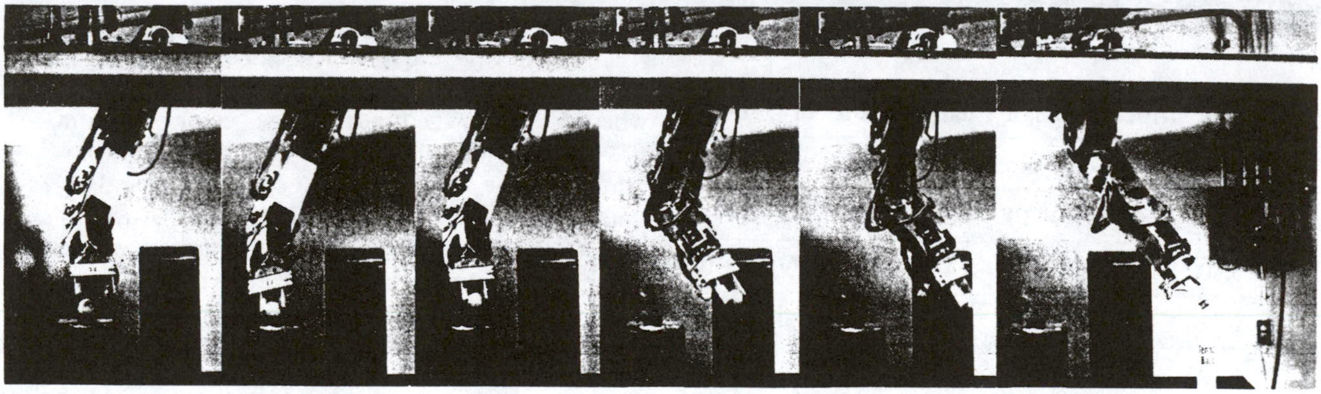

$$\dot{\mathbf{r}}_i^* = \dot{\mathbf{r}}_i^o(t) + G_i\{\mathbf{r}_i^o(t) - \mathbf{r}_i(t)\}, \quad (28)$$

where $\dot{\mathbf{r}}_i^*$ is the velocity command for the ith manipulation variable, $\mathbf{r}_i^o(t)$ is the desired trajectory, and G_i is the scalar feedback coefficient. The values of G_i are determined experimentally. The joint velocity command $\dot{\theta}^*$ is calculated according to Eq. (7) as follows:

$$\dot{\theta}^* = \mathbf{J}_1^{\#}\dot{\mathbf{r}}_1^* + (\mathbf{I} - \mathbf{J}_1^{\#}\mathbf{J}_1)\dot{\mathbf{r}}_2^*. \quad (29)$$

The input voltage is determined according to Eq. (25) as follows:

$$\mathbf{V} = \mathbf{Q}_U\dot{\theta}^* + \mathbf{R}_U \operatorname{sign} \dot{\theta}^* + \mathbf{S}_U\mathbf{C}(\theta), \quad (30)$$

where the acceleration term is neglected.

The desired trajectory is represented by the following equations:

$$\mathbf{r}_1^o(t) = \mathbf{r}_1(t_o) + (t - t_o)(0.0 \quad 0.1 \quad 0.0)^T \text{ (m)}, \quad (31)$$

$$\mathbf{r}_2^o(t) = \boldsymbol{\theta}_r, \quad (32)$$

where $\boldsymbol{\theta}_r \in R^7$ is a constant joint angle, called the *reference arm posture,* which was chosen intuitively so that the elbow of UJIBOT may stay away from the obstacle. Operator intervention of this kind would be a simple but effective method to utilize the global judgment of human operators in obstacle avoidance problems. In the case of complicated obstacles, it would be desirable to choose a time-functional, reference-arm posture. Figure 8 shows the reference-arm posture $\boldsymbol{\theta}_r$.

5.3. Experimental Results and Discussion

Figure 9 shows the motion of UJIBOT when the second term of the right-hand side of Eq. (29) was ne-

Fig. 11. Motion of UJIBOT, projected onto the x_3-x_1 plane. A. Without provisions for the obstacle ($G_1 = 3.0$ 1/s, $G_2 = 0.0$ 1/s). B. With provisions for the obstacle ($G_1 = 3.0$ 1/s, $G_2 = 0.3$ 1/s).

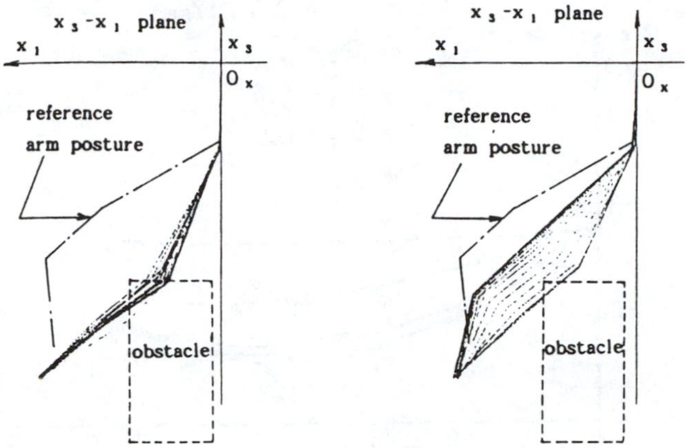

Fig. 12. Errors of the first manipulation variable, $e_1 = \|r_1^o(t) - r_1(t)\|$. A. G_1 changes, $G_2 = 0.3$ 1/s. B. $G_1 = 3.0$ 1/s, G_2 changes.

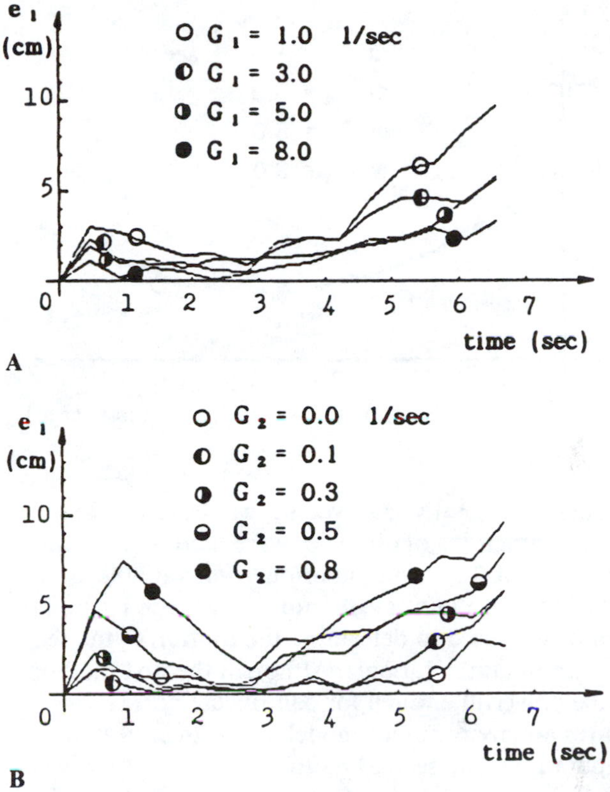

glected: *resolved motion rate control* according to Whitney (1969). UJIBOT came into collision with the obstacle.

Figure 10 shows the motion of UJIBOT when the second term of Eq. (29) was considered. UJIBOT successfully avoided the collision with the obstacle by utilizing redundancy. Figure 11 shows the motions projected onto the x_3-x_1 plane every 0.47 s. The chained lines represent the reference-arm posture. The broken lines mean the obstacle. It was found that the elbow of UJIBOT was lifted by the reference-arm posture and avoided the obstacle while the end effector was tracing the given trajectory. These experimental results proved that inverse kinematics based on the order of priority can be implemented in actual control systems for redundant robot manipulators and can be effective in utilizing redundancy. It took 47 ms of sampling time for real-time computation of Eqs. (28)–(30) using a minicomputer NOVA 03 with a floating-point processing unit.

Experiments were also performed to investigate the relationship between the feedback coefficients and the control performance. Figures 12 and 13 show the errors of the first and the second manipulation variables: that is, $e_1 \triangleq \|r_1^o(t) - r_1(t)\|$ and $e_2 \triangleq \|r_2^o - r_2(t)\|$, when G_1 and G_2 changed. From Figs. 12 and 13, the following facts were clarified concerning the feedback coefficients G_1 and G_2:

1. To increase G_1 as long as it guarantees stability means to reduce e_1 and hardly disturbs e_2.
2. Having G_2 greater than a certain value is fruitless in improving e_2.
3. Increasing G_2 disturbs e_1 to some extent, but this could be overcome by considering a more precise dynamical model.

6. Conclusion

The concept of task priority was introduced into the inverse kinematic problem of redundant manipulators, and an inverse kinematic solution was derived taking into account the order of priority, which can be regarded as an instantaneously optimal solution suitable for real-time redundancy control.

Numerical simulations and experiments were performed in order to verify the effectiveness and the implementation of the solution for redundancy control problems.

We confirmed that dividing a task into subtasks with the order of priority is a remedy for overcoming the degeneracy of degrees of freedom. The obstacle-

Fig. 13. Errors of the second manipulation variable, $e_2 = \|r_2^o - r_2(t)\|$. *A.* G_1 *changes,* $G_2 = 0.3$ *1/s. B.* $G_1 = 3.0$ *1/s,* G_2 *changes.*

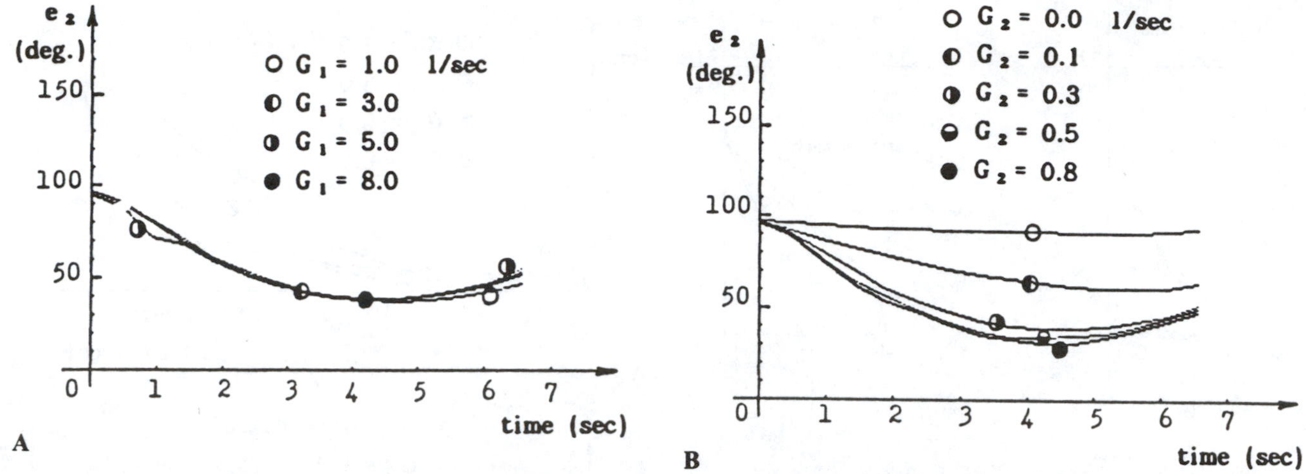

avoidance problem was solved in two ways: (1) the potential functions of obstacles were used to determine the motion of the second manipulation variable, and (2) the reference joint angle intuitively given by an operator was used to determine the motion of the second manipulation variable. Although the first method may be powerful enough for complicated obstacles, it requires an environment model and a great deal of computation. The second method is easier, as it gives the reference information for avoiding obstacles and decreases the computational amount, although it requires the intervention of an operator.

References

Benati, M., Morasso, P., and Tagliasco, V. 1982. The inverse kinematic problem for anthromorphic manipulator arms. *J. Dyn. Sys., Meas., Contr.* 104:110–113.

Boullion, T. L., and Odell, P. L. 1971. *Generalized inverse matrices.* New York: Wiley-Interscience.

Freund, E. 1977 (Tokyo). Path control for a redundant type of industrial robot. *Proc. 7th Int. Symp. Industr. Robots,* pp. 107–114.

Hanafusa, H., Yoshikawa, T., and Nakamura, Y. 1978. Control of articulated robot arms with redundancy. *Preprints 21st Joint Automatic Contr. Conf. in Japan,* pp. 237–238. (in Japanese)

Hanafusa, H., Yoshikawa, T., and Nakamura, Y. 1981 (Kyoto). Analysis and control of articulated robot arms with redundancy. *Preprints 8th Triennial IFAC World Congress* 14:78–83.

Hanafusa, H., Yoshikawa, T., and Nakamura, Y. 1983. Redundancy analysis of articulated robot arms and its utilization for tasks with priority. *Trans. Society of Instruments and Contr. Engineers* 19:421–426. (in Japanese)

Khatib, O. 1985 (St. Louis). Real-time obstacle avoidance for manipulators and mobile robots. *Proc. 1985 Int. Conf. Robotics and Automation,* pp. 500–505.

Khatib, O., and Le Maitre, J.-F. 1978 (Udine, Italy). Dynamic control of manipulators operating in a complex environment. *Proc. 3rd Int. CISM-IFToMM Symp.,* pp. 267–282.

Klein, C. A. 1984 (Kyoto). Use of redundancy in the design of robotic systems. *Preprints 2nd Int. Symp. Robotics Res.,* pp. 58–66.

Klein, C. A., and Huang, C. H. 1983. Review of pseudoinverse control for use with kinematically redundant manipulators. *IEEE Trans. Sys., Man, Cyber.* SMC-13:245–250.

Konstantinov, M. S., Markov, M. D., and Nenchev, D. N. 1981 (Tokyo). Kinematic control of redundant manipulators. *Proc. 11th Int. Symp. Industr. Robots,* pp. 561–568.

Liegeois, A. 1977. Automatic supervisory control of the configuration and behavior of multibody mechanisms. *IEEE Trans. Sys., Man, Cyber.* SMC-7:868–871.

Luh, J. Y. S., Walker, M. W., and Paul, R. P. C. 1980. Resolved acceleration control of mechanical manipulators. *IEEE Trans. Automatic Contr.* 25:468–474.

Maciejewski, A. A., and Klein, C. A. 1985. Obstacle avoidance for kinematically redundant manipulators in dynamically varying environments. *Int. J. Robotics Res.* 4(3):109–117.

Nakamura, Y., and Hanafusa, H. 1986. Inverse kinematic solutions with singularity robustness for robot manipulator control. *J. Dyn. Sys., Meas., Contr.* 108:163–171.
Nakano, E., and Ozaki, S. 1974 (Tokyo). Cooperative control of a pair of anthropomorphous manipulators—MELARM. *Proc. 4th Int. Symp. Industr. Robots,* pp. 250–260.
Uchiyama, M. 1979, Study on dynamic control of artificial arms—part 1. *Trans. Japanese Society of Mechanical Engineers* C-45:314–322. (in Japanese)
Whitney, D. E. 1969. Resolved motion rate control of manipulators and human prostheses. *IEEE Trans. Man-Machine Sys.* MMS-10:47–53.
Whitney, D. E. 1972. The mathematics of coordinated control of prostheses and manipulators. *J. Dyn. Sys., Meas., Contr.* 94:303–309.

Redundancy Resolution of Manipulators through Torque Optimization

JOHN M. HOLLERBACH, MEMBER, IEEE, AND KI C. SUH

Abstract—**Methods for resolving kinematic redundancies of manipulators by the effect on joint torque are examined. When the generalized inverse is formulated in terms of accelerations and incorporated into the dynamics, the effect of redundancy resolution on joint torque can be directly reflected. One method chooses the joint acceleration null-space vector to minimize joint torque in a least squares sense; when the least squares is weighted by allowable torque range, the joint torques tend to be kept within their limits. Contrasting methods employing only the pseudoinverse with and without weighting by the inertia matrix are presented. The results show an unexpected stability problem during long trajectories for the null-space methods and for the inertia-weighted pseudoinverse method, but more seldom for the unweighted pseudoinverse method. Evidently, a whiplash action develops over time that thrusts the endpoint off the intended path, and extremely high torques are required to overcome these natural movement dynamics.**

I. Introduction

INCREASING interest in manipulator redundancy is a direct consequence of perceived limitations of current six-degree-of-freedom (DOF) robots. On the one hand, singularity regions of six-DOF rotary manipulators occupy such a significant portion of the work space as to render them functionally only five-DOF manipulators [1]. On the other hand, work space obstacles may sufficiently constraint movement as to effectively reduce the degrees of freedom.

Most analysis of redundant arms has proceeded independent of consideration of any particular mechanism. Hollerbach [2], [3] proposed a seven-DOF kinematic design with a spherical shoulder joint to eliminate singularities and to improve work space. Yoshikawa [4] proposed a four-DOF wrist to overcome the usual problem of wrist singularity. In practice, very few redundant manipulators have been constructed. At research laboratories seven-DOF arms of anthropomorphic geometry include a tendon-driven torque-controlled robot [5] and the UJIBOT, driven by dc servo motors [6]. An eight-DOF redundant sheep-shearing robot has been discussed in [7]. Commercially, seven-DOF robots have been produced by the Robotics Research Corporation, also of anthropomorphic geometry, and by Cybotech, whose P-15 robot has a yaw axis in the forearm [8]. In addition, a number of six-DOF robots have been mounted on linear tracks, but it is difficult to count this seventh degree of freedom into a fine motion control of the endpoint.

Thus, while the incidence of redundant arms in research labs and industry is gradually increasing, present research in redundant arms has progressed primarily at a theoretical and simulation level, in advance and expectation of available mechanical hardware.

A. Kinematic Resolution of Redundancy

The vast majority of research into the control of redundant arms has involved the instantaneous resolution of the redundancy at the velocity level through use of the pseudoinverse $J^\dagger$ of the Jacobian matrix J. If $\dot{x}$ is the six-dimensional velocity vector of the hand and θ is the $n > 6$-dimensional vector of joint angles, then

$$\dot{x} = J\dot{\theta} \tag{1}$$

$$\dot{\theta} = J^\dagger \dot{x} + (I - J^\dagger J)\dot{\phi} \tag{2}$$

$$J^\dagger = J^T(JJ^T)^{-1} \tag{3}$$

where $\dot{\phi}$ is an arbitrary joint velocity vector and $(I - J^\dagger J)\dot{\phi}$ is its projection into the null space of J, corresponding to a self-motion of the linkage that does not move the end effector. The pseudoinverse is also known as the Moore–Penrose generalized inverse. The attractiveness of this approach is twofold. First, the pseudoinverse is one of the types of generalized inverse that has a least squares property [9], in the present minimizing $\dot{\theta}^T\dot{\theta}$. Presumably, any joint is prevented from moving too fast, leading to a more controllable motion [10]. It is also presumed that squared velocities are approximately related to kinetic energy, which would then also be approximately minimized [10], [11].

Second, the redundancy available beyond that required for the tip motion is succinctly characterized by the null space of the Jacobian, which may be freely utilized to assist in the realization of some chosen objective. Liegeois [12] developed a general formation that tends to minimize a position-dependent scalar performance criterion $p = g(\theta)$. The null-space vector $\dot{\phi} = k\partial g/\partial\theta$, where k is an arbitrary constant, projects the gradient of p onto the joint motion in such a way as to reduce p through subsequent motion:

$$\dot{\theta} = J^\dagger \dot{x} + (I - J^\dagger J)k\frac{\partial g}{\partial \theta}. \tag{4}$$

Manuscript received February 21, 1986; revised September 2, 1986. This work was supported in part by the AFWAL/XRPM Defense Small Business Innovation Research Program under Grant F331615-83-C-5115, awarded to Scientific Systems Inc., and in part by the Defense Advanced Research Projects Agency under Office of Naval Research Contracts N00014-77-C-0389 and N00014-80-C-0505 and by the System Development Foundation, awarded to the MIT Artificial Intelligence Laboratory.

J. M. Hollerbach is with the Artificial Intelligence Laboratory, Massachusetts Institute of Technology, 545 Technology Square, Cambridge, MA 02139.

K. C. Suh was with the Massachusetts Institute of Technology, Cambridge, MA. He is now with AT&T Bell Laboratories, Lincroft, NJ, USA.

IEEE Log Number 8714014.

Reprinted from *IEEE J. Robot. Automat.*, vol. RA-3, no. 4, pp. 308–316, Aug. 1987.

Liegeois demonstrated a way of avoiding joint limits by minimizing the scalar function

$$p=\sum_{i=1}^{n}\left(\frac{\theta_i-\theta_i^{\text{mid}}}{\theta_i^{\text{mid}}-\theta_i^{\text{max}}}\right)^2 \tag{5}$$

where $\theta_i^{\max}$ and $\theta_i^{\min}$ are the upper and lower joint angle limits for joint i and $\theta_i^{\text{mid}} = (\theta_i^{\min} + \theta_i^{\max})/2$.

The null-space vector has also been used in singularity avoidance. Yoshikawa [13] proposed minimizing a dexterity measure w, called by him the manipulatability measure, given by

$$w=\sqrt{\det (JJ^T)}.$$

Since at a singularity $w = 0$, then the scalar function $p = -w(\theta)$ instantaneously maximizes w and tends to keep the arm away from singularities. Klein [14] compared various dexterity measures including w and decided instead on the minimum singular value.

At certain singularities a nonredundant manipulator can actually execute a self-motion, since, for example, a six-DOF robot has an excess freedom with regard to the five-DOF endpoint motion possible at the singularity. The null-space vector corresponding to this self-motion is useful if the endpoint variables are partitioned into high-priority variables $\dot{x}_h$, which must be realized, and low-priority variables $\dot{x}_l$, which are sacrificed to avoid the singularity but which should be realized insofar as is possible [15]. For example, in spray painting the control of rotation about the spray direction is not as important as the control of the other positioning variables and could be sacrificed at a singularity. From the relations

$$\dot{x}_h=J_h\dot{\theta} \tag{7}$$

$$\dot{x}_l=J_l\dot{\theta} \tag{8}$$

$$\dot{\theta}=J_h^{\dagger}\dot{x}_h+(I-J_h^{\dagger}J_h)\phi, \tag{9}$$

the null-space vector left over after realizing the high-priority variables is found by substituting (9) into (8):

$$\phi=[J_l(I-J_h^{\dagger}J_h)]^{\dagger}(\dot{x}_l-J_lJ_h^{\dagger}\dot{x}_h). \tag{10}$$

Substituting into (9) yields

$$\dot{\theta}=J_h^{\dagger}\dot{x}_h+[J_l(I-J_h^{\dagger}J_h)]^{\dagger}(\dot{x}_l-J_lJ_h^{\dagger}\dot{x}_h) \tag{11}$$

where use has been made of the identity $B[CB]^{\dagger} = [CB]^{\dagger}$, with $B = (I - J_h^{\dagger}J_l)$ a hermetian and idempotent matrix [16].

The null-space vector can aid obstacle avoidance [16]. Suppose x_0 is some point on the arm closest to an obstacle, J_0 is the Jacobian such that $\dot{x}_0 = J_0\dot{\theta}$, and $\dot{x}_0$ is a movement of this point away from the obstacle that would be desirable. In a procedure like that above, ϕ is found by substituting (1) into this relation:

$$\phi=[J_0(I-J^{\dagger}J)]^{\dagger}(\dot{x}_0-J_0J^{\dagger}\dot{x}). \tag{12}$$

After substitution and simplification, the joint rates that avoid the obstacle are

$$\dot{\theta}=J^{\dagger}\dot{x}+[J_0(I-J^{\dagger}J)]^{\dagger}(\dot{x}_0-J_0J^{\dagger}\dot{x}). \tag{13}$$

One problem with the Moore–Penrose inverse is that its application is nonconservative [17]. Repetitive motions planned with the pseudoinverse alone do not return at a given tip point to the same joint configuration. Baillieul [18] has proposed an extended Jacobian method, which may be used to make a repetitive motion conservative. If n_J is a vector from the one-dimensional null space of J, obtained by taking cross products of columns of J, and $g(\theta)$ is a scalar function to be minimized, then define

$$G(\theta)=\frac{\partial g(\theta)}{\partial\theta}\cdot n_J. \tag{14}$$

The extended Jacobian J_{ex} is then defined by

$$J_{\text{ex}}\dot{\theta}=\begin{bmatrix}\dot{x}\\0\end{bmatrix}\qquad J_{\text{ex}}=\begin{bmatrix}J\\ \dfrac{\partial G}{\partial\theta}\end{bmatrix}. \tag{15}$$

Assuming the manipulator is nonsingular, this relation may be inverted directly to find $\dot{\theta}$. It is possible to recast this method in terms of a generalized inverse with null-space vector.

An entirely different mechanism from the use of a null-space vector to realize desired performance characteristics is the weighted pseudoinverse. The generalized inverse $J_w^{\dagger}$ that instantaneously minimizes the cost $\dot{\theta}^TW\dot{\theta}$ is [10]

$$J_w^{\dagger}=W^{-1}J^T(JW^{-1}J^T)^{-1} \tag{16}$$

where W is a matrix of weights. Whitney [10] proposed an alternative method for enforcing high and low priority of hand variables through appropriate selection of weights. Konstantinov *et al.* [19] used weights to avoid joint limits.

Most research involving the pseudoinverse deals with the instantaneous kinematics of motion, that is to say, motion that is locally optimized by incremental movement from the current arm state. The Jacobian has also been used in dexterity measures to aid global trajectory planning and the avoidance of singularities. Uchiyama *et al.* [20] parameterized joint angles by polynomials and optimized manipulatability w integrated across the whole trajectory by gradient adjustment of polynomial coefficients. Nakamura and Hanafusa [21] used Pontryagin's maximum principle to minimize a cost function combining manipulatability w, and $\dot{\theta}^T\dot{\theta}$, interpreted by them as energy, for global determination of trajectory.

B. Kinetic Resolution of Redundancy

The previous redundancy resolution schemes are purely kinematic, despite the presumption that since the sum of squares of joint velocities is minimized by the pseudoinverse, then the kinetic energy is approximately minimized. True minimization of kinetic energy would only be realized if the generalized inverse were weighted with the inertia matrix so that the relation to kinetic energy of an unweighted pseudoinverse is not clear. The kinetic energy is

$$T=\frac{1}{2}\dot{\theta}^TH\dot{\theta} \tag{17}$$

where H is the inertia matrix. Minimizing the kinetic energy T is just Whitney's criterion with the weighting matrix $W = H$ in (16).

To incorporate the generalized inverse into dynamics, the pseudoinverse must be formulated in terms of accelerations. Khatib [22], [23] was one of the first researchers to do this, in his case using the inertia-weighted pseudoinverse:

$$\ddot{x} = J\ddot{\theta} + \dot{J}\dot{\theta} \tag{18}$$

$$\ddot{\theta} = J_H^{\dagger}(\ddot{x} - \dot{J}\dot{\theta}). \tag{19}$$

The dynamic equations in closed form are [24]

$$\tau = H\ddot{\theta} + \dot{\theta} \cdot C \cdot \dot{\theta} + g \tag{20}$$

where previously undefined terms are C, the matrix of Coriolis and centrifugal coefficients, and g, the gravity vector. Hence the joint torques are given by

$$\tau = HJ_H^{\dagger}(\ddot{x} - \dot{J}\dot{\theta}) + \dot{\theta} \cdot C \cdot \dot{\theta} + g. \tag{21}$$

Vukobratovic and Kircanski [25] broadened the method of Khatib to include energetic models of hydraulic and electromagnetic motors and applied the resultant generalized inverse to velocities.

While kinetic resolution of redundancy incorporates dynamics, the resulting formulation is only indirectly related to torque production at the joints. A major reason for attempting to reflect directly the effect of redundancy on joint torque is to avoid exceeding torque limits. Past work has been concerned with characterizing the endpoint accelerations that depend on the utilization of the redundancy and on torque limits, rather than on an actual application to trajectory planning. Yoshikawa [26], [27] defined a dynamic manipulability ellipsoid, derived from a pseudoendpoint acceleration $\ddot{\tilde{x}} = \ddot{x} - \dot{J}\dot{\theta}$ related to a pseudotorque $\tilde{\tau} = \tau - \dot{\theta}\cdot C\cdot\dot{\theta} - g$ by

$$\ddot{\tilde{x}} = JH^{-1}\tilde{\tau}. \tag{22}$$

The derivation proceeds at zero velocity by a kind of normalization of the torques according to range. The upper τ^+ and lower τ^- joint torque bounds are presumed equal, and each pseudojoint torque is normalized as $\hat{\tau}_i = \tilde{\tau}_i/(\tau_i^+ - |g_i|)$. Ignoring bounds on acceleration, then

$$\ddot{\tilde{x}} = J\tilde{H}^{-1}\hat{\tau} \tag{23}$$

where $\tilde{H} = WH$ and $W = \text{diag}\,(1/(\tau_i^+ - |g_i|))$ is a diagonal weighting matrix. The dynamic manipulability ellipsoid consists of the set of $\ddot{\tilde{x}}$ that are derivable from a unit input $\hat{\tau} \le 1$, obtained from the dot product

$$\ddot{\tilde{x}}^T(\tilde{H}J^{\dagger})^T\tilde{H}J^{\dagger}\ddot{\tilde{x}} \le 1. \tag{24}$$

The dynamic manipulability index w is derived from this relation and is an extension of (6) through the inclusion of inertia:

$$w = \sqrt{\det\,(J(\tilde{H}^T\tilde{H})^{-1}J^T)}. \tag{25}$$

This index has not yet been applied to redundancy resolution.

Khatib [28] presented a different normalization for torque range in a procedure similar to Yoshikawa's. Defining

$$\hat{\tau}_i = \min\,(|\tau_i^- - g_i|,\ |\tau_i^+ - g_i|),$$

the weighting matrix is $W = \text{diag}\,(1/\hat{\tau}_i)$. Khatib applied this procedure to adjust a two-link manipulator's link lengths and masses to optimize the isotropic nature of endpoint acceleration throughout the workspace.

This paper reports on a method for instantaneously minimizing torque loading at the joints in a least squares sense by specification of a null-space vector applied to generalized inverse accelerations and derived from manipulator dynamics. This method is compared to use of a straightforward pseudoinverse and to use of an inertia-weighted pseudoinverse with regard to the effect on joint torques. Portions of this research have been previously reported [29], [30].

II. The General Inverse and Torque Optimization

Consider an n degree of freedom manipulator whose task is described by m hand variables x, where $m < n$. Assume that the upper and the lower torque limits of the joints are τ^+ and τ^-, respectively, where for simplicity the limits are assumed motion-independent. Then given a desired hand trajectory $x(t)$, we would like to find the set of joint torques that results in $x(t)$ and at the same time reduces actuator demands. One way of reducing the actuator demands is to place the joint torques closest to the midpoint of the joint torque limits $\frac{1}{2}(\tau^+ + \tau^-)$. In the following, a method is presented for finding this set of joint torques using the null space of the Jacobian matrix.

To incorporate dynamics, the redundancy is resolved at the acceleration level rather than at the velocity level. Using the pseudoinverse, the inverse kinematics readily follows from (18):

$$\ddot{\theta} = J^{\dagger}(\ddot{x} - \dot{J}\dot{\theta}) + (I - J^{\dagger}J)\ddot{\phi} \tag{26}$$

where the null-space vector $(I - J^{\dagger}J)\ddot{\phi}$ now appears relative to (19). Substituting (26) into (20),

$$\tau = \tilde{\tau} + H(I - J^{\dagger}J)\ddot{\phi} \tag{27}$$

where

$$\tilde{\tau} = HJ^{\dagger}(\ddot{x} - \dot{J}\dot{\theta}) + \dot{\theta} \cdot C \cdot \dot{\theta} + g. \tag{28}$$

The goal is to place τ closest to $\frac{1}{2}(\tau^+ + \tau^-)$ in a least squares sense, that is to say, to minimize

$$\left\| \tau - \frac{\tau^+ + \tau^-}{2} \right\|^2. \tag{29}$$

Substituting (27) into (29) and defining $\hat{\tau}^+ = \tau^+ - \tilde{\tau}$ and $\hat{\tau}^- = \tau^- - \tilde{\tau}$, the goal is recast as finding the vector $\ddot{\phi}$ to minimize

$$\left\| H(I - J^{\dagger}J)\ddot{\phi} - \frac{\hat{\tau}^+ + \hat{\tau}^-}{2} \right\|^2. \tag{30}$$

This least squares problem can also be solved by the generalized inverse [9], yielding

$$\ddot{\phi} = [H(I - J^{\dagger}J)]^{\dagger}\,\frac{\hat{\tau}^+ + \hat{\tau}^-}{2}. \tag{31}$$

This result can be extended to incorporate the idea that the available torque range is smaller for some joints than for others, through the use of weighted least squares [9]. Given a weighting matrix W, we now want to minimize

$$\left\| \tau - \frac{\tau^+ + \tau^-}{2} \right\|_W^2 = \left(\tau - \frac{\tau^+ + \tau^-}{2} \right)^T W \left(\tau - \frac{\tau^+ + \tau^-}{2} \right). \tag{32}$$

The solution to this is given by

$$\ddot{\phi} = [W^{1/2} H(I - J^\dagger J)]^\dagger \left(W^{1/2} \frac{\hat{\tau}^+ + \hat{\tau}^-}{2} \right) \tag{33}$$

where by definition $W^{1/2}W^{1/2} = W$. Torque range can be incorporated into the least squares by a diagonal matrix $W = \text{diag}\,(1/\tau_i^+ - \tau_i^-)^2)$. Weighting the lease squares by the magnitude of the torque range seems a better procedure than the weighting schemes of Khatib and Yoshikawa, which are biased towards zero.

III. Simulation

The various algorithms for redundancy resolution at the acceleration level are demonstrated by simulation. 1) The *unweighted pseudoinverse algorithm* derives the joint torques from (27) and (28):

$$\tau = HJ^\dagger(\ddot{x} - \dot{J}\dot{\theta}) + \dot{\theta} \cdot C \cdot \dot{\theta} + g. \tag{34}$$

2) The *inertia-weighted pseudoinverse algorithm* is obtained from (21):

$$\tau = HJ_H^\dagger(\ddot{x} - \dot{J}\dot{\theta}) + \dot{\theta} \cdot C \cdot \dot{\theta} + g. \tag{35}$$

3) The *unweighted null-space algorithm* derives the joint torques from (27), (28), and (31):

$$\tau = HJ^\dagger(\ddot{x} - \dot{J}\dot{\theta}) + \dot{\theta} \cdot C \cdot \dot{\theta} + g + H[H(I - J^\dagger J)]^\dagger \frac{\hat{\tau}^+ + \hat{\tau}^-}{2}. \tag{36}$$

4) The *weighted null-space algorithm* is derived from (27), (28), and (33):

$$\tau = HJ^\dagger(\ddot{x} - \dot{J}\dot{\theta}) + \dot{\theta} \cdot C \cdot \dot{\theta} + g + H[W^{1/2} H(I - J^\dagger J)]^\dagger \left(W^{1/2} \frac{\hat{\tau}^+ + \hat{\tau}^-}{2} \right). \tag{37}$$

Methods 3 and 4 again make use of the identity $B[CB]^\dagger = [CB]^\dagger$, where $B = (I - J^\dagger J)$ is hermetian and idempotent [16].

The simulated manipulator is a planar rotary manipulator with three links (Fig. 1). The parameters of the arm are link lengths $l_1 = l_2 = l_3 = 1.0$ m and masses $m_1 = m_2 = m_3 = 10.0$ kg; link inertias are modeled by thin uniform rods. The simulated movements are straight-line Cartesian paths starting and ending with zero velocity; the joint velocities start at zero initially. They are bang-bang movements in the endpoint acceleration, with equal and constant acceleration and deceleration periods in x and y directions of magnitude 1 m/s^2. The total duration of the movement is varied to give trajectories of different lengths. Tip orientation is not constrained, giving one-degree-of-freedom redundancy. For simplicity gravity is ignored, and the upper and lower torque limits are set equal in magnitude yielding $\tau^+ + \tau^- = \mathbf{0}$. Torque limits for joints 1–3 are set at 54, 24, and 6 N·m, respectively, derived from ratios of mass moments at each joint when the arm is perfectly straight to ensure appropriate relative scaling. For the weighted null-space algorithm, the diagonal elements of the weighting matrix τ were set as the inverse squares of these torque limits.

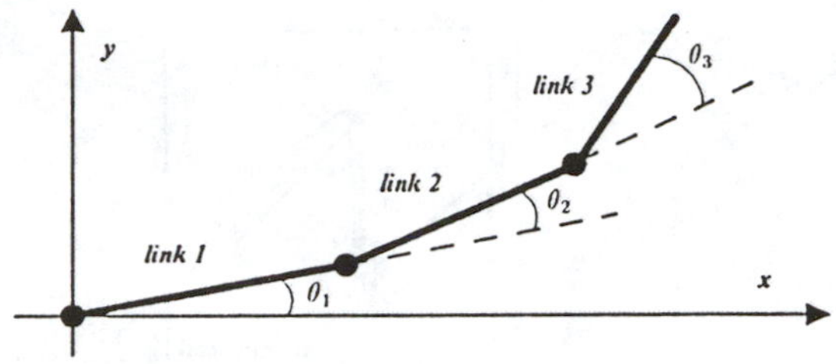

Fig. 1. Geometry of three-link planar manipulator.

The pseudoinverse was found using the singular value decomposition of a matrix [31], via a routine taken from [32]. After the joint torques were found, a fourth-order Runge–Kutta algorithm was used to find the next joint velocities and angles with an integration time interval of 1 ms. Programs were written in Fortran 77 in single precision and were run on the VAX 11/750 computer running 4.2 BSD UNIX.

IV. Results

Performance of the unweighted null-space and pseudoinverse algorithms are compared in Figs. 2–6 for representative trajectories. For the short movement of Fig. 2, the null-space algorithm seems to give a substantial reduction in required joint torques. The most dramatic decrease is in the joint 1 torque, which is much smaller than for the unweighted pseudoinverse algorithm. While the other joint torques in the present example are decreased less, the joint 3 torque is pulled away from the perilously close torque limit of six. For the intermediate length movement of Fig. 3, which is 2.5 times as long as in Fig. 2, the unweighted null-space algorithm still outperforms the unweighted pseudoinverse algorithm, although near the movement midrange and end the joint 2 torques are somewhat larger though still within the torque range.

For the long movement of Fig. 4, which is four times as long as in Fig. 2, the unweighted null-space algorithm unexpectedly encounters stability problems near the movement end. The torques required to keep the manipulator endpoint on the planned trajectory become extremely high, occasioned by the joint alignment of links 2 and 3 coupled with large velocities and accelerations in all the joints (Fig. 5). Evidently, the unweighted null-space algorithm needs a long enough path for this situation to develop. Indeed, nearly any long path led to similar instability problems, since joint alignment and high velocities inevitably followed. This happened less often in the case of the unweighted pseudoinverse

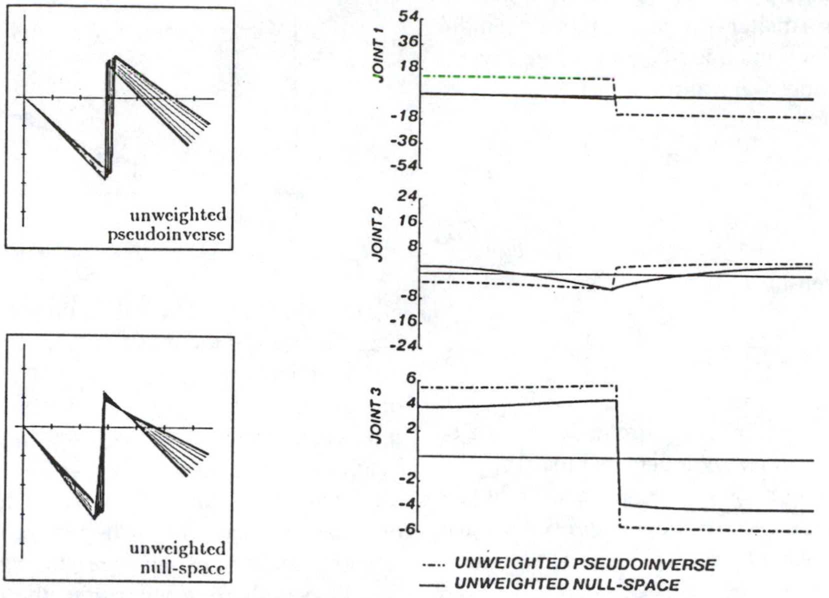

Fig. 2. Motion and torque profiles for pseudoinverse and null-space algorithms for short movement of length 0.2 m in both x and y directions. Initial configuration is $\theta_1 = -45°$, $\theta_2 = 135°$, $\theta_3 = -135°$. Two left boxes show ten successive configurations of manipulator at equal temporal spacings along trajectory; axes tick marks are at 0.25-m spacing. Three joint torque profiles on right show relative performances of two algorithms. Horizontal time axis is scaled from movement start to end.

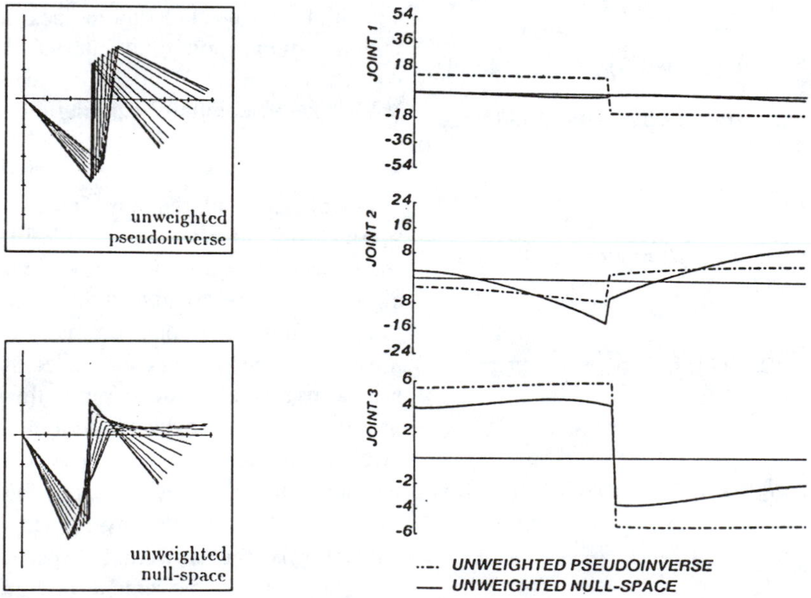

Fig. 3. Medium-length movement of 0.5 m in both x and y directions, with same starting configuration as in Fig. 2.

algorithm. One such example is shown in Fig. 6, where the unweighted pseudoinverse algorithm shows an instability, but the unweighted null-space algorithm does not. Nevertheless, in general, though the unweighted null-space algorithm instantaneously gives smaller torques than the unweighted pseudoinverse algorithm, globally, the unweighted null-space algorithm degrades unacceptably in performance with many longer trajectories.

Ordinarily, one would expect a proximal joint such as joint 1 to have a larger actuator and a wider torque range than the distal joints since it is required to exert more torque. Yet in the unweighted torque optimization all joints are treated equally; the sum of squares of joint torques is minimized by reducing the largest torque, which is the joint 1 torque, by generating high accelerations at all the joints. If the torque optimization is done relative to the allowable torque range at each joint, the instability problem may be overcome.

To test this idea, performance of the weighted null-space algorithm is examined for the same trajectory as in Fig. 2. It can be seen that the joint 3 torque is now much smaller at the expense of more torque at joints 1 and 2, but all remain well within their torque ranges (Fig. 7). Unfortunately, the

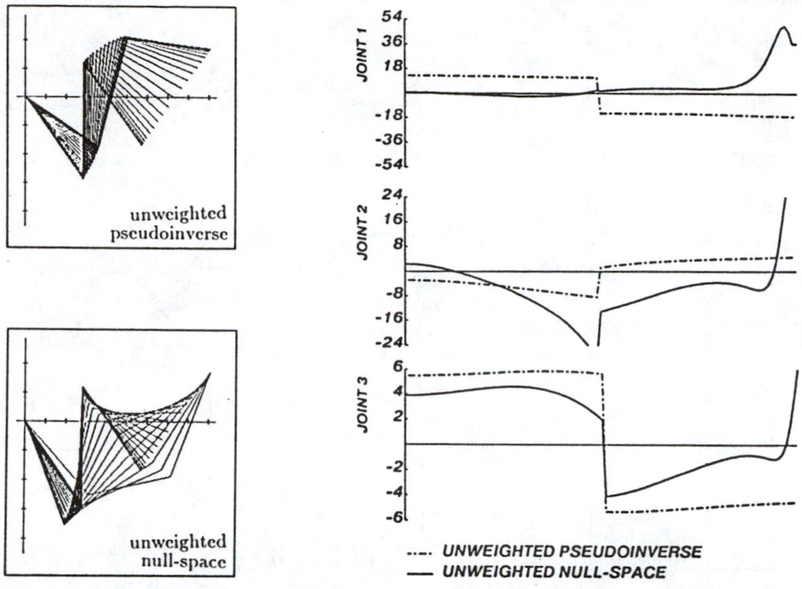

Fig. 4. Long movement of 0.83 m in both x and y directions, with same starting configuration as in Fig. 2. Motion profiles on left show 20 successive manipulator configurations at equal temporal spacings.

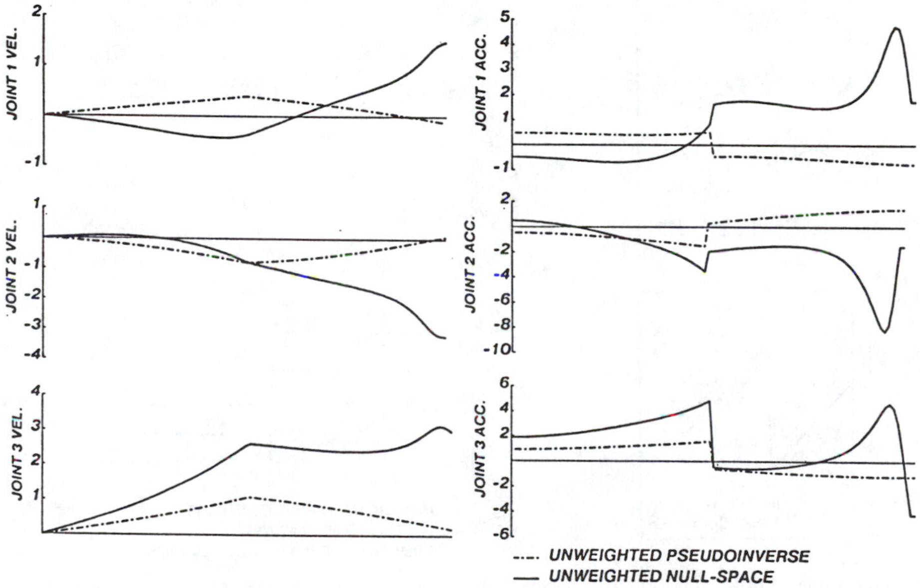

Fig. 5. Joint velocity (rad/s) and joint acceleration (rad/s^2) profiles for movement of Fig. 4.

weighted null-space algorithm also yields stability problems for the long movement of Fig. 4, so that the hoped-for elimination of instability with weights is not realized. There were even movements where the instability was shown only by the weighted case, although the reverse has not yet been observed. The observed characteristics of the instability in the weighted cases were identical to those of the nonweighted cases.

Performance of the inertia-weighted pseudoinverse algorithm is also shown in Fig. 7. Here and elsewhere the joint torques generally fell between the unweighted pseudoinverse algorithm and the unweighted null-space algorithm. The inertia-weighted pseudoinverse algorithm also suffered from instabilities in longer movements. In most of the trajectories studied, whenever the null-space algorithm was unstable, the inertia-weighted pseudoinverse algorithm was unstable as well.

V. Discussion

The goal of this paper was to develop a local algorithm for reducing actuator requirements and avoiding torque limits, and the methods examined include use of a null-space acceleration vector, this vector weighted by torque range, and an inertia-weighted pseudoinverse. These kinetic methods all lead to stability problems, even though locally they reduce actuator torques. Globally, only the unweighted pseudoinverse algorithm, which by contrast is a kinematic method, is generally well-behaved, although it too shows instabilities.

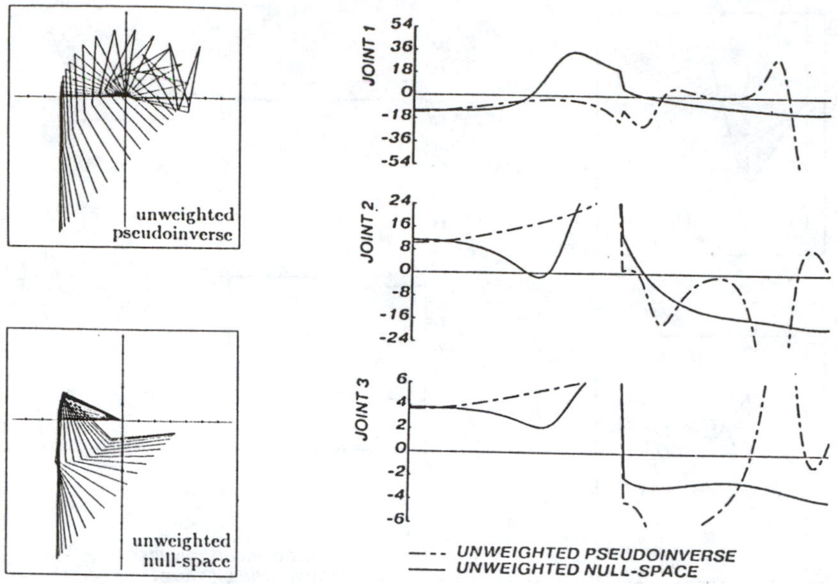

Fig. 6. Long movement of length 1.82 m in both x and y directions but with starting configuration of $\theta_1 = 180°$, $\theta_2 = 90°$, $\theta_3 = 0°$.

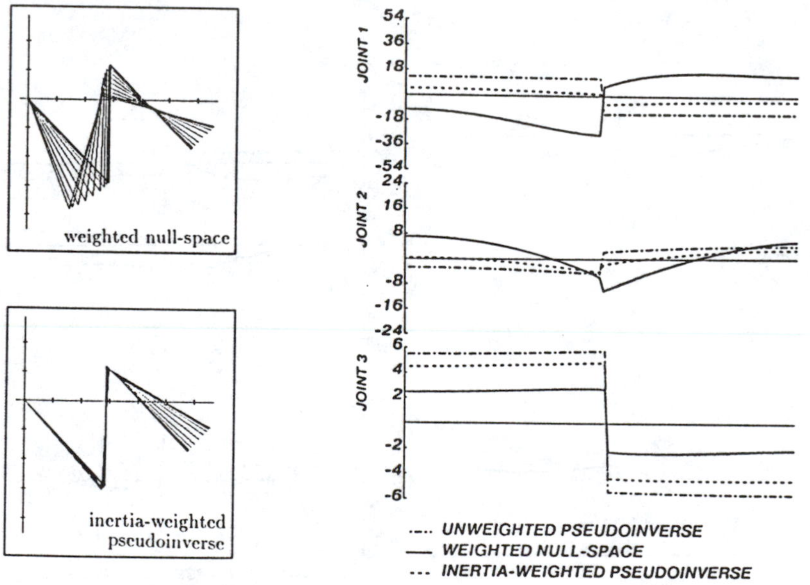

Fig. 7. Motion and torque profiles for short movement of Fig. 2 for weighted null-space and inertia-weighted pseudoinverse algorithms.

It seems that local tampering with the energetics of movement has led to global disaster. The torque profiles in Fig. 4 are representative of the problem. High accelerations in the distal joints to high joint velocities, and at the point where the third joint is straight these high velocities induce a whipping action that strongly thrusts the endpoint off the path. Substantial torques are required to overcome these natural movement dynamics and to maintain the planned endpoint trajectory.

Insofar as torque optimization tends to reduce the torques towards zero, the situation may be analogous to a free-swinging pendulum. In Fig. 8 the manipulator has been started at the same position as the previous plots, but with gravity now acting along the original straight-line motion with a magnitude that yields an initial acceleration equal to the acceleration step in the simulated trajectories. The manipulator is allowed to fall freely under the influence of this gravity. The free-swinging pendulum exhibits a similar whipping action when the third joint straightens out. Thus the kinetic methods do indeed lead to utilization of the natural movement dynamics but ultimately to the detriment of the movement goals with which the natural movement dynamics are incompatible.

Whether the kinetic methods may be modified to avoid the instabilities is a subject of continuing research. One possibility is to weight the local optimization criterion with a kinematic term to avoid high-velocity buildup. For staying with torque bounds, linear programming rather than least squares may be more satisfactory. The broader question is whether any local

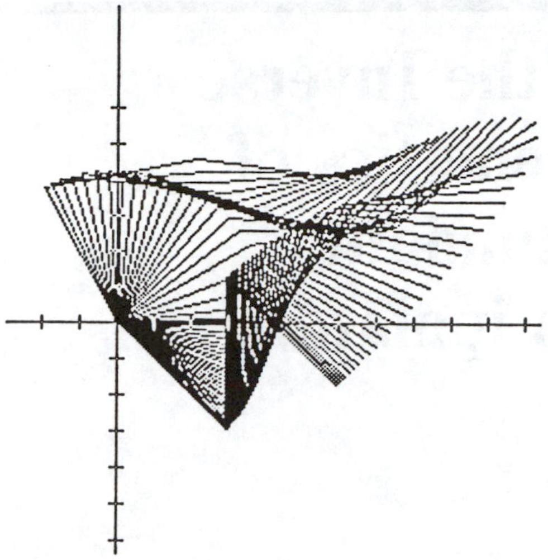

Fig. 8. Free-swinging pendulum, which is acting under influence of gravity directed along former straight-line motion of Fig. 4.

algorithm can every be completely successful, or whether ultimately only a global resolution of redundancy can be guaranteed problem-free.

Acknowledgment

We wish to acknowledge discussions with Roy Featherstone into causes of the instability, and additionally one of us (JMH) acknowledges the many useful discussion concerning redundant arms with Roger Brockett and John Baillieul.

References

[1] R. P. Paul and C. H. Stevenson, "Kinematics of robot wrists," *Int. J. Robotics Res.*, vol. 2, no. 1, pp. 31-38, 1983.

[2] J. M. Hollerbach, "Optimum kinematic design for a seven degree of freedom manipulator," in *Robotics Research: The Second International Symposium*, H. Hanafusa and H. Inoue, Eds. Cambridge, MA: MIT Press, 1985, pp. 215-222.

[3] ——, "Evaluation of redundant manipulators derived from the PUMA Geometry," in *ASME Winter Annual Meeting: Robotics and Manufacturing Automation*, PED-vol. 15, Miami Beach, FL, Nov. 17-22, 1985, pp. 187-192.

[4] T. Yoshikawa, "Manipulability of robotic mechanisms," *Int. J. Robotics Res.*, vol. 4, no. 2, pp. 3-9, 1985.

[5] K. Takase, H. Inoue, and K. Sato, "The design of an articulated manipulator with torque control ability," in *Proc. 4th Int. Symp. Industrial Robotics*, 1974, pp. 261-270.

[6] Y. Nakamura, "Kinematical studies on the trajectory control of robot manipulators," Ph.D. dissertation, Kyoto Univ., Japan, June 1985.

[7] J. P. Trevelyan, P. D. Kovesi, and M. C. H. Ong, "Motion control for a sheep shearing robot," in *Robotics Research: The First International Symposium*, M. Brady an R. Paul, Eds. Cambridge, MA: MIT Press, 1984, pp. 175-190.

[8] F. L. Litvin, T. Costopoulos, V. Perenti Castelli, M. Shaheen, and Y. Yukishige, "Configurations of robot's manipulators and their identification, and the execution of prescribed trajectories. Part 2: Investigations of manipulators having five, seven, and eight degrees of freedom," *ASME J. Mechanisms, Transmissions, and Automation in Design*, vol. 107, pp. 179-188, 1985.

[9] A. Ben-Israel and T. N. E. Greville, *Generalized Inverses: Theory and Applications.* New York: Krieger, 1980.

[10] D. E. Whitney, "Resolved motion rate control of manipulators and human prostheses," *IEEE Trans. Man-Machine Syst.*, vol. MMS-10, pp. 47-53, 1969.

[11] ——, "The mathematics of coordinated control of prosthetic arms and manipulators," *ASME J. Dynamic Systems, Meas., Control*, pp. 303-309, 1972.

[12] A. Liegeois, "Automatic supervisory control of the configuration and behavior of multibody mechanisms," *IEEE Trans. Syst., Man, Cybern.*, vol. SMC-7, pp. 868-871, 1977.

[13] T. Yoshikawa, "Analysis and control of robot manipulators with redundancy," in *Robotics Research: The First International Symposium*, M. Brady and R. Paul, Eds. Cambridge, MA: MIT Press, 1984, pp. 735-748.

[14] C. A. Klein, "Use of redundancy in the design of robotic systems," in *Robotics Research: The Second International Symposium*, H. Hanafusa and H. Inoue, Eds. Cambridge, MA: MIT Press, 1985, pp. 207-214.

[15] H. Hanafusa, T. Yoshikawa, and Y. Nakamura, "Analysis and control of articulated robot arms with redundancy," in *Prep. 8th IFAC World Congress*, pp. XIV-78-83, Aug. 1981.

[16] A. A. Maciejewski and C. A Klein, "Obstacle avoidance for kinematically redundant manipulators in dynamically varying environments," *Int. J. Robotics Res.*, vol. 4, no. 33, pp. 109-117, 1985.

[17] C. A. Klein and C. H. Huang, "Review of pseudoinverse control for use with kinematically redundant manipulators," *IEEE Trans. Syst., Man, Cybern.*, vol. SMC-13, pp. 245-250, 1983.

[18] J. Baillieul, "Kinematic programming alternatives for redundant manipulators," in *Proc. IEEE Conf. Robotics and Automation*, St. Louis, MO, 1985, pp. 722-728.

[19] M. S. Konstantinov, M. D. Markov, and D. N. Nenchev, "Kinematic control of redundant manipulators," in *Proc. 11th Int. Symp. on Industrial Robots*, Tokyo, Japan, 1981, pp. 561-568.

[20] M. Uchiyama, K. Shimizu, and K. Hakomori, "Performance evaluation of manipulators using the Jacobian and its application to trajectory planning," in *Robotics Research: The Second International Symposium*, H. Hanafusa and H. Inoue, Eds. Cambridge, MA: MIT Press, 1985, pp. 447-456.

[21] Y. Nakamura and H. Hanafusa, "Task priority based redundancy control of robot manipulators," in *Robotics Research: The Second International Symposium*, H. Hanafusa and H. Inoue, Eds. Cambridge, MA: MIT Press, 1985, pp. 155-162.

[22] O. Khatib, "Commande dynamique dans l'espace operationnel des robots manipulateurs en presence d'obstacles," Docteur Ingenieur Thesis, L'Ecole Nationale Superieure de l'Aeronautique et de l'Espace, 1980.

[23] ——, "Dynamic control of manipulators in operational space," in *Proc. 6th IFToMM Congress on Theory of Machines and Mechanisms*, New Delhi, 1983, pp. 1123-1131.

[24] J. M. Hollerbach, "Dynamic scaling of manipulator trajectories," *ASME J. Dynamic Syst., Meas. Contr.*, vol. 106, pp. 102-106, 1984.

[25] M. Vukobratovic and M. Kircanski, "A dynamic approach to nominal trajectory synthesis for redundant manipulators," *IEEE Trans. Syst., Man, Cybern.* vol. SMC-14, pp. 580-586, 1984.

[26] T. Yoshikawa, "Dynamic manipulability of robot manipulators," in *Proc. IEEE Conf. Robotics and Automation*, St. Louis, MO, 1985, pp. 1033-1038.

[27] T. Yoshikawa, "Analysis and design of articulated robot arms from the viewpoint of dynamic manipulability," in *Robotics Research: the Third International Symposium*, O. Faugeras and G. Giralt, Eds. Cambridge, MA: MIT Press, 1986, pp. 273-280.

[28] O. Khatib, "The operational space formulation in the analysis, design, and control of robot manipulators," in *Robotics Research: the Third International Symposium*, O. Faugeras and G. Giralt, Eds. Cambridge, MA: MIT Press, 1986, pp. 263-270.

[29] J. Baillieul, J. M. Hollerbach, and R. Brockett, "Programming and control of kinematically redundant manipulators," in *Proc. 23rd IEEE Conf. Decision and Control*, Las Vegas, NV, 1984, pp. 768-774.

[30] J. M. Hollerbach and K. C. Suh, "Redundancy resolution of manipulators through torque optimization," in *Proc. IEEE Int. Conf. Robotics and Automation*, St. Louis, MO, 1985, pp. 1016-1021.

[31] G. Strang, *Linear Algebra and Its Applications.* New York: Academic, 1980, pp. 137-144.

[32] J. J. Dongarra, C. B. Moler, J. R. Bunch, and G. W. Stewart, *LINPAK User's Guide*, Philadelphia, PA: SIAM, 1979.

Daniel R. Baker
Charles W. Wampler II
Mathematics Department
General Motors Research Laboratories
Warren, Michigan 48090

On the Inverse Kinematics of Redundant Manipulators

Abstract

Many conventional nonredundant manipulators have singular configurations, near which some small motions of the end-effector require excessive and physically unrealizable joint speeds. Consequently, the usable workspace of the manipulator is effectively reduced. It has been proposed that high joint speeds could be avoided by introducing redundant joints and using an appropriate kinematic inversion algorithm. For a very general class of kinematic inversion algorithms, the theorems of this paper state some fundamental relations between the properties of the algorithm and its ability to resolve such problems. These results have practical implications in the design of controllers for redundant manipulators, especially when real-time sensory input is used to modify the manipulator's trajectory.

1. Introduction

Most conventional robot manipulators are kinematically nonredundant, that is, each one has the minimum number of joints required for a certain class of motions. Thus, six-degree-of-freedom manipulators are applied to tasks requiring the general positioning and orienting of the end-effector. Likewise, five-degree-of-freedom manipulators are commonly used when the tool must be positioned with one axis pointed in a given direction; that is, the rotation of the tool about the pointing axis is irrelevant, as in spray painting or screw insertion. When manipulators are applied to the tasks of orienting or pointing, there are certain configurations, called *singularities,* where a change of orientation in some direction cannot be produced. Near such configurations, small angular velocities of the end-effector in some directions require physically unrealizable joint speeds, thereby leading to severe inaccuracy in the resultant motion. In recent years, there has been a growing interest in the use of redundant manipulators to reduce joint speeds and work around obstacles. This paper discusses some limitations, based on topological arguments, concerning the extent to which inverse kinematic algorithms for the problems of pointing and orienting can be successful. These limits can help guide the development of useful inverse kinematic solutions for redundant manipulators. In addition, several results are obtained that have implications toward inverse kinematics for pure positioning tasks.

It is a well-established result that any serial-link manipulator has singular configurations associated with orientational tasks. A proof for the nonredundant case of a three-axis wrist was given by Paul and Stevenson (1983), who discussed in detail the effect of twist angles on the location of the singularities. The existence of singularities for any serial-link manipulator, including redundant ones, has been shown by Hollerbach (1985), using a physical argument, and by Gottlieb (1986), using a topological argument. Hollerbach's argument is based on the fact that the joint axes of all the rotational joints of a serial-link manipulator can be oriented parallel to a common plane, at which point it is impossible to produce incremental rotations of the end-effector about an axis perpendicular to that plane. A similar argument holds for the pointing problem, where in addition to orienting all rotational axes parallel to the same plane, the last rotational axis can always be turned until the end-effector pointing axis is also parallel to the plane. Then

The International Journal of Robotics Research,
Vol. 7, No. 2, March/April 1988,

the pointing axis cannot be incrementally rotated about an axis perpendicular to the plane.

In planning paths for a nonredundant manipulator, the only way to avoid singularities is to reduce the workspace by prohibiting orientations near the singularities. This becomes an additional burden on the operator of the robot, who must arrange the workcell so that manipulations near singular orientations are not required. This process is often tedious and iterative, especially when the workspace is already reduced by obstacles or when the singular orientations are coupled to the position of the arm. The problem is even more difficult when allowances must be made for variations in workpiece placement as permitted by the use of vision systems and other sensor input.

In contrast to the nonredundant case, an n-degree-of-freedom redundant manipulator applied to an m-degree-of-freedom task has an $(n - m)$-dimensional family of joint solutions for any given task. Consequently, although singular configurations exist, there is usually a continuum of alternative nonsingular configurations that yield the same end-effector position and orientation. The problem is to utilize the redundant freedoms to produce the desired end-effector motions without inducing excessive joint speeds.

This paper is composed as follows. First, the issues at hand are introduced by means of an illustrative example, which is followed by a discussion of current methods for constructing inverse kinematic solutions. These are divided into two types according to whether or not they require advance knowledge of the manipulator's path. Then, a precise definition is given for a class of methods dubbed "tracking algorithms," which includes most methods appearing in the literature of the type that require only local path information. Five theorems concerning the properties of these methods are stated, and their implications for manipulator design and control are discussed. Finally, a section of conclusions is followed by appendixes containing detailed proofs of the theorems.

2. Background

Basic to a discussion of manipulator kinematics are the notions of joint space and operational space. Denote the joint space of an n-joint, serial-link manipulator as T^n, which consists of n-tuples $\theta = (\theta_1, \ldots, \theta_n)$. Each θ_i represents either a point on a circle or a point on some line segment $[a_i, b_i]$. Circles denote positions of rotational joints without angle stops, and line segments denote displacements of translational joints or rotational joints with angle stops. Also, let X denote the operational space, which is some subset of a vector space. The exact form of X depends on the task; most commonly it consists of three-dimensional position and orientation, but it could also be three-dimensional position with pointing, planar position and orientation, or simply position or orientation alone.

As just indicated, there are three orientation spaces that are relevant to robotics. The simplest of these is orientation in the plane, which has three equivalent representations: a single rotation angle, a unit vector in the plane, or position along the unit circle. Clearly, this space is covered by a single rotational joint, and there is no trouble in finding a singularity-free inverse kinematic solution. Next, the space of pointing directions in three-space is two-dimensional and can be represented by a unit vector in three-space or, equivalently, by the surface of a sphere. In contrast, the configuration space of a manipulator composed of two rotational joints is a torus. The mismatch between the topological properties of a sphere and a torus prevents the construction of a singularity-free inverse kinematic mapping. As is made explicit below, this result extends to redundant manipulators. Finally, the third relevant orientation space is the three-dimensional space of all rigid-body rotations, each of which may be represented by a right-handed set of three mutually orthogonal unit vectors or by the corresponding direction cosine matrix. This space is also topologically distinct from the configuration space of any manipulator, thus giving rise to singularity problems. Consequently, a topological analysis of the kinematics of manipulators can yield some fundamental conclusions that depend on very few assumptions about the specifics of the mechanism.

To illustrate the problem, let us look at the specific task of pointing. For many robotics applications, such as automatic welding or spray painting, the full orientation of the wrist in the manipulator is not necessary; all that matters is the pointing direction of the painter

Fig. 1. A. Nonredundant and B. redundant pointing mechanisms.

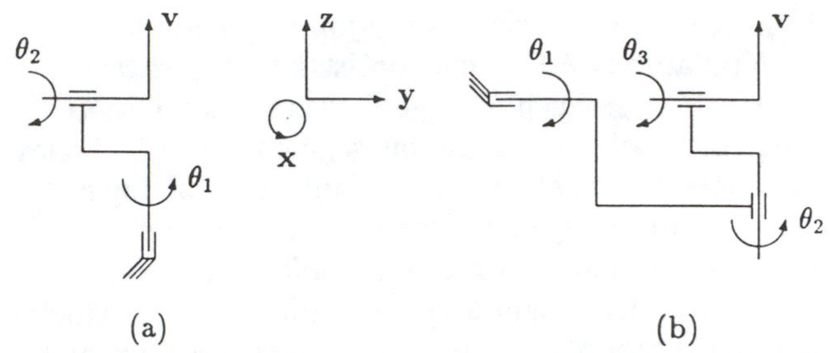

or welding gun. This effectively reduces the operational space for the manipulator wrist from the three-dimensional space of orientations to the surface of the sphere, which has only two degrees of freedom. Each point on the sphere describes a pointing direction of the end-effector. This means that a wrist with three joints will have one redundant joint for this task.

Figure 1A shows a nonredundant pointing mechanism with two rotational joints whose position is described by two angles, θ_1 and θ_2, taken as zero in the configuration shown. The function f that gives the pointing direction $\mathbf{v}$ corresponding to each joint configuration of the mechanism (θ_1, θ_2) is

$$\begin{aligned} \mathbf{v} &= f(\theta_1, \theta_2) \\ &= \cos\theta_1 \sin\theta_2 \mathbf{x} + \sin\theta_1 \sin\theta_2 \mathbf{y} + \cos\theta_2 \mathbf{z}, \end{aligned} \tag{1}$$

where $\mathbf{x}$, $\mathbf{y}$, $\mathbf{z}$ is a set of fixed, mutually orthogonal unit vectors. A joint configuration is singular if the Jacobian of f does not have maximal rank. Since the range of f is the surface of the sphere ($\mathbf{v} \cdot \mathbf{v} = 1$), and not all of three-dimensional space, the maximal rank is 2. In this case, the singular configurations arise when the pointing vector $\mathbf{v}$ is aligned with $\pm\mathbf{z}$; i.e., when $\theta_2 = 0$ or $\theta_2 = \pi$.

Tracking difficulties arise when a path comes too close to one of these singular configurations. As the Jacobian of f starts to become more and more singular, higher and higher velocities in the joint space are necessary to obtain a fixed velocity in the operational space. Suppose, for example, that we want $\mathbf{v}$ to track a great circle that passes very close by the "north pole" ($\theta_2 = 0$) but does not go through it. As the path passes by the pole, the angle θ_1 is forced to make a jump of about π rad, and the closer the path is to the pole the quicker this jump in θ_1 occurs. This means that the mechanism can be forced to move at arbitrarily high speeds in order to track a path of constant velocity.

The singularity at the pole can be avoided by going to a three-joint manipulator and taking advantage of the redundancy in joints. Figure 1B shows such a manipulator. The joint space is described now by three angles $(\theta_1, \theta_2, \theta_3)$, and, as in the above case, there is a function $f(\theta_1, \theta_2, \theta_3)$ that assigns the pointing direction to a joint configuration. As mentioned earlier, every manipulator, regardless of the extent of its redundancies, will have singular configurations, which for this mechanism occur when the entire mechanism lies in a plane, as pictured in Figure 1B. Nevertheless, the three-joint manipulator has nondegenerate alternative configurations for pointing to the north pole. One set of these can be generated by turning θ_2 so as to twist part of the apparatus out of the page. Once this has been done, it becomes possible to track paths that come arbitrarily close to the north pole without having jumps in the velocity of the mechanism.

One might be tempted to bar the mechanism from planar configurations by introducing a constraint function of the form $h(\theta_1, \theta_2, \theta_3) = 0$, but this may produce artificial singularities. For example, the constraint $\theta_2 = \pi/2$ eliminates planar configurations, but this makes the mechanism kinematically equivalent to a nonredundant two-joint wrist with singularities when $\mathbf{v}$ aligns with $\mathbf{y}$. The question arises as to whether there is *any* constraint or, indeed, any general approach, that will yield bounded joint speeds for all paths.

To answer such a question, we will define a class of kinematic inversion methods, to be called "tracking algorithms," that depend only on the current manipulator configuration and the local properties of the path. This enables the manipulator to alter its path as it moves, based on sensor measurements or other real-time information. Before the definition of a tracking algorithm is made precise, it is useful to introduce some notation and to review current methods of planning paths for redundant manipulators.

3. Types of Inverse Kinematic Algorithms

Recall that we denote the joint space of the manipulator as T^n and its operational space as X. Then the forward kinematics of the manipulator is represented by a differentiable function, $f: T^n \to X$, which for each set of joint coordinates, $\theta \in T^n$, gives the corresponding point in the operational space, $\mathbf{x} = f(\theta) \in X$, assumed by the manipulator. Accordingly, for a given trajectory in the operational space, $\mathbf{x}(t)$, an inverse kinematic algorithm produces a corresponding joint-space trajectory, $\theta(t)$, such that $f(\theta(t)) = \mathbf{x}(t)$.

For didactic purposes, inverse kinematic algorithms can be divided into two categories: (1) methods based on complete path information, and (2) methods based only on local path information. Methods of the first type compute a joint-space trajectory from a complete description of the task in operational space. Given limits on the actuator torques and joint speeds, or other constraints, these methods find a joint trajectory that optimizes a performance criterion, such as completion time or average kinetic energy (Nakamura and Hanafusa 1985; Johnson and Gilbert 1985; Bobrow, Dubowsky, and Gibson 1983). This usually produces joint trajectories with acceptable joint speeds, but these methods are limited to off-line planning and do not allow real-time path corrections based on sensor measurements. In contrast, an inversion method based on local path information gives joint paths that depend only on the local behavior of the operational space path. Accordingly, sensor-based path corrections can be incorporated in real-time, but problems of high joint speeds can arise and the paths will, in general, be suboptimal compared to those derived using complete information. Nevertheless, sensor feedback is critical to many tasks, so this paper concentrates on the methods that use only local path information.

Most inversion methods based on local path information use repeated solutions of the differential relations between joint space coordinates and operational space coordinates. That is, if $\dot{\mathbf{x}}$ denotes velocity in the operational space and $J(\theta) = \partial f / \partial \theta$ is the Jacobian matrix of f, then a solution $\dot{\theta}$ of the equation

$$J\dot{\theta}(t) = \dot{\mathbf{x}}(t) \tag{2}$$

is given for all t during the motion. (Higher derivatives also appear in some formulations.) As the velocity equation is solved repeatedly in real-time, $\dot{\mathbf{x}}$ can be modified according to sensor measurements or interactive human commands. A variety of methods has been proposed to solve this equation for redundant manipulators, each differing in the manner in which a unique $\dot{\theta}$ is chosen among the many solutions to this equation. The most common approaches are outlined below.

One proposal for resolving the redundancy, originally due to Whitney (1969), is to find the joint speeds that minimize $\dot{\theta}^T \dot{\theta}$ subject to Eq. (2). This solution can be written in terms of the Moore-Penrose generalized inverse J^+ of the Jacobian matrix as

$$\dot{\theta} = J^+\dot{\mathbf{x}}. \tag{3}$$

Similar strategies include finding joint speeds to minimize kinetic energy or, as investigated by Hollerbach and Suh (1985), finding joint accelerations to minimize joint torques. Although these methods are appealing for their generality, several shortcomings have become apparent. First, both of these methods have the generally undesirable property that repetitive end-effector motions do not necessarily yield repetitive joint motions (Klein and Huang 1983). Worse, although the motion is pointwise optimal, the manipulator configuration can blunder into a region near a singularity of J, where the minimum norm is unacceptably large (Hollerbach and Suh 1985).

To avoid singularities, obstacles, and joint limits, several researchers (e.g., Liégeois 1977; Klein 1985) have proposed replacing the minimum-norm criterion with the gradient minimization of potential fields. That is, one uses

$$\dot{\theta} = J^+\dot{\mathbf{x}} - (I - J^+J)\nabla H, \tag{4}$$

where $H(\theta)$ is some potential function designed to drive the manipulator from undesirable regions of the joint space. Examples of proposed potential functions include the so-called manipulability index (Yoshikawa 1985) and inverse squares of distances to obstacles or limit stops (Khatib 1985). With such a criterion, repetitive end-effector motions tend to be produced by joint motions that converge to a repetitive cycle. However, it is difficult to construct potential functions that guarantee that singularities will be avoided at all times without introducing large gradients that in themselves induce large joint speeds. These issues have not been thoroughly studied, and a full analysis is complicated by the fact that the path traversed in joint space depends on the speed of the path in operational space.

Another class of methods that use only local path information is those that resolve the redundancy by introducing additional constraint functions of the form $h(\boldsymbol{\theta}) = 0$. One could choose such functions directly, or one could choose them indirectly by requiring that the manipulator minimize some side criterion, such as the potential functions described above. Baillieul (1985) and Chang (1986) have given formulations for converting a minimization criterion into constraint functions. Essentially, the constraint is that the projection of the gradient of the minimization criterion onto the null space of the Jacobian of f must be zero. Thus, the joint paths will be those to which the potential field method converges for slow motions. An example where the constraint function was chosen directly can be found in Stanišić and Pennock (1985), where the constraint is designed to keep a four-joint wrist out of degenerate mechanical configurations. Once the constraint is in functional form, one must solve the combined set of equations

$$f(\boldsymbol{\theta}(t)) = \mathbf{x}(t), \qquad h(\boldsymbol{\theta}(t)) = 0. \tag{5}$$

These can be solved numerically for each point along a path (Chang 1986), or, if f and h are differentiable, an initial solution to Eq. (5) can be propagated along a path by solving the differential equation

$$J_e\dot{\theta} = \begin{pmatrix} \dot{\mathbf{x}} \\ 0 \end{pmatrix}, \tag{6}$$

where, following the coinage of Baillieul (1985), J_e is the *extended Jacobian matrix* defined as

$$J_e(\theta) = \begin{pmatrix} J(\theta) \\ \partial h(\theta)/\partial\theta \end{pmatrix}. \tag{7}$$

The condition for the existence of a unique solution $\dot{\theta}(t)$ in Eq. (6) is that the extended Jacobian matrix be nonsingular along the entire path.

A final inversion method of the local type is based on a single inverse function giving the joint coordinates for each point in some subset X_0 of the operational space. In constructing such a function, one may use redundancy to reduce joint speeds and avoid known obstacles. It will be shown that, for both redundant and nonredundant manipulators, such inverse kinematic functions cannot always be defined on the whole operational space. For example, this is true for the two-joint pointing mechanism, where an inverse kinematic function is well defined only on subsets of the workspace that do not include the north and south poles. This fact is reflected in the following definition.

Definition 1.

An inverse kinematic function for f: $T^n \rightarrow X$ on $X_0 \subset X$ is a continuous function of the form g: $X_0 \rightarrow T^n$ with $f(g(\mathbf{x})) = \mathbf{x}$ for all $\mathbf{x} \in X_0$.

If $\mathbf{x}(t)$ is a path in X_0, the joint trajectory $\boldsymbol{\theta}(t) = g(\mathbf{x}(t)) \subset T^n$ will track it. Thus, it is clear that once the inverse function has been constructed off-line, sensor-based motions can be executed on-line. In practice, one might require that the inverse function g be differentiable one or more times so that the smoothness of the path is compatible with machine capabilities. For example, if the joint speeds must be continuous but the joint accelerations can be changed instantaneously, g should be at least once differentiable. As such, continuity is a minimal requirement for inverse functions to be useful in path tracking.

In building an inverse function, it is important to know whether it is possible to define it on the entire operational space. In general, this will depend on detailed facts about the forward kinematic function f, but in a later section we shall show several examples where the impossibility of doing so can be concluded using only the topological properties of the joint and operational spaces. An example of the inverse function method has been given by Wampler (1987).

The extended Jacobian method and the use of inverse kinematic functions turn out to have some important structural similarities, which will be made more precise in Section 5.

4. Definition of a Tracking Algorithm

Having discussed the various possibilities that are available to users of redundant manipulators to do kinematic inversion, we will now try to draw some general conclusions about what one can and cannot do with these methods. Our focus will be the methods based on local path information, and it will be convenient for the purposes of analysis to give an axiomatic description of these methods. The purpose of these axioms is twofold. They should be as general as possible so that the conclusions that one draws from them are applicable to as large a class of inversion methods as possible. They should also be specific enough to allow one to draw meaningful conclusions. These considerations motivate the following definitions. Their form is admittedly somewhat abstract and mathematical, but this has been done to attain maximal generality. The reader may find the explanatory remarks following Definition 3 helpful for motivation.

Definition 2.

Define a *path* $\gamma(t)$ to be a continuous vector-valued function on some interval $[a, b]$ of the real line, which is differentiable at all except a finite number of points $t_1, \ldots, t_n \in [a, b]$, and at the points t_i both right- and left-sided derivatives of γ exist, although they need not be equal. Call $\gamma(t)$ a *closed path* if it starts and ends at the same point.

Let X be some subset of a vector space, hereafter referred to as the operational space. Let T^n denote the joint space of an n-joint, open-chain manipulator, and suppose that $f\colon T^n \rightarrow X$ is a differentiable function. The function f can be thought of as representing the forward kinematics of the manipulator, although the proofs of the results in the next section make use of nothing more than its differentiability.

Definition 3.

A *tracking algorithm* for the function f on some subset $X_0 \subset X$ consists of the following:

1. A bound V on the speeds of the paths in X_0 that can be tracked. The case $V = \infty$ is allowed.
2. For every $\mathbf{x}_0 \in X_0$, there is a nonempty set of *allowable configurations* contained in $f^{-1}(\mathbf{x}_0) = \{\theta | f(\theta) = \mathbf{x}_0\} \subset T^n$. The allowable configurations are the joint configurations that one can use to track paths in X_0. Let $\theta_0 \in f^{-1}(\mathbf{x}_0)$ be an allowable configuration, and let $\mathbf{x}(t)$ be a path through $\mathbf{x}_0$ with speeds bounded as in part 1. The tracking algorithm assigns to $\mathbf{x}(t)$ a *tracking path* $\theta(t)$ of allowable configurations in T^n, through θ_0, such that $f(\theta(t)) = \mathbf{x}(t)$.
3. Suppose $\mathbf{x}_1(t)$ and $\mathbf{x}_2(t)$ are two paths in X_0 that agree on some time interval $[t_1, t_2]$. Then, if the tracking paths $\theta_1(t)$ and $\theta_2(t)$ in T^n agree at some time $t_0 \in [t_1, t_2]$, they must agree on all of $[t_1, t_2]$.
4. The tracking algorithm is time invariant; i.e., if $\theta(t)$ tracks the path $\mathbf{x}(t)$, then $\theta(T + t)$ will track the path $\mathbf{x}(T + t)$ for any fixed time interval T.

Two of the methods foı kinematic inversion that have already been described, i.e., the extended Jacobian method and the use of inverse kinematic functions, satisfy the axioms of a tracking algorithm with $V = \infty$. Tracking paths can be determined for paths with arbitrarily high speeds, although machine limitations may make it impossible to execute them.

When using potential fields to do the kinematic inversion, it may be necessary to set some finite bound on the speed of trackable paths. For example, excessive speeds in the operational space could cause the tracking path to climb a potential hill and run into a singularity. Bounding speeds does not automatically imply that an inversion method based on potential fields will satisfy axiom 2. It must be shown that, for some class of allowable configurations, paths with speeds bounded by a given V can be tracked indefinitely without running into singularities. If this is not the case, the usefulness of the method will be severely limited.

The size of the set of allowable configurations representing a given point in X_0 will vary with the tracking algorithm being used. When an inverse kinematic function $g\colon X_0 \rightarrow T^n$ is used, with $f(g(\mathbf{x})) = \mathbf{x}$, there is only one allowable configuration, $g(\mathbf{x})$, representing each point $\mathbf{x}$ in the operating space. In contrast, in the extended Jacobian method, there may be more than one allowable configuration that satisfies Eq. (5) at a given time. More generally, the allowable configurations might be the set of all nonsingular configurations that reach **x**, or some subset thereof. It might, for example, be necessary to exclude certain nonsingular configurations from the allowable ones because paths through them come too close to singular configurations at some later time.

Axiom 3 expresses the uniqueness of the inverse solution, once the joint configuration at some time t along the path has been specified. It also implies that the tracking path in any small neighborhood of the current manipulator position is determined only by the behavior of the path to be tracked in the corresponding neighborhood of the operational space. This is in contrast to the aforementioned methods that try to minimize some global property of the path. Axiom 3 allows real-time sensor-based path corrections.

Axiom 4 states that the tracking algorithm does not change as a function of time. Every time a given path in the operational space is tracked, starting at a given joint configuration, the same tracking path results.

We conclude this section by remarking that there is one very desirable property of tracking algorithms that has not been included as an axiom. There should be some uniform bound on the joint speeds of the tracking path, given in terms of a bound on the speeds of the path in X_0. None of the kinematic inversion methods already described satisfy this property automatically. For example, for the nonredundant pointing mechanism described in Fig. 1A, there is an inverse kinematic function defined on the set X_0 consisting of the sphere minus the north and south poles. Still, in order to track paths with bounded speeds near one of the poles, arbitrarily high joint speeds are necessary. This effectively limits the subset of the sphere on which such a tracking algorithm can be used to a smaller subset of X_0, i.e., one bounded away from the poles.

The theorems in the following section will state explicit limitations on the sizes of the sets X_0 on which certain types of tracking algorithms can be defined. Since the effective workspace of the manipulator can be smaller than the set X_0, these theorems are to be understood as upper bounds on the size of this workspace. Ultimately the test of such an algorithm will also entail an analysis of the ratios of joint speeds to speeds in the operational space.

5. The Main Results

This section contains a discussion of five theorems and their implications for designing tracking algorithms. The proofs of the theorems are given in Appendixes I and II.

The first question to be considered in this section will be that of the existence of inverse kinematic functions defined on the whole operational space, i.e., the case when $X_0 = X$. Two important cases to be considered are the pointing problem, when X is the surface of a sphere, and the orientation problem, when X is the space of orientations of three-dimensional space.

Let S^2 denote the sphere. The number 2 in this notation underscores the fact that there are two degrees of freedom on the surface of the sphere or, equivalently, that the sphere is a two-dimensional manifold. $SO(3)$ denotes the space of orientations in three-dimensional space, a three-dimensional manifold. A point in $SO(3)$ can be represented by some direction cosine matrix, $C \in SO(3)$.

Theorem 1: Let f: $T^n \rightarrow S^2$ be any continuous function. Then no continuous function g: $S^2 \rightarrow T^n$ exists such that $f(g(\mathbf{v})) = \mathbf{v}$ for every $\mathbf{v} \in S^2$.

Theorem 2: Let f: $T^n \rightarrow SO(3)$ be any continuous function. Then no continuous function g: $SO(3) \rightarrow T^n$ exists such that $f(g(C)) = C$ for every $C \in SO(3)$.

We remark that, although the same method is used to prove both of these theorems, neither of them is an immediate consequence of the other. A proof of Theorem 2, different from the one to be given in this paper, has been given by Gottlieb (1986), but his method of proof does not work for the pointing problem of Theorem 1. In fact, Theorem 2 will be shown to be a consequence of a stronger result, which will also be used in the proof of Theorem 5 below.

Because the number of joints, n, in the joint space T^n of the above theorems is arbitrary, they imply that no inverse kinematic functions can be defined on the whole of these two operational spaces, no matter how great the redundancy.

Note that the theorems use no more specific information about the function f than its continuity. No information beyond this is necessary for the proofs. As a result, no alterations in the construction of an open-chain manipulator will be able to help resolve the problem.

These theorems do not condemn the proposed method to failure; they simply place explicit limitations on what can be accomplished in this fashion. That is, inverse functions can only be defined on proper subsets of S^2 or $SO(3)$. Because of mechanical interferences, a manipulator often can only reach a subset of its ideal workspace, so an inverse function on this subset is all that is required. For example, in the case of a wrist mechanism for pointing, there are some pointing directions in which the end-effector must collide with the forearm to which it is attached. After eliminating these directions from consideration, it might be possible to construct an inverse function on the remaining workspace.

A corollary to Theorems 1 and 2 shows that the same limitations exist when the operational space of the manipulator is a product space containing either $SO(3)$ or S^2 as a factor. This occurs, for example, when the operational space is the six-dimensional space of positions and orientations. Points in this space can be identified as pairs, $(C, \mathbf{x})$, where C gives the orientation and $\mathbf{x}$ is a translation.

Corollary: Let $\tilde{f}$: $T^n \rightarrow X \times Y$ be any continuous function, where the space X is either S^2 or $SO(3)$, and the space Y is arbitrary. Then there is no continuous function $\tilde{g}$: $X \times Y \rightarrow T^n$ with $\tilde{f}(\tilde{g}(\mathbf{x}, \mathbf{y})) = (\mathbf{x}, \mathbf{y})$ for all $(\mathbf{x}, \mathbf{y}) \in X \times Y$.

Proof: Define the projection function π: $X \times Y \rightarrow X$ by $\pi(\mathbf{x}, \mathbf{y}) = \mathbf{x}$.

Suppose the function $\tilde{g}$ of the corollary existed, and let g: $X \rightarrow T^n$ be given by $g(\mathbf{x}) = \tilde{g}(\mathbf{x}, \mathbf{y}_0)$ for some fixed $\mathbf{y}_0 \in Y$. For the function $f = \pi \circ \tilde{f}$: $T^n \rightarrow X$ one then has $f(g(\mathbf{x})) = \mathbf{x}$, in contradiction to Theorems 1 and 2, and this proves the corollary. ■

It is reasonable at this point to ask whether tracking algorithms can be defined on all of S^2 or $SO(3)$ if one is willing to use methods other than inverse kinematic functions. This question contains a hidden difficulty, because it is possible that other inversion methods are equivalent to the use of inverse kinematic functions; i.e., they determine the same tracking paths, even though this fact is not evident from the specific implementation of the algorithm. It turns out that the answer to this problem also makes clear what price must be paid if one is to resort to other inversion methods. To make this explicit, we introduce a new definition.

Definition 4.

Call a tracking algorithm on some subset $X_0 \subset X$ *cyclic* if every closed path in X_0 is tracked only by closed paths in the joint space T^n.

If an algorithm is noncyclic, then repeated motions around a closed path in X_0 may produce unexpected configurations of the manipulator. At every new repetition of the path, the initial position of the manipulator can differ from its position at the start of the previous repetition, resulting in an entirely different tracking path. This can produce problems if, for example, one needs to know that the manipulator respects limits on the joint angles or avoids certain obstacles while executing a given motion. It will no longer suffice to perform the motion only once to ascertain that such constraints will not be violated.

Theorem 3: A tracking algorithm on some subset $X_0 \subset X$ is equivalent to one induced by continuous inverse kinematic functions defined on X_0 if and only if it is cyclic.

Although the proof of Theorem 3 is given in the appendixes, we note briefly how to construct an inverse function g associated to a cyclic tracking algorithm. Choose a starting point $\mathbf{x}_0 \in X_0$ and an allowable configuration $\boldsymbol{\theta}_0 \in f^{-1}(\mathbf{x}_0)$. The function g is determined once $\mathbf{x}_0$ and $\boldsymbol{\theta}_0$ have been chosen. For a point $\mathbf{x}_1 \in X_0$, define $g(\mathbf{x}_1)$ by choosing a path $\mathbf{x}(t)$ between $\mathbf{x}_0$ and $\mathbf{x}_1$, and a tracking path $\boldsymbol{\theta}(t)$ for $\mathbf{x}(t)$ starting at $\boldsymbol{\theta}_0$. Define $g(\mathbf{x}_1)$ to be the value of the tracking path $\boldsymbol{\theta}_1$ upon reaching the point $\mathbf{x}_1$ in the operational space. This definition of $g(\mathbf{x}_1)$ makes sense only if it is independent of the choice of path between $\mathbf{x}_0$ and $\mathbf{x}_1$, and it turns out that, if the tracking algorithm is cyclic, this is indeed the case.

We remark that the above construction will determine a different inverse function for each allowable configuration $\boldsymbol{\theta}_0$ in $f^{-1}(\mathbf{x}_0)$.

Theorem 3, in conjunction with Theorems 1 and 2, states that no cyclic tracking algorithm can be defined on all of S^2 or $SO(3)$. In this context, it seems natural to ask if the same limitations might also apply to the extended Jacobian method. In particular, one would like to know if it is possible to choose some constraint function $h(\theta)$ that avoids singularities over the whole operational space. According to Theorem 3, if this is to be possible, the resulting tracking algorithm will have to track some closed path in the operational space with an open joint trajectory.

In answering the above question, it will be convenient to analyze the extended Jacobian method from the more general perspective of local inverse functions. Rather than expressing the joint space trajectory *implicitly,* using the equations

$$f(\theta(t)) = \mathbf{x}(t), \qquad h(\theta(t)) = 0,$$

one can try to *explicitly* solve these equations, obtaining an inverse function $g_0(\mathbf{x})$ satisfying the equations

$$f(g_0(\mathbf{x})) = \mathbf{x}, \qquad h(g_0(\mathbf{x})) = 0.$$

Given a point $\mathbf{x}_0$ and a joint configuration θ_0, with $f(\theta_0) = \mathbf{x}_0$, $h(\theta_0) = 0$, g_0 will always be defined in some neighborhood U of x_0, with $g_0(\mathbf{x}_0) = \theta_0$, as long as the extended Jacobian J_e at θ_0 is nonsingular. Furthermore, the inverse function g_0 will produce the same tracking paths in its domain of definition as the extended Jacobian method. We refer to g_0 as a *local* inverse function, because its domain of definition may turn out to be quite small, and, as such, it determines tracking paths only in a small area of the operational space.

Definition 5.

Define an *open ϵ-ball* about some point $\mathbf{x}_0$ in the space X_0 to be the set of all points in X_0 that are a distance less than ϵ from $\mathbf{x}_0$. If the specific point $\mathbf{x}_0$ or the value ϵ are not important, we will refer simply to an *open ball* in X_0.

Note that the domain of g_0 can always be chosen to contain an open ball about $\mathbf{x}_0$, because this guarantees that paths starting at $\mathbf{x}_0$ can be tracked at least some distance using g_0, no matter what direction they go in.

Definition 6.

Let U be an open ball about some point $\mathbf{x}_0 \in X$ and suppose $g_0\colon U \to T^n$ is a continuous function with the property that $f(g_0(\mathbf{x})) = \mathbf{x}$ for all $\mathbf{x} \in U$. Then g_0 will be called a *local inverse function* at $\mathbf{x}_0$.

Definition 7.

A tracking algorithm on $X_0 \subset X$ is determined by local inverse functions if the following hold:

1. For every $\mathbf{x}_0 \in X_0$ and every allowable configuration θ_0 with $f(\theta_0) = \mathbf{x}_0$, there is a local inverse function g_0 at $\mathbf{x}_0$ such that $g_0(\mathbf{x}_0) = \theta_0$.
2. Let $\mathbf{x}_0$, θ_0, and g_0 be as in part 1. If $\theta(t)$ is a tracking path for some path $\mathbf{x}(t)$ in X_0, with $\theta(t_0) = \theta_0$ at some time t_0, then $\theta(t) = g_0(\mathbf{x}(t))$ as long as $\mathbf{x}(t)$ remains in the domain of g_0.

From the above discussion, it follows that any tracking algorithm using the extended Jacobian method to track paths will be given by local inverse functions. Readers may at this point protest that local inverse functions would be a clumsy and undesirable way to implement the extended Jacobian method, and they would be correct in doing so. The purpose of these definitions is only to clarify the relationship between the extended Jacobian method and cyclic algorithms and, by so doing, to show some limitations on what can be accomplished with such methods. Theorem 3 states that cyclic tracking algorithms can always be implemented via inverse functions, and analogous statements hold for local inverse functions as well.

Definition 8.

Call a tracking algorithm on X_0 *locally cyclic* if every point in X_0 is contained in some open ball in which closed paths are always tracked by closed paths in the joint space T^n.

The following corollary is a direct consequence of Theorem 3.

Corollary: A tracking algorithm on $X_0 \subset X$ is determined by local inverse functions if and only if it is locally cyclic.

One can now ask if every locally cyclic tracking algorithm must be cyclic. If so, then Theorems 1 and 2 will imply that no tracking algorithms can be defined on all of S^2 or $SO(3)$ that are determined by local inverse functions. The answer to this question turns out to depend on topological properties of the space X_0.

Definition 9.

The set X_0 is called *simply connected* if every closed path in X_0 can be continuously deformed through closed curves in X_0 to the constant path at some point $\mathbf{x}_0 \in X_0$.

A subset of the plane is simply connected if it has no holes. If there is a hole, then any closed path that goes around the hole cannot be deformed to a constant path. The same is not true in higher dimensions. A solid ball with a spherical hole taken from its interior is still simply connected. The sphere S^2 is also simply connected, and one can show, although it is not obvious, that $SO(3)$ is not simply connected.

Theorem 4: Suppose X_0 is simply connected. If a tracking algorithm on X_0 is determined by local inverse functions, then it is cyclic and given by inverse functions defined on all of X_0.

Since S^2 is simply connected, it then follows that no locally cyclic tracking algorithm can be defined on all of S^2. Unfortunately, the argument breaks down for $SO(3)$ because $SO(3)$ is not simply connected. In Appendix I a stronger version of Theorem 2 will be used to get around this fact and prove the same result for $SO(3)$. The results are summarized in the following theorem.

Theorem 5: For f: $T^n \rightarrow X$ no tracking algorithm determined by local inverse functions can be defined on all of $X = S^2$ or $X = SO(3)$.

Corollary: For any serial-link manipulator, the extended Jacobian method cannot be defined without singularities on all of S^2 or $SO(3)$, regardless of the choice of constraint functions $h(\boldsymbol{\theta})$.

Corollary: No tracking algorithm determined by local inverse functions can be defined on an operational space that is a product space $X \times Y$, where X is either S^2 or SO(3) and Y is arbitrary.

The first of these corollaries follows immediately from the theorem, because any extended Jacobian inversion that is free of singularities determines local inverse functions. The second corollary is proven in an analogous fashion to the corollary to Theorems 1 and 2. If $\tilde{f}$: $T^n \rightarrow X \times Y$ describes the forward kinematics, let f: $T^n \rightarrow X$ be given by $f = \pi \circ \tilde{f}$. We can think of X as a subset of $X \times Y$ by choosing a point $\mathbf{y}_0$ and identifying X with $X \times \{\mathbf{y}_0\}$. In this way a locally cyclic tracking algorithm for $\tilde{f}$: $T^n \rightarrow X \times Y$ determines one also for f: $T^n \rightarrow X$, and this is impossible by Theorem 5.

Theorem 5 not only rules out the extended Jacobian method for tracking on all of S^2 or $SO(3)$, but in light of the corollary to Theorem 3 it also gives a necessary condition that any tracking algorithm must satisfy to do so: such a tracking algorithm cannot be locally cyclic. Consequently, at some point in the operational space every open ball, no matter how small, contains closed paths that are tracked by open paths. Thus, there must be a sequence of closed paths in the operational space whose lengths tend to zero, which are tracked by open paths in the joint space. An example of such an algorithm, which is defined for the three-joint pointing mechanism of Fig. 1B, is given in Appendix III. Although the method is of theoretical interest, it requires that all three joints have unlimited ranges in order to track paths, and this makes it impractical to implement.

An immediate consequence of Theorem 5 is that there must be a defect in the algorithm published by Stanišić and Pennock (1985) for orienting a four-axis wrist. In their algorithm, a constraint is imposed to prevent the wrist from entering a mechanically degenerate configuration, and thus it is claimed that the wrist will be able to continuously orient the end-effector over all of $SO(3)$. However, as previously discussed for the three-joint pointing mechanism, such a constraint may produce artificial singularities. Although the authors showed that the Jacobian J is always full rank, their method will fail unless the extended Jacobian J_e (see Eq. (6)) is also always nonsingular. In light of Theorem 5, we must conclude that this is not the case.

Problems of positioning or orienting in two-dimensional space or problems of positioning in three-dimensional space have operational spaces that contain no factors of S^2 or $SO(3)$ in them. As already remarked in Section 2, the orientation problem in two-dimensional space can be solved with one revolute joint. For the positioning problems, however, limitations do exist on the size of the sets X_0 on which cyclic or locally cyclic tracking algorithms can be defined. Brockett (1986) has obtained results of this type, and other results by the authors are planned for a forthcoming report.

6. Summary and Discussion

Some fundamental results concerning a broad class of inverse kinematic methods for redundant serial-link manipulators have been obtained. This class of algorithms, called tracking algorithms, includes inverse function methods, minimum-norm methods, methods involving the side optimization of a potential function, and the extended Jacobian method. The results obtained are of two types: (a) Theorems 1, 2, and 5 concern the possibility of defining an inverse kinematic algorithm for pointing or orienting the end-effector in three-dimensional space, and (b) Theorems 3 and 4 concern the cyclic behavior of tracking algorithms, without reference to specific operational spaces. The main results are restated briefly as follows.

First, Theorems 1 and 2 state that no continuous inverse function can be defined on all of S^2 or SO(3), which are the operational spaces for three-dimensional pointing and orienting, respectively. A corollary extends the result to operational spaces that include either of these as a factor: for example, the operational space for positioning and orienting the end-effector simultaneously. Theorem 5 and its corollaries state similar results for the extended Jacobian method.

Theorem 3 establishes the equivalence of continuous inverse kinematic functions and cyclic tracking algorithms, which are algorithms that always track closed paths in the operational space with closed paths in the joint space. This has several practical implications for the design of an inverse kinematic method. On the one hand, in light of Theorems 1 and 2, it shows that if one requires a cyclic tracking algorithm for three-dimensional pointing or orienting, then the operational space must be restricted. On the other hand, in light of Theorem 5, it shows that a tracking algorithm defined on all of S^2 or $SO(3)$ cannot even be locally cyclic; that is, there are arbitrarily short closed paths in the operational space that are tracked by open paths in the joint space. Yet another implication is that if cyclic behavior is desired on any operational space, one may as well restrict one's attention to inverse function methods, because any other approach is equivalent.

Finally, Theorem 4 partially answers a question originally raised by Baillieul (1985), which is whether or not the addition of constraint functions to resolve redundancy will yield cyclic behavior. Theorem 4 states that this will be the case if the algorithm is defined on a simply connected subset of the operational space. Otherwise, the answer depends on more detailed knowledge of the manipulator and the constraint functions. Theorems 1 and 4 together imply Theorem 5 for the pointing problem. Since they are not specific to the tasks of pointing or orienting, Theorems 3 and 4 are expected to be useful in the study of positional workspaces as well.

Although the results of this paper are very general, it may be instructive to note the limitations of the theorems. First, the results apply only to manipulators whose kinematics are representable by a continuous function from joint space to operational space of the form $f: T^n \to X$. This is true for all open-chain manipulators having only revolute and prismatic joints, but caution must be used in applying the theorems to closed-chain manipulators. If such a manipulator has only revolute and prismatic joints, it can be considered as an open-chain manipulator subject to constraint functions enforcing the loop closures. Consequently, the theorems are applicable to such a mechanism. However, it is possible to build a closed-chain manipulator containing ball-and-socket joints, in which case additional analysis would be required. It is also possible that the kinematic relation between actuator angles and end-effector motion is nonholonomic, as in the case of a wheeled mechanism or a ball-and-socket joint driven by friction wheels.

Additionally, statements concerning tracking algorithms must be interpreted strictly in terms of the definition given for such algorithms. An important class of kinematic inversion methods that are not tracking algorithms are those methods based on complete path information, which as explained above do not satisfy axiom 3. Another potentially useful approach is approximate kinematic inversion methods, which might reduce joint speeds by introducing slight modifications of the path in operational space, thereby violating axiom 2 (Wampler 1986; Nakamura and Hanafusa 1986). However, the axioms are the minimal properties an algorithm needs in order to exactly track sensor-guided paths, so the implications of the theorems are very broad in their applicability.

7. Conclusions

The problem of sensor-guided path tracking for tasks involving pointing or orienting the end-effector of a conventional manipulator leads to a dilemma: the tracking algorithm cannot be both cyclic and defined on the whole operational space. Furthermore, if cyclic behavior is required from a tracking algorithm on any operational space, that is, if it is to track closed paths in the operational space with closed joint space paths, it must be equivalent to one using inverse functions. Thus, the development of methods for constructing such inverses is a logical avenue for future research.

Tracking algorithms which use additional constraint functions to invert the kinematics on redundant manipulators will produce cyclic behavior, if the subset of the operational space in which paths can be tracked is simply connected. A generalization of this fact shows that, for pointing and orienting problems, such algorithms cannot be defined on the entire workspace. These results also provide a necessary condition on any tracking algorithm for pointing or orienting tasks, if it is to be defined on the entire workspace: there must be paths of arbitrarily short length that are tracked by open paths in the joint space.

Appendix I: Proofs of Theorems 3, 4, and 5

In this appendix we prove Theorems 3, 4, and 5. Theorem 5 depends on Theorem 1 (for the case when $X = S^2$), and a generalization of Theorem 2, hereafter referred to as Theorem 2′ (for the case when $X = SO(3)$). To better motivate this version of Theorem 2, we will start with the proof of Theorem 5, and state Theorem 2′ only after having said something about why Theorem 2 is insufficient for our purposes.

Theorems 1 and 2′ are proven with techniques from a field of mathematics called algebraic topology, which is concerned with proving the existence or nonexistence of continuous functions having various special properties, such as those of the functions in these theorems. Since these proofs involve concepts that will be foreign to many readers, they are given separately in Appendix II, along with a list of the facts, with references, that are used.

Theorem 3: A tracking algorithm on some subset $X_0 \subset X$ is equivalent to one induced by continuous inverse kinematic functions defined on X_0 if and only if it is cyclic.

Proof: Suppose first that the tracking algorithm is determined using an inverse kinematic function g: $X_0 \rightarrow T^n$. Then for a closed path $\mathbf{x}(t)$ the tracking path $g(\mathbf{x}(t))$ is also closed, so the tracking algorithm is cyclic.

Conversely, suppose that the tracking algorithm is cyclic. The proof for this case is divided into two stages: first we give a construction for the inverse function, and then we prove its continuity. Before constructing the associated inverse function g, we make a definition. ■

Definition 10.

The set X_0 is *path connected* if any two points in X_0 can be connected by a path lying entirely in X_0. If X_0 is not path connected, then it is the disjoint union of its *path components,* each of which is path connected.

Assume first that the set X_0 on which the tracking algorithm is defined is path connected. Choose a point

$\mathbf{x}_0 \in X_0$ and an allowable configuration $\boldsymbol{\theta}_0 \in f^{-1}(\mathbf{x}_0)$. The function g is uniquely determined once $\mathbf{x}_0$ and $\boldsymbol{\theta}_0$ have been chosen. For any point $\mathbf{x}_1 \in X_0$, define $g(\mathbf{x}_1)$ as follows. First choose a path $\mathbf{x}(t)$ between $\mathbf{x}_0$ and $\mathbf{x}_1$, and let $\boldsymbol{\theta}(t)$ be the tracking path of $\mathbf{x}(t)$, starting at $\boldsymbol{\theta}_0$. Then let $g(\mathbf{x}_1)$ be the value $\boldsymbol{\theta}_1$ of the tracking path upon reaching $\mathbf{x}_1$ in the operational space. This definition makes sense only if it is independent of the choice of path $\mathbf{x}(t)$.

To show that this is indeed the case, suppose that $\mathbf{x}(t)$ and $\tilde{\mathbf{x}}(t)$ are two different paths between $\mathbf{x}_0$ and $\mathbf{x}_1$, and let $\mathbf{y}(t)$ be a path from $\mathbf{x}_1$ to $\mathbf{x}_0$. We shall adopt the convention that a path consisting of shorter paths, traversed sequentially, will be denoted using the names of the shorter paths in the same sequence. For example, the path $\mathbf{xy\tilde{x}}(t)$ goes from $\mathbf{x}_0$ to $\mathbf{x}_1$ via $\mathbf{x}(t)$, returns to $\mathbf{x}_0$ via $\mathbf{y}(t)$, and then goes to $\mathbf{x}_1$ again via $\tilde{\mathbf{x}}(t)$. Choose a tracking path for $\mathbf{xy\tilde{x}}(t)$, starting at $\boldsymbol{\theta}_0$. When the tracking path reaches $\mathbf{x}_1$ the first time, it will take the value $g(\mathbf{x}_1) = \boldsymbol{\theta}_1^x$ determined by the path $\mathbf{x}(t)$. When it returns to $\mathbf{x}_0$ again after the second leg of the trip, it will again take the value $\boldsymbol{\theta}_0$, since the tracking algorithm is cyclic. Thus, as it moves out again toward $\mathbf{x}_1$, this time along $\tilde{\mathbf{x}}(t)$, it will arrive at $\mathbf{x}_1$ assuming the value of $g(\mathbf{x}_1) = \boldsymbol{\theta}_1^{\tilde{x}}$ determined by $\tilde{\mathbf{x}}(t)$. However, $\mathbf{y\tilde{x}}(t)$ is a closed path, starting and ending at $\mathbf{x}_1$, with a tracking path that starts at $\boldsymbol{\theta}_1^x$ and ends at $\boldsymbol{\theta}_1^{\tilde{x}}$. Again, since the tracking algorithm is cyclic, $\boldsymbol{\theta}_1^x = \boldsymbol{\theta}_1^{\tilde{x}}$ and $g(\mathbf{x}_1)$ is well defined.

If X_0 has several path components, a choice of starting point $\mathbf{x}_0$ and configuration $\boldsymbol{\theta}_0$ must be made for each of them. The inverse function g is then defined as above on each path component.

All that remains is to show the continuity of the function g, and to do this we use the following lemma:

Lemma 1: Given a sequence of points $\mathbf{x}_n \rightarrow \mathbf{x}$ in X_0, there is a subsequence $\{\mathbf{y}_n\} \subset \{\mathbf{x}_n\}$ and a path $\mathbf{y}(t)$ in X_0 with $\mathbf{y}(1/n) = \mathbf{y}_n$ and $\mathbf{y}(0) = \mathbf{x}$.

Proof of Lemma 1: Recall that paths were defined to have at most a finite number of discontinuities in their derivatives, and that one-sided derivatives must exist everywhere. The path $\mathbf{y}(t)$ to be constructed will be everywhere differentiable.

Since $\mathbf{x}_n \rightarrow \mathbf{x}$, one can choose the subsequence $\{\mathbf{y}_n\}$ so that the distance between $\mathbf{y}_n$ and $\mathbf{x}$ is less than $1/n^2$, and there is a path $\gamma_n(t)$ with

$$\gamma_n(1/n) = \mathbf{y}_n,$$
$$\gamma_n(1/(n+1)) = \mathbf{y}_{n+1},$$
$$\dot{\gamma}_n(1/n) = \dot{\gamma}_n(1/(n+1)) = 0,$$

and the length of $\gamma_n(t)$ between $\mathbf{y}_n$ and $\mathbf{y}_{n+1}$ is also less than $1/n^2$. The path $\mathbf{y}(t)$ is made by joining all the paths $\gamma_n(t)$ together, end to end. This path will be differentiable for $t > 0$ if each of the paths $\gamma_n(t)$ is. Furthermore, $\lim_{t\rightarrow 0} \mathbf{y}(t) = \mathbf{x}$.

To complete the proof of Lemma 1, we must show that if we define $\mathbf{y}(0) = \mathbf{x}$ then $\mathbf{y}(t)$ will be differentiable at $t = 0$. Since, for $1/(n+1) \leq t \leq 1/n$,

$$|\mathbf{y}(t) - \mathbf{y}(0)| = |\mathbf{y}(t) - \mathbf{x}| \leq |\mathbf{y}(t) - \mathbf{y}(n)| + |\mathbf{y}(n) - \mathbf{x}| \leq 1/n^2 + 1/n^2,$$

it follows that

$$\lim_{t\rightarrow 0} \left| \frac{\mathbf{y}(t) - \mathbf{y}(0)}{t} \right| \leq \lim_{n\rightarrow\infty} \frac{2/n^2}{1/(n+1)} = 0.$$

Thus, $\dot{\mathbf{y}}(0) = 0$, and the lemma is proved.■

We now prove that the function g as defined is continuous at every point $\mathbf{x} \in X_0$. For any sequence of points $\mathbf{x}_n \rightarrow \mathbf{x}$ in X_0, Lemma 1 implies that there is a path $\mathbf{y}(t)$ through a subsequence $\{\mathbf{y}_n\} \subset \{\mathbf{x}_n\}$ with $\mathbf{y}(0) = \mathbf{x}$, $\mathbf{y}(1/n) = \mathbf{y}_n$. If $\boldsymbol{\theta}(t)$ is the tracking path for $\mathbf{y}(t)$, with $\boldsymbol{\theta}(0) = g(\mathbf{x})$, then $\boldsymbol{\theta}(1/n) = g(\mathbf{y}_n)$. Since $1/n$ tends to zero, the continuity of $\boldsymbol{\theta}(t)$ implies that $g(\mathbf{y}_n)$ tends to $g(\mathbf{x})$.

If g were not continuous at $\mathbf{x}$, then there would exist a sequence $\mathbf{x}_n \rightarrow \mathbf{x}$ with $g(\mathbf{x}_n)$ bounded away from $g(\mathbf{x})$. Since this contradicts the above, g must be continuous at $\mathbf{x}$, and, since $\mathbf{x}$ was an arbitrary point in X_0, g must be continuous everywhere in X_0. This completes the proof of Theorem 3.■

Remark: In proving the continuity of g, we have assumed that, when the distances between $\mathbf{y}_n$ and $\mathbf{y}_{n+1}$ tend to zero, there are paths $\gamma_n(t)$ in X_0, between $\mathbf{y}_n$ and $\mathbf{y}_{n+1}$, whose lengths also tend to zero. Although this is certainly the case for all interesting subsets X_0 (e.g., for all manifolds with boundary), it is possible to construct pathological examples of subsets for which

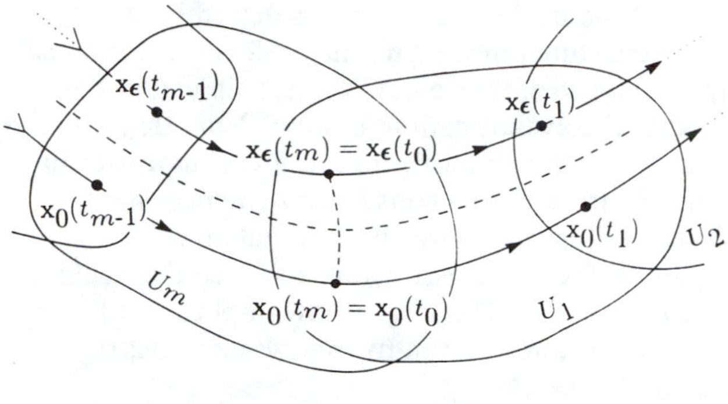

Fig. 2. A section of the deformation of closed paths $x_s(t)$, for $0 \leq s \leq \epsilon$, *and the sets* $U_1, \ldots, U_m$ *on which the local inverses are defined.*

this is not true. In such cases, the continuity proof holds with respect to a different topology on X_0, namely the one determined as above by path length.

We turn now to the question dealt with in Theorem 4—when is a locally cyclic tracking algorithm cyclic? Lemma 3 below provides the main tool for analyzing such problems. The following observation will be used in the proof of Lemma 3.

Lemma 2: Suppose a locally cyclic tracking algorithm is defined on X_0, and let g_1 and g_2 be local inverse functions defined on open balls U_1 and U_2 in X_0. If for some $\mathbf{x}_0 \in U_1 \cap U_2$, $g_1(\mathbf{x}_0) = g_2(\mathbf{x}_0)$, then g_1 and g_2 agree on the entire path component of $U_1 \cap U_2$ containing $\mathbf{x}_0$.

Proof: Let $\mathbf{x}_1$ be in the path component of $U_1 \cap U_2$ containing $\mathbf{x}_0$, and choose a path $\mathbf{x}(t)$ between $\mathbf{x}_0$ and $\mathbf{x}_1$. Then both $g_1(\mathbf{x}(t))$ and $g_2(\mathbf{x}(t))$ are tracking paths for $\mathbf{x}(t)$, with the same starting point $g_1(\mathbf{x}_0) = g_2(\mathbf{x}_0)$. By axiom 2 of tracking algorithms, the two tracking paths must agree everywhere. In particular, $g_1(\mathbf{x}_1) = g_2(\mathbf{x}_1)$, which proves the lemma.■

Definition 11.

Suppose $\mathbf{x}_0(t)$ and $\mathbf{x}_1(t)$ are two closed paths in X_0 defined on some time interval $[0, T]$. A *deformation of* $\mathbf{x}_0(t)$ *to* $\mathbf{x}_1(t)$ is a continuous function $\mathbf{x}$: $[0, 1] \times [0, T] \rightarrow X_0$ with $\mathbf{x}(0, t) = \mathbf{x}_0(t)$ and $\mathbf{x}(1, t) = \mathbf{x}_1(t)$. The deformation paths $\mathbf{x}_s(t) = \mathbf{x}(s, t)$ are also required to be closed for every $s \in [0, 1]$.

Lemma 3: Given a locally cyclic tracking algorithm on X_0, let $\mathbf{x}_s(t)$ be a deformation between closed paths in X_0. Then $\mathbf{x}_0(t)$ is tracked by a closed (resp. open) path in T^n if and only if the same holds for $\mathbf{x}_1(t)$ as well.

Proof: The reader will find it helpful to refer to Fig. 2 while reading this proof. One can track the path $\mathbf{x}_0(t)$ by choosing a sequence of open balls, $U_1, \ldots, U_m$, with local inverse functions g_j: $U_j \rightarrow T^n$ defined, and a sequence of points, $0 = t_0 < \cdots < t_m = T$, such that $\mathbf{x}_0(t) \in U_j$ for $t \in [t_{j-1}, t_j]$. The path is tracked using g_1 on $[t_0, t_1]$, then using g_2 on $[t_1, t_2]$, and so on, until, at the end, g_m is used on $[t_{m-1}, t_m]$. An argument using uniform continuity (see, for example, Theorem XI.4.6 of Dugundji (1968)) then shows that $\mathbf{x}_s(t) \in U_j$ for $t \in [t_{j-1}, t_j]$ and all s in some interval $[0, \epsilon]$.

In fact, the same sequence of local inverse functions can be used to track each $\mathbf{x}_s(t)$ for all $s \in [0, \epsilon]$. To see this, one must check that if $g_j(\mathbf{x}_0(t_j)) = g_{j+1}(\mathbf{x}_0(t_j))$ then

$$g_j(\mathbf{x}_s(t_j)) = g_{j+1}(\mathbf{x}_s(t_j)) \tag{I.1}$$

for all $s \in [0, \epsilon]$, since this is the condition necessary to extend the tracking path for $\mathbf{x}_s(t)$ from one inverse function to the next. Since $\mathbf{x}_s(t_j)$ is continuous in s, it follows that $\mathbf{x}_0(t_j)$ and $\mathbf{x}_s(t_j)$ are in the same path component of $U_j \cap U_{j+1}$ for all $s \in [0, \epsilon]$. Equation (I.1) now holds by Lemma 2.

Lemma 2 also implies that $g_1(\mathbf{x}_0(0)) = g_m(\mathbf{x}_0(T))$ if and only if $g_1(\mathbf{x}_\epsilon(0)) = g_m(\mathbf{x}_\epsilon(T))$, and this implies that the tracking paths determined by the functions g_j for $\mathbf{x}_0$ and $\mathbf{x}_\epsilon$ will both be either closed or open.

One can now iterate this process, showing each time that the tracking paths are either both open or both closed for a sequence of values $0 < s_1 = \epsilon < \cdots < s_l = 1$. This fact for $s_l = 1$ proves the lemma.■

Theorem 4 now follows directly from Lemma 3 and Theorem 3.

Theorem 4: Suppose X_0 is simply connected. If a tracking algorithm on X_0 is determined by local inverse functions, then it is cyclic and given by inverse functions defined on all of X_0.

Fig. 3. The construction of the function $g(\mathbf{y})$, with $f(g(\mathbf{y})) = h(\mathbf{y})$.

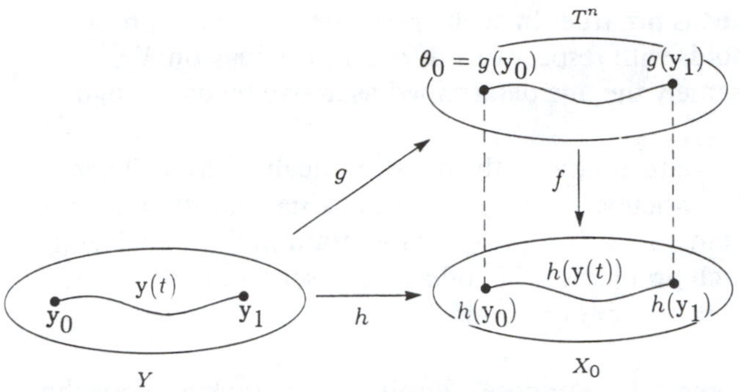

Proof: If the tracking algorithm is determined by local inverse functions, then it is locally cyclic. If X_0 is simply connected, then every closed path can be deformed to a constant path at a point. Note that if the tracking algorithm is locally cyclic, a constant path in X_0 must be tracked by a constant (and hence closed) path in T^n. Lemma 3 now implies that all closed paths in X_0 are tracked by closed paths in T^n, so the tracking algorithm is cyclic. Theorem 3 now implies that the tracking algorithm is given by inverse functions defined on all of X_0.■

Theorem 5: No tracking algorithm determined by local inverse functions can be defined on all of S^2 or $SO(3)$.

Proof: As noted in the previous section, since S^2 is simply connected, Theorems 4 and 1 imply that no tracking algorithm determined by local inverse functions can be defined on all of S^2.■

The analogous proof for $SO(3)$ would use Theorems 4 and 2, but, unfortunately, this strategy does not work, because $SO(3)$ is not simply connected. To deal with this, we switch to the Euler parameter (quaternion) representation of orientations. It turns out that this space *is* simply connected, but there are two Euler parameter representations for each orientation. Accordingly, we will introduce generalizations of Theorems 4 and 2 that allow us to move the problem from the space of orientations to the space of Euler parameters.

Let $\mathbf{R}^4$ be four-dimensional Euclidean space, and let $S^3 \subset \mathbf{R}^4$ be the subset of unit vectors in $\mathbf{R}^4$. S^3 is called the three-dimensional sphere to distinguish it from S^2, the subset of unit vectors in three-dimensional space. (Note that the dimensions of these spheres are always one less than the dimensions of the Euclidean spaces in which they live.) As in S^2, all closed paths in S^3 can be deformed to constant paths; i.e., S^3 is simply connected.

This space is introduced because there is a way of associating a rotation in $SO(3)$ to every unit vector in S^3. The components of the vector in S^3 representing a given rotation are referred to as the Euler parameters of the rotation. Let $\mathbf{v} = (v_1, v_2, v_3, v_4)$ be a unit vector, and let $\mathbf{w} = (v_1, v_2, v_3)$. If $|\mathbf{w}|$ denotes the Euclidean norm of $\mathbf{w}$, it follows that $|\mathbf{w}|^2 + v_4^2 = 1$, so $|\mathbf{w}| = \sin(\theta/2)$ and $v_4 = \cos(\theta/2)$ for some θ, with $0 \leq \theta \leq 2\pi$. We define $p(\mathbf{v}) \in SO(3)$ to be the rotation through the angle θ about the vector $\mathbf{w}$ in three-dimensional space. This defines a function p: $S^3 \rightarrow SO(3)$. One can show that every rotation in $SO(3)$ is represented by exactly two vectors in S^3 under this correspondence, and that these two vectors are antipodal to each other on S^3; i.e., $p(\mathbf{v}) = p(-\mathbf{v})$.

Now we state Theorems 2′ and 4′. Theorem 2′ is proven in Appendix II, and Theorem 4′ is proven below.

Theorem 2′: Let f: $T^n \rightarrow SO(3)$ be any continuous function. Then no continuous function $\tilde{g}$: $S^3 \rightarrow T^n$ exists such that $f(\tilde{g}(\mathbf{v})) = p(\mathbf{v})$, for every $\mathbf{v} \in S^3$.

Note that if an inverse function g: $SO(3) \rightarrow T^n$ as in Theorem 2 could be constructed, then the function $\tilde{g}(\mathbf{v}) = g(p(\mathbf{v}))$ would contradict Theorem 2′, so Theorem 2 follows as an immediate consequence of Theorem 2′.

Theorem 4′: Suppose a tracking algorithm determined by local inverse functions is given on X_0, and let h: $Y \rightarrow X_0$ be a continuous map from some simply connected space Y to X_0. Then there is a continuous function g: $Y \rightarrow T^n$ such that $f(g(\mathbf{y})) = h(\mathbf{y})$ for all $\mathbf{y} \in Y$. Furthermore, any path in X_0 of the form $h(\mathbf{y}(t))$, for some path $\mathbf{y}(t)$ in Y, will be tracked by the path $g(\mathbf{y}(t))$ in T^n.

In the special case where $Y = X_0$ and h is the identity function, Theorem 4′ reduces to Theorem 4.

Proof of Theorem 4′: The proof of this theorem parallels that of Theorem 3 very closely. Pick a starting point $\mathbf{y}_0 \in Y$ (see Fig. 3) along with an allowable configuration $\theta_0 \in T^n$ with $f(\theta_0) = h(\mathbf{y}_0)$. For $\mathbf{y}_1 \in Y$, pick a path $\mathbf{y}(t)$ starting at $\mathbf{y}_0$ and ending at $\mathbf{y}_1$. Let $g(\mathbf{y}_1)$ be the endpoint of the unique tracking path, chosen to start at θ_0, of the path $h(\mathbf{y}(t))$. As in the proof of Theorem 3, we must show that this definition is independent of the choice of path between $\mathbf{y}_0$ and $\mathbf{y}_1$. Since Y is simply connected, any closed path $\tilde{\mathbf{y}}(t)$ in Y can be deformed to the constant path at some point. It follows that the same is true for the closed path $h(\tilde{\mathbf{y}}(t))$ in X_0, since the function h will map the deformation in Y to one in X_0. Thus, by Lemma 3, any tracking path for $h(\tilde{\mathbf{y}}(t))$ must be closed. The proof that the value of $g(\mathbf{y}_1)$ is independent of the choice of path in Y between $\mathbf{y}_0$ and $\mathbf{y}_1$ follows now exactly as the corresponding proof in Theorem 3, and we omit the details.

The continuity of the function g follows from the fact that we can describe g locally as the composite $g(\mathbf{y}) = g_0(h(\mathbf{y}))$, where g_0: $U_0 \rightarrow T^n$ is a local inverse function. Theorem 4′ is now proven.■

The proof of Theorem 5 can now be completed. From problem G4 of Chapter 1 of Spanier (1966), S^n is simply connected for $n > 1$. Given a tracking algorithm determined by local inverse functions, defined on all of $X = SO(3)$, apply Theorem 4′ with $Y = S^3$ and h being the function p: $S^3 \rightarrow SO(3)$. Theorem 4′ now guarantees the existence of the function $\tilde{g}$: $S^3 \rightarrow T^n$ which, according to Theorem 2′, does not exist. Thus the tracking algorithm cannot be defined on all of $SO(3)$.

Appendix II: Proofs of Theorems 1 and 2′

Algebraic topology deals with methods for reducing questions about continuous functions to algebraic questions that can be answered more easily. We start by describing some of these algebraic tools.

Let Y be some space, e.g., S^2, $SO(3)$, or T^n. Using the topological structure of Y, one can construct a vector space of algebraic invariants, $H^*(Y)$, called the cohomology of Y with real coefficients (or just the cohomology of Y, for short). The construction of these invariants is rather complicated, and the reader is referred to Spanier (1966) for details.

If X is some other space, then one can associate to every continuous function f: $X \rightarrow Y$ a linear transformation f^*: $H^*(Y) \rightarrow H^*(X)$. Note that, when going from f to f^*, the functions change direction. Questions about the continuous function f can often be reformulated in terms of the linear transformation f^*. Because algebraic questions about finite-dimensional vector spaces are often easier to answer than the corresponding questions about continuous functions, this can be a valuable simplification of the problem.

We now give a short catalogue of the properties of these invariants that we will need to prove Theorems 1 and 2′. These facts can be found in any standard text on algebraic topology (e.g., Chapters 4 and 5 of Spanier 1966).

1. The vector space $H^*(Y)$ is a direct sum of subspaces $H^i(Y)$, for $i = 0, \ldots, n$, where n is the dimension of the space Y. Each subspace $H^i(Y)$ contains information about the ith-dimensional structure of the space Y, and we say that an element $y \in H^i(Y)$ is an i-dimensional element. The linear transformation f^* is actually a direct sum of linear transformations f^i: $H^i(Y) \rightarrow H^i(X)$ on these subspaces.
2. A product, called the cup product and written $y_1 \cup y_2$, is also defined on the vector space $H^*(Y)$. If $y_1 \in H^i(Y)$ and $y_2 \in H^j(Y)$, then $y_1 \cup y_2 \in H^{i+j}(Y)$.
3. A linear transformation f^* induced by a continuous function f commutes with the cup products:
$$f^*(y_1 \cup y_2) = f^*(y_1) \cup f^*(y_2).$$
4. If Z is yet another space and g: $Y \rightarrow Z$ is a continuous function, then for the composition function $g \circ f$: $X \rightarrow Z$ given by $g \circ f(\mathbf{x}) = g(f(\mathbf{x}))$,
$$(g \circ f)^* = f^* \circ g^*: H^*(Z) \rightarrow H^*(X).$$
5. If h: $X \rightarrow X$ is the identity (i.e., $h(\mathbf{x}) = \mathbf{x}$), then h^*: $H^*(X) \rightarrow H^*(X)$ is also the identity linear transformation.
6. Properties 4 and 5 imply that, for functions g: $X \rightarrow Y$ and f: $Y \rightarrow X$, if $f(g(\mathbf{x})) = \mathbf{x}$, then $g^* \circ f^*(x) = x$ for all $x \in H^*(X)$.

We now give an explicit description of the cohomology $H^*(T^n)$ of the joint space T^n. The computation of this cohomology, as well as of the cohomologies of S^2, S^3, and $SO(3)$, is a fairly standard application of the methods found, for example, in Chapters 4–6 of Spanier (1966). $H^1(T^n)$ is a k-dimensional vector space if there are exactly k rotational joints whose factors in T^n are circles. The other factors, which are line segments, correspond either to translational joints or to rotational joints with angle stops. The cohomology $H^*(T^n)$ remains unchanged by the presence of these factors. Let $t_1, \ldots, t_k$ be a basis for $H^1(T^n)$.

For $j \geqslant 1$, a basis for $H^j(T^n)$ is given by the set of all j-fold products, $t_{i_1} \cup \cdots \cup t_{i_j}$, where $i_1 < \cdots < i_j$. Note that this fact is consistent with the dimension rules given in property 2 for cup products. Since each t_i is in $H^1(T^n)$, a j-fold product of them must be in $H^j(T^n)$. The dimension of $H^j(T^n)$ is thus the binomial coefficient of k things taken j at a time.

Theorem 1: Let f: $T^n \rightarrow S^2$ be any continuous function. Then no continuous function g: $S^2 \rightarrow T^n$ exists such that $f(g(\mathbf{v})) = \mathbf{v}$ for every $\mathbf{v} \in S^2$.

Theorem 2′: Let f: $T^n \rightarrow SO(3)$ be any continuous function. Then no continuous function $\tilde{g}$: $S^3 \rightarrow T^n$ exists such that $f(\tilde{g}(\mathbf{v})) = p(\mathbf{v})$ for every $\mathbf{v} \in S^3$.

Proof of Theorem 1: Suppose a continuous g: $S^2 \rightarrow T^n$ exists, and

$$f(g(\mathbf{v})) = \mathbf{v} \qquad \text{for all } \mathbf{v} \in S^2. \tag{II.1}$$

The idea of the proof is to use Eq. (II.1) to deduce something about the cohomology $H^*(S^2)$, which is known to be false. This will prove that Eq. (II.1) cannot hold.

We do this by noting that, from property 6, $g^*(f^*(x)) = x$ for all $x \in H^*(S^2)$. Since $f^*(x) \in H^*(T^n)$, it follows using our basis for $H^j(T^n)$, $j \geqslant 1$, that $f^*(x)$ is a sum of products of one-dimensional elements, say

$$f^*(x) = \sum_{i_1 < \cdots < i_j} a_{i_1 \ldots i_j} t_{i_1} \cup \cdots \cup t_{i_j},$$

if $x \in H^j(S^2)$.

Since the t_i are one-dimensional, so are the $g^*(t_i)$, by property 1. By property 3, it now follows that

$$x = g^*(f^*(x)) = \sum_{i_1 < \cdots < i_j} a_{i_1 \ldots i_j} g^*(t_{i_1}) \cup \cdots \cup g^*(t_{i_j})$$

is also a sum of products of one-dimensional elements in $H^1(S^2)$.

The above conclusion is valid for all $x \in H^j(S^2)$ and for all $j \geqslant 1$. If we can display elements of some $H^j(S^2)$ that cannot be written in this fashion, this will provide the contradiction we are looking for. To do this, we now give a description of the vector space $H^*(S^2)$.

$H^j(S^2) = \mathbf{0}$, except for $j = 0$ or $j = 2$. These two subspaces are one-dimensional. Since $H^1(S^2) = \mathbf{0}$, it follows that no nonzero element in any $H^j(S^2)$ can be a product of one-dimensional elements. Since there are nonzero elements in $H^2(S^2)$, Theorem 1 is proven.■

Proof of Theorem 2′: The cohomology $H^*(SO(3))$ is given as a special case of Proposition 10.2 of Borel (1953). $H^j(SO(3)) = \mathbf{0}$, except for $j = 0$ or $j = 3$. These two subspaces are one-dimensional. The cohomology $H^*(S^3)$ is isomorphic to $H^*(SO(3))$, and, in fact, p^*: $H^*(SO(3)) \rightarrow H^*(S^3)$ is an isomorphism, where p: $S^3 \rightarrow SO(3)$ is the projection function defined in Appendix I.

As in the proof of Theorem 1, if $f(\tilde{g}(\mathbf{v})) = p(\mathbf{v})$, then

$$\tilde{g}^*(f^*(x)) = p^*(x) \tag{II.2}$$

for all $x \in H^*(SO(3))$. Since $f^*(x)$ must be a sum of products of one-dimensional elements, so must be $p^*(x)$ from Eq. (II.2). Since, as already noted, p^* is an isomorphism, it follows that all non-zero-dimensional elements of $H^*(S^3)$ must be sums of products of one-dimensional elements.

Appendix III. A Noncyclic Tracking Algorithm for Pointing

Together, Theorem 5 and the corollary to Theorem 3 show that any tracking algorithm defined on all of S^2, the operational space for pointing in three-dimen-

Fig. 4. Constraint regions for the noncyclic pointing algorithm. The pattern repeats over the entire plane.

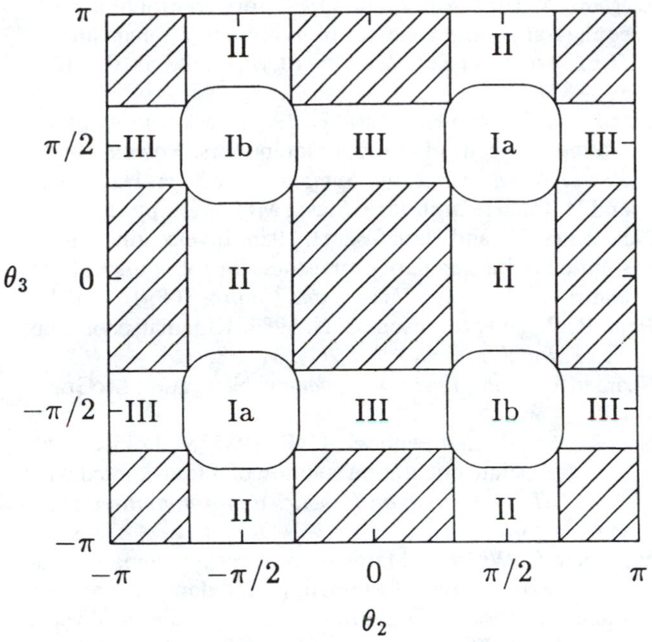

sional space, must be locally noncyclic. For any such algorithm, there is at least one point in the operational space where every open ball, no matter how small, contains closed paths that are tracked by open joint space paths. To show that this is not a vacuous statement, we demonstrate that at least one such algorithm exists, given as follows. For the three-joint mechanism of Fig. 1B, let the angles θ_1, θ_2, and θ_3 be zero for the configuration shown, with positive rotations in the directions indicated. Also, let **v** be the pointing direction. The algorithm is parameterized by two constants a, b, with $0 < a, b < 1$. Starting from the configuration $\mathbf{v} = \mathbf{y}$, with $\boldsymbol{\theta} = (0, \pi/2, \pi/2)$, the algorithm is given by the constraints

$$\begin{aligned} \dot\theta_1 &= 0 \quad && \text{if } |\sin\theta_2 \sin\theta_3| > ab, \\ \dot\theta_2 &= 0 && \text{if } |\sin\theta_2 \sin\theta_3| \le ab, |\sin\theta_2| \ge a, \\ \dot\theta_3 &= 0 && \text{if } |\sin\theta_2 \sin\theta_3| \le ab, |\sin\theta_3| > b. \end{aligned}$$

These constraints are active in the regions labeled I(a) or I(b), II, and III, respectively, in Fig. 4. With this algorithm, the mechanism will never enter the remaining area, where $|\sin\theta_2| < a$, $|\sin\theta_3| \le b$. Consequently, the mechanism never encounters the degenerate configurations at $\sin\theta_2 = \sin\theta_3 = 0$, which is a necessary, but not sufficient, condition for the method to succeed. It can be shown that the method does indeed track all paths in S^2, and that the ratio of joint speeds to operational speed, $\|\dot{\boldsymbol{\theta}}\|/\|\dot{\mathbf{v}}\|$, is bounded.

Path tracking that occurs entirely within one of the constraint regions is cyclic; the noncyclic behavior of the algorithm is observed for paths that enter multiple regions. Since $\mathbf{y} \cdot \mathbf{v} = \sin\theta_2 \sin\theta_3$, regions I(a) and I(b) map to caps of the sphere in the **y** and −**y** directions, and their boundaries are circles parallel to the (**x**, **z**)-plane. Regions II and III both map to the band between the caps. Local inverse functions do not exist for points on the circles $\mathbf{y} \cdot \mathbf{v} = \pm ab$, because of the switching of constraints. For any small closed loop around a point on these boundaries, θ_1 advances on the portion of the loop where $|\mathbf{y} \cdot \mathbf{v}| < ab$, but the return portion of the path is in region I(a) or I(b), where θ_1 is fixed. Consequently, the joint space path is open, and the operational space path is an arbitrarily small closed path, so the algorithm is locally noncyclic, as expected.

In addition to its noncyclic behavior, this algorithm also has the undesirable property that all paths can be tracked only if the joint angles are unlimited. This problem is not limited to paths involving large motions; multiple passes around a small loop of the type just described will incrementally advance θ_1 with each revolution until its limit is reached. It is not clear whether there is a mechanism and a corresponding noncyclic algorithm that can track all paths on S^2 while obeying limits on the joint angles.

References

Baillieul, J. 1985 (St. Louis). Kinematic programming alternatives for redundant manipulators. *Proc. IEEE Int. Conf. on Robotics and Automation,* pp. 722–728.

Bobrow, J. E., Dubowsky, S., and Gibson, J. S. 1983 (San Francisco). On the optimal control of robotic manipulators with actuator constraints. *Proc. Automatic Control Conf.* Vol. 3, pp. 782–787.

Borel, A. 1953. Sur la cohomologie des espaces fibres principaux et des espaces homogènes de groupes de Lie compacts. *Ann. of Math.* 57:115–207.

Brockett, R. W. 1986 (San Francisco). Inverse kinematic programming for redundant mechanisms. Unpublished lecture at IEEE Int. Conf. on Robotics and Automation.

Chang, P. H. 1986 (San Francisco). A closed-form solution for the control of manipulators with kinematic redundancy. *Proc. IEEE Int. Conf. on Robotics and Automation.* Vol. 3, pp. 9–14.

Dugundji, J. 1968. *Topology.* Boston: Allyn and Bacon.

Gottlieb, D. H. 1986 (San Francisco). Topology and robots. *Proc. IEEE Int. Conf. on Robotics and Automation.* Vol. 3, pp. 1689–1691.

Hollerbach, J. M. 1985. Optimum kinematic design for a seven degree of freedom manipulator. In *Robotics Research — The 2nd Intl. Symposium,* eds. H. Hanafusa and H. Inoue, Cambridge, Mass.: MIT Press, pp. 215–222.

Hollerbach, J. M., and Suh, K. C. 1985 (St. Louis). Redundancy resolution of manipulators through torque optimization. *Proc. IEEE Int. Conf. on Robotics and Automation,* pp. 1016–1021.

Johnson, D. W., and Gilbert, E. G. 1985 (Ft. Lauderdale). Minimum time path planning in the presence of obstacles. *Proc. 24th IEEE Conf. on Decision and Control.* Vol. 3, pp. 1748–1753.

Khatib, O. 1985 (St. Louis). Real-time obstacle avoidance for manipulators and mobile robots. *Proc. IEEE Int. Conf. on Robotics and Automation,* pp. 500–505.

Klein, C. A. 1985. Use of redundancy in the design of robotic systems. In *Robotics Research — The 2nd Intl. Symposium,* eds. H. Hanafusa and H. Inoue, Cambridge, Mass.: MIT Press, pp. 207–214.

Klein, C. A., and Huang, C-H. 1983. Review of pseudoinverse control for use with kinematically redundant manipulators. *IEEE Trans. on Sys., Man, and Cybernetics* SMC-13-3: 245–250.

Liégeois, A. 1977. Automatic supervisory control of the configuration and behavior of multibody mechanisms. *IEEE Trans. on Sys., Man, and Cybernetics* SMC-7-12: 868–871.

Nakamura, Y., and Hanafusa, H. 1985. Task priority based redundancy control of robot manipultors. *Robotics Research — The 2nd Intl. Symposium,* eds. H. Hanafusa and H. Inoue, Cambridge, Mass.: MIT Press, pp. 155–162.

Nakamura, Y., and Hanafusa, H. 1986. Inverse kinematic solutions with singularity robustness for robot manipulator control. *J. Dyn. Sys., Meas., and Control* 108: 163–171.

Paul, R. P., and Stevenson, C. N. 1983. Kinematics of robot wrists. *Int. J. Robotics Res.* 2(1): 31–38.

Spanier, E. 1966. *Algebraic Topology.* New York: McGraw-Hill.

Stanišić, M. M., and Pennock, G. R. 1985 (St. Louis). A non-degenerate orientation solution of a four-jointed wrist. *Proc. IEEE Int. Conf. on Robotics and Automation,* pp. 998–1003.

Wampler, C. W. 1986. Manipulator inverse kinematic solutions based on vector formulations and damped least-squares methods. *IEEE Trans. on Sys., Man, and Cybernetics* 16(1): 93.

Wampler, C. W. 1987 (Raleigh, NC). Inverse kinematic functions for redundant manipulators. *Proc. IEEE Int. Conf. on Robotics and Automation.* Vol. 2, pp. 610–617.

Whitney, D. E. 1969. Resolved motion rate control of manipulators and human prostheses. *IEEE Trans. on Man-Machine Systems* MMS-10-2: 47–53.

Yoshikawa, T. 1985. Manipulability of robotic mechanisms. *Robotics Research — The 2nd Intl. Symposium,* eds. H. Hanafusa and H. Inoue, Cambridge, Mass.: MIT Press, pp. 439–446.

Part 11
Dynamically Dexterous Robots

D. E. KODITSCHEK
Yale University

WHILE COMPUTERS have been built that play chess better than almost every human being, we have yet to build a machine as capable as a toddler of walking up the stairs or grabbing a cup. The elusive foundations of such dynamical dexterity evidently transcend the purview of any established discipline and surely belong at the center of robotics research. Yet the study of dynamically dexterous machines remains a lamentably under-represented area of research in robotics. No doubt, this is partially because of the perceived difficulty of the problem area. Surely, vagaries of funding for basic research in an infant discipline forced to earn its keep too early is at least as much to blame. Nevertheless, there is a growing community of researchers whose fascination for building and understanding machines that can balance, strike purposively, and coordinate multiple incompletely actuated degrees of freedom has begun to lay the foundations of a scientific framework in this area.

Part 11 offers four representative papers from the growing literature concerning dynamically dexterous robots. These selections provide some measure of the great diversity of effort presently found in the field. Each concerns a different task: hopping, ping-pong, walking, and juggling. Each places an emphasis on distinct behavioral objectives: control via natural resonance, effective game strategy, passive actuation and sensing, stability of motion. Each introduces different tools: energy considerations, AI blackboard architectures, linearized analysis, and unimodal return maps. Yet, as these introductory remarks are intended to suggest, there is a more-than-causal connection between such apparently disparate themes.

To begin with, each paper is centrally concerned with the control of collisions. From the earliest days of the field, robotics researchers have understood the importance of such a concern (for instance, refer to Goertz, 1963, as quoted by Paul [1]). But it seems fair to say that the preponderance of the last decade's research in robotics has focused on the constituents of manipulation—trajectory tracking, force and impedance control, quasi-static grasping, and compliant motion—to the exclusion of their confluence in accomplishing some complete task-level behavior. In contrast, dynamical dexterity demands an integrated view of and effective passage between no-contact, impact, and compliant manipulation conditions. Controlling collisions takes center place when a task demands that ground strike the body (juggling, ping-pong) or the body strike ground (hopping, walking).

Perhaps in consequence of the dynamical environment facing the robots discussed in these papers, almost all (the notable exception being the ping-pong machine) share a reliance on the natural motion of an appropriately tuned mechanical system to achieve some desired behavior. This is in marked contrast to the prevailing paradigm in robotics, which postulates a reference trajectory generated by some off-line intelligence that the "dumb" controller then forces the robot to track. Although natural motion has been proposed as a means of effecting and controlling tasks in more traditional branches of robotics—for example, artificial potential fields for obstacle avoidance [2], [3], or sensorless manipulation strategies for assembly [4]—the study of dynamically dexterous machines may represent the only area within the field where this point of view has predominated. Perhaps the striking success of the machines described here may invite some wider reconsideration of the virtues of natural motion in intelligent controllers.

And this introduces the most important feature shared by all of the papers in this part. Each enjoys an association with successful empirical study. Tractable models of friction, impact, and deformation being fraught with inaccuracy, there seems to be no substitute for experimentation in the study of dynamical dexterity. One might say that simulation studies in this challenging realm are doomed to succeed [5]. Indeed, at least two of the papers below suggest how lessons learned in making real systems work provide valuable corrective insight for a priori programmatic methodologies.

RAIBERT'S HOPPING MACHINES

The first systematic work in this task domain (and still arguably the most important by any number of measures) is the pioneering research of M. H. Raibert, whose careful experimental studies verify the correctness of his elegant control strategies for legged locomotion [6]. Perhaps the most compelling feature of these control algorithms is their minimal dependence on conventionally conceived higher level intelligence (and, in particular, on planned reference trajectories as mentioned above). His machines rely instead upon the intrinsic dynamical characteristics of actuators and masses.

The paper selected for inclusion in this collection, "Dynamic Stability and Resonance in a One-Legged Hopping Machine," contains one of Raibert's earliest written presentations of his ideas and deserves archival status as reflecting a seminal moment in the history of the field. Here one finds the first suggestion that so apparently complicated a behavior as running or hopping might be decomposed into properly coordinated but intrinsically simple vertical and horizontal subtasks. Moreover, here, perhaps for the first time in the robotics literature, one finds the suggestion that an abstract behavior such as "hop in place with a uniform height" might be parsimoniously encoded and readily effected in terms of restoring energy to a lossy spring. Finally, having been written before any physical experimentation had begun, the paper offers an instructive glimpse of the author's developing insights into how his machine should work.

The essential idea—harnessing a mechanical resonance to achieve the desired vertical hopping mode—comes through loud and clear. Horizontal control in Raibert's later reports is decomposed into a balance/velocity component distinct from the attitude stabilizer [6], [7]. In this paper, these functions are combined within the balance controller. Moreover, the balance-control scheme proposed here—preplanned tabular lookup and interpolation based on off-line simulation—is subsequently abandoned in favor of an elegant, symmetry-based feedback algorithm [7], [8]. One presumes that some of these modifications may well have been occasioned by empirical lessons along the way to getting a physical machine up and running. Intriguingly, the paper also introduces a suggestive line of thinking that would replace the simulation data in the balance controller's lookup table with measured input-output values accumulated over the course of hopping. Such memory-based learning approaches to motor control have recently become very popular in robotics.*

Following Raibert's pioneering investigations, one can begin to see a growing interest in statically unstable gaits in the legged-robot literature [10], [11], [12]. By and large, however, this work retains the traditional reliance upon an a priori intelligently (conventionally equated with open loop) planned reference trajectory for each limb that the "dumb" control system then forces the machine to track. Despite the gradual emergence of legged locomotion as an acceptable sub-field of robotics it seems fair to say that no real competitors have emerged over the past decade since Raibert's original machines appeared. They stand as the first (and, arguably, are still the only) legged machines capable of dynamic locomotion and active balance.

Raibert has subsequently published widely popularized versions of these ideas [7], scientifically broader assessments of their importance and applicability [6], [8], and a variety of technical articles of greater clarity. But this early contribution documents what might be considered one of the few conceptual breakthroughs that have yet occurred in the young field of robotics.

* One of the most cogent (and empirically committed) proponents of this point of view, Atkeson [9], has pursued juggling studies along these lines that will be mentioned herein.

Andersson's Ping-Pong Player

Probably the most dramatic example to date of an autonomous dynamically dexterous machine was contributed by R. L. Andersson of AT & T Bell Laboratories. His ping-pong machine was arguably the first robot ever built that achieved a level of mechanical skill in any way comparable to that of a human being. Moreover, this impressive system was largely comprised of very unimpressive off-the-shelf industrial components—the notable exception to be found in some of the computational hardware [13]. The paper included, "Understanding and Applying a Robot Ping-Pong Player's Expert Controller," has been selected as offering one of Andersson's most concentrated efforts to explain how this complex and intricately engineered system actually functioned.

Apart from the sheer enormity of effort expended by a lone engineer in creating such a machine, there is little exceptional in the materials or models and one is immediately led to consider the system architecture, the focus of the paper. A low-level controller forces the manipulator to track a spline-based reference trajectory [14] whose parameters are selected (and continuously refined) by a higher-level process. The higher level employs a blackboard architecture to optimize (or, more accurately, satisfy and sub-optimize) a collection of performance criteria. These criteria are applied to outcomes predicted by the robot's internal models of the flight and collision dynamics. A comparison with Raibert's work is instructive: There is no natural motion at all in this system and, while the high-level strategist bears close resemblance to the balance scheme Raibert proposes, these are both quite distant from the symmetry-based feedback computations that the working hopping machines actually employed. Thus, in some sense, Andersson's approach seems diametrically opposed to that of Raibert.

Andersson eschews the feasibility and hence the value of formal analytical study in this task domain. Indeed, given the complexity of how he views the task and consequent interrelated functioning of his C calls, it seems unlikely that any simple representation of his robot's operating principles can be forthcoming from even the most careful reading of the code. In consequence, it may be difficult for other researchers to extract from his creation a design methodology applicable to other settings. In his own view, two central precepts emerge from this work. The first concerns the importance of controlling from a continuous sensory data stream: Andersson holds that ballplayers ought to keep their eyes on the ball. The second promotes the virtues of an exception handler: In a more classical setting this idea might have been cast in terms of predictive or look-ahead controllers (although these have traditionally been used to cancel out external disturbances in the feedforward rather than feedback path).

Unquestionably, the ping-pong playing robot must be adjudged a feat of empirical engineering brilliance. Robot practitioners will want to understand whether this bespeaks some fundamental design advance embodied in the creation or simply the considerable talents of the creator. The paper in this collection presents a well-written and succinct contribution toward an appraisal of that question.

McGeer's Passive Dynamical Walker

An emerging point of view in robotics concerns the value of passive elements that enable a mechanical system to funnel key aspects of an environment into a more desirable state with no explicit effort of thought or action. Mason, arguably the central proponent of this approach, has developed sensorless manipulation strategies in the quasi-static task domain [4], [15]. In the domain of dynamical tasks, T. McGeer has further extended this idea by building dynamically stable walking machines that employ neither sensors nor actuators! Beyond the intrinsic interest of these mechanisms and their impressive locomotive capabilities McGeer's papers serve as models of scientific method in the field of robotics. The paper selected for inclusion in this part, "Passive Dynamic Walking," offers a detailed analytical study of one simple machine along with carefully presented experimental evidence to support the validity of the model.

One of the chief strengths of this paper is the methodical development of the model from one-degree-of-freedom wheel to one-degree-of-freedom-with-impact rimless wheel to two-degree-of-freedom-with-impact biped. The biped model, a fourth-order nonlinear difference equation, describes the evolution of the robot's state at successive foot strikes. Its fixed points represent the steady state gaits and their local stability properties determine whether or not they should be experimentally observable. The nonlinear character of the dynamics is such that the fixed points are numerically identified and the eigenvalues of the linearized approximation are computed as well. The domain of attraction is estimated numerically. These predictions compare well with experimental data.

Of course, McGeer does not seriously suggest that passively actuated machines in themselves will find useful applications. Rather, he argues that sound engineering design (both from the point of view of control and energetics) promotes the construction of active machines on platforms whose intrinsic mechanical properties are well suited to the task. This same theme has been implicit [7], and explicit [16] in Raibert's appeal to actively corrected mechanical resonances. Evidently, a reliance on passive sensing or actuation bespeaks an appeal to natural motion in commanding and effecting robot tasks.

Bühler and Koditschek's Juggler

Attempts to build juggling machines no doubt reach back to antiquity, but the first significant contemporary interest in this area must be attributed to Claude Shannon who theorized about and experimented with the timing and spatial patterns characteristic of sustainable juggling policies [6]. Several research groups have attempted to build juggling robots over the last decade. The first programmable machine capable of juggling one (and two) freely falling masses subject to the earth's gravitational field with negligible countervailing frictional or buoyant forces over a sustained period of time was developed by M. Bühler and D. E. Koditschek four years ago. The paper, "From Stable to Chaotic Juggling: Theory, Simulation, and Experiments," takes its place in a series of contributions seeking to advance a more general methodology for robot planning and control in a dynamical environment. This particular selection, because it makes contact with issues arising from the discussion of all three previously discussed papers, may serve as a convenient means of summarizing some of the common threads running through the dynamically dexterous robot literature.

The working algorithm described in this paper conforms to Andersson's general precept of keeping an eye on the ball: The robot is forced to track a nonlinear mirror reflection of the ball's trajectory and the resulting impacts achieve the specified juggling behavior. It is interesting to note that this scheme replaced a theoretically motivated algorithm based on feedback only with respect to the ball's state at impact after the latter failed to produce the desired empirical behavior [18]. Retrospective numerical analysis similar to that described in McGeer's paper showed, in contrast, that the domain of attraction of the original linearized scheme was smaller than the robot's sensory resolution.

The paper is most notable for deriving verifiable and unforeseen empirical predictions from a subsequent body of theory developed to explain the success of the working algorithm. In particular, the appearance of experimentally recorded period-doubling bifurcations in the machine's steady-state behavior implies that the effective dynamics are strongly nonlinear as the authors' earlier analysis had suggested. Moreover, they imply that the stability mechanism underlying the successful juggling behavior is identical to one discovered in an analysis of Raibert's vertical hopping controller [17]. Finally, the latter correspondence would not be coincidental, since the juggling algorithms employed were explicitly motivated by Raibert's use of total energy as a means of encoding a stable bounce. Thus, this paper may offer the first clear evidence of the more general applicability of Raibert's original ideas.

In contrast, it is worth noting that a subsequent spatial

juggling machine developed by Atkeson [9] used a learning scheme remarkably similar in concept to that proposed in the Raibert paper. This tabular lookup and interpolator was designed to correct a programmed linear approximation to the feedforward model of paddle and flight dynamics. Presumably, analysis of the uncorrected effective closed-loop dynamics of this machine might reveal an empirically realizable linear stability mechanism with as large a domain of attraction as in McGeer's work. In a fascinating recent series of experiments, Atkeson and colleagues have demonstrated stable juggling behavior using open-loop actuator strategies [19], lending another instance within the domain of dynamical dexterity to sensorless manipulation, and moving even closer to McGeer's point of view.

Bühler and Koditschek marshal persuasive physical evidence in this paper to suggest that their juggler's successful operation and considerable resistance to adverse perturbations may be attributed to a recently discovered nonlinear stability mechanism. Does this unimodal return map truly describe the stability properties of the vertical hopping mode in Raibert's machines? Even if so, is this relatively complicated mechanism really more useful than the linear dynamics demonstrably harnessed by McGeer? Has it yet been clearly demonstrated that these approaches to task planning and control—employing the natural modes of a mechanical system—are competitive with or even outperform the more conventional paradigm employed to such good effect by Andersson? Future contributions in the area of dynamically dexterous robotics may offer answers to questions such as these.

References

[1] R. P. Paul, *Robot Manipulators: Mathematics Programming and Control,* Cambridge, MA: MIT Press, 1981.

[2] O. Khatib, "Real time obstacle avoidance for manipulators and mobile robots," *Int. J. Robotics Res.,* vol. 5, no. 1, pp. 90–99, 1986.

[3] D. E. Koditschek and E. Rimon, "Robot navigation functions on manifolds with boundary," *Adv. in Appl. Math.,* vol. 11, pp. 412–442, 1990.

[4] M. Erdmann and M. T. Mason, "An exploration of sensorless manipulation," in *Proc. IEEE Int. Conf. Robotics and Automation,* pp. 1569–1574, April 1986.

[5] L. L. Whitcomb, personal communication, 1991.

[6] M. H. Raibert, *Legged Robots that Balance,* Cambridge, MA: MIT Press, 1986.

[7] M. H. Raibert and I. E. Sutherland, "Machines that walk," *Scientific American,* vol. 248, no. 1, pp. 44–53, Jan. 1983.

[8] M. H. Raibert, "Symmetry in running," *Science,* vol. 231, pp. 1292–1294, March 1986.

[9] E. W. Aboaf, S. M. Drucker, and C. G. Atkeson, "Task-level robot learning: juggling a tennis ball more accurately," *Proc. IEEE Int. Conf. Robotics and Automation,* pp. 1290–1295, May 1989.

[10] J. Furusho and M. Masubuchi, "Control of a dynamical biped locomotion," in *Study on Mechanisms and Control of Bipeds,* H. Miura and I. Shimoyama, Eds. Tokyo: University of Tokyo, 1987, pp. 116–127.

[11] H. Miura and I. Shimoyama, "Dynamic walk of a biped," *Int. J. Robotics Res.,* vol. 3, pp. 60–74, 1984.

[12] H. Miura, I. Shimoyama, M. Mitsuishi, and H. Kimura, "Dynamical walk of quadruped robot (Collie-1)," *Int. Symp. Robotics Res.,* pp. 317–324, 1985.

[13] R. L. Andersson, "Real-time gray scale video processing using a moment-generating chip," *IEEE J. Robotics and Automation,* pp. 79–85, 1985.

[14] R. L. Andersson, "Aggressive trajectory generator for a robot ping-pong player," in *Proc. Int. Conf. Robotics and Automation,* pp. 188–193, 1988.

[15] R. H. Taylor and M. T. Mason, "Sensor-based manipulation planning as a game with nature," in *Proc. 4th Int. Symp. Robotics Res.,* Aug. 1987.

[16] C. M. Thompson and M. H. Raibert, "Passive dynamic running," in *Experimental Robotics I,* V. Hayward and O. Khatib, Ed. Berlin: Springer-Verlag, 1990, pp. 74–83.

[17] D. E. Koditschek and M. Bühler, "Analysis of a simplified hopping robot," *Int. J. Robotics Res.,* vol. 10, no. 6, pp. 587–605, Dec. 1991.

[18] M. Bühler, D. E. Koditschek, and P. J. Kindlmann, "A family of robot control strategies for intermittant dynamical environments," *IEEE Cont. Syst. Magazine,* vol. 10, pp. 16–22, Feb. 1990.

[19] S. Schaal, C. G. Atkeson, and S. Botros, "What should be learned," in *Proc. Seventh Yale Workshop in Adaptive and Learning Systems,* May 1992, pp. 199–204.

Dynamic Stability and Resonance in a One-Legged Hopping Machine

M. H. RAIBERT
Computer Science Department and Robotics Institute,
Carnegie-Mellon University, Pittsburgh, Pennsylvania

Abstract—**A one-legged hopping machine is modeled and studied in order to focus on the important locomotion problems of balance, resonance, and dynamic control, while avoiding the problem of coordinating legs. An ideal mechanical hopping system and its differential equations of motion are presented. The control of this system is decomposed into a vertical hopping problem and a horizontal balance problem. A total vertical energy measure is used to control uniformity of hopping height when there are mechanical losses and irregular terrains. Balance and control of translations are achieved by a new table look-up method that extends previous work on computing dynamics with tables. Ongoing work to implement a physical planar hopping machine is discussed.**

Introduction

ONE NEED only watch a few slow-motion instant replays on Saturday afternoon sports television to be amazed by the variety and complexity of ways a human can carry, swing, toss, glide, and otherwise propel his body through space. Orientation, balance, and control are maintained at all times without apparent effort, while the ball is dunked, the bar is jumped, or the base is stolen. And such spectacular performance is not confined to the sports arena—behavior observable at any local playground is equally impressive from a mechanical engineering, sensory motor integration, or computer science point of view. The final wonder comes when we observe the one-year-old infant's first wobbly steps with the knowledge that running and jumping will soon be learned and added to the repertoire.

Despite these accomplishments in using his own legs to locomote, man is still at a primitive stage in the development and construction of vehicles that have the mobility, agility, versatility, and energy advantages of legged devices. Since animals large and small use legs for locomotion they are able to move quickly and reliably over flat, hilly, and mountainous terrain, or through forest, swamp, marsh, and jungle. But we have neither theory nor understanding of legged systems to explain the behavior we see in biological systems, nor to build machines with similar properties. As a result our mobility is generally limited by dependence on roads, runways, the beasts of burden, and our own two feet.

The present attack on this situation was motivated by two observations:

- Most biological walkers larger than insects depend on active balance and dynamic stability as basic operating principles, yet existing man-made walking machines rely solely on static techniques for support.
- Biological systems have elastic muscles and tendons that participate in resonant mechanical oscillations during locomotor behavior [1], [8], whereas man-made walkers usually have rigid legs with no suspension.

A k-legged locomotion system has $(2k-1)!$ nonsingular gates. For $k=1$, $(2k-1)!=1$ [7]. Therefore, since only one gait is possible for a one-legged device, the gait matrix is particularly simple and of little interest, viz: [1, 0]. Instead we address the problem of controlling dynamic motions to maintain an upright balanced posture during translation, and the problem of managing energy flow among kinetic energy of the moving masses, potential energy of the elevated masses, and the elastic elements of the actuation system or environment. We believe that these principles, balance and resonance, and their underlying computational processes are responsible for the robustness, versatility, and surprising mobility seen in the locomotion of biological organisms. Therefore we will force these important problems front and center, where they can get adequate attention.

Background

Aside from study of posture, which does not involve movement of the feet, one can characterize previous work on balance in terms of the fraction of the locomotion cycle that feet are in contact with the ground. A number of workers have examined the case where at least one foot is on the ground at all times—walking. Frank and Vukobratović modeled a passively balanced biped with massless legs that could balance in 3-space without sensors [2], [3]. Feet with finite area were used, so the point of support could change within a step. They characterized this model as synchronous, since the time of stepping was predetermined by an internal clock. Gubina, Hemami, and McGhee have looked at the asynchronous case for a planar biped [4]. They used a linearized feedback control law that combined continuous and discrete methods to control both step size and timing. Golliday and Hemami extended such studies to models having leg mass.

Matsuoka has formulated a massless leg model for locomotion in which the foot is on the ground for a very

Reprinted with permission from *4th Symp. on the Theory and Practice of Robots and Manipulators*, Warsaw: Polish Scientific Publishers.

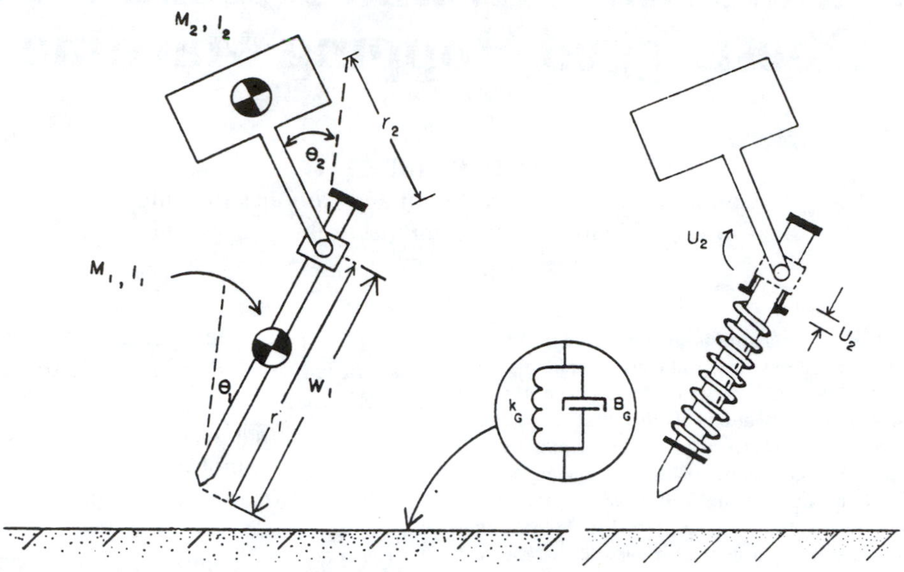

Fig. 1. Diagram of ideal one-legged hopping mechanism.

short time—hopping [5]. His analysis uses difference equations that model the rapid contraction and stretching of the leg with impulsive forces. Matsuoka has also constructed a physical hopping machine based on these principles that can balance in the plane if gravity is reduced to about .2g [6].

The present work falls somewhere between these approaches. During a cycle, the model spends about equal time in stance and in flight. This is a direct result of using a more realistic model of the leg that reflects the interplay between kinetic and potential energy occurring when a mechanically elastic system bounces on the ground. Control is asynchronous and the leg's mass is included.

A Simple Model

During walking, running, and hopping, legs do two things. They store energy in springy muscle and tendon when they are stretched, (returning some of it when they shorten), and they move back and forth to provide balance and to propel. We have modeled this behavior rather directly as a leg that changes length under the influence of a large spring, and a simple rotary hip that moves the leg back and forth. Figure 1 shows the mechanical system under study.

A leg of mass m_1, moment of inertia I_1 is connected to a body of mass m_2 moment of inertia I_2. The joint between these links permits linear sliding motion and rotation. A control torque $U_1(t)$ is applied to the rotary joint. The linear motion is controlled by a spring of stiffness K_S and rest length w_0 in series with a linear position source of length $U_2(t)$. The linear actuator adjusts the zero setting of the spring and therefore the acting between the leg and body. A mechanical stop limits the extension of the spring so that beyond its rest length it becomes very stiff and viscous, i.e., for $w_1 > w_0 + U_2$, $K_S => K_{S2}, B_{S2}$.

The ground is modeled as a two-dimensional spring of stiffness K_G plus damper B_G that interacts with the hopper when the position of the foot is less than zero. The zero setting of the horizontal component is reset each time the foot touches the ground. Equations of motion for this mechanical system are:

$$\ddot{y}_1 = \ddot{y}_0 - r_1\big(\ddot{\theta}_1 \operatorname{Sin}(\theta_1) + \dot{\theta}_1^2 \operatorname{Cos}(\theta_1)\big) \tag{1}$$

$$\ddot{x}_1 = \ddot{x}_0 - r_1\big(\ddot{\theta}_1 \operatorname{Cos}(\theta_1) - \dot{\theta}_1^2 \operatorname{Sin}(\theta_1)\big) \tag{2}$$

$$\begin{aligned}\ddot{y}_2 = \ddot{y}_0 &+ \ddot{w}_1 \operatorname{Cos}(\theta_1) - \dot{w}_1\ddot{\theta}_1 \operatorname{Sin}(\theta_1)\\ &- w_1\dot{\theta}_1^2 \operatorname{Cos}(\theta_1)\\ &- r_2\big(\ddot{\theta}_2 \operatorname{Sin}(\theta_2) + \dot{\theta}_2^2 \operatorname{Cos}(\theta_2)\\ &- 2\dot{w}_1\dot{\theta}_1 \operatorname{Sin}(\theta_1)\end{aligned} \tag{3}$$

$$\begin{aligned}\ddot{x}_2 = \ddot{x}_0 &+ \ddot{w}_1 \operatorname{Sin}(\theta_1) + \dot{w}_1\ddot{\theta}_1 \operatorname{Cos}(\theta_1)\\ &- w_1\theta_1^2 \operatorname{Sin}(\theta_1)\\ &+ r_2\big(\ddot{\theta}_2 \operatorname{Sin}(\theta_2) - \dot{\theta}_2^2 \operatorname{Sin}(\theta_2)\big)\\ &+ 2\dot{w}_1\dot{\theta}_1 \operatorname{Cos}(\theta_1)\end{aligned} \tag{4}$$

$$m_1\ddot{y}_1 = F_y - F_T \operatorname{Cos}(\theta_1) + F_N \operatorname{Sin}(\theta_1) - m_1 g \tag{5}$$

$$m_1\ddot{x}_1 = F_x - F_T \operatorname{Sin}(\theta_1) - F_N \operatorname{Cos}(\theta_1) \tag{6}$$

$$\begin{aligned}I_1\ddot{\theta}_1 = &-F_x r_1 \operatorname{Cos}(\theta_1) + F_y r_1 \operatorname{Sin}(\theta_1)\\ &- F_N(w_1 - r_1) - U_1(t)\end{aligned} \tag{7}$$

$$m_2\ddot{y}_2 = F_T \operatorname{Cos}(\theta_1) - F_N \operatorname{Sin}(\theta_1) - m_2 g \tag{8}$$

$$m_2\ddot{x}_2 = F_T \operatorname{Sin}(\theta_1) + F_N \operatorname{Cos}(\theta_1) \tag{9}$$

$$\begin{aligned}I_2\ddot{\theta}_2 = &F_T r_2 \operatorname{Sin}(\theta_2 - \theta_1)\\ &- F_N r_2 \operatorname{Cos}(\theta_2 - \theta_1) + U_1(t)\end{aligned} \tag{10}$$

Eliminating x_1, y_1, x_2, y_2, and F_N:

$$
\begin{aligned}
&\mathrm{Cos}\,(\theta_1)(m_2 W w_1 + l_1)\ddot{\theta}_1 \\
&\quad + m_2 r_2 W\,\mathrm{Cos}\,(\theta_2)\ddot{\theta}_2 + m_2 W \ddot{x}_0 \\
&\quad + m_2 W\,\mathrm{Sin}\,(\theta_1)\ddot{w}_1 \\
&\quad = W m_2\big(\dot{\theta}_1^2 W\,\mathrm{Sin}\,(\theta_1) - 2\dot{\theta}_1 \dot{w}_1\,\mathrm{Cos}\,(\theta_1) \\
&\quad + r_2 \dot{\theta}_2^2\,\mathrm{Sin}\,(\theta_2) + r_1 \dot{\theta}_1^2\,\mathrm{Sin}\,(\theta_1)\big) \\
&\quad - r_1 F_x\,\mathrm{Cos}\,(\theta_1)2 + \mathrm{Cos}\,(\theta_1) \\
&\quad \cdot\big(r_1 F_y\,\mathrm{Sin}\,(\theta_1) - U_1(t)\big) + F_K W\,\mathrm{Sin}\,(\theta_1)
\end{aligned} \tag{11}
$$

$$
\begin{aligned}
&-\mathrm{Sin}(\theta_1)(m_2 W w_1 + l_1)\ddot{\theta}_1 - m_2 r_2' W\,\mathrm{Sin}\,(\theta_2)\ddot{\theta}_2 \\
&\quad + m_2 W \ddot{y}_0 + m_2 W\,\mathrm{Cos}\,(\theta_1)\ddot{w}_1 \\
&\quad = W m_2\big(\dot{\theta}_1^2 W\,\mathrm{Cos}\,(\theta_1) + 2\dot{\theta}_1 \dot{w}_1\,\mathrm{Sin}\,(\theta_1) \\
&\quad + r_2 \dot{\theta}_2^2\,\mathrm{Cos}\,(\theta_2) + r_1 \dot{\theta}_1^2\,\mathrm{Cos}\,(\theta_1) - g\big) \\
&\quad + r_1 F_x\,\mathrm{Cos}\,(\theta_1)\,\mathrm{Sin}\,(\theta_1) - \mathrm{Sin}\,(\theta_1) \\
&\quad \cdot\big(r_1 F_y\,\mathrm{Sin}\,(\theta_1) - U_1(t)\big) \\
&\quad + F_K W\,\mathrm{Cos}\,(\theta_1)
\end{aligned} \tag{12}
$$

$$
\begin{aligned}
&\mathrm{Cos}\,(0_1)(m_1 r_1 W - l_1)\ddot{\theta}_1 + m_1 W \ddot{x}_0 \\
&\quad = W\big(m_1 r_1 \dot{\theta}_1^2\,\mathrm{Sin}\,(\theta_1) - F_K\,\mathrm{Sin}\,(\theta_1) + F_x\big) \\
&\quad - \mathrm{Cos}\,(\theta_1)\big(F_y r_y\,\mathrm{Sin}\,(\theta_1) \\
&\quad - F_x r_1\,\mathrm{Cos}\,(\theta_1) - U_1(t)\big)
\end{aligned} \tag{13}
$$

$$
\begin{aligned}
&-\mathrm{Sin}\,(\theta_1)(m_1 r_1 W - l_1)\ddot{\theta}_1 + m_1 W \ddot{y}_0 \\
&\quad = W\big(m_1 r_1 \dot{\theta}_1^2\,\mathrm{Cos}\,(\theta_1) - F_K\,\mathrm{Cos}\,(\theta_1) + F_y - m_1 g\big) \\
&\quad - \mathrm{Sin}(\theta_1)\big(F_y r_1\,\mathrm{Sin}\,(\theta_1) \\
&\quad - F_x r_1\,\mathrm{Cos}\,(\theta_1) - U_1(t)\big)
\end{aligned} \tag{14}
$$

$$
\begin{aligned}
&-\mathrm{Cos}(\theta_2 - \theta_1) l_1 r_2 \ddot{\theta}_1 + l_2 W \ddot{\theta}_2 \\
&\quad = W(F_K r_2\,\mathrm{Sin}\,(\theta_2 - \theta_1) + U_1(t)) \\
&\quad - r_2\,\mathrm{Cos}\,(\theta_2 - \theta_1) \\
&\quad \cdot\big(r_1 F_y\,\mathrm{Sin}\,(\theta_1) - r_1 F_x\,\mathrm{Cos}\,(\theta_1) - U_1(t)\big)
\end{aligned} \tag{15}
$$

where $W = w_1 - r_1$.

$$
F_K = \begin{cases} K_S(w_0 - w_1 + U_2) & \text{for } (w_0 - w_1 + U_2) > 0 \\ K_{S2}(w_0 - w_1 + U_2) - B_{S2}\dot{w}_1 & \text{otherwise} \end{cases} \tag{16}
$$

$$
F_x = \begin{cases} K_G(x_0 - x_{y_0=0}) - B_G \dot{x}_0 & \text{for } y_0 < 0 \\ 0 & \text{otherwise} \end{cases} \tag{17}
$$

$$
F_y = \begin{cases} K_G y_0 - B_G \dot{y}_0 & \text{for } y_0 < 0 \\ 0 & \text{otherwise} \end{cases} \tag{18}
$$

Since a closed form solution to these differential equations does not exist, a numerical algorithm is used to determine the system's behavior as a function of time. All the simulations discussed below are accomplished by using such numerical solutions in conjunction with the controllers being studied. The numerical constants used throughout the paper are given in the Appendix.

CONTROL

There are two sets of behaviors that are central to our study of planar hopping: vertical hopping and horizontal balance. The vertical part involves generating a stable series of resonant oscillations that take the system of the ground and control the height of each jump. The horizontal part must manipulate the system so that it remains balanced during hopping, and achieves the correct lateral position and translation on the ground.

Vertical Control: Hopping

$U_2(t)$ can excite the spring-mass system to achieve hopping. Starting from rest this actuator lengthens when its spring is under compression to do positive work on the system, or shortens to do negative work. No work is done if this actuator changes length, either lengthens or shortens, when the spring is unloaded. Therefore, lengthening the actuator at the bottom of each hop and shortening during flight causes the total energy in the system to increase on each hop. On the other hand, shortening the actuator at the bottom of each hop and lengthening during flight causes the total energy in the system to decrease, eventually to zero. The height of hopping should be unaffected by variations and irregularities in the terrain altitude, since total vertical energy does not depend on the location of the ground.

Figure 2 plots the vertical position of the foot and the body during a climbing period and during stable hopping. Starting at rest, the system executes a positive work cycle on each hop until the vertical energy is increased to the correct value. Also shown in this figure is the vertical control signal $U_2(t)$. This actuator lengthens with a quadratic trajectory until the total vertical energy is the desired value, or until the actuator is at maximum length. Shortening occurs during landing by letting the foot just touch the ground until the actuator is fully shortened.

If there were no losses in the system it would not be necessary to use this linear actuator once the desired height of hopping was achieved. However, since there are losses it is necessary to replace the energy lost on each hop to get level hopping. The vertical controller injects or extracts the right amount of energy to achieve a desired hopping height based on a measure of total vertical en-

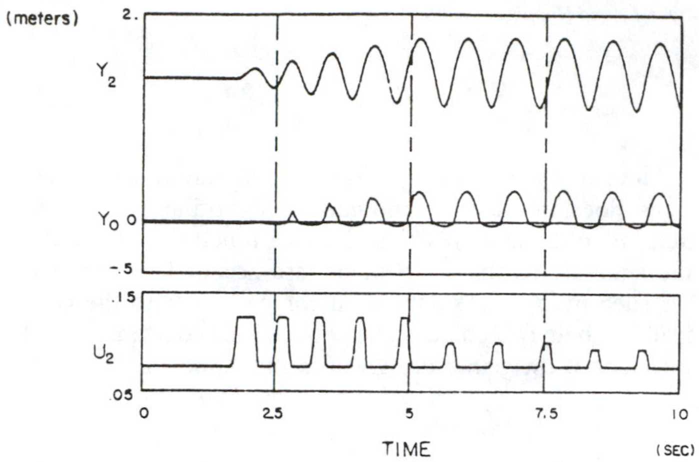

Fig. 2. Vertical hopping. Starting from rest, total energy is increased until desired hopping height is attained.

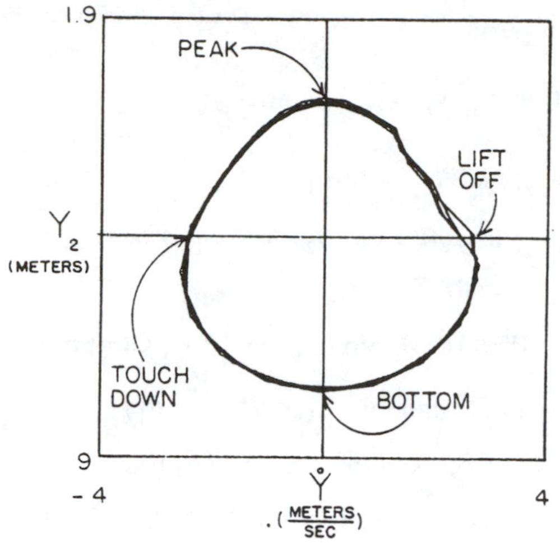

Fig. 3. Phase plot of vertical hopping. Four control events are indicated where plot crosses axes.

ergy made at the bottom of each hop:

$$\begin{aligned} E_{\text{VERT}} &= \text{PE}_g(m_1) + \text{PE}_g(m_2) + \text{KE}(m_1) \\ &\quad + \text{KE}(m_2) + \text{PE}_e(m_1) + \text{PE}_e(m_2) \\ &= m_1 g x_1 + m_2 g x_2 + .5 m_1 \dot{x}_1^2 \\ &\quad + .5 m_2 \dot{x}_2^2 + .5 K_2 (w_0 - w_1 + U_2)^2 \\ &\quad + .5 K_G \min(y_0, 0)^2 \end{aligned} \tag{19}$$

where

PE_g = gravitational potential energy

PE_e = elastic potential energy

KE = kinetic energy

The vertical control cycle can also be described in the phase plane. If we plot body velocity on the abscissa and body altitude with respect to its resting altitude on the ordinate we obtain a phase plot. (See Fig. 3). Note the parabolic trajectory during flight caused by the constant gravitational acceleration, and the circular trajectory during stance caused by the spring. Four events shown in this figure were chosen as control points to synchronize controller sequencing with ongoing behavior of the hopping system. These events are lift-off—the moment the foot leaves the ground, peak—the point in flight when vertical velocity changes from positive to negative, touch-down—when foot first touches the ground, and bottom—when vertical velocity changes from negative to positive. Use of these events as control switching points allows a simple and reliable asynchronous implementation.

Figure 4a is a polar plot of the vertical energy cycle, where the radius is the magnitude of total energy, and the angle is Arctan $(y_2/\dot{y}_2)$. A lossless system would produce a perfectly circular total energy line. Here it can be seen that the primary losses in the system occur when the foot strikes the ground, and when the spring stretches past its rest length to hit the mechanical stop. The injection of energy at bottom can also be seen in the lower right quadrant where the radius increases. Figure 4b plots energy for the data shown in Fig. 3. The spiral shows the start-up period during which the system is pumped up.

It is interesting that the time of leg shortening during a steady-state hop cycle can be manipulated to optimize a variety of criteria. If this action is taken at lift-off, clearance of the foot during flight is optimized. If shortening occurs at the peak height, then the time between actuations is maximized. If shortening occurs upon touch-down, then the ground impact forces are minimized. This last strategy is normally used by humans if they are asked to hop in place on a flat floor.

Horizontal Control: Balance

Torques $U_1(t)$ applied to the hip can be used to control the horizontal behavior of the system in two ways. Torques applied while the foot is in contact with the ground cause rather direct modifications of the body's angular momentum. Torques applied while the system is in flight can be used to manipulate the position of the foot with respect to the projection of the system's center of gravity. In an effort to separate the issues, only the second of these methods is explored here.

The leg must be positioned before the foot touches the ground. There are three levels of control here: an upper level that changes the desired angle of the body with respect to the vertical to control the rate of horizontal movements; a middle level that manipulates the leg landing angle to control balance; a lower level that generates torques at the hip to achieve the desired leg angle.

Our scheme begins positioning the leg at peak, and continues to update its position until just before touch-down. Since the state should be changing gradually during this period, the leg angle changes gradually once it is near

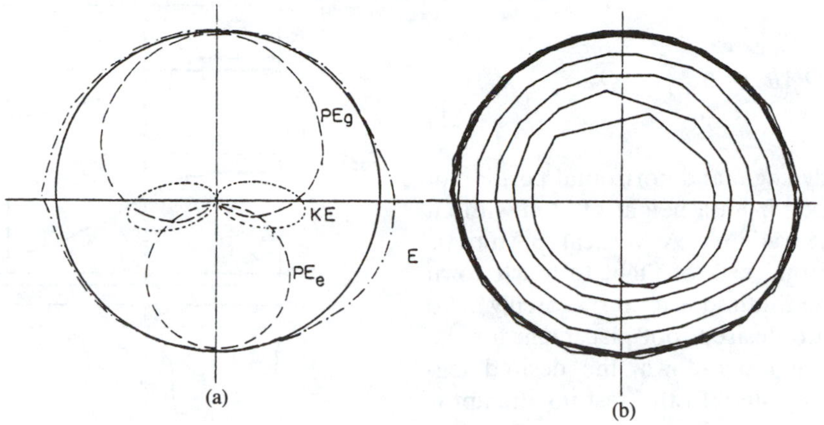

Fig. 4. Polar plot of vertical energy. Radius is magnitude of energy, angle is Arctan $(y_2/\dot{y}_2)$. (a) Broken down into KE, PE_g, and PE_e. (b) Starting from rest, total energy is increased to desired level.

the correct angle. Once the foot touches the ground, hip torques are controlled to keep $\theta_2 - \theta_1$ constant. Hip torques to achieve positioning of the leg are generated from a linear PD controller:

$$U_1(t) = K_P(\theta_1 - \theta_{1d}) + K_v(\dot{\theta}_1 - \dot{\theta}_{1d}) \quad (20)$$

The leg is positioned before each touch-down so that the state at next take-off is optimized. A means of generating these leg angles is presented next.

State Space Response Method (SSRM)

Balance experiments were conducted using an entirely new method of control that extends Raibert's State Space Memory Method [9]. The State Space Response Method (SSRM) is a means of solving the complete nonlinear dynamics for a hopping system without having to do lengthy calculations or numerical integrations in real time. This is a dynamic inverse method in that knowledge of the relationship between desired system state and available control actions is used to choose an appropriate control. We motive this approach by casting the balance problem for a hopping mechanism in the following terms:

> Given complete state information at the time of touch-down, what leg-body angle maintained throughout stance will move the system to minimize the state error at the time of next take-off?

The State Space Response Method is a way of predicting future state, (i.e., response) based on present state and control. We define a multidimensional vector space Γ such that each dimension of Γ corresponds to one element of an augmented system state vector, X. The elements of X are n state variables and i control variables. For each point in Γ there is a future state vector that indicates the state of the system at a specific time in the future—the augmented state vector points to a future state vector.

If such a vector field were accessed at touch-down, and if the future state vectors in the field belonged to the time of next take-off, and if the i control signals were completely determined by their values at take-off, then control could be achieved as follows: At each touch-down the current state vector exclusive of the i control elements determined an i-subfield of Γ. By minimizing a performance index over this subfield a best future state vector is found. Its location in the subfield indicates the best choice for the i control variables.

If each dimension of Γ is quantized, this approach can be implemented with a finite memory. For $N = n + i$ variables, each quantized to M values, there are M^N hyper-regions each storing n values. The values stored in each hyper-region may be obtained from simulation, from evaluating closed form solutions, or form measurements made on a behaving system. In any case, the data must be included for all state variables that are to be controlled.

Note that during the response interval, (i.e., the time between accessing Γ and the time of interest), the i control variables must behave in a regular fashion. Similar initial values must map into similar control signals. This condition is satisfied if control variables remain constant, but that is not required.

There are many unresolved issues concerning appropriate choices of M, methods of approximating the tabled data, and interpolation. For the present, work body angle θ_2, body angular rate θ_2, horizontal velocity $\dot{x}_2$, and leg angle θ_1 are the variables that index a 4-dimensional space. Only θ_1 is a control variable, so $i = 1$. Each dimension of the memory is quantized to eight levels, $M = 8$, requiring that $nM^N = 12{,}288$ values be stored. To compensate for such a course quantization, linear interpolation is performed sequentially on each dimension of the memory.

Use of the memory requires that a search be performed for the hyper-region containing the state vector that minimizes a performance index. The present experiments employ a quadratic index:

$$Pl = Q_1(\theta_2 - \theta_{2d})^2 + Q_2\left(\dot{\theta}_2 - \dot{\theta}_{2d}\right)^2 + Q_3(\dot{x}_2 - \dot{x}_{2d})^2 \tag{21}$$

Figure 5 plots the body angle and horizontal position of the hopper where dropped from a height of .2 m with an initial body angle of .5 radians. A vertical posture is attained in about 6 s. From peak of flight to touch-down the State Space Response Method is used repeatedly at 10 ms intervals to recalculate desired foot placement for the impending landing. Upon touch-down, the desired leg-body angle is no longer adjusted. In this test no attempt is made to control horizontal position, x_2.

Figure 6 shows a lateral translation in which x_2 was controlled indirectly by controlling $\dot{x}_2$. This is accomplished by purposely unbalancing the body. When the system hops with an inclined body and leg, the ground force has a horizontal component in the direction of desired motion. When these data are displayed in motion picture form it can be seen that the behavior is quite precise and stable.

The State Space Response Method requires computing to search for the state that minimizes the performance index, and to interpolate once the correct vector is found. No time is spent evaluating equations of motion directly. However, more work must be done before we know the range of applications to which such methods can be applied. The state vectors stored in the state-space response memory are obtained by simulating a large set of landing/take-off cycles with systematically varied initial conditions. Though it has not yet been attempted, memory data could be obtained by taking measurements from a physical system engaged in hopping. Such a learning controller would benefit from experience gained in practice, and could adapt to changes in the mechanical characteristics of the system.

A Physical Implementation

Figure 7 is a diagram of a physical apparatus under construction that will be used in experiments to verify the ideas presented here.

The important features of this mechanism are a body, a leg, a spring, two pneumatic cylinders, two internal bearings, and three air bearings. Motion of the entire mechanism is constrained to a plane by the three air bearings and a flat inclined table (Fig. 7b). The main assembly of body-cylinders-bearings is permitted to slide vertically along the leg's axis, thereby compressing the spring, and also to rotate about an axis perpendicular to the planar constraint. These linear and angular motions of the assembly are instrumented or measurement.

The entire device is free to translate and rotate with respect to the support surface that simulates the ground. Other sensory information is required to provide state

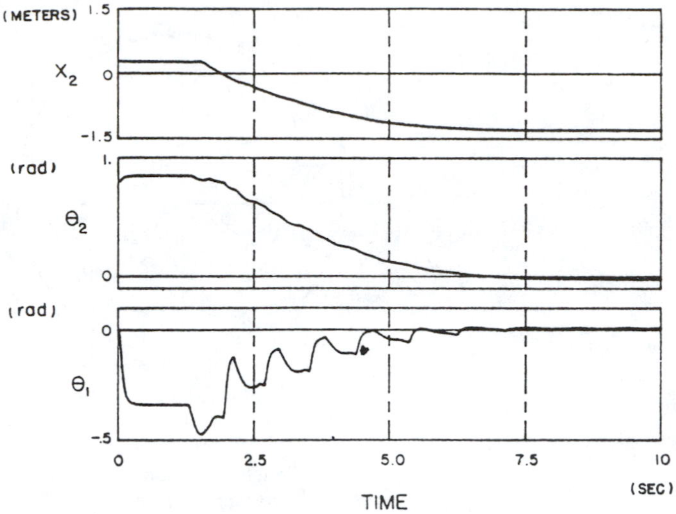

Fig. 5. Balance using State Space Response Method.

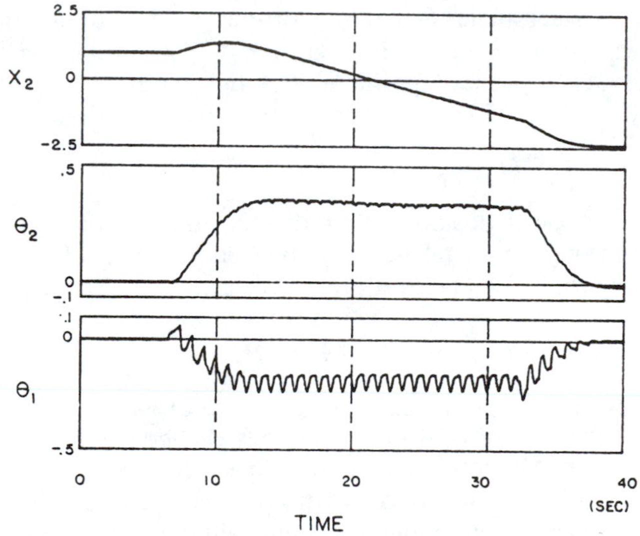

Fig. 6. A lateral step controlled by State Space Response Method.

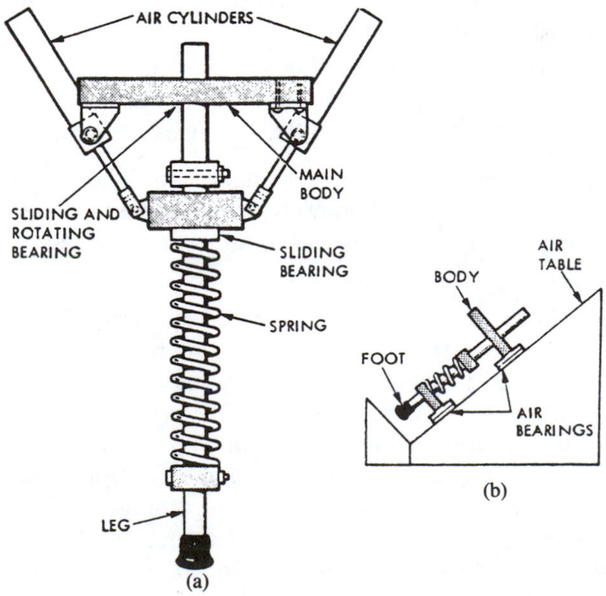

Fig. 7. Physical hopper now under construction.

feedback for these body variables. The basic mode of operation is as follows:

- Simultaneous in-phase rhythmic activation of the two air cylinders causes resonant excitation of the spring-mass, i.e., spring-body system. A mechanical restriction on extension of the spring coupled with this resonant oscillation procedures hopping motions.
- While the device is in flight, differential control of the cylinders causes the leg to rotate relative to the body, thereby selecting a new point for foot placement on subsequent landing. The point thus chosen is determined by considering information from the sensors and the system dynamics.

This mechanical linkage is quite simple, requiring no extraordinary mechanical engineering considerations. However, its performance will depend intimately on the availability of appropriate sensory information adequate computing power to interpret that information, and fast control algorithms that embrace dynamics.

SUMMARY

A one-legged hopping machine is proposed as a simple model for dynamically stable locomotion. The problem of controlling such a machine is decomposed into a vertical part and a horizontal part. A total energy measure is used to control the vertical resonant oscillations of a spring mass system. Horizontal balance during hopping in place and lateral translations is achieved with a new method for calculating dynamics called the State Space Response Method. Simulations show successful control of hopping height, recovery from extreme body angles, and control of horizontal velocity to achieve translations. Work is underway to test these ideas on a physical system.

APPENDIX: SIMULATION PARAMETERS

$m_1 = 1$ kg	$m_2 = 10$ kg
$I_1 = 1\ \text{kg}\cdot\text{m}^2$	$I_2 = 10\ \text{kg}\cdot\text{m}^2$
$r_1 = .5$ m	$r_2 = .4$ m
$w_0 = 1$ m	
$K_p = 900$	$K_v = 60$
$K_s = 10^3$ N/m	
$K_{s2} = 10^5$ N/m	$B_{s2} = 125\ \text{N}\cdot\text{s/m}$
$K_G = 10^4$ N/m	$B_G = 75\ \text{N}\cdot\text{s/m}$
$Q_1 = 5.$	$Q_2 = Q_3 = 1$

ACKNOWLEDGMENT

I am indebted to Ivan Sutherland for finding ways to support this work, both intellectually and financially.

REFERENCES

[1] R. McN. Alexander, and A. Vernon, "The mechanics of hopping by kangaroos (Macropodidas)," *J. Zool., Lond.*, No. 1.77, pp. 265–303, 1975.

[2] A. A. Frank, "An approach to the dynamic analysis and synthesis of biped locomotion machines," *Med. Biolog. Engg.*, vol. 8, pp. 465–476, 1970.

[3] A. A. Frank and M. Vukobratović, "On the synthesis of biped locomotion machines," in *Proc. 8th Int. Conf. Med. Biolog. Engg.*, Evanston, IL, 1969.

[4] F. Gubina, H. Hemami, R. B. McGhee, "On the dynamic stability of biped locomotion," *IEEE Trans. Biomed. Engg.*, vol. BME-21, No. 2, pp. 102–108, 1974.

[5] K. Matsuoka, "A model of repetitive hopping movements in man," in *Proc. Fifth World Cong. Theory Mach. and Mechanisms*, pp. 1168–1171, 1979.

[6] K. Matsuoka, "A mechanical model of repetitive hopping movements," (in Japanese), *Biomechanisms*, vol. 5, pp. 251–258, 1980.

[7] R. B. McGhee, "Some finite state aspects of legged locomotion," *Math. Biosciences*, no. 2, pp. 67–84, 1968.

[8] T. A. McMahon and P. R. Greene, "The influence of track compliance on running," *J. Biomechanics*, vol. 12, pp. 893–904, 1979.

[9] M. H. Raibert, "A model for sensorimotor control and learning," *Biol. Cybern.*, vol. 29, pp. 29–36, 1978.

* Initial funding for this research was provided by the Jet Propulsion Laboratory under NASA contract NAS-7, and by the California Institute of Technology's Computer Science Department. This research is now sponsored by the Cybernetics Technology Office of the Defense Advanced Research Projects Agency under contract MDA903-81-C-0130.

UNDERSTANDING AND APPLYING A ROBOT PING-PONG PLAYER'S EXPERT CONTROLLER

Russell L. Andersson

AT&T Bell Laboratories
Crawford's Corner Road (**Rm. 4B607**)
Holmdel, NJ 07733

ABSTRACT

Our robot ping-pong system attains good performance in a complex non-linear environment subject to tight temporal constraints. We will show how the expert controller combines approximate estimates and feedback to arrive at a suitable task plan, emphasizing feasibility rather than strict optimality. We will discuss the characteristics of the expert controller that cause it to work, and their relationshi· to other tasks.

1. INTRODUCTION

Conventional robot systems operate slowly and methodically, often playing back a pretaught sequence of positions. To investigate the construction of fast, intelligent robot systems, we built a robot ping-pong player able to play against humans.

An important development of the work was an "expert controller" to choose the robot's response to incoming volleys; the expert controller must account not only for the task constraints of ping-pong (to create a valid return), but it must also consider the capabilities of the robot. The robot's small size and less than overwhelming speed ruled out brute-force approaches to plan generation. Without a subtle plan, the robot was destined to fail — early efforts showed this conclusively. The expert controller constructs a plan by choosing values for discretionary, redundant, task and robot degrees of freedom. After an initial description of the ping-pong problem's characteristics and the computing architecture, we will describe the expert controller's organization and operation.

Having done that, we will comment at length on the issue: "what makes it work?" An expert controller requires several essential features to work, for example, task-level redundant degrees of freedom and internal feedback. In the long term, as industrial robots must perform increasingly complex tasks, the discretionary authority required will also increase, making the expert controller's dynamic planning ability mandatory. Although expert controllers ought operate at both the task level and the robot level, in the near term expert controllers might be profitably embedded to control solely robot parameters. As a result of the expert controller's planning ability, we can apply the same mechanical robot to more complex tasks, and achieve higher performance (and lower cost) from it.

2. BACKGROUND

In this section, we will describe the robot ping-pong task: the physical environment, a short description of strategy, and the computer configuration. Additional detail about every aspect of the system may be found in Andersson [2].

2.1 Robot Ping-Pong.

The robot plays ping-pong according to international standard robot ping-pong rules proposed by Billingsley [4]. The game has been modified to be playable by moderate-sized immobile robots, by scaling down and restricting the area that must be covered by the robots (Figure 1). To compensate for this, the paddle has roughly half the surface area of a person's.

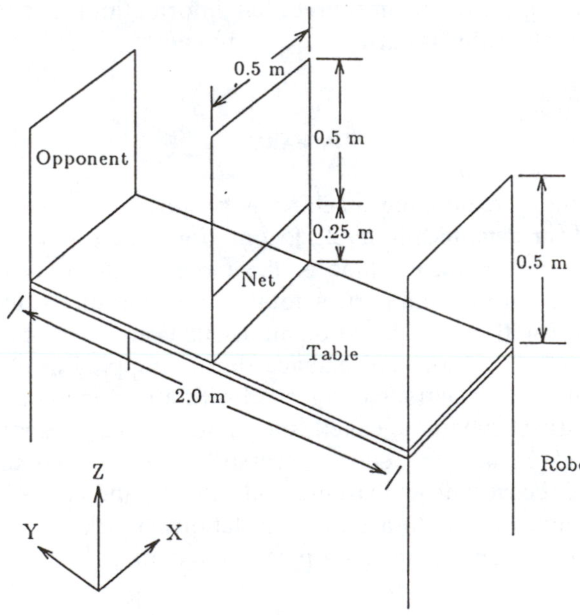

Figure 1. Robot Ping-Pong Table. The players, background, and table are black. The ball must travel sequentially through the square frame at one end, through the center frame, then bounce once and travel through the frame at the table's other end.

The table emphasizes control, not speed: the maximum ball speed is approximately 10 meters per second (10 m/sec). For anything but an ideal player, the limit is about 8 m/sec (20 miles per hour). The minimum ball speed is 3.5 to 4 m/sec. The table is 2 meters long, resulting in maximum available reaction times of 0.4 to 0.8 seconds. Approximately 0.1 second is lost to camera latency.

Reprinted from *Proc. IEEE ROBOT '89*, Scottsdale, AZ, pp. 1284–1289, 1989.

and accumulating enough frames (5) to get a rudimentary estimate of the ball trajectory, leaving little time for robot motion.

2.1.1 Aerodynamics. While in flight, the ping-pong ball is acted on by three significant forces: gravity, air drag, and the Magnus (spin) Effect. We have chosen the following aerodynamic model with reasonable results. Given the velocity vector $\vec{v}$, and the spin vector $\vec{w}$, the ball accelerates at $\vec{a}$:

$$\vec{a} = -C_d |\vec{v}| \vec{v} + C_m |\vec{v}| \vec{w} \times \vec{v} - \vec{g} \quad (1)$$

The vector $\vec{g}$ is the acceleration of gravity; C_d, the drag coefficient; and C_m, the Magnus Effect coefficient.

2.1.2 Bouncing and Hitting. To be able to predict the ultimate trajectory of the ball as early as possible, the system must be able to predict the bounce off the table. With a bit of physics, we can formulate closed-form equations for the ball's final spin and velocity [2] as a function of the table's coefficients of restitution and friction.

We must also know how to hit the ball to create a specific final velocity. The paddle strike resembles a table bounce, but the multilayered rubber paddle surface dramatically affects the bounce. The physics are not at all understood, but we have a plausible model.

Of course, we must not just predict the hit, but cause a desired result. We can find the correct (moving) coordinate system to cause the desired result with a small iterative procedure.

2.2 The Ping-Pong Task.

The kinematics of the robot ping-pong table are much more constrained than the human one. On a human table it is nearly always possible to hit the ball a bit harder with more top-spin, or hit the ball a bit softer with a higher arc, and still have a valid shot.

Returns on a robot table must be tightly constrained in velocity. Simple ball placement strategies such as "aim for the middle of the far frame" just do not work. If the incoming shot is high, there is no choice but to generate a high-speed return to get under the wire on your own side; quickly moving balls are not able to bounce very high before they cross the other end of the table. Balls just clearing the near end of the table must be returned along high-arc lob trajectories. Normally, one would like to aim horizontally straight across from the incoming shot, displaced somewhat towards the center of the far frame, to maximize the probability of a good shot, while minimizing the ball's velocity perpendicular to the face of the paddle. (This strategy reduces the range the robot must cover, and makes the hit less dependent on the paddle surface.) If the incoming shot is wide and near a side frame marker, there is no choice but to aim for the opposing corner, because otherwise the return will strike the side marker on the robot's end of the table.

As may be seen, ping-pong requires substantial strategy to select a suitable return (even without attempting to beat an opponent). The robot's strategy for a shot may be captured by "free variables," which, once chosen, completely define the rest of the stroke (though with effort). The free variables reflect the expert controller's latitude in defining the robot ping-pong task:

1. `hit plane depth` — distance back from the edge of the table at which the ball will be hit,
2. `pad_con_xgen` and `pad_con_ygen` — the position on the paddle's surface where the paddle will hit the ball,
3. `target_alpha`, `target_vy`, `target_vz` — the components of the ball's velocity after the paddle hits it (`target_vx` is encoded in `target_alpha` as `atan (target_vx/target_vy)`),
4. `stick_angle` — roughly, the roll angle of the paddle, but particularly of the stick that supports the paddle, and
5. `settle_time[6]` — the period of time each joint's arrival at its striking velocity will precede contact with the ball.

2.3 Computer System.

The robot ping-pong system demands many CPU cycles for implementation. We have implemented the system on a distributed network of 68020 microprocessors. The task partitions naturally into pieces that run on each processor (Figure 2). The **rtd**, **chief**, **sai**, and **sp2000** programs perform administrative tasks we will not discuss. The processors are connected by the S/Net [1], a high-bandwidth, low-latency inter-processor connect. A multiprocessing and multitasking operating system called MEGLOS runs on the S/Net processors, supporting real-time programming [7]. A micro-VAX II host acts as a file server and software development machine.

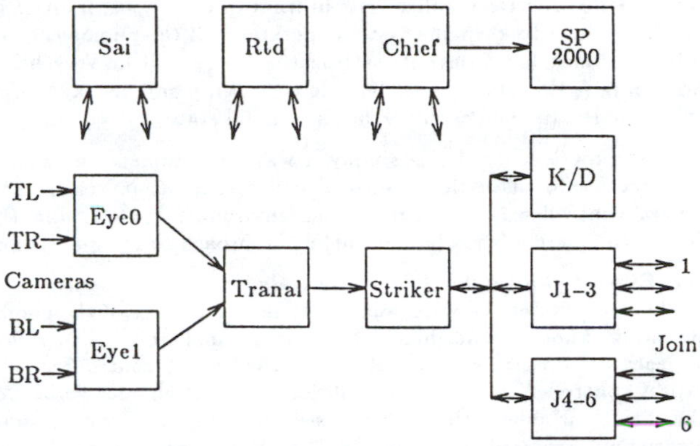

Figure 2. System Block Diagram.

The flow of data through the system begins at the twin vision processors, **eye0** and **eye1**. Each processor contains two real-time vision systems constituting a stereo pair. Both processors operate synchronously with the 60 Hz TV scan, putting an upper bound on the amount of processing they may do. After the vision system and its software driver have accurately located the ball's image in each frame, a stereo algorithm combines this information with camera calibration data to produce the three-dimensional location of the ball.

The position information is sent to the trajectory analyzer, **tranal**, which fits the data to trajectories, with appropriate segmentation at bounce and hit points. **Tranal** sends the position, velocity, and time at which the ball will cross the robot's end of the table to the real-time expert controller, **striker**. **Striker** contains both the expert controller and the low-level robot controller. After a robot motion is planned, it is sent to slave servo processors which cause the motion to occur [3].

3. EXPERT CONTROLLER

What is an expert controller? Syntactically, it is the juxtaposition of an "expert system" and a "robot controller." Semantically, it is a way to increase the intelligence of a robot controller, while maintaining real-time response. But unlike true Artificial Intelligence (AI) systems, our intent is to communicate design intent, not discover it. From both practical and theoretical standpoints, we think it is most immediately relevant to understand how to describe and implement intelligent real-time strategies, which humans can often think up off the cuff, rather than spend a lot of effort trying to generate such (undescribable) strategies automatically. As machine capabilities increase, automatic strategy generation will gain importance.

One particularly simple way to understand the expert controller begins with the complaint of Brooks [5], who pointed out that most programs contain "kludge factors" which greatly affect the results. For example, early vision programs were plagued by thresholds. Sometimes, a program would work only on certain trial data sets, because otherwise the kludge factors would be inappropriate. Brooks rightfully regarded kludge factors as odious hacks; he advocated the detailed confession and documentation of each. Such documentation can be a valuable source of information about the real functioning of an algorithm.

Rather than try to eradicate the kludge factors, an expert controller attempts to institutionalize them, such that they can be scrutinized, reasoned about, and adjusted. The successors to those kludge factors are the "free variables" previously described. As an abstract view of our system, one might picture a black box (the robot controller) with many knobs (one for each free variable). A gnome, (the expert controller), sits in front of the knobs, frantically tweaking them to keep the system operating. If the gnome can see only the black box's output, setting the knobs will be very hard. Accordingly, the gnome must be able to watch many display panels reflecting the (no longer) black box's internal state.

Our expert controller is simply a way to automate the gnome and accelerate his reflexes into the electronic time scale. The expert controller is a programming environment to enable the gnome to describe (thus implement) his approach to knob-twisting.

3.1 Expert Controller Fundamentals.

The real system is more complex than our picture of the gnome tweaking knobs. The black box, or normal robot controller, appears as a series of subroutines to the expert controller. The expert controller calls different subroutines as it chooses values for the free variables; the subroutines perform the conventional algorithms required to generate the details of the plan. For example, subroutines compute the intersection of the ball's trajectory with several possible hit depths, compute the resulting trajectory from striking the ball in a certain direction, or compute the arm's inverse kinematics — generating joint angles corresponding to a particular paddle pose. Both the expert controller and the robot controller are implemented in the programming language "C" with a blackboard structure permitting data-driven control flow.

The expert controller's most important task is to choose values for the free variables. In the simplest case, we use a generalized form of table lookup to choose values. The table lookup process relies on a specialized data structure, the "model," which we use to store both symbolic and numeric information for arbitrary data types. By representing as much information as possible in models, we simplify future supervisory AI programs as well as the expert controller. Library routines present a uniform interface to the data structures; humans create (simple) models with a mouse-driven program.

We often need to combine information from several different sources to generate a value. Although theoretically possible, a single large model with all inputs may require excessive memory. Instead, we can construct several smaller models, one for each independent information source, then combine the results together. We must decide upon an appropriate method to do so.

3.1.1 Selecting among alternatives. The first type of decision we will consider is how to select among several alternatives, given any number of continuous or discrete input variables. For example, we might decide whether to make a forehand or backhand shot (though our system always uses forehand).

We begin picking an alternative by subjectively evaluating all applicable input data over the possible alternatives. Models are used to map input data to one or more figures of merit, usually between ± 2. Not all figures of merit need be generated all the time, in keeping with the symbolic/numeric nature of models. Failure to produce a figure of merit indicates a "don't care" with respect to that alternative and input.

Unlike other work which tries to assign true semantics to the figures of merit [6], we claim that such efforts are largely extraneous in this context. In a system devoted to synthesis, the semantics may be: "Is it a good idea to try to hit the ball when it is dropping at 1.4 m/sec?" It is hard to make a probabilistic estimate of the impact of this variable on the system. The real probability is obscured by the effect of many other possibly unmeasurable variables.

Our approach is to make the evaluators display the right characteristics, to include as many factors as possible, then let the group statistics and robustness of the system operate. There is no *right* answer. In practice, we pick figures of merit for special cases with intuition and hand analysis, then perform spline interpolation between them. By analyzing the system's actual performance, we uncover any gross mistakes.

A `decide` routine chooses the alternative with the highest figure of merit, obtained by summing the evaluators. `Decide` is fast, predictable, and useful: it captures the essential intent of the specifier. It does not try to attach too much significance and accuracy to data which is inherently loosely specified. `Decide` directly creates symbolic information for the blackboard.

3.1.2 Continuous variables. To select a value for a continuous variable, we compute optimal values from restricted views of the problem (and correspondingly small numbers of inputs). For example, we might compute an optimal return speed from the incoming speed, and another return speed from the ball's height. Some data sources may be generated only when other specific input data appears on the blackboard. We then `combine` the values together to get a consensus.

The evaluators used to choose values for continuous outputs produce not figures of merit, but values in the output space. They are the semantic equivalent of "If `x`, then do `y`" or "If `z` increase `x` by 5."

The `combine` operator is a simple (possibly weighted) average. The weights can express such semantics as: "I definitely want `x`=39." The weights are often generated at the same time as the desired output by the same model, but separate meta-models can alternatively specify the relative weighting of the models.

3.2 Overall Organization.

From a task standpoint, the expert controller begins with an "initial planning stage" to create initial values for the free variables and begin operation, once the first data point appears. Subsequently, a loop waits for additional sensor data, then implements a "temporal updating" stage to update the free variables. Both stages apply the `decide` and `combine` operators.

Parallel to this control structure lies another one: the exception handler. Whenever the expert controller detects an error, perhaps an unreachable robot configuration or velocity, it invokes the exception handler. The exception handler will reconfigure the blackboard to cause an attempted recovery from the error. Should the recovery not succeed, the exception handler will resume operating such that it may consider the prior attempt.

At the implementation level, the expert system is tree structured, with blocks of code as the leaves. Control flow may be redirected to the root of the tree; control then switches automatically to a particular leaf selected by the contents of the blackboard. This mechanism supports model-driven control flow alteration and exception recovery.

3.3 Overview of Temporal Updating.

We do not create a new plan from scratch with each new sensor datum, but instead update the old one. The temporal updating process has two primary objectives: to change the plan in response to new sensor data, and to propagate information about the later stages of processing up to the earlier ones, and so achieve a more globally optimum solution to the problem. The temporal updating process has an asset that the initial planning process does not: estimates of program variables and the relationships among them. This data facilitates local linearizations and provides hysteresis.

As the expert controller generates an (initial or updated) plan, it calls upon "evaluator" models. For example, one evaluator comments on the X coordinate at which the ball will cross the far end of the table. The evaluators immediately point out errors (such as an illegal shot), but especially provide data on the plan's quality (such as "we're getting close to hitting the ball off the end of the table").

Global optimality is a loosely defined criterion: we need not have any formal measure. Instead, we try to ensure that no problem becomes severe enough to endanger the task, or at least, that if problems are to become worse, they all do so at roughly the same rate, so that no individual problem occurs prematurely. The evaluators enable the comparison of the severity of different problems.

Specialized knob-twisting routines called "tuners" reside at the heart of the temporal updating process. The tuners are run once each temporal updating iteration, after new sensor data arrives. The tuners adjust the program's free variables in response to global program conditions. Rather than have a single massive tuner, we partition the task into tuner modules operating on individual or related free variables. Partitioning makes the system more manageable, and the granularity it imposes is necessary during exception handling.

A tuner alters its output variables to attempt to correct real or anticipated problems. The tuner does not have to examine all possible problems, only those it believes it can solve. Low-level numeric calculations within each tuner propose the modifications to the free variables. Each correction is described by what amounts to a rule, enabled and disabled by symbolic situation data on the blackboard. (Only a fraction of a tuner runs on any given pass, accelerating execution.) At the end of each tuner, a `combine` generates the new value for the free variable.

4. WHY IT WORKS

The preceding discussion has aimed to outline the general features of the robot ping-pong player's expert controller. For more information, the reader may consult [2].

In this section, we will look at the major factors contributing to the system's successful operation. Preliminary to this, we will describe the system's performance.

4.1 Performance.

The robot system can play complete games against human opponents, even beating its creator. Opponents conditioned to the human game find the robot's small, control-oriented table frustrating. (The larger human ping-pong table favors speed over control.)

The longest volley recorded had 21 hits by the author and by the robot (each, not total) — reflecting well on the performance of both. The robot is far from perfect, as most volleys are considerably shorter (as they are in human games). Shots dropping just over the edge of the table or close to the end wire frame are returned less reliably. Chop or side-spins reduce reliability, though they are tolerated in moderation. Shots that are hard to return reliably are hard to generate reliably. Most of the robot's missed shots hit the top of the wire frame at the robot's end of the table or go off the far (human) end of the table.

Sometimes a significant discrepancy between the expected post-bounce trajectory and the actual trajectory becomes visible too late to correct. The control system tries its best but realizes it can not plan a return, then, refusing to give up, it goes ahead with the last valid plan. The robot can be frozen by hard shots in awkward places — it never generates a successful plan.

In most cases, the control software succeeds at generating a plan: it is surprised by its mistakes. We will encounter several reasons for the control software's success in the remainder of the section, as well as consider why the errors occur.

4.2 Task Redundancy.

The robot ping-pong task's inherent degrees of freedom simplified the implementation. The redundant degrees of freedom offer an alternative means to solve hard problems in a particular degree of freedom — by solving an easier problem with a different degree of freedom. Since tight constraints indicate a risky plan, by using alternate degrees of freedom the task becomes more robust as well. For example, we slow down to navigate a narrow roadway.

Ping-pong has easily found degrees of freedom, such as the direction and velocity of the return, but it is not a special case. We believe that most tasks have equivalent degrees of freedom. Rather than ignorantly choosing fixed values for them in a robot program, we should identify and exploit them.

The extra degrees of freedom indicate fundamental robustness in the task: the free variable corresponding to the ball's left/right placement can be chosen arbitrarily as long as it does not violate any constraints. Even if we do not precisely achieve the value we intend, as long as we do not violate the constraints, we will have a successful operation. The planner exploits this by avoiding the edges of the table, consistent with ball and robot dynamics constraints. The robot ping-pong player does not have any substantial accuracy in its ball return trajectories (compared to the internal plan), but it still succeeds by avoiding problem areas.

4.3 Continuous Improvement.

The continuous sensor data stream was crucial to the development of a working system. In the most general sense, we can say that the continual perception of the environment made the system work. Our particular application virtually required continual perception, because the motion must be started long before accurate data is available. (The robot starts moving when the ball is approximately half way to the net on the opponent's side of the table.) However, all future robots should continuously perceive their environment, much as batch computing gave way to interactive computing.

Continuous perception implies that the robot constantly works with a more accurate model of the environment. In ping-pong, the ever-lengthening temporal window enabled better trajectory predictions. Even a robot watching a stationary object can use additional observations to increase the measurement accuracy. Additionally, the robot becomes sensitive to external changes in the environment, due to equipment operation or malfunction, or due to external agents such as humans or other robots.

As new data arrives, the ping-pong system improves its plan, using internal feedback of the plan's quality. This self-evaluation process was critical to system operation, especially when (as early in system development) the sensor data changes drastically. By continually modifying the plan, we can prevent exceptions from happening in the first place. The presence of the data from the previous plan provides estimates to simplify the updating calculations — contributing to the system's ability to operate in real-time. Also, the prior plan provides the hysteresis needed to prevent decisions from changing from iteration to iteration, which would senselessly expend actuator performance. As the system's planning capabilities improved, it appeared to become lazier, taking the slow but sure shot instead of the riskier and more dramatic one.

4.4 The Importance of Good Physics.

The expert controller creates values for the free variables in an inherently fuzzy way. The fuzziness arises from approximate calculations, model data from humans, quantization errors in models, and from our data-source combination method. We must tolerate fuzziness, because there may not be a right answer, and excessive preoccupation with trying to find the "right answer" can produce catatonia in development or at run time.

To counteract this fuzziness, we have used good physics directly where ever possible. For example, we use physics to estimate the incoming ball's trajectory, to compute the desired paddle velocity, to compute the ball's outgoing flight, and to compute the robot configuration from a ball position and paddle orientation. If we used inherently fuzzy processes for all these stages, as might be proposed by a neural net advocate, the errors would accumulate to the level of nonsense.

We can count on fuzzier processes to generate quick, practical solutions to poorly specified problems. But we can not rely on this approach exclusively, because it lacks absolute accuracy. By combining the approach with good solid physics as a foundation, we can overcome its limitations.

4.5 Applying Control Theory.

Given that the expert controller is some sort of control system, an obvious question is whether or not control theory can analyze it to prove that it works. The question is generally asked by control theorists, who would like to put the system into a nice bin.

Unfortunately, the system resists such analysis because of its complexity and non-linear nature. The system has an essentially infinite number of states, because each note on the blackboard switches in or out some piece of code from the system's "transfer function" (doubling the number of states, thus transfer functions). In addition, the models contribute completely arbitrary functions via lookup tables, not to mention the non-linearities associated with a robot with six rotary joints.

Furthermore, we can't even answer a few fundamental issues: what the problem is, what the right answer is, or that the sensor and actuator system suffices to solve the (unknown) problem (it doesn't). Choosing tractable answers to these questions amounts to sacrificing system performance to the extent that the formalism disagrees with reality. The prospects for a complete analysis are bleak.

We can analyze restricted views of the system, for example, a tuner's proposed correction to a specified problem in the absence of any other effects. The object of the analysis is to determine whether or not the controlled variable converges smoothly (from one side) to a final value. For restricted views, control theory turns out largely not to be necessary: the tuners do not have temporal dynamics of their own. My experience is that the stability of a tuner is either correct by design or so convoluted with other effects as to be intractable.

It would be nice to be able to better quantify the system's operation in the presence of multiple interfering effects. We have not proposed any strong semantics for what should happen with conflicting pressures, but let the convenient `combine` operator try to muddle through.

We wind up having to claim that the system generates "reasonable" solutions to the problems it encounters. When faced with conflicting demands, the `combine` operator takes a mean value. It refuses to make the big mistake by opting for an extremal value. Muddling through maximizes the chance that subsequent data or other parts of the program will be able to solve the problem.

Anecdotal evidence from ping-pong suggests that strong conflicts result from underlying semantic problems, such that no local solution is possible. Instead, the real solution lies elsewhere, with other strategies. For example, the ping-pong system can encounter a situation where the firing elevation of the ball should be increased and decreased at the same time. Rather than trying to find some optimal elevation, the ball must be hit harder instead.

From this discussion, we propose that the lesson is that some details ought not matter too much. If they do, the desired semantics ought be reconsidered. A robust system will result.

4.6 Exception Handler.

The exception handler adds robustness and the ability to step back a level to the system. One part of the initial planning process makes a minor trajectory-planning mistake at a noticeable rate. Rather than causing a disfunctional system and immediately forcing an investigation, the mistake is reliably and repeatedly corrected by the exception handler. Consequently, we can even decide to ignore the problem, saving us the trouble of adding the small additional knowledge to correct the problem.

On a macroscopic scale, the robot can often return balls deflected by the net, probably more often than a (this) human. By funnelling exceptions through common code, we simplify the entire program, and provide a means to examine the overall situation. This "independent viewpoint" enhances program reliability, since possibly buggy experimental strategies do not have to monitor their own performance. For example, the exception handler will reassign problems that are not being solved by the currently assigned solution mechanism.

4.7 Why It Does Not Work.

Since the robot's mistakes are unexpected by the software and the robot tracks its trajectory accurately, poor models of the aerodynamics and bounce dynamics or suboptimal sensor data is suggested. The long-range camera pair has excellent accuracy. The close-up pair has known systematic errors at low speed close to the camera because the exact degree of camera saturation can not be easily predicted. Although the close-up pair's static accuracy is high, small systematic position and velocity-dependent shifts cause inaccurate spin estimates (for the close-up pair).

In general, we believe that to substantially improve performance we would need a better aerodynamic model, a better trajectory fitting technique, and a better bounce model. Improving the models would require an independent spin-measurement technique, many trials, off-line analysis, and more computation at run-time. Some evidence suggests aerodynamic instabilities at low speed — knuckle-balling.

Retrospectively, the close-up cameras should have been placed to emphasize viewing the ball after the bounce, rather than maintaining a continuous view of the ball. By improving the prospects for watching the ball after the bounce, we reduce the importance of an accurate table bounce prediction, as long as we can respond in time. Unfortunately, this technique will not help the challenging bounce from the paddle.

Let us re-emphasize that robot systems should rely on their sensing wherever possible, rather than rely on blind prediction. This comment is in keeping with the ping-pong player's fundamental approach of continuous observation. A system with continuous sensing, instead of observation and prediction, will be more robust. Without infinitely fast robots, some prediction will always be required in time-critical tasks; faster robots are advantageous because less prediction is required.

4.8 Succeeding By Occasionally Failing.

Paradoxically, we must conclude that the system works because it is willing to fail. The same theme repeats at several levels. At the lowest level, we admit we may not generate "perfect" results from models, but we use physics to detect this. At an intermediate level, we may not always combine data sources effectively, but, in addition to the chance that physics may bail us out, the exception handler can be brought to bear to attempt alternative solutions. Ultimately, we can lose a point with grace.

By admitting the possibility of failure, we avoid 90/10 effects where 90 percent of the effort goes into the last 10 percent of the cases. We avoid the spiral of an ever more complex implementation which takes longer and longer to write and debug. Especially, we prevent the system from being slowed to inoperability by time-consuming exact solutions. The resulting system is more robust as well, since it is more prepared to deal with results other than those planned.

Of course, the prospect of fallible robots sounds dangerous. Even well-constructed robots must argue points in ping-pong. I think failure is inherent and must be accepted and accounted for as systems become more complex. This viewpoint bodes ill for those who expect perfection from the Star Wars (SDI) project.

As the final commentary on failure, I remark on the high failure rate of one extraordinarily complex and capable system — a human. Why is it that a tremendously capable human can lock his keys in his office or trip over a crack in the sidewalk? Perhaps occasional failure is a prerequisite for success.

5. FUTURE DIRECTIONS

Robot ping-pong has clearly visible redundant degrees of freedom. We believe that this is the rule rather than the exception: that essentially all tasks have redundant degrees of freedom available to be controlled by a sophisticated robot controller.

We can begin improving controller capabilities by adding internal feedback to extant robot controllers — by designing them to examine the quality of the motions they generate such that they can automatically adjust speeds, transition and acceleration regions, and free space trajectories with human-specified guidelines. Exception handling mechanisms could begin responding to joint limit and tracking errors.

Once the underlying mechanisms are in place for basic robot-related functions, the same mechanisms may be applied to task-level processing as well. Explicitly redundant robot manipulator configurations will hasten the need for this transition if the redundancies are to be exploited in the task domain, say to avoid crashing an elbow, as well as in the bare robot domain. By continually combining both task and robot information, we can achieve both the high performance and robustness needed for cost-effective application.

6. SUMMARY

An expert controller contains several essential features: exploitation of task redundancy, continual integration of new sensor data, continual improvement of the task plan, fast approximate methods to generate values, physics-based models to ensure accuracy, global exception handling, flexible internal architecture, and robustness to failure. At the broadest scale, we hope the reader will take away at least the following two messages.

First, that robots must continuously sense their environment, using the sensor data to constantly improve their task plan. Sensing systems should be designed from the beginning for operation in a dynamic environment. The continual flow of sensor data pushes system designs towards more robustness and higher performance.

Second, that we must design for failure: to admit its possibility and to design systems to compensate for it. Concentrating our effort on producing exactly correct answers is inappropriate, as the eventual execution will not be exact. An error-tolerant system can be faster to design and run, and will be more robust in use.

REFERENCES

[1] S.R. Ahuja, "S/Net: A High Speed Interconnect for Multiple Computers," IEEE Journal of Selected Areas in Communication, Vol. SAC-1, No. 5, p. 751–756, November 1983.

[2] R.L. Andersson, "A Robot Ping-Pong Player: Experiment in Real-Time Intelligent Control," The MIT Press, Cambridge, MA, 1988.

[3] R.L. Andersson, "Aggressive Trajectory Generator for a Robot Ping-Pong Player," IEEE International Conference on Robotics and Automation, Philadelphia, PA, p. 188–193, April 1988.

[4] J. Billingsley, "Machineroe joins new title fight," Practical Robotics, p. 14–16, May/June 1984.

[5] R.A. Brooks, "Self Calibration of Motion and Stereo Vision for Mobile Robots," Fourth International Symposium on Robotics Research, The MIT Press, p. 277–286 (see Appendix), August, 1988.

[6] B.G. Buchanan, E.H. Shortliffe, "Rule-Based Expert Systems," Addison-Wesley, 1984.

[7] R.D. Gaglianello, H.P. Katseff, "Meglos: An Operating System for a Multiprocessor Environment," Proceedings of the Fifth International Conference on Distributed Computing Systems, May 1985.

Tad McGeer
School of Engineering Science
Simon Fraser University
Burnaby, British Columbia, Canada V5A 1S6

Passive Dynamic Walking

Abstract

There exists a class of two-legged machines for which walking is a natural dynamic mode. Once started on a shallow slope, a machine of this class will settle into a steady gait quite comparable to human walking, without active control or energy input. Interpretation and analysis of the physics are straightforward; the walking cycle, its stability, and its sensitivity to parameter variations are easily calculated. Experiments with a test machine verify that the passive walking effect can be readily exploited in practice. The dynamics are most clearly demonstrated by a machine powered only by gravity, but they can be combined easily with active energy input to produce efficient and dextrous walking over a broad range of terrain.

1. Static vs. Dynamic Walking

Research on legged locomotion is motivated partly by fundamental curiousity about its mechanics, and partly by the practical utility of machines capable of traversing uneven surfaces. Increasing general interest in robotics over recent years has coincided with the appearance of a wide variety of legged machines. A brief classification will indicate where our own work fits in. First one should distinguish between *static* and *dynamic* machines. The former maintain static equilibrium throughout their motion. This requires at least four legs and, more commonly, six. It also imposes a speed restriction, since cyclic accelerations must be limited in order to minimize inertial effects. Outstanding examples of static walkers are the Odex series (Russell 1983) and the Adaptive Suspension Vehicle (Waldron 1986). Dynamic machines, on the other hand, are more like people; they can have fewer legs than static machines, and are potentially faster.

The International Journal of Robotics Research,
Vol. 9, No. 2, April 1990,

2. Dynamics vs. Control

Our interest is in dynamic walking machines, which for our purposes can be classified according to the role of active control in generating the gait. At one end of the spectrum is the biped of Mita et al. (1984), whose motion is generated entirely by linear feedback control. At the end of one step, joint angles are commanded corresponding to the end of the next step, and the controller attempts to null the errors. There is no explicit specification of the trajectory between these end conditions. Yamada, Furusho, and Sano (1985) took an approach that also relies on feedback, but in their machine it is used to track a fully specified trajectory rather than just to close the gap between start and end positions. Meanwhile the stance leg is left free to rotate as an inverted pendulum, which, as we shall discuss, is a key element of passive walking. Similar techniques are used in biped walkers by Takanishi et al. (1985), Lee and Liao (1988), and Zheng, Shen, and Sias (1988).

By contrast the bipeds of Miura and Shimoyama (1984) generate their gait by feedforward rather than feedback; joint torque schedules are precalculated and played back on command. Again the stance leg is left free. However, the "feedforward" gait is unstable, so small feedback corrections are added to maintain the walking cycle. Most significantly, these are *not* applied continuously (i.e., for tracking of the nominal trajectory). Instead the "feedforward" step is treated as a processor whose output (the end-of-step state) varies with the input (the start-of-step state). Thus the feedback controller responds to an error in tracking by modifying initial conditions for subsequent steps, and so over several steps the error is eliminated. In this paper you will see analysis of a similar process. Raibert (1986) has developed comparable concepts but with a more pure implementation, and applied them with great success to running machines having from one to four legs.

All of these machines use active control in some form to generate the locomotion pattern. They can be

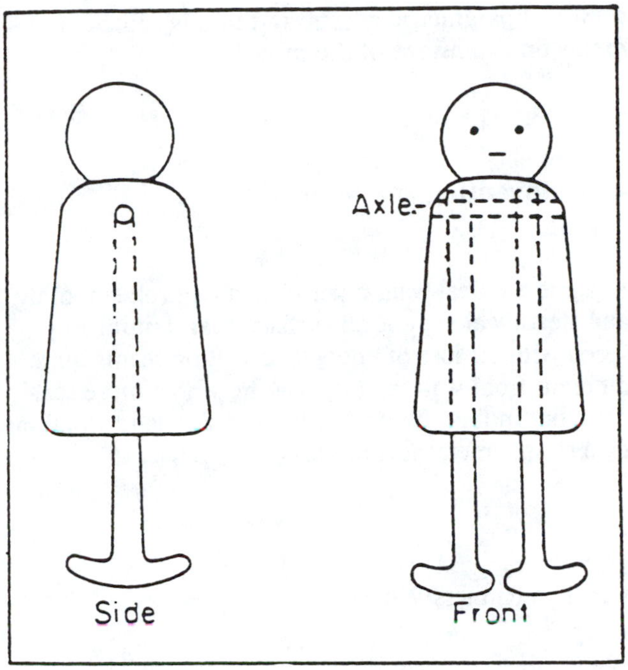

Fig. 1. A bipedal toy that walks passively down shallow inclines. [Reprinted from (McMahon 1984).]

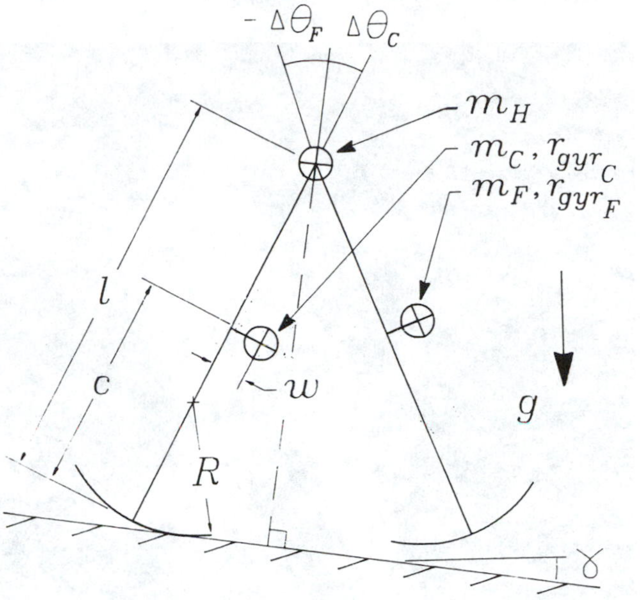

Fig. 2. General arrangement of a 2D biped. It includes legs of arbitrary mass and inertia, semicircular feet, and a point mass at the hip.

ordered according to the style of implementation, ranging from continuous active feedback to once-per-step adjustment of an actively generated but nevertheless fixed cycle. This paper discusses a machine at the extreme end of the spectrum: gravity and inertia alone generate the locomotion pattern, which we therefore call "passive walking."

3. Motivation for Passive Walking

The practical motivation for studying passive walking is, first, that it makes for mechanical simplicity and relatively high efficiency. (The specific resistance of our test biped is ≈ 0.025 in a human-like walk.) Second, control of speed and direction is simplified when one doesn't have to worry about the details of generating a substrate motion. Moreover, the simplicity promotes understanding. Consider an analogy with the development of powered flight. The Wrights put their initial efforts into studying gliders, as did their predecessors Cayley and Lilienthal. Once they had a reasonable grasp of aerodynamics and control, adding a powerplant was only a small change. (In fact, their engine wasn't very good even for its day, but their other strengths ensured their success.) Adding power to a passive walker involves a comparably minor modification (McGeer 1988).

Actually passive walkers existed long before contemporary research machines. My interest was stimulated by the bipedal toy shown in Figure 1; it walks all by itself down shallow slopes, while rocking sideways to lift its swing foot clear of the ground. A similar quadruped toy walks on level ground while being pulled by a dangling weight. I learned of such toys through a paper by Mochon and McMahon (1980), who showed how walking could be generated, at least in large measure, by passive interaction of gravity and inertia. Their "ballistic walking" model is especially helpful for understanding knee flexion, which is discussed toward the end of the paper.

Our discussion here is based on the model shown in Figure 2, which is no more than two stiff legs pin-jointed at the hip. It can be regarded as a two-dimensional version of the toy; its dynamics in the longitudinal plane are similar, but it doesn't rock sideways. This simplifies the motion, but it left us with new problems in building a test machine: how to keep the

Fig. 3. Biped used for experiments on two-dimensional gravity-powered walking. The outer legs are connected by a crossbar and alternate like crutches with the center leg. The feet are semicircular and have roughened rubber treads. Toe-stubbing is prevented by small motors that fold the feet sideways during the swing phase. Apart from that, this machine, like the toy, walks in a naturally stable limit cycle requiring no active control. Leg length is 50 cm, and weight is 3.5 kg.

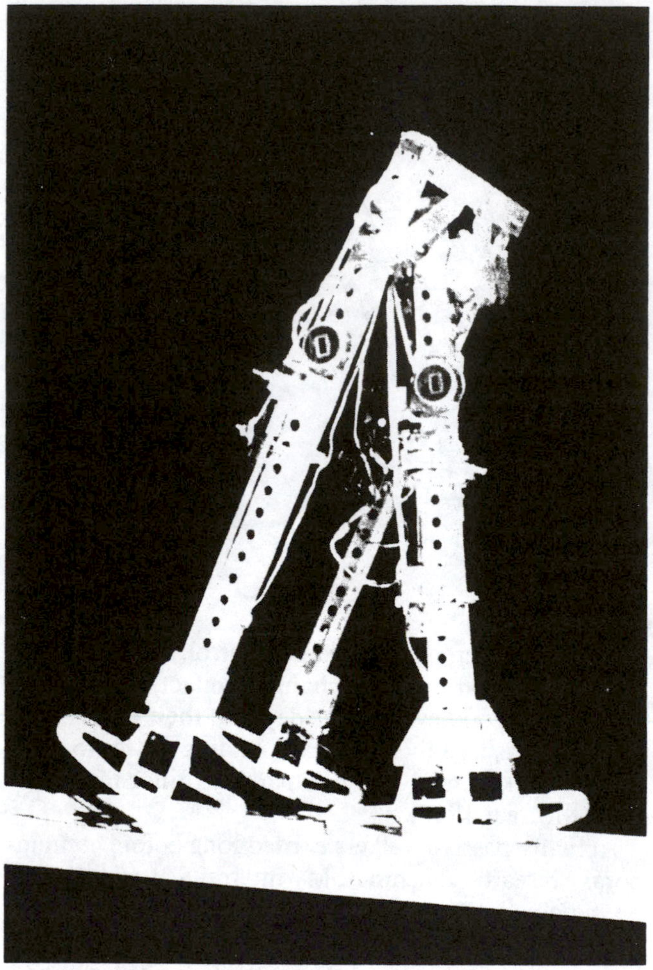

motion two-dimensional, and how to clear the swing foot. Figure 3 is a photo of our solution. Two-dimensionality is enforced by building the outer leg as a pair of crutches. Foot clearance is by either of two methods. Occasionally we use a checkerboard pattern of tiles, which in effect retract the ground beneath the swing feet. However, usually it is more convenient to shorten the legs; thus the machine can lift its feet via small motors driving leadscrews.

The discussion here begins with two elementary models to illustrate the energetics and dynamics of passive walking. Next follow analyses of cyclic walking and stability for the general model of Figure 2, and comparison with experimental results. Then comes a survey of parametric effects, and finally some comments on extensions of the model.

4. Reinventing the Wheel

Imagine an ideal wagon wheel that can roll smoothly and steadily along a level surface, maintaining any speed without loss of energy. Its rolling seems quite different from walking (and on the whole more sensible!), but in fact rolling can be transformed to walking by a simple metamorphosis.

4.1. The Rimless Wheel

Following Margaria (1976), remove the rim from the wagon wheel as in Figure 4, leaving, in effect, a set of legs. Unlike the original wheel, this device cannot roll steadily on a level surface; instead it loses some speed each time a new leg hits the ground. We treat each of these collisions as inelastic and impulsive. In that case the wheel conserves angular momentum about the impact point, and the loss in speed can be calculated as follows. Immediately before the collision the angular momentum is

$$H^- = (\cos 2\alpha_0 + r_{gyr}^2)\, ml^2\Omega^-. \tag{1}$$

(Note that the wheel's radius of gyration is normalized by leg length l.) Immediately after the collision, the angular momentum is simply

$$H^+ = (1 + r_{gyr}^2)\, ml^2\Omega^+. \tag{2}$$

Equating these implies that

$$\frac{\Omega^+}{\Omega^-} = \frac{\cos 2\alpha_0 + r_{gyr}^2}{1 + r_{gyr}^2} \equiv \eta. \tag{3}$$

All of our analysis is cast in dimensionless terms, with mass m, leg length l, and time $\sqrt{l/g}$ providing the base units. To define what amounts to a dimensionless pendulum frequency (7),

$$\sigma^2 \equiv \frac{1}{1 + r_{gyr}^2}. \tag{4}$$

Then

$$\eta = 1 - \sigma^2(1 - \cos 2\alpha_0). \tag{5}$$

It follows from (3) that over a series of k steps

$$\Omega_k \sim \eta^k. \tag{6}$$

Hence on a level surface the rimless wheel will decelerate exponentially. However, on a downhill grade, say with slope γ, the wheel can recoup its losses and so establish a steady rolling cycle. The equilibrium speed can be calculated from the differential equation for rotation about the stance foot. Over the range of angles used in walking, linearization for small angles is entirely justified, so the equation can be written as

$$\ddot{\theta} - \sigma^2\theta \approx \sigma^2\gamma. \tag{7}$$

Here θ is measured from the surface normal; thus the wheel has an unstable equilibrium at $\theta = -\gamma$.

One "step" begins with $\theta = -\alpha_0$, and rolling is cyclic if the initial speed, say Ω_0, repeats from one step to the next. Repetition of Ω_0 implies that each step must end with rotational speed Ω_0/η (6). Thus the steady step has the following initial and final states:

$$\begin{aligned} \theta(0) &= -\alpha_0 & \Omega(0) &= \Omega_0 \\ \theta(\tau_0) &= \alpha_0 & \Omega(\tau_0) &= \Omega_0/\eta \end{aligned} \tag{8}$$

Both Ω_0 and the steady step period τ_0 can be evaluated by applying these boundary conditions to the equation of motion (7). The results are

$$\Omega_0 = \sqrt{\frac{4\gamma\alpha_0\sigma^2\eta^2}{1 - \eta^2}}, \tag{9}$$

$$e^{\sigma\tau_0} = \frac{\gamma + \alpha_0 + \Omega_0/\sigma\eta}{\gamma - \alpha_0 + \Omega_0/\sigma}. \tag{10}$$

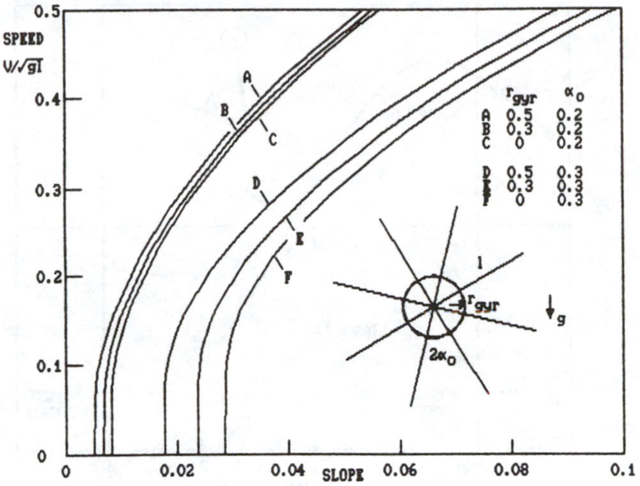

Fig. 4. Removing the rim from a wagon wheel allows a simple illustration of walking energetics. On a level surface a rimless wheel would grind to a halt; but on an incline it can establish a steady rolling cycle somewhat comparable to walking of a passive biped (see Fig. 6). The speed, here in units of $\sqrt{gl}$, is a function of the slope, the inter-leg angle $2\alpha_0$, and the radius of gyration r_{gyr} about the hub.

For small α_0 the dimensionless *forward* speed is then

$$V \approx \frac{2\alpha_0}{\tau_0}. \tag{11}$$

Figure 4 includes plots of this speed as a function of slope for various rimless wheels. These may be compared with speed versus slope in passive walking of a biped, which is plotted in Figure 6. The plots are qualitatively similar and quantitatively comparable. As an example, take $\alpha_0 = 0.3$, which is typical of human walking. Our test machine achieves this stride on a slope of 2.5%, and its forward speed is then 0.46 *m/s*, or 0.21 in units of $\sqrt{gl}$. This speed would be matched by a rimless wheel having $r_{gyr} \approx 0.5$.

Of course the wheel need not always roll at its steady speed. However, it will converge to that speed following a perturbation. In fact, small perturbations decay according to

$$\Omega_k - \Omega_0 \sim (\eta^2)^k \tag{12}$$

We call this decay the "speed mode," which also appears in bipedal walking. Incidentally, it is interesting to observe that convergence on a downhill slope (12) is twice as fast as deceleration on the level (6).

Fig. 5. Walks like a biped; rolls like a wheel. It's a "synthetic wheel," made with two legs, two semicircular feet, and a pin joint at the hip. On each step the free foot swings forward to synthesize a continuous rim. If support is transferred from trailing to leading foot when the leg angles are equal and opposite, then the walking cycle can continue without loss of energy. The period of the cycle (here normalized by the pendulum frequency of the swing leg) is independent of the step length.

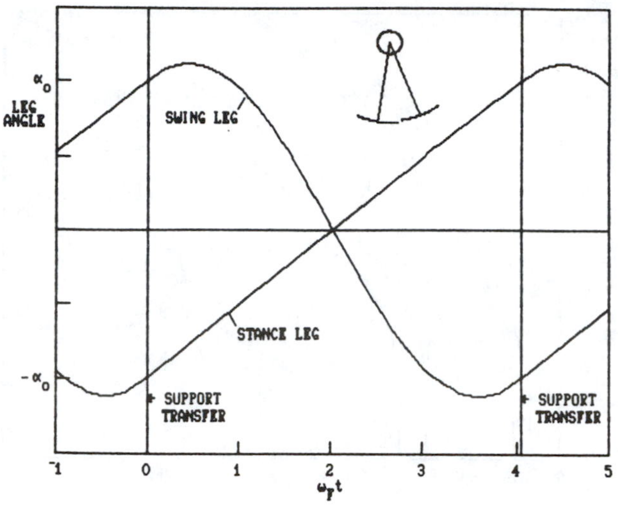

4.2. The Synthetic Wheel

In view of the poor energetics that result, it seems that rim removal is not a very progressive modification of an ordinary wheel. But perhaps improvement might be realized by making cuts elsewhere. In particular, imaging splitting the rim halfway between each spoke. Then discard all but two of the spokes, leaving a pair of legs with big semicircular feet as shown in Figure 5. Put a pin joint at the hip, and ask the following question: could the dynamics be such that while one leg is rolling along the ground, the other swings forward in just the right way to pick up the motion where the first leg leaves off?

In fact, one can devise a solution quite easily. Figure 5 shows a cycle that will continue indefinitely on level ground. The legs start with opposite angles $\pm\alpha_0$ and equal rotational speeds Ω_0, as they did in the original wagon wheel. The appropriate value for α_0 depends on Ω_0, as we will show in a moment. Since the stance leg (subscripted C for "contact") is a section of wheel, it rolls along the ground at constant speed. (Here we presume that the wheel has a large point mass at the hip; otherwise motion of the swing leg would disturb the steady rolling.) The hip (like the hub of an ordinary wheel) therefore translates steadily parallel to the ground, and so the second leg (F for "free") swings as an unforced pendulum. Then following the paths shown in Figure 5 the legs will reach angles $\pm\alpha_0$ with speeds again equal to Ω_0. At that instant, support can roll seamlessly from one rim to the next. Thus a continuous rim has been synthesized from two small pieces.

Of course something must be done to clear the free leg as it swings forward, but we shall deal with that problem later. Also the cycle works only if the step angle α_0 is correct for the speed Ω_0; the relation is derived by matching boundary conditions. First, by inspection of the stance trajectory in Figure 5, Ω_0 must satisfy

$$\Omega_0 = \frac{2\alpha_0}{\tau_0}. \tag{13}$$

τ_0 is determined by the swing leg, which behaves as a pendulum and so follows a sinusoidal trajectory. Its formula is

$$\Delta\theta_F(\tau) = \alpha_0 \cos \omega_F \tau + \frac{\Omega_0}{\omega_F} \sin \omega_F \tau \tag{14}$$

The sinusoid passes through $\Delta\theta_F = -\alpha_0$, with speed Ω_0, when

$$\omega_F \tau_0 = 4.058 \tag{15}$$

Thus the step period for a synthetic wheel is about $\frac{2}{3}$ of the period for a full pendulum swing. Notice that this period is independent of α_0. The speed (13) is then

$$\Omega_0 = \frac{2\omega_F \alpha_0}{4.058} \tag{16}$$

Thus to change the speed of a synthetic wheel, change α_0 (i.e., the length of the step), while τ_0 remains constant —determined solely by the leg's inertial properties and gravity g.

Of course the synthetic wheel is contrived for convenient analysis, but the results have broader application. Our test machine has similar behavior. Figure 6 shows that its step period is fairly insensitive to step length; moreover $\tau_0 \approx 2.8$ in units of $\sqrt{l/g}$ and $\omega_F \approx 1.39$, so $\omega_F \tau_0 \approx 3.9$, which is quite close to the value given by (15). However, (15) is not so accurate for human walking; my own pendulum period, measured by standing on a ledge and dangling one leg, is ≈ 1.4 s,

Fig. 6. Step period τ_0 (in units of $\sqrt{l/g}$, here 0.226 s) and angle α_0 in passive walking of our test biped. Bars show experimental data from trials such as that plotted in Fig. 7. Continuous curves show analytical results, with uncertainty bands carried forward from measurement of the biped's parameters as listed on the plot. The analysis matches experiment poorly if no allowance is made for rolling friction (T_C), but reasonable consistency is realized if the friction is taken to be 0.007 mgl. With this level of friction, walking is calculated to be unstable on slopes less than 0.8%.

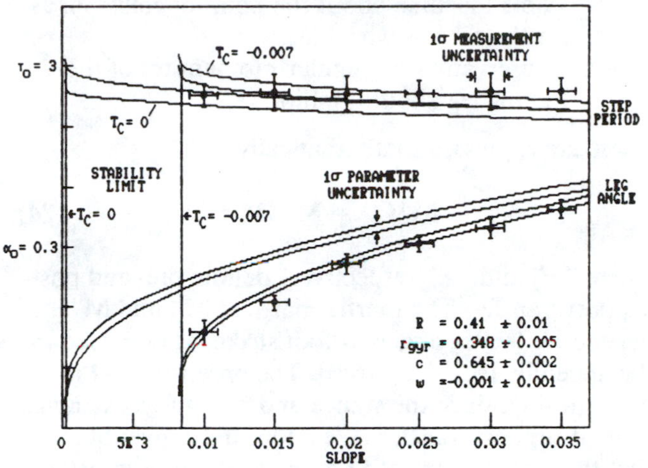

and my step period varies from ≈0.55 s at low speed to ≈0.50 s at high speed. Thus the ratio of step period to pendulum period is ≈0.36 to ≈0.39, as opposed to 0.65 for the wheel (15). I have significantly more life in my stride than either the wheel or our test biped!

While Figure 5 shows that a synthetic wheel is possible in theory, can its cycle actually be realized in practice? This boils down to an issue of stability, and it turns out that the synthetic wheel is just as stable as an ordinary wheel. That is, both have neutrally stable "speed modes"; in the case of a synthetic wheel, a change in speed leads to a new α_0 according to (16). In addition, the synthetic wheel has two other "step-to-step" modes that prove to be stable; these are discussed in section 8.

5. Steady Walking of a General 2D Biped

The biped of Figure 2 is similar to the synthetic wheel but allows for broader variation of parameters. Adjustable parameters include

1. Foot radius R
2. Leg radius of gyration r_{gyr}
3. Leg center of mass height c
4. Fore/aft center of mass offset w
5. Hip mass fraction m_H

We will examine the effects of varying each parameter.

Our calculations of gait and stability rely on step-to-step (S-to-S) analysis, which is explained as follows. At the start of a step, say the kth, the legs have equal and opposite angles $\pm\alpha_k$ and rotational speeds Ω_{Ck}, Ω_{Fk}. The motion proceeds roughly as for the synthetic wheel, with appropriate arrangements to prevent toe-stubbing at mid-stride. The step ends when leg angles are again equal and opposite, but in general different (say α_{k+1}) from α_k. At that point the swing foot hits the ground, and the leg speeds change instantaneously to their initial values for the following step Ω_{Ck+1}, Ω_{Fk+1}. We will formulate equations relating $(\alpha_k, \Omega_{Ck}, \Omega_{Fk})$ to $(\alpha_{k+1}, \Omega_{Ck+1}, \Omega_{Fk+1})$. We will then use these to find naturally repeating initial conditions (i.e., passive cycles) and to examine how perturbations evolve from step to step.

5.1. Start- to End-of-Step Equations

During the step the machine is supported on one foot, and its state is specified by the two leg angles θ_C, θ_F, and the speeds Ω_C, Ω_F. In general the equations of motion are nonlinear in these variables, but since in walking the legs remain near the vertical and the speeds remain small ($\ll \sqrt{g/l}$), linearization is justified. Thus the linearized equations are

$$\mathbf{M}_0 \begin{bmatrix} \dot{\Omega}_C \\ \dot{\Omega}_F \end{bmatrix} \equiv \mathbf{M}_0 \dot{\mathbf{\Omega}} = \mathbf{T} \qquad (17)$$

The inertia matrix $\mathbf{M}_0$ is derived in the appendix (65). $\mathbf{T}$ lists the torques about the stance foot and hip. It can include friction or control inputs, but for our machine the principal torque is gravitational:

$$\mathbf{T}_g = \mathbf{K}[\Delta\boldsymbol{\theta}_{SE} - \Delta\boldsymbol{\theta}] \qquad (18)$$

The stiffness matrix $\mathbf{K}$ and equilibrium position $\Delta\boldsymbol{\theta}_{SE}$ are also derived in the appendix [eqs. (52), (53), (54)].

$\Delta\boldsymbol{\theta}$ is the rotation from the surface normal. For small γ, $\Delta\boldsymbol{\theta}_{SE}$ can be written as

$$\Delta\boldsymbol{\theta}_{SE} = \Delta\boldsymbol{\theta}_W + \mathbf{b}\gamma \tag{19}$$

By way of explanation you can imagine that if the feet were points (i.e., zero foot radius) then the equilibrium position would be legs-vertical, which means that both elements of $\mathbf{b}$ would be -1. However with nonzero foot radius the stance leg would have to rotate past the vertical to put the overall mass center over the contact point, so $b_1 < -1$. Also moving the legs' mass centers fore and aft from the leg axes (i.e., nonzero w, Fig. 2) would make for nonvertical equilibrium even on level ground, and this is handled by $\Delta\boldsymbol{\theta}_W$.

Putting the gravitational torque (18) into the equations of motion (17) leaves a fourth-order linear system. This can be solved to jump from the start-of-step to any later time, via a 4×4 transition matrix $\mathbf{D}$:

$$\begin{bmatrix} \Delta\boldsymbol{\theta}(\tau_k) \\ \boldsymbol{\Omega}(\tau_k) \end{bmatrix} = \mathbf{D}(\tau_k) \begin{bmatrix} \Delta\boldsymbol{\theta}_k - \Delta\boldsymbol{\theta}_{SE} \\ \boldsymbol{\Omega}_k \end{bmatrix} + \begin{bmatrix} \Delta\boldsymbol{\theta}_{SE} \\ \mathbf{0} \end{bmatrix}. \tag{20}$$

The elements of $\mathbf{D}$ for a given $\mathbf{M}_0$, $\mathbf{K}$, and τ_k are calculated by standard methods for linear systems. For transition to the end-of-step, τ_k must be chosen so that the elements of $\Delta\boldsymbol{\theta}(\tau_k)$ are equal and opposite. Thus define

$$\boldsymbol{\lambda} = \begin{bmatrix} -1 \\ 1 \end{bmatrix}. \tag{21}$$

Then

$$\Delta\boldsymbol{\theta}_k = \boldsymbol{\lambda}\alpha_k, \tag{22}$$

$$\Delta\boldsymbol{\theta}(\tau_k) = -\boldsymbol{\lambda}\alpha_{k+1}. \tag{23}$$

5.2. Support Transfer

When (23) is satisfied, foot strike occurs, which as for the rimless wheel we treat as an inelastic collision. In this case two conditions apply:

1. Conservation of angular momentum of the whole machine about the point of collision, as for the rimless wheel (3).
2. Conservation of angular momentum of the trailing leg about the hip.

These are expressed mathematically as

$$\mathbf{M}^{+}\boldsymbol{\Omega}^{+} = \mathbf{M}^{-}\boldsymbol{\Omega}^{-}, \tag{24}$$

where "$-$" and "$+$" respectively denote pre- and post-support transfer. The inertia matrices $\mathbf{M}^-$ and $\mathbf{M}^+$ depend on the leg angles at foot strike, as discussed in the appendix [eqs. (65), (68)]. The ordering of $\boldsymbol{\Omega}$ may be confusing, since the stance and swing legs exchange roles at support transfer. We adopt the convention that the first element of $\boldsymbol{\Omega}$ refers to the *post*-transfer stance leg. Hence the pre-transfer indexing must be flipped:

$$\boldsymbol{\Omega}^- = \begin{bmatrix} 0 & 1 \\ 1 & 0 \end{bmatrix} \boldsymbol{\Omega}(\tau_k) \equiv \mathbf{F}\boldsymbol{\Omega}(\tau_k). \tag{25}$$

From (24), then, the initial speeds for step $k+1$ are

$$\boldsymbol{\Omega}_{k+1} = \mathbf{M}^{+-1}\mathbf{M}^{-}\mathbf{F}\boldsymbol{\Omega}(\tau_k) \equiv \boldsymbol{\Lambda}\boldsymbol{\Omega}(\tau_k). \tag{26}$$

5.3. "S-to-S" Equations

Combining the start- to end-of-step equation (20) with the foot strike conditions [eqs. (22), (23), (26)] produces the S-to-S equations. It is convenient to break $\mathbf{D}$ in (20) into 2×2 submatrices, so the system is written as

$$-\boldsymbol{\lambda}\alpha_{k+1} = \mathbf{D}_{\theta\theta}[\boldsymbol{\lambda}\alpha_k - \Delta\boldsymbol{\theta}_{SE}] + \mathbf{D}_{\theta\Omega}\boldsymbol{\Omega}_k + \Delta\boldsymbol{\theta}_{SE}, \tag{27}$$

$$\boldsymbol{\Omega}_{k+1} = \boldsymbol{\Lambda}\mathbf{D}_{\Omega\theta}[\boldsymbol{\lambda}\alpha_k - \Delta\boldsymbol{\theta}_{SE}] + \boldsymbol{\Lambda}\mathbf{D}_{\Omega\Omega}\boldsymbol{\Omega}_k. \tag{28}$$

Bear in mind that while formulation of this set has been simple, evaluation is not quite so straightforward. Given initial conditions (α_k, $\boldsymbol{\Omega}_k$), the time τ_k (which determines $\mathbf{D}$) at which the leg angles are next equal and opposite must be determined. Then α_{k+1} (27)

Fig. 7. Step period and angle measured at heel strike in trials of the test biped. The machine was started by hand on a 5.5-m ramp inclined at 2.5% downhill, and after a few steps settled into a fairly steady gait. Dots show the mean values, and bars the scatter recorded over six trials. Lengths are normalized by leg length l, and periods by $\sqrt{l/g}$. "C" denotes start-of-stance on the center leg; "O" on the outer legs.

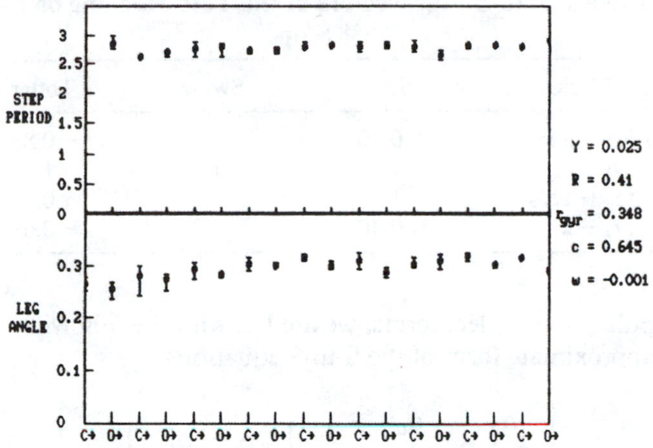

determines Λ (26), which in turn allows evaluation of $\mathbf{\Omega}_{k+1}$ (28).

5.4. Solution for the Walking Cycle

For cyclic walking, initial conditions must repeat from step to step:

$$\alpha_{k+1} = \alpha_k = \alpha_0, \quad (29)$$

$$\mathbf{\Omega}_{k+1} = \mathbf{\Omega}_k = \mathbf{\Omega}_0. \quad (30)$$

Imposing these conditions on S-to-S equations (27) and (28) leads to a compact solution for the walking cycle. First solve for $\mathbf{\Omega}_0$ using (28):

$$\mathbf{\Omega}_0 = [\mathbf{I} - \Lambda \mathbf{D}_{\Omega\Omega}]^{-1} \Lambda \mathbf{D}_{\Omega\theta}[\lambda\alpha_0 - \Delta\boldsymbol{\theta}_{SE}]. \quad (31)$$

Then substitute for $\mathbf{\Omega}_0$ in (27); the end result can be written as follows. Define

$$\mathbf{D}'(\alpha_0, \tau_0) = \mathbf{D}_{\theta\theta} + \mathbf{D}_{\theta\Omega}[\mathbf{I} - \Lambda \mathbf{D}_{\Omega\Omega}]^{-1} \Lambda \mathbf{D}_{\Omega\theta}. \quad (32)$$

Then (27) can be written as

$$\begin{aligned} \mathbf{0} &= [\mathbf{D}' - \mathbf{F}]\lambda\alpha_0 - [\mathbf{D}' - \mathbf{I}]\Delta\boldsymbol{\theta}_{SE} \\ &= [\mathbf{D}' - \mathbf{F}]\lambda\alpha_0 - [\mathbf{D}' - \mathbf{I}][\Delta\boldsymbol{\theta}_W + \mathbf{b}\gamma]. \quad (33) \end{aligned}$$

The last line follows from (19). This is the steady-cycle condition, with two equations in three variables α_0, τ_0, and γ. Usually we specify α_0 and solve for the other two. Equation (33) then has either two solutions or none. If two, then one cycle has $\omega_F\tau_0 < \pi$, and is invariably unstable. The other corresponds to a synthetic-wheel-like cycle with $\omega_F\tau_0$ between π and $3\pi/2$. This is the solution of interest. Thus we use the synthetic-wheel estimate (15) to start a Newton's method search for τ_0, and if a solution exists, then convergence to five significant figures requires about five iterations.

6. Steady Walking of the Test Machine

The analytical solution for steady walking is compared with experiment in Figure 6. Experimental data were obtained by multiple trials as shown in Figure 7; in each trial our test biped was started by hand from the top of a ramp, and after a few steps it settled into the steady gait appropriate for the slope in use. (Notice, incidentally, that the rate of convergence is a measure of stability; we will pursue this analytically in the next section.) Several trials were done on each slope, and means and standard deviations for α_0 and τ_0 were calculated over all data except those from the first few steps of each trial.

Our predictive ability to compare with analysis was limited by uncertainty in the machine's parameters. Each leg's center-of-mass height c was measured to about 1 mm by balancing each leg on a knife edge; w to about 0.5 mm by hanging each leg freely from the hip; and r_{gyr} to about 2.5 mm by timing pendulum swings. The center and outer legs were ballasted to match within these tolerances and differed in mass by only a few grams (i.e., $\approx 0.001\ m$).

When these parameters were put into (33) we found (as Figure 6 indicates) that the observed cadence was slower than predicted. We suspected that the discrepancy was caused by rolling friction on the machine's rubber-soled feet. Hence we added a constant to the first element of $\mathbf{T}$ in (17), which modified $\Delta\boldsymbol{\theta}_W$ in (19). A value of -0.007 brought the analytical results into line with observation. In dimensional terms this is equivalent to 25 gm-force applied at the hip. We could not devise an independent measurement of rolling friction, but this figure is credible.

However, after adjusting friction for a reasonable fit, we still found a discrepancy in that the observed step period is less sensitive to speed than the analysis indicates. We therefore suspected that the inelastic model for foot strike was imprecise, and in fact there was a bit of bouncing at high speed. But in any case the residual discrepancy is small, so our current S-to-S equations (27) and (28) apparently form a sound basis for further investigation, in particular of stability and parametric variations.

Here the efficiency of biped walking should be noted. For comparison with other modes of transport, we measure efficiency by specific resistance:

$$SR \equiv \frac{\text{resistive force}}{\text{weight}} = \frac{\text{mechanical work done}}{\text{weight} \times \text{distance travelled}}$$

For a vehicle powered by descending a slope, *SR* is just equal to γ, or about 2.5% for our machine using $\alpha_0 \approx 0.3$, which is comfortable for humans. This figure would be terrible for a car, but it is very good in comparison with other rough-terrain vehicles such as multilegged crawlers and bulldozers (Waldron 1984).

7. Linearized Step-to-Step Equations

A cyclic solution is a necessary but not quite sufficient condition for practical passive walking. The walking cycle must also be *stable.* Stability can be assessed by linearizing eqs. (27) and (28) for small perturbations on the steady gait. The transition matrix **D** and the support transfer matrix $\mathbf{\Lambda}$ can be approximated for this purpose as follows:

$$\mathbf{D}(\tau_k) \approx \mathbf{D}(\tau_0) + \frac{\partial \mathbf{D}}{\partial \tau}(\tau_k - \tau_0), \tag{34}$$

$$\mathbf{\Lambda}(\alpha_{k+1}) \approx \mathbf{\Lambda}(\alpha_0) + \frac{\partial \mathbf{\Lambda}}{\partial \alpha}(\alpha_{k+1} - \alpha_0). \tag{35}$$

After substituting these into (27) and (28) and manipulating to collect terms, we are left with the following approximate form of the S-to-S equations:

Table 1. S-to-S Eigenvectors of the Test Machine on a 2.5% Slope

Mode	Speed	Swing	Totter
Eigenvalue, z	0.70	−0.05	−0.83
$\alpha - \alpha_0$	1	1	1
$\Omega_C - \Omega_{C0}$	1.1	1.1	0.12
$\Omega_F - \Omega_{F0}$	0.30	7.3	−0.03

$$\begin{bmatrix} \alpha_{k+1} - \alpha_0 \\ \Omega_{Ck+1} - \Omega_{C0} \\ \Omega_{Fk+1} - \Omega_{F0} \\ \tau_k - \tau_0 \end{bmatrix} = \mathbf{S} \begin{bmatrix} \alpha_k - \alpha_0 \\ \Omega_{Ck} - \Omega_{C0} \\ \Omega_{Fk} - \Omega_{F0} \end{bmatrix}. \tag{36}$$

The formula for **S** is given in the appendix [eqs. (69), (70), and (72)]. The important point here is that **S** is the transition matrix of a standard linear difference equation, so the eigenvalues of its upper 3×3 block indicate stability [the equation for $(\tau_k - \tau_0)$ is ancillary]. If all have magnitude less than unity, then the walking cycle is stable; the smaller the magnitude, the faster the recovery from a disturbance. (If, however, the walking cycle is unstable, then linearized S-to-S equations are helpful for design of a stabilizing control law. This was in essence the approach of Miura and Shimoyama (1984). See also McGeer (1989) for active stabilization of a passive cycle.)

8. S-to-S Modes of the Test Machine

Results of stability analysis for our test biped are listed in Table 1. Similar results are found over a wide range of parameter variations, so the modes can be considered typical for passive walking.

The "speed mode" is analogous to the transient behavior of a rimless wheel, with z corresponding to η^2 in (12). The eigenvalue is linked to energy dissipation at support transfer. Thus for the synthetic wheel, $z = 1$ in this mode (i.e., stability is neutral with respect to speed change), as we noted earlier.

The "swing mode" is so named because the eigenvector is dominated by Ω_F. The eigenvalue of this mode is usually small, and in fact reduces to zero in the synthetic wheel, which means that a "swing" perturbation is eliminated immediately at the first support transfer. Physically this occurs because the post-transfer speed of the synthetic wheel's legs is determined entirely by the momentum of the large hip mass; the pretransfer Ω_F is irrelevant.

Finally, the "totter mode" is distinguished by a negative eigenvalue, which means that perturbations alternate in sign from one step to the next. Again the synthetic wheel offers an easily understood special case. Since the wheel cannot dissipate energy, it retains its initial rolling speed for all time. If the initial step angle is not appropriate to the initial speed as specified by (13), then the angle must accommodate through some sort of transient, which turns out to be the totter mode. The eigenvalue can be found analytically by generalizing the swing trajectory formula (14) for $\alpha_k \neq \alpha_0$:

$$-\alpha_{k+1} = \alpha_k \cos \omega_F \tau_k + \frac{\Omega_0}{\omega_F} \sin \omega_F \tau_k. \tag{37}$$

Meanwhile the generalized stance trajectory [cf. (13)] is

$$\alpha_{k+1} = -\alpha_k + \Omega_0 \tau_k. \tag{38}$$

Now differentiate eqs. (37) and (38) with respect to α_k:

$$-\frac{d\alpha_{k+1}}{d\alpha_k} = \cos \omega_F \tau_k + (-\alpha_k \omega_F \sin \omega_F \tau_k + \Omega_0 \cos \omega_F \tau_k) \frac{d\tau_k}{d\alpha_k} \tag{39}$$

$$\frac{d\alpha_{k+1}}{d\alpha_k} = -1 + \Omega_0 \frac{d\tau_k}{d\alpha_k} \tag{40}$$

Solving for the derivatives and evaluating at τ_0 gives

$$\frac{d\alpha_{k+1}}{d\alpha_k} = -\frac{4 \cos \omega_F \tau_0 - \omega_F \tau_0 \sin \omega_F \tau_0}{2 + 2 \cos \omega_F \tau_0 - \omega_F \tau_0 \sin \omega_F \tau_0} = -0.20, \tag{41}$$

$$\frac{d\tau_k}{d\alpha_k} = \frac{1}{\Omega_0} \frac{2 - 2 \cos \omega_F \tau_0}{2 + 2 \cos \omega_F \tau_0 - \omega_F \tau_0 \sin \omega_F \tau_0} = \frac{0.80}{\Omega_0}. \tag{42}$$

It follows from (41) that perturbations converge exponentially according to

$$\alpha_k - \alpha_0 \sim -0.2^k \quad \text{in the totter mode} \tag{43}$$

In summary then, and with some oversimplification for clarity, the "speed" mode in passive walking is a monotonic convergence to the speed appropriate for the slope in use. The "swing" mode is a rapid adjustment of the swing motion to a normal walking pattern. The "totter" mode is an oscillatory attempt to match step length with forward speed. An arbitrary perturbation will excite all three modes simultaneously. Of the three, the totter mode differs most between the synthetic wheel and our test biped ($z = -0.2$ vs. -0.83), and our parametric surveys indicate that it bears watching. We will show examples of parameter choices that make the totter mode unstable.

9. Larger Perturbations

The preceding analysis applies only for small perturbations, but an obvious question is, "How small is small?" There is a limit: if the machine were started just by standing the legs upright and letting go, it would fall over rather than walk! Thus starting requires a bit more care. In general, the machine could be released with arbitrary leg angles and speeds, and it would be nice to know the "convergence zone" in the four-dimensional state space. Unfortunately the boundary of a 4D volume is rather difficult to map, so we have restricted attention to finding the maximum tolerable perturbation in the initial Ω_C or Ω_F. This is done by serial evaluation of the nonlinear S-to-S equations (27) and (28). The results show that Ω_F can vary over wide limits (e.g., <0 to $>+150\%$ of the cyclic value). However, Ω_C is sensitive; many passive walkers can only tolerate an error of a few percent in the starting speed. Still this is not so exacting as it sounds; manual starting (by various techniques) is quite natural, and success is achievable with little or no practice. Furthermore, the boundaries of the convergence zone are sharp, and if the machine starts even barely inside the edge it settles to the steady cycle very nearly as suggested by the small-perturbation analysis.

An alternative to the convergence-zone measure of robustness is resistance to jostling. Thus we calculated transients produced by disrupting the steady gait with a horizontal impulse, applied at the hip just as the legs passed through $\Delta\theta = 0$. As an intuitive standard of reference, we compared the walker with a similar machine resting with legs *locked* at $\pm\alpha_0$. It turns out that a passive walker can tolerate a useful fraction of the impulse required to topple the locked machine; the level varies widely with choice of machine parameters, but 25% (forward or backward) is representative. The calculation has to be done carefully, since bands of tolerable and intolerable impulse magnitudes are interspersed. Twenty-five percent is a typical upper bound for the first tolerable band. More aggressive jostling would have to be countered by active control.

The jostling calculation was particularly helpful in dispelling the inclination to associate robustness with rapid convergence from small perturbations. In fact, a biped with a slowly convergent or even slightly unstable totter mode may well tolerate a stronger midstride jostle than a machine with better totter stability. Thus a practical biped designer might be willing to accept a requirement for "weak" active stabilization of the steady cycle in return for better resistance to knockdowns.

10. Effect of Parameter Variations

We now embark on a brief survey of the effect of design variables on walking performance. Our purpose is twofold: first, to outline the designer's options, and second, to illustrate that walking cycles can be found over a wide range of variations on the 2D biped theme.

10.1. Scale

Before discussing variation of dimensionless parameters we should note the effect of scale (i.e., changing

Fig. 8. Step period and slope for passive walking of bipeds having various foot sizes. The step angle is specified to be 0.3, which is typical in human walking, and the mass center and inertia are held constant while R is varied. The slope (and so the resistance) is zero with R = l, as for the synthetic wheel.

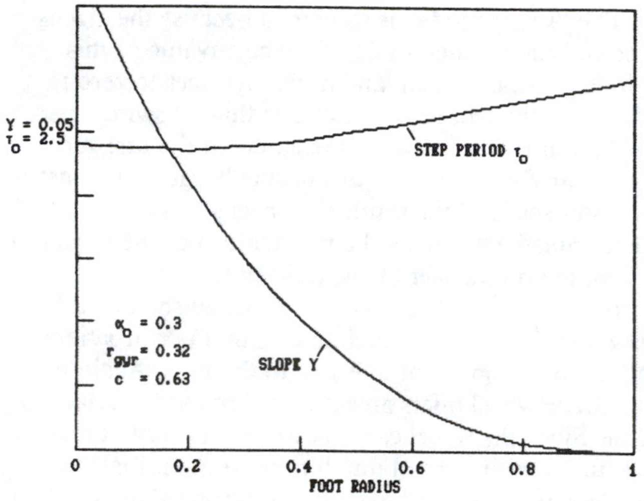

m, l, or g). Quite simply, changing m scales the forces but doesn't change the gait. Changing l scales the step period by $1/\sqrt{l}$, and the speed by $\sqrt{l}$. Changing g scales both period and speed by $\sqrt{g}$. A noteworthy consequence was experienced by the lunar astronauts in 1/6th Earth's gravity. Apparently they had a sensation of being in slow motion, and indeed walking could achieve only 40% of normal speed. Rather than accept that, they hopped instead (McMahon 1984).

10.2. Foot Radius

Figure 8 shows an example of the effect of foot radius on steady walking. Most notable is the improvement in efficiency as the foot changes from a point to a section of wheel. Thus with $R = 0$ we have a biped that, like the rimless wheel, needs a relatively steep slope, and with $R = 1$ we have a synthetic wheel rolling on the level.

Figure 9 shows the locus of S-to-S eigenvalues as a function of R. The speed eigenvalue increases from $z \approx 0.2$ with a point foot to $z = 1$ (i.e., neutral speed stability) with a wheel; this is associated with the improvement in efficiency and is consistent with (12). Meanwhile the totter and swing modes are well separated with a point foot, but coupled with mid-sized

Fig. 9. Locus of the three step-to-step eigenvalues for the walking cycles plotted in Fig. 8, with foot radius as the parameter. In this example one eigenvalue (for the "totter mode") lies outside the unit circle if the feet are very small; this indicates that passive walking is unstable. However, with other choices for c and r_{gyr}, walking is stable even on point feet.

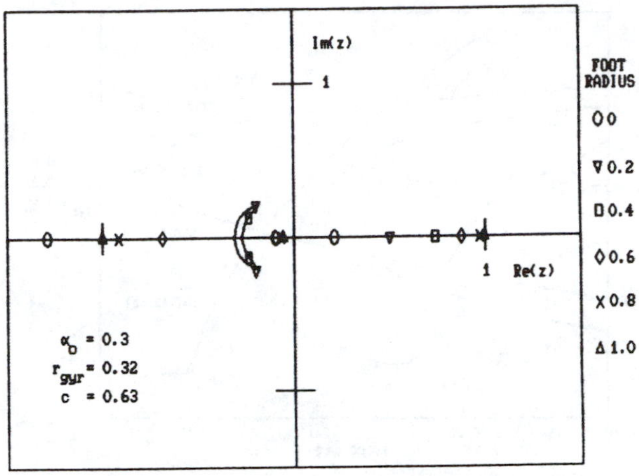

feet so that the two eigenvalues form a complex pair. Then with larger foot radius they separate again, and $R = 1$ puts the totter eigenvalue near $z = -1$. [A true synthetic wheel has $z = -0.2$ (43), but its hip mass is much larger than specified in this example.] The big-footed bipeds in this example thus have relatively weak totter stability, but it still turns out that their resistance to jostling is better than with small feet.

We should mention that the semicircular foot is a mathematical convenience rather than a physical necessity; doubtless other arrangements are feasible. For example, a flat foot could be used on which support would transfer impulsively from heel to toe at mid-stride. Walking would be less efficient than with a curved foot, but apart from that we would expect a similar passive gait. For comparison with a circular foot, the key feature is translation of the support point during stance, which is $2\,\alpha_0 R$. Thus a human, with heel-to-ball distance of $\approx 0.2\,l$ and a typical stride angle of $\alpha_0 = 0.3$, has an "equivalent radius" of ≈ 0.3.

10.3. Leg Inertia and Height of the Mass Center

Leg radius of gyration and center-of-mass height have similar effects, as illustrated in Figures 10 and 11. Increasing r_{gyr}, or raising c with r_{gyr} held constant, lengths the pendulum period and so slows the cadence (15). If the mass center is too close to the hip, then the

Fig. 10. Variation of the walking cycle with c, the height of the leg's mass center. |z| is the magnitude of the totter mode eigenvalue; here it indicates that passive walking is unstable for bipeds with c too low or too high. (The kinks appear where z branches into the complex plane, as in Fig. 9.) The high-c problem arises because of excessively long swing-pendulum periods, while the low-c problem is caused by inefficient support transfer. The latter can be remedied by adding mass at the hip. In this example, m_H is zero.

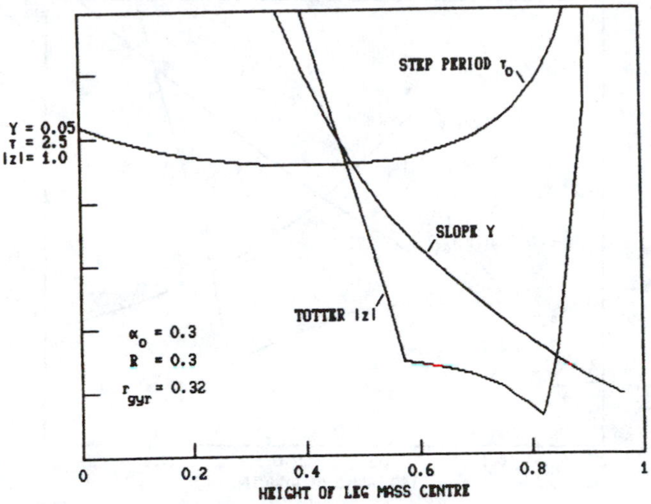

swing leg can't come forward in time to break the fall of the stance leg, and the walking cycle vanishes. On the other hand, lowering r_{gyr} or c causes a different problem: support transfer becomes inefficient according to (5). [Note that α_0 in (5) is the angle subtended by the feet, *at the overall mass center*.] If the efficiency is too poor, then the cycle becomes unstable or even vanishes entirely, as in the example of Figure 11. However, the situation can be retrieved by adding mass at the hip, which raises the overall mass center and thus improves efficiency. A human has about 70% of body mass above the hip; this is sufficient to make support transfer efficient regardless of leg properties.

10.4. Hip Mass

Figure 12 illustrates the effect of adding point mass at the hip. Efficiency improves, and it turns out that jostling resistance improves as well. Still more advantage can be gained if the mass is in the form of an extended torso. For example, the torso can be held in a backward recline, reacting against the stance leg; the reaction provides a braking torque that allows a steep descent. Analysis of this scheme and other roles of the torso is reported by McGeer (1988).

Fig. 11. Leg inertia and c have similar effects on the walking cycle, as indicated by comparing this plot with Fig. 10. The effects here are also mediated by the swing pendulum frequency and by the efficiency of support transfer.

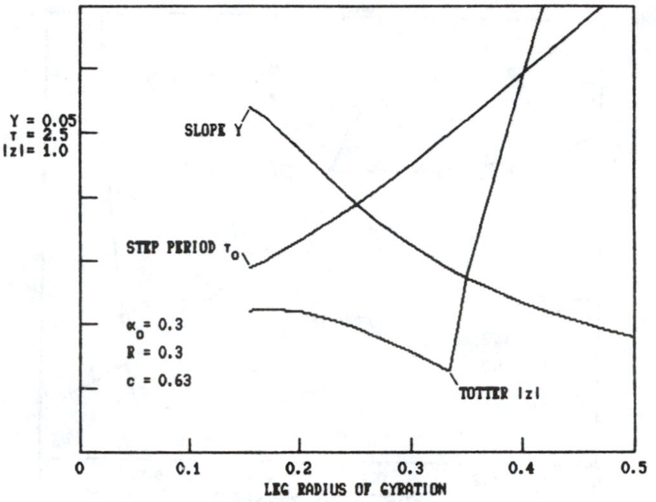

Fig. 12. Our test biped is just a pair of legs, but passive walking also works while carrying a "payload" at the hip. Actually the added mass improves efficiency of the walking cycle.

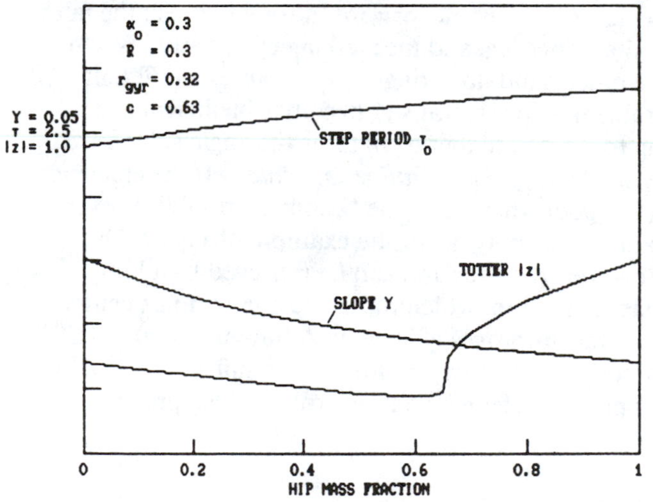

Fig. 13. Passive walking is forgiving of most parameter variations, but even a small amount of friction in the hip joint can destroy the cycle. However, walking can be restored by moving the legs' mass centers backward from the leg axes ($w < 0$, Fig. 18) as long as the friction is only moderate. For this example we have used a viscous model for friction, which is measured by the damping ratio for pendulum oscillations of the swing leg.

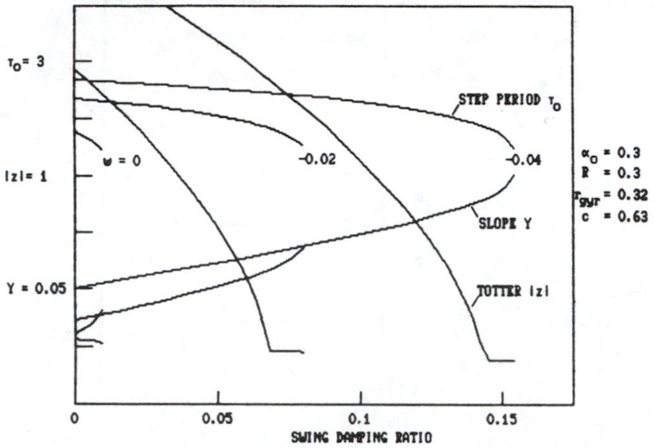

10.5. Hip Damping and Mass Offset

The last few examples have indicated that passive walking is robust with respect to parametric variations. However, it is not universally tolerant, and the hip joint is particularly sensitive. As shown in Figure 13, introducing only a small amount of friction makes the cyclic solution vanish. Consequently, in our test biped we use ball bearings on the hip axle, and these keep the friction acceptably low. However, alternative mechanical arrangements, for instance those made by bone and cartilage, may not be quite as good. Fortunately, compensation can be made by lateral offset (w) of the legs' mass centers from the axes. Thus, as Figure 13 indicates, if the joint has significant friction, then the designer should shift the leg mass backward from the line between hip and foot center of curvature (i.e., $w < 0$).

Actually w is a powerful remedy for a variety of similar ailments (McGeer 1988; 1989). But like any powerful medicine it must be treated with some care. Figure 14 shows that for any given set of machine parameters, w must be set within narrow limits if passive walking is to work. Experiments bring the point home; changing w by only a few millimeters has a very noticeable effect on the feel of manual starting. The power of w makes it attractive not only as a design parameter but also as a *dynamically adjustable* control variable; it might be used, for example, to modulate the gait from one step to the next.

10.6. Leg Mismatch

To close this section on parametric effects, we present a curiosity that may have some implication for the study of gait pathology. Since our biped's legs could not be matched precisely, we were concerned about the sensitivity to differences between them. By way of

Fig. 14. Experiments confirm the powerful effect of mass offset w on the walking cycle. (w was adjusted by shifting the feet relative to the legs.) With each setting we did a series of trials as in Fig. 7. α_0 and τ_0 are well matched by calculations if T_C is taken to be -0.007, but then walks that were in fact sustained for the full length of three six-foot tables are calculated to be unstable. Taking T_C to be zero gives a better match to the observed stability, but leaves relatively large discrepancies in α_0 and τ_0. We suspect that our support transfer model is imprecise, but in any case the sensitivity to w is clear.

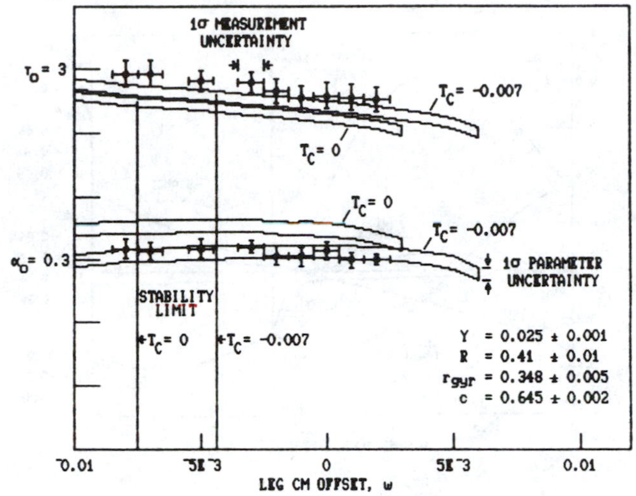

investigation we calculated walking cycles with legs of different mass. The steady-walking solution (33) doesn't apply with mismatched legs, so instead we did serial calculations with S-to-S equations (27) and (28) until a steady gait emerged. With a small mismatch, the cycle repeats in two steps (as would be expected). But with larger mismatch, the stable cycle repeats in four steps, as plotted in Figure 15. Here, then, is an example of frequency jumping, which is not uncommon in nonlinear systems.

11. Fully Passive Walking

We mentioned earlier that our 2D, stiff-legged model allows simple analytical treatment of walking, but it forces resort to inelegant methods of clearing the swing foot. Fully passive foot clearance would be preferable; we will discuss two options.

11.1. Rocking

The toy of Figure 1, and its quadrupedal cousins, clear swing feet by lateral rocking. The frontal view shows

Fig. 15. The stable cycle calculated for a biped having legs with a 10% mass mismatch. m_C is the mass of the stance leg. Each step is similar to that of a synthetic wheel, but the cycle repeats over four steps rather than one. (With smaller mismatch the cycle repeats in two steps.)

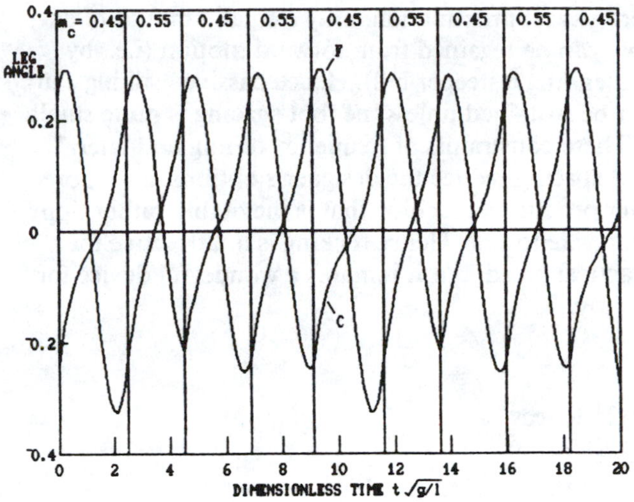

the key design feature, which is that the feet are approximately circular in lateral as well as longitudinal section, and have approximately coincidental centers of curvature. The center of curvature is above the center of gravity, which makes lateral rocking a pendulum oscillation.

We have analyzed the combination of rocking with 2D synthetic wheel dynamics, under the approximation of zero yaw (i.e., that the hip axle remains normal to the direction of motion). This is exactly true for quadrupedal toys and seems fairly accurate for the biped. The rocking frequency must be tuned to the swing frequency, such that half a rocking cycle is completed in one step. Hence the ratio of swing to rocking frequencies, from (15), should be

$$\frac{\omega_F}{\omega_R} = \frac{4.058}{\pi}. \tag{44}$$

If this condition is satisfied, then rocking and swinging naturally phase-lock at the first support transfer, and thereafter walking is non-dissipative. However, if the condition is not satisfied, then the phasing must be "reset" on each step; this entails an energy loss. If the mistuning is too large, then there is no cycle at all.

An additional problem arises if the feet have nonzero lateral separation. Then support transfer bleeds energy out of the rocking motion via the rimless-wheel mechanism (3). But it turns out that, at least in the zero yaw approximation, only a small amount of energy can be regained from forward motion (i.e., by descending a steeper hill). Hence passive walking cannot be sustained unless the foot spacing is quite small.

These constraints of frequency tuning and lateral foot spacing reduce the designer's options, and moreover produce a machine that is inevitably rather tippy from side to side. Hence rocking is unattractive for a practical biped, but it remains a wonderful device for the toy.

11.2. Knees

Of course we humans rock as we walk, but only for lateral balance; for foot clearance we rely on knee flexion. Mochon and McMahon (1980) demonstrated that this motion might well be passive. The key result of the study is that if a straight stance leg and knee-jointed swing leg are given initial conditions in an appropriate range, then they will shift ballistically from start- to end-of-step angles. Recently we have found that support transfer can then regenerate the start-of-step conditions, thus producing a closed passive cycle.

A full discussion of passive walking with knees must be left for a future report, but here we will review one example as shown in Figure 16. The model is still a 2D biped, but with pin-jointed knees and mechanical stops, as in the human knee, to prevent hyperextension. As in ballistic walking, the stance knee is *specified* to remain locked against the stop throughout the step, while the swing knee is initially free and flexes passively as plotted in Figure 16. Then it re-extends and hits the stop. (As always we treat the collision as impulsive and inelastic.) Thereafter both knees remain locked until foot strike.

Parametric studies of this knee-jointed model have produced these results:

1. Passive cycles are found over parameter ranges more limited than those in stiff-legged walking, but still quite broad.
2. With proper choice of parameters, naturally arising torques keep the stance and swing knees locked through the appropriate phases of the cycle.

Fig. 16. Our test biped has rigid legs, but passive walking also works with knees. The parameters here are quoted according to the convention of Fig. 18 and are similar to those of a human (including $m_H = 0.676$). (A) Flexure of the swing knee takes care of foot clearance (although just barely). Meanwhile a naturally generated torque holds the stance knee locked against a mechanical stop. (ϵ_K is the angle from the locked knee to the foot's center of curvature.) (B) The motion is obviously similar to human walking, but it is slower; the period of this cycle would be ≈ 0.85 s for a biped with 1-m leg length. There is also a faster cycle, but we prefer the slower because it turns out to be stable and more efficient.

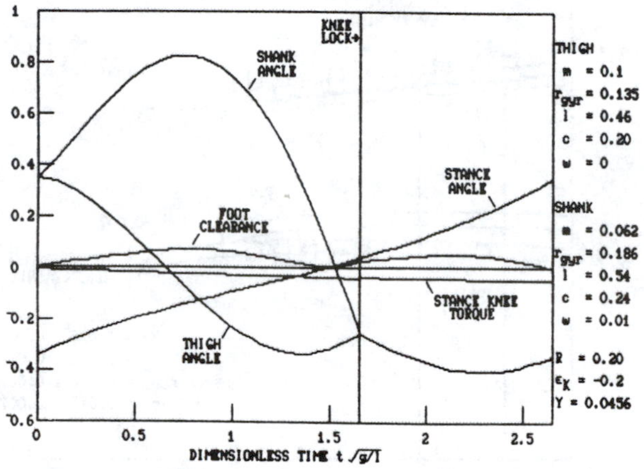

A

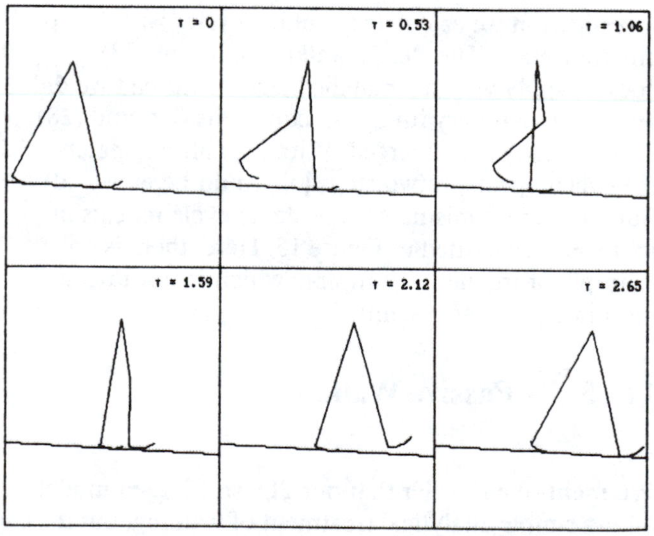

B

3. A passive cycle with knee joints is possible only if the foot is displaced forward relative to the leg, as shown in Figure 16. The implication is that humans might have difficulty if their feet didn't stick out in front of their legs.
4. A large fraction of the swing leg's kinetic energy is dissipated at knee lock, and therefore

knee-jointed walking is less efficient than stiff-legged walking. To keep the penalty small, the swing leg's energy must be small compared with the total energy of motion. This implies that most of the mass should be above the hip.

On the last point it might be argued that foot clearance with knees seems inefficient, whereas *active* retraction can in principle consume no energy at all. However, in practice economical food retraction may be hard to achieve. Our test biped has a "fundamental" dissipation (from Figure 6) of ≈ 0.2 J per step, while the tiny motors that lift its feet waste ≈ 3 J per step! Some goes into friction in the leadscrews, and some into switching transistors. At this rate, passive knee flexion is a bargain, and moreover a small active intervention at the right time may reduce the knee-locking loss substantially.

12. Action on Passive Walking

Having surveyed the physics of passive walking, our next objective must be exploitation of the effect to make machines with some practical capability. We see developments proceeding as follows:

1. "Powered" passive walking on shallow up- and downhill slopes in 2D
2. Step-to-step gait modulation over unevenly-spaced footholds in 2D
3. Walking on steep slopes, stairs, and 2D rough terrain
4. Lateral balance and steering

We are presently building a machine to test the first two developments. It will be similar to the first biped, but with 1-m leg length and a torso. Power will be provided by pushing with the trailing leg as it leaves the ground. (This is analogous to plantar flexion in humans, but the implementation is quite different.) Gait modulation is to be done by applying torques between the legs and torso. McGeer (1988) presents the analytical basis for this machine, as well as alternative options for applying power and control. S-to-S treatment is a key principle; thus control laws look at the state of the machine only once per step. This approach proves attractively simple, both mathematically and mechanically.

We have yet to proceed to the next problem of walking with steep changes in elevation. One promising strategy is to seek passive fore/aft leg swinging that can proceed in synchrony with actively cycled variation in leg length. This approach works on shallow slopes, with the length adjustments serving as a source of energy (McGeer 1988). However, further results await reformulation of equations of motion (17) and (18), since the linearization used here is invalid on steep slopes.

Finally, we expect that lateral balance will have to be done actively. One possibility is to control as when standing still (i.e., by leaning in appropriate directions). Another scheme more in line with the "S-to-S principle" is once-per-step adjustment of lateral foot placement; this has been analyzed by Townsend (1985). Turning, at least at low rates, can be done by the same mechanism.

Much of this discussion and development strategy can also apply to running. Running is of practical interest because walking has a fundamental speed limit of order $\sqrt{gl}$; at higher speeds centrifugal effect would lift the stance leg off the ground. It turns out that by adding a torsional spring at the hip and translational springs in each leg, the model of Figure 2 becomes a passive runner, with the same features of simplicity, efficiency, and ease of control that make passive walking attractive. Passive running has been described by McGeer (1989), and Thompson and Raibert (1989) have independently found similar behavior with both monoped and biped models. Investigation thus far has been limited to shallow slopes; as in walking, we have yet to do the steep-terrain and 3D investigations.

We hope that passive models will provide much insight into the dynamics and control of legged machines, but ultimately we must admit that this is only the easy part of the design problem. Legged locomotion is not competitive on smooth terrain; therefore practical machines must be capable of finding footholds and planning paths through difficult surroundings. Yet contemporary robots are hard pressed even to pick their way along paved highways and across flat floors! Thus it appears that the demands of a legged automaton must stimulate research for some years to come.

Fig. 17. Notation for an N-link, two-dimensional open chain with rolling support.

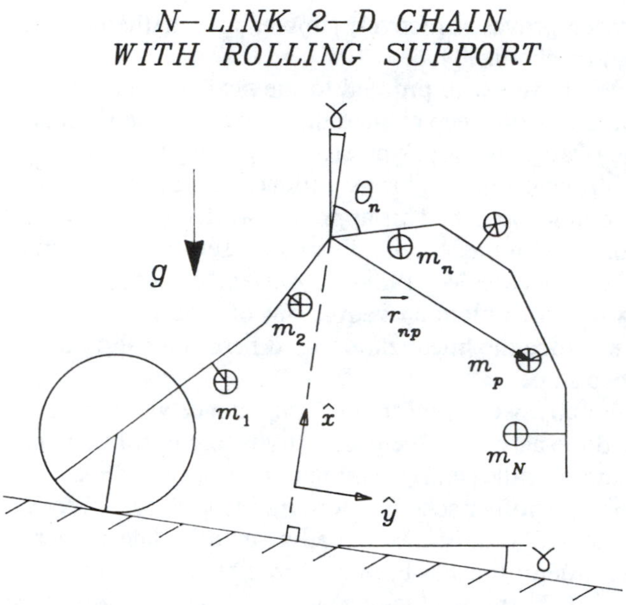

Appendix A. Dynamics of an N-Link Chain

Although we are interested in a machine with only two rigid links, there is little extra effort involved in generalizing to N links, and the more general result is needed for a knee-jointed machine like that in Figure 16. The dynamics can be expressed in N second-order equations of the form

$$\dot{H}_n = T_n, \tag{45}$$

where H_n is the angular momentum and T_n the torque about the nth joint. In the case of $n = 1$, the "joint" is the instantaneous point of contact. Simplification will be afforded by subtracting equations for successive joints, so that (45) becomes

$$\dot{H}_n - \dot{H}_{n+1} = T_n - T_{n+1}. \tag{46}$$

The torque is produced by gravity, and amounts to

$$T_{g_n} = \sum_{p=n}^{N} m_p \bar{r}_{np} \times \bar{g}; \tag{47}$$

Figures 17 and 18 provide the necessary notation. Then

$$T_{g_n} - T_{g_{n+1}} = \left(m_n \bar{r}_{nn} + \sum_{p=n+1}^{N} m_p \bar{l}'_n \right) \times \bar{g} \tag{48}$$

where $\bar{r}_{nn}$ is the vector from joint n to the mass center of link n, and $\bar{l}'_n$ the vector from joint n to joint $n + 1$. For link 1

$$\bar{r}_{11} = (c_1 - R)\hat{x}_1 + w_1 \hat{y}_1 + R\hat{x}, \tag{49}$$

$$\bar{l}'_1 = (l_1 - R)\hat{x}_1 + R\hat{x}. \tag{50}$$

$\hat{x}_1$ and $\hat{y}_1$ are unit vectors fixed in link 1, as illustrated by Figure 18.

For walking analysis we linearize the torque for small $\Delta\theta$, γ. Thus from (18) and (19) we have

$$\mathbf{T}_g \equiv \begin{bmatrix} T_{g_1} - T_{g_2} \\ T_{g_2} - T_{g_3} \\ \vdots \\ T_{g_N} \end{bmatrix} = \mathbf{K}[\Delta\boldsymbol{\theta}_W + \mathbf{b}\gamma - \Delta\boldsymbol{\theta}], \tag{51}$$

where $(\mathbf{K}\Delta\boldsymbol{\theta}_W)_n$ is the value of (48) at $\Delta\boldsymbol{\theta} = 0$, $\gamma = 0$; $(\mathbf{Kb})_n$ is the derivative of (48) with respect to γ, evaluated at $\Delta\boldsymbol{\theta} = 0$, $\gamma = 0$; and $\mathbf{K}_{nm}$ is the derivative of (48) with respect to θ_m, evaluated at $\Delta\boldsymbol{\theta} = 0$, $\gamma = 0$. For a two-link chain with a point hip mass m_H, the results are as follows:

$$\mathbf{K}\Delta\boldsymbol{\theta}_W = g \begin{bmatrix} m_1 w_1 \\ -m_2 w_2 \end{bmatrix} \tag{52}$$

$$\mathbf{Kb} = g \begin{bmatrix} m_1 c_1 + (m_2 + m_H) l_1 \\ -m_2 c_2 \end{bmatrix} \tag{53}$$

$$\mathbf{K} = -g \begin{bmatrix} m_1(c_1 - R) + (m_2 + m_H)(l_1 - R) & 0 \\ 0 & -m_2 c_2 \end{bmatrix} \tag{54}$$

Note that according to the convention of Figure 17, a biped with matched legs has

$$m_2 = m_1, \tag{55}$$

$$c_2 = l_1 - c_1, \tag{56}$$

$$w_2 = -w_1. \tag{57}$$

Fig. 18. Notation for individual links of the chain.

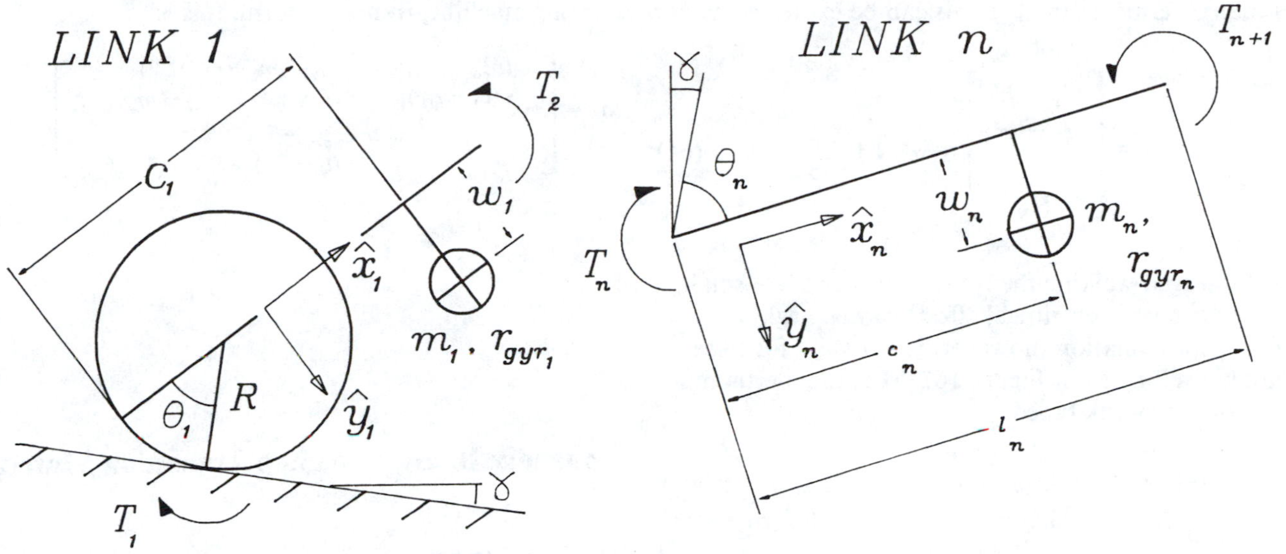

Now to work on the angular momentum terms in the equations of motion (46). H_n is

$$H_n = \sum_{p=n}^{N} m_p r^2_{gyr_p} \Omega_p + \sum_{p=n}^{N} m_p \bar{r}_{np} \times \bar{V}_p. \tag{58}$$

Therefore

$$H_n - H_{n+1} = m_n r^2_{gyr_n} \Omega_n + m_n \bar{r}_{nn} \times \bar{V}_n + \sum_{p=n+1}^{N} m_p \bar{l}'_n \times \bar{V}_p. \tag{59}$$

Equation (58) should be recognized as the angular momentum about a *stationary* point that is instantaneously coincident with joint n. Therefore, for the equations of motion (46) only Ω_n and $\bar{V}_n$ are differentiated:

$$\dot{H}_n - \dot{H}_{n+1} = m_n r^2_{gyr_n} \dot{\Omega}_n + m_n \bar{r}_{nn} \times \dot{\bar{V}}_n + \sum_{p=n+1}^{N} m_p \bar{l}'_n \times \dot{\bar{V}}_p. \tag{60}$$

The kinematics of the chain are such that

$$\bar{V}_n = \Omega_n \times \bar{r}_{nn} + \sum_{p=1}^{n-1} \Omega_p \times \bar{l}'_p, \tag{61}$$

where the Ω_n vector is directed into the page of Figure 17. This is differentiated for (60) as follows:

$$\dot{\bar{V}}_n = \dot{\Omega}_n \times \bar{r}_{nn} - \bar{r}_{nn}\Omega_n^2 + \sum_{p=1}^{n-1} \dot{\Omega}_p \times \bar{l}'_p - \sum_{p=1}^{n-1} \bar{l}'_p \Omega_p^2 + R\Omega_1^2 \hat{x}. \tag{62}$$

The last term comes from differentiating $\bar{r}_{11}$ (49), and accounts for rolling of the contact point. Substituting into (60) leaves (after substantial but straightforward simplification)

$$\begin{aligned}
\dot{H}_n - \dot{H}_{n+1} = & \sum_{p=1}^{n-1} \bar{l}'_p \cdot \left(m_n \bar{r}_{nn} + \sum_{k=n+1}^{N} m_k \bar{l}'_n \right) \dot{\Omega}_p \\
& + \left(m_n r^2_{gyr_n} + m_n |\bar{r}_{nn}|^2 + \sum_{k=n+1}^{N} m_k |\bar{l}'_n|^2 \right) \dot{\Omega}_n \\
& + \bar{l}'_n \cdot \sum_{p=n+1}^{N} \left(m_p \bar{r}_{pp} + \sum_{k=p+1}^{N} m_k \bar{l}'_p \right) \dot{\Omega}_p \\
& + \sum_{p=1}^{n-1} \bar{l}'_p \times \left(m_n \bar{r}_{nn} + \sum_{k=n+1}^{N} m_k \bar{l}'_n \right) \Omega_p^2 \\
& - \bar{l}'_n \times \sum_{p=n+1}^{N} \left(m_p \bar{r}_{pp} + \sum_{k=p+1}^{N} m_k \bar{l}'_p \right) \Omega_p^2 \\
& - R\hat{y} \cdot \left(m_n \bar{r}_{nn} + \sum_{k=n+1}^{N} m_k \bar{l}'_n \right) \Omega_1^2.
\end{aligned} \tag{63}$$

In matrix terms, as in (17), this can be expressed as

$$\dot{\mathbf{H}} = \begin{bmatrix} \dot{H}_1 - \dot{H}_2 \\ \dot{H}_2 - \dot{H}_3 \\ \vdots \\ \dot{H}_N \end{bmatrix} = \mathbf{M}\dot{\boldsymbol{\Omega}} + \mathbf{C}\boldsymbol{\Omega}^2. \tag{64}$$

In stiff-legged walking the rotational speeds are small, so the $\boldsymbol{\Omega}^2$ term (centrifugal effect) can be neglected. (This approximation, however, is *not* valid for knee-jointed walking as in Figure 16). **M** is the inertia matrix; for a two-link biped

$$\mathbf{M} = \begin{bmatrix} m_1(r^2_{gyr_1} + |\bar{r}_{11}|^2) + (m_2 + m_H)|\bar{l}'_1|^2 & m_2\,\bar{l}'_1 \cdot \bar{r}_{22} \\ m_2\bar{l}'_1 \cdot \bar{r}_{22} & m_2(r^2_{gyr_2} + |\bar{r}_{22}|^2) \end{bmatrix}. \tag{65}$$

In equation of motion (17), **M** is linearized; thus $\mathbf{M}_0$ is found by evaluating (65) with $\Delta\boldsymbol{\theta} = 0$. However, for support transfer $\mathbf{M}^+$ (24) is evaluated with $\Delta\boldsymbol{\theta} = \boldsymbol{\lambda}\alpha$ (22).

Support transfer also involves the pre-transfer inertia matrix $\mathbf{M}^-$, which is calculated as follows. Equation (59) still holds for the angular momentum, but (61) is no longer correct for the velocity, since prior to transfer the chain is rolling about link N rather than link 1. Thus recasting (61) gives

$$\bar{V}_n^- = \boldsymbol{\Omega}_n^- \times (\bar{r}_{nn} - \bar{l}'_n) + \sum_{p=n+1}^{N} \bar{l}'_p \times \boldsymbol{\Omega}_p^- + R(\hat{y} + \hat{y}_N)\boldsymbol{\Omega}_N^-. \tag{66}$$

Inserting this into (59) and collecting terms leaves

$$\begin{aligned} H_n - H_{n+1} = {} & m_n(r^2_{gyr_n} + |\bar{r}_{nn}|^2 - \bar{l}'_n \cdot \bar{r}_{nn})\boldsymbol{\Omega}_n^- \\ & - \sum_{p=n+1}^{N} \left(\left[m_n\bar{r}_{nn} + \sum_{k=n+1}^{p} m_k\bar{l}'_n \right] \right. \\ & \left. \cdot\, \bar{l}'_p - m_p\bar{l}'_n \cdot \bar{r}_{pp} \right) \boldsymbol{\Omega}_p^- \\ & + R\left(m_n\bar{r}_{nn} + \sum_{p=n+1}^{N} m_p\bar{l}'_n \right) \\ & \cdot\, (\hat{x} + \hat{x}_N)\boldsymbol{\Omega}_N^-. \end{aligned} \tag{67}$$

For our machine, in matrix form, this is

$$\mathbf{M}^- = \begin{bmatrix} m_1(r^2_{gyr_1} - \bar{r}_{11} \cdot (\bar{l}'_1 - \bar{r}_{11})) & (m_1\bar{r}_{11} + (m_2 + m_H)\bar{l}'_1) \cdot (R(\hat{x} + \hat{x}_2) - \bar{l}_2) + m_2\bar{r}_{22} \cdot \bar{l}'_1 \\ 0 & m_2(r^2_{gyr_2} - \bar{r}_{22} \cdot (\bar{l}_2 - \bar{r}_{22}) + R\bar{r}_{22} \cdot (\hat{x} + \hat{x}_2)) \end{bmatrix}. \tag{68}$$

Appendix B. Step-to-Step Transition Matrix

The S-to-S transition (36) has been derived by McGeer (1988); here we present only the final formula.

$$\mathbf{S} = \mathbf{S}_1^{-1}\mathbf{S}_2 \tag{69}$$

where

$$\mathbf{S}_1 = \left[\begin{array}{c|cc|c} -\boldsymbol{\lambda} & \begin{matrix} 0 \\ 0 \end{matrix} & \begin{matrix} 0 \\ 0 \end{matrix} & -\dfrac{\partial \mathbf{D}_{\theta\theta}}{\partial \tau}[\boldsymbol{\lambda}\alpha_0 - \Delta\boldsymbol{\theta}_{SE}] - \dfrac{\partial \mathbf{D}_{\theta\Omega}}{\partial \tau}\boldsymbol{\Omega}_0 \\ \hline -\dfrac{\partial \boldsymbol{\Lambda}}{\partial \alpha}(\mathbf{D}_{\Omega\theta}[\boldsymbol{\lambda}\alpha_0 - \Delta\boldsymbol{\theta}_{SE}] + \mathbf{D}_{\Omega\Omega}\boldsymbol{\Omega}_0) & \begin{matrix} 1 \\ 0 \end{matrix} & \begin{matrix} 0 \\ 1 \end{matrix} & -\boldsymbol{\Lambda}\left(\dfrac{\partial \mathbf{D}_{\Omega\theta}}{\partial \tau}[\boldsymbol{\lambda}\alpha_0 - \Delta\boldsymbol{\theta}_{SE}] + \dfrac{\partial \mathbf{D}_{\Omega\Omega}}{\partial \tau}\boldsymbol{\Omega}_0\right) \end{array}\right], \tag{70}$$

$$\mathbf{S}_2 = \begin{bmatrix} \mathbf{D}_{\theta\theta}\boldsymbol{\lambda} & \mathbf{D}_{\theta\Omega} \\ \boldsymbol{\Lambda}\mathbf{D}_{\Omega\theta}\boldsymbol{\lambda} & \boldsymbol{\Lambda}\mathbf{D}_{\Omega\Omega} \end{bmatrix}. \tag{71}$$

With the torque given by (18), the time derivative of **D** is

$$\frac{\partial \mathbf{D}}{\partial \tau} = \begin{bmatrix} \mathbf{D}_{\Omega\theta} & \mathbf{D}_{\Omega\Omega} \\ -\mathbf{M}_0^{-1}\mathbf{K}\mathbf{D}_{\Omega\theta} & -\mathbf{M}_0^{-1}\mathbf{K}\mathbf{D}_{\Omega\Omega} \end{bmatrix}. \tag{72}$$

Appendix C. Explanation of Symbols

Roman

- **b** derivative of static equilibrium *w.r.t.* slope (19), (53)
- c Figs. 2, 18
- **D** start- to end-of-step transition matrix (20)
- $\mathbf{D}_{\theta\theta}$, $\mathbf{D}_{\theta\Omega}$, $\mathbf{D}_{\Omega\theta}$, $\mathbf{D}_{\Omega\Omega}$ submatrices of start- to end-of-step transition matrix (27), (28)
- $\mathbf{D}'$ (32)
- **F** index exchanger (25)
- $\bar{g}$ gravitational acceleration, Fig. 2
- H angular momentum
- **I** 2×2 identity matrix
- **K** stance stiffness matrix (54)
- l leg length, Figs. 2, 18
- $\bar{l}$ joint-to-joint vector (50), Fig. 18
- **M** inertia matrix (65), (68)
- $\mathbf{M}_0$ **M** for $\Delta\theta = 0$ (17), (65)
- m mass, Figs. 2, 17
- m_H hip mass fraction
- R foot radius, Fig. 2
- $\bar{r}$ joint-to-mass center vector (49), Fig. 17
- r_{gyr} radius of gyration
- **S** step-to-step transition matrix (36), (69)
- T torque
- V linear velocity
- w offset from leg axis to leg mass center (Fig. 2)
- $\hat{x}$, $\hat{y}$ unit vectors, Figs. 17, 18
- z eigenvalue of step-to-step equations (36)

Greek

- α leg angle at support transfer (22), Figs. 4, 5
- γ slope, Fig. 2
- $\Delta\theta$ angle from surface normal, Fig. 2
- η coefficient of restitution (5)
- $\Delta\theta_{SE}$ static equilibrium position (19)
- $\Delta\theta_W$ static equilibrium on level ground (52)
- $\mathbf{\Lambda}$ support transfer matrix (26)
- $\boldsymbol{\lambda}$ support transfer vector (21)
- σ pendulum frequency (4)
- τ dimensionless time $t\sqrt{g/l}$
- Ω angular speed
- ω_F swing pendulum frequency
- ω_R lateral rocking frequency

Sub- and superscripts

- $+$ immediately after support transfer
- $-$ immediately before support transfer
- 0 steady cycle conditions
- C stance leg
- F swing leg
- g gravitational effect
- k step index

References

Lee, T-T., and Liao, J-H. 1988 (Philadelphia). Trajectory planning and control of a 3-link biped robot. *Proc. 1988 IEEE Int. Conf. Robot. Automat.* New York: IEEE, pp. 820–823.

Margaria, R. 1976. *Biomechanics and Energetics of Muscular Exercise.* Oxford, U.K.: Clarendon Press.

McGeer, T. 1988. Stability and control of two-dimensional biped walking. Technical report CSS-IS TR 88-01. Simon Fraser University, Centre for Systems Science, Burnaby, B.C.

McGeer, T. 1989. Passive bipedal running. Technical report CSS-IS TR 89-02. Simon Fraser University, Centre for Systems Science, Burnaby, B.C.

McMahon, T. 1984. Mechanics of locomotion. *Int. J. Robot. Res.* 3(2):4–28.

Mita, T., Yamaguchi, T., Kashiwase, T., and Kawase, T. 1984. Realisation of a high speed biped using modern control theory. *Int. J. Control* 40(1):107–119.

Miura, H., and Shimoyama, I. 1984. Dynamic walking of a biped. *Int. J. Robot. Res.* 3(2):60–74.

Mochon, S., and McMahon, T. 1980. Ballistic walking: An improved model. *Math. Biosci.* 52:241–260.

Raibert, M. 1986. *Legged Robots That Balance.* Cambridge, Mass.: MIT Press.

Russell, M. 1983. Odex 1: The first functionoid. *Robot. Age* 5:12–18.

Takanishi, A., Ishida, M., Yamazaki, Y., and Kato, I. The realisation of dynamic walking by the biped walking robot WL-10RD. *Proc. Int. Conf. Adv. Robot.* Tokyo: Robotics Society of Japan, pp. 459–466.

Thompson, C. and Raibert, M. (in press). Passive dynamic running. In Hayward, V., and Khatib, O. (eds.): *Int. Symposium of Experimental Robotics*. New York: Springer-Verlag.

Townsend, M. 1985. Biped gait stabilisation via foot placement. *J. Biomech.* 18(1):21–38.

Waldron, K., Vohnout, V., Perry, A., and McGhee, R. 1984. Configuration design of the adaptive suspension vehicle. *Int. J. Robot. Res.* 3(2):37–48.

Waldron, K. 1986. Force and motion management in legged locomotion. *IEEE J. Robot. Automat.* 2(4):214–220.

Yamada, M., Furusho, J., and Sano, A. 1985. Dynamic control of a walking robot with kick action. *Proc. Int. Conf. Adv. Robot.* Tokyo: Robotics Society of Japan, pp. 405–412.

Zheng, Y-F., Shen, J., and Sias, F. 1988 (Philadelphia). A motion control scheme for a biped robot to climb sloping surfaces. *Proc. 1988 IEEE Int. Conf. Robot. Automat.* New York: IEEE, pp. 814–816.

From Stable to Chaotic Juggling: Theory, Simulation, and Experiments

M. Bühler and D. E. Koditschek [1]

Center for Systems Science
Yale University, Department of Electrical Engineering

Abstract

Robotic tasks involving intermittent robot-environment interactions give rise to return maps defining discrete dynamical systems that are, in general, strongly nonlinear. In our work on robotic juggling, we encounter return maps for which a global stability analysis has heretofore proven intractable. At the same time, local linear analysis has proven inadequate for any practical purposes.

In this paper we appeal to recent results of dynamical systems theory to derive strong predictions concerning the global properties of a simplified model of our planar juggling robot. In particular, we find that certain lower order local (linearized) stability properties determine the essential global (nonlinear) stability properties, and that successive increments in the controller gain settings give rise to a cascade of stable period doubling bifurcations that comprise a "universal route to chaos." The theoretical predictions are first verified via simulation and subsequently corroborated by experimental data from the juggling robot.

1 Introduction

We have built the "plane-juggler," a one degree of freedom robot that juggles two degree of freedom bodies — pucks falling (otherwise) freely on a frictionless plane inclined into the earth's gravitational field [3]. We have developed the rudiments of a "geometric language" — a family of "mirror laws" that map puck states into desired robot states — capable of translating certain abstract goals (juggling one and two pucks, catching) into robot control laws whose closed loop behavior has been shown experimentally to accomplish these tasks in a stable and robust manner [2]. We have proven, as well, that for the task of juggling one puck the mirror law is correct by resort to a local stability analysis of linearized closed loop model [3, 1]. Unfortunately, this local analysis and its intrinsically weak conclusions have been of little use in predicting the physical consequences of different gain settings [1] and have convinced us of the practical necessity of a global stability analysis. However, even for the simpler line-juggler that gives rise to a *scalar* return map, a global stability analysis with standard mathematical techniques is exceedingly difficult. This paper marshals an array of theoretical tools for the global analysis of a large class of scalar maps — among them our line-juggler map — that gives considerable promise of narrowing the gap between our analysis and practice in general. We capitalize on a "by-product" of the recently burgeoning study of bifurcations, chaos, and sensitive dependence on initial conditions in qualitative dynamical systems theory [8, 11, 5, 7, 6]. Specifically, we find that certain (seemingly unrestrictive) sufficient conditions for "reading off" from the derivative at a fixed point the essential global stability properties of an entire dynamical system are met by a simplified model of our closed loop system, the line-juggler. The strong predictions concerning the global properties are validated by a gratifying correspondence between theory and experiment. In addition, the analytical predictions seem to be relevant, as well, to the plane juggler as is strongly suggested by experimental data. Moreover, the coincidence of our systems' stability mechanisms with these special cases may be shared by the underlying stability mechanism of Raibert's hoppers [10, 9]. This coincidence, if physically intrinsic, holds great promise for advancing the science of robotics in intermittent dynamical environments.

[1] This work has been supported in part by PMI Motion Technologies, the North American Philips Laboratories, INMOS Corporation and the the National Science Foundation under a Presidential Young Investigator Award held by the second author.

2 The Line-Juggler Model

This section, devoted to the illustrative one degree of freedom case — the line-juggler — portrays more simply than can the two degree of freedom experimental system the modeling process as well as the underlying ideas in our new feedback control law, or "mirror algorithm." First, we derive the open loop model for the line-juggler. The specification of a feedback control law will then give rise the closed loop model, a scalar return map of puck impact velocities that is analyzed in the following section. Both the modeling and the control law generalize in a straightforward fashion to the two degree of freedom juggler. For a detailed discussion of these issues and the complete derivation of the plane-juggler model and mirror algorithm refer to [1, 3, 2].

2.1 The open loop model

The simple model of the one degree of freedom line-juggler is displayed in Figure 1. A puck falls freely in the gravitational field toward a prismatic robot actuator. The robot's and puck's positions are denoted by r and b, respectively. The task — the vertical one-juggle — is to force the puck into a stable periodic trajectory with specified impact velocity (and thus apex position). Since the robot can only provide intermittent impacts to the puck it makes sense to examine the discrete map between puck states at those interactions as a function of the robot's in-

Reprinted from *Proc. IEEE ROBOT '90*, Cincinnati, OH, pp. 1976–1981, May 1990.

puts. For now, we will ignore the robot's dynamics and assume it capable of applying arbitrary inputs to the puck during these recurring interactions. We can now examine the following control problem: Given a sequence of desired puck states — the task — find a sequence of robot control inputs that achieves it.

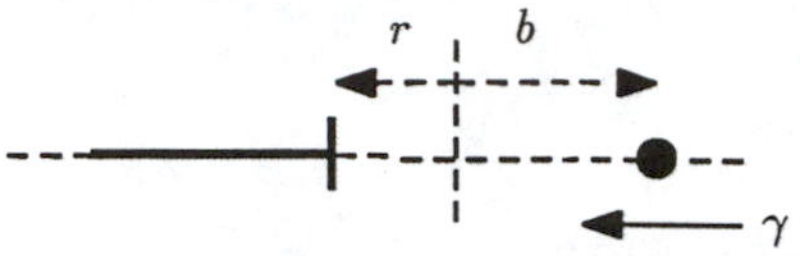

Figure 1: The Line-Juggler Model

First we construct the discrete model that relates two successive puck states $w = (b, \dot{b})$ just before impact as a function of the robot control inputs. This process consists of modeling the puck-robot impacts and the puck's flight dynamics.

For the impact model we make the common assumption that the elastic impact can be modeled accurately by a coefficient of restitution law and that the robot's velocity $\dot{r}$ during impact remains unchanged. Assuming the puck and the robot are moving toward each other, $(\dot{b} - \dot{r}) < 0$, then the puck velocity just after impact $\dot{b}'$ is related to the velocity just before impact $\dot{b}$ via

$$\dot{b}' = -\alpha\dot{b} + (1+\alpha)\dot{r} = \mathbf{c}(\dot{b}, \dot{r}), \quad (1)$$

where $\alpha \in (0,1)$ denotes the coefficient of restitution. Neglecting friction during flight, the puck's flight dynamics are given by

$$\begin{bmatrix} b(t) \\ \dot{b}(t) \end{bmatrix} = \begin{bmatrix} b' + \dot{b}'t - \frac{1}{2}\gamma t^2 \\ \dot{b}' - \gamma t \end{bmatrix}. \quad (2)$$

Here $w' = (b', \dot{b}')$ denotes the initial conditions, the puck states just after impact, and γ the gravitational constant. As the impacts are modeled to be instantaneous, the puck position during an impact remains unchanged, $b' = b$. If we now combine the impact model (1) with the flight model (2) and select the time of flight and the robot velocity at impact as our robot control inputs, we obtain the discrete map between successive puck impacts as a function of the two robot control inputs,

$$f(b, \dot{b}, u_1, u_2) = \begin{bmatrix} b + \mathbf{c}(\dot{b}, u_2)u_1 - \frac{1}{2}\gamma u_1^2 \\ \mathbf{c}(\dot{b}, u_2) - \gamma u_1 \end{bmatrix}. \quad (3)$$

2.2 The Mirror Law

The vertical one juggle task can now be specified as a sequence of desired puck states just before impact. Selecting $w^* = \left(b^*, \dot{b}^*\right)$ as the desired constant set point of (3) indicates that the impact should always occur at the position b^* with the velocity just before impact $\dot{b}^*$. If w^* is a fixed point of the closed loop dynamics, then the velocity just after impact must be $-\dot{b}^*$, and this "escape velocity" leads to a free flight puck trajectory whose apex occurs at the height $b_{apex} = b^* + \frac{\dot{b}^{*2}}{2\gamma}$. Thus, a constant w^* "encodes" a periodic puck trajectory which passes forever through a specified apex point, b_{apex}.

Successful control of the vertical one juggle task is achieved via a new class of feedback algorithms termed "mirror algorithms" [3]. Suppose the robot tracks exactly the continuous "distorted mirror" trajectory of the puck,

$$r = -\kappa_{10} b,$$

where κ_{10} is a constant. In this case, impacts between the two do occur only when $(r, b) = (0, 0)$ with robot velocity

$$\dot{r} = -\kappa_{10}\dot{b}. \quad (4)$$

For simplicity we will assume that the desired impact position is always selected to be $b^* = 0$. Any other impact position can be achieved by shifting the coordinate frame for robot and puck to that position. Now solving the fixed point condition $\dot{b}' = \mathbf{c}(\dot{b}^*, \dot{r}(\dot{b}^*)) = -\dot{b}^*$ for κ_{10} using (1) and (4), yields a choice of that constant, $\kappa_{10} = (1-\alpha)/(1+\alpha)$ which ensures a return of the puck to the original height. Thus a properly tuned "distortion constant," κ_{10} will maintain a correct puck trajectory in its proper periodic course.

The ability to maintain the vertical one-juggle — fixed point condition — with such a simple mirror control law is an encouraging first step, but still impractical, as it is not stable. The second idea at work which will assure stability is borrowed from Marc Raibert [10], who also uses the total energy for controlling hopping robots. In the absence of friction, the desired steady state periodic puck trajectory is completely determined by its total vertical energy,

$$\eta(w) = \frac{1}{2}\dot{b}^2 + \gamma b.$$

This suggests the addition to the the original mirror trajectory,

$$r = -\kappa_1(w)b; \qquad \kappa_1(w) \triangleq \kappa_{10} + \kappa_{11}[\eta(w^*) - \eta(w)], \quad (5)$$

of a term which "servos" around the desired steady state energy level. Thus, implementing a mirror algorithm is an exercise in robot trajectory tracking wherein the reference trajectory is a function of the puck's state.

2.3 The closed loop model

We will now show that all impacts between the one degree of freedom robot and the puck under the mirror law (5) must occur at the desired position $b = b^* = 0$. The impact conditions $r - b = 0, \dot{r} - \dot{b} > 0$ translate via equation (5) and its derivative into

$$[1 + \kappa_1(w)]b = 0, \quad (6)$$

$$(1 + \kappa_1(w))\dot{b} < 0. \quad (7)$$

Here we exploit $\dot{\kappa}_1 = -\kappa_{11}\dot{\eta} = 0$ as we ignore friction. As noted earlier, we restrict ourselves to puck imact velocities pointing down in the gravitational field toward the robot, $\dot{b} < 0$. Therefore, condition (7) is equivalent to $1 + \kappa_1(w) > 0$ and now the only solution to (6) is the desired puck robot impact position $b = 0$.

To construct the closed loop impact map we must now evaluate the effective robot control inputs at impact. The time of flight and the robot impact velocity is

$$u_1 = \frac{2}{\gamma}\dot{b}' \quad \text{and} \quad u_2 = \dot{r} = -\kappa_1\dot{b}.$$

Substituting these robot control inputs in (3) we obtain the scalar map of puck impact velocities just before impact at the invariant impact position $b^* = 0$,

$$f(\dot{b}) = \dot{b}\left(1 - \beta(\dot{b}^2 - \dot{b}^{*2})\right), \qquad (8)$$

where $\beta = \kappa_{11} \cdot (1+\alpha)/2$.

In an effort to synthesize a more realistic closed loop map than (8) that will serve us in predicting experimental results we now include coulomb friction between puck and sliding surface. Furthermore, in order to prevent the puck from falling off the sliding plane, we incline the juggler in the gravitational field away from the vertical by an angle δ, which also has the effect of decreasing gravitational acceleration. Now the dynamics of the puck in isolation are described by

$$\ddot{b} = -\gamma\cos\delta - sgn(\dot{b})\mu_{fric}\gamma\sin\delta.$$

Here μ_{fric} denotes the friction coefficient for dry friction. Proceeding now analogously to the frictionless case, we apply the same mirror law (5) and obtain the closed loop impact map corresponding to (8),

$$f(\dot{b}) = \zeta\dot{b}\left(1 - \beta(\dot{b}^2 - \dot{b}^{*2})\right), \qquad (9)$$

where β is defined as above and $\zeta = \zeta(\mu_{fric})$.

Note that the impact maps (8) and (9) are only defined for positive velocities after impact. This restricts our previous domain $\dot{b} < 0$ for the vertical puck velocity just before impact for both cases without and with friction to the domain W defined by

$$\bar{\dot{b}} < \dot{b} < 0 \quad \text{where} \quad \bar{\dot{b}} \triangleq -[\dot{b}^{*2} + \frac{1}{\beta}]^{\frac{1}{2}}.$$

3 The Stability Properties of Unimodal Maps

We now sketch our theoretical tools derived from the Singer-Guckenheimer theory (for a complete derivation, see [4]) and describe their relevance to the present application. Bifurcation plots are generated on the computer as an illustration of the theoretical statements and for purposes of comparison with the experimental data presented in Section 4.

3.1 The Singer-Guckenheimer Theory

3.1.1 S-unimodal Maps

Singer and Guckenheimer stated their results for a very particular class of functions which preserve the unit interval, called *normal S-unimodal* maps. These functions increase strictly towards a unique maximum and strictly decrease over the remainder of the interval. Moreover, they have a negative *Schwarzian Derivative* [11] except at the maximum. Rather than seeking global asymptotic stability of a fixed point, we will content ourselves with *essential global asymptotic stability*. This property holds when the set of initial conditions that fails to converge to the fixed point has measure zero. Note that for all engineering purposes essential global asymptotic stability is indistinguishable from global asymptotic stability.

Singer showed that normal S-unimodal maps can have at most one attracting periodic orbit [11]. Guckenheimer showed that the domain of attraction of such attracting orbits includes the entire unit interval with the possible exception of a zero measure set [8]. Thus, an asymptotically stable orbit of a normal S-unimodal map is essentially globally asymptotically stable. Although these strong results are stated in terms of the apparently restrictive class of normal S-unimodal maps, they extend as well to all differentiable conjugates. Namely, say that g is a *smooth S-unimodal* map if there is some normal S-unimodal map, f, to which g is differentiably conjugate — i.e. there exists a smooth and smoothly invertible function, h such that $g = h \circ f \circ h^{-1}$. It is straightforward to show that an attracting orbit of a smooth S-unimodal map is essentially globally asymptotically stable [4].

Smooth S-unimodal maps form a sufficiently large family that this theory appears to have broad engineering applicability. For example, we demonstrate below that the line-juggler map falls within this class. Moreover, we have shown that simplified models of Raibert's hopping robots give rise to smooth S-unimodal maps as well [9].

3.1.2 Bifurcations of Unimodal Maps

If we now consider *one-parameter families* of unimodal maps, then certain strong and essentially universal (i.e. independent of the particular parametrized family) properties hold true. Namely we can expect predictable structural changes in the qualitative dynamics pertaining to almost identically related values of the parameter, entirely independent of the details of the particular family. Let g_μ be a unimodal map for each μ in some real interval, $\mathcal{M} \subset \mathbb{R}$. A particular value, μ_0, is said to be a *bifurcation point* if there is no neighborhood of μ_0 in $\mathcal{M}$ such that g_μ is conjugate to g_{μ_0} when μ is in that neighborhood. Intuitively, the qualitative behavior of the dynamics changes around a bifurcation point.

Now suppose that $\{g_\mu\}$ is a family of smooth S-unimodal maps. If there is an interval of values $\mathcal{M} \triangleq (\mu_0, \mu_\infty)$ of μ for which $g_{\mu_0}(c) = c$ and $g_{\mu_\infty}(c) = 1$, then we shall say that g_μ is a *full family* [7]. A full family must exhibit an accumulating cascade of period doubling bifurcations: i.e., from an asymptotically stable period one orbit until μ_1, to an asymptotically stable period two orbit until μ_2, an asymptotically stable period four orbit until μ_3, and so on. Thus, unimodal families give rise to theoretically determined global bifurcation diagrams. A typical such bifurcation diagram is displayed in Figure 2 as taken straight out of [5].

Unimodal period doublings have a universal structure. Denote as μ_n those points in the bifurcation diagram where there is a bifurcation from length 2^{n-1} to 2^n. Then the ratios $\frac{\mu_n - \mu_{n-1}}{\mu_{n+1} - \mu_n}$ converge to some universal number $\delta = 4.66920...$ *regardless of the family or the details of the parametrization* [5]. We will show later that our line-juggler satisfies these conditions and indeed, from the simulated bifurcations diagram we can verify this universal number.

3.1.3 Practical Summary

In the fortunate case of encountering a smooth S-unimodal return map, f, the job of determining the essential global limit behavior of its associated dynamics reduces to simple algebra and calculus. After finding the fixed points of the Nth iterate

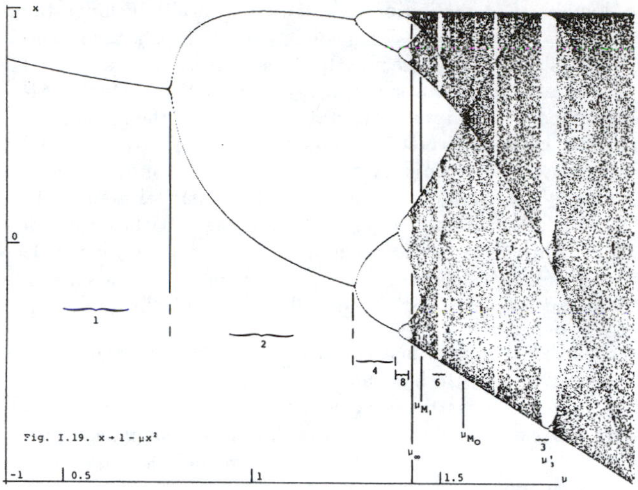

Figure 2: Bifurcation Diagram for $f_\mu = 1 - \mu x^2$ as Shown in Collet and Eckmann

of f (algebra), we compute the magnitude of its first derivative (calculus) at a fixed point. If that magnitude is less than unity then we may expect *all* experiments performed upon the corresponding physical apparatus (with gains set to the appropriate values) to result in periodic steady state behavior exhibiting no more than N distinct states.

If, moreover, we encounter a one-parameter family of smooth S-unimodal maps, and the family is full, then appropriate adjustments of the parameter will afford any conceivable stable periodic behavior. Eventually, when the period, N, gets to be a sufficiently large number, our ability to distinguish periodic from "chaotic" steady state behavior will be compromised by the imprecision of our measurements.

3.2 Applications to the Line-Juggler Return Map

We now explore the implications of the preceding theory for our particular dynamics. In order to be directly applicable to our physical apparatus, we will use the models that include friction in the following sections. The case without friction can readily be recovered by setting $\zeta = 1$.

The original map for successive impact velocities (9) has a fixed point at

$$\dot{b}_{fp} = -\sqrt{\dot{b}^{*2} + \frac{1 - 1/\zeta}{\beta}} \tag{10}$$

and a unique minimum at $\dot{b}_c = \frac{1}{\sqrt{3}}\bar{\dot{b}}$ with $f(\dot{b}_c) = \zeta\frac{2}{\beta}\dot{b}_c^3$. Moreover, it is not hard to see that f is smooth S-unimodal [4]. Thus from the preceeding results we immediately have

Theorem 1 ([4]) *The mirror algorithm for the line-juggler results in a successful vertical one juggle which is essentially globally asymptotically stable as long as*

$$0 < \beta < \frac{2/\zeta - 1}{\dot{b}^{*2}}.$$

It is easy to verify that g_μ is also a *full family*: For $\mu \in (\mu_0, \mu_\infty) = (\frac{3}{2}, \frac{3}{2}\sqrt{3})$, we obtain $g_{\mu_0}(c) = c$ and $g_{\mu_\infty}(c) = 1$. Now we know that after the fixed point $\dot{b}_{fp}$ becomes unstable, the map f will exhibit period doublings that will eventually lead to chaotic behavior, as predicted before. This is confirmed in Figure 3, which show the bifurcation diagram (obtained via simulation) for our specific line-juggler map (with friction). The ratio $\frac{\mu_n - \mu_{n-1}}{\mu_{n+1} - \mu_n}$ is evaluated for the first three bifurcations, $n = 2$, directly from the two figures, as 4.6 with an accuracy of about 0.3 due to the large β stepsize. This value is close to the expected limiting value for $n \to \infty$ of $\delta = 4.66920\ldots$.

Given the settings $\mu_{fric} = 0.16, \alpha = 0.7, \zeta = 0.9115$, and $\dot{b}^* = -125$, one can predict from $f'(\dot{b}_{fp}) = -1$ the first β-bifurcation value $7.6 \cdot 10^{-5}$. The corresponding value of $\kappa_{11} = 2\beta/(1+\alpha) = 8.9 \cdot 10^{-5}$ is confirmed accurately from Figure 3.

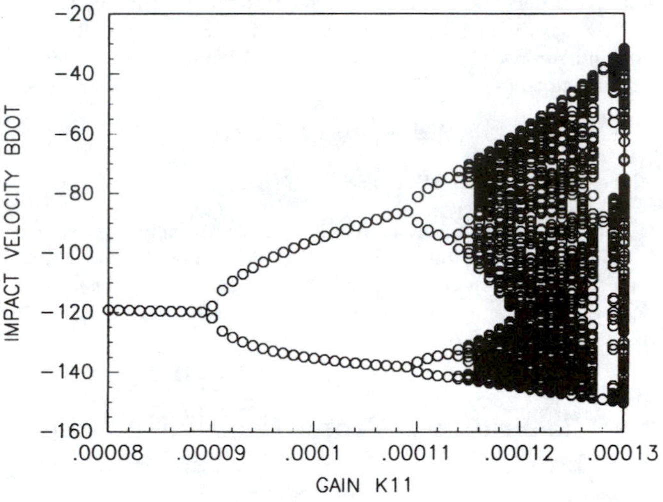

Figure 3: Line-Juggler with Friction: Simulated Bifurcation Diagram

4 Experimental Results

In this section we present experimental data to validate the models developed in Section 2 and the theoretical predictions by comparisons with simulated data from Section 3.2. In Section 4.1 devoted to the line-juggler, we will use the theoretical insights presented in Section 3.1 for scalar return maps to predict the dynamical behavior of the physical apparatus. The correspondence between simulated and experimental data is gratifying. We are able to predict and verify experimentally the transients of the stable fixed points as well as higher period stability properties, specifically bifurcations to stable period two and period four orbits. In Section 4.2 we back with experiments our speculations [4] about the applicability of the theory to the two degree of freedom juggler.

4.1 Line-Juggler Experiments

All our past experiments were based on the plane-juggler, which allows for planar puck motion which is controlled by impacts with a revolute motor, as depicted in Figure 4. The one-juggle is implemented by restricting the puck motion to a vertical line. The resulting closed loop impact map, when using the proper mirror algorithm for the motor, is identical to the one-juggle map (9) [4].

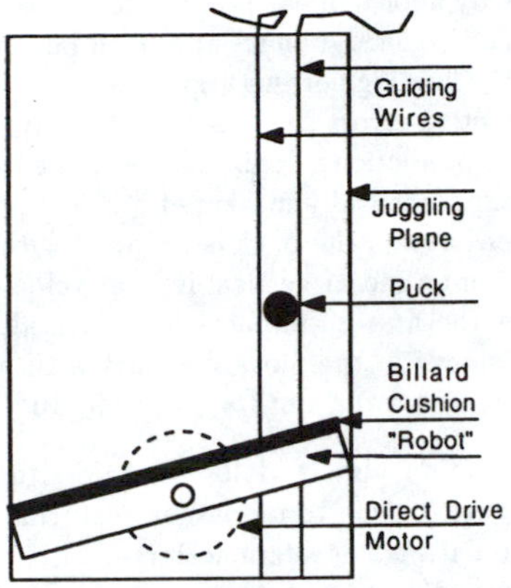

Figure 4: The Line-Juggler Implementation

We have shown in the previous section that the local dynamical behavior is essentially global. The data in Figure 5 confirm that the transients can be predicted by recourse to local linear analysis of the scalar impact map. Evaluating the derivative of (9) at the fixed point (10) for the four gain settings $\kappa_{11} = 3/5/7/9 \cdot 10^{-5}$ shown in the figure, we predict locally an overdamped, critically damped, underdamped and an unstable response, respectively. This behavior is confirmed even from large initial conditions ("globally") on the juggling apparatus. When inspecting the transient for the last gain setting $\kappa_{11} = 9 \cdot 10^{-5}$ closely we see that it maintains a small oscillation, the predicted onset of instability. The fixed point in the presence of friction (10) depends on the gain setting κ_{11}. Figure 5 confirms as well our ability to predict the steady state values for the vertical impact velocities with less than 3% error.

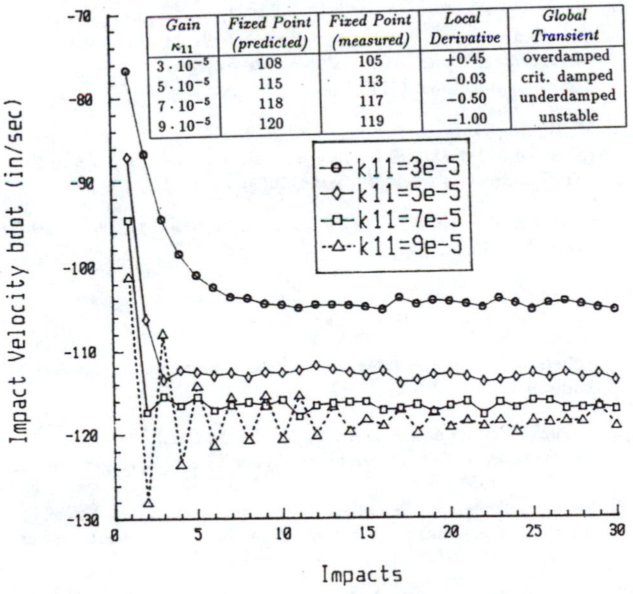

Gain κ_{11}	*Fixed Point (predicted)*	*Fixed Point (measured)*	*Local Derivative*	*Global Transient*
$3 \cdot 10^{-5}$	108	105	+0.45	overdamped
$5 \cdot 10^{-5}$	115	113	−0.03	crit. damped
$7 \cdot 10^{-5}$	118	117	−0.50	underdamped
$9 \cdot 10^{-5}$	120	119	−1.00	unstable

Figure 5: Line-Juggler Transients: Experimental Data

When increasing the gain beyond the predicted first κ_{11} bifurcation value of $8.99 \cdot 10^{-5}$, we expect to see a stable period two orbit. Visually, this shows the puck oscillating between two different juggling heights. Indeed, for $\kappa_{11} = 11 \cdot 10^{-5}$, Figure 6 shows the divergence from the unstable fixed point of f (120.8 in/sec) towards a stable period two — a stable fixed point of $f \circ f$.

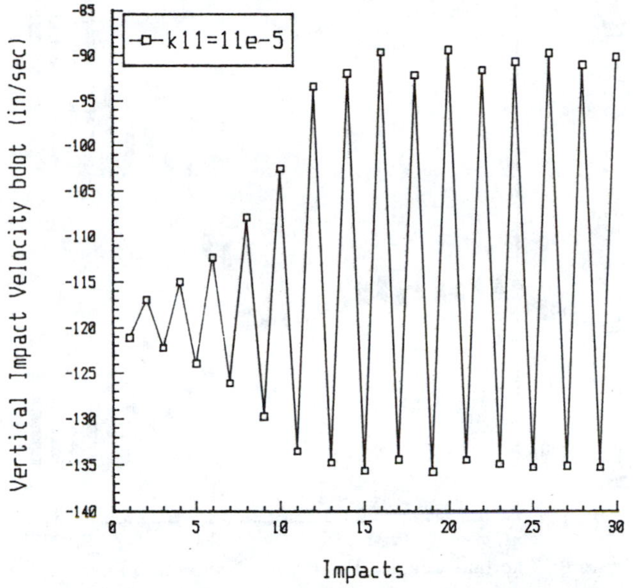

Figure 6: A Transient Toward Period Two: Experimental Data

The strength of our analytical predictions is demonstrated further when increasing the gain to $\kappa_{11} = 12.25 \cdot 10^{-5}$: an observable period four trajectory emerges, as shown in Figure 7. As the gain increases, the higher period orbits become more and more sensitive to perturbations. This together with the slower transient recovery from these perturbations causes the period four orbit to appear and disappear at unregular intervals, and explains the fact that we have not seen higher period orbits than period four.

This most satisfying correspondence between analytical predictions and experimental data is summarized in Figure 8: the experimentally acquired bifurcation diagram for the line-juggler. It coincides with good accuracy with the model prediction in Figure 3 up to the second bifurcation. This plot was acquired in an

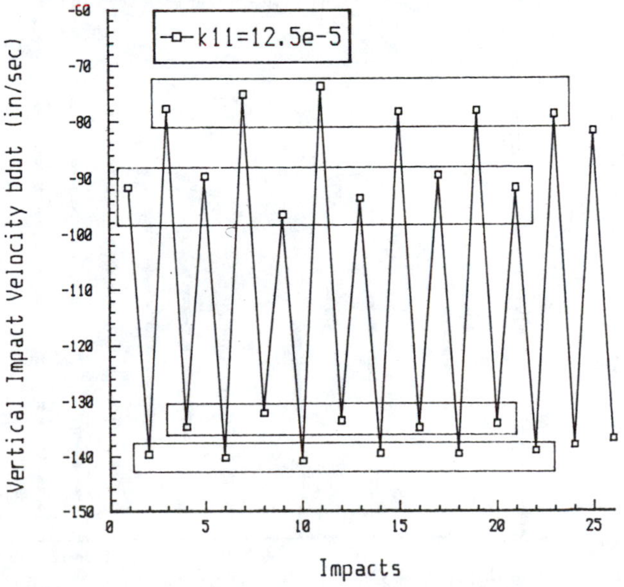

Figure 7: A Period Four Sequence: Experimental Data

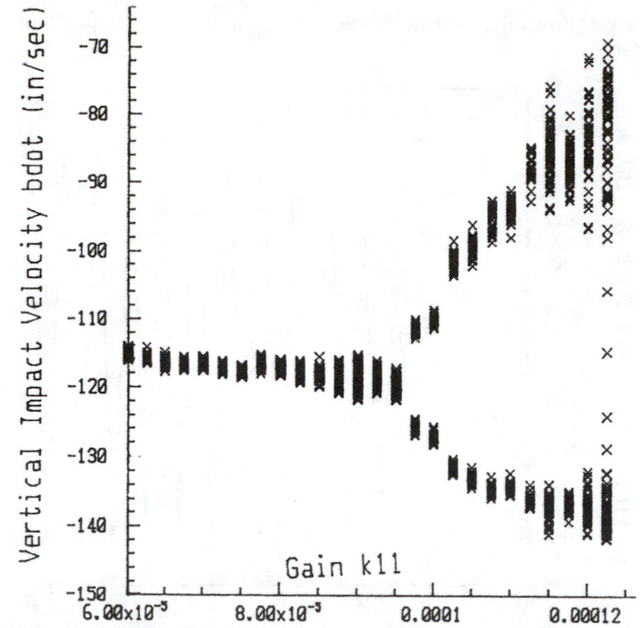

Figure 8: The Line-Juggler Bifurcation Diagram: Experimental Data

analogous fashion to the simulated plot. The puck was dropped at a height corresponding to $\dot{b} = -100$ in/sec. We discarded the first 20 impacts to assure steady state juggling and logged the following 50 impacts. For the last four gain settings we logged the following 100 impacts since the spread of impact velocities increased. This was repeated for the κ_{11} gain range in $0.25 \cdot 10^{-5}$ increments.

4.2 Planar Juggler Experiments

We now remove the guiding wires in Figure 4 to obtain the original plane-juggler. The one-juggle mirror algorithm can be generalized to accomplish the vertical one-juggle task on the plane-juggler [4]. However, the complexity of the new closed loop system admits only a local linear proof of correctness. After showing that the plane-juggler dynamics at a fixed horizontal position are identical to the line-juggler, we argue that even in the presence of horizontal position errors the line-juggler analysis should serve us for qualitative predictions. This can be verified in Figure 9 which displays an experimentally measured bifurcation diagram for the planar juggler. Again, the puck is dropped with an initial height which results in an initial vertical impact velocity of about -100 in/sec. After the first 20 impacts have passed, we log the next 100 impacts shown in the plot. We start with an initial value of $6 \cdot 10^{-5}$ and record a run until $\kappa_{11} = 12.5 \cdot 10^{-5}$.

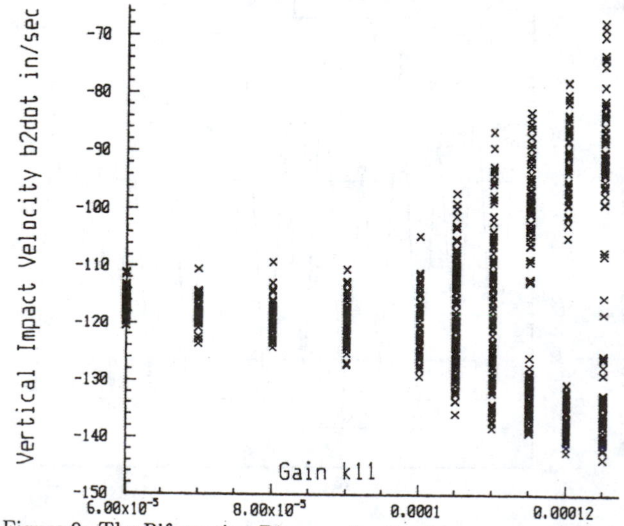

Figure 9: The Bifurcation Diagram for the Unconstrained (2dof) Juggler: Experimental Data

This bifurcation plot is almost identical to that of the line-juggler in Figure 8. Due to the larger perturbations present, the spread of the data is considerable larger. However, we can still identify a bifurcation to a stable period two orbit after the value of $\kappa_{11} = 10 \cdot 10^{-5}$, which is close to the value $\kappa_{11} = 8.99 \cdot 10^{-5}$ predicted from the theory of the simpler system.

Acknowledgements

We would like to thank John Guckenheimer for his kind tutorial efforts on our behalf.

References

[1] M. Bühler, D. E. Koditschek, and P. J. Kindlmann. A Simple Juggling Robot: Theory and Experimentation. In V. Hayward and O. Khatib, editors, *International Symposium on Experimental Robots.* Springer-Verlag, Montréal, Canada, Jun 1989.

[2] M. Bühler, D. E. Koditschek, and P. J. Kindlmann. Planning and Control of Robotic Juggling Tasks. In H. Miura, editor, *Fifth International Symposium on Robotics Research*, pages 270–281, Tokyo, Japan, 1989. MIT Press.

[3] M. Bühler, D. E. Koditschek, and P. J. Kindlmann. A family of robot control strategies for intermittent dynamical environments. *IEEE Control Systems Magazine*, 10(2):16–22, Feb 1990.

[4] Martin Bühler. *Robotic Tasks with Intermittent Dynamics.* PhD thesis, Yale University, May 1990.

[5] P. Collet and J. P. Eckmann. *Iterated Maps on the Interval as Dynamical Systems.* Birkhäuser, Boston, 1980.

[6] Robert L. Devaney. *Introduction to Chaotic Dynamical Systems.* Addison Wesley, Reading, MA, 1987.

[7] J. Guckenheimer and P. Holmes. *Nonlinear Oscillations, Dynamical Systems, and Bifurcations of Vector Fields.* Springer-Verlag, New York, 1983.

[8] John Guckenheimer. Sensitive dependence to initial conditions for one dimensional maps. *Communications in Mathematical Physics*, (70):133–160, 1979.

[9] D. E. Koditschek and M. Bühler. Analysis of a simplified hopping robot. *The International Journal of Robotics Research*, (to appear).

[10] Marc H. Raibert. *Legged Robots That Balance.* MIT Press, Cambridge, MA, 1986.

[11] David Singer. Stable orbits and bifurcations of maps of the interval. *SIAM J. Applied Mathematics*, 35(2):260–267, Sep 1978.

Author Index

Subject Index

E

F

G

N

O

P

R

S

T

U

V

W

Editors' Biographies

Mark W. Spong (S'81–M'81–SM'89) was born in Warren, OH on November 5, 1952. He received the B.A. degree (magna cum laude, Phi Beta Kappa) in mathematics and physics from Hiram College in 1975, the M.S. degree in mathematics from New Mexico State University in 1977, and the M.S. and D.Sc. degrees in systems science and mathematics from Washington University in St. Louis in 1979 and 1981, respectively.

After spending one year at Lehigh University and two years at Cornell University, he joined the University of Illinois at Urbana-Champaign in August, 1984. Dr. Spong is currently Professor of General Engineering, Professor of Electrical and Computer Engineering, Research Professor in the Coordinated Science Laboratory, and Director of the Department of General Engineering Robotics and Automation Laboratory, which he founded in January 1987.

Dr. Spong's research interests are in robotics and nonlinear control systems. He has published over 80 articles and is coauthor with M. Vidyasagar of the textbook *Robot Dynamics and Control* (New York: John Wiley & Sons, 1989). Dr. Spong is a past Associate Editor of the *IEEE Control Systems Magazine* and of the *IEEE Transactions on Automatic Control*.

Frank L. Lewis (S'78–M'81–SM'86) was born in Würzburg, Germany, subsequently studying in Chile and Scotland. He obtained the B.S. degree in physics/electrical engineering and the M.E.E. at Rice University in 1971. He spent six years in the U.S. Navy, serving as navigator aboard the frigate USS Trippe (FF-1075), and Executive Officer and Acting Commanding Officer aboard USS Salinan (ATF-161). In 1977, he received the M.S. degree in aeronautical engineering from the University of West Florida. In 1981, he obtained the Ph.D. degree at The Georgia Institute of Technology in Atlanta, where he was employed from 1981 to 1990. In 1990 he was awarded the Moncrief-O'Donnell endowed chair at the Automation and Robotics Research Institute of The University of Texas at Arlington.

Dr. Lewis has studied the geometric properties of the Riccati equation and implicit systems: his current interests include robotics, nonlinear systems, and manufacturing process control. He is the author of *Optimal Control*, *Optimal Estimation*, and *Applied Optimal Control and Estimation*, and coauthor of *Aircraft Control and Simulation* and *Control of Robot Manipulators* (to appear). Dr. Lewis is an Associate Editor of *Circuits, Systems, and Signal Processing*, and the recipient of an NSF Research Initiation Grant, a Fulbright Research Award, the American Society of Engineering Education F. E. Terman Award, three Sigma Xi Research Awards, and the UTA Halliburton Research Award.

Chaouki Abdallah (S'81–M'88) is a native of Lebanon. He received the B.E. degree in electrical engineering in 1981 from Youngstown State University, OH, the M.S. degree in 1982 and the Ph.D. degree in electrical engineering in 1988 from The Georgia Institute of Technology, Atlanta.

Between 1983 and 1985, he was with SAWTEK Inc., Orlando, FL, where he was a member of the technical staff. Since September 1988 he has been an Assistant Professor of Electrical and Computer Engineering at the University of New Mexico, Albuquerque.

Dr. Abdallah was exhibit chairman of the 1990 International Conference on Acoustics, Speech, and Signal Processing (ICASSP) in Albuquerque, New Mexico. His research interests are in the areas of robotics, nonlinear and robust control, digital control, and neural networks. He is a member of Sigma Xi and Tau Beta Pi.